The Elements

Element	Symbol	Atomic number	Molar mass/ $(g\ mol^{-1})$	Element	Symbol	Atomic number	Molar mass/ $(g\ mol^{-1})$
Actinium	Ac	89	227.03	Mercury	Hg	80	200.59
Aluminum	Al	13	26.98	Molybdenum	Mo	42	95.94
Americium	Am	95	241.06	Neodymium	Nd	60	144.24
Antimony	Sb	51	121.75	Neon	Ne	10	20.18
Argon	Ar	18	39.95	Neptunium	Np	93	237.05
Arsenic	As	33	74.92	Nickel	Ni	28	58.71
Astatine	At	85	210	Niobium	Nb	41	92.91
Barium	Ba	56	137.34	Nitrogen	N	7	14.01
Berkelium	Bk	97	249.08	Nobelium	No	102	255
Beryllium	Be	4	9.01	Osmium	Os	76	190.2
Bismuth	Bi	83	208.98	Oxygen	O	8	16.00
Bohrium	Bh	107	–	Palladium	Pd	46	106.4
Boron	B	5	10.81	Phosphorus	P	15	30.97
Bromine	Br	35	79.91	Platinum	Pt	78	195.09
Cadmium	Cd	48	112.40	Plutonium	Pu	94	239.05
Calcium	Ca	20	40.08	Polonium	Po	84	210
Californium	Cf	98	251.08	Potassium	K	19	39.10
Carbon	C	6	12.01	Praseodymium	Pr	59	140.91
Cerium	Ce	58	140.12	Promethium	Pm	61	146.92
Cesium	Cs	55	132.91	Protactinium	Pa	91	231.04
Chlorine	Cl	17	35.45	Radium	Ra	88	226.03
Chromium	Cr	24	52.01	Radon	Rn	86	222
Cobalt	Co	27	58.93	Rhenium	Re	75	186.2
Copper	Cu	29	63.54	Rhodium	Rh	45	102.91
Curium	Cm	96	247.07	Rubidium	Rb	37	85.47
Dubnium	Db	105	–	Ruthenium	Ru	44	101.07
Dysprosium	Dy	66	162.50	Rutherfordium	Rf	104	–
Einsteinium	Es	99	254.09	Samarium	Sm	62	150.35
Erbium	Er	68	167.26	Scandium	Sc	21	44.96
Europium	Eu	63	151.96	Seaborgium	Sg	106	–
Fermium	Fm	100	257.10	Selenium	Se	34	78.96
Fluorine	F	9	19.00	Silicon	Si	14	28.09
Francium	Fr	87	223	Silver	Ag	47	107.87
Gadolinium	Gd	64	157.25	Sodium	Na	11	22.99
Gallium	Ga	31	69.72	Strontium	Sr	38	87.62
Germanium	Ge	32	72.59	Sulfur	S	16	32.06
Gold	Au	79	196.97	Tantalum	Ta	73	180.95
Hafnium	Hf	72	178.49	Technetium	Tc	43	98.91
Hassium	Hs	108	–	Tellurium	Te	52	127.60
Helium	He	2	4.00	Terbium	Tb	65	158.92
Holmium	Ho	67	164.93	Thallium	Tl	81	204.37
Hydrogen	H	1	1.008	Thorium	Th	90	232.04
Indium	In	49	114.82	Thulium	Tm	69	168.93
Iodine	I	53	126.90	Tin	Sn	50	118.69
Iridium	Ir	77	192.2	Titanium	Ti	22	47.90
Iron	Fe	26	55.85	Tungsten	W	74	183.85
Krypton	Kr	36	83.80	Uranium	U	92	238.03
Lanthanum	La	57	138.91	Vanadium	V	23	50.94
Lawrencium	Lr	103	257	Xenon	Xe	54	131.30
Lead	Pb	82	207.19	Ytterbium	Yb	70	173.04
Lithium	Li	3	6.94	Yttrium	Y	39	88.91
Lutetium	Lu	71	174.97	Zinc	Zn	30	65.37
Magnesium	Mg	12	24.31	Zirconium	Zr	40	.22
Manganese	Mn	25	54.94				
Meitnerium	Mt	109	–				
Mendelevium	Md	101	258.10				

Inorganic Chemistry

Inorganic Chemistry

Third edition

D. F. Shriver

Morrison Professor of Chemistry
Northwestern University, Evanston, Illinois

P. W. Atkins

Professor of Chemistry, University of Oxford
and Fellow of Lincoln College

W. H. Freeman and Company
New York

Cover Illustration: Janet Hamlin

Library of Congress Cataloging-in-Publication Data

Shriver, D. F. (Duward F.), 1934–

    Inorganic chemistry.—3d ed. / Duward F. Shriver, Peter Atkins.

        p. cm.

Includes index.

ISBN: 0716736241 (EAN: 9780716736240)

I. Chemistry, Inorganic. I. Atkins, P. W. (Peter William), 1 940—

    II. Title.

QD154.5.S57 1999b

546—dc2   98-52143

                                               CIP

Printed in the United States of America

Sixth printing

W. H. Freeman and Company

41 Madison Avenue

New York, NY 10010

Preface

As with previous editions, our aim is to introduce the thriving discipline of inorganic chemistry. Inorganic chemistry deals with the properties of over a hundred elements. These elements range from highly reactive metals, such as sodium, to noble metals, such as gold. They include the nonmetals, which range from the aggressive oxidizing agent fluorine to unreactive gases such as helium. This variety is one of the great attractions of the subject. To provide mastery and appreciation of the broad sweep of the subject, the text focuses on trends in reactivity, structure, and properties of the elements and their compounds in relation to their position in the periodic table. These general periodic trends provide a foundation for an initial understanding.

Inorganic compounds vary from ionic solids, which can be described by simple applications of classical electrostatics, to covalent compounds and metals, which are best described by models that have their origin in quantum mechanics. For the rationalization and interpretation of most inorganic properties we use qualitative models that are based on quantum mechanics, such as the properties of atomic orbitals and their use to form molecular orbitals. Similar bonding models should already be familiar from introductory chemistry courses. Theory has contributed greatly to our understanding of inorganic chemistry, and qualitative models of bonding and reactivity clarify and systematize the subject. Nevertheless, inorganic chemistry, like organic chemistry and biochemistry, is essentially an experimental subject. The ultimate authority consists of observations and measurements, such as the identities of products of a reaction, structures, thermodynamic properties, spectroscopic signatures, and measurements of reaction rates. Large areas of inorganic chemistry remain unexplored, so new and often unusual inorganic compounds are constantly being synthesized. The exploratory inorganic syntheses continue to enrich the field with compounds that give us new perspectives on structure, bonding, and reactivity.

In addition to its intellectual attractions, inorganic chemistry has considerable practical impact and touches on all other branches of science. The chemical industry is strongly dependent on inorganic chemistry. Inorganic chemistry is essential to the formulation and improvement of modern materials such as catalysts, semiconductors, light guides, nonlinear optical devices, superconductors, and advanced ceramic materials. The environmental impact of inorganic chemistry is also huge. In this connection, the extensive role of metal ions in plants and animals led to the thriving area of bioinorganic chemistry. These current topics are mentioned throughout the book and developed more thoroughly in later chapters.

In preparing this new edition, we have refined the presentation, logical organization, and visual representation. In fact, much of the book has been rewritten, hundreds of illustrations have been redrawn, and we have reorganized the text. We have written with the student in mind, and have added new pedagogical features and have enhanced others.

The brief introductions to the chapters set the scene for the material that follows. We are all too well aware that the process of reading a text can quickly become a passive undertaking, without significant comprehension or retention. To turn this into a more active learning experience, we provide brief summaries—which we think of as 'bottom lines'—at the end of most subsections throughout the text. These summaries are designed to engage the reader in actively thinking about the material and to facilitate review of the material. They do not pretend to be complete; their aim is simply to be helpful.

There are more *Exercises* at the end of each chapter. These *Exercises* are intended to consolidate the understanding of the material; we provide brief answers at the end of the book. The *Exercises* are worked out in detail and thoroughly explained in S.H. Strauss's accompanying *Guide to solutions for inorganic chemistry*. Students who used the previous editions of Strauss's *Guide* were enthusiastic about its clear and helpful perspective. Because we believe that the *Guide* greatly enriches the text by showing how the principles we describe are applied in practice, we have inserted a small **marginal icon** to indicate where the *Guide* elaborates a particular point (other than explaining an approach to an *Exercise*). The *Problems* at the ends of chapters are intended to be more searching, and many require reference to current literature.

The topics in Part 1, *Fundamentals*, have been rearranged and largely rewritten to provide better flow of thought and improved clarity. Part 2, *Systematic chemistry of the elements*, now starts with hydrogen and then progresses through the periodic table. To avoid interrupting this flow, the chapter on main-group organometallic compounds has been moved to Part 3. The combination of descriptive material and principles in the first two parts of the book provides a solid foundation in inorganic chemistry. Each of the descriptive chapters in Part 2 presents chemistry within the framework of the principles covered in Part 1. This type of rationalization is necessary to bring the mass of observations into a coherent context.

In Part 3, *Advanced topics*, we present background material on the achievements in many areas of current research, including a more detailed treatment of electronic spectra and discussions of rates and mechanisms of reactions, organometallic chemistry, catalysis, solid state chemistry, and bioinorganic chemistry. These chapters have been extensively revised to make them more accessible and to provide information on recent advances. The chapter on bioinorganic chemistry, which has been largely rewritten, now more closely mirrors the topics of current bioinorganic research.

The text is supported by a CD-ROM that has been compiled under the direction of Karl Harrison, University of Oxford. Almost all the numbered molecular structures will be found in a three-dimensional viewable, rotatable form on this CD-ROM together with many of the basic types of crystal structures. We consider it enormously important to develop a sense of the three-dimensional character of inorganic compounds, and know that the information on the CD-ROM will be of great value in this regard. Also on the CD-ROM will be found almost all the illustrations from the text in full color and in formats suitable for downloading into presentation software. We hope our colleagues will make full use of these images in their lectures.

Further support is available on our two websites, www.oup.co.uk/best.textbooks/ichem3e and www.whfreeman.com/chemistry. For more detailed information than we have space to provide in the text itself, we refer readers to a number of compendia. Two outstanding sources are *Chemistry of the elements*, by N.N. Greenwood and A. Earnshaw (Butterworth-Heinemann, Oxford, 1997), and *Advanced inorganic chemistry*, by F.A. Cotton and G.

Wilkinson (Wiley, New York, 1988). The *Encyclopedia of inorganic chemistry*, edited by R.B. King (Wiley, New York, 1994), is an excellent source for further information, as are *Comprehensive organometallic chemistry*, edited by G. Wilkinson, F.G.A. Stone, and E.W. Abel (Elsevier, Oxford, 1987 and 1995); and *Comprehensive coordination chemistry*, edited by G. Wilkinson, R.D. Gillard, and J. McCleverty (Pergamon Press, Oxford, 1987). Inorganic industrial processes are summarized in *Industrial inorganic chemistry*, by W. Buchner, R. Schliebs, G. Winter, and K.H. Buchel (VCH, Deerfield Beach, 1989). Two multi-volume compendia of industrial chemistry are *Kirk–Othmer encyclopedia of chemical technology* (Wiley-Interscience, 1991 *et seq.*), and *Ullmann's encyclopedia of industrial chemistry* (VCH, Deerfield Beach, 1985 *et seq.*).

We have taken particular care to ensure that the text is free of errors. This is difficult in a rapidly changing field, where today's knowledge is soon replaced by tomorrow's. We acknowledge below all those colleagues who so unstintingly gave their time and expertise to a careful reading of a variety of draft chapters, and would like to thank all those readers—too numerous to mention here by name—who, with great good will, wrote unsolicited letters to us full of helpful advice.

Cooper Langford is not an author of this edition, but his contribution to the first two editions has a lasting impact on the book's content and style. Yet again, we wish to thank our publishers for the understanding and assistance they have provided at all stages of the intricate and time-consuming task of producing a book of such structural complexity as this. We owe a considerable debt to their patience, wisdom, and understanding.

Evanston   D.F.S.
Oxford   P.W.A.

Acknowledgements

As with the second edition we have benefited from the suggestions and expertise of others. Many instructors and students gave us useful suggestions for this revision, as did several of the foreign language translators and our colleagues. In particular we acknowledge the help of:

Dr H C Aspinall, University of Liverpool

Dr Philip J Bailey, University of Edinburgh

Dr Ronald Bailey, Rensselaer Polytechnic Institute, NY

Professor D W Bruce, University of Exeter

Dr Clive Buckley, North East Wales Institute

Dr William Byers, University of Ulster

Professor Martin Cowie, University of Alberta, Canada

Dr Peter J Cragg, University of Brighton

Professor Dainis Dakternieks, Deakin University, Victoria, Australia

Dr Melinda J Duer, University of Cambridge

Dr Dennis Edwards, University of Bath

Professor J Evans, University of Southampton

Professor David Fenton, University of Sheffield

Professor R O Gould, University of Edinburgh

Dr E M Green, University of Greenwich

Dr Malcolm A Halcrow, University of Cambridge

Dr D A House, University of Canterbury

Professor Michael D Johnson, New Mexico State University, NM

Professor Silvia Jurisson, University of Missouri, MO

Professor Martin L Kirk, University of New Mexico, NM

Dr Ken Kite, University of Exeter

Dr W Levason, University of Southampton

Professor D E Linn Jr, IPFW, Fort Wayne, IN

Dr Debbie Mans, University of Paisley

Dr Caroline Martin, University of Cambridge

Dr A G Osborne, University of Exeter

Dr A W Parkins, Kings College, London

Professor A Pidcock, University of Central Lancashire

Professor D W H Rankin, University of Edinburgh

Dr P R Raithby, University of Cambridge

Dr J M Rawson, University of Cambridge

Dr James L Reed, Clark Atlanta University, GA

Dr Roger Reeve, University of Sunderland

Dr Lee Roecker, Berca College, KY

Dr David Rosseinsky, University of Exeter

Professor Peter Sadler, University of Edinburgh

Dr Ian Salter, University of Exeter

Professor Robert Slade, University of Surrey

Dr A W Sleight, Oregon State University, OR

Professor Lothar Stahl, University of North Dakota, ND

Professor Claire Tessler, University of Akron, OH

Dr George M H van de Velde, University of Twente, Netherlands

Professor J H C van Hooff, Eindhoven University of Technology, Eindhoven, Netherlands

Professor Mark Weller, University of Southampton

Professor A R West, University of Aberdeen

Professor Mark Wicholas, Western Washington University, WA

Professor R G Williams, Albuquerque, NM

Dr Mark J Winter, University of Sheffield

Dr L J Wright, University of Auckland

Summary of contents

Contents

Part 1   Foundations

The seven chapters of this part of the text lay the foundations of inorganic chemistry. The first four chapters develop an understanding of the structures of atoms, molecules, and solids in terms of quantum theory. Because all models of bonding are based on atomic properties, atomic structure is described in Chapter 1. The following chapter develops a description of the simplest bonding model, ionic bonding, in terms of the structures and properties of ionic solids. Chapter 3 likewise develops a description of the properties of the covalent bond by presenting molecular structure in terms of increasingly sophisticated theories. Chapter 4 shows how intuitive ideas on symmetry can be made into precise arguments, and then used to discuss the bonding, physical properties, and vibrations of molecules.

The next two chapters introduce two fundamental reaction types. Chapter 5 describes the reactions of acids and bases in which the reaction takes place by the transfer of a proton or by the sharing of electron pairs. We see that many reactions can be expressed as one type or the other, and the introduction of these reaction types helps to systematize inorganic chemistry. Chapter 6 introduces another major class of chemical reactions, those proceeding by oxidation and reduction, and shows how electrochemical data can be used to systematize a large class of reactions.

Chapter 7 brings these principles together, by treating the coordination compounds formed by the d-block metals. Here we see the role of symmetry in determining the electronic structures of molecules, and meet some elementary ideas about how reactions take place.

1 Atomic structure

The origin of the elements

The structure of hydrogenic atoms

Many-electron atoms

Further reading

Exercises

Problems

This chapter introduces our current understanding of the origin and nature of the matter in our solar system. It then discusses the atomic properties of the elements and summarizes how atomic structure is rationalized in terms of the behavior of electrons in atoms. We review trends in atomic parameters such as radius and the spacing of energy levels, and rationalize these trends in terms of the results from quantum theory. The concepts of quantum theory are introduced qualitatively, with emphasis on pictorial representations rather than mathematical rigor. Some of the atomic parameters encountered in earlier courses, such as atomic and ionic radii, ionization energy, electron affinity, and electronegativity, will be reviewed. In later chapters we shall see how useful these parameters are for organizing observed trends in the physical and chemical properties and the structures of inorganic compounds.

The observation that the universe is expanding has led to the current view that about 15 billion years ago it was concentrated into a point-like region which exploded in an event called the **Big Bang**. With initial temperatures immediately after the Big Bang thought to be about 10^9 K, the fundamental particles produced in the explosion had too much kinetic energy to bind together in the forms we know today. However, the universe cooled as it expanded, the particles moved more slowly, and they soon began to adhere together under the influence of a variety of forces. In particular, the **strong force**, a short-range but powerful attractive force between nucleons (protons and neutrons), bound these particles together into nuclei. As the temperature fell still further, the **electromagnetic force**, a relatively weak but long-range force between electric charges, bound electrons to nuclei to form atoms.

The properties of the only subatomic particles that we need to consider in chemistry are summarized in Table 1.1. The 110 or so known elements that are formed from these subatomic particles are distinguished by their **atomic number**, Z, the number of protons in the nucleus of an atom of the element. Many elements have a number of **isotopes**, which are atoms with the same atomic number but different atomic masses. These isotopes are distinguished by the **mass number**, A, the total number of protons and neutrons in the nucleus; A is also sometimes termed more appropriately the 'nucleon number'. Hydrogen, for instance, has three isotopes. In each case, $Z = 1$, indicating that the nucleus contains one proton. The most abundant isotope has $A = 1$, denoted ^{1}H: its nucleus consists of a lone proton. Far

Table 1.1 Subatomic particles of relevance to chemistry

Particle	Symbol	Mass/u*	Mass number	Charge/e†	Spin
Electron	e^-	5.486×10^{-4}	0	-1	$\frac{1}{2}$
Proton	p	1.0073	1	$+1$	$\frac{1}{2}$
Neutron	n	1.0087	1	0	$\frac{1}{2}$
Photon	γ	0	0	0	1
Neutrino	ν	c. 0	0	0	$\frac{1}{2}$
Positron	e^+	5.486×10^{-4}	0	$+1$	$\frac{1}{2}$
α particle	α	[$_2^4$He^{2+} nucleus]	4	$+2$	0
β particle	β	[e^- ejected from nucleus]	0	-1	$\frac{1}{2}$
γ photon	γ	[electromagnetic radiation from nucleus]	0	0	1

*Masses are expressed in atomic mass units, **u**, with $1\ \text{u} = 1.6605 \times 10^{-27}$ kg.
†The elementary charge e is 1.602×10^{-19} C.

less abundant (only 1 atom in 6000) is deuterium, with $A = 2$. This mass number indicates that, in addition to a proton, the nucleus contains one neutron. The formal designation of deuterium is ^{2}H, but it is commonly denoted D. The third, short-lived, radioactive isotope of hydrogen is tritium (^{3}H, or T). Its nucleus consists of one proton and two neutrons. In certain cases it is helpful to display the atomic number of the element as a left suffix; so the three isotopes of hydrogen would then be denoted $_1^1$H, $_1^2$H, and $_1^3$H.

The origin of the elements

If current views are correct, by about 2 h after the start of the universe the temperature had fallen so much that most of the matter was in the form of H atoms (89 per cent) and He atoms (11 per cent). In one sense, not much has happened since then for, as Fig. 1.1 shows, hydrogen and helium remain the most abundant elements in the universe. However, nuclear reactions have formed a wide assortment of other elements and have immeasurably enriched the variety of matter in the universe.

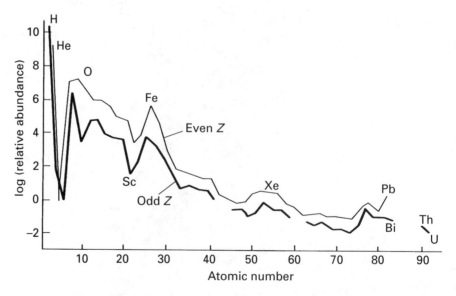

1.1 The abundances of the elements in the universe. Elements with odd Z are less stable than their neighbors with even Z. The abundances refer to the number of atoms of each element relative to Si taken as 10^6.

1.1 Nucleosynthesis of the light elements

The earliest stars resulted from the gravitational condensation of clouds of H and He atoms. The compression of these clouds under the influence of gravity gave rise to high temperatures and densities within them, and fusion reactions began as nuclei merged together. The earliest nuclear reactions are closely related to those now being studied in connection with the development of controlled nuclear fusion.

Energy is released when light nuclei fuse together to give elements of higher atomic number. For example, the nuclear reaction in which an α particle (a ^{4}He nucleus, consisting of two protons and two neutrons) fuses with a carbon-12 nucleus to give an oxygen-16 nucleus and a γ-ray photon (γ) is

$$^{12}_{6}C + ^{4}_{2}\alpha \longrightarrow ^{16}_{8}O + \gamma$$

This reaction releases 7.2 MeV.[1] Nuclear reactions are very much more energetic than normal chemical reactions because the strong force is much stronger than the electromagnetic force that binds electrons to atoms. Whereas a typical chemical reaction might release about 10^3 kJ mol^{-1}, a nuclear reaction typically releases a million times more energy, about 10^9 kJ mol^{-1}. In this nuclear equation, the **nuclide**, a nucleus of specific atomic number Z and mass number A, is designated $^{A}_{Z}E$, where E is the chemical symbol of the element. Note that, in a balanced nuclear equation, the mass numbers of the reactants sum to the same value as the mass numbers of the products ($12 + 4 = 16$). The atomic numbers sum likewise ($6 + 2 = 8$) provided an electron, e$^-$, when it appears as a β particle, is denoted $_{-1}^{0}$e and a positron, e$^+$, is denoted $^{0}_{1}$e. A positron is a positively charged version of an electron: it has zero mass number and a single positive charge. When it is emitted, the mass number of the nuclide is unchanged but the atomic number decreases by 1 because the nucleus has lost one positive charge. Its emission is equivalent to the conversion of a proton in the nucleus into a neutron: $^{1}_{1}p \longrightarrow ^{1}_{0}n + e^+ + \nu$. A neutrino, ν, is electrically neutral and has a very small (possibly zero) mass.

Elements of atomic number up to 26 were formed inside stars. Such elements are the products of the nuclear fusion reactions referred to as 'nuclear burning'. The burning reactions, which should not be confused with chemical combustion, involved H and He nuclei and a complicated fusion cycle catalyzed by C nuclei. (The stars that formed in the earliest stages of the evolution of the cosmos lacked C nuclei and used non-catalyzed H-burning reactions.) Some of the most important nuclear reactions in the cycle are

Proton (p) capture by carbon-12:	$^{12}_{6}C + ^{1}_{1}p \longrightarrow ^{13}_{7}N + \gamma$
Positron decay accompanied by neutrino (ν) emission:	$^{13}_{7}N \longrightarrow ^{13}_{6}C + e^+ + \nu$
Proton capture by carbon-13:	$^{13}_{6}C + ^{1}_{1}p \longrightarrow ^{14}_{7}N + \gamma$
Proton capture by nitrogen-14:	$^{14}_{7}N + ^{1}_{1}p \longrightarrow ^{15}_{8}O + \gamma$
Positron decay, accompanied by neutrino emission:	$^{15}_{8}O \longrightarrow ^{15}_{7}N + e^+ + \nu$
Proton capture by nitrogen-15:	$^{15}_{7}N + ^{1}_{1}p \longrightarrow ^{12}_{6}C + ^{4}_{2}\alpha$

The net result of this sequence of nuclear reactions is the conversion of four protons (four ^{1}H nuclei) into an α particle (a ^{4}He nucleus):

$$4\,^{1}_{1}p \longrightarrow ^{4}_{2}\alpha + 2\,e^+ + 2\,\nu + 3\,\gamma$$

---

[1] An electronvolt (1 eV) is the energy required to move an electron through a potential difference of 1 V. It follows that 1 eV $= 1.602 \times 10^{-19}$ J, which is equivalent to 96.48 kJ mol^{-1}; 1 MeV $= 10^6$ eV.

The reactions in the sequence are rapid at temperatures between 5 and 10 MK (where 1 MK = 10^6 K). Here we have another contrast between chemical and nuclear reactions, because chemical reactions take place at temperatures a hundred-thousand times lower. Moderately energetic collisions between species can result in chemical change, but only highly vigorous collisions can overcome the barrier to activation typical of most nuclear processes.

Heavier elements are produced in significant quantities when hydrogen burning is complete and the collapse of the star's core raises the density there to 10^8 kg m^{-3} (about 10^5 times the density of water) and the temperature to 100 MK. Under these extreme conditions, **helium burning** becomes viable. The low abundance of beryllium in the present-day universe is consistent with the observation that $^8_4$Be formed by collisions between α particles goes on to react with more α particles to produce the more stable carbon nuclide, $^{12}_6$C:

$$^8_4\text{Be} + ^4_2\alpha \longrightarrow ^{12}_6\text{C} + \gamma$$

Thus, the helium-burning stage of stellar evolution does not result in the formation of beryllium as a stable end product; for similar reasons, it also results in low concentrations of lithium and boron. The nuclear reactions leading to these three elements are still uncertain, but they may result from the fragmentation of C, N, and O nuclei by collisions with high-energy particles.

Elements can also be produced by nuclear reactions such as neutron (n) capture accompanied by proton emission:

$$^{14}_7\text{N} + ^1_0\text{n} \longrightarrow ^{14}_6\text{C} + ^1_1\text{p}$$

This reaction still continues in our atmosphere as a result of the impact of cosmic rays and contributes to the steady-state concentration of radioactive carbon-14 on Earth.

The high abundance of iron and nickel in the universe is consistent with their having the most stable of all nuclei. This stability is expressed in terms of the **binding energy**, which represents the difference in energy between the nucleus itself and the same numbers of individual protons and neutrons. This binding energy is often presented in terms of a difference in mass between the nucleus and its individual protons and neutrons, for, according to Einstein's theory of relativity, mass and energy are related by $E = mc^2$, where c is the speed of light. Therefore, if the mass of a nucleus differs from the total mass of its components by $\Delta m = m_{\text{nucleons}} - m_{\text{nucleus}}$, then its binding energy is

$$E_{\text{bind}} = (\Delta m)c^2 \tag{1}$$

A positive binding energy corresponds to a nucleus that has a lower, more favorable, energy (and lower mass) than its constituent nucleons. The binding energy of ^{56}Fe, for example, is the difference in energy between the ^{56}Fe nucleus and 26 protons and 30 neutrons.

Figure 1.2 shows the binding energy (expressed as an energy per nucleon by dividing the total binding energy by the number of nucleons) for all the elements, and we see that iron (and nickel) occur at the maximum of the curve, showing that their nucleons are bound more strongly than in any other nuclide. More difficult to discern from the curve is the alternation of binding energies as the atomic number varies from even to odd; there is a corresponding alternation in cosmic abundances, with nuclides of even atomic number being marginally more abundant than those of odd atomic number.

> Nuclear mass number and overall charge are conserved in nuclear reactions; a large binding energy signifies a stable nucleus. The lighter elements were formed by nuclear reactions in stars formed from primeval hydrogen and helium.

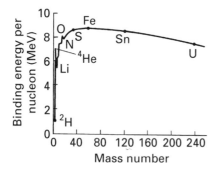

1.2 Nuclear binding energies. The greater the binding energy, the more stable the nucleus. The most stable nuclide is $^{56}_{26}$Fe.

1.2 The nucleosynthesis of heavy elements

Because nuclei close to iron are the most stable, heavier elements are produced by a variety of energy-consuming processes. These processes include the capture of free neutrons, which are not present in the earliest stages of stellar evolution but are produced later in reactions such as

$$\ce{^{23}_{10}Ne} + \ce{^{4}_{2}\alpha} \longrightarrow \ce{^{26}_{12}Mg} + \ce{^{1}_{0}n}$$

Under conditions of intense neutron flux, as in a supernova (the explosion of a star), a given nucleus may capture a succession of neutrons and become a progressively heavier isotope. However, there comes a point at which the nucleus will eject an electron from the nucleus as a β particle (a high-velocity electron, e^-). Because β decay leaves the mass number of the nuclide unchanged but increases its atomic number by 1 (the nuclear charge increases by 1 unit when an electron is ejected), a new element is formed. An example is

Neutron capture: $\qquad \ce{^{98}_{42}Mo} + \ce{^{1}_{0}n} \longrightarrow \ce{^{99}_{42}Mo} + \gamma$

Followed by β decay accompanied

by neutrino emission: $\qquad \ce{^{99}_{42}Mo} \longrightarrow \ce{^{99}_{43}Tc} + e^- + \nu$

The **daughter nuclide**, the product of a nuclear reaction ($\ce{^{99}_{43}Tc}$, an isotope of technetium, in this example), can absorb another neutron, and the process can continue, gradually building up the heavier elements.

> Heavier nuclides are formed by processes that include neutron capture and subsequent β decay.

Example 1.1 Balancing equations for nuclear reactions

Synthesis of heavy elements occurs in neutron capture reactions believed to take place in the interior of cool 'red giant' stars. One such reaction is the conversion of $\ce{^{68}_{30}Zn}$ to $\ce{^{69}_{31}Ga}$ by neutron capture to form $\ce{^{69}_{30}Zn}$, which then undergoes β decay. Write balanced nuclear equations for this process.

Answer Neutron capture increases the mass number of a nuclide by 1 but leaves the atomic number (and hence the identity of the element) unchanged:

$$\ce{^{68}_{30}Zn} + \ce{^{1}_{0}n} \longrightarrow \ce{^{69}_{30}Zn} + \gamma$$

The excess energy is carried away as a photon. β decay, the loss of an electron from the nucleus, leaves the mass number unchanged but increases the atomic number by 1. Because zinc has atomic number 30, the daughter nuclide has $Z = 31$, corresponding to gallium. Therefore, the nuclear reaction is

$$\ce{^{69}_{30}Zn} \longrightarrow \ce{^{69}_{31}Ga} + e^-$$

In fact, a neutrino is also emitted, but this cannot be inferred from the data as a neutrino is effectively massless and electrically neutral.

Self-test 1.1 Write the balanced nuclear equation for neutron capture by $\ce{^{80}_{35}Br}$.

1.3 The classification of the elements

Some substances that we now recognize as chemical elements have been known since antiquity: they include carbon, sulfur, iron, copper, silver, gold, and mercury. The alchemists and their immediate successors, the early chemists, had added about another 18 elements by 1800. By that time, the precursor of the modern concept of an element had been formulated as a substance that consists of only one type of atom. (Now, of course, by 'type' of atom we mean an atom with a particular atomic number.) By 1800 a variety of experimental techniques were available for converting oxides and other compounds into elements. These techniques were considerably enhanced by the introduction of electrolysis. The list of elements grew rapidly in the later nineteenth century. This growth was in part a result of the development of atomic spectroscopy, in which thermally excited atoms of a particular element are observed to emit electromagnetic radiation with a unique pattern of frequencies. These spectroscopic observations made it much easier to detect previously unknown elements.

(a) Patterns and periodicity

A useful broad division of elements is into **metals** and **nonmetals**. Metallic elements (such as iron and copper) are typically lustrous, malleable, ductile, electrically conducting solids at about room temperature. Nonmetals are often gases (oxygen), liquids (bromine), or solids that do not conduct electricity appreciably (sulfur). The chemical implications of this classification should already be clear from introductory chemistry:

1 Metallic elements combine with nonmetallic elements to give compounds that are typically hard, non-volatile solids (for example, sodium chloride).
2 When combined with each other, the nonmetals often form volatile molecular compounds (such as phosphorus trichloride).
3 When metals combine (or simply mix together) they produce alloys that have most of the physical characteristics of metals.

A more detailed classification of the elements is the one devised by D.I. Mendeleev in 1869; this scheme is familiar to every chemist as the **periodic table**. Mendeleev arranged the known elements in order of increasing atomic weight (molar mass). This arrangement resulted in families of elements with similar chemical properties, which he arranged into the groups of the periodic table. For example, the following formulas for the compounds of some elements with hydrogen suggest that the elements belong to two different groups:

$$CH_4 \qquad\qquad NH_3$$
$$SiH_4 \qquad\qquad PH_3$$
$$GeH_4 \qquad\qquad AsH_3$$
$$SnH_4 \qquad\qquad SbH_3$$

Other compounds of these elements show family similarities, as in the formulas CF_4 and SiF_4 in the first group, and NF_3 and PF_3 in the second.

Mendeleev concentrated on the chemical properties of the elements. At about the same time Lothar Meyer in Germany was investigating their physical properties, and found that similar values repeated periodically with increasing atomic weight. A classic example is shown in Fig. 1.3, where the molar volume of the element (its volume per mole of atoms) in its normal form is plotted against atomic number.

Mendeleev provided a spectacular demonstration of the usefulness of the periodic table by correctly predicting the general chemical properties, such as the numbers of bonds they form, of unknown elements corresponding to gaps in his original periodic table. The same process of inference from periodic trends is still used by inorganic chemists to rationalize trends in the physical and chemical properties of compounds and to suggest the synthesis of previously unknown compounds. For instance, by recognizing that carbon and silicon are in the same family, the existence of alkenes ($R_2C\!=\!CR_2$) suggests that $R_2Si\!=\!SiR_2$ ought to exist too. Compounds with silicon–silicon double bonds (disilaethenes) do indeed exist, but it was not until 1981 that inorganic chemists succeeded in isolating a stable member of the family.

The elements are broadly divided into metals and nonmetals according to their physical and chemical properties; the organization of elements into the form resembling the modern periodic table is accredited to Mendeleev.

(b) The modern periodic table

The general structure of the modern periodic table will be familiar from previous chemistry courses (Fig. 1.4), and the following is a review. The elements are listed in order of atomic

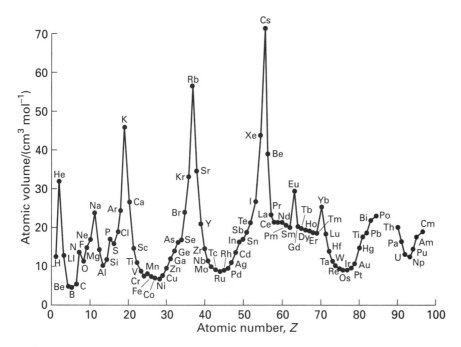

1.3 The periodicity of molar volume with atomic number.

number, not atomic weight, for the atomic number tells us the number of electrons in the atom and is therefore a more fundamental quantity. The horizontal rows of the table are called **periods** and the vertical columns are called **groups**.[2] We often use the group number to designate the general position of an element as in 'gallium is in Group 13'; alternatively, the lightest element in the group is used to designate the group, as in 'gallium is a member of the boron group'. The members of the same group as a given element are called the **congeners** of that element. Thus, sodium and potassium are congeners of lithium.

The periodic table is divided into four **blocks** which are labeled as in Fig. 1.4. The members of the *s*- and *p*-blocks are collectively called the **main-group elements**, and the *d*-block

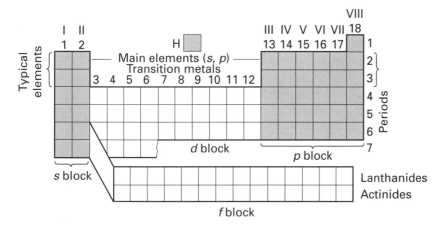

1.4 The general structure of the periodic table. The tinted areas denote the main-group elements. Compare this template with the complete table inside the front cover for the identities of the elements that belong to each block.

....

[2] The numbering system used for groups in Fig. 1.4 follows the IUPAC recommendation (IUPAC is the International Union of Pure and Applied Chemistry, which recommends nomenclature, symbols, units, and sign conventions). The volume that outlines the conventions for the periodic table and inorganic substances is *Nomenclature of inorganic chemistry*, Blackwell Scientific, Oxford (1990), and is known colloquially as the 'Red Book' on account of its distinctive red cover.

elements (often with the exception of Group 12, zinc, cadmium, and mercury) are also referred to collectively as the **transition elements**. The f-block elements are divided into the lighter series (atomic numbers 57–71) called the **lanthanides** and the heavier series (atomic numbers 89–103) called the **actinides**. The **representative elements** are the members of the first three periods of the main group elements (from hydrogen to argon).

The numbering system of the groups is still in contention. In the illustration we show both the traditional numbering of the main groups (with the roman numerals from I to VIII) and the current IUPAC recommendations, in which the groups of the s-, d-, and p-blocks are numbered from 1 through 18. The groups of the f-block are not numbered because there is little similarity between the lanthanides and the corresponding actinides in the period below.

> *The periodic table is divided into periods and groups; the groups belong to four major blocks; the main-group elements are those in the s- and p-blocks.*

The structure of hydrogenic atoms

The organization of the periodic table is a direct consequence of periodic variations in the electronic structure of atoms. Initially, we consider **hydrogenic atoms**, which have only one electron and so are free of the complicating effects of electron–electron repulsions. Hydrogenic atoms include ions such as He^+ and C^{5+} (found in stellar interiors) as well as the hydrogen atom itself. Then we use the concepts these atoms introduce to build up an approximate description of the structures of **many-electron atoms**, which are atoms with more than one electron.[3]

1.4 Some principles of quantum mechanics

Because the electronic structures of atoms must be expressed in terms of quantum mechanics, we need to review some of the concepts and terminology of this description of matter. A fundamental concept of quantum mechanics is that *matter has wave-like properties*. This attribute is normally not evident for macroscopic objects but it dominates the nature of subatomic particles such as the electron.

An electron is described by a **wavefunction**, ψ, which is a mathematical function of the position coordinates x, y, and z and of the time t. We interpret the wavefunction by using the **Born interpretation**, according to which the *probability* of finding the particle in an infinitesimal region of space is proportional to the square ψ^2 of the wavefunction.[4] According to this interpretation, there is a high chance of finding the particle where ψ^2 is large, and the particle will not be found where ψ^2 is zero (Fig. 1.5). The quantity ψ^2 is called the **probability density** of the particle. It is a density in the sense that the product of ψ^2 and the infinitesimal volume element $d\tau = dxdydz$ is proportional to the probability of finding the electron in that volume element. The probability is *equal* to $\psi^2 d\tau$ if the wavefunction is **normalized** in the sense that

$$\int \psi^2 \, d\tau = 1 \qquad (2)$$

where the integration is over all the space accessible to the electron. This expression simply states that the total probability of finding the electron somewhere must be 1. Any wavefunction can be made to fulfil this condition by multiplication by a **normalization**

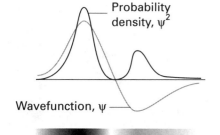

Probability
density, ψ^2

Wavefunction, ψ

Probability
density

1.5 The Born interpretation of the wavefunction is that its square is a probability density. There is zero probability density at a node. In the lower part of the illustration, the probability density is indicated by the density of shading.

---

[3] IUPAC favors the name 'polyelectron atoms'.

[4] If the wavefunction is complex, in the sense of having real and imaginary parts, the probability is proportional to the square modulus, $\psi^*\psi$, where ψ^* is the complex conjugate of ψ. For simplicity, we shall usually assume that ψ is real and write all formulas accordingly.

constant, N, a numerical constant which ensures that the integral in eqn 2 is indeed equal to 1.

The important implication of the Born interpretation is that, with its emphasis on the probability of finding particles in various regions rather than precise predictions of their locations, quantum mechanics does away with the classical concept of an orbit.

Like other waves, wavefunctions in general have regions of positive and negative amplitude. However, this sign has no direct physical significance. When we want to interpret a wavefunction, we should focus on its magnitude, not whether it happens to be positive or negative. The sign of the wavefunction, however, is of crucial importance when two wavefunctions spread into the same region of space, for then a positive region of one wavefunction may add to a positive region of the other wavefunction to give a region of enhanced amplitude. This enhancement is called **constructive interference** (Fig. 1.6a). It means that, where the two wavefunctions spread into the same region of space, such as occurs when two atoms are close enough to form a bond, there may be a significantly enhanced probability of finding the particles in that region. Conversely, a positive region of one wavefunction may be canceled by a negative region of the second wavefunction (Fig. 1.6b). This **destructive interference** between wavefunctions will greatly lessen the probability that a particle will be found in that region. As we shall see, the interference of wavefunctions is of great importance in the explanation of chemical bonding. To keep track of the relative signs of different regions of a wavefunction in illustrations we shall label regions of opposite sign with dark and light shading (sometimes white in the place of light shading).

The wavefunction for a particle is found by solving the **Schrödinger equation**, a partial differential equation proposed by Erwin Schrödinger in 1926. When this equation is solved for a free particle it is found that there is no restriction on the energy, so it can exist with all possible energies. In contrast, when the equation is solved for a particle that is confined to a small region of space or is bound to an attractive center (like an electron in an atom), it is found that acceptable solutions can be obtained only for certain energies. We speak of the energy as being **quantized**, meaning that it is confined to discrete values. Later we shall see that certain other properties (for instance, angular momentum) are also quantized. This quantization of physical observables is of the most profound importance in chemistry for it endows atoms and molecules with stability and governs the bonds they can form.

> *The probability of finding an electron at a given location is proportional to the square of the wavefunction there. Wavefunctions generally have regions of positive and negative amplitude, and may undergo constructive or destructive interference with one another. The energy of a bound or confined particle is quantized.*

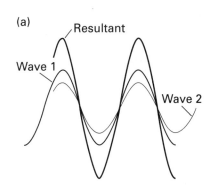

(a)

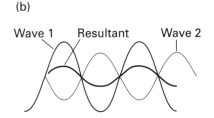

(b)

1.6 Wavefunctions interfere where they spread into the same region of space. (a) If they have the same sign in a region, they interfere constructively and the total wavefunction has an enhanced amplitude in the region. (b) If the wavefunctions have opposite signs, then they interfere destructively, and the resulting superposition has a reduced amplitude.

1.5 Atomic orbitals

The wavefunctions of an electron in a hydrogenic atom are called **atomic orbitals**. Hydrogenic atomic orbitals are central to a large part of the interpretation of inorganic chemistry, and we shall spend some time describing their shapes and significance.

(a) Hydrogenic energy levels

The wavefunctions obtained by solving the Schrödinger equation for hydrogenic atoms are specified by giving the values of three numbers called **quantum numbers**. These quantum numbers are designated n, l, and m_l: n is called the **principal quantum number**, l is the **orbital angular momentum quantum number** (or 'azimuthal quantum number'), and m_l is called the **magnetic quantum number**. Each quantum number labels a quantized physical property of the electron: n labels the quantized energy, l labels the quantized orbital angular momentum, and m_l labels the quantized orientation of the angular momentum.

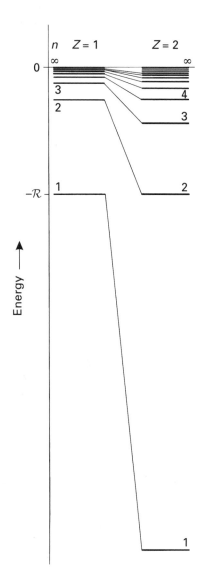

1.7 The quantized energy levels of an H atom ($Z = 1$) and an He$^+$ ion ($Z = 2$). The energy levels of a hydrogenic atom are proportional to Z^2.

The allowed energies are determined solely by the principal quantum number, n, and for a hydrogenic atom of atomic number Z are given by

$$E = -\frac{hcZ^2\mathscr{R}}{n^2} \qquad n = 1, 2, \ldots \tag{3}$$

The zero of energy corresponds to the electron and nucleus widely separated and stationary; the energies given by this expression are all negative, signifying that the atom has a lower energy than a widely separated stationary electron and nucleus. The constant $\mathscr{R}$ is a collection of fundamental constants called the **Rydberg constant**:[5]

$$\mathscr{R} = \frac{m_e e^4}{8h^3 c\varepsilon_0^2} \tag{4}$$

Its numerical value is 1.097×10^5 cm^{-1}, corresponding to 13.6 eV. The $1/n^2$ dependence of the energy leads to a rapid convergence of energy levels at high (less negative) energies (Fig. 1.7). The zero of energy, which occurs when $n = \infty$, corresponds to an infinite separation of a stationary nucleus and electron, and therefore to ionization of the atom. Above this zero of energy, the electron is unbound and may travel with any velocity and hence possess any energy.

> *The wavefunction of an electron in a hydrogenic atom is defined by the three quantum numbers n, l, and m_l. The energy of the bound electron is determined by the principal quantum number alone and is given by the expression in eqn 3: note the variation of the energy with Z^2 and $1/n^2$.*

(b) Shells, subshells, and orbitals

In a hydrogenic atom, all orbitals with the same value of n correspond to the same energy and hence are said to be **degenerate**. The principal quantum number therefore defines a series of **shells** of the atom, or sets of orbitals with the same value of n and hence (in a hydrogenic atom) with the same energy.

The orbitals belonging to each shell are classified into **subshells** distinguished by a quantum number l. This quantum number determines the magnitude of the orbital angular momentum of the electron around the nucleus through the formula $\{l(l + 1)\}^{1/2}\hbar$, where $\hbar = h/2\pi$. For a given principal quantum number n, the quantum number l can have the values $l = 0, 1, \ldots, n - 1$, giving n different values in all. Thus, the shell with $n = 2$ consists of two subshells of orbitals, one with $l = 0$ and the other with $l = 1$; the former corresponds to zero orbital angular momentum around the nucleus and the latter to $2^{1/2}\hbar$. It is common to refer to each subshell by a letter:

l:	0	1	2	3	4	$\cdots$
	s	p	d	f	g	$\cdots$

It follows that there is only one subshell (an s subshell) in the shell with $n = 1$, two subshells in the shell with $n = 2$ (the s and p subshells), three in the shell with $n = 3$ (the s, p, and d subshells), four when $n = 4$ (the s, p, d, and f subshells), and so on. The orbital angular momentum of an electron in one of these subshells increases along the series from s to f. For most purposes in chemistry we need consider only s, p, d, and f subshells.

A subshell with quantum number l consists of $2l + 1$ individual orbitals. These orbitals are distinguished by the magnetic quantum number, m_l, which can take the $2l + 1$ values $m_l = l, l - 1, l - 2, \ldots, -l$. The quantum number m_l specifies the component of orbital angular momentum on an arbitrary axis (commonly designated z) passing through the

---
[5] The fundamental constants in this expression are given inside the back cover.

nucleus, and limits its values to $m_l\hbar$. In classical terms, m_l denotes the orientation of the orbit occupied by the electron, with $m_l = +l$ corresponding to counterclockwise rotation in the xy-plane (viewed from above), $m_l = -l$ corresponding to clockwise rotation in the same plane, and $m_l = 0$ corresponding to a more 'polar' orbit. Thus, a d subshell of an atom consists of five individual atomic orbitals that are distinguished by the values $m_l = +2, +1, 0, -1, -2$.

The practical conclusion for chemistry from these remarks is that there is only one orbital in an s subshell ($l = 0$), the one with $m_l = 0$: this orbital is called an **s orbital**. There are three orbitals in a p subshell ($l = 1$), with quantum numbers $m_l = +1, 0, -1$; they are called **p orbitals**. The five orbitals of a d subshell ($l = 2$) are called **d orbitals**, and so on.

> *Orbitals belong to subshells that in turn belong to shells; all orbitals of the same shell have the same value of n, those belonging to a given subshell also have the same value of l and are distinguished by the value of m_l.*

(c) Electron spin

In addition to the three quantum numbers required to specify the spatial distribution of an electron in a hydrogenic atom, two more quantum numbers are needed to define the state of an electron completely. These additional quantum numbers relate to the intrinsic angular momentum of an electron, its **spin**. This evocative name suggests that an electron can be considered as having an angular momentum arising from a spinning motion, rather like the daily rotation of a planet as it travels in its annual orbit around the sun. However, spin is a purely quantum mechanical property and differs considerably from its classical namesake.

Spin is described by two quantum numbers, s and m_s. The former is the analog of l for orbital motion, but it is restricted to the single, unchangeable value of $\frac{1}{2}$. The magnitude of the spin angular momentum is given by the expression $\{s(s+1)\}^{1/2}\hbar$, so for an electron this magnitude is fixed at $\frac{1}{2}3^{1/2}\hbar$ for any electron. The second quantum number, the **spin magnetic quantum number**, m_s, may take only two values, $+\frac{1}{2}$ (counterclockwise spin, imagined from above) and $-\frac{1}{2}$ (clockwise spin). These quantum numbers specify the orientation of the spin with respect to a chosen axis, and the component of spin angular momentum around an axis is limited to the values $\pm\frac{1}{2}\hbar$. Classically, we can picture these two spin states as the rotation of an electron on its axis either clockwise or counterclockwise. The two states are often represented by the two arrows ↑ ('spin-up', $m_s = +\frac{1}{2}$) and ↓ ('spin-down', $m_s = -\frac{1}{2}$) or by the Greek letters α and β, respectively.

Because the spin state of an electron must be specified if the state of the atom is to be specified fully, it is common to say that the state of an electron in a hydrogenic atom is characterized by four quantum numbers, namely n, l, m_l, and m_s (the fifth quantum number, s, is fixed at $\frac{1}{2}$).

> *The intrinsic spin angular momentum of an electron is defined by the two quantum numbers s and m_s; the latter can have one of two values. Four quantum numbers are needed to define the state of an electron in a hydrogenic atom.*

(d) The radial shapes of hydrogenic orbitals

The expressions for some of the hydrogenic orbitals are shown in Table 1.2. Because the Coulomb potential of the nucleus is spherically symmetric (it is proportional to Z/r and independent of orientation relative to the nucleus), the orbitals are best expressed in terms of the spherical polar coordinates defined in Fig. 1.8. In these coordinates, the orbitals all have the form

$$\psi_{nlm_l} = R_{nl}(r)Y_{lm_l}(\theta, \phi) \tag{5}$$

This formula and the entries in the table may look somewhat complicated, but they express the simple idea that a hydrogenic orbital can be written as the product of a function R of the

Table 1.2 Hydrogenic orbitals

(a) Radial wavefunctions
$$R_{nl}(r) = f(r)\,(Z/a_0)^{3/2}\mathrm{e}^{-\rho/2}$$
where a_0 is the Bohr radius $(0.53\,\text{Å})$ and $\rho = 2Zr/na_0$

n	l	$f(r)$
1	0	2
2	0	$(1/2\sqrt{2})(2-\rho)$
2	1	$(1/2\sqrt{6})\rho$
3	0	$(1/9\sqrt{3})(6-6\rho+\rho^2)$
3	1	$(1/9\sqrt{6})(4-\rho)\rho$
3	2	$(1/9\sqrt{30})\rho^2$

(b) Angular wavefunctions
$$Y_{l,m_l}(\theta, \phi) = (1/4\pi)^{1/2}y(\theta, \phi)$$

l	m_l	$y(\theta, \phi)$
0	0	1
1	0	$3^{1/2}\cos\theta$
1	± 1	$\mp(3/2)^{1/2}\sin\theta\,\mathrm{e}^{\pm i\phi}$
2	0	$(5/4)^{1/2}(3\cos^2\theta - 1)$
2	± 1	$\mp(15/4)^{1/2}\cos\theta\sin\theta\,\mathrm{e}^{\pm i\phi}$
2	± 2	$(15/8)^{1/2}\sin^2\theta\,\mathrm{e}^{\pm 2i\phi}$

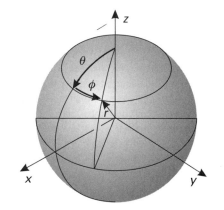

1.8 Spherical polar coordinates: r is the radius, θ (theta) the colatitude, and ϕ (phi) the azimuth.

radius and a function Y of the angular coordinates. The **radial wavefunction**, R, determines the variation of the orbital with distance from the nucleus. The **angular wavefunction**, Y, expresses the orbital's angular shape. Most of the time we shall use pictorial representations and not the expressions themselves. The locations where the radial wavefunction passes through zero are called **radial nodes**.[6] The planes on which the angular wavefunction passes through zero are called **angular nodes** or **nodal planes**. We shall see examples shortly.

The variations of the wavefunction with radius are shown in Fig. 1.9 and Fig. 1.10. A $1s$ orbital, the wavefunction with $n = 1$, $l = 0$, and $m_l = 0$, decays exponentially with distance from the nucleus and never passes through zero. All orbitals decay exponentially at sufficiently great distances from the nucleus, but some orbitals oscillate through zero close to the nucleus, and thus have one or more radial nodes before beginning their final exponential decay. An orbital with quantum numbers n and l in general has $n - l - 1$ radial nodes independent of the value of m_l. This oscillation is evident in the $2s$ orbital, the orbital with $n = 2$, $l = 0$, and $m_l = 0$, which passes through zero once and hence has one radial node. A $3s$ orbital passes through zero twice and so has two radial nodes. A $2p$ orbital (one of the three orbitals with $n = 2$ and $l = 1$) has no radial nodes because its radial wavefunction does not pass through zero anywhere. However, a $2p$ orbital, like *all* orbitals other than s orbitals, is zero at the nucleus.[7] Although an electron in an s orbital may be found at the nucleus, an electron in any other type of orbital will not be found there. We shall soon see that this apparently minor detail, which is a consequence of the absence of orbital angular momentum when $l = 0$, is one of the key concepts for understanding the periodic table.

1.9 The radial wavefunctions of the $1s$, $2s$, and $3s$ hydrogenic orbitals. Note that the number of radial nodes is 0, 1, and 2, respectively. Each orbital has a nonzero amplitude at the nucleus (at $r = 0$); the amplitudes have been adjusted to match at $r = 0$.

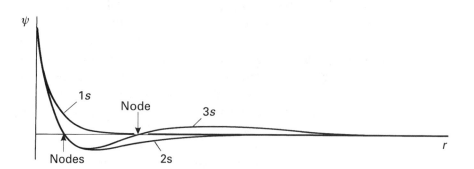

1.10 The radial wavefunctions of the $2p$ and $3p$ hydrogenic orbitals. Note that the number of radial nodes is 0 and 1, respectively. Each orbital has zero amplitude at the nucleus (at $r = 0$).

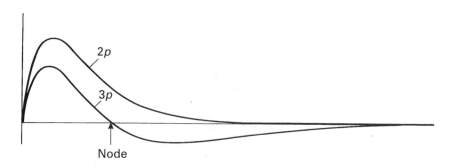

---

[6] All orbitals approach zero exponentially at large distances from the nucleus but, as they do not pass *through* zero at infinity, these zeros are not nodes.

[7] The zero at the nucleus is not a radial node because the radial wavefunction does not pass *through* zero. The wavefunction looks as though it is poised to pass through zero at $r = 0$; however, only non-negative values of the radius are physically significant.

> An s orbital has a nonzero amplitude at the nucleus; all other orbitals (those with $l > 0$) vanish at the nucleus.

(e) The radial distribution function

The Coulombic force that binds the electron is centered on the nucleus, so it is often of interest to know the probability of finding an electron at a given distance from the nucleus regardless of its direction. This information enables us to judge how tightly the electron is bound. The total probability of finding the electron in a spherical shell of radius r and thickness dr is the integral of $\psi^2 d\tau$ over all angles. This result is often written Pdr and, for a spherical wavefunction (one that is independent of angle),[8]

$$P = 4\pi r^2 \psi^2 \tag{6}$$

The function P is called the **radial distribution function**. If we know the value of P at some radius r (which we can find once we know ψ), then we can state the probability of finding the electron somewhere in a shell of thickness dr at that radius simply by multiplying P by dr. In general, a radial distribution function for an orbital in a shell of principal quantum number n has $n - l$ peaks, the outermost peak being the highest.

Because a $1s$ orbital decreases exponentially with distance from the nucleus, and r^2 increases, the radial distribution function of a $1s$ orbital goes through a maximum (Fig. 1.11). Therefore, there is a distance at which the electron is most likely to be found. In general, this distance decreases as the nuclear charge increases (because the electron is attracted more strongly to the nucleus). This most probable distance increases as n increases because, the higher the energy, the more likely it is that the electron will be found far from the nucleus. The most probable distance of an electron from the nucleus in the lowest energy state of a hydrogenic atom is at the point where P goes through a maximum. For a $1s$ electron in a hydrogenic atom of atomic number Z this maximum occurs at

$$r_{max} = \frac{a_0}{Z} \tag{7}$$

We see that the most probable distance of a $1s$ electron decreases as the atomic number increases.

> A radial distribution function gives the probability that an electron will be found at a given distance from the nucleus, regardless of the direction.

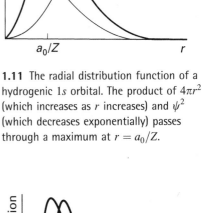

1.11 The radial distribution function of a hydrogenic $1s$ orbital. The product of $4\pi r^2$ (which increases as r increases) and ψ^2 (which decreases exponentially) passes through a maximum at $r = a_0/Z$.

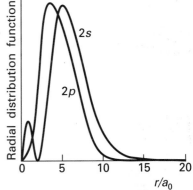

1.12 The radial distribution functions of hydrogenic orbitals. Although the $2p$ orbital is *on average* closer to the nucleus (note where its maximum lies), the $2s$ orbital has a high probability of being close to the nucleus on account of the inner maximum.

Example 1.2 Interpreting radial distribution functions

Figure 1.12 shows the radial distribution functions for $2s$ and $2p$ hydrogenic orbitals. Which orbital gives the electron a greater probability of close approach to the nucleus?

Answer The radial distribution function of a $2p$ orbital approaches zero near the nucleus faster than that of a $2s$ electron. This difference is a consequence of the fact that a $2p$ orbital has zero amplitude at the nucleus on account of its orbital angular momentum. Thus, the $2s$ electron has a greater probability of close approach to the nucleus.

Self-test 1.2 Which orbital, $3p$ or $3d$, gives an electron a greater probability of being found close to the nucleus?

---

[8]   The corresponding expression for a nonspherical orbital, one with $l > 0$, is $P = r^2 R^2$, where R is the radial wavefunction for the orbital.

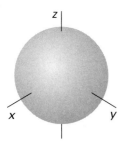

1.13 The spherical boundary surface of an *s* orbital.

(f) The angular shapes of atomic orbitals

An *s* orbital has the same amplitude at a given distance from the nucleus whatever the angular coordinates of the point of interest: that is, an *s* orbital is spherically symmetrical. The orbital is normally represented by a spherical surface with the nucleus at its center. The surface is called the **boundary surface** of the orbital, and defines the region of space within which there is a high (typically 75 per cent) probability of finding the electron. The boundary surface of any *ns* orbital is spherical (Fig. 1.13).

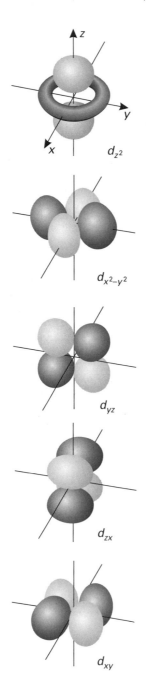

d_{z^2}

$d_{x^2-y^2}$

d_{yz}

d_{zx}

d_{xy}

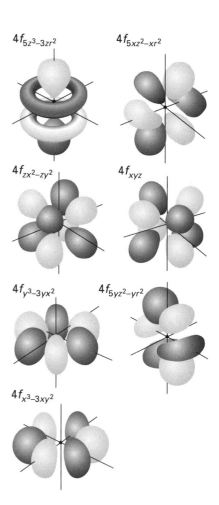

$4f_{5z^3-3zr^2}$ $4f_{5xz^2-xr^2}$

$4f_{zx^2-zy^2}$ $4f_{xyz}$

$4f_{y^3-3yx^2}$ $4f_{5yz^2-yr^2}$

$4f_{x^3-3xy^2}$

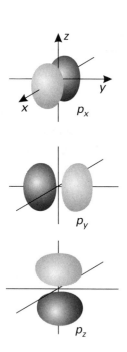

p_x

p_y

p_z

1.14 The boundary surfaces of *p* orbitals. Each orbital has one nodal plane running through the nucleus. For example, the nodal plane of the p_z orbital is the *xy*-plane. The lightly shaded lobe has a positive amplitude; the more darkly shaded one is negative.

1.15 One representation of the boundary surfaces of the *d* orbitals. Four of the orbitals have two perpendicular nodal planes that intersect in a line passing through the nucleus. In the d_{z^2} orbital, the nodal surface forms two cones that meet at the nucleus.

1.16 One representation of the boundary surfaces of the *f* orbitals. Other representations (with different shapes) are also sometimes encountered.

All orbitals with $l > 0$ have amplitudes that vary with the angle. In the most common graphical representation, the boundary surfaces of the three p orbitals of a given shell are identical apart from the fact that their axes lie parallel to each of the three different cartesian axes centered on the nucleus, and each one possesses an angular node passing through the nucleus (Fig. 1.14). This representation is the origin of the labels p_x, p_y, and p_z, which are alternatives to the m_l values as labels for individual orbitals.[9]

As in Fig. 1.14, we depict the positive and negative amplitudes of the wavefunction by different shading: positive amplitude will be shown by light gray or white and negative by dark gray. Each p orbital has a single nodal plane. A p_z orbital, for instance, is proportional to $\cos \theta$ (see Table 1.2), so its wavefunction vanishes everywhere on the plane corresponding to $\theta = 90°$ (the xy-plane). An electron will not be found anywhere on a nodal plane. A nodal plane cuts through the nucleus and separates the regions of positive and negative sign of the wavefunction.

The boundary surfaces and labels we use for the d and f orbitals are shown in Fig. 1.15 and Fig. 1.16, respectively. Note that a typical d orbital has two nodal planes which intersect at the nucleus, and a typical f orbital has three nodal planes.

> *The boundary surface of an orbital indicated the region within which the electron is most likely to be found; orbitals with the quantum number l have l nodal planes.*

Many-electron atoms

As remarked at the start of the chapter, a 'many-electron atom' is an atom with more than one electron, so even He, with two electrons, is technically a many-electron atom. The exact solution of the Schrödinger equation for an atom with N electrons would be a function of the $3N$ coordinates of all the electrons. There is no hope of finding exact formulas for such complicated functions; however, as computing power has increased it has become possible to perform numerical computations that provide ever more precise energies and probability densities. The price of numerical precision, though, is the loss of the ability to visualize the solutions. For most of inorganic chemistry we rely on the **orbital approximation**, in which each electron occupies an atomic orbital that resembles those found in hydrogenic atoms.[10]

1.6 Penetration and shielding

It is quite easy to account for the electronic structure of the helium atom in its **ground state**, its state of lowest energy. According to the orbital approximation, we suppose that both electrons occupy an atomic orbital that has the same spherical shape as a hydrogenic $1s$ orbital, but with a more compact radial form: because the nuclear charge of helium is greater than in hydrogen, the electrons are drawn in toward the nucleus more closely than is the one electron of an H atom. The ground-state **configuration** of an atom is a statement of the orbitals its electrons occupy in its ground state. For helium, with two electrons in the $1s$ orbital, the ground-state configuration is denoted $1s^2$.

As soon as we come to the next atom in the periodic table, lithium ($Z = 3$), we encounter several major new features. The ground-state configuration is *not* $1s^3$. This configuration is forbidden by a fundamental feature of nature known as the **Pauli exclusion principle**:

- No more than two electrons may occupy a single orbital and, if two do occupy a single orbital, then their spins must be paired.

By 'paired' we mean that one electron spin must be ↑ and the other ↓; the pair is denoted ↑↓. Another way of expressing the principle is to note that, because an electron in an atom is

---

[9] Specifically, the p_x, p_y, and p_z orbitals are linear combinations of the orbitals labeled by m_l: p_z is identical to p_0, p_x is $(p_{-1} - p_{+1})/2^{1/2}$, and p_y is $i(p_{-1} + p_{+1})/2^{1/2}$.

[10] When we say that an electron 'occupies' an atomic orbital, we mean that it is described by the corresponding wavefunction.

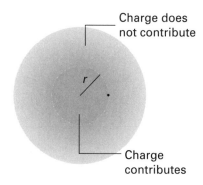

1.17 The electron indicated by the dot at the radius r experiences a repulsion from the total charge within the sphere of radius r; charge outside that radius has no net effect

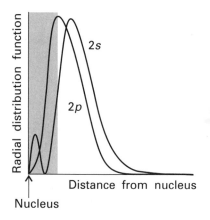

1.18 The penetration of a 2s electron through the inner core is greater than that of a 2p electron because the latter vanishes at the nucleus. Therefore, the 2s electrons are less shielded than the 2p electrons.

described by four variable quantum numbers, *no two electrons can have the same four quantum numbers*. The Pauli principle was introduced originally to account for the absence of certain transitions in the spectrum of atomic helium.

The proposed ground-state configuration of He, $1s^2$, is feasible (with the two electrons paired). However, the configuration of Li cannot be $1s^3$, for that would require all three electrons to occupy the same orbital, which is forbidden. Therefore, the third electron must occupy an orbital of the next higher shell, the shell with $n = 2$. The question that now arises is whether the third electron occupies a 2s orbital or one of the three 2p orbitals. To answer this question, we need to examine the energies of the two subshells and the effect of the other electrons in the atom. Although 2s and 2p orbitals have the same energy in a hydrogenic atom, spectroscopic data and detailed calculations show that that is not the case in a many-electron atom.

In the orbital approximation we treat the repulsion between electrons in an approximate manner by supposing that the electronic charge is spherically distributed around the nucleus. Then each electron moves in the attractive field of the nucleus plus this average repulsive charge distribution. According to classical electrostatics, the field that arises from a spherical distribution of charge is equivalent to the field generated by a single point charge at the center of the distribution. The magnitude of the point charge is equal to the total charge within a sphere with a radius equal to the distance of the location of interest to the center of the distribution (Fig. 1.17). In this approximation, an electron is assumed to experience an **effective nuclear charge**,[11] $Z_{eff}e$, determined by the total electronic charge with a sphere of radius equal to the mean distance of the electron from the nucleus. This effective nuclear charge depends on the values of n and l of the electron of interest, because electrons in different shells and subshells have different radial distribution functions. The reduction of the true nuclear charge to the effective nuclear charge is called **shielding**. The effective nuclear charge is sometimes expressed in terms of the true nuclear charge, Ze, and an empirical **shielding parameter**, σ, by writing $Z_{eff} = Z - \sigma$.

The closer to the nucleus that an electron can penetrate in an atom, the closer the value of Z_{eff} to Z itself, because the electron is repelled less by the other electrons present in the atom. With this point in mind, consider a 2s electron in the Li atom. There is a nonzero probability that the 2s electron **penetrates**, that is, has a nonzero probability of being found inside, the 1s shell and experiences the full nuclear charge. A 2p electron does not penetrate through the core so effectively because it has a nodal plane through the nucleus; therefore, it is more fully shielded from the nucleus by the electrons of the core (Fig. 1.18). We can conclude that a 2s electron has a lower energy (is bound more tightly) than a 2p electron, and therefore that the ground-state electron configuration of Li is $1s^2 2s^1$. The ground-state electron configuration of Li is commonly denoted $[He]2s^1$, where [He] denotes the atom's helium-like $1s^2$ core.

The pattern of energies in lithium, with 2s lower than 2p, is a general feature of many-electron atoms. This pattern can be seen from Table 1.3, which gives the values of Z_{eff} for a number of valence-shell atomic orbitals in the ground-state electron configuration of atoms. The typical trend in effective nuclear charge is an increase across a period, for in most cases the increase in nuclear charge in successive groups is not canceled by the additional electron. The values in the table also confirm that an s electron in the outermost shell of the atom is generally less shielded than a p electron of that shell. So, for example, $Z_{eff} = 5.13$ for a 2s electron in an F atom, whereas for a 2p electron $Z_{eff} = 5.10$, a lower value. Similarly, the effective nuclear charge is larger for an electron in an np orbital than for one in an nd orbital.

As a result of penetration and shielding, the order of energies in many-electron atoms is typically

$$ns < np < nd < nf$$

---

[11] In common usage, Z_{eff} itself is termed the 'effective nuclear charge'.

Table 1.3 Effective nuclear charges, Z_{eff}

	H							He
Z	1							2
$1s$	1.00							1.69
	Li	Be	B	C	N	O	F	Ne
Z	3	4	5	6	7	8	9	10
$1s$	2.69	3.68	4.68	5.67	6.66	7.66	8.65	9.64
$2s$	1.28	1.91	2.58	3.22	3.85	4.49	5.13	5.76
$2p$			2.42	3.14	3.83	4.45	5.10	5.76
	Na	Mg	Al	Si	P	S	Cl	Ar
Z	11	12	13	14	15	16	17	18
$1s$	10.63	11.61	12.59	13.57	14.56	15.54	16.52	17.51
$2s$	6.57	7.39	8.21	9.02	9.82	10.63	11.43	12.23
$2p$	6.80	7.83	8.96	9.94	10.96	11.98	12.99	14.01
$3s$	2.51	3.31	4.12	4.90	5.64	6.37	7.07	7.76
$3p$			4.07	4.29	4.89	5.48	6.12	6.76

Source: E. Clementi and D.L. Raimondi, *Atomic screening constants from SCF functions*, IBM Research Note NJ-27 (1963).

because, in a given shell, *s* orbitals are the most penetrating and *f* orbitals are the least penetrating. The overall effect of penetration and shielding is depicted in the energy-level diagram for a neutral atom shown in Fig. 1.19.

Figure 1.20 summarizes the energies of the orbitals through the periodic table. The effects are quite subtle, and the order of the orbitals depends strongly on the numbers of electrons

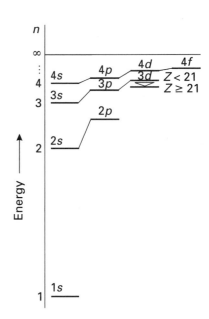

1.19 A schematic diagram of the energy levels of a many-electron atom with $Z < 21$ (as far as calcium). There is a change in order for $Z \geq 21$ (from scandium onward). This is the diagram that justifies the building-up principle (Section 1.7), with up to two electrons being allowed to occupy each orbital.

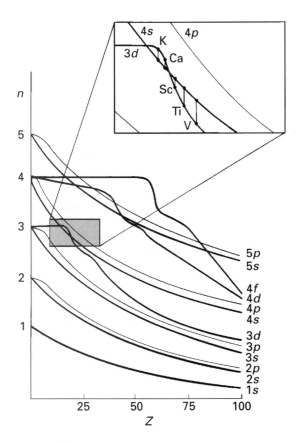

1.20 A more detailed portrayal of the energy levels of many-electron atoms in the periodic table. The inset shows a magnified view of the order near $Z = 20$, where the $3d$ elements begin.

present in the atom. For example, the effects of penetration are very pronounced for $4s$ electrons in K and Ca, and in these atoms the $4s$ orbitals lie lower in energy than the $3d$ orbitals. However, from Sc through Zn, the $3d$ orbitals in the neutral atom lie close to but lower than the $4s$ orbitals. In atoms from Ga ($Z = 31$) onward, the $3d$ orbitals lie well below the $4s$ orbital in energy, and the outermost electrons are unambiguously those of the $4s$ and $4p$ subshells.

Once we know the effective nuclear charge we can write approximate forms of the atomic orbitals and begin to make estimates of their extent and other properties. This was first done by J.C. Slater, who devised a set of rules for estimating the value of Z_{eff} for an electron in any atom, and using the value to write down an approximate wavefunction for any atomic orbital. Slater's rules have been superseded by more accurate calculated values (those given in Table 1.3). Some points to note from the values in the table are that across each period Z_{eff} for the outermost electrons increases in line with the atomic number. However, Z_{eff} for an electron in an s orbital is greater than that for the corresponding p orbital of the same atom. An additional point is that Z_{eff} for the outermost electrons of Period 3 elements are only slightly greater than those of Period 2 elements even though the nuclear charge itself is considerably greater.

> For a given principal quantum number, electrons approach the nucleus less closely as l increases and, as a result of the combined effects of penetration and shielding, the order of energy levels in a many-electron atom is $s < p < d < f$.

1.7 The building-up principle

The ground-state electron configurations of many-electron atoms are determined experimentally by spectroscopy and are summarized in Appendix 1. To account for them, we need to take into account both the effects of penetration and shielding on the energies of the orbitals and the role of the Pauli principle. The **building-up principle** (which is also called by its German name, the *Aufbau* principle, and is described below) is a procedure that leads to plausible ground-state configurations. It is not infallible, but it is an excellent starting point for the discussion. Moreover, as we shall see, it provides a theoretical framework for understanding the structure and implications of the periodic table.

(a) Ground-state electron configurations

In the building-up principle, orbitals of the neutral atoms are treated as being occupied in the order determined in part by the principal quantum number and in part by penetration and shielding:

Order of occupation: $1s\ 2s\ 2p\ 3s\ 3p\ 4s\ 3d\ 4p \cdots$

Each orbital can accommodate up to two electrons. Thus, the three orbitals in a p subshell can accommodate a total of six electrons and the five orbitals in a d subshell can accommodate up to ten electrons. The ground-state configurations of the first five elements are therefore expected to be

H	He	Li	Be	B
$1s^1$	$1s^2$	$1s^2 2s^1$	$1s^2 2s^2$	$1s^2 2s^2 2p^1$

This order agrees with experiment. When more than one orbital is available for occupation—such as when the $2p$ orbitals begin to be filled in B and C—we adopt **Hund's rule**:

- When more than one orbital has the same energy, electrons occupy separate orbitals and do so with parallel spins ($\uparrow\uparrow$).

The occupation of separate orbitals (such as a p_x orbital and a p_y orbital) can be understood in terms of the weaker repulsive interactions between electrons that occupy different regions of space (electrons in different orbitals) than between those that occupy the same region of space (the electrons occupy the same orbital). The requirement of parallel spins for electrons that do occupy different orbitals is a consequence of a quantum mechanical effect called **spin correlation**, the tendency of two electrons with parallel spins to stay apart from one another and hence to repel each other less.

It follows from the building-up principle that the ground-state configuration of C is $1s^2 2s^2 2p_x^1 2p_y^1$ (it is arbitrary which of the p orbitals is occupied first because they are degenerate: it is common to adopt the alphabetical order p_x, p_y, p_z) or, more briefly, $1s^2 2s^2 2p^2$. If we recognize the helium-like core ($1s^2$), an even briefer notation is $[He]2s^2 2p^2$, and we can think of the electronic structure of the atom as consisting of two paired $2s$ electrons and two parallel $2p$ electrons surrounding a closed helium-like core. The electron configurations of the remaining elements in the period are similarly

C $[He]2s^2 2p^2$ N $[He]2s^2 2p^3$ O $[He]2s^2 2p^4$ F $[He]2s^2 2p^5$ Ne $[He]2s^2 2p^6$

The $2s^2 2p^6$ configuration of neon is another example of a **closed shell**, a shell with its full complement of electrons. The configuration $1s^2 2s^2 2p^6$ is denoted [Ne] when it occurs as a core.

Example 1.3 Accounting for trends in effective nuclear charge

Refer to Table 1.3. Suggest a reason why the increase in Z_{eff} for a $2p$ electron is smaller between N and O than between C and N given that the configurations of the three atoms are C: $[He]2s^2 2p^2$, N: $[He]2s^2 2p^3$, and O: $[He]2s^2 2p^4$.

Answer On going from C to N, the additional electron occupies an empty $2p$ orbital. On going from N to O, the additional electron must occupy a $2p$ orbital that is already occupied by one electron. It therefore experiences a stronger repulsion and the increase in nuclear charge is more completely canceled than between C and N.

Self-test 1.3 Account for the larger increase in effective nuclear charge for a $2p$ electron on going from B to C compared with a $2s$ electron on going from Li to Be.

The ground-state configuration of Na is obtained by adding one more electron to a neon-like core, and is $[Ne]3s^1$, showing that it consists of a single electron outside a core. Now a similar sequence begins again, with the $3s$ and $3p$ orbitals complete at argon, with configuration $[Ne]3s^2 3p^6$, denoted [Ar]. Because the $3d$ orbitals are so much higher in energy, this configuration is effectively closed. Moreover, the $4s$ orbital is next in line for occupation, so the configuration of K is analogous to that of Na, with a single electron outside a noble-gas core: specifically, it is $[Ar]4s^1$. The next electron, for Ca, also enters the $4s$ orbital, giving $[Ar]4s^2$, the analog of Mg. However, it is observed that the next element, scandium, accommodates the added electron into a $3d$ orbital, and the d-block of the periodic table begins.

In the d-block, the d orbitals of the atoms are in the process of being occupied (according to the formal rules of the building-up principle). However, the energy levels in Figs 1.19 and 1.20 are for individual atomic orbitals and do not fully take into account interelectronic repulsions. For most of the d-block, the spectroscopic determination of ground states (and detailed computation) shows that it is advantageous to occupy *higher* energy orbitals (the $4s$ orbitals). The explanation is that the occupation of orbitals of higher energy can result in a reduction in the repulsions between electrons that would occur if the lower-energy $3d$

orbitals were occupied. It is essential to consider all contributions to the energy of a configuration, not merely the one-electron orbital energies.[12] Spectroscopic data show that the ground-state configurations of d-block atoms are of the form $3d^n4s^2$, with the $4s$ orbitals fully occupied despite individual $3d$ orbitals being lower in energy.

An additional feature is that, in some cases, a lower total energy may be obtained by forming a half-filled or filled d subshell even though that may mean moving an s electron into the d subshell. Therefore, close to the center of the d-block the ground-state configuration is likely to be d^5s^1 and not d^4s^2 (as for Cr). Close to the right of the d block the configuration is likely to be $d^{10}s^1$ rather than d^9s^2 (as for Cu). A similar effect occurs in the f-block, where f orbitals are being occupied, and a d electron may be moved into the f subshell so as to achieve an f^7 or an f^{14} configuration, with a net lowering of energy. For instance, the ground-state electron configuration of Gd is $[Xe]4f^75d^16s^2$ and not $[Xe]4f^65d^26s^2$.

The complication of the orbital energy not being a reliable guide to the total energy disappears when the $3d$ orbital energies fall well below that of the $4s$ orbitals, for then the competition is less subtle. The same is true of the cations of the d-block elements, where the removal of electrons reduces the complicating effects of electron–electron repulsions: all d-block cations have d^n configurations; the Fe^{2+} ion, for instance, has a d^6 configuration outside an argon-like closed shell. For the purposes of chemistry, the electron configurations of the d-block ions are more important than those of the neutral atoms. In later chapters (starting in Chapter 7), we shall see the great significance of the configurations of the d-metal ions, for the subtle modulations of their energies is the basis of important properties of their compounds.

> *For a given value of n, the occupation of atomic orbitals follows the order ns, np, (n − 1)d, (n − 2)f. Degenerate orbitals are occupied singly before being doubly occupied. Certain modifications of the order of occupation occur for d and f orbitals.*

....

Example 1.4 Deriving an electron configuration

Give the ground-state electron configurations of the Ti atom and the Ti^{3+} ion.

Answer For the atom, we add $Z = 22$ electrons in the order specified above, with no more than two electrons in any one orbital. This results in the configuration Ti $1s^22s^22p^63s^23p^64s^23d^2$, or $[Ar]4s^23d^2$, with the two $3d$ electrons in different orbitals with parallel spins. However, because the $3d$ orbitals lie below the $4s$ orbitals once we are past Ca, it is appropriate to reverse the order in which they are written. The configuration is therefore reported as $[Ar]3d^24s^2$. The configuration of the cation is obtained formally by removing *first* the s electrons and *then* as many d electrons as required. We must remove three electrons in all, two s electrons and one d electron. The configuration of Ti^{3+} is therefore $[Ar]3d^1$.

....

Self–test 1.4 Give the ground-state electron configurations of Ni and Ni^{2+}.

....

(b) The format of the periodic table

The modern form of the periodic table reflects the underlying electronic structure of the elements. We can now see, for instance, that the blocks of the table indicate the type of subshell currently being occupied according to the building-up principle. Each period of the table corresponds to the completion of the s and p subshells of a given shell. The period

....

[12] For a further discussion of this point, see L.G. Vanquickenborne, K. Pierloot, and D. Devoghel, *J. Chem. Educ.* **71**, 469–71 (1994).

number is the value of the principal quantum number n of the shell currently being occupied in the main groups of the table.

The group numbers are closely related to the number of electrons in the **valence shell**, the outermost shell of the atom. However, the precise relation depends on the group number G and the numbering system adopted. In the '1–18' numbering system recommended by IUPAC:

Block	Number of electrons in valence shell
s, d	G
p	$G - 10$

For the purpose of this expression, the 'valence shell' of a d-block element consists of the ns and $(n - 1)d$ orbitals, so a scandium atom has three valence electrons (two $4s$ and one $3d$ electron). The number of valence electrons for the p-block element selenium (Group 16) is $16 - 10 = 6$. Alternatively, in the roman numeral system, the group number is equal to the number of s and p valence electrons for the s- and p-blocks. Thus, selenium belongs to Group VI; hence it has six valence (s and p) electrons. Thallium belongs to Group III, so it has three valence s and p electrons.

> The blocks of the periodic table reflect the identity of the orbitals that are occupied last in the building-up process. The period number is the principal quantum number of the valence shell. The group number is related, as set out above, to the number of valence electrons.

1.8 Atomic parameters

Certain characteristic properties of atoms, particularly their sizes and the energies associated with the removal and addition of electrons, show periodic variations with atomic number. These atomic properties are of considerable importance for accounting for the chemical properties of the elements. A knowledge of the variation enables chemists to rationalize observations and predict likely chemical and structural behavior without resort to tabulated data for each element.

(a) Atomic and ionic radii

One of the most useful atomic properties of an element is the size of its atoms and ions. As we shall see in later chapters, geometrical considerations are central to the structures of many solids and individual molecules, and the distance of electrons from an atom's nucleus correlates well with the energy needed to remove them in the process of ion formation.

The quantum theory of the atom does not result in precise atomic or ionic radii because at large distances the wavefunction of the electrons falls off exponentially with increasing distance from the nucleus. However, despite this lack of a precise radius, we can expect atoms with numerous electrons to be larger, in some sense, than atoms that have only a few electrons. Such considerations have led chemists to propose a variety of definitions of atomic radius on the basis of empirical considerations.

The **metallic radius** of a metallic element is defined as half the experimentally determined distance between the nuclei of nearest-neighbor atoms in the solid ((1), but see Section 2.7 for a refinement of this definition). The **covalent radius** of a nonmetallic element is similarly defined as half the internuclear separation of neighboring atoms of the same element in a molecule (2). Multiple bonds are shorter than single bonds. The periodic trends in metallic and

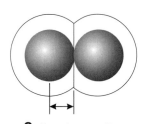

1 Metallic radius

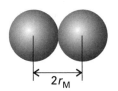

2 Covalent radius

Table 1.4 Atomic radii (in ångströms)*

Li	Be													B	C	N	O	F
1.57	1.12													0.88	0.77	0.74	0.66	0.64
Na	Mg													Al	Si	P	S	Cl
1.91	1.60													1.43	1.18	1.10	1.04	0.99

K	Ca	Sc	Ti	V	Cr	Mn	Fe	Co	Ni	Cu	Zn	Ga	Ge	As	Se	Br
2.35	1.97	1.64	1.47	1.35	1.29	1.37	1.26	1.25	1.25	1.28	1.37	1.53	1.22	1.21	1.17	1.14
Rb	Sr	Y	Zr	Nb	Mo	Tc	Ru	Rh	Pd	Ag	Cd	In	Sn	Sb	Te	I
2.50	2.15	1.82	1.60	1.47	1.40	1.35	1.34	1.34	1.37	1.44	1.52	1.67	1.58	1.41	1.37	1.33
Cs	Ba	Lu	Hf	Ta	W	Re	Os	Ir	Pt	Au	Hg	Tl	Pb	Bi		
2.72	2.24	1.72	1.59	1.47	1.41	1.37	1.35	1.36	1.39	1.44	1.55	1.71	1.75	1.82		

*The values refer to coordination number 12 (see Section 2.7). 1 Å = 100 pm.
Source: A.F. Wells, *Structural inorganic chemistry*, 5th edn. Clarendon Press, Oxford (1984).
Metallic radii are given for metals, single-bond covalent radii for the rest.

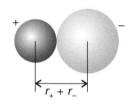

3 Ionic radius

covalent radii can be discerned in Table 1.4 and are illustrated in Fig. 1.21. We shall refer to metallic and covalent radii jointly as **atomic radii**.[13] The **ionic radius** of an element is related to the distance between the nuclei of neighboring cations and anions (**3**). An arbitrary decision had to be taken on how to apportion the cation–anion distance between these two ions. In one common scheme, the radius of the O^{2-} ion is taken to be $1.40\ \text{Å}$ (see Section 2.10 for a refinement of this definition). For example, the ionic radius of Mg^{2+} is obtained by subtracting $1.40\ \text{Å}$ from the internuclear separation of neighboring Mg^{2+} and O^{2-} ions in solid MgO. Some ionic radii are listed in Table 1.5.

 The data in Table 1.4 show that *atomic radii increase down a group*, and within the *s*- and *p*-blocks *they decrease from left to right across a period*. These trends are readily interpreted in terms of the electronic structure of the atoms. On descending a group, the valence electrons are found in orbitals of successively higher principal quantum number and hence larger mean radius. Across a period, the valence electrons enter orbitals of the same shell; however, the increase in effective nuclear charge across the period draws in the electrons and

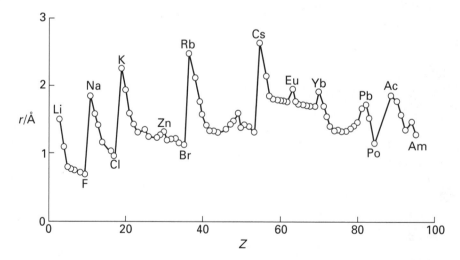

1.21 The variation of atomic radii through the periodic table. Note the contraction of radii following the lanthanides in Period 6. Metallic radii have been used for the metallic elements and covalent radii have been used for the nonmetallic elements.

[13] Atoms may be linked by single, double, and triple bonds, and different bond orders result in different covalent radii. The covalent radii in Fig. 1.21 refer to single bonds between atoms.

Table 1.5 Ionic radii (in ångströms)*

Li$^+$	Be^{2+}	B^{3+}			N^{3-}	O^{2-}	F$^-$
0.59(4) 0.76(6)	0.27(4)	0.12(4)			1.71	1.35(2) 1.38(4) 1.40(6) 1.42(8)	1.28(2) 1.31(4) 1.33(6)
Na$^+$	Mg^{2+}	Al^{3+}			P^{3-}	S^{2-}	Cl$^-$
0.99(4) 1.02(6) 1.16(8)	0.49(4) 0.72(6) 0.89(8)	0.39(4) 0.53(6)			2.12	1.84(6)	1.67(6)
K$^+$	Ca^{2+}	Ga^{3+}			As^{3-}	Se^{2-}	Br$^-$
1.38(6) 1.51(8) 1.59(10) 1.60(12)	1.00(6) 1.12(8) 1.28(10) 1.35(12)	0.62(6)			2.22	1.98(6)	1.96(6)
Rb$^+$	Sr^{2+}	In^{3+}	Sn^{2+}	Sn^{4+}		Te^{2-}	I$^-$
1.49(6) 1.60(8) 1.73(12)	1.16(6) 1.25(8) 1.44(12)	0.79(6) 0.92(8)	0.93(8)	0.74(6)		2.21(6)	20.6(6)
Cs$^+$	Ba^{2+}	Tl^{3+}					
1.67(6) 1.74(8) 1.88(12)	1.49(6) 1.56(8) 1.75(12)	0.88(6)					

*Numbers in parentheses are the coordination number of the ion. 1 Å = 100 pm.
Source: R.D. Shannon, *Acta Crystallogr.* **A32**, 751 (1976). The radius of the six-coordinate NH$_4^+$ ion is approximately 1.46 Å.

results in progressively more compact atoms (Table 1.4 and Fig. 1.21). The general increase in radius down a group and decrease across a period should be remembered as they correlate well with trends in many chemical properties.

Period 6 shows an interesting and important modification to these otherwise general trends. We see from Fig. 1.21 that the metallic radii in the third row of the *d*-block are very similar to those in the second row, and not significantly larger as might be expected given their considerably larger numbers of electrons. For example, the radii of molybdenum ($Z = 42$) and tungsten ($Z = 74$) are 1.40 Å and 1.41 Å, respectively, despite the latter having 32 more electrons. The reduction of radius below that expected on the basis of a simple extrapolation down the group is called the **lanthanide contraction**. The name points to the origin of the effect. The elements in the third row of the *d*-block (Period 6) are preceded by the elements of the first row of the *f*-block, the lanthanides, in which the 4*f* orbitals are being occupied. These orbitals have poor shielding properties.[14] The repulsions between electrons being added on crossing the *f*-block fail to compensate for the increasing nuclear charge, so Z_{eff} increases from left to right across a period. The dominating effect of the latter is to draw in all the electrons, and hence to result in a more compact atom.

A similar contraction is found in the elements that follow the *d*-block. For example, although there is a substantial increase in atomic radius between boron and aluminum (from 0.88 Å for B to 1.43 Å for Al), the atomic radius of gallium (1.53 Å) is only slightly greater than that of aluminum. As for the lanthanide contraction, this effect can be traced to the poor shielding characteristics of the *d* electrons of the elements earlier in the period. Although small variations

---

[14] A careful analysis (see D.R. Lloyd, *J. Chem. Educ.* **53**, 502 (1986)) supports the view that the poor shielding abilities of *f* electrons stem from their radial distribution, not, as is sometimes expressed, from their highly angular shapes.

due to greater $e^\ominus$-$e^\ominus$ repulsions

·Anions are larger than parent atoms

· Cations are smaller than parent atoms

in atomic radii may seem of little importance, in fact atomic radius plays a central role in the chemical properties of the elements, and small changes can have profound consequences.

A general feature apparent from Table 1.5 is that *all anions are larger than their parent atoms and all cations are smaller* (in some cases, markedly so). The increase in radius of an atom on anion formation is a result of the greater electron–electron repulsions that occur in an anion compared to those in the neutral atom. The smaller radius of a cation compared with its parent atom is a consequence not only of the reduction in electron–electron repulsions that follows electron loss, but also of the fact that cation formation typically results in the loss of the valence electrons. That loss often leaves behind only the much more compact core of the atom. Once these gross differences are taken into account, the variation in ionic radii through the periodic table mirrors that of the atoms.

> *Atomic radii increase down a group and, within the s- and p-blocks, decrease from left to right across a period. The lanthanide contraction is a decrease in atomic radius for elements following the f-block. All anions are larger than their parent atoms and all cations are smaller.*

(b) Ionization energy

The ease with which an electron can be removed from an atom is measured by its **ionization energy**, I, the minimum energy needed to remove an electron from a gas-phase atom:

$$A(g) \longrightarrow A^+(g) + e^-(g) \qquad I = E(A^+, g) - E(A, g) \qquad (8)$$

The **first ionization energy**, I_1, is the ionization energy of the least tightly bound electron of the neutral atom, the **second ionization energy**, I_2, is the ionization energy of the resulting cation, and so on.

Ionization energies are conveniently expressed in **electronvolts** (eV), where 1 eV is the energy acquired by an electron when it falls through a potential difference of 1 V. Because this energy is equal to $e \times (1\ \text{V})$, it is easy to deduce that 1 eV is equivalent to 96.485 kJ mol^{-1}.

The ionization energy of the H atom is 13.6 eV, so to remove an electron from an H atom is equivalent to dragging the electron through a potential difference of 13.6 V. In thermodynamic calculations it is often more convenient to use the **ionization enthalpy**, the standard enthalpy of the process in eqn 8, typically at 298 K. The molar ionization enthalpy is larger by $\frac{5}{2}RT$ than the ionization energy.[15] However, because RT is only 2.5 kJ mol^{-1} (corresponding to 0.026 eV) at room temperature and ionization energies are of the order of 10^2–10^3 kJ mol^{-1} (1–10 eV), the difference between ionization energy and enthalpy can often be ignored. In this book we express ionization energies in electronvolts and ionization enthalpies in kilojoules per mole.

To a large extent, the first ionization energy of an element is determined by the energy of the highest occupied orbital of its ground-state atom.[16] First ionization energies vary systematically through the periodic table (Table 1.6 and Fig. 1.22), being smallest at the lower left (near cesium) and greatest near the upper right (near helium). The variation follows the pattern of effective nuclear charge already mentioned in connection with the building-up principle, and (as Z_{eff} itself shows) there are some subtle modulations arising from the effect

---

[15] This difference stems from the change in temperature (at constant pressure) from $T = 0$ (assumed implicitly for I) to the temperature T (typically 298 K) to which the enthalpy value refers, and the replacement of 1 mol of gas particles by 2 mol at the latter temperature.

[16] This approximation is valid if the electrons remaining after ionization do not adjust their spatial distributions significantly. For the ionization of He, for example, the ionization energy is the difference in energy between He and He$^+$ and, as the electron in He$^+$ is much more tightly bound than it is in He, the ionization energy cannot be ascribed solely to the one-electron orbital energy of the neutral atom.

Table 1.6 First and second (and some higher) ionization energies of the elements (in electronvolts*)

H							He
13.60							24.59
							54.51
Li	Be	B	C	N	O	F	Ne
5.32	9.32	8.30	11.26	14.53	13.62	17.42	21.56
75.63	18.21	25.15	24.38	29.60	35.11	34.97	40.96
122.4	153.85	37.93	47.88	47.44	54.93	62.70	63.45
		259.30					
Na	Mg	Al	Si	P	S	Cl	Ar
5.14	7.64	5.98	8.15	10.48	10.36	12.97	15.76
47.28	15.03	18.83	16.34	19.72	23.33	23.80	27.62
71.63	80.14	28.44	33.49	30.18	34.83	39.65	40.71
		119.96					
K	Ca	Ga	Ge	As	Se	Br	Kr
4.34	6.11	6.00	7.90	9.81	9.75	11.81	14.00
31.62	11.87	20.51	15.93	18.63	21.18	21.80	24.35
45.71	50.89	30.71	34.22	28.34	30.82	36.27	36.95
Rb	Sr	In	Sn	Sb	Te	I	Xe
4.18	5.69	5.79	7.34	8.64	9.01	10.45	12.13
27.28	11.03	18.87	14.63	18.59	18.60	19.13	21.20
40.42	43.63	28.02	30.50	25.32	27.96	33.16	32.10
Cs	Ba	Tl	Pb	Bi	Po	At	Rn
3.89	5.21	6.11	7.42	7.29	8.42	9.64	10.74
25.08	10.00	20.43	15.03	16.69	18.66	16.58	
35.24	37.51	29.83	31.94	25.56	27.98	30.06	
	Ra						
	5.28						
	10.15						
	34.20						

*To convert to **kJ mol**$^{-1}$, multiply by 96.485. See Appendix 1 for a longer list.
Source: C.E. Moore, *Atomic energy levels*, NBS Circular 467 (1949–58).

of electron–electron repulsions within the same subshell. Ionization energies also correlate strongly with atomic radii, for elements that have small atomic radii generally have high ionization energies. The explanation of the correlation is that in a small atom an electron is close to the nucleus and experiences a strong Coulombic attraction.

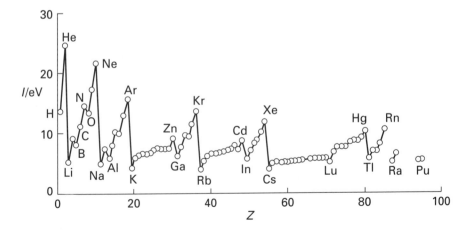

1.22 The variation of first ionization energies through the periodic table.

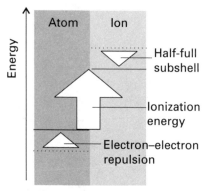

O [He]$2s^2 2p^4$ O$^+$ [He]$2s^2 2p^3$

1.23 The energy changes accompanying the ionization of an oxygen atom. Two factors contribute to the decrease between nitrogen and oxygen. One is the repulsion between the two electrons in a $2p$ orbital in the neutral atom. The other is the relatively low energy of the half-filled $2p$ subshell of the ionized atom.

Some differences in ionization energy can be explained quite readily. An example is the observation that the first ionization energy of boron is smaller than that of beryllium despite the former's higher nuclear charge. This anomaly is readily explained by noting that, on going to boron, the outermost electron occupies a $2p$ orbital and hence is less strongly bound than if it had entered a $2s$ orbital. As a result, the value of I_1 falls back to a lower value. The decrease between nitrogen and oxygen has a slightly different explanation. The configurations of the two atoms are

$$\text{N } [\text{He}]2s^2 2p_x^1 2p_y^1 2p_z^1 \qquad \text{O } [\text{He}]2s^2 2p_x^2 2p_y^1 2p_z^1$$

We see that, in an O atom, two electrons are present in a single $2p$ orbital. They are so close together that they repel each other strongly, and this strong repulsion offsets the greater nuclear charge. Another contribution to the difference is the lower energy of the O$^+$ ion on account of its $2s^2 2p^3$ configuration: as we have seen, a half-filled subshell has a relatively low energy (Fig. 1.23).

Example 1.5 Accounting for a variation in ionization energy

Account for the decrease in first ionization energy between phosphorus and sulfur.

Answer The ground-state configurations of the two atoms are

$$\text{P } [\text{Ne}]3s^2 3p_x^1 3p_y^1 3p_z^1 \qquad \text{S } [\text{Ne}]3s^2 3p_x^2 3p_y^1 3p_z^1$$

As in the analogous case of N and O, in the ground-state configuration of S, two electrons are present in a $3p$ orbital. They are so close together that they repel each other strongly, and this increased repulsion offsets the effect of the greater nuclear charge of S compared with P. As in the difference between N and O, the half-filled subshell of S$^+$ also contributes to the lowering of energy of the ion and hence to the smaller ionization energy.

Self-test 1.5 Account for the decrease in first ionization energy between fluorine and chlorine.

When building fluorine and neon, the last electrons enter orbitals that are already half full, and continue the trend in ionization energy from oxygen. The higher values of the ionization energies of these two elements reflect the high value of Z_{eff}. The value of I_1 falls back sharply from neon to sodium as the outermost electron occupies the next shell with an increased principal quantum number and hence further from the nucleus.

Another pattern in the ionization energies of the elements of considerable importance in inorganic chemistry is that *successive ionizations of a species require higher energies*. Thus, the second ionization energy of an element (the energy needed to remove an electron from the cation E$^+$) is higher than its first ionization energy, and its third ionization energy is higher still. The explanation is that, the higher the positive charge of a species, the greater the energy needed to remove an electron from the species. The difference in ionization energy is greatly magnified when the electron is removed from an inner shell of the atom (as is the case for the second ionization energy of lithium and any of its congeners) because the electron must then be extracted from a compact orbital in which it interacts strongly with the nucleus. The first ionization energy of lithium, for instance, is 5.3 eV, but its second ionization energy is 75.6 eV, more than ten times greater.

The pattern of successive ionization energies down a group is far from simple. Figure 1.24 shows the first, second, and third ionization energies of the boron group (Group 13/III). Although they lie in the expected order $I_1 < I_2 < I_3$, the shape of each curve does not suggest any simple trend. The lesson to be drawn is that, whenever an argument hangs on trends in small differences in ionization energies, it is always best to refer to actual numerical values rather than to guess a likely outcome.

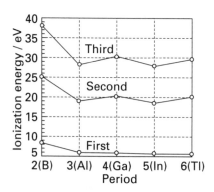

1.24 The first, second, and third ionization energies of the elements of Group 13/III. Successive ionization energies increase, but there is no clear pattern of ionization energies down the group.

First ionization energies vary systematically through the periodic table (Table 1.6 and Fig. 1.22), being smallest at the lower left (near cesium) and greatest near the upper right (near helium). Successive ionizations of a species require higher energies.

(c) Electron affinity

The **electron-gain enthalpy**, $\Delta_{eg}H^{\ominus}$, is the standard enthalpy change per mole of atoms when a gas-phase atom gains an electron:

$$A(g) + e^-(g) \longrightarrow A^-(g) \qquad \Delta_{eg}H^{\ominus}$$

Electron gain may be either exothermic or endothermic. Although the electron-gain enthalpy is the thermodynamically appropriate term, much of inorganic chemistry is discussed in terms of a closely related property, the **electron affinity**, E_a, of an element (Table 1.7), which is the energy difference

$$E_a = E(A, g) - E(A^-, g) \qquad (9)$$

At $T = 0$, the electron affinity is the negative of the electron-gain enthalpy.[17] A positive electron affinity indicates that the ion A^- has a lower, more negative, and hence more favorable, energy than the neutral atom A.[18] As for ionization energies and enthalpies, we shall quote electron affinities in electronvolts and electron-gain enthalpies in kilojoules per mole.

The electron affinity of an element is determined in large part by the energy of the lowest unfilled (or half-filled) orbital of the ground-state atom. This orbital is the second of the two **frontier orbitals** of an atom, its partner being the highest filled atomic orbital. The frontier orbitals are the site of many of the changes in electron distributions when bonds form, and we shall see more of their importance as the text progresses. An element has a high electron affinity if the additional electron can enter a shell where it experiences a strong effective nuclear charge. This is the case for elements toward the top right of the periodic table, as we have already explained. Therefore, elements close to fluorine (specifically oxygen, nitrogen,

Table 1.7 Electron affinities of the main-group elements (in electronvolts[*])

H							He
0.754							−0.5
Li	Be	B	C	N	O	F	Ne
0.618	⩽0	0.277	1.263	−0.07	1.461	3.399	−1.2
					−8.75		
Na	Mg	Al	Si	P	S	Cl	Ar
0.548	⩽0	0.441	1.385	0.747	2.077	3.617	−1.0
					−5.51		
K	Ca	Ga	Ge	As	Se	Br	Kr
0.502	+0.02	0.30	1.2	0.81	2.021	3.365	−1.0
Rb	Sr	In	Sn	Sb	Te	I	Xe
0.486	+0.05	0.3	1.2	1.07	1.971	3.059	−0.8

[*]To convert to $kJ\,mol^{-1}$, multiply by 96.485.
The first values refer to the formation of the ion X^- from the neutral atom X; the second value to the formation of X^{2-} from X^-.
Source: H. Hotop and W.C. Lineberger, *J. Phys. Chem. Ref. Data* **14**, 731 (1985).

....

[17] Specifically, $\Delta_{eg}H^{\ominus} = -E_a - \frac{5}{2}RT$. The contribution $\frac{5}{2}RT$ is commonly ignored.

[18] Be alert to the fact that some people identify electron affinities with electron-gain enthalpies, so, somewhat irrationally, a positive electron affinity then indicates that A^- has a higher, less favorable, energy than A.

and chlorine but not the noble gases) can be expected to have the highest electron affinities. The second electron-gain enthalpy, the enthalpy change for the attachment of a second electron to an initially neutral atom, is invariably positive because the electron repulsion outweighs the nuclear attraction.

| *Electron affinities are highest for elements near fluorine in the periodic table.*

Example 1.6 Accounting for the variation in electron affinity

Account for the large decrease in electron affinity between lithium and beryllium despite the increase in nuclear charge.

Answer The electron configurations of the two atoms are $[He]2s^1$ and $[He]2s^2$. The additional electron enters the $2s$ orbital of lithium, but it must enter the $2p$ orbital of beryllium, and hence is much less tightly bound. In fact, the nuclear charge is so well shielded in beryllium that electron gain is endothermic.

Self-test 1.6 Account for the decrease in electron affinity between carbon and nitrogen.

(d) Electronegativity

The **electronegativity**, χ (chi), of an element is the power of an atom of the element to attract electrons to itself when it is part of a compound. If an atom has a strong tendency to acquire electrons, it is said to be highly **electronegative** (like the elements close to fluorine). If it has a tendency to lose electrons (like the alkali metals), it is said to be **electropositive**. Electronegativities have numerous applications, and will be used in arguments throughout this text: they include a description of bond energies, the predictions of polarities of bonds and molecules, and the rationalization of the types of reactions that substances undergo.

The electronegativity of an element has been defined in many different ways and its precise interpretation is still the subject of debate.[19] Similarly, there have been a variety of attempts to formulate a quantitative scale of electronegativity. Linus Pauling's original formulation (which results in the values called χ_P in Table 1.8 and illustrated in Fig. 1.25) draws on concepts relating to the energetics of bond formation which will be dealt with in Chapter 3. A definition more in the spirit of this chapter in the sense that it is based on the properties of atoms was proposed by Robert Mulliken. He observed that, if an atom has a high ionization energy, I, and a high electron affinity, E_a, then it will be likely to acquire rather than lose electrons when it is part of a compound, and hence be classified as highly electronegative. Conversely, if its ionization energy and electron affinity are both low, then the atom will tend to lose electrons rather than gain them, and hence be classified as electropositive. These observations motivate the definition of the **Mulliken electronegativity**, χ_M, as the average value of the ionization energy and the electron affinity of the element:

$$\chi_M = \tfrac{1}{2}(I + E_a) \tag{10}$$

If both I and E_a are high, then the electronegativity is high; if both are low, then the electronegativity is low.

The complication in the definition of the Mulliken electronegativity is that the ionization energy and electron affinity in the definition relate to the **valence state**, the electron configuration the atom is supposed to have when it is part of a molecule. Hence, some calculation is required because the ionization energy and electron affinity to be used in calculating χ_M are mixtures of values for various actual states of the atom. We need not go

[19] To get the flavor of this debate see: R.G. Pearson, *Acc. Chem. Res.* **23**, 1 (1990), and L.C. Allen, *Acc. Chem. Res.* **23**, 23 (1990).

Table 1.8 Pauling (*italics*), χ_P, and Mulliken electronegativities, χ_M

H							He
2.20							
3.06							5.5
Li	Be	B	C	N	O	F	Ne
0.98	*1.57*	*2.04*	*2.55*	*3.04*	*3.44*	*3.98*	
1.28	1.99	1.83	2.67	3.08	3.22	4.43	4.60
Na	Mg	Al	Si	P	S	Cl	Ar
0.93	*1.31*	*1.61*	*1.90*	*2.19*	*2.58*	*3.16*	
1.21	1.63	1.37	2.03	2.39	2.65	3.54	3.36
K	Ca	Ga	Ge	As	Se	Br	Kr
0.82	*1.00*	*1.81*	*2.01*	*2.18*	*2.55*	*2.96*	*3.0*
1.03	1.30	1.34	1.95	2.26	2.51	3.24	2.98
Rb	Sr	In	Sn	Sb	Te	I	Xe
0.82	*0.95*	*1.78*	*1.96*	*2.05*	*2.10*	*2.66*	*2.6*
0.99	1.21	1.30	1.83	2.06	2.34	2.88	2.59
Cs	Ba	Tl	Pb	Bi			
0.79	*0.89*	*2.04*	*2.33*	*2.02*			

Source: Pauling values: A.L. Allred, *J. Inorg. Nucl. Chem.* **17**, 215 (1961); L.C. Allen and J.E. Huheey, *ibid.*, **42**, 1523 (1980); Mulliken values: L.C. Allen, *J. Am. Chem. Soc.* **111**, 9003 (1989). The Mulliken values have been scaled to the range of the Pauling values.

into the calculation, but the resulting values given in Table 1.8 may be compared with the Pauling values. The two scales are approximately in line. One reasonably reliable conversion between the two is

$$\chi_P = 1.35\chi_M^{1/2} - 1.37 \qquad (11)$$

Because the elements near fluorine are the ones with both high ionization energies and appreciable electron affinities, these elements have the highest Mulliken electronegativities. Because χ_M depends on atomic energy levels—and in particular on the location of the frontier orbitals (Fig. 1.26)—the electronegativity of an element is high if the two frontier orbitals of its atoms are low in energy.

Various alternative 'atomic' definitions of electronegativity have been proposed. A widely used scale, suggested by A.L. Allred and E. Rochow, is based on the view that electronegativity is determined by the electric field at the surface of an atom. As we have seen, an electron in an

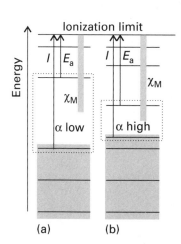

1.26 The interpretation of the electronegativity and polarizability of an element in terms of the energies of the frontier orbitals (the highest filled and lowest unfilled atomic orbitals). The electronegativity is high when the frontier orbitals lie low in energy. The polarizability is high when the frontier orbitals are close together. (a) Low χ, low α; (b) higher χ, higher α.

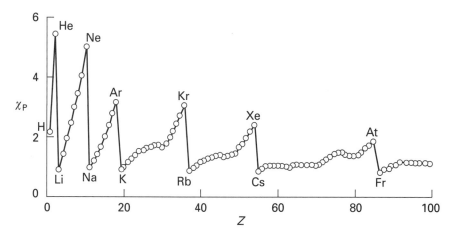

1.25 The variation of Pauling electronegativity through the periodic table.

atom experiences an effective nuclear charge Z_{eff}. The Coulombic potential at the surface of such an atom is proportional to Z_{eff}/r, and the electric field there is proportional to Z_{eff}/r^2. In the **Allred–Rochow definition** of electronegativity, χ_{AR} is assumed to be proportional to this field, with r taken to be the covalent radius of the atom:

$$\chi_{AR} = 0.744 + \frac{0.3590 Z_{eff}}{(r/\text{Å})^2} \tag{12}$$

The numerical constants have been chosen so that values comparable to Pauling electronegativities are obtained. According to the Allred–Rochow definition, elements with high electronegativity are those with high effective nuclear charge and the small covalent radius: these elements lie close to fluorine. The Allred–Rochow values parallel closely those of the Pauling electronegativities and are useful for discussing the electron distributions in compounds.

> *The electronegativity of an element is the power of an atom of the element to attract electrons when it is part of a compound; according to the Mulliken definition, the electronegativity of an element is the mean of the ionization energy and electron affinity and hence the mean energy of the frontier orbitals of the atom.*

(e) Polarizability

The **polarizability**, α, of an atom is its ability to be distorted by an electric field (such as that of a neighboring ion).[20] An atom or ion (most commonly, an anion) is highly polarizable if its electron distribution can be distorted readily, which is the case if unfilled atomic orbitals lie close to the highest-energy filled orbitals. That is, the polarizability is likely to be high if the separation of the frontier orbitals is small and the polarizability will be low if the separation of the frontier orbitals is large (see Fig. 1.26). Closely separated frontier orbitals are typically found for large, heavy atoms and ions, such as the atoms and ions of the heavier alkali metals and the heavier halogens, so these atoms and ions are the most polarizable. Small, light atoms, such as the atoms and ions near fluorine, typically have widely spaced energy levels, so these atoms an ions are least polarizable.

> *A polarizable atom or ion is one with frontier orbitals that lie close in energy; large, heavy atoms and ions tend to be highly polarizable.*

---

[20] The polarizability is related to the empirical quantity called *hardness*; see Section 5.12.

Further reading

P.A. Cox, *The elements: their origin, abundance, and distribution.* Oxford University Press (1989). A concise and readable account.

G.P. Wulfsberg, Periodic table trends in the properties of the elements. In *Encyclopedia of inorganic chemistry* (ed. R.B. King), Vol. 6, pp. 3079–91. Wiley, New York (1994).

R.J. Puddephatt and P.K. Monaghan, *The periodic table of the elements.* Oxford University Press (1986). An elementary survey of the structure of the periodic table and the trends in properties of elements.

P.W. Atkins, *Physical chemistry.* Oxford University Press and W.H. Freeman & Co, New York (1998). Chapters 11 and 12 give an introduction to the principles of atomic structure.

P.W. Atkins, *Quanta: a handbook of concepts.* Oxford University Press (1991). A collection of nonmathematical accounts of the concepts of quantum chemistry arranged like an encyclopedia.

J. Emsley, *The elements.* Oxford University Press (1998). A very useful collection of data and information in a handy format.

D.M.P. Mingos, *Essential trends in inorganic chemistry.* Oxford University Press (1998). An overview of structure and bonding trends throughout the periodic table.

Exercises

1.1 Write balanced equations for the following nuclear reactions (show emission of excess energy as a photon of electromagnetic radiation, γ): (a) $^{14}N + {}^4He$ to produce ^{17}O; (b) $^{12}C + p$ to produce ^{13}N; (c) $^{14}N + n$ to produce ^3H and ^{12}C. (The last reaction produces a steady state concentration of radioactive ^3H in the upper atmosphere.)

1.2 One possible source of neutrons for the neutron-capture processes mentioned in the text is the reaction of ^{22}Ne with α particles to produce ^{25}Mg and neutrons. Write the balanced equation for the nuclear reaction.

1.3 Without consulting reference material, draw the form of the periodic table with the numbers of the groups and the periods and identify the *s*-, *p*-, and *d*-blocks. Identify as many elements as you can. (As you progress through your study of inorganic chemistry, you should learn the positions of all the *s*-, *p*-, and *d*-block elements and associate their positions in the periodic table with their chemical properties.)

1.4 What is the ratio of the energy of a ground-state He$^+$ ion to that of a Be^{3+} ion?

1.5 The ionization energy of H is 13.6 eV. What is the difference in energy between the $n = 1$ and $n = 6$ levels?

1.6 What is the relation of the possible angular momentum quantum numbers to the principal quantum number?

1.7 How many orbitals are there in a shell of principal quantum number n? (Hint: begin with $n = 1, 2$, and 3 and see if you can recognize the pattern.)

1.8 Using sketches of 2*s* and 2*p* orbitals, distinguish between (a) the radial wavefunction, (b) the radial distribution function, and (c) the angular wavefunction.

1.9 Compare the first ionization energy of calcium with that of zinc. Explain the difference in terms of the balance between shielding with increasing numbers of *d* electrons and the effect of increasing nuclear charge.

1.10 Compare the first ionization energies of strontium, barium, and radium. Relate the irregularity to the lanthanide contraction.

1.11 The second ionization energies of some Period 4 elements are

Ca	Sc	Ti	V	Cr	Mn
11.87	12.80	13.58	14.65	16.50	15.64 eV

Identify the orbital from which ionization occurs and account for the trend in values.

1.12 Give the ground-state electron configurations of (a) C, (b) F, (c) Ca, (d) Ga^{3+}, (e) Bi, (f) Pb^{2+}.

1.13 Give the ground-state electron configurations of (a) Sc, (b) V^{3+}, (c) Mn^{2+}, (d) Cr^{2+}, (e) Co^{3+}, (f) Cr^{6+}, (g) Cu, and (h) Gd^{3+}.

1.14 Give the ground-state electron configurations of (a) W, (b) Rh^{3+}, (c) Eu^{3+}, (d) Eu^{2+}, (e) V^{5+}, (f) Mo^{4+}.

1.15 Account for the trends, element by element, across Period 3 in (a) ionization energy, (b) electron affinity, and (c) electronegativity.

1.16 Account for the fact that the two Group 5 elements niobium (Period 5) and tantalum (Period 6) have the same metallic radii.

1.17 Discuss the trend in electronegativities across Period 2 from lithium to fluorine. Can you account for the difference in details from the trend in ionization energy?

1.18 Identify the frontier orbitals of a Be atom.

1.19 Compare the broad trends in ionization energy, atomic radius, and electonegativity. Account for the parallels.

Problems

1.1 Show that an atom with the configuration ns^2np^6 is spherically symmetrical. Is the same true of an atom with the configuration ns^2np^3?

1.2 According to the Born interpretation, the probability of finding an electron in a volume element $d\tau$ is proportional to $\psi^*\psi d\tau$. (a) What is the most probable location of an electron in an H atom in its ground state? (b) What is its most probable distance from the nucleus, and why is this different? (c) What is the most probable distance of a $2s$ electron from the nucleus?

1.3 The ionization energies of rubidium and silver are respectively 4.18 eV and 7.58 eV. Calculate the ionization energies of an H atom with its electron in the same orbitals as in these two atoms and account for the differences in values.

1.4 When 58.4 nm radiation from a helium discharge lamp is directed on a sample of krypton, electrons are ejected with a velocity of 1.59×10^6 m s^{-1}. The same radiation ejects electrons from rubidium atoms with a velocity of 2.45×10^6 m s^{-1}. What are the ionization energies (in eV) of the two elements?

1.5 Survey the early and modern proposals for the construction of the periodic table. You should consider attempts to arrange the elements on helices and cones as well as the more practical two-dimensional surfaces. What, in your judgement, are the advantages and disadvantages of the various arrangements?

1.6 The decision about which elements should be identified as belonging to the f-block has been a matter of some controversy. A view has been expressed by W.B. Jensen (*J. Chem. Educ.* **59**, 634 (1982)). Summarize the controversy and Jensen's arguments.

1.7 The natural abundances of adjacent elements in the periodic table usually differ by a factor of ten or more. Explain this phenomenon.

1.8 Balance the following nuclear reaction: $^{246}_{96}Cm + ^{12}_{6}C \rightarrow$ _____ $+ 5\,^1_0n$

1.9 Explain how you would determine, using data you would look up in tables, whether the nuclear reaction in Problem 1.8 corresponds to a release of energy or not.

1.10 Consider the process of shielding in atoms, using Be as an example. What is being shielded? What is it shielded from? What is doing the shielding?

1.11 In general, ionization energies increase across a period from left to right. Explain why the second ionization energy of chromium is higher, not lower, than that of manganese.

1.12 Draw pictures of the two d orbitals in the xy-plane as flat projections in the plane of the paper. Label each drawing with the appropriate mathematical function, and include a labeled pair of cartesian coordinate axes. Label the orbital lobes correctly with $+$ and $-$ signs.

1.13 At various times the following two proposals have been advanced for the elements to be included in Group 3: (a) Sc, Y, La, Ac; (b) Sc, Y, Lu, Lr. Because ionic radii strongly influence the chemical properties of the metallic elements, it might be thought that ionic radii could be employed as one criterion for the periodic arrangement of the elements. Using this criterion describe which of these sequences is preferred.

2

The structures of simple solids

The packing of spheres

The structures of metals

Ionic solids

Further reading

Exercises

Problems

This chapter surveys the patterns adopted by atoms and ions in simple solids and explores the reasons why one arrangement may be preferred to another. We begin with the simplest model, in which atoms are represented by spheres and the structure of the solid is the outcome of stacking the spheres together densely. This close-packed arrangement provides a good description of many metals and a useful starting point for the discussion of alloys and ionic solids. We then see that the structures of many solids can be expressed in terms of a few commonly occurring structural types. This classification greatly simplifies the task of remembering structures of solids. Furthermore, some structural types are more likely when the bonding in the solid has partially covalent character and some of the trends in structural type therefore correlate with the positions of the constituent atoms in the periodic table. The final part of the chapter describes some of the parameters that are used to rationalize the trends in structure. These arguments also systematize the discussion of the thermal stabilities and solubilities of ionic solids.

The most elementary kind of chemical bonding is **metallic bonding** in which the atoms of a single element lose one or more electrons to a common sea. The strength of the bonding stems from the attraction between these electrons and the cations left behind. The familiar properties of metals stem from the nature of the bonding: metals are malleable and ductile because the electrons can adjust rapidly to relocation of the cations, and they are lustrous because the electrons can respond almost freely to an incident wave of electromagnetic radiation and reflect it. Because metallic bonding depends on electron loss to a common sea, it is characteristic of elements with low ionization energies, those on the left of the periodic table, through the *d*-block, and into part of the *p*-block close to the *d*-block.

Not much more complex than metallic bonding is **ionic bonding**, in which ions of different elements are held together in rigid, symmetrical arrays as a result of the attraction between their opposite charges. Ionic bonding also depends on electron loss, so it is found typically in compounds of metals with electronegative elements. However, there are plenty of exceptions: not all compounds of metals are ionic and some compounds of nonmetals (such as ammonium nitrate) are ionic.

The packing of spheres

The most significant feature of metals and ionic compounds is the arrangement adopted by the atoms and ions that make up a crystal, and initially we concentrate on this aspect. The arrangement of atoms or ions in simple structures can often be represented by different arrangements of hard spheres. The spheres used to describe metallic solids represent neutral atoms because each cation is still surrounded by its full complement of electrons. The spheres used to describe ionic solids represent the cations and anions because there has been an actual transfer of electrons from one type of atom to the other.

2.1 Unit cells and the description of crystal structure

A crystal of an element or compound can be regarded as constructed from regularly repeating **asymmetric units**, which may be atoms, molecules, or ions. The **space lattice** is the pattern formed by the points that represent the locations of these structural elements. More formally, the space lattice is a three-dimensional, infinite array of points, each of which is surrounded in an identical way by its neighboring points, and which defines the basic structure of the crystal. In some cases an asymmetric unit may be centered on the lattice point, but that is not necessary. The **crystal structure** itself is obtained by associating an identical asymmetric unit with each lattice point. A **unit cell** of the crystal is an imaginary parallel-sided region from which the entire crystal can be built up by purely translational displacements (Fig. 2.1).[1] There is a wide range of choices in the selection of a unit cell, as the two-dimensional example in the illustration shows, but it is generally preferable to choose the smallest cell that exhibits the greatest symmetry. Thus, the unit cell in Fig. 2.1a is preferred to that shown in Fig. 2.1b. A **primitive unit cell** has a lattice point at each vertex and nowhere else.

Many metallic and ionic solids can be regarded as constructed from entities such as atoms and ions that may be represented as hard spheres. If there are no directional covalent bonds, electrically neutral spheres are free to pack together as closely as geometry allows and hence adopt a **close-packed structure** in which there is the least waste of space. In close-packed structures of identical atoms each sphere has 12 neighbors, the greatest number that geometry allows, and the spheres occupy all but 26 per cent of the available space. When directional bonding is important, the resulting structures are no longer close-packed and the **coordination number** (C.N.) of an atom is less than 12. The attractions and repulsions between the charged spheres used to represent ions also have a profound effect on the packing pattern, but nevertheless we shall see that close-packed structures are often a good starting point for the discussion of these solids too.

> *A unit cell is a subdivision of a crystal that, when stacked together without rotation or reflection, reproduces the crystal. The structure of solids may be described in terms of the stacking of hard spheres representing the atoms or ions. In a close-packed structure, there is least waste of space.*

2.2 The close packing of spheres

We can picture the formation of close-packed structures of identical spheres by laying close-packed two-dimensional layers on top of each other. The first layer is started by placing a sphere in the indentation between two touching spheres, so making a triangle (**1**). The layer is then formed by continuing this process of laying spheres together in the indentations between those already in place. A complete close-packed layer consists of spheres in contact,

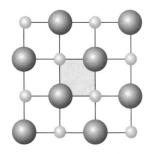

(a)

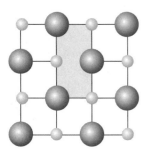

(b)

2.1 A two-dimensional solid and two choices of unit cell. The entire crystal is reproduced by translational displacements of either unit cell, but (a) is generally preferred because it displays the maximum symmetry of the structure, whereas (b) does not.

1 Close packing

---

[1]   Three-dimensional structures are often difficult to visualize on a two-dimensional page, but crystal models help to clarify the arrangements. Animation also clarifies three-dimensional structures. An excellent source of animated structures is the CD-ROM *Solid State Resources, JCE: Software*, University of Wisconsin, Department of Chemistry, 1101 University Ave., Madison, WI 53706-1396.

with each sphere having six nearest neighbors in the plane (Fig. 2.2). This arrangement is shown by the dark gray spheres in the illustration.

The second layer (mid gray) is formed by placing spheres in the dips of the first layer. The third layer (light gray) can be laid in either of two ways and hence can give rise to either of two **polytypes**, or structures that are the same in two dimensions (in this case, in the planes) but different in the third. The coordination number is 12 for each polytype.[2]

In one polytype, the spheres of the third layer lie directly above the spheres of the first. This ABAB . . . pattern of layers, where A denotes layers that have spheres directly above each other and likewise for B, gives a structure with a hexagonal unit cell, and hence is said to be **hexagonally close-packed** (hcp, Fig. 2.3a). In the second polytype, the spheres of the third layer are placed above the *gaps* in the first layer. The second layer covers half the holes in the first layer and the third layer lies above the remaining holes. This arrangement results in an ABCABC . . . pattern, where C denotes layers that have spheres directly above each other but above the spheres of neither the A nor the B layers. This pattern corresponds to a structure with a cubic unit cell (Fig. 2.3b). Hence it is **cubic close-packed** (ccp) or, more specifically, **face-centered cubic** (fcc; the origin of this name will become clear later).

> *The close-packing of identical spheres can result in a variety of polytypes, of which hexagonal and cubic close-packed structures are the most common.*

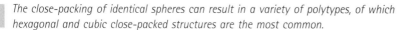

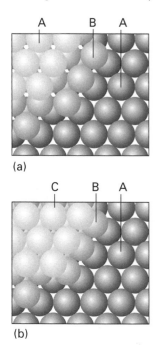

(a)

(b)

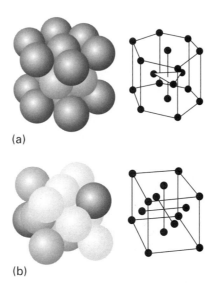

(a)

(b)

2.2 The formation of two close-packed polytypes. (a) The third layer reproduces the first, giving an ABA structure. (b) The third layer lies above the gaps in the first layer, giving an ABC structure. The different shadings identify the different layers of identical spheres.

2.3 (a) The hexagonal (hcp) unit of the ABAB . . . close-packed solid and (b) the cubic (fcc) unit of the ABCABC . . . polytype. The tints of the spheres correspond to the layers shown in Fig. 2.2.

2.3 Holes in close-packed structures

The feature of a close-packed structure modeled by a collection of hard spheres that enables us to extend the concept to describe structures more complicated than elemental metals is

---

[2] Later we shall see that many different polytypes can be formed; those described here are two very important special cases.

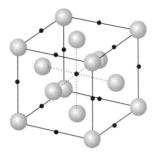

2.4 The locations of octahedral holes relative to the atoms in an fcc structure. The holes take names from the relative positions of lattice points around them.

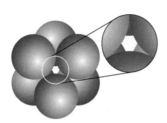

2 Octahedral hole

the existence of two types of **hole**, or unoccupied space. The unoccupied space in a close-packed structure amounts to 26 per cent of the total volume of a sample.[3] This figure is obtained by considering a ccp unit cell composed of spheres of radius r. The side of such a cell is $8^{1/2}r$, so its volume is $8^{3/2}r^3$. There are the equivalent of four complete spheres inside the unit cell, so the total volume they occupy is $4(\frac{4}{3}\pi r^3)$. The occupied fraction is therefore $16\pi/8^{3/2}3$, or 0.740. The type and distribution of holes are important because many structures, including those of some alloys and many ionic compounds, can be regarded as formed from an expanded close-packed arrangement in which additional atoms or ions occupy all or some of the holes.

An **octahedral hole** lies between two oppositely directed planar triangles of spheres in adjoining layers (**2**). If there are N atoms in the close-packed structure, there are N octahedral holes. The distribution of these holes in an fcc arrangement is shown in Fig. 2.4. This illustration also shows that the hole has local octahedral symmetry in the sense that it is surrounded by six nearest-neighbor spheres with their centers at the corners of an octahedron. If each hard sphere has radius r, then we show in the example below that each octahedral hole can accommodate another hard sphere with a radius no larger than $0.414r$.

Example 2.1 Calculating the size of an octahedral hole

Calculate the maximum radius of a sphere that may be accommodated in an octahedral hole in a close-packed solid composed of spheres of radius r.

Answer The structure of a hole, with the top spheres removed, is shown in (**3**). If the radius of a sphere is r and that of the hole is r_h, it follows from the Pythagorean theorem that

$$(r + r_h)^2 + (r + r_h)^2 = (2r)^2$$

and therefore that

$$(r + r_h)^2 = 2r^2 \qquad \text{or} \qquad r + r_h = 2^{1/2}r$$

That is,

$$r_h = (2^{1/2} - 1)r = 0.414r$$

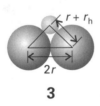

3

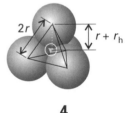

4

Self-test 2.1 Show that the maximum radius of a sphere that can fit into a tetrahedral hole (see below) is $r_h = ((3/2)^{1/2} - 1)r$; base your calculation on (**4**).

A **tetrahedral hole**, T (**5**), is formed by a planar triangle of touching spheres capped by a single sphere lying in the dip they form. There are two types of tetrahedral hole in any close-packed solid: in one the apex of the tetrahedron is directed up (T) and in the other the apex points down (T'). There are N tetrahedral holes of each type and $2N$ tetrahedral holes in all. In a close-packed structure of spheres of radius r, a tetrahedral hole can accommodate another

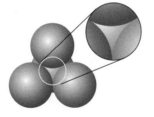

5 Tetrahedral hole

---

[3] The space represented by the holes is not empty in a real solid because electron density does not end as abruptly as the hard-sphere model suggests.

hard sphere of radius no greater than $0.225r$ (see Self-test 2.1). The location of tetrahedral holes in an fcc lattice is shown in Fig. 2.5; it can be seen from the illustration that each hole has four nearest-neighbor spheres arranged at the corners of a tetrahedron. Larger spheres may be accommodated in the holes if the close-packing of the parent structure is relaxed.

▌ *The structures of many solids can be discussed in terms of nearly close-packed structures in which tetrahedral or octahedral holes are occupied by other atoms. An octahedral hole can accommodate spheres of radius $0.414r$ and each type of tetrahedral hole can accommodate a sphere of radius $0.225r$.*

2.5 The locations of tetrahedral holes relative to the atoms in an fcc structure. There are two types of hole corresponding to tetrahedra in different orientations.

The structures of metals

X-ray diffraction studies (Box 2.1) reveal that many metallic elements have close-packed structures, indicating that their atoms have only a weak tendency toward directed covalency (Table 2.1). One consequence of this close packing is that metals often have high densities, because most mass is packed into the smallest volume. Indeed, the elements deep in the *d*-block, near iridium and osmium, include the densest solids known under normal conditions of temperature and pressure. A feature of almost all hcp metals is that they are distorted from pure hard-sphere packing: in most cases the interlayer spacing is slightly less than expected.

Table 2.1 The crystal structures adopted by some metallic elements at 25°C and 1 bar

Crystal structure	Elements
Hexagonal close-packed (hcp)	Be, Cd, Co, Mg, Ti, Zn
Cubic close-packed (fcc)	Ag, Al, Au, Ca, Cu, Ni, Pb, Pt
Body-centered cubic (bcc)	Ba, Cr, Fe, W, alkali metals
Primitive cubic (cubic-P)	Po

2.4 Polytypism

Which common close-packed polytype—hcp or fcc—a metal adopts depends on the details of the electronic structure of the element, the extent of interaction between second-nearest neighbors, and the residual effects of some directional character in their bonds. Indeed, a close-packed structure need not be either of the common ABAB . . . or ABCABC . . . polytypes. An infinite range of polytypes can in fact occur, for the close-packed layers may stack in a more complex manner.

Cobalt is an example of more complex polytypism. Above 500 °C, cobalt is fcc, but it undergoes a transition when cooled. The structure that results is a randomly stacked set (ABACBABABC . . .) of close-packed layers. In some samples of cobalt (and of SiC too), the polytypism is not random, for the sequence of planes repeats after several hundred layers. It is difficult to account for this behavior in terms of valence forces. The long-range repeat may be a consequence of a spiral growth of the crystal that requires several hundred turns before a stacking pattern is repeated.

▌ *Complex polytypism is sometimes observed for close-packed structures.*

2.5 Structures that are not close-packed

Not all metals are close-packed, and some other packing patterns use space nearly as efficiently. Even metals that are close-packed may undergo a phase transition to a less closely packed structure when they are heated and their atoms undergo large-amplitude vibrations.

Box 2.1 **X-ray diffraction**

X-ray diffraction is the most widely used and least ambiguous method for the precise determination of the positions of atoms in molecules and solids. X-ray structure determinations play a much more prominent role in inorganic chemistry than in organic chemistry because inorganic molecules and solids are structurally more diverse. Thus structural inferences from spectroscopic data frequently suffice for organic molecules, but spectroscopy is less successful for the unambiguous characterization of new inorganic compounds. Additionally, the bonding in inorganic molecules is more varied than in organic molecules, so inorganic chemists depend on bond distance and bond angle information to infer the nature of bonds.

A typical X-ray diffractometer (Fig. B2.1) consists of an X-ray source with a fixed wavelength, a mount for a single crystal of the compound being investigated, and an X-ray detector. The positions of the detector and the crystal, which is typically as small as

The pictorial display of an X-ray structure often has the appearance of Fig. B2.2. This type of computer-generated drawing is called an **ORTEP diagram** (an acronym for Oak Ridge Thermal Ellipsoid Program). The ORTEP diagram depicts the bond distances and bond angles of the molecule in the solid. In addition, the atoms are displayed as ellipsoids that indicate the direction and amplitude of their thermal motion. Since the restoring force for bond angle bending is generally less than for bond stretching, the ellipsoids are generally elongated in directions perpendicular to the bonds, as clearly shown by the N atoms of the CN^- ligands in the illustration.

Further reading

G.H. Stout and L.H. Jensen, *X-ray structure determination: a practical guide.* Wiley, New York (1989).

A. Domenico and I. Hargitai (ed.), *Accurate molecular structures: their determination and importance.* Oxford University Press (1992).

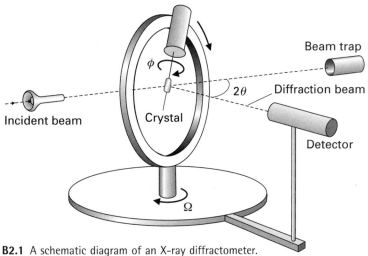

B2.1 A schematic diagram of an X-ray diffractometer.

0.2 mm on a side, are controlled by a computer. For certain orientations of the crystal relative to the X-ray beam the crystal diffracts the X-rays at a fixed angle and the intensity is measured when the detector is placed in the direction of this diffracted beam. Under computer control, the detector is scanned through each reflection while the intensities are recorded and stored. It is common to collect data on the intensities and positions of over 1000 reflections, and to obtain more than ten observed reflections for each structural parameter to be determined (positions of atoms and a range of locations associated with their thermal motion). A trial structure is chosen, either by means of a 'direct methods' program or by hints from the diffraction data together with a knowledge of physically reasonable arrangements of atoms. This structural model is refined by systematic shifts in the atom positions until satisfactory agreement between observed and calculated X-ray diffraction intensities is obtained.

B2.2 An ORTEP diagram of $[Re(CN)_7]^{4-}$ in $K_4[Re(CN)_7]\cdot 2H_2O$.

One common structure is the **body-centered cubic** (cubic-I or bcc) structure in which there is a sphere at the center of a cube with spheres at each corner (Fig. 2.6). Metals with this structure have a coordination number of 8. Although a bcc structure is less closely packed than the ccp and hcp structures (for which the coordination number is 12), the difference is not very great because the central atom has six second-nearest neighbors only 15 per cent further away. This arrangement leaves 32 per cent of the space unfilled compared with 26 per cent in the close-packed structures.

The least common metallic structure is the **primitive cubic** (cubic-P) structure (Fig. 2.7), in which spheres are located at the corners of a cube. The coordination number of a cubic-P structure is only 6. One form of polonium (α-Po) is the only example of this structure among the elements under normal conditions. Solid mercury, however, has a closely related structure: it is obtained from the simple cubic arrangement by stretching the cube along one of its body diagonals. Although bismuth normally has a layer structure, it converts to a cubic-P structure under pressure, and then to a bcc structure at higher pressures.

Metals that have structures more complex than those described so far can sometimes be regarded as slightly distorted versions of simple structures. Zinc and cadmium, for instance, have almost hcp structures, but the planes of close-packed atoms are separated by a slightly greater distance than in pure hcp. This difference suggests stronger bonding between the atoms in the plane: the bonding draws these atoms together and, in doing so, squeezes out the atoms of the neighboring layers.

A common non-close-packed structure is body-centered cubic; a primitive cubic structure is occasionally observed. Metals that have structures more complex than those described so far can sometimes be regarded as slightly distorted versions of simple structures.

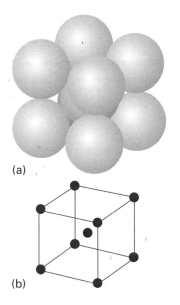

(a)

(b)

2.6 (a) A body-centered cubic (bcc) unit cell and (b) its lattice-point representation.

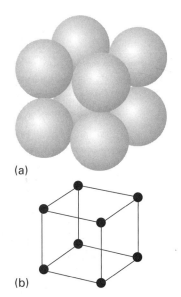

(a)

(b)

2.7 (a) A primitive cubic unit cell and (b) its lattice-point representation.

2.6 Polymorphism of metals

The low directionality of the bonds that metal atoms may form accounts for the wide occurrence of **polymorphism**, the ability to adopt different crystal forms, under different conditions of pressure and temperature. Iron, for example, shows several solid–solid phase transitions as it is heated and the atoms adopt new packing arrangements. It is often but not universally found that the most closely packed phases are thermodynamically favored at low temperatures and the less closely packed structures are favored at high temperatures.

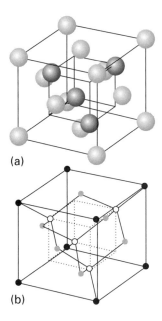

(a)

(b)

2.8 The structure of α-tin. Note its relation to the fcc unit cell in Fig. 2.5, with an additional Sn atom in half the tetrahedral holes. This structure is also that of diamond, silicon, and germanium.

The polymorphs of metals are generally (but not always systematically) labeled α, β, γ, ... with increasing temperature. Some metals revert to a low-temperature form at higher temperatures. Iron, for example, is polymorphic: α-Fe, which is bcc, occurs up to 906 °C, γ-Fe, which is fcc, occurs up to 1401 °C, and then α-Fe occurs again up to the melting point at 1530 °C. β-Fe, which is hcp, is formed at high pressures. A substance typically adopts a denser arrangement at high pressures because the Gibbs energy of a phase with a low molar volume (high density) increases more slowly with pressure than that of the phase with a high molar volume.[4] Therefore, at sufficiently high pressures the Gibbs energy of the denser phase will lie below that of the less dense phase, and the transition to the dense phase becomes spontaneous.

The room-temperature polymorph of tin is white tin (β-Sn): it undergoes a transition to gray tin (α-Sn) below 14.2 °C, but the conversion occurs at an appreciable rate only after prolonged exposure to a much lower temperature. Gray tin has a diamond-like structure (Fig. 2.8). The structure of white tin is unusual in that each atom has four nearest neighbors that are more distant than in gray tin, as would be expected for the high-temperature form. However, white tin is the appreciably denser polymorph (7.31 g cm^{-3} compared with 5.75 g cm^{-3} for gray tin). The explanation is that in white tin the second-nearest neighbors are closer than in gray tin, so overall the solid is more compact. A further point of interest, which shows that crystal structure can influence chemical properties through its influence on kinetics, is that, when tin dissolves in concentrated hydrochloric acid, white tin forms tin(II) chloride whereas gray tin forms tin(IV) chloride.

The bcc structure is common at high temperatures for metals that are close-packed at low temperatures because the increased amplitude of atomic vibrations in the hotter solid demands a less close-packed structure. For many metals (among them calcium, titanium, and manganese), the transition temperature is above room temperature. For others (among them lithium and sodium), the transition temperature is below room temperature. It is also found empirically that a bcc structure is favored by a small number of valence electrons per orbital. This observation suggests that a dense electron sea is needed to draw cations together into the close-packed arrangement and that the alkali metals do not have enough valence electrons for such close packing to be achieved.

> Polymorphism is a common consequence of the low directionality of metallic bonding. A bcc structure is common at high temperatures for metals that are close-packed at low temperatures because the increased amplitude of atomic vibrations in the hotter solid demand a less close-packed structure.

2.7 Atomic radii of metals

An informal definition of the atomic radius of a metallic element was given in Section 1.8 as half the distance between the centers of neighboring atoms in the solid. However, it is found that this distance generally increases with the coordination number of the lattice. For example, the same atom in lattices with different coordination numbers may appear to have different radii, and an atom of an element with coordination number 12 appears bigger than one with coordination number 8. In an extensive study of internuclear separations in a wide variety of polymorphic elements and alloys, V. Goldschmidt found that the average relative radii are as shown in Table 2.2.

It is desirable to put all elements on the same footing when comparing trends in their characteristics—that is, when comparing the *intrinsic* properties of their atoms rather than the properties that stem from their environment. Therefore, it is common to adjust the empirical internuclear separation to the value that would be expected if the element were in fact close-packed (with coordination number 12). Thus, the empirical atomic radius of Na is

Table 2.2 The variation of radius with coordination number

Coordination number	Relative radius
12	1
8	0.97
6	0.96
4	0.88

---

4 This difference follows from the thermodynamic relation $dG_m = V_m dp$, which shows that the molar Gibbs energy, G_m, is more responsive to pressure if the molar volume, V_m, is large.

1.85 Å, but that is for a structure in which the coordination number is 8. To adjust to 12-coordination we multiply this radius by $1/0.97 = 1.03$ and obtain 1.91 Å as the radius that an Na atom would have if it were in a close-packed structure.

Goldschmidt radii of the elements were in fact the ones listed in Table 1.5 as 'metallic radii' and used in the discussion of the periodicity of atomic radius (Section 1.8). The essential features of that discussion to bear in mind now are as follows:

- Atomic radii generally increase down a group.
- Atomic radii generally decrease from left to right across a period.

As remarked in Section 1.8, atomic radii reveal the presence of the lanthanide contraction in Period 6, with atomic radii of the elements that follow the lanthanides found to be smaller than simple extrapolation from earlier periods would suggest. As remarked there, this contraction can be traced to the poor shielding effect of f electrons. A similar contraction occurs across each row of the d-block.

> The Goldschmidt correction converts atomic radii to the value they would have in a close-packed structure. Atomic radii generally increase down a group and decrease from left to right across a period.

2.8 Alloys

An **alloy** is a blend of metals prepared by mixing the molten components and then cooling the mixture. Alloys may be homogeneous solid solutions, in which the atoms of one metal are distributed randomly among the atoms of the other, or they may be compounds with a definite composition and internal structure.

Solid solutions are sometimes classified as either substitutional or interstitial. A **substitutional solid solution** is a solid solution in which atoms of the solute metal occupy some of the locations of the solvent metal atoms (Fig. 2.9a). An **interstitial solid solution** is a solid solution in which the solute atoms occupy the interstices (the holes) between the solvent atoms (Fig. 2.9b). However, this distinction is not particularly fundamental, because interstitial atoms often lie in a definite array (Fig. 2.9c), and hence can be regarded as a substitutional version of another structure. A better viewpoint is that a solid solution is a new structure, and that its relation to the host structure may be largely coincidental. Some of the classic examples of alloys are brass (up to 40 per cent zinc in copper), bronze (a metal other than zinc or nickel in copper; casting bronze, for instance, is 10 per cent tin and 5 per cent lead), and stainless steel (over 12 per cent chromium in iron).

(a) Substitutional solid solutions

Substitutional solid solutions are generally formed if three criteria are fulfilled:

1 The atomic radii of the elements are within about 15 per cent of each other.
2 The crystal structures of the two pure metals are the same, for this indicates that the directional forces between the two types of atom are compatible with each other.
3 The electropositive characters of the two components are similar, for otherwise compound formation would be more likely.

Thus, although sodium and potassium are chemically similar and have bcc structures, the atomic radius of sodium (1.91 Å) is 19 per cent smaller than that of potassium (2.35 Å), and the two metals do not form a solid solution. On the other hand, copper and nickel, two neighbors late in the d-block, have similar electropositive character, similar crystal structures (both fcc), and similar atomic radii (Ni 1.25 Å, Cu 1.28 Å, only 2.3 per cent different), and form a continuous series of solid solutions, ranging from pure nickel to pure copper. Zinc, copper's other neighbor in Period 4, has a similar atomic radius (1.37 Å, 7 per cent larger), but it is hcp, not fcc. In this instance, zinc and copper are partially miscible and form solid solutions over only a limited concentration range.

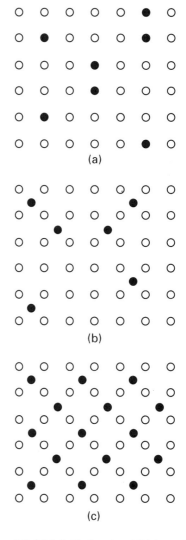

2.9 (a) Substitutional and (b) interstitial alloys. (c) In some cases an interstitial alloy may be regarded as a substitutional alloy derived from another lattice.

Substitutional solid solutions may be formed if the criteria set out above are fulfilled.

(b) Interstitial solid solutions of nonmetals

Interstitial solid solutions are formed by solute nonmetals (such as boron and carbon) that have atoms small enough to inhabit the interstices in the solvent structure. The small atoms enter the host solid with preservation of the crystal structure of the original metal; there is either a simple whole-number ratio of metal and interstitial atoms (as in Fe_3C), or the small atoms are distributed randomly in the available holes in the metal. The former substances are true compounds and the latter are solid solutions.

Considerations of size can help to decide whether the formation of a solid solution is likely. Thus, the largest solute atom that can enter a close-packed solid without distorting the structure appreciably is one that just fits an octahedral hole, which as we have seen has radius 0.414r (if the structure is modeled in terms of hard spheres). On geometrical grounds and with no reconstruction of the crystal structure, to accommodate hard-sphere H, B, C, or N atoms, the atomic radii of the host metal atoms must be no less than 0.89 Å, 2.13 Å, 1.86 Å, or 2.19 Å, respectively. The observation that the Period 4 metals close to nickel (which have atomic radii close to 1.3 Å) form an extensive series of interstitial solid solutions with boron, carbon, and nitrogen must indicate that specific bonding occurs between the host and the interstitial atom. Hence, as electronegativity considerations should lead us to expect, these substances are best regarded as examples of compounds of the nonmetals, and they are treated as such in Part 2.

Interstitial solid solutions may be formed by solute nonmetals with atoms small enough to inhabit the interstices in the parent structure.

(c) Intermetallic compounds

In contrast to the interstitial solid solutions of metals and nonmetals, where geometrical considerations are consistent with intuition, there is a class of solid solutions formed between two metals that are better regarded as actual compounds despite the similarity of the electronegativities of the metals. For instance, when some liquid mixtures of metals are cooled, they form phases with definite structures that are often unrelated to the parent structure. These phases are called **intermetallic compounds**. They include β-brass (CuZn) and compounds of composition $MgZn_2$, Cu_3Au, and Na_5Zn_{21}. Chemical formulas such as these show the *limiting* compositions at the boundaries of phase diagrams and are not necessarily applicable to a specific sample of the compound.

An intermetallic compound is a compound formed between metals.

Ionic solids

Ionic solids, such as sodium chloride and potassium nitrate, are often recognized by their brittleness for, instead of there being an adaptable, mobile electron sea, the electrons available from cation formation are pinned down to form a neighboring anion. They also commonly have moderately high melting points and many are soluble in polar solvents, particularly water. However, there are exceptions: calcium fluoride, CaF_2, for example, is a high-melting ionic solid, but it is insoluble in water. Ammonium nitrate, NH_4NO_3, is ionic, but melts at 170 °C (and is treacherously unstable). The existence of anomalies like MgO and NH_4NO_3 points to the need for a more fundamental definition of an 'ionic solid'.

An indication that a solid is ionic is provided by X-ray diffraction, for the coordination number of an ionic lattice is generally low: this low value is consistent with the low densities of ionic solids. However, although coordination number helps to distinguish metallic bonding from ionic, it does not distinguish between ionic and covalent bonding: covalent solids also have low coordination numbers (diamond, for example, has coordination number 4), and so we need a more fundamental criterion.

The classification of a solid as ionic is based on comparison of its properties with those of the **ionic model**, which treats the solid as an assembly of oppositely charged spheres that interact primarily by Coulombic forces (together with repulsions between the complete shells of the ions in contact). If the thermodynamic properties of the solid calculated on this model agree with experiment, then the solid may be ionic. However, it should be noted that many examples of coincidental agreement with the ionic model are known, so numerical agreement alone does not imply ionic bonding.

As we did for metals, we start by describing some common ionic structures in terms of the packing of hard spheres, but in this case the spheres have different sizes and opposite charges. After that, we shall see how to rationalize the structures in terms of the energetics of crystal formation. The structures we shall describe were obtained using X-ray diffraction and were among the first inorganic solids to be examined in this way.

> *The ionic model treats a solid as an assembly of oppositely charged spheres that interact primarily by Coulombic forces. If the thermodynamic properties of the solid calculated on this model agree with experiment, then the solid may be ionic.*

2.9 Characteristic structures of ionic solids

The ionic structures described in this section are prototypes of a wide range of solids. For instance, although the rock-salt structure takes its name from a mineral form of NaCl, it is characteristic of numerous other solids (Table 2.3). Many of the structures we describe can be regarded as derived from arrays in which the anions (sometimes the cations) stack together in fcc or hcp patterns and the counterions (the ions of opposite charge) occupy the octahedral or tetrahedral holes in the lattice. Throughout the following discussion, it will be helpful to refer back to Figs 2.4 and 2.5 to see how the structure being described is related to the hole patterns shown there. The close-packed layer usually needs to expand in order to accommodate the counterions, but this expansion is often a minor perturbation of the anion arrangement. Hence, the close-packed structure is often a good starting point for the discussion of ionic structures.

Table 2.3 Compounds with particular crystal structures

Crystal structure	Example*
Antifluorite	K_2O, K_2S, Li_2O, Na_2O, Na_2Se, Na_2S
Cesium chloride	**CsCl**, CaS, TlSb, CsCN, CuZn
Fluorite	**CaF₂**, UO_2, $BaCl_2$, HgF_2, PbO_2,
Nickel arsenide	**NiAs**, NiS, FeS, PtSn, CoS
Perovskite	**CaTiO₃**, $BaTiO_3$, $SrTiO_3$
Rock salt	**NaCl**, LiCl, KBr, RbI, AgCl, AgBr, MgO, CaO, TiO, FeO, NiO, SnAs, UC, ScN
Rutile	**TiO₂**, MnO_2, SnO_2, WO_2, MgF_2, NiF_2
Sphalerite (zinc blende)	**ZnS**, CuCl, CdS, HgS, GaP, InAs
Wurtzite	**ZnS**, ZnO, BeO, MnS, AgI,[†] AlN, SiC, NH_4F

*The substance in bold type is the one that gives its name to the structure.
[†]Silver iodide is also found with a sphalerite structure, which is metastable.

(a) The rock-salt structure

The **rock-salt structure** is based on an fcc array of bulky anions in which the cations occupy all the octahedral holes (Fig. 2.10). Alternatively, it can be viewed as a structure in which the

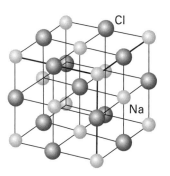

2.10 The rock-salt structure. Note its relation to the fcc structure in Fig. 2.4 with an anion in each octahedral hole. Alternatively, regard the fcc lattice as the location of anions, in which case the cations occupy the octahedral holes.

anions occupy all the octahedral holes in an fcc array of cations. It can be seen from the diagram that each ion is surrounded by an octahedron of six counterions. The coordination number of each type of ion is therefore 6, and the structure is said to have (6, 6)-coordination. In this notation, the first number in parentheses is the coordination number of the cation, and the second number is the coordination number of the anion.

To visualize the local environment of an ion in the rock-salt structure, we should note that the six nearest neighbors of the central ion of the cell shown in Fig. 2.10 lie at the centers of the faces of the cell and form an octahedron around the central ion. All six neighbors have a charge that is opposite to that of the central ion. The 12 second-nearest neighbors of the central ion, those next further away, are at the centers of the edges of the cell, and all have the same charge as the central ion. The eight third-nearest neighbors are at the corners of the unit cell, and have a charge opposite to that of the central ion.

When assessing the number of ions of each type in a unit cell we must take into account that any ions that are not fully inside the cell are shared by neighboring cells:

1 An ion in the *body* of a cell belongs entirely to that cell and counts as 1.
2 An ion on a *face* is shared by two cells and contributes $\frac{1}{2}$ to the cell in question.
3 An ion on an *edge* is shared by four cells and hence contributes $\frac{1}{4}$.
4 An ion at a *vertex* is shared by eight cells that share the vertex, and so contributes $\frac{1}{8}$.

In the unit cell shown in Fig. 2.10, there are the equivalent of four Na^+ ions and four Cl^- ions. Hence, each unit cell contains four NaCl formula units.

> The rock-salt structure consists of an fcc array of anions in which the cations occupy all the octahedral holes (or vice versa).

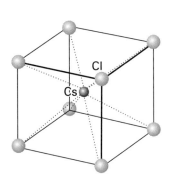

2.11 The cesium-chloride structure. Note that the corner ions, which are shared by eight cells, are surrounded by eight nearest-neighbor atoms. The cation occupies a cubic hole.

(b) The cesium-chloride structure

Much less common than the rock-salt structure is the **cesium-chloride structure** (Fig. 2.11), which is possessed by CsCl, CsBr, and CsI as well as some other compounds formed of ions of similar radii to these, including NH_4Cl (see Table 2.3). The cesium-chloride structure has a cubic unit cell with each vertex occupied by an anion and a cation occupying the 'cubic hole' at the cell center (or vice versa). The coordination number of both types of ion is 8. Their radii are so similar that this energetically highly favorable (8, 8) form of packing is feasible, with numerous counterions adjacent to a given ion.

> The cesium-chloride structure has a cubic unit cell with each vertex occupied by an anion and a cation at the cell center (or vice versa).

(c) The sphalerite structure

The **sphalerite structure** (Fig. 2.12), which is also known as the **zinc-blende structure** takes its name from a mineral form of ZnS. It is based on an expanded fcc anion lattice, but now the cations occupy one type of tetrahedral hole. Each ion is surrounded by four neighbors, so the structure has (4, 4)-coordination.

> The sphalerite structure consists of an expanded fcc anion lattice with cations occupying one type of tetrahedral hole.

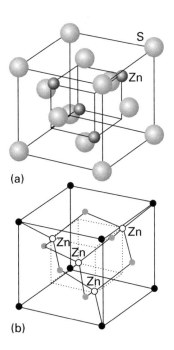

2.12 The sphalerite (zinc-blende) structure. Note its relation to the fcc structure in Fig. 2.5, with half the tetrahedral holes occupied by Zn^{2+} ions.

Example 2.2 Counting the number of ions in a unit cell

How many ions are there in the unit cell shown in the sphalerite structure shown in Fig. 2.12?

Answer An ion in the body of the cell belongs entirely to that cell and counts as 1. An ion on the face is shared by two adjacent unit cells and counts $\frac{1}{2}$. An ion on an edge is shared by four cells and so counts $\frac{1}{4}$. An ion at a vertex is shared by eight cells and so counts $\frac{1}{8}$. In the sphalerite structure, the count is

Location (share)	Number of cations	Number of anions	Contribution
Body (1)	4	0	4
Face ($\frac{1}{2}$)	0	6	3
Edge ($\frac{1}{4}$)	0	0	0
Vertex ($\frac{1}{8}$)	0	8	1
Total:	4	4	8

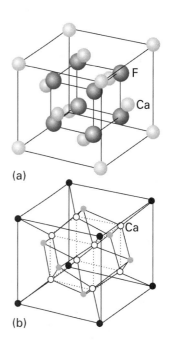

(a)

(b)

Note that there are four cations and four anions in the unit cell. This ratio is consistent with the chemical formula ZnS.

Self-test 2.2 Count the ions in the cesium-chloride unit cell shown in Fig. 2.11.

(d) The fluorite and antifluorite structures

The **fluorite structure** takes its name from its exemplar fluorite (CaF_2). In it, the Ca^{2+} cations lie in an expanded fcc array and the F^- anions occupy the two types of tetrahedral hole (Fig. 2.13).[5] The **antifluorite structure** is the inverse of the fluorite structure in the sense that the locations of cations and anions are reversed. The latter structure is shown by some alkali metal oxides, including K_2O. In it, the cations (which are twice as numerous as the anions) occupy both types of tetrahedral hole of the fcc array.[6]

In the fluorite structure, the anions in their tetrahedral holes have four nearest neighbors. The cation site is surrounded by a cubic array of eight anions. From an alternative viewpoint, the anions form a primitive cubic lattice and the cations occupy half the cubic holes in this lattice. Note the relation of this structure to the cesium-chloride structure, in which all the cubic holes are occupied. From either viewpoint, the lattice has (8, 4)-coordination, which is consistent with there being twice as many anions as cations. The coordination in the antifluorite structure is the opposite, namely (8, 4).

> In the fluorite structure, cations occupy half the cubic holes of a primitive cubic array of anions. Alternatively, the anions occupy both types of tetrahedral hole in an expanded fcc lattice of cations. In the antifluorite structure, the roles of the cations and anions are reversed.

2.13 The fluorite structure. This structure has an fcc array of cations and all the tetrahedral holes are filled with anions. Alternatively, view the structure as a simple cubic array of anions with cations in half the cubic holes. (a) and (b) are two different representations of the same structure.

(e) The wurtzite structure

The **wurtzite structure** (Fig. 2.14) takes its name from another polymorph of zinc sulfide. It differs from the sphalerite structure in being derived from an expanded hexagonally close-packed anion array rather than a face-centered cubic array, but as in sphalerite the cations occupy one type of tetrahedral hole. This wurtzite structure, which has (4, 4)-coordination, is possessed by ZnO, AgI, and one polymorph of SiC as well as several other compounds (Table 2.3). The local symmetries of the cations and anions are identical toward their nearest neighbors in wurtzite and sphalerite but differ at the second-nearest neighbors.

> The wurtzite structure is derived from an expanded hexagonally close-packed anion array with cations occupying one type of tetrahedral hole.

(f) The nickel-arsenide structure

The **nickel-arsenide structure** (NiAs, Fig. 2.15) is also based on an expanded, distorted hcp anion array, but the Ni atoms now occupy the octahedral holes and each As atom lies at the center of a trigonal prism of Ni atoms. This structure is adopted by NiS, FeS, and a number of other sulfides. The nickel-arsenide structure is typical of **MX** compounds that contain polarizable cations in combination with polarizable anions, which suggests that it is favored

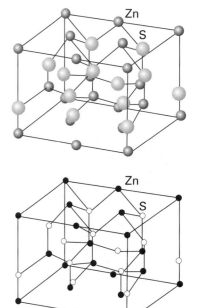

2.14 The wurtzite structure, which is derived from the hcp structure (Fig. 2.3a).

---

[5]   Fluorite is so called because it melts and flows on being heated with a mineralogist's blowpipe, and so can be distinguished from gemstones.

[6]   Recall that, if there are N atoms, then there are $2N$ tetrahedral holes.

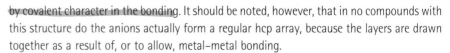

by covalent character in the bonding. It should be noted, however, that in no compounds with this structure do the anions actually form a regular hcp array, because the layers are drawn together as a result of, or to allow, metal–metal bonding.

> The nickel-arsenide structure is based on an expanded, distorted hcp anion array, with the cations occupying the octahedral holes.

(g) The rutile structure

The **rutile structure** (Fig. 2.16) takes its name from rutile, a mineral form of titanium(IV) oxide, TiO_2. It is also an example of an hcp anion lattice, but now the cations occupy only half the octahedral holes. This arrangement results in a structure that reflects the strong tendency of a Ti atom to acquire octahedral coordination. Each Ti atom is surrounded by six O atoms and each O atom is surrounded by three Ti atoms; hence the rutile structure has (6,3)-coordination. The principal ore of tin, cassiterite (SnO_2), has the rutile structure, as do a number of fluorides (Table 2.3).

> The rutile structure consists of an hcp anion lattice with cations occupying half the octahedral holes.

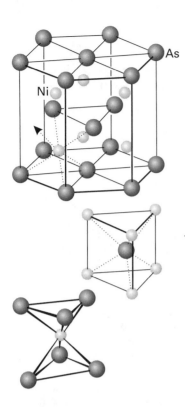

2.15 The nickel-arsenide structure, another structure derived from the hcp structure (Fig. 2.3a). Note the prismatic and trigonal-antiprismatic local symmetries of the As and Ni atoms, respectively.

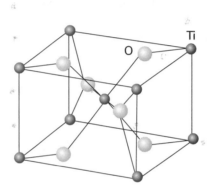

2.16 The rutile structure. Rutile itself is one polymorph of TiO_2.

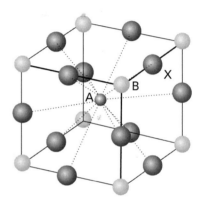

2.17 The perovskite structure, ABX_3. Perovskite itself is $CaTiO_3$.

(h) The perovskite structure

The mineral perovskite $CaTiO_3$ has a structure that is the prototype of many **ABX_3** solids (Table 2.3), particularly oxides. In its ideal form, the perovskite structure (Fig. 2.17) is cubic with the A atoms surrounded by 12 X atoms and the B atoms surrounded by 6 X atoms. The sum of the charges on the A and B ions must be 6, but that can be achieved in several ways ($A^{2+}B^{4+}$ and $A^{3+}B^{3+}$ among them), including the possibility of mixed oxides of formula $A(B_{0.5}B'_{0.5})O_3$, as in $La(Ni_{0.5}Ir_{0.5})O_3$. The perovskite structure is closely related to the materials that show interesting electrical properties, such as piezoelectricity, ferroelectricity, and high-temperature superconductivity (see Chapter 18).

> The perovskite structure is cubic with the A atoms (of ABX_3) surrounded by 12 X atoms and the B atoms surrounded by 6 X atoms.

Example 2.3 Interpreting a prototype structure

What is the coordination number of a Ti atom in perovskite?

Answer We need to imagine eight of the unit cells in Fig. 2.17 stacked together with a Ti atom shared by them all. A local fragment of the structure is shown in Fig. 2.18; we can see at once that there are six O atoms around the central Ti atom; so the coordination number of Ti in perovskite is 6.

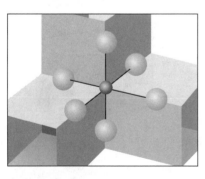

2.18 The local coordination environment of a Ti atom in perovskite.

Self-test 2.3 What is the coordination number of a Ti atom in rutile?

(i) Spinels

Spinel itself is $MgAl_2O_4$, and spinels in general have the formula AB_2O_4. The **spinel structure** consists of an fcc array of O^{2-} ions in which the A cations occupy one-eighth of the tetrahedral holes and the B cations occupy the octahedral holes (Fig. 2.19). Spinels are sometimes denoted $A[B_2]O_4$, the square brackets denoting the species that occupies the octahedral holes. Examples of compounds that have spinel structures include some simple *d*-block oxides, such as Fe_3O_4, Co_3O_4, and Mn_3O_4. Note that in these structures A and B are the same element. There are also **inverse spinels**, in which the cation distribution is $B[AB]O_4$. Spinels and inverse spinels are discussed again in Section 18.15.

> *The spinel structure, AB_2O_4, consists of an fcc array of O^{2-} ions in which the A cations occupy one-eighth of the tetrahedral holes and the B cations occupy the octahedral holes.*

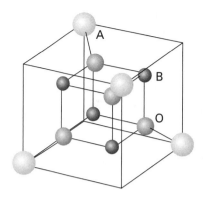

2.19 The spinel structure, AB_2O_4, consists of an fcc array of oxide ions in which the A cations occupy one-eighth of the tetrahedral holes and the B cations occupy the octahedral holes.

2.10 The rationalization of structures

The thermodynamic properties of such ionic solids can be treated very simply in terms of the ionic model. However, a model of a solid in terms of charged spheres interacting Coulombically is likely to be quite crude, and we should expect significant departures from its predictions because many solids are more covalent than ionic. Even conventional 'good' ionic solids, such as the alkali metal halides, have significant covalent character. Nevertheless, the ionic model provides an attractively simple and effective scheme for correlating many properties.

(a) Ionic radii

A difficulty that confronts us at the outset is the meaning of the term **ionic radius**. As remarked in Section 1.8, it is necessary to apportion the single internuclear separation of nearest-neighbor ions between the two different species (for example, an Na^+ ion and a Cl^- ion in contact).[7] The most direct way to solve the problem is to make an assumption about the radius of one ion, and then to use that value to compile a set of self-consistent values for all other ions. The O^{2-} ion has the advantage of being found in combination with a wide range of elements. It is also reasonably unpolarizable, so its size does not vary much as the identity of the accompanying cation is changed. In a number of compilations, therefore, the values are based on $r(O^{2-}) = 1.40\ \text{Å}$ (140 pm). This value not only produces a consistent set of radii but also satisfies a number of theoretical criteria proposed by Linus Pauling. However, this value is by no means sacrosanct: a set of values compiled by Goldschmidt was based on $r(O^{2-}) = 1.32\ \text{Å}$.

The lesson at this stage is that for certain purposes (such as for predicting the sizes of unit cells) ionic radii can be helpful, but they are reliable only if they are all based on the same fundamental choice (such as the value 1.40 Å for O^{2-}). *If values of ionic radii are used from different sources, it is essential to verify that they are based on the same convention.*

An additional complication that was first noted by Goldschmidt is that ionic radii increase with coordination number (Fig. 2.20, as happens with the atomic radius too). Hence, when comparing ionic radii, we should compare like with like, and use values for a single coordination number (typically 6).

The problems of the early workers have only partly been resolved by developments in X-ray diffraction. It is now possible to measure the electron density between two neighboring ions,

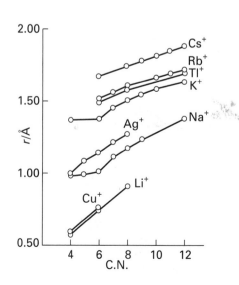

2.20 The variation of ionic radius with coordination number.

---

[7] A good survey of radii and their determination is given by R.D. Shannon in *Encyclopedia of inorganic chemistry* (ed. R.B. King), p. 929. Wiley, New York (1994).

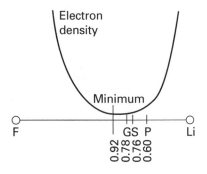

2.21 The variation in electron density along the Li–F axis in LiF. The point P indicates the Pauling radius of the ions, G the original (1927) Goldschmidt radius, and S the Shannon radius.

locate the minimum, and identify that with the boundary between the two ions. However, as we see from Fig. 2.21, the electron density passes through a very broad minimum, and its exact location may be very sensitive to experimental uncertainties and to the identities of the two neighbors. That being so, and in the general spirit of inorganic chemistry, it is still probably more useful to express the sizes of ions in a self-consistent manner than to seek calculated values of individual radii in certain combinations. Very extensive lists of self-consistent values, which have been compiled by analyzing X-ray data on thousands of compounds, particularly oxides and fluorides, are now available, and some were given in Table 1.5.

The general trends for ionic radii are the same as for atomic radii. Thus:

1 Ionic radii increase on going down a group. (The lanthanide contraction, Section 1.8, restricts the increase among the heaviest ions of the $5d$ metal ions.)
2 The radii of ions of the same charge decrease across a period.
3 When an ion can occur in environments with different coordination numbers, its radius increases as the coordination number increases.
4 If an element can form cations with different charge numbers, then for a given coordination number its ionic radius decreases with increasing charge number.
5 Because a positive charge indicates a reduced number of electrons, and hence a more dominant nuclear attraction, cations are usually smaller than anions for elements with similar atomic numbers.

> *Ionic radii are defined in terms of a choice for one species of ion. Ionic radii increase down a group, decrease across a period, increase with coordination number, and decrease with increasing charge number.*

(b) The radius ratio

A parameter that figures quite widely in the literature of inorganic chemistry, particularly in introductory texts, is the **radius ratio**, ρ, of the ions. The radius ratio is the ratio of the radius of the smaller ion $(r_<)$ to that of the larger $(r_>)$:

$$\rho = \frac{r_<}{r_>} \tag{1}$$

In many cases, $r_<$ is the cation radius and $r_>$ is the anion radius. The minimum radius ratio that can tolerate a given coordination number is then calculated by considering the geometrical problem of packing together spheres of different sizes. The results are listed in Table 2.4. It is argued that, if the radius ratio falls below the minimum given, then ions of opposite charge will not be in contact and ions of like charge will touch. According to a simple electrostatic argument, the lower coordination number, in which the contact of oppositely charged ions is restored, then becomes favorable. As the ionic radius of the M^+ ion increases, more anions can pack around it, as we have seen for CsCl with $(8,8)$-coordination compared with NaCl and its $(6,6)$-coordination.

In practice, the radius ratio is most reliable when the cation coordination number is 8, less reliable with 6-coordinate cations, and unreliable for 4-coordinate cations.[8] The radius ratio rules are probably an over-elaboration of the qualitative observation that large cations generally have large coordination numbers.

> *The radius ratio gives an indication of the likely coordination number of a compound: the higher the ratio, the greater the coordination number.*

(c) Structure maps

Granted that the use of radius ratios is uncertain, it is still possible to make progress with the rationalization of structures by collecting enough information empirically and looking for a

Table 2.4 The radius ratio rule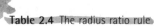

Coordination number	Radius ratio	Diagram
8	>0.7	
6	0.4–0.7	
4	0.2–0.4	
3	0.1–0.2	

---

8 L.C. Nathan, *J. Chem. Educ.* **62**, 215 (1985).

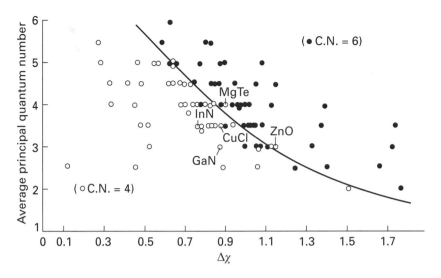

2.22 A structure map for compounds of formula MX. A point is defined by the electronegativity difference, $\Delta\chi$, between anion and cation and the average principal quantum number n. Its location in the map indicates the coordination number expected for that pair of properties. From E. Mooser and W.B. Pearson, *Acta Crystallogr.* **12**, 1015 (1959).

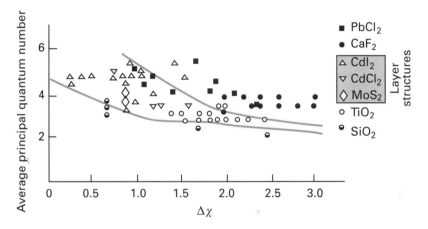

2.23 A structure map for compounds of formula MX_2. From E. Mooser and W.B. Pearson, *Acta Crystallogr.* **12**, 1015 (1959).

pattern in it. This attitude has motivated the compilation of **structure maps**. A structure map is an empirically compiled map that depicts the dependence of crystal structure on the electronegativity difference, $\Delta\chi$ (Section 1.8d), between the elements involved and the average principal quantum number of the valence shells of the two atoms.[9]

The ionic character of a bond increases with $\Delta\chi$, so moving from left to right along the horizontal axis of a structure map corresponds to an increase in ionic character in the bonding. The principal quantum number is an indication of the radius of an ion, so moving up the vertical axis corresponds to an increase in the average radius of the ions. Because atomic energy levels also become closer as the atom expands, the polarizability of the atom increases too (Section 1.8e). Consequently, the vertical axis of a structure map corresponds to increasing size and polarizability of the bonded atoms.

Figure 2.22 is an example of a structure map for **MX** compounds and Fig. 2.23 is an example of one for MX_2 compounds. We see from the former that the structures we have been discussing fall in quite distinct regions of the map. Elements with large $\Delta\chi$ have (6,6)-

9   E. Mooser and W.B. Pearson, *Acta Crystallogr.* **12**, 1015 (1959).

coordination, such as is found in the rock-salt structure; elements with small $\Delta\chi$ (and hence where there is the expectation of covalence) have a lower coordination number. In terms of a structure map representation, GaN is in a more covalent region of Fig. 2.22 than ZnO because $\Delta\chi$ is appreciably smaller.

> A structure map is a representation of the variation in crystal structure with the character of the bonding, the electronegativity difference horizontally and the average polarizability vertically.

Example 2.4 Using a structure map

What type of crystal structure should be expected for magnesium sulfide, MgS?

Answer The electronegativities of magnesium and sulfur are 1.3 and 2.6 respectively, so $\Delta\chi = 1.3$. The average principal quantum number is 3 (both elements are in Period 3). The point $\Delta\chi = 1.3$, $n = 3$ lies just in the coordination number 6 region of the structure map in Fig. 2.22. This location is consistent with the observed rock-salt structure of MgS.

Self-test 2.4 Predict the coordination number for the cation and anion in rubidium chloride, RbCl.

2.11 The energetics of ionic bonding

A compound tends to adopt the crystal structure that corresponds to lowest Gibbs energy. Therefore, if for the process

$$M^+(g) + X^-(g) \longrightarrow MX(s) \qquad \Delta G^{\ominus} = \Delta H^{\ominus} - T\Delta S^{\ominus}$$

the change in Gibbs energy, $\Delta G^{\ominus}$, is more negative for the formation of a structure A rather than B, then the transition from B to A is spontaneous under the prevailing conditions, and we can expect the solid to be found with structure A.

The process of solid formation from the gas of ions is so exothermic that at and near room temperature the contribution of the entropy may be neglected (this neglect is rigorously true at $T = 0$). Hence, discussions of the thermodynamic properties of solids normally focus, initially at least, on changes in enthalpy. That being so, we look for the structure that is formed most exothermically and identify it as the thermodynamically most stable form.

(a) Lattice enthalpy

The **lattice enthalpy**, $\Delta H_L^{\ominus}$, is the standard molar enthalpy change accompanying the formation of a gas of ions from the solid:

$$MX(s) \longrightarrow M^+(g) + X^-(g) \qquad \Delta H_L^{\ominus}$$

Because lattice disruption is always endothermic, lattice enthalpies are always positive and their positive signs are normally omitted from their numerical values. If entropy considerations are neglected, the most stable crystal structure of the compound is the structure with the greatest lattice enthalpy under the prevailing conditions.

Lattice enthalpies are determined from enthalpy data by using a **Born–Haber cycle**, such as that shown in Fig. 2.24. In this special case of a thermodynamic cycle, a cycle (or closed path) of steps is formed that includes lattice formation as one stage. The standard enthalpy of decomposition of a compound into its elements is the negative of its standard enthalpy of formation, $\Delta_f H^{\ominus}$; the standard enthalpy of lattice formation is the negative of the lattice enthalpy. For a solid element, the standard enthalpy of atomization, $\Delta_{atom} H^{\ominus}$, is the standard enthalpy of sublimation, $\Delta_{sub} H^{\ominus}$, as in the process

$$K(s) \longrightarrow K(g) \qquad \Delta_{sub} H^{\ominus} = +89 \text{ kJ mol}^{-1}$$

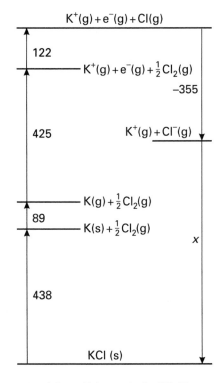

$K^+(g) + e^-(g) + Cl(g)$

122

$K^+(g) + e^-(g) + \frac{1}{2}Cl_2(g)$

-355

425   $K^+(g) + Cl^-(g)$

$K(g) + \frac{1}{2}Cl_2(g)$

89   $K(s) + \frac{1}{2}Cl_2(g)$

x

438

KCl (s)

2.24 A Born–Haber cycle for KCl. The lattice enthalpy is equal to $-x$.

For a gaseous element, the standard enthalpy of atomization is the standard enthalpy of dissociation, as in

$$Cl_2(g) \longrightarrow 2\,Cl(g) \qquad \Delta_{dis}H^{\ominus} = +244 \text{ kJ mol}^{-1}$$

The standard enthalpy of ionization is the ionization enthalpy (for the formation of cations), $\Delta_{ion}H^{\ominus}$, and the electron-gain enthalpy (for anions), $\Delta_{eg}H^{\ominus}$. Two examples are

$$K(g) \longrightarrow K^+(g) + e^-(g) \qquad \Delta_{ion}H^{\ominus} = +425 \text{ kJ mol}^{-1}$$
$$Cl(g) + e^-(g) \longrightarrow Cl^-(g) \qquad \Delta_{eg}H^{\ominus} = -355 \text{ kJ mol}^{-1}$$

The value of the lattice enthalpy—the only unknown in a well-chosen cycle—is found from the requirement that the sum of the enthalpy changes round a complete cycle is zero (because enthalpy is a state property).[10]

Example 2.5 Using a Born–Haber cycle to determine a lattice enthalpy

Calculate the lattice enthalpy of KCl(s) using a Born–Haber cycle and the following information:

Step	$\Delta H^{\ominus}/(\text{kJ mol}^{-1})$
Sublimation of K(s)	+89
Ionization of K(g)	+425
Dissociation of $Cl_2(g)$	+244
Electron gain by Cl(g)	−355
Formation of KCl(s)	−438

Answer The required cycle is shown in Fig. 2.24. The sum of the enthalpy changes around the cycle is 0, so

$$-\Delta H_L^{\ominus} + 719 \text{ kJ mol}^{-1} = 0$$

Therefore, $\Delta H_L^{\ominus} = 719 \text{ kJ mol}^{-1}$.

Self-test 2.5 Calculate the lattice enthalpy of magnesium bromide from the following data:

Step	$\Delta H^{\ominus}/(\text{kJ mol}^{-1})$
Sublimation of Mg(s)	+148
Ionization of Mg(g) to $Mg^{2+}(g)$	+2187
Vaporization of $Br_2(l)$	+31
Dissociation of $Br_2(g)$	+193
Electron gain by Br(g)	−331
Formation of $MgBr_2(s)$	−524

Once the lattice enthalpy is known, it can be used to judge the character of the bonding in the solid. If the value calculated on the assumption that the lattice consists of ions interacting Coulombically is in good agreement with the measured value, then it may be appropriate to adopt a largely ionic model of the compound. A discrepancy indicates a degree of covalence. As mentioned earlier, it is important to remember that numerical coincidences can be misleading in this assessment.

---

[10] When the lattice enthalpy is known from calculation, a Born–Haber cycle may be used to determine the value of another elusive quantity, the electron-gain enthalpy (and hence the electron affinity, Section 1.8c).

> Lattice enthalpies are determined from enthalpy data by using a Born–Haber cycle. If entropy considerations are neglected, the most stable crystal structure of the compound is the structure with the greatest lattice enthalpy under the prevailing conditions.

(b) Coulombic contributions to lattice enthalpies

To calculate the lattice enthalpy of a supposedly ionic solid we need to take into account several contributions to its energy, including the attractions and repulsions between the ions. The mean kinetic energy of the vibrating atoms (apart from its contribution to the zero-point energy of the crystal) can be neglected if it is supposed that the solid is at the absolute zero of temperature.

The total Coulombic potential energy of a crystal is the sum of the individual Coulomb potential energy terms of the form

$$V_{AB} = \frac{(z_A e)(z_B e)}{4\pi\varepsilon_0 r_{AB}} \tag{2}$$

for ions of charge numbers z_A and z_B (with cations having positive charge numbers and anions negative charge numbers) separated by a distance r_{AB}.[11] The sum over all the pairs of ions in the solid may be carried out for any crystal structure, although in practice it converges very slowly because nearest neighbors contribute a large negative term, second-nearest neighbors an only slightly weaker positive term, and so on. The overall result is that the attraction between the cations and anions predominates and yields a favorable (negative) contribution to the energy of the solid.

For example, in a uniformly spaced one-dimensional line of alternating cations and anions with $z_A = +z$ and $z_B = -z$ (Fig. 2.25), the interaction of one ion with all the others is proportional to

$$-\frac{2z^2}{d} + \frac{2z^2}{2d} - \frac{2z^2}{3d} + \frac{2z^2}{4d} - \cdots = -\frac{2z^2}{d}\left(1 - \tfrac{1}{2} + \tfrac{1}{3} - \tfrac{1}{4} + \cdots\right)$$
$$= -\frac{2z^2}{d}\ln 2$$

The factor of 2 comes from the fact that the same ions occur on both sides of the central ion. As in this case, it is found that, apart from the explicit appearance of the ion charge numbers, the value of the sum depends only on the type of the lattice and on a single scale parameter which may be taken as the separation of the centers of nearest neighbors, d. We may write

$$V = -\frac{e^2}{4\pi\varepsilon_0} \times \frac{z^2}{d} \times 2\ln 2 \tag{3}$$

The first factor is a collection of fundamental constants. The second is specific to the identities of the ions and to the scale of the lattice. The third term, $2\ln 2 = 1.386$, characterizes the symmetry of the lattice (in this case a straight line of ions) and is the simplest example of a **Madelung constant**, $\mathscr{A}$. In simple solids, the Madelung constant is specific to the crystal type and independent of the interionic distances. The same expression applies both to a cation and to an anion in the lattice, so the potential energy of one formula unit is also given by the same expression (not twice V because we must not count ion–ion interactions twice).

In general, the total potential energy per mole of formula units in an arbitrary crystal structure is

$$V = \frac{N_A e^2}{4\pi\varepsilon_0}\left(\frac{z_A z_B}{d}\right)\mathscr{A} \tag{4}$$

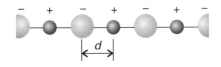

2.25 The basis of the calculation of the Madelung constant for a line of alternating cations and anions with lattice constant d.

---

[11] The fundamental constant ε_0 is the vacuum permittivity; see inside back cover. In a precise calculation, the actual permittivity of the medium should be used.

where N_A is the Avogadro constant (in this case, the number of formula units per mole) and z_A and z_B are the charge numbers of the ions: because the charge number of cations is positive and that of anions is negative, V is negative overall, corresponding to a lowering in potential energy relative to the gas of widely separated ions.

Some computed values of the Madelung constant for a variety of lattices are given in Table 2.5. The general trend is for the values to increase with coordination number. This trend reflects the fact that a large contribution comes from nearest neighbors, and such neighbors are more numerous when the coordination number is large. The values for the rock-salt structure (C.N. 6) and for the cesium-chloride structure (C.N. 8) illustrate the trend. However, a high coordination number does not necessarily mean that the interactions are stronger in the cesium-chloride structure, for the potential energy also depends on the scale of the lattice. Thus, d may be so large in lattices with ions big enough to adopt eightfold coordination that the separation of the ions overcomes the small increase in Madelung constant and results in a less favorable potential energy.

Another secondary contribution to the lattice enthalpy arises from the **van der Waals interactions** between the ions and molecules, the weak intermolecular interactions that are responsible for the formation of condensed phases of neutral species. An important and sometimes dominant contribution of this kind is the **dispersion interaction** (the 'London interaction'). The dispersion interaction arises from the transient fluctuations in electron density (and, consequently, instantaneous electric dipole moment) on one molecule driving a fluctuation in electron density (and dipole moment) on a neighboring molecule, and the attractive interaction between these two instantaneous electric dipoles. The potential energy of this interaction is inversely proportional to the sixth power of the separation, so in the crystal as a whole we can expect it to vary as the sixth power of the scale of the lattice:

$$V = -\frac{N_A C}{d^6} \tag{5}$$

The constant C depends on the substance. For ions of low polarizability, this contribution is only about 1 per cent of the Coulombic contribution and is ignored in elementary lattice enthalpy calculations of ionic solids.

> The Madelung constant reflects the role of the geometry of the lattice on the strength of the net Coulomb interaction; other contributions to the lattice energy include van der Waals interactions, particularly the dispersion interaction.

(c) Repulsions arising from overlap

When two closed-shell ions are in contact, another contribution to the total potential energy is the repulsion arising from the overlap of their electron distributions. Because orbitals decay exponentially toward zero at large distances from the nucleus, and repulsive interactions depend on the overlap of orbitals, it is plausible that their contribution to the potential energy has the form

$$V = +N_A C' e^{-d/d^*} \tag{6}$$

where C' and d^* are constants. We shall see in a moment that C' cancels and that its value need not be known. The constant d^* can be estimated from measurements of compressibilities, which reflect the increase in potential energy that occurs when ions are pressed together by an applied force. Although the values of d^* measured in this way actually span a range, it is often found that setting $d^* = 0.345$ Å gives reasonable agreement with experiment.

> The repulsion between atoms in a crystal is often assumed to decrease exponentially with increasing separation.

Table 2.5 Madelung constants*

Structural type	$\mathscr{A}$
Cesium chloride	1.763
Fluorite	2.519
Rock salt	1.748
Rutile	2.408
Sphalerite	1.638
Wurtzite	1.641

*The values given are for the geometric factor described in the text. Some sources cite values that include the charge numbers of the ions (so, for instance, the value for CaF_2 is quoted as 5.039), and it is necessary to verify the definition before using them.

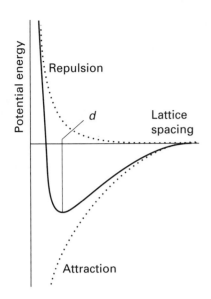

2.26 Contributions to the potential energy of ions in a crystal. The attractive interaction scales as $1/d$, where d represents the dimensions of the lattice. The repulsive interaction has a shorter range, but becomes strong very rapidly once the ions are in contact. The total potential energy, the sum of these two contributions, passes through a minimum that corresponds to the equilibrium dimension of the lattice.

(d) The Born–Mayer equation

The contribution of the attractive and repulsive components to the total potential energy of a solid is depicted by the curve in Fig. 2.26. The total potential energy passes through a minimum at the equilibrium lattice parameter of the solid, which can be found by looking for a minimum in the expression

$$V = \frac{N_A e^2}{4\pi\varepsilon_0}\left(\frac{z_A z_B}{d}\right)\mathscr{A} + N_A C' e^{-d/d^*} \tag{7}$$

The minimum potential energy is obtained where $\mathrm{d}V/\mathrm{d}d = 0$, which occurs when

$$N_A C' e^{-d/d^*} = -\frac{N_A e^2}{4\pi\varepsilon_0}\left(\frac{z_A z_B d^*}{d^2}\right)\mathscr{A}$$

Substitution of this relation into the preceding one results in the **Born–Mayer equation**:

$$V = \frac{N_A z_A z_B e^2}{4\pi\varepsilon_0 d}\left(1 - \frac{d^*}{d}\right)\mathscr{A} \tag{8}$$

The experimental scale of the lattice may now be used for d. At $T = 0$, where there is no contribution from the mean kinetic energy, we can identify this minimum potential energy as the molar internal energy of the crystal relative to the energy of the widely separated ions. It follows that the negative of V may be identified as the lattice enthalpy (strictly, its value at $T = 0$ with the zero-point vibrational energy of the lattice ignored).

As remarked previously, the agreement between the experimental lattice enthalpy and the value calculated using the ionic model of the solid (in practice, from the Born–Mayer equation) provides a measure of the extent to which the solid is ionic. Some calculated and measured lattice enthalpies are given in Table 2.6.

As a general guide, the ionic model is reasonably valid if the difference in electronegativity, $\Delta\chi$, for the elements is more than about 2, and the bonding is predominantly covalent if $\Delta\chi$ is less than about 1. However, it should be remembered that the electronegativity criterion ignores the role of polarizability of the ions. Thus, the alkali metal halides give quite good agreement with the ionic model, the best with the least polarizable halide ions (F^-) formed from the highly electronegative F atom, and the worst with the highly polarizable halide ions (I^-) formed from the less electronegative I atom. It is not always clear whether it is the electronegativity of the atoms or the polarizability of the resultant ions that should be used as a criterion. The worst agreement with the ionic model is for polarizable cation–polarizable

Table 2.6 Measured and calculated lattice enthalpies

Compound	$\Delta H_L^{\ominus}/(\mathbf{kJ\,mol^{-1}})$		Calc/expt (per cent)
	(calc)	(expt)	
LiF[a]	1033	1037	99.6
LiCl[a]	845	852	99.2
LiBr[a]	798	815	97.9
LiI[a]	740	761	97.2
CsF[a]	748	750	99.7
CsCl[b]	652	676	96.4
CsBr[b]	632	654	96.6
CsI[b]	601	620	96.9
AgF[a]	920	969	94.9
AgCl[a]	833	912	91.3
AgBr[a]	816	900	90.7

[a]Rock-salt structure; [b]cesium-chloride structure.
Source: D. Cubicciotti, *J. Chem. Phys.* **31**, 1646 (1951).
The calculated values use a more complete ionic model that includes terms beyond those in the Born–Mayer equation.

anion combinations, which are likely to be substantially covalent. Here again, though, the difference between the electronegativities of the parent elements is small and it is not clear whether electronegativity or polarizability provides the better criterion.

> The Born–Meyer equation is used to estimate lattice enthalpy for an ionic lattice. The ionic model is reasonably valid if the electronegativity difference for the neutral atoms is more than about 2, and the bonding is predominantly covalent if this difference is less than about 1.

(e) The Kapustinskii equation

The Russian chemist A.F. Kapustinskii noticed that, if the Madelung constants for a number of structures are divided by the number of ions per formula unit, n, then approximately the same value is obtained for them all. Moreover, he also noticed that the value so obtained increases with the coordination number. Therefore, because ionic radius also increases with coordination number, the variation in $\mathscr{A}/nd$ from one structure to another can be expected to be quite small. This observation led Kapustinskii to propose that there exists a hypothetical rock-salt structure that is energetically equivalent to the true structure of any ionic solid. That being so, the lattice enthalpy can be calculated by using the Madelung constant and the appropriate ionic radii for $(6,6)$-coordination. The resulting expression is called the **Kapustinskii equation**:

$$\Delta H_{\mathrm{L}}^{\ominus} = -\frac{nz_{\mathrm{A}}z_{\mathrm{B}}}{d}\left(1 - \frac{d^*}{d}\right)\mathscr{K} \tag{9}$$

In this equation, $d = r_{\mathrm{A}} + r_{\mathrm{B}}$, $\mathscr{K} = 1.21$ MJ Å mol^{-1}.

The Kapustinskii equation can be used to ascribe numerical values to the 'radii' of nonspherical molecular ions, for their values can be adjusted until the calculated value of the lattice enthalpy matches that obtained experimentally from the Born–Haber cycle. The self-consistent set of parameters obtained in this way are called **thermochemical radii** (Table 2.7). They may be used to estimate lattice enthalpies, and hence enthalpies of formation, of a wide range of compounds.

> The Kapustinskii equation is used to estimate lattice enthalpies of ionic compounds and to give a measure of the thermochemical radii of the constituent ions.

Table 2.7 The thermochemical radii of ions, $r/\text{Å}$

Main-group elements

BeF_4^{2-}	BF_4^-	CO_3^{2-}	NO_3^-	OH^-	
2.45	2.28	1.85	1.89	1.40	
		CN^-	NO_2^-	O_2^{2-}	
		1.82	1.55	1.80	
			PO_4^{3-}	SO_4^{2-}	ClO_4^-
			2.38	2.30	2.36
			AsO_4^{3-}	SeO_4^{2-}	
			2.48	2.43	
			SbO_4^{3-}	TeO_4^{2-}	IO_4^-
			2.60	2.54	2.49
					IO_3^-
					1.82

Complex ions / d-Metal oxoanions

$[TiCl_6]^{2-}$	$[IrCl_6]^{2-}$	$[SiF_6]^{2-}$	$[GeCl_6]^{2-}$	CrO_4^{2-}	MnO_4^-
2.48	2.54	1.94	2.43	2.40	2.40
$[TiBr_6]^{2-}$	$[PtCl_6]^{2-}$	$[GeF_6]^{2-}$	$[SnCl_6]^{2-}$	MoO_4^{2-}	
2.61	2.59	2.01	2.47	2.54	
$[ZrCl_6]^{2-}$			$[PbCl_6]^{2-}$		
2.47			2.48		

Source: A.F. Kapustinskii, *Q. Rev. Chem. Soc.* **10**, 283 (1956). 1 Å = 100 pm.

Example 2.6 Using the Kapustinskii equation

Estimate the lattice enthalpy of potassium nitrate, KNO_3.

Answer To use the Kapustinskii equation we need the number of ions per formula unit ($n = 2$), their charge numbers $z_{K^+} = +1$ and $z_{NO_3^-} = -1$, and the sum of their thermochemical radii ($1.38\ \text{Å} + 1.89\ \text{Å} = 3.27\ \text{Å}$). Then, with $d^* = 0.345\ \text{Å}$,

$$\Delta H_L^\ominus = -\frac{2(+1)(-1)}{3.27\ \text{Å}} \times \left(1 - \frac{0.345\ \text{Å}}{3.27\ \text{Å}}\right) \times 1.21\ \text{MJ Å mol}^{-1}$$

$$= 662\ \text{kJ mol}^{-1}$$

Self-test 2.6 Estimate the lattice enthalpy of calcium sulfate, $CaSO_4$.

2.12 Consequences of lattice enthalpies

The Born–Mayer equation shows that, for a given lattice type (a given value of $\mathscr{A}$), the lattice enthalpy increases with increasing ion charge numbers (as $|z_A z_B|$). The lattice enthalpy also increases as the scale d of the lattice decreases. Energies that vary as the **electrostatic parameter**, ξ (xi),

$$\xi = \frac{z^2}{d} \tag{10}$$

where z is an ion charge number and d is a scale length, are widely adopted in inorganic chemistry as indicative that an ionic model is appropriate.[12] In this chapter we consider three consequences of lattice enthalpy and its correlation with the electrostatic parameter.

(a) Thermal stabilities of ionic solids

The particular aspect we consider here is the temperatures needed to bring about thermal decomposition of carbonates. Magnesium carbonate, for instance, decomposes when heated to about $300\,°C$, whereas calcium carbonate decomposes only if the temperature is raised to over $800\,°C$. In general, it is found that large cations stabilize large anions (and vice versa). In particular, the decomposition temperatures of thermally unstable compounds (such as carbonates) increase with cation radius.

The stabilizing influence of a large cation on an unstable anion can be explained in terms of trends in lattice enthalpies. First, we note that the decomposition temperatures of solid inorganic compounds can be discussed in terms of their Gibbs energies of decomposition into specified products. In many cases it is sufficient to consider only the corresponding reaction enthalpy, for the reaction entropy is almost constant for the substances being compared. Moreover, as we shall see, the calculations center on differences of lattice enthalpy between the solid reactants and the solid products, for these differences dominate the reaction enthalpy. The differences in lattice enthalpy in turn may be discussed, at least in a general way, in terms of the Kapustinskii equation. Hence the variation in stability may be correlated with the variation of the electrostatic parameter ξ.

Consider the thermal decomposition of carbonates:

$$MCO_3(s) \longrightarrow MO(s) + CO_2(g)$$

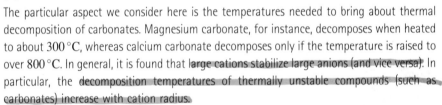

The experimental observation that the decomposition temperature increases with increasing cation radius can be expressed in terms of the reaction Gibbs energy: the temperature at which $\Delta_r G^\ominus$ becomes negative and the reaction favorable increases with cation radius. It

---

[12] The correlation of properties with ξ is a very useful guide for many properties, but we should not take it as proof of the dominance of charge–charge interactions. Often there is a correlation of a property with a wide range of parameters such as z^2/r^2 or z/r: indeed, a correlation can often be found with almost any expression with charge number in the numerator and radius in the denominator.

Table 2.8 Decomposition data on carbonates*

	Mg	Ca	Sr	Ba
$\Delta G^{\ominus}/(\text{kJ mol}^{-1})$	+ 48.3	+130.4	+183.8	+218.1
$\Delta H^{\ominus}/(\text{kJ mol}^{-1})$	+100.6	+178.3	+234.6	+269.3
$\Delta S^{\ominus}/(\text{J K}^{-1}\text{ mol}^{-1})$	+175.0	+160.6	+171.0	+172.1
$\theta/°C$	300	840	1100	1300

*Data are for the reaction

$$MCO_3(s) \rightarrow MO(s) + CO_2(g) \quad \Delta X^{\ominus}(298\text{ K})$$

θ is the temperature required to reach 1 bar of CO_2, and has been estimated from the 298 K reaction enthalpy and entropy. The experimental values reported in the literature for calcium (and rubidium) are unreliable because the samples were probably damp.

follows from the thermodynamic relation $\Delta_r G^{\ominus} = \Delta_r H^{\ominus} - T\Delta_r S^{\ominus}$ that the decomposition temperature is reached when

$$T = \frac{\Delta_r H^{\ominus}}{\Delta_r S^{\ominus}} \quad \Longleftarrow \text{ Temperature at which decomposition is reached} \tag{11}$$

The decomposition entropy is almost constant for all carbonates because in each case it is dominated by the formation of gaseous carbon dioxide. Hence, we can expect that, the higher the reaction enthalpy, the higher the decomposition temperature. This trend can be verified by reference to the experimental data in Table 2.8.

The standard enthalpy of decomposition depends in part on the difference between the lattice enthalpy of the decomposition product, MO, and that of the parent carbonate, MCO_3. The overall reaction enthalpy is positive (decomposition is endothermic), but it is less strongly positive if the lattice enthalpy of the oxide is markedly greater than that of the carbonate. Hence, the decomposition temperature will be low for oxides that have relatively high lattice enthalpies compared with their parent carbonates. The compounds for which this is true are composed of small, highly charged cations, such as Mg^{2+}.

Figure 2.27 suggests why a small cation increases the lattice enthalpy of an oxide more than that of a carbonate. The illustration shows that the relative change in scale of the lattice is large when a compound composed of a small cation and a large anion becomes an oxide. The change in scale is relatively small when the parent compound has a large cation initially.

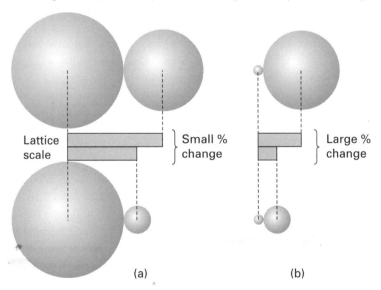

2.27 A greatly exaggerated representation of the change in lattice parameter for cations of different sizes. (a) When the anion changes size (when CO_3^{2-} decomposes to O^{2-} and CO_2, for instance) and the cation is large, the lattice scale changes by a relatively small amount. (b) If the cation is small, the relative change in lattice scale is large, and decomposition is thermodynamically more favorable.

As the illustration shows in an exaggerated way, when the cation is very big, the change in size of the anion barely affects the scale of the lattice. Therefore, with a given unstable polyatomic anion, the lattice enthalpy difference is more favorable to decomposition when the cation is small than when it is large.

The difference in lattice enthalpy between MO and MCO_3 is magnified by a large charge number on the cation. As a result, thermal decomposition of a carbonate will occur at lower temperatures if it contains a highly charged cation. One consequence of this dependence on charge number is that alkaline earth carbonates (M^{2+}) decompose at lower temperatures than the corresponding alkali metal carbonates (M^+).

> *Cations in high oxidation states are stabilized by large anions. Large cations stabilize large anions (and vice versa). The decomposition temperatures of thermally unstable compounds increase with cation radius. The higher the reaction enthalpy, the higher the decomposition temperature.*

Example 2.7 Assessing the dependence of stability on ionic radius

Present an argument to account for the fact that, when they burn in oxygen, lithium forms the oxide Li_2O but sodium forms the peroxide Na_2O_2.

Answer Because the small Li^+ ion results in Li_2O having a more favorable lattice enthalpy than Na_2O, the decomposition reaction $M_2O_2 \rightarrow M_2O + \frac{1}{2}O_2$ is thermodynamically more favorable for Li_2O_2 than for Na_2O_2.

Self-test 2.7 Predict the order of decomposition temperatures of alkaline earth metal sulfates in the reaction $MSO_4(s) \rightarrow MO(s) + SO_3(g)$.

(b) The stabilities of oxidation states

A similar argument can be used to account for the general observation that high oxidation states are stabilized by small anions.[13] In particular, fluorine has a greater ability compared with the other halogens to stabilize the high oxidation states of metals. Thus, the only known halides of Ag(II), Co(III), and Mn(IV) are the fluorides. Another sign of the decrease in stability of the heavier halides of metals in high oxidation states is that the iodides of Cu(II) and Fe(III) decompose on standing at room temperature. The oxygen atom is a very effective species for drawing out the high oxidation states of atoms because not only is it small, but it can accept up to two electrons.

To explain these observations, we consider the reaction

$$MX + \tfrac{1}{2}X_2 \longrightarrow MX_2$$

where X is a halogen. Our aim is to show why this reaction is most strongly spontaneous for X = F. If we ignore entropy contributions, we must show that the reaction is most exothermic for fluorine.

One contribution to the reaction enthalpy is the conversion of $\frac{1}{2}X_2$ to X^-. Despite fluorine having a lower electron affinity than chlorine, this step is more exothermic for X = F than for X = Cl on account of the lower bond enthalpy of F_2 compared with that of Cl_2. The lattice enthalpies, however, play the major role. In the conversion of MX to MX_2, the charge number of the cation increases from +1 to +2, so the lattice enthalpy increases. As the radius of the anion increases, though, this difference in the two lattice enthalpies diminishes, and the exothermic contribution to the overall reaction decreases too. Hence, both the lattice energy and the X^- formation enthalpy lead to a less exothermic reaction as the halogen changes from

---

[13] Oxidation states and oxidation numbers are introduced in Section 3.1; they should already be familiar from introductory chemistry. For monatomic species, the oxidation number is the same as the charge number.

F to I. Provided entropy factors are similar, we expect an increase in thermodynamic stability of MX relative to MX_2 on going from $X = F$ to $X = I$ down Group 17/VII.

> *Cations with high oxidation states are stabilized by small anions.*

(c) Solubility

Lattice enthalpies play a role in solubilities, but one that is much more difficult to analyze than for reactions. One general rule that is quite widely obeyed is that compounds that contain ions with widely different radii are generally soluble in water. Conversely, the least water-soluble salts are those of ions with similar radii. That is, in general, *difference in size favors solubility in water*. It is found empirically that an ionic compound MX tends to be most soluble when the radius of M^+ is smaller than that of X^- by about 0.8 Å.

Two familiar series of compounds illustrate these trends. In gravimetric analysis, Ba^{2+} is used to precipitate SO_4^{2-}, and the solubilities of the Group 2 sulfates decrease from $MgSO_4$ to $BaSO_4$. In contrast, the solubility of the alkaline earth metal hydroxides increases down the group: $Mg(OH)_2$ is the sparingly soluble 'milk of magnesia' but $Ba(OH)_2$ is used as a soluble hydroxide for preparation of solutions of OH^-. The first case shows that a large anion requires a large cation for precipitation. The second case shows that a small anion requires a small cation for precipitation.

Before attempting to rationalize the observations, we should note that the solubility of an ionic compound depends on the Gibbs energy of the process

$$MX(s) \longrightarrow M^+(aq) + X^-(aq)$$

In the process, the interactions responsible for the lattice enthalpy of MX are replaced by hydration (and by solvation in general) of the ions. However, the exact balance of enthalpy and entropy effects is delicate and difficult to assess, particularly because the entropy change also depends on the degree of order of the *solvent* molecules that is brought about by the presence of the dissolved solute. The data in Fig. 2.28 suggest that enthalpy considerations are important in some cases at least, for the graph shows that there is a correlation between

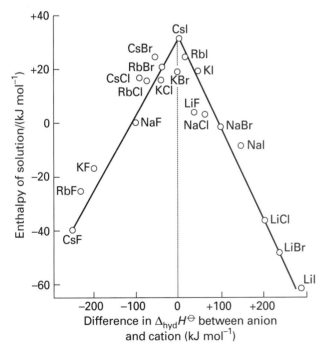

2.28 The correlation between enthalpies of solution of halides and the differences between the hydration enthalpies of the ions. Dissolving is most exothermic when the difference is large.

the enthalpy of solution of a salt and the difference in hydration enthalpies of the two ions. If the cation has a larger hydration enthalpy than its anion partner (reflecting the difference in their sizes) or vice versa, the dissolution of the salt is exothermic (reflecting the favorable solubility equilibrium).

The variation in enthalpy can be explained using the ionic model. The lattice enthalpy is inversely proportional to the distance between the ions:

$$\Delta H_L^{\ominus} \propto \frac{1}{r_+ + r_-} \tag{12a}$$

whereas the hydration enthalpy is the sum of individual ion contributions:

$$\Delta_{hyd} H^{\ominus} \propto \frac{1}{r_+} + \frac{1}{r_-} \tag{12b}$$

If the radius of one ion is small, the term in the hydration enthalpy for that ion will be large. However, in the expression for the lattice enthalpy one small ion cannot make the denominator of the expression small by itself. Thus, one small ion can result in a large hydration enthalpy but not necessarily lead to a high lattice enthalpy, so ion size asymmetry can result in exothermic dissolution. However, if both ions are small, then both the lattice enthalpy and the hydration enthalpy may be large, and dissolution might not be very exothermic.

Compounds that contain ions with widely different radii are generally soluble in water. Conversely, the least water-soluble salts are those of ions with similar radii.

Example 2.8 Accounting for trends in the solubility of *s*-block compounds

What is the trend in the solubilities of the Group 2 metal carbonates?

Answer The CO_3^{2-} anion has a large radius and is of the same charge type (2) as the cations M^{2+} of the Group 2 elements. The least soluble carbonate of the group is predicted to be that of the largest cation, Ra^{2+}. The most soluble is expected to be the carbonate of the smallest cation, Mg^{2+} (beryllium has too much covalent character in its bonding for it to be included in this analysis). Although magnesium carbonate is more soluble than radium carbonate, it is still only sparingly soluble: its solubility constant (its solubility-product constant) is only 3×10^{-8}.

Self-test 2.8 Which can be expected to be more soluble in water, $NaClO_4$ or $KClO_4$?

Further reading

Some introductory texts on solid-state inorganic chemistry at about the level of this text are:

U. Müller, *Inorganic structural chemistry*. Wiley, New York (1993). Chapters 1–5 introduce prototype structures and bonding in ionic solids.

D.M. Adams, *Inorganic solids*. Wiley, New York (1974).

A.R. West, *Solid state chemistry and its applications*. Wiley, New York (1984).

M.F.C. Ladd, *Chemical bonding in solids and fluids*. Ellis Harwood (1994).

P.A. Cox, *The electronic structure and chemistry of solids*. Oxford University Press (1987).

M.F.C. Ladd, *Structure and bonding in solid state chemistry*. Wiley, New York (1979).

The standard reference book, which surveys the structures of a huge number of elements and compounds, is

A.F. Wells, *Structural inorganic chemistry*. Clarendon Press, Oxford (1984).

Two very useful thoughtful introductory texts on the application of thermodynamic arguments to inorganic chemistry are:

W.E. Dasent, *Inorganic energetics*. Cambridge University Press (1982).

D.A. Johnson, *Some thermodynamic aspects of inorganic chemistry*. Cambridge University Press (1982).

Also of interest are:

J.K. Burdett, *Chemical bonding in solids*. Oxford University Press (1995).

P.A. Cox, *Transition metal oxides*. Oxford University Press (1992).

Exercises

2.1 Which of the following schemes for the repeating pattern of close-packed planes are not ways of generating close-packed lattices? (a) ABCABC … ; (b) ABAC … ; (c) ABBA … ; (d) ABCBC … ; (e) ABABC … ; (f) ABCCB … .

2.2 Draw one layer of close-packed spheres. On this layer mark the positions of the centers of the B layer atoms using the symbol $\otimes$ and, with the symbol $\bigcirc$ mark the positions of the centers of the C layer atoms of an fcc lattice.

2.3 (a) Distinguish between the terms polymorph and polytype. (b) Give an example of each.

2.4 The interstitial alloy tungsten carbide, WC, has the rock-salt structure. Describe it in terms of the holes in a close-packed structure.

2.5 Depending on temperature, RbCl can exist in either the rock-salt or cesium-chloride structure. (a) What is the coordination number of the anion and cation in each of these structures? (b) In which of these structures will Rb have the larger apparent radius?

2.6 The ReO_3 structure is cubic with Re at each corner of the unit cell and one O atom on each unit cell edge midway between the Re atoms. Sketch this unit cell and determine (a) the coordination number of the cation and anion and (b) the identity of the structure type that would be generated if a cation were inserted in the center of the ReO_3 structure.

2.7 Sketch the *s*- and *p*-blocks of the periodic table and mark the boxes for the elements that contribute monatomic cations and anions to solids that are described well by the ionic model. Identify the elements.

2.8 Consider the structure of rock salt. (a) What are the coordination numbers of the anion and cation? (b) How many Na^+ ions occupy second-nearest neighbor locations of an Na^+ ion? (c) Pick out the closest-packed plane of Cl^- ions. (Hint: this hexagonal plane will be perpendicular to a threefold axis.)

2.9 Consider the structure of cesium chloride. (a) What is the coordination number of the anion and cation? (b) How many Cs^+ ions occupy second-nearest neighbor locations of a Cs^+ ion?

2.10 (a) How many Cs^+ ions and how many Cl^- ions are in the CsCl unit cell? (b) How many Zn^{2+} ions and how many S^{2-} ions are in the sphalerite unit cell?

2.11 Confirm that in rutile (TiO_2, Fig 2.16) the stoichiometry is consistent with the structure.

2.12 Figure 2.17 shows the perovskite structure of $CaTiO_3$. Confirm that the stoichiometry is consistent with the structure.

2.13 Imagine the construction of an MX_2 structure from the bcc CsCl structure by removal of half the Cs^+ ions to leave tetrahedral coordination around each Cl^-. What is this MX_2 structure?

2.14 Given the following data for the length of a side of the unit cell for compounds that crystallize in the rock-salt structure, determine the cation radii: MgSe (5.45 Å), CaSe (5.91Å), SrSe (6.23 Å), BaSe (6.62 Å). (To determine the Se^{2-} radius, assume that the Se^{2-} ions are in contact in MgSe.)

2.15 Use the structure map in Fig. 2.22 to predict the coordination numbers of the cations and anions in (a) LiF, (b) RbBr, (c) SrS, and (d) BeO. The observed coordination numbers are (6, 6) for LiF, RbBr, and SrS, and (4, 4) for BeO. Propose a possible reason for the discrepancies.

2.16 (a) Calculate the enthalpy of formation of the hypothetical compound KF_2 assuming a CaF_2 structure. Use the Born–Mayer equation to obtain the lattice enthalpy and estimate the radius of K^{2+} by extrapolation of trends in Table 1.5. Ionization enthalpies and electron gain enthalpies can be obtained from Tables 1.6 and 1.7. (b) What factor prevents the formation of this compound despite the favorable lattice enthalpy?

2.17 The Coulombic attraction of nearest-neighbor cations and anions accounts for the bulk of the lattice enthalpy of an ionic compound. With this fact in mind, estimate the order of increasing lattice enthalpy for the following solids, all of which crystallize in the rock-salt structure: (a) MgO, (b) NaCl, (c) LiF. Give your reasoning.

2.18 Which one of each of the following pairs of isostructural compounds is likely to undergo thermal decomposition at lower

temperature? Give your reasoning. (a) $MgCO_3$ and $CaCO_3$ (decomposition products $MO + CO_2$). (b) CsI_3 and $N(CH_3)_4I_3$ (both compounds contain I_3^-; decomposition products $MI + I_2$; the radius of $N(CH_3)_4^+$ is much bigger than that of Cs^+).

2.19 Which member of each pair is likely to be the more soluble in water: (a) $SrSO_4$ or $MgSO_4$, (b) NaF or $NaBF_4$?

Problems

2.1 In the structure of MoS_2, the S atoms are arranged in close-packed layers that repeat themselves in the sequence AAA .... . The Mo atoms occupy holes with C.N. = 6. Show that each Mo atom is surrounded by a trigonal prism of S atoms.

2.2 Show that the maximum fraction of the available volume occupied by hard spheres on various lattices is (a) simple cubic: 0.52, (b) bcc: 0.68, (c) fcc: 0.74.

2.3 The common oxidation number for an alkaline earth metal is +2. Aided by the Born–Mayer equation for lattice enthalpy and a Born–Haber cycle, show that CaCl is an exothermic compound. Use a suitable analogy to estimate an ionic radius for Ca^+. The sublimation enthalpy of Ca(s) is $176\,kJ\,mol^{-1}$. Show that an explanation for the nonexistence of CaCl can be found in the enthalpy change for the reaction

$$2CaCl(s) \rightarrow Ca(s) + CaCl_2(s)$$

2.4 For a close-packed structure (such as solid Xe, mp $-112\,°C$), give the number of nearest-neighbor atoms, the number of tetrahedral holes per atom, and the number of octahedral holes per atom.

2.5 Recommend a specific cation for the quantitative precipitation of carbonate ion in water. Justify your recommendation.

2.6 Metallic sodium adopts a body-centered cubic structure at $25\,°C$ and 1 atm. The density of sodium under these conditions is $0.97\,g\,cm^{-3}$. What is the unit cell edge length?

2.7 If you look up the ionic radius of Na^+ in a table of ionic radii, several values appear. This is not because different scientists use different ways to estimate ionic radii. It is due to a more fundamental reason. Explain.

2.8 There are two common polymorphs of zinc sulfide: cubic and hexagonal. Based on an analysis of Madelung constants alone, predict which polymorph should be the more stable? Assume that the Zn-S distances in the two polymorphs are identical.

2.9 Explain why lattice energy calculations based on the Born–Mayer equation duplicate the experimentally determined values to within 1 per cent for LiCl but only to within 10 per cent for AgCl. Both compounds have the rock-salt structure. (b) Choose a pair of compounds containing 2+ and two ions that might be expected to show similar behavior.

3

Molecular structure and bonding

The interpretation of structures and reactions in inorganic chemistry is often based on semiquantitative models. Pictorial representations, such as Lewis structures, are widely used and can be very helpful for correlating bonding and structure; we review some of the most useful aspects. In addition, semiquantitative pictorial models for molecular bonding can be achieved by using valence-bond and molecular orbital concepts, and much of this chapter introduces this useful formalism. We shall see this interplay between qualitative models, experiment, and calculation in this chapter and throughout the remainder of the text.

A great deal of inorganic chemistry depends on being able to relate the chemical properties of compounds to their electronic structures. The most elementary discussion of covalent bonding, the topic of this chapter, is in terms of shared pairs of electrons. That approach was introduced by G.N. Lewis in 1916. Since that time our understanding of bonding has been greatly enriched both experimentally and theoretically, particularly through the development of molecular orbital theory. In this chapter we review Lewis's elementary but useful theory and show how modern theories capture some of its spirit but go far beyond it.

Lewis structures: a review

Lewis proposed that a **covalent bond** is formed when two neighboring atoms share an electron pair. A single pair of shared electrons is denoted A—B; double bonds (A=B) and triple bonds (A≡B) consist of two and three shared pairs of electrons, respectively. Unshared pairs of valence electrons on atoms are called **lone pairs**. Although lone pairs do not contribute directly to the bonding, they do influence the shape of the molecule and its chemical properties.

3.1 The octet rule

Lewis found that he could account for the existence of a wide range of molecules by proposing the **octet rule**: *each atom shares electrons with neighboring atoms to achieve a total of eight valence electrons*. As we saw in Section 1.7, a closed-shell, noble-gas configuration is achieved when eight electrons occupy the s and p subshells of the valence shell. One exception is the hydrogen atom, which fills its valence shell, the $1s$ orbital, with two electrons.

The octet rule provides a simple way of constructing a **Lewis structure**, a diagram that shows the pattern of bonds and lone pairs in a molecule. In most cases we can construct a Lewis structure in three steps.

1 Decide on the number of electrons that are to be included in the structure by adding together the numbers of all the valence electrons provided by the atoms.

Each atom provides all its valence electrons (thus, H provides one electron and O, with the configuration $[He]2s^2 2p^4$, provides six). Each negative charge on an ion corresponds to an additional electron; each positive charge corresponds to one electron less.

2 Write the chemical symbols of the atoms in the arrangement showing which are bonded together.

In most cases we know the arrangement or can make an informed guess. The less electronegative element is usually the central atom of a molecule, as in CO_2 and SO_4^{2-}, but there are plenty of well-known exceptions (H_2O and NH_3 among them).

3 Distribute the electrons in pairs so that there is one pair of electrons between each pair of atoms bonded together, and then supply electron pairs (to form lone pairs or multiple bonds) until each atom has an octet.

Each bonding pair is then represented by a single line. The net charge of a polyatomic ion is assumed to be possessed by the ion as a whole, not by a particular individual atom.

Example 3.1 Writing a Lewis structure

Write a Lewis structure for the BF_4^- ion.

Answer The atoms supply $3 + (4 \times 7) = 31$ valence electrons; the single negative charge of the ion reflects the presence of an additional electron. We must therefore accommodate 32 electrons in 16 pairs around the five atoms. One solution is (**1**). The negative charge is ascribed to the ion as a whole, not to a particular individual atom.

Self-test 3.1 Write a Lewis structure for the PCl_3 molecule*

Table 3.1 gives examples of Lewis structures of some common molecules and ions. Except in simple cases, a Lewis structure does not portray the shape of the species, but only the pattern of

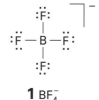

1 BF_4^-

Table 3.1 Lewis structures of some common molecules*

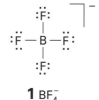

*Only representative resonance structures are given. Shapes are indicated only for diatomic and triatomic molecules. Octet expansion is treated on p. 70.

bonds and lone pairs: it shows the topology of the links, not the geometry of the molecule. For example the BF_4^- ion is actually tetrahedral (**2**), not planar, and PF_3 is trigonal pyramidal (**3**).

> A covalent bond is a shared electron pair; atoms share electron pairs until they have acquired an octet of valence electrons.

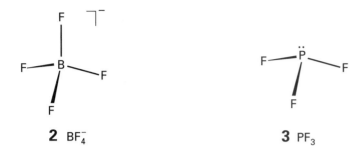

2 BF_4^- **3** PF_3

(a) Resonance

A single Lewis structure is often an inadequate description of the molecule. The Lewis structure of O_3 (**4**), for instance, suggests incorrectly that one OO bond is different from the other, whereas in fact they have identical lengths (1.28 Å) intermediate between those of typical single O—O and double O=O bonds (1.48 Å and 1.21 Å, respectively). This deficiency of the Lewis description is overcome by introducing the concept of **resonance**, in which the actual structure of the molecule is taken to be a superposition of all the feasible Lewis structures corresponding to a given atomic arrangement.

Resonance is indicated by a double-headed arrow:

$$:\ddot{O}—\ddot{O}=\ddot{O} \longleftrightarrow \ddot{O}=\ddot{O}—\ddot{O}:$$

It should be pictured as a *blending* of structures, not a flickering alternation between them. In quantum mechanical terms, the electron distribution of each structure is represented by a wavefunction, and the actual wavefunction, ψ, of the molecule is the superposition of the individual wavefunctions for each structure:[1]

$$\psi = \psi(O—O=O) + \psi(O=O—O)$$

The first of the contributing wavefunctions, for instance, represents an electron distribution in which there is a double bond between the right-hand pair of atoms in O_3. The wavefunction is written as a superposition with equal contributions from each structure because the two structures have identical energies. The blended structure of two or more Lewis structures is called a **resonance hybrid**. Note that resonance occurs between structures that differ only in the allocation of electrons; resonance does not occur between structures in which the atoms themselves lie in different positions. For instance, there is no resonance between the structures SOO and OSO.

Resonance has two main effects:

1 Resonance averages the bond characteristics over the molecule.
2 The energy of a resonance hybrid structure is lower than that of any single contributing structure.

The energy of the O_3 resonance hybrid, for instance, is lower than that of either individual structure alone. Resonance is most important when there are several structures of identical

$$:\ddot{O}—\ddot{O}=\ddot{O}$$

4 O_3

---

[1] This wavefunction is not normalized. We shall often omit normalization constants from linear combinations so as to clarify their structure. The wavefunctions themselves are formulated in the valence-bond theory, which is described later.

5 BF_3

energy that can be written to describe the molecule, as for O_3. In such cases, all the structures of the same energy contribute equally to the overall structure.

Structures with different energies may also contribute to an overall resonance hybrid but, in general, the greater the energy difference between two Lewis structures, the less the higher energy structure contributes. The BF_3 molecule, for instance, could be regarded as a resonance hybrid of the structures shown in (**5**), but the first structure dominates even though the octet is incomplete. Consequently, BF_3 is regarded *primarily* as having that structure with a small admixture of double-bond character. In contrast, for the NO_3^- ion (**6**), the last three structures dominate, and we treat the ion as having partial double-bond character.

> *Resonance between Lewis structures lowers the calculated energy of the molecule and models the bonding character by distributing it over the molecule. Lewis structures with similar energies provide the greatest resonance stabilization.*

6 NO_3^-

(b) Formal charge

The decision about which Lewis structure is likely to have lowest energy and hence contribute predominantly to resonance can be put on a simple quantitative footing by assessing the **formal charge**, f, on each atom. The formal charge of an atom in a Lewis structure is the charge it would have if the bonding were perfectly covalent, with each shared pair of electrons shared equally between the two bonded atoms. That is, each atom is assumed to 'own' one electron of a bonding pair. Each lone pair of electrons belongs wholly to the atom on which the lone pair resides. The formal charge is then the net charge of the atom based on this 'purely covalent, perfect sharing' model of the bonding and is a measure of the extent to which the atom has gained or lost electrons in reaching the Lewis structure. More formally,

$$f = V - L - \tfrac{1}{2}P \tag{1}$$

where V is the number of valence electrons on the parent atom, L is the number of lone pair electrons in the Lewis structure of the molecule, and P is the number of shared electrons. Thus the formal charge is the difference between the number of valence electrons in the free atom and the number that the atom has in the molecule supposing it to own one electron of each shared pair and both electrons of any lone pair (Fig. 3.1). The formal charge represents (in some idealized sense) the number of electrons that an atom gains or loses when it enters into perfect covalent bonding with other atoms. The sum of the formal charges on a Lewis structure is equal to the total charge of the species (and is zero for an electrically neutral molecule).

It is commonly the case that the lowest energy structure is the one with: (1) the lowest formal charges on the atoms; and (2) the structure in which the more electronegative element is assigned a formal negative charge and the less electronegative element is assigned a formal positive charge.

> *The formal charge is the charge an atom would have if electron pairs were shared equally. Lewis structures with low formal charges typically have the lowest energy.*

Example 3.2 Writing resonance structures

Write resonance structures for an NO_2F molecule and identify the dominant structures.

Answer Four Lewis structures and their formal charges are shown in (**7**). It is very unlikely that a low energy will be achieved with a positive charge on an F atom or on the N atom, so the two structures with N═O bonds are most likely to dominate in the resonance.

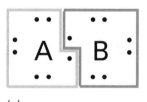

(a)

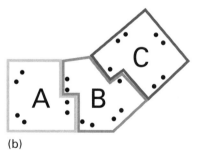

(b)

3.1 A pictorial representation of the calculation of formal charge. The lines show how the bonding electrons and lone-pair electrons are apportioned to each atom (a) in a diatomic molecule with Lewis structure A—B and (b) in a triatomic molecule with Lewis structure A═B—C. The formal charge on each atom is the difference between the number of electrons obtained in this way and the number in the free, neutral atom.

$$\underset{:F:\ 0}{\overset{-1\quad +2\quad -1}{:O-N-O:}} \qquad \underset{:F:\ 0}{\overset{-1\quad +1\quad 0}{:O-N=O}} \qquad \underset{:F:\ 0}{\overset{0\quad +1\quad -1}{O=N-O:}} \qquad \underset{:F:\ +1}{\overset{-1\quad +1\quad -1}{:O-N-O:}}$$

7 NO_2F

Self-test 3.2 Write resonance structures for the NO_2^- ion.

(c) Oxidation number

Formal charge is a parameter derived by exaggerating the covalent character of a bond. The **oxidation number**, ω, is a parameter obtained by exaggerating the ionic character of a bond. It can be regarded as the charge that an atom would have if the more electronegative atom in a bond acquired the two electrons of the bond completely. Each O atom in a compound therefore becomes an oxide ion (provided F is not present), and hence is ascribed an oxidation number of -2. Likewise, the exaggerated ionic structure of NO_3^- is treated as $N^{5+}(O^{2-})_3$, so the oxidation number of nitrogen in this compound is $+5$. When an element is ascribed a particular oxidation number, it is said to be in a specific **oxidation state**. Thus, when nitrogen has the oxidation number $+5$ it is said to be in the $+5$ oxidation state and denoted either N(V), with a Roman numeral denoting the oxidation number, or N($+5$). These conventions may be used even if the oxidation number is negative, so oxygen is in the -2 oxidation state, denoted O(-2) or O($-$II), in most of its compounds.

In practice, oxidation numbers (and the corresponding states) are assigned by applying a set of rules (Table 3.2). These rules reflect the consequences of electronegativity for the 'exaggerated ionic' structures of compounds and match the increase in the degree of

Table 3.2 The determination of oxidation number*

		Oxidation number
1.	The sum of the oxidation numbers of all the atoms in the species is equal to its total charge	
2.	For atoms in their elemental form	0
3.	For atoms of Group 1	$+1$
	For atoms of Group 2	$+2$
	For atoms of Group 13/III (except B)	$+3(EX_3), +1(EX)$
	For atoms of Group 14/IV (except C, Si)	$+4(EX_4), +2(EX_2)$
4.	For hydrogen	$+1$ in combination with nonmetals
		-1 in combination with metals
5.	For fluorine	-1 in all its compounds
6.	For oxygen	-2 unless combined with F
		-1 in peroxides (O_2^{2-})
		$-\frac{1}{2}$ in superoxides (O_2^-)
		$-\frac{1}{3}$ in ozonides (O_3^-)
7.	Halogens	-1 in most compounds, unless the other elements include oxygen or more electronegative halogens

*To determine an oxidation number, work through the following rules in the order given. Stop as soon as the oxidation number has been assigned. These rules are not exhaustive, but they are applicable to a wide range of common compounds.

oxidation that we would expect as the number of oxygen atoms in a compound increase (as in going from NO to NO_3^-). This aspect of oxidation numbers is taken further in Chapter 6.

Example 3.3 Assigning the oxidation number of an element

What is the oxidation number of (a) N in the azide ion, N_3^-, (b) Mn in the permanganate ion, MnO_4^-?

Answer We work through the steps set out in Table 3.2, in the order given. (a) The charge of the species is -1; all three atoms are of the same element, so $3\omega(N) = -1$, which implies that $\omega(N) = -\frac{1}{3}$. (b) The sum of the oxidation numbers of all the atoms is -1, so $\omega(Mn) + 4\omega(O) = -1$. Because $\omega(O) = -2$, it follows that $\omega(Mn) = -1 - 4(-2) = +7$. That is, MnO_4^- is a compound of Mn(VII).

Self-test 3.3 What is the oxidation number of (a) O in O_2^+, (b) P in PO_4^{3-}?

8 PCl_5

9 SF_6

(d) Hypervalence

The elements of Period 2, Li through Ne, obey the octet rule quite well, but elements of later periods show deviations from it. For example, the bonding in PCl_5 requires the P atom to have 10 electrons in its valence shell, one pair for each P—Cl bond (**8**). Similarly, in SF_6 the S atom must have 12 electrons if each F atom is to be bound to the central S atom by an electron pair (**9**). Species of this kind, which demand the presence of more than an octet of electrons around at least one atom, are called **hypervalent**. Species for which the resonance structures include expanded octets but which do not *necessarily* have more than eight valence electrons are not regarded as hypervalent: thus, SO_4^{2-} is not hypervalent even though some contributing Lewis structures have 12 electrons in the valence shell of S. In certain compounds, an atom may be located in an environment in which it has more than its normal coordination number but in which the bonding is not necessarily a result of shared electron pairs. Examples are the eightfold coordination of C in Be_2C and sixfold coordination in $[Co_6C(CO)_{15}]^{2-}$. Such species are best regarded as **hypercoordinate** rather than hypervalent. Five- and six-coordinate B and C are also known in covalent cage and cluster compounds and in extended solid-state metal borides and carbides (Chapter 10).

The traditional explanation of hypervalence (for SF_6, for instance) and of octet expansion in general (for certain Lewis structures of SO_4^{2-}, for instance) invokes the availability of low-lying unfilled d orbitals, which can accommodate the additional electrons. According to this explanation, a P atom can accommodate more than eight electrons if it uses its vacant $3d$ orbitals. In PCl_5, for instance, at least one $3d$ orbital must be used. The rarity of hypervalence in Period 2 is then ascribed to the absence of $2d$ orbitals in these elements. However, a more compelling reason for the rarity of hypervalence in Period 2 may be the geometrical difficulty of packing more than four atoms around a single small central atom, and may in fact have little to do with the availability of d orbitals. Recent computations (of the kind described later) suggest that the traditional explanation overemphasizes the role of d orbitals in hypervalent compounds. In Section 3.12b, for instance, we shall see how to account for hypervalent compounds without making use of d orbitals.

> *Hypervalence, bonding to more atoms than conventional octet formation allows, and octet expansion in resonance structures occur for elements following Period 2.*

3.2 Structure and bond properties

Certain properties of bonds are approximately the same in different compounds of the elements. Thus, if we know the strength of an O—H bond in H_2O, then with some confidence we can use the same value if we are interested in the strength of the O—H bond in CH_3OH. At this stage we confine our attention to two of the most important characteristics of a bond: its length and its strength.

(a) Bond lengths

The **equilibrium bond length** in a molecule is the internuclear separation of the two bonded atoms.[2] A wealth of useful and accurate bond length information is available in the literature, most of it obtained by X-ray diffraction on solids. Equilibrium bond lengths of molecules in the gas phase are usually determined by infrared or microwave spectroscopy, or by electron diffraction. Some typical values are shown in Table 3.3.

To a reasonable first approximation, equilibrium bond lengths can be partitioned into contributions from each atom of the bonded pair. The contribution of an atom to a covalent bond is called the **covalent radius** of the element (**10**) (Table 3.4). We can use covalent radii to predict, for example, that the length of a P—N bond is $1.10\ \text{Å} + 0.74\ \text{Å} = 1.84\ \text{Å}$; experimentally, this bond length is close to $1.8\ \text{Å}$ in a number of compounds. Experimental bond lengths should be used whenever possible, but covalent radii are useful for making cautious estimates when experimental data are not available.

Covalent radii vary through the periodic table in much the same way as metallic and ionic radii (Section 1.8a), for the same reasons, and are smallest close to F. Sums of covalent radii are approximately equal to the separation of nuclei when the cores of the two atoms are in contact: the valence electrons draw the two atoms together until the repulsion between the cores starts to dominate. A covalent radius expresses the closeness of approach of *bonded* atoms; the closeness of approach of *nonbonded* atoms in neighboring molecules that are in contact is expressed in terms of the **van der Waals radius** of the element, which is the internuclear separation when the *valence* shells of the two atoms are in nonbonding contact (**11**). Van der Waals radii are of paramount importance for the packing of molecular compounds in crystals, the conformations adopted by small but flexible molecules, and the shapes of biological macromolecules.

(b) Bond strength

A convenient thermodynamic measure of the strength of an **AB** bond is the **bond dissociation enthalpy**, $\Delta H^{\ominus}(\text{A}-\text{B})$, the standard reaction enthalpy for the process

$$\text{A}-\text{B(g)} \longrightarrow \text{A(g)} + \text{B(g)}$$

Table 3.3 Bond lengths

	$R_e/\text{Å}$
H_2^+	1.06
H_2	0.74
HF	0.92
HCl	1.27
HBr	1.41
HI	1.60
N_2	1.09
O_2	1.21
F_2	1.44
Cl_2	1.99
I_2	2.67

Source: G. Herzberg, *Spectra of diatomic molecules.* Van Nostrand, Princeton (1950).

Table 3.4 Covalent radii* ($r_{cov}/\text{Å}$)

H			
0.37			
C	N	O	F
0.77 (1)	0.74 (1)	0.66 (1)	0.64
0.67 (2)	0.65 (2)	0.57 (2)	
0.60 (3)			
Si	P	S	Cl
1.18 (1)	1.10	1.04 (1)	0.99
		0.95 (2)	
Ge	As	Se	Br
1.22 (1)	1.21	1.17	1.14
	Sb	Te	I
	1.41	1.37	1.33

*Values are for single bonds except where otherwise stated (in parentheses).

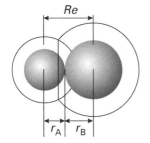

10 Covalent radius

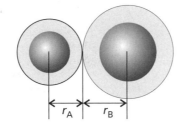

11 Van der Waals radius

---

[2] When the molecular potential energy curve has been introduced (Section 3.4), we shall see that the equilibrium bond length is the internuclear separation corresponding to the minimum of the curve.

The **mean bond enthalpy**, B, is the average bond dissociation enthalpy taken over a series of A—B bonds in different molecules (Table 3.5).

Mean bond enthalpies can be used to estimate reaction enthalpies. However, thermodynamic data on actual species should be used whenever possible in preference to mean values because the latter can be misleading. For instance, the Si—Si bond enthalpy ranges from 226 kJ mol^{-1} in Si_2H_6 to 322 kJ mol^{-1} in $Si_2(CH_3)_6$. The values in Table 3.5 are best considered as data of last resort: they may be used to make rough estimates of reaction enthalpies when enthalpies of formation or actual bond enthalpies are unavailable.

> *The strength of a bond is measured by its dissociation enthalpy. Mean bond enthalpies may be used with caution to make estimates of reaction enthalpies.*

Example 3.4 Making estimates using mean bond enthalpies

Estimate the reaction enthalpy for the production of $SF_6(g)$ from $SF_4(g)$ given that the mean bond enthalpies of F_2, SF_4, and SF_6 are $+158$, $+343$, and $+327 \text{ kJ mol}^{-1}$, respectively, at $25 \, ^\circ C$.

Answer The reaction is

$$SF_4(g) + F_2(g) \longrightarrow SF_6(g)$$

1 mol F—F bonds and 4 mol S—F bonds (in SF_4) must be broken, corresponding to an enthalpy change of $158 + 4 \times 343 = +1530 \text{ kJ}$. Then 6 mol S—F bonds (in SF_6) must be formed, corresponding to an enthalpy change of $6 \times (-327) = -1962 \text{ kJ}$. The net enthalpy change is therefore

$$\Delta H^{\ominus} = +1530 \text{ kJ} - 1962 \text{ kJ} = -432 \text{ kJ}$$

Hence, the reaction is strongly exothermic. The experimental value for the reaction is -434 kJ.

Self-test 3.4
Estimate the enthalpy of formation of H_2S from S_8 (a cyclic molecule) and H_2.

Table 3.5 Mean bond enthalpies* $(B/\text{kJ mol}^{-1})$

	H	C	N	O	F	Cl	Br	I	S	P	Si
H	436										
C	412	348 (1)									
		612 (2)									
		518 (a)									
N	388	305 (1)	163 (1)								
		613 (2)	409 (2)								
		890 (3)	946 (3)								
O	463	360 (1)	157	146 (1)							
		743 (2)		497 (2)							
F	565	484	270	185	155						
Cl	431	338	200	203	254	242					
Br	366	276				219	193				
I	299	238				210	178	151			
S	338	259			496	250	212		264		
P	322(1)									201	
										480(3)	
Si	318			466							226

*Values are for single bonds except where otherwise stated (in parentheses); (a) denotes aromatic.

(c) Bond enthalpy trends in the p-block

The trend in bond enthalpies in the p-block can be summarized as follows:

- For an element E that has no lone pairs, the E—X bond enthalpy decreases down the group.

For example:

	$B/(\text{kJ mol}^{-1})$
C—C	347
Si—C	301
Ge—C	242

Another general trend is as follows:

- For an element that has lone pairs, the bond enthalpy decreases down the group, but the value for an element at the head of a group is anomalous and smaller than the bond enthalpy of an element in Period 3.

Two examples are:

	$B/(\text{kJ mol}^{-1})$			$B/(\text{kJ mol}^{-1})$
N—O	163		C—Cl	326
P—O	368		Si—Cl	401
As—O	330		Ge—Cl	339
			Sn—Cl	314

The relative weakness of single bonds between Period 2 elements that have lone pairs is often ascribed to the closeness of the lone pairs on neighboring atoms and the repulsion between them.

A number of features of the p-block can be interpreted (and remembered) with the aid of arguments based on bond enthalpies. For example, the bond dissociation enthalpy of gaseous BO is 788 kJ mol^{-1} whereas the B—O single bond enthalpy is 523 kJ mol^{-1}. It follows that the boron–oxygen bond in BO must be at least double and perhaps triple. Although acyclic alkanes with less than four C atoms are thermodynamically stable with respect to decomposition into their elements, their silicon analogs the silanes, Si_nH_{2n+2}, are unstable with respect to Si(s) and $H_2(g)$. This difference is a consequence of the weakness of the Si—H bond compared to that of the H—H bond. Third, because the Cl—Cl bond of Cl_2 is weaker than the H—H bond of H_2, and an Si—Cl bond is stronger than an Si—H bond, compounds of formula Si_nCl_{2n+2} can be expected to exist, and indeed are known for n as high as 10.

The trend in bond strengths for atoms with lone pairs is illustrated by the P—P single bond, for which $B = 200 \text{ kJ mol}^{-1}$, compared with the N—N single bond, for which $B = 165 \text{ kJ mol}^{-1}$. However, nitrogen forms much stronger multiple bonds than phosphorus. This difference can account for the fact that phosphorus is found as P_4 molecules (12) whereas nitrogen is found as N_2 molecules (:N≡N:). The considerable difference in single and triple bond enthalpies of nitrogen also explains the rarity of **catenation**, the formation of chains of atoms of the same element, in its compounds. Thus, hydrazine, H_2N-NH_2, is quite strongly endoergic (its standard Gibbs energy of formation is $+149 \text{ kJ mol}^{-1}$) and analogs of higher alkanes are unknown.

Similar differences are found in Group 16/VI. The S—S single bond enthalpy is larger than the O—O single bond enthalpy (263 kJ mol^{-1} and 142 kJ mol^{-1}, respectively). However, the bond enthalpies of the doubly bonded species O_2 is significantly larger than that of S_2 (498 kJ mol^{-1} and 431 kJ mol^{-1}, respectively). This difference explains why elemental sulfur forms rings or chains with S—S single bonds, whereas oxygen exists as diatomic molecules.

12 P_4

Similarly, sulfur catenation leads to polysulfides of formula $[S—S—S]^{2-}$ and $[S—S—S—S—S]^{2-}$, but polyoxygen anions beyond O_2^{2-} are unstable.

One enlightening application of bond-enthalpy arguments concerns **subvalent compounds**, or molecules in which fewer bonds are formed than valence rules allow, such as PH_2. This compound is thermodynamically stable with respect to dissociation into the constituent atoms; its instability lies in its tendency to decompose as follows:[3]

$$3\,PH_2(g) \longrightarrow 2\,PH_3(g) + \tfrac{1}{4}P_4(s)$$

The origin of the spontaneity of this reaction is the strength of the P—P bonds in solid phosphorus (P_4). There are the same number (six) of P—H bonds in the reactants as there are in the products, but the reactants have no P—P bonds.

> For an element E that has no lone pairs, the E—X bond enthalpy decreases down the group. For an element that has lone pairs, the E—X bond enthalpy typically increases between Periods 2 and 3 and then decreases down the group.

(d) Electronegativity and bond enthalpy

The concept of electronegativity, χ, was introduced in Section 1.8d, where we saw that it is the power of an atom of the element to attract electrons to itself when it is part of a compound. The greater the difference in electronegativity between two elements A and B, the greater the ionic character of the A—B bond.

Linus Pauling's original formulation drew on concepts relating to the energetics of bond formation. He argued that the excess energy Δ of an A—B bond over the average energy of A—A and B—B bonds can be attributed to the presence of an ionic contributions to the covalent bonding. He defined the difference in electronegativity as

$$|\chi_P(A) - \chi_P(B)| = 0.102(\Delta/(kJ\,mol^{-1}))^{1/2} \tag{2a}$$

where

$$\Delta = B(A—B) - \tfrac{1}{2}\{B(A—A) + B(B—B)\} \tag{2b}$$

with $B(X—Y)$ the mean X—Y bond enthalpy. Thus, if the A—B bond enthalpy differs markedly from the average of the nonpolar A—A and B—B bonds, it is presumed that there is a substantial ionic contribution to the wavefunction, and hence a large difference in electronegativity between the two atoms.

There are complications with this definition, because Pauling electronegativities increase with increasing oxidation number of the element. The values in Table 1.8 are for the maximum oxidation number of the element concerned. Pauling electronegativities are useful for estimating bond enthalpies between elements of different electronegativity and for a qualitative assessment of the polarities of bonds.

> The Pauling definition of electronegativity is based on bond enthalpies; it is useful for estimating bond enthalpies and for assessing the polarities of bonds.

3.3 The VSEPR model

The **valence-shell electron pair repulsion model** (VSEPR model) of molecular shape is a simple extension of Lewis's ideas and is surprisingly successful for predicting the shapes of polyatomic molecules. The theory stems from suggestions made by Nevil Sidgwick and

---

[3] In terms of the language introduced in Section 6.8, this reaction is an example of a disproportionation in which $3\,P(II) \rightarrow 2\,P(III) + P(0)$.

Herbert Powell in the years up to 1940 and later extended and put into a more modern context by Ronald Gillespie and Ronald Nyholm.[4]

(a) The basic shapes

The primary assumption of the VSEPR model is that regions of enhanced electron density, by which we mean bonding pairs, lone pairs, or the concentrations of electrons associated with multiple bonds, take up positions as far apart as possible so that the repulsions between them are minimized. For instance, four such regions of electron density will lie at the corners of a regular tetrahedron, five will lie at the corners of a trigonal bipyramid, and so on (Table 3.6). As an example, an SF_6 molecule, with six single bonds around the central S atom, is predicted (and found) to be octahedral (**13**), and a PCl_5 molecule, with five single bonds, is predicted (and found) to be trigonal-bipyramidal (**14**).

Although the arrangement of regions of electron density, both bonding regions and regions associated with lone pairs, governs the shape of the molecule, the *name* of the shape is determined by the arrangement of *atoms*, not the arrangement of the regions of electron density (Table 3.7). For instance, the NH_3 molecule has four electron pairs that are disposed tetrahedrally, but as one of them is a lone pair the molecule itself is classified as trigonal-pyramidal. One apex of the pyramid is occupied by the lone pair. Similarly, H_2O has a tetrahedral arrangement of its electron pairs but, as two of the pairs are lone pairs, the molecule is reported as angular.

According to the VSEPR model, a multiple bond is treated as just another region of enhanced electron density. Consequently, a linear structure is predicted for O=C=O, on the grounds that the C atom has two regions of electron density (the two double bonds), which adopt a linear arrangement. This approach does away with any worry about which resonance structure to consider. Thus, the Lewis structure of SO_4^{2-} in which the S atom has an octet and all SO bonds are single and an expanded-octet structure with two SO double bonds and two SO single bonds are both predicted to be tetrahedral.

Apart from the fact that the VSEPR model is highly successful, there is no firm evidence for the assumptions about the distribution of electron density on which it is based. Moreover, some basic shapes have repulsions that are not much lower than those in alternative arrangements, and a molecule may adopt one of these alternatives if there are other contributions to the energy that result in a lower energy overall. For example, a square-pyramidal arrangement of bonds is only slightly higher in electron-repulsion energy than a trigonal-bipyramidal arrangement, and there are several examples of the former (**15**). Similarly, the basic shapes for seven regions of electron density are less readily predicted than

Table 3.6 The basic arrangement of regions of electron density according to the VSEPR model

Number of electron regions	Arrangement
2	Linear
3	Trigonal planar
4	Tetrahedral
5	Trigonal bipyramidal
6	Octahedral

Table 3.7 The description of molecular shapes

Linear	HCN, CO_2
Angular	H_2O, O_3, NO_2^-
Trigonal planar	BF_3, SO_3, NO_3^- CO_3^{2-}
Trigonal pyramidal	NH_3, SO_3^{2-}
Tetrahedral	CH_4, SO_4^{2-}, NSF_2^*
Square planar	XeF_4
Square pyramidal	$Sb(Ph)_5$
Trigonal bipyramidal	$PCl_5(g)$, SOF_4^*
Octahedral	SF_6, PCl_6^-, $IO(OH)_5^*$

*Approximate shape.

13 SF_6 **14** PCl_5 **15** $[InCl_5]^{2-}$

[4] For an excellent introduction to modern attitudes to VSEPR theory, see R.J. Gillespie and I. Hargittai, *The VSEPR model of molecular geometry*. Allyn & Bacon, Needham Heights (1991). An advanced discussion of the role of electron density distributions in molecules and their role in the determination of shape has been given by R.F.W. Bader, *Atoms in molecules*. Clarendon Press, Oxford (1990). See also R.F.W. Bader, R.J. Gillespie, and P.J. MacDougall, *J. Am. Chem. Soc.* **110**, 7329 (1988).

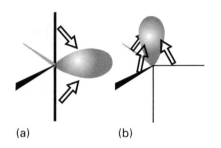

(a)   (b)

3.2 In the VSEPR model of molecular shape, a lone pair in (a) the equatorial position of a trigonal-bipyramidal arrangement interacts strongly with two bonding pairs, but (b) in the axial position it interacts strongly with three bonding pairs. The former arrangement is generally of lower energy.

others, partly because so many different conformations correspond to similar energies. Lone pairs are stereochemically less influential when they belong to heavy p-block elements. The SeF_6^{2-} and $TeCl_6^{2-}$ ions, for instance, are octahedral despite the presence of a lone pair on the Se and Te atoms. Lone pairs that do not influence the molecular geometry are said to be **stereochemically inert**.

> In the VSEPR model, regions of enhanced electron density are supposed to take up positions as far apart as possible, and the shape of the molecule is identified by referring to the locations of the atoms in the resulting structure.

(b) Modifications of the basic shapes

Once the number of electron regions has been used to identify the basic shape of a molecule, minor adjustments are made by taking into account the differences in electrostatic repulsion between bonding regions and lone pairs. These repulsions are generally assumed to lie in the order

lone pair/lone pair > lone pair/bonding region > bonding region/bonding region

In elementary accounts, the greater repelling effect of a lone pair is explained by supposing that the lone pair is on average closer to the nucleus than a bonding pair and therefore repels other electron pairs more strongly. However, the true origin of the difference is obscure. An additional detail about this order of repulsions is that, given the choice between an axial and an equatorial site for a lone pair in a trigonal-bipyramidal array, the lone pair occupies the equatorial site because it is then repelled less by the bonding pairs (Fig. 3.2).

Example 3.5 Using VSEPR theory to predict shapes

Predict the shape of an SF_4 molecule.

Answer First, we write the Lewis structure of SF_4 (**16**). This structure has five electron pairs around the central atom, and they adopt a trigonal-bipyramidal arrangement. Repulsion from the one lone pair is minimized if it occupies an equatorial site, for then it interacts with the two axial bonding pairs strongly, rather than an axial site, when it would interact strongly with three equatorial bonding pairs. The S–F bonds then bend away from the lone pair to give a molecular shape that resembles a see-saw, with the axial bonds forming the 'plank' of the see-saw and the equatorial bonds the 'pivot' (**17**).

Self-test 3.5 Predict the shape of an XeF_2 molecule.

The angle between the O—H bonds in H_2O decreases slightly from its tetrahedral value (109.5°) as the two lone pairs move apart. This decrease is in agreement with the observed HOH bond angle of 104.5°. A similar effect accounts for the HNH bond angle of 107° in NH_3.

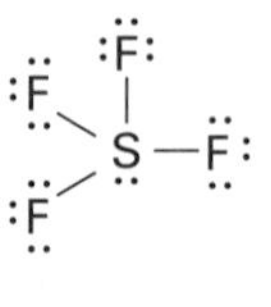

16 SF_4

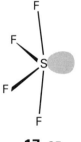

17 SF_4

It also accords neatly with the regular tetrahedral shape of NH_4^+, in which all four bonds are equivalent.

> Lone pairs repel other pairs more strongly than bonding pairs do.

Valence-bond theory

The **valence-bond theory** (VB theory) of bonding was the first quantum mechanical theory of bonding to be developed, and can be regarded as a way of expressing Lewis's concepts in terms of wavefunctions. The computational techniques involved have been largely superseded by molecular orbital theory, but much of the language and some of the concepts of VB theory still remain.

3.4 The hydrogen molecule

The two-electron wavefunction for two widely separated H atoms is $\psi = \phi_A(1)\phi_B(2)$, where ϕ_J is an H$1s$ orbital on atom J. When the atoms are close, it is not possible to know whether it is electron 1 that is on A or electron 2. An equally valid description is therefore $\psi = \phi_A(2)\phi_B(1)$, in which electron 2 is on A and electron 1 is on B. When two outcomes are equally probable, quantum mechanics instructs us to describe the true state of the system as a superposition of the wavefunctions for each possibility, so a better description of the molecule than either wavefunction alone is the linear combination

$$\psi = \phi_A(1)\phi_B(2) + \phi_A(2)\phi_B(1) \tag{3}$$

This function is the (unnormalized) VB wavefunction for an H—H bond. The formation of the bond can be pictured as due to the high probability that the two electrons will be found between the two nuclei and hence will bind them together. More formally, the wave pattern represented by the term $\phi_A(1)\phi_B(2)$ interferes constructively with the wave pattern represented by the contribution $\phi_A(2)\phi_B(1)$, and there is an enhancement in the value of the wavefunction in the internuclear region. For technical reasons stemming from the Pauli principle, only electrons with paired spins can be described by a wavefunction of the type written in eqn 3, so only paired electrons can contribute to a bond in VB theory.

The electron distribution described by the wavefunction in eqn 3 is called a **σ bond**. A σ bond has cylindrical symmetry around the internuclear axis, and the electrons in it have zero orbital angular momentum about that axis. The **molecular potential energy curve** for H_2, the curve showing the variation of the energy of the molecule with internuclear separation, is calculated by changing the internuclear separation R and evaluating the energy at each selected separation (Fig. 3.3). The energy falls below that of two separated H atoms as the two atoms are brought within bonding distance and each electron is free to migrate to the other atom. However, the resulting lowering of energy is counteracted by an increase in energy from the Coulombic repulsion between the two positively charged nuclei. This positive contribution to the energy becomes large as R becomes small. Consequently, the total potential energy curve passes through a minimum and then climbs to a strongly positive value at small internuclear separations. The depth of the minimum of the curve is denoted D_e. The deeper the minimum, the more strongly the atoms are bonded together. The steepness of the well shows how rapidly the energy of the molecule rises as the bond is stretched or compressed. The steepness therefore governs the vibrational frequency of the molecule (Section 4.8).

> In the valence-bond theory, the wavefunction of an electron pair is formed by superimposing the wavefunctions for the separated fragments of the molecule, as in eqn 3. A molecular potential energy curve shows the variation of the molecular energy with internuclear separation.

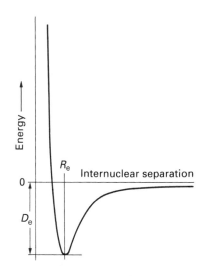

3.3 A molecular potential energy curve showing how the total energy of a molecule (with nuclei stationary) varies as the internuclear separation is changed. The zero of energy is the energy of the free atoms.

3.5 Homonuclear diatomic molecules

The same description can be applied to more complex molecules, such as **homonuclear diatomic molecules**, which are diatomic molecules in which both atoms belong to the same element. Nitrogen, N_2, is an example. To construct the valence-bond description of N_2, we consider the valence-electron configuration of each atom, which from Section 1.7 we know to be N $2s^2 2p_x^1 2p_y^1 2p_z^1$. It is conventional to take the z-axis to be the internuclear axis, so we can imagine each atom as having a $2p_z$ orbital pointing towards a $2p_z$ orbital on the other atom (Fig. 3.4), with the $2p_x$ and $2p_y$ orbitals perpendicular to the axis. A σ bond is then formed by spin pairing between the two electrons in the opposing $2p_z$ orbitals. Its spatial wavefunction is still given by eqn 3, but now ϕ_A and ϕ_B stand for the two $2p_z$ orbitals.

The remaining $2p$ orbitals cannot merge to give σ bonds as they do not have cylindrical symmetry around the internuclear axis. Instead, the electrons in them merge to form two **π bonds**. A π bond arises from the spin pairing of electrons in two p orbitals that approach side-by-side (Fig. 3.5). The bond is so called because, viewed along the internuclear axis, it resembles a pair of electrons in a p orbital. More precisely, an electron in a π bond has one unit of orbital angular momentum about the internuclear axis.

There are two π bonds in N_2, one formed by spin pairing in two neighboring $2p_x$ orbitals and the other by spin pairing in two neighboring $2p_y$ orbitals. The overall bonding pattern in N_2 is therefore a σ bond plus two π bonds (Fig. 3.6), which is consistent with the Lewis structure : N≡N : for nitrogen.

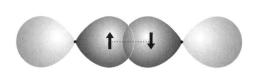

3.4 The formation of a σ bond in VB theory is pictured as arising from the pairing of electrons in the two contributing atomic orbitals and the latter's overlap.

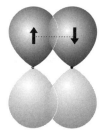

3.5 The formation of π bond in VB theory when two unpaired electrons in neighboring orbitals pair and the orbitals overlap.

> To form a VB description of a diatomic molecule, electrons in atomic orbitals of the same symmetry but on neighboring atoms are paired to form σ and π bonds.

3.6 Polyatomic molecules

Each σ bond in a polyatomic molecule is formed by the spin pairing of electrons in any neighboring atomic orbitals with cylindrical symmetry about the relevant internuclear axis. Likewise, π bonds are formed by pairing electrons that occupy neighboring atomic orbitals of the appropriate symmetry.

Consider the VB description of H_2O. The valence-electron configuration of an O atom is $2s^2 2p_x^2 2p_y^1 2p_z^1$. The two unpaired electrons in the $O2p$ orbitals can each pair with an electron in an $H1s$ orbital, and each combination results in the formation of a σ bond (each bond has cylindrical symmetry about the respective O–H internuclear axis). Because the $2p_y$ and $2p_z$ orbitals lie at 90° to each other, the two σ bonds also lie at 90° to each other (Fig. 3.7). We can predict, therefore, that H_2O should be an angular molecule, which it is. However, the theory predicts a bond angle of 90°, whereas the actual bond angle is 104.5°. Similarly, to

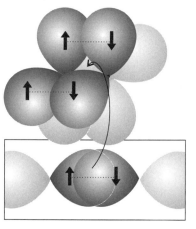

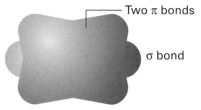

Two π bonds

σ bond

3.6 The valence-bond description of nitrogen, in which two electrons form a σ bond, and another two pairs form π bonds. In linear molecules, where the x- and y-axes are not specified, the electron density of π bonds is cylindrically symmetrical around the internuclear axis.

predict the structure of an ammonia molecule, we start by noting that the valence-electron configuration of an N atom given previously suggests that three H atoms can form bonds by spin pairing with the electrons in the three half-filled $2p$ orbitals. The latter are perpendicular to each other, so we predict a trigonal pyramidal molecule with a bond angle of 90°. The molecule is indeed trigonal pyramidal, but the experimental bond angle is 107°. The origin of this discrepancy is discussed below.

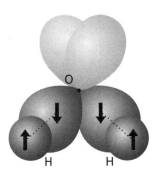

3.7 The valence-bond description of a water molecule. There are two σ bonds, each formed by spin pairing of electrons in $O2p$ orbitals and $H1s$ orbitals. The bond angle according to this primitive model is 90°.

(a) Promotion

An apparent deficiency of valence-bond theory is its inability to account for carbon's tetravalence, its ability to form four bonds. The ground-state configuration of C is $2s^2 2p_x^1 2p_y^1$, which suggests that a carbon atom should be capable of forming only two bonds, not four. This deficiency is overcome by allowing for **promotion, the excitation of an electron to an orbital of higher energy.** Although electron promotion requires an investment of energy, it is worthwhile if that energy can be more than recovered from the greater strength or number of bonds that it allows to be formed. Promotion is not a 'real' process in which an atom somehow becomes excited and then forms bonds: it is a contribution to the overall energy change that occurs when bonds form.

In carbon, for example, the promotion of a $2s$ electron to a $2p$ orbital can be thought of as leading to the configuration $2s^1 2p_x^1 2p_y^1 2p_z^1$, with four unpaired electrons in separate orbitals. These electrons may pair with four electrons in orbitals provided by four other atoms (such as four $H1s$ orbitals if the molecule is CH_4), and hence form four σ bonds. Although energy was required to promote the electron, it is more than recovered by the atom's ability to form four bonds in place of the two bonds of the unpromoted atom. Promotion, and the formation of four bonds, is a characteristic feature of carbon (and of its congeners in Group 14/IV) because the promotion energy is quite small: the promoted electron leaves a doubly occupied $2s$ orbital and enters a vacant $2p$ orbital, hence significantly relieving the electron–electron repulsion it experiences in the former.

> In polyatomic molecules, each bond is described by a wavefunction like that in eqn 3; promotion may occur if the outcome is to achieve more bonds and a lower overall energy.

(b) Hybridization

The description of the bonding in AB_4 molecules of Group 14/IV is still incomplete because it appears to imply the presence of three σ bonds of one type (formed from ϕ_B and ϕ_{A2p} orbitals) and a fourth σ bond of a distinctly different character (formed from ϕ_B and ϕ_{A2s}) whereas all the evidence (bond lengths, strengths, shape) point to the equivalence of all four A—B bonds.

This problem is overcome by realizing that the electron density distribution in the promoted atom is equivalent to the electron density in which each electron occupies a **hybrid orbital** formed by interference between the $A2s$ and $A2p$ orbitals. The origin of the hybridization can be appreciated by thinking of the four atomic orbitals, which are waves centered on a nucleus, as being like ripples spreading from a single point on the surface of a lake: the waves interfere destructively and constructively in different regions, and give rise to four new shapes.

The specific linear combinations that give rise to four equivalent hybrid orbitals are

$$h_1 = s + p_x + p_y + p_z \qquad h_2 = s - p_x - p_y + p_z$$
$$h_3 = s - p_x + p_y - p_z \qquad h_4 = s + p_x - p_y - p_z \tag{4}$$

As a result of the interference between the component orbitals, each hybrid orbital consists of a large lobe pointing in the direction of one corner of a regular tetrahedron (Fig. 3.8). The angle between the axes of the hybrid orbitals is the tetrahedral angle, $\arccos(-1/3) = 109.47°$. Because each hybrid is built from one s orbital and three p orbitals, it is called an sp^3 **hybrid orbital.**

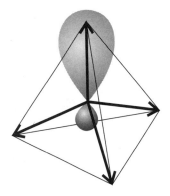

3.8 The four equivalent sp^3 tetrahedral hybrid orbitals. Each one points toward the vertex of a regular tetrahedron.

It is now easy to see how the valence-bond description of an AB_4 molecule leads to a tetrahedral molecule containing four equivalent A—B bonds. Each hybrid orbital of the promoted A atom contains a single unpaired electron; an electron in ϕ_B can pair with each one, giving rise to a σ bond pointing in a tetrahedral direction. Because each sp^3 hybrid orbital has the same composition, all four σ bonds are identical apart from their orientation in space.

A further feature of hybridization is that a hybrid orbital has pronounced directional character, in the sense that it has an enhanced amplitude in the internuclear region. This directional character arises from the constructive interference between the s orbital and the positive lobes of the p orbitals. As a result of the enhanced amplitude in the internuclear region, the bond strength is greater than for an s or p orbital alone. This increased bond strength is another factor that helps to repay the promotion energy.

Hybrid orbitals of different compositions are used to match different molecular geometries and to provide a basis for their VB description. For example, sp^2 hybridization is used to reproduce the electron distribution needed for trigonal planar molecules, such as on B in BF_3 and N in NO_3^-, and sp hybridization reproduces a linear distribution. Table 3.8 gives the hybrids needed to match the geometries of a variety of electron distributions.

> Hybrid orbitals are formed in VB theory when a distribution of electron density needs to be modeled by using atomic orbitals on a given atom.

(c) Isolobality

The concept of hybridization can be used to identify analogies in the structures of apparently unrelated molecules. Thus, we may view $N(CH_3)_3$ as derived from NH_3 by substitution of a CH_3 fragment for each H atom. In current terminology, the structurally analogous fragments are said to be **isolobal**, and the relationship is expressed by the symbol $\overleftrightarrow{\circlearrowleft}$. Two fragments are isolobal if their highest-energy orbitals have the same symmetry (such as the σ symmetry of the H1s and a Csp^3 hybrid orbital), similar energies, and the same electron occupation (one in each case in H1s and Csp^3). The origin of the name is the lobe-like shape of a hybrid orbital in a molecular fragment. As will become clear, the concept of isolobality is a helpful device for rationalizing observations.

Table 3.8 Some hybridization schemes

Coordination number	Arrangement	Composition
2	Linear	sp, pd, sd
	Angular	sd
3	Trigonal planar	sp^2, p^2d
	Unsymmetrical planar	spd
	Trigonal pyramidal	pd^2
4	Tetrahedral	sp^3, sd^3
	Irregular tetrahedral	spd^2, p^3d, pd^3
	Square planar	p^2d^2, sp^2d
5	Trigonal bipyramidal	sp^3d, spd^3
	Tetragonal pyramidal	sp^2d^2, sd^4, pd^4, p^3d^2
	Pentagonal planar	p^2d^3
6	Octahedral	sp^3d^2
	Trigonal prismatic	spd^4, pd^5
	Trigonal antiprismatic	p^3d^3

Source: H. Eyring, J. Walter, and G.E. Kimball, *Quantum chemistry.* Wiley, New York (1944).

A simple valence-bond viewpoint can be adopted for many of the applications of isolobality.[5] This viewpoint allows us to identify families of isolobal fragments, such as

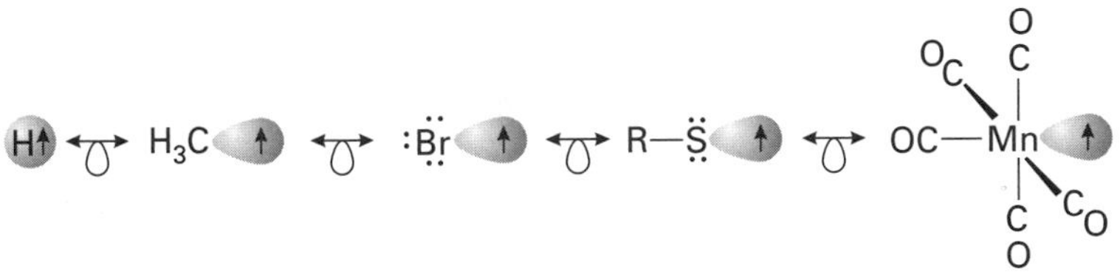

where, in this case, the arrow represents a single electron (other isolobal fragments may have two paired electrons in the corresponding orbitals). The recognition of this family permits us to anticipate by analogy with H—H that molecules such as H_3C—Br and $(OC)_5Mn$—CH_3 can be formed. It is also possible to identify isolobal fragments with two available singly occupied orbitals:

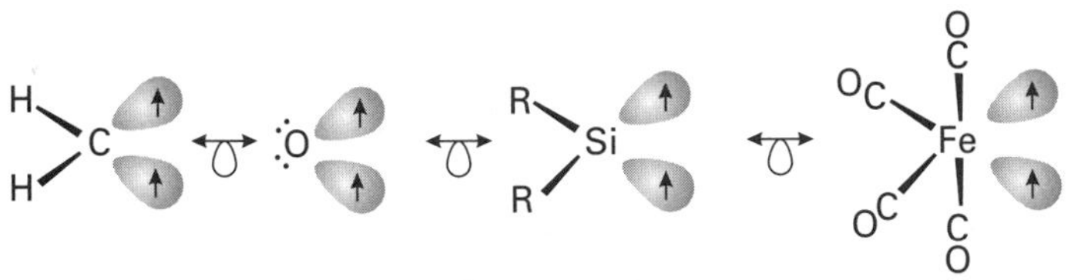

Some three-orbital isolobal fragments can also be identified:

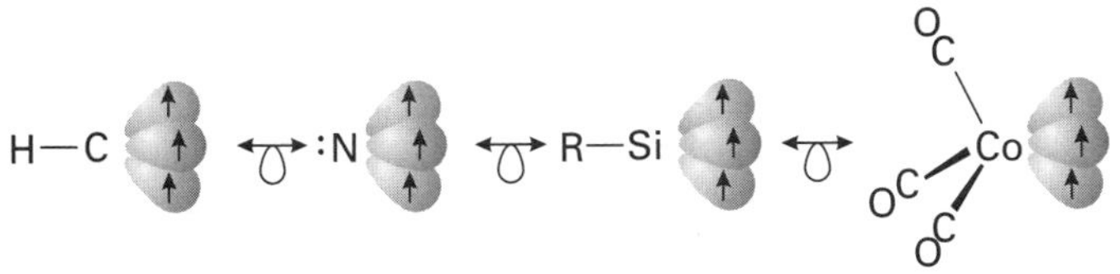

The existence of these families suggests that we should expect to encounter molecules such as cyclo-C_4H_8, $O(CH_3)_2$, and $N(CH_3)_3$, which may all be built from these and similar fragments. These species are all known. The two complexes $Co_4(CO)_{12}$ (**18**) and $Co_3(CO)_9CH$ (**19**) are two more examples. However, isolobal analogies must be used with care, for they may also tempt us to postulate the existence of $(OC)_5Mn$—O—$Mn(CO)_5$ and $(OC)_4Fe$=$Fe(CO)_4$, but both are unknown. Isolobal analogies provide useful correlations and hints for the synthesis of new molecules, but they are no substitute for experimental facts.

Groups of isolobal molecular fragments can be used to suggest patterns of bonding.

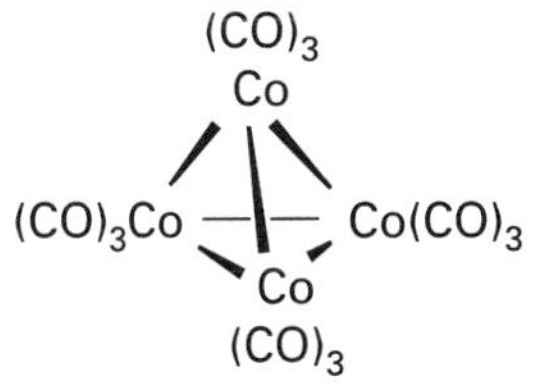

Molecular orbital theory

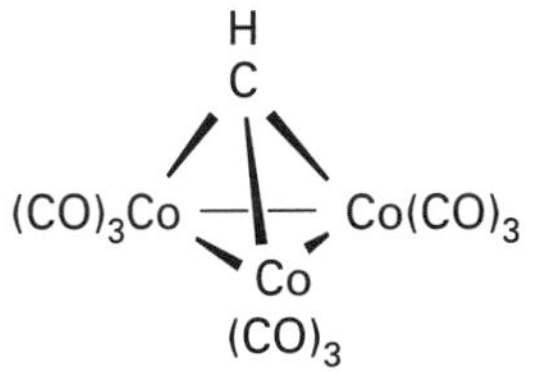

We shall now generalize the *atomic* orbital description of atoms in a very natural way to **molecular orbitals**. In contrast to valence-bond theory, these orbitals describe how electrons spread over all the atoms in a molecule and bind them together. In the spirit of this chapter, we continue to treat the concepts qualitatively and to give a sense of how inorganic chemists discuss the electronic structures of molecules.

3.7 An introduction to the theory

Almost all calculations on inorganic molecules are now carried out within the framework of **molecular orbital theory** (MO theory). However, certain concepts of VB theory are

18 $Co_4(CO)_{12}$

19 $Co_3(CO)_9CH$

---

[5] In molecular orbital terms, isolobality can be used in a localized orbital description.

sometimes imported into qualitative discussions. We begin by considering homonuclear diatomic molecules and diatomic ions. The concepts these species introduce are readily extended to **heteronuclear diatomic molecules**, which are molecules built from two atoms of different elements. They are also easily extended, as we see later, to polyatomic molecules and solids composed of huge numbers of atoms and ions. In parts of this section we shall include molecular fragments in the discussion, such as the SF diatomic group in the SF_6 molecule or the OO diatomic group in H_2O_2, for similar concepts also apply to pairs of atoms bound together as parts of larger molecules.

(a) The approximations of the theory

As in the description of the electronic structures of atoms, we set out by making the **orbital approximation**, in which we assume that the wavefunction, Ψ, of the N electrons in the molecule can be written as a product of N one-electron wavefunctions, ψ: $\Psi = \psi(1)\psi(2)\cdots\psi(N)$. The interpretation of this expression is that electron 1 is described by the wavefunction $\psi(1)$, electron 2 by the wavefunction $\psi(2)$, and so on. These one-electron wavefunctions are the **molecular orbitals** of the theory. As for atoms, the square of a one-electron wavefunction gives the probability distribution for that electron in the molecule: an electron in a molecular orbital is likely to be found where the orbital has a large amplitude, and will not be found at all at any of its nodes.

The next approximation is motivated by noticing that, when an electron is close to the nucleus of one atom, its wavefunction closely resembles an atomic orbital of that atom. For instance, when an electron is close to the nucleus of an H atom in a molecule, its wavefunction is like a $1s$ orbital of that atom. Therefore we may suspect that we can construct a reasonable first approximation to the molecular orbital by superimposing atomic orbitals contributed by each atom. This modeling of a molecular orbital in terms of contributing atomic orbitals is called the **linear combination of atomic orbitals** (LCAO) approximation. A 'linear combination' is a sum with various weighting coefficients.

In the most elementary form of MO theory, only the valence-shell atomic orbitals are used to form molecular orbitals. Thus, the molecular orbitals of H_2 are approximated by using two hydrogen $1s$ orbitals, one from each atom:

$$\psi = c_A\phi_A + c_B\phi_B \tag{5}$$

In this case the **basis set**, the atomic orbitals ϕ from which the molecular orbital is built, consists of two H$1s$ orbitals, one on atom A and the other on atom B. The coefficients c in the linear combination show the extent to which each atomic orbital contributes to the molecular orbital: the greater the value of c^2, the greater the contribution of that orbital to the molecular orbital.

The linear combination that gives the lowest energy for the H_2 molecule has equal contributions from each $1s$ orbital ($c_A^2 = c_B^2$). As a result, electrons in this orbital are equally likely to be found near each nucleus. Specifically, the coefficients are $c_A = c_B = 1$, and

$$\psi_+ = \phi_A + \phi_B \tag{6}$$

The combination that corresponds to the next higher energy orbital also has equal contributions from each $1s$ orbital ($c_A^2 = c_B^2$), but the coefficients have opposite signs ($c_A = +1$, $c_B = -1$):

$$\psi_- = \phi_A - \phi_B \tag{7}$$

The relative signs of coefficients in LCAOs play a very important role in determining the energies of the orbitals. As we shall see, the relative signs determine whether atomic orbitals interfere constructively or destructively in different regions of the molecule and hence lead to an accumulation or reduction of electron density in those regions.

Two more preliminary points should be noted. We see from this discussion that *two* molecular orbitals may be constructed from *two* atomic orbitals. In due course, we shall see the importance of the general point that *N* molecular orbitals can be constructed from a basis set of *N* atomic orbitals. For example, if we use all four valence orbitals on each O atom in O_2, then from the total of eight atomic orbitals we can construct eight molecular orbitals. Secondly, as in atoms, the Pauli exclusion principle implies that each molecular orbital may be occupied by up to two electrons; if two electrons are present, their spins must be paired. Thus, in a diatomic molecule constructed from two Period 2 atoms and in which there are eight molecular orbitals available for occupation, up to 16 electrons may be accommodated before all the molecular orbitals are full.

The general pattern of the energies of molecular orbitals formed from *N* atomic orbitals is that one molecular orbital lies below that of the parent atomic energy levels, one lies higher in energy than they do, and the remainder are distributed between these two extremes.

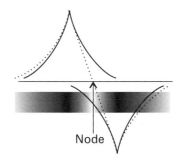

3.9 The enhancement of electron density in the internuclear region arising from the constructive interference between the atomic orbitals on neighboring atoms.

> *N molecular orbitals are constructed by forming linear combinations of N atomic orbitals; there is a high probability of finding electrons in atomic orbitals that have large coefficients in the linear combination. Each molecular orbital can be occupied by up to two paired electrons.*

(b) Bonding and antibonding orbitals

The orbital ψ_+ is an example of a **bonding orbital**. It is so called because the energy of the molecule is lowered relative to that of the separated atoms if this orbital is occupied by electrons. In elementary discussions of the chemical bond, the bonding character of ψ_+ is ascribed to the constructive interference between the two atomic orbitals and the enhanced amplitude this causes between the two nuclei (Fig. 3.9). An electron that occupies ψ_+ has an enhanced probability of being found in the internuclear region, and can interact strongly with both nuclei. Hence **orbital overlap**, the spreading of one orbital into the region occupied by another, leading to enhanced probability of electrons being found in the internuclear region, is taken to be the origin of the strength of bonds.[6]

The orbital ψ_- is an example of an **antibonding orbital**. It is so called because, if it is occupied, the energy of the molecule is higher than for the two separated atoms. The greater energy of an electron in this orbital arises from the destructive interference between the two atomic orbitals, which cancels their amplitudes and gives rise to a nodal plane between the two nuclei (Fig. 3.10). Electrons that occupy ψ_- are largely excluded from the internuclear region and are forced to occupy energetically less favorable locations. It is generally true that the energy of a molecular orbital in a polyatomic molecule is higher the more internuclear nodes it has. The increase in energy reflects an increasingly complete exclusion of electrons from the regions between nuclei.

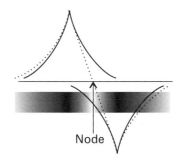

Node

3.10 The destructive interference that leads to a nodal surface in an antibonding molecular orbital if the overlapping orbitals have opposite phases.

The energies of the two molecular orbitals in H_2 are depicted in Fig. 3.11, which is an example of a **molecular orbital energy level diagram**. An indication of the size of the energy gap between the two molecular orbitals is the observation of a spectroscopic absorption in H_2 at 11.4 eV (in the ultraviolet at 109 nm) which can be ascribed to the transition of an electron from the bonding orbital to the antibonding orbital. The dissociation energy of H_2 is 4.5 eV, which gives an indication of the location of the bonding orbital relative to the separated atoms.

The Pauli exclusion principle limits to two the number of electrons that can occupy any molecular orbital, and requires that those two electrons be paired ($\uparrow\downarrow$). The exclusion principle

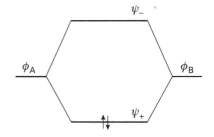

3.11 The molecular orbital energy level diagram for H_2 and analogous molecules.

---

[6] For a straightforward, readable discussion of conventional wisdom on the origin of the strengths of bonds according to molecular orbital theory, see Y. Jean and F. Volatron, *An introduction to molecular orbitals* (ed. J.K. Burdett). Oxford University Press, New York (1993).

is the origin of the importance of the electron pair in bond formation in MO theory just as it is in VB theory: in the context of MO theory, two electrons is the maximum number that can occupy an orbital that contributes to the stability of the molecule. The H_2 molecule, for example, has a lower energy than that of the separated atoms because two electrons can occupy the orbital ψ_+ and both can contribute to the lowering of its energy (as shown in Fig. 3.11). A weaker bond can be expected if only one electron is present in a bonding orbital, but, nevertheless, H_2^+ is known as a transient gas-phase ion; its dissociation energy is 2.6 eV. Three electrons are less effective than two electrons because the third electron must occupy the antibonding orbital ψ_- and hence destabilize the molecule. With four electrons, the antibonding effect of two electrons in ψ_- overcomes the bonding effect of two electrons in ψ_+. There is then no net bonding. It follows that a four-electron molecule with only $1s$ orbitals available for bond formation, such as He_2, cannot be expected to be stable relative to dissociation into its atoms.

> *A bonding orbital arises from the constructive interference of neighboring atomic orbitals; an antibonding orbital arises from their destructive interference, as indicated by a node between the atoms.*

3.8 Homonuclear diatomic molecules

Although the structures of diatomic molecules can be calculated effortlessly by using commercial software packages, the validity of these calculations needs to be monitored by appeal to experiment. Moreover, insight can often be achieved into molecular structure by drawing on experimental information. One of the most direct portrayals of electronic structure is obtained from photoelectron spectroscopy.

(a) Photoelectron spectra

Although ultraviolet absorption spectra have proved very useful for the analysis of the electronic structure of diatomic molecules more complex than H_2, a more direct portrayal of the molecular orbital energy levels can often be achieved by using **ultraviolet photoelectron spectroscopy** (UV-PES) in which electrons are ejected from the orbitals they occupy in molecules and their energies determined. In this technique, a sample is irradiated with 'hard' (high-frequency) ultraviolet radiation (typically 21.2 eV radiation from excited He atoms) and the kinetic energies of the **photoelectrons**, the ejected electrons, are measured. Because a photon of frequency ν has energy $h\nu$, if it expels an electron with ionization energy I from the molecule, the kinetic energy, E_K, of the photoelectron will be

$$E_K = h\nu - I \tag{8}$$

The lower in energy the electron lies initially (that is, the more tightly it is bound in the molecule), the greater its ionization energy and hence the lower its kinetic energy after it is ejected (Fig. 3.12). Because the peaks in a photoelectron spectrum correspond to the various kinetic energies of photoelectrons ejected from different orbitals of the molecule, the spectrum gives a vivid portrayal of the molecular orbital energy levels of a molecule.

The UV-PES spectrum of N_2 is shown in Fig. 3.13. We see that the photoelectrons have a series of discrete ionization energies close to 15.6 eV, 16.7 eV, and 18.8 eV. This pattern of energies strongly suggests a shell structure for the arrangement of electrons in the molecule. All the values are close to, but greater than, the ionization energy of the atom (14.5 eV). Because the ionization energy corresponds to the removal of a valence electron, the shell structure suggests that, when a molecule forms, the valence electrons arrange themselves into the molecular analog of atomic shells. The electrons in these shells differ only slightly in the strengths with which they are bound to the molecule. The lines corresponding to the lowest ionization energy in N_2 (close to 15.6 eV) are due to photoelectrons ejected from the occupied molecular orbital with highest energy in the molecule (the orbital in which an electron is bound most weakly). The higher ionization energies of 16.7 eV and 18.8 eV must

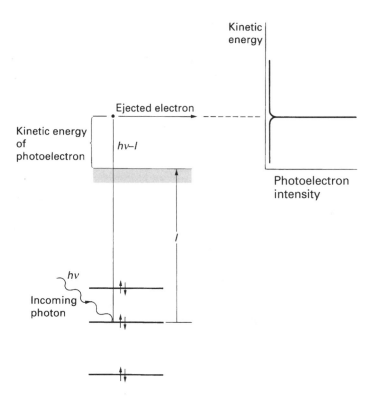

3.12 The photoelectron experiment. The incoming photon has energy $h\nu$; when it ejects an electron from an orbital of ionization energy I, the electron acquires a kinetic energy $h\nu - I$. The spectrometer detects the numbers of photoelectrons with different kinetic energies.

indicate the presence of molecular orbitals with successively lower energies (in which the electrons are bound more strongly), and indicate the existence of a ladder-like array of orbital energies in the molecule. There may also be more deeply lying orbitals, but the 21.2 eV ultraviolet photons cannot eject electrons from them, so they are not observed in the photoelectron spectrum.

The detailed structure of each group of lines in a photoelectron spectrum is a result of the ionized molecule being left in a vibrationally excited state after it is ionized. Some of the energy of the incoming photon is used to excite the vibration, so less energy is available for the kinetic energy of the ejected electron. Each quantum of vibration that is excited results in a corresponding reduction in kinetic energy of the photoelectron. As a result, the photoelectrons appear at a series of kinetic energies separated by the energy of each vibrational excitation that has occurred.

The vibrational structure in a photoelectron spectrum can be very helpful in assigning the origin of a spectral line. For example, extensive vibrational structure occurs when the ejected electron comes from an orbital in which it exerts an appreciable force on the nuclei, for the loss of the electron has a pronounced effect on the forces experienced by the nuclei. If the orbital is neutral in bonding character, photoejection leaves the force field largely undisturbed and there is little vibrational structure.

Photoelectron spectra observe the ionization energies of electrons from different molecular orbitals and give a portrayal of the energies of the orbitals.

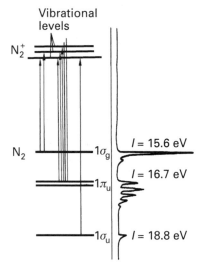

3.13 The UV photoelectron spectrum of N_2. The fine structure in the spectrum arises from the excitation of vibrations in the cation formed by photoejection.

(b) The orbitals

Our task now is to see how molecular orbital theory can account for the features revealed by photoelectron spectroscopy and the other techniques, principally absorption spectroscopy,

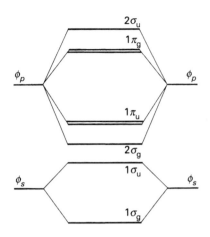

3.14 The molecular orbital energy level diagram for the later Period 2 homonuclear diatomic molecules. This diagram should be used for O_2 and F_2 molecules.

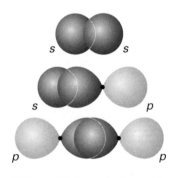

3.15 σ orbitals can be formed in various ways, which include s,s overlap, s,p overlap, and p,p overlap, with the p orbitals directed along the internuclear axis in each case.

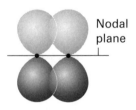

3.16 Two p orbitals can overlap to form a π orbital. The orbital has a nodal plane passing through the internuclear axis: this view shows the plane from the side.

that are used to study diatomic molecules. As with H_2, the starting point in the theoretical discussion is the **minimal basis set**, the smallest set of atomic orbitals from which useful molecular orbitals can be built. In Period 2 diatomic molecules, the minimal basis set consists of the one valence s orbital and three valence p orbitals on each atom, giving eight atomic orbitals in all. As remarked previously, N atomic orbitals can be used to construct N molecular orbitals. We shall now show how the minimal basis set of eight valence-shell atomic orbitals (four from each atom) is used to construct eight molecular orbitals. Then we shall use the Pauli principle to predict the ground-state electron configurations of the molecules.

The energies of the atomic orbitals that form the basis set are shown on either side of the molecular orbital diagram in Fig. 3.14. We can form **σ orbitals** by allowing overlap between atomic orbitals that have cylindrical symmetry around the internuclear axis, which is conventionally labeled z. The notation σ signifies that the orbital has cylindrical symmetry. The orbitals in Fig. 3.15 are σ orbitals. Atomic orbitals that can form σ orbitals include the $2s$ and $2p_z$ orbitals on the two atoms, as we show in Fig. 3.15. From these four orbitals (the $2s$ and the $2p_z$ orbitals on atom A and the corresponding orbitals on atom B) with cylindrical symmetry we can construct four σ molecular orbitals. Two of the molecular orbitals will be bonding and two will be antibonding. Their energies will resemble those shown in Fig. 3.14, but it is difficult to predict the precise locations of the central two orbitals: the orbitals are labeled $1\sigma, 2\sigma, \cdots$ starting with the orbital of lowest energy.

The remaining two $2p$ orbitals on each atom, which have a nodal plane through the z-axis, overlap to give **π orbitals** (Fig. 3.16). Bonding and antibonding π orbitals can be formed from the mutual overlap of the two $2p_x$ orbitals and, also, from the mutual overlap of the two $2p_y$ orbitals. This pattern of overlap gives rise to the two pairs of doubly degenerate energy levels shown in Fig. 3.14.

The procedure we have adopted for the description of diatomic molecules in terms of molecular orbitals can be summarized as follows:

1 From a basis set of N atomic orbitals, N molecular orbitals are constructed. In Period 2, $N = 8$.
2 The eight orbitals can be classified by symmetry into two sets: there are four σ orbitals and four π orbitals.
3 The four π orbitals form one doubly degenerate pair of bonding orbitals and one doubly degenerate pair of antibonding orbitals.
4 The four σ orbitals span a range of energies, one being strongly bonding and another strongly antibonding, with the remaining two σ orbitals lying between these extremes.
5 To establish the actual location of the energy levels, it is necessary to use electronic absorption spectroscopy, photoelectron spectroscopy, or detailed computation.

Photoelectron spectroscopy and detailed computation (the numerical solution of the Schrödinger equation for the molecules) let us build the orbital energy schemes shown in Fig. 3.17. As we see there, for Li_2 through N_2 the arrangement of orbitals is that shown in Fig. 3.18, whereas for O_2 and F_2 the order of the $2\sigma_g$ and $1\pi_u$ orbitals is reversed and the array is that shown in Fig. 3.14. The reversal of order can be traced to the increasing separation of the $2s$ and $2p$ orbitals that occurs on going to the right across Period 2. A general principle of quantum mechanics is that the mixing of wavefunctions is strongest if their energies are similar. Therefore, as the s and p energy separation increases, the molecular orbitals become more purely s-like and p-like. When the s,p energy separation is small, each molecular orbital is a more extensive mixture of s and p character on each atom.

If we are considering species containing two neighboring d-block atoms, as in Hg_2^{2+} and $[Cl_4ReReCl_4]^{2-}$, we should also allow for the possibility of forming bonds from d orbitals. The d_{z^2} orbital has cylindrical symmetry with respect to the internuclear (z) axis, and hence can contribute to the σ orbitals that are formed from s and p_z orbitals. The d_{yz} and d_{zx} orbitals

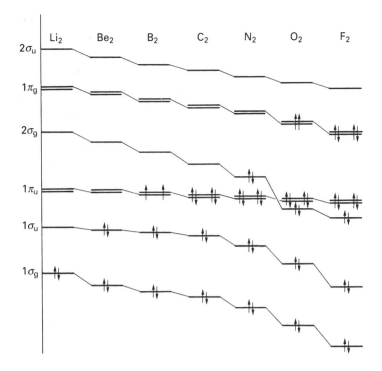

3.17 The variation in orbital energies for Period 2 homonuclear diatomic molecules as far as F_2.

both look like p orbitals when viewed along the axis, and hence can contribute to the π orbitals formed from p_x and p_y. The new feature is the role of $d_{x^2-y^2}$ and d_{xy}, which have no counterpart in the orbitals discussed up to now. These two orbitals can overlap with matching orbitals on the other atom to give rise to doubly degenerate pairs of bonding and antibonding **δ orbitals** (Fig. 3.19). As we shall see in Chapter 7, δ orbitals are important for the discussion of bonds between d-metal atoms and lead to descriptions of some species in terms of quadruple bonds, as in $[Cl_4Re \equiv ReCl_4]^{2-}$.

For homonuclear diatomics, it is convenient (particularly for spectroscopic discussions) to signify the symmetry of the molecular orbitals with respect to their behavior under inversion through the center of the molecule. The operation of inversion consists of starting at an

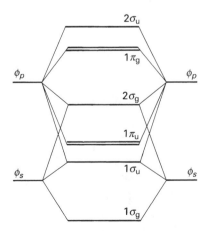

3.18 The alternative ordering of orbital energies found in homonuclear diatomic molecules from Li_2 to N_2.

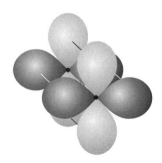

3.19 The formation of δ orbitals from d-orbital overlap. The orbital has two mutually perpendicular nodal planes that intersect along the internuclear axis.

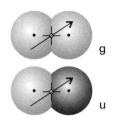

20 σ_g and σ_u orbitals

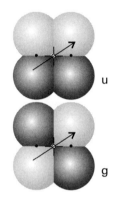

21 π_g and π_u orbitals

arbitrary point in the molecule, traveling in a straight line to the center of the molecule, and then continuing an equal distance out on the other side of the center. The orbital is designated **g** (for *gerade*, even) if it is identical under inversion, and **u** (for *ungerade*, odd) if it changes sign. Thus, a bonding σ orbital is **g** and an antibonding σ orbital is **u** (**20**). On the other hand, a bonding π orbital is **u** and an antibonding π orbital is **g** (**21**). These labels have been included in Figs 3.14, 3.17, and 3.18. Note that the σ_g orbitals are numbered separately from the σ_u orbitals, and similarly for the π orbitals.

> *Orbitals are classified as σ, π, or δ according to their symmetry about the internuclear axis, and (in centrosymmetric species) as g or u according to their symmetry with respect to inversion.*

(c) The building-up principle for molecules

We use the building-up principle in conjunction with the energy level diagram in the same way as for atoms. The order of occupation of the orbitals is the order of increasing energy as depicted in Figs 3.14 or 3.18. Each orbital can accommodate up to two paired electrons. If more than one orbital is available for occupation (because they happen to have identical energies, as in the case of pairs of π orbitals), then the orbitals are occupied separately. In that case, the electrons in the half-filled orbitals adopt parallel spins ($\uparrow\uparrow$), just as is required by Hund's rule for atoms.

With very few exceptions, these rules lead to the actual ground-state configuration of the Period 2 diatomic molecules. For example, the electron configuration of N_2, with 10 valence electrons, is

$$N_2: \quad 1\sigma_g^2 1\sigma_u^2 1\pi_u^4 2\sigma_g^2$$

Molecular orbital configurations are written like those for atoms: the orbitals are listed in order of increasing energy, and the number of electrons in each one is indicated by a superscript. Note that π^4 is shorthand for the occupation of two different π orbitals.

....

Example 3.6 Writing electron configurations of diatomic molecules

Give the ground-state electron configurations of the oxygen molecule, O_2, the superoxide ion, O_2^-, and the peroxide ion, O_2^{2-}.

Answer The O_2 molecule has 12 valence electrons. The first ten electrons recreate the N_2 configuration except for the reversal of the order of the $1\pi_u$ and $2\sigma_g$ orbitals (see Fig. 3.17). Next in line for occupation are the doubly degenerate $1\pi_g$ orbitals. The last two electrons enter these orbitals separately, and have parallel spins. The configuration is therefore

$$O_2: \quad 1\sigma_g^2 1\sigma_u^2 2\sigma_g^2 1\pi_u^4 1\pi_g^2$$

The O_2 molecule is interesting because the lowest energy configuration has two unpaired electrons in different π orbitals. Hence, O_2 is paramagnetic (tends to move into a magnetic field). The next two electrons can be accommodated in the $1\pi_g$ orbitals, giving

$$O_2^-: \quad 1\sigma_g^2 1\sigma_u^2 2\sigma_g^2 1\pi_u^4 1\pi_g^3$$

$$O_2^{2-}: \quad 1\sigma_g^2 1\sigma_u^2 2\sigma_g^2 1\pi_u^4 1\pi_g^4$$

We are assuming that the orbital order does not change; this might not be the case.

....

Self-test 3.6 Write the valence-electron configuration for S_2^{2-} and Cl_2^-.

....

The **highest occupied molecular orbital** (HOMO) is the molecular orbital that, according to the building-up principle, is occupied last. The **lowest unoccupied molecular orbital** (LUMO) is the next higher molecular orbital. In Fig. 3.17, the HOMO of F_2 is $1\pi_g$ and its LUMO is $2\sigma_u$; for N_2 the HOMO is $2\sigma_g$ and the LUMO is $1\pi_g$. We shall increasingly see that these **frontier orbitals** play special roles in structural and kinetic studies. The term SOMO, denoting a **singly occupied molecular orbital**, is sometimes encountered, and is of crucial importance for the properties of radical species.

> *The building-up principle is used to predict the ground-state electron configurations by accommodating electrons in the array of molecular orbitals summarized in Figs 3.14 or 3.18 and recognizing the constraints of the Pauli principle.*

3.9 Heteronuclear diatomic molecules

The molecular orbitals of heteronuclear diatomic molecules differ from those of homonuclear diatomic molecules in having unequal contributions from each atomic orbital. Each molecular orbital has the form

$$\psi = c_A\phi_A + c_B\phi_B + \cdots \qquad (9)$$

as in homonuclear molecules. The unwritten orbitals include all the other orbitals of the correct symmetry for forming σ or π bonds, but which typically make a smaller contribution than the two valence-shell orbitals we are considering. In contrast to orbitals for homonuclear species, the coefficients c_A and c_B are not necessarily equal in magnitude. If c_A^2 is larger than c_B^2, the orbital is composed principally of ϕ_A and an electron that occupies the molecular orbital is more likely to be found near atom **A** than atom **B**. The opposite is true for a molecular orbital in which c_B^2 is larger than c_A^2.

(a) Molecular orbitals built from atoms of different elements

The greater contribution to a bonding molecular orbital normally comes from the more electronegative atom: the bonding electrons are then likely to be found close to that atom and hence be in an energetically favorable location. The extreme case of a **polar covalent bond**, a covalent bond formed by an electron pair that is unequally shared by the two atoms, is an **ionic bond** in which one atom gains complete control over the electron pair. The less electronegative atom normally contributes more to an antibonding orbital (Fig. 3.20). That is, antibonding electrons are more likely to be found in an energetically unfavorable location, close to the less electronegative atom.

A second difference between homonuclear and heteronuclear diatomic molecules stems from the energy mismatch in the latter between the two sets of atomic orbitals. We have already remarked that two wavefunctions interact less strongly as their energies diverge. This dependence on energy separation implies that the lowering of energy as a result of the overlap of atomic orbitals on different atoms in a heteronuclear molecule is less pronounced than in a homonuclear molecule, in which the orbitals have the same energies. However, we cannot necessarily conclude that A—B bonds are weaker than A—A bonds, because other factors (which include orbital size and closeness of approach) are also important. The CO molecule, for example, which is isoelectronic with N_2, has an even higher bond enthalpy (1070 kJ mol^{-1}) than N_2 (946 kJ mol^{-1}).

> *Heteronuclear diatomic molecules are polar. Bonding electrons tend to be found on the more electronegative atom and antibonding electrons on the less electronegative atom.*

(b) Hydrogen fluoride

As an illustration of these general points, consider a simple heteronuclear diatomic molecule, HF. The valence orbitals available for molecular orbital formation are the $1s$ orbital of H and

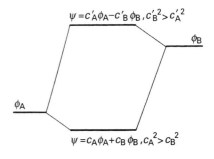

3.20 When two atomic orbitals with different energies overlap, the lower molecular orbital is primarily composed of the lower atomic orbital, and vice versa. Moreover, the shift in energies of the two levels is less than if they had had the same energy in the atoms.

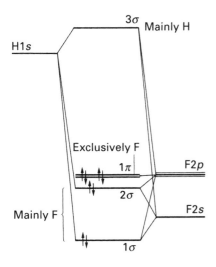

3.21 The molecular orbital energy level diagram for HF. The relative positions of the atomic orbitals reflect the ionization energies of the atoms.

the $2s$ and $2p$ orbitals of F; there are $1 + 7 = 8$ valence electrons to accommodate in the molecular orbitals.

The σ orbitals for HF can be constructed by allowing an $H1s$ orbital to overlap the $2s$ and $2p_z$ orbitals of F (z being the internuclear axis). These three atomic orbitals combine to give three σ molecular orbitals of the form

$$\psi = c_1 \phi_{H1s} + c_2 \phi_{F2s} + c_3 \phi_{F2p_z}$$

This procedure leaves the $F2p_x$ and $F2p_y$ orbitals unaffected as they have π symmetry and there are no valence H orbitals of that symmetry. The π orbitals are therefore **nonbonding orbitals**, or orbitals that are neither bonding nor antibonding in character and which (in a diatomic molecule) are confined to a single atom.

The resulting energy level diagram is shown in Fig. 3.21. The 1σ bonding orbital is predominantly $F2s$ in character (in accord with the high electronegativity of fluorine) because that orbital lies so low in energy and plays the major role in the bonding orbitals it forms. The 2σ orbital is largely nonbonding and confined mainly to the F atom. The bulk of its density lies on the opposite side of the F atom to the H atom, so it barely participates in bonding. The 3σ orbital is antibonding, and principally $H1s$ in character: the $1s$ orbital has a relatively high energy (compared with the fluorine orbitals) and hence contributes predominantly to the high-energy antibonding molecular orbital.

Two of the eight electrons enter the 1σ orbital, forming a bond between the two atoms. Six more enter the 2σ and 1π orbitals; these two orbitals are largely nonbonding and confined mainly to the F atom. All the electrons are now accommodated, so the configuration of the molecule is $1\sigma^2 2\sigma^2 1\pi^4$. One important feature to note is that all the electrons occupy orbitals that are predominantly on the F atom. It follows that we can expect the HF molecule to be polar, with a partial negative charge on the F atom, which is found experimentally. The observed dipole moment is $1.91\,\mathrm{D}$.[7] Unfortunately, simple MO theory is not very successful when used to calculate molecular dipole moments, and even the qualitative analysis of the contributions to dipole moments is fraught with difficulty.

In HF, *the bonding orbital is more concentrated on the* F *atom and the antibonding orbital is more concentrated on the* H *atom.*

(c) Carbon monoxide

The molecular orbital energy level diagram for carbon monoxide (and the isoelectronic CN^- ion) is a somewhat more complicated example than HF because both atoms have $2s$ and $2p$ orbitals that can participate in the formation of σ and π orbitals. The energy level diagram is shown in Fig. 3.22. The ground-state configuration is

$$CO: \quad 1\sigma^2 2\sigma^2 1\pi^4 3\sigma^2$$

The HOMO in CO is 3σ, which is a largely nonbonding lone pair on the C atom. The LUMO is the doubly degenerate pair of antibonding π orbitals, with mainly $C2p$ orbital character (Fig. 3.23). This combination of frontier orbitals is very significant, and we shall see that it is one reason why metal carbonyls are such a characteristic feature of the chemical properties of the d elements (Chapter 16). In metal carbonyls, the HOMO lone pair orbital of CO participates in the formation of a σ bond, and the LUMO antibonding π orbital participates in the formation of π bonds to the metal atom.

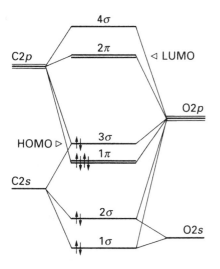

3.22 The molecular orbital energy level diagram of CO.

---

[7] The SI unit for reporting dipole moments, the product of charge in coulombs and distance in meters, is coulomb-meter (C m). However, it proves more convenient to adopt the non-SI Debye unit, D, which was originally defined in terms of electrostatic units. The conversion is $1\,\mathrm{D} = 3.336 \times 10^{-30}\,\mathrm{C\,m}$. A dipole consisting of charges e and $-e$ separated by $1\,\text{Å}$ (100 pm) has a dipole moment of 4.8 D. In most cases, dipole moments are somewhat smaller since they arise from partial charges; typical values are close to 1 D.

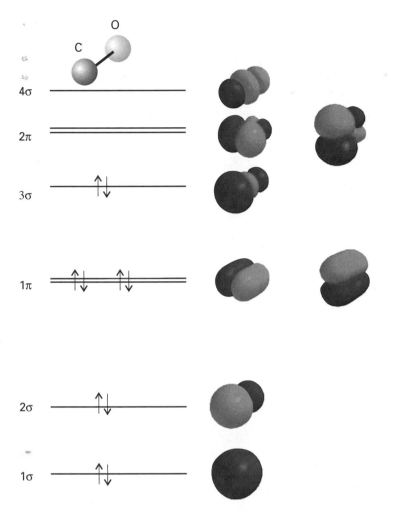

3.23 The molecular orbitals of CO as calculated by an *ab initio* procedure.

Example 3.7 Accounting for the structure of a heteronuclear diatomic molecule

The halogens form compounds among themselves. One of these 'interhalogen' compounds is iodine monochloride, ICl, in which the order of orbitals is $1\sigma, 2\sigma, 3\sigma, 1\pi, 2\pi, 4\sigma$ (from calculation). What is its ground-state electron configuration?

Answer First, we identify the atomic orbitals that are to be used to construct molecular orbitals: these are the $3s$ and $3p$ valence-shell orbitals of Cl and the $5s$ and $5p$ valence-shell orbitals of I. As for Period 2 elements, an array of σ and π orbitals can be constructed, and is shown in Fig. 3.24. The bonding orbitals are predominantly Cl in character (because that is the more electronegative element) and the antibonding orbitals are predominantly I in character. There are $7 + 7 = 14$ valence electrons to accommodate, which results in the ground-state electron configuration $1\sigma^2 2\sigma^2 3\sigma^2 1\pi^4 2\pi^4$.

Self-test 3.7 Describe the structure of the hypochlorite ion, ClO⁻, in terms of molecular orbitals.

Although the difference in electronegativity between C and O is large, the experimental value of the electric dipole moment of the CO molecule (0.1 D) is small. Moreover, the negative end of the dipole is on the C atom despite that being the less electronegative atom.

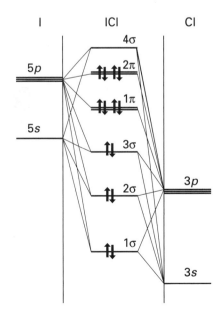

3.24 A schematic molecular orbital energy level diagram for the interhalogen molecule ICl as calculated by an *ab initio* procedure.

This odd situation stems from the fact that the lone pairs and bonding pairs have a complex distribution. It is wrong to conclude that, because the bonding electrons are mainly on the O atom, O is the negative end of the dipole, for this ignores the balancing effect of the lone pair on the C atom. The inference of polarity from electronegativity is particularly unreliable when antibonding orbitals are occupied.

> The HOMO of a carbon monoxide molecule is an almost nonbonding σ orbital largely localized on C; the LUMO is an antibonding π orbital.

3.10 Bond properties in the molecular orbital formalism

We now bring the discussion full circle, and show that, although seemingly very different, molecular orbital theory does elucidate many features of the Lewis description of molecules. We have already seen the origin of the importance of the electron pair: two electrons is the maximum number that can occupy a bonding orbital and hence contribute to a chemical bond. We now extend this concept by introducing the concept of 'bond order'.

(a) Bond order

The **bond order**, b, identifies a shared electron pair as counting as a 'bond' and an electron pair in an antibonding orbital as an 'antibond' between two atoms. More precisely, the bond order is defined as

$$b = \tfrac{1}{2}(n - n^*) \tag{10}$$

where n is the number of electrons in bonding orbitals and n^* is the number in antibonding orbitals. For example, N_2 has the configuration $1\sigma_g^2 1\sigma_u^2 1\pi_u^4 2\sigma_g^2$ and, because $1\sigma_g$ and $1\pi_u$ orbitals are bonding but $1\sigma_u$ is antibonding, $b = \tfrac{1}{2}(2 + 4 + 2 - 2) = 3$. A bond order of 3 corresponds to a triply bonded molecule, which is in line with the Lewis structure :N≡N:. The high bond order is reflected in the high bond enthalpy of the molecule ($+946\ \text{kJ mol}^{-1}$), one of the highest for any molecule.

The bond order of the isoelectronic CO molecule is also 3, in accord with the analogous Lewis structure :C≡O:. However, this method of assessing bonding is primitive, especially for heteronuclear species. For instance, inspection of the computed molecular orbitals in Fig. 3.23 suggests that 2σ and 3σ are best regarded as nonbonding orbitals largely localized on O and C, and hence should really be disregarded in the calculation of b. The resulting bond order is unchanged by this modification. The lesson is that the definition of bond order provides a useful indication of the multiplicity of the bond, but any interpretation of contributions to b needs to be done in the light of guidance from the properties of computed orbitals.

Electron loss from N_2 leads to the formation of the transient species N_2^+, in which the bond order is reduced from 3 to 2.5. This reduction in bond order is accompanied by a corresponding decrease in bond strength (to $855\ \text{kJ mol}^{-1}$) and increase of the bond length from $1.09\ \text{Å}$ for N_2 to $1.12\ \text{Å}$ for N_2^+. The bond order of F_2 is 1, which is consistent with the Lewis structure :F—F: and the conventional description of the molecule as having a single bond.

The definition of bond order allows for the possibility that an orbital is only singly occupied. The bond order in O_2^-, for example, is 1.5, because three electrons occupy the $1\pi_g$ antibonding orbitals. Isoelectronic molecules and ions have the same bond order, so F_2 and O_2^{2-} both have bond order 1, and N_2, CO, and NO^+ all have bond order 3.

> The bond order assesses the net number of bonds between two atoms in the molecular orbital formalism.

(b) Bond correlations

The strengths and lengths of bonds correlate quite well with each other and with the bond order:

- Bond enthalpy for a given pair of atoms increases as bond order increases.
- Bond length decreases as bond order increases.

These trends are illustrated in Fig. 3.25 and Fig. 3.26. The strength of the dependence varies with the elements. In Period 2 it is relatively weak in CC bonds, with the result that a C=C double bond is less than twice as strong as a C—C single bond. This difference has profound consequences in organic chemistry, particularly for the reactions of unsaturated compounds. It implies, for example, that it is energetically favorable (but kinetically slow in the absence of a catalyst) for ethene and ethyne to polymerize: in this process, C—C single bonds form at the expense of the appropriate numbers of multiple bonds.

Familiarity with carbon's properties, however, must not be extrapolated without caution to the bonds between other elements. An N=N double bond (409 kJ mol^{-1}) is more than twice as strong as an N—N single bond (163 kJ mol^{-1}), and an N≡N triple bond (945 kJ mol^{-1}) is more than five times as strong. It is on account of this trend that NN multiply bonded compounds are stable relative to polymers or three-dimensional compounds having only single bonds. The same is not true of phosphorus, where the P—P, P=P, and P≡P bond enthalpies are 200 kJ mol^{-1}, 310 kJ mol^{-1}, and 490 kJ mol^{-1}, respectively. For this element, single bonds are stable relative to the matching number of multiple bonds. Thus, phosphorus exists in a variety of solid forms in which P—P single bonds are present, including the tetrahedral P_4 molecules of white phosphorus, and not as P_2 molecules.

The two correlations with bond order taken together imply that:

- For a given pair of elements, the bond enthalpy increases as bond length decreases.

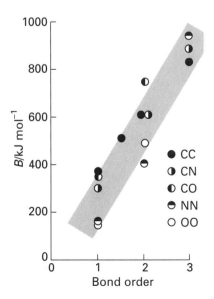

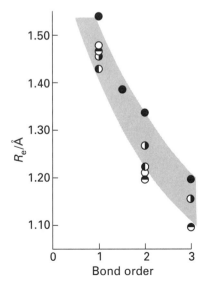

3.25 The correlation of bond strength and bond order.

3.26 The correlation of bond length and bond order. The key for the points is the same as in Fig. 3.25.

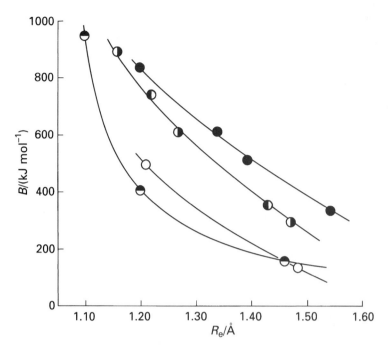

3.27 The correlation of bond length and bond strength. The key for the points is the same as in Fig. 3.25.

This correlation is illustrated in Fig. 3.27. It is a useful feature to bear in mind when considering the stabilities of molecules, because bond lengths may be readily available from independent sources.

For a given pair of elements, the bond enthalpy increases as bond length decreases.

Molecular orbitals of polyatomic molecules

Molecular orbital theory can be used to discuss in a uniform manner the electronic structures of triatomic molecules, finite groups of atoms, and solids. In each case the molecular orbitals resemble those of diatomic molecules, the only important difference being that the orbitals are built from a more extensive basis set of atomic orbitals. As remarked earlier, a key point to bear in mind is that from N atomic orbitals it is possible to construct N molecular orbitals.

We saw in Section 3.8 that the general structure of molecular orbital energy level diagrams can be derived by grouping the orbitals into different sets, the σ and π orbitals, according to their shapes. The same procedure is used in the discussion of the molecular orbitals of polyatomic molecules. However, as their shapes are more complex than diatomic molecules, we need a more powerful approach. The discussion of polyatomic molecules will therefore be carried out in two stages. In this chapter we use intuitive ideas about molecular shape to construct molecular orbitals. In the next chapter we discuss the shapes of molecules and the use of their symmetry characteristics in the construction of molecular orbitals and for the discussion of other properties. That chapter will rationalize the procedures presented here.

3.11 The construction of molecular orbitals

We shall use the simplest polyatomic species H_3^+ and H_3 to identify some of the central features of all polyatomic molecules. Although H_3^+ and H_3 might seem far removed from real inorganic chemistry, the orbitals we are about to derive occur widely, as in NH_3 and, in a more disguised form, in other molecules, such as BF_3. The transitory gas-phase ion H_3^+ has been detected spectroscopically in the auroras of Jupiter, Saturn, and Uranus.[8] Tracer experiments have also provided evidence that H_3^+ may also occur as a reaction intermediate in solution.

A major question in connection with polyatomic molecules is what controls their shapes. We shall therefore begin by considering two possibilities for H_3^+ and H_3, namely linear and triangular, and look for reasons why one shape may have a lower energy than the other.

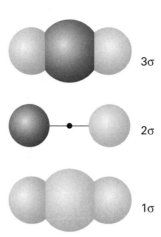

3.28 The three molecular orbitals of a linear H_3 molecule formed from overlap of the H1s orbitals. Throughout this chapter, white and shades of gray denote opposite signs of the wavefunction.

(a) Linear trihydrogen

The simplest basis set for the H_3 molecule and its ions consists of a $1s$ orbital on each of the three H atoms, H_A, H_B, and H_C. We denote these orbitals $1s_A$, $1s_B$, and $1s_C$. As always, N molecular orbitals can be formed from N atomic orbitals, and in this case $N = 3$. The three molecular orbitals of the linear H_3 species (both H_3 and any of its ions) are combinations of the three atomic orbitals are shown in Fig. 3.28: we can expect one combination to be strongly bonding, one strongly antibonding, and the third in between these two extremes. We shall continue to omit normalization constants because then the form of the orbitals is much clearer; where we judge it appropriate, we give the normalized forms in footnotes.

A specific calculation (which we shall not describe in detail) shows that the most strongly bonding of the three combinations is

$$1\sigma = \phi_{1s_A} + 2^{1/2}\phi_{1s_B} + \phi_{1s_C}$$

This orbital has a low energy because it is bonding between H_A and H_B and between H_B and H_C. (It is also bonding between H_A and H_C but, as these are so far apart, this contribution is less important.) The orbital is designated σ because it has cylindrical symmetry around the molecular axis. The next higher orbital is also a σ orbital:

$$2\sigma = \phi_{1s_A} - \phi_{1s_C}$$

This molecular orbital has no contribution from the central atom and, because the outer two orbitals are so far apart, there is negligible interaction between them. The orbital is an example of a nonbonding orbital, and neither lowers nor raises the energy of the molecule when it is occupied. The third molecular orbital that can be formed from the basis set is another σ orbital:

$$3\sigma = \phi_{1s_A} - 2^{1/2}\phi_{1s_B} + \phi_{1s_C}$$

This orbital is antibonding between both neighboring pairs of atoms and therefore has the highest energy of the three σ orbitals.[9]

The three σ orbitals introduce one feature that should always be kept in mind:

- The energies of molecular orbitals generally increase as the number of nodes between neighboring atoms increases.

The physical reason for this trend is that electrons are progressively excluded from the internuclear regions as the number of nodes increases.

A linear trihydrogen species has three molecular orbitals with 0, 1, and 2 nodal surfaces; two electrons occupy the fully bonding orbital.

....

[8] R. McNab, The spectroscopy of H_3^+. *Adv. Chem. Phys.* **89**, 1 (1995).

[9] The normalized forms of these three orbitals, if we ignore overlap, are

$$1\sigma = \tfrac{1}{2}\{\phi_{1s_A} + 2^{1/2}\phi_{1s_B} + \phi_{1s_C}\}$$
$$2\sigma = \tfrac{1}{2^{1/2}}\{\phi_{1s_A} - \phi_{1s_C}\}$$
$$3\sigma = \tfrac{1}{2}\{\phi_{1s_A} - 2^{1/2}\phi_{1s_B} + \phi_{1s_C}\}$$

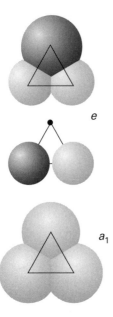

3.29 The three molecular orbitals of an equilateral triangular H_3 molecule formed from overlap of the $H1s$ orbitals.

(b) Triangular trihydrogen

Now suppose that H_3 (or its ions) is an equilateral triangle. Its three molecular orbitals have the same form as in the linear molecule, but the orbitals previously called 2σ and 3σ now have the same energy. It is not immediately obvious that 2σ and 3σ are degenerate, but symmetry arguments of the kind to be introduced in Chapter 4 can be used to show that they do in fact have exactly the same energy. That their energies are the same can be seen at least to be plausible by the following argument. In the triangular molecule, A and C are next to each other (Fig. 3.29), so 2σ is antibonding between them and hence higher in energy than in the linear molecule. On the other hand, the 3σ orbital is bonding between A and C (but still antibonding between A and B and between C and B), so its energy is lower than before. The two orbitals therefore converge in energy as the molecule bends, and when all three H–H separations are the same the energies are found to be equal.

It is no longer strictly appropriate to use the label σ in the triangular molecule because that label applies to a linear molecule. However, it is often convenient to continue to use the notation σ (and in other molecules π too) when we are concentrating on the *local* form of an orbital, its shape relative to the internuclear axis between two neighboring atoms. The correct procedure for labeling orbitals in polyatomic molecules according to their symmetry is described in Chapter 4. For our present purposes all we need know is the following:

- a, b denote a nondegenerate orbital.
- e denotes a doubly degenerate orbital.
- t denotes a triply degenerate orbital.

Subscripts and superscripts are sometimes added to these letters, as in a_1, b'', e_g, and t_2 because it is sometimes necessary to distinguish different a, b, e, and t orbitals according to a more detailed analysis of their symmetries. These details are explained in Section 4.5.

With these rules in mind, the orbital we have called 1σ is labeled a. The orbitals 2σ and 3σ are doubly degenerate and so are labeled e. Therefore, the molecular orbitals of the equilateral triangular molecule are

$$a = \phi_{1s_A} + \phi_{1s_B} + \phi_{1s_C}$$
$$e = \begin{cases} \phi_{1s_A} - \phi_{1s_C} \\ \phi_{1s_A} - 2\phi_{1s_B} + \phi_{1s_C} \end{cases}$$

It is possible to compute the energies of these three orbitals as the bond angle is changed from $180°$ (linear) to $60°$ (equilateral triangular) and the σ orbitals of the linear molecule evolve into the a and e orbitals of the equilateral triangular molecule. The resulting diagram is called a **correlation diagram** (Fig. 3.30). We shall see that such diagrams play an important role in understanding the shapes, spectra, and reactions of polyatomic molecules. Note that, whereas the a orbital is bonding, the e orbitals have mixed bonding and antibonding character between different pairs of atoms, and their net character is antibonding.

> *Equilateral triangular trihydrogen has three molecular orbitals: one is fully bonding, the other two are degenerate and have net antibonding character; two electrons occupy the fully bonding orbital.*

(c) Electron configurations in H_3^+

Now that the form of the molecular orbitals is available and their relative energies are known, it is possible to deduce the electron configuration of the two-electron molecular ion H_3^+ for any bond angle.

The two electrons occupy the orbital with the lowest energy. The resulting configuration for linear H_3^+ will be $1\sigma^2$; for equilateral triangular H_3^+ it is a^2. It is possible to decide which

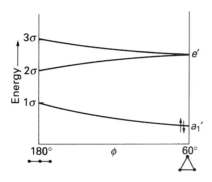

3.30 The orbital correlation diagram for an H_3 molecule shows how the energies of the orbitals change as the linear molecule becomes equilateral triangular.

geometry has the lower energy only by detailed calculation. However, the fact that a is bonding between all three atoms whereas 1σ is bonding only between A and B and between B and C suggests that the lower energy is obtained for the equilateral triangular arrangement. This is in fact the case, and spectroscopic data and numerical computation indicate that H_3^+ is an equilateral triangular species with configuration a^2.

An important point, which can be seen in Fig. 3.29, is that the a orbital spreads over all the atoms. The same is true of the 1σ orbital in the linear molecule. Therefore, in each case *the two electrons bind the entire cluster together*. In other words, molecular orbitals are **delocalized orbitals**, in the sense that they spread over several atoms and are not, in general, localized between two atoms. The H_3^+ molecule-ion is in fact the simplest example of a **three-center, two-electron bond** (a 3c,2e bond) in which three nuclei are bound by only two electrons. This type of bond is represented by the three-spoke pattern shown in (**22**). There should be nothing particularly puzzling about 3c,2e bonding, for it is a natural outcome of the formation of delocalized molecular orbitals.

22 3c,2e bond

> The only occupied orbital in triangular H_3^+ is fully bonding, and is designated a three-center, two-electron bond.

3.12 Polyatomic molecules in general

The photoelectron spectrum of NH_3 (Fig. 3.31) indicates some of the features that a theory of the structure of polyatomic molecules must elucidate. The spectrum shows two bands. The one with the lower ionization energy (in the region of 11 eV) has considerable vibrational structure. This structure indicates that the orbital from which the electron is ejected plays a considerable role in the determination of the molecule's shape. The broad band in the region of 16 eV arises from electrons that are bound more tightly.

(a) The formation of molecular orbitals

The features that have been introduced in connection with trihydrogen are present in all polyatomic molecules. In each case, we write the molecular orbital of a given symmetry (such as the σ orbitals of a linear molecule) as a sum of *all* the orbitals that can overlap to form orbitals of that symmetry:

$$\psi = \sum_i c_i \phi_i \qquad (11)$$

In this linear combination the ϕ_i are atomic orbitals (usually, the valence orbitals of each atom in the molecule) and the index i runs over all the atomic orbitals that have the appropriate symmetry. From N atomic orbitals we can construct N molecular orbitals. Then:

1 The greater the number of nodes in a molecular orbital, the greater the antibonding character and the higher the orbital energy.
2 Interactions between non-nearest neighbor atoms are weakly bonding (lower the energy slightly) if the orbital lobes on these atoms have the same sign (and interfere constructively). They are weakly antibonding if the signs are opposite (and interfere destructively).
3 Orbitals constructed from lower energy atomic orbitals lie lower in energy. (So atomic s orbitals typically produce lower energy molecular orbitals than atomic p orbitals of the same shell.)

For instance, to account for the features in the photoelectron spectrum of NH_3, we need to build molecular orbitals that will accommodate the eight valence electrons in the molecule. Each molecular orbital is the combination of seven atomic orbitals: the three H1s orbitals, the

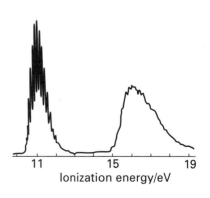

3.31 The UV photoelectron spectrum of NH_3 using helium 21 eV radiation.

Ionization energy/eV

N2s orbital, and the three N2p orbitals. It is possible to construct seven molecular orbitals from these seven atomic orbitals. The form of the orbitals is shown in Fig. 3.32.

The formal rules for the construction of the orbitals are described in Chapter 4, but it is possible to obtain a sense of their origin by imagining viewing the NH_3 molecule along its threefold axis (designated z). The $N2p_z$ and N2s orbitals both have cylindrical symmetry about that axis. If the three H1s orbitals are superimposed with the same sign relative to each other (that is, so that all have the same tint in the diagram), then they match this cylindrical symmetry. It follows that we can form molecular orbitals of the form

$$\psi = c_1\phi_{N2s} + c_2\phi_{N2p_z} + c_3\{\phi_{H1s_A} + \phi_{H1s_B} + \phi_{H1s_C}\}$$

From these *three* orbitals (the specific combination of H1s orbitals counts as a single orbital), it is possible to construct three molecular orbitals (with different values of the coefficients c). The orbital with no nodes between the N and H atoms is the lowest in energy, that with a node between all the NH neighbors is the highest in energy, and the third orbital lies between the two. The three orbitals are nondegenerate and are labeled $1a$, $2a$, and $3a$ in order of increasing energy.

The $N2p_x$ and $N2p_y$ orbitals have π symmetry with respect to the z-axis, and can be used to form orbitals with combinations of the H1s orbitals that have a matching symmetry. For example, one such superposition will have the form

$$\psi = c_1\phi_{N2p_x} + c_2\{\phi_{H1s_A} - \phi_{H1s_B}\}$$

As can be seen from Fig 3.32, the signs of the H1s orbital combination match those of the $N2p_x$ orbital. The N2s orbital cannot contribute to this superposition, so only *two*

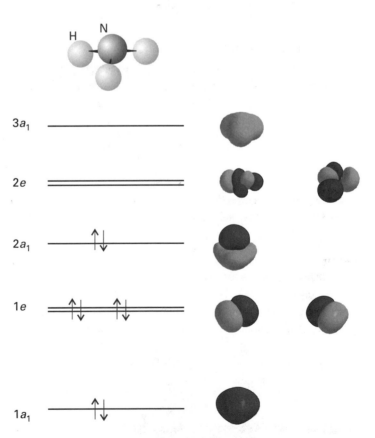

3.32 The composition of the molecular orbitals of NH_3. The signs of the orbitals are represented by tinting.

combinations can be formed, one without a node between the N and H orbitals and the other with a node. The two orbitals differ in energy, the former being lower. A similar combination of orbitals can be formed with the $N2p_y$ orbital, and it turns out (by the symmetry arguments that we use in Chapter 4) that the two orbitals are degenerate with the two we have just described. The combinations are examples of e orbitals (because they form doubly degenerate pairs), and are labeled $1e$ and $2e$ in order of increasing energy.

The general form of the molecular orbital energy level diagram is shown in Fig. 3.33. The actual location of the orbitals (particularly the relative positions of the a set and the e set), can be found only by detailed computation or by identifying the orbitals responsible for the photoelectron spectrum. We have indicated the probable assignment of the 11 eV and 16 eV peaks, which fixes the locations of two of the occupied orbitals. The third occupied orbital is out of range of the 21 eV radiation used to obtain the spectrum.

The photoelectron spectrum is consistent with the need to accommodate eight electrons in the orbitals. The electrons enter the molecular orbitals in increasing order of energy, starting with the orbital of lowest energy, and taking note of the requirement of the exclusion principle that no more than two electrons can occupy any one orbital. The first two electrons enter $1a$ and fill it. The next four enter the doubly degenerate $1e$ orbitals and fill them. The last two enter the $2a$ orbital, which calculations show is almost nonbonding and localized on the N atom. The resulting overall ground-state electron configuration is therefore $1a^2 1e^4 2a^2$. No antibonding orbitals are occupied, so the molecule has a lower energy than the separated atoms. The conventional description of NH_3 as a molecule with a lone pair is also mirrored in the configuration: the HOMO is $2a$, which is largely confined to the N atom and makes only a small contribution to the bonding. We have already seen that lone pair electrons play a considerable role in determining the shapes of molecules. The extensive vibrational structure in the 11 eV band of the photoelectron spectrum is consistent with this observation, for photoejection of a $2a$ electron removes the effectiveness of the lone pair and the shape of the ionized molecule is considerably different from that of the NH_3 itself. Photoionization therefore results in extensive vibrational structure in the spectrum.

Molecular orbitals are formed from linear combinations of atomic orbitals of the same symmetry. Their energies can be determined experimentally from gas-phase photoelectron spectra and interpreted in terms of the pattern of orbital overlap.

3.33 The molecular orbital energy level diagram of NH_3 when the molecule has its observed bond angle (107°) and bond lengths.

(b) Hypervalence in the context of molecular orbitals

A slightly more complicated example, but one that makes an important point, is the octahedral molecule SF_6. A simple basis set that is adequate to illustrate the point we want to make consists of the valence-shell s and p orbitals of the S atom and one p orbital of each of the six F atoms and pointing toward the S atom. We use the $F2p$ orbitals rather than the $F2s$ orbitals because they match the S orbitals more closely in energy. From these ten atomic orbitals it is possible to construct ten molecular orbitals. Calculations[10] indicate that four of the orbitals are bonding and four are antibonding; the two remaining orbitals are nonbonding (Fig. 3.34).[11]

There are 12 electrons to accommodate. The first two can enter $1a$ and the next six can enter $1t$. The remaining four fill the nonbonding pair of orbitals, resulting in $1a^2 1t^6 e^4$. As we see, none of the antibonding orbitals ($2a$ and $2t$) is occupied. The theory, therefore, accounts for the formation of SF_6, with four bonding orbitals and two nonbonding orbitals occupied.

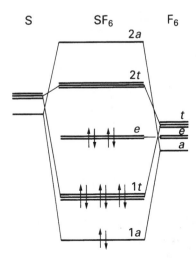

3.34 A schematic molecular orbital energy level diagram for SF_6.

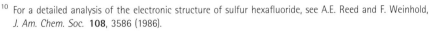

[10] For a detailed analysis of the electronic structure of sulfur hexafluoride, see A.E. Reed and F. Weinhold, *J. Am. Chem. Soc.* **108**, 3586 (1986).

[11] The notation for the orbitals has been simplified from that appropriate to an octahedral molecule: see Section 4.5 for the full labels.

The important point is that the bonding in SF_6 can be explained without using $S3d$ orbitals to expand the octet. This does not mean that d orbitals cannot participate in the bonding, but it does show that they are not *necessary* for bonding six F atoms to the central S atom. The limitation of valence-bond theory is the assumption that each atomic orbital on the central atom can participate in the formation of only one bond. Molecular orbital theory takes hypervalence into its stride by having available plenty of orbitals, not all of which are antibonding. Therefore the question of when hypervalence can occur appears to depend on factors other than d-orbital availability.

> The delocalization of molecular orbitals means that an electron pair can contribute to the bonding of more than two atoms.

(c) Electron deficiency

The formation of molecular orbitals by combining several atomic orbitals accounts effortlessly for the existence of **electron-deficient compounds**, which are compounds for which there are not enough electrons for a Lewis structure to be written. This point can be illustrated most easily with diborane, B_2H_6 (**23**). There are only 12 valence electrons, but at least seven electron pairs are needed to bind eight atoms together if Lewis's model is correct. The problem is solved effortlessly by molecular orbital theory. The eight atoms of this molecule contribute a total of 14 valence orbitals (four orbitals from each B atom, making eight, and one orbital each from the six H atoms). These 14 atomic orbitals can be used to construct 14 molecular orbitals. About seven of these molecular orbitals will be bonding or nonbonding, which is more than enough to accommodate the 12 valence electrons provided by the atoms. In MO terms, there is nothing mysterious about the existence of this molecule.

A slightly different viewpoint is achieved by concentrating on the two BHB fragments and the molecular orbitals that can be formed by combining the atomic orbitals contributed by the three atoms (Fig. 3.35). This is the first of several examples where we construct molecular orbitals from the orbitals of molecular fragments. This approach is a half-way stage between the atomic orbitals and the final molecular orbitals, and is very useful for drawing analogies between molecules. The bonding orbital spanning these three atoms can accommodate two electrons and pull the molecule together, exactly as in H_3^+. Because the bridging bond is formed from two electrons in a molecular orbital built from orbitals on three atoms, it is another example of a (3c,2e) bond. That two electrons can bind two pairs of atoms (BH and HB) in this way is not particularly remarkable, but it is an insurmountable difficulty for the Lewis approach. Electron deficiency is in fact well developed not only in boron (where it was first clearly recognized), but also in carbocations and a variety of other classes of compounds that we encounter later in the text.

> The existence of electron-deficient species is explained by the delocalization of the bonding influence of electrons over several atoms.

(d) Localization

A striking feature of the Lewis approach to chemical bonding is its accord with chemical instinct, for it identifies something that can be called 'an A—B bond'. Both O—H bonds in H_2O, for instance, are treated as localized, equivalent, structures, because each one consists of an electron pair shared between O and H. This feature appears to be absent from molecular orbital theory, for molecular orbitals are delocalized, and the electrons that occupy them bind all the atoms together, not just a specific pair of neighboring atoms. The concept of an A—B bond as existing independently of other bonds in the molecule, and of being transferable from one molecule to another, appears to have been lost. However, we shall now show that the molecular orbital description is mathematically almost equivalent to a localized description of the overall electron distribution. The demonstration hinges on the fact that linear

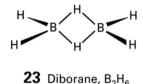

23 Diborane, B_2H_6

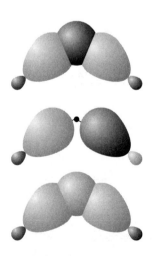

3.35 The molecular orbitals formed by two B orbitals and one H orbital on an atom lying between them, as in B_2H_6. Two electrons occupy the bonding combination and hold all three atoms together.

combinations of molecular orbitals can be formed that result in the same *overall* electron distribution, but the individual orbitals are distinctly different.[12]

Consider the H_2O molecule. The two occupied bonding orbitals of the delocalized description, $1a_1$ and $1b_2$, are shown in Fig. 3.36. If we form the sum $1a_1 + 1b_2$, the negative half of $1b_2$ cancels half the $1a_1$ orbital almost completely, leaving a localized orbital between O and the other H. Likewise, when we form the difference $1a_1 - 1b_2$, the other half of the $1a_1$ orbital is canceled almost completely, so leaving a localized orbital between the other pair of atoms. That is, by taking sums and differences of delocalized orbitals, localized orbitals are created (and vice versa). Because these are two equivalent ways of describing the same overall electron population, one description cannot be said to be better than the other. Hence it is reasonably well justified to use the localized description of a molecule when chemical evidence suggests that it is appropriate.

Table 3.9 suggests when it is appropriate to select a delocalized description or a localized description. In general, a delocalized description is needed for dealing with global properties of the entire molecule. Such properties include electronic spectra (UV and visible transitions), photoionization spectra, ionization and electron attachment energies, and reduction potentials. In contrast, a localized description is most appropriate for dealing with properties of a fragment of a total molecule. Such properties include bond strength, bond length, bond force constant, and some aspects of reactions (such as acid–base character). The localized description is more appropriate because it focuses attention on the distribution of electrons in and around a particular bond.

> Localized and delocalized descriptions of bonds are mathematically equivalent, but one description may be more suitable for a particular property, as summarized in Table 3.9.

(e) Localized bonds and hybridization

The localized molecular orbital description of bonding can be taken a stage further by invoking the concept of hybridization. Strictly speaking, hybridization belongs to VB theory, but it is commonly invoked in simple qualitative descriptions of molecular orbitals.

We have seen that in general a molecular orbital is constructed from all atomic orbitals of the appropriate symmetry. However, it is sometimes convenient to form a mixture of orbitals on one atom (the O atom in H_2O, for instance), and then to use these hybrid orbitals to construct localized molecular orbitals. In H_2O, for instance, each O—H bond can be regarded

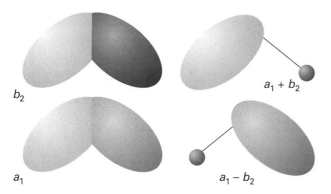

b_2 $a_1 + b_2$

a_1 $a_1 - b_2$

3.36 The two occupied $1a_1$ and $1b_2$ orbitals of the H_2O molecule and their sum $1a_1 + 1b_2$ and difference $1a_1 - 1b_2$. In each case we form an almost fully localized orbital between a pair of atoms.

Table 3.9 A general indication of the properties for which localized and delocalized descriptions are appropriate

Localized appropriate	Delocalized appropriate
Bond strengths	Electronic spectra
Force constants	Photoionization
Bond lengths	Electron attachment
Brønsted acidity*	Magnetism
VSEPR description of molecular geometry	Walsh description of molecular geometry
	Standard potentials†

*Chapter 5.   †Chapter 6.

[12] The justification of this remark is that a many-electron wavefunction that satisfies the Pauli principle can be written as a determinant. Linear combinations of the rows or columns of any determinant can be taken without affecting its value.

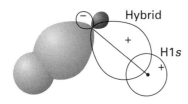

3.37 The formation of localized O—H orbitals in H_2O by the overlap of hybrid orbitals on the O atom and H1s orbitals. The hybrid orbitals are a close approximation to the sp^3 hybrids shown in Fig. 3.8.

as formed by the overlap of an H1s orbital and a hybrid orbital composed of O2s and O2p orbitals (Fig. 3.37).

We have already seen that the mixing of s and p orbitals on a given atom results in hybrid orbitals that have a definite direction in space, as in the formation of tetrahedral hybrids (eqn 4 in Section 3.6). Once the hybrid orbitals have been selected, a localized molecular orbital description can be constructed. For example, four bonds in CF_4 can be formed by building bonding and antibonding localized orbitals by overlap of each hybrid and one F2p orbital directed toward it. Similarly, to describe the electron distribution of BF_3, we could consider each localized B—F σ orbital as formed by the overlap of an sp^2 hybrid with an F2p orbital. A localized orbital description of a PCl_5 molecule would be in terms of five P—Cl σ bonds formed by overlap of each of the five trigonal-bipyramidal sp^3d hybrid orbitals with a 3p orbital of a Cl atom. Similarly, where we wanted to form six localized orbitals in a regular octahedral arrangement, we would need two d orbitals: the resulting six sp^3d^2 hybrids point in the required directions.

Hybrid atomic orbitals are sometimes used in the discussion of localized molecular orbitals.

3.13 Molecular shape in terms of molecular orbitals

The VSEPR model is based on a localized picture of electron distributions. However, in molecular obital theory the electrons responsible for bonding are delocalized over the entire molecule. Current *ab initio* and semiempirical molecular orbital calculations, which are widely performed with commercial software, are able to predict the shapes of even quite complicated molecules with high reliability. Nevertheless, there is still a need to understand the qualitative factors that contribute to the shape of a molecule within the framework of MO theory.

A simple pictorial approach to the task of analyzing molecular shape in terms of delocalized molecular orbitals was devised by A.D. Walsh in a classic series of papers published in 1953. Walsh's approach to the discussion of the shape of an H_2X triatomic molecule (such as BeH_2 and H_2O) is illustrated in Fig. 3.38. The illustration shows an example of a **Walsh diagram**, a graph showing the dependence of orbital energy on molecular geometry. The Walsh diagram for an H_2X molecule is constructed by considering how the composition and energy of each molecular orbital changes as the bond angle is varied from 90° to 180°. The diagram is in fact just a more elaborate version of the correlation diagram that we illustrated for an H_3 molecule or an H_3^+ ion in Section 3.11 (Fig. 3.30).

The molecular orbitals to consider in the angular molecule are[13]

$$\psi_{a_1} = c_1\phi_{2s} + c_2\phi_{2p_z} + c_3\phi_+$$
$$\psi_{b_1} = \phi_{2p_x} \tag{12}$$
$$\psi_{b_2} = c_4\phi_{2p_y} + c_5\phi_-$$

The linear combinations ϕ_+ and ϕ_- are illustrated in Fig. 3.39. There are three a_1 orbitals and two b_2 orbitals; the lowest energy orbitals of each type are shown on the left of Fig. 3.40. In the linear molecule, the molecular orbitals are

$$\psi_{\sigma_g} = c_1\phi_{2s} + c_2\phi_+$$
$$\psi_{\pi_u} = \phi_{2p_x} \text{ and } \phi_{2p_z} \tag{13}$$
$$\psi_{\sigma_u} = c_3\phi_{2p_y} + c_4\phi_-$$

The lowest energy molecular orbital in 90° H_2X is the one labeled $1a_1$, which is built from the overlap of the $X2p_z$ orbital (the $2p_x$ orbital on X) with the ϕ_+ combination of H1s orbitals. As

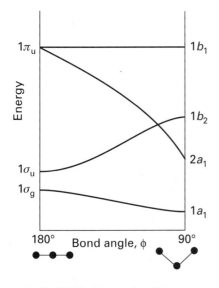

3.38 The Walsh diagram for XH_2 molecules. Only the bonding and nonbonding orbitals are shown.

[13] We continue to use the letters a and b to label nondegenerate orbitals, and will explain their full significance in Chapter 4.

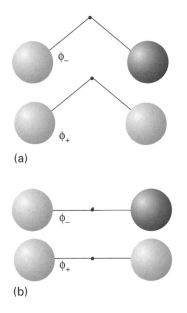

(a)

(b)

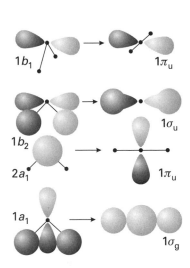

3.39 The combinations of H1s orbitals that are used to construct molecular orbitals in (a) angular, (b) linear XH$_2$ molecules.

3.40 The composition of the molecular orbitals of an XH$_2$ molecule at the two extremes of the correlation diagram shown in Fig. 3.38.

the bond angle changes to 180° the energy of this orbital increases in part because the H–H overlap decreases and in part because the loss of X2p_z character decreases the overlap with ϕ_+ (Fig. 3.40). The energy of the 1b_2 orbital is lowered because the H1s orbitals move into a better position for overlap with the X2p_y orbital. The weakly antibonding H–H contribution is also decreased. The biggest change occurs for the 2a_1 orbital. It has considerable X2s character in the 90° molecule, but correlates with a pure X2p_z orbital in the 180° molecule. Hence, it shows a steep rise in energy as the bond angle increases. The 1b_1 orbital is a nonbonding X2p orbital perpendicular to the molecular plane in the 90° molecule and remains nonbonding in the linear molecule. Hence, its energy barely changes with angle.

The principal feature that determines whether or not the molecule is angular is whether 2a_1 is occupied. This is the orbital that has considerable X2s character in the angular molecule but not in the linear molecule. Hence, a lower energy is achieved if, when it is occupied, the molecule is angular. The shape adopted by an H$_2$X molecule therefore depends on the number of electrons that occupy the orbitals.

The simplest XH$_2$ molecule in Period 2 is the transient gas-phase BeH$_2$ molecule,[14] in which there are four valence electrons. These four electrons occupy the lowest two molecular orbitals. If the lowest energy is achieved with the molecule angular, then that will be its shape. We can decide whether the molecule is likely to be angular by accommodating the electrons in the lowest two orbitals corresponding to an arbitrary bond angle in Fig. 3.38. We then note that the HOMO (the highest occupied molecular orbital) decreases in energy on going to the left of the diagram and that the lowest total energy is obtained when the molecule is linear. Hence, BeH$_2$ is predicted to be linear and to have the configuration 1$\sigma^2$2σ^2. In CH$_2$, which has two more electrons than BeH$_2$, three of the molecular orbitals must be occupied. In this case, the lowest energy is achieved if the molecule is angular and has configuration 1$a_1^2$2$a_1^2$1b_2^2.

---

[14] BeH$_2$ normally exists as a polymeric solid, with four-coordinate beryllium atoms.

In general, any XH_2 molecule with from five to eight valence electrons is predicted to be angular. The observed bond angles are

BeH_2	BH_2	CH_2	NH_2	OH_2
180	131	136	103	105

These experimental observations are qualitatively in line with Walsh's approach, but for quantitative predictions we have to turn to detailed molecular orbital calculations.

Example 3.8 Using a Walsh diagram to predict a shape

Predict the shape of an H_2O molecule on the basis of a Walsh diagram for an XH_2 molecule.

Answer We choose an intermediate bond angle along the horizontal axis of the XH_2 diagram in Fig. 3.38 and accommodate eight electrons. The resulting configuration is $1a_1^2 2a_1^2 1b_2^2 1b_1^2$. The $2a_1$ orbital is occupied, so we expect the nonlinear molecule to have a lower energy than the linear molecule.

Self-test 3.8 Is any XH_2 molecule, in which X denotes an atom of a Period 3 element, expected to be linear? If so, which?

Walsh applied his approach to molecules other than compounds of hydrogen, but the correlation diagrams soon become very complicated. His approach represents a valuable complement to the VSEPR model because it traces the influences on molecular shapes of the occupation of orbitals spreading over the entire molecule and concentrates less on localized repulsions between pairs of electrons. Correlation diagrams like those introduced by Walsh are frequently encountered in contemporary discussions of the shapes of complex molecules, and we shall see a number of examples in later chapters. They illustrate how inorganic chemists can sometimes identify and weigh competing influences by considering two extreme cases (such as linear and $90°$ XH_2 molecules), and then rationalize the fact that the state of a molecule is a compromise intermediate between the two extremes.

> In the Walsh model, the shape of a molecule is predicted on the basis of the occupation of molecular orbitals that, in a correlation diagram, show a strong dependence on bond angle.

The molecular orbital theory of solids

The molecular orbital theory of small molecules can be extended to account for the properties of solids, which are aggregations of a virtually infinite number of atoms. This approach is strikingly successful for the description of metals, for it may be used to explain their characteristic luster, their good electrical and thermal conductivity, and their malleability. All these properties stem from the ability of the atoms to contribute electrons to a common sea. The luster and electrical conductivities stem from the mobility of these electrons, either in response to the oscillating electric field of an incident ray of light or to a potential difference. The high thermal conductivity is also a consequence of electron mobility, because an electron can collide with a vibrating atom, pick up its energy, and transfer it to another atom elsewhere in the solid. The ease with which metals can be mechanically deformed is another aspect of electron mobility, because the electron sea can quickly readjust to a deformation of the solid and continue to bind the atoms together.

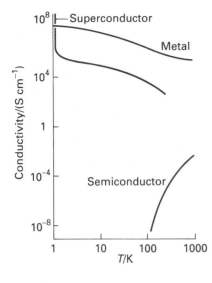

3.41 The variation of the electrical conductivity of a substance with temperature is the basis of its classification as a metallic conductor, a semiconductor, or a superconductor.

Electronic conduction is also a characteristic of semiconductors. The criterion for distinguishing between a metal and a semiconductor is the temperature dependence of the electric conductivity (Fig. 3.41):[15]

- A **metallic conductor** is a substance with an electric conductivity that *decreases* with increasing temperature.

- A **semiconductor** is a substance with an electric conductivity that *increases* with increasing temperature.

It is also generally the case (but not the criterion for distinguishing them) that the conductivities of metals at room temperature are higher than those of semiconductors. A solid **insulator** is a substance with a very low electrical conductivity. However, when that conductivity can be measured, it is found to increase with temperature, like that of a semiconductor. For some purposes, therefore, it is possible to disregard the classification 'insulator' and to treat all solids as either metals or semiconductors. **Superconductors** are a special class of materials that have zero electrical resistance below a critical temperature.

A metallic conductor is a substance with an electric conductivity that decreases with increasing temperature; a semiconductor is a substance with an electric conductivity that increases with increasing temperature.

3.14 Molecular orbital bands

The central idea underlying the description of the electronic structure of solids is that the valence electrons supplied by the atoms spread through the entire structure. This concept is expressed more formally by making a simple extension of MO theory in which the solid is treated like an indefinitely large molecule.[16] The description in terms of delocalized electrons can also be used to describe nonmetallic solids. We shall therefore begin by showing how metals are described in terms of molecular orbitals. Then we shall go on to show that the same principles can be applied, but with a different outcome, to ionic and molecular solids.

---

[15] The *resistance*, R, of a sample is measured in ohms, Ω. The inverse of the resistance is called the *conductance*, G, and is measured in siemens, S, where $1\ S = 1\ \Omega^{-1}$. The resistance of a sample increases with its length, l, and decreases with its cross-sectional area, A, and we write $R = \rho l/A$, where ρ is the *resistivity* of the substance. The units of resistivity are ohm-meter $(\Omega\,m)$. The *conductivity*, σ, is the reciprocal of the resistivity, and its units are siemens per meter $(S\,m^{-1})$.

[16] In solid state physics, this approach is called the 'tight-binding approximation'.

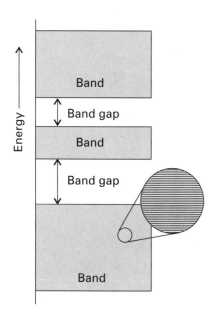

3.42 The electronic structure of a solid is characterized by a series of bands of orbitals which are separated by gaps in the energy for which there are no orbitals.

(a) Band formation by orbital overlap

The overlap of a large number of atomic orbitals leads to molecular orbitals that are closely spaced in energy and so form a virtually continuous **band** (Fig. 3.42). Bands are separated by **band gaps**, which are values of the energy for which there are no molecular orbitals.

The formation of bands can be understood by considering a line of atoms, and supposing that each atom has an s orbital that overlaps the s orbitals on its immediate neighbors (Fig. 3.43). When the line consists of only two atoms, there is a bonding and an antibonding molecular orbital. When a third atom joins them, there are three orbitals. The central orbital of the set is nonbonding and the outer two are at low energy and high energy, respectively. As more atoms are added, each one contributes an atomic orbital, and hence one more molecular orbital is formed. When there are N atoms in the line, there are N molecular orbitals. The orbital of lowest energy has no nodes between neighboring atoms. The orbital of highest energy has a node between every pair of neighbors. The remaining orbitals have successively $1, 2, \ldots$ internuclear nodes and a corresponding range of energies between the two extremes.

The total width of the band, which remains finite even as N approaches infinity (as shown in Fig. 3.44), depends on the strength of the interaction between neighboring atoms. The greater the strength of interaction (in broad terms, the greater the degree of overlap between neighbors), the greater the energy separation of the no-node orbital and the all-node orbital. However, whatever the number of atomic orbitals used to form the molecular orbitals, there is only a finite spread of orbital energies (as depicted in Fig. 3.44). It follows that the separation in energy between neighboring orbitals must approach zero as N approaches infinity, for otherwise the range of orbital energies could not be finite. That is, a band consists of a countable number but near-continuum of energy levels.

The band just described is built from s orbitals and is called an **s band**. If there are p orbitals available, a **p band** can be constructed from their overlap as shown in Fig. 3.45. Because p orbitals lie higher in energy than s orbitals of the same valence shell, there is often an energy gap between the s band and the p band (Fig. 3.46). However, if the bands span a wide range of energy and the atomic s and p energies are similar (as is often the case), then the two bands overlap. The d **band** is similarly constructed from the overlap of d orbitals.

The overlap of atomic orbitals in solids gives rise to bands separated by gaps.

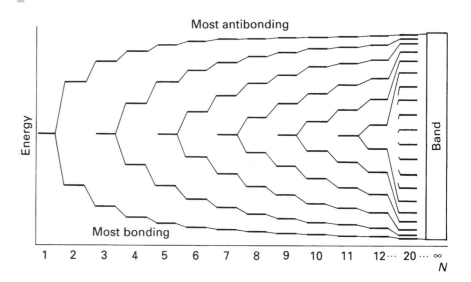

3.44 The energies of the orbitals that are formed when N atoms are brought up to form a line.

3.43 A band can be thought of as formed by bringing up atoms successively to form a line of atoms. N atomic orbitals give rise to N molecular orbitals.

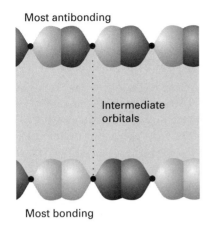

Most antibonding

Intermediate orbitals

Most bonding

3.45 An example of a p band in a one-dimensional solid.

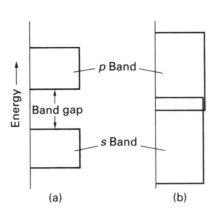

(a)   (b)

3.46 (a) The s and p bands of a solid and the gap between them. Whether there is in fact a gap depends on the separation of the s and p orbitals of the atoms and the strength of interaction between the atoms. (b) If the interaction is strong, the bands are wide and may overlap.

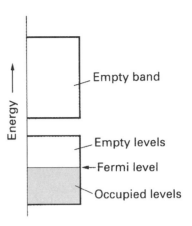

Empty band

Empty levels

Fermi level

Occupied levels

3.47 If each of the N atoms supplies one s electron, then at $T = 0$ the lowest $\frac{1}{2}N$ are occupied, and the Fermi level lies near the center of the band.

(b) The Fermi level

At $T = 0$, electrons occupy the individual molecular orbitals of the bands in accord with the building-up principle. If each atom supplies one s electron, then at $T = 0$ the lowest $\frac{1}{2}N$ are occupied. The highest occupied orbital at $T = 0$ is called the **Fermi level**; it lies near the center of the band (Fig. 3.47).

At temperatures above $T = 0$, the population, P, of the orbitals is given by the **Fermi–Dirac distribution**, which is a version of the Boltzmann distribution that takes into account the requirement that no more than two electrons can occupy any level. The distribution has the form

$$P = \frac{1}{e^{(E-\mu)/kT} + 1} \tag{14}$$

where μ is the **chemical potential**, which in this context is the energy of the level for which $P = \frac{1}{2}$. The chemical potential depends on the temperature, and at $T = 0$ is equal to the energy of the Fermi level. As the temperature is increased, the chemical potential rises above the Fermi level because electrons start to occupy higher states and the level at which $P = \frac{1}{2}$ becomes higher in energy. The shape of the Fermi–Dirac distribution is shown in Fig. 3.48. At high energies ($E \gg \mu$), the 1 in the denominator of the Fermi–Dirac distribution can be neglected, and the populations for $T > 0$ resemble a Boltzmann distribution in as much as they decline exponentially with increasing energy:

$$P \approx e^{-(E-\mu)/kT} \tag{15}$$

When the band is not completely full, the electrons close to the Fermi level can easily be promoted to nearby empty levels. As a result, they are mobile, and can move relatively freely through the solid. The substance is an electronic conductor. To understand this mobility, we can think of the individual orbitals in a band as being standing waves. Standing waves are superpositions of traveling waves corresponding to motion in opposite directions. In the absence of a potential difference, the two directions of travel are degenerate and are equally

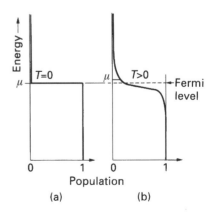

(a)   (b)

Population

3.48 The shape of the Fermi distribution (a) at $T = 0$ and (b) at $T > 0$. The population decays exponentially at energies well above the Fermi level.

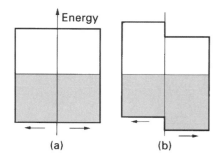

3.49 Another way of representing the bands is to draw the orbitals for motion to the right and motion to the left separately. (a) In the absence of a potential difference applied across the metal, the corresponding orbitals are degenerate. However, when a field is applied (b), one 'half band' has a lower energy than the other. It is more heavily populated, and the net result is a flow of electrons.

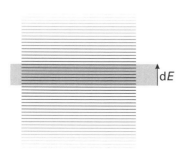

3.50 The density of states is the number of energy levels in an infinitesimal region of the band divided by the width of the infinitesimal region (dE).

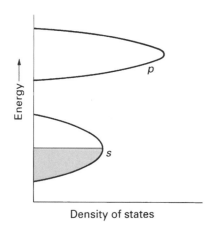

3.51 A typical density of states in a metal.

populated up to the Fermi level (Fig. 3.49). However, when a potential difference is applied, electrons traveling in one direction have a different energy from those traveling in the opposite direction, and the two sets of orbitals are no longer equally populated. Consequently, there are now more electrons traveling in one direction than in the other, and an electric current flows through the solid.

We have seen that the criterion of metallic conduction is the decrease of electrical conductivity with increasing temperature. This behavior is the opposite of what we might expect if the conductivity were governed by a Boltzmann distribution of electrons. The competing effect can be identified once we recognize that the ability of an electron to travel smoothly through the solid in a conduction band depends on the uniformity of the arrangement of the atoms. An atom vibrating vigorously at a site is equivalent to an impurity that disrupts the orderliness of the orbitals. This decrease in uniformity reduces the ability of the electron to travel from one edge of the solid to the other, so the conductivity of the solid is less than at $T = 0$. If we think of the electron as described by a wave propagating through the solid, then we would say that it was 'scattered' by the impurity (the atomic vibration). This **carrier scattering** increases with increasing temperature as the lattice vibrations increase, and the increase accounts for the observed inverse temperature dependence of the conductivity of metals.

> The occupation of orbitals in a band is given by the Fermi–Dirac distribution. The decreate in conductivity of metals as the temperature is raised is due to carrier scattering.

(c) Densities of states

The number of energy levels in a range divided by the width of the range is called the **density of states**, ρ (Fig. 3.50). The density of states is not uniform across a band because the energy levels are packed together more closely at some energies than at others. This variation is apparent even in one dimension, for—compared with its edges—the center of the band is relatively sparse in orbitals (as can be seen in Fig. 3.44). In three dimensions, the variation of density of states is more like that shown in Fig. 3.51, with the greatest density of states near the center of the band and the lowest density at the edges. The reason for this behavior can be traced to the number of ways of producing a particular linear combination of atomic orbitals. There is only one way of forming a fully bonding molecular orbital (the lower edge of the band) and only one way of forming an fully antibonding orbital (the upper edge). However, there are many ways (in a three-dimensional array of atoms) of forming a molecular orbital with an energy in the interior of the band.

The density of states is zero in the band gap—there are no energy levels in the gap. In certain special cases, though, a full band and an empty band might coincide in energy, but with a zero density of states at their conjunction (Fig. 3.52). Solids with this band structure are called **semimetals**. Because they have only a few electrons that can act as carriers, semimetals are characterized by a low metallic conductivity. One important example is graphite, which is a semimetal in directions parallel to the sheets of carbon atoms.[17]

> The density of states is not uniform across a band: in most cases, the states are densest close to the center of the band.

(d) Photoelectron and X-ray analysis of bands

Evidence for the existence of bands and a map of their densities of states can be obtained experimentally with photoelectron spectroscopy in much the same way as for discrete molecules. The densities of states of discrete molecules consist of a series of widely separated narrow peaks, each spike corresponding to the energy of a discrete molecular orbital. These

---

[17] This use of the term semimetal should be distinguished from its use as a synonym for metalloid.

peaks appear in the photoelectron spectrum as photoelectrons with discrete ionization energies.

Analogous information about solids can be obtained from the analysis of **X-ray emission bands**. In this technique, electrons are ejected (by electron bombardment) from the inner closed shells of the atoms and X-rays are emitted as electrons from the valence bands fall into the resulting vacancies (Fig. 3.53). Because a valence electron can originate from any occupied level in the band, the X-ray emission that accompanies its transition covers a range of frequencies. The greatest emission intensities are obtained when there are many states in the valence band with similar energies, and the weakest intensities are obtained when there are only a few states with a particular energy. Hence, it follows that the shape of the X-ray emission band is an indication of the variation of the density of states across the band. The intensity profile does not match the band profile exactly, however, because allowance must be made for the differences between the ease with which the incoming photon may eject an electron from different types of orbital (more specifically, the differences in transition probabilities).

> X-ray emission bands give information about the densities of states of the occupied parts of bands.

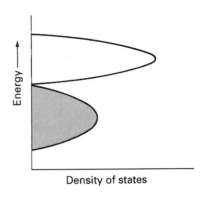

3.52 The density of states typical of a semimetal.

(e) Special features of one-dimensional solids

In recent years, a series of solids have been investigated that have a partially filled band formed by a linear chain of metal atoms. An example is $K_2Pt(CN)_4Br_{0.3} \cdot 3H_2O$, commonly called 'KCP' (Fig. 3.54). These **one-dimensional solids** are not quite as simple as has been implied by the discussion so far, for a theorem due to Rudolph Peierls states that, at $T = 0$, no one-dimensional solid is a metal!

The origin of Peierls' theorem can be traced to a hidden assumption in the discussion so far: we have supposed that the atoms lie in a line with a regular separation. However, the actual spacing in a one-dimensional solid (and any solid) is determined by the distribution of the electrons, not vice versa, and there is no guarantee that the state of lowest energy is a solid with a regular lattice spacing. In fact, in a one-dimensional solid at $T = 0$, there always exists

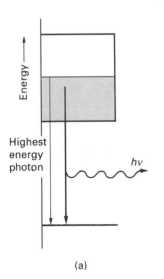

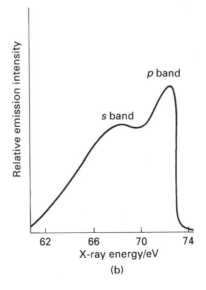

3.53 (a) The formation of an X-ray emission band and (b) a typical example (aluminum).

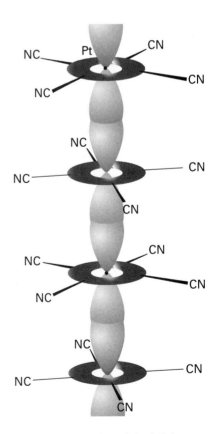

3.54 A representation of the infinite chain structure of KCP, $K_2Pt(CN)_4Br_{0.3} \cdot 3H_2O$, and a schematic illustration of its d band.

(a)

(b)

3.55 The formation of a Peierls distortion: the energy of the line of atoms with alternating bond lengths (b) is lower than that of the uniformly spaced atoms (a).

a distortion, a **Peierls distortion**, which leads to a lower energy than in the perfectly regular solid.

An idea of the origin and effect of a Peierls distortion can be obtained by considering a one-dimensional solid of N atoms and N valence electrons (Fig. 3.55). Such a line of atoms distorts to one that has long and short alternating bonds. Although the longer bond is energetically unfavorable, the strength of the short bond more than compensates for the weakness of the long bond, and the net effect is a lowering of energy below that of the regular solid. Now, instead of the electrons near the Fermi surface being free to move through the solid, they are trapped between the longer-bonded atoms (these electrons have antibonding character, and so are found outside the internuclear region between strongly bonded atoms). The Peierls distortion introduces a band gap in the center of the original conduction band, and the filled orbitals are separated from the empty orbitals. Hence, the distortion results in a semiconductor or insulator, not a metallic conductor.

The conduction band in KCP is a d band formed principally by overlap of platinum's $5d_{z^2}$ orbitals. The small proportion of bromine in the compound, which is present as Br^-, removes a small number of electrons from this otherwise full d band, so turning it into a conduction band. Indeed, at room temperature, doped KCP is a lustrous bronze color with its highest conductivity along the axis of the Pt chain. However, below 150 K the conductivity drops sharply on account of the onset of a Peierls distortion. At higher temperatures, the motion of the atoms averages the distortion to zero, the separation is regular (on average), the gap is absent, and the solid is metallic.

> *A Peierls distortion ensures that no one-dimensional solid is a metallic conductor below a critical temperature.*

(f) Insulators

A solid is an insulator if enough electrons are present to fill a band completely and there is a considerable energy gap before an empty orbital becomes available (Fig. 3.56). In a sodium chloride crystal, for instance, the N Cl^- ions are nearly in contact and their $3s$ and $3p$ valence orbitals overlap to form a narrow band consisting of $4N$ levels. The Na^+ ions are also nearly in contact and also form a band. The electronegativity of chlorine is so much greater than that of sodium that the chlorine band lies well below the sodium band, and the band gap is about 7 eV. A total of $8N$ electrons are to be accommodated (seven from each chlorine atom, one from each sodium atom). These $8N$ electrons enter the lower chlorine band, fill it, and leave the sodium band empty. Because kT is equivalent to 0.03 eV at room temperature, very few electrons occupy the orbitals of the sodium band.

We normally think of an ionic or molecular solid as consisting of discrete ions or molecules. According to the picture we have just described, though, it appears that they should be regarded as having a band structure. The two pictures can be reconciled, because it is possible to show that a *full* band is equivalent to a sum of localized electron densities. In sodium chloride, for example, a full band built from Cl orbitals is equivalent to a collection of discrete Cl^- ions. As with molecules, the delocalized band picture is needed for description of spectra where processes involve one electron at a time, such as photoelectron spectra and X-ray spectra.

> *A solid insulator is a semiconductor with a large band gap.*

Energy →

3.56 The structure of a typical insulator: there is a significant gap between the filled and empty bands.

3.15 Semiconduction

The characteristic physical property of a semiconductor is that its electrical conductivity increases with increasing temperature. At room temperature, the conductivities of semiconductors are typically intermediate between those of metals and insulators (in the region of 10^3 S cm^{-1}). The dividing line between insulators and semiconductors is a matter of the size of the band gap (Table 3.10); the conductivity itself is an unreliable criterion because, as the temperature is increased, a given substance may have a low, intermediate, or high conductivity. The values of the band gap and conductivity that are taken as indicating semiconduction rather than insulation depend on the application being considered.

(a) Intrinsic semiconductors

In an **intrinsic semiconductor**, the band gap is so small that the Fermi–Dirac distribution results in some electrons populating the empty upper band (Fig. 3.57). This occupation of the conduction band introduces negative carriers into the upper level and positive holes into the lower, and as a result the solid is conducting. A semiconductor at room temperature generally has a much lower conductivity than a metallic conductor because only very few electrons and holes can act as charge carriers. The strong, increasing temperature dependence of the conductivity follows from the exponential Boltzmann-like temperature dependence of the electron population in the upper band.

It follows from the exponential form of the population of the conduction band that the conductivity of a semiconductor should show an Arrhenius-like temperature dependence of the form

$$\sigma = \sigma_0 e^{-E_a/kT} \tag{16}$$

The relation of the activation energy E_a to the band gap is established by deciding how $E - \mu$, which appears in the high-temperature form of the Fermi–Dirac distribution (eqn 15), depends on E_g.

In a simple picture of the band structure, μ (the energy at which $P = \frac{1}{2}$) is approximately half way between the upper and lower bands (Fig. 3.58), and the energy of the lowest level of the upper band E_- is related to μ and E_g by

$$E_- - \mu \approx \tfrac{1}{2}E_g$$

It follows that the conductivity should follow the expression

$$\sigma = \sigma_0 e^{-E_g/2kT} \tag{17}$$

That is, the conductivity of a semiconductor can be expected to be Arrhenius-like with an activation energy equal to half the band gap, $E_a \approx \frac{1}{2}E_g$. This is found to be the case in practice.

> The band gap in a semiconductor controls the temperature dependence of the conductivity through an Arrhenius-like expression.

Example 3.9 Determining the band gap from the temperature dependence of the conductivity

The conductance, G, of a sample of germanium varied with temperature as indicated below. Estimate the value of E_g.

T/K	312	354	420
G/S	0.0847	0.429	2.86

Answer From eqn 17 we see that the analysis is similar to that used to obtain the activation energy of a chemical reaction. Because the conductance is proportional to the conductivity, $G \propto \sigma$, we can write

$$G = G_0 e^{-E_g/2kT}$$

Taking logarithms gives

$$\ln G = \ln G_0 - \frac{E_g}{2kT}$$

Table 3.10 Some typical band gaps at 25°C

Material	E_g/eV
Carbon (diamond)	5.47
Silicon carbide	3.00
Silicon	1.12
Germanium	0.66
Gallium arsenide	1.42
Indium arsenide	0.36

Source: S.A. Schwartz, *Kirk–Othmer encyclopedia of chemical technology.* Wiley-Interscience, New York (1982). Vol. 20, p. 601.

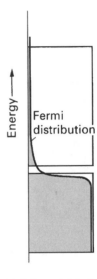

3.57 In an intrinsic semiconductor, the band gap is so small that the Fermi distribution results in some electrons populating the empty upper band.

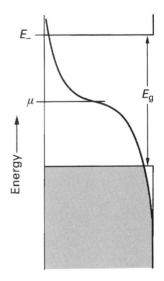

3.58 The relation between the Fermi distribution and the band gap.

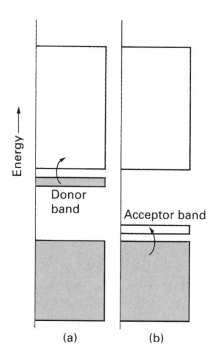

3.59 The band structure in (a) an n-type semiconductor and (b) a p-type semiconductor.

Therefore, a plot of $\ln G$ against $1/T$ should yield a straight line of slope $-E_g/2k$. The data give a slope of -4.26×10^3, and because $k = 8.614 \times 10^{-5}$ eV K^{-1}, we find $E_g = 0.73$ eV.

Self-test 3.9 What is the conductance of the sample at 370 K?

(b) Extrinsic semiconductors

The number of electron carriers can be increased if atoms with more electrons than the parent element can be introduced by the process called **doping**. Remarkably low levels of dopant concentration are needed—only about one atom per 10^9 of the host material—so it is essential to achieve very high purity of the parent element initially.

If arsenic atoms are introduced into a silicon crystal, one additional electron will be available for each dopant atom that is substituted. Note that the doping is *substitutional* in the sense that the dopant atom takes the place of an Si atom. If the donor atoms, the As atoms, are far apart from each other, their electrons will be localized and the **donor band** will be very narrow (Fig. 3.59a). Moreover, the foreign atom levels will lie at higher energy than the valence electrons of the host lattice. The filled dopant band is commonly near the empty band of the lattice. For $T > 0$, some of its electrons will be thermally promoted into the empty conduction band. In other words, thermal excitation will lead to the transfer of an electron from an As atom into the empty orbitals on a neighboring Si atom. From there it will be able to migrate through the lattice in the molecular orbitals formed by Si–Si overlap. This process gives rise to **n-type semiconductivity**, the 'n' indicating that the charge carriers are negative electrons.

An alternative substitutional procedure is to dope the silicon with atoms of an element with fewer valence electrons per atom, such as gallium. A dopant atom of this kind effectively introduces holes into the solid. More formally, the dopant atoms form a very narrow, empty **acceptor band** that lies above the full Si band (Fig. 3.59b). At $T = 0$ the acceptor band is empty, but at higher temperatures it can accept thermally excited electrons from the Si valence band. By doing so, it introduces holes into the latter and hence allows the remaining electrons in the band to be mobile. Because the charge carriers are now effectively positive holes in the lower band, this type of semiconductivity is called **p-type semiconductivity.**

Several d-metal oxides, including ZnO and Fe_2O_3, are n-type semiconductors. In their case, the property is due to nonstoichiometry and a small deficit of O atoms. The electrons that should occupy the localized O atomic orbitals (giving a very narrow oxide band, essentially localized individual O^{2-} ions) occupy a previously empty conduction band formed by the metal orbitals. The conductivity decreases when the solid is heated in oxygen because the deficit of O atoms is replaced, and as the atoms are added electrons are withdrawn from the conduction band.

p-type semiconduction is observed for some low oxidation number d-metal chalcogenides and halides, including Cu_2O, FeO, FeS, and CuI. In these nonstoichiometric compounds, the loss of electrons is equivalent to the oxidation of some of the metal atoms, with the result that holes appear in the metal-ion band. The conductivity increases when these compounds are heated in oxygen because more holes are formed in the metal-ion band as oxidation progresses.

> *p-type semiconductors are doped with atoms that remove electrons from the valence band; n-type semiconductors are doped with atoms that supply electrons to the conduction band.*

3.16 Superconduction

Until 1987, the only known superconductors (which included metals, some oxides, and some halides) needed to be cooled to below about **20 K** before they became superconducting. However, in 1987 the first **high-temperature superconductors** (HTSC) were discovered;

their superconduction is well established at **120 K** and spasmodic reports of superconduction at even higher temperatures have appeared. We will not consider these high-temperature materials at this stage (see Chapter 18), but sketch the ideas behind the mechanism of low-temperature superconduction.

The central concept of low-temperature superconduction is the existence of a **Cooper pair**, a pair of electrons that exists on account of the two electrons' indirect interaction via the nuclei of the atoms in the lattice. Thus, if one electron is in a particular region of a solid, the nuclei there move toward it to give a distorted local structure (Fig. 3.60). Because that local distortion is rich in positive charge, it is favorable for a second electron to join the first. Hence, there is a virtual attraction between the two electrons, and they move together as a pair. The local distortion can be easily disrupted by thermal motion of the ions, so the virtual attraction occurs only at very low temperatures.

A Cooper pair undergoes less scattering than an individual electron as it travels through the solid, because the distortion caused by one electron can attract back the other electron should it be scattered out of its path in a collision. This has been likened to the difference between the motion of a herd of cattle, with members of the herd that are deflected from their path by boulders in their way, and a team of cattle yoked together, which will travel forward largely regardless of obstacles. Because the Cooper pair is stable against scattering, it can carry charge freely through the solid, and hence give rise to superconduction.

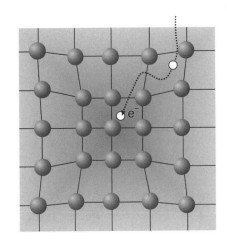

3.60 The formation of a Cooper pair. One electron distorts the crystal lattice, and the second electron has a lower energy if it goes to that region. This effectively binds the two electrons into a pair.

| *Conventional superconduction occurs as a result of the formation of Cooper pairs by interactions between the electrons and the atoms of the solid.*

Further reading

Y.-J.F. Volatron and J.K. Burdett, *An introduction to molecular orbitals.* Oxford University Press (1993). A good overview of bonding in molecules with emphasis on the fragment orbital approach.

R.L. DeKock and H.B. Gray, *Chemical structure and bonding.* Benjamin/Cummings, Menlo Park (1980).

T.A. Albright and J.K. Burdett, *Orbital interactions in chemistry.* Wiley, New York (1985). This text has a thorough discussion of isolability.

J.N. Murrell, S.F.A. Kettle, and J.M. Tedder, *The chemical bond.* Wiley, New York (1978).

T.A. Albright and J.K. Burdett, *Problems in molecular orbital theory.* Oxford University Press (1992).

More thorough discussions are given in the following:

B. Webster, *Chemical bonding theory.* Blackwell Scientific, Oxford (1990).

D.M.P. Mingos, *Essential trends in inorganic chemistry.* Oxford University Press (1998). An overview of inorganic chemistry from the perspective of structure and bonding.

Exercises

3.1 Write Lewis electron dot structures for (a) $GeCl_3^-$, (b) FCO_2^-, (c) CO_3^{2-}, (d) $AlCl_4^-$, and (e) FNO. Where more than one resonance structure is important, give examples of all major contributors.

3.2 Construct Lewis structures of typical resonance contributions of (a) ONC^- and (b) NCO^- and assign formal charges to each atom. Which resonance structure is likely to be the dominant contribution in each case?

3.3 (a) Write Lewis structures for the major resonance forms of NO_2^-. (b) Assign formal charges. (c) Assign oxidation numbers to the atoms. (d) Describe whether oxidation numbers or formal charges are appropriate for the following applications: (i) predominant resonance Lewis dot structure among several resonance forms, (ii) determining whether there

is the possibility of nitrogen being oxidized or reduced, (iii) determining the physical charge on the nitrogen atom.

3.4 Write Lewis structures for (a) XeF_4, (b) PF_5, (c) BrF_3, (d) $TeCl_4$, (e) ICl_2^-.

3.5 What shapes would you expect for the species (a) SO_3, (b) SO_3^{2-}, (c) IF_5?

3.6 Solid phosphorus pentachloride is an ionic solid composed of PCl_4^+ cations and PCl_6^- anions, but the vapor is molecular. What are the shapes of the ions in the solid?

3.7 Use the covalent radii in Table 3.4 to calculate the bond lengths in (a) CCl_4 (1.77 Å), (b) $SiCl_4$ (2.01 Å), (c) $GeCl_4$ (2.10 Å). (The values in parentheses are experimental bond lengths and are included for comparison.)

3.8 Given that $B(\text{Si}{=}\text{O})$ is 640 kJ mol^{-1}, show that bond enthalpy considerations predict that silicon–oxygen compounds are likely to contain networks of tetrahedra with Si—O single bonds and not discrete molecules with Si=O double bonds.

3.9 The common forms of nitrogen and phosphorus are $N_2(g)$ and $P_4(s)$, respectively. Account for the difference in terms of the single and multiple bond enthalpies.

3.10 Use the data in Table 3.5 to calculate the standard enthalpy of the reaction $2H_2(g) + O_2(g) \rightarrow 2H_2O(g)$. The experimental value is -484 kJ. Account for the difference between the estimated and experimental values.

3.11 Predict the standard enthalpies of the reactions

(a) $S_2^{2-}(g) + \frac{1}{4}S_8(g) \rightarrow S_4^{2-}(g)$

(b) $O_2^{2-}(g) + O_2(g) \rightarrow O_4^{2-}(g)$

using mean bond enthalpy data. Assume that the unknown species O_4^{2-} is a singly bonded chain analog of S_4^{2-}.

3.12 Use molecular orbital diagrams to determine the number of unpaired electrons in (a) O_2^-, (b) O_2^+, (c) BN, and (d) NO^-.

3.13 Use Fig. 3.17 to write the electron configurations of (a) Be_2, (b) B_2, (c) C_2^-, and (d) F_2^+ and sketch the form of the HOMO in each case.

3.14 Determine the MO bond orders of (a) S_2, (b) Cl_2, and (c) NO^- from their molecular orbital configurations and compare the values with the bond orders determined from Lewis structures. (NO has orbitals like O_2.)

3.15 What are the expected changes in bond order and bond distance that accompany the following ionization processes? (a) $O_2 \rightarrow O_2^+ + e^-$, (b) $N_2 + e^- \rightarrow N_2^-$, (c) $NO \rightarrow NO^+ + e^-$.

3.16 (a) How many independent linear combinations are possible for four $1s$ orbitals? (b) Draw pictures of the linear combinations of $H1s$ orbitals for a hypothetical linear H_4 molecule. (c) From a consideration of the number of nonbonding and antibonding interactions, arrange these molecular orbitals in order of increasing energy.

3.17 (a) Construct the form of each molecular orbital in linear $[\text{HHeH}]^{2+}$ using $1s$ basis atomic orbitals on each atom and considering successive nodal surfaces. (b) Arrange the MOs in increasing energy. (c) Indicate the electron population of the MOs. (d) Should $[\text{HHeH}]^{2+}$ be stable in isolation or in solution? Explain your reasoning.

3.18 Based on the MO discussion of NH_3 in the text, find the average NH bond order in NH_3 by calculating the net number of bonds and dividing by the number of NH groups.

3.19 From the relative atomic orbital and molecular orbital energies depicted in Fig. 3.34, describe the character as mainly F or mainly S for the frontier orbitals e (the HOMO) and $2t$ (the LUMO) in SF_6. Explain your reasoning.

3.20 Classify the hypothetical species (a) square H_4^{2+}, (b) bent O_3^{2-} as electron-precise or electron-deficient. Explain your answer and decide whether either of them is likely to exist.

3.21 Identify (a) the hydrogen–nitrogen molecule or molecular fragment that is isolobal with CH_3^-, (b) the hydrogen–boron molecule or molecular fragment that is isolobal with an O atom, (c) a nitrogen-containing species that is isolobal with $[\text{Mn(CO)}_5]^-$.

3.22 (a) Draw a simple band picture to distinguish a metallic conductor from a semiconductor. (b) Explain how the temperature dependence of the electrical conductivity can be used to distinguish a metallic conductor from a semiconductor. (c) Can the temperature dependence of the conductivity be used to distinguish an insulator from a semiconductor?

3.23 Decide whether the following systems are likely to be n-type or p-type semiconductors: (a) arsenic-doped germanium, (b) gallium-doped germanium, (c) silicon-doped germanium.

3.24 The promotion of an electron from the valence band into the conduction band in pure titanium(IV) oxide by light absorption requires a wavelength of less than 350 nm. Calculate the energy gap in electronvolts between the valence and conduction bands.

3.25 When titanium(IV) oxide is heated in hydrogen, a blue color develops, indicating light absorption in the red. Does reduction of Ti(IV) to Ti(III) correspond to n-doping or p-doping?

3.26 Gallium arsenide is a semiconductor that is widely used for the construction of the red-light-emitting displays and is under development for advanced central processor chips in supercomputers. If gallium arsenide (GaAs) is doped with selenium (on the As site), is it n-doped or p-doped?

3.27 Cadmium sulfide, CdS, is used as a photoconductor in light meters. The band gap is about 2.4 eV. What is the greatest wavelength of light that can promote an electron from the valence band to the conduction band in cadmium sulfide?

3.28 The band gap of silicon as determined from optical absorption spectra is about 1.12 eV. Calculate the ratio of the conductivities at 373 K and 273 K.

Problems

3.1 Use the concepts from Chapter 1, particularly the effects of penetration and shielding on the radial wavefunction, to account for the variation of single bond covalent radii with position in the periodic table.

3.2 Develop an argument based on bond enthalpies for the importance of SiO bonds in substances common in the Earth's crust in preference to SiSi or SiH bonds. How and why does the behavior of silicon differ from that of carbon?

3.3 When an He atom absorbs a photon to form the excited configuration $1s^1 2s^1$ (here called He*) a weak bond forms with another He atom to give the diatomic molecule HeHe*. Construct a molecular orbital description of the bonding in this species.

3.4 Construct an approximate molecular orbital energy diagram for a hypothetical planar form of NH_3. You may refer to Appendix 4 to determine the form of the appropriate orbitals on the central N atom and on the triangle of H_3 atoms. From a consideration of the atomic energy levels, place the N and H_3 orbitals on either side of a molecular orbital energy-level diagram. Then use your judgement about the effect of bonding and antibonding interactions and energies of the parent orbitals to construct the molecular orbital energy levels in the center of your diagram and draw lines indicating the contributions of the atomic orbitals to each molecular orbital. Atomic orbital energy level data are $H1s = -13.6$ eV, $N2s = -26.0$ eV, $N2p = -13.4$ eV.

3.5 (a) Use an extended Hückel molecular orbital program[18] or input and output from such a program supplied by your instructor, to construct a molecular orbital energy level diagram to correlate the MO (from the output) and AO (from the input) energies and indicate the occupancy of the MOs (in the manner of Fig. 3.17) for one of the following molecules: HF (bond length 0.92 Å), HCl (1.27 Å), or CS (1.53 Å); (b) Use the output to sketch the form of the occupied orbitals, showing signs of the AO lobes by shading and their amplitudes by means of size of the orbital.

3.6 Perform an extended Hückel MO calculation of H_3 by using the H energy given in Problem 3.4 and H–H distances from NH_3 (N–H length 1.02 Å, HNH bond angle 107°) and then carry out the same type of calculation for NH_3. Use energy data for N2s and N2p orbitals from Problem 3.4. From the output, plot the molecular orbital energy levels with proper symmetry labels and correlate them with the N orbitals and H_3 orbitals of the appropriate symmetries. Compare the results of this calculation with the qualitative description in Problem 3.4.

3.7 Assign the lines in the UV photoelectron spectrum of CO shown in Fig. 3.61 and predict the appearance of the UV photoelectron spectrum of the SO molecule.

---
[18] Suitable programs are QCMP001, from QCPE, Chemistry Department, Indiana University, Bloomington, IN; CACAO, by C. Maelli and D.M. Proserpio, *J. Chem. Educ.* **67**, 399 (1990); PLOT3D, by J.A. Bertrand and M.R. Johnson, School of Chemistry, Georgia Institute of Technology, Atlanta, GA.

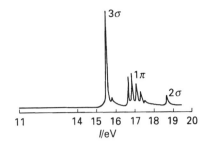

Fig. 3.61 The ultraviolet photoelectron spectrum of CO using 21 eV radiation.

3.8 The K-shell X-ray emission spectrum of aluminum oxide is illustrated in Fig. 3.53. It is so called because emission arises from valence-band electrons falling into a vacancy in the K-shell (an alternative name for the $n = 1$ shell) that has been produced by bombardment. The corresponding absorption arises from promotion of an electron from the K-shell into the conduction band. What is the band gap energy in Al_2O_3? Is alumina an insulator or a semiconductor? Are energy levels dense near the band edges or near the band centers? Which peak gives an indication of the distribution of levels derived mainly from O orbitals?

3.9 The electrical conductivity of bismuth is 9.1×10^5 S m^{-1} at 273 K, 6.4×10^5 S m^{-1} at 373 K, and 7.8×10^5 S m^{-1} at 573 K. What type of material is bismuth? Bismuth melts at 271 °C.

3.10 The conductivity of VO increases sharply with increasing temperature up to 125 K, and reaches 1×10^{-4} S m^{-1}. At about 125 K the conductivity rises abruptly to 1×10^2 S m^{-1} and then declines slowly to about 5×10^1 S m^{-1} near 400 K. How would you classify (a) the low-temperature and (b) the high-temperature forms of VO?

3.11 The neutral NH_2 molecular fragment is bent. Should the first excited state of NH_2 be more bent or more linear than the ground state? Explain. (*Hint*. You should consult Fig. 3.38.)

3.12 The band gap of pure Si as determined from optical absorption spectra is 1.14 eV. Calculate the ratio of the electrical conductivities of Si at 100 °C and 0 °C.

3.13 Predict the effect on the HOMO–LUMO gap in SF_6 if all six S—F bonds are stretched by the same amount. (*Hint*. You should consult Fig. 3.34.)

3.14 Refer to Fig. 3.27. One might incorrectly conclude from this figure that the P—P single bond is weaker than the shorter N—N single bond. What is wrong with this conclusion?

3.15 When NiO, a poor conductor at 25 °C, is heated in air to 800 °C, its conductivity increases dramatically. Is the behavior of NiO more like that of FeO or ZnO? Why? (*Hint*. Consult Appendix 1.)

4

Molecular symmetry

The concept of symmetry is of the greatest importance in inorganic chemistry. Symmetry helps to determine the physical properties of a molecule and provides hints about how reactions might occur. In this chapter we explore some of the consequences of molecular symmetry, and refine that concept into the powerful concepts of group theory. We shall see that symmetry considerations can be used to construct molecular orbitals, to discuss electronic structure, and to simplify the discussion of molecular vibrations.

The systematic treatment of symmetry is called **group theory**. Group theory is a rich and powerful subject, but we shall confine our use of it at this stage to the classification of molecules, constructing molecular orbitals, and analyzing molecular vibrations and their selection rules. We shall also see that it is possible to draw some general conclusions about the properties of molecules without doing any calculations at all. A lot of group theory is straightforward common sense but, because it makes the analysis of symmetry systematic, it can also be used to draw conclusions when the consequences of symmetry are not immediately obvious.

An introduction to symmetry analysis

Our initial aim is to define the symmetries of molecules precisely, not just intuitively, and to provide a scheme for specifying and reporting these symmetries. It will become clear in later chapters that symmetry analysis is one of the most pervasive techniques in inorganic chemistry.

4.1 Symmetry operations and symmetry elements

A fundamental concept of group theory is the **symmetry operation**, which is an action, such as a rotation through a certain angle, that leaves the molecule apparently unchanged. An example is the rotation of an H_2O molecule by $180°$ (but not any smaller angle) around the bisector of the HOH angle. Associated with each symmetry operation there is a **symmetry element**, a point, a line, or a plane with respect to which the symmetry operation is performed. The most important symmetry operations and their corresponding elements are listed in Table 4.1. All these operations leave at least one point of the molecule unmoved, just as a rotation of a sphere leaves its center unmoved, and hence they are the operations of **point-group symmetry**.

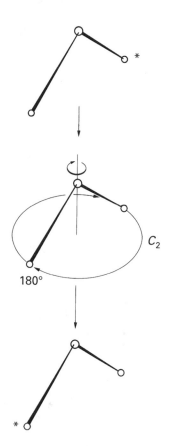

4.1 An H_2O molecule may be rotated through any angle about the bisector of the HOH bond angle, but only a rotation of 180°, C_2, leaves it apparently unchanged.

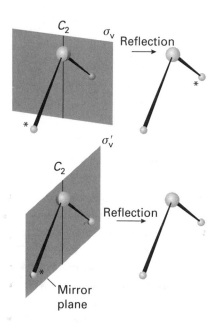

4.3 The two vertical mirror planes σ_v and σ'_v in H_2O and the corresponding operations. Both planes cut through the C_2 axis.

Table 4.1 Important symmetry operations and symmetry elements

Symmetry element	Symmetry operation	Symbol
	Identity*	E
n-Fold symmetry axis	Rotation by $2\pi/n$	C_n
Mirror plane	Reflection	σ
Center of inversion	Inversion	i
n-Fold axis of improper rotation†	Rotation by $2\pi/n$ followed by reflection perpendicular to rotation axis	S_n

*The symmetry element can be thought of as the whole of space.
†Note the equivalences $S_1 = \sigma$ and $S_2 = i$.

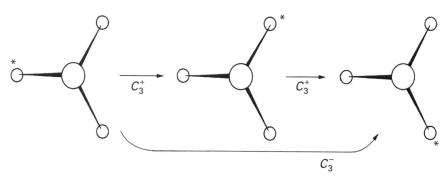

4.2 A threefold rotation and the corresponding C_3 axis in NH_3. There are two rotations associated with this axis, one through 120° (C_3^+) and the other through −120° (C_3^-).

The **identity operation**, E, leaves the whole molecule unchanged. Every molecule has at least this operation, and some have only this operation, so we need it if we are to classify all molecules according to their symmetry. The rotation of an H_2O molecule by 180° around a line bisecting the HOH angle is a symmetry operation, so the H_2O molecule possesses a 'twofold' rotation axis, C_2 (Fig. 4.1). In general, an **n-fold rotation** is a symmetry operation if the molecule appears unchanged after rotation by 360°/n. The corresponding symmetry element is a line, the **n-fold rotation axis**, C_n, about which the rotation is performed. The trigonal-pyramidal NH_3 molecule has a threefold rotation axis, denoted C_3, but there are two operations associated with this axis, one a clockwise rotation by 120° and the other a counterclockwise rotation through the same angle (Fig. 4.2). The two operations are denoted C_3^+ and C_3^-, respectively. Notice that C_3^+ and C_3^- are identical, so there is only one rotation associated with a C_2 axis (as in H_2O).

The **reflection** of an H_2O molecule in either of the two planes shown in Fig. 4.3 is a symmetry operation; the corresponding symmetry element—the plane of the mirror—is a **mirror plane**, σ. The H_2O molecule has two mirror planes that intersect at the bisector of the HOH angle. Because the planes are 'vertical', in the sense of being parallel to the rotational axis of the molecule, they are labeled with a subscript v, as in σ_v and σ'_v. The C_6H_6 molecule also has a mirror plane σ_h in the plane of the molecule. The subscript h signifies that the plane is 'horizontal' in the sense that the principal rotational axis of the molecule is perpendicular to it. This molecule also has two more sets of three mirror planes that intersect the sixfold axis (Fig. 4.4). The symmetry elements (and the associated operations) are denoted σ_v for the planes that pass through the C atoms of the ring and σ_d for the planes that bisect the angle between the C atoms. The d denotes 'dihedral' and signifies that the plane bisects the angle between two C_2 axes (the C—H axes).

To understand the **inversion** operation, we need to imagine that each atom is projected in a straight line through a single point and then out to an equal distance on the other side

(Fig. 4.5). If we consider CO_2, for instance, with the point at the center of the molecule (at the C nucleus), then the operation exchanges the two O atoms. In an octahedral molecule such as SF_6, with the point at the center of the molecule, diametrically opposite pairs of atoms at the apices of the octahedron are interchanged. The symmetry element, the point through which the projections are made, is called the **center of inversion, i.** The center of inversion of CO_2 lies at the C nucleus; that of the SF_6 molecule lies at the nucleus of the S atom. There need not be an atom at the center of inversion: an N_2 molecule has a center of inversion midway between the two nitrogen nuclei. An H_2O molecule does not possess a center of inversion. No tetrahedral AB_4 molecule has a center of inversion; for example, $Ni(CO)_4$ has no center of inversion. Although an inversion and a twofold rotation may sometimes achieve the same effect (Fig. 4.6), that is not the case in general and the two operations should be clearly distinguished.

An **improper rotation** is a composite operation (and one of the most difficult to identify in a molecule). It consists of a rotation of the molecule through a certain angle around an axis

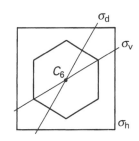

4.4 Some of the symmetry elements of the benzene ring. There is one horizontal reflection plane (σ_h) and two sets of vertical reflection plane (σ_v and σ_d); one example of each is shown.

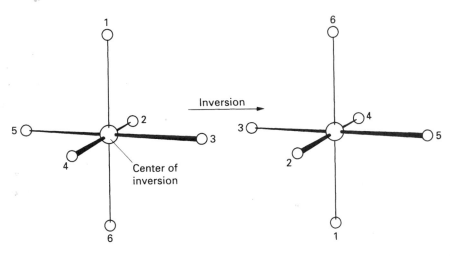

4.5 The inversion operation and the center of inversion i in SF_6.

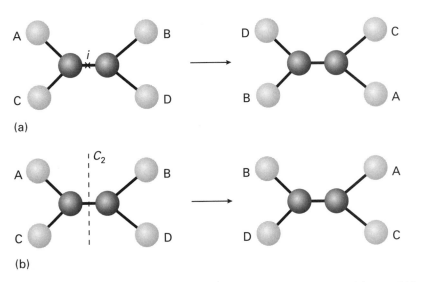

4.6 Care must be taken not to confuse (a) an inversion operation with (b) a twofold rotation. Although the two operations may sometimes appear to have the same effect, that is not the case in general.

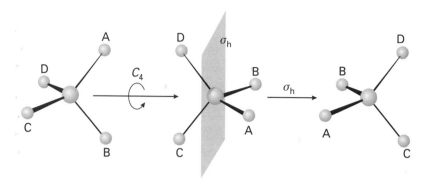

4.7 A fourfold axis of improper rotation S_4 in the CH_4 molecule.

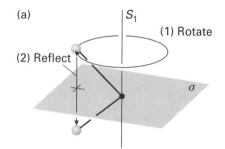

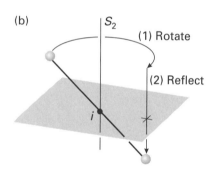

4.8 (a) An S_1 axis is equivalent to a mirror plane and (b) an S_2 axis is equivalent to a center of inversion.

followed by a reflection in the plane perpendicular to that axis (Fig. 4.7). A rotation can actually be carried out; an improper rotation cannot actually be carried out because it turns a left-handed object (a left hand, for instance) into a right-handed object (a right hand). The illustration shows a fourfold improper rotation of a tetrahedral CH_4 molecule. In this case, the operation consists of a 90° rotation about an axis bisecting two HCH bond angles, followed by a reflection through a plane perpendicular to the rotation axis. Neither the 90° operation nor the reflection alone is a symmetry operation for CH_4. In each case we can see that the molecule has been moved after either one of the operations has been applied, but their *overall* effect is a symmetry operation because we cannot tell that the molecule has been moved if we close our eyes until after both operations have been applied. This fourfold improper rotation is denoted S_4. The symmetry element, the **improper-rotation axis**, S_n (S_4 in the example), is the corresponding combination of an n-fold rotational axis and a perpendicular mirror plane.

An S_1 axis, a rotation through a full 360° followed by a reflection in the horizontal plane, is equivalent to a horizontal reflection alone, so S_1 and σ_h are the same; the symbol σ_h is generally used rather than S_1. Similarly, an S_2 axis, a rotation through 180° followed by a reflection in the horizontal plane, is equivalent to an inversion, i (Fig. 4.8), and the symbol i is employed rather than S_2.

> With each symmetry operation there is associated a symmetry element; symmetry elements include a symmetry axis, a mirror plane, a center of inversion, and an improper-rotation axis.

Example 4.1 Identifying a symmetry element

Which conformation of a CH_3CH_3 molecule has an S_6 axis?

Answer We need to find a conformation that leaves the molecule looking the same after a $360°/6 = 60°$ rotation followed by a reflection in a plane perpendicular to that axis. The conformation and axis are shown in (**1**); this 'staggered' conformation of the molecule also happens to be the conformation of lowest energy, but the energy cannot be predicted by symmetry arguments alone.

Self-test 4.1 Identify a C_3 axis of an NH_4^+ ion. How many of these axes are there in the ion?

4.2 The point groups of molecules

To assign a molecule to a particular point group, we make a list of symmetry elements it possesses and compare that list with the list that defines each point group. The symmetry elements that define each group are summarized in Table 4.2. For example, if a molecule has only the identity element (CHBrClF is an example; (**2**)), we list its elements as E alone and look for the group that has only this element. Because the group labeled C_1 has only the element

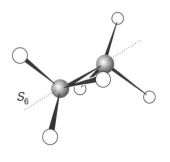

1 S_6 axis

Table 4.2 The composition of some common groups*

Point group	Symmetry elements	Shape	Examples
C_1	E		SiBrClFI
C_2	E, C_2		H_2O_2
C_s	E, σ		NHF_2
C_{2v}	$E, C_2, \sigma_v, \sigma_v$		H_2O, SO_2Cl_2
C_{3v}	$E, 2C_3, 3\sigma_v$		NH_3, PCl_3, $POCl_3$
$C_{\infty v}$	$E, C_2, 2C_\phi, \ldots \infty \sigma_v$		CO, HCl, OCS
D_{2h}	$E, C_2(x,y,z), \sigma(xy, yz, zx), i$		N_2O_4, B_2H_6
D_{3h}	$E, C_3, 3C_2, 3\sigma_v, \sigma_h, S_3$		BF_3, PCl_5
D_{4h}	$E, C_4, C_2, 2C_2', 2C_2'', i, S_4, \sigma_h, 2\sigma_v, 2\sigma_d$		XeF_4, *trans*-MA_4B_2
$D_{\infty h}$	$E, C_\infty, \ldots, \infty \sigma_v, i, S_\infty, \ldots, \infty C_2$		H_2, CO_2, C_2H_2
T_d	$E, 3C_2, 4C_3, 6\sigma_d, 4S_4$		CH_4, $SiCl_4$
O_h	$E, 6C_2, 4C_3, 3C_4, 4S_6, 3S_4, i, 3\sigma_h, 6\sigma_d$		SF_6

*Not all the elements of each group are listed, but enough are listed for unambiguous assignments to be made.

E, the CHBrClF molecule belongs to that group. The molecule CH_2BrCl belongs to a slightly richer group: it has the elements E (all groups have that element) and a mirror plane. The group of elements (E, σ) is called C_s, so the CH_2BrCl molecule belongs to that group. This procedure can be continued, and molecules assigned to the group that matches the symmetry

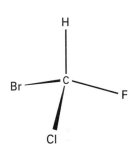

2 CHBrClF

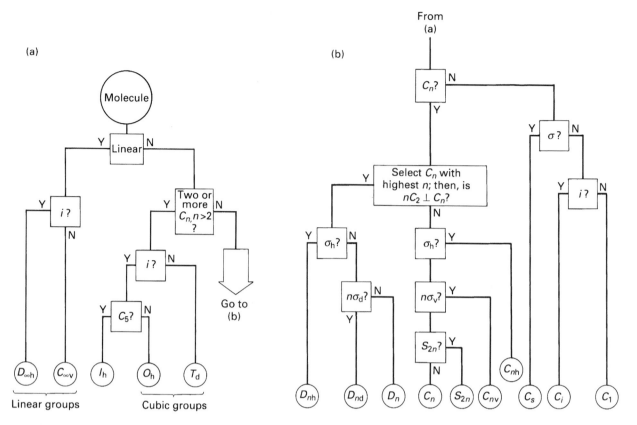

4.9 The decision tree for identifying a molecular point group. After passing though part (a), go to part (b) if necessary. The symbols at each decision point refer to the symmetry elements (not the corresponding operations).

elements they possess. Some of the more common groups and their names are listed in Table 4.2. The assignment of a molecule to its group depends on identifying the symmetry elements it possesses and then referring to the table. In practice, the shapes in the table give a very good clue to the identity of the group to which the molecule belongs, at least in simple cases. The flowchart in Fig. 4.9 can be used to assign most common point groups systematically by answering the questions at each decision point of the tree.

Example 4.2 Identifying the point group of a molecule

To what point groups do H_2O and NH_3 belong?

Answer Work through Fig. 4.9. The symmetry elements are shown in Fig. 4.10. (a) H_2O possesses the identity (E), a twofold rotation axis (C_2), and two vertical mirror planes (σ_v and σ'_v). The set of elements ($E, C_2, \sigma_v, \sigma'_v$) corresponds to the group C_{2v}. (b) NH_3 possesses the identity (E), a threefold axis (C_3), and three vertical mirror planes ($3\sigma_v$). The set of elements ($E, C_3, 3\sigma_v$) identifies the group as C_{3v}.

Self-test 4.2 Identify the point groups of (a) BF_3, a trigonal-planar molecule, and (b) the tetrahedral SO_4^{2-} ion.

It is very useful to be able to recognize immediately, simply at a glance, the point groups of some common molecules. Linear molecules with a center of symmetry (H_2, CO_2, HC≡CH; (**3**)) belong to the point group $D_{\infty h}$. A molecule that is linear but has no center of symmetry (HCl, OCS, NNO; (**4**)) belongs to the point group $C_{\infty v}$. Tetrahedral (T_d) and octahedral (O_h) molecules (Fig. 4.11) have more than one principal axis of symmetry: a tetrahedral CH_4 molecule, for instance, has four C_3 axes, one along each C—H bond. A closely related group,

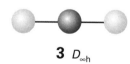

3 $D_{\infty h}$

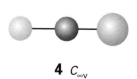

4 $C_{\infty v}$

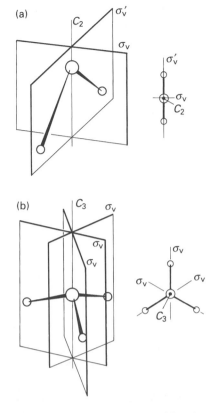

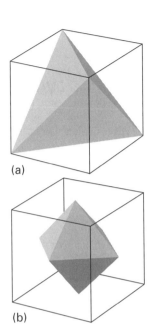

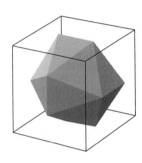

4.10 The symmetry elements of (a) H_2O and (b) NH_3. The diagrams on the right are views from above and summarize the diagrams to their left.

4.11 Shapes with the symmetries of the groups (a) T_d, the tetrahedron, (b) O_h, the octahedron. Both are closely related to the symmetries of a cube.

4.12 The regular icosahedron of point group I_h and its relation to a cube.

the icosahedral group I_h characteristic of the icosahedron, has 12 fivefold axes (Fig. 4.12). The icosahedral group is important for boron compounds and for the C_{60} fullerene molecule.

The distribution of molecules over the various point groups is very uneven. Some of the most common groups for molecules are the low-symmetry groups C_1 and C_s, the groups for a number of polar molecules C_{2v} (as in SO_2) and C_{3v} (as in NH_3), and the highly symmetrical tetrahedral and octahedral groups. There are many linear molecules, which belong to the groups $C_{\infty v}$ (HCl, OCS) and $D_{\infty h}$ (Cl_2 and CO_2), and a number of planar-trigonal molecules, D_{3h} (such as BF_3 (**5**)), trigonal-bipyramidal molecules (such as PCl_5 (**6**)), which are also D_{3h}, and square-planar molecules, D_{4h} (**7**). So-called 'octahedral' molecules with two substituents opposite each other, as in (**8**), are also D_{4h}. The last example shows that the point-group

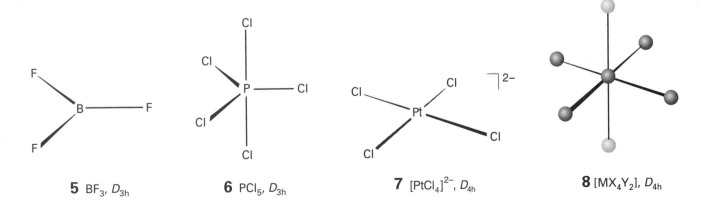

5 BF_3, D_{3h}

6 PCl_5, D_{3h}

7 $[PtCl_4]^{2-}$, D_{4h}

8 $[MX_4Y_2]$, D_{4h}

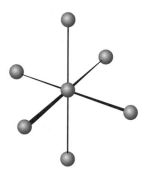

9 [MX$_6$], O_h

classification of a molecule is more precise than the casual use of the term 'octahedral' or 'tetrahedral'. For instance, a molecule may loosely be called octahedral even if it has six different groups attached to the central atom. However, it belongs to the octahedral point group O_h only if all six groups are identical (**9**).

> The point groups of molecules are identified by noting the symmetry elements of the molecule and comparing the list of elements with the elements that define each group.

Applications of symmetry

An important application of symmetry in inorganic chemistry is to the construction and labeling of molecular orbitals (see Sections 4.5–4.7). However, there are several other applications of the group classification of molecules. One application is to use the group to decide whether a molecule is polar or chiral. In most cases we do not need to use the sledgehammer of group theory to crack the nut of deciding whether a molecule has these characteristics, but these examples give an impression of the approach that can be adopted when the result is not obvious.

4.3 Polar molecules

A **polar molecule** is a molecule with a permanent electric dipole moment. There are certain symmetry elements that rule out a permanent electric dipole in molecules, or forbid it to lie in certain orientations. First, *a molecule cannot be polar if it has a center of inversion.* Inversion implies that a molecule has matching charge distributions at all diametrically opposite points about a center. This symmetry rules out a dipole moment. Second, and for the same reason, a dipole moment cannot lie perpendicular to any mirror plane or axis of rotation that the molecule may possess. For example, a mirror plane demands identical atoms on either side of the plane, so there can be no dipole moment across the plane. Similarly, a symmetry axis implies the presence of identical atoms at points related by the corresponding rotation, which rules out a dipole moment perpendicular to the axis. In summary:

1 A molecule cannot be polar if it has a center of inversion.
2 A molecule cannot have an electric dipole moment perpendicular to any mirror plane.
3 A molecule cannot have an electric dipole moment perpendicular to any axis of rotation.

Some molecules have a symmetry axis that rules out a dipole moment in one plane and another symmetry axis or mirror plane that rules it out in another direction. The two or more symmetry elements jointly forbid the presence of a dipole moment in *any* direction. For example, any molecule that has a C_n axis and either a C_2 axis perpendicular to that C_n axis, or a σ_h plane perpendicular to the axis, cannot have a dipole moment in any direction. Any molecule belonging to a D point group is of this kind, so any such molecule must be nonpolar; a BF$_3$ molecule (D_{3h}) is therefore nonpolar. Likewise, molecules belonging to the tetrahedral, octahedral, and icosahedral groups have several perpendicular rotation axes that rule out dipoles in all three directions, so such molecules must be nonpolar; hence, SF$_6$ (O_h) and CCl$_4$ (T_d) are nonpolar.

> A molecule cannot be polar if it belongs to any of the following point groups: (1) any group that includes a center of inversion; (2) any of the groups D and their derivatives; (3) the cubic groups (T, O), the icosahedral group (I), and their modifications.

Example 4.3 Judging whether a molecule can be polar

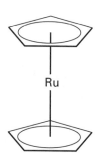

10 Ruthenocene, Ru(Cp)$_2$

The ruthenocene molecule (**10**), is a pentagonal prism with the Ru atom sandwiched between two C$_5$H$_5$ rings. Is it polar?

Answer We should decide whether the point group is D or cubic, because in neither case can it have a permanent electric dipole. Reference to Fig. 4.9 shows that a pentagonal prism belongs to the point group D_{5h}. Therefore, the molecule must be nonpolar.

Self-test 4.3 A conformation of the ferrocene molecule that lies 4 kJ mol^{-1} (0.04 eV) above the lowest energy configuration is a pentagonal antiprism (**11**). Is it polar?

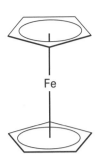

11 Ferrocene (higher energy conformation), $Fe(Cp)_2$

4.4 Chiral molecules

A **chiral molecule** (from the Greek word for 'hand') is a molecule that cannot be superimposed on its own mirror image. A hand is chiral in the sense that the mirror image of a left hand is a right hand, and the two hands cannot be superimposed. Provided they are sufficiently long-lived to be observed, chiral molecules are **optically active**, which means that they can rotate the plane of polarized light. A chiral molecule and its mirror image partner are called **enantiomers** (from the Greek word for 'both'). Enantiomeric pairs of molecules rotate the plane of polarization of light by equal amounts in opposite directions.

The group-theoretical criterion of chirality is that a molecule should not have an improper rotation axis, S_n. A molecule with such an axis cannot be chiral. Groups in which S_n is present include D_{nh} (which include S_n), D_{nd}, and some of the cubic groups (specifically, T_d and O_h). Therefore, a molecule such as CH_4 or $Ni(CO)_4$ belonging to the group T_d is not chiral. That a 'tetrahedral' carbon atom leads to optical activity (as in CHClFBr) should serve as another reminder that group theory is stricter in its terminology than casual conversation: a molecule such as CHBrClF belongs to the group C_1, not to the group T_d. A CHBrClF molecule is tetrahedral in a colloquial sense but it is not group-theoretically tetrahedral.

When judging chirality, it is important to be alert for improper-rotation axes that are present in disguise. Thus, we have seen that a mirror plane alone is an S_1 axis and a center of inversion is equivalent to an S_2 axis. Therefore, molecules with either a mirror plane or a center of inversion have improper-rotation axes and cannot be chiral. Molecules with neither a center of inversion nor a mirror plane (and hence with neither S_1 nor S_2 axes) are usually chiral, but it is important to verify that a higher-order improper-rotation axis is not also present. For instance, the quaternary ammonium ion (**12**) has neither a mirror plane (S_1) nor an inversion center (S_2), but it does have an S_4 axis (it belongs to the group S_4, in fact) and so it is not chiral.

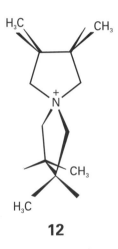

12

> A molecule is not chiral if (1) it possesses an improper rotation axis, (2) it belongs to the groups D_{nh} or D_{nd} (but it may be chiral if it belongs to the groups D_n), (3) it belongs to T_d or O_h.

Example 4.4 Judging whether a molecule is chiral

The ion $[Cr(ox)_3]^{3-}$, where ox denotes the oxalate ion $O_2CCO_2^{2-}$, has the structure shown as (**13**). Is it chiral?

Answer We begin by identifying the point group. Working through the chart in Fig. 4.9 shows that the ion belongs to the point group D_3. This group consists of the elements $(E, C_3, 3C_2)$ and hence does not contain an improper-rotation axis (either explicitly or in a disguised form). The complex ion is chiral and hence, as it is long-lived, optically active.

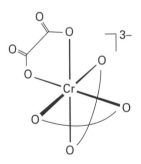

13 $[Cr(ox)_3]^{3-}$, D_3

Self-test 4.4 Is the skew form of H_2O_2 (**14**) chiral?

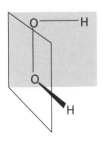

14

The symmetries of orbitals

We shall now see in more detail the significance of the labels used for molecular orbitals introduced in Sections 3.11 and 3.12, and gain more insight into their construction. The discussion will continue to be informal and pictorial, its aim being to give an elementary introduction to group theory, but not the details of the specific calculations involved. The objective here is to show how to identify the symmetry label of an orbital from a drawing like those in Appendix 4 and, conversely, to appreciate the significance of a symmetry label. The arguments later in the book are all based on simply 'reading' molecular orbital diagrams qualitatively.[1]

4.5 Character tables and symmetry labels

Molecular orbitals of diatomic molecules (and linear polyatomic molecules) are labeled σ, π, and so on. These labels refer to the symmetries of the orbitals with respect to rotations around the principal symmetry axis of the molecule. The same designations can be used for specifying orbitals by noting the *local* symmetry with respect to the axis of a given bond, so we can speak, for instance, of σ and π orbitals in benzene. A σ orbital does not change sign under a rotation through any angle about the internuclear axis, a π orbital changes sign when rotated by 180°, and so on (Fig. 4.13). This labeling of orbitals according to their behavior under rotations can be generalized and extended to the global symmetries of nonlinear polyatomic molecules, where there may be reflections and inversions to take into account as well as rotations.

The labels σ and π can be assigned to individual *atomic* orbitals in a linear molecule (and, locally, in nonlinear molecules too). For example, we often speak of an individual p_z orbital as having σ symmetry with respect to an internuclear axis. This ability to classify individual atomic orbitals is important, because, as we remarked in Chapter 3, only atomic orbitals that have the same symmetry type contribute to a given molecular orbital. For example, an s orbital of σ symmetry on one atom and a p_x orbital of π symmetry on its neighbor cannot contribute to the same molecular orbital in a linear molecule.

Symmetry labels may also be ascribed to linear combinations of atomic orbitals on symmetry-related atoms. Thus, a symmetry label can be ascribed to the combination $\phi_{A1s} + \phi_{B1s} + \phi_{C1s}$ of the three H1s orbitals in NH_3, and only the orbitals of the same symmetry type on the N atom will have a nonzero net overlap with this combination. The combinations of atomic orbitals of a specified symmetry from which molecular orbitals are built are called **symmetry-adapted linear combinations** (SALCs) of atomic orbitals.

The labels σ and π are based on the rotational symmetry of orbitals with respect to an internuclear axis. The more elaborate labels a, a_1, e, e_g, and so on, which are ascribed to molecular orbitals in nonlinear molecules, are based on the behavior of the orbitals under all the symmetry operations of the relevant molecular point group. The label is assigned by referring to the **character table** of the group, a table that characterizes the different symmetry types possible in a point group. Thus, when we assign the labels σ and π, we use

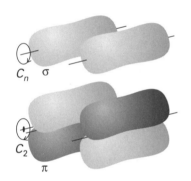

4.13 The σ and π classification of orbitals is based on their symmetries with respect to rotations about an axis. A σ orbital is unchanged by any rotation; a π orbital changes sign when rotated by 180° about the internuclear axis.

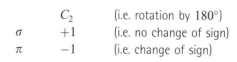

	C_2	
σ	+1	(i.e. rotation by 180°)
π	−1	(i.e. no change of sign)
		(i.e. change of sign)

---

[1] For details of the calculations on which the illustrations are based, and for the construction of combinations of specific symmetry, refer to *Further information 3*.

Table 4.3 The C_{3v} character table*

	E	$2C_3$	$3\sigma_v$			
A$_1$	1	1	1	z	x^2+y^2, z^2	
A$_2$	1	1	-1			R_z
E	2	-1	0	(x,y)	$(x^2-y^2,xy),\ (zx,yz)$	(R_x, R_y)

*(a,b) denotes a degenerate pair of orbitals, the characters in the table refer to the symmetry of the pair jointly. The symbol R_q denotes a rotation around the axis q.

This table is a fragment of the character table for a linear molecule. The entry $+1$ indicates that an orbital remains the same and the entry -1 indicates that it changes sign under the operation C_2.

The entries in a complete character table are derived by using the formal techniques of group theory and are called **characters**, χ. These numbers characterize the essential features of each symmetry type in a way that will be illustrated as we go on.[2] We shall illustrate these essential features by considering the C_{3v} character table (Table 4.3). The character tables for a selection of other groups are given in Appendix 3.

The columns in a character table are labeled with the symmetry operations of the group. If there are several operations of the same kind (technically, of the same *class*), then they are collected together into a single column. In the C_{3v} character table, for instance, there are two threefold rotations (C_3^+ and C_3^-) about a single axis, as indicated by the heading $2C_3$.

The rows, which summarize the characteristic symmetries of orbitals (and, as we shall see, other entities too), are labeled with the **symmetry type** (the generalizations of σ and π). The symmetry types label the 'irreducible representations' of a group, which (in a technical sense that we shall not pursue) are the primitive types of symmetry that may occur in any molecule that belongs to the specific group. For instance, in linear molecules the primitive symmetry types are Σ, Π, etc. By convention, in general (and with the exception of the groups of linear molecules) the symmetry types are given upper case roman letters (such as A$_1$ and E) but the orbitals to which they apply are labeled with the lower case italic equivalents (so an orbital of symmetry type A$_1$ is called an a_1 orbital). Note that care must be taken to distinguish the identity element E (italic, a column heading) from the symmetry label E (roman, a row label).

The symmetries of various functions (including individual atomic orbitals) centered on the midpoint of a molecule (the 'point' of the point group) to which a row of characters refers are indicated by the letters (such as xy) on the right of the table. When we are interested in orbitals, these letters indicate their angular variation. Thus, xy denotes a d_{xy} orbital centered on the midpoint of the molecule. An s orbital appears in a slightly disguised form as $x^2+y^2+z^2$, or some other similar totally symmetrical expression. The parentheses around pairs of functions indicate that they must be treated jointly as a pair. For example, we treat (x,y), and by extension, the orbitals (p_x, p_y) centered on the midpoint, as an indivisible pair in molecules of symmetry C_{3v}, such as NH$_3$.

The entry in the column headed by the identity operation E gives the degeneracy of the orbitals (the number of orbitals of the same energy). Thus, in a C_{3v} molecule, any orbital with a symmetry label a_1 or a_2 (Fig. 4.14) must be nondegenerate and have a character of 1 in the column headed E. Conversely, if we know that we are dealing with a nondegenerate orbital in a C_{3v} molecule, then we also know that its symmetry type must be either A$_1$ or A$_2$ and therefore that the orbital will be labeled either a_1 or a_2. Similarly any doubly degenerate pair of orbitals in C_{3v} must be labeled e and have a character 2 in the column labeled E (Fig. 4.15).

2   For an account of the generation and use of character tables without too much mathematical background, see P.W. Atkins, *Physical chemistry*. Oxford University Press and W.H. Freeman & Co, New York (1998). For a more rigorous introduction, see P.W. Atkins and R.S. Friedman, *Molecular quantum mechanics*. Oxford University Press (1997).

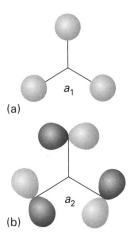

4.14 The combinations of H1s orbitals that are used to build molecular orbitals in NH$_3$. (a) An a_1 orbital is formed by the overlap of the combination shown with N2s and N2p_z orbitals. (b) None of the valence orbitals of nitrogen has the appropriate symmetry to form an a_2 orbital.

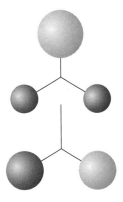

4.15 The combination of H1s orbitals that are used to form e orbitals in NH$_3$. They overlap the p_x and p_y orbitals on the N atom.

Because there are no characters with the value 3 in the column headed E, we know at a glance that there can be no triply degenerate orbitals in a C_{3v} molecule.

> Only atomic orbitals, or combinations of atomic orbitals, of the same symmetry type can contribute to a molecular orbital of a given symmetry type. The character of the identity operation is the degeneracy of a set of orbitals.

Example 4.5 Using a character table to judge degeneracy

Can there be triply degenerate orbitals in BF_3?

Answer First, identify the point group; which in this case is D_{3h}. Reference to the character table for this group (Appendix 3) shows that, because no character exceeds 2 in the column headed E, the maximum degeneracy is 2. Therefore, none of its orbitals can be triply degenerate.

Self-test 4.5 The SF_6 molecule is octahedral. What is the maximum possible degree of degeneracy of its orbitals?

4.6 The interpretation of character tables

We have seen that the entries under the identity operation, E, indicate the degeneracy of the orbitals. The entries in the other columns indicate the behavior of an orbital or a set of orbitals under the corresponding operations. Broadly speaking, the interpretation of the characters is as follows:[3]

Character	Significance
$+1$	the orbital is unchanged
-1	the orbital changes sign
0	the orbital undergoes a more complicated change

It follows that, in simple cases at least, we can identify the symmetry label of the orbital by comparing the changes that occur to an orbital under each operation and then comparing the resulting $+1$ or -1 with the entries in a row of the character table for the relevant point group.

The principal complication in this interpretation is that, for the rows labeled E or T (which refer to sets of doubly and triply degenerate orbitals, respectively), the characters are the *sums* of the characters summarizing the behavior of the individual orbitals of the degenerate set. Thus, if one member of a doubly degenerate pair remains unchanged under a symmetry operation but the other changes sign, then the entry is reported as $\chi = 1 - 1 = 0$. We now see why the first column tells us the degeneracy: it is the sum of 1 for each orbital in the degenerate set (because each orbital remains unchanged under the identity operation), so a nondegenerate orbital gives $\chi = 1$, a doubly degenerate orbital gives $\chi = 1 + 1 = 2$, and so on.

As a specific example, consider a $2p_x$ orbital on the O atom of H_2O. Because H_2O belongs to the point group C_{2v}, we know by referring to the C_{2v} character table (see margin) that the labels available for the orbitals are A_1, A_2, B_1, and B_2. We can decide on the appropriate label for $2p_x$ by noting that under a 180° rotation (C_2) the orbital changes sign (Fig. 4.16), so it must be either B_1 or B_2 as only these two symmetry types have character -1 under C_2. The $2p_x$ orbital also changes sign under the reflection σ'_v, which identifies it as B_1. As we shall see, any molecular orbital built from this atomic orbital will also be a b_1 orbital. Similarly, $2p_y$ changes sign under C_2 but not under σ'_v and so contributes to b_2 orbitals.

C_{2v} (2mm)	E	C_2	$\sigma_v(xz)$	$\sigma'_v(yz)$		$h = 4$
A_1	1	1	1	1	z	x^2, y^2, z^2
A_2	1	1	-1	-1	R_z	xy
B_1	1	-1	1	-1	x, R_y	xz
B_2	1	-1	-1	1	y, R_x	yz

---

[3] We say 'broadly speaking' because the precise details depend on the group.

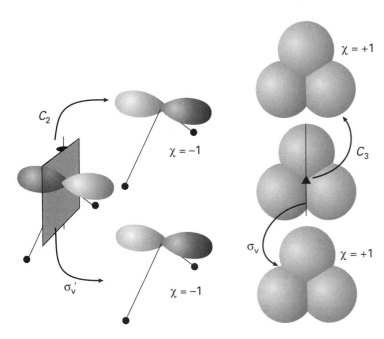

4.16 In a C_{2v} molecule such as H_2O, the $2p_x$ orbital on the central atom changes under a C_2 rotation, which signifies that it is a B orbital. That it also changes sign under the reflection σ'_v identifies it as B_1.

4.17 The combination $\phi_1 = \phi_A + \phi_B + \phi_C$ of the three $H1s$ orbitals in the C_{3v} molecule NH_3 remains unchanged under a C_3 rotation and under any of the vertical reflections.

A slightly more complicated example is the symmetry classification of the combination $\phi_1 = \phi_{A1s} + \phi_{B1s} + \phi_{C1s}$ of the three $H1s$ orbitals in the C_{3v} molecule NH_3 (Fig. 4.17). Because ϕ_1 is nondegenerate, it is either A_1 or A_2. It remains unchanged under a C_3 rotation and under any of the vertical reflections, so its characters are

E	$2C_3$	$3\sigma_v$
1	1	1

Comparison with the C_{3v} character table in Table 4.3 shows that ϕ_1 is of symmetry type A_1 and therefore that it contributes to a_1 molecular orbitals in NH_3. An extensive collection of symmetry-adapted linear combinations of orbitals is shown in Appendix 4, and it is usually a simple matter to identify the symmetry type of a combination of orbitals by comparing it with the diagrams provided there.

▍ *The entries in a character table display the behavior of atomic orbitals, and linear combinations of orbitals, under the symmetry operations of the group.*

Example 4.6 Identifying the symmetry type of orbitals

Identify the symmetry type of the orbital $\psi = \phi - \phi'$ in the C_{2v} molecule NO_2, where ϕ is a $2p_x$ orbital on one O atom and ϕ' a $2p_x$ orbital on the other O atom.

Answer The combination is shown in Fig. 4.18. Under a C_2 rotation, ψ changes into itself, implying a character of $+1$. Under the reflection σ_v both orbitals change sign, so $\psi \rightarrow -\psi$, implying a character of -1. Under σ'_v, ψ also changes sign, so the character for this operation is also -1. The characters are therefore

E	C_2	σ_v	σ'_v
1	1	-1	-1

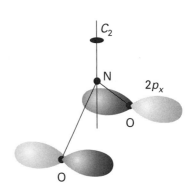

4.18 The combination of $O2p_x$ orbitals referred to in Example 4.6.

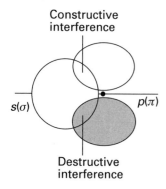

4.19 An s orbital (with σ symmetry) has zero net overlap with a p orbital (with π symmetry) because the constructive interference between the parts of the atomic orbitals with the same sign exactly matches the destructive interference between the parts with opposite signs.

These values match the characters of the A_2 symmetry label, so ψ can contribute to an a_2 orbital.

Self-test 4.6 Identify the symmetry type of the combination $\phi_{A1s} - \phi_{B1s} + \phi_{C1s} - \phi_{D1s}$ in a square-planar array of H atoms.

4.7 The construction of molecular orbitals

Molecular orbitals are constructed from symmetry-adapted linear combinations of atomic orbitals of the same symmetry type. Thus, in a linear molecule with the z-axis as the internuclear axis, an s orbital and a p_z orbital both have σ symmetry and so may combine to form σ molecular orbitals. On the other hand, an s orbital and a p_x orbital have different symmetries (σ and π, respectively) and hence cannot contribute to the same molecular orbital (Fig. 4.19). As can be seen from the illustration, the physical interpretation for this difference is that the contribution from the region of constructive interference is canceled by the contribution from the region of destructive interference.

The analogous argument for a nonlinear molecule can be illustrated by considering NH_3 again. We have seen that the combination ϕ_1 has A_1 symmetry (see Appendix 4). The N2s and N2p_z orbitals also have the same set of characters, as shown explicitly to the right of the character table for C_{3v}, so these two atomic orbitals also have A_1 symmetry. Because they have the same symmetry as the ϕ_1 SALC, the N2s and N2p_z orbitals and the ϕ_1 SALC all contribute to a_1 molecular orbitals. The resulting molecular orbitals have the form

$$\psi_{a_1} = c_1 \phi_{N2s} + c_2 \phi_{N2p_z} + c_3 \phi_1 \tag{1}$$

Only three such linear combinations are possible (because the H combination ϕ_1 counts as a single orbital), and they are labeled $1a_1$, $2a_1$, and $3a_1$ in order of increasing energy (the order of increasing number of internuclear nodes).

We have also seen (and can confirm by referring to Appendix 4) that the symmetry-adapted combinations ϕ_2 and ϕ_3 have E symmetry in C_{3v}. The character table shows that the same is true of the N2p_x and N2p_y orbitals, and this identification is confirmed by noting that jointly the two 2p orbitals behave exactly like ϕ_2 and ϕ_3 (Fig. 4.20). It follows that ϕ_2 and ϕ_3

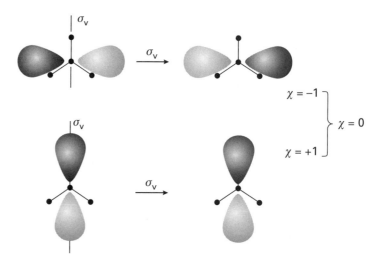

4.20 An N2p_x orbital in NH_3 changes sign under a σ_v reflection but an N2p_y orbital is left unchanged. Hence the degenerate pair jointly has character 0 for this operation. The plane of the paper is the xy-plane.

can combine with these two $2p$ orbitals and form doubly degenerate bonding and antibonding e orbitals of the molecule. These orbitals have the form

$$\psi_e = \{c_1\phi_{\mathrm{N}2p_x} + c_2\phi_2, \qquad c_3\phi_{\mathrm{N}2p_y} + c_4\phi_3\} \tag{2}$$

The bonding pair is labeled $1e$ and the antibonding pair is labeled $2e$.

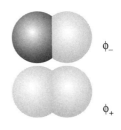

15

Example 4.7 Identifying symmetry-adapted orbitals

The H$1s$ orbitals in H_2O (of point group C_{2v}) form two symmetry-adapted linear combinations $\phi_+ = \phi_{\mathrm{A}1s} + \phi_{\mathrm{B}1s}$ and $\phi_- = \phi_{\mathrm{A}1s} - \phi_{\mathrm{B}1s}$ (**15**). What symmetry labels do they have? With what O orbitals will they overlap to form molecular orbitals?

Answer Under C_2, ϕ_+ does not change sign but ϕ_- does; their characters are $+1$ and -1, respectively. Under the reflections, ϕ_+ does not change sign; ϕ_- changes sign under σ_v, so that its character is -1 for this operation. The characters are therefore

	E	C_2	σ_v	σ_v'
ϕ_+	1	1	1	1
ϕ_-	1	-1	-1	1

This table identifies their symmetry types as A_1 and B_2, respectively. The same conclusion could have been obtained more directly by referring to Appendix 3. According to the right of the character table, the O$2s$ and O$2p_z$ orbitals also have A_1 symmetry; O$2p_y$ has B_2 symmetry. The linear combinations that can be formed are therefore

$$\psi_{a_1} = c_1\phi_{\mathrm{O}2s} + c_2\phi_{\mathrm{O}2p_z} + c_3\phi_+$$
$$\psi_{b_2} = c_4\phi_{\mathrm{O}2p_y} + c_5\phi_- \tag{3}$$

The three a_1 orbitals are bonding, intermediate, and antibonding in character according to the relative signs of the three coefficients. Similarly, depending on the relative signs of the two coefficients, one of the two b_2 orbitals is bonding and the other is antibonding.

Self-test 4.7 What is the symmetry label of the symmetry-adapted linear combination $\phi = \phi_{\mathrm{A}1s} + \phi_{\mathrm{B}1s} + \phi_{\mathrm{C}1s} + \phi_{\mathrm{D}1s}$ in CH_4, where $\phi_{\mathrm{J}1s}$ is an H$1s$ orbital atom J?

A symmetry analysis has nothing to say about the energies of orbitals other than to identify degeneracies. To calculate the energies, and even to arrange the orbitals in order, it is necessary to resort to quantum mechanics; to assess them experimentally it is necessary to turn to techniques such as photoelectron spectroscopy. In simple cases, though, we can use the general rules set out in Section 3.12a to judge the relative energies of the orbitals. For example, in NH_3, the $1a_1$ orbital, being composed of the low-lying N$2s$ orbital, will lie lowest in energy and its antibonding partner, $3a_1$, will probably lie highest. The $1e$ bonding orbital is next after $1a_1$, and is followed by the $2a_1$ orbital, which is largely nonbonding. This qualitative analysis leads to the energy level scheme shown in Fig. 4.21. These days, there is no difficulty in using one of the widely available software packages to calculate the energies of the orbitals directly by either an *ab initio* or semiempirical procedure. The energies given in Fig. 4.21 have in fact been calculated by using an *ab initio* package. Nevertheless, the ease of achieving computed values should not be seen as a reason for disregarding the understanding of the energy level order that comes from investigating the structures of the orbitals.

> The general procedure for constructing a molecular orbital scheme for a reasonably simple molecule can be summarized as follows:[4]

[4] A more pictorial procedure is given by S. K. Dhar, *J. Coord. Chem.* **29**, 17 (1993).

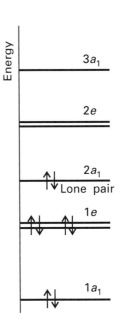

4.21 A schematic molecular orbital energy level diagram for NH_3 and an indication of its ground-state electron configuration. The forms of the orbitals themselves are shown in Fig. 3.32.

1 Assign a point group to the molecule.

2 Look up the shapes of the symmetry-adapted linear combinations in Appendix 4.

3 Arrange the SALCs of each molecular fragments in increasing order of energy, first noting whether they stem from s, p, or d orbitals (and put them in the order $s < p < d$), and then their number of internuclear nodes.

4 Combine SALCs of the same symmetry type from the two fragments, and from N SALCs form N molecular orbitals.

5 Estimate the relative energies of the molecular orbitals from considerations of overlap and relative energies of the parent orbitals, and draw the levels on a molecular orbital energy level diagram (showing the origin of the orbitals).

6 Confirm, correct, and revise this qualitative order by carrying out a molecular orbital calculation using commercial software.

The symmetries of molecular vibrations

Molecular vibrations are small periodic distortions from the equilibrium geometry of molecules. Their excitation by infrared (IR) radiation is the basis of IR spectroscopy and their excitation by inelastic collisions with visible and ultraviolet photons is the basis of Raman spectroscopy. The interpretation of IR and Raman spectra is greatly simplified by taking into account the symmetry of the molecule in question, and in this section we see a little of the reasoning involved.

4.8 Vibrating molecules: the modes of vibration

First, we review some general principles of molecular vibrations from the viewpoint of quantum theory.

(a) Vibrational energy levels

We may think of a bond in a molecule as a spring, and stretching the spring through a distance x produces a restoring force, F. For small displacements from equilibrium, the restoring force is proportional to the displacement, and we write $F = -kx$. The constant of proportionality, k, is the **force constant** of the bond: the stiffer the bond, the greater the force constant. A particle that experiences this type of restoring force is called a **harmonic oscillator**. The solutions of the Schrödinger equation for a harmonic oscillator show that the allowed energies are

$$E_v = (v + \tfrac{1}{2})\hbar\omega \qquad \omega = \left(\frac{k}{m}\right)^{1/2} \qquad v = 0, 1, 2, \ldots \tag{4}$$

where m is the **effective mass** of the oscillator. For a diatomic molecule consisting of atoms of masses m_A and m_B both atoms move during the vibration and the effective mass is given by

$$\frac{1}{m} = \frac{1}{m_A} + \frac{1}{m_B} \tag{5}$$

This expression is plausible for, if A is very heavy (in the sense $m_A \gg m_B$, so that $m \approx m_B$), then only B moves appreciably during the vibration and the vibrational energy levels are determined largely by m_B, the mass of the lighter atom alone. The energy levels given by eqn 4 are illustrated in Fig. 4.22. The frequency ω is high when the force constant is large (a stiff bond) and the mass of the oscillator is low (only light atoms are moved during the molecular vibration).

There are two important features to note about the energy levels of a harmonic oscillator. One is that there is a **zero-point energy**, the lowest possible energy, which in this case is

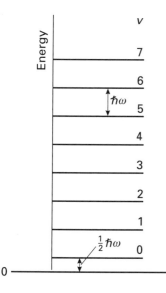

4.22 The energy levels of a harmonic oscillator. Note that all levels are equally spaced, and that the zero-point energy is half the spacing. The spacing increases as the force constant (stiffness) of the bond increases and as the effective mass of the vibrating molecule decreases.

$E_0 = \frac{1}{2}\hbar\omega$. This minimum, irremovable energy is large if the bond is stiff (k large) and the masses of the vibrating atoms are small (m small). The second feature to note is the separation of any two neighboring levels (for example, the separation of the levels with $v = 0$ and $v = 1$), which is $\hbar\omega$ for any v. This separation is large if the bond is stiff and the masses of the vibrating atoms are small. Vibrational transition energies are normally expressed in terms of the **wavenumber**, $\tilde{\nu}$, the reciprocal of the wavelength, with $\tilde{\nu} = \omega/2\pi c$.[5] Typical values of $\tilde{\nu}$ lie in the range from $500\ cm^{-1}$ to $3500\ cm^{-1}$ and the corresponding transitions lie in the infrared region of the electromagnetic spectrum. Table 4.4 shows some typical values of vibrational wavenumbers of diatomic molecules and the force constants derived from them.

> Molecules vibrate with a frequency that increases with increasing force constant (rigidity of the bond) and decreasing effective mass.

(b) Normal modes of vibration

The vibrations of polyatomic molecules present a much more complex problem than those of diatomic molecules. A diatomic molecule can vibrate in only one mode (its stretching mode) whereas a nonlinear polyatomic molecule of N atoms can vibrate in $3N - 6$ modes.

The number of independent vibrational modes of a polyatomic molecule of N atoms is calculated by recognizing that the motion of each atom may be described in terms of displacements along three perpendicular directions in space. It follows that the complicated motion of the N atoms of a polyatomic molecule can be expressed in terms of $3N$ displacements. Three combinations of these displacements result in the motion of the center of mass of the molecule through space and correspond to translations of the entire molecule (Fig. 4.23). If the molecule is nonlinear, another three combinations of the displacements are needed to specify the rotation of the whole molecule about its center of mass. That leaves $3N - 6$ combinations of displacements that leave the center of mass and the orientation of the molecule unchanged, and which are distortions of the molecule. Linear molecules are a special case because there is no rotation about the axis of the molecule: only two combinations of displacements correspond to a change in orientation of the molecule, so there are $3N - 5$ vibrational modes. Thus, the linear triatomic molecule CO_2 has four modes of vibration ($3 \times 3 - 5 = 4$), the angular triatomic molecule H_2O has three modes ($3 \times 3 - 6 = 3$), and the octahedral SF_6 molecule has fifteen ($3 \times 7 - 6 = 15$).

The $3N - 6$ modes of vibration of a nonlinear molecule (and the $3N - 5$ modes of a linear molecule) can be described in a number of ways. One of the most useful descriptions involves identifying the **normal modes** of the molecule by a combination of symmetry arguments and computation based on the force constants of the bonds. The normal modes of a molecule are collective motions of the atoms that are independent of each other (provided the motion is harmonic) in the sense that if any one is excited, it does not stimulate motion of another normal mode. A representation of the three independent modes of vibration of the angular triatomic molecule H_2O is shown in Fig. 4.24. Each diagram in the illustration represents an extreme point (in classical terms, a turning point) of the vibration, In diagrams of this kind, we show only the relative direction of displacement in only one phase of the vibration; we do not show the matching displacements as the atoms swing back to and then past their original positions. Each mode has its own characteristic frequency, and the corresponding wavenumbers for H_2O are shown in the illustration.

Although normal modes are collective motions of all the atoms in a molecule, in many cases it is possible to identify the vibration as primarily bond-stretching or bond-bending. For

---

Table 4.4 Vibrational wavenumbers and force constants of diatomic molecules

Molecule	$\tilde{\nu}/cm^{-1}$	$k/(N\ m^{-1})$
HCl	2885	4.8×10^2
Cl_2	557	3.2×10^2
Br_2	321	2.4×10^2
CO	2143	1.9×10^4
NO	1876	1.6×10^4

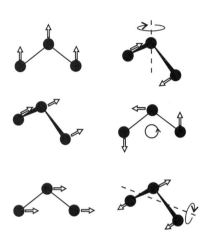

4.23 An illustration of the counting procedure for displacements of the atoms of a nonlinear molecule. There are nine atomic displacements in all. Three combinations of the displacements correspond to translation of the molecule through space, and three combinations correspond to rotations. Hence, three combinations (shown in the next illustration) correspond to vibrations of the molecule that leave its center of mass and orientation unchanged.

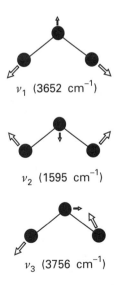

ν_1 (3652 cm^{-1})

ν_2 (1595 cm^{-1})

ν_3 (3756 cm^{-1})

4.24 The three normal modes of H_2O and their wavenumbers.

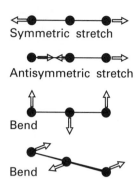

Symmetric stretch

Antisymmetric stretch

Bend

Bend

4.25 The four normal modes of CO_2. The two bending modes have the same frequency. The symmetric stretch leaves the electric dipole moment unchanged (at zero) but the antisymmetric stretch and the bending modes change the electric dipole moment.

Table 4.5 Group wavenumbers of some important groups often present in inorganic molecules

Group	Wavenumber/cm^{-1}
Terminal CN	2200–2000
Terminal CO	2150–1850
Bridging CO	1850–1700
Terminal **MH** of d-block hydride	1950–1750
Superoxo OO	1200–1100
Peroxo OO	920–750
MX in d-block Metal halides (**X** = Cl, Br, I)	450–150
Metal–metal bond	250–150*

example, two of the vibrations shown in Fig. 4.24 (ν_1 and ν_3) have similar wavenumbers because both modes mainly involve the stretching of O—H bonds. Whenever a vibration can be identified as primarily one type of displacement (such as a stretch or a bend) it is a useful approximation to refer to it as a particular type of vibration. Thus, we speak of the O—H stretching region of the spectrum, the C—O stretching region, or the H—O—H bending region.

Table 4.5 shows some of the characteristic wavenumbers of the absorption regions in the IR spectra of inorganic molecules. These characteristic wavenumbers are called **group wavenumbers** (more colloquially, 'group frequencies'). Their observation in a spectrum is a great help in the recognition of the presence of groups of atoms in a molecule and to its overall identification.

> A nonlinear molecule that consists of N atoms has $3N - 6$ modes of vibration; if it is linear, it has $3N - 5$ vibrational modes. The normal modes of a molecule are collective independent motions of the atoms.

4.9 Symmetry considerations

Now we turn to a consideration of how the symmetry of a molecule can assist in the analysis of IR and Raman spectra. It is convenient to consider two aspects of symmetry. One is the information that can be obtained directly by knowing to which point group a molecule as a whole belongs. The other is the additional information that comes from a knowledge of the symmetry type of each normal mode.

The general principles of IR and Raman spectroscopy are set out in Box 4.1. For the purpose of this discussion we need to note that absorption of infrared radiation can occur when a vibration results in a change in the electric dipole moment of a molecule. A Raman transition can occur when the polarizability of a molecule changes during a vibration. Vibrations that can contribute to IR and Raman spectra are called **infrared active** or **Raman active**, respectively. Vibrations that cannot contribute to a spectrum are classified as **inactive**.

(a) Information from the point group: the exclusion rule

It should be quite easy to see intuitively that all three displacements shown in Fig. 4.24 lead to a change in the molecular dipole moment of H_2O. It follows that all three modes are IR active. It is much more difficult to judge if a mode is Raman active, because it is hard to know whether a particular distortion of a molecule results in a change of polarizability. In fact, all three modes of H_2O are Raman active as well as IR active. The difficulty of deciding about the activity is partly overcome by the **exclusion rule**, which is sometimes very helpful:

• If a molecule has a center of inversion, none of its modes can be both IR and Raman active.

(A mode may be inactive in both.) There is no center of inversion in H_2O, so all three vibrational modes may be both IR and Raman active.

As an illustration of the usefulness of the exclusion rule, consider CO_2, a linear molecule with the C atom at the center of inversion. Its four normal modes are shown in Fig. 4.25. One mode involves a symmetrical stretching motion of the two O atoms and is called the **symmetric stretch**. This mode of vibration leaves the electric dipole moment unchanged at zero and so it is IR inactive and may be Raman active (in fact, it is). In another mode, the **antisymmetric stretch**, the C atom moves in a direction opposite to that of the two O atoms. As a result, the electric dipole moment changes from zero in the course of the vibration and the mode is IR active. Because the CO_2 molecule has a center of inversion, it follows from the exclusion rule that this mode cannot be Raman active. The two **bending modes** cause a departure of the dipole moment from zero and are therefore IR active. It follows from the exclusion rule that the two bending modes are Raman inactive.

Box 4.1 Infrared and Raman spectroscopy

Box 4.1 Infrared (IR) and Raman spectroscopy are primarily used to determine energy differences between vibrational states of molecules and solids and to identify a compound by comparison of spectra with those for authentic samples. They may also be used to identify individual structural features, such as a CO group in a molecule or an SO_4^{2-} ion in a compound. The spectra also give information on the symmetry and shapes of simple molecules and the force constants of bonds.

In an IR absorption process (Fig. B4.1), a photon of infrared radiation of frequency ν is absorbed and the molecule or solid is promoted to a higher vibrational state. For this absorption process to occur the energy of the photon must match the separation of vibrational states in the sample. In Raman spectroscopy, a photon of frequency ν_0 is scattered inelastically, giving up a part of its energy and emerging from the sample with a lower frequency $\nu_0 - \nu_i$ (Fig. B4.2), where ν_i is a vibrational frequency of the molecule. The energy lost by the photon is equal to the energy of a vibrational transition in the molecule.

Fourier transform infrared (FTIR) spectra are now widely used on account of their high sensitivity and their digital data handling capability. An FTIR spectrometer makes use of interference effects between two rays from a broad-band source, and detects an interferogram, a signal at the detector that oscillates in time. The mathematical technique of Fourier transformation converts the interferogram into an absorption spectrum as a function of frequency. The sample may be a liquid, gas, or solid. Solid samples are commonly prepared as powdered suspensions in oil or in pressed KBr disks. The spectrum produced by an infrared spectrometer is commonly displayed as percentage transmission against wavenumber.

In a Raman spectrometer, a monochromatic laser source of frequency ν_0 is directed through the sample (Fig. B4.3). The scattered light is collected at right angles to the incident beam, and passed through a monochromator. The latter scans frequencies from near the incident laser frequency to lower frequencies and the output is plotted as scattered intensity against wavenumber. The Raman effect is so feeble that only about one photon in every 10^{12} incident photons is scattered inelastically. However, the use of intense laser radiation and very efficient photomultiplier detectors makes the technique viable.

Raman spectroscopy is somewhat more tedious and costly than IR spectroscopy. Chemists go to the trouble of collecting Raman data because:

1 The combination of IR and Raman data is more conclusive than either one alone for identifying the symmetries of simple molecules.
2 The Raman scattered light is polarized, and this additional information helps to identify the symmetries of molecular vibrations.
3 Raman spectroscopy in aqueous solution and on single crystals is generally much more successful than IR spectroscopy.

Further reading

E.A.V. Ebsworth, D.W.H. Rankin, and S. Cradock, *Structural methods in inorganic chemistry.* Blackwell, Oxford (1991).

K. Nakamoto, *Infrared and Raman spectra of inorganic and coordination compunds.* Wiley, New York (1986).

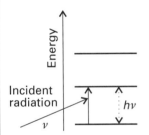

B4.1 Infrared absorption of a photon of frequency ν: This transition is real.

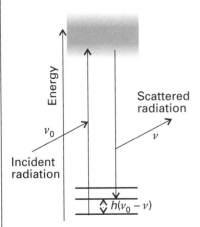

B4.2 Raman scattering can be envisaged as arising from a 'virtual' transition to excited states and the return of the molecule to a different state.

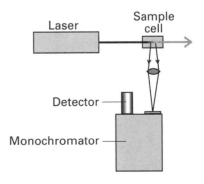

B4.3 A Raman spectrometer. In FT-Raman spectroscopy the monochromator is replaced by an interferometer.

If a molecule has a center of inversion, none of its modes can be both IR and Raman active.

....

Example 4.8 Using the exclusion rule

The symmetrical 'breathing' mode of the SF_6 is Raman active. Can it be detected in the IR spectrum?

Answer The molecule belongs to the point group O_h and has a center of inversion (at the S atom). The exclusion rule applies, so the Raman active breathing mode cannot be IR active.

....

Self-test 4.8 The bending mode of N_2O is active in the IR. Can it be Raman active also?

....

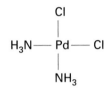

16 *cis*–$[PdCl_2(NH_3)_2]$

17 *trans*–$[PdCl_2(NH_3)_2]$

(b) Information from the symmetries of normal modes

So far, we have remarked that it is often intuitively obvious whether a vibrational mode gives rise to a changing electric dipole and is therefore IR active. When intuition is unreliable (perhaps because the molecule is complex or the mode of vibration difficult to visualize), a symmetry analysis can be used instead. We shall illustrate the procedure by considering the two square-planar species (**16**) and (**17**).[6] The *cis* isomer has C_{2v} symmetry, whereas the *trans* isomer is D_{2h}. Both species have bands in the Pd–Cl stretching region between 200 and 400 cm^{-1}. We know immediately from the exclusion rule that the two modes of the *trans* isomer may not be active in both IR and Raman. However, to decide which modes are IR active and which are Raman active, we may consider the characters of the modes themselves.

The symmetric and antisymmetric stretches of a Pd–Cl group are shown in Fig. 4.26, where the NH_3 group is treated as a single mass point. To decide on the activity of the modes, we need to classify them according to their symmetry types in their respective point groups. The approach is similar to the symmetry analysis of molecular orbitals in terms of SALCs.[7]

The group C_{2v} consists of the operations E, C_2, σ_v, and σ_v'. Consider the symmetric stretch of the *cis* isomer labeled A_1 in Fig. 4.26. The point to notice is that each operation of the group leaves the displacement vectors representing the vibration apparently unchanged. For example, the twofold rotation interchanges two equivalent displacement vectors. It follows that the character of each operation is $+1$:

	E	C_2	σ_v	σ_v'
χ	1	1	1	1

Comparison with the C_{2v} character table identifies the symmetry species of this mode to be A_1. Next consider the antisymmetric mode. The identity E leaves the displacement vectors unchanged; the same is true of σ_v', which lies in the plane containing the two Cl atoms. However, both C_2 and σ_v interchange the two oppositely directed displacement vectors, and so convert the overall displacement into -1 times itself:

	E	C_2	σ_v	σ_v'
χ	1	-1	-1	1

....

[6]   The Pt analogs of these species (and the distinction between them) are of considerable social and practical significance because the *cis* Pt isomer is used as a chemotherapeutic agent against certain types of cancer whereas the *trans* Pt isomer is therapeutically inactive.

[7]   A helpful account of the analogies between symmetry-adapted molecular orbitals and normal modes has been given by J.G. Verkade, *J. Chem. Educ.* **64**, 411 (1987).

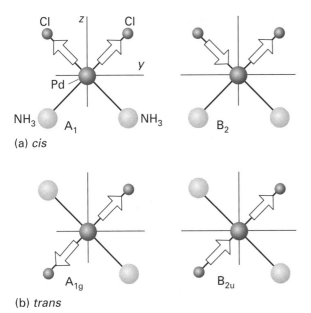

4.26 Some of the Pd–Cl stretching modes of a $[PdCl_2(NH_3)_2]$ square-planar complex. The motion of the Pd atom (which preserves the center of mass of the molecule) is not shown.

A comparison of this set of characters with the C_{2v} character table lets us identify the symmetry type of this mode as B_2. A similar analysis of the *trans* isomer, but using the D_{2h} group, results in the labels A_g and B_{2u} for the symmetric and antisymmetric Pd–Cl stretches, respectively.

....

Example 4.9 Identifying the symmetry of vibrational displacements

The *trans* isomer in Fig. 4.26 has D_{2h} symmetry. Verify that the antisymmetric stretch is of the symmetry type B_{2u}.

Answer The elements of D_{2h} are E, $C_2(x)$, $C_2(y)$, $C_2(z)$, i, $\sigma(xy)$, $\sigma(yz)$, $\sigma(zx)$. Of these, E, $C_2(y)$, $\sigma(xy)$, and $\sigma(yz)$ leave the displacement vectors unchanged and so have characters $+1$. The remaining operations reverse the directions of the vectors, so giving characters of -1:

	E	$C_2(x)$	$C_2(y)$	$C_2(z)$	i	$\sigma(xy)$	$\sigma(yz)$	$\sigma(zx)$
χ	1	−1	1	−1	−1	1	1	−1

Comparison of this set of characters with the D_{2h} character table shows that the symmetry type is B_{2u}.

....

Self-test 4.9 Confirm that the symmetric mode of the *trans* isomer is of symmetry A_{1g}.

To decide whether a vibration produces a change in the dipole moment of the molecule we must consider the symmetry of the electric dipole moment vector:

- A vibrational mode is IR active if it has the same symmetry as a component of the electric dipole vector.

To use this rule, we need to know that the components of the dipole vector have the same symmetries as the functions x, y, and z, which are indicated on the right of the character table. For example, in C_{2v}, z is A_1 and y is B_2. Both A_1, and B_2 vibrations in C_{2v} are therefore

IR active. In D_{2h}, x, y, and z are B_{3u}, B_{2u}, and B_{1u}, respectively, so only the B_{2u} vibration of the *trans* isomer can be IR active.

To judge whether a mode is Raman active we use a similar rule:

- A vibrational mode is Raman active if it has the same symmetry as a component of the molecular polarizability.

To use this rule, we need to know that the components of the polarizability have the same symmetries as the functions x^2, y^2, z^2, xy, yz, and zx. The symmetry species of these products may also be found by referring to the right of the character table for each group. It follows that, in the group C_{2v}, modes of symmetry A_1, A_2, B_1, and B_2 are Raman active. In D_{2h}, on the other hand, only A_g, B_{1g}, B_{2g}, and B_{3g} are Raman active. The experimental distinction between the *cis* and *trans* isomers now emerges. In the Pd–Cl stretching region, the *cis* (C_{2v}) isomer will have two bands in both the Raman and IR spectra. In contrast, the *trans* (D_{2h}) isomer will have one band at a different frequency in each spectrum. The IR spectra of the two isomers are shown in Fig. 4.27.

> *A vibrational mode is IR active if it has the same symmetry as a component of the electric dipole vector; a vibrational mode is Raman active if it has the same symmetry as a component of the molecular polarizability.*

(c) The assignment of molecular symmetry from vibrational spectra

One important application of vibrational spectra is to the identification of molecular symmetry and hence shape. An especially important example arises in metal carbonyls in which CO molecules are bound to a metal atom. Vibrational spectra are especially useful because the CO stretch is responsible for strong characteristic absorptions between 1850 and 2200 cm^{-1}.

The first metal carbonyl to be characterized was the tetrahedral (T_d) molecule Ni(CO)$_4$. The vibrational modes of the molecule that arise from stretching motions of the CO groups are four combinations of the four CO displacement vectors. The problem of recognizing the

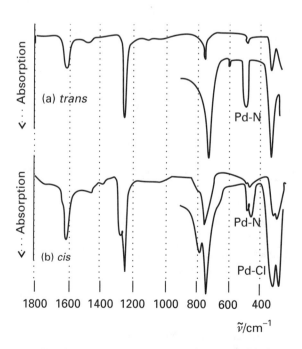

4.27 The IR spectra of *cis*- and *trans*-[PdCl$_2$(NH$_3$)$_2$]. (R. Layton, D.W. Sink, and J.R. Durig, *J. Inorg. Nucl. Chem.* **28**, 1965 (1966).)

appropriate linear combinations is similar to the problem of finding SALCs of atomic orbitals in the construction of molecular orbitals. Reference to Appendix 4 shows that that four atomic s orbitals give one A_1 SALC and three T_2 SALCs. The analogous linear combinations of the CO displacements are depicted in Fig. 4.28.

At this stage we consult the character table for T_d. We see that the combination labeled A_1 transforms like $x^2 + y^2 + z^2$ indicating that it will be Raman active but not IR active. In contrast, x, y, and z and the products xy, yz, and zx transform as T_2, so the T_2 modes are both Raman and IR active. Consequently, a tetrahedral carbonyl is recognized by one IR band and two Raman bands in the CO stretching region.[8]

> The shape of simple molecules and complexes may be inferred from the IR and Raman bands of a spectrum.

Example 4.10 Predicting the IR bands of an octahedral molecule

Consider an AB_6 molecule, such as SF_6. Sketch the form of linear combinations of A–B stretches that have the symmetry types A_{1g} and T_{1u} in the group O_h.

Answer Identify the SALCs that can be constructed from s orbitals in an octahedral arrangement (Appendix 4). These orbitals are the analogs of the stretching displacements of the A—B bonds and the signs represent their relative phases. The resulting linear combinations of stretches are illustrated in Fig. 4.29.

Self-test 4.10 Does the single IR absorption band of SF_6 arise from the A_1 or the T_{1u} mode?

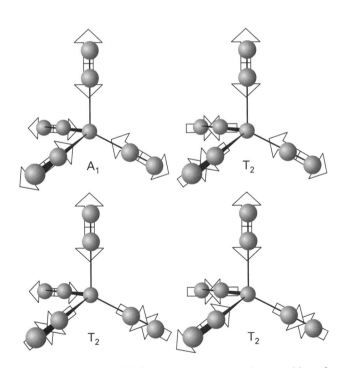

4.28 The modes of $Ni(CO)_4$ that correspond to the stretching of the CO bonds. The A_1 mode is nondegenerate; the T_2 modes are triply degenerate. (Compare the relative phases of the displacements with the relative phases of the SALCs of the same symmetry in Appendix 4.)

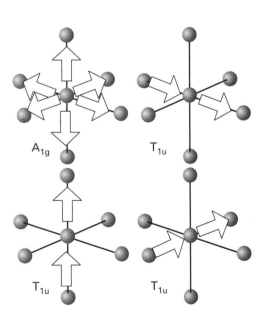

4.29 The A_{1g} and T_{1u} A–B stretching modes of an octahedral AB_6 molecule. The motion of the A atom, which preserves the center of mass of the molecule is not shown (it is stationary in the A_1 mode).

---

[8] The number of IR bands expected for various symmetries of metal carbonyls are given in Table 16.5.

Further reading

R.J. Gillespie, *Molecular geometry*. Van Nostrand-Reinhold, New York (1972). The book describes VSEPR theory in some detail.

R.J. Gillespie and I. Hargittai, *The VSEPR model of molecular geometry*. Allyn & Bacon, Needham Heights (1991). An introductory text, which includes an account of the model's quantum mechanical basis.

J.K. Burdett, *Molecular shapes: theoretical models of inorganic stereochemistry*. Wiley, New York (1980).

Structural techniques

E.A.V. Ebsworth, D.W.H. Rankin, and S. Cradock, *Structural methods in inorganic chemistry*. Blackwell Scientific, Oxford (1991). A general book that covers the principles of spectroscopic techniques such as NMR and vibrational spectroscopy as well as X-ray diffraction.

K. Nakamoto, *Infrared and Raman spectra of inorganic and coordination compounds* (in two volumes). Wiley, New York

(1997). A more specialized text which introduces vibrational spectroscopy and provides a very useful collection of data and its interpretation.

R.S. Drago, *Physical methods for chemists*. Saunders, Philadelphia (1992). A good general reference.

Point groups

Four reasonably elementary accounts of the use of point groups and character tables in chemistry are:

S.F.A. Kettle, *Symmetry and structure*. Wiley, New York (1995).

B.E. Douglas and C.A. Hollingsworth, *Symmetry in bonding and spectra*. Academic Press, New York (1985).

D.C. Harris and M.D. Bertolucci, *Symmetry and spectroscopy*. Oxford University Press (1978).

F.A. Cotton, *Chemical applications of group theory*. Wiley, New York (1990).

Exercises

4.1 Draw sketches to identify the following symmetry elements: (a) a C_3 axis and a σ_v plane in the NH_3 molecule; (b) a C_4 axis and a σ_h plane in the square-planar $[PtCl_4]^{2-}$ ion.

4.2 Which of the following molecules and ions has (1) a center of inversion, (2) an S_4 axis: (a) CO_2, (b) C_2H_2, (c) BF_3, (d) SO_4^{2-}?

4.3 Determine the symmetry elements and assign the point group of (a) NH_2Cl, (b) CO_3^{2-}, (c) SiF_4, (d) HCN, (e) $SiFClBrI$, (f) BF_4^-.

4.4 Determine the symmetry elements of objects with the shape of the boundary surfaces of (a) an s orbital, (b) a p orbital, (c) a d_{xy} orbital, and (d) a d_{z^2} orbital.

4.5 (a) State the symmetry elements that imply that a molecule is nonpolar. (b) Use symmetry criteria to determine whether or not each species in Exercise 4.3 is polar.

4.6 (a) State the symmetry criteria for chirality. (b) Determine whether any of the species in Exercise 4.3 can be optically active.

4.7 (a) Determine the symmetry group appropriate to the SO_3^{2-} ion. (b) What is the maximum degeneracy of a molecular orbital in this ion? (c) If the sulfur orbitals are $3s$ and $3p$, which of them can contribute to molecular orbitals of this maximum degeneracy?

4.8 (a) Determine the point group of the PF_5 molecule. (Use VSEPR, if necessary, to assign geometry.) (b) What is the maximum degeneracy of its molecular orbitals? (c) Which phosphorus $3p$ orbitals contribute to a molecular orbital of this degeneracy?

4.9 In a metal complex $[MH_4L_2]$ (where L is a ligand), four H atoms form a square-planar array around the central atom. (a) Draw the symmetry-adapted combinations of $H1s$ orbitals (it may be helpful to refer to Appendix 4). (b) Determine the point group for the complex and assign the symmetry label of each of these symmetry-adapted orbitals. (c) Which metal d orbitals have the correct symmetry to form molecular orbitals with these H linear combinations?

4.10 (a) How many vibrational modes does an SO_3 molecule have in the plane of the nuclei? (b) How many vibrational modes does it have perpendicular to the molecular plane?

4.11 What are the symmetries of the vibrations of (a) SF_6, (b) BF_3 that are *both* IR and Raman active?

4.12 What are the symmetries of the vibrational modes of a C_{6v} molecule that are neither IR nor Raman active?

Problems

4.1 Consider a molecule IF_3O_2 (with I as the central atom). How many isomers are possible? Which is likely to have the lowest energy? Assign point group designations to each isomer.

4.2 Group theory is often used by chemists as an aid in the interpretation of infrared spectra. For NH_4^+, four stretching modes are possible. There is the possibility that several vibrational modes are degenerate. A quick glance at the character table will tell if degeneracy is possible. (a) In the case of the tetrahedral NH_4^+ ion, is it necessary to consider the possibility of degeneracies? (b) Are degeneracies possible in any of the vibrational frequencies of $NH_2D_2^+$?

4.3 Figure 4.30 shows the energy levels for CH_3^+. What is the point group used for this illustration? What H1s orbital linear combination participates in a_1'? What C orbitals contribute to a_1'? What H linear combinations participate in the bonding e' pair? What C orbitals are e'? Which C orbitals are a_2''? Is any H linear combination a_2''? Now add two H1s orbitals on the z-axis (above and below the plane), modify the linear

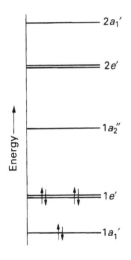

Fig. 4.30 A schematic molecular orbital energy level diagram for planar CH_3^+.

combinations of each symmetry type accordingly, and construct a new a_2'' linear combination. Are there bonding and nonbonding (or only weakly antibonding orbitals) that can accommodate 10 electrons and allow carbon to become hypervalent? (*Hint.* Refer to Appendices 3 and 4.)

4.4 Construct a Walsh diagram correlating the orbitals of a hypothetical square-planar H_4 (D_{4h}) molecule with those of a linear H_4 ($D_{\infty h}$) molecule.

4.5 Construct a Walsh diagram correlating the orbitals of a trigonal-planar XH_3 (D_{3h}) molecule with those of a trigonal-pyramidal XH_3 (C_{3v}) molecule.

4.6 (a) Determine the point group of the most symmetric planar conformation of $B(OH)_3$ and the most symmetric nonplanar conformation of $B(OH)_3$. Assume that the B—O—H bond angles are all $109.5°$ in all conformations. (b) Sketch a conformation of $B(OH)_3$ that is chiral, once again keeping all three of the B—O—H bond angles equal to $109.5°$.

4.7 How many isomers are there for 'octahedral' molecules with the formula MA_3B_3, where A and B are monoatomic ligands? What is the point group of each isomer? Are any of the isomers chiral? Repeat this exercise for molecules with the formula $MA_2B_2C_2$.

4.8 Determine whether the number of IR and Raman active stretching vibrations could be used to determine uniquely whether a sample of gas is BF_3, NF_3, ClF_3.

4.9 How many SALCs of 1s atomic orbitals will a square-planar array of four H atoms produce? Consider that the four H atoms lie directly on the x- and y-axes of a cartesian coordinate system. Draw the SALC with no nodal planes and determine its symmetry type. Draw the SALC with two nodal planes and determine its symmetry type.

4.10 How many planes of symmetry does a benzene molecule possess? What chloro-substituted benzene of formula $C_6H_nCl_{6-n}$ has exactly four total planes of symmetry?

5

Acids and bases

This chapter focuses on the wide variety of species that are classified as acids and bases. The acids and bases described in the first part of the chapter take part in proton transfer reactions. Proton transfer equilibria can be discussed quantitatively in terms of acidity constants, which are a measure of the strength with which species donate protons. In the second part of the chapter, we broaden the definition of acids and bases to include reactions that involve electron pair sharing between a donor and an acceptor. Because of the greater diversity of these species, a single scale of strength is not appropriate. Therefore, we describe two approaches: in one, acids and bases are classified as 'hard' or 'soft'; in the other, thermochemical data are used to obtain a set of parameters characteristic of each species.

The first recognition of the existence of acids and bases was based, hazardously, on criteria of taste and feel: acids were sour and bases felt soapy. A deeper chemical understanding of their properties emerged from Arrhenius's conception of an acid as a compound that produced hydrogen ions in water. The modern definitions, the only ones we consider, are based on a broader range of chemical reactions. The definition due to Brønsted and Lowry focuses on proton transfer, and that due to Lewis is based on the interaction of electron pair acceptor and donor molecules and ions.

Brønsted acidity

Johannes Brønsted in Denmark and Thomas Lowry in England proposed (in 1923) that the essential feature of an acid–base reaction is the transfer of a proton from one species to another. In the context of the Brønsted–Lowry definitions, a proton is a hydrogen ion, H^+. They suggested that any substance that acts as a proton donor should be classified as an **acid**, and any substance that acts as a proton acceptor should be classified as a **base**. Substances that act in this way are now called **Brønsted acids** and **Brønsted bases**, respectively. The definitions make no reference to the environment in which proton transfer occurs, so they apply to proton transfer behavior in any solvent, and even in no solvent at all.

An example of a Brønsted acid is hydrogen fluoride, HF, which can donate a proton to another molecule, such as H_2O, when it dissolves in water:

$$HF(g) + H_2O(l) \longrightarrow H_3O^+(aq) + F^-(aq)$$

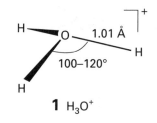

1 H_3O^+

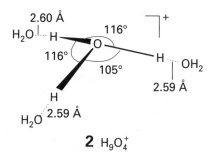

2 $H_9O_4^+$

An example of a Brønsted base is ammonia, NH_3, which can accept a proton from a proton donor:

$$H_2O(l) + NH_3(aq) \longrightarrow NH_4^+(aq) + OH^-(aq)$$

As these two examples show, water is an example of an **amphiprotic** substance, a substance that can act as both a Brønsted acid and a Brønsted base.

When an acid donates a proton to a water molecule, the latter is converted into a **hydronium ion**, H_3O^+. The dimensions of the hydronium ion (**1**) are taken from the crystal structure of $H_3O^+ClO_4^-$. However, the structure H_3O^+ is almost certainly an oversimplified description of the ion in water, for it participates in extensive hydrogen bonding. If a simple formula is required, the hydronium ion in water is probably best represented as $H_9O_4^+$ (**2**). Gas-phase studies of water clusters using mass spectrometry suggest that a cage of 20 H_2O molecules can condense around one H_3O^+ ion in a regular pentagonal dodecahedral arrangement, resulting in the formation of the species $H^+(H_2O)_{21}$.[1] As these structures indicate, the most appropriate description of a proton in water varies according to the environment and the experiment under consideration.

> A Brønsted acid is a proton donor and a Brønsted base is a proton acceptor. A simple representation of a hydrogen ion in water is as the polyatomic hydronium ion, H_3O^+.

5.1 Proton transfer equilibria in water

Proton transfer between acids and bases is fast in both directions, so the dynamic equilibria

$$HF(aq) + H_2O(l) \rightleftharpoons H_3O^+(aq) + F^-(aq)$$
$$H_2O(l) + NH_3(aq) \rightleftharpoons NH_4^+(aq) + OH^-(aq) \tag{1}$$

give a more complete description of the behavior of the acid HF and the base NH_3 in water than the forward reaction alone. The central feature of Brønsted acid–base chemistry in aqueous solution is that of rapidly attained proton transfer equilibrium, and we concentrate on this aspect.

(a) Conjugate acids and bases

The symmetry of each of the two forward and reverse reactions in eqn 1, both of which depend on the transfers of a proton from an acid to a base, is expressed by writing the general **Brønsted equilibrium** as

$$Acid_1 + Base_2 \rightleftharpoons Acid_2 + Base_1 \tag{2}$$

The species $Base_1$ is called the **conjugate base** of $Acid_1$, and $Acid_2$ is the **conjugate acid** of $Base_2$. The conjugate base of an acid is the species that is left after a proton is lost. The conjugate acid of a base is the species formed when a proton is gained. Thus, F^- is the conjugate base of HF and H_3O^+ is the conjugate acid of H_2O. There is no *fundamental* distinction between an acid and a conjugate acid or a base and a conjugate base: a conjugate acid is just another acid and a conjugate base is just another base.

> When a species donates a proton, it becomes the conjugate base; when a species gains a proton, it becomes the conjugate acid. Conjugate acids and bases are in equilibrium in solution.

Example 5.1 Identifying acids and bases

Identify the Brønsted acid and its conjugate base in the following reactions:

(a) $HSO_4^-(aq) + OH^-(aq) \longrightarrow H_2O(l) + SO_4^{2-}(aq)$

(b) $PO_4^{3-}(aq) + H_2O(l) \longrightarrow HPO_4^{2-}(aq) + OH^-(aq)$

---

[1] A.W. Castleman, S. Wei, and Z. Shi, *J. Chem. Phys.* **94**, 3268 (1991).

Answer (a) The hydrogensulfate ion, HSO_4^-, transfers a proton to hydroxide; it is therefore the acid and the SO_4^{2-} ion produced is its conjugate base. (b) The H_2O molecule transfers a proton to the phosphate ion acting as a base; thus H_2O is the acid and the OH^- ion is its conjugate base.

....

Self–test 5.1 Identify the acid, base, conjugate acid, and conjugate base in the reactions (a) $HNO_3(aq) + H_2O(l) \rightarrow H_3O^+(aq) + NO_3^-(aq)$,
(b) $CO_3^{2-}(aq) + H_2O(l) \rightarrow HCO_3^-(aq) + OH^-(aq)$,
(c) $NH_3(aq) + H_2S(aq) \rightarrow NH_4^+(aq) + HS^-(aq)$.

....

(b) The strengths of Brønsted acids

The strength of a Brønsted acid, such as HF, in aqueous solution is expressed by its **acidity constant** (or 'acid ionization constant'), K_a:

$$HF(aq) + H_2O(l) \rightleftharpoons H_3O^+(aq) + F^-(aq) \qquad K_a = \frac{[H_3O^+][F^-]}{[HF]} \qquad (3)$$

In this definition, [X] denotes the numerical value of the molar concentration of the species X (so, if the molar concentration of HF molecules is $0.001 \ mol \, L^{-1}$, then $[HF] = 0.001$).[2] A value $K_a \ll 1$ implies that proton retention by the acid is favored. The experimental value of K_a for hydrogen fluoride in water is 3.5×10^{-4}, indicating that under normal conditions, only a very small fraction of HF molecules are deprotonated in water. The actual fraction deprotonated can be calculated as a function of acid concentration from the numerical value of K_a as described in any introductory chemistry text.[3]

The proton transfer equilibrium characteristic of a base, such as NH_3, in water can also be expressed in terms of an equilibrium constant, the **basicity constant**, K_b:

$$NH_3(aq) + H_2O(l) \rightleftharpoons NH_4^+(aq) + OH^-(aq) \qquad K_b = \frac{[NH_4^+][OH^-]}{[NH_3]} \qquad (4)$$

If $K_b \ll 1$, the base is a weak proton acceptor and its conjugate acid is present in low abundance in solution. The experimental value of K_b for ammonia in water is 1.8×10^{-5}, indicating that under normal conditions, only a very small fraction of NH_3 molecules are protonated in water. As for the acid calculation, the actual fraction of base protonated can be calculated from the numerical value of K_b.

Because water is amphiprotic, a proton transfer equilibrium exists even in the absence of added acids or bases. The proton transfer from one water molecule to another is called **autoprotolysis** (or 'autoionization'). The extent of autoprotolysis and the composition of the solution at equilibrium is described by the water **autoprotolysis constant**, K_w:

$$2H_2O(l) \rightleftharpoons H_3O^+(aq) + OH^-(aq) \qquad K_w = [H_3O^+][OH^-] \qquad (5)$$

The experimental value of K_w is 1.00×10^{-14} at $25 \, °C$, indicating that only a very tiny fraction of water molecules is present as ions in pure water.

An important role for the autoprotolysis constant of a solvent is that it enables us to express the strength of a base in terms of the strength of its conjugate acid. Thus, the value of K_b for the equilibrium in eqn 4 is related to the value of K_a for the equilibrium

$$NH_4^+(aq) + H_2O(l) \rightleftharpoons H_3O^+(aq) + NH_3(aq)$$

....

[2]   In precise work, K_a is expressed in terms of the activity of X, $a(X)$, its effective thermodynamic concentration.

[3]   See, for instance, P.W. Atkins and L.L. Jones, *Chemistry: molecules, matter, and change*. W.H. Freeman & Co, New York (1997).

by

$$K_a K_b = K_w \qquad (6)$$

This relation may be verified by multiplying together the expressions for the acidity constant of NH_4^+ and the basicity constant of NH_3. The implication of eqn 6 is that the larger the value of K_b, the smaller the value of K_a. That is, the *stronger the base, the weaker its conjugate acid*. A further implication of eqn 6 is that the strengths of bases may be reported in terms of the acidity constants of their conjugate acids.

Because molar concentrations and acidity constants span many orders of magnitude, it proves convenient to report them as their common logarithms (logarithms to the base 10) by using

$$pH = -\log[H_3O^+] \qquad pK = -\log K \qquad (7)$$

where K may be any of the constants we have introduced. At $25\,^\circ$C, for instance, $pK_w = 14.00$. It follows from this definition and the relation in eqn 6 that

$$pK_a + pK_b = pK_w \qquad (8)$$

A similar expression applies to the strengths of conjugate acids and bases in any solvent, with pK_w replaced by the appropriate autoprotolysis constant of the solvent, pK_{sol}.

> The strength of a Brønsted acid is measured by its acidity constant, and the strength of a Brønsted base is measured by its basicity constant. The stronger the base, the weaker its conjugate acid.

(c) Strong and weak acids and bases

Table 5.1 lists the acidity constants of some common acids and conjugate acids of some bases. A substance is classified as a **strong acid** if the proton transfer equilibrium lies strongly in favor of donation to water. Thus, a substance with $pK_a < 0$ (corresponding to $K_a > 1$ and usually to $K_a \gg 1$) is a strong acid. Such acids are commonly regarded as being fully deprotonated in solution (but it must never be forgotten that that is only an approximation).

Table 5.1 Acidity constants for species in aqueous solution at $25\,^\circ$C

Acid	HA	A^-	K_a	pK_a
Hydriodic	HI	I^-	10^{11}	-11
Perchloric	$HClO_4$	ClO_4^-	10^{10}	-10
Hydrobromic	HBr	Br^-	10^9	-9
Hydrochloric	HCl	Cl^-	10^7	-7
Sulfuric	H_2SO_4	HSO_4^-	10^2	-2
Hydronium ion	H_3O^+	H_2O	1	0.0
Chloric	$HClO_3$	ClO_3^-	10^{-1}	1
Sulfurous	H_2SO_3	HSO_3^-	1.5×10^{-2}	1.81
Hydrogensulfate ion	HSO_4^-	SO_4^{2-}	1.2×10^{-2}	1.92
Phosphoric	H_3PO_4	$H_2PO_4^-$	7.5×10^{-3}	2.12
Hydrofluoric	HF	F^-	3.5×10^{-4}	3.45
Pyridinium ion	$HC_5H_5N^+$	C_5H_5N	5.6×10^{-6}	5.25
Carbonic	H_2CO_3	HCO_3^-	4.3×10^{-7}	6.37
Hydrogen sulfide	H_2S	HS^-	9.1×10^{-8}	7.04
Boric acid*	$B(OH)_3$	$B(OH)_4^-$	7.2×10^{-10}	9.14
Ammonium ion	NH_4^+	NH_3	5.6×10^{-10}	9.25
Hydrocyanic	HCN	CN^-	4.9×10^{-10}	9.31
Hydrogencarbonate ion	HCO_3^-	CO_3^{2-}	4.8×10^{-11}	10.32
Hydrogenarsenate ion	$HAsO_4^{2-}$	AsO_4^{3-}	3.0×10^{-12}	11.53
Hydrogensulfide ion	HS^-	S^{2-}	1.1×10^{-19}	19
Hydrogenphosphate ion	HPO_4^{2-}	PO_4^{3-}	2.2×10^{-13}	12.67
Dihydrogenphosphate ion	$H_2PO_4^-$	HPO_4^{2-}	6.2×10^{-8}	7.21

*The proton transfer equilibrium is $B(OH)_3(aq) + 2H_2O(l) \rightleftharpoons H_3O^+(aq) + B(OH)_4^-(aq)$.

For example, hydrochloric acid is regarded as a solution of hydronium ions and chloride ions, and a negligible concentration of HCl molecules. A substance with $pK_a > 0$ (corresponding to $K_a < 1$) is classified as a **weak acid**; for such species, the proton transfer equilibrium lies in favor of nonionized acid. Hydrogen fluoride is a weak acid in water, and hydrofluoric acid consists of hydronium ions, fluoride ions, and a high proportion of HF molecules.

A **strong base** is a species that is virtually fully protonated in water. An example is the oxide ion, O^{2-}, which is immediately converted into OH^- ions in water. A **weak base** is only partially protonated in water. An example is NH_3, which is present almost entirely as NH_3 molecules in water, with a small proportion of NH_4^+ ions. The conjugate base of any strong acid is a weak base, because it is thermodynamically unfavorable for such a base to accept a proton.

> An acid or base is classified as either weak or strong depending on the size of its acidity constant. There is a reciprocal relation between the strengths of conjugate acids and bases.

(d) Polyprotic acids

A **polyprotic acid** is a substance that can donate more than one proton. An example is hydrogen sulfide, H_2S, a diprotic acid. For a diprotic acid, there are two successive proton donations and two acidity constants:

$$H_2S(aq) + H_2O(l) \rightleftharpoons HS^-(aq) + H_3O^+(aq) \qquad K_{a1} = \frac{[H_3O^+][HS^-]}{[H_2S]}$$

$$HS^-(aq) + H_2O(l) \rightleftharpoons S^{2-}(aq) + H_3O^+(aq) \qquad K_{a2} = \frac{[H_3O^+][S^{2-}]}{[HS^-]}$$

From Table 5.1, $K_{a1} = 9.1 \times 10^{-8}$ ($pK_{a1} = 7.04$) and $K_{a2} \approx 10^{-19}$ ($pK_{a2} = 19$). The second acidity constant, K_{a2}, is almost always smaller than K_{a1} (and hence pK_{a2} is generally larger than pK_{a1}). The decrease in K_a is consistent with an electrostatic model of the acid in which, in the second deprotonation, a proton must separate from a center with one more negative charge than in the first deprotonation. Because additional electrostatic work must be done to remove the positively charged proton, the deprotonation is less favorable.

The clearest representation of the concentrations of the species that are formed in the successive proton transfer equilibria of polyprotic acids is a **distribution diagram**. In such a diagram, the fraction, $\alpha(X)$, of a specified species X is plotted against the pH. Consider, for instance, the triprotic acid H_3PO_4, which releases three protons in succession to give $H_2PO_4^-$, HPO_4^{2-}, and PO_4^{3-}. The diagram in Fig. 5.1 shows the fraction of each of these species as a function of pH; for the fraction of H_3PO_4 molecules, for instance,

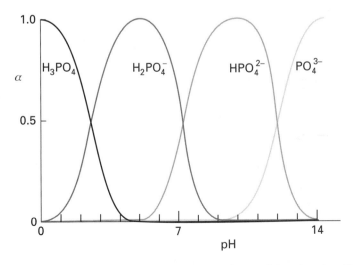

5.1 The distribution diagram for the various forms of the triprotic acid, phosphoric acid, as a function of pH.

$$\alpha(H_3PO_4) = \frac{[H_3PO_4]}{[H_3PO_4] + [H_2PO_4^-] + [HPO_4^{2-}] + [PO_4^{3-}]} \tag{9}$$

The variation of α with pH reveals the relative importance of each acid and its conjugate base at each pH. Conversely, the diagram indicates the pH of the solution that contains a particular fraction of the species. We see, for instance, that if $pH < pK_{a1}$, corresponding to high hydronium ion concentrations, then the dominant species is the fully protonated H_3PO_4 molecule. However, if $pH > pK_{a3}$, corresponding to low hydronium ion concentrations, then the dominant species is the fully deprotonated PO_4^{3-} ion. The intermediate species are dominant at pH values that lie between the relevant pK_as.

> A polyprotic acid loses protons in succession, and successive deprotonations are less favorable; a distribution diagram summarizes how the fraction of each species present depends on the pH of the solution.

5.2 Solvent leveling

An acid that is weak in water may appear strong in a solvent that is a stronger proton acceptor, and vice versa. Indeed, in sufficiently basic solvents (such as liquid ammonia), it may not be possible to arrange a series of acids according to their strengths, because all of them will be fully deprotonated. Similarly, bases that are weak in water may appear strong in a more strongly proton-donating solvent (such as anhydrous acetic acid). It may not be possible to arrange a series of bases according to strength, for all of them will be effectively fully protonated in acidic solvents. We shall now see that the autoprotolysis constant of a solvent plays a crucial role in determining the range of acid or base strengths that can be distinguished for species dissolved in it.

Any acid stronger than H_3O^+ in water donates a proton to H_2O and forms H_3O^+. Consequently, no acid significantly stronger than H_3O^+ can remain protonated in water. No experiment conducted in water can tell us which is the stronger acid between HBr and HI because both transfer their protons essentially completely to give H_3O^+. In effect, solutions of the strong acids HX and HY behave as though they are solutions of H_3O^+ ions regardless of whether HX is intrinsically stronger than HY. Water is therefore said to have a **leveling effect** that brings all stronger acids down to the acidity of H_3O^+. To distinguish the strengths of such acids we can use a less basic solvent. For instance, although HBr and HI have indistinguishable acid strengths in water, in acetic acid HBr and HI behave as weak acids and their strengths can be distinguished. It is found in this way that HI is a stronger proton donor than HBr.

The acid leveling effect can be expressed in terms of the pK_a of the acid. An acid such as HCN dissolved in a solvent, HSol, is classified as strong if $pK_a < 0$, where K_a is the acidity constant of the acid in the solvent Sol:

$$HCN(sol) + HSol(l) \rightleftharpoons H_2Sol^+(sol) + CN^-(sol) \quad K_a = \frac{[H_2Sol^+][CN^-]}{[HCN]} \tag{10}$$

That is, all acids with $pK_a < 0$ (corresponding to $K_a > 1$) display the acidity of H_2Sol^+ when they are dissolved in the solvent HSol.

An analogous effect can be found for bases in water. Any base that is strong enough to undergo complete protonation by water produces an OH^- ion for each molecule of base added. The solution behaves as though it contains OH^- ions. Therefore, we cannot distinguish the proton-accepting power of such bases, and we say that they are leveled to a common strength. Indeed, the OH^- ion is the strongest base that can exist in water because any species that is a stronger proton acceptor immediately forms OH^- ions by proton transfer from water. For this reason, we cannot study NH_2^- or CH_3^- in water by dissolving alkali metal amides or methides because both anions generate OH^- ions and are fully protonated to NH_3 and CH_4:

$$KNH_2(s) + H_2O(l) \longrightarrow K^+(aq) + OH^-(aq) + NH_3(aq)$$
$$Li_4(CH_3)_4(s) + 4H_2O(l) \longrightarrow 4Li^+(aq) + 4OH^-(aq) + 4CH_4(g)$$

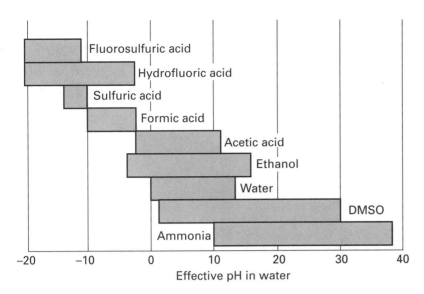

5.2 The acid–base discrimination windows of a variety of solvents. In each case, the width of the window is proportional to the autoprotolysis constant pK of the solvent.

The base leveling effect can be expressed in terms of the pK_b of the base. A base dissolved in HSol is classified as strong if $pK_b < 0$, where K_b is the basicity constant of the base in HSol:

$$NH_3(sol) + HSol(l) \rightleftharpoons NH_4^+(sol) + Sol^-(sol) \qquad K_b = \frac{[NH_4^+][Sol^-]}{[NH_3]} \qquad (11)$$

That is, all bases with $pK_b < 0$ (corresponding to $K_b > 1$) behave like Sol^- in the solvent HSol. Now, because $pK_a + pK_b = pK_{sol}$, this criterion for leveling may be expressed as follows: all bases with $pK_a > pK_{sol}$ behave like Sol^- in the solvent HSol.

It follows from this discussion of acids and bases in a common solvent HSol that, because any acid is leveled if $pK_a < 0$ in HSol and any base is leveled if $pK_a > pK_{sol}$ in the same solvent, then the window of strengths that are not leveled in the solvent is from $pK_a = 0$ to pK_{sol}. For water, $pK_w = 14$. For liquid ammonia, the autoprotolysis equilibrium is

$$2NH_3(l) \rightleftharpoons NH_4^+(am) + NH_2^-(am) \qquad pK_{am} = 33$$

(The 'am' signifies solution in liquid ammonia.) It follows from these figures that acids and bases are discriminated much less in water than they are in ammonia. The discrimination windows of a number of solvents are shown in Fig. 5.2. The window for dimethylsulfoxide (DMSO, $(CH_3)_2SO$) is wide because $pK_{dmso} = 37$. Consequently, DMSO can be used to study a wide range of acids (from H_2SO_4 to PH_3). Water has a narrow window compared to some of the other solvents shown in the illustration. One reason is the high relative permittivity (dielectric constant) of water, which favors the formation of H_3O^+ and OH^- ions.

> *A solvent with a large autoprotolysis constant can be used to discriminate between a wide range of acid and base strengths.*

Periodic trends in Brønsted acidity

From now on we shall concentrate on Brønsted acids and bases in water. The largest class of acids in water consists of species that donate protons from an –OH group attached to a central atom. A donatable proton of this kind is called an **acidic proton** to distinguish it from other protons that may be present in the molecule, such as the nonacidic methyl protons in CH_3COOH.

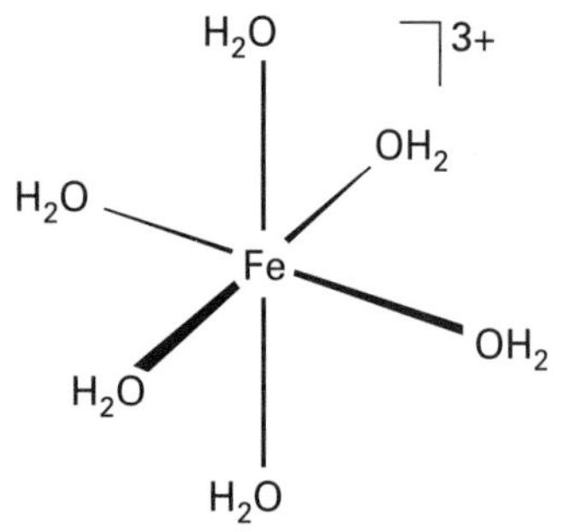

3 $[Fe(OH_2)_6]^{3+}$

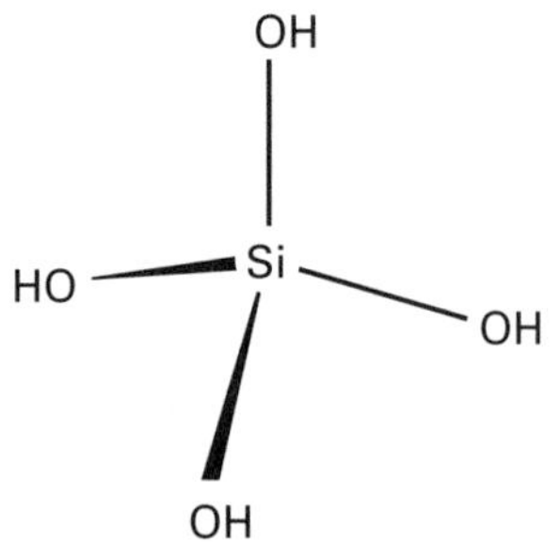

4 $Si(OH)_4$

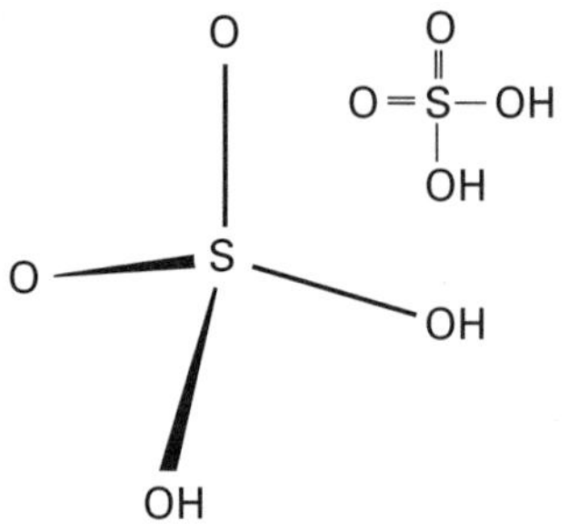

5 H_2SO_4

There are three classes of acids to consider:

1 An **aqua acid**, in which the acidic proton is on a water molecule coordinated to a central metal ion.

$$E(OH_2)(aq) + H_2O(l) \rightleftharpoons E(OH)^-(aq) + H_3O^+(aq)$$

An example is

$$[Fe(OH_2)_6]^{3+}(aq) + H_2O(l) \longrightarrow [Fe(OH_2)_5(OH)]^{2+}(aq) + H_3O^+(aq)$$

The aqua acid, the hexaaquairon(III) ion, is shown as (**3**).

2 A **hydroxoacid**, in which the acidic proton is on a hydroxyl group without a neighboring oxo group ($=O$).

An example is $Si(OH)_4$ (**4**), which is important in the formation of minerals.

3 An **oxoacid**, in which the acidic proton is on a hydroxyl group with an oxo group attached to the same atom.

Sulfuric acid, H_2SO_4 ($O_2S(OH)_2$; (**5**)), is an example of an oxoacid. The three classes of acids can be regarded as successive stages in the deprotonation of an aqua acid:

$$H_2O-E-OH_2 \xrightarrow{-2H^+} HO-E-OH^{2-} \xrightarrow{-H^+} HO-E=O^{3-}$$

aqua acid   hydroxoacid   oxoacid

An example of these successive stages is provided by a *d*-block metal in an intermediate oxidation state, such as Ru(IV):

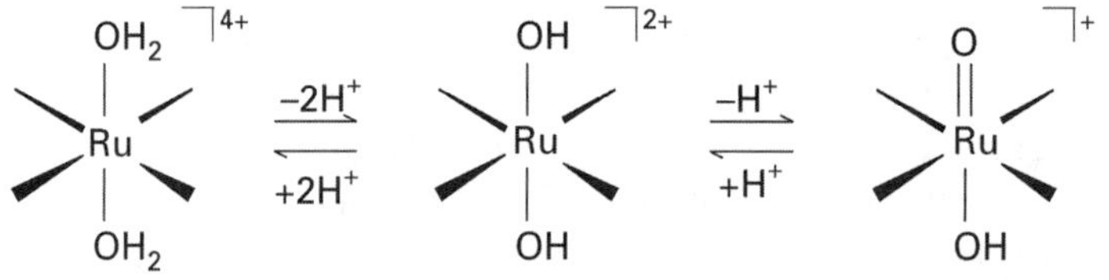

Aqua acids are characteristic of central atoms in low oxidation states, of *s*- and *d*-block metals, and of metals on the left of the *p*-block. Oxoacids are commonly found where the central element is in a high oxidation state. An element from the right of the *p*-block in one of its intermediate oxidation states may also produce an oxoacid ($HClO_2$ is an example).

> Aqua acids, hydroxoacids, and oxoacids are typical of specific regions of the periodic table, as detailed above.

5.3 Periodic trends in aqua acid strength

The strengths of aqua acids typically increase with increasing positive charge of the central metal ion and with decreasing ionic radius. This variation can be rationalized to some extent in terms of an ionic model, in which the metal cation is represented by a sphere carrying z positive charges. The gas-phase pK_a is proportional to the work of removing a proton to infinity from a distance equal to the sum of the ionic radius, r_+, and the diameter of a water molecule, d.[4] Because protons are more easily removed from the vicinity of cations of high charge and small radius, the model predicts that the acidity should increase with increasing z

---

[4] This relation comes from $pK_a \propto \Delta_r G^\ominus$ coupled with the identification of ΔG and electrical work; so $pK_a \propto w_{electrical}$.

and with decreasing r_+, very roughly as the electrostatic parameter $\zeta = z^2/(r_+ + d)$. The trends this model predicts for the gas phase will also apply in solution if the effects of solvation are reasonably constant.

The validity of the ionic model of acid strengths can be judged from Fig. 5.3. Aqua ions of elements that form ionic solids (principally those from the s-block) have pK_as that are quite well described by the ionic model. Several d-block ions (such as Fe^{2+} and Cr^{3+}) lie reasonably near the same straight line, but many ions (particularly those with low pK_a, corresponding to high acid strength) deviate markedly from it. This deviation indicates that the metal ions repel the departing proton more strongly than is predicted by the ionic model. This enhanced repulsion can be rationalized by supposing that the cation's positive charge is not fully localized on the central ion but is delocalized over the ligands and hence is closer to the departing proton. The delocalization is equivalent to attributing covalence to the element-oxygen bond. Indeed, the correlation is worst for ions that are disposed to form covalent bonds.

For the later d-block and the p-block metals (such as Cu^{2+} and Sn^{2+}, respectively), the strengths of the aqua acids are much greater than the ionic model predicts. For these species, covalent bonding is more important than ionic bonding and the ionic model is unrealistic. The overlap between metal orbitals and the orbitals of an oxygen ligand increases from left to right across a period. It also increases down a group, so aqua ions of heavier d-block metals tend to be stronger acids.

> The strengths of aqua acids typically increase with increasing positive charge of the central metal ion and with decreasing ionic radius. Exceptions are commonly due to the effects of covalent bonding.

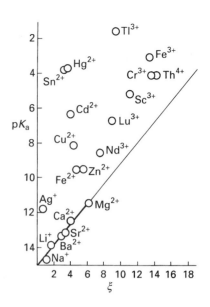

5.3 The correlation between acidity constant and electrostatic parameter ξ of aqua ions. Note that only hard ions with low charge follow the correlation; all others are more acidic than the correlation suggests.

Example 5.2 Accounting for trends in aqua acid strength

Account for the trend in acidity $[Fe(OH_2)_6]^{2+} < [Fe(OH_2)_6]^{3+} < [Al(OH_2)_6]^{3+} \approx [Hg(OH_2)_n]^{2+}$.

Answer The weakest acid is the Fe^{2+} complex on account of its relatively large radius and low charge. The increase of charge to $+3$ increases the acid strength. The greater acidity of Al^{3+} can be explained by its smaller radius. The anomalous ion in the series is the Hg^{2+} complex. This complex reflects the failure of an ionic model, for in this complex there is a large transfer of positive charge to oxygen as a result of covalent bonding.

Self-test 5.2 Arrange the following ions in order of increasing acidity: $[Na(OH_2)_n]^+$, $[Sc(OH_2)_6]^{3+}$, $[Mn(OH_2)_6]^{2+}$, $[Ni(OH_2)_6]^{2+}$.

5.4 Simple oxoacids

The simplest oxoacids are the **mononuclear acids**, which contain one atom of the parent element. They include H_2CO_3, HNO_3, H_3PO_4, and H_2SO_4. These oxoacids are formed by the electronegative elements at the upper right of the periodic table and by other elements in high oxidation states (Table 5.2). One interesting feature in the table is the occurrence of planar $B(OH)_3$, H_2CO_3, and HNO_3 molecules but not their analogs in later periods. As we saw in Chapter 3, π bonding is more important among the Period 2 elements, so their atoms are more likely to be constrained to lie in a plane.

(a) Substituted oxoacids

One or more $-OH$ groups of an oxoacid may be replaced by other groups to give a series of substituted oxoacids, which include fluorosulfuric acid, $O_2SF(OH)$, and aminosulfuric acid, $O_2S(NH_2)OH$ (**6**). Because fluorine is highly electronegative, it withdraws electrons from the central S atom and confers on S a higher effective positive charge. As a result, the substituted acid is a stronger acid than $O_2S(OH)_2$. Another electron acceptor substituent is CF_3, as in the

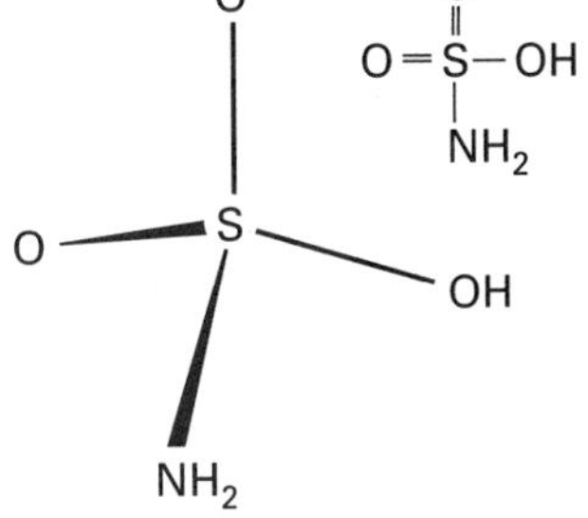

6 $O_2S(NH_2)OH$

Table 5.2 The structure and pK_a values of oxoacids*

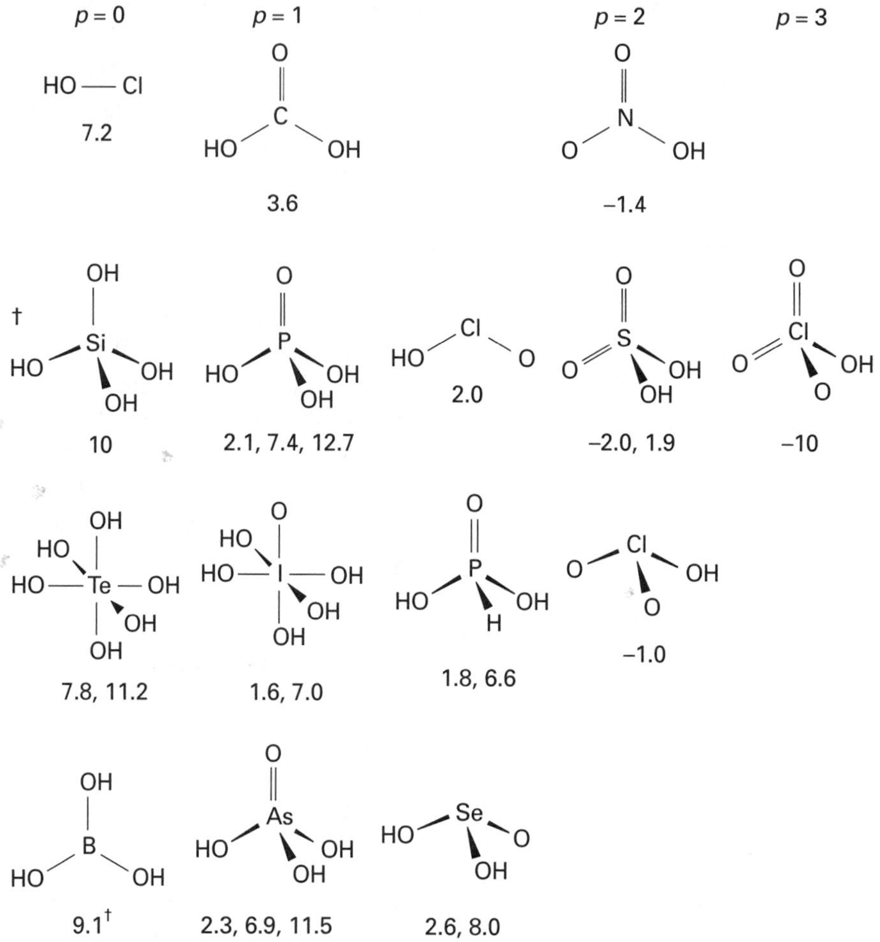

$p = 0$	$p = 1$		$p = 2$	$p = 3$

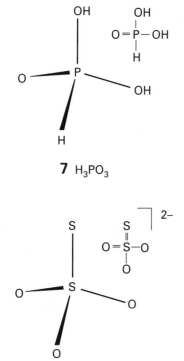

†

strong acid trifluoromethylsulfonic acid, CF_3SO_3H (that is, $O_2S(CF_3)(OH)$). In contrast, the NH_2 group, which has lone pair electrons, can donate electron density to S by π bonding. This transfer of charge reduces the positive charge of the central atom and weakens the acid.

A trap for the unwary is that not all oxoacids follow the familiar structural pattern of a central atom surrounded by OH and O groups. Occasionally an H atom is attached directly to the central atom, as in phosphorous acid, H_3PO_3. Phosphorous acid is in fact only a diprotic acid, for the substitution results in a P—H bond (**7**) and consequently a nonacidic proton. This structure is consistent with NMR and vibrational spectra, and the structural formula is $OPH(OH)_2$. Substitution for an oxo group (as distinct from a hydroxyl group) is another example of a structural change that can occur. An important example is the thiosulfate ion, $S_2O_3^{2-}$ (**8**), in which an S atom replaces an O atom of a sulfate ion.

> Substituted oxoacids have strengths that may be rationalized in terms of the electron withdrawing power of the substitutent; in a few cases, a nonacidic H atom is attached directly to the central atom of an oxoacid.

(b) Pauling's rules

The observed strengths of mononuclear oxoacids can be systematized by using two empirical rules devised by Linus Pauling:

footnotes

* p is the number of non-protonated O atoms.

† See Table 5.1

7 H_3PO_3

8 $S_2O_3^{2-}$

1 For the oxoacid $O_pE(OH)_q$, $pK_a \approx 8 - 5p$.

2 The successive pK_a values of polyprotic acids (those with $q > 1$), increase by 5 units for each successive proton transfer.

Neutral hydroxoacids with $p = 0$ have $pK_a \approx 8$, acids with one oxo group have $pK_a \approx 3$, and acids with two oxo groups have $pK_a \approx -2$. For example, sulfuric acid, $O_2S(OH)_2$, has $p = 2$ and $q = 2$, and $pK_{a1} \approx -2$ (signifying a strong acid). Similarly, pK_{a2} is predicted to be $+3$, which is reasonably close to the experimental value of 1.9.

The success of these simple rules may be gauged by inspection of Table 5.2, in which acids are grouped according to p, the number of oxo groups. That the estimates are good to about ± 1 is pleasantly surprising. The variation in strengths down a group is not large, and the complicated and perhaps canceling effects of changing structures allow the rules to work moderately well. The more important variation across the periodic table from left to right and the effect of change of oxidation number are taken into account by the number of oxo groups characteristic of the molecular acids. In Group 15/V, the oxidation number $+5$ requires one oxo group (as in $OP(OH)_3$) whereas in Group 16/VI the oxidation number $+6$ requires two (as in $O_2S(OH)_2$).

> The strengths of a series of oxoacids containing a specific central atom with a variable number of oxo and hydroxyl groups are summarized by the Pauling rules as set out above.

(c) Structural anomalies

An interesting use of the Pauling rules is to detect structural anomalies. For example, carbonic acid, $OC(OH)_2$, is commonly reported as having $pK_{a1} = 6.4$, but the rules predict $pK_{a1} = 3$. The anomalously low acidity indicated by the experimental value is the result of treating the concentration of dissolved CO_2 as if it were all H_2CO_3. However, in the equilibrium

$$CO_2(aq) + H_2O(l) \rightleftharpoons OC(OH)_2(aq)$$

only about 1 per cent of the dissolved CO_2 is present as $OC(OH)_2$, so the actual concentration of acid is much less than the concentration of dissolved CO_2. When this difference is taken into account, the true pK_{a1} of H_2CO_3 is about 3.6, as the Pauling rules predict.

The experimental value $pK_{a1} = 1.8$ reported for sulfurous acid, H_2SO_3, suggests another anomaly, this time acting in the opposite direction. In fact, spectroscopic studies have failed to detect the molecule $OS(OH)_2$ in solution, and the equilibrium constant for

$$SO_2(aq) + H_2O(l) \rightleftharpoons H_2SO_3(aq)$$

is less than 10^{-9}. The equilibria of dissolved SO_2 are complex, and a simple analysis is inappropriate. The ions that have been detected include HSO_3^- and $S_2O_5^{2-}$, and there is evidence for an S—H bond in the solid salts of the hydrogensulfite ion.[5]

The pK_a values of aqueous CO_2 and SO_2 call attention to the important point that not all nonmetal oxides react fully with water to form acids. Carbon monoxide is another example: although it is formally the anhydride of formic acid, $HCOOH$, carbon monoxide does not in fact react with water at room temperature to give the acid. The same is true of some metal oxides: OsO_4, for example, can exist as dissolved neutral molecules.

> In certain cases, notably H_2CO_3 and H_2SO_3, a simple molecular formula misrepresents the composition of aqueous solutions of nonmetal oxides.

Example 5.3 Using Pauling's rules

Identify the structural formulas that are consistent with the following pK_a values: H_3PO_4, 2.12; H_3PO_3, 1.80; H_3PO_2, 2.0.

---

[5] Solution spectroscopic studies of the sulfite problem have been carried out by R.E. Connick and his collaborators. See *Inorg. Chem.* **21**, 103 (1982) and **25**, 2414 (1986).

Answer All three values are in the range that Pauling's first rule associates with one oxo group. This observation suggests the formulas $(HO)_3P{=}O$, $(HO)_2HP{=}O$, and $(HO)H_2P{=}O$. The second and the third formulas are derived from the first by replacement of $-OH$ by $-H$ bound to P (as in structure **7**).

Self-test 5.3 Predict the pK_a values of (a) H_3PO_4, (b) $H_2PO_4^-$, (c) HPO_4^{2-}. Experimental values are given in Table 5.2.

5.5 Anhydrous oxides

We have treated oxoacids as derived by deprotonation of their parent aqua acids. It is also useful to take the opposite viewpoint and to consider aqua acids and oxoacids as derived by hydration of the oxides of the central atom. This approach emphasizes the acid and base properties of oxides and their correlation with the location of the element in the periodic table.

(a) Acidic and basic oxides

An **acidic oxide** is one which, on dissolution in water, binds an H_2O molecule and releases a proton to the surrounding solvent:

$$CO_2(g) + H_2O(l) \longrightarrow [OC(OH)_2](aq)$$
$$[OC(OH)_2](aq) + H_2O(l) \rightleftharpoons [O_2C(OH)]^-(aq) + H_3O^+(aq)$$

An equivalent interpretation is that an acidic oxide is an oxide that reacts with an aqueous base (an alkali):

$$CO_2(g) + OH^-(aq) \longrightarrow [O_2C(OH)]^-(aq)$$

A **basic oxide** is an oxide to which a proton is transferred when it dissolves in water:

$$CaO(s) + H_2O(l) \longrightarrow Ca^{2+}(aq) + 2\,OH^-(aq)$$

The equivalent interpretation in this case is that a basic oxide is an oxide that reacts with an acid:

$$CaO(s) + 2\,H_3O^+(aq) \longrightarrow Ca^{2+}(aq) + 3\,H_2O(l)$$

Because acidic and basic oxide character often correlates with other chemical properties, a wide range of properties can in fact be predicted from a knowledge of the character of oxides. In a number of cases the correlations follow from the basic oxides being largely ionic and of acidic oxides being largely covalent. For instance, an element that forms an acidic oxide is likely to form volatile, covalent halides. In contrast, an element that forms a basic oxide is likely to form solid, ionic halides. In short, the acidic or basic character of an oxide is a chemical indication of whether an element should be regarded as a metal or a nonmetal.

> *Metallic elements typically form basic oxides; nonmetallic elements typically form acidic oxides.*

(b) Amphoterism

An **amphoteric oxide** is an oxide that reacts with both acids and bases.[6] Thus, aluminum oxide reacts with acids and alkalis:

$$Al_2O_3(s) + 6\,H_3O^+(aq) + 3\,H_2O(l) \longrightarrow 2\,[Al(OH_2)_6]^{3+}(aq)$$
$$Al_2O_3(s) + 2\,OH^-(aq) + 3\,H_2O(l) \longrightarrow 2\,[Al(OH)_4]^-(aq)$$

---

[6] The word 'amphoteric' is derived from the Greek word for 'both'.

Amphoterism is observed for the lighter elements of Groups 2 and 13/III, as in BeO, Al_2O_3, and Ga_2O_3. It is also observed for some of the d-block elements in high oxidation states, such as TiO_2 and V_2O_5, and some of the heavier elements of Groups 14/IV and 15/V, such as SnO_2 and Sb_2O_5. Figure 5.4 shows the location of elements that in their characteristic group oxidation states have amphoteric oxides. They lie on the frontier between acidic and basic oxides, and hence serve as an important guide to the metallic or nonmetallic character of an element. The onset of amphoterism correlates with a significant degree of covalent character in the bonds formed by the elements, either because the metal ion is strongly polarizing (as for Be) or because the metal is polarized by the oxygen atom attached to it (as for Sb).

An important issue in the d-block is the oxidation number necessary for amphoterism. Figure 5.5 shows the oxidation number for which an element in the first row of the block has an amphoteric oxide. We see that on the left of the block, from titanium to manganese and perhaps iron, oxidation state $+4$ is amphoteric (with higher values on the border of acidic and lower values of the border of basic). On the right of the block, amphoterism occurs at lower oxidation numbers: the oxidation states $+3$ for cobalt and nickel and $+2$ for copper and zinc are fully amphoteric. There is no simple way of predicting the onset of amphoterism. However, it presumably reflects the ability of the metal cation to polarize the oxide ions that surround it—that is, to introduce covalence into the metal–oxygen bond. The degree of covalence typically increases with the oxidation state of the metal.

> The frontier between metals and nonmetals in the periodic table is characterized by the formation of amphoteric oxides; amphoterism also varies with the oxidation state of the element.

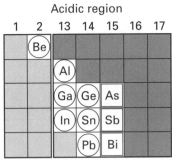

5.4 The elements in circles have amphoteric oxides even in their highest oxidation states. The elements in boxes have acidic oxides in their maximum oxidation states and amphoteric oxides in lower oxidation states.

Example 5.4 Using oxide acidity in qualitative analysis

In the traditional scheme of qualitative analysis, a solution of metal ions is oxidized and then aqueous ammonia is added to raise the pH. The ions Fe^{3+}, Ce^{3+}, Al^{3+}, Cr^{3+}, and V^{3+} precipitate as hydrous oxides. The addition of H_2O_2 and NaOH redissolves the aluminum, chromium, and vanadium oxides. Discuss these steps in terms of the acidities of oxides.

Answer When the oxidation number is $+3$, all the metal oxides are sufficiently basic to be insoluble in a $pH \approx 10$ solution. Aluminum(III) is amphoteric and redissolves in strong base to give aluminate ions, $[Al(OH)_4]^-$. Vanadium(III) and chromium(III) are oxidized by H_2O_2 to give vanadate ions, $[VO_4]^{3-}$, and chromate ions, $[CrO_4]^{2-}$, which are the anions derived from the acidic oxides V_2O_5 and CrO_3, respectively.

Self-test 5.4 If titanium(IV) ions were present in the sample, how would they behave?

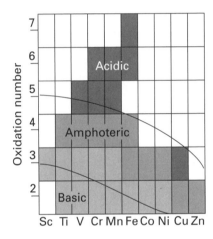

5.5 The influence of oxidation number on the acid–base character of oxides of the elements in the first row of the d-block. Mainly acidic oxidation states are shown darkly shaded. Mainly basic oxidation states are more lightly shaded. Oxidation states in the region bounded by the two curved lines are amphoteric.

5.6 Polyoxo compound formation

One of the most important aspects of the reactivity of acids containing the O–H group is the formation of condensation polymers. Polycation formation from simple aqua cations occurs with the loss of H_3O^+ ions:

$$2\,[Al(OH_2)_6]^{3+}(aq) \longrightarrow [(H_2O)_5Al(OH)Al(OH_2)_5]^{5+}(aq) + H_3O^+(aq)$$

Polyanion formation from oxoanions occurs by protonation of an O atom and its departure as H_2O:

$$2\,[CrO_4]^{2-}(aq) + 2\,H_3O^+(aq) \longrightarrow [O_3CrOCrO_3]^{2-}(aq) + 3\,H_2O(l)$$

The importance of polyoxo anions can be judged by the fact that they account for most of the mass of oxygen in the Earth's crust, for they include almost all silicate minerals. They also

5.6 The structure of the $[AlO_4(Al(OH)_2)_{12}]^{7+}$ ion, with AlO_6 groups represented by octahedra around the central tetrahedron which represents the AlO_4 unit.

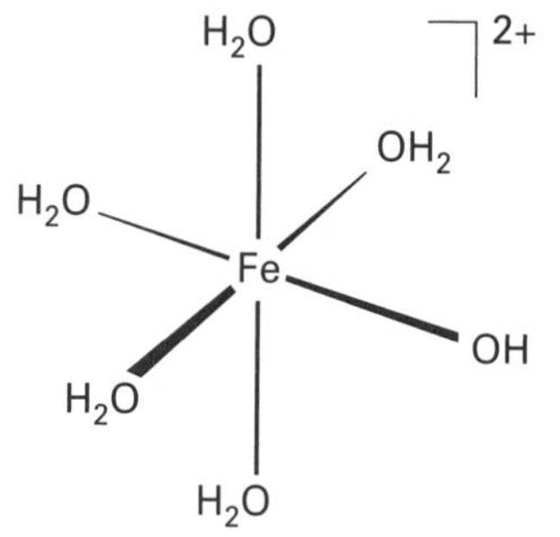

include the phosphate polymers (such as ATP) used for energy storage in living cells. The silicates are so important that they are treated separately (Chapter 10).

> Acids containing the OH group may form condensation polymers; polycation formation from simple aqua cations occurs with the loss of H_3O^+.

(a) Polymerization of aqua ions to polycations

As the pH of a solution is increased, the aqua ions of metals that have basic or amphoteric oxides generally undergo polymerization and precipitation. One application is to the separation of metal ions, because the precipitation occurs quantitatively at a pH characteristic of each metal.

With the exception of Be^{2+} (which is amphoteric), the elements of Groups 1 and 2 have no important solution species beyond the aqua ions $M^+(aq)$ and $M^{2+}(aq)$. In contrast, the solution chemistry of the elements becomes very rich as the amphoteric region of the periodic table is approached. The two most common examples are polymers formed by Fe(III) and Al(III), both of which are abundant in the Earth's crust. In acidic solutions, both form octahedral hexaaquaions, $[Al(OH_2)_6]^{3+}$ and $[Fe(OH_2)_6]^{3+}$. In solutions of $pH > 4$, both precipitate as gelatinous hydrous oxides:

$$[Fe(OH_2)_6]^{3+}(aq) + (3+n)\,H_2O(l) \longrightarrow Fe(OH)_3 \cdot nH_2O(s) + 3\,H_3O^+(aq)$$

$$[Al(OH_2)_6]^{3+}(aq) + (3+n)\,H_2O(l) \longrightarrow Al(OH)_3 \cdot nH_2O(s) + 3\,H_3O^+(aq)$$

The precipitated polymers, which are often of colloidal dimensions, slowly crystallize to stable mineral forms.

Aluminum and iron behave differently at intermediate pH, in the region between the existence of aqua ions and the occurrence of precipitation. Relatively few iron species have been characterized, but known species include the two monomers (**9**) and (**10**), a dimer (**11**), and a polymer containing about 90 Fe atoms. In contrast, Al(III) forms a series of discrete polymeric cations in which the monomer unit consists of a central Al^{3+} ion surrounded tetrahedrally by four O atoms (**12**). A 'simple' polymeric cation of this kind is $[AlO_4(Al(OH)_2)_{12}]^{7+}$, with an average charge of $+0.54$ per Al atom. An impression of the structure is given in Fig. 5.6, where the AlO_6 units are represented as octahedral blocks packed around the central tetrahedron; the larger Fe^{3+} ion does not fit so well in such a structure. This particular Al^{3+} polycation with 13 Al atoms has been shown to inhibit plant growth at one-tenth the concentration of $Al^{3+}(aq)$. It is now thought that polycations may

9 $[Fe(OH_2)_5OH]^{2+}$

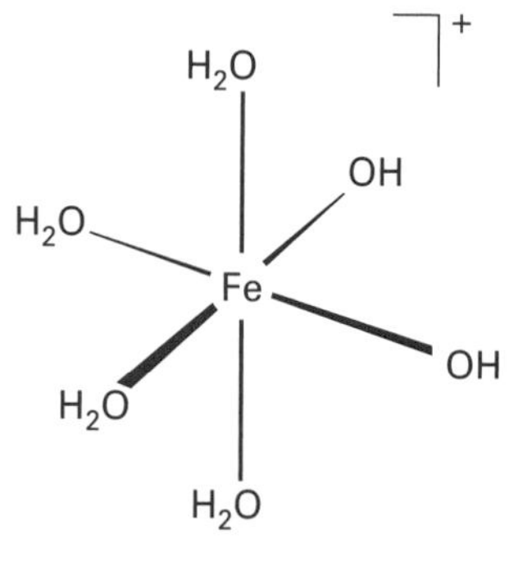

10 $[Fe(OH_2)_4(OH)_2]^+$

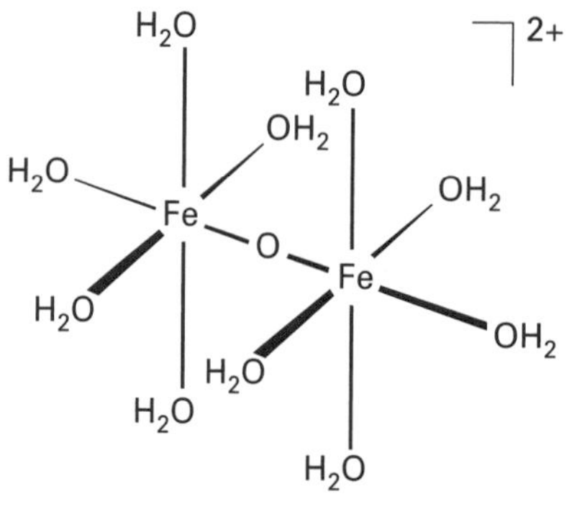

11 $[Fe_2O(OH_2)_{10}]^{2+}$

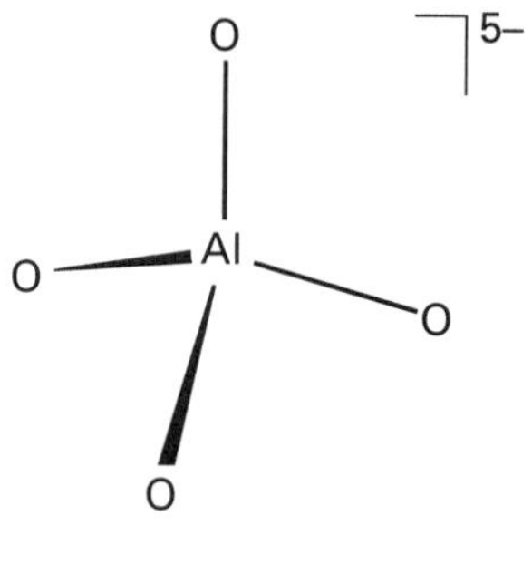

12 $[AlO_4]^{5-}$

be the most important toxic species leached by acid rain into lakes and soils. Studies using [7] ^{27}Al-NMR show that the 13-Al polycation is a prominent species of aluminum in acidic organic soils.[7]

The extensive network structure of aluminum polymers, which are neatly packed in three dimensions, contrasts with the linear polymers of their iron analogs. Aluminum polycations and similar ions are used constructively in water treatment to precipitate anions (such as F^-) that are present as pollutants in effluents from aluminum refining plants.

As the pH is increased, H^+ ions are plucked off these polycations and the net charge on the cations is thereby reduced. The **pH at which the net charge is zero is known as the point of zero charge.** Because both Fe(III) and Al(III) form amphoteric oxides, increasing the pH to a sufficiently high value can lead to the redissolution of their oxides as anions (Fig. 5.7).

> *Aqua ions of metals that have basic oxides generally undergo polymerization and precipitation as the pH of their solution is increased. For amphoteric ions, the precipitate redissolves at high pH.*

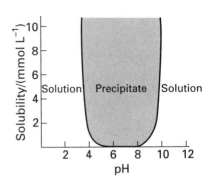

5.7 The variation of the solubility of Al_2O_3 with pH expressed as total concentration of Al. In the extreme acid region the aluminum is present as $[Al(OH_2)_6]^{3+}$. In the extreme basic region it is present as $[Al(OH)_4]^-$.

(b) Polyoxoanions

Polyoxoanions are formed when base is added to aqueous solutions of the early *d*-block ions or oxides in high oxidation states. Such polymerization is important for V(V), Mo(VI), W(VI), and (to a lesser extent) Nb(V), Ta(V), and Cr(VI); see Section 9.7.

A solution formed by dissolving the amphoteric oxide V_2O_5 in a strongly basic solution is colorless, and the dominant species is the tetrahedral $[VO_4]^{3-}$ ion (the analog of the colorless PO_4^{3-} ion). As the pH is increased by the addition of more base, the solution passes through a series of deeper colors from orange to red. This sequence indicates a complicated series of condensations and hydrolyses that yield ions including $[V_2O_7]^{4-}$, $[V_3O_9]^{3-}$, $[V_4O_{12}]^{4-}$, $[HV_{10}O_{28}]^{5-}$, and $[H_2V_{10}O_{28}]^{4-}$ (see Section 9.7). The successive increase in average anionic charge per V atom as the polyanions grow should be noted. The strongly acid solution is pale yellow and contains the hydrated $[VO_2]^+$ ion (**13**). Polyoxoanions are also formed by some nonmetals, but their structures are different from those of their *d*-metal analogs. The common species in solution are rings and chains. As already remarked, the silicates are very important examples of polymeric oxoanions, and we discuss them in detail in Chapter 10. One example of a polysilicate mineral is $MgSiO_3$, which contains an infinite chain of SiO_3^{2-} units. In this section we illustrate some features of polyoxoanions using phosphates as examples.

The simplest condensation reaction, starting with the orthophosphate ion, PO_4^{3-}, is

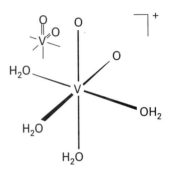

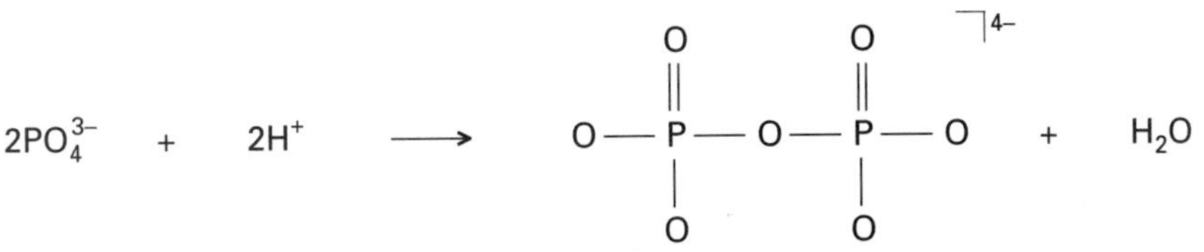

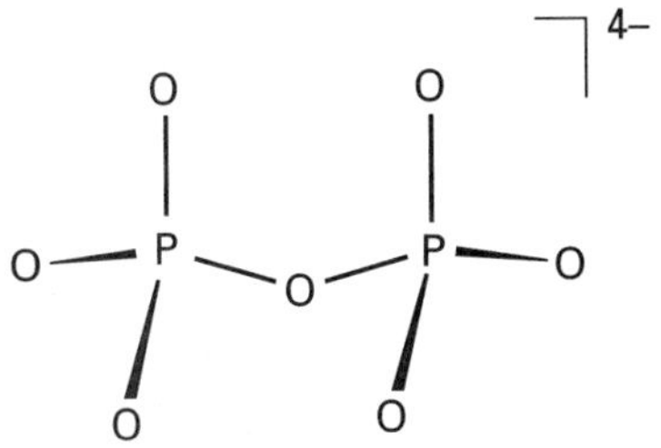

13 $[VO_2(OH_2)_4]^+$

$$2PO_4^{3-} \quad + \quad 2H^+ \quad \longrightarrow \quad \begin{bmatrix} O & & O \\ \parallel & & \parallel \\ O-P-O-P-O \\ | & & | \\ O & & O \end{bmatrix}^{4-} \quad + \quad H_2O$$

14 $[P_2O_7]^{4-}$

The elimination of water consumes protons and reduces the average charge number of each P atom to -2. The diphosphate ion, $[P_2O_7]^{4-}$ (**14**), can be drawn as (**15**) if each phosphate group is represented as a tetrahedron with the O atoms located at the corners. Phosphoric acid can be prepared by hydrolysis of the solid phosphorus(V) oxide, P_4O_{10}. An initial step

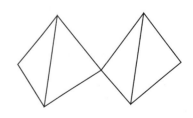

15 $[P_2O_7]^{4-}$

[7] D. Hunter and D.S. Ross, *Science* **251**, 1056 (1991).

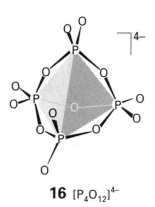

16 $[P_4O_{12}]^{4-}$

using a limited amount of water produces a metaphosphate ion with the formula $[P_4O_{12}]^{4-}$ (**16**). This reaction is only the simplest among many, and the separation of products from the hydrolysis of phosphorus(V) oxide by chromatography reveals the presence of chain species with from one to nine P atoms. Higher polymers are also present and can be removed from the column only by hydrolysis. A diagrammatic representation of a two-dimensional paper chromatogram is shown in Fig. 5.8. The upper spot sequence corresponds to linear polymers and the lower sequence corresponds to rings. Chain polymers of formula P_n with $n = 10$ to 50 can be isolated as mixed amorphous glasses analogous to those formed by silicates and borates.

The biological importance of polyphosphates was mentioned at the outset. At physiological pH values (close to 7.4), the P—O—P bond is unstable with respect to hydrolysis. Consequently, its hydrolysis can serve as a mechanism for the delivery of Gibbs energy. Similarly, the formation of the P—O—P bond is a means of storing Gibbs energy. The key to energy exchange in metabolism is the hydrolysis of adenosine triphosphate, ATP (**17**), to adenosine diphosphate, ADP (**18**):

$$ATP^{4-} + 2\,H_2O \longrightarrow ADP^{3-} + HPO_4^{2-} + H_3O^+$$

for which $\Delta_r G^{\ominus} = -41\ kJ\,mol^{-1}$ at pH $= 7.4$. Energy storage in metabolism depends on the subtle construction of pathways to make ATP from ADP. The energy is utilized metabolically by pathways that have evolved to exploit the delivery of a thermodynamic driving force resulting from the hydrolysis of ATP.

Polyoxoanions are formed when base is added to the oxoacids of the early d-block elements in high oxidation states and the oxoacids of some nonmetals; the latter commonly form rings and chains.

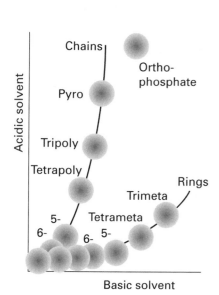

5.8 Two-dimensional paper chromatogram of a complex mixture of phosphates formed by condensation reactions. The sample spot was placed at the lower left; basic solvent separation was used first, followed by acidic solvent perpendicular to the first elution. The latter treatment separates open chain from cyclic phosphates.

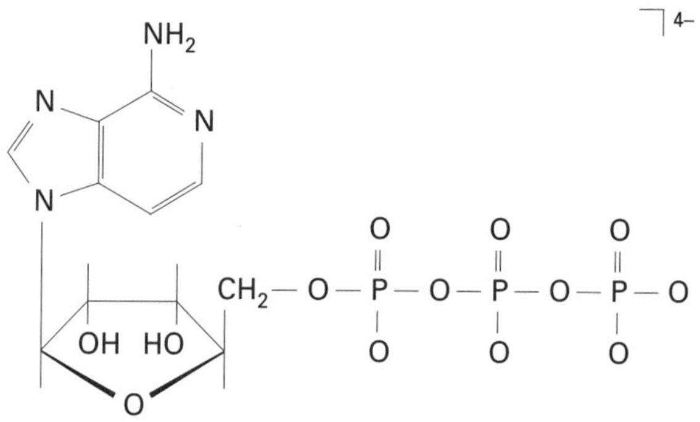

17 ATP^{4-}

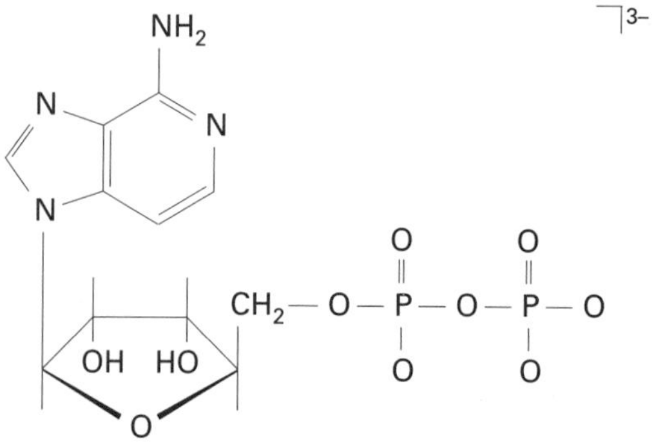

18 ADP^{3-}

Lewis acidity

The Brønsted–Lowry theory of acids and bases focuses on the transfer of a proton between species. While this concept is more general than any that preceded it, it still fails to take into account reactions between substances that show similar features but in which no protons are transferred. This deficiency was remedied by a more general theory of acidity introduced by G.N. Lewis in the same year as Brønsted and Lowry introduced theirs (1923). Lewis's approach became influential only in the 1930s.

A **Lewis acid** is a substance that acts as an electron pair acceptor. A **Lewis base** is a substance that acts as an electron pair donor. We denote a Lewis acid by A and a Lewis base by :B, often omitting any other lone pairs that may be present. The fundamental reaction of Lewis acids and bases is the formation of a **complex** (or adduct), A—B, in which A and :B bond together by sharing the electron pair supplied by the base.[8]

5.7 Examples of Lewis acids and bases

A proton is a Lewis acid because it can attach to an electron pair, as in the formation of NH_4^+ from NH_3. It follows that any Brønsted acid, since it provides protons, exhibits Lewis acidity too.[9] All Brønsted bases are Lewis bases, because a proton acceptor is also an electron pair donor: an NH_3 molecule, for instance, is a Lewis base as well as a Brønsted base. Therefore, the whole of the material presented in the preceding sections of this chapter can be regarded as a special case of Lewis's approach. However, because the proton is not essential to the definition of a Lewis acid or base, a wider range of substances can be classified as acids and bases in the Lewis scheme than can be classified in the Brønsted scheme.

We meet many examples of Lewis acids later, but we should be alert for the following possibilities:

1 A metal cation can bond to an electron pair supplied by the base in a coordination compound.

This aspect of Lewis acids and bases will be treated at length in Chapter 7. An example is the hydration of Co^{2+}, in which the lone pairs of H_2O (acting as a Lewis base) donate to the central cation to give $[Co(OH_2)_6]^{2+}$. The cation is therefore the Lewis acid. We also need to be alert for complex formation in which a cation, the acid, interacts with the π electrons of a base, as in the formation of the complex of Ag^+ and benzene (**19**).

2 A molecule with an incomplete octet can complete its octet by accepting an electron pair.

A prime example is $B(CH_3)_3$, which can accept the lone pair of NH_3 and other donors:

19 $[C_6H_6Ag]^+$

---

[8] The terms Lewis acid and base are used in discussions of the equilibrium properties of reactions. In the context of reaction rates (Section 14.1), an electron pair donor is called a *nucleophile* and an electron acceptor is called an *electrophile*.

[9] Note that the *Brønsted* acid HA is the complex formed by the *Lewis* acid H^+ with the Lewis base A^-. We say that a Brønsted acid 'exhibits' Lewis acidity rather than saying that a Brønsted acid *is* a Lewis acid.

Hence, $B(CH_3)_3$ is a Lewis acid.

3   A molecule or ion with a complete octet may be able to rearrange its valence electrons and accept an additional electron pair.

For example, CO_2 acts as a Lewis acid when it forms HCO_3^- by accepting an electron pair from an O atom in an OH^- ion:

$$
\begin{array}{c}
\overset{\displaystyle :O:}{\underset{\displaystyle :O:}{\overset{\|}{\underset{\|}{C}}}} \quad + \quad :\ddot{O}H^- \quad \longrightarrow \quad \left[\overset{\displaystyle :O:}{\underset{\displaystyle :\ddot{O}:}{\overset{\|}{\underset{|}{C}} - \ddot{O}H}} \right]^-
\end{array}
$$

4   A molecule or ion may be able to expand its valence shell (or simply be large enough) to accept another electron pair.

An example is the formation of the complex $[SiF_6]^{2-}$ when two F^- ions (the Lewis bases) bond to SiF_4 (the acid).

$$
\begin{array}{c}
\underset{F}{\overset{F}{\underset{\diagup}{\overset{|}{Si}}}}\diagdown_F \quad + \quad 2(:F^-) \quad \longrightarrow \quad \left[\underset{F}{\overset{F}{\underset{F\diagup}{\overset{F\diagdown|\diagup F}{S}}}}\diagdown_F \right]^{2-}
\end{array}
$$

This type of Lewis acidity is common for the halides of the heavier *p*-block elements, such as SiX_4, AsX_3, and PX_5 (with X a halogen).

5   A closed-shell molecule may be able to use one of unoccupied antibonding molecular orbitals to accommodate an incoming electron pair.

An example of this behavior is the ability of tetracyanoethene (TCNE, (**20**)) to accept a lone pair into its antibonding π^* orbital, and hence to act as an acid.

> A Lewis acid is an electron pair acceptor; a Lewis base is an electron pair donor.

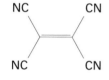

20 TCNE

....

Example 5.5 Identifying Lewis acids and bases

Identify the Lewis acids and bases in the reactions (a) $BrF_3 + F^- \rightarrow [BrF_4]^-$, (b) $(CH_3)_2CO + I_2 \rightarrow (CH_3)_2CO{-}I_2$, (c) $KH + H_2O \rightarrow KOH + H_2$.

Answer (a) The acid BrF_3 adds the base $:F^-$. (b) Acetone (propanone) acts as a base and donates a lone pair of electrons from O into an empty antibonding orbital of the I_2 molecule, which therefore acts as an acid. The I_2 molecule could also donate into the π^* orbital of the CO group, in which case it would be acting as an acid. (c) The ionic hydride complex KH provides the base H^- to displace the acid H^+ from water to give H_2 together with KOH, in which the base OH^- is combined with the very weak acid K^+.

....

Self-test 5.5
Identify the acids and bases in the reactions (a) $FeCl_3 + Cl^- \rightarrow [FeCl_4]^-$, (b) $I^- + I_2 \rightarrow I_3^-$, (c) $[:SnCl_3]^- + (CO)_5MnCl \rightarrow (CO)_5Mn{-}SnCl_3 + Cl^-$.

5.8 Boron and carbon group acids

The planar molecules BX_3 and AlX_3 have incomplete octets, and the vacant p orbital perpendicular to the plane (21) can accept a lone pair from a Lewis base:

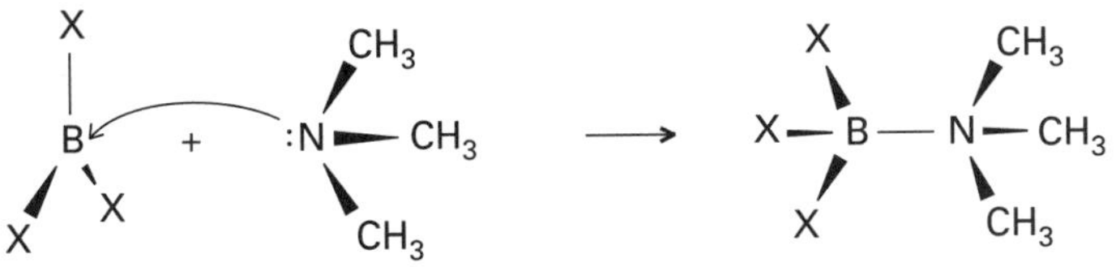

21 AlX_3 and BX_3

The acid molecule becomes pyramidal as the complex is formed and the B—X bonds bend away from their new neighbors.

(a) Boron halides

The order of thermodynamic stability of complexes of $:N(CH_3)_3$ with BX_3 is $BF_3 < BCl_3 < BBr_3$. This order is opposite to that expected on the basis of the relative electronegativities of the halogens: an electronegativity argument would suggest that fluorine, the most electronegative halogen, ought to leave the B atom in BF_3 most electron deficient and hence able to form the strongest bond to the incoming base. The currently accepted explanation is that the halogen atoms in the BX_3 molecule can form π bonds with the empty $B2p$ orbital (22), and that these π bonds must be disrupted to make the acceptor orbital available for complex formation. The small F atom forms the strongest π bonds with the $B2p$ orbital: recall that p–p π bonding is strongest for Period 2 elements, largely on account of the small atomic radii of these elements and the significant overlap of their compact $2p$ orbitals. Thus, the BF_3 molecule has the strongest π bond to be broken when the amine forms an N—B bond.

Boron trifluoride is widely used as an industrial catalyst. Its role there is to extract bases bound to carbon and hence to generate carbocations:

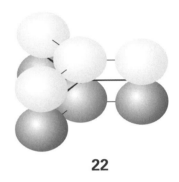

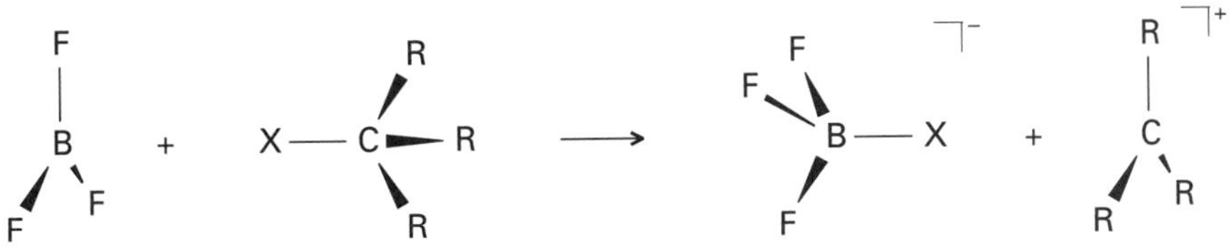

22

Boron trifluoride is a gas, but it dissolves in diethyl ether to give a solution that is convenient to use. This dissolution is also an aspect of Lewis acid character for, as BF_3 dissolves, it forms a complex with the :O atom of a solvent molecule.

> The ability of boron trihalides to act as Lewis acids generally increases in the order $BF_3 < BCl_3 < BBr_3$.

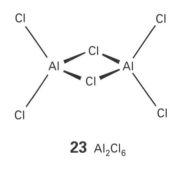

23 Al_2Cl_6

(b) Aluminum halides

Aluminum halides are dimers in the gas phase; aluminum chloride, for example, has molecular formula Al_2Cl_6 in the vapor (23). Each Al atom acts as an acid toward a Cl atom initially belonging to the other Al atom.

Aluminum chloride is widely used as a Lewis acid catalyst for organic reactions. The classic examples are Friedel–Crafts alkylation (the attachment of R^+ to an aromatic ring) and acylation (the attachment of RCO^+). The catalytic cycle is displayed in Fig. 5.9.

> Aluminum halides are dimeric in the gas phase; they are used as catalysts in solution.

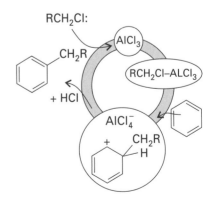

5.9 The catalytic cycle for the action of aluminum chloride in a Friedel–Crafts alkylation reaction.

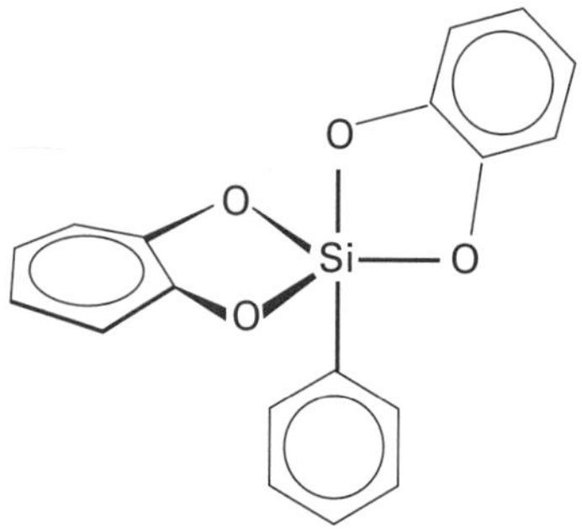

(c) Silicon and tin complexes

Unlike carbon, an Si atom can expand its valence shell (or is simply large enough) to become hypervalent:

Germanium can react similarly. Because the Lewis base F^-, aided by a proton, can displace O^{2-} from Si, hydrofluoric acid is corrosive toward glass (SiO_2). The trend in acidity for SiX_4, which follows the order $SiI_4 < SiBr_4 < SiCl_4 < SiF_4$, and correlates with the increase in the electron-withdrawing power of the halogen from I to F and is the reverse of that for BX_3.

Coordination number 6, as in $[SiF_6]^-$, is not the only state of coordination for silicon above 4. For example, a five-coordinate trigonal bipyramidal structure is possible (**24**).

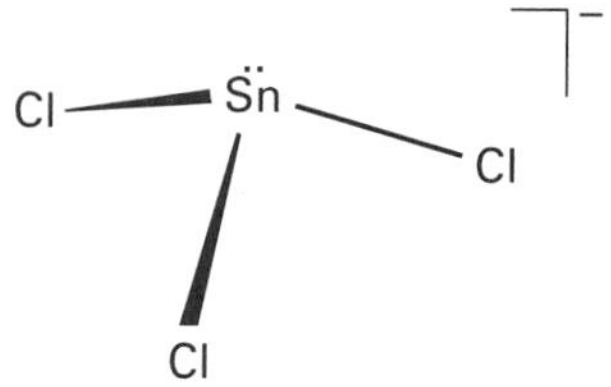

Tin(II) chloride is both a Lewis acid and a Lewis base. As an acid, $SnCl_2$ combines with Cl^- to form the complex $[SnCl_3]^-$ (**25**). The complex retains a lone pair, and it is sometimes more revealing to write its formula as $:SnCl_3^-$. It acts as a base to give metal–metal bonds, as in the complex $(CO)_5Mn–SnCl_3$ (**26**). Compounds containing metal–metal bonds are currently the focus of much attention in inorganic chemistry, as we see later in the text (Chapter 15).

> Silicon and germanium halides act as Lewis acids by becoming five- or six-coordinate; tin(II) chloride is both a Lewis acid and a Lewis base.

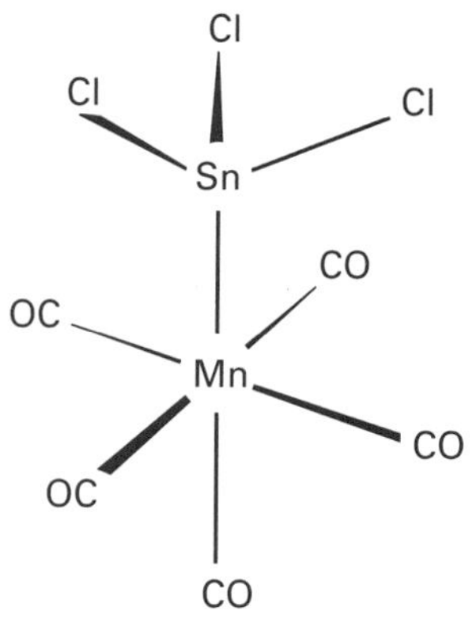

....

Example 5.6 Predicting the relative basicity of compounds

Rationalize the relative Lewis basicities $(H_3Si)_2O < (H_3C)_2O$; (b) $(H_3Si)_3N < (H_3C)_3N$.

Answer Nonmetallic elements in Period 3 and below can expand their valence shells by delocalization of the O or N lone pairs (**27**), so the silyl ether and silyl amine are the weaker Lewis bases in each pair.

....

Self-test 5.6 Given that π bonding between Si and the lone pairs of N is important, what difference in structure between $(H_3Si)_3N$ and $(H_3C)_3N$ do you expect?

....

5.9 Nitrogen and oxygen group acids

The heavier elements of the nitrogen group (Group 15/V) form some of the most important Lewis acids, SbF_5 being one of the most widely studied compounds. This Lewis acid can be used to produce some of the strongest Brønsted acids, as in the reaction

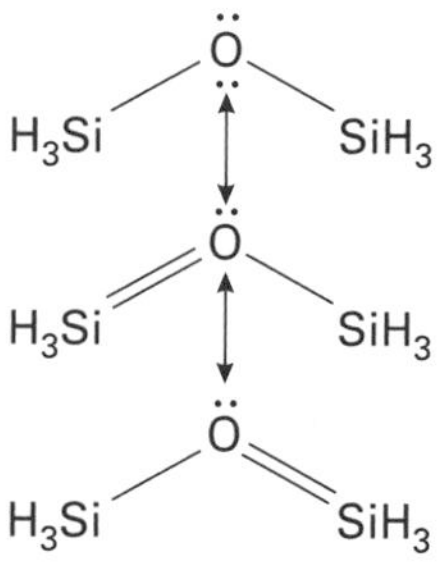

24 $[Si(C_6H_5)(C_6H_4O_2)_2]^-$

25 $[SnCl_3]^-$

26 $(CO)_5MnSnCl_3$

27

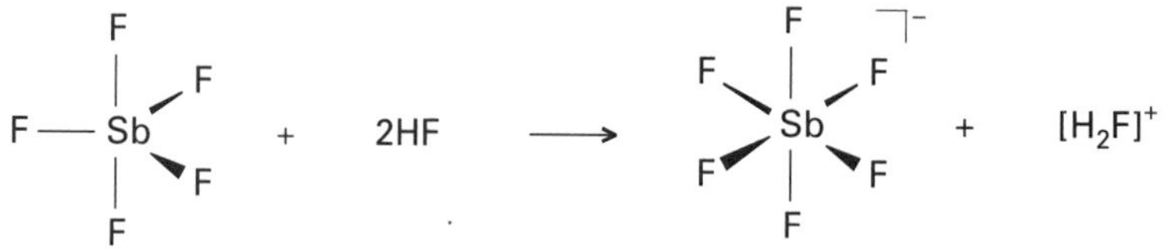

A **superacid** is a mixture that can protonate almost any organic compound. One can be produced by dissolving SbF_5 in a mixture of HSO_3F and SO_3. The simplest of the many reactions occurring in this mixture is

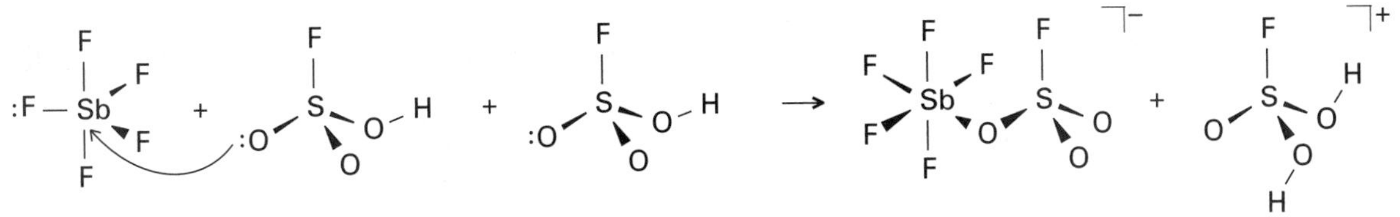

The doubly protonated fluorosulfate acts as a powerful Brønsted acid.

Sulfur dioxide is both a Lewis acid and a Lewis base. Its Lewis acidity is illustrated by the formation of a complex with the Lewis base trimethylamine:

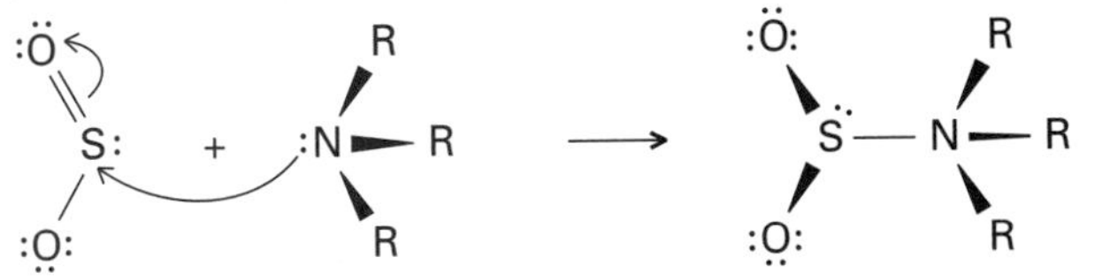

To act as a Lewis base, the SO_2 molecule can donate either its S or its O lone pair to a Lewis acid. When SbF_5 is the acid, the O atom of SO_2 acts as the electron pair donor, but when Ru(II) is the acid, the S atom acts as donor (**28**).[10]

Sulfur trioxide is a strong Lewis acid and a very weak (O donor) Lewis base. Its acidity is illustrated by the reactions

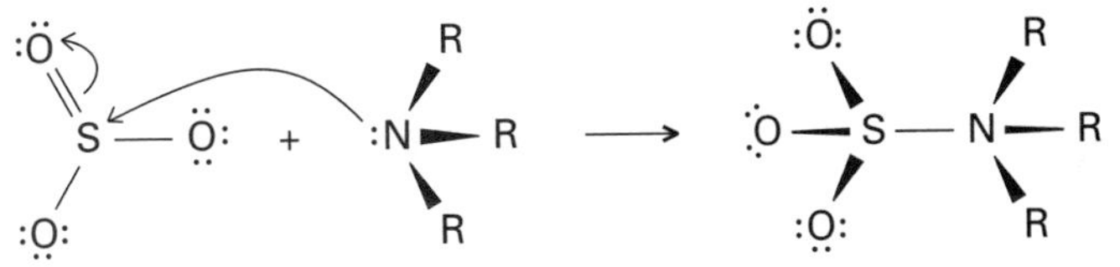

A classic aspect of the acidity of SO_3 is its highly exothermic reaction with water in the formation of sulfuric acid. The resulting problem of having to remove large quantities of heat from the reactor used for the commercial production of the acid is alleviated by using a two-stage process that employs another aspect of the trioxide's Lewis acidity. Prior to dilution, sulfur trioxide is dissolved in sulfuric acid to form the mixture known as *oleum*. This reaction is in fact an example of Lewis acid–base complex formation:

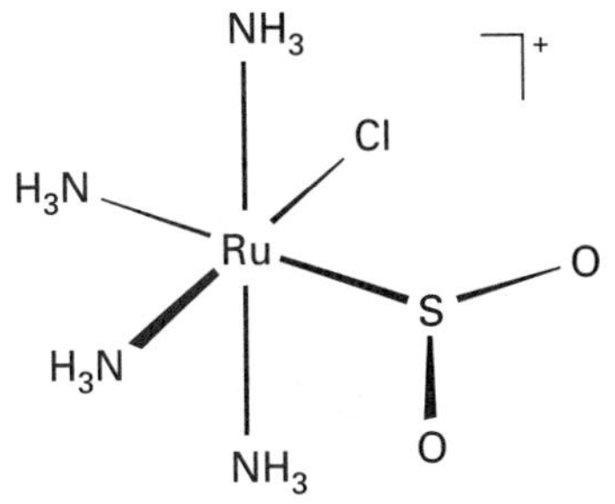

28 $[RuCl(NH_3)_4(SO_2)]^+$

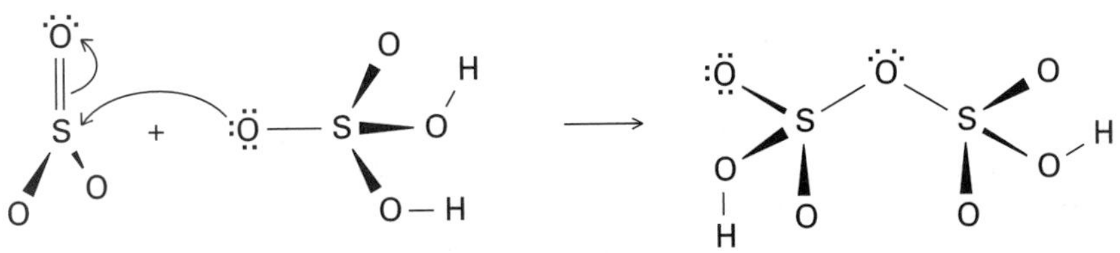

The resulting $H_2S_2O_7$ can then be hydrolyzed in a less exothermic reaction:

$$H_2S_2O_7 + H_2O \longrightarrow 2\,H_2SO_4$$

---

[10] We shall see in Section 5.12 that this preference is an illustration of the concept of hard and soft bases.

> *Oxides and halides of the heavier Group V/15 elements act as Lewis acids; sulfur dioxide and sulfur trioxide act as Lewis acids and weak Lewis bases.*

5.10 Halogen acids

Lewis acidity is expressed in an interesting and subtle way by Br_2 and I_2. Iodine is violet in the solid and gas phases and in non-donor solvents such as trichloromethane. In water, propanone, or ethanol, all of which are Lewis bases, iodine is brown. Visible, UV, and IR spectroscopy of iodine dissolved in trichloromethane that also contains added donor molecules, such as $(CH_3)_2CO$, provide convincing evidence for 1:1 complex formation. The color changes because the solvent–solute complex (which is formed from the lone pair of donor molecule O atoms and a low-lying σ^* orbital of the dihalogen) has a strong optical absorption.

The strong visible absorption spectra of Br_2 and I_2 arise from transitions to low-lying unfilled orbitals. The colors of the species therefore suggest that the empty orbitals may be low enough in energy to serve as acceptor orbitals in Lewis acid–base complex formation.[11]

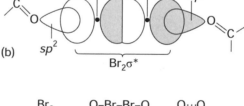

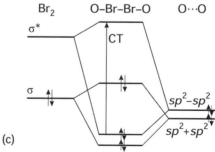

5.10 (a) The structure of $(CH_3)_2COBr_2$ as shown by X-ray diffraction. (b) The orbital overlap responsible for complex formation. (c) A partial molecular orbital energy level diagram for the interaction of the σ and σ^* orbitals of Br_2 with the appropriate symmetry-adapted combinations of sp^2 orbitals on the two O atoms. The charge-transfer band in the near-UV is labeled CT.

....

[11] The terms 'donor–acceptor complex' and 'charge-transfer complex' were at one time used to denote these complexes. However, the distinction between these complexes and the more familiar Lewis acid–base complexes is arbitrary and in the current literature the terms are used more or less interchangeably.

The interaction of Br_2 with the carbonyl group of propanone is shown in Fig. 5.10. The illustration also shows the transition responsible for the new absorption band observed when a complex is formed. The orbital from which the electron originates in the transition is predominantly the lone pair orbital of the base (the ketone). The orbital to which the transition occurs is predominantly the LUMO of the acid (the dihalogen). Thus, to a first approximation, the transition transfers an electron from the base to the acid and is therefore called a **charge-transfer transition**.

The triiodide ion, I_3^-, is an example of a complex between a halogen acid (I_2) and a halide base (I^-). One of the applications of its formation is to render molecular iodine soluble in water so that it can be used as a titration reagent:

$$I_2(s) + I^-(aq) \longrightarrow I_3^-(aq) \qquad K = 725$$

The triiodide ion is one example of a large class of polyhalide ions (Section 12.5).

> Bromine and iodine molecules act as mild Lewis acids.

Systematics of Lewis acids and bases

Lewis acids and bases undergo a variety of characteristic reactions. We review these reactions here, and then show how to characterize the strengths of Lewis acids and bases.

5.11 The fundamental types of reaction

The simplest Lewis acid–base reaction in the gas phase or non-coordinating solvents is **complex formation**:

$$A +:B \longrightarrow A{-}B$$

Three examples are

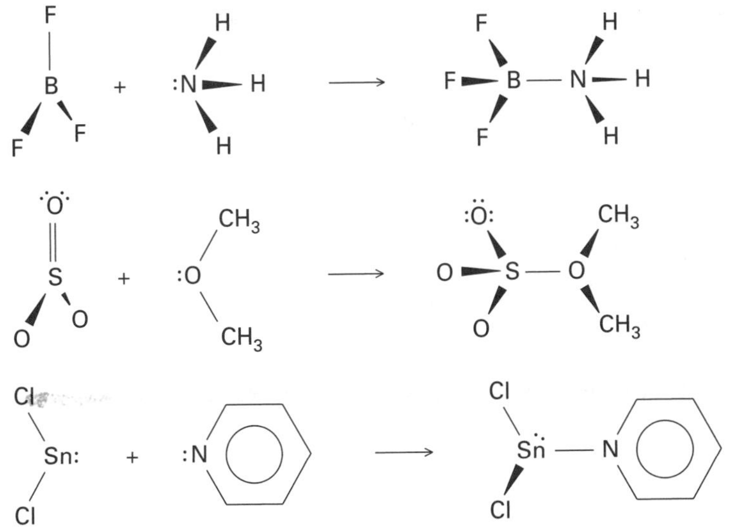

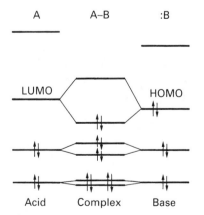

5.11 Localized molecular orbital representation of the interaction between frontier orbitals in the formation of a complex between a Lewis acid A and a Lewis base :B.

All three reactions involve Lewis acids and bases that are independently stable in the gas phase or in solvents that do not form complexes with them. Consequently, the individual species (as well as the complexes) may be studied experimentally.

Figure 5.11 shows the interaction of orbitals responsible for bonding in Lewis complexes. The exothermic character of the formation of the complex stems from the fact that the newly formed bonding orbital is populated by the two electrons supplied by the base whereas the newly formed antibonding orbital is left unoccupied. As a result, there is a net lowering of energy when the bond forms.

(a) Displacement reactions

A **displacement** of one Lewis base by another is a reaction of the form

$$B—A + :B' \longrightarrow B: + A—B'$$

An example is

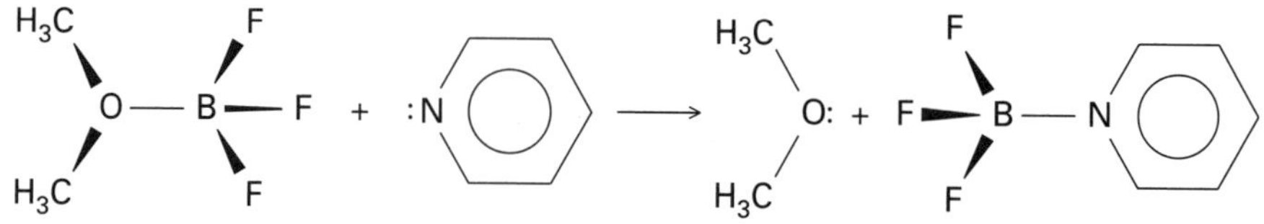

All Brønsted proton transfer reactions are of this type, as in

$$HS^-(aq) + H_2O(l) \longrightarrow S^{2-}(aq) + H_3O^+(aq)$$

In this reaction, the Lewis base H_2O displaces the Lewis base S^{2-} from its complex with the acid H^+. Displacement of one acid by another is also possible, as in the reaction

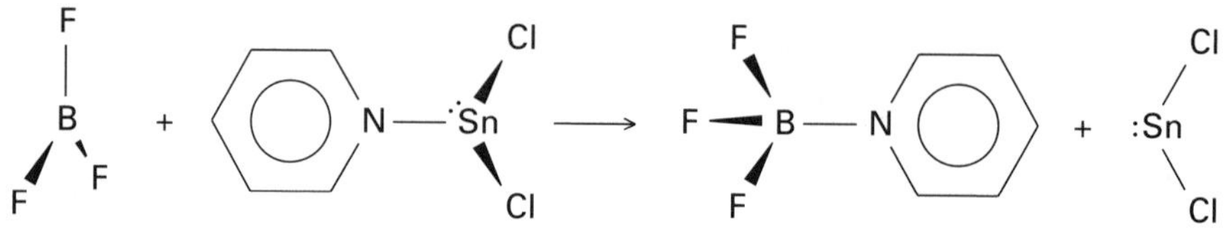

In the context of *d*-metal complexes, a displacement reaction in which one ligand is driven out of the complex and is replaced by another, is generally called a **substitution reaction** (Section 7.8).

> In a displacement reaction, an acid or base drives out another acid or base from a Lewis complex.

(b) Metathesis reactions

A **metathesis reaction** (or 'double displacement reaction') is the interchange of partners:[12]

$$A—B + A'—B' \longrightarrow A—B' + A'—B$$

The displacement of the base :B by :B' is assisted by the extraction of :B by the acid A'. An example is the reaction

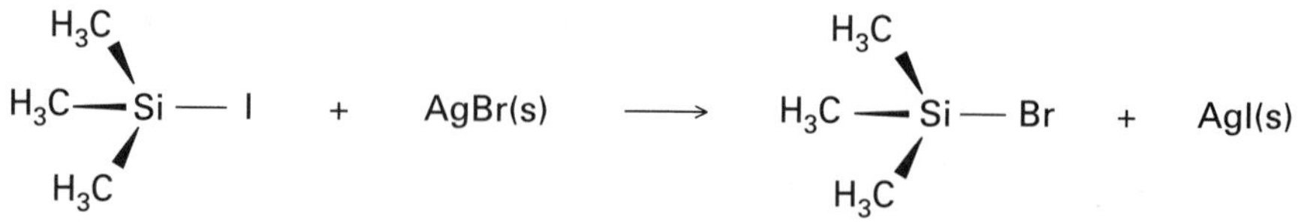

---

[12] The name *metathesis* comes from the Greek word for exchange.

Here the base Br^- displaces I^-, and the extraction is assisted by the formation of an ionic lattice incorporating the acid Ag^+.

> *A metathesis reaction is a displacement reaction assisted by the formation of another complex.*

5.12 Hard and soft acids and bases

The proton (H^+) was the key electron pair acceptor in the discussion of Brønsted acid and base strengths. When considering Lewis acids and bases we must allow for a greater variety of acceptors and hence more factors that influence the interactions between electron pair donors and acceptors in general.

(a) The classification of acids and bases

It proves helpful when considering the interactions of Lewis acids and bases containing elements drawn from throughout the periodic table to consider at least two main classes of substance. The classification of substances as 'hard' and 'soft' acids and bases was introduced by R.G. Pearson; it is a generalization—and a more evocative renaming—of the distinction between two types of behavior which originally were named simply 'class *a*' and 'class *b*', respectively, by Ahrland, Chatt, and Davies.

The two classes are identified empirically by the opposite order of strengths (as measured by the equilibrium constant, K_f, for the formation of the complex) with which they form complexes with halide ion bases:

* Hard acids bond in the order: $I^- < Br^- < Cl^- < F^-$
* Soft acids bond in the order: $F^- < Cl^- < Br^- < I^-$

Figure 5.12 shows the trends in K_f for complex formation with a variety of halide ion bases. The equilibrium constants increase steeply from F^- to I^- when the acid is Hg^{2+}, indicating that Hg^{2+} is a soft acid. The trend is less steep but in the same direction for Pb^{2+}, which indicates that this ion is a borderline soft acid. The trend is in the opposite direction for Zn^{2+}, so this ion is a borderline hard acid. The steep downward slope for Al^{3+} indicates that it is a hard acid.

For Al^{3+}, the binding strength increases as the electrostatic parameter, $\xi = z^2/r$, of the anion increases, which is consistent with an ionic model of the bonding. For Hg^{2+}, the binding strength increases with increasing polarizability of the anion. These two correlations suggest that hard acid cations form complexes in which simple Coulombic interactions are dominant, and that soft acid cations form more complexes in which covalent bonding is important.

A similar classification can be applied to neutral molecular acids and bases. For example, the Lewis acid phenol forms a more stable complex by hydrogen bonding to $(C_2H_5)_2O$: than to $(C_2H_5)_2S$:. This behavior is analogous to the preference of Al^{3+} for F^- over Cl^-. In contrast, the Lewis acid I_2 forms a more stable complex with $(C_2H_5)_2S$:. We can conclude that phenol is hard whereas I_2 is soft.

In general, acids are identified as hard or soft by the thermodynamic stability of the complexes they form, as set out for the halide ions above and for other species as follows:

* Hard acids bond in the order: $R_3P \ll R_3N, R_2S \ll R_2O$
* Soft acids bond in the order: $R_2O \ll R_2S, R_3N \ll R_3P$

It follows from the definition of hardness that

* Hard acids tend to bind to hard bases.
* Soft acids tend to bind to soft bases.

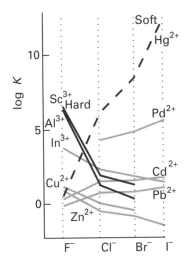

5.12 Trends in stability constants and the classification of cations as hard, borderline, and soft. Borderline ions are indicated by gray lines and may be borderline soft or borderline hard. This diagram, which is adapted from J. Burgess, *Metal ions in solution*, Ellis Horwood, Chichester (1988), emphasizes that there are degrees of hardness and softness.

Table 5.3 The classification of Lewis acids and bases*

Hard	Borderline	Soft
Acids H^+, Li^+, Na^+, K^+ Be^{2+}, Mg^{2+}, Ca^{2+} Cr^{2+}, Cr^{3+} Al^{3+} SO_3, BF_3	Fe^{2+}, Co^{2+} Ni^{2+} Cu^{2+}, Zn^{2+}, Pb^{2+} SO_2, BBr_3	Cu^+, Ag^+, Au^+, Tl^+, Hg^+ Pd^{2+}, Cd^{2+}, Pt^{2+}, Hg^{2+} BH_3
Bases F^-, OH^-, H_2O, NH_3 CO_3^{2-}, NO_3^-, O^{2-} SO_4^{2-}, PO_4^{3-}, ClO_4^-	$\underline{N}O_2^-$, SO_3^{2-}, Br^- N_3^-, N_2 C_6H_5N, $SC\underline{N}^-$	H^-, R^-, $\underline{C}N^-$, $\underline{C}O$, I^- $\underline{S}CN^-$, R_3P, C_6H_6 R_2S

*The underlined element is the site of attachment to which the classification refers.

When species are analyzed with these rules in mind, it is possible to identify the classification summarized in Table 5.3.

> Hard and soft acids and bases are identified empirically by the trends in stabilities of the complexes they form; hard acids tend to bind to hard bases and soft acids tend to bind to soft bases.

(b) The interpretation of hardness

The bonding between hard acids and bases can be described approximately in terms of ionic or dipole–dipole interactions. Soft acids and bases are more polarizable than hard acids and bases, so the acid–base interaction has a more pronounced covalent character.

Although the type of bond formation is a major reason for the distinction between the two classes, we must not forget that there are other contributions to the Gibbs energy of complex formation and hence to the equilibrium constant. Among these contributions are:

1 The rearrangement of the substituents of the acid and base that may be necessary to permit formation of the complex.

2 Steric repulsion between substituents on the acid and the base.

3 Competition with the solvent in reactions in solution.

Later in the chapter we shall see that these additional contributions can have a marked effect on the outcome of a reaction.

> Hard acid–base interactions are predominantly electrostatic; soft acid–base interactions are predominantly covalent.

(c) Chemical consequences of hardness

The concepts of hardness and softness help to rationalize a great deal of inorganic chemistry. For instance, they are useful for choosing preparative conditions and predicting the directions of reactions, and they help to rationalize the outcome of metathesis reactions. However, the concepts must always be used with due regard for other factors that may affect the outcome of reactions. This deeper understanding of chemical reactions will grow in the course of the rest of the book. For the time being we shall limit the discussion to a few straightforward examples.

The classification of molecules and ions as hard or soft acids and bases helps to clarify the terrestrial distribution of the elements described in Chapter 1. The tendency of soft acids to bond to soft bases and of hard acids to bond to hard bases explains certain aspects of the **Goldschmidt classification** of the elements into four types, a scheme widely used in geochemistry. Two of the classes are the **lithophile elements** and the **chalcophile elements**. The lithophile elements, which are found primarily in the Earth's crust (the lithosphere) in silicate minerals, include lithium, magnesium, titanium, aluminum, and chromium (as their

hard acids

cations). These cations are hard, and are found in association with the hard base O^{2-}. On the other hand, the chalcophile elements are often found in combination with sulfide (and ←Soft acids selenide and telluride) minerals, and include cadmium, lead, antimony, and bismuth. These elements (as their cations) are soft, and are found in association with the soft base S^{2-} (or Se^{2-} and Te^{2-}). Zinc cations are borderline hard, but softer than Al^{3+} and Cr^{3+}, and zinc is also often found as its sulfide.

Example 5.7 Explaining the Goldschmidt classification

The common ores of nickel and copper are sulfides. In contrast, aluminum is obtained from the oxide and calcium from the carbonate. Can these observations be explained in terms of hardness?

Answer From Table 5.3 we know that O^{2-} and CO_3^{2-} are hard bases; S^{2-} is a soft base. The table also shows that the cations Ni^{2+} and Cu^{2+} are considerably softer acids than Al^{3+} or Ca^{2+}. Hence the hard-hard and soft-soft rule accounts for the sorting observed.

Self-test 5.7 Of the metals cadmium, rubidium, chromium, lead, strontium, and palladium, which might be expected to be found in aluminosilicate minerals and which in sulfides?

These considerations also help to systematize many reactions in solids and molten salt solutions. These reactions often involve the transfer of a basic anion (typically O^{2-}, S^{2-}, or Cl^-) from one cationic acid center to another. For example, the reaction of CaO with SiO_2 to give the Ca^{2+} salt of the polyanion $[SiO_3^{2-}]_n$ can be regarded as the transfer of the base O^{2-} from the weak acid Ca^{2+} to the stronger acid 'Si^{4+}'. This reaction is a model for slag formation, which is used to remove silicates from the molten iron phase during reduction of iron ores in a blast furnace, and the floating of slag on top of iron is a microcosm of the core/mantle/crust division of the Earth. Similar molten salt and solid reactions are involved in the formation of glass and ceramics. In these reactions, alkali metal oxides or hydroxides transfer a basic O^{2-} ion to the acid silicate center.

Polyatomic anions may contain donor atoms of different hardnesses. For example, the SCN^- ion is a base by virtue of both the harder N atom and the softer S atom. The ion binds to the hard Si atom through N. However, with a soft acid, such as a metal ion in a low oxidation state, the ion bonds through S. Platinum(II), for example, forms Pt-SCN in the complex $[Pt(SCN)_4]^{2-}$.

> *Hard-hard and soft-soft interactions help to systematize complex formation, but must be considered in the light of other possible influences on bonding.*

5.13 Thermodynamic acidity parameters

An important alternative to the hard-soft classification of acids and bases makes use of an approach in which electronic, structural rearrangement, and steric effects are incorporated into a small set of parameters. A leading example of this approach parametrizes the standard reaction enthalpies of complex formation:

$$A(g) + :B(g) \longrightarrow A—B(g) \qquad \Delta_r H^{\ominus}(A—B)$$

It has been found that the values of $\Delta_r H^{\ominus}$ for reactions of this kind can be reproduced by the **Drago–Wayland equation**:

$$-\Delta_r H^{\ominus}(A—B)/(kJ\,mol^{-1}) = E_A E_B + C_A C_B \qquad (12)$$

The parameters E and C were introduced with the idea that they represented 'electrostatic' and 'covalent' factors, respectively, but in fact they must accommodate all factors except

Table 5.4 Drago–Wayland parameters for some acids and bases*

	E	C
Acids		
Antimony pentachloride	15.1	10.5
Boron trifluoride	20.2	3.31
Iodine	2.05	2.05
Iodine monochloride	10.4	1.70
Phenol	8.86	0.90
Sulfur dioxide	1.88	1.65
Trichloromethane	6.18	0.32
Trimethylboron	12.6	3.48
Bases		
Acetone	2.02	4.67
Ammonia	2.78	7.08
Benzene	0.57	1.21
Dimethyl sulfide	0.70	15.26
Dimethylsulfoxide	2.76	5.83
Methylamine	2.66	12.00
p-Dioxane	2.23	4.87
Pyridine	2.39	13.10
Trimethylphosphine	17.2	13.40

*E and C parameters are often reported to give ΔH in kcal mol^{-1}; we have multiplied both by $\sqrt{(4.184)}$ to obtain ΔH in kJ mol^{-1}

solvation. The compounds for which the parameters are listed in Table 5.4 satisfy the equation with an error of less than ± 3 kJ mol^{-1}, as do a much larger number of examples in the original papers.[13]

The Drago–Wayland equation is very successful and useful. In addition to giving the enthalpies of complex formation for over 1500 complexes, these enthalpies can be combined to calculate the enthalpies of displacement and metathesis reactions. Moreover, the equation is useful for reactions of acids and bases in nonpolar, non-coordinating solvents as well as for reactions in the gas phase. The major limitation is that the equation is restricted to substances that can conveniently be studied in the gas phase or non-coordinating solvents; hence, in the main it is limited to neutral molecules.

The standard enthalpies of complex formation are reproduced by the E and C parameters of the Drago–Wayland equation that reflect, in part, the ionic and covalent contributions to the bond in the complex.

5.14 Solvents as acids and bases

Most solvents are either electron-pair acceptors or donors and hence are either Lewis acids or bases. The chemical consequences of solvent acidity and basicity are considerable, for they help to account for the differences between reactions in aqueous and nonaqueous media (see Box 5.1). It follows that a displacement reaction often occurs when a solute dissolves in a solvent, and that the subsequent reactions of the solution are also usually either displacements or metatheses. For example, when antimony pentafluoride dissolves in bromine trifluoride, the following displacement reaction occurs:

$$SbF_5 + BrF_3(l) \longrightarrow [BrF_2]^+ + [SbF_6]^-$$

In the reaction, the strong Lewis acid SbF_5 abstracts F^- from BrF_3. A more familiar example of the solvent as participant in a reaction is in Brønsted theory. In this theory, the acid (H^+) is always regarded as complexed with the solvent, as in H_3O^+ if the solvent is water, and reactions are treated as the transfer of the acid, the proton, from a basic solvent molecule to another base. Only the saturated hydrocarbons among common solvents lack significant Lewis acid or base character.

(a) Basic solvents

Solvents with Lewis base character are common. Most of the well known polar solvents, including water, alcohols, ethers, amines, dimethylsulfoxide (DMSO, $(CH_3)_2SO$), dimethylformamide (DMF, $(CH_3)_2NCHO$), and acetonitrile (CH_3CN), are hard Lewis bases. Dimethylsulfoxide is an interesting example of a solvent that is hard on account of its O donor atom and soft on account of its S donor atom. Reactions of acids and bases in these solvents are generally displacements:

Basic solvents are common; they may form complexes with the solute and participate in displacement reactions.

[13] A good source of E and C parameters is R.S. Drago, N. Wong, C. Bilgrien, and C. Vogel, *Inorg. Chem.* **26**, 9 (1987). For extensions of these concepts, see R.S. Drago, *Applications of electrostatic–covalent models in chemistry.* Surfside Scientific Publications, Gainesville (1996).

Box 5.1 Some useful nonaqueous solvents

A good example of a synthetically useful nonaqueous solvent is tetrahydrofuran, THF (**B1**), a cyclic nonpolar ether which boils at 66 °C. This weak hard-base solvent can readily be dried and deoxygenated by distillation from sodium strips in a nitrogen atmosphere. After use in syntheses (including preparation of air-sensitive compounds) it can be easily removed from a reaction mixture by heating under reduced pressure. For these reasons, it is one of the most widely used solvents for the synthesis of organometallic compounds. The Lewis basicity of the O atom is sometimes important, for it can coordinate to cations. For example, the preparation of substituted metal carbonyls, $[M(CO)_5L]$ (L = phosphine, amine, etc.) can be accomplished by photolysis of $[M(CO)_6]$ in THF to give $[M(CO)_5(thf)]$ and carbon monoxide. This intermediate is then allowed to react with the entering ligand L. Other useful polar aprotic solvents for inorganic substances include methyl cyanide, CH_3CN, and dimethylsulfoxide, $(CH_3)_2SO$ (DMSO).

Ammonia is a strong hard base; this property is very important for its ability to coordinate to d-block acids and protons. It should not be forgotten that, although ammonia is commonly regarded as a base, its own protons can act as centers of Lewis acidity by virtue of their ability to participate in hydrogen bonding. Liquid ammonia (b.p. -33 °C) is readily exploited as a solvent by using a Dewar flask. Despite a somewhat lower relative permittivity ($\varepsilon_r = 22$) than that of water, many salts of alkali metal cations with large anions are reasonably soluble in liquid ammonia. Organic compounds are often more soluble in liquid ammonia than they are in water.

One of the most remarkable reactions in liquid ammonia is the result of dissolution of alkali metals. These solutions are strongly reducing. Electron paramagnetic resonance spectra show that they contain unpaired electrons. The blue color typical of the solutions is the outcome of a very broad optical absorption band in the near IR with a maximum near 1500 nm. The metal is ionized in ammonia solution to give 'solvated electrons';

$$Na(s) + NH_3(l) \longrightarrow Na^+(am) + e^-(am)$$

where 'am' denotes ammonia solution. The blue solutions survive for long times at low temperature but decompose slowly to give hydrogen and sodium amide, $NaNH_2$. The exploitation of the blue solutions to produce compounds called 'electrides' is discussed in Section 9.5.

Liquid hydrogen fluoride (b.p. 19.5 °C) is an acidic solvent with considerable Brønsted acid strength and a relative permittivity comparable to that of water. It is a good solvent for ionic substances. However, as it is both highly reactive and toxic, it presents handling problems, including its ability to etch glass. In practice, hydrogen fluoride is usually contained in polytetrafluoro-ethylene and polychlorotrifluoroethylene vessels.

Although the conjugate base of HF is formally F^-, the ability of HF to form a strong hydrogen bond to F^- means that the conjugate base is better regarded as the bifluoride ion, FHF^-. Many fluorides are soluble in hydrogen fluoride as a result of the formation of this ion; for example

$$LiF(s) + HF(l) \longrightarrow Li^+(hf) + [FHF]^-(hf)$$

where 'hf' denotes solution in hydrogen fluoride. Because HF is quite acidic, protonation of the solute commonly accompanies bifluoride ion formation:

$$CH_3OH(l) + 2 HF(l) \longrightarrow [CH_3OH_2]^+(hf) + [FHF]^-(hf)$$
$$CH_3COOH(l) + 2 HF(l) \longrightarrow [CH_3C(OH)_2]^+(hf) + [FHF]^-(hf)$$

The second reaction is striking since acetic acid, an acid in water, is here acting as a base.

Further reading

W.L. Jolly, *The synthesis and characterization of inorganic compounds*. Waveland Press, Prospect Heights (1991).

J.J. Lagowski (ed.), *The chemistry of nonaqueous solvents*, Vols 1–5. Academic Press, New York (1966–78).

B1 Tetrahydrofuran

Example 5.8 Accounting for properties in terms of the Lewis basicity of solvents

Silver perchlorate, $AgClO_4$, is significantly more soluble in benzene than in alkane solvents. Account for this observation in terms of Lewis acid–base properties.

Answer The π electrons of benzene, a soft base, are available for complex formation with the empty orbitals of the cation Ag^+, a soft acid (recall **19**). The species $[Ag-C_6H_6]^+$ is the complex of the acid Ag^+ with π electrons of the weak base benzene.

Self-test 5.8 Boron trifluoride, BF_3, a hard acid, is often used in the laboratory as a solution in diethyl ether, $(C_2H_5)_2O$:, a hard base. Draw the structure of the complex that results from the dissolution of $BF_3(g)$ in $(C_2H_5)_2O(l)$.

(b) Acidic and neutral solvents

Hydrogen bonding can be regarded as an example of complex formation. The 'reaction' is between A–H (the Lewis acid) and :B (the Lewis base) and gives the complex conventionally denoted A–H$\cdots$B. Hence, many solutes that form hydrogen bonds with a solvent can be regarded as dissolving by virtue of complex formation. A consequence of this view is that an acidic solvent molecule is displaced when proton transfer occurs:

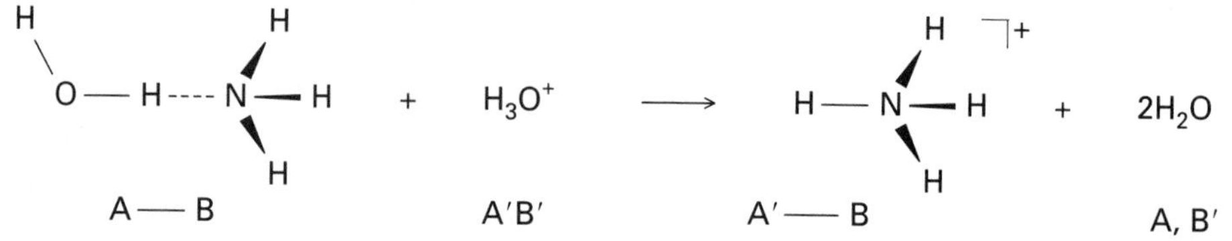

Liquid SO_2 is a good soft acidic solvent for dissolving the soft base benzene. Unsaturated hydrocarbons may act as acids or bases by using their π or π^* orbitals as frontier orbitals. Alkanes with electronegative substitutents, such as haloalkanes (e.g. $CHCl_3$), are significantly acidic at the hydrogen atom.

> Hydrogen bond formation is an example of Lewis complex formation. Other solvents may also show Lewis acid character.

(c) Solvent parameters

A quantitative measure of the basicity of a solvent is the reaction enthalpy for the formation of a complex between the solvent and a reference acid. V. Gutmann selected the strong Lewis acid $SbCl_5$ in 1,2-dichloroethane for this reference role, the relevant reaction being

$$SbCl_5 + :B \longrightarrow Cl_5Sb-B \qquad \Delta_r H^{\ominus}$$

The negative of $\Delta_r H^{\ominus}/(\text{kcal mol}^{-1})$ (kilocalories are used for historical reasons) is called the **donor number** of the solvent. Some representative values are collected in Table 5.5: the higher the donor number, the stronger the Lewis base.

A corresponding parameter to measure solvent acidity, the **acceptor number**, has been introduced. In this case, triethylphosphine oxide, $(C_2H_5)_3PO$:, is used as a reference base and the NMR chemical shift of ^{31}P is measured for the base dissolved in the pure solvent. The zero of the scale is defined as the shift in hexane and the value 100 is ascribed to $SbCl_5$. On this arbitrary basis, numbers similar in magnitude to donor numbers are obtained (Table 5.5). As for donor numbers, the higher the acceptor number, the stronger the Lewis acid.

> Donor numbers and acceptor numbers are sometimes used to report the strengths of bases and acid, respectively. The higher the donor or acceptor number, the stronger the base or the acid.

Table 5.5 Donor and acceptor numbers (D.N. and A.N.) and relative permittivities (dielectric constants, ε_r) at 25°C

Solvent	D.N.	A.N.	ε_r
Acetic acid		52.9	6.2
Acetone	17.0	12.5	20.7
Benzene	0.1	8.2	2.3
Carbon tetrachloride		8.6	2.2
Diethyl ether	19.2	3.9	4.3
Dimethylsulfoxide	29.8	19.3	45
Ethanol	19.0	37.1	24.3
Pyridine	33.1	14.2	12.3
Tetrahydrofuran	20.0	8.0	7.3
Water	18	54.8	81.7

Source: V. Gutmann, *Coordination chemistry in nonaqueous solutions.* Springer-Verlag, Berlin (1968).

Heterogeneous acid–base reactions

Some of the most important reactions involving the Lewis and Brønsted acidity of inorganic compounds occur at solid surfaces. For example, **surface acids**, which are solids with a high surface area and Lewis acid sites, are used as catalysts in the petrochemical industry for the interconversion of hydrocrbons. The surfaces of many materials that are important in the chemistry of soil and natural waters also have Brønsted and Lewis acid sites.[14]

Silica surfaces do not readily produce Lewis acid sites because –OH groups remain tenaciously attached at the surface of SiO_2 derivatives; as a result, Brønsted acidity is dominant. The Brønsted acidity of silica surfaces themselves is only moderate (and comparable to that of acetic acid). However, as we have already remarked, aluminosilicates display strong Brønsted acidity. When surface OH groups are removed by heat treatment, the aluminosilicate surface possesses strong Lewis acid sites.

Surface reactions carried out using the Brønsted acid sites of silica gels are used to prepare thin coatings of a wide variety of organic groups using surface modification reactions such as

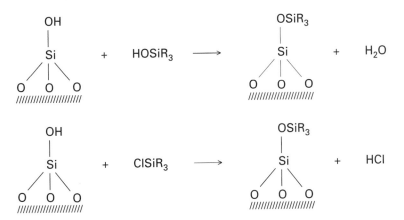

Thus, silica gel surfaces can be modified to have affinities for specific classes of molecules. This procedure greatly expands the range of stationary phases that can be used for chromatography. The surface O–H groups on glass can be modified similarly, and glassware treated in this manner is sometimes used in the laboratory when proton-sensitive compounds are being studied.

The surfaces of many catalytic materials and minerals have Brønsted and Lewis acid sites.

[14] An interesting account is found in G. Sposito, *The surface chemistry of soils.* Oxford University Press (1984).

Further reading

R.P. Bell, *The proton in chemistry*. Cornell University Press, Ithaca (1973). A classic discussion of Brønsted acidity with many examples drawn from organic chemistry.

C.F. Bates, Jr and R.E. Messmer, *The hydrolysis of cations*. Wiley-Interscience, New York (1976). A survey of the acidities and the polymerization of aqua ions.

J. Burgess, *Ions in solution*. Ellis Horwood, Chichester, UK (1988). A readable account of ion solvation with an introduction to acidity and polymerization.

W. Stumm and J.J. Morgan, *Aquatic chemistry*. Wiley-Interscience, New York (1996). The standard text on the chemistry of natural waters.

Two general treatments of Lewis acids and bases are:

R.S. Drago and N.A. Matwiyoff, *Acids and bases*. Heath, Boston (1968).

W.B. Jensen, *The Lewis acid-base concepts*. Wiley, New York (1980).

More specialized books are:

R.G. Pearson, In *Survey of progress in chemistry* (ed. A. Scott), Vol. 1, Chapter 1. Academic Press, New York (1969). This is an account of the hard and soft classification by the originator of the terminology.

V. Gutmann, *Coordination chemistry in nonaqueous solution*. Springer-Verlag, Berlin (1968). This monograph analyzes the role of solvents in detail.

Exercises

5.1 Sketch an outline of the *s*- and *p*-blocks of the periodic table and indicate on it the elements that form (a) strongly acidic oxides and (b) strongly basic oxides, and (c) show the regions for which amphoterism is common.

5.2 Identify the conjugate bases corresponding to the following acids: $[Co(NH_3)_5(OH_2)]^{3+}$, HSO_4^-, CH_3OH, $H_2PO_4^-$, $Si(OH)_4$, HS^-.

5.3 Identify the conjugate acids of the bases C_5H_5N (pyridine), HPO_4^{2-}, O^{2-}, CH_3COOH, $[Co(CO)_4]^-$, CN^-.

5.4 List the bases HS^-, F^-, I^-, and NH_2^- in order of increasing proton affinity.

5.5 Aided by Fig. 5.2 (taking solvent leveling into account), identify which bases from the following lists are (a) too strong to be studied experimentally; (b) too weak to be studied experimentally; or (c) of directly measurable base strength.

 (i) CO_3^{2-}, O^{2-}, ClO_4^-, and NO_3^- in water

 (ii) HSO_4^-, NO_3^-, ClO_4^- in H_2SO_4

5.6 The aqueous solution pK_a values for HOCN, H_2NCN, and CH_3CN are approximately 4, 10.5, and 20 (estimated), respectively. Explain the trend in these –CN derivatives of binary acids and compare them with H_2O, NH_3, and CH_4. Is –CN electron donating or withdrawing?

5.7 The pK_a value of $HAsO_4^{2-}$ is 11.6. Is this value consistent with the two Pauling rules?

5.8 Draw the structures and indicate the charges of the tetraoxoanions of $X = Si$, P, S, and Cl. Summarize and account for the trends in the pK_a values of their conjugate acids.

5.9 Which of the following pairs is the stronger acid? Give reasons for your choice. (a) $[Fe(OH_2)_6]^{3+}$ or $[Fe(OH_2)_6]^{2+}$, (b) $[Al(OH_2)_6]^{3+}$ or $[Ga(OH_2)_6]^{3+}$, (c) $Si(OH)_4$ or $Ge(OH)_4$, (d) $HClO_3$ or $HClO_4$, (e) H_2CrO_4 or $HMnO_4$, (f) H_3PO_4 or H_2SO_4.

5.10 Arrange the oxides Al_2O_3, B_2O_3, BaO, CO_2, Cl_2O_7, SO_3 in order from the most acidic through amphoteric to the most basic.

5.11 Arrange the acids HSO_4^-, H_3O^+, H_4SiO_4, CH_3GeH_3, NH_3, HSO_3F in order of increasing acid strength.

5.12 The ions Na^+ and Ag^+ have similar radii. Which aqua ion is the stronger acid? Why?

5.13 Which of the elements Al, As, Cu, Mo, Si, B, Ti form oxide polyanions and which form oxide polycations?

5.14 When a pair of aqua cations forms an M—O—M bridge with the elimination of water, what is the general rule for the change in charge per M atom on the ion?

5.15 Write a balanced equation for the formation of $P_2O_7^{4-}$ from PO_4^{3-}. Write a balanced equation for the dimerization of the complex $[Fe(OH_2)_6]^{3+}$ to give $[(H_2O)_4Fe(OH)_2Fe(OH_2)_4]^{4+}$

5.16 Write balanced equations for the main reaction occurring when (a) H_3PO_4 and Na_2HPO_4 and (b) CO_2 and $CaCO_3$ are mixed in aqueous media.

5.17 Sketch the *p*-block of the periodic table. Identify as many elements as you can that form Lewis acids in one of their oxidation states and give the formula of a representative Lewis acid for each element.

5.18 For each of the following processes, identify the acids and bases involved and characterize the process as complex formation or acid–base displacement. Identify the species that exhibit Brønsted acidity as well as Lewis acidity.

 (a) $SO_3 + H_2O \rightarrow HSO_4^- + H^+$

 (b) $CH_3[B_{12}] + Hg^{2+} \rightarrow [B_{12}]^+ + CH_3Hg^+$; $[B_{12}]$ designates the Co-macrocycle, vitamin B_{12}.

 (c) $KCl + SnCl_2 \rightarrow K^+ + [SnCl_3]^-$

 (d) $AsF_3(g) + SbF_5(l) \rightarrow [AsF_2]^+[SbF_6]^-(s)$

(e) Ethanol dissolves in pyridine to produce a nonconducting solution.

5.19 Select the compound on each line with the named characteristic and state the reason for your choice.

(a) Strongest Lewis acid:

BF_3 BCl_3 BBr_3

$BeCl_2$ BCl_3

$B(n\text{-}Bu)_3$ $B(t\text{-}Bu)_3$

(b) More basic toward $B(CH_3)_3$:

Me_3N Et_3N

$2\text{-}CH_3C_5H_4N$ $4\text{-}CH_3C_5H_4N$

5.20 Using hard–soft concepts, which of the following reactions are predicted to have an equilibrium constant greater than 1? Unless otherwise stated, assume gas-phase or hydrocarbon solution and $25\,°C$.

(a) $R_3PBBr_3 + R_3NBF_3 \rightleftharpoons R_3PBF_3 + R_3NBBr_3$

(b) $SO_2 + (C_6H_5)_3PHOC(CH_3)_3 \rightleftharpoons (C_6H_5)_3PSO_2 + HOC(CH_3)_3$

(c) $CH_3HgI + HCl \rightleftharpoons CH_3HgCl + HI$

(d) $[AgCl_2]^-(aq) + 2CN^-(aq) \rightleftharpoons [Ag(CN)_2]^-(aq) + 2Cl^-(aq)$

5.21 The molecule $(CH_3)_2N—PF_2$ has two basic atoms, P and N. One is bound to B in a complex with BH_3, the other to B in a complex with BF_3. Decide which is which and state your reason.

5.22 The enthalpies of reaction of trimethylboron with NH_3, CH_3NH_2, $(CH_3)_2NH$, and $(CH_3)_3N$ are -58, -74, -81, and $-74\,kJ\,mol^{-1}$, respectively. Why is trimethylamine out of line?

5.23 With the aid of the table of E and C values, discuss the relative basicity in (a) acetone and dimethylsulfoxide, (b) dimethylsulfide and dimethylsulfoxide. Comment on a possible ambiguity for DMSO.

5.24 Give the equation for the dissolution of SiO_2 glass by HF and interpret the reaction in terms of Lewis and Brønsted acid–base concepts.

5.25 Aluminum sulfide, Al_2S_3, gives off a foul odor characteristic of hydrogen sulfide when it becomes damp. Write a balanced chemical equation for the reaction and discuss it in terms of acid–base concepts.

5.26 Describe the solvent properties which would (a) favor displacement of Cl^- by I^- from an acid center, (b) favor basicity of R_3As over R_3N, (c) favor acidity of Ag^+ over Al^{3+}, (d) promote the reaction $2\,FeCl_3 + ZnCl_2 \rightarrow Zn^{2+} + 2\,[FeCl_4]^-$. In each case, suggest a specific solvent which might be suitable.

5.27 Why are strongly acidic solvents (e.g. SbF_5/HSO_3F) used in the preparation of cations like I_2^+ and Se_8^+, whereas strongly basic solvents are needed to stabilize anionic species such as S_4^{2-} and Pb_9^{4-}?

5.28 The Lewis acid $AlCl_3$ catalysis of the acylation of benzene was described in Section 5.8. Propose a mechanism for a similar reaction catalyzed by an alumina surface.

5.29 Use acid–base concepts to comment on the fact that the only important ore of mercury is cinnabar, HgS, whereas zinc occurs in nature as sulfides, silicates, carbonates, and oxides.

5.30 Write balanced Brønsted acid–base equations for the dissolution of the following compounds in liquid hydrogen fluoride: (a) CH_3CH_2OH, (b) NH_3, (c) C_6H_5COOH.

5.31 Is the dissolution of silicates in HF a Lewis acid–base reaction, a Brønsted acid–base reaction, or both?

5.32 The f-block elements are found as $M(III)$ lithophiles in silicate minerals. What does this indicate about their hardness?

5.33 Consider the reaction forming metasilicates from carbonates:

$$nCaCO_3(s) + nSiO_2(s) \rightarrow [CaSiO_3]_n + nCO_2(g)$$

Identify the stronger acid between SiO_2 and CO_2.

5.34 The ores of titanium, tantalum, and niobium may be brought into solution near $800\,°C$ using sodium disulfate. A simplified version of the reaction is

$$TiO_2 + Na_2S_2O_7 \rightarrow Na_2SO_4 + TiO(SO_4)$$

Identify the acids and bases.

5.35 Sketch the shapes of AsF_5 and its complex with F^- (use VSEPR if necessary) and identify their point groups. What are the point groups of X_3BNH_3 and Al_2Cl_6?

Problems

5.1 In analytical chemistry a standard trick for improving the detection of the stoichiometric point in titrations of weak bases with strong acids is to use acetic acid as a solvent. Explain the basis of this approach.

5.2 In the gas phase, the base strength of amines increases regularly along the series $NH_3 < CH_3NH_2 < (CH_3)_2NH < (CH_3)_3N$. Consider the role of steric effects and the electron-donating ability of CH_3 in determining this order. In aqueous solution, the order is reversed. What solvation effect is likely to be responsible?

5.3 The hydroxoacid $Si(OH)_4$ is weaker than H_2CO_3. Write balanced equations to show how dissolving a solid M_2SiO_4 can lead to a reduction

in the pressure of CO_2 over an aqueous solution. Explain why silicates in ocean sediments might limit the increase of CO_2 in the atmosphere.

5.4 The precipitation of $Fe(OH)_3$ discussed in this chapter is used to clarify waste waters, because the gelatinous hydrous oxide is very efficient at the coprecipitation of some contaminants and the entrapment of others. The solubility constant of $Fe(OH)_3$ is $K_s = [Fe^{3+}][OH^-]^3 \approx 10^{-38}$. Since the autoprotolysis constant of water links $[H_3O^+]$ to $[OH^-]$ by $K_w = [H_3O^+][OH^-] = 10^{-14}$, we can rewrite the solubility constant by substitution as $[Fe^{3+}]/[H^+]^3 = 10^4$. (a) Balance the chemical equation for the precipitation of $Fe(OH)_3$ when iron(III) nitrate is added to water. (b) If $6.6\,kg$ of $Fe(NO_3)_3 \cdot 9H_2O$ is added to $100\,L$ of water, what is the final pH of the solution and the molar concentration of Fe^{3+}(aq), neglecting other forms of dissolved Fe(III)? Give formulas for the two most important Fe(III) species that have been neglected in this calculation.

5.5 The frequency of the symmetrical M–O stretching vibration of the octahedral aqua ions $[M(OH_2)_6]^{2+}$ increases along the series, $Ca^{2+} < Mn^{2+} < Ni^{2+}$. How does this trend relate to acidity?

5.6 An electrically conducting solution is produced when $AlCl_3$ is dissolved in the basic polar solvent CH_3CN. Give formulas for the most probable conducting species and describe their formation using Lewis acid–base concepts.

5.7 The complex anion $[FeCl_4]^-$ is yellow whereas $[Fe_2Cl_6]$ is reddish. Dissolution of $0.1\,mol$ $FeCl_3$(s) in $1\,L$ of either $POCl_3$ or $PO(OR)_3$ produces a reddish solution which turns yellow on dilution. Titration of red solutions in $POCl_3$ with Et_4NCl solutions leads to a sharp color change (from red to yellow) at $1:1$ mole ratio of $FeCl_3/Et_4NCl$. Vibrational spectra suggest that oxochloride solvents form adducts with typical Lewis acids *via* coordination of oxygen. Compare the following two sets of reactions as possible explanations of the observations.

(a) $Fe_2Cl_6 + 2POCl_3 \rightleftharpoons 2[FeCl_4]^- + 2[POCl_2]^+$

$POCl_2^+ + Et_4NCl \rightleftharpoons Et_4N^+ + POCl_3$

(b) $Fe_2Cl_6 + 4POCl_3 \rightleftharpoons [FeCl_2(OPCl_3)_4]^+ + [FeCl_4]^-$

Both equilibria are shifted to products by dilution.

5.8 In the traditional scheme for the separation of metal ions from solution that is the basis of qualitative analysis, ions of Au, As, Sb, and Sn precipitate as sulfides but redissolve on addition of excess ammonium polysulfide. In contrast, ions of Cu, Pb, Hg, Bi, and Cd precipitate as sulfides but do not redissolve. In the language of this chapter, the first group is amphoteric for reactions involving SH^- in place of OH^-. The second group is less acidic. Locate the amphoteric boundary in the periodic table for sulfides implied by this information. Compare this

boundary with the amphoteric boundary for hydrous oxides in Fig. 5.4. Does this analysis agree with describing S^{2-} as a softer base than O^{2-}?

5.9 The compounds SO_2 and $SOCl_2$ can undergo an exchange of radioactively labeled sulfur. The exchange is catalyzed by Cl^- and $SbCl_5$. Suggest mechanisms for these two exchange reactions with the first step being the formation of an appropriate complex.

5.10 In the reaction of *t*-butyl bromide with $Ba(NCS)_2$, the product is 91 per cent S-bound *t*-butyl-SCN. However, if $Ba(NCS)_2$ is impregnated into solid CaF_2, the yield is higher and the product is 99 per cent *t*-butyl-NCS. Discuss the effect of alkaline earth metal salt support on the hardness of the ambidentate nucleophile SCN^-. See T. Kimura, M. Fujita, and T. Ando, *J. Chem. Soc., Chem. Commun.* 1213 (1990).

5.11 Order the following cations in terms of *increasing* Brønsted acidity in water: Sr^{2+}, Ba^{2+}, Hg^{2+}.

5.12 Draw the structures of chloric and chlorous acid and predict their pK_a values using Pauling's rules.

5.13 Pyridine forms a stronger Lewis acid–base complex with SO_3 than with SO_2. However, pyridine forms a weaker complex with SF_6 than with SF_4. Explain the difference.

5.14 Predict whether the equilibrium constants for the following reactions should be greater than 1 or less than 1:

(a) $CdI_2(s) + CaF_2(s) \rightleftharpoons CdF_2(s) + CaI_2(s)$,

(b) $[CuI_4]^{2-}(aq) + [CuCl_4]^{3-}(aq) \rightleftharpoons [CuCl_4]^{2-}(aq) + [CuI_4]^{3-}(aq)$,

(c) $NH_2^-(aq) + H_2O(l) \rightleftharpoons NH_3(aq) + OH^-(aq)$

5.15 For parts (a), (b), and (c) state which of the two solutions has the *lower* pH: (a) $0.1\,M\,Fe(ClO_4)_2$ or $0.1\,M\,Fe(ClO_4)_3$, (b) $0.1\,M\,Ca(NO_3)_2$ or $0.1\,M\,Mg(NO_3)_2$, (c) $0.1\,M\,Hg(NO_3)_2$ or $0.1\,M\,Zn(NO_3)_2$.

5.16 Consider the three manganese oxides MnO, MnO_2, and Mn_2O_7. One of them is acidic, one of them is basic, and one is amphoteric. Which one is which?

6

Oxidation and reduction

A third major class of chemical reactions consists of redox reactions, in which electrons are transferred from one species to another. We analyze their practicality by considering both thermodynamics and kinetics. The discussion begins with a thermodynamic analysis of the conditions needed for some major industrial processes carried out at such elevated temperatures that kinetics are only a secondary consideration. Next, we develop procedures for analyzing redox reactions in solution close to room temperature, where both thermodynamic and kinetic considerations are important. We see that the electrode potentials of electroactive species provide thermodynamic data in a useful form. In particular, we see how to use diagrammatic techniques to summarize trends in the stabilities of various oxidation states, including the influence of pH. Overpotentials are introduced as an approximate way of treating kinetic factors.

A very large class of reactions can be regarded as occurring by the transfer of electrons from one species to another. Electron gain is called **reduction** and electron loss is **oxidation**; the joint process is called a **redox reaction**. The species that supplies electrons is the **reducing agent** (or 'reductant') and the species that removes electrons is the **oxidizing agent** (or 'oxidant').

The transfer of electrons is often accompanied by the transfer of atoms, and it is sometimes difficult to keep track of where electrons have come from and gone to. It is therefore safest—and simplest—to analyze redox reactions according to a set of formal rules expressed in terms of oxidation numbers (Section 3.1c) and not to think in terms of actual electron transfers. Oxidation then corresponds to the increase in oxidation number of an element and reduction corresponds to the decrease in oxidation number. A redox reaction is a chemical reaction in which there are changes in oxidation number of at least one of the elements involved.

Extraction of the elements

The original definition of 'oxidation' was a reaction in which an element reacts with oxygen and is converted to an oxide. 'Reduction' originally meant the reverse reaction, in which an oxide of a metal was converted to the metal. Both terms have been generalized and expressed in terms of electron transfer and changes in oxidation number, but these special cases are still the basis of a major part of chemical industry and laboratory chemistry.

6.1 Elements extracted by reduction

Oxygen has been a component of the atmosphere since photosynthesis became a dominant process over 10^9 years ago, and many metals are found as their oxides. Copper could be extracted from its ores at temperatures attainable in the primitive hearths that became available in about 4000 BC, and the process of 'smelting' was discovered, in which ores are heated with a reducing agent such as carbon. Smelting is still used but, as many important ores of easily reduced metals are sulfides, it is often preceded by conversion of some of the sulfide to an oxide by 'roasting' in air, as in the reaction

$$2\,Cu_2S(s) + 3\,O_2(g) \longrightarrow 2\,Cu_2O(s) + 2\,SO_2(g)$$

It was not until nearly 1000 BC that the higher temperatures could be reached and less readily reduced elements, such as iron, could be extracted. The Iron Age then began. Carbon remained the dominant reducing agent until the end of the nineteenth century, and metals that needed higher temperatures for their production remained unavailable even though their ores were reasonably abundant.

The technological breakthrough in the nineteenth century that resulted in the conversion of aluminum from a rarity into a major construction metal was the introduction of **electrolysis**, the driving of a non-spontaneous reaction (including the reduction of ores) by the passage of an electric current. The availability of electric power also expanded the scope of carbon reduction, for electric furnaces can reach much higher temperatures than carbon-combustion furnaces, such as the blast furnace. Thus, magnesium is a twentieth-century metal even though one of its modes of recovery, the **Pidgeon process**, the electrothermal reduction of the oxide in the reaction

$$MgO(s) + C(s) \xrightarrow{\;\Delta\;} Mg(l) + CO(g)$$

uses carbon as the reducing agent.

> *Metals are obtained from their ores by using a chemical reducing agent at high temperatures or by electrolysis.*

(a) Thermodynamic aspects of extraction

Thermodynamic arguments can be used to identify which reactions are spontaneous (that is, have a natural tendency to occur) under the prevailing conditions, and help us to choose the most economical reducing agents and reaction conditions. The thermodynamic criterion of spontaneity is that, at constant temperature and pressure, the reaction Gibbs energy, $\Delta_r G$, is negative. It is usually sufficient to consider the *standard* reaction Gibbs energy, $\Delta_r G^{\ominus}$, which is related to the equilibrium constant through

$$\Delta_r G^{\ominus} = -RT \ln K \tag{1}$$

Hence, a negative value of $\Delta_r G^{\ominus}$ corresponds to $K > 1$ and therefore to a 'favorable' reaction. It should be noted that equilibrium is rarely attained in commercial processes, and even a process for which $K < 1$ can be viable if the products are swept out of the reaction chamber. In principle, we also need to consider rates when judging whether a reaction is feasible in practice, but reactions are often fast at high temperature, and thermodynamically favorable reactions are likely to occur. A fluid phase (typically a gas) is usually required to facilitate what would otherwise be a sluggish reaction between coarse particles.

To achieve a negative $\Delta_r G^{\ominus}$ for the reduction of a metal oxide with carbon or carbon monoxide, one of the reactions

(a) $C(s) + \frac{1}{2} O_2(g) \to CO(g) \qquad \Delta_r G^{\ominus}(C, CO)$
(b) $\frac{1}{2} C(s) + \frac{1}{2} O_2(g) \to \frac{1}{2} CO_2(g) \qquad \Delta_r G^{\ominus}(C, CO_2)$
(c) $CO(g) + \frac{1}{2} O_2(g) \to CO_2(g) \qquad \Delta_r G^{\ominus}(CO, CO_2)$

must have a more negative $\Delta_r G^{\ominus}$ than a reaction of the form

(d) $x\mathrm{M}(\mathrm{s}\ \mathrm{or}\ \mathrm{l}) + \frac{1}{2}\mathrm{O}_2(\mathrm{g}) \longrightarrow \mathrm{M}_x\mathrm{O}(\mathrm{s})$ $\Delta_\mathrm{r}G^\ominus(\mathrm{M}, \mathrm{M}_x\mathrm{O})$

under the same reaction conditions. If that is so, then one of the reactions

(a − d) $\mathrm{M}_x\mathrm{O}(\mathrm{s}) + \mathrm{C}(\mathrm{s}) \longrightarrow x\mathrm{M}(\mathrm{s}\ \mathrm{or}\ \mathrm{l}) + \mathrm{CO}(\mathrm{g})$
$$\Delta_\mathrm{r}G^\ominus(\mathrm{C}, \mathrm{CO}) \ - \ \Delta_\mathrm{r}G^\ominus(\mathrm{M}, \mathrm{M}_x\mathrm{O})$$

(b − d) $\mathrm{M}_x\mathrm{O}(\mathrm{s}) + \frac{1}{2}\mathrm{C}(\mathrm{s}) \longrightarrow x\mathrm{M}(\mathrm{s}\ \mathrm{or}\ \mathrm{l}) + \frac{1}{2}\mathrm{CO}_2(\mathrm{g})$
$$\Delta_\mathrm{r}G^\ominus(\mathrm{C}, \mathrm{CO}_2) \ - \ \Delta_\mathrm{r}G^\ominus(\mathrm{M}, \mathrm{M}_x\mathrm{O})$$

(c − d) $\mathrm{M}_x\mathrm{O}(\mathrm{s}) + \mathrm{CO}(\mathrm{g}) \longrightarrow x\mathrm{M}(\mathrm{s}\ \mathrm{or}\ \mathrm{l}) + \mathrm{CO}_2(\mathrm{g})$
$$\Delta_\mathrm{r}G^\ominus(\mathrm{CO}, \mathrm{CO}_2) \ - \ \Delta_\mathrm{r}G^\ominus(\mathrm{M}, \mathrm{M}_x\mathrm{O})$$

will have a negative reaction Gibbs energy, and therefore be spontaneous. The relevant information is commonly summarized in an **Ellingham diagram**, which is a graph of $\Delta_\mathrm{r}G^\ominus$ against temperature for each of these reactions (Fig. 6.1).

An understanding of the appearance of an Ellingham diagram can be obtained by noting that

$$\Delta_\mathrm{r}G^\ominus = \Delta_\mathrm{r}H^\ominus - T\Delta_\mathrm{r}S^\ominus \qquad\qquad (2)$$

and using the fact that the enthalpy and entropy of reaction are, to a reasonable approximation, independent of temperature: the slope of a line in an Ellingham diagram should therefore be equal to $-\Delta_\mathrm{r}S^\ominus$ for the relevant reaction. Because the standard entropies of gases are much larger than those of solids, the standard reaction entropy of (d), in which there is a net consumption of gas, is negative, and hence the plot in the Ellingham diagram should have a positive slope, as shown in Fig. 6.2. The kinks in the lines, where the slope of the metal oxidation line changes, are where the metal undergoes a phase change, particularly melting, and the reaction entropy changes accordingly.

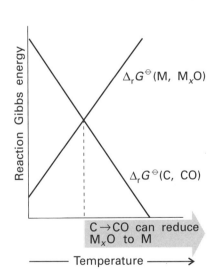

6.1 The variation of the standard reaction Gibbs energies for the formation of a metal oxide and carbon monoxide with temperature. The formation of carbon monoxide from carbon can reduce the metal oxide to the metal at temperatures higher than the point of intersection of the two lines. More specifically, at the intersection K changes from less than 1 to more than 1.

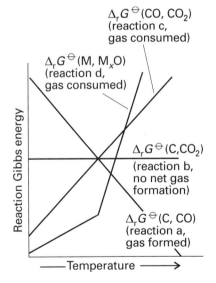

6.2 A fragment of an Ellingham diagram showing the standard reaction Gibbs energy for the formation of a metal oxide and the three carbon oxidation Gibbs energies. The slopes of the curves are determined largely by whether or not there is net gas formation or consumption in the course of the reaction. A phase change can result in a kink in the graph (because the entropy of the substance changes).

The reaction entropy of (a), in which there is a net formation of gas (because 1 mol CO replaces $\frac{1}{2}$ mol O_2), is positive, and its line in the Ellingham diagram therefore has a negative slope. The standard reaction entropy of (b) is close to zero as there is no net change in the amount of gas, so its line is horizontal in the Ellingham diagram. Finally, (c) has a negative reaction entropy because $\frac{3}{2}$ mol of gas molecules is replaced by 1 mol CO_2; hence the line in the diagram has a positive slope.

At temperatures for which the C,CO line lies above the metal oxide lines in Fig. 6.2, $\Delta_r G^{\ominus}(M, M_xO)$ is more negative than $\Delta_r G^{\ominus}(C, CO)$. At these temperatures, $\Delta_r G^{\ominus}(C, CO) - \Delta_r G^{\ominus}(M, M_xO)$ is positive, so the reaction (a – d) is not spontaneous. However, for temperatures for which the C,CO line lies below the metal oxide line, the reduction of the metal oxide by carbon is spontaneous. Similar remarks apply to the temperatures at which the other two carbon oxidation lines lie above or below the metal oxide lines in Fig. 6.2. In summary:

- For temperatures at which the C,CO line lies below the metal oxide line, carbon can be used to reduce the metal oxide and itself is oxidized to carbon monoxide.
- For temperatures at which the C,CO_2 line lies below the metal oxide line, carbon can be used to achieve the reduction, but is oxidized to carbon dioxide.
- For temperatures at which the CO,CO_2 line lies below the metal oxide line, carbon monoxide can reduce the metal oxide to the metal and is oxidized to carbon dioxide.

Figure 6.3 shows an Ellingham diagram for a selection of common metals. In principle, production of all the metals shown in the diagram, even magnesium and calcium, could be accomplished by **pyrometallurgy**, heating with a reducing agent. However, there are severe

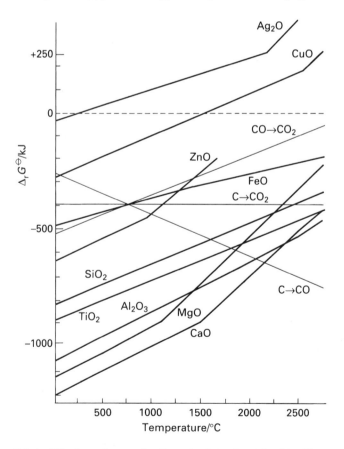

6.3 An Ellingham diagram for the reduction of metal oxides. The standard reaction Gibbs energies are for the formation of the oxides from the metal and for the three carbon oxidations quoted in the text.

practical limitations. Efforts to produce aluminum by pyrometallurgy (most notably in Japan, where electricity is expensive) were frustrated by the volatility of Al_2O_3 at the very high temperatures required. A difficulty of a different kind is encountered in the pyrometallurgical extraction of titanium, where titanium carbide, TiC, is formed instead of the metal. In practice, pyrometallurgical extraction of metals is confined principally to magnesium, iron, cobalt, nickel, zinc, and a variety of ferroalloys (alloys with iron).

Example 6.1 Using an Ellingham diagram

What is the lowest temperature at which ZnO can be reduced to zinc metal by carbon? What is the overall reaction at this temperature?

Answer The C,CO line in Fig. 6.3 lies below the ZnO line at approximately 950 °C; above this temperature, reduction of the metal oxide is spontaneous. The contributing reactions are reaction (a) and

$$Zn(g) + \tfrac{1}{2}O_2(g) \longrightarrow ZnO(s)$$

so the overall reaction is the difference, or

$$C(s) + ZnO(s) \longrightarrow CO(g) + Zn(g)$$

The physical state of zinc is given as a gas because the element boils at 907 °C (the corresponding inflexion in the ZnO line in the Ellingham diagram can be seen in Fig. 6.3).

Self-test 6.1 What is the minimum temperature for reduction of MgO by carbon?

Similar principles apply to reductions using other reducing agents. For instance, an Ellingham diagram can be used to explore whether a metal M' can be used to reduce the oxide of another metal M. In this case, we note from the diagram whether at a temperature of interest the $M'O$ line lies below the MO line, for M' is now taking the place of C. When

$$\Delta_r G^{\ominus} = \Delta_r G^{\ominus}(M', M'O) - \Delta_r G^{\ominus}(M, MO)$$

is negative, where the Gibbs energies refer to the reactions

(a) $M'(s \text{ or } 1) + \tfrac{1}{2}O_2(g) \longrightarrow M'O(s)$ $\Delta_r G^{\ominus}(M', M'O)$
(b) $M(s \text{ or } 1) + \tfrac{1}{2}O_2(g) \longrightarrow MO(s)$ $\Delta_r G^{\ominus}(M, MO)$

the reaction

(a − b) $MO(s) + M'(s \text{ or } 1) \longrightarrow M(s \text{ or } 1) + M'O(s)$

(and its analogs for MO_2, and so on) is feasible. For example, because in Fig. 6.3 the line for magnesium lies above the line for silicon at temperatures below 2200 °C, magnesium may be used to reduce SiO_2 below that temperature. This reaction has in fact been employed to produce low-grade silicon.

> An Ellingham diagram summarizes the temperature dependence of the standard Gibbs energies of formation of metal oxides, and hence may be used to determine the temperature at which reduction by carbon or carbon monoxide becomes spontaneous.

(b) A survey of chemical reductions

Industrial processes for achieving the reductive extraction of metals show a greater variety than the thermodynamic analysis might suggest. An important factor is that the ore and carbon are both solids, and a reaction between two coarse solids is rarely fast. Most processes exploit gas-solid or liquid-solid reactions. Current industrial processes are quite varied in the strategies they adopt to ensure economical rates, exploit materials, and avoid environmental problems. We can explore these strategies by considering three important examples that reflect low, moderate, and extreme difficulty of reduction.

The least difficult reductions include that of copper ores. Roasting and smelting are still widely used in the pyrometallurgical extraction of copper. However, some recent techniques seek to avoid the major environmental problems caused by the production of the large quantity of SO_2 that accompanies roasting. One promising development is the **hydrometallurgical extraction** of copper, the extraction of a metal by reduction of aqueous solutions of its ions, by using H_2 or scrap iron as the reducing agent. In this process, Cu^{2+} ions leached from low-grade ores by acid or bacterial action are reduced by hydrogen in the reaction

$$Cu^{2+}(aq) + H_2(g) \longrightarrow Cu(s) + 2\,H^+(aq)$$

or by a similar reduction using iron. As well as being environmentally relatively benign, this process also allows economic exploitation of lower grade ores.

That extraction of iron is of intermediate difficulty is shown by the fact that the Iron Age followed the Bronze Age. In economic terms, iron ore reduction is the most important application of carbon pyrometallurgy. In a blast furnace (Fig. 6.4), which is still the major source of the element, the mixture of iron ores (Fe_2O_3, Fe_3O_4), coke (C), and limestone ($CaCO_3$) is heated with a blast of hot air. Combustion of coke in this air blast raises the temperature to $2000\,°C$, and the carbon burns to carbon monoxide in the lower part of the furnace. The supply of Fe_2O_3 from the top of the furnace meets the hot CO rising from below. The iron(III) oxide is reduced, first to Fe_3O_4 and then to FeO at 500–$700\,°C$, and the CO is oxidized to CO_2. At the same time, limestone is converted to lime (CaO), so increasing the CO_2 content of the exhaust gases. The final reduction to iron occurs between 1000 and $1200\,°C$ in the central region of the furnace:

$$FeO(s) + CO(g) \longrightarrow Fe(s) + CO_2(g)$$

A sufficient supply of CO is assured by the reaction

$$CO_2(g) + C(s) \longrightarrow 2\,CO(g)$$

which augments the carbon monoxide formed by the incomplete combustion of carbon. This series of gas–solid reactions accomplishes the overall solid–solid reaction between the ore and the coke.

The function of the lime, CaO, formed by the thermal decomposition of calcium carbonate is to combine (in a Lewis acid–base reaction) with the silicates present in the ore to form a molten layer of slag in the hottest (lowest) part of the furnace. Slag is less dense than iron and can be drained away. The iron formed melts at about $400\,°C$ below the melting point of the pure metal on account of the dissolved carbon it contains. The iron, the densest phase, settles to the bottom and is drawn off to solidify into 'pig iron', in which the carbon content is high (about 4 per cent by mass). The manufacture of steel is then a series of reactions in which the carbon content is reduced and other metals are used to form alloys with the iron.

More difficult than the extraction of either copper or iron is the extraction of silicon from its oxide: indeed, silicon is very much an element of the twentieth century. Silicon of 96 to 99 per cent purity is prepared by reduction of quartzite or sand (SiO_2) with high purity coke. The Ellingham diagram shows that the reduction is feasible only at temperatures in excess of about $1500\,°C$. This high temperature is achieved in an electric arc furnace in the presence of excess silica (to prevent the accumulation of SiC):

$$SiO_2(l) + 2\,C(s) \xrightarrow{1500\,°C} Si(l) + 2\,CO(g)$$
$$2\,SiC(s) + SiO_2(l) \longrightarrow 3\,Si(l) + 2\,CO(g)\cdot$$

Very pure silicon (for semiconductors) is made by converting crude silicon to volatile compounds, such as $SiCl_4$. These compounds are purified by exhaustive fractional distillation and then reduced to silicon with pure hydrogen. The resulting semiconductor-grade silicon is

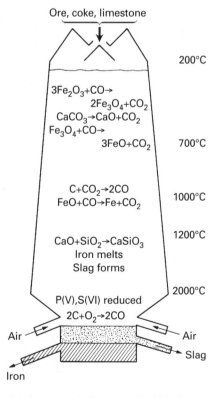

6.4 A schematic diagram of a blast furnace showing the typical composition and temperature profile.

melted and large single crystals are pulled slowly from the melt: this procedure is called the **Czochralski process**.

The Ellingham diagram shows that the direct reduction of Al_2O_3 with carbon becomes feasible only above 2000 °C, which is uneconomically expensive and complicated by features common to high temperature chemistry, such as the volatility of the oxide. However, the reduction can be brought about electrolytically, and all modern production uses the **Hall–Héroult process**, which was invented in 1886 independently by Charles Hall in the United States and Paul Héroult in France.

Electrolytic reduction is a technique for driving a reduction reaction by using an applied potential difference, E. If the reaction Gibbs energy is $\Delta_r G$, thermodynamic arguments lead to the conclusion that

$$E(\text{applied}) \geq \frac{\Delta_r G}{\nu F} \tag{3}$$

where ν is the number of electrons transferred in the reaction (for the chemical equation as written). For a reaction in which $\Delta_r G = +100 \text{ kJ mol}^{-1}$ and $\nu = 1$, the potential difference required to bring about the reaction is $E(\text{applied}) \approx 1$ V or more.

In practice, the bauxite ore used as a source of aluminum is a mixture of the acidic oxide SiO_2 and the amphoteric oxides Al_2O_3 and Fe_2O_3 (plus some TiO_2). The Al_2O_3 is extracted with aqueous sodium hydroxide, which separates the aluminum and silicon oxides from the iron(III) oxide (which needs a more concentrated alkali for significant reaction). Neutralization of the solution with CO_2 results in the precipitation of $Al(OH)_3$, leaving silicates in solution. The aluminum hydroxide is then dissolved in fused cryolite (Na_3AlF_6) and the melt is reduced electrolytically at a steel cathode with graphite anodes. The latter participate in the overall reaction, which is

$$2\,Al_2O_3 + 3\,C \longrightarrow 4\,Al + 3\,CO_2$$

Because the electrode material is consumed, it needs to be continuously replaced. The commercial process uses an electrode potential of about 4.5 V with a current density of about 1 $A\,cm^{-2}$. This current density might not seem much, but when multiplied by the total area of the electrodes, the current required is enormous. Moreover, because the power (in watts) is the product of current and potential ($P = IE$), the power consumption of a typical plant is huge. Consequently, aluminum is often produced where electricity is cheap (in Quebec, for instance) and not where bauxite is mined (in Jamaica, for example).

> *Electrolysis may be used to bring about a non-spontaneous reduction; the minimum potential required is given by eqn 3.*

6.2 Elements extracted by oxidation

The halogens are the most important elements extracted by oxidation. The standard reaction Gibbs energy for the oxidation of Cl^- ions in water

$$2\,Cl^-(aq) + 2\,H_2O(l) \longrightarrow 2\,OH^-(aq) + H_2(g) + Cl_2(g)$$
$$\Delta_r G^{\ominus} = +422 \text{ kJ mol}^{-1}$$

is strongly positive, which suggests that electrolysis is required. The minimum potential difference that can achieve the oxidation of Cl^- is about 2.2 V (because $\nu = 2$ for the reaction as written).

It may appear that there is a problem with the competing reaction

$$2\,H_2O(l) \longrightarrow 2\,H_2(g) + O_2(g) \qquad \Delta_r G^{\ominus} = +414 \text{ kJ mol}^{-1}$$

which can be driven forward by a potential difference of only 1.2 V (in this reaction, $\nu = 4$). However, the rate of oxidation of water is very slow at potentials at which it first becomes favorable thermodynamically. This slowness is expressed by saying that the reduction requires a high **overpotential**, η, the potential that must be applied in addition to the equilibrium value before a significant rate of reaction is achieved.[1] Consequently, the electrolysis of brine produces Cl_2, H_2, and aqueous NaOH, but not much O_2. One problem faced by the industry is therefore to try to develop uses for both Cl_2 and NaOH in a market balance as closely adjusted to the stoichiometry of the reaction as possible. Environmental pressures to decrease the consumption of chlorine have resulted in the exploitation of alternative sources of NaOH, such as

$$Na_2CO_3(aq) + Ca(OH)_2(aq) \longrightarrow CaCO_3(s) + 2\,NaOH(aq)$$

The sodium carbonate used in this process is mined and the calcium hydroxide is formed by roasting calcium carbonate and hydrating the resulting oxide.

Oxygen, not fluorine, is produced if aqueous solutions of fluorides are electrolyzed. Therefore, F_2 is prepared by the electrolysis of an anhydrous mixture of potassium fluoride and hydrogen fluoride, an ionic conductor which is molten above $72\,°C$. The more readily oxidizable halogens, Br_2 and I_2, are obtained by chemical oxidation of the aqueous halides with chlorine.

Because O_2 is available from fractional distillation of air, chemical methods of oxygen production are not necessary (but may become necessary as we colonize other planets). Sulfur is an interesting mixed case. Elemental sulfur is either mined or produced by oxidation of the H_2S that is removed from 'sour' natural gas and crude oil by trapping in 2-hydroxyethylamine ($HOCH_2CH_2NH_2$). The oxidation is accomplished by the **Claus process**, which consists of two stages. In the first, some hydrogen sulfide is oxidized to sulfur dioxide:

$$2\,H_2S + 3\,O_2 \longrightarrow 2\,SO_2 + 2\,H_2O$$

In the second stage, this sulfur dioxide is allowed to react in the presence of a catalyst with more hydrogen sulfide:

$$2\,H_2S + SO_2 \xrightarrow{\text{Oxide catalyst, } 300\,°C} 3\,S + 2\,H_2O$$

The catalyst is typically Fe_2O_3 or Al_2O_3. The Claus process is environmentally benign, for otherwise it would be necessary to burn the toxic hydrogen sulfide to polluting sulfur dioxide.

The only important metals obtained by oxidation are the ones that occur native (that is, as the element). Gold is an example, because it is difficult to separate the granules of metal in low-grade ores by simple 'panning'. The dissolution of gold depends on oxidation, which is favored by complexation with CN^- ions to form $[Au(CN)_2]^-$. This complex is then reduced to the metal by reaction with another reactive metal, such as zinc:

$$2\,[Au(CN)_2]^-(aq) + Zn(s) \longrightarrow 2\,Au(s) + [Zn(CN)_4]^{2-}(aq)$$

Elements obtained by oxidation include the halogens, sulfur, and (in the course of their purification) certain noble metals.

Reduction potentials

The driving reaction in a blast furnace is the oxidation of carbon or carbon monoxide, and these reducing agents are in direct contact with the metal oxide. In electrochemical metallurgy, however, the reducing agent and the species undergoing reduction are physically separated. Indeed, the driving reaction for the non-spontaneous redox reaction (or physical process) is at a distant location, such as a power station.

---

[1] Overpotential is treated more fully in Section 6.4a.

6.3 Redox half-reactions

It is convenient to think of a redox reaction as the sum of two conceptual **half-reactions** in which the electron loss (oxidation) and gain (reduction) are displayed explicitly. In a reduction half-reaction, a substance gains electrons, as in

$$2\,H^+(aq) + 2\,e^- \longrightarrow H_2(g)$$

In an oxidation half-reaction, a substance loses electrons, as in

$$Zn(s) \longrightarrow Zn^{2+}(aq) + 2\,e^-$$

A state is not ascribed to the electrons in the equation of a half-reaction: the division into half-reactions is only conceptual and may not correspond to an actual physical separation of the two processes.

The oxidized and reduced species in a half-reaction constitute a redox **couple**. A couple is written with the oxidized species before the reduced, as in H^+/H_2 and Zn^{2+}/Zn, and typically the phases are not shown. It is usually desirable to adopt the convention of writing all half-reactions as reductions. Because an oxidation is the reverse of a reduction, we can write the second of the two half-reactions above as the reduction

$$Zn^{2+}(aq) + 2\,e^- \longrightarrow Zn(s)$$

The overall reaction is then the difference of the two reduction half-reactions. It may be necessary to multiply each half-reaction by a factor to ensure that the numbers of electrons match. This procedure is similar to that adopted in the discussion of an Ellingham diagram, in which all the reactions are written as oxidations, and the overall reaction is the difference of reactions with matching numbers of O atoms.

> A redox reaction is regarded as the outcome of a reduction and an oxidation half-reaction; all half-reactions are written as reductions, and the overall chemical equation is the difference of the chemical equations for two reduction half-reactions.

(a) Standard potentials

Because the overall chemical equation is the difference of two reduction half-reactions, the standard Gibbs energy of the overall reaction is the difference of the standard Gibbs energies of the two half-reactions. The overall reaction is favorable (in the sense that $K > 1$) in the direction that corresponds to a negative value of the resulting overall $\Delta_r G^{\ominus}$.

Because reduction half-reactions must always occur in pairs in any actual reaction, only the difference in their standard Gibbs energies has any significance. Therefore, we can choose one half-reaction to have $\Delta_r G^{\ominus} = 0$, and report all other values relative to it. By convention, the specially chosen half-reaction is the reduction of hydrogen ions:

$$2\,H^+(aq) + 2\,e^- \longrightarrow H_2(g) \qquad \Delta_r G^{\ominus} = 0 \tag{4}$$

at all temperatures. With this choice, the standard Gibbs energy for the reduction of Zn^{2+} ions, for example, is found by determining experimentally that

$$Zn^{2+}(aq) + H_2(g) \longrightarrow Zn(s) + 2\,H^+(aq) \qquad \Delta_r G^{\ominus} = +147 \text{ kJ mol}^{-1}$$

Then, because the H^+ reduction half-reaction makes zero contribution to the reaction Gibbs energy (according to our convention), it follows that

$$Zn^{2+}(aq) + 2\,e^- \longrightarrow Zn(s) \qquad \Delta_r G^{\ominus} = +147 \text{ kJ mol}^{-1}$$

Standard reaction Gibbs energies may be measured by setting up a galvanic cell (an electrochemical cell in which a chemical reaction is used to generate an electric current) in which the reaction driving the electric current through the external circuit is the reaction of

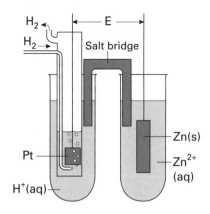

6.5 A schematic diagram of a galvanic cell. The standard cell potential, $E^{\ominus}$, is the potential difference when the cell is not generating current and all the substances are in their standard states.

interest (Fig. 6.5). The potential difference between its electrodes is then measured,[2] and, if desired, converted to a Gibbs energy by using $\Delta_r G = -\nu F E$. Tabulated values—normally for standard conditions[3]—are usually kept in the units in which they were measured, namely volts (V).

The potential that corresponds to the $\Delta_r G^{\ominus}$ of a half-reaction is written $E^{\ominus}$, with

$$\Delta_r G^{\ominus} = -\nu F E^{\ominus} \tag{5}$$

The potential $E^{\ominus}$ is called the **standard potential** (or standard reduction potential, to emphasize that the half-reaction is a reduction). Because $\Delta_r G^{\ominus}$ for the reduction of H^+ is arbitrarily set at zero, the standard potential of the H^+/H_2 couple is also zero at all temperatures:

$$2\,H^+(aq) + 2\,e^- \longrightarrow H_2(g) \qquad E^{\ominus}(H^+, H_2) = 0 \tag{6}$$

Similarly, for the Zn^{2+}/Zn couple, for which $\nu = 2$, it follows from the measured value of $\Delta_r G^{\ominus}$ that at 25 °C:

$$Zn^{2+}(aq) + 2\,e^- \longrightarrow Zn(s) \qquad E^{\ominus}(Zn^{2+}, Zn) = -0.76\ V$$

Because the standard reaction Gibbs energy is the difference of the $\Delta_r G^{\ominus}$ for the half-reactions, $E^{\ominus}$ for an overall reaction is also the difference of the two standard potentials of the reduction half-reactions into which the overall reaction can be divided. Thus, from the half-reactions

(a) $2\,H^+(aq) + 2\,e^- \longrightarrow H_2(g) \qquad E^{\ominus} = 0$

(b) $Zn^{2+}(aq) + 2\,e^- \longrightarrow Zn(s) \qquad E^{\ominus} = -0.76\ V$

it follows that the difference (a − b) is

$$2\,H^+(aq) + Zn(s) \longrightarrow Zn^{2+}(aq) + H_2(g) \qquad E^{\ominus} = +0.76\ V$$

The consequence of the negative sign in eqn 5, is that *a reaction is favorable if $E^{\ominus} > 0$*. Because $E^{\ominus} > 0$ for the reaction above ($E^{\ominus} = +0.76\ V$), we know that zinc has a thermodynamic tendency to reduce H^+ ions under standard conditions (acidic solution, pH = 0, and Zn^{2+} at unit activity). The same is true for any couple with a negative standard potential.

> A reaction is favorable if $E^{\ominus} > 0$, where $E^{\ominus}$ is the difference of the standard potentials corresponding to the half-reactions into which the overall reaction may be divided.

(b) The electrochemical series

A negative standard potential signifies a couple in which the reduced species (the Zn in Zn^{2+}/Zn) is a reducing agent for H^+ ions under standard conditions in aqueous solution. That is, if $E^{\ominus}(Ox, Red) < 0$, then the substance 'Red' is a strong enough reducing agent to reduce H^+ ions (in the sense that $K > 1$).

A short list of $E^{\ominus}$ values at 25 °C is given in Table 6.1. The list is arranged in the order of the **electrochemical series**:

Ox/Red couple with strongly positive $E^{\ominus}$ [Ox is strongly oxidizing]
⋮
Ox/Red couple with strongly negative $E^{\ominus}$ [Red is strongly reducing]

---

[2] In practice, we must ensure that the cell is acting reversibly in a thermodynamic sense, which means that the potential difference must be measured with negligible current flowing.

[3] Standard conditions are all substances at **1 bar** pressure and unit activity. For our purposes, there is negligible difference using **1 atm** in place of **1 bar**. For reactions involving H^+ ions, standard conditions correspond to pH = 0, approximately **1 M** acid. Pure solids and liquids have unit activity.

Table 6.1 Selected standard potentials at 25°C

Couple	$E^{\ominus}/V$
$F_2(g) + 2e^- \rightarrow 2F^-(aq)$	+3.05
$Ce^{4+}(aq) + e^- \rightarrow Ce^{3+}(aq)$	+1.76
$MnO_4^-(aq) + 8H^+(aq) + 5e^- \rightarrow Mn^{2+}(aq) + 4H_2O(l)$	+1.51
$Cl_2(g) + 2e^- \rightarrow 2Cl^-(aq)$	+1.36
$O_2(g) + 4H^+(aq) + 4e^- \rightarrow 2H_2O(aq)$	+1.23
$[IrCl_6]^{2-}(aq) + e^- \rightarrow [IrCl_6]^{3-}(aq)$	+0.87
$Fe^{3+}(aq) + e^- \rightarrow Fe^{2+}(aq)$	+0.77
$[PtCl_4]^{2-}(aq) + 2e^- \rightarrow Pt(s) + 4Cl^-(aq)$	+0.76
$I_3^-(aq) + 2e^- \rightarrow 3I^-(aq)$	+0.54
$[Fe(CN)_6]^{3-}(aq) + e^- \rightarrow [Fe(CN)_6]^{4-}(aq)$	+0.36
$AgCl(s) + e^- \rightarrow Ag(s) + Cl^-(aq)$	+0.22
$2H^+(aq) + 2e^- \rightarrow H_2(g)$	0
$AgI(s) + e^- \rightarrow Ag(s) + I^-(aq)$	−0.15
$Fe^{2+}(aq) + 2e^- \rightarrow Fe(s)$	−0.44
$Zn^{2+}(aq) + 2e^- \rightarrow Zn(s)$	−0.76
$Al^{3+}(aq) + 3e^- \rightarrow Al(s)$	−1.68
$Ca^{2+}(aq) + 2e^- \rightarrow Ca(s)$	−2.87
$Li^+(aq) + e^- \rightarrow Li(s)$	−3.04

An important feature of the electrochemical series is that the reduced member of a couple has a thermodynamic tendency to reduce the oxidized member of any couple that lies above it in the series. Note the classification refers to the thermodynamic aspect of the reaction—its spontaneity, not its rate.

> The oxidized member of a couple is a strong oxidizing agent if $E^{\ominus} > 0$; the reduced member is a strong reducing agent if $E^{\ominus} < 0$.

Example 6.2 Using the electrochemical series

Among the couples in Table 6.1 is the permanganate ion, MnO_4^-, the common analytical reagent used in redox titrations of iron. Which of the ions Fe^{2+}, Cl^-, and Ce^{3+} can permanganate oxidize in acidic solution?

Answer The standard potential of MnO_4^- to Mn^{2+} in acidic solution is +1.51 V. The standard potentials for the listed ions are +0.77, +1.36, and +1.76 V, respectively. It follows that permanganate ions are sufficiently strong oxidizing agents in acidic solution (pH = 0) to oxidize the first two ions, which have less positive standard potentials. Precautions should be taken to avoid the rapid oxidation of Cl^- ions in titrations of iron solutions with permanganate. Permanganate ions cannot oxidize Ce^{3+}, which has a more positive standard potential. It should be noted that the presence of other ions in the solution can modify the potentials and the conclusions. This variation with conditions is particularly important in the case of hydrogen ions, and the influence of **pH** is discussed in Section 6.5.

Self-test 6.2 Another common analytical oxidizing agent is an acidic solution of dichromate ions, $Cr_2O_7^{2-}$, for which $E^{\ominus} = +1.38$ V. Is the solution useful for titration of Fe^{2+}? Could there be a side reaction when Cl^- is present?

(c) The Nernst equation

To judge the tendency of a reaction to run in a particular direction at an arbitrary composition, we need to know the sign and value of $\Delta_r G$ at that composition. For this information, we use the thermodynamic result that

$$\Delta_r G = \Delta_r G^{\ominus} + RT \ln Q \qquad (7)a$$

where Q is the **reaction quotient**[4]

$$a \, Ox_A + b \, Red_B \longrightarrow a' \, Red_A + b' \, Ox_B \qquad Q = \frac{[Red_A]^{a'} [Ox_B]^{b'}}{[Ox_A]^{a} [Red_B]^{b}} \qquad (8)b$$

The reaction is spontaneous under the stated conditions if $\Delta_r G < 0$. This criterion can be expressed in terms of potentials by substituting $E = -\Delta_r G / \nu F$ and $E^{\ominus} = -\Delta_r G^{\ominus} / \nu F$, which gives the **Nernst equation**:

$$E = E^{\ominus} - \frac{RT}{\nu F} \ln Q \qquad (9)$$

A reaction is spontaneous if, under the prevailing conditions, $E > 0$ for then $\Delta_r G < 0$. At equilibrium $E = 0$ and $Q = K$, so eqn 8 implies the following very important relation between the standard potential of a reaction and its equilibrium constant at a temperature T:

$$\ln K = \frac{\nu F E^{\ominus}}{RT} \qquad (10)$$

Table 6.2 The relation between K and $E^{\ominus}$

$E^{\ominus}/V$	K
+2	10^{34}
+1	10^{17}
0	1
−1	10^{-17}
−2	10^{-34}

Table 6.2 lists the value of K that corresponds to potentials in the range −2 to +2 V, with $\nu = 1$ and at 25 °C. The table shows that, although electrochemical data are often compressed into the range −2 to +2 V, that narrow range corresponds to 68 orders of magnitude in the value of the equilibrium constant.

If we regard E as the difference of two potentials, just as $E^{\ominus}$ is the difference of two *standard* potentials, then the potential of each couple can be written as in eqn 8, but with

$$a \, Ox + \nu \, e^- \longrightarrow a' \, Red \qquad Q = \frac{[Red]^{a'}}{[Ox]^{a}} \qquad (11)$$

Notice that the electrons do not appear in the expression for Q.

> The potential of a reaction at an arbitrary composition of the reaction mixture is given by the Nernst equation, eqn 8, and the relation between the standard potential and the equilibrium constant is given by eqn 9.

Example 6.3 Using the Nernst equation

What is the dependence of the potential of the H^+/H_2 couple on pH when the hydrogen pressure is 1.00 bar and the temperature is 25 °C?

Answer The reduction half-reaction is $2 \, H^+(aq) + 2 \, e^- \rightarrow H_2(g)$. We can expect a decrease in hydrogen ion concentration (an increase in pH) to lower the tendency of the reaction to form products and hence to lead to an increase in reaction Gibbs energy and therefore a decrease in potential. The Nernst equation for the couple is

$$E(H^+, H_2) = E^{\ominus}(H^+, H_2) - \frac{RT}{2F} \ln \frac{1}{[H^+]^2}$$

$$= \frac{RT \ln 10}{F} \log[H^+] = -(59 \text{ mV})\text{pH}$$

---

[4] The molar concentrations, [X], in the reaction quotient should be interpreted as molar concentrations relative to a standard molar concentration of 1 mol L^{-1}: that is, they are the numerical value of the molar concentration. For gas-phase reactions, the relative molar concentration of a species is replaced by its partial pressure relative to the standard pressure $p^{\ominus} = 1 \text{ bar}$.

We have used $\ln(1/x) = -\ln x$ and $\ln x = (\ln 10)\log x$. That is, the potential becomes more negative by 59 mV for each unit increase in pH. Note that in neutral solution (pH = 7), $E(H^+, H_2) = -0.41$ V.

....

Self-test 6.3 What is the potential for reduction of MnO_4^- to Mn^{2+} in neutral (pH = 7) solution?

....

6.4 Kinetic factors

We have already remarked that thermodynamic considerations can be used only to analyze the spontaneity of a reaction under the prevailing conditions, its *tendency* to occur. Thermodynamics is silent on the rate at which a spontaneous reaction occurs. In this section we summarize some of the information that is available about the rates of redox processes in solution.

(a) Overpotential

A negative standard potential for a metal–metal ion couple indicates that the metal can reduce H^+ ions or any more positive couple under standard conditions in aqueous solution; it does not assure us that a mechanistic pathway exists for this reduction to be realized. There are no fully general rules that predict when reactions are likely to be fast, for the factors that control them are diverse (as we see in Chapter 14). However, a useful rule-of-thumb (but one with notable exceptions) is that couples with potentials that are lower than the hydrogen potential (at the prevailing pH) by more than about 0.6 V bring about the reduction of hydrogen ions to H_2 at a significant rate. Similarly, potentials that are higher than the potential of the couple $O_2, H^+/H_2O$ at the prevailing pH by more than about 0.6 V can bring about the oxidation of water.

The additional potential of 0.6 V is another example of an overpotential, the potential beyond the zero-current ('equilibrium') potential that must exist before the reaction proceeds at a significant rate. Overpotentials have subtle origins and depend on the details of the reaction mechanism. They may also vary between isotopes. For example, the commercial (electrolytic) concentration of D_2O depends on the greater overpotential for the evolution of D_2 than for the evolution of H_2.

The existence of an overpotential explains why some metals reduce acids but not water itself. Such metals (which include iron and zinc) have negative standard potentials, but these potentials are not low enough to achieve the necessary overpotential for the reduction of H^+ in neutral solution (see the example below). However, the difference $E(H^+, H_2) - E(Fe^{2+}, Fe)$ can be increased if $E(H^+, H_2)$ is made more positive, which can be done by lowering the pH from 7 toward a more acidic value. When the difference in potential exceeds about 0.6 V, the metal brings about reduction at an appreciable rate.

> Significant reduction of one couple by another occurs only if the difference in potentials of two couples exceeds a certain characteristic value known as the overpotential.

....

Example 6.4 Taking overpotential into account

Is iron likely to be rapidly oxidized to $Fe^{2+}(aq)$ by water at 25 °C?

Answer We need to judge the difference in potentials of the couples at pH = 7. The Nernst equation for the hydrogen couple (supposing that the partial pressure of hydrogen is about 1 bar) was derived in Example 6.3 and at pH = 7.0 gives $E \approx -0.41$ V. The Nernst equation for the iron couple is

$$Fe^{2+}(aq) + 2e^- \longrightarrow Fe(s) \qquad E = -0.44\ \text{V} + \tfrac{1}{2}(0.059\ \text{V})\log[Fe^{2+}]$$

If the concentration of Fe^{2+} approaches $1\ mol\ L^{-1}$, then $E \approx -0.44$ V. Although the iron couple is more positive than the hydrogen couple, the difference (0.03 V) is less than the 0.6 V typically required for a significant rate of hydrogen evolution. This small value suggests that the reaction will be slow.

Self-test 6.4 If an exposed surface is maintained, magnesium can undergo rapid oxidation in water at $pH = 7$, $25\,°C$. Is the value of E for this reaction in line with the generalization about overpotential stated in the text?

(b) Electron transfer

In some cases, the overpotential can be interpreted mechanistically. The transfer of an electron in bulk solution often takes place by **outer-sphere electron transfer**, a process in which a minimal change occurs in the coordination spheres of the redox centers (the atoms undergoing a change of oxidation number). This process is depicted in Fig. 6.6: we can envisage it as taking place by the transfer of an electron when the two complexes encounter each other in solution and the contact of their coordination spheres provides a route for the migration of the electron.[5] In such processes, the coordination spheres remain intact and the reaction is accompanied by only small changes in metal–ligand distances. Electron transfer of this kind can be fast, and it is often found that the rate is proportional to the exponential of the difference in the standard potentials of the two couples. That is, the more favorable the

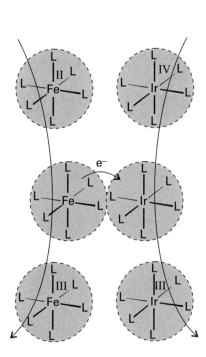

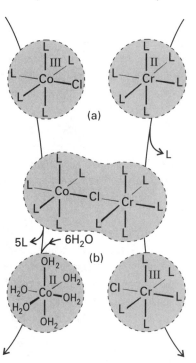

6.6 Schematic diagram of an outer-sphere redox reaction. (a) The reacting ions diffuse together through the solvent and, when they are in contact, an electron is transferred from one to the other. (b) The products then diffuse apart with their original coordination spheres (represented by the shaded spheres) intact. A small readjustment of the metal–ligand bond lengths may occur to accommodate the changes in radius of the central metal ions.

6.7 Schematic diagram of an inner-sphere redox reaction. (a) The reacting ions diffuse together through the solvent and, when they are in contact, a ligand substitution reaction occurs to create a ligand-bridged species. (b) The bridging ligand transfers (with all its electrons, as :Cl⁻, in this example) from one complex to the other, so resulting in a change in oxidation state of the metal atoms in the complexes. In the present example, a substitution-labile Co(II) complex is formed, and it quickly undergoes ligand substitution with the surrounding solvent.

5   The terminology used in this section is explained in Section 7.1.

equilibrium, the faster the reaction. The quantitative form of the dependence is discussed in Section 14.13. One of the techniques that is used to study inorganic reaction mechanisms is outlined in Box 6.1.

The next simplest case is an **inner-sphere electron transfer**, in which the reaction involves a change in the composition of the coordination sphere of a complex. A classic example is the reduction of $[CoCl(NH_3)_5]^{2+}$ by $Cr^{2+}(aq)$, for which the reaction is

$$[CoCl(NH_3)_5]^{2+}(aq) + [Cr(OH_2)_6]^{2+}(aq) + 5\,H_2O(l) + 5\,H^+(aq)$$
$$\longrightarrow [Co(OH_2)_6]^{2+}(aq) + [CrCl(OH_2)_5]^{2+}(aq) + 5\,NH_4^+(aq)$$

Because it is known that replacement of NH_3 or Cl^- at Co(III) is a slow process, the change of ligands on Co must take place after the formation of the more labile Co(II) complex. The observation of Cl^- coordinated to Cr(III) indicates that a bridging Cl atom is involved in the mechanism because the incorporation of Cl^- into the coordination sphere of an aqua Cr(III) ion, $[Cr(OH_2)_6]^{3+}$, is very slow (Fig. 6.7). It is frequently found in this kind of reaction that, the greater the difference in standard potentials of the couples, then the faster the reaction.

A **non-complementary redox reaction** is a reaction in which the changes in oxidation numbers of the oxidizing and reducing agents are unequal. Such reactions are often slow because the reaction cannot be accomplished in a single electron-transfer step. The reaction may be slow if any one of the steps is slow or the concentration of an essential intermediate is low.

Non-complementary reactions are characteristic of the oxidation of p-block oxoanions by d-block ions because the oxidation numbers of the element in such oxoanions often differ by steps of 2 (as in NO_2^-, with oxidation number $+3$, and NO_3^-, with oxidation number $+5$) whereas redox reactions of d-block ions frequently occur in one-electron steps. Even if the net change in oxidation number is 1, more than one step may be involved. For example, the oxidation of sulfite ions, SO_3^{2-} (oxidation number $+4$) to dithionate ions, $S_2O_6^{2-}$ (oxidation number $+5$) by Fe^{3+} ions requires two steps. In the first step, Fe(III) oxidizes a sulfite ion to a radical, and then the reaction is completed by dimerization:

$$2\,(\cdot SO_3^-)(aq) \longrightarrow S_2O_6^{2-}(aq)$$

Single electron-transfer processes may occur either by inner- or outer-sphere mechanisms; non-complementary reactions occur in more than one step.

(c) Empirical generalizations

Inner-sphere mechanisms are common for oxoanions, and a number of trends have been observed. For instance, it is found that *the higher the oxidation number of the parent element in the oxoanion, the slower its reduction*. Two examples of this trend in reaction rates are

$$ClO_4^- < ClO_3^- < ClO_2^- < ClO^-$$

and

$$ClO_4^- < SO_4^{2-} < HPO_4^{2-}$$

The radius of the central atom is also important, and it is found that *the smaller the central atom, the lower the rate of reduction*:

$$ClO_3^- < BrO_3^- < IO_3^-$$

Iodate reactions are fast and equilibrate rapidly enough to be useful in titrations.

A third very useful empirical rule is that *the formation and consumption of common diatomic molecules are slow*. Thus, the formation or consumption of O_2, N_2, and H_2 have complex mechanisms and are usually slow.

The higher the oxidation state and the smaller the radius of an element in an oxoanion, the slower its reduction; diatomic molecule formation and decomposition are generally slow.

Box 6.1 Cyclic voltammetry

In the course of investigations of inorganic reactions, it is often very important to identify the species (aqua ions, complexes, and so on) that are involved in redox reactions and to identify any reaction intermediates. The technique that has become popular with inorganic chemists is *cyclic voltammetry* (Fig. B6.1).

Cyclic voltammetry makes use of a small 'working electrode' of about 1 mm or 2 mm in diameter. The current toward the electrode is limited by the rate at which an electrochemically active species can diffuse toward it. The experiment is conducted in a three-electrode cell in which the potential of the platinum working electrode is held constant with respect to a reference electrode by controlling the current through a counter electrode. The potential is applied to the working electrode in a sawtooth manner. A *cyclic voltammogram* is a plot of the current through the electrode as the potential is increased linearly, then reduced linearly, and so on. Figure B6.2 shows a cyclic voltammogram of a solution containing the redox couple $[Fe(CN)_6]^{3-}/[Fe(CN)_6]^{4-}$ at a platinum electrode.

The experiment begins by setting the potential at a highly cathodic (negative) value (on the left in Fig. B6.2). After a reasonable period of time, all the $[Fe(CN)_6]^{3-}$ near the working electrode will be reduced to $[Fe(CN)_6]^{4-}$. The current will now be very small because the concentration of the oxidized species will be very low near the electrode. A potential scan is then initiated. As the potential approaches the value of $E^{\ominus}$ for the couple, the oxidation of $[Fe(CN)_6]^{4-}$ at the electrode surface begins. As a result, at point A in the illustration, a current begins to flow as some iron(II) complex is oxidized to iron(III). As the value of $E^{\ominus}$ is passed, a condition is approached at which all the complex near the electrode has been oxidized to iron(III). When the oxidizable iron(II) species becomes scarce near the electrode, at Point B, the current falls again as it becomes limited by diffusion of the oxidizable iron(II) complex from

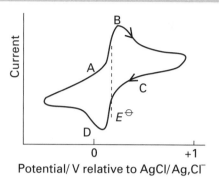

B6.2 A typical cyclic voltammogram. The points are described in the text.

a distance far from the electrode surface. The result is that the current returns to a small value at highly positive potentials (on the right of the illustration). There is a peak in the current–voltage curve near (but not at) $E^{\ominus}$ for the couple.

When the current has reached a low value on the right, the direction of the potential sweep is reversed. As the potential becomes more negative, a current begins to flow in the reverse direction (point C in the illustration). The remainder of the curve is the reverse of the scan in the anodic direction. Notice that the peak current (point B) in the first sweep is to the right (positive) of the peak in the second sweep (point D). These positions reflect the fact that the concentration of the product species becomes small only after $E^{\ominus}$ has been passed. In every case, the peak current will appear at a potential beyond $E^{\ominus}$ in the direction of the sweep of the potential. The two peak current values in the two directions bracket $E^{\ominus}$, so it is easy to recognize the couple responsible for the peaks. In the illustration, $E^{\ominus} = +0.36$ V, which is recognizable as that of the $[Fe(CN)_6]^{4-}/[Fe(CN)_6]^{3-}$ couple.

The difference in potential between the peak current in the two potential sweeps is directly derivable from theory if the electron-transfer reaction at the electrode is rapid and the couple is reversible (in the sense that the concentrations at the surface of the electrode rapidly adopt the values dictated by the Nernst equation). If electron transfer at the electrode is slow, the peak currents are separated by more than the reversible value, so a cyclic voltammogram also provides information about electron-transfer kinetics.

In some cases, the reaction in one direction is simple but the reaction in the reverse direction is complex. For example, oxidation of RCO_2^- produces the radical $RCO_2\cdot$ which decomposes rapidly to $R\cdot$ and CO_2. The result of the decomposition is that the oxidation is irreversible. Neither $R\cdot$ nor CO_2 is reduced at a potential near the standard potential of RCO_2^-. For a couple like this, the 'wave' for oxidation will appear in the cyclic voltammogram, but the reduction wave will be absent in the potential region near $E^{\ominus}$ (Fig. B6.3).

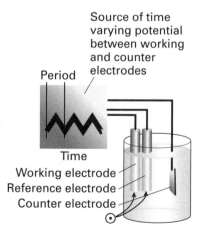

B6.1 The arrangement for a cyclic voltammetry experiment. Each period of the time-varying potential (the first is marked by the vertical lines) corresponds to one complete cycle of a cyclic voltammogram (as in the following two illustrations).

(continued)

Box 6.1 *continued*

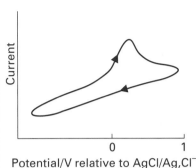

B6.3 A cyclic voltammogram in which the reaction in one direction is simple but that in the reverse direction is complex. The reduction wave is absent.

The electron-transfer couple is characterized by peaks near $E^{\ominus}$ that can be used to identify the electron-transfer reaction and to measure the reduction potential. Whether or not the electron reaction is rapid at the electrode is identified by the separation of the anodic and cathodic peaks. The absence of a peak in one direction shows that the product of the electrode reaction undergoes a reasonably rapid chemical transformation before the potential is swept back in the reverse direction. Often it is possible to recover the peak in the back sweep direction by increasing the rate at which the potential is scanned. This recovery implies that the potential cycle is being completed faster than the irreversible reaction, and hence its onset can be interpreted in terms of the rate of that reaction.

Cyclic voltammetry can be used to identify the specific electron-transfer reaction occurring at the electrode and to decide whether or not it is electrochemically reversible. In fact, considerations beyond those discussed here allow additional inferences to be extracted from cyclic voltammograms. The technique is very powerful for the identification of redox reaction pathways.

Further reading

P.H. Rieger, *Electrochemistry*. Prentice-Hall, Englewood Cliffs (1987).

A.J. Bard and L.R. Faulkner, *Electrochemical methods*. Wiley, New York (1980).

Redox stability in water

An ion or molecule in solution may be destroyed by oxidation or reduction caused by any of the other species present. Therefore, when assessing the stability of a species in solution, we must bear in mind all possible reactants: the solvent, other solutes, the solute itself, and dissolved oxygen. In the following discussion, we shall stress the types of reaction that result from the thermodynamic instability of a solute. We shall also comment on rates, but the trends they show are generally less systematic than those shown by stabilities.

6.5 Reactions with water

Water may act as an oxidizing agent. When it acts in this way it is reduced to H_2:

$$2\,H_2O(l) + 2\,e^- \longrightarrow H_2(g) + 2\,OH^-(aq) \qquad E = -(0.059\text{ V})pH \qquad (12)$$

However, the potential required for the reduction of hydronium ions is the same,

$$2\,H^+(aq) + 2\,e^- \longrightarrow H_2(g) \qquad E = -(0.059\text{ V})pH \qquad (13)$$

and this is the reaction that chemists typically have in mind when they refer to 'the reduction of water'. To derive this expression, we have taken the partial pressure of hydrogen to be 1 bar and have set $\nu = 2$ in the Nernst equation. Water may also act as a reducing agent, when it is oxidized to O_2:

$$O_2(g) + 4\,H^+(aq) + 4\,e^- \longrightarrow 2\,H_2O(l) \qquad E = 1.23\text{ V} - (0.059\text{ V})pH \qquad (14)$$

To derive the pH dependence from the Nernst equation we have assumed that the partial pressure of oxygen is 1 bar and have set $\nu = 4$. The variation of these three potentials with pH is shown in Fig. 6.8 (the additional features shown in this illustration are shown below). Species that can survive in water must have potentials that lie between the limits defined by these processes.

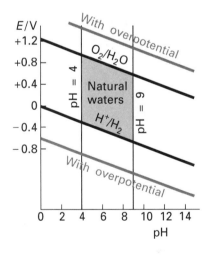

6.8 The stability field of water. The vertical axis is the reduction potential of the redox couple in water: those above the upper line can oxidize water; those below the lower line can reduce water. The gray lines are the boundaries when overpotential is taken into account, and the vertical lines represent the normal range for natural waters. Hence, the gray area is the stability field for natural waters.

(a) Oxidation by water

The reaction of metals with water or aqueous acid is actually the oxidation of the metal by water or hydrogen ions, because the overall reaction is one of the following processes (and their analogs for more highly charged metal ions):

$$M(s) + H_2O(l) \longrightarrow M^+(aq) + \tfrac{1}{2}H_2(g) + OH^-(aq)$$
$$M(s) + H^+(aq) \longrightarrow M^+(aq) + \tfrac{1}{2}H_2(g)$$

These reactions are thermodynamically favorable when M is an s-block metal other than beryllium or a first d-series metal from Group 4 through at least Group 7 (Ti, V, Cr, Mn). A number of other metals undergo similar reactions but with different numbers of electrons being transferred. An example from Group 3 is

$$2\,Sc(s) + 6\,H^+(aq) \longrightarrow 2\,Sc^{3+}(aq) + 3\,H_2(g)$$

When the standard potential for the reduction of a metal ion to the metal is negative, the metal should undergo oxidation in 1 M acid with the evolution of H_2. However, the reaction may be slow, in which case it is appropriate to consider the role of the overpotential, as we have already explained.

Although the reactions of magnesium and aluminum with moist air are spontaneous, both metals can be used for years in the presence of water and oxygen. They survive because they are **passivated**, or protected against reaction, by an impervious film of oxide. Magnesium oxide and aluminum oxide both form a protective skin on the parent metal beneath. A similar passivation occurs with iron, copper, and zinc. The process of 'anodizing' a metal, in which the metal is made an anode (the site of oxidation) in an electrolytic cell, is one in which partial oxidation produces a smooth, hard passivating film on its surface. Anodizing is especially effective for the protection of aluminum.

(b) Reduction by water

The strongly positive potential in eqn 13 shows that acidified water is a poor reducing agent except toward strong oxidizing agents. An example of the latter is $Co^{3+}(aq)$, for which $E^{\ominus}(Co^{3+}, Co^{2+}) = +1.92$ V. It is reduced by water with the evolution of O_2:

$$4\,Co^{3+}(aq) + 2\,H_2O(l) \longrightarrow 4\,Co^{2+}(aq) + O_2(g) + 4\,H^+(aq) \qquad E^{\ominus} = +0.69\ V$$

This $E^{\ominus}$ is very close to the overpotential needed for a significant reaction rate. Because H^+ ions are produced in the reaction, lower acidity favors the oxidation, for lowering the concentration of H^+ ions encourages the formation of the products.

Only a few oxidizing agents (Ag^{2+} is another example) can oxidize water rapidly enough to give appreciable rates of O_2 evolution: the necessary overpotentials are not attained by many reagents. Indeed, standard potentials greater than $+1.23$ V occur for several redox couples that are regularly used in aqueous solution, including Ce^{4+}/Ce^{3+} ($+1.76$ V), the acidified dichromate ion couple $Cr_2O_7^{2-}/Cr^{3+}$ ($+1.38$ V), and the acidified permanganate couple MnO_4^-/Mn^{2+} ($+1.51$ V). The origin of the barrier to reaction is the need to transfer four electrons and to form an oxygen–oxygen double bond.

Given that the rates of redox reactions are often controlled by the slowness with which an oxygen–oxygen bond can be formed, it remains a challenge for inorganic chemists to find good catalysts for O_2 evolution. Some progress has been made using ruthenium complexes. Existing catalysts include the relatively poorly understood coatings that are used in the anodes of cells for the commercial electrolysis of water. They also include the enzyme system that is found in the O_2 evolution apparatus of the plant photosynthetic center. This system is based on manganese with four identifiable levels of oxidation (see Section 19.11). Although Nature is elegant and efficient, it is also complex, and the photosynthetic process is only slowly being elucidated by biochemists and bioinorganic chemists.

Water can act as a reducing agent (that is, be oxidized by other species) provided the overpotential can be exceeded.

(c) The stability field of water

A reducing agent that can reduce water to H_2 rapidly, or an oxidizing agent that can oxidize water to O_2 rapidly, cannot survive in aqueous solution. The **stability field** of water, which is shown tinted in Fig. 6.8, is the range of values of potential and **pH** for which water is thermodynamically stable toward both oxidation and reduction.

The upper and lower boundaries of the stability field are identified by finding the dependence of E on **pH** for the relevant half-reactions. Thus, any species with a potential higher than that given in eqn 13 can oxidize water with the production of O_2. Hence, this expression defines the upper boundary of the stability field. Similarly, the reduction of $H^+(aq)$ to H_2 can be caused by any couple with a potential that is more negative than that given in eqn 12. This expression therefore gives the lower boundary of the stability field.

Couples that are thermodynamically unstable in water lie outside the limits defined by the sloping lines in Fig. 6.8. The couple consists either of too strong a reducing agent (below the H_2 production line) or too strong an oxidizing agent (above the O_2 production line). The stability field in 'natural' water is represented by the addition of two vertical lines at $pH = 4$ and $pH = 9$, which mark the limits on **pH** that are commonly found in lakes and streams. Diagrams like that shown in the illustration are widely used in geochemistry, as we shall see in Section 6.10.

The stability field of water shows the region of pH and reduction potential where couples are neither oxidized by nor reduce hydrogen ions.

6.6 Disproportionation

Because $E^{\ominus}(Cu^+, Cu) = +0.52$ V and $E^{\ominus}(Cu^{2+}, Cu^+) = +0.16$ V, and both potentials lie within the stability field of water, Cu^+ ions neither oxidize nor reduce water. Despite this, Cu(I) is not stable in aqueous solution because it can undergo **disproportionation**, a redox reaction in which the oxidation number of an element is simultaneously raised and lowered. In other words, the element undergoing disproportionation serves as its own oxidizing and reducing agent:

$$2\,Cu^+(aq) \longrightarrow Cu^{2+}(aq) + Cu(s)$$

This reaction is the difference of the following two half-reactions:

$$Cu^+(aq) + e^- \longrightarrow Cu(s) \qquad E^{\ominus} = +0.52 \text{ V}$$
$$Cu^{2+}(aq) + e^- \longrightarrow Cu^+(aq) \qquad E^{\ominus} = +0.16 \text{ V}$$

Disproportionation is spontaneous because $E^{\ominus} = (0.52 \text{ V}) - (0.16 \text{ V}) = +0.36$ V. We can obtain a more quantitative picture of the equilibrium position by employing the relation in eqn 9 with $\nu = 1$ to obtain $K = 1.3 \times 10^6$ at 298 K.

Hypochlorous acid is also subject to disproportionation:

$$5\,HClO(aq) \longrightarrow 2\,Cl_2(g) + ClO_3^-(aq) + 2\,H_2O(l) + H^+(aq)$$

This redox reaction is the difference of the following two half-reactions:

$$4\,HClO(aq) + 4\,H^+(aq) + 4\,e^- \longrightarrow 2\,Cl_2(g) + 4\,H_2O(l) \qquad E^{\ominus} = +1.63 \text{ V}$$
$$ClO_3^-(aq) + 5\,H^+(aq) + 4\,e^- \longrightarrow HClO(aq) + 2\,H_2O(l) \qquad E^{\ominus} = +1.43 \text{ V}$$

So overall $E^{\ominus} = (1.63 \text{ V}) - (1.43 \text{ V}) = +0.20$ V, implying that $K = 3 \times 10^{13}$ at 298 K.

Example 6.5 Assessing the importance of disproportionation

Show that manganate(VI) ions are unstable with respect to disproportionation into Mn(VII) and Mn(II) in acidic aqueous solution.

Answer The overall reaction must involve reduction of one Mn(VI) to Mn(II) balanced by four Mn(VI) oxidizing to Mn(VII). The equation is therefore

$$5\,HMnO_4^-(aq) + 3\,H^+(aq) \longrightarrow 4\,MnO_4^-(aq) + Mn^{2+}(aq) + 4\,H_2O(l)$$

This reaction can be expressed as the difference of the reduction half-reactions

$$HMnO_4^-(aq) + 7\,H^+(aq) + 4\,e^- \longrightarrow Mn^{2+}(aq) + 4\,H_2O(l) \quad E^\ominus = +1.63\text{ V}$$
$$4\,MnO_4^-(aq) + 4\,H^+(aq) + 4\,e^- \longrightarrow 4\,HMnO_4^-(aq) \quad E^\ominus = +0.90\text{ V}$$

The difference of the standard potentials is $+0.73$ V, so the disproportionation is essentially complete $(K = 10^{50}$ at 298 K, because $\nu = 4)$. A practical consequence of the disproportionation is that high concentrations of MnO_4^{2-} ions cannot be obtained in acidic solution but must be prepared in a basic solution.

Self-test 6.5 The standard potentials for the couples Fe^{2+}/Fe and Fe^{3+}/Fe^{2+} are -0.41 V and $+0.77$ V, respectively. Can Fe^{2+} disproportionate under standard conditions?

The reverse of disproportionation is **comproportionation**. In comproportionation, two species with the same element in different oxidation states form a product in which the element is in an intermediate oxidation state. An example is

$$Ag^{2+}(aq) + Ag(s) \longrightarrow 2\,Ag^+(aq) \qquad E^\ominus = +1.18\text{ V}$$

The large positive potential indicates that Ag(II) and Ag(0) are completely converted to Ag(I) in aqueous solution $(K = 1 \times 10^{20}$ at 298 K).

> *Copper(I) and hypochlorous acid are subject to disproportionation in solution; Ag(II) undergoes comproportionation.*

6.7 Oxidation by atmospheric oxygen

The possibility of reaction between the solutes and dissolved O_2 (eqn 13) must be considered when a solution is contained in an open beaker or is otherwise exposed to air. As an example, consider a solution containing Fe^{2+}. The standard potential of the Fe^{3+}/Fe^{2+} couple is $+0.77$ V, which lies within the stability field of water and hence suggests that Fe^{2+} should survive in water. Moreover, we can also infer that the oxidation of metallic iron by $H^+(aq)$ should not proceed beyond Fe(II), because further oxidation to Fe(III) is unfavorable (by 0.77 V) under standard conditions. However, the picture changes considerably in the presence of O_2. In fact, Fe(III) is the most common form of iron in the Earth's crust, and most iron in sediments that have been deposited from aqueous environments is present as Fe(III). The reaction

$$4\,Fe^{2+}(aq) + O_2(g) + 4\,H^+(aq) \longrightarrow 4\,Fe^{3+}(aq) + 2\,H_2O(l)$$

is the difference of the following two half-reactions:

$$O_2(g) + 4\,H^+(aq) + 4\,e^- \longrightarrow 2\,H_2O(l) \qquad E = +1.23\text{ V} - (0.059\text{ V})pH$$
$$Fe^{3+}(aq) + e^- \longrightarrow Fe^{2+}(aq) \qquad E^\ominus = +0.77\text{ V}$$

which implies that $E^\ominus = +0.46$ V and that the oxidation of Fe^{2+} by O_2 is spontaneous (for pH = 0). However, $+0.46$ V is not a large enough overpotential for rapid reaction, and atmospheric oxidation of Fe(II) in aqueous solution is slow in the absence of catalysts. At pH = 7, $E = (0.82\text{ V}) - (0.77\text{ V}) = +0.05$ V and the reaction is less spontaneous and even less likely to occur. As a result, it is possible to use Fe^{2+} aqueous solutions in laboratory procedures without elaborate precautions other than maintaining the pH on the acid side of neutrality.

Example 6.6 Judging the importance of atmospheric oxidation

The oxidation of copper roofs to a substance with a characteristic green color is another example of atmospheric oxidation in a damp environment. Estimate the potential for oxidation of copper by oxygen.

Answer The reduction half-reactions are

$$O_2(g) + 4\,H^+(aq) + 4\,e^- \longrightarrow 2\,H_2O(l) \qquad E = +1.23\ V - (0.059\ V)pH$$
$$Cu^{2+}(aq) + 2\,e^- \longrightarrow Cu(s) \qquad E^{\ominus} = +0.34\ V$$

The difference is $E^{\ominus} = +0.89\ V$ at $pH = 0$ and $+0.48\ V$ at $pH = 7$, so atmospheric oxidation by the reaction

$$2\,Cu(s) + O_2(g) + 4\,H^+(aq) \longrightarrow 2\,Cu^{2+}(aq) + 2\,H_2O(l)$$

is spontaneous when the pH is on the acid side of neutrality. Nevertheless, copper roofs do last for more than a few minutes: their familiar green surface is a passive layer of an almost impenetrable hydrated copper(II) carbonate and sulfate formed from oxidation in the presence of atmospheric CO_2 and SO_2. Near the sea, the copper patina contains substantial amounts of chloride.

Self-test 6.6 The standard potential for the conversion of sulfate ions, $SO_4^{2-}(aq)$, to $SO_2(aq)$ by the reaction $SO_4^{2-}(aq) + 4\,H^+(aq) + 2\,e^- \rightarrow SO_2(aq) + 2\,H_2O(l)$ is $+0.16\ V$. What is the thermodynamically expected fate of SO_2 emitted into fog or clouds?

The diagrammatic presentation of potential data

There are several useful diagrammatic summaries of the relative thermodynamic stabilities of a series of species in which one element exists with a series of different oxidation numbers. We shall describe two of them. *Latimer diagrams* are useful for summarizing quantitative data element by element. *Frost diagrams* are useful for the qualitative portrayal of the relative stabilities of oxidation states. We shall use them frequently in this context in the following chapters to convey the sense of trends in the redox properties of the members of a group. For numerical calculations, however, Latimer diagrams and other explicit tabulations of standard potentials are more convenient.

6.8 Latimer diagrams

In a **Latimer diagram** for an element, the numerical value of the standard potential (in volts) is written over a horizontal line connecting species with the element in different oxidation states. The most highly oxidized form of the element is on the left, and in species to the right the element is in successively lower oxidation states. A Latimer diagram summarizes a great deal of information in a compact form and (as we shall explain) shows the relationships between the various species in a particularly clear and compact manner.

The Latimer diagram for chlorine in acidic solution, for instance, is

$$\underset{+7}{ClO_4^-} \xrightarrow{+1.20} \underset{+5}{ClO_3^-} \xrightarrow{+1.18} \underset{+3}{HClO_2} \xrightarrow{+1.65} \underset{+1}{HClO} \xrightarrow{+1.67} \underset{0}{Cl_2} \xrightarrow{+1.36} \underset{-1}{Cl^-}$$

As in this example, oxidation numbers are sometimes written under (or over) the species. The notation

$$CIO_4^- \xrightarrow{\text{+1.20}} CIO_3^-$$

denotes

$$CIO_4^-(aq) + 2\,H^+(aq) + 2\,e^- \longrightarrow CIO_3^-(aq) + H_2O(l) \qquad E^{\ominus} = +1.20 \text{ V}$$

Similarly,

$$HCIO \xrightarrow{\text{+1.67}} CI_2$$

denotes

$$2\,HCIO(aq) + 2\,H^+(aq) + 2\,e^- \longrightarrow CI_2(g) + 2\,H_2O(l) \qquad E^{\ominus} = +1.63 \text{ V}$$

As in this example, the conversion of a Latimer diagram to a half-reaction often involves balancing elements by including the predominant species present in acidic aqueous solution (H^+ and H_2O). Electric charge is then balanced by the inclusion of the appropriate number of e^-. The standard state for this couple includes the condition that $pH = 0$.

In basic aqueous solution (corresponding to $pH = 14$), the Latimer diagram for chlorine is

$$CIO_4^- \xrightarrow{\text{+0.37}} CIO_3^- \xrightarrow{\text{+0.30}} CIO_2^- \xrightarrow{\text{+0.68}} CIO^- \xrightarrow{\text{+0.42}} CI_2 \xrightarrow{\text{+1.36}} CI^-$$

with a connecting line labelled $+0.89$ from CIO^- to CI^-.

Note that the value for the CI_2/CI^- couple is the same as in acidic solution because its half-reaction does not involve the transfer of protons. Under basic conditions, the predominant species present in solution are OH^- and H_2O, so these species are used to balance the equations for the half-reactions. As an example, the half-reaction for the CIO^-/CI_2 couple in basic solution is

$$2\,CIO^-(aq) + 2\,H_2O(l) + 2\,e^- \longrightarrow CI_2(g) + 4\,OH^-(aq) \qquad E^{\ominus} = +0.42 \text{ V}$$

The standard state for this reaction includes the condition that $pOH = 0$, corresponding to $pH = 14$.

> A Latimer diagram portrays the standard potential of couples diagrammatically: oxidation numbers decrease from left to right and the numerical values of $E^{\ominus}$ in volts are written above the line joining the species involved in the couple.

(a) Nonadjacent species

The Latimer diagram given above includes the standard potential for two nonadjacent species (the couple CIO^-/CI^-). This information is redundant, but it is often included for commonly used couples as a convenience. To derive the standard potential of a nonadjacent couple when it is not listed explicitly we make use of the relation eqn 5 ($\Delta_r G^{\ominus} = -\nu F E^{\ominus}$) and the fact that the overall $\Delta_r G^{\ominus}$ for two successive steps is the sum of the individual values:

$$\Delta_r G^{\ominus} = \Delta_r G^{\ominus\prime} + \Delta_r G^{\ominus\prime\prime}$$

To find the standard potential of the composite process, we convert the individual $E^{\ominus}$ values to $\Delta_r G^{\ominus}$ values by multiplication by the relevant $-\nu F$ factor, add them together, and then convert the sum back to $E^{\ominus}$ for the nonadjacent couple by division by $-\nu F$ for the overall electron transfer:

$$-\nu F E^{\ominus} = -\nu' F E^{\ominus\prime} - \nu'' F E^{\ominus\prime\prime}$$

Because the factors $-F$ cancel and $\nu = \nu' + \nu''$, the net result is

$$E^{\ominus} = \frac{\nu' E^{\ominus\prime} + \nu'' E^{\ominus\prime\prime}}{\nu' + \nu''} \qquad (15)$$

> *The standard potential for a couple involving species with nonadjacent oxidation numbers can be derived from the intermediate values by converting the potentials to standard Gibbs energies; alternatively, use eqn 14 directly.*

Example 6.7 Extracting $E^{\ominus}$ for nonadjacent oxidation states

Use the Latimer diagram to calculate the value of $E^{\ominus}$ for reduction of HClO to Cl^- in aqueous acidic solution.

Answer The oxidation number of chlorine changes from $+1$ to 0 in one step ($\nu' = 1$) and from 0 to -1 ($\nu'' = 1$) in the subsequent step. From the Latimer diagram given in the text above, we can write

$$HClO(aq) + H^+(aq) + e^- \longrightarrow \tfrac{1}{2}Cl_2(g) + H_2O(l) \qquad E^{\ominus\prime} = +1.63 \text{ V}$$
$$\tfrac{1}{2}Cl_2(g) + e^- \longrightarrow Cl^-(aq) \qquad E^{\ominus\prime\prime} = +1.36 \text{ V}$$

We see that $\nu' = 1$ and $\nu'' = 1$, so $\nu = 2$. It follows that the standard potential of the HClO/Cl^- couple is

$$E^{\ominus} = \frac{E^{\ominus\prime} + E^{\ominus\prime\prime}}{2} = \frac{1.63 \text{ V} + 1.36 \text{ V}}{2} = +1.50 \text{ V}$$

Self-test 6.7 Calculate $E^{\ominus}$ for the reduction of ClO_3^- to HClO in aqueous acidic solution.

(b) Disproportionation

Consider the disproportionation

$$2 M^+(aq) \longrightarrow M(s) + M^{2+}(aq)$$

This reaction is spontaneous if $E^{\ominus} > 0$. To analyse this criterion in terms of a Latimer diagram, we express the overall reaction as the difference of two half-reactions:

$$M^+(aq) + e^- \longrightarrow M(s) \qquad E^{\ominus}(R)$$
$$M^{2+}(aq) + e^- \longrightarrow M^+(aq) \qquad E^{\ominus}(L)$$

The designations L and R refer to the relative positions, left and right, of the couples in a Latimer diagram (recall that the more highly oxidized species lies to the left). The standard potential for the overall reaction is

$$E^{\ominus} = E^{\ominus}(R) - E^{\ominus}(L) \qquad (16)$$

which is positive if $E^{\ominus}(R) > E^{\ominus}(L)$. We can conclude that *a species has a thermodynamic tendency to disproportionate into its two neighbors if the potential on the right of the species is higher than the potential on the left.*

An actual example is H_2O_2, which has a tendency to disproportionate into O_2 and H_2O under acid conditions:

$$O_2 \xrightarrow{\;+0.70\;} H_2O_2 \xrightarrow{\;+1.76\;} H_2O$$

To verify this conclusion we could write the two half-reactions:

$$H_2O_2(aq) + 2\,H^+(aq) + 2\,e^- \longrightarrow 2\,H_2O(l) \qquad E^\ominus = +1.76\text{ V}$$
$$O_2(g) + 2\,H^+(aq) + 2\,e^- \longrightarrow H_2O_2(aq) \qquad E^\ominus = +0.70\text{ V}$$

and form the difference:

$$2\,H_2O_2(aq) \longrightarrow 2\,H_2O(l) + O_2(g) \qquad E^\ominus = +1.06\text{ V}$$

Because $E^\ominus > 0$, the disproportionation is spontaneous.

> A species has a thermodynamic tendency to disproportionate into its two neighbors if the potential on the right of the species in a Latimer diagram is higher than the potential on the left.

6.9 Frost diagrams

A **Frost diagram** of an element X is a plot of $NE^\ominus$ for the couple $X(N)/X(0)$ against the oxidation number, N, of the element:

$$X(N) + N\,e^- \longrightarrow X(0) \qquad E^\ominus$$

An example is shown in Fig. 6.9. Because $NE^\ominus$ is proportional to the standard reaction Gibbs energy for the conversion of the species $X(N)$ to the element (explicitly, $NE^\ominus = -\Delta_r G^\ominus/F$, where $\Delta_r G^\ominus$ is the standard reaction Gibbs energy for the half-reaction given above), a Frost diagram can also be regarded as a plot of standard reaction Gibbs energy against oxidation number. It follows that the most stable oxidation state of an element corresponds to the species that lies lowest in its Frost diagram (1).

> A Frost diagram is a plot of $NE^\ominus$ for the couple $X(N)/X(0)$ against the oxidation number N of an element. The most stable oxidation state of a species lies lowest in the diagram.

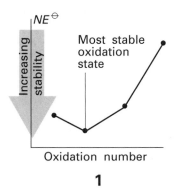

1

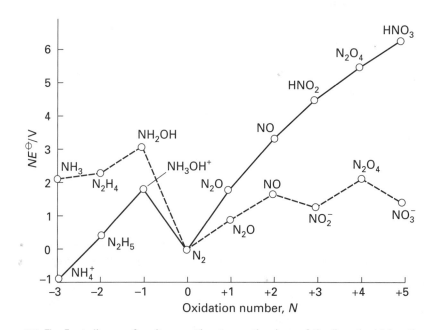

6.9 The Frost diagram for nitrogen: the steeper the slope of the line, the higher the standard potential for the couple. The solid line refers to standard conditions (pH = 0), the dotted line to pH = 14.

Example 6.8 Constructing a Frost diagram

Construct a Frost diagram for oxygen from the Latimer diagram

Answer The oxidation numbers of O are 0, -1, and -2 in the three species. For the change of oxidation number from 0 to -1 (O_2 to H_2O_2), $E^{\ominus} = +0.70$ V and $N = -1$, so $NE^{\ominus} = -0.70$ V. Because the oxidation number of O in H_2O is -2, $N = -2$, and $E^{\ominus}$ for the formation of H_2O is $+1.23$ V, $NE^{\ominus} = -2.46$ V. These results are plotted in Fig. 6.10.

Self-test 6.8 Construct a Frost diagram from the Latimer diagram for Tl:

$$Tl^{3+} \xrightarrow{+1.25} Tl^{+} \xrightarrow{-0.34} Tl$$
$$\underset{+0.72}{\rule{3cm}{0.4pt}}$$

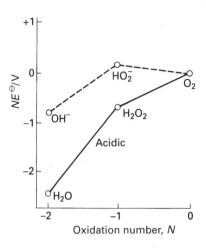

6.10 The Frost diagram for O in acidic solution (full line) and basic solution (broken line).

(a) The interpretation of Frost diagrams

To interpret the qualitative information contained in a Frost diagram it will be useful to keep the following features in mind.

The slope of the line joining any two points in a Frost diagram is equal to the standard potential of the couple formed by the two species the points represent. To see that this is so, consider the two points in Fig. 6.11. One corresponds to the point $(N'E^{\ominus\prime}, N')$ and the other corresponds to the point $(N''E^{\ominus\prime\prime}, N'')$:

$$X(N') + N'\,e^- \longrightarrow X(0) \qquad \Delta_r G^{\ominus\prime} = -N'FE^{\ominus\prime}$$
$$X(N'') + N''\,e^- \longrightarrow X(0) \qquad \Delta_r G^{\ominus\prime\prime} = -N''FE^{\ominus\prime\prime}$$

First, note that the difference of the two half-reactions is

$$X(N') + (N' - N'')\,e^- \longrightarrow X(N'') \qquad \Delta_r G^{\ominus} = -(N' - N'')FE^{\ominus}$$

with

$$\Delta_r G^{\ominus} = \Delta_r G^{\ominus\prime} - \Delta_r G^{\ominus\prime\prime} = -F(N'E^{\ominus\prime} - N''E^{\ominus\prime\prime})$$

It follows from these two equations that

$$E^{\ominus} = \frac{N'E^{\ominus\prime} - N''E^{\ominus\prime\prime}}{N' - N''} \tag{13}$$

Now consider the slope of the straight line joining the two points in the Frost diagram:

$$\text{Slope} = \frac{N'E^{\ominus\prime} - N''E^{\ominus\prime\prime}}{N' - N''}$$

This expression is the same as in eqn 16, so we can conclude that the slope of the line is indeed the standard potential of the couple joined by the straight line.

For an illustration, refer to the oxygen diagram in Fig. 6.10. At the point corresponding to oxidation number -1 (for H_2O_2), $(-1)E^{\ominus} = -0.70$ V, and at oxidation number -2 (for H_2O), $(-2)E^{\ominus} - 2.46$ V. The difference of the two values is -1.76 V. The change in oxidation number of oxygen on going from H_2O_2 to H_2O is -1. Therefore, the slope of the line is $(-1.76 \text{ V})/(-1) = +1.76$ V, in accord with the value for the H_2O_2/H_2O couple in the Latimer diagram.

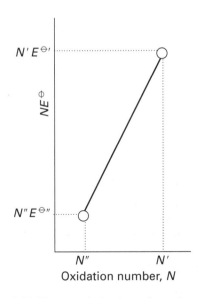

6.11 The general structure of a region of a Frost diagram used to establish the relation betwen the slope of a line and the standard potential of the corresponding couple.

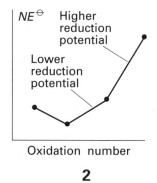

$NE^{\ominus}$ Higher reduction potential

Lower reduction potential

Oxidation number

2

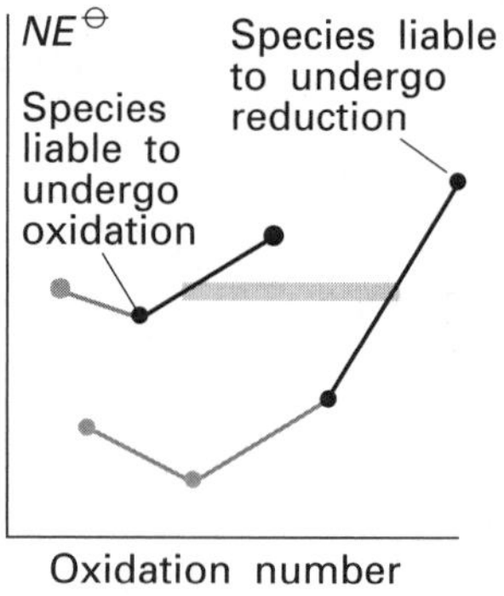

$NE^{\ominus}$

Species liable to undergo reduction

Species liable to undergo oxidation

Oxidation number

3

It follows from this discussion that, the steeper the line joining two points in a Frost diagram, the higher the potential of the corresponding couple (**2**). Therefore, we can infer the spontaneity of the reaction between any two couples by comparing the slopes of the corresponding lines. In particular (**3**):

1 The oxidizing agent in the couple with the more positive slope (the more positive $E^{\ominus}$) is liable to undergo reduction.

2 The reducing agent of the couple with the less positive slope (the less positive $E^{\ominus}$) is liable to undergo oxidation.

For example, the steep slope connecting HNO_3 to lower oxidation numbers in Fig. 6.9 shows that nitric acid is a good oxidizing agent under standard conditions.

> The steeper the line joining two points in a Frost diagram, the higher the potential of the corresponding couple.

(b) Disproportionation and comproportionation

We saw in the discussion of Latimer diagrams that a species is liable to undergo disproportionation if the potential for its reduction from $X(N)$ to $X(N-1)$ is greater than its potential for oxidation from $X(N)$ to $X(N+1)$ (see Section 6.8, where the criterion was discussed in terms of the potentials on the left and right of a species in a Latimer diagram). The same criterion can be expressed in terms of a Frost diagram (**4**):

1 A species in a Frost diagram is unstable with respect to disproportionation if its point lies above the line connecting two adjacent species.

When this criterion is satisfied, the potential for the couple to the left of the species is greater than that for the species on the right. A specific example is NH_2OH in Fig. 6.9, which is unstable with respect to disproportionation into NH_3 and N_2. However, it is one of the intricacies of nitrogen chemistry that slow formation of N_2 often prevents its production. The origin of this rule is illustrated in (**5**), where we show geometrically that the reaction Gibbs energy of a species with intermediate oxidation number lies above the average value for the two terminal species. As a result, there is a tendency for the intermediate species to disproportionate into the two terminal species.

The criterion for comproportionation to be thermodynamically spontaneous is analogous (**6**):

2 Two species will tend to comproportionate into an intermediate species that lies below the straight line joining the terminal species.

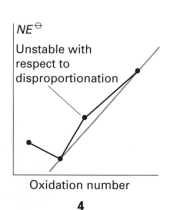

$NE^{\ominus}$

Unstable with respect to disproportionation

Oxidation number

4

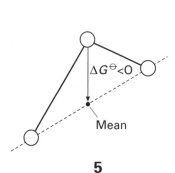

$\Delta G^{\ominus} < 0$

Mean

5

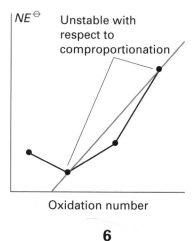

$NE^{\ominus}$

Unstable with respect to comproportionation

Oxidation number

6

A substance that lies below the line connecting its neighbors in a Frost diagram is more stable than they are because their average Gibbs energy is higher (**7**). Accordingly, comproportionation is thermodynamically favorable. The nitrogen in NH_4NO_3, for instance, has two ions with oxidation numbers -3 (NH_4^+) and $+5$ (NO_3^-). Because N_2O, in which the oxidation number of N has the average of these two values, $+1$, lies below the line joining NH_4^+ to NO_3^-, their comproportionation is spontaneous:

$$NH_4^+(aq) + NO_3^-(aq) \longrightarrow N_2O(g) + 2H_2O(l)$$

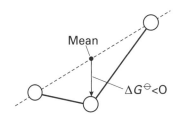

7

The reaction is kinetically inhibited in solution and does not ordinarily occur. However, it does occur in the solid state, when it can be explosively fast: indeed, ammonium nitrate is often used in place of dynamite for blasting rocks.

When the points for three substances lie approximately on a straight line, no one species will be the exclusive product. The three reactions

$$NO(g) + NO_2(g) + H_2O(l) \longrightarrow 2HNO_2(aq) \qquad (rapid)$$
$$3HNO_2(aq) \longrightarrow HNO_3(aq) + 2NO(g) + H_2O(l) \qquad (rapid)$$
$$2NO_2(g) + H_2O(l) \longrightarrow HNO_3(aq) + HNO_2(aq)$$
$$(slow\ and\ normally\ rate\text{-}determining)$$

are examples of this behavior. They are all important in the industrial synthesis of nitric acid by the oxidation of ammonia.

> *Disproportionation is spontaneous if the species lies above a straight line joining its two product species; comproportionation is spontaneous if the intermediate species lies below the straight line joining two reactant species.*

Example 6.9 Using a Frost diagram to judge the stability of ions

Figure 6.12 shows a Frost diagram for manganese. Comment on the stability of Mn^{3+} in aqueous solution.

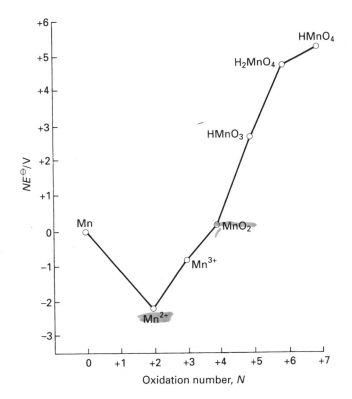

6.12 The Frost diagram for manganese species in acidic solution.

Answer Because Mn^{3+} lies above the line joining Mn^{2+} to MnO_2, it should disproportionate. The disproportionation involves a one-electron transfer, and is therefore likely to be reasonably rapid.

....

Self–test 6.9 What is the oxidation number of Mn in the product when MnO_4^- is used as an oxidizing agent in aqueous acid?

....

6.10 pH dependence

So far, we have discussed Latimer and Frost diagrams only for acidic aqueous solution at $pH = 0$, but they can equally well be constructed for other conditions. As is done in Appendix 2, Latimer diagrams are commonly presented for both $pH = 0$ and $pH = 14$.[6]

(a) Conditional diagrams

'Basic Latimer diagrams' are expressed in terms of reduction potentials at $pH = 14$ (corresponding to $pOH = 0$). The potentials are denoted $E_B^\ominus$ and the dotted line in Fig. 6.9 is a representation of their values in a 'basic Frost diagram'. The important difference from the behavior in acidic solution is the stabilization of NO_2^- against disproportionation: its point in the basic Frost diagram no longer lies above the line connecting its neighbors. The practical outcome is that metal nitrites can be isolated whereas HNO_2 cannot. In some cases, there are marked differences between strongly acidic and basic solutions, as for the phosphorus oxoanions:

$$E^\ominus$$

$$H_3PO_4 \xrightarrow{-0.28} H_3PO_3 \xrightarrow{-0.50} H_3PO_2 \xrightarrow{-0.51} P \xrightarrow{-0.06} PH_3$$

$$E_B^\ominus$$

$$PO_4^{3-} \xrightarrow{-1.12} HPO_3^{2-} \xrightarrow{-1.56} HPO_2^- \xrightarrow{-2.05} P \xrightarrow{-0.89} PH_3$$

This example illustrates an important general point about oxoanions: when their reduction requires removal of oxygen, the reaction consumes H^+ ions, and all oxoanions are stronger oxidizing agents in acidic than in basic solution.

Standard potentials in neutral solution ($pH = 7$) are denoted $E_W^\ominus$. These potentials are particularly useful in biochemical discussions because cell fluids are buffered near $pH = 7$. In that context, a **biochemical standard state** corresponds to $pH = 7$ and is denoted $E^\oplus$ for the potential and $\Delta_r G^\oplus$ for the reaction Gibbs energy. It follows from the Nernst equation that

$$E^\oplus = E^\ominus + 7\nu_{H^+}\left(\frac{RT}{F}\right)\ln 10 \tag{18}$$

where ν_{H^+} is the stoichiometric coefficient of hydrogen ions in the reaction.[7] At 298 K, $7(RT/F)\ln 10 = 0.414$ V.

> *Conditional Latimer and Frost diagrams summarize potential data under the specified conditions of pH; their interpretation is the same as for pH = 0, but oxoanions often display markedly different stabilities. All oxoanions are stronger oxidizing agents in acidic than in basic solution.*

....

[6]   Limited potential data are available for nonaqueous solvents. For example, data are available for liquid ammonia solutions. See W.L. Jolly, *J. Chem. Educ.* **33**, 512 (1956).

[7]   If the hydrogen ions occur as reactants, take ν_{H^+} to be negative.

Example 6.10 Using conditional diagrams

Potassium nitrite is stable in basic solution but, when the solution is acidified, a gas is evolved that turns brown on exposure to air. What is the reaction?

Answer Figure 6.9 shows that the NO_2^- ion in basic solution lies below the line joining NO to NO_3^-; the ion therefore is not liable to disproportionation. On acidification, the HNO_2 point rises and the straightness of the line through NO, HNO_2, and N_2O_4 implies that all three species are present at equilibrium. The brown gas is NO_2 formed from the reaction of NO evolved from the solution with air. In solution, the species of oxidation number +2 (NO) tends to disproportionate. However, the escape of NO from the solution prevents its disproportionation to N_2O and HNO_2.

Self-test 6.10 By reference to Fig. 6.9, compare the strength of NO_3^- as an oxidizing agent in acidic and basic solution.

(b) Pourbaix diagrams

Diagrams can also be used for discussing the general relations between redox activity and Brønsted acidity. In a **Pourbaix diagram**, the regions indicate the conditions of **pH** and potential under which a species is thermodynamically stable.

Figure 6.13 is a simplified Pourbaix diagram for iron, omitting such low concentration species as oxygen-bridged dimers. This diagram is useful for the discussion of iron species in natural water because the total iron concentration is low.

We can see how the diagrams are constructed by considering some of the reactions involved. The reduction half-reaction

$$Fe^{3+}(aq) + e^- \longrightarrow Fe^{2+}(aq) \qquad E^\ominus = +0.77 \text{ V}$$

does not involve H^+ ions, and so (if we ignore the effect of **pH** on the activity of the iron ions) its potential is independent of **pH**, giving a horizontal line on the diagram. If the environment contains a couple with a potential above this line (a more positive, more oxidizing couple), then the oxidized species, Fe^{3+}, will be the major species. Hence, the horizontal line toward the top left of the diagram is a boundary that separates the regions where Fe^{3+} and Fe^{2+} dominate.

Another reaction to consider is

$$Fe^{3+}(aq) + 3 H_2O(l) \longrightarrow Fe(OH)_3(s) + 3 H^+(aq)$$

This reaction is not a redox reaction (there is no change in oxidation number of any element), and the boundary between the region where Fe^{3+} is soluble and the region where $Fe(OH)_3$ precipitates and limits $Fe^{3+}(aq)$ to very low concentration is independent of any redox couples. However, this boundary does depend on **pH**, with $Fe^{3+}(aq)$ favored by low **pH** and the $Fe(OH)_3$ favored by high **pH**. We shall accept that Fe^{3+} is the dominant species in the solution if its concentration exceeds $10 \ \mu\text{mol L}^{-1}$ (a typical freshwater value). The equilibrium concentration of Fe^{3+} varies with **pH**, and the vertical boundary represents the **pH** at which the Fe^{3+} becomes dominant according to this definition.

A third reaction we need to consider is the reduction

$$Fe(OH)_3(s) + 3 H^+(aq) + e^- \longrightarrow Fe^{2+}(aq) + 3 H_2O(l)$$

This reaction depends on the **pH**, as we can see by noting that H^+ ions appear in the half-reaction, and as is shown explicitly by writing the Nernst equation:

$$\begin{aligned} E &= E^\ominus - (0.059 \text{ V}) \log \frac{[Fe^{2+}]}{[H^+]^3} \\ &= E^\ominus - (0.059 \text{ V}) \log[Fe^{2+}] - (0.177 \text{ V})\text{pH} \end{aligned} \qquad (19)$$

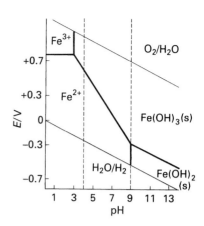

6.13 A simplified Pourbaix diagram for some important naturally occurring compounds of iron. The broken vertical lines represent the normal pH range in natural waters.

The potential falls linearly as the pH increases, as shown in Fig. 6.13. The line has been drawn for a Fe^{2+} concentration of 10 $\mu mol\,L^{-1}$. The regions of potential and pH that lie above this line correspond to conditions in which the oxidized species ($Fe(OH)_3$) is stable; the regions below it correspond to conditions under which the reduced species (Fe^{2+}) is stable. We see that a stronger oxidizing agent is required to oxidize and precipitate Fe^{2+} ions in acidic media than in basic media.

It is a general rule for coupled redox–acid reactions that the presence of H^+ on the left of the half-reaction for a redox couple implies a downward slope of a boundary in a Pourbaix diagram. Conversely, H^+ on the right implies an upward slope of a boundary.

The vertical line at $pH = 9$ divides the regions in which either the reactants or the products are stable in the reaction

$$Fe^{2+}(aq) + 2\,H_2O(l) \longrightarrow Fe(OH)_2(s) + 2\,H^+(aq)$$

This reaction is not a redox reaction, and the vertical line is drawn at a pH that corresponds to an Fe^{2+} equilibrium concentration of 10 $\mu mol\,L^{-1}$.

Another line separates the regions in which $Fe(OH)_2$ and $Fe(OH)_3$ are stable:

$$Fe(OH)_3(s) + H^+(aq) + e^- \longrightarrow Fe(OH)_2(s) + H_2O(l)$$

The potential of this reduction depends on the pH, but when we write the Nernst equation

$$E = E^{\ominus} - \frac{RT}{F}\ln\frac{1}{[H^+]} = E^{\ominus} - (0.059\ V)pH \tag{20}$$

we see that it has a less steep slope than for the $Fe(OH)_3/Fe^{2+}$ couple (because the number of H^+ ions involved is smaller than in the latter).

Finally, we have drawn on the diagram the two sloping lines that act as boundaries for the stability field of water (recall Fig. 6.8). As we saw earlier, any couples present with potentials more positive than the upper line will oxidize water to O_2, and any with potentials more negative than the lower line will reduce it to H_2. All the couples we need consider when discussing the iron redox reactions in water therefore lie within the stability field.

> A Pourbaix diagram is a map of conditions of potential and pH under which species are stable in water.

(c) Natural waters

The chemistry of natural waters can be rationalized by using Pourbaix diagrams of the kind we have just constructed. Thus, where fresh water is in contact with the atmosphere, it is saturated with O_2, and many species may be oxidized by this powerful oxidizing agent ($E^{\ominus} = +1.23\ V$). More fully reduced forms are found in the absence of oxygen, especially where there is organic matter to act as a reducing agent. The major acid system that controls the pH of the medium is the $CO_2/H_2CO_3/HCO_3^-/CO_3^{2-}$ diprotic system, where atmospheric CO_2 provides the acid and dissolved carbonate minerals provide the base. Biological activity is also important, because respiration consumes oxygen and releases CO_2. This acidic oxide reduces the pH and hence makes the potential more negative. The reverse process, photosynthesis, consumes CO_2 and releases O_2. This consumption of acid raises the pH and makes the potential less negative. The condition of typical natural waters—their pH and the potentials of the redox couples they contain—is summarized in Fig. 6.14.

From Fig. 6.13 we see that Fe^{3+} can exist in water if the environment is oxidizing; hence, where O_2 is plentiful and the pH is low (below 4), iron will be present as Fe^{3+}. Because few natural waters are so acidic, Fe^{3+} is very unlikely to be found in them. The iron in insoluble Fe_2O_3 can enter solution as Fe^{2+} if it is reduced, which occurs when the condition of the water lies below the sloping boundary in the diagram. We should observe that, as the pH rises, Fe^{2+} can form only if there are strong reducing couples present, and its formation is very unlikely in

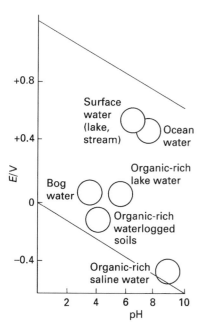

6.14 The stability field of water showing regions typical of various natural waters.

oxygen-rich water. If we compare Figs 6.13 and 6.14, we see that iron will be reduced and dissolved in the form of Fe^{2+} in bog waters and organic-rich waterlogged soils (pH near 4.5 and E near $+0.03$ V and -0.1 V, respectively).

It is instructive to analyze a Pourbaix diagram in conjunction with an understanding of the physical processes that occur in water. As an example, consider a lake where the temperature gradient tends to prevent vertical mixing. At the surface, the water is fully oxygenated and the iron must be present in particles of the insoluble $Fe(OH)_3$; these particles will tend to settle. At greater depth, the O_2 content is low. If the organic content or other sources of reducing agents are sufficient, the oxide will be reduced and iron will dissolve as Fe^{2+}. The Fe(II) ions will then diffuse toward the surface where they encounter O_2 and be oxidized to insoluble $Fe(OH)_3$ again.

Pourbaix diagrams are useful for discussing the relative stabilities of metal-containing species in natural waters.

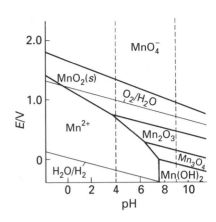

6.15 A section of the Pourbaix diagram for manganese. The broken vertical lines represent the normal pH range in natural waters.

....

Example 6.11 Using a Pourbaix diagram

Figure 6.15 is part of a Pourbaix diagram for manganese. Identify the environment in which the solid MnO_2 or its corresponding hydrous oxides are important.

Answer Manganese dioxide is thermodynamically favored only if $E > +0.6$ V. Under mildly reducing conditions, the stable species is $Mn^{2+}(aq)$. Thus, MnO_2 is important in well aerated waters (near the air–water boundary) where E approaches the value for the O_2/H_2O couple.

....

Self-test 6.11 Use Figs 6.13 and 6.14 to evaluate the possibility of finding $Fe(OH)_3(s)$ in a waterlogged soil.

....

The effect of complex formation on potentials

The formation of metal complexes (see Chapter 7) affects standard potentials because the ability of the complex ML_6 to accept or release an electron is different from that of the corresponding aqua ion with $L = H_2O$. An example is

$$[Fe(OH_2)_6]^{3+}(aq) + e^- \longrightarrow [Fe(OH_2)_6]^{2+}(aq) \qquad E^\ominus = +0.77 \text{ V}$$

compared with

$$[Fe(CN)_6]^{3-}(aq) + e^- \longrightarrow [Fe(CN)_6]^{4-}(aq) \qquad E^\ominus = +0.36 \text{ V}$$

We see that the hexacyanoferrate(III) complex is more resistant to reduction than the aqua complex. To analyze this difference, we treat the reduction of the cyano complex as the sum of the following three reactions:

   (a) $[Fe(CN)_6]^{3-}(aq) + 6 H_2O(l) \longrightarrow [Fe(OH_2)_6]^{3+}(aq) + 6 CN^-(aq)$
   (b) $[Fe(OH_2)_6]^{3+}(aq) + e^- \longrightarrow [Fe(OH_2)_6]^{2+}(aq)$
   (c) $[Fe(OH_2)_6]^{2+}(aq) + 6 CN^-(aq) \longrightarrow [Fe(CN)_6]^{4-}(aq) + 6 H_2O(l)$

The replacement of aqua ligands by cyano ligands stabilizes the hexacyanoferrate(III) complex more than it stabilizes the hexacyanoferrate(II) complex, so the net contribution of reactions 1 and 2 is to favor the reactants in the overall reaction

$$[Fe(CN)_6]^{3-}(aq) + e^- \longrightarrow [Fe(CN)_6]^{4-}(aq)$$

As a result, the hexacyanoferrate(III) complex is less susceptible to reduction than the hexaaquairon(III) complex.

> *The formation of a more stable complex when the metal has the higher oxidation number favors oxidation and makes the reduction potential more negative. The formation of a more stable complex when the metal has the lower oxidation number favors reduction and the reduction potential becomes more positive.*

Example 6.12 Assessing the effect of complexation on potential

In general, CN^- forms complexes that are thermodynamically more stable than those formed by Br^-. Which complex, $[Ni(CN)_4]^{2-}$ or $[NiBr_4]^{2-}$, is expected to have the more positive potential for reduction to Ni(s)?

Answer The reduction half-reaction is

$$[NiX_4]^{2-}(aq) + 2\,e^- \longrightarrow Ni(s) + 4\,X^-(aq)$$

The more stable the complex, the higher its resistance to reductive decomposition. Because $[Ni(CN)_4]^{2-}$ is more stable, it is more difficult to reduce than $[NiBr_4]^{2-}$, and therefore it will have the more negative standard potential.

Self-test 6.12 Which couple has the higher standard potential, $[Ni(en)_3]^{2+}/Ni$ or $[Ni(NH_3)_6]^{2+}/Ni$?

Further reading

T.W. Swaddle, *Inorganic chemistry*. Academic Press (1997). This book provides a good introductory account of practical aspects of environmental chemistry and extractive metallurgy. It also treats redox reactions in solution, including kinetics, electrolysis, and corrosion.

A.J. Bard, R. Parsons, and R. Jordan, *Standard potentials in aqueous solution*. M. Dekker, New York (1985). A critical collection of potential data. Comments on rates are also included.

S.G. Bartsch, Standard electrode potentials and temperature coefficients in water at 298.15. *J. Phys. Chem. Ref. Data* **18**, 1 (1989).

D.M. Stanbury, Potentials for radical species. *Adv. Inorg. Chem.* **33**, 69 (1989).

I. Barin, *Thermochemical data of pure substances*, Vols 1 and 2. VCH, Weinheim (1989). A comprehensive and reliable source of thermodynamic data as a function of temperature for inorganic substances.

J. Emsley, *The elements*. Oxford University Press (1998). A very useful collection of data on the elements, including standard potentials.

P.H. Rieger, *Electrochemistry*. Prentice-Hall, Englewood Cliffs (1987). A good introduction to electrochemical methods and their interpretation.

M. Pourbaix, *Atlas of electrochemical equilibria in aqueous solution*. Pergamon Press, Oxford (1966). The source of Pourbaix diagrams for many elements.

R.M. Garrels and C.L. Christ, *Solutions, minerals, and equilibria*. Harper and Row, New York (1965). A standard text on aqueous geochemistry, which makes use of Pourbaix diagrams throughout and analyzes the implications of chemical equilibria in geology.

W. Stumm and J.J. Morgan, *Aquatic chemistry*. Wiley, New York (1996). The standard reference on natural water chemistry.

D.C. Harris, *Introduction to analytical chemistry*. W.H. Freeman, New York (1991). Chapters 14 and 16 are good introductions to electrochemical cells and electrochemical measurements.

R.B. Jordan, *Reaction mechanisms of inorganic and organometallic systems*. Oxford University Press (1991).

R.G. Wilkins, *Kinetics and mechanism of reactions of transition metal complexes*. VCH, New York (1991).

Exercises

6.1 Consult the Ellingham diagram in Fig. 6.3 and determine if there are any conditions under which aluminum might be expected to reduce MgO. Comment on these conditions.

6.2 Standard potential data suggest that gaseous oxygen rather than gaseous chlorine should be the anode product in electrolysis of aqueous NaCl. In fact, only traces of oxygen are produced. Suggest an explanation.

6.3 Using data from Appendix 2, suggest chemical reagents which would be suitable for carrying out the following transformations and write balanced equations: (a) oxidation of HCl to chlorine gas, (b) reducing $Cr^{(III)}(aq)$ to $Cr^{(II)}(aq)$, (c) reducing $Ag^+(aq)$ to $Ag(s)$, (d) reducing I_2 to I^-.

6.4 Use standard potential data from Appendix 2 as a guide to write balanced equations for the reactions that might be expected for each of the following species in aerated aqueous acid. If the species is stable, write no reaction. (a) Cr^{2+}, (b) Fe^{2+}, (c) Cl^-, (d) HClO, (e) $Zn(s)$.

6.5 Use the information in Appendix 2 to write balanced equations for the reactions, including disproportionations, that can be expected for each of the following species in aerated acidic aqueous solution: (a) Fe^{2+}, (b) Ru^{2+}, (c) $HClO_2$, (d) $Br_2(l)$.

6.6 Write the Nernst equation for (a) the reduction of O_2:

$$O_2(g) + 4H^+(aq) + 4e^- \rightarrow 2H_2O(l)$$

(b) the reduction of $Fe_2O_3(s)$:

$$Fe_2O_3(s) + 6H^+(aq) + 6e^- \rightarrow 2Fe(s) + 3H_2O(l)$$

and in each case express the formula in terms of **pH**. What is the potential for the O_2 reduction at **pH** = 7 and $p(O_2) = 0.20$ bar (the partial pressure of oxygen in air)?

6.7 Given the following potentials in basic solution

$$CrO_4^{2-}(aq) + 4H_2O(l) + 3e^- \rightarrow Cr(OH)_3(s) + 5OH^-(aq)$$

$$E^\ominus = -0.11\ V$$

$$[Cu(NH_3)_2]^+(aq) + e^- \rightarrow Cu(s) + 2NH_3(aq)\quad E^\ominus = -0.10\ V$$

and assuming that a reversible reaction can be established on a suitable catalyst, calculate $E^\ominus$, $\Delta_r G^\ominus$, and K for the reduction of CrO_4^{2-} and $[Cu(NH_3)_2]^+$ in basic solution. Comment on why $\Delta_r G^\ominus$ and K are so different between the two cases despite the values of $E^\ominus$ being so similar.

6.8 Answer the following questions using the Frost diagram in Fig. 6.16. (a) What are the consequences of dissolving Cl_2 in aqueous basic solution? (b) What are the consequences of dissolving Cl_2 in aqueous acid? (c) Is the failure of $HClO_3$ to disproportionate in aqueous solution a thermodynamic or a kinetic phenomenon?

6.9 Use standard potentials as a guide to write equations for the main net reaction which you would predict in the following experiments: (a) N_2O is bubbled into aqueous NaOH solution, (b) zinc metal is added to aqueous sodium triiodide, (c) I_2 is added to excess aqueous $HClO_3$.

6.10 Characterize the condition of acidity or basicity which would most favor the following transformations in aqueous solution: (a) $Mn^{2+} \rightarrow MnO_4^-$, (b) $ClO_4^- \rightarrow ClO_3^-$, (c) $H_2O_2 \rightarrow O_2$, (d) $I_2 \rightarrow 2I^-$.

6.11 Comment on the likelihood that the following reactions occur by a simple outer-sphere electron transfer, simple atom transfer, or a multistep mechanism:

(a) $HIO(aq) + I^-(aq) \rightarrow I_2(aq) + OH^-(aq)$,

(b) $[Co(phen)_3]^{3+}(aq) + [Cr(bipy)_3]^{2+}(aq) \rightarrow$

$$[Co(phen)_3]^{2+}(aq) + [Cr(bipy)_3]^{3+}(aq),$$

(c) $IO_3^-(aq) + 8I^-(aq) + 6H^+(aq) \rightarrow 3I_3^-(aq) + 3H_2O(l)$.

6.12 Use the Latimer diagram for chlorine to determine the potential for reduction of ClO_4^- to Cl_2. Write a balanced equation for this half-reaction.

6.13 Calculate the equilibrium constant of the reaction $Au^+(aq) + 2CN^-(aq) \rightarrow [Au(CN)_2]^-(aq)$ from the standard potentials

$$Au^+(aq) + e^- \rightarrow Au(s)\quad E^\ominus = +1.69\ V$$

$$[Au(CN)_2]^-(aq) + e^- \rightarrow Au(s) + 2CN^-(aq)\quad E^\ominus = -0.60\ V$$

6.14 Use Fig. 6.14 to find the approximate potential of an aerated lake at **pH** = 6. With this information and Latimer diagrams from Appendix 2, predict the species at equilibrium for the elements (a) iron, (b) manganese, (c) sulfur.

6.15 The species Fe^{2+} and H_2S are important at the bottom of a lake where O_2 is scarce. If the **pH** = 6, what is the maximum value of E characterizing the environment?

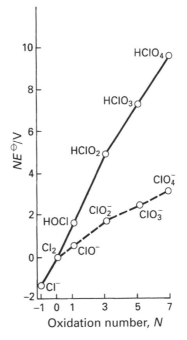

Fig. 6.16 A Frost diagram for chlorine. The solid line refers to standard conditions (pH = 0) and the broken line to pH = 14.

6.16 The ligand edta forms stable complexes with hard acid centers. How will complexation with edta affect the reduction of M^{2+} to the metal in the first d series?

6.17 In Fig. 6.13, which of the boundaries depend on the choice of Fe^{2+} concentration as 10^{-5} mol L^{-1}?

Problems

6.1 Using standard potential data, suggest why permanganate is not a suitable oxidizing agent for the quantitative estimation of Fe^{2+} in the presence of HCl but becomes so if sufficient Mn^{2+} and phosphate ion are added to the solution? (*Hint.* Phosphate forms a complex with Fe^{3+}, thereby stabilizing it.)

6.2 R.A. Binstead and T.J. Meyer have described (*J. Am. Chem. Soc.* **109**, 3287 (1987)) the reduction of $[Ru^{IV}O(bipy)_2(py)]^{2+}$ to $[Ru^{III}(OH)(bipy)_2(py)]^{2+}$. The study revealed that the rate of the reaction is strongly influenced by the change of the solvent from H_2O to D_2O. What does this suggest about the mechanism of the reaction? Comment on the relation of this result to the cases of simple electron transfer and simple atom transfer described in the chapter.

6.3 It is often found that O_2 is a slow oxidizing agent. Suggest a mechanistic explanation that takes into account the two standard potentials.

$$O_2(g) + 4\,H^+(aq) + 4\,e^- \rightarrow 2\,H_2O(l) \qquad E^\ominus = +1.23\ \text{V}$$

$$O_2(g) + 2\,H^+(aq) + 2\,e^- \rightarrow H_2O_2(aq) \qquad E^\ominus = +0.70\ \text{V}$$

6.4 A number of anaerobic bacteria utilize oxidizing agents other than O_2 as an energy source. Some important examples include SO_4^{2-}, NO_3^-, and Fe^{3+}. One half-reaction is

$$FeO(OH)(s) + HCO_3^-(aq) + 2\,H^+(aq) + e^- \rightarrow FeCO_3(s) + 2\,H_2O(l)$$

for which $E^\ominus = +1.67$ V. What mass of iron gives the same standard reaction Gibbs energy as 1.00 g of oxygen?

6.5 Many of the tabulated data for standard potentials have been determined from thermochemical data rather than direct electrochemical measurements of cell potentials. Carry out a calculation to illustrate this approach for the half-reaction

$$Sc_2O_3(s) + 3\,H_2O(l) + 6\,e^- \rightarrow 2\,Sc(s) + 6\,OH^-(aq)$$

	Sc^{3+} (aq)	OH^- (aq)	$H_2O(l)$
$\Delta_f H^\ominus$/kJ mol^{-1}	-614.2	-230.0	-285.8
$S_m^\ominus$/J K^{-1} mol^{-1}	-255.2	-10.75	$+69.91$

6.18 How could a cyclic voltammogram be used to determine whether a metal ion in aqueous solution is present in a complex with ligands other than water?

6.19 Explain why water with high concentrations of dissolved carbon dioxide and open to atmospheric oxygen is very corrosive toward iron.

	$Sc_2O_3(s)$	$Sc(s)$
$\Delta_f H^\ominus$/kJ mol^{-1}	-1908.7	0
$S_m^\ominus$/J K^{-1} mol^{-1}	$+77.0$	$+34.76$

6.6 The reduction potential of an ion such as OH^- can be strongly influenced by the solvent. (a) From the review article by D.T. Sawyer and J.L. Roberts, *Acc. Chem. Res.* **21**, 469 (1988), describe the magnitude of the change in the $OH \cdot /OH^-$ potential on changing the solvent from water to acetonitrile, CH_3CN. (b) Suggest a qualitative interpretation of the difference in solvation of the OH^- ion in these two solvents.

6.7 Reactions that involve atom transfer may be expressed either in terms of the redox language of changes in oxidation number or from the standpoint of nucleophilic substitution. Describe, from these two viewpoints, the mechanism of the reaction of NO_2^- with ClO^- in water to produce NO_3^- and Cl^-. See D.W. Johnson and D.W. Margerum, *Inorg. Chem.* **30**, 4845 (1991).

6.8 Balance the following redox reaction in acid solution: $MnO_4^- + H_2SO_3 \rightarrow Mn^{2+} + HSO_4^-$. Predict the qualitative **pH** dependence on the net potential for this reaction (i.e. increases, decreases, remains the same).

6.9 Draw a *Frost* diagram for mercury in acid solution given the following Latimer diagram:

$$Hg^{2+} \xrightarrow{\ 0.911\ } Hg_2^{2+} \xrightarrow{\ 0.796\ } Hg$$

Comment on the tendency of any of the species to act as an oxidizing agent, a reducing agent, or to undergo disproportionation.

6.10 From the following Latimer diagram, calculate the net $E^\ominus$ value for the following reaction

$$O_2 \xrightarrow{\ -0.125\ } HO_2 \xrightarrow{\ 1.510\ } H_2O_2 \qquad 2\,HO_2 \rightarrow O_2 + H_2O_2$$

Given your value of $E^\ominus$, comment on the thermodynamic tendency of HO_2 to undergo disproportionation.

6.11 Using the following aqueous acid solution reduction potentials $E^\ominus$ $(Pd^{2+}/Pd) = +0.915$ V and $E^\ominus([PdCl_4]^{2-}/Pd) = +0.600$ V, calculate the equilibrium constant for the following reaction in 1 M aqueous HCl:

$$Pd^{2+}(aq) + 4Cl^-(aq) \rightleftharpoons [PdCl_4]^{2-}(aq)$$

6.12 Calculate the half-cell potential at $25\,°C$ for the reduction of MnO_4^-(aq) to MnO_2(s) in aqueous solution at **pH** $= 9.00$ and 1 M MnO_4^-(aq) $E^\ominus(MnO_4^-/MnO_2) = +1.69$ V.

d-Metal complexes

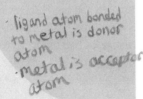

Metal complexes, in which a single central metal atom or ion is surrounded by several atoms or ions, play an important role in inorganic chemistry, especially for elements of the d-block. In this chapter, we introduce the common structural arrangements for ligands around a central metal atom. We then discuss the nature of the ligand–metal bonding in terms of two theoretical models. We start with the simple but useful crystal-field theory, which is based on an electrostatic model of the bonding, and then progress to the more sophisticated ligand-field theory. Both theories invoke a parameter, the ligand-field splitting parameter, to correlate spectroscopic and magnetic properties. The same parameter also helps to systematize the discussion of the stabilities of complexes and their rates of reaction.

In the context of *d*-metal chemistry, the term **complex** means a central metal atom or ion surrounded by a set of ligands. A **ligand** is an ion or molecule that can have an independent existence. An example of a complex is $[Co(NH_3)_6]^{3+}$, in which the Co^{3+} ion is surrounded by six NH_3 ligands. We shall use the term **coordination compound** to mean a neutral complex or an ionic compound in which at least one of the ions is a complex. Thus, $[Ni(CO)_4]$ and $[Co(NH_3)_6]Cl_3$ are both coordination compounds. A complex is a combination of a Lewis acid (the central metal atom) with a number of Lewis bases (the ligands). The atom in the Lewis base ligand that forms the bond to the central atom is called the **donor atom**, because it donates the electrons used in bond formation. Thus, O is the donor atom when H_2O acts as a ligand. The metal atom or ion, the Lewis acid in the complex, is the **acceptor atom**. This chapter focusses on complexes that contain *d*-metal atoms or ions, but *s*- and *p*-metal ions also form complexes (see Chapter 9).

The principal features of the geometrical structures of *d*-metal complexes were identified by the Swiss chemist Alfred Werner (1866–1919), whose training was in organic stereochemistry. Werner combined the interpretation of optical and geometrical isomerism, patterns of reactions, and conductance data in work that remains a model of how to use physical and chemical evidence effectively and imaginatively.[1] The striking colors of

------

[1] G.B. Kauffman gives a fascinating account of the history of structural coordination chemistry in *Inorganic coordination compounds.* Wiley, New York (1981). Translations of Werner's key papers are available in G.B. Kauffman, *Classics in coordination chemistry; I Selected papers of Alfred Werner.* Dover, New York (1968).

many coordination compounds, which reflect their electronic structures, were a mystery to Werner. This characteristic was clarified only when the description of electronic structure in terms of orbitals was applied to the problem in the period from 1930 to 1960.

The geometrical structures of *d*-metal complexes, which we consider first, can now be determined in many more ways than Werner had at his disposal. When single crystals of a compound can be grown, X-ray diffraction (Box 2.1) gives precise shapes, bond distances, and angles. Nuclear magnetic resonance (see *Further information 2*) can be used to study complexes with lifetimes longer than microseconds. Very short-lived complexes, those with lifetimes comparable to diffusional encounters in solution (a few nanoseconds), can be studied by vibrational and electronic spectroscopy. It is possible to infer the geometries of complexes with long lifetimes in solution (such as the classic complexes of Co(III), Cr(III), and Pt(II) and many 4*d*, 5*d*, and organometallic compounds) by analyzing patterns of reactions and isomerism. This method was originally exploited by Werner, and it still teaches us much about the synthetic chemistry of the compounds as well as helping to establish their structures.

Structures and symmetries

In an **inner–sphere complex**, the ligands are attached directly to the central metal atom or ion; the **primary coordination sphere** consists of these directly attached ligands. The number of ligands in the coordination sphere is called the **coordination number** (C.N.) of the complex. As in solids, a wide range of coordination numbers can occur, and the origin of the structural richness and chemical diversity of complexes is the ability of the coordination number to range up to 12.

Although we shall concentrate on inner-sphere complexes throughout this chapter, we should keep in mind that complex cations can associate electrostatically with anionic ligands (and, by other weak interactions, with solvent molecules) without displacement of the ligands already present (**1**). The product of this association is called an **outer-sphere complex** or an **ion pair**. For $[Mn(OH_2)_6]^{2+}$ and SO_4^{2-} ions, for instance, the equilibrium concentration of the outer-sphere complex $\{[Mn(OH_2)_6]^{2+}SO_4^{2-}\}$ can, depending on the concentration, exceed that of the inner-sphere complex $[Mn(OH_2)_5SO_4]$ in which the SO_4^{2-} ligand is directly attached to the metal ion. It is worth remembering that most methods of measuring complex formation equilibria do not distinguish outer-sphere from inner-sphere complex formation but simply detect the sum of all bound ligands. Outer-sphere complexation should be suspected whenever the metal and ligand have opposite charges.

The coordination number of a *d*-metal atom or ion is not always evident from the composition of the solid, for solvent molecules and species that are potentially ligands may simply fill spaces within the structure and not have any direct bonds to the metal ion. For example, X-ray diffraction shows that $CoCl_2 \cdot 6H_2O$ contains the neutral complex $[Co(Cl)_2(OH_2)_4]$ and two uncoordinated H_2O molecules occupying well defined positions in the crystal. Such additional solvent molecules are called **lattice solvent** or **solvent of crystallization**.

1 $[Mn(OH_2)_6]SO_4$

7.1 Constitution

Three factors determine the coordination number of a complex:

1 The size of the central atom or ion.
2 The steric interactions between the ligands.
3 Electronic interactions.

In general, the large radii of atoms and ions in Periods 5 and 6 favor higher coordination numbers for the complexes of these elements. For similar steric reasons, bulky ligands often result in low coordination numbers. High coordination numbers are most common on the left of a row of the d-block, where the atoms have larger radii. They are especially common when the metal ion has only a few d electrons, for a small number of d electrons means that the metal ion can accept more electrons from Lewis bases; one example is $[Mo(CN)_8]^{4-}$. Lower coordination numbers are found on the right of the block, particularly if the ions are rich in d electrons; an example is $[PtCl_4]^{2-}$. Such atoms are less able to accept electrons from any Lewis bases that are potential ligands. Low coordination numbers occur if the ligands can form multiple bonds with the central metal, as in MnO_4^- and CrO_4^{2-}, for now the electrons provided by each ligand tend to exclude the attachment of more ligands.

(a) Low coordination numbers

The best known complexes with C.N. = 2 that are formed in solution under ordinary laboratory conditions are linear species of the Group 11 and 12 ions, such as Cu(I), Ag(I), Au(I), and Hg(II); see *Summary chart 1*. The complex $[AgCl_2]^-$, which is responsible for the dissolution of solid silver chloride in aqueous solutions containing excess Cl^- ions, is one example. The toxic complex $[Hg(CH_3)_2]$ (**2**), which is formed in nature by bacterial action on $Hg^{2+}(aq)$, is another example. A series of linear Au(I) complexes of formula LAuX, where X is a halogen and L is a neutral Lewis base such as a substituted phosphine, R_3P, or thioether, R_2S, are also known. The thioether ligand in $[(R_2S)AuCl]$ is easily displaced by the stronger donor SR^-.[2] Two-coordinate complexes often gain additional ligands to form three- or four-coordinate complexes.

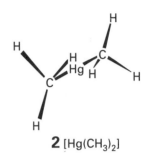

2 $[Hg(CH_3)_2]$

	11	**12**
Cu(I)	X — Cu — X ⌐− X = Cl, Br	
Ag(I)	H_3N — Ag — NH_3 ⌐+	
Au(I)	R_3P — Au — PR_3 ⌐+	H_3C — Hg — CH_3

Summary chart 1 Linear complexes.

A formula that suggests two-coordination in a solid compound in some cases conceals a polymer with a higher coordination number. For example, the salt $K[Cu(CN)_2]$ contains a chain-like anion with three-coordinate Cu atoms (**3**). Three-coordination is rare among d-metal complexes, and MX_3 compounds with X a halogen are usually chains or networks with a higher coordination number and shared ligands.

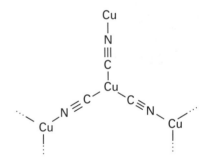

3 $[Cu(CN)_2]^-$

> *Two-coordinate complexes are found for Cu^+ and Ag^+; these complexes readily accommodate more ligands if they are available. It is important to be alert for complexes with coordination numbers higher than their empirical formulas suggest.*

(b) Four-coordination

Four-coordination is found in an enormous number of compounds. Tetrahedral complexes of approximately T_d symmetry (**4**) are favored over higher coordination numbers when the

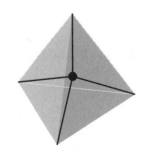

4 Tetrahedral complex, T_d

---

[2] Gold(I) complexes are exploited in the treatment of rheumatoid arthritis, and the interaction of Au(I) with thiol groups of proteins is believed to be involved. See S.J. Lippard (ed.), *Platinum, gold, and other chemotherapeutic agents. Chemistry and biochemistry*, ACS Symposium Series, No. 209. American Chemical Society, Washington, DC (1983).

central atom is small and the ligands are large (such as Cl^-, Br^-, and I^-), for then ligand–ligand repulsions override the energy advantage of forming more metal–ligand bonds.[3] Tetrahedral complexes are common for oxoanions of metal atoms on the left of the *d*-block in high oxidation states, such as $[CrO_4]^{2-}$. The halide complexes of M^{2+} ions on the right of the $3d$ series, such as $[NiBr_4]^{2-}$, are generally tetrahedral. See *Summary chart 2* for examples of each type of complex.

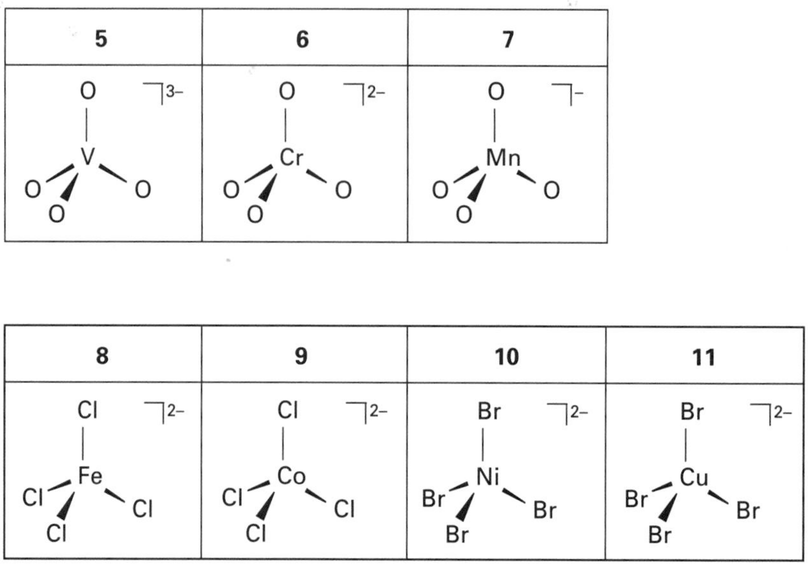

5	6	7
$\begin{array}{c} O \\ \| \\ O \diagup V \diagdown O \\ \| \\ O \end{array}$ $\rceil^{3-}$	$\begin{array}{c} O \\ \| \\ O \diagup Cr \diagdown O \\ \| \\ O \end{array}$ $\rceil^{2-}$	$\begin{array}{c} O \\ \| \\ O \diagup Mn \diagdown O \\ \| \\ O \end{array}$ $\rceil^{-}$

8	9	10	11
$\begin{array}{c} Cl \\ \| \\ Cl \diagup Fe \diagdown Cl \\ \| \\ Cl \end{array}$ $\rceil^{2-}$	$\begin{array}{c} Cl \\ \| \\ Cl \diagup Co \diagdown Cl \\ \| \\ Cl \end{array}$ $\rceil^{2-}$	$\begin{array}{c} Br \\ \| \\ Br \diagup Ni \diagdown Br \\ \| \\ Br \end{array}$ $\rceil^{2-}$	$\begin{array}{c} Br \\ \| \\ Br \diagup Cu \diagdown Br \\ \| \\ Br \end{array}$ $\rceil^{2-}$

Summary chart 2 Tetrahedral complexes.

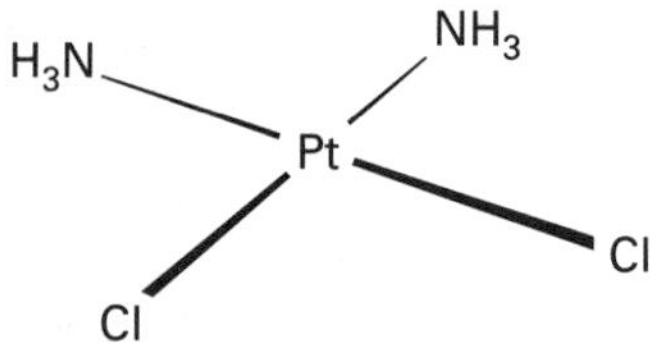

Werner studied a series of four-coordinate Pt(II) complexes formed by the reactions of $PtCl_2$ with NH_3 and HCl. For a complex of formula MX_2L_2, only one isomer is expected if the species is tetrahedral, but two isomers are expected if the species is square-planar, (**5a**) and (**5b**). Because Werner was able to isolate *two* nonelectrolytes of formula $[PtCl_2(NH_3)_2]$, he concluded that they could not be tetrahedral and were, in fact, square-planar (**6**). The complex with like ligands on adjacent corners of the square is called a *cis* isomer (**5a**) and the complex with like ligands opposite is the *trans* isomer (**5b**). The existence of different spatial arrangements of the same ligands is called **geometrical isomerism**. Geometrical isomerism is far from being of only academic interest: platinum complexes are used in cancer chemotherapy, and it is found that only *cis*-Pt(II) complexes can bind to the bases of DNA for long enough to be effective.

> Tetrahedral complexes are favored if the central atom is small or the ligands large; they are common for oxoanions of metal atoms on the left of the *d*-block in high oxidation states, and for halogeno complexes of M^{2+} ions on the right of the $3d$ series. Square-planar complexes may exhibit geometrical isomerism.

Example 7.1 Identifying isomers from chemical evidence

Use the reactions in Fig. 7.1 to show how the *cis* and *trans* geometries may be assigned.

Answer The *cis* isomer reacts with Ag_2O to lose Cl^-, and the product adds one oxalic acid molecule ($H_2C_2O_4$) at neighboring positions. The *trans* isomer loses Cl^-, but the product cannot displace the two OH^- ligands with only one $H_2C_2O_4$ molecule. A reasonable explanation is that the $H_2C_2O_4$ molecule cannot reach across the square plane to bridge two *trans* positions. This conclusion is supported by X-ray crystallography.

$\begin{array}{c} H_3N \\ \diagdown \\ \quad Pt \quad \diagup NH_3 \\ \diagup \quad \diagdown \\ Cl \qquad Cl \end{array}$

5a *cis*-[PtCl$_2$(NH$_3$)$_2$]

$\begin{array}{c} Cl \quad \diagup NH_3 \\ \diagdown Pt \diagup \\ \diagup \quad \diagdown \\ H_3N \qquad Cl \end{array}$

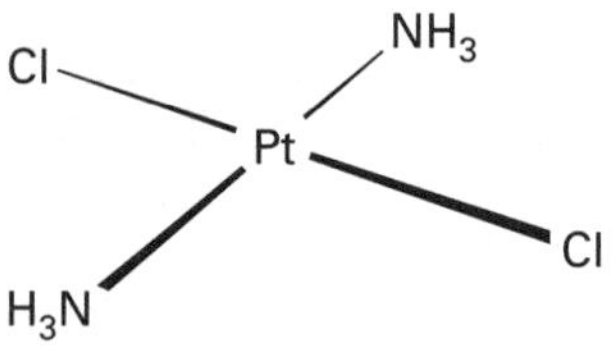

5b *trans*-[PtCl$_2$(NH$_3$)$_2$]

6 Square-planar complex, D_{4h}

---

[3] Four-coordinate *s*- and *p*-block complexes with no lone pairs on the central atom, such as $[BeCl_4]^{2-}$, $[BF_4]^-$, and $[SnCl_4]$, are almost always tetrahedral.

Self-test 7.1 The two square-planar isomers of [PtBrCl(PR$_3$)$_2$] (where PR$_3$ is a trialkylphosphine) have different ^{31}P–NMR spectra, Fig. 7.2. One (A) shows a single ^{31}P group of lines; the other (B) shows two distinct ^{31}P resonances, each one being similar to the single resonance region of A. Which isomer is *cis* and which is *trans*?

Square-planar complexes of the 3*d* metals are typically formed by metal ions with *d*^{8} configurations (for example, Ni^{2+}) in combination with ligands that can form π bonds by accepting electrons from the metal atom, as in [Ni(CN)$_4$]$^{2-}$ (see *Summary chart 3*). The four-coordinate *d*^{8} complexes of the elements belonging to the second and third rows of the *d*-block (4*d*^{8} and 5*d*^{8}), such as those formed by Rh$^+$, Ir$^+$, Pd^{2+}, Pt^{2+}, Au^{3+}, are almost invariably square-planar. Square-planar geometry can also be forced on a central atom by complexation with a ligand that contains a rigid ring of four donor atoms, much as in the formation of five-coordinate porphyrin complexes described below.

> Square-planar complexes in the first row of the *d*-block are typically observed for metal atoms and ions with *d*^{8} configurations and π-acceptor ligands; four-coordinate *d*^{8} complexes of the elements belonging to the second and third rows of the *d*-block are almost invariably square-planar regardless of the π donor or acceptor character of the ligand.

9	10	11
	$\begin{bmatrix} NC & & CN \\ & Ni & \\ NC & & CN \end{bmatrix}^{2-}$ Ni(II)*	
$\begin{matrix} Me_3P & & Cl \\ & Rh & \\ Me_3P & & PMe_3 \end{matrix}$ Rh(I)	$\begin{bmatrix} Cl & & Cl \\ & Pd & \\ Cl & & Cl \end{bmatrix}^{2-}$ Pd(II)	
$\begin{matrix} Me_3P & & Cl \\ & Ir & \\ OC & & PMe_3 \end{matrix}$ Ir(I)	$\begin{bmatrix} H_3N & & NH_3 \\ & Pt & \\ H_3N & & NH_3 \end{bmatrix}^{2+}$ Pt(II)	$\begin{bmatrix} Cl & & Cl \\ & Au & \\ Cl & & Cl \end{bmatrix}^{-}$ Au(III)

Summary chart 3 Planar complexes.

(c) Five-coordination

Five-coordinate complexes, which are less common than four- or six-coordinate complexes in the *d*-block, are either square-pyramidal or trigonal-pyramidal. However, distortions from these ideal geometries are common.[4] A trigonal-bipyramidal shape minimizes ligand–ligand repulsions, but steric constraints on polydentate ligands can favor a square-pyramidal structure. For instance, square-pyramidal five-coordination is found among the biologically

---

[4] E.L. Muetterties and L.J. Guggenberger, *J. Am. Chem. Soc.* **96**, 1748 (1974), identified a series of five-coordinate structures that show a smooth transition from ideal trigonal-bipyramidal to ideal square-pyramidal. The compounds in question ranged from (trigonal-bipyramidal) [CdCl$_5$]$^{3-}$, [P(C$_6$H$_5$)$_5$], [Co(C$_6$H$_7$NO)$_5$]$^{2+}$, [Ni(CN)$_5$]$^{3-}$, [Nb(NC$_5$H$_{10}$)$_5$], to [Sb(C$_6$H$_5$)$_5$] (square-pyramidal).

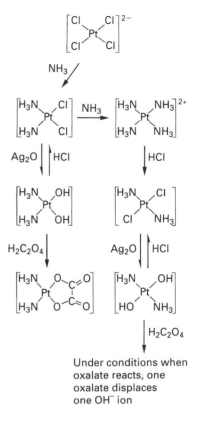

7.1 The preparation of *cis*- and *trans*-diamminedichloroplatinum(II) and a chemical method for distinguishing the isomers.

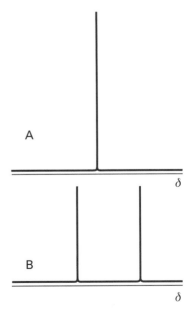

7.2 Idealized NMR spectra of the isomeric phosphine complexes of Pt(II).

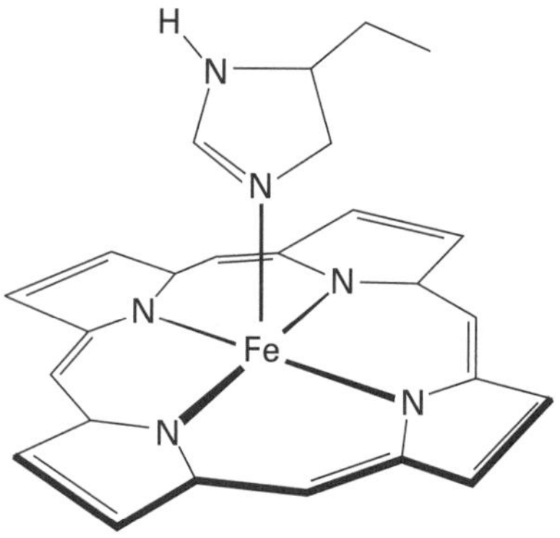

7

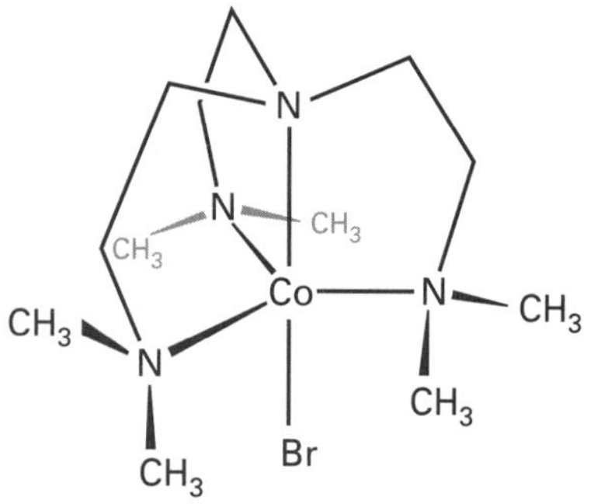

important porphyrins, where the ligand ring enforces a square-planar structure and a fifth ligand attaches above the plane. Structure (**7**) shows the active center of myoglobin, the oxygen transport protein; the location of the Fe atom above the plane of the ring is important to its function, as we shall see in Section 19.3. In some cases, five-coordination is induced by a polydentate ligand containing a donor atom that can bind to an axial location of a trigonal bipyramid, with its remaining donor atoms reaching down to the three equatorial positions (**8**).

The energies of the various geometries of five-coordinate complexes often differ little from one another. The delicacy of this balance is underlined by the fact that $[Ni(CN)_5]^{3-}$ can exist as both square-pyramidal (**9a**) and trigonal-bipyramidal (**9b**) conformations in the same crystal. In solution, trigonal-bipyramidal complexes with monodentate ligands are often highly fluxional, so a ligand that is axial at one moment becomes equatorial at the next moment: the conversion from one stereochemistry to another may occur by a **Berry pseudorotation** (Fig. 7.3). The neutral complex $[Fe(CO)_5]$, for instance, is trigonal-bipyramidal in the crystal; however, in solution the ligands exchange their axial and equatorial positions at a rate that is fast on an NMR timescale but slow on an IR timescale.

> In the absence of polydentate ligands that enforce the geometry, the energies of the various geometries of five-coordinate complexes differ little from one another and such complexes are often fluxional.

8

(d) Six-coordination

Six-coordination is the most common arrangement for electronic configurations ranging from d^0 to d^9. For example, complexes formed by M^{3+} ions of the 3*d* series are usually octahedral (**10**). A few examples representative of the wide range of six-coordinate complexes

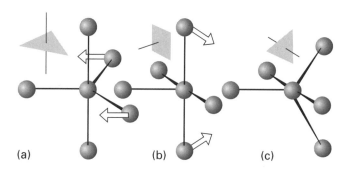

7.3 A Berry pseudorotation in which (a) a trigonal-bipyramidal $[Fe(CO)_5]$ distorts into (b) a square-pyramidal isomer and then (c) becomes trigonal-bipyramidal again, but with two initially equatorial carbonyls now axial. An example of a complex of this kind is $[Fe(CO)_5]$.

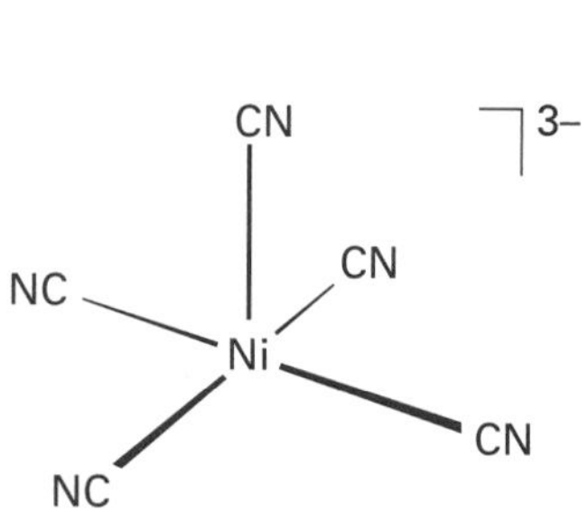

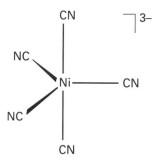

9a $[Ni(CN)_5]^{3-}$ (square-pyramidal conformation)

9b $[Ni(CN)_5]^{3-}$ (trigonal-bipyramidal conformation)

10 Octahedral complex, O_h

that can occur are $[Sc(OH_2)_6]^{3+}$ (d^0), $[Cr(NH_3)_6]^{3+}$ (d^3), $[Mo(CO)_6]$ (d^6), $[Fe(CN)_6]^{3-}$ (d^5), and $[RhCl_6]^{3-}$ (d^6). Even some halides of the f-block elements can display six-coordination, but higher coordination numbers, especially 8 and 9, are more common with these large cations.

Almost all six-coordinate complexes are octahedral, at least in the colloquial sense of the term. A *regular* octahedral (O_h) arrangement of ligands is especially important in six-coordination not only because it is found for many complexes of formula ML_6 but also because it is the starting point for discussions of complexes of lower symmetry, such as those shown in Fig. 7.4. The simplest distortion from O_h symmetry is tetragonal (D_{4h}), and occurs when two ligands along one axis differ from the other four. For the d^9 configuration (particularly for Cu^{2+} complexes) a tetragonal (D_{4h}) distortion may occur even when all ligands are identical. Rhombic (D_{2h}) and trigonal (D_{3d}) distortions also occur. Trigonal distortion gives rise to a large family of structures that are intermediate between regular octahedral and trigonal-prismatic (D_{3h}).

Trigonal-prismatic complexes (**11**) are rare, but have been found in solid MoS_2 and WS_2; the trigonal prism is also the shape of several complexes of formula $[M(S_2C_2R_2)_3]$ ((**12**), see *Summary chart 4*). Trigonal-prismatic d^0 complexes such as $[Zr(CH_3)_6]^{2-}$, have also been isolated. Such structures require either very small σ-donor ligands or favorable ligand–ligand interactions that can constrain the complex into a trigonal-prismatic shape; such ligand–ligand interactions are often provided by ligands that contain sulfur atoms, which can form covalent bonds to each other.[5]

> The overwhelming majority of six-coordinate complexes are octahedral or have shapes that are small distortions of octahedral.

11 Trigonal-prismatic complex, D_{3h}

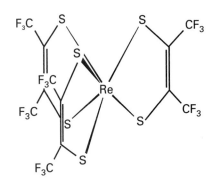

12 $[Re(S_2C_2(CF_3)_2)_3]$

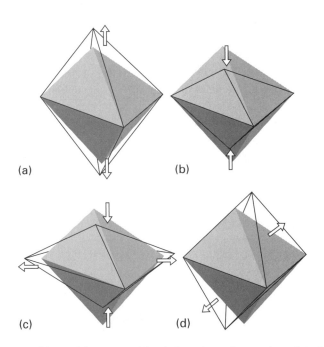

7.4 (a) and (b) Tetragonal (D_{4h}) distortions of a regular octahedron, (c) rhombic (D_{2h}), and (d) trigonal (D_{3d}) distortions. The last can lead to a trigonal prism (D_{3h}) by a further 60° rotation of the faces containing the arrows.

5   The structures of six-coordinate trigonal-prismatic zirconium and hafnium complexes are described and interpreted in P.M. Morse and G.S. Girolami, *J. Am. Chem. Soc.* **111**, 4114 (1989).

13 Pentagonal-bipyramidal complex, D_{5h}

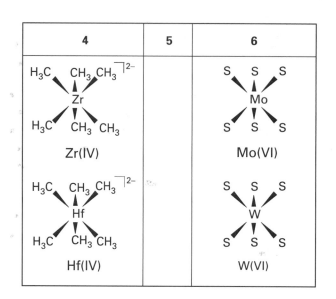

	4	5	6
	H₃C⟍CH₃⟍CH₃ ⌉²⁻ Zr H₃C⟋CH₃⟋CH₃ Zr(IV)		S⟍S⟍S Mo S⟋S⟋S Mo(VI)
	H₃C⟍CH₃⟍CH₃ ⌉²⁻ Hf H₃C⟋CH₃⟋CH₃ Hf(IV)		S⟍S⟍S W S⟋S⟋S W(VI)

Summary chart 4 Trigonal-prismatic complexes.

14 Capped octahedral complex

(e) Higher coordination numbers

Seven-coordination is encountered for a few $3d$ complexes and many more $4d$ and $5d$ complexes, where the larger central atom can accommodate more than six ligands. It resembles five-coordination in the similarity in energy of its various geometries. These limiting 'ideal' geometries include the pentagonal bipyramid (**13**), a capped octahedron (**14**), and a capped trigonal prism (**15**); in each of the latter two, the seventh capping ligand occupies one face. There are a number of intermediate structures, and interconversions are facile. Examples include $[Mo(CNR)_7]^{2+}$, $[ZrF_7]^{3-}$, $[TaCl_4(PR_3)_3]$, and $[ReOCl_6]^{2-}$ from the d-block and $[UO_2(OH_2)_5]^{2+}$ from the f-block. A method to force seven-coordination rather than six on the lighter elements is to synthesize a ring of five donor atoms (**16**) which then occupy the equatorial positions, leaving the axial positions free to accommodate two more ligands.

Stereochemical non-rigidity is also shown in eight-coordination, for such complexes may be square-antiprismatic (**17**) in one crystal but dodecahedral (**18**) in another. Two examples of complexes with these geometries are shown as (**19**) and (**20**), respectively.

Nine-coordination is important in the structures of f-block elements, for their relatively large ions can act as host to a large number of ligands. An example is $[Nd(OH_2)_9]^{3+}$. The MCl_3 solids, with M ranging from La to Gd, achieve high coordination numbers through metal–halide–metal bridges. An example of nine-coordination in the d-block is $[ReH_9]^{2-}$ (**21**), which has small enough ligands for this coordination number to be feasible.

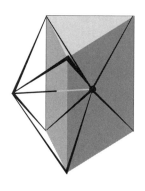

15 Capped trigonal prism

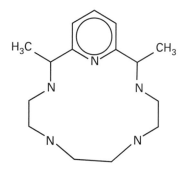

16

17 Square-antiprismatic complex

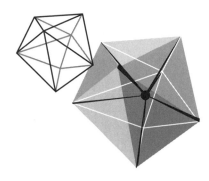

18 Dodecahedral complex

Coordination numbers 10 and 12 are encountered in complexes of the f-block M^{3+} ions. Examples include $[Ce(NO_3)_6]^{2-}$ (**22**), which is formed in the reaction of Ce(IV) salts with nitric acid. Each NO_3^- ligand is bonded to the metal atom by two O atoms. An example of a ten-coordinate complex is $[Th(ox)_4(OH_2)_2]^{4-}$, in which each oxalate ion ligand (ox, $C_2O_4^{2-}$) provides two O atoms. These high coordination numbers are rare with d-block M^{3+} ions.

> *Larger atoms and ions, particularly those of the f-block, tend to form complexes with high coordination numbers; nine-coordination is particularly important in the f-block.*

(f) Polymetallic complexes

A considerable amount of attention has been given recently to the synthesis of **polymetallic complexes**, which are complexes that contain more than one metal atom (Fig. 7.5). In some cases, the metal atoms are held together by bridging ligands, in others there are direct metal–metal bonds, and in yet others there are both types of link. The term **metal cluster** is usually reserved for polymetallic complexes in which there are direct metal–metal bonds that form triangular or larger closed structures. When no metal–metal bonds are present, polymetallic complexes are referred to as **cage complexes** (or cage compounds).[6]

Cage complexes may be formed with a wide variety of anionic ligands. For example, two Cu^{2+} ions can be held together with acetate-ion bridges (Fig. 7.5a). Figure 7.5b shows an example of a cubic structure formed from four Fe atoms bridged by RS^- ligands. This type of structure is of great biological importance, for it is involved in a number of biochemical redox reactions (see Section 19.6).

With the advent of modern structural techniques, such as automated X-ray diffractometers and multinuclear NMR, many polymetallic clusters containing metal–metal bonds have been discovered and have given rise to an active area of research. A simple example is the

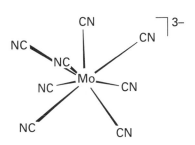

19 $[Mo(CN)_8]^{3-}$

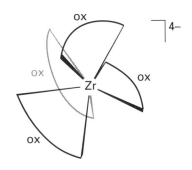

20 $[Zr(ox)_4]^{4-}$

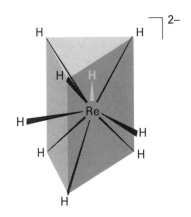

21 $[ReH_9]^{2-}$, D_{3h}

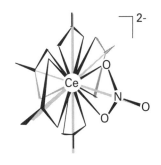

22 $[Ce(NO_3)_6]^{2-}$

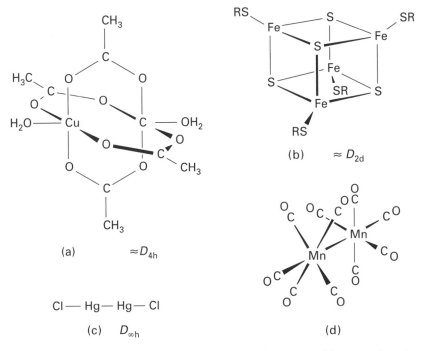

7.5 Representative types of polymetallic complexes. (a) The copper(II) acetate dimer in which there is negligible metal–metal bonding. (b) A synthetic Fe—S complex which models biochemically important electron-transfer agents. (c) Mercury(I) chloride, with a definite Hg—Hg bond. (d) the complex $[Mn_2(CO)_{10}]$ which is held together by an Mn—Mn bond.

---

6 The term 'cage compound' has a variety of meanings in inorganic chemistry, and it is important to keep them distinct. For example, another use of the term is as a synonym of a clathrate compound (an inclusion compound, in which a species is trapped in a cage formed by molecules of another species).

mercury(I) cation Hg_2^{2+}, and complexes derived from it, such as $[Hg_2Cl_2]$ (Fig. 7.5c). A metal cluster containing CO ligands is illustrated in Fig. 7.5d. More details about *d*-block cage and cluster compounds will be found in Sections 9.9 and 16.10.

> Polymetallic complexes contain more than one metal atom and are classified as metal clusters, if they contain **M–M** bonds, or as cage complexes, if they contain ligand-bridged metal atoms.

7.2 Representative ligands and nomenclature

We outline here a few key ideas of nomenclature and introduce a number of common ligands. More detailed guidance is given in *Further information 1*. We begin by considering complexes that contain only **monodentate ligands**, or 'one-toothed' ligands, that have only one point of attachment to the metal atom, as distinct from **polydentate ligands**, which are 'many-toothed' ligands that have more than one point of attachment.

(a) Nomenclature

Table 7.1 gives the names and formulas of a number of common simple ligands. Table 7.2 gives the Greek prefixes that are used to denote the number of each type of ligand in a complex.

Memorize

Table 7.1 Typical ligands and their names

Name	Formula	Abbreviation	Classification*
Acetylacetonato	$(CH_3COCHCOCH_3)^-$	acac	B(O)
Ammine	NH_3		M(N)
Aqua	OH_2		M(O)
2,2-Bipyridine	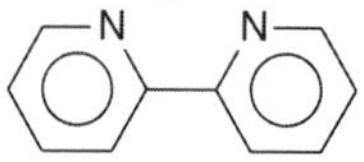	bpy	B(N)
Bromo	Br^-		M(Br)
Carbonato	CO_3^{2-}		M(O) or B(O)
Carbonyl	CO		M(C)
Chloro	Cl^-		M(Cl)
Cyano	CN^-		M(C)
Diethylenetriamine	$NH(C_2H_4NH_2)_2$	dien	T(N)
Ethylenediamine	$H_2NCH_2CH_2NH_2$	en	B(N)
Ethylenediaminetetreacetato	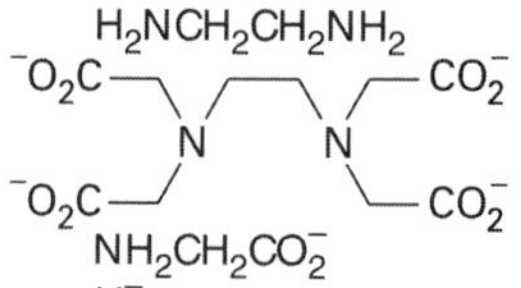 ^-O_2C～N～N～CO_2^- / ^-O_2C～CO_2^-	edta	S(N,O)
Glycinato	$NH_2CH_2CO_2^-$	gly	B(N,O)
Hydrido	H^-		M
Hydroxo	OH^-		M(O)
Maleonitriledithiolato	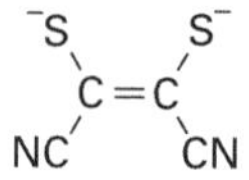 S^-～S^- / $C=C$ / NC～CN	mnt	B(S)
Nitrilotriacetato	$N(CH_2CO_2^-)_3$	nta	Te(N,O)
Nitro, nitrito-*N*	NO_2^-		M(N)
Oxo	O^{2-}		M
Oxalato	$C_2O_4^{2-}$	ox	B(O)
Nitrito	NO_2^-		M(O)
Tetraazacyclotetradecane	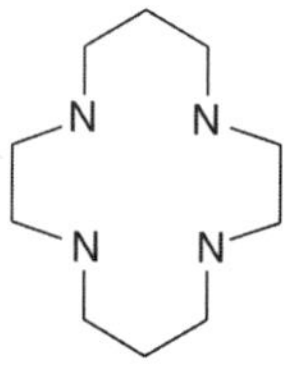	cyclam	Te(N)
Thiocyanato	SCN^-		M(S)
Isothiocyanato	SCN^-		M(N)
2,2′,2″-Triaminotriethylamine	$N(C_2H_4NH_2)_3$	trien	Te(N)

*M: monodentate, B: bidentate, T: tridentate, Te: tetradentate, S: sexidentate. The letters in parentheses identify the donor atoms.

Table 7.2 Prefixes used for naming complexes

Prefix	Meaning
mono-	1
di-, bis-	2
tri-, tris-	3
tetra-, tetrakis-	4
penta-	5
hexa-	6
hepta-	7
octa-	8
nona-	9
deca-	10
undeca-	11
dodeca-	12

Complexes are named with their ligands in alphabetical order (ignoring any numerical prefixes). The ligand names are followed by the name of the metal with either its oxidation number in parentheses, as in hexaamminecobalt(III) for $[Co(NH_3)_6]^{3+}$, or with the overall charge on the complex specified in parentheses, as in hexaamminecobalt(3+). The suffix -ate is added to the name of the metal (sometimes in its Latin form) if the complex is an anion, as in the name hexacyanoferrate(II) for $[Fe(CN)_6]^{4-}$.

The number of a particular type of ligand in a complex is indicated by the prefixes mono-, di-, tri-, and tetra-. The same prefixes are used to state the number of metals atoms if more than one is present in a complex, as in octachlorodirhenate(III), $[Re_2Cl_8]^{2-}$ (**23**). Where confusion with the names of ligands is likely, perhaps because the name already includes a prefix, as with ethylenediamine (en, $H_2NCH_2CH_2NH_2$) the alternative prefixes bis-, tris-, and tetrakis- are used, with the ligand name in parentheses. For example, dichloro- is unambiguous but tris(ethylenediamine) shows more clearly that there are three en ligands, as in tris(ethylenediamine)cobalt(II), $[Co(en)_3]^{2+}$. Ligands that bridge two metal centers are denoted by a prefix μ- added to the name of the relevant ligand, as in μ-oxo-bis(pentamminechromium(III)) (**24**).

The formula of a complex should be enclosed within square brackets whether it is charged or not; however, in casual usage, neutral complexes and oxoanions are often written without brackets, as in Ni(CO)$_4$ for tetracarbonylnickel(0)[7] and MnO$_4^-$ for tetraoxomanganate(VII) ('permanganate'). The metal symbol is given first, then the anionic ligands, and finally the neutral ligands, as in $[CoCl_2(NH_3)_4]^+$ for tetraamminedichlorocobalt(III). (This order is sometimes varied to clarify which ligand is involved in a reaction.) Polyatomic ligand formulas are sometimes written in an unfamiliar sequence (as for OH$_2$ in $[Fe(OH_2)_6]^{2+}$ for hexaaquairon(II)) to place the donor atom adjacent to the metal atom and so help to make the structure of the complex clear.

The names of ligands are summarized in Table 7.1.

(b) Ambidentate ligands

A ligand with different potential donor atoms is called **ambidentate**. An example is the thiocyanate ion (NCS$^-$) which can attach to a metal atom either by the N atom, to give isothiocyanato complexes, or by the S atom, to give thiocyanato complexes. Another example of an ambidentate ligand is NO$_2^-$: as M—NO$_2^-$ the ligand is nitro and as M—ONO$^-$ it is nitrito.[8]

The existence of ambidentate ligands gives rise to the possibility of **linkage isomerism**, in which the same ligand may link through different atoms. This type of isomerism accounts for the red and yellow isomers of the formula $[Co(NO_2)(NH_3)_5]^{2+}$. The red compound has a nitrito Co—O link (**25**). The yellow isomer, which forms from the unstable red form on standing, has a nitro Co—N link (**26**).

An ambidentate ligand gives rise to the possibility of linkage isomerism.

(c) Chelating ligands

Polydentate ligands can produce a **chelate** (from the Greek for claw), a complex in which a ligand forms a ring that includes the metal atom. An example is the bidentate ligand ethylenediamine, which forms a five-membered ring when both N atoms attach to the same metal atom (**27**). The hexadentate ligand ethylenediaminetetraacetic acid as its anion (edta) can attach at six points and can form an elaborate complex with five five-membered

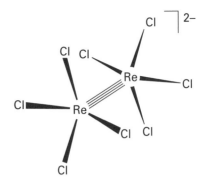

23 $[Re_2Cl_8]^{2-}$, D_{4h}

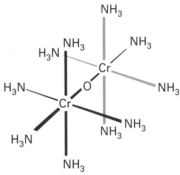

24 μ-Oxo-bis(pentaamminechromium(III))

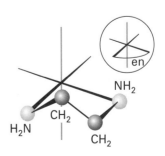

25 Nitrito ligand

26 Nitro ligand

27 Ethylenediamine ligand (en)

---

[7] When assigning oxidation numbers in carbonyl complexes, CO is ascribed a net oxidation number of 0.

[8] Many other interesting examples, including transient linkage isomerism, are given in the article, Ambidentate ligands, the schizophrenics of coordination chemistry, by J.L. Burmeister, *Coord. Chem. Rev.* **105**, 77 (1990).

28 [Co(edta)]⁻

29

30

31 Bite angle

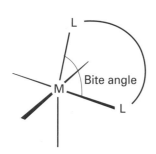

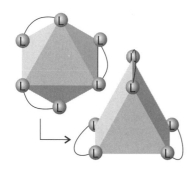

32

rings (**28**). This ligand is used to trap metal ions, such as Ca^{2+} ions in 'hard' water. Table 7.1 includes some of the most common chelating ligands.

Example 7.2 Naming complexes

Name the complexes (a) *trans*-$[PtCl_2(NH_3)_4]^{2+}$, (b) $[Ni(CO)_3(py)]$, (c) $[Cr(edta)]^-$.

Answer (a) The complex has two anionic ligands (Cl^-) and four neutral ligands (NH_3); hence the oxidation number of platinum must be +4. According to the alphabetical order rules, the name of the complex is *trans*-tetraamminedichloroplatinum(IV). (b) The ligands CO and py (pyridine) are neutral, so the oxidation number of nickel must be 0. It follows that the name of the complex is tricarbonylpyridinenickel(0). (c) This complex contains the hexadentate edta⁴⁻ ion as the sole ligand. The four negative charges of the ligand result in a complex with a single negative charge if the central metal ion is Cr^{3+}. The complex is therefore ethylenediaminetetraacetatochromate(III).

Self–test 7.2

Write the formulas of the following complexes: (a) *cis*-diaquadichloroplatinum(II); (b) diamminetetra(isothiocyanato)chromate(III); (c) tris(ethylenediamine)rhodium(III).

In a chelate formed from a saturated organic ligand, such as ethylenediamine, a five-membered ring can fold into a conformation that preserves the tetrahedral angles within the ligand and yet still achieve an L–M–L angle of 90°, the angle typical of octahedral complexes. Six-membered rings may be favored sterically or by electron delocalization through their π orbitals. The bidentate β-diketones, for example, coordinate as the anions of their enols in six-membered ring structures (**29**). An important example is the acetylacetonato anion, $CH_3COCHCOCH_3^-$ ((**30**); acac). Because biochemically important amino acids can form five- or six-membered rings, they also chelate readily.

The degree of strain in a chelating ligand is often expressed in terms of the **bite angle**, the L–M–L angle in the chelate ring (**31**). A chelating ligand that permits only a small bite angle is one of the main causes of distortion from octahedral toward trigonal-prismatic geometry in six-coordinate complexes (**32**).

> Polydentate ligands can form chelates; a bidentate ligand with a small bite angle can result in distortions from standard structures.

(d) The template effect

A **condensation reaction** is a reaction in which a bond is formed between two molecules, and a small molecule—often H_2O—is eliminated. A metal ion such as Ni(II) can be used to assemble a group of ligands which then undergo a condensation reaction among themselves to form a macrocyclic ligand. This phenomenon, which is called the **template effect**, can be applied to produce a surprising variety of macrocyclic ligands. An example of a **template synthesis**, a synthesis that uses the template effect at some stage in a reaction sequence, starts with the coordination of four 1,2-dicyanobenzene molecules to a metal ion, and the ligands then react to give a square-planar Ni(II) complex:

The origin of the template effect may be either kinetic or thermodynamic. For example, the condensation may stem either from the increase in the rate of the reaction between coordinated ligands (on account of their proximity or electronic effects) or from the added stability of the chelated ring product.[9]

> A template synthesis of a macrocyclic ligand makes use of the template effect, the assembly of a macrocyclic ligand from smaller ligands attached to a central metal atom.

7.3 Isomerism and chirality

The three-dimensional character of octahedral complexes results in a much richer variety of isomers than in the essentially two-dimensional square-planar complexes. In particular, octahedral complexes may show both geometrical and optical isomerism.

(a) Geometrical isomerism in six-coordination

One type of geometrical isomerism in six-coordinate complexes resembles that in square-planar complexes. For example, the two Y ligands of an $[MX_4Y_2]$ complex may be placed on adjacent octahedral positions to give a *cis* isomer (33) or on diametrically opposite positions to give a *trans* isomer (34). The *trans* isomer has D_{4h} symmetry and the *cis* isomer has C_{2v} symmetry if the ligands themselves are highly symmetric.

A further type of geometrical isomerism is obtained in complexes of composition $[MX_3Y_3]$. For example, one of the products of the oxidation of Co(II) in the presence of nitrite ions and ammonia is the yellow nonelectrolyte triamminetrinitrocobalt(III), $[Co(NO_2)_3(NH_3)_3]$. There are two ways of arranging the ligands in $[MX_3Y_3]$ complexes. In one isomer, three X ligands lie in one plane and three Y ligands lie in a perpendicular plane (35). This complex is designated the *mer* isomer (for meridional) because each set of ligand can be regarded as lying on a meridian of a sphere. In the second isomer, all three X (or Y) ligands are adjacent and occupy the corners of one triangular face of the octahedron (36); this complex is designated the *fac* isomer (for facial). An approach to the synthesis of specific isomers is summarized in Box 7.1.

> Octahedral complexes of formula $[MX_4Y_2]$ exist as cis and trans geometrical isomers; octahedral complexes of formula $[MX_3Y_3]$ exist as mer and fac geometrical isomers.

33 *cis*-$[CoCl_2(NH_3)_4]^+$

34 *trans*-$[CoCl_2(NH_3)_4]^+$

35 *mer*-$[Co(NO_2)_3(NH_3)_3]$

36 *fac*-$[Co(NO_2)_3(NH_3)_3]$

....

Example 7.3 Identifying types of isomerism

When the four-coordinate square-planar complex $[IrCl(PMe_3)_3]$ (where PMe_3 is trimethylphosphine) reacts with Cl_2, two six-coordinate products of formula $[IrCl_3(PMe_3)_3]$ are formed by a reaction known as 'oxidative addition'. ^{31}P-NMR spectra indicate one P environment in one of these isomers and two in the other. What isomers are possible?

....

[9] An account of the origins of template effect in the case of Pt(II), in which relatively slow reactions of complex formation assist in the elucidation of the details, has been given by D.J. Sheeran and K.B. Mertes, *J. Am. Chem. Soc.* **112**, 1055 (1990).

Box 7.1 **The synthesis of specific isomers**

The synthesis of specific isomers often requires subtle changes in synthetic conditions. For example, the most stable Co(II) complex in ammoniacal solutions of Co(II) salts, $[Co(NH_3)_6]^{2+}$, is only slowly oxidized to $[Co(NH_3)_6]^{3+}$ by air. As a result, a variety of complexes containing other ligands as well as NH_3 can be prepared by bubbling air through a solution containing ammonia and a Co(II) salt. Starting with ammonium carbonate yields $[CoCO_3(NH_3)_4]^+$, in which CO_3^{2-} is a bidentate ligand that occupies two adjacent coordination positions. The complex *cis*-$[Co(NH_3)_4L_2]$ can be prepared by decomposition of the CO_3^{2-} ligand in acidic solution. When concentrated hydrochloric acid is used, the violet *cis*-$[CoCl_2(NH_3)_4]Cl$ compound (**B1**) can be isolated:

$$[Co(CO_3)(NH_3)_4]^+(aq) + 2H^+(aq) + 3Cl^-(aq) \longrightarrow$$
$$\text{*cis*-}[CoCl_2(NH_3)_4]Cl(s) + H_2CO_3(aq)$$

In contrast, reaction with a mixture of HCl and H_2SO_4 gives the bright green *trans*-$[CoCl_2(NH_3)_4]Cl$ isomer (**B2**).

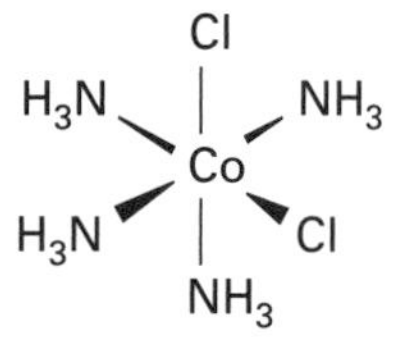

B1

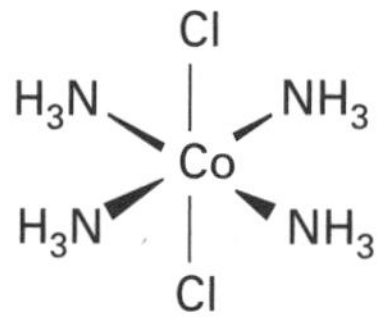

B2

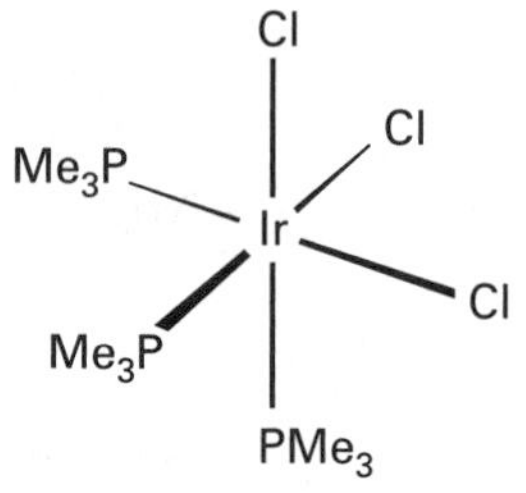

---

37 *fac*-[IrCl$_3$(PMe$_3$)$_3$]

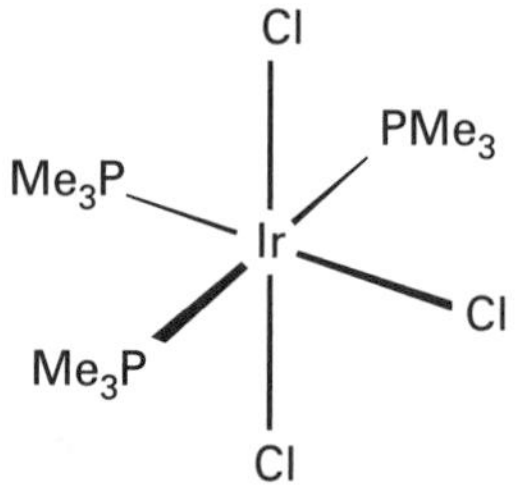

38 *mer*-[IrCl$_3$(PMe$_3$)$_3$]

Answer Because the complexes have the formula $[MX_3Y_3]$, we expect meridional and facial isomers. Structures (**37**) and (**38**) show the arrangement of the three Cl^- ions in the *fac* and *mer* isomers, respectively. All P atoms are equivalent in the *fac* isomer and two environments exist in the *mer* isomer.

Self-test 7.3 When the anion of the amino acid glycine, $H_2NCH_2CO_2^-$ (gly$^-$) reacts with Co(III) oxide, both the N and an O atom of gly$^-$ coordinate and two Co(III) nonelectrolyte *mer* and *fac* isomers of $[Co(gly)_3]$ are formed. Sketch the two isomers.

(b) Chirality and optical isomerism

A **chiral complex** is a complex that is not superimposable on its own mirror image. The existence of a pair of chiral complexes that are each other's mirror image (like a right and left hand), and which have lifetimes that are long enough for them to be separable, is called **optical isomerism**. The two mirror-image isomers jointly make up an **enantiomeric pair**. Optical isomers are so called because they are optically active, in the sense that one enantiomer rotates the plane of polarized light in one direction and the other rotates it through an equal angle in the opposite direction.

As an example of optical isomerism, consider the products of the reaction of Co(III) and ethylenediamine. The product includes a pair of dichloro complexes, one of which is violet and the other green; they are, respectively, the *cis* and *trans* isomers of dichlorobis(ethylenediamine)cobalt(III), $[CoCl_2(en)_2]^+$. (The reaction also results in the formation of a yellow complex, tris(ethylenediamine)cobalt(III), $[Co(en)_3]^{3+}$.) As can be seen from Fig. 7.6, the *cis* isomer cannot be superimposed on its mirror image. It is therefore chiral and hence (because the complexes are long-lived) optically active. The *trans* isomer has a mirror plane and can be superimposed on its mirror image; it is achiral and optically inactive.

The formal criterion of chirality is the absence of an axis of improper rotation (S_n, an *n*-fold axis in combination with a horizontal mirror plane, Section 4.4). The existence of such a symmetry element is implied by the presence of either a mirror plane through the central atom (which is equivalent to an S_1 axis) or a center of inversion (which is equivalent to an S_2 axis), and if either of these elements is present the complex is achiral. As remarked in Section 4.4, we must also be alert for higher-order axes of improper rotation (particularly S_4), because the presence of any S_n axis implies achirality (the absence of chirality). An example of a species with an S_4 axis but not an S_1 or S_2 axis was given as structure **12** in Chapter 4.

Example 7.4 Recognizing chirality

Which of the complexes (a) $[\mathrm{Cr(edta)}]^-$, (b) $[\mathrm{Ru(bipy)_3}]^{2+}$, (c) $[\mathrm{PtCl(dien)}]^+$ are chiral?

Answer The complexes are shown schematically in (**39**), (**40**), and (**41**). Neither (**39**) nor (**40**) has a mirror plane or a center of inversion; so both are chiral (they also have no higher S_n axis); (**41**) has a plane of symmetry and hence is achiral. (Although the CH_2 groups in a dien ligand are not in the mirror plane, they oscillate rapidly above and below it.)

Self–test 7.4 Which of the complexes (a) *cis*-$[\mathrm{CrCl_2(ox)_2}]^{3-}$, (b) *trans*-$[\mathrm{CrCl_2(ox)_2}]^{3-}$, (c) *cis*-$[\mathrm{RhH(CO)(PR_3)_2}]$ are chiral?

The absolute configuration of a chiral complex is described by imagining a view along a threefold rotation axis of a regular octahedron and noting the handedness of the helix formed by the ligands (Fig. 7.7). Right rotation of the helix is then designated Δ and left rotation Λ. The designation of the absolute configuration must be distinguished from the experimentally determined direction in which an isomer rotates polarized light: some Λ compounds rotate in

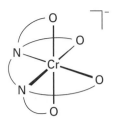

39a

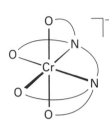

39b

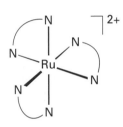

40a

40b

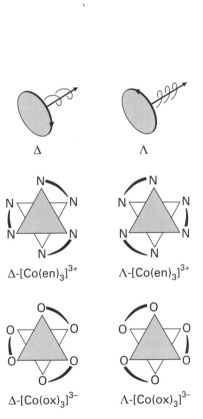

7.6 (a) and (b) Enantiomers of *cis*-$[\mathrm{CoCl_2(en)_2}]$ and (c) the achiral *trans* isomer. The curves represent the CH_2CH_2 bridges in the en ligands. The mirror plane for testing whether there is an S_1 axis is also shown.

7.7 Absolute configurations of $[\mathrm{M(L-L)_3}]$ complexes; Δ is a right-hand screw and Λ is a left-hand screw, as is indicated in the diagrams at the top of the figure by the direction that a screw would turn when being driven in the direction shown.

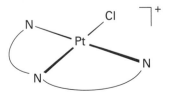

41 $[\mathrm{PtCl(dien)}]^+$

one direction, others rotate in the opposite direction, and the direction may change with wavelength. The isomer that rotates to the right (when viewed into the oncoming beam) at a specified wavelength is designated the *d*-isomer, or the (+)-isomer; the one rotating the plane to the left is designated the *l*-isomer, or the (−)-isomer.

> Complexes that lack an S_n axis are chiral and, if sufficiently long-lived, they are optically active.

(c) The resolution of enantiomers

Optical activity is the only physical manifestation of chirality for a compound with a single chiral center. However, as soon as more than one chiral center is present, other physical properties—such as solubility and melting points—are affected because they depend on the strengths of intermolecular forces, which are different between different isomers (just as there are different forces between a given nut and bolts with left- and right-handed threads). One method of separating a pair of enantiomers into the individual isomers is therefore to prepare **diastereomers**. As far as we need be concerned, diastereomers are isomeric compounds that contain two chiral centers, one being of the same absolute configuration in both components and the other being enantiomeric between the two components. An example of diastereomers is provided by the two salts of an enantiomeric pair of cations, A, with an optically pure anion, B, and hence of composition [Δ-A][Δ-B] and [Λ-A][Δ-B]. Because diastereomers differ in physical properties (such as solubility), they are separable by conventional techniques. The procedure is summarized in Box 7.2.

> The formation of diastereomers may be used to separate chiral complexes.

Box 7.2 A typical chiral resolution procedure

A classical chiral resolution procedure begins with the isolation of a naturally optically active species from a biochemical source (many naturally occurring compounds are chiral). A convenient compound is *d*-tartaric acid (**B3**), a carboxylic acid obtained from grapes. This molecule is a chelating ligand for complexation of antimony, so a convenient resolving agent is the potassium salt of the singly charged 'antimony *d*-tartrate' anion (**B4**). This anion is used for the resolution $[Co(NO_2)_2(en)_2]^+$ as follows.

The enantiomeric mixture of the cobalt(III) complex is dissolved in warm water and a solution of potassium antimony *d*-tartrate is added. The mixture is cooled immediately to induce crystallization. The less soluble diastereomer $\{l\text{-}[Co(NO_2)_2(en)_2]\}\{d\text{-}[SbOC_4H_4O_6]\}$ separates as fine yellow crystals. The filtrate is reserved for isolation of the *d*-enantiomer. The solid diastereomer is ground with water and

sodium iodide. The sparingly soluble compound $l\text{-}[Co(NO_2)_2(en)_2]I$ separates, leaving sodium antimony tartrate in the solution. The *d*-isomer is obtained from the filtrate by precipitation of the bromide salt.

Further reading

G. Pass and H. Sutcliffe, *Practical inorganic chemistry*. Chapman & Hall, London (1974).

W.L. Jolly, *The synthesis and characterization of inorganic compounds*. Waveland Press, Prospect Heights (1991).

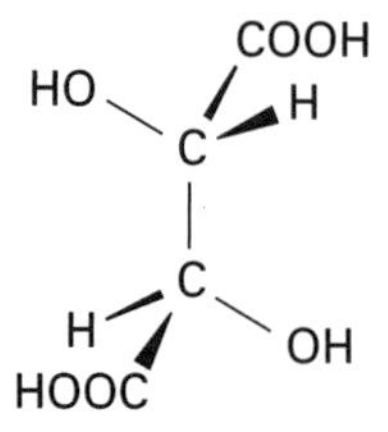

B3 *d*-Tartaric acid

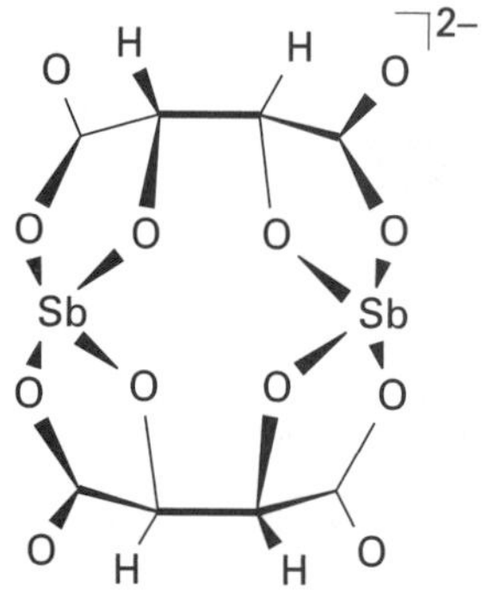

B4 $[Sb_2(d\text{-}C_4H_4O_6)_2]^{2-}$

Bonding and electronic structure

An early theory of the electronic structure of complexes was developed to account for the properties of *d*-metal ions in ionic crystals. In this **crystal-field theory**, a ligand lone pair is modeled as a point negative charge (or as the partial negative charge of an electric dipole) that repels electrons in the *d* orbitals of the central metal ion. This approach concentrates on the resulting splitting of the *d* orbitals into groups of different energies. It then uses that splitting to account for the number of unpaired electrons on the ion and for the spectra, stability, and magnetic properties of complexes. The crystal-field approach is simple and readily visualized; however, it ignores covalent interactions between the ligand and the central metal ion, and the approach has been superseded by **ligand-field theory**. This theory focusses on the overlap of *d* orbitals with ligand orbitals to form molecular orbitals. The qualitative variation in the separation of orbitals associated primarily with the metal atom is the same as that of crystal-field theory, but ligand-field theory provides a better understanding of the origin of the energy separation.

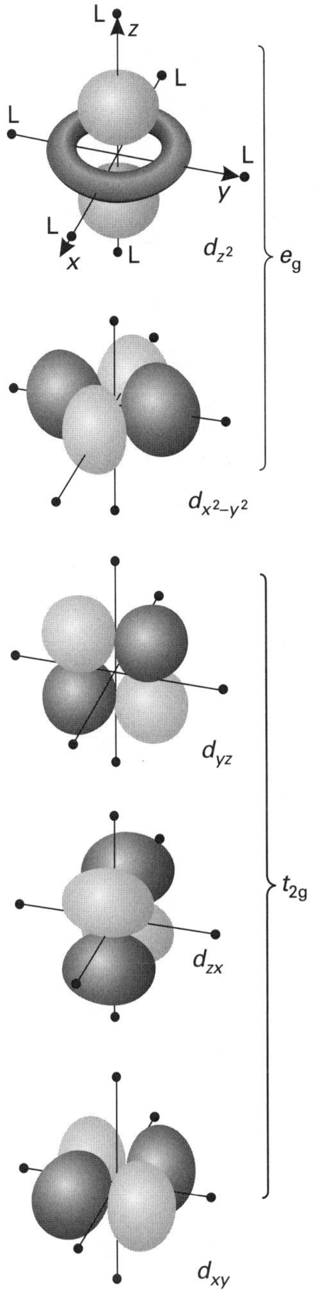

7.4 Crystal-field theory

In the model of an octahedral complex used in crystal-field theory, six ligands are placed on the cartesian axes centered on the metal ion. The ligands interact strongly with the central metal ion, and the stability of the complex stems in large part from this interaction. There is a much smaller secondary effect arising from the fact that electrons in different *d* orbitals interact with the ligands to different extents. Although this differential interaction is little more than about 10 per cent of the overall metal–ligand interaction, it has major consequences for the properties of the complex and is the principal focus of this section.

(a) Ligand-field splitting parameters

Electrons in the two *d* orbitals pointing directly along the cartesian axes and directly at the ligands, namely d_{z^2} and $d_{x^2-y^2}$ (which are jointly of symmetry type e_g in O_h), are repelled more strongly by negative charge on ligands than electrons in the three *d* orbitals that point between the ligands, namely, d_{xy}, d_{yz}, and d_{zx} (symmetry type, t_{2g}). Group theory shows that the e_g orbitals are doubly degenerate (although this is not readily apparent from drawings), and that the t_{2g} orbitals are triply degenerate (Fig. 7.8). This simple model leads to an energy level diagram in which the t_{2g} orbitals lie below the e_g orbitals (Fig. 7.9). The separation of the orbitals is called the **ligand-field splitting parameter**, Δ_O (where the subscript O signifies an octahedral crystal field).[10]

The simplest property that can be interpreted by crystal-field theory is the absorption

7.8 The orientation of the five *d* orbitals with respect to the ligands of an octahedral complex.

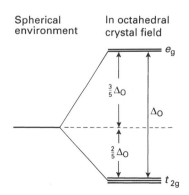

Spherical environment | In octahedral crystal field

e_g

$\frac{3}{5}\Delta_O$

Δ_O

$\frac{2}{5}\Delta_O$

t_{2g}

7.9 The energies of the *d* orbitals in an octahedral crystal field. Note that the mean energy remains unchanged relative to the energy of the *d* orbitals in a spherically symmetric environment (such as in a free atom).

---

[10] Strictly, in the context of crystal-field theory, the ligand-field splitting parameter should be called the 'crystal-field splitting parameter', but we use the former name to avoid a proliferation of names.

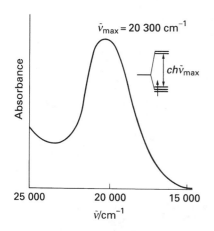

7.10 The optical absorption spectrum of $[Ti(OH_2)_6]^{3+}$.

spectrum of a one-electron complex. Figure 7.10 shows the optical absorption spectrum of the d^1 hexaaquatitanium(III) ion, $[Ti(OH_2)_6]^{3+}$. Crystal-field theory assigns the first absorption maximum at $20\,300\ cm^{-1}$ to the transition $e_g \leftarrow t_{2g}$. (In keeping with spectroscopic notation, the higher-energy orbital is shown first.) We can identify $20\,300\ cm^{-1}$ with Δ_O for the complex. It is more complicated to obtain values of Δ_O for complexes with more than one d electron because the energy of a transition then depends not only on orbital energies (which we wish to know) but also on the repulsion energies between the several electrons present. This aspect is treated more fully in Chapter 13, and the results from the analyses described there have been used to obtain the values of Δ_O in Table 7.3.

The ligand-field splitting parameter varies systematically with the identity of the ligand. The empirical evidence for this trend was the observation, by the Japanese chemist R. Tsuchida, that there are certain regularities in the absorption spectra as the ligands of a complex are varied. For instance, in the series of complexes $[CoX(NH_3)_5]^{n+}$ with $X = I^-$, Br^-, Cl^-, H_2O, and NH_3, the colors range from deep purple (for $X = I^-$) through pink (for Cl^-) to yellow (with NH_3). This observation indicates that there is an increase in the energy of the lowest-energy electronic transition (and therefore in Δ_O) as the ligands are varied along the series. Moreover, this variation is quite general, for the same order of ligands is followed regardless of the identity of the metal ion.

On the basis of these observations, Tsuchida proposed that ligands could be arranged in a **spectrochemical series**, in which the members are arranged in order of increasing energy of transitions that occur when they are present in a complex:

$$I^- < Br^- < S^{2-} < \underline{S}CN^- < Cl^- < \underline{NO_3^-} < N_3^- < F^- < OH^-$$
$$< C_2O_4^{2-} < H_2O < \underline{N}CS^- < CH_3CN < py < NH_3 < en < bipy$$
$$< phen < \underline{NO_2^-} < PPh_3 < \underline{C}N^- < CO$$

(The donor atom in an ambidentate ligand is underlined.) For example, the series indicates that the optical absorption of a hexacyano complex will occur at much higher energy than that of a hexachloro complex of the same metal.

The values of Δ_O also depend in a systematic way on the metal ion, and it is not in general possible to say that a particular ligand exerts a strong or a weak ligand field without considering the metal ion too. In this connection, the most important trends to keep in mind are that

1 Δ_O increases with increasing oxidation number.
2 Δ_O increases down a group.

The variation with oxidation number reflects the smaller size of more highly charged ions and the consequent smaller metal–ligand distances. The second factor reflects the improved

Table 7.3 Ligand-field splitting parameters Δ_O of ML_6 complexes*

	Ions	Ligands				
		Cl^-	H_2O	NH_3	en	CN^-
d^3	Cr^{3+}	13.7	17.4	21.5	21.9	26.6
d^5	Mn^{2+}	7.5	8.5		10.1	30
	Fe^{3+}	11.0	14.3			(35)
d^6	Fe^{2+}		10.4			(32.8)
	Co^{3+}		(20.7)	(22.9)	(23.2)	(34.8)
	Rh^{3+}	(20.4)	(27.0)	(34.0)	(34.6)	(45.5)
d^8	Ni^{2+}	7.5	8.5	10.8	11.5	

*Values are in multiples of $1000\ cm^{-1}$; entries in parentheses are for low-spin complexes.
Source: H.B. Gray, *Electrons and chemical bonding.* Benjamin, Menlo Park (1965).

metal–ligand bonding of the more expanded $4d$ and $5d$ orbitals compared with the compact $3d$ orbitals. The spectrochemical series for metal ions is approximately:

$$Mn^{2+} < Ni^{2+} < Co^{2+} < Fe^{2+} < V^{2+} < Fe^{3+} < Co^{3+} < Mn^{4+}$$
$$< Mo^{3+} < Rh^{3+} < Ru^{3+} < Pd^{4+} < Ir^{3+} < Pt^{4+}$$

> In the presence of an octahedral crystal field, the d orbitals split into a lower-energy triply degenerate (t_{2g}) and a higher-energy doubly degenerate set (e_g) separated by an energy Δ_O. The ligand-field splitting parameter increases along the spectrochemical series of ligands and varies with the identity and charge of the metal atom.

(b) Ligand-field stabilization energies

Because there are three t_{2g} and two e_g orbitals, the t_{2g} orbitals lie $\frac{2}{5}\Delta_O$ below the average energy and the e_g orbitals lie $\frac{3}{5}\Delta_O$ above that average (Fig. 7.9). Therefore, relative to the average energy, the energy of the t_{2g} orbitals is $-0.4\Delta_O$ and that of the e_g orbitals is $+0.6\Delta_O$. It follows that the net energy of a $t_{2g}^x e_g^y$ configuration relative to the average energy of the orbitals, which is called the **ligand-field stabilization energy** (LFSE), is

$$LFSE = (-0.4x + 0.6y)\Delta_O \tag{1}$$

The values of the LFSE for different configurations are given in Table 7.4. The LFSE is generally only a few tens of per cent of the total energy of complex formation, so the LFSE should be viewed as a modulation of the overall interaction between the metal atom and the ligands. The latter metal–ligand interactions increase in strength from left to right across a period on account of the decrease in radius of the M^{2+} ions along that series.

> The ligand-field stabilization energy, eqn 1, is a measure of the net energy of occupation of the d orbitals relative to their mean energy.

Table 7.4 Ligand-field stabilization energies (absolute values)*

d^n	Example	Octahedral					Tetrahedral	
			N	LFSE			N	LFSE
d^0	Ca^{2+}, Sc^{3+}		0	0			0	0
d^1	Ti^{3+}		1	0.4			1	0.6
d^2	V^{3+}		2	0.8			2	1.2
d^3	Cr^{3+}, V^{2+}		3	1.2			3	0.8
		Strong-field			Weak-field			
d^4	Cr^{2+}, Mn^{3+}	2	1.6		4	0.6	4	0.4
d^5	Mn^{2+}, Fe^{3+}	1	2.0		5	0	5	0
d^6	Fe^{2+}, Co^{3+}	0	2.4		4	0.4	4	0.6
d^7	Co^{2+}	1	1.8		3	0.8	3	1.2
d^8	Ni^{2+}	2	1.2				2	0.8
d^9	Cu^{2+}	1	0.6				1	0.4
d^{10}	Cu^+, Zn^{2+}	0	0				0	0

*N is the number of unpaired electrons; LFSE is in units of Δ_O for octahedra or Δ_T for tetrahedra; the calculated relation is $\Delta_T \approx \frac{4}{9}\Delta_O$.

(c) Weak-field and strong-field limits

To infer the ground-state electron configurations of d-metal complexes, we use the d-orbital energy level diagram shown in Fig. 7.9 as a basis for applying the building-up principle. As usual, we look for the lowest energy configuration subject to the Pauli exclusion principle (a maximum of two electrons in an orbital) and (if more than one degenerate orbital is available) to the requirement that electrons first occupy separate orbitals and do so with parallel spins. We begin by considering complexes formed by $3d$ elements.

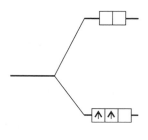

42 d^2

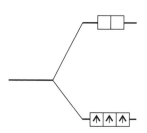

43 d^3

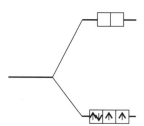

44 Strong-field d^4

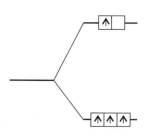

45 Weak-field d^4

The first three $3d$ electrons of a $3d^n$ complex occupy separate t_{2g} nonbonding orbitals, and do so with parallel spins. For example, the ions Ti^{2+} and V^{2+} have electron configurations $3d^2$ and $3d^3$, respectively. The $3d$ electrons occupy the lower t_{2g} orbitals ((**42**) and (**43**), respectively) and the complexes are stabilized by $2 \times 0.4\Delta_O = 0.8\Delta_O$ (Ti^{2+}) and $3 \times 0.4\Delta_O = 1.2\Delta_O$ (V^{2+}).

The next electron needed for the $3d^4$ ion Cr^{2+} may enter one of the t_{2g} orbitals and pair with the electron already there (**44**). However, if it does so, it experiences a strong Coulombic repulsion, which is called the **pairing energy**, P. Alternatively, the electron may occupy one of the e_g orbitals (**45**). Although the pairing penalty is avoided, the orbital energy is higher by Δ_O. In the first case (t_{2g}^4), the LFSE is $1.6\Delta_O$, the pairing energy is P, and the net stabilization is $1.6\Delta_O - P$. In the second case ($t_{2g}^3 e_g^1$), the LFSE is $3 \times 0.4\Delta_O - 0.6\Delta_O = 0.6\Delta_O$, and there is no pairing energy to consider. Which configuration is adopted depends on which of $1.60\Delta_O - P$ and $0.60\Delta_O$ is the larger.

If $\Delta_O < P$, which is called the **weak-field case**, a lower energy is achieved if the upper orbital is occupied. The resulting configuration is $t_{2g}^3 e_g^1$. If $\Delta_O > P$, which is called the **strong-field case**, a lower energy is achieved by occupying only the lower orbitals despite the cost of the pairing energy. The resulting configuration is now t_{2g}^4. For example, $[Cr(OH_2)_6]^{2+}$ has the ground-state configuration $t_{2g}^3 e_g^1$ whereas $[Cr(CN)_6]^{4-}$, with relatively strong-field ligands (as given by the spectrochemical series), has the configuration t_{2g}^4.

The ground-state electron configurations of $3d^1$, $3d^2$, and $3d^3$ complexes are unambiguous because there is no competition between LFSE and pairing energy: the configurations are t_{2g}^1, t_{2g}^2, and t_{2g}^3, respectively. A competition is possible for $3d^n$ complexes in which $n = 4$ or 5, because a strong field favors the occupation of the lower orbitals (and gives rise to t_{2g}^n configurations) whereas a weak field allows electrons to escape the pairing energy by occupying the upper orbitals. Then the configurations will be $t_{2g}^3 e_g^1$ and $t_{2g}^3 e_g^2$. Because in the latter case all the electrons occupy different orbitals, they will have parallel spins.

When alternative configurations are possible, the species with the smaller number of parallel electron spins is called a **low-spin complex**, and the species with the greater number of parallel electron spins is called a **high-spin complex**. A $3d^4$ complex is likely to be low-spin if the crystal field is strong but high-spin if the field is weak. The same applies to $3d^5$ complexes (see Table 7.4). High- and low-spin configurations are also found for $3d^6$ and $3d^7$ complexes. In these cases, a strong crystal field results in the low-spin configurations t_{2g}^6 (no unpaired electrons) and $t_{2g}^6 e_g^1$ (one unpaired electron), respectively, and a weak crystal field results in the high-field configurations $t_{2g}^4 e_g^2$ (four unpaired electrons) and $t_{2g}^5 e_g^2$ (three unpaired electrons), respectively.

The strength of the crystal field (as measured by the value of Δ_O) and the spin-pairing energy (as measured by P) depend on the identity of both the metal and the ligand, so it is not possible to specify a universal point in the spectrochemical series at which a complex changes from high spin to low spin. However, low-spin complexes commonly occur for ligands that are very high in the spectrochemical series (such as CN^-) in combination with $3d$-metal ions. High-spin complexes are common for ligands that are very low in the series (such as F^-) in combination with $3d$-metal ions. For octahedral d^n complexes with $n = 1$–3 and 8–10 there is no ambiguity about the configuration (Table 7.4), and the designations high-spin and low-spin are not employed.

Now consider complexes of $4d$ and $5d$ metals. For these metals, values of Δ_O are typically higher than for $3d$ metals. Consequently, complexes of these metals generally have electron configurations that are characteristic of strong crystal fields and typically have low spin. An example is the $4d^4$ complex $[RuCl_6]^{2-}$, which has a t_{2g}^4 configuration, corresponding to a strong crystal field, despite Cl^- being low in the spectrochemical series. Likewise, $[Ru(ox)_3]^{3-}$ has the low-spin configuration t_{2g}^5 whereas $[Fe(ox)_3]^{3-}$ has the high-spin configuration $t_{2g}^3 e_g^2$.

The ground-state configuration of a complex reflects the relative values of the ligand-field splitting parameter and the pairing energy; for 3d^n species with n = 4–7, high-spin and low-spin complexes occur in the weak-field and strong-field cases, respectively. Complexes of 4d and 5d metals are typically low-spin species.

(d) Magnetic measurements

The experimental distinction between high-spin and low-spin complexes is based on the determination of their magnetic properties. Complexes are classified as **diamagnetic** if they tend to move out of a magnetic field and **paramagnetic** if they tend to move into a magnetic field. The two can be distinguished experimentally as explained in Box 7.3. The extent of the paramagnetism of a complex is commonly reported in terms of the magnetic dipole moment it possesses: the higher the magnetic dipole moment of the complex, the greater the paramagnetism of the sample.

In a free atom or ion, both the orbital and the spin angular momenta give rise to a magnetic moment and contribute to the paramagnetism. When the atom or ion is part of a complex, any orbital angular momentum may be eliminated—the technical term is **quenched**—as a result of the interactions of the electrons with their nonspherical environment. However, the electron spin angular momentum survives, and gives rise to **spin-only paramagnetism**, which is characteristic of many *d*-metal complexes. The spin-only magnetic moment, μ, of a complex with total spin quantum number S is

$$\mu = 2\{S(S+1)\}^{1/2}\mu_B \qquad (2)$$

Box 7.3 **The measurement of magnetic susceptibility**

The attraction to or repulsion from a magnetic field can be measured by the change in the apparent weight of a sample when the magnetic field is turned on in a 'Gouy balance' (Fig. B7.1). The observed change in apparent weight of the sample is the net outcome of a paramagnetic term related to unpaired electrons and a diamagnetic term common to all matter. If unpaired electrons are present, the paramagnetic term is commonly much larger than the diamagnetic contribution. The force on the sample is proportional to the magnetic field gradient, so the sample usually protrudes from the field (so that it is subject to a field gradient). The measurement yields a value for the magnetic susceptibility that must be corrected for diamagnetism by using measurements on samples with similar total electron population but without unpaired electrons.

The modern procedure for measuring magnetic susceptibility makes use of the solid-state device known as a 'superconducting quantum interference device' (SQUID). A SQUID takes advantage of the quantization of magnetic flux and the property of current loops in superconductors that, as a part of the circuit, include a weakly conducting link through which electrons must tunnel. The current that flows in the loop in a magnetic field depends on the value of the magnetic flux, and SQUIDs can be exploited as very sensitive magnetometers.

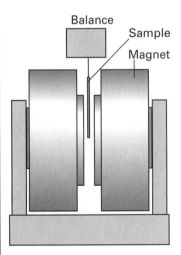

B7.1 A Gouy balance, which is used to measure the magnetic susceptibilities of substances. The change in apparent weight when the electromagnet is turned on is proportional to the magnetic susceptibility of the sample. The latter is determined by the number of unpaired electrons in the sample.

Further reading

W.E. Hatfield, Magnetic measurements. In *Solid state chemistry* (ed. A.K. Cheetham and P. Day). Oxford University Press (1987).

L.N. Mulay and I.L. Mulay, Static magnetic techniques and applications. *Techniques of physical chemistry* **IIIB**, 133 (1989).

where μ_B is the collection of fundamental constants known as the **Bohr magneton**:

$$\mu_B = \frac{e\hbar}{2m_e} \tag{3}$$

Its value is 9.274×10^{-24} J T^{-1}. Because each unpaired electron has a spin quantum number of $\frac{1}{2}$, it follows that $S = \frac{1}{2}N$, where N is the number of unpaired electrons; therefore

$$\mu = \{N(N+2)\}^{1/2}\mu_B \tag{4}$$

A measurement of the magnetic moment of a *d*-block complex can usually be interpreted in terms of the number of unpaired electrons it contains, and hence the measurement can be used to distinguish between high-spin and low-spin complexes. For example, magnetic measurements on a d^6 complex permit us to distinguish between a high-spin $t_{2g}^4 e_g^2$ ($N = 4$, $\mu = 4.90\mu_B$) configuration and a low-spin t_{2g}^6 ($S = 0$, $\mu = 0$) configuration.

The spin-only magnetic moments for the configurations $t_{2g}^n e_g^m$ are listed in Table 7.5 and compared there with experimental values for a number of 3*d* complexes. For most 3*d* complexes (and some 4*d* complexes), experimental values lie reasonably close to spin-only predictions, so it becomes possible to identify correctly the number of unpaired electrons and assign the ground-state configuration. For instance, $[Fe(OH_2)_6]^{3+}$ is paramagnetic with a magnetic moment of $6.3\mu_B$. As shown in Table 7.5, this value is reasonably close to the value for five unpaired electrons, which implies a high-spin $t_{2g}^3 e_g^2$ configuration.

> *Magnetic measurements can be used to determine the number of unpaired spins in a complex and hence to identify its ground-state configuration.*

Table 7.5 Calculated spin-only magnetic moments

Ion	N	S	μ/μ_B	
			Calculated	Experiment
Ti^{3+}	1	$\frac{1}{2}$	1.73	1.7–1.8
V^{3+}	2	1	2.83	2.7–2.9
Cr^{3+}	3	$\frac{3}{2}$	3.87	3.8
Mn^{3+}	4	2	4.90	4.8–4.9
Fe^{3+}	5	$\frac{5}{2}$	5.92	5.9

Example 7.5 Inferring an electron configuration from a magnetic moment

The magnetic moment of an octahedral Co(II) complex is $4.0\mu_B$. What is its electron configuration?

Answer A Co(II) complex is d^7. The two possible configurations are $t_{2g}^5 e_g^2$ (high spin) with three unpaired electrons or $t_{2g}^6 e_g^1$ (low spin) with one unpaired electron. The spin-only magnetic moments (Table 7.5) are $3.87\mu_B$ and $1.73\mu_B$, respectively. Therefore, the only consistent assignment is the high-spin configuration $t_{2g}^5 e_g^2$.

Self-test 7.5 The magnetic moment of the complex $[Mn(NCS)_6]^{4-}$ is $6.06\mu_B$. What is its electron configuration?

The interpretation of magnetic susceptibility measurements is sometimes less straightforward than Example 7.5 might suggest. For example, the potassium salt of $[Fe(CN)_6]^{3-}$ has $\mu = 2.3\mu_B$, which is between the spin-only values for one and two unpaired electrons ($1.7\mu_B$ and $2.8\mu_B$, respectively). In this case, the spin-only assumption has failed and the orbital magnetic contribution is substantial.

For orbital angular momentum to contribute, and hence for the paramagnetism to differ significantly from the spin-only value, there must be an unfilled or half-filled orbital similar in energy to that of the orbitals occupied by the unpaired spins. If that is so, the electrons can make use of the available orbital to circulate through the complex and hence generate orbital angular momentum and an orbital contribution to the total magnetic moment (Fig. 7.11). Departure from spin-only values is generally large for low-spin d^5 and for high-spin $3d^6$ and $3d^7$ complexes.

> *Departure from spin-only values due to orbital contributions is generally large for low-spin d^5 and for high-spin $3d^6$ and $3d^7$ complexes.*

(e) Thermochemical correlations

The concept of ligand-field stabilization energy helps to explain the double-humped hydration enthalpies of $3d$-metal M^{2+} ions (Fig. 7.12). The nearly linear increase across a period shown by the filled circles represents the increasing strength of the bonding between H_2O ligands and the central metal ion as the ionic radii decrease from left to right across the period. The wave-like deviation of hydration enthalpies from a straight line reflects the variation in the ligand-field stabilization energies. As Table 7.4 shows, the LFSE increases from d^1 to d^3, decreases again to d^5, then rises to d^8. (As we see in the next section, d^9 is a special case.) The filled circles in Fig. 7.12 were calculated by subtracting the high-spin LFSE from $\Delta_{hyd}H$ by using the spectroscopic values of Δ_O in Table 7.4. We see that the LFSE calculated from spectroscopic data nicely accounts for the additional ligand binding energy for the complexes shown in the illustration.

> *The experimental variation in hydration enthalpies reflects a combination of the variation in radii of the ions (the linear trend) and the variation in LFSE (the wave-like variation).*

Example 7.6 Using the LFSE to account for thermochemical properties

The oxides of formula MO, which all have octahedral coordination of the metal ions, have the following lattice enthalpies:

CaO	TiO	VO	MnO	
3460	3878	3913	3810	kJ mol^{-1}

Account for the trends in terms of the LFSE.

Answer The general trend across the d-block is the increase in lattice enthalpy from CaO (d^0) to MnO (d^5) as the ionic radii decrease. Both Ca^{2+} and Mn^{2+} have an LFSE of zero. Because O^{2-} is a weak-field ligand, TiO (d^2) has an LFSE of $0.8\Delta_O$ and VO (d^3) has an LFSE of $1.2\Delta_O$ (Table 7.4). It follows that the greater lattice enthalpies of TiO and VO arise from the ligand-field stabilization energy.

Self-test 7.6 Account for the variation in lattice enthalpy of the solid fluorides in which each metal ion is surrounded by an octahedral array of F^- ions: MnF_2 (2780 kJ mol^{-1}), FeF_2 (2926 kJ mol^{-1}), CoF_2 (2976 kJ mol^{-1}), NiF_2 (3060 kJ mol^{-1}), and ZnF_2 (2985 kJ mol^{-1}).

7.5 The electronic structures of four-coordinate complexes

Second only in abundance to octahedral complexes are the tetrahedral and square-planar four-coordinate complexes. The same kind of arguments within the context of crystal-field theory can be applied to these species as for octahedral complexes, but we have to take into account the different order of the d-orbital energies.

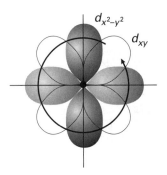

7.11 If there is a low-lying orbital of the correct symmetry, the applied field may induce the circulation of the electrons in a complex and hence generate orbital angular momentum. This diagram shows the way in which circulation may arise when the field is applied perpendicular to the xy-plane (perpendicular to this page).

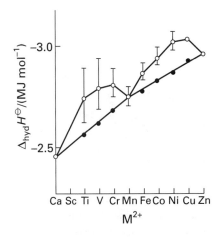

7.12 The hydration enthalpy of M^{2+} ions of the first row of the d-block. The straight lines show the trend when the ligand-field stabilization energy has been subtracted from the observed values. Note the general trend to greater hydration enthalpy (more exothermic hydration) on crossing the period from left to right.

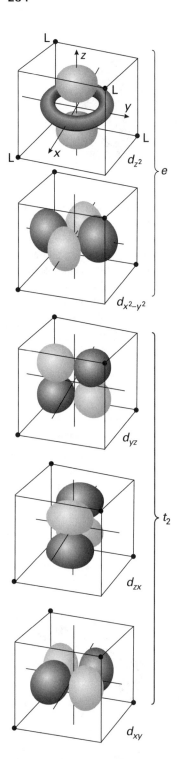

7.13 The effect of a tetrahedral crystal field on a set of *d* orbitals is to split them into two sets; the *e* pair (which point less directly at the ligands) lie lower in energy than the t_2 triplet.

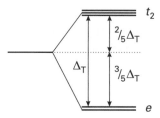

7.14 The orbital energy level diagram used in the application of the building-up principle in a crystal-field analysis of a tetrahedral complex.

Table 7.6 Values of Δ_T for representative tetrahedral complexes

Complex	Δ_T/cm^{-1}
VCl_4	9010
$[CoCl_4]^{2-}$	3300
$[CoBr_4]^{2-}$	2900
$[CoI_4]^{2-}$	2700
$[Co(NCS)_4]^{2-}$	4700

(a) Tetrahedral complexes

A tetrahedral crystal field splits *d* orbitals into two sets, one of which is doubly degenerate and the other triply degenerate (Fig. 7.13): the doubly degenerate *e* set lies *below* the triply degenerate t_2 set (Fig. 7.14).[11] This difference can be understood from a detailed analysis of the spatial arrangement of the orbitals, for the two *e* orbitals point between the positions of the ligands and their partial negative charges whereas the three t_2 orbitals point more directly toward the ligands. A secondary difference is that the ligand-field splitting parameter, Δ_T, in a tetrahedral complex is less than Δ_O, as might be expected for complexes with fewer ligands (but in fact, $\Delta_T < \frac{1}{2}\Delta_O$, which is harder to explain). Hence, only weak-field tetrahedral complexes are common, and they are the only ones we need consider.

Ligand-field stabilization energies can be calculated in exactly the same way as for octahedral complexes, the only differences being the order of occupation (*e* before t_2) and the contribution of each orbital to the total energy ($-\frac{3}{5}\Delta_T$ for an *e* electron and $+\frac{2}{5}\Delta_T$ for a t_2 electron). Some calculated values are given in Table 7.4 and the experimental values of Δ_T for a number of complexes are collected in Table 7.6. As can be inferred from Table 7.4, the configurations of d^n complexes are $e^1, e^2, e^2t_2^1, \ldots, e^2t_2^3, e^3t_2^3, e^4t_2^3, \ldots, e^4t_2^6$ as *n* increases from 1 to 10.

> In a tetrahedral complex, the *e* orbitals lie below the t_2 orbitals; only the weak-field case need be considered.

(b) Tetragonal and square-planar complexes

Copper(II), which has nine *d* electrons, forms six-coordinate complexes that usually depart considerably from O_h symmetry, and often show pronounced tetragonal distortions. Low-spin d^7 six-coordinate complexes may show a similar distortion, but they are less common.

It is convenient to take the regular octahedron as a starting point for the explanation of the existence of six-coordinate complexes with geometries that are distorted toward square-planar. A tetragonal distortion corresponding to extension along the *z*-axis and compression on the *x*- and *y*-axes, reduces the energy of the $e_g(d_{z^2})$ orbital and increases the energy of the $e_g(d_{x^2-y^2})$ orbital (Fig. 7.15). Therefore, if one, two, or three electrons occupy the e_g orbitals (as in d^7, d^8, and d^9 complexes) a tetragonal distortion may be energetically advantageous. For example, in a d^9 complex (with configuration that would be $t_{2g}^6 e_g^3$ in O_h), such a distortion leaves two electrons stabilized and one destabilized. A similar distortion can be expected when there are one, two, or three e_g electrons in the complex, for in each case the distortion lowers the total energy.

The distortion of a high-spin d^8 ($t_{2g}^6 e_g^2$) complex may be large enough to encourage the two e_g electrons to pair in the d_{z^2} orbital. This distortion may go as far as the total loss of the

---

[11] Because there is no center of inversion in a tetrahedral complex, the orbital designation does not include the parity label g or u.

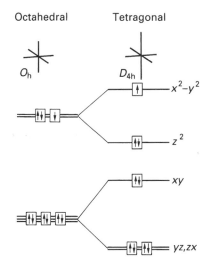

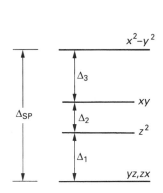

7.15 The effect of tetragonal distortions (compression along x and y and extension along z) on the energies of d orbitals. The electron occupation is for a d^9 complex.

7.16 The orbital splitting parameters for a square-planar complex.

ligands on the z-axis and the formation of the d^8 square-planar complexes, such as those found for Rh(I), Ir(I), Pt(II), Pd(II), and Au(III). The common occurrence of square-planar $4d^8$ and $5d^8$ complexes correlates with the high values of the ligand-field splitting parameter in these two series, which gives rise to a high ligand-field stabilization of the low-spin square-planar complexes. In contrast, $3d$-metal complexes such as $[NiX_4]^{2-}$, with X a halogen, are generally tetrahedral because the ligand-field splitting parameter is generally quite small for members of this series. Only when the ligand is high in the spectrochemical series is the LFSE large enough to result in the formation of a square-planar complex, as for example with $[Ni(CN)_4]^{2-}$.

The d-orbital splitting for square-planar complexes is shown in Fig. 7.16. The sum of the three distinct orbital splittings is denoted Δ_{SP}. This sum is greater than Δ_O; simple theory predicts that $\Delta_{SP} = 1.3\Delta_O$ for complexes of the same metal and ligands with the same M—L bond lengths.

> *A tetragonal distortion can be expected when there are one, two, or three e_g electrons in the complex; in a $4d^8$ or $5d^8$ complex the distortion may result in the formation of a square-planar species.*

(c) The Jahn–Teller effect

The tetragonal distortions just described represent specific examples of the **Jahn–Teller effect**:

- If the ground electronic configuration of a nonlinear complex is orbitally degenerate, the complex will distort so as to remove the degeneracy and achieve a lower energy.

An octahedral d^9 complex is orbitally degenerate because the $e_g(d_{x^2-y^2})$ and $e_g(d_{z^2})$ orbitals have the same energy and the single electron can occupy either one of them. A tetragonal distortion results in the two orbitals having different energies, and the energy of the resulting complex is lower than in the undistorted complex. A low-spin octahedral d^8 complex is degenerate because the electrons may occupy the $e_g(d_{x^2-y^2})$ orbital and the $e_g(d_{z^2})$ orbital in a variety of ways with paired spins. A tetragonal distortion removes the orbital degeneracy, with d_{z^2} commonly lower than $d_{x^2-y^2}$, and the configuration $d_{z^2}^2$ has the lowest energy.

Distortion to square-planar, with the electrons paired in d_{z^2}, can be regarded as an extreme case of the Jahn–Teller effect.

The Jahn–Teller effect identifies an unstable geometry; it does not predict the preferred distortion. The examples we have cited involve the elongation of two axial bonds and the compression of the four bonds that lie in a plane. The alternative distortion, compression along an axis and elongation in a plane, would also remove the degeneracy. Which distortion occurs in practice is a matter of energetics, not symmetry. However, because the axial elongation weakens only two bonds whereas elongation in the plane would weaken four, axial elongation is more common than axial compression.

The Jahn–Teller distortion can hop from one orientation to another and give rise to the **dynamic Jahn–Teller effect**. For example, below 20 K, the EPR spectrum of $[Cu(OH_2)_6]^{2+}$ shows a static distortion (more precisely, one effectively stationary on the timescale of the resonance experiment). However, above 20 K the distortion apparently disappears because it hops more rapidly than the timescale of the EPR observation.

> *If the ground electronic configuration of a nonlinear complex is orbitally degenerate, the complex will distort so as to remove the degeneracy and achieve a lower energy.*

7.6 Ligand–field theory

Crystal-field theory provides a simple conceptual model and can be used to interpret spectra and thermochemical data by appealing to empirical values of Δ_O. On closer inspection, however, the theory is defective because it treats ligands as point charges or dipoles and does not take into account the overlap of ligand and metal orbitals.

Ligand-field theory, which is an application of molecular orbital theory that concentrates on the *d*-orbitals of the central metal atom, provides a more substantial framework for understanding Δ_O. The strategy for describing the molecular orbitals of a metal complex follows procedures similar to those described in Chapter 3 for bonding in polyatomic molecules: the valence orbitals on the metal and ligand are used to form symmetry-adapted linear combinations (SALCs), and then—using empirical energy and overlap considerations—the relative energies of the molecular orbitals are estimated. The estimated order can be verified and positioned more precisely by comparison with experimental data (particularly optical absorption and photoelectron spectroscopy).[12]

(a) σ Bonding

We begin the systematic discussion of ligand-field theory by imagining an octahedral complex in which each ligand has a single valence orbital directed toward the central metal atom; each of these orbitals has local σ symmetry with respect to the M–L axis. Examples of such ligands include the isolobal NH_3 molecule and F^- ion.

In a strictly octahedral (O_h) environment, the metal orbitals divide by symmetry into four sets (Fig. 7.17 and Appendix 4):

Metal orbital	Symmetry label	Degeneracy
s	a_{1g}	1
p_x, p_y, p_z	t_{1u}	3
d_{xy}, d_{yz}, d_{zx}	t_{2g}	3
$d_{x^2-y^2}, d_{z^2}$	e_g	2

---

[12] High-resolution PES data are available only for complexes in the gas phase, and are therefore restricted to uncharged complexes. For an introduction, see S.F.A. Kettle, *Physical inorganic chemistry*. Spektrum, Oxford (1996).

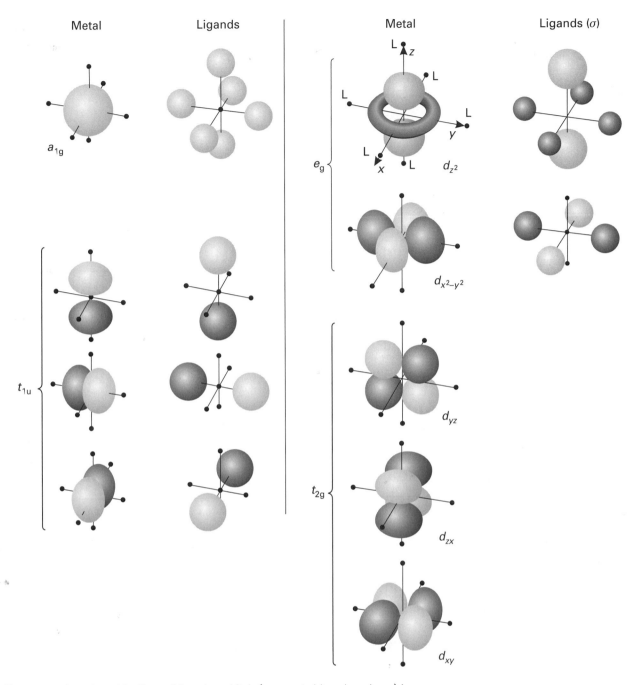

7.17 Symmetry-adapted combinations of ligand σ orbitals (represented here by spheres) in an octahedral complex. For symmetry-adapted orbitals in other point groups, see Appendix 4.

Six symmetry-adapted linear combinations of the six ligand σ orbitals can also be formed. These combinations can be taken from Appendix 4 and are also shown in Fig. 7.17. One (unnormalized) ligand combination is a nondegenerate a_{1g} linear combination:

$$a_{1g}: \qquad \sigma_1 + \sigma_2 + \sigma_3 + \sigma_4 + \sigma_5 + \sigma_6$$

where σ_i denotes a σ orbital on ligand $i = 1, 2, \ldots, 6$. Three form a t_{1u} set:

$$t_{1u}: \qquad \sigma_1 - \sigma_3, \qquad \sigma_2 - \sigma_4, \qquad \sigma_5 - \sigma_6$$

The remaining two form an e_g pair:

$$e_g: \qquad \sigma_1 - \sigma_2 + \sigma_3 - \sigma_4, \qquad 2\sigma_6 + 2\sigma_5 - \sigma_1 - \sigma_2 - \sigma_3 - \sigma_4$$

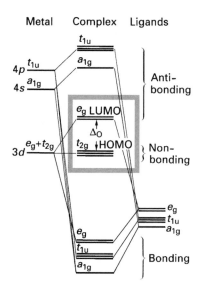

Metal   Complex   Ligands

7.18 Molecular orbital energy levels of a typical octahedral complex. The frontier orbitals are inside the gray box.

These six combinations account for all the ligand orbitals of σ symmetry: there is no combination of ligand σ orbitals that has the symmetry of the metal t_{2g} orbitals, so the latter do not participate in σ bonding.[13]

Molecular orbitals are formed by combining SALCs and metal atomic orbitals of the same symmetry type. For example, the (unnormalized) form of an a_{1g} molecular orbital is $c_M\psi_{Ms} + c_L\psi_{La_{1g}}$. The values of the coefficients are determined by a molecular orbital calculation and the probability of finding an electron in the metal atomic orbitals is proportional to c_M^2; the probability of finding an electron in the ligand atomic orbitals is proportional to c_L^2. The metal and ligand a_{1g} orbitals overlap to give two molecular orbitals (one bonding and one antibonding), the doubly degenerate metal and ligand e_g orbitals overlap to give four molecular orbitals (two degenerate bonding, two degenerate antibonding), and the triply degenerate metal and ligand t_{1u} orbitals overlap to give six molecular orbitals (three degenerate bonding, three degenerate antibonding). There are therefore six bonding combinations in all and six antibonding combinations. The three triply degenerate metal t_{2g} orbitals remain nonbonding and fully localized on the metal atom. Calculations of the resulting energies (adjusted to agree with a variety of spectroscopic data) result in the molecular orbital energy level diagram shown in Fig. 7.18.

It should be recalled from Chapter 3 that the greatest contribution to the molecular orbital of lowest energy is from atomic orbitals of lowest energy. For NH_3, F^-, and most other ligands, the ligand σ orbitals are derived from atomic orbitals with energies that lie well below those of the metal d orbitals. As a result, the six bonding molecular orbitals of the complex are mainly ligand-orbital in character in the sense that $c_L^2 > c_M^2$. These six bonding orbitals can accommodate the 12 electrons provided by the six ligand lone pairs. In a sense, therefore, the electrons provided by the ligands are largely confined to the ligands in the complex, just as the crystal-field theory presumes. However, the coefficients of the d orbitals in these bonding molecular orbitals are not zero, so 'ligand electrons' do in fact leak on to the central metal atom.

The number of electrons to accommodate in addition to those supplied by the ligands depends on the number of d electrons, n, supplied by the metal atom or ion. These additional electrons occupy the nonbonding d orbitals (the t_{2g} orbitals) and the antibonding combination (the upper e_g orbitals) of the d orbitals and ligand orbitals. The t_{2g} orbitals are confined to the metal atom and the antibonding e_g orbitals are largely metal-ion in character too, so the n electrons supplied by the central ion remain largely on the metal ion. In summary, the frontier orbitals of the complex are the nonbonding entirely metal t_{2g} orbitals and the antibonding e_g orbitals (mainly metal in character). The octahedral ligand-field splitting parameter, Δ_O, in this approach is the HOMO–LUMO separation. In ligand-field theory, the molecular orbitals involved are largely, but not completely, confined to the metal atom. Crystal-field theory exaggerates that confinement by supposing that the d electrons are strictly confined to the metal.

Once the molecular orbital energy level diagram has been established, the building-up principle can be used to arrive at the ground-state electron configuration of the complex. First, we note that the six bonding molecular orbitals accommodate the 12 electrons supplied by the ligands. The remaining n electrons of a d^n complex are accommodated in the nonbonding t_{2g} orbitals and the antibonding e_g orbitals. Now the story is essentially exactly the same as for crystal-field theory: the types of complex that are obtained (high-spin or low-spin, for instance) depending on the relative values of Δ_O and the pairing energy P. The only difference from the discussion in Section 7.4 is that the qualitative molecular orbital theory permits us to identify more deeply the origin and magnitude of the splitting of the orbitals shown in the box in Fig. 7.18.

---

[13] The normalization constants (with overlap neglected) are $N(a_{1g}) = 1/6^{1/2}$, $N(t_{1u}) = 1/2^{1/2}$ for all three orbitals, and $N(e_g) = 1/2$ and $1/12^{1/2}$, respectively.

> In ligand-field theory, the building-up principle is used in conjunction with a molecular-orbital energy level diagram constructed by noting the symmetries of the d orbitals and linear combinations of ligand orbitals.

Example 7.7 Using a photoelectron spectrum to obtain information about a complex

The photoelectron spectrum of gas-phase $[Mo(CO)_6]$ is shown in Fig. 7.19. Use the spectrum to infer the energies of the molecular orbitals of the complex.

Answer Twelve electrons are provided by the six CO ligands (treated as :CO); they enter the bonding orbitals and result in the configuration $a_{1g}^2 t_{1u}^6 e_g^4$. The oxidation number of molybdenum, Group 6, is 0, so Mo provides a further six valence electrons. The ligand and metal valence electrons are distributed over the orbitals shown in the box in Fig. 7.18 and, as CO is a strong-field ligand, the ground-state electron configuration of the complex is expected to be low-spin: $a_{1g}^2 t_{1u}^6 e_g^4 t_{2g}^6$. The HOMOs are the three t_{2g} orbitals that are largely confined to the Mo atom, and their energy can be identified by ascribing the peak of lowest ionization energy (close to 8 eV) to them. The group of ionization energies around 14 eV are probably due to the Mo–CO σ bonding orbitals. However, 14 eV is close to the ionization energy of CO itself, so the variety of peaks at that energy also arise from bonding orbitals in CO.

Self-test 7.7 Suggest an interpretation of the photoelectron spectrum of $[Fe(C_5H_5)_2]$ and $[Mg(C_5H_5)_2]$ shown in Fig. 7.20.

(b) π Bonding

So far, we have considered only metal–ligand σ interactions. If the ligands in a complex have orbitals with local π symmetry with respect to the M–L axis (as two of the p orbitals of a halide ligand have), they may form bonding and antibonding π molecular orbitals with the metal orbitals (Fig. 7.21). We shall illustrate the effect of π bonding by a simplified argument based on π bonding in a single M–X fragment; in a fuller treatment, SALCs would be constructed from all the available π orbitals on all the ligand atoms. The important point is that when ligand π orbitals are taken into account, the combinations that can be formed from them include SALCs of t_{2g} symmetry. These combinations have net overlap with the metal t_{2g} orbitals, which are therefore no longer purely nonbonding on the metal atom. Depending on the relative energies of the ligand and metal orbitals, the energies of the t_{2g} molecular orbitals lie above or below the energies they had as nonbonding atomic orbitals, so the HOMO–LUMO gap (that is, Δ_O) is decreased or increased, respectively.

To explore the role of π bonding, we need two of the general principles described in Chapter 3. First, we shall make use of the idea that, when atomic orbitals overlap strongly, they mix strongly: the resulting bonding molecular orbitals are significantly lower in energy and the antibonding molecular orbitals are significantly higher in energy than the atomic orbitals. Second, we note that atomic orbitals with similar energies interact strongly, whereas those of very different energies mix only slightly even if their overlap is large.

A π-**donor ligand** is a ligand that, before any bonding is considered, has *filled* orbitals of π symmetry around the M–L axis. The energies of these full π orbitals are usually close to, but somewhat lower than, those of the metal d orbitals. It is also commonly the case that ligands of this kind have no low-energy vacant π orbitals, so we need consider only the full orbitals when considering the effects of π bonding in the complex. Such ligands include Cl^-, Br^-, and H_2O. Because the full π orbitals of π-donor ligands lie lower in energy than the partially filled d orbitals of the metal, when they form molecular orbitals with the metal t_{2g} orbitals, the bonding combination lies lower than the ligand orbitals and the antibonding combination lies

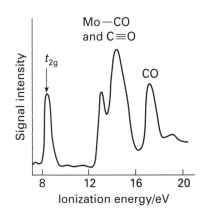

7.19 The He(II) (30.4 nm) photoelectron spectrum of $[Mo(CO)_6]$. With six electrons from Mo and twelve from :CO, the configuration indicated by Fig. 7.18 is $a_{1g}^2 t_{1u}^6 e_g^4 t_{2g}^6$. (From B.R. Higgenson, D.R. Lloyd, P. Burroughs, D.M. Gibson, and A.F. Orchard, *J. Chem. Soc. Faraday II* **69**, 1659 (1973).)

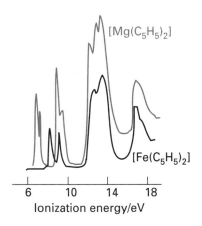

7.20 Photoelectron spectra of ferrocene and magnesocene.

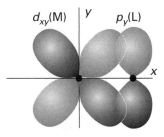

7.21 The π overlap that may occur between a ligand p orbital perpendicular to the M–L axis and a metal d_{xy} orbital.

M   ML$_6$ L

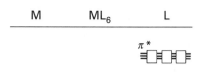

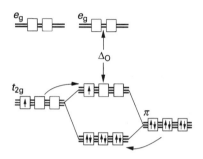

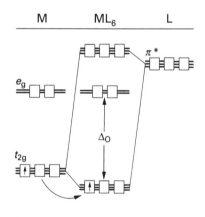

7.22 The effect of π bonding on the ligand-field splitting parameter. Ligands that act as π donors decrease Δ_O.

M   ML$_6$ L

7.23 Ligands that act as π acceptors increase Δ_O.

above the energy of the d orbitals of the free metal ion (Fig. 7.22). The electrons supplied by the ligand lone pairs occupy and fill the bonding combinations, leaving the electrons originally in the d orbitals of the central metal atom to occupy the antibonding t_{2g} orbitals. The net effect is that the hitherto nonbonding metal-ion t_{2g} orbitals become antibonding and hence are raised closer in energy to the mainly metal antibonding e_g orbital. It follows that strong π-donor ligands decrease Δ_O.

A **π-acceptor ligand** is a ligand that has filled π orbitals at (usually) lower energies than metal t_{2g} orbitals; it also has empty π orbitals that are available for occupation. Typically, the π acceptor orbitals are vacant antibonding orbitals on the ligand, as in CO and N$_2$, and these orbitals lie above the metal d orbitals in energy. For example, the π^* orbital of CO has its largest amplitude on the C atom and has the correct symmetry for overlap with the t_{2g} orbitals of the metal. In contrast, the full bonding π orbital of CO is low in energy and is largely localized on the O atom (because that is the more electronegative atom). As a result, the π-donor character of CO is very low and in most (if not all) d-metal carbonyl complexes CO is a net π acceptor.

Because the π-acceptor orbitals on most ligands are higher in energy than the metal d orbitals, they form molecular orbitals in which the bonding t_{2g} combinations are largely of metal d orbital character (Fig. 7.23). These bonding combinations lie slightly lower in energy than the d orbitals themselves. The net result is that Δ_O is increased by the π-acceptor interaction.

We can now put the role of π bonding in perspective. The order of ligands in the spectrochemical series is partly that of the strengths with which they can participate in M—L σ bonding. For example, both CH$_3^-$ and H$^-$ are very high in the spectrochemical series because they are very strong σ donors. However, when π bonding is significant, it has a strong influence on Δ_O: π-donor ligands decrease Δ_O and π-acceptor ligands increase Δ_O. This effect is responsible for CO (a strong π acceptor) being high on the spectrochemical series and for OH$^-$ (a strong π donor) being low in the series. The overall order of the spectrochemical series may be interpreted in broad terms as dominated by π effects (with a few important exceptions), and in general the series can be interpreted as follows:

$$- \text{ increasing } \Delta_O \longrightarrow$$
$$\pi \text{ donor} < \text{weak } \pi \text{ donor} < \text{no } \pi \text{ effects} < \pi \text{ acceptor}$$

Representative ligands that match these classes are

$$I^- < Br^- < Cl^- < F^- < H_2O < NH_3 < PR_3 < CO$$
$$\pi \text{ donor} < \text{weak } \pi \text{ donor} < \text{no } \pi \text{ effects} < \pi \text{ acceptor}$$

Notable examples of where the σ bonding effect dominates include CH$_3^-$ and H$^-$, which are neither π-donor nor π-acceptor ligands.

> π-*Donor ligands decrease Δ_O and π-acceptor ligands increase Δ_O; the spectrochemical series is largely a consequence of the effects of π bonding when such bonding is feasible.*

Reactions of complexes

The reactions of d-metal complexes are usually studied in solution. The solvent molecules compete for the central metal ion, and the formation of a complex with another ligand is a **substitution reaction**, a reaction in which an incoming group displaces a ligand (in this case, a solvent molecule) already present. The incoming group is called the **entering group** and the displaced ligand is the **leaving group**. We normally denote the leaving group as X and the entering group as Y. Then a substitution reaction is the Lewis displacement reaction

$$MX + Y \longrightarrow MY + X$$

Both the thermodynamics and the kinetics of complex formation lead to an understanding of the reaction of complexes. We start with thermodynamic considerations and then survey some general trends in the kinetics of ligand substitution. A more thorough account of the kinetics of reactions is given in Chapter 14.

7.7 Coordination equilibria

A specific example of a coordination equilibrium is the reaction of Fe(III) with SCN^- to give $[Fe(NCS)(OH_2)_5]^{2+}$, a red complex used to detect either iron(III) or the thiocyanate ion:[14]

$$[Fe(OH_2)_6]^{3+}(aq) + NCS^-(aq) \rightleftharpoons [Fe(NCS)(OH_2)_5]^{2+}(aq) + H_2O(l)$$

$$K_f = \frac{[Fe(NCS)(OH_2)_5^{2+}]}{[Fe(OH_2)_6^{3+}][NCS^-]} \tag{5}$$

The equilibrium constant, K_f, is the **formation constant** of the complex. The concentration of H_2O does not appear because it is taken to be constant in dilute solution and is absorbed into K_f. A ligand for which K_f is large is one that binds more tightly than H_2O. A ligand for which K_f is small may not be a weak ligand in an absolute sense, but merely weaker than H_2O.

(a) Formation constants

The discussion of stabilities is more involved when more than one ligand may be replaced. In the series from $[Ni(OH_2)_6]^{2+}$ to $[Ni(NH_3)_6]^{2+}$, for instance, there are as many as six steps even if *cis-trans* isomerism is ignored. For the general case of the complex ML_n, the **stepwise formation constants** are

$$M + L \rightleftharpoons ML \qquad K_1 = \frac{[ML]}{[M][L]}$$

$$ML + L \rightleftharpoons ML_2 \qquad K_2 = \frac{[ML_2]}{[ML][L]} \tag{6}$$

$$\cdots$$

$$ML_{n-1} + L \rightleftharpoons ML_n \qquad K_n = \frac{[ML_n]}{[ML_{n-1}][L]}$$

These stepwise constants are the ones to consider when seeking to understand the relations between structure and reactivity. When we want to calculate the concentration of the final product (the complex ML_n) we use the **overall formation constant**, β_n:

$$\beta_n = \frac{[ML_n]}{[M][L]^n} \tag{7}$$

The overall formation constant is the product of the stepwise constants:

$$\beta_n = K_1 K_2 K_3 \ldots K_n \tag{8}$$

The inverse of K_f, the **dissociation constant**, K_d, is also sometimes useful:

$$ML \rightleftharpoons M + L \qquad K_d = \frac{[M][L]}{[ML]} \tag{9}$$

Because K_d has the same form as K_a for acids, its use facilitates comparisons between metal complexes and Brønsted acids. The values of K_d and K_a can be tabulated together if the proton is considered to be simply another cation, with HF, for instance, the complex formed from the Lewis acid H^+ and the Lewis base F^- playing the role of a ligand.

---

[14] In expressions for equilibrium constants, square brackets denote the numerical value of the molar concentration of a species.

> The stabilities of complexes are expressed in terms of their formation constants; formation constants express the strengths of species as ligands relative to the strength of H_2O as a ligand.

(b) Trends in successive formation constants

The magnitude of the formation constant is a direct reflection of the sign and magnitude of the Gibbs energy of formation (because $\Delta_r G^{\ominus} = -RT \ln K_f$). It is commonly observed that stepwise formation constants lie in the order $K_1 > K_2 > K_3 \ldots > K_n$. This general trend can be explained quite simply by considering the decrease in the number of the ligand H_2O molecules available for replacement in the formation step, as in

$$M(OH_2)_5L + L \longrightarrow M(OH_2)_4L_2 + H_2O$$

compared with

$$M(OH_2)_4L_2 + L \longrightarrow M(OH_2)_3L_3 + H_2O$$

This sequence reduces the number of H_2O ligands available for replacement as n increases. Conversely, the increase in the number of bound L groups increases the importance of the reverse of these reactions as n increases. Therefore, provided the reaction enthalpy is largely unaffected, the equilibrium constants lie progressively in favor of the reactants as n increases. Physically, the decrease in the stepwise formation constants reflects the unfavorable entropy change as free ligands are progressively immobilized by coordination to the metal. That such a simple explanation is more or less correct is illustrated by data for the successive complexes in the series from $[Ni(OH_2)_6]^{2+}$ to $[Ni(NH_3)_6]^{2+}$ (Table 7.7). The enthalpy changes for the six successive steps are known to vary by less than 2 kJ mol^{-1} from 16.7 to 18.0 kJ mol^{-1}.

A reversal of the relation $K_n < K_{n+1}$ is usually an indication of a major change in the electronic structure of the complex as more ligands are added. An example is that the tris(bipyridine) complex of Fe(II), $[Fe(bipy)_3]^{2+}$, is strikingly stable compared with the bis complex, $[Fe(bipy)_2(OH_2)_2]^{2+}$. This observation can be correlated with the change from a weak field $t_{2g}^4 e_g^2$ in the bis complex (note the presence of weak-field H_2O ligands) to a strong field t_{2g}^6 configuration in the tris complex. A contrasting example is the anomalously low value of K_3/K_2 (approximately $\frac{1}{7}$) for the halogeno complexes of Hg(II). This decrease is too large to be explained statistically and suggests a major change in the nature of the complex, such as the onset of four-coordination:

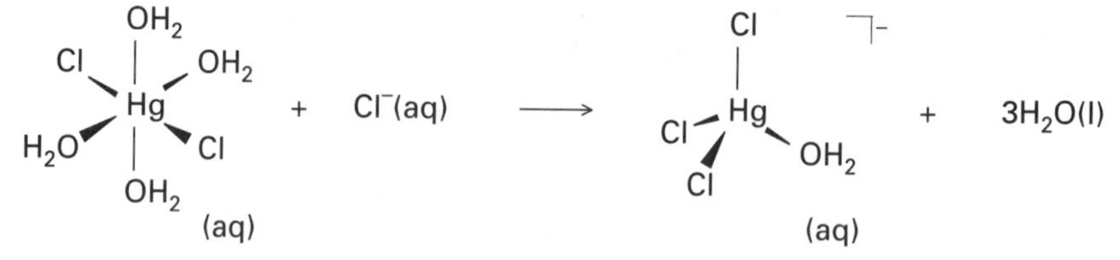

Table 7.7 Formation constants of Ni(II) ammines, $[Ni(NH_3)_n(OH_2)_{6-n}]^{2+}$

n	pK_f	K_n/K_{n-1}	
		Experimental	Statistical*
1	−2.72		
2	−2.17	0.28	0.42
3	−1.66	0.31	0.53
4	−1.12	0.29	0.56
5	−0.67	0.35	0.53
6	−0.03	0.2	0.42

*Based on ratios of numbers of ligands available for replacement, with the reaction enthalpy assumed constant.

> Stepwise formation constants typically lie in the order $K_n < K_{n+1}$, as expected statistically; deviations from this order are a sign of a major change in structure.

Example 7.8 Interpreting irregular successive formation constants

The formation of cadmium complexes with Br^- exhibit the successive equilibrium constants $\log K_1 = 1.56$, $\log K_2 = 0.54$, $\log K_3 = 0.06$, $\log K_4 = 0.37$. Suggest an explanation of why K_4 is larger than K_3.

Answer The anomaly suggests a structural change. Aqua complexes are usually six-coordinate whereas halo complexes of M^{2+} ions are commonly tetrahedral. The reaction of the complex with three Br^- groups to add the fourth is

$$[CdBr_3(OH_2)_3]^-(aq) + Br^-(aq) \longrightarrow [CdBr_4]^{2-}(aq) + 3\,H_2O(l)$$

This step is favored by the release of three molecules of water from the relatively restricted coordination sphere environment. The result is an increase in K.

Self-test 7.8 A square-planar four-coordinate Fe(II) porphyrin (P) complex may add two further ligands (L) axially. The maximally coordinated complex $[FePL_2]$ is low-spin, whereas $[FePL]$ is high-spin. Account for an increase of the second formation constant with respect to the first.

(c) The chelate effect

When K_1 for a bidentate chelate ligand, such as ethylenediamine, is compared with the value of β_2 for the corresponding bisammine complex, it is found that the former is generally larger:

$$[Cu(OH_2)_6]^{2+} + en \rightleftharpoons [Cu(en)(OH_2)_4]^{2+} + 2\,H_2O$$

$$\log K_1 = 10.6 \qquad \Delta_r H^{\ominus} = -54\;kJ\,mol^{-1} \qquad \Delta_r S^{\ominus} = +23\;J\,K^{-1}\,mol^{-1}$$

$$[Cu(OH_2)_6]^{2+} + 2\,NH_3 \rightleftharpoons [Cu(NH_3)_2(OH_2)_4]^{2+} + 2\,H_2O$$

$$\log \beta_2 = 7.7 \qquad \Delta_r H^{\ominus} = -46\;kJ\,mol^{-1} \qquad \Delta_r S^{\ominus} = -8.4\;J\,K^{-1}\,mol^{-1}$$

Two similar Cu—N bonds are formed in each case, yet the formation of the chelate is distinctly more favorable. The **chelate effect** is this greater stability of chelated complexes compared with their non-chelated analogs.

The chelate effect can be traced primarily to differences in reaction entropy between chelated and non-chelated complexes in dilute solutions. The chelation reaction results in an increase in the number of independent molecules in solution. In contrast, the non-chelating reaction produces no net change (compare the two chemical equations above). The former therefore has the more positive reaction entropy and hence is the more favorable process. The reaction entropies measured in dilute solution support this interpretation.

The chelate effect is of great practical importance. The majority of reagents used in complexometric titrations in analytical chemistry are multidentate chelates like $edta^{4-}$. Most biochemical metal binding sites are chelating ligands. When a formation constant is measured as 10^{12} to 10^{25} it is generally a sign that the chelate effect is in operation.

The entropy advantage of chelation extends beyond bidentate ligands. It accounts in part for the great stability of complexes containing the tetradentate porphyrin ligand and the hexadentate $edta^{4-}$ ligand.

> The chelate effect is the greater stability of a complex containing a coordinated polydentate ligand compared with a complex containing the equivalent number of analogous monodentate ligands.

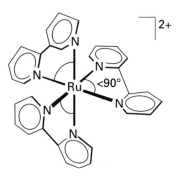

46   **47** bipy   **48** phen

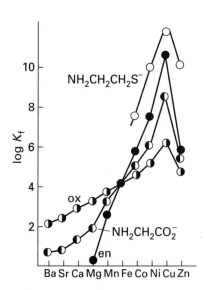

49 [Ru(bipy)₃]²⁺

(d) Steric effects and electron delocalization

Steric effects have an important influence on formation constants. They are particularly important in chelate formation because ring completion may be difficult geometrically. Chelate rings with five members are generally very stable, as we saw in Section 7.2. Six-membered rings are reasonably stable and may be favored if their formation results in electron delocalization.

Complexes containing chelating ligands with delocalized electronic structures may be stabilized by electronic effects in addition to the entropy advantages of chelation. For example, diimine ligands (**46**), such as bipyridine (**47**) and phenanthroline (**48**), are constrained to form five-membered rings with the metal. The great stability of their complexes is probably a result of their ability to act as π acceptors as well as σ donors, and to form π bonds by overlap of the full metal *d* orbitals and the empty ring π^* orbitals. This bond formation is favored by electron population in the metal t_{2g} orbitals, which allows the metal atom to act as a π donor and donate electron density to the ligand rings. An example is the complex [Ru(bipy)₃]²⁺ (**49**). The small bite angle imposed by these ligands distorts the complex from octahedral symmetry.

> The stability of chelates involving diimine ligands is a result of the chelate effect in conjunction with the ability of the ligands to act as π acceptors as well as σ donors.

(e) The Irving–Williams series

Figure 7.24 is obtained when $\log K_f$ is plotted for complexes of the M^{2+} ions of the first *d* series. This variation is summarized by the **Irving–Williams series** for the order of formation constants. For M^{2+} cations:

$$Ba^{2+} < Sr^{2+} < Ca^{2+} < Mg^{2+} < Mn^{2+} < Fe^{2+} < Co^{2+} < Ni^{2+} < Cu^{2+} < Zn^{2+}$$

The order is relatively insensitive to the choice of ligands.

In the main, the increase in stability progresses from metal ions of large radius to those of smaller radius, which suggests that the Irving–Williams series reflects electrostatic effects However, beyond Mn^{2+} there is a sharp increase in the value of K_f for Fe(II), d^6; Co(II), d^7; Ni(II), d^8; and Cu(II), d^9. These ions experience an additional stabilization proportional to the ligand-field stabilization energies (Table 7.4). However, there is one important exception: the stability of Cu(II) complexes is greater than that of Ni(II) despite the fact that Cu(II) has an additional antibonding e_g electron. This anomaly is a consequence of the stabilizing influence of the Jahn–Teller effect, which results in strong binding of the four ligands in the plane of the tetragonally distorted Cu(II) complex, and that stabilization enhances the value of K_f. The distorted axial positions are correspondingly more weakly bound.

> The Irving–Williams series summarizes the relative stabilities of complexes formed by M^{2+} ions of the first *d* series.

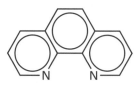

7.24 The variation of formation constants for the M^{2+} ions of the Irving–Williams series.

7.8 Rates and mechanisms of ligand substitution

Rates of reaction are as important as equilibria in coordination chemistry. The numerous isomers of the ammines of Co(III) and Pt(II), which were so important to the development of the subject, could not have been isolated if ligand substitutions and interconversion of the isomers had been fast. But what determines whether one complex will survive for long periods while another will undergo rapid reaction?

(a) Lability and inertness

Complexes that are thermodynamically unstable but survive for long periods (at least a minute) are called **inert**. Complexes that undergo more rapid equilibration are called **labile**.

Figure 7.25 shows the characteristic lifetimes of octahedral complexes of the important aqua metal ions. We see a range of lifetimes starting at about 1 ns, which is approximately the time it takes for a molecule to diffuse one molecular diameter in solution. At the other end of the scale are lifetimes in years. Even so, the illustration does not show the longest times that could be considered, which are comparable to geological eras.

There are a number of generalizations that help us to anticipate the lability of the complexes we are likely to meet:[15]

1 All complexes of s-block ions except the smallest (Be^{2+} and Mg^{2+}) are very labile.

2 Across the first d series, complexes of d-block **M(II)** ions are moderately labile, with distorted Cu(II) complexes among the most labile. However, d^6 complexes with high field ligands are an exception; for example, $[Fe(CN)_6]^{4-}$ and $[Fe(phen)_3]^{2+}$. Complexes of **M(III)** ions are distinctly less labile.

3 Complexes of low oxidation number d^{10} ions (Zn^{2+}, Cd^{2+}, and Hg^{2+}) are highly labile.

4 Strong-field d^3 and d^6 octahedral complexes of the first series (such as Cr(III) an Co(III) complexes, respectively) are generally inert; all others are generally labile.

5 In the first d series, the least labile **M(II)** and **M(III)** ions are those with greatest LFSE.

6 Inertness is quite common among the complexes of the second and third d series, which reflects the high LFSE and strength of the metal–ligand bonding.

7 The **M(III)** ions of the f-block are all very labile.

> The distinction between labile and inert complexes can be summarized by a number of generalizations, as set out in the list above.

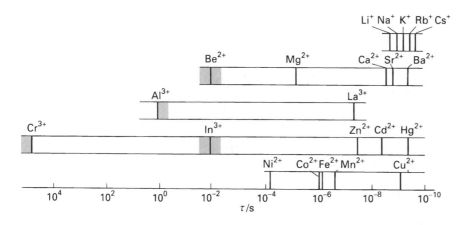

7.25 Characteristic lifetimes for exchange of water molecules in aqua complexes.

---

[15] These features are discussed more fully in Chapter 14.

(b) Associative reactions

The most revealing information about the mechanisms of ligand substitution reactions comes from studies of the rates of substitution reactions as the ligand is varied for a given metal ion.

Reactions of square-planar complexes of Pt(II) are first-order in the complex, first-order in the entering group (Y), and second-order overall.[16] The rate constants vary widely with the choice of the entering group. For example, trans-$[PtCl_2(py)_2]$ reacts with various entering groups (Y) to give trans-$[PtClY(py)_2]$ and the rate constant (in units of $L\,mol^{-1}\,s^{-1}$) varies from 4.7×10^{-4} for $Y = NH_3$, through 3.7×10^{-3} for Br^- and 0.107 for I^-, to 6.00 for thiourea, a factor of 10^4. The corresponding change in the rate constants of reactions in which the leaving group (X) is changed from Cl^- to I^- is a factor of only 3.5. In general, the effect of changing the entering group greatly exceeds that of changing the leaving group. This observation suggests that the entering group is the major factor determining the overall activation energy. This behavior is expected in an **associative substitution**, a reaction in which the coordination number of the activated complex[17] is greater than that of the initial complex. Associative substitution is a typical feature of square-planar complexes because it is reasonably easy for an entering group to enter the relatively uncrowded region of the central metal atom and to form a bond with it. The observation that cis or trans isomers substitute with retention of the original stereochemistry suggests that the activated complex is approximately a trigonal bipyramid:

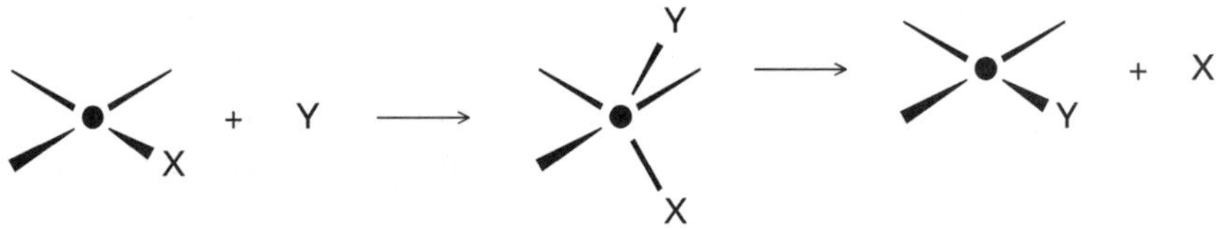

It is difficult to make generalizations about rates of associative substitutions because both the initial complex and the entering group must be specified.[18] However, one generalization that can be made is that, among the d^8 systems with similar entering and leaving groups, most are comparatively inert. Typically, lability decreases in the series $Ni(II) > Pd(II) \gg Pt(II) \approx Rh(I)$.

> In an associative reaction, the coordination number of the activated complex is greater than that of the initial complex; the rates of such reactions, which are typical of square-planar complexes, depend strongly on the identity of the entering group.

(c) Dissociative substitutions

Octahedral Ni(II) complex formation reactions are representative of reactions of octahedral complexes. They are first-order in $Ni(aq)^{2+}$, first order in the entering group, and second-order overall. However, in contrast to the square-planar case, there is only a comparatively small change in the rate constant as the entering group is varied: for example, there is only a factor of 10 difference between the rate constants for the formation of $[Ni(SO_4)(OH_2)_4]$ and $[Ni(OH_2)_4(phen)]^{2+}$. The higher rate constant of SO_4^{2-} substitution arises from the electrostatic attraction between Ni^{2+} and SO_4^{2-} ions, which increases the probability of their encountering each other in solution (this topic is discussed in detail in Section 14.6).

---

[16] In some cases, attack by abundant water solvent results in a labile intermediate that reacts with Y in a first-order reaction.

[17] Here the term 'complex' has the meaning characteristic of activated complex theory (transition state theory).

[18] One of the most interesting generalizations that can be made is that the ligand trans to the leaving group plays an important role in determining reactivity. This 'trans effect' is discussed in Section 14.4.

The principal factor determining the activation energy for the replacement of a ligand in such an octahedral complex is the breaking of the bond between the metal atom and the leaving group. The entering group plays only a minor role in the substitution reaction. Reactions of this kind are called **dissociative substitutions**: they are reactions in which the activated complex has a lower coordination number than the initial complex. We would expect dissociative reactions to be characteristic of octahedral complexes because the metal atom is relatively crowded and the leaving group needs to make room for the entering group. The pathway can be written as follows:

In this example, the dotted lines represent weak bonds.

The dissociative character of a substitution reaction is identified from the insensitivity of the rate to variation of the entering group. Judged in this way, many octahedral substitutions seem more dissociative than associative. This conclusion is helpful because the lability of an octahedral complex can be described without specifying the entering group: lability is a property of the complex itself. It is observed, for instance, that among the aqua metal ions shown in Fig. 7.25, the ones that form 'weak' bonds on account of their relatively low charge and large ionic radius are more labile than those with higher charge and smaller radius. This trend is consistent with a dissociative mechanism of substitution, which involves the breaking of a bond.

In a dissociative reaction, the coordination number of the activated complex is lower than that of the initial complex; the rates of such reactions, which are typical of many octahedral complexes, depend only weakly on the identity of the entering group.

Further reading

G. Wilkinson, R.D. Gillard, and J. McCleverty (ed.), *Comprehensive coordination chemistry*. Pergamon Press, Oxford (1987 *et seq.*). Volume 1 of this set provides a general introduction and succeeding volumes cover the particulars for each metallic element.

D.M.P. Mingos (ed.), *Structure and bonding*. Vol. 87: *Structures and electronic paradigms in cluster chemistry*. Springer (1997). This volume discusses cluster chemistry throughout the periodic table, but emphasizes metal clusters.

A.F. Wells, *Structural inorganic chemistry*. Oxford University Press (1984). This book is a good single-volume reference for information on the structures of compounds of the metals.

Ullman's encyclopedia of industrial chemistry. VCH, Weinheim (1985-96); *Kirk-Othmer Encyclopedia of chemical technology*. Wiley-Interscience, New York (1991-7). These volumes provide general information on the metallic elements, with emphasis on mineral sources, methods of extraction, and applications.

J.C. Bailar, Jr, H.J. Emeleus, R. Nyholm, and A.F. Trotman-Dickenson (ed.), *Comprehensive inorganic chemistry*, Vols 1–5. Pergamon Press, Oxford (1973). These volumes provide much useful information on the chemical properties of the metallic elements.

S.F.A. Kettle, *Physical inorganic chemistry: a coordination chemistry approach*. Oxford University Press (1998). (First published by Spectrum (1996).)

G.T. Seaborg and W.D. Loveland, *The elements beyond uranium*. Wiley-Interscience, New York (1990). This primer for the transuranium elements describes their discovery and provides a good overview of their separation, detection, and chemical properties.

J. Katz, G. Seaborg, and L. Morss, *The chemistry of the actinide elements*. Longman, London (1986). A comprehensive two-volume account of the actinides.

A.E. Martell and R.D. Hancock, *Metal complexes in aqueous solution*. Plenum Press (1996).

Exercises

7.1 Preferably without using reference material, write out the 3*d* elements in their arrangement in the periodic table. Indicate the metal ions that commonly form tetrahedral complexes of formula $[MX_4]^{2-}$ where X^- is a halide ion.

7.2 (a) On a chart of the *d*-block elements in their periodic table arrangement, identify the elements and associated oxidation numbers that form square-planar complexes. (b) Give formulas for three examples of square-planar complexes.

7.3 (a) Sketch the two structures that describe most six-coordinate complexes. (b) Which one of these is rare? (c) Give formulas for three different *d*-metal complexes that have the more common six-coordinate structure.

7.4 Name and draw structures of the following complexes: (a) $[Ni(CO)_4]$; (b) $[Ni(CN)_4]^{2-}$; (c) $[CoCl_4]^{2-}$; (d) $[Ni(NH_3)_6]^{2+}$.

7.5 Draw the structures of representative complexes that contain the ligands (a) en, (b) ox^{2-}, (c) phen, and (d) $edta^{4-}$.

7.6 Draw the structure of (a) a typical square-planar four-coordinate complex; (b) a typical trigonal prismatic six-coordinate complex; (c) a typical complex of coordination number 2. Name each complex.

7.7 Give formulas for (a) pentaamminechlorocobalt(III) chloride; (b) hexaaquairon(3+) nitrate; (c) *cis*-dichlorobis(ethylenediamine)ruthenium(II); (d) μ-hydroxobis[pentaamminechromium(III)] chloride.

7.8 Name the octahedral complex ions (a) *cis*-$[CrCl_2(NH_3)_4]^+$, (b) *trans*-$[Cr(NCS)_4(NH_3)_2]^-$, and (c) $[Co(C_2O_4)(en)_2]^+$. Is the oxalato complex *cis* or *trans*?

7.9 Draw all possible isomers of (a) octahedral $[RuCl_2(NH_3)_4]$, (b) square-planar $[IrH(CO)(PR_3)_2]$, (c) tetrahedral $[CoCl_3(OH_2)]^-$, (d) octahedral $[IrCl_3(PEt_3)_3]$, and (e) octahedral $[CoCl_2(en)(NH_3)_2]^+$.

7.10 The compound Na_2IrCl_6 reacts with triphenylphosphine in diethylene glycol under an atmosphere of CO to give *trans*-$[IrCl(CO)(PPh_3)_2]$, known as Vaska's compound. Excess CO produces a five-coordinate species and treatment with $NaBH_4$ in ethanol gives $[IrH(CO)_2(PPh_3)_2]$. Draw and name the three complexes.

7.11 Which of the following complexes are chiral? (a) $[Cr(ox)_3]^{3-}$, (b) *cis*-$[PtCl_2(en)]$, (c) *cis*-$[RhCl_2(NH_3)_4]^+$, (d) $[Ru(bipy)_3]^{2+}$, (e) $[Co(edta)]^-$, (f) *fac*-$[Co(NO_2)_3(dien)]$, (g) *mer*-$[Co(NO_2)_3(dien)]$. Draw the enantiomers of the complexes identified as chiral and identify the plane of symmetry in the structures of the achiral complexes.

7.12 One pink solid has the formula $CoCl_3 \cdot 5NH_3 \cdot H_2O$. A solution of this salt is also pink and rapidly gives **3 mol** AgCl on titration with silver nitrate solution. When the pink solid is heated, it loses **1 mol** H_2O to give a purple solid with the same ratio of NH_3:Cl:Co. The purple solid releases two of its chlorides rapidly; then, on dissolution and after titration with $AgNO_3$, releases one of its chlorides slowly. Deduce the structures of the two octahedral complexes and draw and name them.

7.13 The hydrated chromium chloride that is available commercially has the overall composition $CrCl_3 \cdot 6H_2O$. On boiling a solution, it becomes violet and has a molar electrical conductivity similar to that of $[Co(NH_3)_6]Cl_3$. In contrast, $CrCl_3 \cdot 5H_2O$ is green and has a lower molar conductivity in solution. If a dilute acidified solution of the green complex is allowed to stand for several hours, it turns violet. Interpret these observations with structural diagrams.

7.14 The complex first denoted β-$[PtCl_2(NH_3)_2]$ was identified as the *trans* isomer. (The *cis* isomer was denoted α.) It reacts slowly with solid Ag_2O to produce $[Pt(NH_3)_2(OH_2)_2]^{2+}$. This complex does not react with ethylenediamine to give a chelated complex. Name and draw the structure of the diaqua complex.

7.15 The 'third isomer' (neither α nor β, see Exercise 7.14) of composition $PtCl_2 \cdot 2NH_3$, is an insoluble solid which, when ground with $AgNO_3$, gives a solution containing $[Pt(NH_3)_4](NO_3)_2$ and a new solid phase of composition $Ag_2[PtCl_4]$. Give the structures and names of each of the three Pt(II) compounds.

7.16 Phosphine and arsine analogs of $[PtCl_2(NH_3)_2]$ were prepared in 1934 by Jensen. He reported zero dipole moments for the β isomers, where the β designation represents the product of a synthetic route analogous to that of the ammines. Give the structures of the complexes.

7.17 Describe the differences in the ^{31}P-NMR spectra of the *cis* and *trans* isomers referred to in Exercise 7.16.

7.18 (a) Indicate what parameter of NMR spectra might be used to distinguish the *cis* and *trans* isomers of $[W(CO)_4(P(CH_3)_3)_2]$. (b) What features of NMR spectra would distinguish a complex $[M(CO)_3PR_3)_2]$ of trigonal bipyramid geometry with the phosphine ligands in the axial positions from one with the phosphine ligands in the trigonal plane?

7.19 In *trans*-$[W(CO)_4(PR_3)_2]$, the alkylphosphine ligands define the *z*-axis of the coordinate system. Sketch the symmetry-adapted linear combination of σ orbitals from the two P atoms that can combine with the metal d_{z^2} orbital. Show the bonding and antibonding orbitals that may be formed.

7.20 Determine the configuration (in the form $t_{2g}^m e_g^n$ or $e^m t_2^n$, as appropriate), the number of unpaired electrons, and the ligand-field stabilization energy as a multiple of Δ_O or Δ_T for each of the following complexes using the spectrochemical series to decide, where relevant, which are likely to be strong-field and which weak-field. (a) $[Co(NH_3)_6]^{3+}$, (b) $[Fe(OH_2)_6]^{2+}$, (c) $[Fe(CN)_6]^{3-}$, (d) $[Cr(NH_3)_6]^{3+}$, (e) $[W(CO)_6]$, (f) tetrahedral $[FeCl_4]^{2-}$, and (g) tetrahedral $[Ni(CO)_4]$.

7.21 Both H^- and $P(C_6H_5)_3$ are ligands of similar field strength high in the spectrochemical series. Recalling that phosphines act as π acceptors, is π-acceptor character required for strong-field behavior? What orbital factors account for the strength of each ligand?

7.22 Estimate the spin-only contribution to the magnetic moment for each complex in Exercise 7.20.

7.23 Solutions of the complexes $[Co(NH_3)_6]^{2+}$, $[Co(H_2O)_6]^{2+}$ (both O_h), and $[CoCl_4]^{2-}$ are colored. One is pink, another is yellow, and the third is blue. Considering the spectrochemical series and the relative magnitudes Δ_T and Δ_O try to assign each color to one of the complexes.

7.24 Interpret the variation, including the overall trend across the $3d$ series, of the following values of oxide lattice enthalpies (in $kJ\ mol^{-1}$). All the compounds have the rock-salt structure. CaO (3460); TiO (3878); VO (3913); MnO (3810); FeO (3921); CoO (3988); NiO (4071).

7.25 A neutral macrocyclic ligand with four donor atoms produces a red diamagnetic low-spin d^8 complex of Ni(II) if the anion is the weakly coordinating perchlorate ion. When perchlorate is replaced by two thiocyanate ions, SCN^-, the complex turns violet and is high-spin with two unpaired electrons. Interpret the change in terms of structure.

7.26 Bearing in mind the Jahn–Teller theorem, predict the structure of $[Cr(OH_2)_6]^{2+}$.

7.27 The spectrum of $d^1\ Ti^{3+}$(aq) is attributed to a single electronic transition $e_g \leftarrow t_{2g}$. The band shown in Fig. 7.10 is not symmetrical and suggests that more than one state is involved. Suggest how to explain this observation using the Jahn–Teller theorem.

7.28 The rate constants for the formation of $[CoX(NH_3)_5]^{2+}$ from $[Co(NH_3)_5OH_2]^{3+}$ for $X = Cl^-$, Br^-, N_3^-, and SCN^- differ by no more than a factor or two. What is the mechanism of the substitution?

7.29 If a substitution process is associative, why may it be difficult to characterize an aqua ion as labile or inert?

Problems

7.1 Air oxidation of Co(II) carbonate and aqueous ammonium chloride gives a pink chloride salt with a ratio of 4 NH_3 : Co. On addition of HCl to a solution of this salt, a gas is rapidly evolved and the solution slowly turns violet on heating. Complete evaporation of the violet solution yields $CoCl_3 \cdot 4NH_3$. When this compound is heated in concentrated HCl, a green salt can be isolated which analyzes as $CoCl_3 \cdot 4NH_3 \cdot HCl$. Write balanced equations for all the transformations occuring after the air oxidation. Give as much information as possible concerning the isomerism occurring and give the basis of your reasoning. If you know that the form of $[Co(en)_2Cl_2]^+$ that is resolvable into enantiomers is violet, is that helpful?

7.2 By considering the splitting of the octahedral orbitals as the symmetry is lowered, draw the symmetry-adapted linear combinations and the molecular orbital energy level diagram for σ bonding in a trans-$[ML_4X_2]$ complex. Assume that the ligand X is lower in the spectrochemical series than L.

7.3 By referring to Appendix 4, draw the appropriate symmetry-adapted linear combinations and the molecular orbital diagram for σ bonding in a square-planar complex. The group is D_{4h}. Take note of the small overlap of the ligand with the d_{z^2} orbital. What is the effect of π bonding?

7.4 When cobalt(II) salts are oxidized by air in a solution containing ammonia and sodium nitrite, a yellow solid, $[Co(NO_2)_3(NH_3)_3]$, can be isolated. In solution it displays a conductivity small enough to be attributed to impurities only. Werner treated it with HCl to give a complex which, after a series of further reactions, he identified as trans-$[CoCl_2(NH_3)_3(OH_2)]^+$. It required an entirely different route to prepare cis-$[CoCl_2(NH_3)_3(OH_2)]^+$. Is the yellow substance fac or mer? What assumption must you make in order to arrive at a conclusion?

7.5 The reaction of $[ZrCl_4(dppe)]$ (dppe is a bidentate phosphine ligand) with $Mg(CH_3)_2$ gives $[Zr(CH_3)_4(dppe)]$. NMR spectra indicate that all methyl groups are equivalent. Draw octahedral and trigonal prism structures for the complex and show how the conclusion from NMR supports the trigonal prism assignment. (P.M. Morse and G.S. Girolami, J. Am. Chem. Soc. **111**, 4114 (1989).)

7.6 Write balanced equations for the formation of the triaza (three N atoms) and tetraaza (four N atoms) macrocycles that can be formed by a template synthesis in which o-aminobenzaldehyde self-condenses in the presence of Ni(II). (See G.A. Melson and D.H. Busch, J. Am. Chem. Soc. **86**, 4830 (1964).)

7.7 The resolving agent d-cis-$[Co(NO_2)_2(en)_2]Br$ can be converted to the soluble nitrate by grinding in water with $AgNO_3$. Outline the use of this species for resolving a racemic mixture of the d and l enantiomers of $K[Co(edta)]$. (The l-$[Co(edta)]^-$ enantiomer forms the less soluble diastereomer. See F.P. Dwyer and F.L. Garvan, Inorg. Synth. **6**, 192 (1965).)

7.8 Figures 7.15 and 7.16 show the relation between the frontier orbitals of octahedral and square-planar complexes. Construct similar diagrams for two-coordinate linear complexes.

7.9 In a fused magma liquid from which silicate minerals crystallize, the metal ions can be four coordinate. In olivine crystals, the M(II) coordination sites are octahedral. Partition coefficients, which are defined as $K_p = [M(II)]_{olivine}/[M(II)]_{melt}$, follow the order Ni(II) > Co(II) > Fe(II) > Mn(II). Account for this in terms of ligand-field theory. (See I.M. Dale and P. Henderson, 24th Int. Geol. Congress, Sect. **10**, 105 (1972).)

7.10 The equilibrium constants for the successive reactions of ethylenediamine with Co^{2+}, Ni^{2+}, and Cu^{2+} are as follows.

$$[M(OH_2)_6]^{2+} + en \rightleftharpoons [M(en)(OH_2)_4]^{2+} + 2H_2O \qquad K_1$$

$$[M(en)(OH_2)_4]^{2+} + en \rightleftharpoons [M(en)_2(OH_2)_2]^{2+} + 2H_2O \qquad K_2$$

$$[M(en)_2(OH_2)_2]^{2+} + en \rightleftharpoons [M(en)_3]^{2+} + 2H_2O \qquad K_3$$

Ion	$\log K_1$	$\log K_2$	$\log K_3$
Co^{2+}	5.89	4.83	3.10
Ni^{2+}	7.52	6.28	4.26
Cu^{2+}	10.55	9.05	−1.0

Discuss whether these data support the generalizations in the text about successive formation constants and the Irving–Williams series. How do you account for the very low value of K_3 for Cu^{2+}?

7.11 Sketch all possible isomers of $[Co(en)_2(NO_2)_2]$, including optical isomers, in which the en ligands are bidentate. One of the possible isomers is shown in (**50**). Do not draw two different orientations of the same isomer.

7.12 For each of the following pairs of complexes, circle the one that has the larger LFSE:

(a) $[Cr(OH_2)_6]^{2+}$ or $[Mn(OH_2)_6]^{2+}$, (b) $[Mn(OH_2)_6]^{2+}$ or $Fe(OH_2)_6]^{3+}$, (c) $[Fe(OH_2)_6]^{3+}$ or $[Fe(CN)_6]^{3-}$, (d) $[Fe(CN)_6]^{3}$ or $[Ru(CN)_6]^{3-}$, (e) tetrahedral $[FeCl_4]^{2-}$ or tetrahedral $[CoCl_4]^2$.

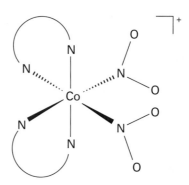

Fig. 50 *cis*-[bis(ethylenediamine)dinitrocobalt(III)] ion

Part 2

Systematic chemistry of the elements

The chapters of this part illustrate the utility of the periodic table for the mastery of descriptive chemistry. We focus in this part on periodic trends in the properties of the elements. Some physical and chemical properties vary in a simple manner down a group or across a period, but often the trend is less straightforward. In addition, we shall see that many trends can be rationalized in terms of the models of bonding, structure, and reactivity that were developed in Part 1. It is important to appreciate, though, that explanations presented in inorganic chemistry range from sound deductions supported by data and calculation to speculations based on insight. Some of the explanations will be overthrown by new experimental results or more accurate theories. Facts relating to structures will, for the most part, survive.

The chapters of this part proceed systematically through the periodic table. The opening chapter describes the chemistry of the unique element hydrogen. Chapter 9 then surveys the systematics of the metallic elements. These elements are the most numerous, for they span the periodic table from Group 1 to the last glimmers of metallic behavior in Group 16/VI. The remaining chapters of Part 2 deal with the p-block elements, particularly the nonmetals. Although fewer in number, the nonmetallic elements display a wide range of chemical properties and play a major role in all aspects of chemistry.

8

Hydrogen

Despite its simplicity, hydrogen has a very rich chemistry; it was in fact at the forefront of much of the discussion in Part 1, most importantly in the discussion of acids and bases. Here we summarize some of the properties of the element and its binary compounds. The latter range from solid ionic compounds and solids with metallic properties to molecular compounds. We shall see that many of the reactions of these compounds are understandable in terms of their ability to supply H^- or H^+ ions, and that it is often possible to anticipate when one tendency is likely to dominate.

Hydrogen is the most abundant element in the universe and the fifteenth most abundant in the Earth. The partial depletion of hydrogen from the Earth reflects its volatility during the formation of the planet. Hydrogen is one of the most important elements in the Earth's crust, where it is found in minerals, the oceans, and in all living things.

Because a hydrogen atom has only one electron, it might be thought that the element's chemical properties will be mundane, but this is far from the case. Hydrogen has richly varied chemical properties, for despite its single electron under certain circumstances it can bond to more than one atom simultaneously. Moreover, it ranges in character from a strong Lewis base (as the hydride ion, H^-) to a strong Lewis acid (as the hydrogen cation, H^+, the proton).

The element

Hydrogen does not fit neatly into the periodic table. It is sometimes placed at the head of the alkali metals in Group 1, with the rationale that hydrogen and the alkali metals have only one valence electron. That position, however, is not entirely satisfactory in terms of either the chemical or physical properties of the elements. In particular, hydrogen is not a metal under normal conditions. To achieve a metallic form, hydrogen must be subjected to very high pressures. The necessary pressures have not yet been achieved in the laboratory, but currently attainable pressures would appear to be approaching the values required. Hydrogen may be metallic at the core of Jupiter.[1] Less frequently, hydrogen is placed above the halogens in Group 17/VII, with the rationale that, like the halogens, it requires one electron to

---

1 W.J. Nellis, B.T. Weir, and A.C. Mitchell, *Science* **273**, 936 (1996).

complete its valence shell, and, as we shall see, there are indeed some parallels between its properties and those of the halogens.

8.1 Nuclear properties

There are three isotopes of hydrogen: hydrogen itself (^{1}H), deuterium (D, ^{2}H), and tritium (T, ^{3}H); tritium is radioactive. The three isotopes are unique in having different names, but that reflects the significant differences in their masses and the chemical properties that stem from mass, such as the rates of bond-cleavage reactions. The distinct properties of the isotopes make them useful as **tracers**, which are isotopes that can be followed through a series of reactions, such as by infrared (IR) spectroscopy or NMR. Tritium is sometimes preferred as a tracer because it can be detected by its radioactivity, which is a far more sensitive probe than spectroscopy.

The lightest isotope, ^{1}H—which is very occasionally called protium—is by far the most abundant. Deuterium has variable natural abundance with an average value of about 0.016 per cent. Neither ^{1}H nor ^{2}H is radioactive, but tritium decays by the loss of a β particle to yield a rare but stable isotope of helium:

$$^3_1H \longrightarrow {}^3_2He + \beta^-$$

The half-life for this decay is 12.4 y. Tritium's abundance of 1 in 10^{21} hydrogen atoms in surface water reflects a steady state between its production by bombardment of cosmic rays on the upper atmosphere and its loss by radioactive decay. However, the low natural abundance of the isotope is augmented artificially because it has been manufactured on an undisclosed scale for use in thermonuclear weapons. This synthesis uses the neutrons from a fission reactor with Li as a target:

$$^1_0n + {}^6_3Li \longrightarrow {}^3_1H + {}^4_2He$$

The physical and chemical properties of isotopically substituted molecules are usually very similar. However, the same is not true when D is substituted for H, for the mass of the substituted atom is doubled. For instance, Table 8.1 shows that the differences in boiling points and bond enthalpies are easily measurable for H_2 and D_2. The difference in boiling point between H_2O and D_2O reflects the greater strength of the $O \cdots D$—O hydrogen bond compared with that of the $O \cdots H$—O bond.

Reaction rates are often measurably different for processes in which E—H and E—D bonds are broken, made, or rearranged. The detection of this **kinetic isotope effect** can often help to support a proposed reaction mechanism. Kinetic isotope effects are frequently observed when an H atom is transferred from one atom to another in an activated complex. The effect on rate when this atom transfer occurs is called the **primary isotope effect**, and D atom transfer can be as much as ten times slower than H atom transfer. For example, the electrochemical reduction of $H^+(aq)$ to $H_2(g)$ occurs with a substantial isotope effect, with H_2 being liberated more rapidly. A practical consequence of the difference in rates of formation of H_2 and D_2 is that D_2O may be concentrated electrolytically. Substantial **secondary isotope effects** may occur when H is not transferred. For example, the rate of hydrolysis of the Cr—NCS link in *trans*-$[Cr(NCS)_4(NH_3)_2]^-$ is twice as fast when the ligands are ND_3 than when they are NH_3 even though no N—H bonds are broken in the reaction. This effect is ascribed to the change in the strength of the N—H $\cdots$ O hydrogen bond between the complex and the solvent and the effect of this bonding on the ease with which the NCS$^-$

Table 8.1 The effect of deuteration on physical properties

	H_2	D_2	H_2O	D_2O
Normal boiling point/°C	−252.8	−249.7	100.00	101.42
Mean bond enthalpy/(kJ mol^{-1})	436.0	443.3	463.5	470.9

ligand is able to depart from the Cr(III) site. The rate is greater for D than for H because N—D···O bonds are stronger than N—H···O bonds.

Because the frequencies of transitions between vibrational levels depend on the masses of atoms, they are strongly influenced by substitution of D for H. The heavier isotope results in the lower frequency.[2] Inorganic chemists often take advantage of this isotope effect by observing the IR spectra of isotopomers (molecules differing in their isotopic composition) to determine whether a particular infrared transition involves significant motion of hydrogen.

Another important property of the hydrogen nucleus is its spin, $I = \frac{1}{2}$, which is widely utilized in NMR (the nuclear spins of D and T are 1 and $\frac{1}{2}$, respectively). Proton NMR is useful for detecting the presence of H atoms in a compound and for identifying the atoms to which the hydrogen is bound by observation of the chemical shift or the spin–spin coupling to another nucleus.

> The three hydrogen isotopes H, D, and T have large differences in their atomic masses and different nuclear spin, which give rise to easily observed changes in IR, Raman, and NMR spectra of molecules containing these isotopes.

8.2 Hydrogen atoms and ions

The H atom has a high ionization energy (1310 kJ mol^{-1}, 13.6 eV) and a low but positive electron affinity (77 kJ mol^{-1}). The Pauling electronegativity of hydrogen is 2.2. This value is similar to those of B, C, and Si, so E—H bonds involving these elements are not expected to be very polar.

Hydrogen forms compounds with metals that are referred to as 'metal hydrides'; however, only the compounds of the highly electropositive metals in Groups 1 and 2 can be regarded as containing the hydride ion, H^{-}. When combined with the highly electronegative elements from the right of the periodic table, the E—H bond is best regarded as covalent and polar, with the H atom carrying the very small partial positive charge. As can be seen by inspection of Fig. 8.1, strong acids have strongly deshielded proton NMR signals. In accord with its electronegativity, hydrogen is normally assigned oxidation number -1 when in combination with metals (as in NaH and AlH$_3$) and $+1$ when in combination with nonmetals (as in NH$_3$ and HCl).

When a high-voltage discharge is passed through hydrogen gas at low pressure, the molecules dissociate, ionize, and recombine to form a plasma that contains spectroscopically observable amounts of H, H^{+}, H$_2^{+}$, and H$_3^{+}$. The free hydrogen cation (H^{+}, the proton) has a

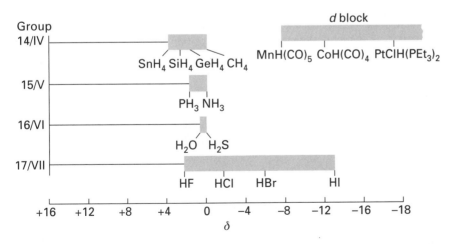

8.1 Typical ^{1}H-NMR chemical shifts. The tinted boxes group families of elements together.

---

[2] Recall from quantum mechanical treatments of the simple harmonic oscillator that the vibrational frequency of an oscillator of effective mass μ and force constant k is proportional to $(k/\mu)^{1/2}$.

very high charge/radius ratio and so it is not surprising to find that it is a very strong Lewis acid. In the gas phase it readily attaches to other molecules and atoms; for example, it attaches to He to form HeH^+. In condensed phases, H^+ is always found in combination with a Lewis base, and its ability to transfer between Lewis bases gives it the special role in chemistry that we explored in detail in Chapter 5.

The molecular cations H_2^+ and H_3^+ have only a transitory existence in the gas phase and are unknown in solution. As remarked in Section 3.11, H_3^+ has been detected in the interstellar medium and in the auroras of Uranus, Jupiter, and Saturn. Their electronic structures were described in Section 3.11, where we saw that spectroscopic data indicate that $H_3^+(g)$ is an equilateral triangle. The H_3^+ ion is the simplest example of a three-center, two-electron bond (a 3c,2e-bond) in which three nuclei are bonded by only two electrons.

> In combination with metals hydrogen is often regarded as a hydride; hydrogen compounds with elements of similar electronegativity have low polarity.

8.3 Properties and reactions of dihydrogen

The stable form of elemental hydrogen under normal conditions is *dihydrogen*, H_2, more informally and henceforth plain 'hydrogen'. The H_2 molecule has a high bond enthalpy (436 kJ mol^{-1}) and a short bond length (0.74 Å). Because it has so few electrons, the forces between neighboring H_2 molecules are weak, and at 1 atm the gas condenses to a liquid only when cooled to 20 K.

(a) Production

Molecular hydrogen is not present in significant quantities in the Earth's atmosphere or in underground gas deposits, but it is produced in huge quantities to satisfy the needs of industry. The main commercial process for the production of hydrogen is currently *steam reforming*, the catalyzed reaction of water and hydrocarbons (typically methane from natural gas) at high temperatures:

$$CH_4(g) + H_2O(g) \xrightarrow{1000\,°C} CO(g) + 3\,H_2(g)$$

A similar reaction, but with coke as the reducing agent, is sometimes called the *water-gas reaction*:

$$C(s) + H_2O(g) \xrightarrow{1000\,°C} CO(g) + H_2(g)$$

This reaction was once a primary source of H_2 and it may become important again when natural hydrocarbons are depleted. Both reactions are generally followed by a second reaction, often called the *shift reaction*, in which water is reduced to hydrogen by reaction with carbon monoxide:

$$CO(g) + H_2O(g) \longrightarrow CO_2(g) + H_2(g)$$

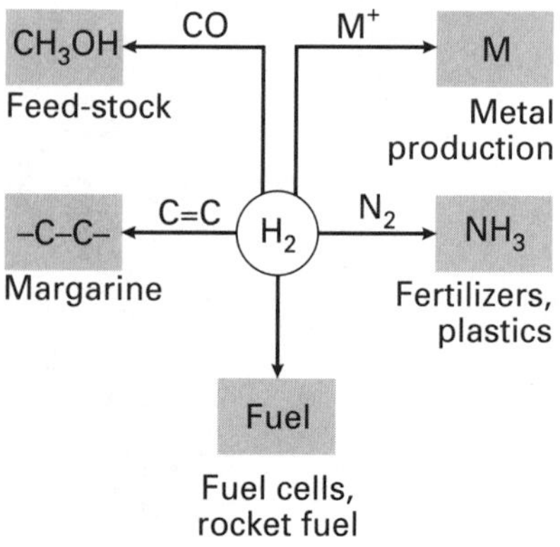

Hydrogen production is often integrated with chemical processes that require H_2 as a feed-stock. As shown in Chart 8.1, a major use of hydrogen is direct combination with N_2 to produce NH_3, the primary source of nitrogen-containing chemicals, plastics, and fertilizers. Another major chemical, methanol, is produced from the catalytic combination of H_2 and CO.

Because of its high specific enthalpy,[3] hydrogen is an excellent fuel for large rockets. The more general use of hydrogen as a fuel has been analyzed seriously since the early 1970s when petroleum prices rose sharply. Strategies have been devised for a 'hydrogen economy' in

Chart 8.1

(diagram around H_2:)

CH$_3$OH — CO — M$^+$ → M
Feed-stock / Metal production

–C–C– — C=C — H_2 — N$_2$ → NH$_3$
Margarine / Fertilizers, plastics

Fuel
Fuel cells, rocket fuel

---

3 Specific enthalpy is the enthalpy of combustion of a sample divided by the mass of the sample; the specific enthalpy of hydrogen is 142 kJ g^{-1}; that of a typical hydrocarbon is 50 kJ g^{-1}.

which hydrogen is the primary fuel. One scenario is to collect solar energy with photovoltaic cells and employ the resulting electricity to electrolyze water. The hydrogen from this electrolysis would be a form of stored solar energy that could be burned as a fuel when needed or used as a feed-stock in chemical processes. The current prices of petroleum, natural gas, and coal are too low to make such a hydrogen economy viable, but the idea might have its day.[4] Quite aside from the eventual scarcity of fossil fuels, when hydrogen burns it produces water, which is not as significant a greenhouse gas as CO_2, so a hydrogen economy would not necessarily lead to the global warming that now seems to confront us.[5] The availability of photochemically produced H_2 would also eliminate the CO_2 resulting from steam reforming and the shift reaction. A first step toward a hydrogen economy might therefore be the use of electrolytic or photochemical hydrogen in the chemical and petrochemical industries.

> Much of the hydrogen for industry is produced by high-temperature reaction of H_2O with CH_4 or a similar reaction with coke. Hydrogen's major uses are in the processing of hydrocarbon fuels and synthesis of ammonia.

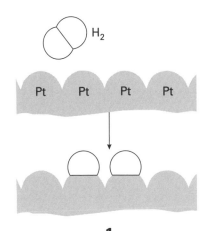

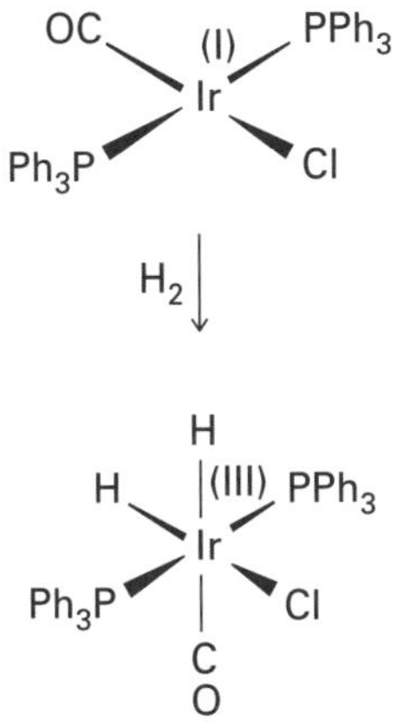

1

(b) Reactions of hydrogen

Molecular hydrogen reacts slowly with most other elements, partly because of its high bond enthalpy and hence high activation energy for reaction. Under special conditions, however, the reactions are rapid. These conditions include:

1 The activation of the molecule by homolytic dissociation on a metal surface or a metal complex.
2 Heterolytic dissociation by a surface or a metal ion.
3 Initiation of a radical chain reaction.

Two examples of homolytic dissociation are chemisorption on platinum (**1**) and coordination to the Ir atom in a complex (**2**). The former accounts for the use of finely divided platinum metal to catalyze the hydrogenation of alkenes and of platinum as the electrode material used in the electrolytic reduction of H^+. The overpotential for H_2 formation is much smaller at a platinum electrode, on which dissociative chemisorption occurs, than at a mercury electrode, where chemisorption is unfavorable. Similarly, H_2 readily bonds to the Ir atom in complexes such as $[IrClCO(PPh_3)_2]$ to give a complex containing two hydrido (H^-) ligands:

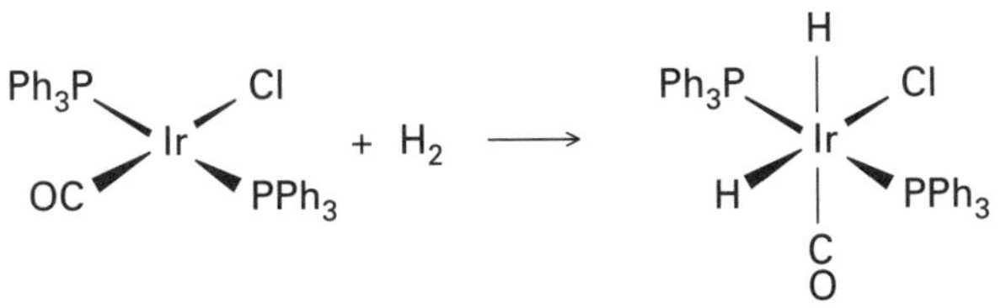

2

This type of reaction is an example of **oxidative addition**, in which the metal center undergoes a formal increase in oxidation number when the reactant adds to it. Oxidative

[4] Research on the generation of hydrogen by advanced photochemical methods is described in M. Grätzel (ed.), *Energy resources through photochemistry and catalysis*. Academic Press, New York (1983).

[5] A greenhouse gas is a species in the atmosphere that is relatively transparent to incoming ultraviolet and visible radiation from the Sun, but absorbs outgoing infrared radiation. It acts rather like the glass in a greenhouse, and the atmosphere becomes warmed.

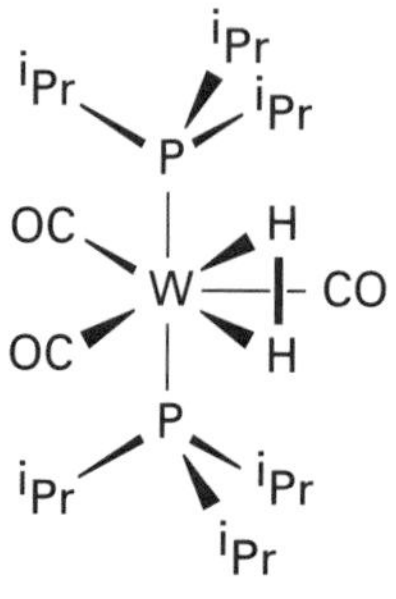

3

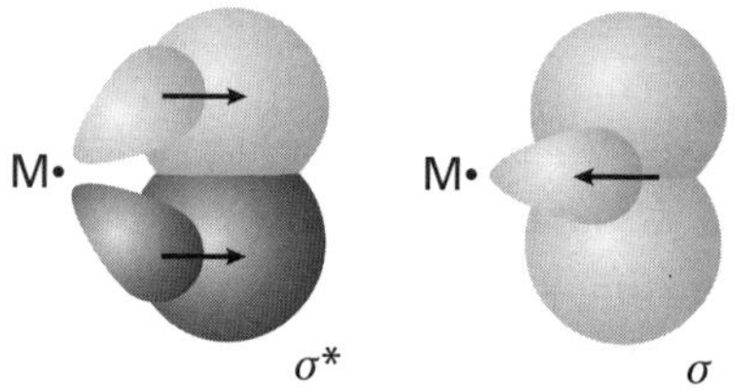

4

addition of H_2 is common for Rh(I), Ir(I), and Pt(0) complexes. There are many d-block hydrido complexes, some of which are formed by oxidative addition, and they are discussed in more detail in Chapter 16.

An H_2 molecule can coordinate to a metal atom without cleavage of the H—H bond.[6] The first such compound to be made was $[W(CO)_3(H_2)(P^iPr_3)_2]$ (**3**), where ^{i}Pr denotes isopropyl, $-CH(CH_3)_2$, and well over a hundred compounds of this type are now known. The significance of these compounds is that they provide examples of species intermediate between molecular H_2 and a dihydrido complex. Molecular orbital calculations indicate that the bonding may be considered in terms of donation of electron density from the H—H bond into a vacant metal d orbital and a simultaneous back π bonding into the σ^* orbital of H_2 (**4**). This bonding pattern is similar to that of CO and ethene with metal atoms (Sections 7.6 and 17.3). The back-donation of electron density from the metal to a σ^* orbital on H_2 is consistent with the observation that the H—H bond is cleaved when the metal center is made electron-rich by highly basic ligands. No dihydrogen complexes are known for the early d-block (Groups 3, 4, and 5), the f-block, or the p-block metals.

An example of heterolytic dissociation is the reaction of H_2 with a ZnO surface, which appears to produce a Zn(II)-bound hydride and an O-bound proton:

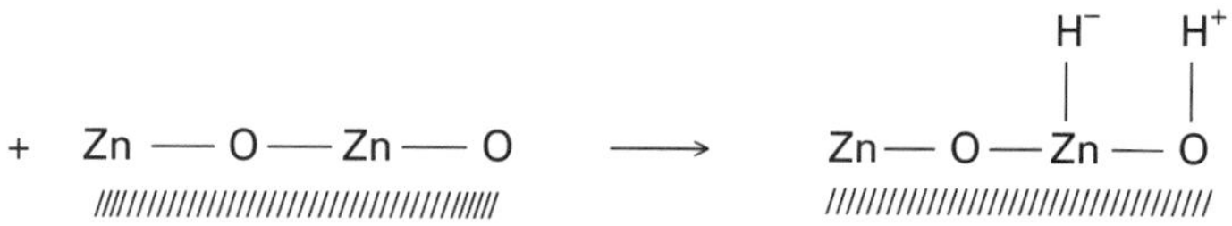

$$H_2 + \quad Zn—O—Zn—O \quad \longrightarrow \quad Zn—O—Zn—O$$

This reaction is thought to be involved in the catalytic hydrogenation of carbon monoxide to methanol

$$CO(g) + 2\,H_2(g) \xrightarrow{\ Cu/Zn\ } CH_3OH(g)$$

which is carried out on a very large scale worldwide. Another example is the sequence

$$H_2(g) + Cu^{2+}(aq) \longrightarrow [CuH]^+(aq) + H^+(aq) \longrightarrow Cu(s) + 2\,H^+(aq)$$

which is important in the hydrometallurgical reduction of Cu^{2+} (Section 6.1). The $[CuH]^+$ intermediate has only a transitory existence.

6   An informative account of this discovery and its implications is given by G. Kubas, *Acc. Chem. Res.* **21**, 120 (1988) and the detection of dihydrogen complexes by NMR is described by R.H. Crabtree, *Acc. Chem. Res.* **23**, 95 (1990).

Radical chain mechanisms account for the thermally or photochemically initiated reactions between H_2 and the halogens. Initiation is by thermal or photochemical dissociation of the dihalogen molecules to give atoms that act as radical chain carriers:

Initiation: $Br_2 \xrightarrow{\Delta \text{ or } h\nu} Br\cdot + \cdot Br$

Propagation: $Br\cdot + H_2 \longrightarrow HBr + H\cdot$

     $H\cdot + Br_2 \longrightarrow HBr + Br\cdot$

The activation energy for radical attack is low because a new bond is formed as one bond is lost, so once initiated the formation and consumption of radicals is self-sustaining and the production of HBr is very rapid. Chain termination occurs when the radicals recombine:

Termination: $H\cdot + \cdot H \longrightarrow H_2$ $Br\cdot + \cdot Br \longrightarrow Br_2$

Termination becomes more important toward the end of the reaction when the concentrations of H_2 and Br_2 are low.

> Molecular hydrogen is activated by dissociation on a metal surface or by formation of a metal complex. Radical chain reactions are common in the reaction of hydrogen with oxygen and halogens.

The classification of compounds of hydrogen

Figure 8.2 summarizes the classification of the binary compounds of hydrogen and their distribution through the periodic table. The main point of devising this classification is to emphasize the principal trends in properties. In fact, there is a gradation of structural types, and some elements form compounds with hydrogen that do not fall strictly into any one category. The three classes of binary hydrogen compounds that we shall consider are:

1 **Molecular compounds**, binary compounds of an element and hydrogen in the form of individual, discrete molecules.
2 **Saline hydrides**, non-volatile, electrically non-conducting, crystalline solids.
3 **Metallic hydrides**, nonstoichiometric, electrically conducting solids.

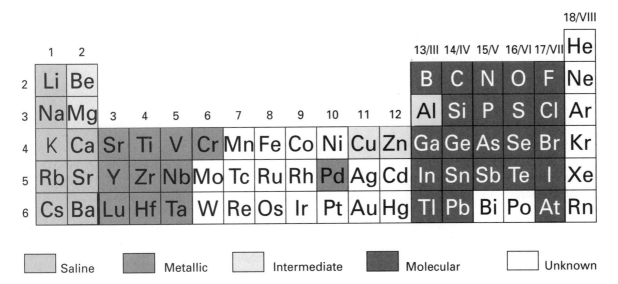

8.2 Classification of the binary hydrogen compounds of the *s*-, *p*-, and *d*-block elements. Although some *d*-block elements, such as iron and ruthenium, do not form binary hydrides, they do form metal complexes containing the hydride ligand.

8.4 Molecular compounds

Molecular compounds of hydrogen are common for the electronegative elements of Groups 13/III through 17/VII, and examples include B_2H_6, CH_4, NH_3, H_2O, and HF.

(a) Nomenclature and classification

The systematic names of the molecular hydrogen compounds are formed from the name of the element and the suffix -ane, as in phosphane for PH_3. The more traditional names, though, such as phosphine and hydrogen sulfide (H_2S, sulfane) are still widely used (Table 8.2). The nonsystematic names ammonia and water are universally used rather than their systematic names azane and oxidane.

It is convenient to classify molecular compounds of hydrogen into three categories:

1 In an **electron-precise compound**, all valence electrons of the central atom are engaged in bonds.
2 In an **electron-deficient compound**, there are too few electrons to be able to write a Lewis structure for the molecule.
3 In an **electron-rich compound**, there are more electron pairs on the central atom than are needed for bond formation.

The hydrocarbons, such as methane and ethane, are electron-precise; so too are silane, SiH_4, and germane, GeH_4. All these molecules are characterized by the presence of two-center, two-electron bonds (2c,2e bonds) and the absence of lone pairs on the central atom.

Diborane, B_2H_6 (**5**), is an example of an electron-deficient compound. Its Lewis structure would require at least 14 valence electrons to bind the eight atoms together, but the molecule has only 12 valence electrons. The simple explanation of its structure is the presence of B—H—B 3c,2e bonds acting as bridges between the two B atoms, so two electrons can help to bind three atoms. Electron-deficient hydrogen compounds are common for boron and aluminum.

Electron-rich compounds are formed by the elements in Groups 15/V through 17/VII. Ammonia, NH_3, with one lone pair on nitrogen, and water, H_2O, with two lone pairs on oxygen, are examples. These types of molecules generally exhibit Lewis basicity. For example, the electron-rich compound trimethylamine, $N(CH_3)_3$, reacts with boron trifluoride to form the Lewis acid–base complex trimethylamine trifluoroborane, $(H_3C)_3NBF_3$. The hydrogen halides, HF, HCl, HBr, and HI, are another important group of electron-rich compounds.

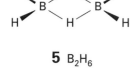

5 B_2H_6

Table 8.2 Some molecular hydrogen compounds

Group	Formula	Traditional name	IUPAC name
13/III	B_2H_6	Diborane	Diborane(6)
14/IV	CH_4	Methane	Methane
	SiH_4	Silane	Silane
	GeH_4	Germane	Germane
	SnH_4	Stannane	Stannane
15/V	NH_3	Ammonia	Azane
	PH_3	Phosphine	Phosphane
	AsH_3	Arsine	Arsane
	SbH_3	Stibine	Stibane
16/VI	H_2O	Water	Oxidane
	H_2S	Hydrogen sulfide	Sulfane
	H_2Se	Hydrogen selenide	Sellane
	H_2Te	Hydrogen telluride	Tellane
17/VII	HF	Hydrogen fluoride	Hydrogen fluoride
	HCl	Hydrogen chloride	Hydrogen chloride
	HBr	Hydrogen bromide	Hydrogen bromide
	HI	Hydrogen iodide	Hydrogen iodide

> *Molecular compounds of hydrogen are classified as electron-rich, electron-precise, or electron-deficient, as outlined above.*

(b) General aspects of properties

The shapes of the electron-precise and electron-rich compounds can all be predicted by the VSEPR rules (Section 3.3). Thus, CH_4 is tetrahedral (**6**), NH_3 is pyramidal (**7**), H_2O is angular (**8**), and HF is (necessarily) linear. However, the simple VSEPR rules do not indicate the considerable change in bond angle between NH_3 and its heavier analogs or between H_2O and its analogs in Group 16/VI. As noted in Table 8.3, the bond angles in NH_3 and H_2O are slightly less than the tetrahedral angle, but for their heavier analogs the bond angle is as small as 90°.

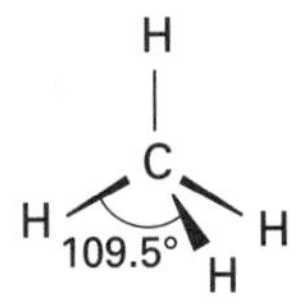

An important consequence of the simultaneous presence of highly electronegative atoms (N, O, and F) and lone pairs in the electron-rich compounds is the possibility of forming hydrogen bonds. A **hydrogen bond** consists of an H atom between atoms of more electronegative nonmetallic elements. This definition includes the widely recognized N—H···N and O—H···O hydrogen bonds but excludes the B—H—B bridges in boron hydrides because boron is not more electronegative than hydrogen. The definition also excludes the W—H—W links present in $[(OC)_5WHW(CO)_5]^-$, for tungsten is a metal.[7] The concept of hydrogen bonding originated from the observation of unusually high relative permittivities of water and similar fluids.[8] Striking evidence for hydrogen bonding is provided by the trends in boiling points, which are unusually high for the strongly hydrogen-bonded molecules water, ammonia, and hydrogen fluoride (Fig. 8.3). Some of the most persuasive evidence for hydrogen bonding comes from structural data, such as the open network structure of ice (Fig. 8.4), the existence of chains in solid HF (**9**) that survive partially even in

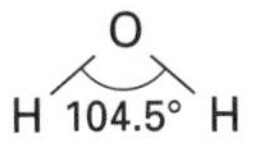

Table 8.3 Bond angles (in degrees) for Group 15/V and 16/VI hydrogen compounds

NH_3	106.6	H_2O	104.5
PH_3	93.8	H_2S	92.1
AsH_3	91.8	H_2Se	91
SbH_3	91.3	H_2Te	89

Source: A.F. Wells, *Structural inorganic chemistry.* Oxford University Press (1984).

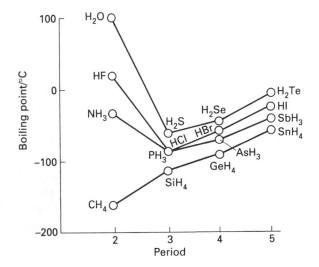

8.3 Normal boiling points of *p*-block binary hydrogen compounds.

6 CH_4, T_d

7 NH_3, C_{3v}

8 H_2O, C_{2v}

9 $(HF)_n$

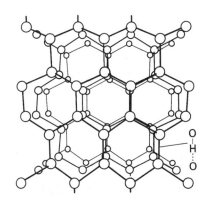

8.4 The structure of ice. The large spheres represent the O atoms; the H atoms lie on the lines joining O atoms.

---

[7] Moreover, even though tungsten is a metal, it is more electronegative than hydrogen on the Pauling scale.

[8] W.M. Latimer and W.H. Rhodebush, *J. Am. Chem. Soc.* **42**, 1635 (1912). It took many years after this publication for the full significance of hydrogen bonding to be appreciated.

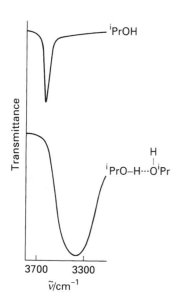

8.5 Infrared spectra of 2-propanol (isopropanol). In the upper curve, 2-propanol is present as unassociated molecules in dilute solution. In the lower curve the pure alcohol is associated through hydrogen bonds. The association lowers the frequency and broadens the O—H stretching absorption band. (From N.B. Colthup, L.H. Daly, and S.E. Wiberley, *Introduction to infrared and Raman spectroscopy*. Academic Press, New York (1975).)

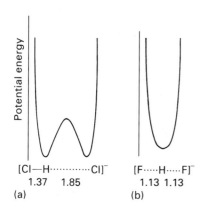

8.6 The variation of the potential energy with the position of the proton between two atoms in a hydrogen bond. (a) The double-minimum potential characteristic of a weak hydrogen bond. (b) The single-minimum potential characteristic of the strong hydrogen bond in [FHF]⁻.

Table 8.4 Comparison of hydrogen bond enthalpies with the corresponding E—H covalent bond enthalpies ($kJ\,mol^{-1}$)

	Hydrogen bond ($\cdots$)		Covalent bond (—)
HS—H $\cdots$ SH$_2$	7	S—H	363
H$_2$N—H $\cdots$ NH$_3$	17	N—H	386
HO—H $\cdots$ OH$_2$	22	O—H	464
F—H $\cdots$ F—H	29	F—H	565
HO—H $\cdots$ Cl$^-$	55	Cl—H	428
F $\cdots$ H $\cdots$ F$^-$	165	F—H	565

the vapor, and the location of hydrogen atoms in solids by diffraction and NMR techniques.[9] As indicated by boiling point trends, H_2S, PH_3, HCl, and the heavier *p*-block hydrogen compounds do not form strong hydrogen bonds.

Although hydrogen bonds are much weaker than conventional bonds (Table 8.4), they have important consequences for the properties of the electron-rich hydrogen compounds of Period 2, including their densities, viscosities, vapor pressures, and acid–base characters. Hydrogen bonding is readily detected by the shift to lower wavenumber and broadening of E—H stretching bands in infrared spectra (Fig. 8.5). The majority of hydrogen bonds are weak. In these cases, the H atom is not midway between the two nuclei, even when the heavier linked atoms are identical. For example, the $[ClHCl]^-$ ion is linear but the H atom is not midway between the Cl atoms (Fig. 8.6). In contrast, the bifluoride ion, $[FHF]^-$, has a strong hydrogen bond and the H atom lies midway between the F atoms; the F—F separation (2.26 Å) is significantly less than twice the van der Waals radius of the F atom (2×1.35 Å).

The structures of hydrogen-bonded complexes have been observed in the gas phase through the use of microwave spectroscopy.[10] To a first approximation, the lone-pair orientation of electron-rich compounds implied by VSEPR theory shows good agreement with the HF orientation (Fig. 8.7). For example, HF is oriented along the threefold axis of NH_3 and PH_3, out of the H_2O plane in its complex with H_2O, and off the HF axis in the HF dimer. X-ray single crystal structure determinations often show the same patterns, as for example in the structure of ice and in solid HF, but packing forces in solids may have a strong influence on the orientation of the relatively weak hydrogen bond.

One of the most interesting manifestations of hydrogen bonding is the structure of ice. There are seven different phases but only one at lower pressures. The familiar low-pressure phase of ice crystallizes in a hexagonal unit cell with each O atom surrounded tetrahedrally by four others. These O atoms are held together by hydrogen bonds with O—H $\cdots$ O and O $\cdots$ H—O bonds largely randomly distributed through the solid. The resulting structure is quite open, which accounts for the density of ice being lower than that of water. When ice melts, the network of hydrogen bonds partially collapses.

Water can also form **clathrate hydrates**, consisting of hydrogen-bonded cages of water molecules surrounding foreign molecules or ions. One example is the clathrate hydrate of composition $Cl_2 \cdot (H_2O)_{7.25}$ (Fig. 8.8). In this structure, the cages with O atoms defining their corners consist of 14-faced and 12-faced polyhedra in the ratio 3:2. These O atoms are held together by hydrogen bonds and Cl_2 molecules occupy the interiors of the 14-faced polyhedra. Similar clathrate hydrates are formed at elevated pressures and low temperatures with Ar, Xe, and CH_4, but in these cases all the polyhedra appear to be occupied. Aside from

[9] W.C. Hamilton and J.A. Ibers, *Hydrogen bonding in solids*. W.A. Benjamin, New York (1968); J. Emsley, *Chem. Soc. Rev.* **9**, 91 (1990).

[10] A.C. Legon and D.J. Millen, *Acc. Chem. Res.* **20**, 39 (1987), describe the structures of gas-phase hydrogen-bonded dimers and the simple interpretation of their electronic origin.

their interesting structures—which illustrate the organization that can be enforced by hydrogen bonding—clathrate hydrates are often used as models for the way in which water appears to become organized around nonpolar groups, such as those in proteins. Methane clathrate hydrates occur in the Earth at high pressures, and it is estimated that huge quantities of natural gas are trapped in these formations.

Some ionic compounds form clathrate hydrates in which the anion is incorporated into the framework by hydrogen bonding. This type of clathrate is particularly common with the very strong hydrogen bond acceptors F^- and OH^-. One such example is $N(CH_3)_4F \cdot 4H_2O$.

> *Hydrogen compounds of electronegative nonmetals with at least one lone pair often associate through hydrogen bonds; water, ice, and clathrate hydrates are major examples of this association.*

8.5 Saline hydrides

The Group 1 hydrides have rock-salt structures, and the Group 2 hydrides have crystal structures like those of some heavy metal halides (Table 8.5). These structures (and their chemical properties) are the basis for classifying hydrogen compounds of the s-block elements other than beryllium as saline. The ionic radius of H^- determined from X-ray diffraction varies from 1.26 Å in LiH to 1.54 Å in CsH. This wide variability reflects the loose control that the single charge of the proton has on its two surrounding electrons and the resulting high compressibility of H^-.

The saline hydrides are insoluble in common nonaqueous solvents but they do dissolve in molten alkali halides. Electrolysis of these molten-salt solutions produces hydrogen gas at the anode (the site of oxidation):

$$2\,H^-(melt) \longrightarrow H_2(g) + 2\,e^-$$

This reaction provides chemical evidence for the existence of H^-. The reaction of saline hydrides with water is dangerously vigorous:

$$NaH(s) + H_2O(l) \longrightarrow H_2(g) + NaOH(aq)$$

Indeed, finely divided sodium hydride can ignite if it is left exposed to humid air. Such fires are difficult to extinguish because even carbon dioxide is reduced when it comes into contact with hot metal hydrides (water, of course, forms even more flammable hydrogen); they may be blanketed with an inert solid, such as silica sand. The reaction between metal hydrides and water is employed in the laboratory to remove traces of water from solvents and inert gases such as nitrogen and argon:

$$CaH_2(s) + 2\,H_2O(g) \longrightarrow Ca(OH)_2(s) + 2\,H_2(g)$$

Calcium hydride is preferred in this application because it is the cheapest of the saline hydrides, and is available in granular form, which is easy to handle. Large amounts of water should not be removed from a solvent in this manner because the strongly exothermic reaction evolves flammable hydrogen.

The absence of convenient solvents limits the use of saline hydrides as reagents, but this problem is partially overcome by the availability of commercial dispersions of finely divided

Table 8.5 Structures of s-block hydrides

Compound	Crystal structure
LiH, NaH, KH, RbH, CsH	Rock salt
MgH₂	Rutile
CaH₂, SrH₂, BaH₂	Distorted PbCl₂

Source: A.F. Wells, *Structural inorganic chemistry.* Oxford University Press (1984).

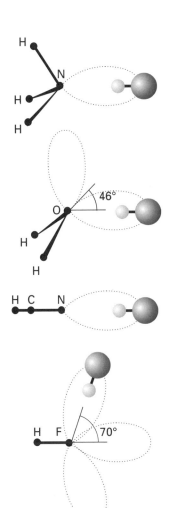

8.7 The orientation of lone pairs as indicated by VSEPR theory (left) compared with the structure of the gas-phase hydrogen-bonded complex formed with HF (right). (After A.C. Legon and D.J. Millen, *Acc. Chem. Res.* **20**, 39 (1987).)

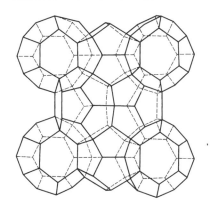

8.8 The cages of water molecules in clathrate hydrates, such as $Cl_2 \cdot (H_2O)_{7.25}$. Each intersection is the location of an O atom, and H atoms lie on the lines joining these O atoms. Two of the central 14-hedra are about 80 per cent occupied by Cl_2.

NaH in oil.[11] Alkali metal hydrides are convenient reagents for making other hydride compounds. For example, reaction with a trialkylboron compound yields a hydride complex which is soluble in polar organic solvents and therefore a useful reducing agent and source of hydride ions:

$$NaH(s) + B(C_2H_5)_3(et) \longrightarrow Na[HB(C_2H_5)_3](et)$$

where et denotes solution in diethyl ether.

> *Hydrogen compounds of electropositive metals may be regarded as metal hydrides, M^+H^-; accordingly, they liberate H_2 in contact with acids and transfer H^- to electrophiles.*

8.6 Metallic hydrides

Nonstoichiometric metallic hydrides are formed by all the d-block metals of Groups 3, 4, and 5 and for the f-block elements (Fig. 8.9). However, the only hydride in Group 6 is CrH, and no hydrides are known for the unalloyed metals of Groups 7, 8, and 9. Claims have been made, though, that hydrogen dissolves in iron at very high pressure, and that iron hydride is abundant at the center of the Earth. The region of the periodic table covered by Groups 7 through 9 is sometimes referred to as the **hydride gap** because few if any stable binary metal–hydrogen compounds are formed by these elements.

The Group 10 metals, especially nickel and platinum, are often used as hydrogenation catalysts in which surface hydride formation is implicated (Section 17.5). However, somewhat surprisingly, at moderate pressures only palladium forms a stable bulk phase; its composition is PdH_x, with $x < 1$. Nickel forms hydride phases at very high pressures but platinum does not

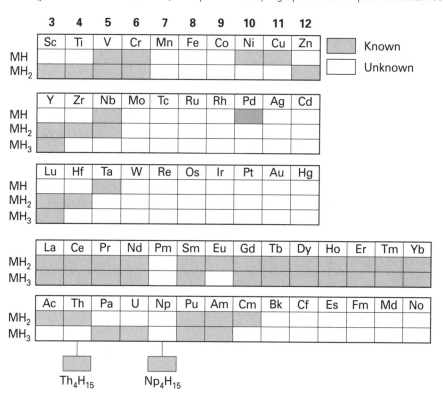

8.9 Hydrides formed by d- and f-block elements. The formulas are limiting stoichiometries based on the structural type. In many cases they are not attained. For example, PdH_x never attains $x = 1$. (From G.G. Libowitz, *Solid state chemistry of the binary metal hydrides.* Benjamin (1965).)

---

[11] Even more finely divided and reactive alkali metal hydrides can be prepared from the metal alkyl and hydrogen: P.A.A. Klusener, L. Brandsma, H.D. Verkruijsse, P. von Rague Schleyer, T. Friedl, and R. Pi, *Angew. Chem., Int. Ed. Engl.* **25**, 465 (1986).

form any at all. Apparently the Pt—H bond enthalpy is sufficiently great to disrupt the H—H bond but not to offset the loss of Pt—Pt bonding, which would occur upon formation of a bulk platinum hydride. In agreement with this interpretation, the enthalpies of sublimation, which reflect M—M bond enthalpies, increase in the order Pd (378 kJ mol^{-1}) < Ni (430 kJ mol^{-1}) < Pt (565 kJ mol^{-1}).

Most metallic hydrides are metallic conductors (hence their name) and have variable composition. For example, at 550 °C a compound ZrH$_x$ exists over a composition range from ZrH$_{1.30}$ to ZrH$_{1.75}$; it has the fluorite structure (Fig. 2.13) with a variable number of anion sites unoccupied. The variable stoichiometry and metallic conductivity of these hydrides can be understood in terms of a model in which the band of delocalized orbitals responsible for the conductivity accommodates electrons supplied by arriving H atoms. In this model, the H atoms as well as the metal atoms take up equilibrium positions in this electron sea. The conductivities of metallic hydrides typically vary with hydrogen content, and this variation can be correlated with the extent to which the conduction band is filled or emptied as hydrogen is added or removed. Thus, CeH$_{2-x}$ is a metallic conductor, whereas CeH$_3$ (which has a full conduction band) is an insulator.

A striking property of many metallic hydrides is the high mobility of hydrogen at slightly elevated temperatures. This mobility is utilized in the ultrapurification of H$_2$ by diffusion through a palladium–silver alloy tube (Fig. 8.10). The high mobility of the hydrogen they contain, and their variable composition, also make the metallic hydrides potential hydrogen storage media. Thus the intermetallic compound LaNi$_5$ forms a hydride phase with a limiting composition LaNi$_5$H$_6$, and at this composition it contains a greater density of hydrogen than liquid H$_2$. A less expensive system with the composition FeTiH$_x$ ($x < 1.95$) is now commercially available for low-pressure hydrogen storage and it has been tested as a source of energy in vehicle trials.

> No stable binary metal hydrides are known for the metals in Groups 7 through 9. Metallic hydrides have metallic conductivity and in many the hydrogen is very mobile.

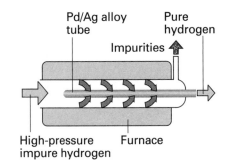

8.10 Schematic diagram of a hydrogen purifier. Because of a pressure differential and the mobility of H atoms in palladium, hydrogen diffuses through the palladium–silver alloy tube as H atoms but impurities do not.

Example 8.1 Correlating the classification and properties of hydrogen compounds

Classify the compounds PH$_3$, CsH, B$_2$H$_6$ and discuss their probable physical properties. For the molecular compounds specify their subclassification (electron-deficient, electron-precise, or electron-rich).

Answer The compound CsH is a compound of a Group 1 element, and so it can be expected to be a saline hydride typical of the *s*-block metals. It is an electrical insulator with the rock-salt structure. As with the hydrogen compounds of other *p*-block elements, the hydrides PH$_3$ and B$_2$H$_6$ are molecular with low molar masses and high volatilities. They are in fact gases under normal conditions. The Lewis structure indicates that PH$_3$ has a lone pair on the phosphorus atom and that it is therefore an electron-rich molecular compound. As remarked in the text (see also Section 3.12), diborane is an electron-deficient compound.

Self-test 8.1 Give balanced equations (or NR, for no reaction) for (a) Ca + H$_2$, (b) NH$_3$ + BF$_3$, (c) LiOH + H$_2$.

The synthesis and reactions of hydrogen compounds

Many of the chemical properties of hydrogen compounds can be rationalized by noting the periodic trends in their thermodynamic stabilities and their polarities (specifically, whether the partial charge on hydrogen is $H^{\delta+}$ or $H^{\delta-}$). Furthermore, the less predictable but important radical chain reactions play a major role in the gas-phase reactions of hydrogen with oxygen and the halogens.

8.7 Stability and synthesis

A negative Gibbs energy of formation is a clue that the direct combination of hydrogen and an element may be the preferred synthetic route for a hydrogen compound. When a compound is thermodynamically unstable with respect to its elements, an indirect synthetic route from other compounds can often be found, but each step in the indirect route must be thermodynamically favorable. Our principal objective in this section is to convey the intuition that suggests which reactions are likely to be thermodynamically favorable.

(a) Thermodynamic aspects

The standard Gibbs energies of formation of the hydrogen compounds of s- and p-block elements reveal a regular variation in stability (Table 8.6). With the possible exception of BeH_2, for which good data are not available, all the s-block hydrides are exoergic ($\Delta_f G^{\ominus} < 0$) and therefore thermodynamically stable with respect to their elements at room temperature. The trend is erratic in Group 13/III, in that only AlH_3 is exoergic. In all the other groups of the p-block, the hydrogen compounds of the first members of the groups (CH_4, NH_3, H_2O, and HF) are exoergic but the hydrogen compounds become progressively less stable down the group. This general trend in stability is largely a reflection of the decreasing E—H bond strength down a group in the p-block (Fig. 8.11). The weak bonds formed by the heavier elements are often attributed to poor overlap between the relatively compact H$1s$ orbital and the more diffuse s and p orbitals of the heavier atoms in these groups. The heavier members become more stable on going from Group 14/IV across to the halogens. For example, SnH_4 is highly endoergic whereas HI is barely so.

> *Compound formation between H_2 and Group 1 and 2 metals except Be is thermodynamically favorable. On going down a group in the p-block, the element-hydrogen bond strengths decrease and the hydrides of the heavier elements become thermodynamically unstable.*

Bond enthalpy

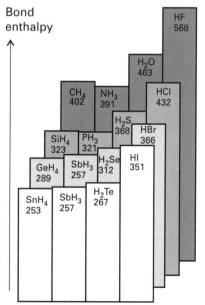

8.11 Average bond enthalpies (in kilojoules per mole, at 298 K).

Table 8.6 Standard Gibbs energy of formation, $\Delta_f G^{\ominus}/(kJ\ mol^{-1})$, of binary s- and p-block hydrogen compounds at 25°C

Period	Group 1 I	2 II	13 III	14 IV	15 V	16 VI	17 VII
2	LiH(s) −68.4	BeH$_2$(s) (+20)	B$_2$H$_6$(g) +86.7	CH$_4$(g) −50.7	NH$_3$(g) −16.5	H$_2$O(l) −237.1	HF(g) −273.2
3	NaH(s) −33.5	MgH$_2$(s) −35.9	AlH$_3$(s) (−1)	SiH$_4$(g) +56.9	PH$_3$(g) +13.4	H$_2$S(g) −33.6	HCl(g) −95.3
4	KH(s) (−36)	CaH$_2$(s) −147.2	Ga$_2$H$_6$(s) >0	GeH$_4$(g) +113.4	AsH$_3$(g) +68.9	H$_2$Se(g) +15.9	HBr(g) −53.5
5	RbH(s) (−30)	SrH$_2$(s) (−141)		SnH$_4$(g) +188.3	SbH$_3$(g) +147.8	H$_2$Te(g) >0	HI(g) +1.7
6	CsH(s) (−32)	BaH$_2$(s) (−140)					

Data from *J. Phys. Chem. Ref. Data* **11**, Supplement 2 (1982). Values in parentheses are based on $\Delta_f H^{\ominus}$ data from this source and entropy contributions estimated by the method of W.M. Latimer, p. 359 of *Oxidation potentials.* Prentice Hall, Englewood Cliffs, NJ (1952).

(b) Synthesis

There are three common methods for synthesizing binary hydrogen compounds:

1 Direct combination of the elements:

$$2E + H_2(g) \longrightarrow 2EH$$

2 Protonation of a Brønsted base:

$$E^- + H_2O(aq) \longrightarrow EH + OH^-$$

3 Metathesis (double replacement) of a halide or pseudohalide with a hydride:[12]

$$E^+H^- + EX \longrightarrow E^+X^- + EH$$

In such general equations, the symbol **E** could also denote an element with higher valence, with corresponding changes of detail in the formulas and stoichiometric numbers.

An example of a direct synthesis is the reaction of H_2 with an element. This method is used commercially for the synthesis of compounds that have negative Gibbs energies of formation, including NH_3 and the hydrides of lithium, sodium, and calcium. However, in some cases forcing conditions (high pressure, high temperature, and a catalyst) are necessary to overcome the unfavorable kinetics. The high temperature used for the lithium reaction is an example: it melts the metal and hence helps to break up the surface layer of hydride that would otherwise passivate it. This inconvenience is avoided in many laboratory preparations by adopting one of the alternative synthesis routes, which may also be used to prepare compounds with positive Gibbs energies of formation.

An example of protonation of a Brønsted base, such as a nitride ion, is

$$Li_3N(s) + 3H_2O(l) \longrightarrow 3LiOH(aq) + NH_3(g)$$

Lithium nitride is too expensive for the reaction to be suitable for the commercial production of NH_3, but it is very useful in the laboratory for the preparation of ND_3. The success of the reaction depends on the proton donor being stronger than the conjugate acid of the N^{3-} anion (NH_3 in this case). Water is a sufficiently strong acid to protonate the very strong base N^{3-}, but a stronger acid, such as H_2SO_4, is required to protonate the weak base Cl^-:

$$NaCl(s) + H_2SO_4(l) \longrightarrow NaHSO_4(s) + HCl(g)$$

A synthesis by a metathesis reaction is

$$LiAlH_4 + SiCl_4 \longrightarrow LiAlCl_4 + SiH_4$$

This reaction involves (at least formally) the exchange of Cl^- ions for H^- ions in the coordination sphere of the Si atom. Hydrides of the more electropositive elements (LiH, NaH, and $[AlH_4]^-$) are the most active H^- sources. The favorite sources are often the $[AlH_4]^-$ and $[BH_4]^-$ ions in salts such as $LiAlH_4$ ('lithium aluminum hydride'; more formally lithium tetrahydridoaluminate) and $NaBH_4$ ('sodium borohydride', formally sodium tetrahydridoborate) that are soluble in ether solvents (such as $CH_3OC_2H_4OCH_3$) which solvate the alkali metal ion. Of these anion complexes, $[AlH_4]^-$ is much the strongest hydride donor.

The general routes to hydrogen compounds are direct reaction of H_2 and the element, protonation of nonmetal anions, and metathesis between a hydride source and a halide or pseudohalide.

[12] A pseudohalide is a diatomic or polyatomic ion that resembles a halide ion chemically. Pseudohalides include SCN^- and CH_3O^-; see Section 12.3.

8.8 Reaction patterns of hydrogen compounds

Three types of reaction can lead to the scission of an E—H bond:

1 Heterolytic cleavage by hydride transfer:

$$E\text{—}H \longrightarrow E^+ + :H^-$$

2 Homolytic cleavage:

$$E\text{—}H \longrightarrow E\cdot + H\cdot$$

3 Heterolytic cleavage by proton transfer:

$$E\text{—}H \longrightarrow :E^- + H^+$$

A free H^- or H^+ ion is unlikely to be formed in reactions that take place in solution. Instead, the ion is transferred via a complex that involves a hydrogen bridge.

Detailed experiments are required to establish which of these three primary processes occurs in practice. For example, any of them can lead to the addition of EH across a double bond:

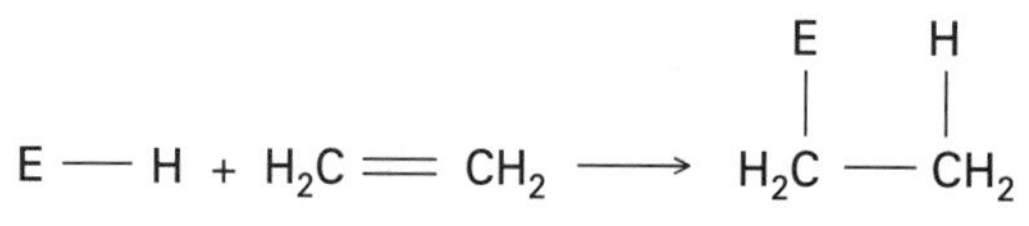

$$E\text{—}H + H_2C = CH_2 \longrightarrow \begin{array}{cc} E & H \\ | & | \\ H_2C & \text{—} CH_2 \end{array}$$

However, there are some patterns of behavior that at least suggest which one is likely to occur and these are discussed in the next section.

(a) Heterolytic cleavage and hydridic character

Compounds with **hydridic character** react vigorously with Brønsted acids by the transfer of an H^- ion (Process 1) with the evolution of H_2:

$$NaH(s) + H_2O(l) \longrightarrow NaOH(aq) + H_2(g)$$

A compound that takes part in this reaction with a weak proton donor (water, in this example) is said to be 'strongly hydridic'. A compound that requires a strong proton donor is classified as 'weakly hydridic' (an example is germane, GeH_4). Hydridic character is most pronounced toward the left of a period where the element is most electropositive (in the s-block) and decreases rapidly after Group 13/III. For example, the saline hydrides of Groups 1 and 2 are strongly hydridic, as demonstrated by their vigorous reaction with water or alcohols. In addition to protolysis, the saline hydrides display hydridic behavior, by transferring an H^-, in the following synthetically useful reactions:

1 Metathesis with a halide, such as the reaction of finely divided lithium hydride with silicon tetrachloride in dry ether (et):

$$4\,LiH(s) + SiCl_4(et) \longrightarrow 4\,LiCl(s) + SiH_4(g)$$

2 Addition to a Lewis acid:

$$LiH(s) + B(CH_3)_3(g) \longrightarrow Li[BH(CH_3)_3](et)$$

3 Reaction with a proton source, to produce H_2:

$$NaH(s) + CH_3OH(l) \longrightarrow NaOCH_3(s) + H_2(g)$$

Note that the H atom of an OH group, but not the hydrogen bonded to the C atom, is sufficiently acidic to undergo this reaction.

Combinations of these reactions are also common. For example the following reaction may be thought of as metathesis (to form B_2H_6) to which $2H^-$ may then add (to form BH_4^-):

$$4\,LiH(s) + BF_3(et) \longrightarrow Li[BH_4](et) + 3\,LiF(s)$$

> *Hydrides of active metals, such as LiH, are strongly hydridic: they react vigorously with proton sources, including relatively weak proton sources such as water and alcohols, to liberate H_2 and form a salt, and they may displace halides to form anionic hydride complexes, such as BH_4^-.*

(b) Homolytic cleavage and radical character

Homolytic cleavage appears to occur readily for the hydrogen compounds of some p-block elements, especially the heavier elements. For example, the use of a radical initiator greatly facilitates the reaction of trialkylstannanes, R_3SnH, with haloalkanes, R—X, as a result of the formation of $R_3Sn\cdot$ radicals:

$$R_3SnH + R'X \longrightarrow R'H + R_3SnX$$

The tendency toward radical reactions increases toward the heavier elements in each group, and Sn–H compounds are in general more prone to radical reactions than Si—H compounds. The ease of homolytic E—H bond cleavage correlates with the decrease in E—H bond strength down a group.

The order of reactivity for various haloalkanes with trialkylstannanes is

$$RF < RCl < RBr < RI$$

Thus fluoroalkanes do not react with R_3SnH, chloroalkanes require heat, photolysis, or chemical radical initiators, and bromoalkanes and iodoalkanes react spontaneously at room temperature. This trend indicates that the initiation step is halogen abstraction.

> *Homolytic cleavage of an E—H bond to produce a radical E· and H_2 occurs most readily for the hydrides of the heavy p-block elements.*

(c) Heterolytic cleavage and protonic character

Compounds reacting by deprotonation are said to show **protic behavior**: in other words, they are Brønsted acids. We saw in Section 5.1 that Brønsted acid strength increases from left to right across a period in the p-block and down a group. One striking example of this trend is the increase in acidity from CH_4 to HF.

> *Hydrogen attached to an electronegative element has protic character and the compound is typically a Brønsted acid.*

....

Example 8.2 Using hydrogen compounds in synthesis

Suggest a procedure for synthesizing lithium tetraethoxyaluminate from $Li[AlH_4]$ and reagents and solvents of your choice.

Answer The reaction of the slightly acidic compound ethanol with the strongly hydridic lithium tetrahydroaluminate should yield the desired alkoxide and hydrogen. The reaction might be carried out by dissolving $LiAlH_4$ in tetrahydrofuran and slowly dropping ethanol into this solution:

$$LiAlH_4(thf) + 4\,C_2H_5OH(l) \longrightarrow Li[Al(OEt)_4](thf) + 4\,H_2(g)$$

This type of reaction should be carried out slowly under a stream of inert gas (N_2 or Ar) to dilute the H_2, which is explosively flammable.

....

Self-test 8.2 Suggest a way of making triethylmethylstannane from triethylstannane and a reagent of your choice.

....

The electron-deficient hydrides of the boron group

The *boranes*, the binary hydrogen compounds of boron, were first isolated in pure form by the German chemist Alfred Stock. All the boron hydrides burn with a characteristic green flame and several of them ignite explosively on contact with air. We have already seen that the simplest member of the series, diborane (B_2H_6), is electron-deficient and that its structure can be described in terms of 2c,2e and 3c,2e bonds (Section 3.12).

8.9 Diborane

In this section, we explore some of the chemical properties of diborane for comparison with other molecular hydrogen compounds. The structures, bonding, and reactions of the higher boron hydrides are covered in Chapter 10.

(a) Synthesis

As shown in Table 8.6, diborane is an endoergic compound at 25 °C, so direct combination of the elements is not feasible as a synthetic route. Stock's entry into the world of boron hydrides was the protolysis of magnesium boride, but this route has been superseded by syntheses that give better yields of specific compounds. Thus, the simplest stable boron hydride, diborane, B_2H_6, may be prepared in the laboratory by metathesis of a boron halide with either $LiAlH_4$ or $LiBH_4$ in ether:

$$3\,LiEH_4 + 4\,BF_3 \longrightarrow 2\,B_2H_6 + 3\,LiEF_4 \qquad (E = B, Al)$$

Both $LiBH_4$ and $LiAlH_4$, like LiH, are good reagents for the transfer of H^-, but they are generally preferred over LiH and NaH because they are soluble in ethers. The synthesis is carried out with the strict exclusion of air (typically, in a vacuum line, because diborane ignites on contact with air; see Box 8.1). Diborane decomposes very slowly at room temperature, forming higher boron hydrides and a non-volatile and insoluble yellow solid, the boron chemist's counterpart of the organic chemist's 'black tar'.

Stock prepared six different boron hydrides,[13] which fell into two classes. One class had the formula B_nH_{n+4} and the other, which was richer in hydrogen and less stable, had the formula B_nH_{n+6}. Examples include pentaborane(11), B_5H_{11}, tetraborane(10), B_4H_{10}, and pentaborane(9), B_5H_9. The nomenclature, in which the number of B atoms is specified by a prefix and the number of H atoms is given in parentheses, should be noted. Thus, the systematic name for diborane is diborane(6); however, as there is no diborane(8), the simpler term 'diborane' is almost always used.

All the boranes are colorless, diamagnetic substances. They range from gases (B_2 and B_4 hydrides), through volatile liquids (B_5 and B_6 hydrides), to the sublimable solid $B_{10}H_{14}$.

> *Diborane may be synthesized by metathesis between a boron halide, such as BF_3, and a hydride source, such as $LiAlH_4$; many of the higher boranes can be prepared by the partial pyrolysis of diborane.*

(b) Oxidation

All the boron hydrides are flammable, and several of the lighter ones, including diborane, react spontaneously with air, often with explosive violence and a green flash (an emission from an

---

[13] Stock had good but inconclusive evidence for a seventh boron hydride, B_6H_{12}, which (together with many other boron hydrides) was characterized only after his death. In 1933 he wrote an interesting account of his research, which involved the development of vacuum line apparatus for handling these highly air-sensitive compounds. See *Hydrides of boron and silicon*. Cornell University Press, Ithaca (1957).

Box 8.1 Chemical vacuum lines

Chemical vacuum line techniques provide a method for handling gases or volatile liquids without exposure to the air. They enable a wide range of operations necessary for the preparation and characterization of such air-sensitive compounds to be performed.

A vacuum line, such as the one in Fig. B8.1, is evacuated by a high-vacuum pumping system. Once the apparatus is evacuated, condensable gases can be moved from one site into another by low-temperature condensation. For example, a sample of gas from a storage bulb on the right might be condensed onto a solid reactant in reaction vessel on the left. Products of different volatility can be separated by passing the vapors through U-traps. The first trap in the series is held at the highest temperature (perhaps $-78\,°C$, by means of a carbon dioxide/acetone slush bath surrounding the trap) to capture the least volatile component; the next trap might be held at $-196\,°C$ by means of a Dewar flask of liquid nitrogen; noncondensable gases such as H_2 might be pumped away.

As well as excluding air and enabling the manipulation of vapors with ease, the vacuum line provides a closed system in which gaseous reactants or products can be measured quantitatively. For instance, a sample of gas in the bulb could be allowed to expand into the adjacent manometer, where its pressure is measured. If the system has been previously calibrated, the amount of gas present can be calculated using the gas laws. The identity and purity of a compound can be determined on the vacuum line by measuring its vapor pressure at one or more fixed temperatures, and comparison with tabulated vapor pressures. NMR, IR, and other spectroscopic cells can also be attached to the vacuum line and loaded by condensation of a vapor.

Glass vacuum systems such as the one illustrated here are widely used for volatile hydrogen compounds, organometallic compounds, and halides. Since elemental fluorine and highly reactive fluorides corrode glass, they are generally handled in a vacuum line constructed from nickel tubing, with metal valves in place of the stopcocks, and electronic pressure transducer gauges in place of the mercury-filled manometers shown here.

Further reading

R.J. Angelici, *Synthesis and techniques in inorganic chemistry.* Saunders, Philadelphia (1977).

R.J. Errington, *Advanced practical inorganic and metal organic chemistry.* Chapman & Hall, London (1997).

D.F. Shriver and M.A. Drezdzon, *The manipulation of air-sensitive compounds.* Wiley, New York (1986).

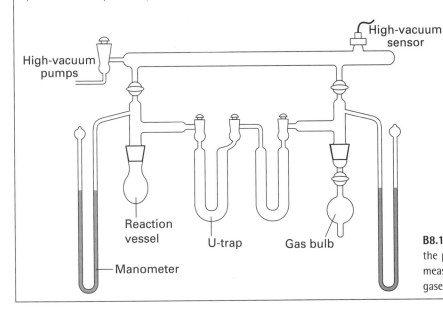

B8.1 A simple chemical vacuum line suitable for the preparation, characterization, and quantitative measurement of condensable gases, particularly gases that are sensitive to air and moisture.

excited state of the reaction intermediate BO). The final product of the reaction is the hydrated oxide:

$$B_2H_6(g) + 3\,O_2(g) \longrightarrow 2\,B(OH)_3(s)$$

Air oxidation is fairly general for the *p*-block hydrides. Only HF and H_2O do not burn in air; some (including B_2H_6) ignite spontaneously as soon as they come into contact with air.

The lighter boranes are readily hydrolyzed by water:

$$B_2H_6(g) + 6\,H_2O(l) \longrightarrow 2\,B(OH)_3(aq) + 6\,H_2(g)$$

As described below, BH_3 is a Lewis acid, and the mechanism of this hydrolysis reaction involves coordination of H_2O acting as a Lewis base. Molecular hydrogen forms through the combination of the partially positively charged H atom on O with the partially negatively charged H atom on B.

All the boron hydrides are flammable, sometimes explosively, and many of them are susceptible to hydrolysis.

(c) Lewis acidity

As implied by the mechanism of hydrolysis, diborane and many other light boron hydrides act as Lewis acids and they are cleaved by reaction with Lewis bases. Two different cleavage patterns have been observed, namely, symmetric cleavage and unsymmetric cleavage.

In **symmetric cleavage**, B_2H_6 is broken symmetrically into two BH_3 fragments, each of which forms a complex with a Lewis base:[14]

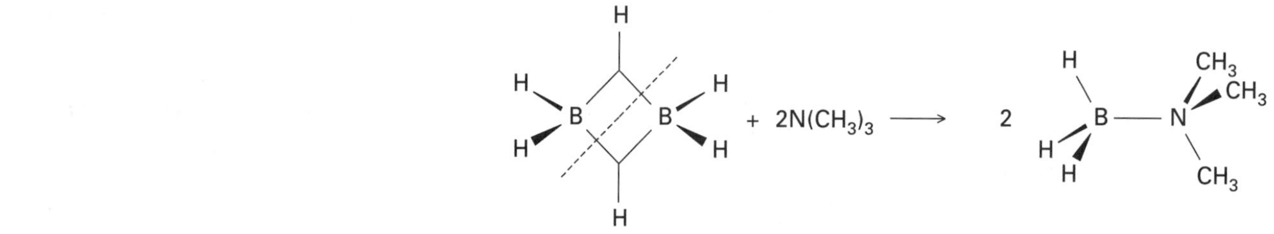

Many complexes of this kind exist. They are interesting partly because they are isoelectronic with hydrocarbons. For instance, the product of the reaction above is isoelectronic with 2,2-dimethylpropane (neopentane, $C(CH_3)_4$). Stability trends indicate that BH_3 is a soft Lewis acid, as illustrated by the following reaction, in which BH_3 transfers to the soft S donor atom and the harder Lewis acid, BF_3, combines with the hard N donor atom:

$$H_3BN(CH_3)_3 + F_3BS(CH_3)_2 \longrightarrow H_3BS(CH_3)_2 + F_3BN(CH_3)_3$$

The direct reaction of diborane and ammonia results in **unsymmetrical cleavage**, which is cleavage leading to an ionic product:

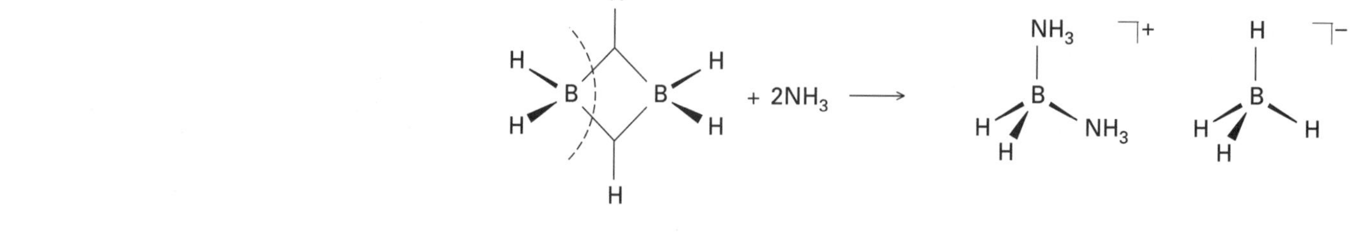

---

[14] Wherever possible, we use the symbol

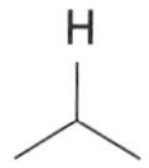

to depict a 3c,2e bond. Unlike in the representation of the structures of organic molecules by lines, there is no atom at the intersection of the three spokes. In large structures, where this symbol would clutter the diagram, we simply connect the central H atom to its neighbors by lines, as in

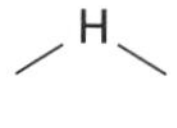

Unsymmetrical cleavage of this kind is generally observed when diborane and a few other boron hydrides react with strong, sterically uncrowded bases at low temperatures. The steric repulsion is such that only two small ligands can attack one B atom in the course of the reaction.

> *Soft and bulky Lewis bases* (L) *cleave diborane symmetrically, giving* H_3BL; *more compact and hard Lewis bases cleave the hydrogen bridge unsymmetrically, giving* $[H_2BL_2][BH_4]$. *Although it reacts with many hard Lewis bases, diborane is best regarded as a soft Lewis acid.*

(d) Hydroboration

An important component of a synthetic chemist's repertoire of reactions is **hydroboration**, the addition of H—B across a multiple bond:

$$H_3B\text{—}OR_2 + H_2C\text{=}CH_2 \xrightarrow{\text{ether}} CH_3CH_2BH_2 + R_2O$$

From the viewpoint of an organic chemist, the C—B bond in the primary product of hydroboration is an intermediate stage in the stereospecific formation of C—H or C—OH bonds, to which it can be converted. From the viewpoint of the inorganic chemist, the reaction is a convenient method for the preparation of wide variety of organoboranes. The hydroboration reaction is one of a class of reactions in which E—H adds across the multiple bond; hydrosilylation (Section 8.12b) is another important example. Much of the driving force for these types of reaction stems from the strength of the C—H bond compared with those of B—H and Si—H.

> *Hydroboration, the reaction of diborane with alkenes in ether solvent, produces organoboranes that are useful intermediates in synthetic organic chemistry.*

8.10 The tetrahydroborate ion

Diborane reacts with alkali metal hydrides to produce salts containing the tetrahydroborate ion:

$$B_2H_6 + 2\,LiH \longrightarrow 2\,LiBH_4$$

Because of the sensitivity of diborane and LiH to water and oxygen, the synthesis must be carried out in the absence of air and in a nonaqueous solvent such as the short-chain polyether $CH_3OCH_2CH_2OCH_3$. We can view this reaction as another example of the Lewis acidity of BH_3 toward the strong Lewis basicity of H^-. The BH_4^- ion is isoelectronic with CH_4 and NH_4^+, and the three show the following variation in chemical properties as the electronegativity of the central atom increases:

	BH_4^-	CH_4	NH_4^+
Character:	hydridic	—	protic

CH_4 is neither acidic nor basic under the conditions prevailing in aqueous solution.

Alkali metal tetrahydroborates are very useful laboratory and commercial reagents. They are often used as a mild source of H^- ions, as general reducing agents, and as precursors for most boron–hydrogen compounds. Most of these reactions are carried out in polar nonaqueous solvents. The preparation of diborane mentioned previously,

$$3\,LiEH_4 + 4\,BF_3 \longrightarrow 2\,B_2H_6 + 3\,LiEF_4 \qquad (E = B, Al)$$

is one example. Although BH_4^- is thermodynamically unstable with respect to hydrolysis, the reaction is very slow at high **pH** and some synthetic applications have been devised in water.

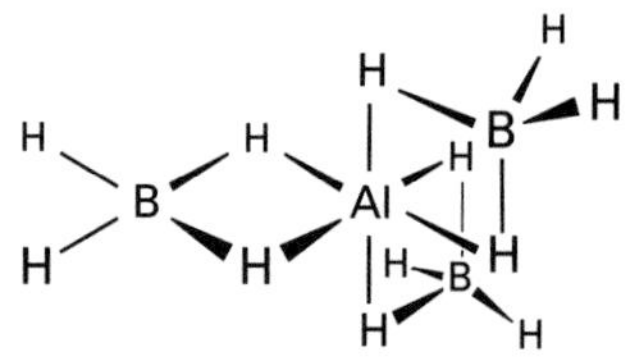

10 [Al(BH$_4$)$_3$]

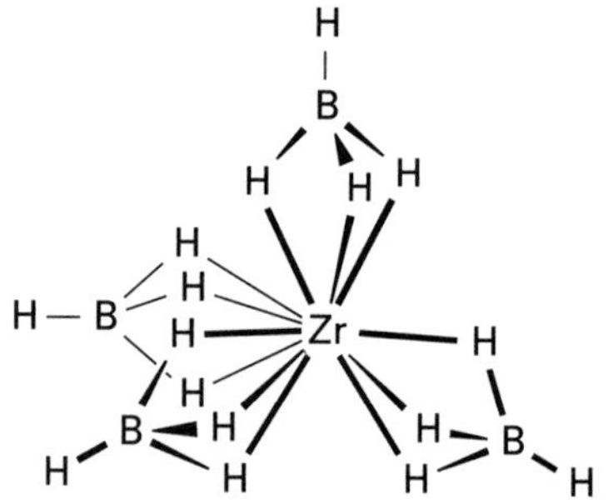

11 [Zr(BH$_4$)$_4$]

For example, germane can be prepared by dissolving GeO_2 and KBH_4 in aqueous potassium hydroxide and then acidifying the solution:[15]

$$HGeO_3^-(aq) + BH_4^-(aq) + 2H^+(aq) \longrightarrow GeH_4(g) + B(OH)_3(aq)$$

Aqueous BH_4^- also can serve as a simple reducing agent, as in the reduction of aqua ions such as Ni^{2+} or Cu^{2+} to the metal or metal boride. With halo complexes of $4d$ and $5d$ elements that also have stabilizing ligands such as phosphines, tetrahydroborate ions can be used to introduce a hydride ligand by a metathesis reaction in a nonaqueous solvent:

$$RuCl_2(PPh_3)_3 + NaBH_4 + PPh_3 \xrightarrow{\text{benzene/ethanol}} RuH_2(PPh)_4 + \text{other products}$$

It is probable that many of these metathesis reactions proceed via a transient BH_4^- complex. Indeed, many borohydride complexes are known, especially with highly electropositive metals: they include $Al(BH_4)_3$ (**10**), which contains a diborane-like double hydride bridge, and $Zr(BH_4)_4$ (**11**), in which triple hydride bridges are present. We see from these examples that many compounds can be described in terms of 3c, 2e bonds.

| *The tetrahydroborate ion is a useful intermediate for the preparation of metal hydride complexes and borane adducts of formula H_3BL.*

.....

Example 8.3 Predicting the reactions of boron–hydrogen compounds

By means of a chemical equation, indicate the products resulting from the interaction of equal amounts of $[HN(CH_3)_3]Cl$ with $LiBH_4$ in tetrahydrofuran (THF).

Answer The interaction of the hydridic BH_4^- ion with the protonic $[HN(CH_3)_3]^+$ ion will evolve hydrogen to produce trimethylamine and BH_3. In the absence of other Lewis bases, the BH_3 would coordinate to THF; however the stronger Lewis base trimethylamine is produced in the initial reactions, so the overall reaction will be

$$[HN(CH_3)_3]Cl + LiBH_4 \longrightarrow H_2 + H_3BN(CH_3)_3 + LiCl$$

The boron in the $H_3BN(CH_3)_3$ product has four tetrahedrally attached groups.

.....

Self-test 8.3 Write a plausible equation for the reaction of $LiBH_4$ with propene in ether solvent and a 1:1 stoichiometry, and another equation for its reaction with ammonium chloride in THF with the same stoichiometry.

.....

8.11 The hydrides of aluminum and gallium

The hydrides of indium and thallium are very unstable. Pure Ga_2H_6 has been prepared only recently[16] but derivatives have been known for some time. The range of binary aluminum hydrides is much more limited than that of boron. The metathesis of their halides with LiH leads to lithium tetrahydridoaluminate, $LiAlH_4$, or the analogous tetrahydrogallate, $LiGaH_4$:

$$4LiH + ECl_3 \xrightarrow{\text{ether}} LiEH_4 + 3LiCl \qquad (E = Al, Ga)$$

The direct reaction of lithium, aluminum, and hydrogen leads to the formation of either $LiAlH_4$ or Li_3AlH_6, depending on the conditions of the reaction. Their formal analogy with halo complexes such as $[AlCl_4]^-$ and $[AlF_6]^{3-}$ should be noted.

.....

[15] The detailed procedure is given by W.L. Jolly, *The synthesis and characterization of inorganic substances*, p. 496. Waveland Press, Prospect Heights, IL (1991).

[16] For a description of this experimental *tour de force*, see A.J. Downs and C.R. Pulham, The hunting of gallium hydrides. *Adv. Inorg. Chem.* **41**, 171 (1994).

The $[AlH_4]^-$ and $[GaH_4]^-$ ions are tetrahedral, and are much more hydridic than $[BH_4]^-$. The latter property is consistent with the higher electronegativity of boron compared with aluminum and gallium. They are also much stronger reducing agents; $LiAlH_4$ is commercially available and widely used as a strong hydride source and as a reducing agent. With the halides of many nonmetallic elements $[AlH_4]^-$ serves as a hydride source in metathesis reactions, such as the reaction of lithium tetrahydridoaluminate with silicon tetrachloride in tetrahydrofuran solution (denoted thf here) to produce silane:[17]

$$LiAlH_4(thf) + SiCl_4(thf) \longrightarrow LiAlCl_4(thf) + SiH_4(g)$$

The general rule in this important type of reaction is that H^- *transfers from the element of lower electronegativity* (Al in the example) *to the element of greater electronegativity* (Si in the example). That tetrahydridoaluminate is a more powerful hydride source than tetrahydroborate is in line with the lower electronegativity of aluminum in comparison with that of boron. For example, $NaAlH_4$ reacts violently with water, but basic aqueous solutions of $NaBH_4$ are useful in synthetic chemistry, as described above.

Under conditions of controlled protolysis, both AlH_4^- and GaH_4^- lead to complexes of aluminum or gallium hydride:

$$LiEH_4 + [(CH_3)_3NH]Cl \longrightarrow (CH_3)_3N{-}EH_3 + LiCl + H_2 \qquad (E = Al, Ga)$$

In striking contrast to BH_3 complexes, these complexes will add a second molecule of base to form five-coordinate complexes of aluminum or gallium hydride (**12**):

$$(CH_3)_3N{-}EH_3 + N(CH_3)_3 \longrightarrow ((CH_3)_3N)_2EH_3 \qquad (E = Al, Ga)$$

This behavior is consistent with the trend for Period 3 and higher *p*-block elements to form five- and six-coordinate hypervalent compounds (Section 3.1).

Aluminum hydride, AlH_3, is a solid that is best regarded as saline, like the hydrides of the *s*-block metals. Unlike CaH_2 and NaH, which are more readily available commercially, AlH_3 does not find many applications in the laboratory. The alkylaluminum hydrides, such as $Al_2(C_2H_5)_4H_2$ (**13**), are well known molecular compounds and contain Al—H—Al 3c,2e bonds. Hydrides of this kind are used to couple alkenes, the initial step being the addition of Al—H across the C=C double bond, as in hydroboration.

> $LiAlH_4$ and $LiGaH_4$ are useful precursors of MH_3L_2 complexes; $LiAlH_4$ also is used as a source of H^- ions in the preparation of metalloid hydrides, such as SiH_4. The alkyl aluminum hydrides are used to couple alkenes.

12 $[(CH_3)_3N]_2AlH_3$, R = CH_3

13 $Al_2(C_2H_5)_4H_2$

Electron-precise hydrides of the carbon group

The electron-precise hydrides of the carbon group (Group 14/IV) have no lone pairs of electrons and so are not Lewis bases. Their lack of lone pairs also means that they are not associated by hydrogen bonding. Moreover, the C atom in hydrocarbons has no vacant, energetically accessible orbitals, and so is not a Lewis acid. The numerous hydrocarbons are best regarded from the viewpoint of organic chemistry, and this section concentrates mainly on the *silanes*, the silicon hydrides. The extensive range of metals and metalloids bound to hydrocarbon ligands are discussed in Chapters 15 and 16.

8.12 Silanes

The silanes, with their greater number of electrons and stronger intermolecular forces, are less volatile than their hydrocarbon analogs: whereas propane, C_3H_8, boils at $-44\,°C$ and is a gas under normal conditions, its analog trisilane, Si_3H_8, is a liquid and boils at $53\,°C$. The chemical properties of silanes are less fully documented than those of the alkanes and other

---

[17] This reaction was first encountered in Section 8.7.

hydrocarbons. This difference is partly due to the fact that there have been few practical or theoretical reasons to prepare many of them, and their greater reactivity makes their investigation more demanding. Stable unsaturated analogs of alkenes, alkynes, and arenes are unknown for silanes containing only silicon and hydrogen.

(a) Synthesis

Partly because he happened to use magnesium boride contaminated with magnesium silicide, Stock also found himself studying the silicon hydrides. He identified four silicon analogs of the alkanes, specifically SiH_4, Si_2H_6, Si_3H_8, and Si_4H_{10}. Separation of these products by modern gas chromatography suggests that Si_4H_{10} in fact consists of a mixture of a linear and a branched isomer analogous to butane and methylpropane. Indeed, gas chromatography suggests that there are silicon analogs of all the straight and branched-chain alkanes, at least up to Si_9H_{20}.

The silanes are thermally less stable than the hydrocarbons. Thus, silanes are cracked at moderate temperatures, when they form a mixture of silane itself and higher silanes:

$$Si_2H_6 \xrightarrow{400\,°C} H_2 + SiH_4 + \text{higher silanes}$$

Complete decomposition occurs above about 500 °C, when the silane decomposes to silicon and hydrogen. These **pyrolysis reactions**, reactions in which a compound is degraded by heat, have considerable technological utility because silane is used as a source of pure crystalline silicon by the semiconductor industry:

$$SiH_4 \xrightarrow{500\,°C} Si(s, \text{crystalline}) + 2\,H_2$$

In this process a thin film of silicon can be laid down on a heated substrate.[18] A related reaction, the decomposition of silane in an electric discharge, yields amorphous silicon:

$$SiH_4(g) \xrightarrow{\text{electric discharge}} Si(s, \text{amorphous}) + 2\,H_2(g)$$

Amorphous silicon is used in photovoltaic devices, such as power sources for pocket calculators. 'Amorphous silicon' is in fact a misnomer, for this material contains a significant proportion of hydrogen, which IR spectroscopy shows to be bound by Si—H bonds. The amorphous structure arises because Si—H bonds prevent the formation of the ordered Si—Si bonds that are responsible for the diamond-like structure of pure crystalline silicon.

Stock's protonation of a silicide is no longer of much importance in the preparation of Si—H bonds. Instead, the laboratory preparation is generally based on the metathesis reactions of Si—Cl or Si—Br compounds with $LiAlH_4$:

$$4\,R_3SiCl + LiAlH_4 \xrightarrow{\text{ether}} 4\,R_3SiH + LiAlCl_4$$

Although useful in the laboratory, this method is too expensive for the commercial production of silane. Instead, the commercial production of silane employs the less expensive reagents HCl and either silicon or iron silicide, which form trichlorosilane, $HSiCl_3$. The trichlorosilane is then heated in the presence of a catalyst, such as aluminum chloride, to yield silane:

$$4\,SiHCl_3 \xrightarrow{\Delta} SiH_4 + 3\,SiCl_4$$

This reaction is endoergic and becomes more so at higher temperatures $(\Delta_r S^{\ominus} = -58.5\ \mathrm{J\,K^{-1}\,mol^{-1}})$, but it is possible to drive the reaction forward by removing the highly volatile SiH_4. Silane itself is also prepared commercially by the reduction of SiO_2 with aluminum under a high pressure of hydrogen in a molten salt mixture of NaCl and $AlCl_3$. An idealized equation for this reaction is

$$6\,H_2(g) + 3\,SiO_2(s) + 4\,Al(s) \longrightarrow 3\,SiH_4(g) + 2\,Al_2O_3(s)$$

[18] J.M. Jasinski and S.M. Gates, *Acc. Chem. Res.* **24**, 9 (1991).

Silane is used in the production of semiconductor devices such as solar cells and in the hydrosilylation of alkenes; it is prepared commercially by the high-pressure reaction of hydrogen, silicon dioxide, and aluminum.

(b) Reactions

The silanes are generally more reactive than their hydrocarbon analogs, and ignite spontaneously on contact with air. They also react explosively with fluorine, chlorine, and bromine. Silane itself is also a reducing agent in aqueous solution. For example, when silane is bubbled through an oxygen-free aqueous solution containing Fe^{3+}, it reduces the iron to Fe^{2+}.

Bonds between silicon and hydrogen are not readily hydrolyzed in neutral water, but the reaction is rapid in strong acid or in the presence of traces of base. Likewise, alcoholysis is accelerated by catalytic amounts of alkoxide:

$$SiH_4 + 4\,ROH \xrightarrow{OR^-} Si(OR)_4 + 4\,H_2$$

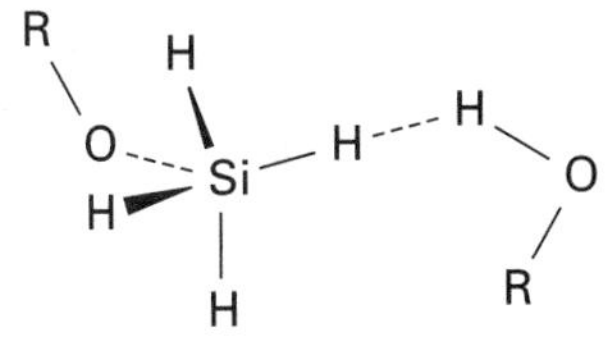

Kinetic studies indicate that the reaction proceeds through a structure in which OR^- attacks the Si atom while H_2 is being formed via a kind of $H \cdots H$ hydrogen bond between hydridic and protic hydrogen atoms (**14**). Although the silicon tetrahalides form many stable complexes with amines and similar donors, the proposed intermediate cannot be isolated because silane is a far weaker Lewis acid than the tetrahalides. This difference in Lewis acidity is reasonable in view of the low electronegativity of hydrogen relative to the halogens, which results in a more positive Si atom and therefore a stronger Lewis acid site in $SiCl_4$ than in SiH_4.

The silicon analog of hydroboration is **hydrosilylation**, the addition of Si—H across the multiple bonds of alkenes and alkynes. This reaction, which is used in both industrial and laboratory syntheses, can be carried out under conditions (300 °C or UV irradiation) that produce a radical intermediate. In practice it is usually performed under far milder conditions by using a platinum complex as catalyst:

$$CH_2{=}CH_2 + SiH_4 \xrightarrow{H_2PtCl_6,\ isopropanol} CH_3CH_2SiH_3$$

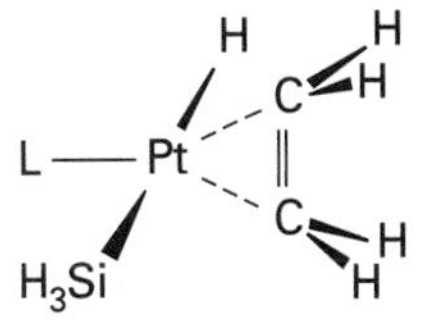

The current view is that this reaction proceeds through an intermediate in which both the alkene and silane are attached to the metal atom (**15**).

Silane is a reducing agent. With a platinum complex as catalyst it adds across carbon-carbon double bonds (hydrosilylation) and forms $Si(OR)_4$ with alcohols.

14

15

8.13 Germane, stannane, and plumbane

The decreasing stability of the hydrides on going down the group severely limits the accessible chemical properties of stannanes and plumbane. Germane, GeH_4, and stannane, SnH_4, can be synthesized by the reaction of the appropriate tetrachloride with $LiAlH_4$ in THF solution. Plumbane has been synthesized in trace amounts by the protolysis of a magnesium/lead alloy. The presence of alkyl or aryl groups stabilizes the hydrides of all three elements. For example, trimethylplumbane, $(CH_3)_3PbH$, begins to decompose at −30 °C, but it manages to survive for several hours at room temperature.

Thermal stability decreases from germane to stannane and plumbane.

Electron-rich compounds of Groups 15/V to 17/VII

We consider here the production of the commercially important electron-rich hydrogen compounds. We also discuss the mechanism for the reaction of hydrogen and oxygen to form water, because it is interesting mechanistically and presents a serious safety problem whenever hydrogen gas escapes into the atmosphere. The Brønsted acidity of the hydrogen halides has already been discussed in Chapter 5 and will not be covered here.

8.14 Ammonia

Ammonia, NH_3, is produced in huge quantities worldwide for use as a fertilizer and as a primary source of nitrogen in the production of many chemicals. Virtually only one method, the **Haber process**, is used for the entire global production. In this process, nitrogen and hydrogen combine at high temperature (450 °C) and pressure (100 atm):

$$N_2(g) + 3 H_2(g) \longrightarrow 2 NH_3(g)$$

This process is the direct combination of N_2 and H_2 carried out over a promoted iron catalyst; the promoters required include SiO_2, MgO, and other oxides. The reaction is carried out at high temperature and in the presence of a catalyst to overcome the kinetic inertness of N_2, and at high pressure to overcome the thermodynamic effect of an unfavorable equilibrium constant at the operating temperature.

So novel and great were the chemical and engineering problems arising from the then (early twentieth century) uncharted area of large-scale high-pressure technology, that two Nobel Prizes were awarded. One went to Fritz Haber (in 1918), who developed the chemical process. The other went to Carl Bosch (in 1931), the chemical engineer who designed the first plants to realize Haber's process. The Haber–Bosch process has had major impact on civilization because ammonia is the primary source of most nitrogen-containing compounds (Chart 8.2), including fertilizers and most commercially important compounds of nitrogen. More details on the synthesis and individual reactions of ammonia may be found in Section 11.4.

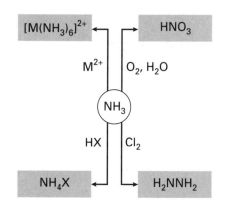

Chart 8.2

> *The high-pressure and temperature-catalyzed combination of N_2 and H_2 produces ammonia, which is a fertilizer and the starting material for other nitrogen-containing chemicals.*

8.15 Phosphine, arsine, and stibine

In contrast to the commanding role that ammonia plays in nitrogen chemistry, the highly poisonous hydrides of the heavier nonmetallic elements of Group 15/V (particularly phosphine, PH_3, and arsine, AsH_3) are of minor importance in the chemistry of their respective *p*-block elements. Both phosphine and arsine are used in the semiconductor industry to dope silicon or to prepare other semiconductor compounds such as GaAs by chemical vapor deposition. These thermal decomposition reactions reflect the positive Gibbs energy of formation of these hydrides.

The commercial synthesis of PH_3 is the disproportionation of white phosphorus in basic solution:

$$P_4(s) + 3 OH^-(aq) + 3 H_2O(l) \longrightarrow PH_3(g) + 3 H_2PO_2^-(aq)$$

Arsine and stibine may be prepared by the protolysis of compounds that contain an electropositive metal in combination with arsenic or antimony:

$$Zn_3E_2(s) + 6 H^+(aq) \longrightarrow 2 EH_3(g) + 3 Zn^{2+}(aq) \qquad (E = As, Sb)$$

Phosphine and arsine are poisonous gases that readily ignite in air, but the much more stable organic analogs PR_3 and AsR_3 (R = alkyl or aryl groups) are widely used as ligands in metal coordination chemistry. In contrast with the hard donor properties of ammonia and alkylamine ligands, the organophosphines and organoarsines, such as $P(C_2H_5)_3$ and $As(C_6H_5)_3$, are soft ligands and therefore are often incorporated into metal complexes having central metal atoms in low oxidation states. The stability of these complexes correlates with the soft acceptor nature of the metals in low oxidation states, and the stability of soft-donor, soft-acceptor combinations (Section 5.12).

It is evident from Fig. 8.3 that PH_3, AsH_3, and SbH_3 are subject to little if any hydrogen bonding with themselves; however, PH_3 and AsH_3 can be protonated by strong acids, such as HI. All the Group 15/V hydrides are pyramidal, but an interesting change in bond angle occurs down the series:

NH_3, 107.8° PH_3, 93.6° AsH_3, 91.8° SbH_3, 91.3°

The large change in bond angle and basicity from NH_3 to PH_3 is in harmony with the idea that the N—H bonds and lone pair of NH_3 can be regarded as sp^3-hybrid orbitals, whereas the lone pair in PH_3 appears to have much more s-orbital character and the P—H bonds correspondingly more p-orbital character.

Unlike liquid ammonia, liquid phosphine, arsine, and stibine do not associate through hydrogen bonding. Their much more stable alkyl and aryl analogs are useful soft ligands.

8.16 Water

By far the most important binary compound with hydrogen in this group is water, H_2O. The bonding and properties of water have been discussed extensively in a variety of contexts already. Its standard Gibbs energy of formation is strongly negative ($-237.1 \text{ kJ mol}^{-1}$), so the compound is thermodynamically stable with respect to its elements. The chemical properties of water include its role as an excellent solvent for many ionic compounds and for substances with which it can form hydrogen bonds. It acts as a Lewis base, and metal cations in water are normally present with about six H_2O molecules attached as ligands via the O atoms. Water is a mild reducing agent, but it needs powerful oxidizing agents before it can release electrons. It is a stronger oxidizing agent. These features were discussed in Section 6.10 in the context of the stability field of water, and that section should be consulted for a full discussion.

The mechanism of reaction of hydrogen gas with oxygen gas to produce water has been extensively studied. A hint of its complexity is evident from the pressure dependence of the rate of the reaction (Fig. 8.12). This diagram shows that at 550 °C an increase in total pressure can turn an explosive reaction into a smooth reaction and that a further pressure increase leads to an explosion again. It is now recognized that the complexity of the reaction stems from the coexistence of a branched-chain mechanism as well as a simple chain propagation step. In the simple chain, a radical carrier (·OH) is consumed with the production of another carrier (·H),

$$\cdot OH + H_2 \longrightarrow H_2O + \cdot H$$

In a branching chain, more than one radical carrier is produced when one radical reacts:

$$\cdot H + O_2 \longrightarrow \cdot OH + \cdot O \cdot$$
$$\cdot O \cdot + H_2 \longrightarrow \cdot OH + \cdot H$$

Under normal reaction conditions chain carriers are scavenged by collisions with the walls of the reactor or other chain-terminating gas-phase encounters. However, the branching steps allow the rate of carrier production to outstrip quenching, producing a cascade of radical carriers, an increase in rate, and an explosion. Hydrogen in air poses a serious explosion hazard because of the very wide range of hydrogen partial pressures under which its combustion is explosive.

Water is unique as a solvent for ionic compounds; it is a mild reducing agent and a more powerful oxidizing agent; it acts as a Lewis base. The formation of water from its elements occurs by a radical chain mechanism.

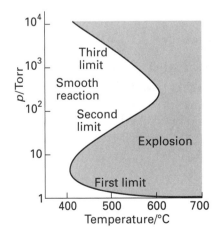

8.12 The variation of the character of the $H_2 + O_2$ reaction with pressure and temperature.

8.17 Hydrogen sulfide, selenide, and telluride

The decrease in E—H bond strength down the group from H_2S to H_2Te (see Fig. 8.11) is reflected in a sharp change in the Gibbs energy of formation to more positive values; indeed, both H_2S and H_2Te are unstable with respect to decomposition into the elements. All the H_2E compounds can be prepared by protonation of their metal salts:

$$Na_2E(s) + 2\,H^+(aq) \longrightarrow 2\,Na^+(aq) + H_2E(g)$$

The Brønsted basicity of H_2S and heavier members of this series is negligible in aqueous solution, so a nonaqueous solvent and a strong acid are needed to distinguish their basicities.

In addition to the simple hydrogen compounds of formula EH_2, a series of hydrogen polysulfides may be prepared by the protonation of a metal polysulfide:

$$Na_2S_n + 2\,H^+(aq) \longrightarrow 2\,Na^+(aq) + H_2S_n \qquad (n = 4 \text{ to } 6)$$

Hydrogen polysulfide molecules consist of zig-zag chains of S atoms that are capped at the ends by H atoms.

Element-hydrogen bond strengths and basicities of EH_2 decrease sharply from $E = O$ to S and more slowly from S to Te. Sulfur, selenium, and tellurium, unlike oxygen, display a strong tendency to catenate.

8.18 Hydrogen halides

The hydrogen halides may be formed by direct reaction of the elements in a radical chain reaction, such as that for the reaction between Br_2 and H_2. With the lighter halogens (F_2 and Cl_2), the reaction is explosive under a wide range of conditions. However, all commercial HF and most HCl is synthesized by the protonation of halide ions:

$$CaF_2(s) + H_2SO_4(s) \longrightarrow CaSO_4(s) + 2\,HF(g)$$

This reaction cannot be used for the preparation of HI because concentrated sulfuric acid oxidizes the product to iodine; it can be used to prepare HBr if the concentration of H_2SO_4 is carefully controlled.

With the exception of HF, which is a weak acid in water, all the hydrogen halides are strong acids. Their Brønsted acidic character has been treated in general terms in Chapter 5.

Hydrogen halides are prepared by the action of a strong acid on the halide salt.

Further reading

G.A. Jeffrey, *An introduction to hydrogen bonding*. Oxford University Press (1997).

R.B. King, *Inorganic chemistry of the main group elements*, Chapter 1, pp. 1–20. VCH, Weinheim (1994).

M. Kakiuchi, Hydrogen inorganic. In *Encyclopedia of inorganic chemistry* (ed. R.B. King), Vol. 3, pp. 1444–71. Wiley, New York (1994).

Kirk–Othmer encyclopedia of chemical technology. Wiley-Interscience, New York (1991 *et seq.*).
In particular, see articles on hydrogen, deuterium, and tritium.

J. Emsley, Very strong hydrogen bonds. *Chem. Soc. Rev.* **9**, 91 (1980).

G.A. Jeffrey and W. Saenger, *Hydrogen bonding in biological structures*. Springer-Verlag, New York (1991).

Exercises

8.1 Assign oxidation numbers to the elements in (a) H_2S, (b) KH, (c) $[ReH_9]^{2-}$, (d) H_2SO_4, (e) $H_2PO(OH)$.

8.2 Write balanced chemical equations for three major industrial preparations of hydrogen gas. Propose two different reactions that would be convenient for the preparation of hydrogen in the laboratory.

8.3 Preferably without consulting reference material, construct the periodic table, identify the elements, and (a) indicate positions of salt-like, metallic, and molecular hydrides, (b) add arrows to indicate trends in $\Delta_f G^\ominus$ for the hydrogen compounds of the p-block elements, (c) delineate the areas where the molecular hydrides are electron-deficient, electron-precise, and electron-rich.

8.4 Name and classify the following hydrogen compounds: (a) BaH_2, (b) SiH_4, (c) NH_3, (d) AsH_3, (e) $PdH_{0.9}$, (f) HI.

8.5 Identify the compounds from Exercise 8.4 that provide the most pronounced example of the following chemical characteristics and give a balanced equation that illustrates each of the characteristics: (a) hydridic character, (b) Brønsted acidity, (c) variable composition, (d) Lewis basicity.

8.6 Divide the compounds in Exercise 8.4 into those that are solids, liquids, or gases at room temperature and pressure. Which of the solids are likely to be good electrical conductors?

8.7 Use Lewis structures and the VSEPR model to predict the shapes of H_2Se, P_2H_4, and H_3O^+ and to assign point groups. Assume a skew structure for P_2H_4.

8.8 Identify the reaction that is most likely to give the highest proportion of HD and give your reasoning: (a) $H_2 + D_2$ equilibrated over a platinum surface, (b) $D_2O + NaH$, (c) electrolysis of HDO.

8.9 Identify the compound in the following list that is most likely to undergo radical reactions with alkyl halides, and describe the reason for your choice: H_2O, NH_3, $(CH_3)_3SiH$, $(CH_3)_3SnH$.

8.10 Arrange H_2O, H_2S, and H_2Se in order of (a) increasing acidity, (b) increasing basicity toward a hard acid such as the proton.

8.11 Describe the three different common methods for the synthesis of binary hydrogen compounds and illustrate each one with a balanced chemical equation.

8.12 Give balanced chemical equations for practical laboratory methods of synthesizing (a) H_2Se, (b) SiD_4, and (c) $Ge(CH_3)_2H_2$ from $Ge(CH_3)_2Cl_2$, and for the commercial synthesis of (d) SiH_4 from elemental silicon and HCl.

8.13 Does B_2H_6 survive in air? If not, write the equation for the reaction. Describe a step-by-step procedure for transferring B_2H_6 quantitatively from a gas bulb at 200 Torr into a reaction vessel containing diethyl ether.

8.14 What is the trend in hydridic character of $[BH_4]^-$, $[AlH_4]^-$, and $[GaH_4]^-$? Which is the strongest reducing agent? Give the equations for the reaction of $[GaH_4]^-$ with excess 1 M HCl(aq).

8.15 Given $NaBH_4$, a hydrocarbon of your choice, and appropriate ancillary reagents and solvents, give formulas and conditions for the synthesis of (a) $B(C_2H_5)_3$, (b) Et_3NBH_3.

8.16 Write balanced chemical equations for the formation of pure silicon from crude silicon via silane.

8.17 Describe the important physical differences and a chemical difference between each of the hydrogen compounds of the p-block elements in Period 2 and their counterparts in Period 3.

8.18 What type of compound is formed between water and krypton at low temperatures and elevated krypton pressure? Describe the structure in general terms.

8.19 Sketch the approximate potential energy surfaces for the hydrogen bond between H_2O and the Cl^- ion and contrast this with the potential energy surface for the hydrogen bond in $[FHF]^-$.

Problems

8.1 What is the expected infrared stretching wavenumber of gaseous ^3HCl given that the corresponding value for ^1HCl is 2991 cm^{-1}?

8.2 Consult *Further information 2* and then sketch the qualitative splitting pattern and relative intensities within each set for the ^1H- and ^{31}P-NMR spectra of PH_3.

8.3 (a) Sketch a qualitative molecular orbital energy level diagram for the HeH^+ molecule-ion and indicate the correlation of the molecular orbital levels with the atomic energy levels. The ionization energy of H is 13.6 eV and the first ionization energy of He is 24.6 eV. (b) Estimate the relative contribution of H1s and He1s orbitals to the bonding orbital and predict the location of the partial positive charge of the polar molecule. (c) Why do you suppose that HeH^+ is unstable on contact with common solvents and surfaces?

8.4 Borane exists as the molecule B_2H_6 and trimethylborane exists as a monomer $B(CH_3)_3$. In addition, the molecular formulas of the compounds of intermediate compositions are observed to be $B_2H_5(CH_3)$, $B_2H_4(CH_3)_2$, $B_2H_3(CH_3)_3$, and $B_2H_2(CH_3)_4$. Based on these facts, describe the probable structures and bonding in this series of mixed alkyl hydrides.

8.5 Observations on a complex in which HD replaces H_2 were used to show that H_2 is bound in $[W(CO)_3(P^iPr_3)_2(H_2)]$ without H—H bond rupture (G. Kubas, R.R. Ryan, B.I. Swanson, P.J. Vergamini, and H.J. Wasserman, *J. Am. Chem. Soc.* **106**, 451 (1984)). If the dihydrogen complex has an H—H stretch at 2695 cm^{-1}, what would be the expected wavenumber of the HD complex? What should be the pattern of the ^1H-NMR signal arising from coupling with D in the HD complex?

8.6 Hydrogen bonding can influence many reactions, including the rates of O_2 binding and dissociation from metalloproteins (G.D. Armstrong and A.G. Sykes, *Inorg. Chem.* **25**, 3135 (1986)). Describe the evidence for (or against) hydrogen bonding with O_2 in the metalloproteins hemerythrin, myoglobin, and hemocyanin.

8.7 (a) Compare the structures of borane and aluminum hydride as described in this text with gallane (see reference in footnote 16 of this chapter). (b) Compare the compounds formed from diborane and digallane with trimethylamine and rationalize any differences.

8.8 Spectroscopic evidence has been obtained for the existence of $[Ir(C_5H_5)(H_3)(PR_3)]^+$, a complex in which one ligand is formally H_3^+. Devise a plausible molecular orbital scheme for the bonding in the complex, assuming that an angular H_3 unit occupies one coordination site and interacts with the e_g and t_{2g} orbitals of the metal. An alternative formulation of the structure of the complex, however, is as a trihydro species with very large coupling constants (see *J. Am. Chem. Soc.* **113**, 6074 (1991) and the references therein, and especially *J. Am. Chem. Soc.*

112, 909 and 920 (1990)). Review the evidence for this alternative formulation.

8.9 Correct the faulty statements in the following description of hydrogen compounds. 'Hydrogen, the lightest element, forms thermodynamically stable compounds with all the nonmetals and most metals. The isotopes of hydrogen have mass numbers of 1, 2, and 3, and the isotope of mass number 2 is radioactive. The structures of the hydrides of the Group 1 and 2 elements are typical of ionic compounds because the H^- ion is compact and has a well-defined radius. The structures of the hydrogen compounds of the nonmetals are adequately described by VSEPR theory. The compound $NaBH_4$ is a versatile reagent because it has greater hydridic character than the simple Group 1 hydrides such as NaH. Heavy element hydrides such as the tin hydrides frequently undergo radical reactions, in part because of the low E—H bond energy. The boron hydrides are called electron-deficient compounds because they are easily reduced by hydrogen.'

9

The metals

The metallic elements are the most numerous of the elements, and their chemical properties are central to both industry and contemporary research. We shall see many systematic trends in properties of the metals within each of the blocks of the periodic table, and some striking differences between these blocks. In this chapter we survey the periodic trends in their chemical properties. One of the most important is the trend in stabilities of the oxidation states within each block. These stabilities are closely related to the ease of recovery of metals from their ores and to the ways in which the various metals and their compounds are handled in the laboratory. Other important properties include ligand preference and the arrangement of ligands around metal ions in complexes. It will be seen that simple binary compounds, such as metal halides and oxides, often follow systematic trends; however, when the metal is in a low oxidation state, interesting variations such as metal–metal bonding may be encountered.

All the elements in the s-, d-, and f-blocks of the periodic table are metals, and seven of the thirty p-block elements (aluminum, gallium, indium, thallium, tin, lead, and bismuth) are usually considered to be metals (Fig. 9.1). We say 'usually' because the diagonal borderline between metals and nonmetals in the p-block is ill-defined, and the metalloids germanium (Ge) and polonium (Po) are also sometimes described as metals.

General properties of metals

Most metals have high electric and thermal conductivities and are malleable and ductile. However, there is a wide range of personalities within this general uniformity. One aspect of this diversity is the strength of the cohesion between atoms as indicated by the range of enthalpies of vaporization (Fig. 9.2). A practical application of the low enthalpies of vaporization in Groups 1 and 12 (the alkali metals, and zinc, cadmium, and mercury) is the use of sodium and mercury in electric discharge lamps, such as fluorescent lamps and street lamps. The metals with the highest enthalpies of vaporization are found in the middle of the 4d and 5d series. Tungsten, with the highest value for any element, is used as a filament in incandescent lights because it volatilizes only very slowly at high temperatures.

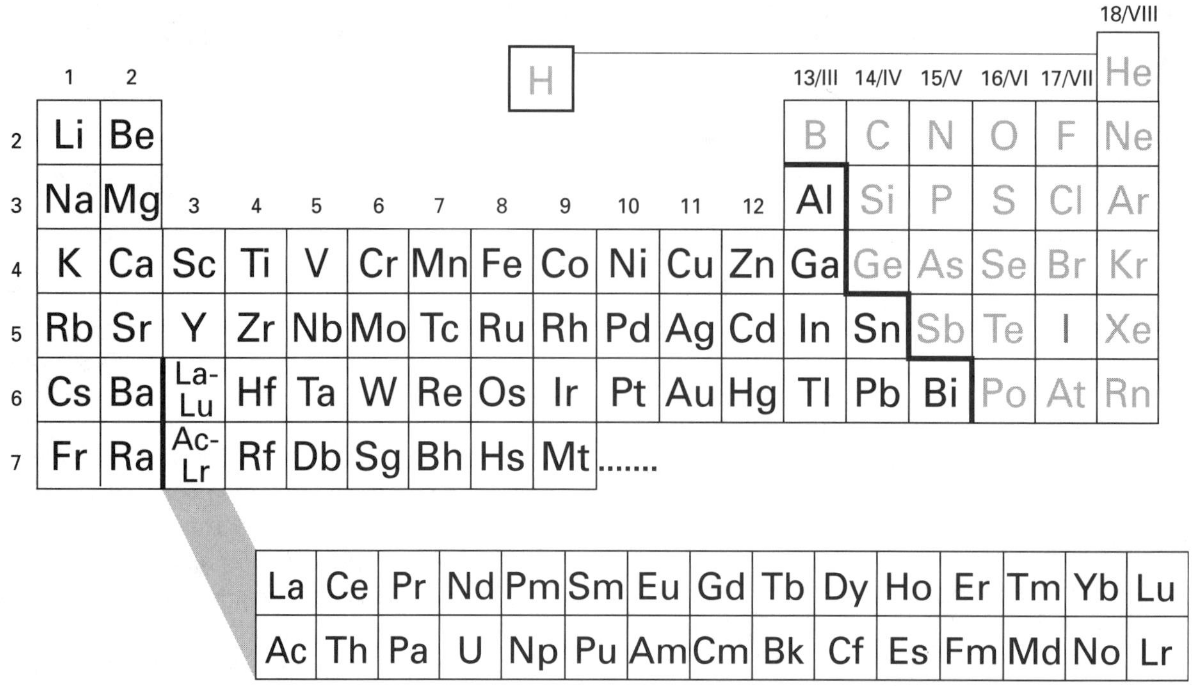

9.1 The distribution of metallic elements (shown bold) in the periodic table.

Two characteristic chemical properties of many metals are the formation of basic oxides and hydroxides when the metal is in the +1 or +2 oxidation state and the formation of simple (hydrated) cations in acidic aqueous solution. Most metals react with oxygen. However, the thermodynamic spontaneity and rate of this reaction vary greatly from cesium, which ignites on contact with air, to metals such as aluminum and iron that survive in air. The latter metals are commercially useful because they react only slowly with the atmosphere at

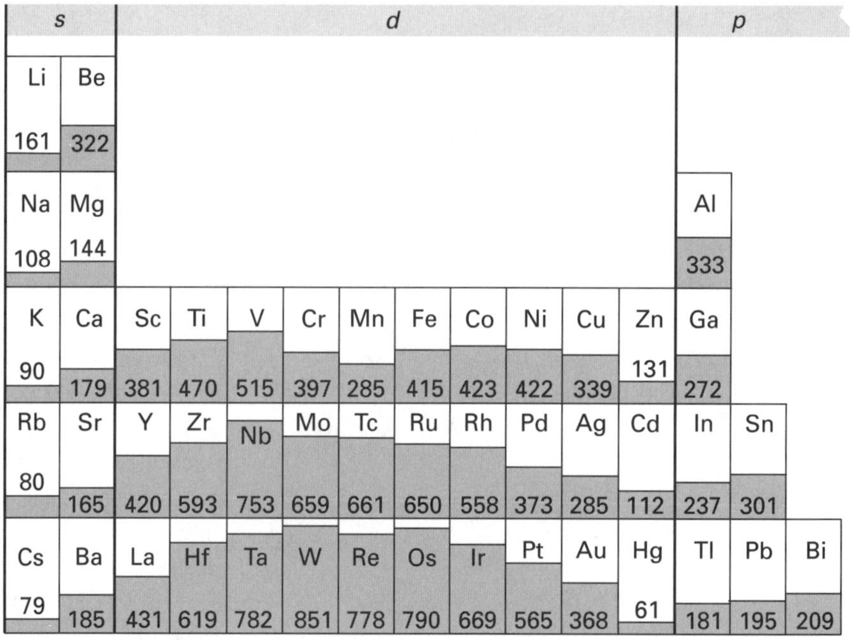

9.2 The enthalpies of vaporization (in kilojoules per mole) for the metallic elements in the s-, d-, and p-blocks.

ordinary temperatures on account of their passivation by a thin protective layer of oxide that forms on their surface. The only metals that have no tendency to form oxides under standard conditions are the small group of noble metals in the lower right of the d-block (Fig. 9.3), the most familiar examples being gold and platinum. The elements in the middle of the d-block have highly varied chemical properties on account of their wide range of accessible oxidation states and ability to form a multitude of complexes (Chapter 7).

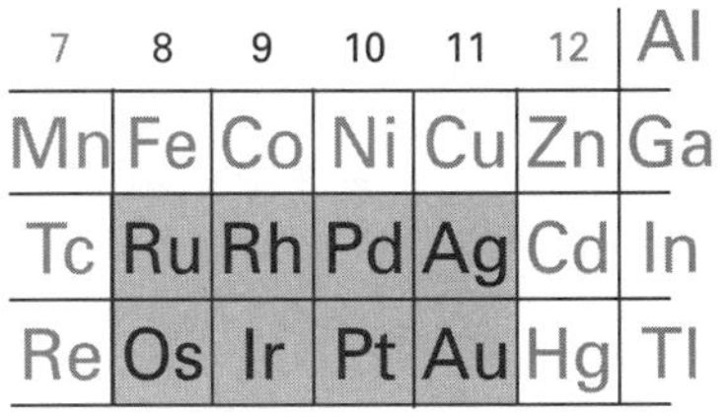

The traditional view of the chemical properties of the metallic elements is from the perspective of ionic solids and complexes that contain a single central metal ion. However, with the development of improved techniques for the determination of structure it has been recognized that there are also many d-metal compounds that have metal–metal (M—M) bond distances comparable to or shorter than those in the pure metals. These findings have stimulated additional X-ray structural investigations of compounds that might contain M—M bonds, and have inspired research into syntheses designed to broaden the range of such cluster compounds. Examples of metal (and nonmetal) cluster compounds are now known in every block of the periodic table (Fig. 9.4), but they are most numerous in the d-block.

Figure 9.4 shows how cluster compounds may be classified according to ligand type. Although this classification is not accurate in detail for all cluster compounds, it provides information on the principal ligand types and some insight into the bonding. For example, alkyllithium compounds (Chapter 15) often have alkyl groups attached to a metal cluster by two-electron multicenter bonds. Clusters of the early d-block elements and the lanthanides generally contain σ-donor ligands such as Br^-. These ligands can fill some of the low-lying orbitals of these electron-poor metal atoms by σ- and π-electron donation. In contrast, electron-rich metal clusters on the right of the d-block generally contain π-acceptor ligands such as CO, which remove some of the electron density from the metal. Many p-block elements do not require ligands to complete the valence shells of the individual atoms in clusters and can exist as **naked clusters**. The ionic clusters Pb_5^{2-} and Sn_9^{4-} are two examples of naked clusters formed by metals.

Although the chemical properties of the metallic elements are very rich, each block of elements displays moderately consistent trends in the stability of the oxidation states, the tendency to form complexes, and the nature of the halides and chalcogenides of the various oxidation states.

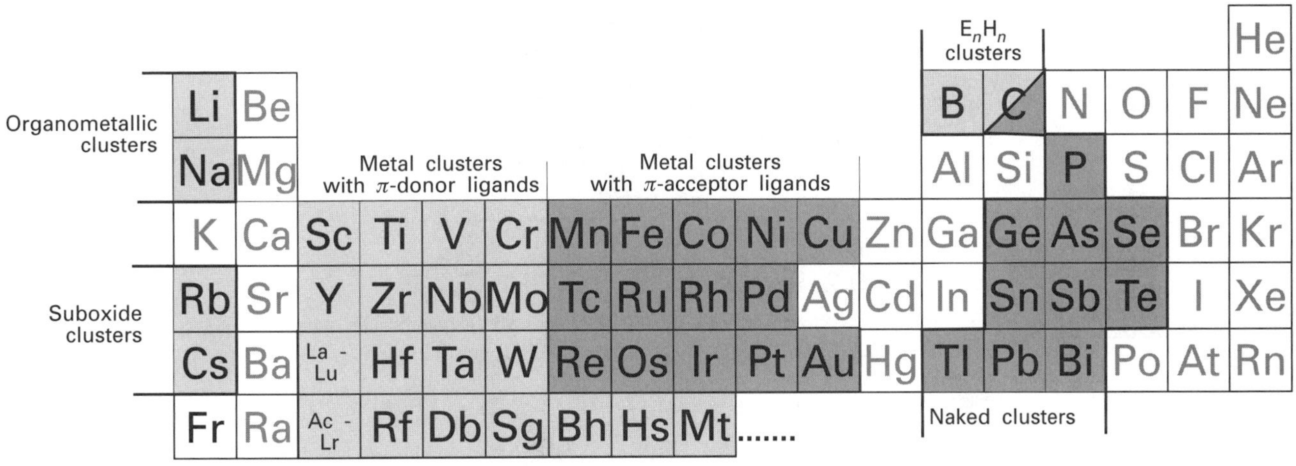

9.3 The location of the noble metals in the periodic table.

9.4 The major classes of cluster-forming elements. Note that carbon is in two classes (for example, it occurs as C_8H_8 and C_{60}). (Adapted from D.M.P. Mingos and D.J. Wales, *Introduction to cluster chemistry*. Prentice Hall, Englewood Cliffs (1990).)

The s-block metals

Alkali metal (Group 1) and alkaline earth metal (Group 2) cations are commonly found in minerals and natural waters, and some are important constituents of biological fluids such as blood. That the atoms of these metals have only poor control over their valence electrons is shown by their low ionization energies and enthalpies of vaporization (Fig. 9.2). As a consequence of both properties the metals are strong reducing agents, and all the Group 1 metals and calcium through barium in Group 2 react rapidly with water to liberate hydrogen. The cheaper metals (lithium, sodium, potassium, and calcium) are often used in the laboratory and industry as powerful reducing agents for chemical reactions in nonaqueous solvents, such as liquid ammonia. The characteristic oxidation numbers of the s-block elements are the same as their group numbers: $+1$ for the alkali metals and $+2$ for the alkaline earth metals.

Until the 1950s few simple complexes of the alkali metal ions were recognized; however, the synthesis of polydentate ligands containing hard donor atoms such as oxygen and nitrogen has led to the discovery of many complexes of Group 1 and 2 metal ions. It has also been found that when water and air are excluded, some unusual compounds with the metals in negative oxidation states can be prepared. Examples include the sodides, which contain Na^-.

9.1 Occurrence and isolation

The abundance of Group 1 and 2 metals in the Earth's crust spans a broad range from calcium (the fifth most abundant metal), through sodium (sixth), magnesium (seventh), and potassium (eighth), to the relatively rare metals cesium and beryllium (Fig. 9.5). We saw in Section 1.1 that the low abundances of lithium and beryllium arise from the details of nucleosynthesis. The low abundances of the heavy alkali and alkaline earth metals are associated with the decline in nuclear binding energies of elements beyond iron.

Table 9.1 lists the principal natural sources and methods of isolation of the commercially important Group 1 and 2 metals. Because these elements are all highly reducing, their recovery requires relatively expensive technology, such as the electrolysis of molten salts, the use of other alkali metals as reducing agents, or high-temperature metal reduction.

> Sodium, potassium, magnesium, and calcium are abundant in the Earth's crust, but recovery of the metals is energy-intensive.

Table 9.1 Mineral sources and methods of recovery of some commercially important s-block metals

Metal	Natural source	Method of recovery
Lithium	Spodumene, $LiAl(SiO_3)_2$	Electrolysis of molten $LiCl + KCl$
Sodium	Rock salt, $NaCl$, sea water and brines	Electrolysis of molten $NaCl$
Potassium	Silvite, KCl brines	Action of sodium on KCl at $850°C$
Beryllium	Beryl, $Be_3AlSi_6O_{18}$	Electrolysis of molten $BeCl_2$
Magnesium	Dolomite, $CaMg(CO_3)_2$	$2MgCaO_2(l) + FeSi(l) \xrightarrow{1150°C}$ $Mg(g) + Fe(l) + Ca_2SiO_4(l)$
Calcium	Limestone, $CaCO_3$	Electrolysis of molten $CaCl_2$

9.2 Redox reactions

The standard potentials of the alkali and alkaline earth metals (Table 9.2) suggest that they are all capable of being oxidized by water:

$$\text{Group 1}: \quad M(s) + H_2O(l) \longrightarrow M^+(aq) + OH^-(aq) + H_2(g)$$

$$\text{Group 2}: \quad M(s) + 2H_2O(l) \longrightarrow M^{2+}(aq) + 2OH^-(aq) + H_2(g)$$

This reaction is so rapid and exothermic for sodium and its heavier congeners that the hydrogen is ignited. The vigor of these reactions is associated with the low melting points of the metals, because once molten a clean metal surface is more readily exposed and rapid

Group 1

Li	Na	K	Rb	Cs
1.30	4.36	4.23	1.85	0.20

Group 2

Be	Mg	Ca	Sr	Ba
0.30	4.45	4.71	2.65	2.58

9.5 The crustal abundances of Group 1 and 2 elements. The abundances are given as their logarithms (to the base 10) of grams of metal per 10^3 kg of sample. Because the vertical scale is logarithmic, the differences are much greater than they appear.

reaction ensues. In Group 2, both beryllium and magnesium are protected from further oxidation by a thin oxide coating and therefore survive in the presence of water and air.

The standard potentials of the alkali metals are surprisingly uniform (and close to -3 V). This uniformity stems from compensating contributions in the thermodynamic cycle into which the reduction half-reaction may be analyzed (Fig. 9.6). The enthalpies of sublimation and ionization both decrease down the group (making oxidation more favorable); however, this trend is counteracted by a smaller enthalpy of hydration as the radii of the ions increase (making oxidation less favorable).

The Group 1 and 2 metals are strong reducing agents.

Table 9.2 Standard potentials for the s-block elements ($E^{\ominus}/V$)

Group 1		Group 2	
Li	−3.04	Be	−1.97
Na	−2.71	Mg	−2.36
K	−2.94	Ca	−2.87
Rb	−2.92	Sr	−2.90
Cs	−3.06	Ba	−2.92

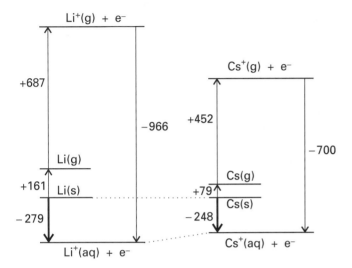

9.6 Thermochemical cycles (enthalpy changes in kilojoules per mole) for the oxidation half-reaction $M(s) \rightarrow M^+(aq) + e^-$ for lithium and cesium. Note that the large difference in the enthalpy of formation of $M^+(g)$ is compensated by a similar difference in the M^+ hydration enthalpies.

9.3 Binary compounds

The standard potentials for couples in aqueous solution often give an indication of the tendency of the s-block elements to form compounds. However, the interactions of the cations with anions in the solid state may differ significantly from their interaction with H_2O molecules, so standard potentials are not always a good guide. For example, despite the relative uniformity of the standard potentials down a group, the only stable nitride of the Group 1 metals is lithium nitride, Li_3N. The chemical individuality of the s-block elements also appears, as we shall see, in their reactions with oxygen.

Trends in chemical properties are much simpler for metal halides, and examples are known of all binary combinations of an alkali metal and a halogen. Most of the alkali halides have a $(6,6)$-coordinated rock-salt structure (Section 2.9), but CsCl, CsBr, and CsI have the more closely packed, $(8,8)$-coordinated cesium-chloride structure. At high pressures the halides of sodium, potassium, and rubidium undergo a transition from the rock-salt to the more closely packed cesium-chloride structure.

The great majority of the oxides and sulfides of the s-block metals display the group oxidation number ($+1$ for Group 1, $+2$ for Group 2). However, the possibility of catenation (chain formation) of the anion gives a rich range of compounds. For example, Na_2O_2 contains the peroxide ion, O_2^{2-}, and KO_2 contains the superoxide ion, O_2^-. In addition, the larger Group 2 metal ions form peroxides. The stabilization of the large peroxide and superoxide anions by the larger alkali metal and alkaline earth metal cations was used in Section 2.12 as an example of how large cations can help to stabilize large anions.

A distinctive aspect of the alkali metal cations is the high solubility in water of most of their simple salts. The major exceptions to this rule are compounds in which the larger cations, K^+ through Cs^+, are combined with large anions: this property was also mentioned in Section 2.12 as an example of the variation of lattice enthalpies with ionic radius. For instance, the heavy alkali metal perchlorates are much less soluble than the light alkali metal perchlorates: whereas the molar concentration of saturated $CsClO_4(aq)$ is $0.09\ mol\ L^{-1}$, that of $LiClO_4(aq)$ is $4.5\ mol\ L^{-1}$. Tetraphenylborate salts of potassium and the heavier alkali metals are even less soluble in water. Similarly, the heavier alkaline earth metal cations form insoluble salts with large dinegative ions: a familiar example is hydrated calcium sulfate, the major constituent of plaster. The trend toward lower solubility in water down the group is very striking: $MgSO_4$ is highly soluble but the molar solubility of $CaSO_4 \cdot 2H_2O$ is $5 \times 10^{-2}\ mol\ L^{-1}$ and that of $BaSO_4$ is only $10^{-5}\ mol\ L^{-1}$.

X-ray structure determinations show that the Li^+ ion is often found in sites ranging from four- to six-coordination. Examples include Li_2O, in which it is four-coordinate (in the antifluorite structure, Section 2.9), and LiF, in which it is six-coordinate (in the rock-salt structure). The larger alkali metal ions display even more variation in their common coordination number.

The structures of beryllium compounds are distinctive on account of the element's tendency toward covalent bond formation, which in turn arises from its small size and high polarizing power. A recurring motif in beryllium compounds is a tetrahedral unit with Be at its center. For example, the low-temperature form of BeO has the wurtzite structure (Section 2.9), with tetrahedral coordination of the Be atom. Were it not for the high toxicity of beryllium compounds, ceramic BeO might find many applications because it displays the uncommon combination of very low electrical conductivity with high thermal conductivity. In contrast with BeO, the simple oxides of all the larger alkaline earth metal ions are six-coordinate in the rock-salt structure.

> The smallest Group 2 ion, Be^{2+}, is typically found with four-coordination; the larger ions typically have six-coordination.

9.4 Complex formation

The *s*-block metal ions in their group oxidation states are chemically hard. Therefore, most of the complexes they form arise from Coulombic interactions with small, hard donors, such as those possessing O or N atoms. In general, the smaller the ion and the greater the charge, the more stable is the complex. For example, Be^{2+} and Mg^{2+} complexes with hard ligands are often stable with respect to decomposition, but complexes of the other *s*-block metal ions are not. For these complexes, lability parallels stability. Indeed, only for $Be^{2+}(aq)$ and $Mg^{2+}(aq)$ is the formation of simple complexes slower than the rate of encounter in solution.

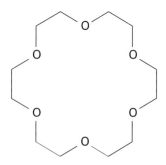

1 18–crown–6

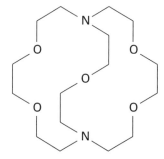

2 2.2.1 crypt

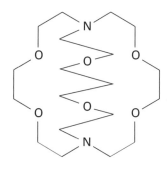

3 2.2.2 crypt

The most notable complexes of the cations of Group 1 and the heavier Group 2 metals (calcium through barium) are formed by polydentate ligands. Monodentate ligands are only weakly bound on account of the weak Coulombic interactions and lack of significant covalent bonding by these ions. Crown ethers such as 18-crown-6 (**1**) form complexes with alkali metal ions that survive indefinitely in nonaqueous solution. Bicyclic cryptand ligands, such as 2.2.1 crypt (**2**) and 2.2.2 crypt (**3**), form complexes with alkali metals that are even more stable, and they can survive even in aqueous solution. These ligands are sterically selective for a particular metal ion, the dominant factor being the fit between the cation and the cavity in the ligand that accommodates it (Fig. 9.7).

The Group 2 cations form complexes with the crown and crypt ligands. The most stable complexes are formed with charged polydentate ligands, such as the analytically important ethylenediaminetetraacetate ion (($^-O_2C)_2NCH_2CH_2N(CO_2^-)_2$, edta). The formation constants of edta complexes of the alkali metal complexes lie in the order $Ca^{2+} > Mg^{2+} > Sr^{2+} > Ba^{2+}$. The analysis of this odd trend is quite involved and beyond the scope of the present discussion. In the solid state, the structure of the Mg^{2+} edta complex is seven-coordinate (**4**), with H_2O filling a coordination site. The calcium complex is either seven- or eight-coordinate, depending on the counterion, with one or two H_2O molecules serving as ligands. A large number of complexes of Ca^{2+} and Mg^{2+} occur naturally, the most familiar being chlorophyll (see Section 19.11).

Compounds of beryllium show properties consistent with a greater covalent character than those of its congeners, and some of its complexes with ordinary ligands are quite stable. For example, basic beryllium acetate (beryllium oxoacetate, $Be_4O(O_2CCH_3)_6$) consists of a central O atom surrounded by a tetrahedron of four Be atoms, which in turn are bridged by acetate ions (**5**). It may be prepared by the reaction of acetic acid with beryllium carbonate:

$$4\,BeCO_3(s) + 6\,CH_3COOH(l) \longrightarrow 4\,CO_2(g) + 3\,H_2O(l) + Be_4O(O_2CCH_3)_6(s)$$

Basic beryllium acetate is a colorless, sublimable, molecular compound; it is soluble in chloroform, from which it can be recrystallized.

> Cyclic or bicyclic polydentate ligands are employed to form stable complexes of Group 1 and 2 ions.

Example 9.1 Explaining trends in *s*-block chemistry

Use simple bonding models to explain why barium forms a peroxide but beryllium does not.

Answer Large anions are generally stabilized by large cations (Section 2.12). Therefore, barium peroxide should be more stable than beryllium peroxide. In fact the peroxide compound forms spontaneously on exposure of barium to air. Beryllium forms only BeO.

Self-test 9.1 Choose the most appropriate ligand for each of the two groups of metal ions and give an approximate order of stability within each group. Ligand: 2.2.2 crypt, edta, OH^-; metal cation: (a) Li^+, Na^+, Rb^+, Cs^+; (b) Cu^{2+}, Fe^{3+}.

9.5 Metal-rich oxides, electrides, and alkalides

Special conditions are needed to form compounds of the *s*-block metals in which the elements occur with oxidation numbers lower than their group numbers. These remarkable compounds are formed only when air, water, and other oxidizing agents are rigorously excluded. For example, a series of metal-rich oxides is formed by the reaction of rubidium or cesium with a limited supply of oxygen. These compounds, which are classified in Fig. 9.4 as cluster compounds, are dark, highly reactive metallic conductors with seemingly strange formulas, such as Rb_6O, Rb_9O_2, Cs_4O_4, and Cs_7O. A clue to the explanation of these constitutions is that Rb_9O_2 consists of O atoms surrounded by octahedra of six Rb atoms, with

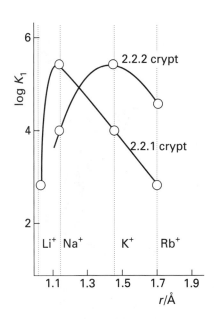

9.7 The stability of complexes of metals with cryptand ligands shown as a plot of the logarithm of the formation constant in water versus cation radius. Note that the smaller 2.2.1 crypt favors complex formation with Na^+ and the larger 2.2.2 crypt favors K^+.

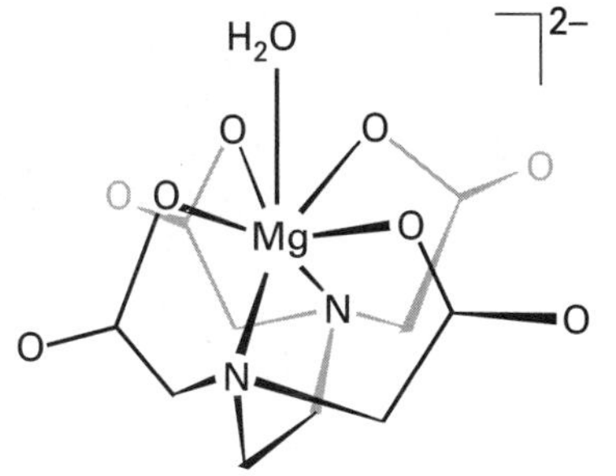

4 $[Mg(edta)(OH_2)]^{2-}$

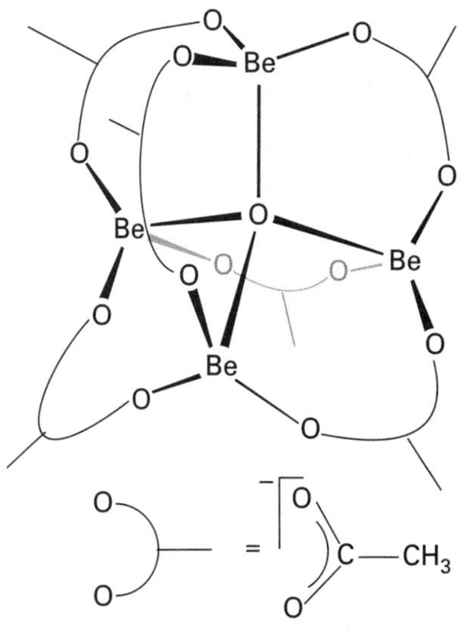

5 $[Be_4O(O_2CCH_3)_6]$

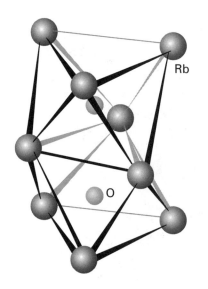

9.8 The structure of Rb_9O_2. Each O atom in this cluster is surrounded by an octahedron of Rb atoms.

two neighboring octahedra sharing faces (Fig. 9.8). These clusters are thought to owe their existence to Coulombic M^+O^{2-} interactions together with weaker M—M bonds delocalized over the metal system. The metallic conduction of the compounds suggests that the valence electrons are delocalized beyond the individual Rb_9O_2 clusters.

Another interesting set of compounds has been characterized in the course of studying sodium in liquid ammonia. Sodium dissolves in pure, anhydrous liquid ammonia (without hydrogen evolution) to give solutions that are deep blue when dilute. The color of these **metal–ammonia solutions** originates from the tail of a strong absorption band that peaks in the near infrared. Other electropositive metals with low enthalpies of sublimation, including calcium and europium, dissolve in liquid ammonia to give solutions with a color that is independent of the metal. A variety of experiments indicate that the absorption responsible for the color arises from transitions of an electron in a cavity formed by a group of NH_3 molecules and with energy levels that resemble those of a particle in a spherical well.

The dissolution of a sodium in liquid ammonia to give a very dilute solution is represented by the equation

$$Na(s) \longrightarrow Na^+(am) + e^-(am)$$

where 'am' denotes solution in ammonia. These solutions survive for long periods at the temperature of boiling ammonia ($-33\,°C$) and in the absence of air. However, they are only metastable, and their decomposition is catalyzed by some d-block compounds:

$$Na^+(am) + e^-(am) + NH_3(l) \longrightarrow NaNH_2(am) + \tfrac{1}{2}H_2(g)$$

In more concentrated solutions, $e^-(am)$ associates with the cation, and the concentrated solutions have a bronze appearance. Optical spectra and electrical conductivity measurements indicate that the electrons are delocalized throughout the solution, much as they are in a metal.

The blue metal–ammonia solutions are excellent reducing agents. For example, the unusual Ni(I) complex $[Ni_2(CN)_6]^{4-}$ may be prepared by the reduction of Ni(II) with potassium in liquid ammonia:

$$2\,K_2[Ni(CN)_4] + 2\,K^+(am) + 2\,e^-(am) \longrightarrow K_4[Ni_2(CN)_6](am) + 2\,KCN$$

The reaction is performed in the absence of air in a vessel cooled to the boiling point of ammonia.

Alkali metals also dissolve in ethers and alkylamines to give solutions with absorption spectra that depend on the alkali metal. The metal dependence suggests that the spectrum is associated with charge transfer from an *alkalide ion*, M^- (such as a sodide ion, Na^-), to the solvent. Further evidence for the occurrence of alkalide ions is the diamagnetism associated with the species assigned as M^-, which would have the spin-paired s^2 valence electron configuration. Another observation in agreement with this interpretation is that, when sodium–potassium alloy is dissolved, the metal-dependent band is the same as for solutions of sodium itself. When ethylenediamine is used as a solvent (denoted 'en'), the metal-independent band is not observed, so the dissolution equation is written as

$$2\,Na(s) \longrightarrow Na^+(en) + Na^-(en)$$
$$NaK(l) \longrightarrow K^+(en) + Na^-(en)$$

The complexation of a cation with a cryptand can be used to prepare solid sodides, such as $[Na(2.2.2)]^+[Na]^-$, where (2.2.2) denotes the cryptand ligand. X-ray structure determination reveals the presence of $[Na(2.2.2)]^+$ and Na^- ions, with the latter located in a cavity of the crystal with an apparent radius larger than that of I^-. The preparation of sodides and other alkalides demonstrates the powerful influence of solvents and complexing agents on the chemical properties of metals. It is even possible to crystallize solids containing solvated electrons, the so-called **electrides**, and to obtain X-ray crystal structures of these compounds.

Figure 9.9 , for example, shows the inferred position of the maxima of the electron density in such a solid.

The s-block elements also form a range of organometallic compounds that are useful in organic and inorganic synthesis. Two familiar examples are the Grignard reagents, such as CH_3MgBr, and methyllithium, $Li_4(CH_3)_4$. These compounds are treated at length in Chapter 15.

> Group 1 and 2 metals form metal-rich oxides and dissolve in liquid ammonia; when oxidizing agents are absent or limited, they form metal-rich compounds with cryptands, many of which are electronic conductors.

The d-block metals

Most of the d-block metals are much more rigid than the metals in Groups 1 and 2. This property, together with moderate rates of oxidation in air, accounts for the wide use of iron, copper, and titanium in the construction of vehicles and buildings. Another contrast with the s-block metals is the wide range of oxidation states that many of the d-block metals display, which leads to a rich and interesting chemistry. The d-block metals also form a much more extensive range of coordination compounds (Chapter 7) and organometallic compounds (Chapter 16). Their range of oxidation states accounts for interesting electronic properties of many solid compounds of the d-block elements (Chapter 18), their ability to participate in catalysis (Chapter 17), and their subtle and interesting role in biochemical processes (Chapter 19). In this section we focus on trends in the stabilities of oxidation states within the d-block and on the properties of some representative compounds.

9.6 Occurrence and recovery

The elements on the left of the 3d series occur in nature primarily as metal oxides or as metal cations in combination with oxoanions (Table 9.3). Of these elements, titanium is the most

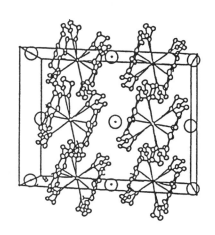

9.9 An ORTEP diagram of the crystal structure of $[Cs(18\text{-crown-}6)_2]^+e^-$. The circled dots mark the sites of highest electron density, and so indicate the locations of the 'anion' e^-. (From S.B. Dawes, D.L. Ward, R.H. Huang, and J.L. Dye, *J. Am. Chem. Soc.* **108**, 3534 (1986).)

Table 9.3 Mineral sources and methods of recovery of some commercially important d-block metals

Metal	Principal minerals	Method of recovery	Note
Titanium	Ilmenite, $FeTiO_3$ Rutile, TiO_2	$TiO_2 + 2C + 2Cl_2 \rightarrow TiCl_4 + 2CO$ followed by reduction of $TiCl_4$ with Na or Mg	
Chromium	Chromite, $FeCr_2O_4$	$FeCr_2O_4 + 4C \rightarrow Fe + 2Cr + 4CO$	(a)
Molybdenum	Molybdenite, MoS_2	$2MoS_2 + 7O_2 \rightarrow 2MoO_3 + 4SO_2$ followed by either: $MoO_3 + 2Fe \rightarrow Mo + Fe_2O_3$ or: $MoO_3 + 3H_2 \rightarrow Mo + 3H_2O$	 (b)
Tungsten	Scheelite, $CaWO_4$ Wolframite, $FeMn(WO_4)_2$	$CaWO_4 + 2HCl \rightarrow WO_3 + CaCl_2 + H_2O$ followed by $2WO_3 + 6H_2 \rightarrow 2W + 6H_2O$	
Manganese	Pyrolusite, MnO_2	$MnO_2 + C \rightarrow Mn + CO_2$	(c)
Iron	Hematite, Fe_2O_3 Magnetite, Fe_3O_4 Limonite, $FeO(OH)$	$Fe_2O_3 + 3CO \rightarrow 2Fe + 3CO_2$	
Cobalt	Cobaltite, $CoAsS$ Smaltite, $CoAs_2$ Linnaeite, Co_3S_4	Byproduct of copper and nickel production	
Nickel	Pentlandite, $(Fe,Ni)_6S_8$	$2NiS + 2O_2 \rightarrow 2Ni + 2SO_2$	(d)
Copper	Chalcopyrite, $CuFeS_2$ Chalcocite, Cu_2S	$2CuFeS_2 + 2SiO_2 + 5O_2 \rightarrow 2Cu + 2FeSiO_3$ $+ 4SO_2$	

(a) The iron–chromium alloy is used directly for stainless steel.
(b) The iron–molybdenum alloy is used in cutting steel.
(c) The reaction is carried out in a blast furnace with Fe_2O_3 to produce alloys.
(d) NiS is formed by melting the mineral and separated by physical processes. NiO is used in a blast furnace with iron oxides to produce steels. Nickel is purified by electrolysis or the Mond process via $Ni(CO)_4$.

difficult to reduce, and it is widely produced by heating TiO_2 with Cl_2 and carbon to produce $TiCl_4$, which is then reduced by molten magnesium at about $1000\,°C$ in an inert-gas atmosphere. The oxides of chromium, manganese, and iron are reduced with carbon (see Section 6.1), a much cheaper reagent. To the right of iron in the $3d$ series, cobalt, nickel, and copper occur mainly as sulfides and arsenides, which is consistent with the increasingly soft Lewis acid character of the dipositive ions toward the right of the series. Copper is used in large quantities for electrical conductors, and electrolysis is used to refine crude copper so as to achieve the high purity needed for high electrical conductivity.

The difficulty of reducing the early $4d$ and $5d$ metals molybdenum and tungsten is apparent from Table 9.3. It reflects the tendency of these elements to have stable high oxidation states, as discussed later in this section. The platinum metals (Ru and Os, Rh and Ir, and Pd and Pt), which are found at the lower right of the d-block, occur as sulfide and arsenide ores, usually in association with larger quantities of copper, nickel, and cobalt. They are collected from the sludge that forms during the electrolytic refinement of copper and nickel.

> *Sulfide minerals of chemically soft metals such as Cu are partially oxidized to obtain the metal; the more electropositive and chemically hard metals occur as oxides, and are extracted by reduction.*

9.7 High oxidation states

As a result of the convention of assigning a negative oxidation number to any nonmetal in combination with a metal, high formal oxidation numbers may be encountered, such as Re(VII) in $[ReH_9]^{2-}$ and W(VI) in $W(CH_3)_6$. However, these compounds are not oxidizing agents in the usual sense, and they are discussed together with other organometallic compounds in Chapter 16. We shall confine this discussion to compounds containing electronegative ligands such as the halogens, oxygen, and sulfur.

(a) Oxidation states across the $3d$ series

The group oxidation number (in the 1–18 system) can be achieved in elements that lie toward the left of a d-block but not by elements on the right. For example, scandium, yttrium, and lanthanum in Group 3 are found in aqueous solution only with oxidation number $+3$, and the great majority of their complexes contain the metals in this oxidation state. The group oxidation number is never achieved after Group 9 (Co, Rh, and Ir). This limit on the maximum oxidation number correlates with the increase in noble character from left to right across each period in the d-block.

The trend in thermodynamic stability of the group oxidation states of the $3d$ metals is clearly illustrated by the Frost diagram in Fig. 9.10. We see that the group oxidation states of scandium, titanium, and vanadium fall in the lower part of the diagram. This location indicates that the metal and any species in intermediate oxidation states are readily oxidized to the group oxidation state. In contrast, species in the group oxidation state for chromium and manganese ($+6$ and $+7$, respectively) lie in the upper part of the diagram. This location indicates that they are very susceptible to reduction. The illustration also shows that the group oxidation number is not achieved in Groups 8 through 11 of Period 4 (iron, cobalt, nickel, copper, and zinc).

The binary compounds of the $3d$ elements with halogens and oxygen also illustrate the trend in stability of the group oxidation state. The earliest metals can achieve their group oxidation states in compounds with chlorine (for example, $ScCl_3$ and $TiCl_4$), but the more strongly oxidizing halogen fluorine is necessary to achieve the group oxidation state of vanadium (Group 5) and chromium (Group 6), which form VF_5 and CrF_6, respectively. Beyond Group 6 in Period 4, even fluorine cannot bring out the group oxidation state, and MnF_7 and FeF_8 have never been prepared. Oxygen brings out the group oxidation state for many metals

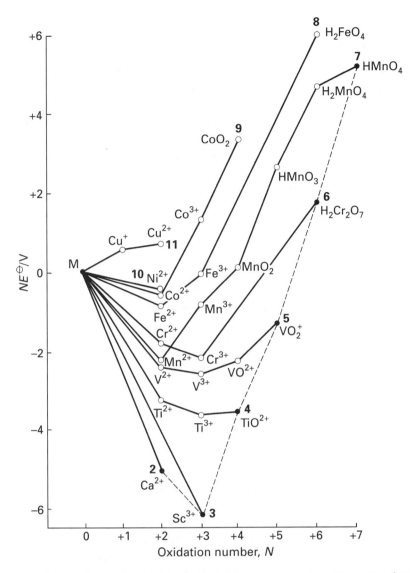

9.10 A Frost diagram for the first series of d-block elements in acidic solution (pH = 0). The bold numbers designate the group numbers and the broken line connects species in their group oxidation states.

more readily than does fluorine because fewer O atoms than F atoms are needed to achieve the same oxidation number. For example, the group oxidation state of $+7$ for manganese is achieved in permanganate salts such as potassium permanganate, $KMnO_4$. Claims for the existence of FeO_4, however, are strongly disputed.

As can be inferred from the Frost diagram in Fig. 9.10, chromate, CrO_4^{2-}, permanganate, MnO_4^-, and ferrate, FeO_4^{2-} (see Box 9.1), are strong oxidizing agents and become stronger from CrO_4^{2-} to FeO_4^{2-}. This trend is another illustration of the decreasing stability of the maximum attainable oxidation state for Groups 6, 7, and 8. Yet another example of the greater difficulty of oxidizing an element to the right of chromium to its maximum oxidation state is that the air oxidation of MnO_2 in molten potassium hydroxide does not take manganese to its group oxidation state but instead yields the deep green compound potassium manganate, K_2MnO_4, which contains Mn(VI). The disproportionation of MnO_4^{2-} in acidic aqueous solution yields manganese(IV) oxide, MnO_2, and the deep purple permanganate ion, MnO_4^-, containing Mn(VII):

$$3\,MnO_4^{2-}(aq) + 4\,H^+(aq) \longrightarrow 2\,MnO_4^-(aq) + MnO_2(s) + 2\,H_2O(l)$$

Box 9.1 The preparation of dichromate

To prepare the Cr(VI) oxoanion, chromite ore, $FeCr_2O_4$, which contains Fe(II) and Cr(III), is dissolved in molten potassium hydroxide and oxidized with atmospheric oxygen:

$$FeCr_2O_4(s) + 6\,KOH(l) + \tfrac{5}{2}O_2(g) \longrightarrow$$
$$K_2FeO_4(melt) + 2\,K_2CrO_4(melt) + 3\,H_2O(g)$$

In this process, both the iron and chromium are converted to the +6 state, FeO_4^{2-} and CrO_4^{2-}, so chromium reaches its group oxidation state but iron does not. Dissolution in water followed by filtration leads to a solution of the two oxoanions. These ions can be separated by taking advantage of the greater oxidizing power of Fe(VI) relative to that of Cr(VI) in acidic solution. Acidification leads to the reduction of FeO_4^{2-} and the conversion of CrO_4^{2-} to dichromate $Cr_2O_7^{2-}$; the latter conversion is a simple acid–base reaction, and not a redox reaction:

$$2\,FeO_4^{2-}(aq) + 4\,H^+(aq) \longrightarrow$$
$$2\,Fe^{3+}(aq) + 3\,O_2(g) + 2\,H_2O(l) \qquad \text{redox}$$

$$2\,CrO_4^{2-}(aq) + 2\,H^+(aq) \longrightarrow$$
$$O_3CrOCrO_3^{2-}(aq) + H_2O(l) \qquad \text{acid–base}$$

The group oxidation number of iron (+8) has not been achieved in any solvent.

> The group oxidation state may be achieved in elements that lie toward the left of the d-block but not by elements on the right.

(b) Oxidation states down a group

In Groups 4 through 10, the highest oxidation state of an element becomes more stable on descending a group, with the greatest change in stability occurring between the first two rows of the d-block. This trend is illustrated for Group 6 in Fig. 9.11. Note that the Mo(VI) and W(VI) species lie below Cr(VI) in $H_2Cr_2O_7$, indicating that the maximum oxidation state is more stable for molybdenum and tungsten. The relative positions of Cr(VI), Mo(VI), and W(VI) in the diagram also illustrate the remark about the change between the first and second rows of the d-block.

The increasing stability of high oxidation states for the heavier d-block metals can be seen in the formulas of their halides (Table 9.4), and the limiting formulas MnF_4, TcF_6, and ReF_7 show the greater ease of oxidizing the 4d and 5d metals. The hexafluorides of the heavier d-block elements have been prepared from Group 6 through Group 10 (as in PtF_6). In keeping with the stability of high oxidation states for the heavier metals, WF_6 is not a significant oxidizing agent. However, the oxidizing character of the hexafluorides increases to the right, and PtF_6 is so potent that it can oxidize O_2 to O_2^+:

$$O_2(g) + PtF_6(s) \longrightarrow [O_2][PtF_6](s)$$

The ability to achieve the highest oxidation state does not correlate with the ease of oxidation of the bulk metal to an intermediate oxidation state. For example, elemental iron is susceptible to oxidation by $H^+(aq)$ under standard conditions,

$$Fe(s) + 2\,H^+(aq) \longrightarrow Fe^{2+}(aq) + H_2(g) \qquad (E^\ominus = +0.44\ V)$$

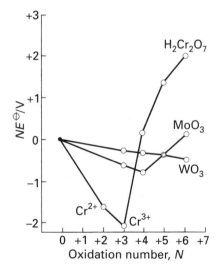

9.11 A Frost diagram for the chromium group in the d-block (Group 6) in acidic solution (pH = 0).

Table 9.4 Highest oxidation state d-block binary halides*

Group							
4	5	6	7	8	9	10	11
TiI_4	VF_5	CrF_5†	MnF_4	$FeBr_3$	CoF_3	NiF_4	$CuBr_2$
ZrI_4	NbI_5	MoF_6	$TcCl_6$	RuF_6	RhF_6	PdF_4	AgF_3
HfI_4	TaI_5	WBr_6	ReF_7	OsF_6	IrF_6	PtF_6	$AuCl_6$

*The formulas show the least electronegative halide that brings out the highest oxidation state.
†CrF_6 exists for several days at room temperature in a passivated Monel container.

However, as already noted, no oxidizing agent has been found that will take it to its group oxidation state in solution. The two heavier metals in Group 8 (ruthenium and osmium) are not oxidized by H^+ in acidic aqueous solution:

$$Os(s) + 2H_2O(l) \longrightarrow OsO_2(s) + 2H_2(g) \qquad (E^\ominus = -0.65\text{ V})$$

However, they can be oxidized by oxygen to the +8 state, as RuO_4 and OsO_4:

$$Os(s) + 2O_2(g) \longrightarrow OsO_4(s)$$

Ruthenium tetroxide and osmium tetroxide are low melting, highly volatile toxic molecular compounds that are used as selective oxidizing agents. For example, osmium tetroxide (as well as the permanganate ion) is used to oxidize alkenes to diols:

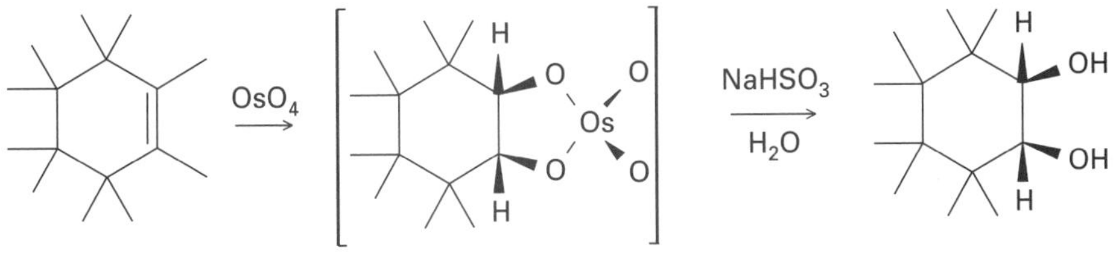

$$C_6H_{10} \qquad\qquad\qquad\qquad\qquad\qquad\qquad\qquad C_6H_{10}(OH)_2$$

The Frost diagrams for the positive oxidation states of molybdenum and (especially) tungsten are quite flat (Fig. 9.11). This flatness indicates that neither element exhibits the marked tendency to form **M(III)** that is so characteristic of chromium. Mononuclear complexes of molybdenum and tungsten with oxidation numbers +2, +3, +4, +5, or +6 are common. We shall see in more detail later that **M—M** bonded dinuclear or polynuclear complexes of molybdenum and tungsten with oxidation numbers +2 and +3 are numerous.

> In Groups 4 through 10, the highest oxidation state of an element becomes more stable
> on descending a group, with the largest change in stability occurring between the first two
> rows of the d-block.

(c) Structural trends down a group

As may be anticipated from considerations of atomic and ionic radii, the $4d$ and $5d$ elements often have higher coordination numbers than their $3d$ congeners. This trend is illustrated in Table 9.5 for the fluoro and cyano complexes of the early d-block metals. Note that with the small F^- ligand, the early $3d$ metals tend to form six-coordinate complexes but that the larger $4d$ and $5d$ metals in the same oxidation state tend to form seven- and nine-coordinate complexes. The octacyanomolybdate complex illustrates the tendency toward high coordination numbers with compact ligands. The same complex is readily reduced electrochemically or chemically:

$$[Mo(CN)_8]^{3-}(aq) + e^- \longrightarrow [Mo(CN)_8]^{4-}(aq) \qquad (E^\ominus = +0.73\text{ V})$$

> The 4d and 5d elements often have higher coordination numbers than their 3d congeners.

Table 9.5 Coordination numbers of some early d-block fluoro and cyano complexes

	Complex (coordination number) for group		
	3	4	5
$3d$	$[NH_4]_3[ScF_6]$ (6)	$Na_2[TiF_6]$ (6)	$K[VF_6]$ (6); $K_2[V(CN)_7]\cdot 2H_2O$ (7)
$4d$	$NaYF_9$ (9)	$Na_3[ZrF_7]$ (7)	$K_2[NbF_7]$ (7); $K_5[Nb(CN)_8]$ (8)
$5d$	$NaLaF_9$ (9)	$Na_3[HfF_7]$ (7)	$K_3[TaF_8]$ (8)

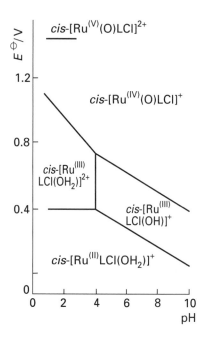

9.12 A Pourbaix diagram for *cis*-[RuLCl(OH₂)]²⁺ and related species. The ligand L is quadridentate, and its structure is shown in (6). (Adapted from C.-K. Li, W.-T. Tang, C.-M. Chi, K.-Y. Wong, R.-J. Wang, and T.C.W. Mak, *J. Chem. Soc., Dalton Trans.* 1909 (1991).)

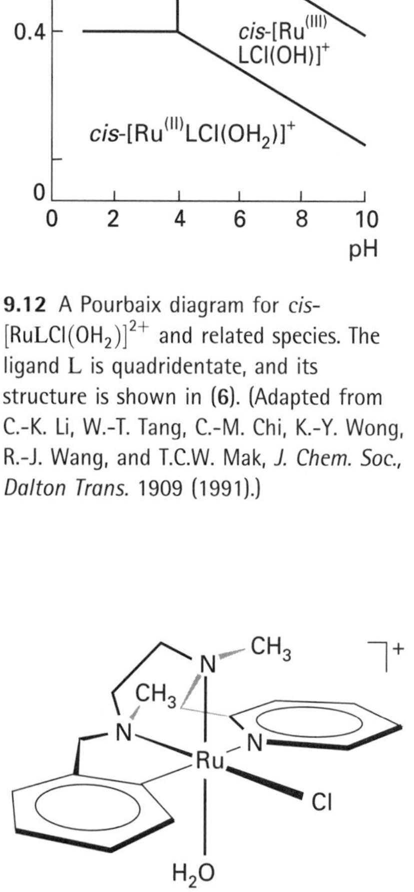

(d) Mononuclear oxo complexes

Because oxygen is readily available in an aqueous environment, in the atmosphere, and as the donor atom in many organic molecules, it is not surprising that oxygen-containing ligands play a major role in the chemistry of the metallic elements. This is particularly true of the chemically hard, high oxidation state metals on the left of the *d*-block. Our focus in this section is on the oxo ligand and its ability to bring out the high oxidation states of metals. Another important issue is the relation between oxo and aqua ligands through acid–base equilibria, and the relation of the oxo ligand to ligands that are isoelectronic with it.

Metals in high oxidation states typically occur as oxoanions in aqueous solution, such as permanganate, MnO_4^-, which contains manganese(VII), and chromate, CrO_4^{2-}, which contains chromium(VI). The existence of these oxoanions contrasts with the existence of simple aqua ions for the same metals in lower oxidation states, such as $[Mn(OH_2)_6]^{2+}$ for manganese(II) and $[Cr(OH_2)_6]^{2+}$ for chromium(II).

The formation of an oxo ligand rather than an aqua ligand is favored by high **pH**. The effect of **pH** is quite easy to rationalize: the OH^- ions in solution tend to remove protons from the aqua ligands. The occurrence of aqua complexes for low oxidation state metal cations can be rationalized by noting the relatively small electron-withdrawing effect that these cations exert on the O atoms of the water molecule. As a result, the aqua ligands are only weak proton donors. A metal ion in a high oxidation state depletes the electron density on the attached O atoms and thereby increases the Brønsted acidity of H_2O and OH^- ligands.

The combined influence of **pH** and oxidation state is conveniently summarized by Pourbaix diagrams (Section 6.10). The example shown in Fig. 9.12 involves a complex containing a stabilizing tetradentate ligand, L, which results in a similar coordination geometry for a wide range of conditions. The aqua complex *cis*-[RuLCl(OH₂)]²⁺ (**6**) in solution at pH = 2 is stable up to +0.40 V. Just above this potential, simple oxidation (that is, electron removal) occurs to give *cis*-[RuLCl(OH₂)]³⁺. Under even more oxidizing conditions (at +0.95 V) and pH = 2, both oxidation and deprotonation occur and result in the formation of an Ru(IV) oxo species, *cis*-[RuLCl(O)]⁺. At an even higher potential (about +1.4 V), further oxidation yields *cis*-[RuLCl(O)]²⁺. As we have remarked, the influence of more basic conditions is to deprotonate an H_2O ligand and thus to favor hydroxo or oxo complexes. For example, at pH = 8, the Ru(III) species is deprotonated to the hydroxo complex *cis*-[RuLCl(OH)]⁺, which in turn is converted to the oxo-Ru(IV) complex at a lower potential than for the Ru(III)–Ru(IV) transformation at pH = 2.

A list of simple complexes containing oxo ligands is given in Table 9.6. Many complexes are known that contain the vanadyl moiety,[1] VO^{2+}, with vanadium in its penultimate oxidation state (+4). Vanadyl complexes generally contain four additional ligands and are square-pyramidal (**7**). Many of these d^1 complexes are blue (as a result of a *d–d* transition) and may take on a weakly bound sixth ligand *trans* to the oxo ligand. The vanadyl V—O bond length (1.58 Å) in [VO(acac)₂] is short compared with the four V—O bond lengths to the acac ligand (1.97 Å). The short bond distances in vanadyl complexes, together with high VO stretching

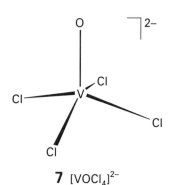

6 *cis*-[RuLCl(OH₂)]⁺

7 [VOCl₄]²⁻

Table 9.6 Some common monoxo and dioxo complexes

Group	Element	Structure	Formula
5	V(IV),d^1	Square-pyramidal	$[V(O)(acac)_2]$, $[V(O)Cl_4]^{2-}$
	V(V),d^0	*cis*-Octahedral	$[V(O)_2(OH_2)_4]^+$
6	Mo(VI),d^0; W(VI),d^0	Tetrahedral	$[M(O)_2(Cl)_2]$
7, 8	Re(V),d^2; Os(VI),d^2	*trans*-Octahedral	$[Re(O)_2(py)_4]^+$, $[Re(O)_2(CN)_4]^{3-}$, $[Os(O)_2Cl_4]^{2-}$

---

[1] Although the term moiety strictly means one-half of an entity, it is widely used in the looser sense of one of the parts into which an entity may be divided.

wavenumbers (940–980 cm^{-1}) provide strong evidence for VO multiple bonding where lone pairs on the oxygen ligand are donated to the central vanadium. The d–p π-orbital overlap involved in a multiple bond is illustrated in Fig. 9.13; $4p$–p overlap may also make a significant contribution. This strong multiple bonding with the oxygen appears to be responsible for the *trans* influence of the oxo ligand, which disfavors attachment of the ligand *trans* to O.[2]

Vanadium in its highest oxidation state forms an extensive series of oxo compounds, many of which are the polyoxo species discussed later. The simplest oxo complex, $[V(O)_2(OH_2)_4]^+$, exists in the acidic solution formed when the sparingly soluble vanadium pentoxide, V_2O_5, dissolves in water. This pale yellow complex has a *cis* geometry (**8**). Here again we see the *trans* influence of an oxo ligand. A *cis* geometry minimizes competition between the oxo ligands for π bonding to the vanadium atom. As shown in Fig. 9.14, the two O^{2-} ligands in the *trans* configuration can π bond with two d orbitals; in the *cis* configuration the metal atom has only one d orbital in common with the π orbitals on the O atoms.

In contrast with the *cis* structure of the d^0 complex $[V(O)_2(OH_2)_4]^+$, many *trans*-dioxo complexes are known for the d^2 metal centers Re(V) and Os(VI) (**9**); some examples are given in Table 9.6. This geometry is probably favored because the *trans* configuration leaves a vacant low-energy orbital that can be occupied by the two d electrons. According to this explanation, the avoidance of the destabilizing effect of the d^2 electrons more than offsets the disadvantage of the ligands having to compete for the same d orbital in the *trans*-oxo geometry.

> The conversion of an aqua ligand to an oxo ligand is favored by a high *pH* and by a high oxidation state of the central metal atom. In vanadium complexes with the oxidation state +4 or +5, the site *trans* to the oxo ligand may be vacant or occupied by a weak ligand.

(e) Nitrido and alkylidyne complexes

The ligands N^{3-} and RC_2^- are isolobal with O^{2-}, and like O^{2-} they form strong bonds to the metal and weaken the *trans* metal–ligand bond. The highly reactive compound

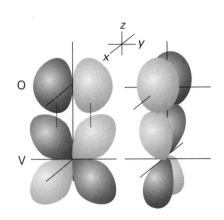

9.13 The d–p π-bonding between oxygen and vanadium in the vanadyl moiety, VO^{2+}. It is conventional to think of the nonmetal oxygen as having oxidation number -2 and vanadium as having oxidation number $+4$. From this point of view, both electron pairs for the two π bonds are donated from the oxide ligand to the metal d_{zx} and d_{yz} orbitals.

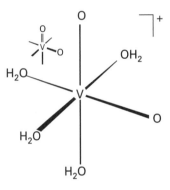

8 $[V(O)_2(OH_2)_4]^+$

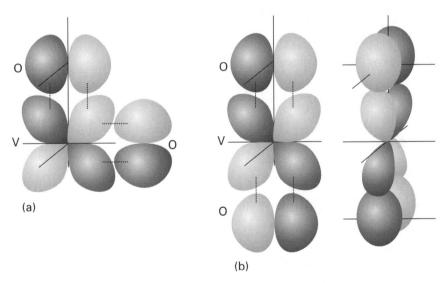

(a)

(b)

9.14 Comparison of the competition between *cis* and *trans* oxo ligands attached to vanadium. (a) In the *cis* configuration the metal atom has only one d orbital in common with the π orbitals on the O atoms. (b) In the *trans* configuration the metal atom has two d orbitals in common with the π orbitals on the two O atoms.

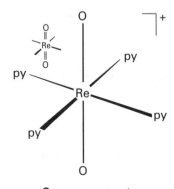

9 $[Re(O)_2(py)_4]^+$

----

2 The *trans* influence of the oxo ligand has been reviewed by E.M. Shustrovitch and M.A. Pori-Koshits, *Coord. Chem. Rev.* **17**, 1 (1975). For a concise summary of the effect and theoretical justification see I.B. Bersuker, *Electronic structure and properties of transition metal compounds*, p. 473. Wiley, New York (1996).

10 [M(N)X$_4$]$^-$

11 [Ta$_2$(N)Br$_8$]$^{3-}$

(Me$_3$CO)$_3$W≡W(OCMe$_3$)$_3$ cleaves the C≡N bond in phenylcyanide to give both a nitrido and an alkylidyne ligand:

$$(Me_3CO)_3W\equiv W(OCMe_3)_3 + PhC\equiv N \longrightarrow (Me_3CO)_3W\equiv CPh + (Me_3CO)_3W\equiv N$$

The remarkable feature of this reaction is that the C≡N bond is broken despite its considerable strength (890 kJ mol^{-1}), so the W≡C and W≡N bond enthalpies must be substantial.[3]

Both nitrido and alkylidyne complexes are known to have short M—N and M—C bonds, which suggests the existence of multiple metal–ligand bonds (as with their oxo counterparts). Another analogous feature with the oxo ligand is that nitrido and alkylidyne ligands often weaken the *trans* metal–ligand bond. The order of this influence is RC≡ > N≡ > O≡; this weakening is attributed to competition for back π donation into a common set of metal *d* orbitals.

Many nitrido complexes are known for the elements in Groups 5 through 8. They are most numerous for molybdenum and tungsten in Group 6, rhenium in Group 7, and ruthenium and osmium in Group 8. Square-pyramidal nitrido complexes of formula [MNX$_4$]$^-$ are known for M(VI), with M = Mo, Re, Ru, and Os, and X = F, Cl, Br, and (in some cases) I. These square-pyramidal structures (**10**) have short M≡N bonds (1.57 Å to 1.66 Å). In a few cases, a ligand is attached in the sixth site, but the resulting bond is long and weak, as with the oxo complexes discussed earlier. Examples of M≡N≡M species are known, such as [Ta$_2$NBr$_8$]$^{3-}$ (**11**), but nitrogen analogs of the polyoxometallates described below are as yet unknown.

> *Multiple M≡L bonds are common with high oxidation state metals of the early d-metal series, and these bonds weaken the bonds to the trans ligands.*

(f) Polyoxometallates

A **polyoxometallate** is an oxoanion containing more than one metal atom. The H$_2$O ligand created by protonation of an oxo ligand at low pH can be eliminated from the central metal atom and thus lead to the condensation of mononuclear oxometallates. A familiar example is the reaction of a basic chromate solution, which is yellow, with excess acid to form the oxo-bridged dichromate ion, which is orange:

$$2\,CrO_4^{2-}(aq) + 2\,H^+(aq) \longrightarrow Cr_2O_7^{2-}(aq) + H_2O(l)$$

In highly acidic solution, longer chain oxo-bridged Cr(VI) species are formed. The tendency for Cr(VI) to form polyoxo species is limited by the fact that the O tetrahedra link only through vertices: edge and face bridging would result in too close an approach of the metal centers. In contrast, it is found that five- and six-coordinate metal oxo complexes, which are common with 4*d* and 5*d* metals, can share oxo ligands between either vertices or edges. These structural possibilities lead to a richer variety of polyoxometallates than found with the 3*d* metals.

Chromium's neighbors in Groups 5 and 6 form six-coordinate polyoxo complexes (Fig. 9.15). In Group 5 the polyoxometallates are most numerous for vanadium, which

4	5	6	7
Ti	V$^{(IV,V)}$	Cr$^{(VI)}$	Mn
Zr	Nb$^{(V)}$	Mo$^{(VI)}$	Tc
Hf	Ta$^{(V)}$	W$^{(VI)}$	Re

9.15 Elements in the *d*-block that form polyoxometallates. The shaded elements form the greater variety of polyoxometallates.

---

[3] Metathesis of tungsten–tungsten triple bonds with acetylenes and nitriles to give alkylidyne and nitrido complexes. R.R. Schrock, M.L. Listmann and L.G. Sturgeoff, *J. Am. Chem. Soc.* **104**, 4291 (1982).

forms many V(V) complexes and a few V(IV) or mixed oxidation state V(IV)–V(V) polyoxo complexes. In short, polyoxometallate formation is most pronounced in Groups 5 and 6 for vanadium(V), molybdenum(VI), and tungsten(VI).

It is often convenient to represent the structures of the polyoxometallate ions by polyhedra, with the metal atom understood to be in the center and O atoms at the vertices. For example, the sharing of O atom vertices in the dichromate ion, $Cr_2O_7^{2-}$, may be depicted in either the traditional way (**12**) or in the polyhedral representation (**13**). Similarly, the important M_6O_{19} structure of $[Nb_6O_{19}]^{8-}$, $[Ta_6O_{19}]^{8-}$, $[Mo_6O_{19}]^{2-}$, and $[W_6O_{19}]^{2-}$, is depicted by the conventional or polyhedral structures shown in Fig. 9.16. The structures for this series of polyoxometallates contain terminal O atoms (those projecting outward from a single metal) and two types of bridging O atoms: two-metal bridges, M—O—M, and one 'hypercoordinated' O atom in the center of the structure that is common to all six metal atoms. The structure consists of six MO_6 octahedra, each sharing an edge with four neighbors. The overall symmetry of the M_6O_{19} array is O_h. Another example of a polyoxometallate is $[W_{12}O_{40}(OH)_2]^{10-}$ (**14**). As shown by this formula, the polyoxoanion is protonated. Proton transfer equilibria are common for the polyoxometallates, and may occur together with condensation and fragmentation reactions.[4]

Polyoxometallate anions can be prepared by carefully adjusting pH and concentrations.[5] For example, polyoxomolybdates and polyoxotungstates are formed by acidification of solutions of the simple molybdate or tungstate:

$$6\,[MoO_4]^{2-}(aq) + 10\,H^+(aq) \rightleftharpoons [Mo_6O_{19}]^{2-}(aq) + 5\,H_2O(l)$$

$$8\,[MoO_4]^{2-}(aq) + 12\,H^+(aq) \rightleftharpoons [Mo_8O_{26}]^{4-}(aq) + 6\,H_2O(l)$$

In addition to the large number of polyoxomolybdates and polyoxotungstates, there is a large class of **heteropolyoxometallates**, such as the molybdates and tungstates, that also incorporate phosphorus, arsenic, and other heteroatoms. For example, $[PMo_{12}O_{40}]^{3-}$ contains a PO_4^{3-} tetrahedron that shares O atoms with surrounding octahedral MoO_6 groups (**15**). Many different heteroatoms can be incorporated into this structure, and the general formulation is $[X(n+)Mo_{12}O_{40}]^{(8-n)-}$ where $(n+)$ represents the oxidation state of the heteroatom, which may be As(V), Si(IV), Ge(IV), or Ti(IV). An even broader range of heteroatoms is observed with the analogous tungsten heteropolyoxoanion. Heteropolyoxomolybdates and

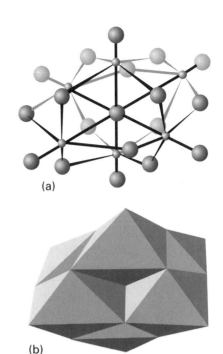

(a)

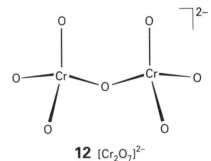

(b)

9.16 (a) Conventional representation and (b) polyhedral representation of the six edge-shared octahedra as found in $[M_6O_{19}]^{2-}$.

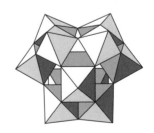

12 $[Cr_2O_7]^{2-}$

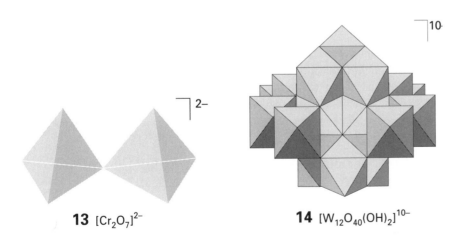

13 $[Cr_2O_7]^{2-}$ **14** $[W_{12}O_{40}(OH)_2]^{10-}$ **15** $[PMo_{12}O_{40}]^{3-}$

4   V.W. Day and W.G. Klemperer, Metal oxide chemistry in solution: the early transition metal polyoxyanions. *Science* **228**, 533 (1985).

5   For excellent summaries of this area, see M.T. Pope, *Heteropoly and isopoly oxometallates.* Springer, Berlin (1983), and *Comprehensive coordination chemistry*, Vol. 3. Pergamon Press, Oxford (1987), or M.T. Pope, Polyoxoanions. In *Encyclopedia of inorganic chemistry* (ed. R.B. King), p. 3361. Wiley, New York (1994).

tungstates can undergo one-electron reduction with no change in structure but with the formation of a deep blue color. The color appears to arise from the excitation of the added electron from Mo(V) or W(V) to an adjacent Mo(VI) or W(VI) site.

| In their highest oxidation states, metals in Groups 5 and 6 readily form polyoxometallate and heteropolyoxometallate species.

9.8 Intermediate oxidation states

Most +1 d-block metal cations (M^+) disproportionate (to M and M^{2+}) because the bonds in the solid metal are so strong. However, the more noble metals, copper, silver, and gold, form many salts containing M^+; for silver, Ag^+ is the most important oxidation state. For other d metals, the +2 oxidation state is generally the lowest one to consider in aqueous solution and in combination with hard ligands. Exceptions to this behavior are mainly confined to metal–metal bonded and organometallic compounds, such as the metal carbonyls, $Ni(CO)_4$ and $Mo(CO)_6$, discussed in Chapter 16, in which the metal has oxidation number 0.

(a) Oxidation state +2 of the 3d metals

The dipositive aqua ions, $M^{2+}(aq)$ (specifically, the octahedral complexes $[M(OH_2)_6]^{2+}$), play an important role in the chemistry of the 3d metals. Many of these ions are colored as a result of d–d transitions in the visible region of the spectrum. For example, $Mn^{2+}(aq)$ is pale pink, $Fe^{2+}(aq)$ is pale green, $Co^{2+}(aq)$ is pink, $Ni^{2+}(aq)$ is green, and $Cu^{2+}(aq)$ is blue.

The +2 oxidation state becomes increasingly common from left to right across the period. For example, among the early members of this series, $Sc^{2+}(aq)$ (Group 3) is unknown and $Ti^{2+}(aq)$ (Group 4) is not readily accessible. For Groups 5 and 6, $V^{2+}(aq)$ and $Cr^{2+}(aq)$ are thermodynamically unstable with respect to oxidation by H^+:

$$2\,V^{2+}(aq) + 2\,H^+(aq) \longrightarrow 2\,V^{3+}(aq) + H_2(g) \qquad (E^{\ominus} = +0.26\ V)$$

However, the slowness of H_2 evolution makes it possible to work with solutions of these dipositive ions in the absence of air, and as a result they are useful reducing agents. Beyond chromium (for Mn^{2+}, Fe^{2+}, Co^{2+}, Ni^{2+}, and Cu^{2+}), the +2 state is stable with respect to reaction with water, and only Fe^{2+} is oxidized by air. The only stable aqua ions of copper and nickel are M^{2+}. The same is nearly true for cobalt, because $Co^{3+}(aq)$ is reduced to $Co^{2+}(aq)$ by water:

$$4\,Co^{3+}(aq) + 2\,H_2O(l) \longrightarrow 4\,Co^{2+}(aq) + O_2(g) + 4\,H^+(aq) \quad (E^{\ominus} = +0.69\ V)$$

The trend toward increased stability of the lower oxidation states in the middle and right of the d-block can be understood from the general increase in ionization energies from left to right across a period in the d-block (Section 1.8).

Water is an unsafe environment for many metal ions, as it can serve as an oxidizing agent. A wider range of M^{2+} ions is therefore known in solids than in aqueous solution. For example, $TiCl_2$ can be prepared by the reduction of $TiCl_4$ with hexamethyldisilane:

$$(CH_3)_3SiSi(CH_3)_3(l) + TiCl_4(l) \longrightarrow TiCl_2(s) + 2\,(CH_3)_3SiCl(l)$$

Table 9.7 Structures of dihalides*

	Ti	V	Cr	Mn	Fe	Co	Ni	Cu
F	R	R†	R†	R	R	R	R†	
Cl	L	L	R†	L	L	L	L	L
Br	L	L	L	L	L	L	L	L
I	L	L	L	L	L	L	L	L

*R = rutile, L = layered (CdI_2, $CdCl_2$ or related structures).
†The structure is distorted from the ideal type.
Source: Adapted from A.F. Wells, Structural inorganic chemistry. Oxford University Press (1984).

but it is oxidized by both water and air. With the exception of $ScCl_2$, these dihalides are well described by structures containing isolated M^{2+} ions (Table 9.7). As we see from the table, the fluorides have a rutile structure (Section 2.9) and most of the heavier halides are layered; in both cases the metal is in an octahedral site. This shift in structure is understandable because the rutile structure is associated with ionic bonding and the layered structures are associated with more covalent bonding. The metal–metal bonded $ScCl_2$ and related M—M bonded dihalides of the $4d$ and $5d$ elements are discussed in Section 9.9.

The dipositive ions of the $3d$ metals form four-coordinate complexes with halide ions, such as the brown–yellow complex $[NiBr_4]^{2-}$. The absorption intensities of these tetrahedral complexes are greater than those of the simple octahedral hexaaqua complexes because the lack of a center of symmetry in the tetrahedral array means that the $d-d$ transition is allowed. The structures of $[CuX_4]^{2-}$ complexes are distorted because the tetrahedral d^9 configuration is orbitally degenerate and therefore susceptible to Jahn–Teller distortion (Section 7.5). The magnetic moments of the $3d$ tetrahalo complexes demonstrate that they are all high-spin, which is in line with the small ligand-field splitting in tetrahedral complexes.

Monoxides are known for many of the $3d$ metals (Table 9.8). They have the rock-salt structure characteristic of ionic solids, but their properties, which will be discussed in more detail in Chapter 18, indicate significant deviations from the simple ionic $M^{2+}O^{2-}$ model. For example, TiO has metallic conductivity and FeO is always deficient in iron. The early d-block monoxides are strong reducing agents. Thus, TiO is easily oxidized by water or oxygen, and MnO is a convenient oxygen scavenger, which is used in the laboratory to remove oxygen impurity in inert gases down to the parts-per-billion range.

> Oxidation state $+2$ is common for the $3d$ metals from the middle to the right of the block. Toward the left, the $3d$ series M^{2+} aqua ions are increasingly susceptible to oxidation by air and by water.

Table 9.8 d-block MO compounds

	Group						
	4	5	6	7	8	9	10
Rock-salt structure (shaded)	Ti Zr	V		Mn	Fe	Co	Ni Pd*

*PtS structure (square-planar, four-coordinate metal atom).
Source: A.F. Wells, *Structural inorganic chemistry*, p. 537. Oxford University Press (1984).

Example 9.2 Judging trends in redox stability in the d-block

On the basis of trends in the $3d$ series, suggest possible M^{2+} aqua ions for use as reducing agents, and write a balanced chemical equation for the reaction of one of these ions with O_2 in acidic solution.

Answer Because oxidation state $+2$ is most stable for the late $3d$ elements, strong reducing agents include ions of the metals on the left of the $3d$ series: such ions include $Ti^{2+}(aq)$, $V^{2+}(aq)$, and $Cr^{2+}(aq)$. The ion $Fe^{2+}(aq)$ is only weakly reducing. The ions $Co^{2+}(aq)$, $Ni^{2+}(aq)$, and $Cu^{2+}(aq)$ are not oxidizable in water. The Latimer diagram for iron indicates that Fe^{3+} is the only accessible higher oxidation state of iron in acid solution:

$$Fe^{3+} \xrightarrow{+0.77} Fe^{2+} \xrightarrow{-0.44} Fe$$

The chemical equation is then

$$4\,Fe^{2+}(aq) + O_2(g) + 4\,H^+(aq) \longrightarrow 4\,Fe^{3+}(aq) + 2\,H_2O(l)$$

16 [Ru(OH₂)(NH₃)₅]²⁺

17 [RuCl₂(PPh₃)₃]

18 [Re₂Cl₈]²⁻, D_{4h}

Self-test 9.2 Refer to the appropriate Latimer diagram in Appendix 2 and identify the oxidation state and formula of the species that is thermodynamically favored when an acidic aqueous solution of V^{2+} is exposed to oxygen.

(b) Oxidation states +2 of the 4d and 5d elements

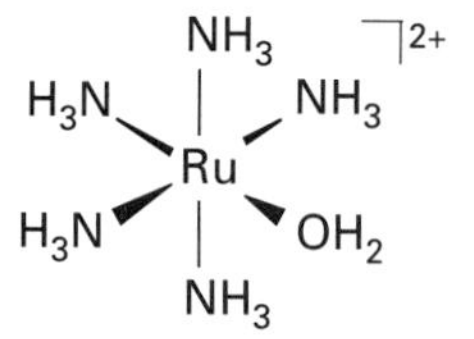

In contrast to the 3d metals, the 4d and 5d metals only rarely form simple $M^{2+}(aq)$ ions. A few examples have been characterized, including $[Ru(OH_2)_6]^{2+}$, $[Pd(OH_2)_4]^{2+}$, and $[Pt(OH_2)_4]^{2+}$. However, the 4d and 5d metals do form many M(II) complexes with ligands other than H_2O; they include the very stable d^6 octahedral complexes, such as (**16**), and the much rarer square-pyramidal d^6 complexes, such as (**17**), that form with bulky ligands.[6] Palladium(II) and platinum(II) form many square-planar d^8 complexes, such as $[PtCl_4]^{2-}$, and they will be discussed in Section 9.10. The Ru(II) complex in (**16**) is obtained by reducing $RuCl_3 \cdot 3H_2O$ with zinc in the presence of ammonia; it is a useful starting material for the synthesis of a range of ruthenium(II) pentaammine complexes that have π-acceptor ligands, such as CO, as the sixth ligand:

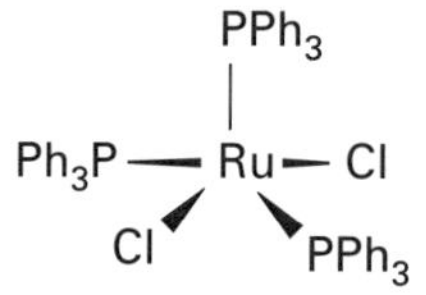

$$[Ru(NH_3)_5(OH_2)]^{2+}(aq) + L(aq) \longrightarrow [Ru(NH_3)_5L]^{2+}(aq) + H_2O(l)$$

$$(L = CO, N_2, N_2O)$$

These ruthenium and the related osmium pentaammine species are strong π-donors, and consequently the complexes they form with the π-acceptors CO and N_2 are stable. The combination of an ammine ligand and a 4d or 5d metal ion results in a strong ligand field, so the configuration is t_{2g}^6. The electrons in two of the t_{2g} orbitals have π symmetry with respect to the M—CO bond and so can back-donate into that π-acceptor ligand.[7]

> M(II) *complexes with σ-donor ligands are common for the 3d metals but* M(II) *complexes of 4d and 5d metals are less common; they generally contain π-acceptor ligands.*

9.9 Metal–metal bonded d-metal compounds

A general feature of the low oxidation state metals from the early part of the d-block is their formation of metal–metal bonds (Fig. 9.4).[8] These clusters are stabilized by π-donor ligands, such as halides and alkoxides. They may involve M—M bonds within a discrete molecular cluster or in extended solid state compounds. When no bridging ligands are present in a cluster, as in $[Re_2Cl_8]^{2-}$ (**18**), the presence of a metal–metal bond is unambiguous. Although this compound may be prepared by the reduction of ReO_4^- by a conventional reducing agent (such as zinc and a dilute acid), it is best prepared by the reduction of ReO_4^- with benzyl chloride.[9] When bridging ligands are present, careful observations and measurement (typically of bond lengths and magnetic properties) are needed to identify direct metal–metal bonding. We shall see in Chapter 16 that there is an extensive range of organometallic metal cluster compounds of the middle to late d-block elements that are stabilized by π-acceptor ligands, particularly CO.

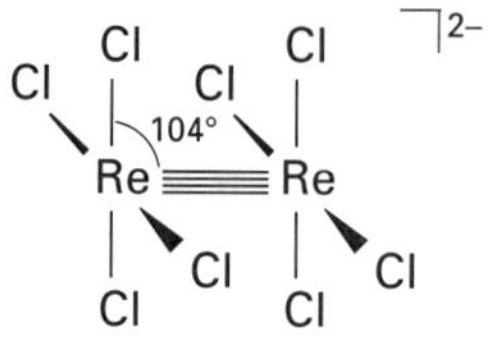

The bonding patterns in most metal cluster compounds are so intricate that metal–metal bond strengths cannot be determined with great precision. Some evidence, however, such as the stability of compounds and the magnitudes of M—M force constants, indicates that

[6] P.R. Hoffman and K.G. Caulton, *J. Am. Chem. Soc.* **97**, 4221 (1975).

[7] The t_{2g} designation applies strictly to an O_h complex, but the terminology is used here to facilitate the discussion. The precise orbital designations for the C_{4v} group of the complex are b_2 and e, and the configuration is $b_2^2 e^4$. The e electrons form the π bond.

[8] F.A. Cotton and R.A. Walton, *Multiple bonds between metal atoms.* Oxford University Press, New York (1983); M.H. Chisholm, The $\sigma^2\pi^4$ triple bond between molybdenum and tungsten atoms: developing the chemistry of an inorganic functional group. *Angew. Chem., Intl. Edn. Engl.* **25**, 21 (1986); M.H. Chisholm (ed.) *Early transition metal clusters with π-donor ligands.* VCH, Weinheim (1995).

[9] T.J. Border and R.A. Walton, *Inorg. Synth.* **23**, 116 (1985).

there is an increase in **M**—**M** bond strength down a group, perhaps on account of the greater spatial extension of *d* orbitals in heavier atoms. This trend may be the reason why there are so many more metal–metal bonded compounds for the 4*d* and 5*d* metals than for their 3*d* counterparts. Figure 9.2 shows that for the bulk metals, the metal–metal bonds in the *d*-block are strongest in the 4*d* and 5*d* series, and this feature carries over into their compounds. In contrast, element–element bonds weaken down a group in the *p*-block.

Not all the early *d*-block metal–metal bonded compounds are discrete clusters. Many extended metal–metal bonded compounds exist, such as the multiple chain scandium subhalides, Sc_7Cl_{12} and Sc_5Cl_6 (Fig. 9.17) and layered compounds such as ZrCl (Fig. 9.18). In the latter compound, the Zr atoms are within bonding distance in two adjacent layers of metal atoms sandwiched between Cl^- layers. We have already seen that zirconium and scandium adopt their group oxidation states (+3 and +4, respectively) when exposed to air and moisture, so these metal–metal bonded compounds are prepared out of contact with air and moisture. For example, ZrCl is prepared by reducing zirconium tetrachloride with zirconium metal in a sealed tantalum tube at high temperatures:

$$3\,Zr(s) + ZrCl_4(g) \xrightarrow{600-800°C} 4\,ZrCl(s)$$

Discrete clusters—as distinct from the extended **M**—**M** bonded solid state compounds just described—are often soluble, and can be manipulated in solution. The π-donor ligands in these clusters typically occur in one of several locations, namely, in a terminal position (**19**), bridging two metals (**20**), or bridging three metals (**21**). Clusters may be linked by bridging halides or chalcogenides in the solid state. For example, the solid compound of formula $MoCl_2$ consists of octahedral Mo clusters linked by Cl bridges. It may be prepared in a sealed glass tube by the reaction of $MoCl_5$ in a mixture of molten $NaAlCl_4$ and $AlCl_3$, together with aluminum metal as the reducing agent:

$$MoCl_5(s) + Al(s) \xrightarrow{NaAlCl_4/AlCl_3(l),\ 200°C} MoCl_2(s) + AlCl_3(l)$$

The compound can withstand oxidation under mild conditions. When treated with hydrochloric acid it forms the anionic cluster $[Mo_6Cl_{14}]^{2-}$. This cluster contains an octahedral array of Mo atoms with a Cl atom bridging each triangular face and a terminal Cl atom on each Mo vertex (**22**). These terminal Cl atoms can be replaced by other halogens, alkoxides, and phosphines. An analogous series of tungsten cluster compounds is known. Once formed, these molybdenum and tungsten compounds can be handled in air and water at room temperature on account of their kinetic barriers to decomposition. The oxidation number of the metal is +2, so the metal valence electron count is $(6-2) \times 6 = 24$. One-electron oxidation and reduction of the clusters is possible.

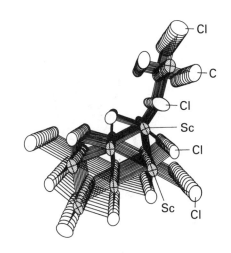

9.17 The structure of Sc_7Cl_{10} showing the single chain of Sc atoms at the top and multiple chains at the bottom. (From J.D. Corbett, *Acc. Chem. Res.* **14**, 239 (1981).)

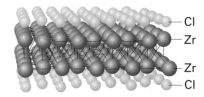

9.18 The structure of ZrCl consists of layers of metal atoms in graphite-like hexagonal nets.

Cl
|
M

19

Cl
/ \
M — M

20

Cl
/|\
M — M
  M

21

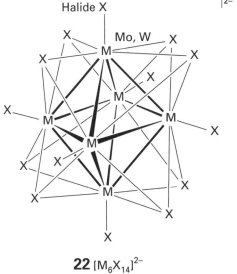

22 $[M_6X_{14}]^{2-}$

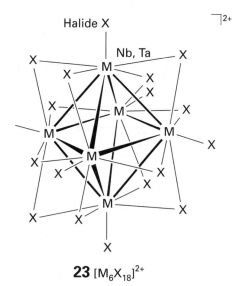

23 $[M_6X_{18}]^{2+}$

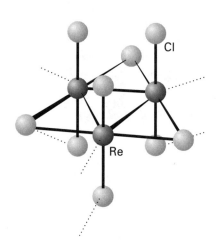

9.19 The structure of $ReCl_3$ in the solid state. The dotted bonds correspond to interaction with a Cl atom belonging to an adjacent cluster.

Similar octahedral clusters are known for niobium and tantalum in Group 5, and for zirconium in Group 4. The cluster $[Nb_6Cl_{12}L_6]^{2+}$ and its tantalum analog have an octahedral framework with edge-bridging Cl atoms and six terminal ligands (**23**). The metal valence electron count in this case is 16, but these clusters can be oxidized in several steps. One example from Group 4 is the cluster $[Zr_6Cl_{18}C]^{4-}$, which has the same metal and Cl atom array together with a C atom in the center of the octahedron.

Rhenium trichloride, which consists of Re_3Cl_9 clusters linked by weak halide bridges (Fig. 9.19), provides the starting material for the preparation of a series of three-metal clusters that have been studied thoroughly. As with molybdenum dichloride, the intercluster bridges in the solid state can be broken by reaction with potential ligands. For example, treatment of Re_3Cl_9 with Cl^- ions produces the discrete complex $[Re_3Cl_{12}]^{3-}$. Neutral ligands, such as alkylphosphines, can also occupy these coordination sites and result in clusters of the general formula $Re_3Cl_9L_3$.

Many compounds containing bonds between two different metals are known. Some of their common structural motifs are an ethane-like structure (**24**), an edge-shared bioctahedron (**25**), a face-shared bioctahedron (**26**), and a tetragonal prism, which we have already encountered with $[Re_2Cl_8]^{2-}$ (**18**). We shall focus on bonding in the last of these structural types, in which the bond order may be considered as lying between 1 and 4.

Figure 9.20 shows that a σ bond between two metals can arise from the overlap of a d_{z^2} orbital from each metal, π bonds can arise from the overlap of d_{zx} or d_{yz} orbitals (two such π bonds are possible), and a δ bond can be formed from the overlap of two face-to-face d_{xy} orbitals on two different metal atoms (the remaining $d_{x^2-y^2}$ orbital is used in the **M**—**L** σ bonds). A quadruple **M**—**M** bond results when all the bonding orbitals are occupied and the configuration is $\sigma^2\pi^4\delta^2$ (Fig. 9.21). Evidence for quadruple bonding comes from the observation that $[Re_2Cl_8]^{2-}$ has an eclipsed array of Cl ligands, which is sterically unfavorable. It is argued that the δ bond, which is formed only when the d_{xy} orbitals are confacial, locks the complex in the eclipsed configuration. Another well known quadruply bonded compound is molybdenum(II) acetate (**27**), which is prepared by heating the molybdenum(0) compound $Mo(CO)_6$ with acetic acid:

$$2\,Mo(CO)_6 + 4\,CH_3COOH \longrightarrow Mo_2(O_2CCH_3)_4 + 4\,H_2 + 12\,CO$$

The quadruply bonded molybdenum acetato complex is an excellent starting material for other Mo—Mo compounds. For example, the quadruply bonded chloro complex is obtained when the acetato complex is treated with concentrated hydrochloric acid at below room temperature:

$$Mo_2(OCCH_3)_4 + 4\,H^+(aq) + 8\,Cl^-(aq) \longrightarrow [Mo_2Cl_8]^{4-}(aq) + 4\,CH_3COOH(aq)$$

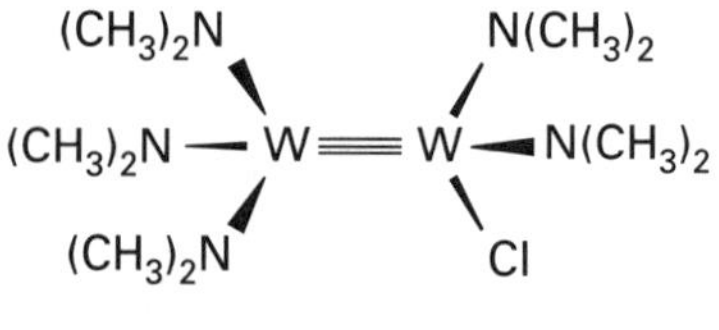

24 $[((CH_3)_2N)_3WWCl(N(CH_3)_2)_2]$

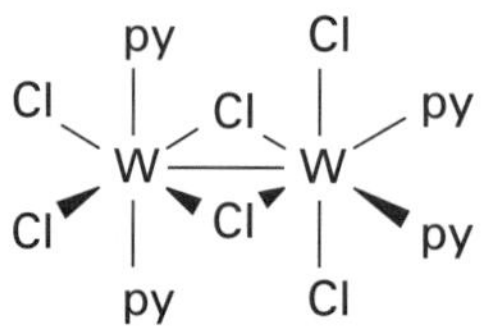

25 $[W_2(py)_4Cl_6]$

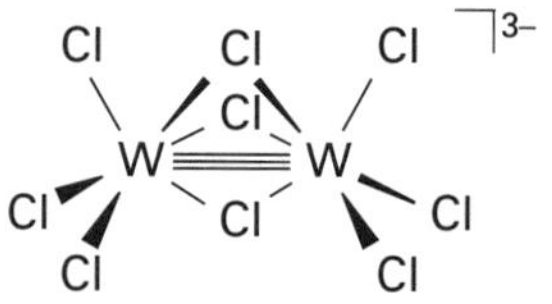

26 $[W_2Cl_9]^{3-}$

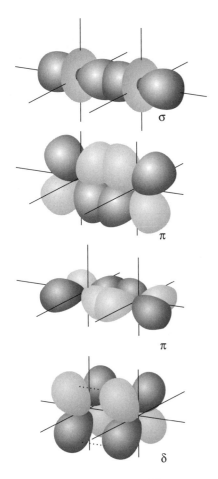

9.20 The origin of σ, π, and δ interactions between two *d*-block metal atoms situated along the *z*-axis. Only the bonding combinations are shown.

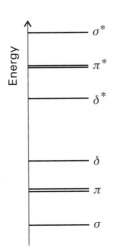

9.21 Approximate molecular orbital energy level scheme for the M—M interactions in tetragonal-prismatic two-metal clusters.

27 [Mo$_2$(CH$_3$CO$_2$)$_4$]

As shown in Table 9.9, triply bonded **M—M** systems arise for tetragonal-prismatic complexes when δ and δ^* orbitals are both occupied. These complexes are more numerous than the quadruply bonded complexes and, because δ bonds are weak, **M≡M** bond lengths are often similar to those of quadruply bonded systems. Many triply bonded systems also include bridging ligands (Table 9.9). The table shows that singly occupied δ or δ^* orbitals lead to a formal bond order of $3\frac{1}{2}$. Once the δ and δ^* orbitals are fully occupied, successive occupation of the two higher lying π^* orbitals leads to further reduction in the bond order from $2\frac{1}{2}$ to 1.

Like carbon–carbon multiple bonds, metal–metal multiple bonds are centers of reaction. However, the variety of structures resulting from the reactions of metal–metal multiply bonded compounds is more diverse than for organic compounds.[10] For example,

$$Cp(OC)_2Mo \equiv Mo(CO)_2Cp \quad + \quad HI \quad \longrightarrow \quad Cp(OC)_2Mo \overset{H}{\underset{I}{\diamond}} Mo(CO)_2Cp$$

[10] For several reviews see M.H. Chisholm (ed.), *Reactivity of metal–metal bonds*, ACS Symposium Series 155. American Chemical Society, Washington DC (1981). R.E. McCarley, T.R. Ryan, and C.M.C. Torardi, p. 41; R.A. Walton, p. 207, M.D. Curtis *et al*., p. 221, and A.F. Dyke *et al*., p. 259.

where Cp denotes the cyclopentadienyl group, C_5H_5. In this reaction, HI adds across a triple bond but both the H and I bridge the metal atoms; the outcome is quite unlike the addition of HX to an alkyne, which results in a substituted alkene. The reaction product can be regarded as containing a 3c,2e M—H—M bridge and an I atom bonded by two conventional 2c,2e bonds, one to each Mo atom.

Table 9.9 Examples of metal–metal bonded tetragonal-prismatic complexes

Complex	Configuration	Bond order	M—M bond length/Å
(structure, $4-$)	$\sigma^2\pi^4\delta^2$	4	2.11
(structure, $3-$)	$\sigma^2\pi^4\delta^1$	3.5	2.17
(structure, $2-$)	$\sigma^2\pi^4$	3	2.22
(structure, $-$)	$\sigma^2\pi^4\delta^2\delta^*\pi^{*2}$	2.5	2.27
(structure)	$\sigma^2\pi^4\delta^2\delta^{*2}\pi^{*2}$	2	2.26

Table 9.9 continued

Complex	Configuration	Bond order	M—M bond length/Å
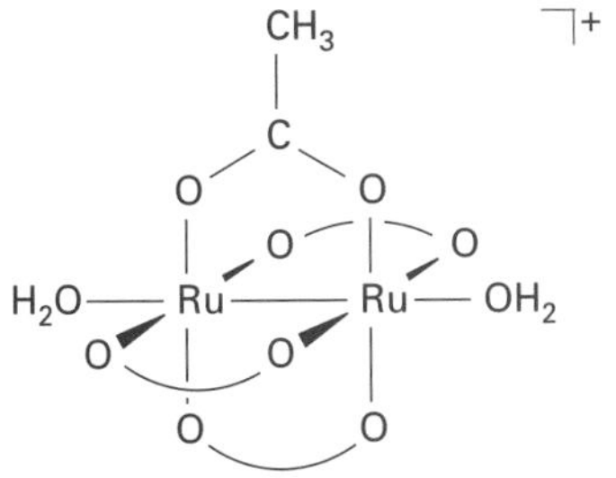	$\sigma^2\pi^4\delta^2\delta^{*2}\pi^{*3}$	1.5	2.32
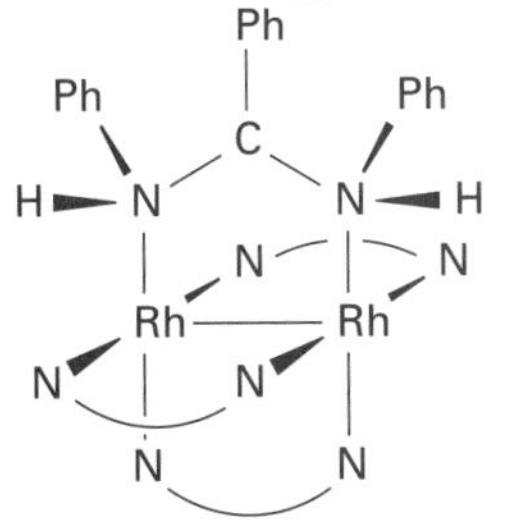	$\sigma^2\pi^4\delta^2\delta^{*2}\pi^{*4}$	1	2.39

Source: F.A. Cotton and G. Wilkinson, *Advanced inorganic chemistry.* Wiley, New York (1988) and F.A. Cotton. *Chem. Soc. Rev.* **12**, 35 (1983).
*When multiple bridging ligands are present only one is shown in detail.

It is possible to build larger metal clusters by addition to a metal–metal multiple bond. For example, [Pt(PPh$_3$)$_4$] loses two triphenylphosphine ligands when it adds to the Mo–Mo triple bond, resulting in a three-metal cluster:

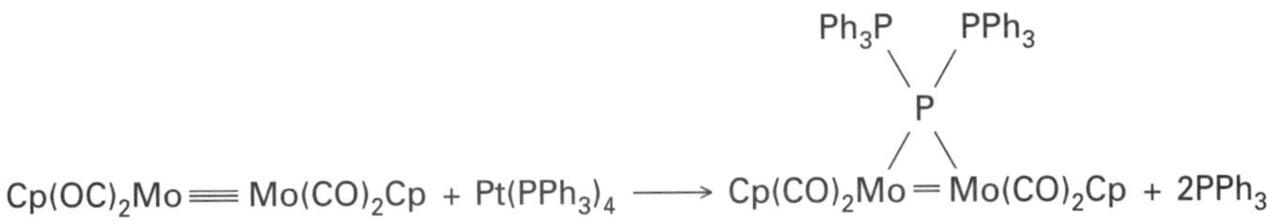

Cp(OC)$_2$Mo $\equiv$ Mo(CO)$_2$Cp + Pt(PPh$_3$)$_4$ $\longrightarrow$ Cp(CO)$_2$Mo $=$ Mo(CO)$_2$Cp + 2PPh$_3$

> Metal–metal bonds with bond orders up to 4 are formed by the early 4*d* and 5*d* metals in low oxidation states and in conjunction with π-donor ligands. The multiple bonds are susceptible to attack by H$^+$ or electron-rich metal complexes; the latter yield larger metal cluster compounds.

Example 9.3 Deciding probable structures of metal halides

List the major structural classes of *d*-metal halides and decide the most likely class for (a) MnF$_2$, (b) WCl$_2$, (c) RuF$_6$, and (d) FeI$_2$.

Answer The difluorides of the 3*d* metals (MnF$_2$) have the simple rutile structure characteristic of ionic **AB$_2$** compounds; heavier halides (FeI$_2$) are commonly layered. The chlorides, bromides, and iodides of the low oxidation state early 4*d* and 5*d* metals have metal–metal bonds (WCl$_2$ is in this class; in fact it contains a W$_6$ cluster). The hexahalides are molecular (RuF$_6$).

Self-test 9.3 Describe the probable structure of the compound formed when Re$_3$Cl$_9$ is dissolved in a solvent containing PPh$_3$.

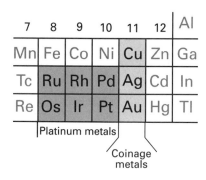

9.22 The location of the platinum and coinage metals in the periodic table.

9.10 Noble character

Metals on the right of the *d*-block are resistant to oxidation. This resistance is most evident for silver, gold, and the 4*d* and 5*d* metals in Groups 8 through 10 (Fig. 9.22). The latter are referred to as the **platinum metals** because they occur together in platinum-bearing ores. In recognition of their traditional use, copper, silver, and gold are referred to as the **coinage metals**. Gold occurs as the metal; silver, gold, and the platinum metals are also recovered in the electrolytic refining of copper. The prices of the individual platinum metals vary widely because they are recovered together but their consumption is not proportional to their abundance. Rhodium is by far the most expensive metal in this group because it is widely used in industrial catalytic processes and in automotive catalytic converters (Chapter 17). Rhodium is about 40 times more costly than the less catalytically useful metal palladium even though they occur in similar abundance.

Copper, silver, and gold are not susceptible to oxidation by hydrogen ions under standard conditions, and this noble character accounts for their use, together with platinum, in jewelry and ornaments. *Aqua regia*, a 3:1 mixture of concentrated hydrochloric and nitric acids, is an old but effective reagent for the oxidation of gold and platinum. Its function is twofold: the NO_3^- ions provide the oxidizing power and the Cl^- ions act as complexing agents. The overall reaction is

$$Au(s) + 4\,H^+(aq) + NO_3^-(aq) + 4\,Cl^-(aq) \longrightarrow$$
$$[AuCl_4]^-(aq) + NO(g) + 2\,H_2O(l)$$

The active species in solution are thought to be Cl_2 and NOCl, which are generated in the reaction

$$3\,HCl(aq) + HNO_3(aq) \longrightarrow Cl_2(aq) + NOCl(aq) + 2\,H_2O(l)$$

Oxidation state preferences are erratic in Group 11. For copper, the +1 and +2 states are most common, but for silver +1 is typical and for gold +1 and +3 are common. The simple aqua ions $Cu^+(aq)$ and $Au^+(aq)$ undergo disproportionation in aqueous solution:

$$2\,Cu^+(aq) \longrightarrow Cu(s) + Cu^{2+}(aq)$$
$$3\,Au^+(aq) \longrightarrow 2\,Au(s) + Au^{3+}(aq)$$

Complexes of Cu(I), Ag(I) and Au(I) are often linear. For example, $[H_3NAgNH_3]^+$ forms in aqueous solution and linear $[XAgX]^-$ complexes have been identified by X-ray crystallography. The currently preferred explanation for the tendency toward linear coordination is the similarity in energy of the outer *ns*, *np*, and $(n-1)d$ orbitals, which permits the formation of collinear *spd* hybrids (Fig. 9.23).

The soft character of Cu^+, Ag^+ and Au^+, which also results from the relatively small energy difference between the frontier orbitals of these ions, is illustrated by their affinity order, which is $I^- > Br^- > Cl^-$. Complex formation, as in the formation of $[Cu(NH_3)_2]^+$ and $[AuI_2]^-$, provides a means of stabilizing the +1 oxidation state of these metals in aqueous solution. Many tetrahedral complexes are also known for Cu(I), Ag(I), and Au(I).

Square-planar complexes are common for the platinum metals and gold in oxidation states that yield the d^8 electronic configuration, which include Rh(I), Ir(I), Pd(II), Pt(II), and Au(III). An example is $[Pt(NH_3)_4]^{2+}$. Characteristic reactions for these complexes are ligand substitution (Section 7.8) and, except for gold(III) complexes, **oxidative addition**. In the latter reaction, a

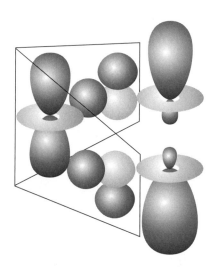

9.23 The hybridization of *s*, p_z, and d_{z^2} with the choice of phases shown here produces a pair of collinear orbitals that can be used to form strong σ bonds.

ligand **XY** is cleaved and the components **X** and **Y** fill two coordination sites to give a six-coordinate complex:

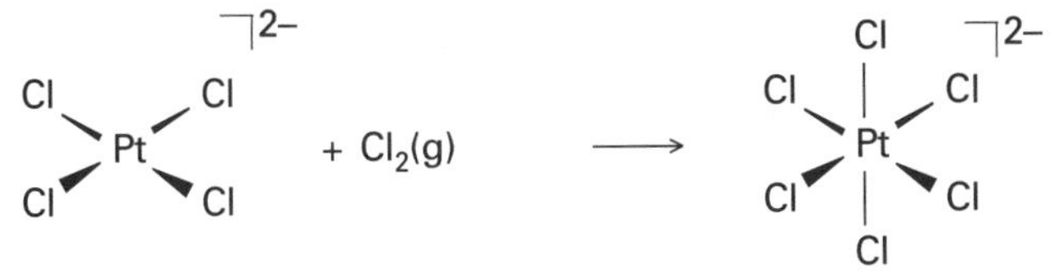

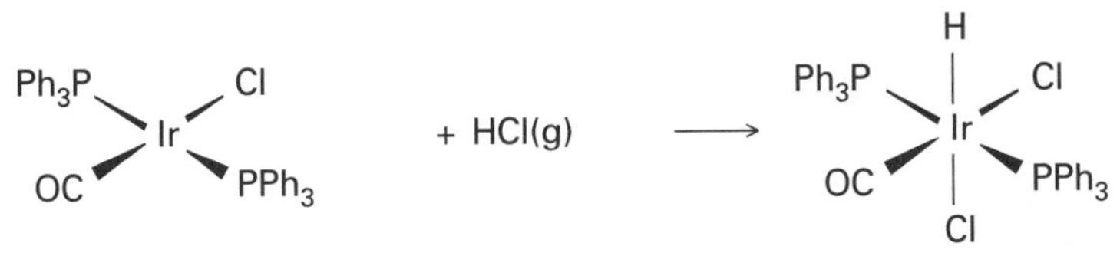

The classification of the reaction as oxidative addition stems from the formal increase in oxidation number of the metal center (from Pt(II) to Pt(IV), and from Ir(I) to Ir(III), in the two examples above) upon addition of a nonmetal species of formula **XY**, such as H_2, CH_3I, or HCl. Because negative oxidation numbers are assigned to nonmetal atoms attached to a metal, the attachment of **X** and **Y** leads to an increase of 2 in the oxidation number of the metal. This type of reaction is frequently implicated in the mechanisms of catalytic reactions of platinum complexes (see Section 17.3).

The properties of platinum complexes with organic ligands are discussed in Chapter 16. An interesting series of Pt(I) and Pd(I) complexes has been found in which two M—M bonded metal atoms are bridged by phosphine ligands (**28**). These compounds undergo reactions in which a two-electron ligand inserts into the M—M bond:

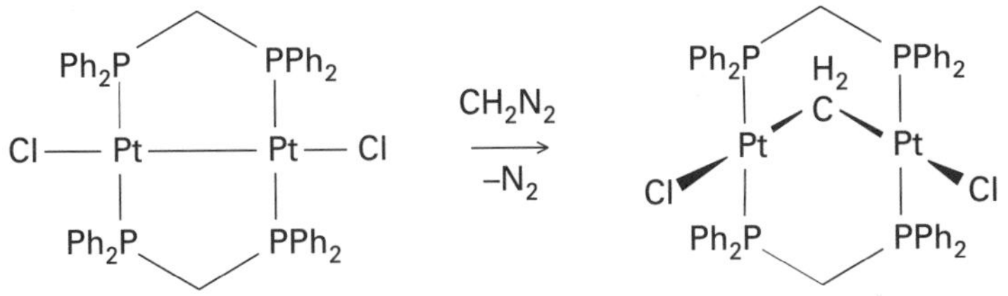

28   **29**

In this example, the CH_2 group inserts into the M—M bond and the resulting compound (**29**), which is informally called an 'A-frame complex', now contains two square-planar Pt(II) groups bridged by CH_2 as well as the phosphine ligands.

> *Metals on the right of the d-block prefer low oxidation states and soft ligands.*

9.11 Metal sulfides and sulfide complexes

Sulfur is softer and less electronegative than oxygen; it therefore has a broader span of oxidation states and a stronger affinity for the softer metals on the right of the d-block. For example, zinc(II) sulfide is readily precipitated when Zn^{2+}(aq) is added to an aqueous solution containing H_2S and NH_3 (to adjust the pH) but Sc^{3+}(aq) gives $Sc(OH)_3$(s) instead. The covalent contribution to the lattice enthalpy of all the softer metal sulfides cannot be overcome in water, and in general the sulfides are only very sparingly soluble. Another striking feature of sulfur is its tendency to form catenated sulfide ions. Thus, in combination with alkali metals a broad set of compounds containing S_n^{2-} ions can be prepared, and the smaller polysulfides can serve as chelating ligands toward d-block metal ions.

(a) Monosulfides

As with the d-metal monoxides, the monosulfides are most common in the first series of the d-block (Table 9.10). In contrast to the monoxides, however, most of the monosulfides have the nickel-arsenide structure (Fig. 2.15). The different structures are consistent with the rock-salt structure of the monoxides being favored by the ionic (harder) cation–anion combinations. The nickel-arsenide structure is favored by more covalent (softer) combinations and is found only when there is appreciable metal–metal bonding and correspondingly short metal–metal distances.

Monosulfides are formed by most 3d-metal ions, and their nickel-arsenide structure is attributed to covalent character in the bonding.

Table 9.10 Structures of d-block MS compounds*

	Group						
	4	5	6	7	8	9	10
Nickel-arsenide structure (shaded)	Ti	V		Mn†	Fe	Co	Ni
Rock-salt structure (unshaded)	Zr	Nb					

*Metal monosulfides not shown here for Group 6; some of the heavier metals have more complex structures.
†MnS has two polymorphs; one has a rock-salt structure, the other has a wurtzite structure.
Source: A.F. Wells, *Structural inorganic chemistry*, p. 752. Oxford University Press (1984).

(b) Disulfides

The disulfides of the d metals fall into two broad classes (Table 9.11). One class consists of layered compounds with the CdI_2 or MoS_2 structure and the other of compounds containing discrete S_2^{2-} groups.

The layered disulfides are built from a sulfide layer, a metal layer, and then another sulfide layer (Fig. 9.24). These sandwiches stack together in the crystal with sulfide layers in one slab adjacent to a sulfide slab in the next. Clearly, this crystal structure is not in harmony with a simple ionic model and its formation is a sign of covalence in the bonds between the soft sulfide ion and d-metal cations. The metal ion in these layered structures is surrounded by six S atoms. Its coordination environment is octahedral in some cases (PtS_2, for instance) and trigonal-prismatic in others (MoS_2). The layered MoS_2 structure is favored by S—S bonding as indicated by short S—S distances within each of the MoS_2 slabs. The common occurrence of the trigonal-prismatic structure in many of these compounds is in striking contrast to isolated metal complexes, where the octahedral arrangement of ligands is by far the most common.

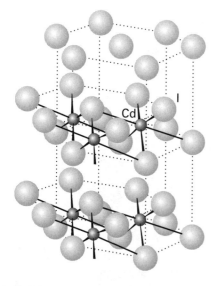

9.24 The CdI_2 structure adopted by many disulfides. The latter have adjacent sulfide layers in place of the I atoms.

Table 9.11 Structures of d-block MS_2 compounds*

	Group							
	4	5	6	7	8	9	10	11
Layered (shaded)	Ti			Mn	Fe	Co	Ni	Cu
Pyrite or marcasite (unshaded)	Zr	Nb	Mo		Ru	Rh		
	Hf	Ta	W	Re	Os	Ir	Pt	

*Metals not shown either do not form disulfides or have disulfides with complex structures.
Source: A.F. Wells, *Structural inorganic chemistry*, p. 757. Oxford University Press (1984).

Some of the layered metal sulfides readily undergo **intercalation reactions** in which ions or molecules penetrate between adjacent sulfide layers, often with accompanying redox reactions:

$$0.6\,Na(am) + TaS_2(s) \longrightarrow Na_{0.6}TaS_2(s)$$

In this reaction, sodium dissolved in liquid ammonia gives up an electron to a vacant band in TaS_2, and the Na^+ ion worms its way into positions between the sulfide layers. Similar intercalation reactions are common for graphite (Section 10.8) and various d-block oxides and sulfides (Section 18.7). The intercalation reaction between lithium and TiS_2 has been investigated for use in lightweight automobile batteries that can be recharged rapidly because the solid-state reaction involves no essential structural change.

Compounds containing discrete S_2^{2-} ions adopt the pyrite (Fig. 9.25) or marcasite structure. The stability of the formal S_2^{2-} ion in metal sulfides is much greater than that of the O_2^{2-} ion in peroxides, and there are many more metal sulfides in which the anion is S_2^{2-} than there are peroxides.

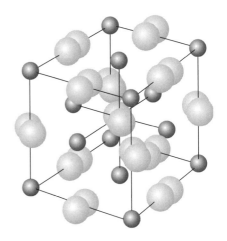

9.25 The structure of pyrite, FeS_2.

> 4d and 5d metals often form disulfides with alternating layers of metal ions and sulfide ions. The 3d metals in oxidation state +2 generally form sulfides containg discrete S_2^{2-} ions.

Example 9.4 Contrasting the structures of two different d-block disulfides

Contrast the structures of MoS_2 and FeS_2 and explain their existence in terms of the oxidation states of the metal ions.

Answer The point to be decided is whether the metals are likely to be present as $M(IV)$, in which case the two S atoms would be present as S^{2-} ions. If $M(IV)$ is not readily inaccessible, the metal might be present as $M(II)$, and the S atoms present as the S–S bonded species S_2^{2-}. Because it is a fair reducing agent, S^{2-} will only be found with a metal ion in an oxidation state that is not easily reduced. As with many of the d metals in Periods 5 and 6, molybdenum is easily oxidized to Mo(IV); therefore Mo(IV) can coexist with S^{2-}. Molybdenum(IV) sulfide (MoS_2) has the layered structure typical of metal disulfides. Iron is readily oxidized to Fe(II) but not to Fe(IV). Therefore, Fe(IV) cannot coexist with S^{2-}. The compound is therefore likely to contain Fe(II) and S_2^{2-}. The mineral name for FeS_2 is *pyrite*; its common name, 'fools' gold', indicates its misleading color.

Self-test 9.4 Molybdenum(IV) sulfide is a very effective lubricant. Present a plausible reason for this property.

(c) Sulfido complexes

The coordination chemistry of sulfur is quite different from that of oxygen. Much of this difference is connected with the ability of sulfur to catenate (form chains) and the preference of sulfur for metal centers that are not highly oxidizing. Research on Fe—S cluster complexes has flourished as a result of the discovery that they are present in electron-transfer and nitrogen-fixing enzymes (Section 19.6).[11] The structure of one such model compound, $[Fe_4S_4(SR)_4]^{2-}$, was shown in Fig. 7.4. It is readily prepared from simple starting materials in the absence of air:

$$4\,FeCl_3 + 4\,HS^- + 6\,RS^- + 4\,CH_3O^-$$

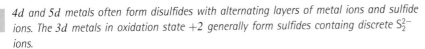

$$\xrightarrow{\text{methanol}} [Fe_4S_4(SR)_4]^{2-} + RS—SR + 12\,Cl^- + 4\,CH_3OH$$

[11] R. Cammock and A.G. Sykes (ed.), *Adv. Inorg. Chem.* **38** (1992).

30 $[Mo_2(S_2)_6]^{2-}$

31 $[MoS(S_4)_2]^{2-}$

The observation that this reaction is successful with many different R groups, and its good yield, indicate that the Fe_4S_4 cage is thermodynamically more stable than other possibilities. The HS^- ion provides the sulfide ligands for the cage, the RS^- ion serves both as a ligand and a reducing agent, and the CH_3O^- ion acts as a base. The cubic cluster contains Fe and S atoms on alternate corners, so each S atom bridges three Fe atoms. Each of the thiolate groups, RS^-, occupies a terminal position on an Fe atom. The cluster remains intact upon one-electron reduction to $[Fe_4S_4(SR)_4]^{3-}$. Analogous clusters are implicated in the redox reactions of the electron-transfer protein ferredoxin.

Simple thiometallate complexes such as $[MoS_4]^{2-}$ can be synthesized easily by passing H_2S gas through a strongly basic aqueous solution of the molybdate or tungstate ion:

$$[MoO_4]^{2-}(aq) + 4\,H_2S(g) \longrightarrow [MoS_4]^{2-}(aq) + 4\,H_2O(l)$$

These tetrathiometallate anions are building blocks for the synthesis of complexes containing more metal atoms. For example, they will coordinate to many dipositive metal ions, such as Co^{2+} and Zn^{2+}:

$$Co^{2+}(aq) + 2\,[MoS_4]^{2-}(aq) \longrightarrow [S_2MoS_2CoS_2MoS_2]^{2-}(aq)$$

The polysulfides such as S_2^{2-} and S_3^{2-}, which are formed by addition of elemental sulfur to a solution of ammonium sulfide, can also act as ligands. An example is $[Mo_2(S_2)_6]^{2-}$ (**30**), which is formed from ammonium polysulfide and MoO_4^{2-}; it contains side-bonded S_2^{2-} ligands. The larger polysulfides bond to metal atoms forming chelate rings, as in $[MoS(S_4)_2]^{2-}$ (**31**), which contains chelating S_4 ligands.

Binary disulfides of the early d-block metals often have a layered structure, whereas Fe^{2+} and many of the later d-block metal disulfides contain discrete S_2^{2-} ions. Chelating polysulfide ligands are common in metal–sulfur coordination compounds of the 4d and 5d metals. Sulfur prefers metal centers that are not highly oxidizing.

The elements of Group 12

We shall see in this section that Group 12 displays trends in oxidation states that contrast strongly with those just described for the other d-block metals. For instance, the noble character—the resistance to oxidation—that developed across the d-block is suddenly lost at Group 12. We shall see that there is a connection between the greater ease of oxidation of these metals and an abrupt lowering of d-orbital energies at the end of the d-block.

9.12 Occurrence and recovery

Zinc is by far the most abundant element in Group 12. It ranks twenty-third in crustal abundance, just ahead of copper. Cadmium and mercury are much less abundant: they are even less abundant than most of the lanthanides.

Sulfides are the principal ores for the elements in the group, with zinc and cadmium often occurring together (Table 9.12). Zinc sulfide is roasted in air to produce the oxide:

$$ZnS(s) + \tfrac{3}{2}O_2(g) \longrightarrow ZnO(s) + SO_2(g)$$

Table 9.12 Occurrence and methods of recovery of the Group 12 metals

Metal	Principal minerals	Method of recovery
Zinc	Sphalerite, ZnS	$ZnS + \tfrac{3}{2}O_2 \rightarrow ZnO + SO_2$ followed by $2\,ZnO + C \rightarrow 2\,Zn + CO_2$
Cadmium	Traces in zinc ores	
Mercury	Cinnabar, HgS	$HgS + O_2 \rightarrow Hg + SO_2$

The oxide is then reduced in a blast furnace charged with carbon. The reduction occurs primarily by CO in a hot portion of the blast furnace, as was shown in Fig. 6.4 for the reduction of iron. When cadmium and zinc occur together, the sulfide ore is roasted in air to produce a mixture of the metal sulfates and oxides. This mixture is dissolved in sulfuric acid and reduced. The separation is based on the greater ease of reduction of Cd^{2+} compared with Zn^{2+}.

Mercury occurs in the bright red mineral cinnabar, HgS, which was at one time used as a pigment (vermilion) by artists, a practice which has been discontinued because of the toxicity of mercury. Elemental mercury is recovered from the sulfide by roasting in air:

$$HgS(s) + O_2(g) \longrightarrow Hg(l) + SO_2(g)$$

Zinc and mercury are removed from their sulfide ores by oxidation of sulfide to SO_2.

9.13 Redox reactions

Zinc and cadmium are much more readily oxidized than their neighbors, copper and silver. This difference is apparent in the standard potentials, which are much lower for Zn^{2+} (-0.76 V) and Cd^{2+} (-0.40 V) than for Cu^{2+} ($+0.34$ V) and Ag^+ ($+0.80$ V). The origin of the difference between the two groups is the lower sublimation enthalpy, and to a lesser extent the lower ionization enthalpy, of Group 12 elements compared with those of Group 11: $\Delta_{sub}H^{\ominus}$ is 338 kJ mol^{-1} for Cu but only 131 kJ mol^{-1} for Zn (Fig. 9.2). This decrease in sublimation enthalpy in turn reflects a decrease in metal–metal bond strength from Group 11 to 12. A variety of evidence suggests that the weaker metal–metal bonds arise from the lack of d-orbital contribution to the bonding in Group 12. This decrease in d–d bonding correlates with the decrease in energy of the d orbitals beyond Group 12, as was illustrated in Fig. 1.19.

Another difference from other d-block elements is the substantial change in chemical properties between the lightest d-block element (Zn) and its congeners (Cd and Hg). For example, mercury is much less electropositive than zinc and cadmium and is alone in having a $+1$ oxidation state (the Hg_2^{2+} ion) that survives in aqueous solution. Partly as a result of this stability, Hg_2^{2+} is a far more important species than the rare and much more readily oxidized species Zn_2^{2+} and Cd_2^{2+}.

The chemical, spectroscopic, and X-ray structural evidence that Hg(I) occurs as the dinuclear cation Hg_2^{2+} provided the first example of a metal–metal bonded species. Much later Cd_2^{2+} was identified in $Cd_2(AlCl_4)_2$, and there is spectroscopic evidence for the formation of Zn_2^{2+} in the reaction of zinc metal with molten $ZnCl_2$. The Cd_2^{2+} compound $Cd_2(AlCl_4)_2$ is well characterized. It appears to owe its existence in part to the presence of the large anion, which helps to stabilize the large cation. The synthesis of the Cd_2^{2+} compound is performed in a strictly nonaqueous molten salt medium:

$$CdCl_2(l) + 2\,AlCl_3(l) + Cd(s) \longrightarrow Cd_2[AlCl_4]_2(s)$$

Cadmium(I) rapidly disproportionates in the presence of water:

$$Cd_2^{2+}(aq) \longrightarrow Cd(s) + Cd^{2+}(aq)$$

The substantial hydration enthalpy of Cd^{2+} provides a driving force for this reaction. The situation with mercury(I) ions is not radically different because they have a measurable equilibrium constant for disproportionation:

$$Hg_2^{2+}(aq) \rightleftharpoons Hg(l) + Hg^{2+}(aq) \qquad K = 6.0 \times 10^{-3}$$

This reaction is driven to the right by ligands that strongly complex or precipitate Hg^{2+}: for instance, the addition of CN^- yields a stable Hg(II) complex:

$$Hg_2^{2+}(aq) + 2\,CN^-(aq) \longrightarrow Hg(l) + Hg(CN)_2(aq)$$

and OH^- gives mercury(II) oxide:

$$Hg_2^{2+}(aq) + 2\,OH^-(aq) \longrightarrow Hg(l) + HgO(aq) + H_2O(l)$$

The ease of reduction of the Group 12 cations decreases from Hg^{2+} to Zn^{2+}.

9.14 Coordination chemistry

Mercury(II) forms some linear two-coordinate complexes, such as $Hg(CN)_2$, $Hg(CH_3)_2$, and $O-Hg-O$ in solid mercury(II) oxide, HgO. Zinc and cadmium usually exhibit higher coordination numbers, ranging from 4 to 6.[12] The coordination of zinc in ZnO is tetrahedral, and that of cadmium in CdO is octahedral, in the rock-salt structure. Both the reduced tendency toward linear coordination and the harder character of Zn^{2+} compared with Cu^+, and of Cd^{2+} compared with Ag^+, can be traced to the much larger energy difference between filled d orbitals and empty s and p orbitals that occurs at Group 12.

The increase in amphoterism of lower oxidation states on going from left to right across Period 4 was illustrated in Fig. 5.5. In Group 12 we find that Zn^{2+} is amphoteric:

$$Zn(OH)_2(s) + 2\,H^+(aq) \longrightarrow Zn^{2+}(aq) + 2\,H_2O(l)$$
$$Zn(OH)_2(s) + OH^-(aq) \longrightarrow Zn(OH)_3^-(aq)$$

Cadmium hydroxide also reacts with hydroxide ions, but it requires more concentrated base to dissolve. Mercury(II) hydroxide is not amphoteric.

In keeping with their closed-subshell (d^{10}) electronic configurations, the Group 12 cations Zn^{2+}, Cd^{2+}, and Hg^{2+} are colorless in aqueous solution. However, some of their heavy halides and chalcogenides are colored. For example, $ZnCl_2$ and ZnI_2 are colorless, but only the chlorides of cadmium and mercury are colorless; CdI_2 is yellow and HgI_2 occurs as two polymorphs, one of which is yellow and the other red. The presence of heavy halides shifts the energy of these charge-transfer transitions from the ultraviolet (for the colorless materials) to the visible region of the spectrum. Compounds that owe their color to charge-transfer transitions are very useful in practice (for pigments) because their absorptions are more intense than those of $d-d$ transitions.

Zn(II) and Cd(II) usually form tetrahedral complexes; Hg(II) forms linear $XHgX$ with CN^- and similar strong donors.

The p-block metals

To provide comparisons and contrasts with the chemical properties of the other metallic elements, we shall describe the redox stability and some coordination chemistry of the p-block elements aluminum, gallium, indium, and thallium in Group 13/III, of tin and lead in Group 14/IV, and of bismuth in Group 15/V. Further information on these elements in relation to the p-block nonmetals is presented in Chapters 11 and 12.

In contrast to the elements in the d-block, the heavier p-block metals favor low oxidation states. This trend is illustrated in Fig. 9.26 for the Group 13/III metals, where the maximum oxidation state is easily achieved for gallium but not for thallium. The latter favors oxidation state +1. This tendency to favor an oxidation state 2 less than the group oxidation number also occurs in Groups 14/IV (Fig. 9.27) and 15/V and is an example of the **inert pair effect**. There is no simple explanation for this effect: it is probably best attributed to the low $M-X$ bond enthalpies for the heavy p-block elements and the fact that it requires less energy to oxidize an element to a low oxidation state than to a higher oxidation state. This energy has to

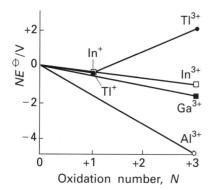

9.26 Frost diagrams for the Group 13/III p-block metals in acidic solution.

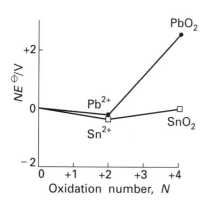

9.27 Frost diagrams for the Group 14/IV p-block metals in acidic solution.

[12] One major exception to this rule is that zinc, cadmium, and mercury all form linear alkyl compounds, MR_2, Section 15.3.

be supplied by the formation of ionic or covalent bonds, so the high oxidation state may be inaccessible if bonding to a particular element is weak.

The most commonly encountered oxidation states for the three heaviest elements in these groups are Tl(I), Pb(II) and Bi(III), and compounds containing these elements in their group oxidation states, Tl(III), Pb(IV), and Bi(V), are easily reduced. The highly oxidizing character of the heavy p-block metals in their group oxidation states is also reflected in the limited range of binary compounds of that state of the metal. For example, TlX_3 exists for $X = F$, Cl, and Br, but TlI_3 contains the Tl^+ cation in combination with the I_3^- anion. Similarly, PbF_4 is known, $PbCl_4$ does not survive slightly above room temperature, and the Pb(IV) bromides and iodides are unknown.

The heavier p-block metals favor low oxidation states.

9.15 Occurrence and recovery

The abundances of the p-block metals in the Earth's crust vary widely from aluminum, which is the third most abundant element (behind oxygen and silicon), to bismuth, which is the heaviest element with a stable isotope. Gallium, which is more abundant than lithium, boron, lead, and many other familiar elements, is an expensive element because it is widely dispersed in aluminum- and iron-containing minerals. Gallium is difficult to recover because the chemical properties of Ga^{3+}, Al^{3+}, and Fe^{3+} are similar owing to their similar radii and acid–base properties. The recovery of these elements is summarized in Table 9.13. The lighter, more electropositive metals aluminum and gallium are recovered from oxides. The energy demands are high for the high-temperature electrolytic recovery of the most electropositive element, aluminum (Section 7.1), but the price is moderated by the abundance of bauxite and the economies of large-scale production.

The least abundant elements considered here are thallium and bismuth. The low abundance of these three elements is in harmony with the general observation of lower nuclear binding energies with increasing atomic number. For reasons related to the details of its nuclear structure, lead does not follow this trend, and is more abundant than the lanthanide elements and germanium. As with mercury and cadmium, lead is environmentally hazardous because it is highly toxic. Unfortunately, lead, mercury, and cadmium are uniquely suited for and used in a variety of consumer products, ranging from batteries to electrical switches.

Most of the aluminum on Earth is distributed in clays and aluminosilicate minerals that are not economically attractive sources of the metal. The primary ore for aluminum is bauxite, a hydrated oxide. Gallium, which occurs as a trace component of bauxite, is produced as a byproduct of aluminum refining. The heavier and chemically softer p-block elements in Group 13/III (indium, and thallium) together with germanium in Group 14/IV, are recovered

Table 9.13 Occurrence and methods of recovery of commercially important metals in Groups 13/III to 15/V

Metal	Principal minerals	Method of recovery
Group 13		
Aluminum	Bauxite, $Al_2O_3 \cdot xH_2O$	Electrolytic (Hall process)
Gallium	Traces in aluminum and zinc ores	
Group 14		
Tin	Cassiterite, SnO_2	$SnO_2 + C \rightarrow Sn + CO_2$
Lead	Galena, PbS	$PbS + \frac{3}{2}O_2 \rightarrow PbO + SO_2$
		followed by $2\,PbO + C \rightarrow 2\,Pb + CO_2$
Group 15		
Bismuth	Traces in zinc, copper, and lead sulfide ores	

as byproducts of the refining of sulfide ores of more abundant elements. Bismuth is sometimes recovered from the minerals bismuthinite, Bi_2S_3, and bismite, Bi_2O_3, but like the other heavy *p*-block metals most is recovered during copper, zinc, or lead smelting.

> Among the p-block elements, silicon and aluminum are highly abundant; thallium and bismuth are the least abundant.

9.16 Group 13/III

The Group 13/III metals have a silver luster and erratic variations in melting points down the group: Al (660 °C), Ga (30 °C), In (157 °C), and Tl (303 °C). The low melting point of gallium is reflected in the unusual structure of the metal, which contains Ga_2 units that persist into the melt. Gallium, indium, and thallium are all mechanically soft metals.

(a) The group oxidation state (+3)

Although direct reaction of aluminum or gallium with a halogen yields a halide, both these electropositive metals also react with HCl or HBr gas, and this is usually a more convenient route:

$$2\,Al(s) + 6\,HCl(g) \xrightarrow{100°C} 2\,AlCl_3(s) + 3\,H_2(g)$$

In the laboratory, a 'hot tube' reactor (one is illustrated in Fig. B12.1) is often used. The halides of both elements are available commercially, but it is common to synthesize them in the laboratory when a material free of hydrolysis products is required.

Because the F^- ion is so small, the fluorides AlF_3 and GaF_3 are hard solids that have much higher melting points and sublimation enthalpies than the other halides. Their high lattice enthalpies also result in their having very limited solubility in most solvents, and they do not act as Lewis acids to simple donor molecules. In contrast, the heavier halides are soluble in a wide variety of polar solvents and are excellent Lewis acids. However, despite their low reactivity toward most donors, AlF_3 and GaF_3 form salts of the type Na_3AlF_6 and Na_3GaF_6 which contain octahedral $[MF_6]^{3-}$ complex ions.

The Lewis acidities of the halides reflect the relative chemical hardness of the Group 13/III elements. Thus, toward a hard Lewis base (such as ethyl acetate, which is hard on account of its O donor atoms), the Lewis acidities of the halides weaken as the softness of the acceptor element increases: so the Lewis acidities fall in the order:

$$BCl_3 > AlCl_3 > GaCl_3$$

In contrast, toward a soft Lewis base (such as dimethylsulfide, which is soft on account of its S atom), the Lewis acidities strengthen as the softness of the acceptor element increases:

$$GaX_3 > AlX_3 > BX_3 \qquad (X = Cl \text{ or } Br)$$

In keeping with the general tendency toward higher coordination numbers for the heavier *p*-block elements, halides of aluminum and its heavier congeners may take on more than one Lewis base and become hypervalent:

$$AlCl_3 + N(CH_3)_3 \longrightarrow Cl_3AlN(CH_3)_3$$
$$Cl_3AlN(CH_3)_3 + N(CH_3)_3 \longrightarrow Cl_3Al(N(CH_3)_3)_2$$

The most stable form of Al_2O_3, α-alumina, is a very hard and refractory material. In its mineral form it is known as *corundum* and as a gemstone it is *sapphire*. The blue of the latter arises from a charge-transfer transition from Fe^{2+} to Ti^{4+} ion impurities. The structure of α-alumina and gallia, Ga_2O_3, consists of an hcp array of O^{2-} ions with the metal ions occupying two-thirds of the octahedral holes in an ordered array. Ruby is α-alumina in which a small fraction of the Al^{3+} ions are replaced by Cr^{3+}. The Cr(III) is red rather than the characteristic violet of $[Cr(OH_2)_6]^{3+}$ or the green of Cr_2O_3 because, when Cr^{3+} substitutes for the smaller

Al^{3+} ion, the O ligands are compressed about Cr^{3+}. This compression increases the ligand-field splitting parameter and shifts the first spin-allowed d-d band to higher energies.

Dehydration of aluminum hydroxide at temperatures below $900\,°C$ leads to the formation of γ-alumina, a metastable polycrystalline form with a very high surface area. Partly on account of its surface acid and base sites, this material is used as a solid phase in chromatography and as a heterogeneous catalyst and catalyst support (Section 17.5).

> *Boron, aluminum, and gallium all favor the +3 oxidation state, and their trihalides are Lewis acids.*

(b) Low oxidation state gallium, indium, and thallium

Oxidation states of aluminum lower than +3 are rarely encountered. Thus, AlCl(g) is encountered only at very high temperatures and Al^+ is not stable in solid compounds. The oxidation state +1 increases in stability down the group and gallium forms compounds such as the solids GaI and $Ga(AlCl_4)$. Halides of Ga(I) and Ga(II), such as GaCl and $GaCl_2$, can be prepared by a comproportionation reaction in which the Ga(III) halide is heated with gallium metal:

$$2\,GaX_3 + Ga \xrightarrow{\Delta} 3\,GaX_2 \qquad (X = Cl, Br, \text{ or } I \text{ but not } F)$$

The formula $GaCl_2$ is deceiving, for this solid and most other apparently divalent salts do not contain Ga(II); instead they are mixed oxidation state compounds containing Ga(I) and Ga(III). Mixed oxidation state compounds are also known for the heavier metals, such as $InCl_2$ and $TlBr_2$. The presence of M^{3+} ions is indicated by the existence of $[MX_4]^-$ complexes in these salts with short M—X distances, and the presence of M^+ ions is indicated by longer and less regular separation from the halide ions. There is in fact only a fine line between the formation of a mixed oxidation state ionic compound and the formation of a compound that contains M—M bonds. For example, mixing $GaCl_2$ with a solution of $[N(CH_3)_4]Cl$ in a nonaqueous solvent yields the compound $[N(CH_3)_4]_2[Cl_3Ga—GaCl_3]$, in which the anion has an ethane-like structure with a Ga—Ga bond.

Gallium(I) has much in common with indium(I); for example, both disproportionate when dissolved in water:

$$3\,MX(s) \longrightarrow 2\,M(s) + M^{3+}(aq) + 3\,X^-(aq) \qquad (M = Ga, In)$$

On the other hand, Tl^+ is stable with respect to disproportionation in water because Tl^{3+} is difficult to achieve, as discussed above in connection with the inert pair effect.

We have seen that the d-block elements are soft Lewis acids in their low oxidation states; however, the trend for the heavy p-block elements is just the opposite. The relative affinities for hard and soft donors suggest that Tl^+ is harder than Tl^{3+}. However, the Tl^+ ion is in fact still borderline between hard and soft. For example, it is transported into cells along with the hard K^+ ion. Like the alkali metal hydroxides, TlOH is soluble in water; unlike K^+, however, thallium(I) chloride and thallium(I) bromide and sulfide are insoluble. The transport through cell walls and reactions with soft donors appears to account for the observation that thallium is highly toxic to mammals.

The monohalides GaX, InX, and TlX are known for X = Cl, Br, and I. Under ordinary conditions the Tl(I) halides are insulators, as is typical of ionic compounds. However, at high pressures a new phase is formed with a significant electrical conductivity which decreases with increasing temperature. This behavior signifies metallic conduction (see the introduction preceding Section 3.14).

> *The +1 oxidation state becomes progressively more stable from gallium to thallium.*

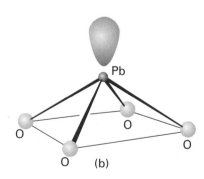

32 $[SnCl_3]^-$

33 $[(SnCl)_2(Pt(SnCl_3)_2)_3]$

Example 9.5 Proposing reactions of Group 13/III halides

Propose balanced chemical equations (or indicate no reaction) for reactions between (a) $AlCl_3$ and $(C_2H_5)_3NGaCl_3$ in toluene, (b) $(C_2H_5)_3NGaCl_3$ and GaF_3 in toluene, and (c) TlCl and NaI in water.

Answer (a) Al(III) is a stronger and harder Lewis acid than Ga(III); therefore, the following reaction can be expected:

$$AlCl_3 + (C_2H_5)_3NGaCl_3 \longrightarrow (C_2H_5)_3NAlCl_3 + GaCl_3$$

(b) No reaction, because GaF_3 has a very high lattice enthalpy and thus is not a good Lewis acid. (c) Tl(I) is chemically borderline soft, so it combines with the softer I^- ion rather than Cl^-:

$$TlCl(s) + NaI(aq) \longrightarrow TlI(s) + NaCl(aq)$$

Like silver halides, Tl(I) halides have low solubility in water, so the reaction will probably proceed very slowly.

Self-test 9.5 Propose, with reasons, the chemical equation (or indicate no reaction) for reactions between (a) $(CH_3)_2SAlCl_3$ and $GaBr_3$ and (b) TlCl and formaldehyde (CH_2O) in acidic aqueous solution. (*Hint:* formaldehyde is easily oxidized to CO_2 and H^+.)

9.17 Tin and lead

Aqueous and nonaqueous solutions of tin(II) salts are useful mild reducing agents, but they must be stored under an inert atmosphere because air oxidation is spontaneous and rapid:

$$Sn^{2+}(aq) + \tfrac{1}{2}O_2(g) + 2H^+(aq) \longrightarrow Sn^{4+}(aq) + H_2O(l) \qquad (E^\ominus = +1.08 \text{ V})$$

Tin dihalides and tetrahalides are both well known. The tetrachloride, bromide, and iodide are molecular compounds, but the tetrafluoride has a structure consistent with it being an ionic solid because the small F^- ion permits a six-coordinate structure. Lead tetrafluoride also has a structure consistent with it being an ionic solid but, as a manifestation of the inert pair effect, $PbCl_4$ is an unstable compound that decomposes into $PbCl_2$ and Cl_2 at room temperature. Lead tetrabromide and tetraiodide are unknown, so the dihalides dominate the halogen chemistry of lead. The arrangement of halogen atoms around the central metal atom in the dihalides of tin and lead often deviates from simple tetrahedral or octahedral coordination, and is attributed to the presence of a stereochemically active lone pair. The tendency to achieve the distorted structure is more pronounced with the small F^- ion, and less distorted structures are observed with larger halides.

Both Sn(IV) and Sn(II) form a variety of complexes. Thus, $SnCl_4$ forms complex ions such as $[SnCl_5]^-$ and $[SnCl_6]^{2-}$ in acidic solution. In nonaqueous solution, a variety of donors interact with the moderately strong Lewis acid $SnCl_4$ to form complexes such as *cis*-$SnCl_4(OPMe_3)_2$. In aqueous and nonaqueous solutions Sn(II) forms trihalo complexes, such as $[SnCl_3]^-$, where the pyramidal structure indicates the presence of a stereochemically active lone pair (**32**). The $[SnCl_3]^-$ ion can serve as a soft donor to *d*-metal ions. One unusual example of this ability is the red cluster compound $Pt_3Sn_8Cl_{20}$, which is trigonal-bipyramidal (**33**).

The oxides of lead are very interesting from both fundamental and technological standpoints. In the red form of PbO, the Pb(II) ions are four-coordinate (Fig. 9.28), but the O^{2-} ions around the Pb(II) lie in a square. As for the halides, this structure can be rationalized by the presence of a stereochemically active lone pair on the metal atom. Lead also forms mixed oxidation state oxides. The best known is 'red lead', Pb_3O_4, which contains Pb(IV) in an octahedral environment and Pb(II) in an irregular six-coordinate environment. The assignment of different oxidation numbers to the lead in these two sites is based on the shorter Pb—O

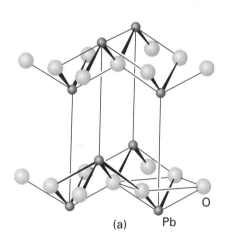

9.28 (a) The structure of PbO. (b) The square-pyramidal array of PbO showing the possible orientation of a stereochemically active lone pair of electrons.

Box 9.2 **The lead–acid battery**

The chemistry of the lead–acid battery is noteworthy because, as well as being the most successful rechargeable battery, it illustrates the role of both kinetics and thermodynamics in the operation of cells.

In its fully charged state, the active material on the cathode is PbO_2 and at the anode it is lead metal; the electrolyte is dilute sulfuric acid. One feature of this arrangement is that the lead-containing reactants and products at both electrodes are insoluble. When the cell is producing current, the reaction at the cathode is the reduction of Pb(IV) as PbO_2 to Pb(II), which in the presence of sulfuric acid is deposited on the electrode as insoluble $PbSO_4$:

$$PbO_2(s) + HSO_4^-(aq) + 3H^+(aq) + 2e^- \longrightarrow$$
$$PbSO_4(s) + 2H_2O(l)$$

At the anode, lead is oxidized to Pb(II), which is also deposited as the sulfate:

$$Pb(s) + SO_4^{2-}(aq) \longrightarrow PbSO_4(s) + 2e^-$$

The overall reaction is

$$PbO_2(s) + 2HSO_4^-(aq) + 2H^+(aq) + Pb(s) \longrightarrow$$
$$2PbSO_4(s) + 2H_2O(l)$$

The potential difference of about 2 V is remarkably high for a cell in which an aqueous electrolyte is used, and exceeds by far the potential for the oxidation of water to O_2, which is 1.23 V. The success of the battery hinges on the high overpotentials (and hence low rates) of oxidation of H_2O on PbO_2 and of reduction of H_2O on lead.

---

distances for the atom identified as Pb(IV). The maroon form of lead(IV) oxide, PbO_2, crystallizes in the rutile structure. This oxide is a component of the cathode of a lead-acid battery (Box 9.2).

> Oxidation states +2 and +4 are possible for many tin and lead compounds; however, Pb(IV) is highly oxidizing so the tetrachloride is unstable and the tetrabromide and tetraiodide are unknown. Sn(II) has a stereochemically active lone pair, so its trihalide anions are pyramidal and can serve as ligands to soft metal centers.

9.18 Bismuth

The chemical properties of bismuth illustrate the inert pair effect in a striking manner. For instance, Bi gives up its full complement of five valence electrons with great difficulty, and most of its compounds are bismuth(III). Bismuth(III) can be regarded as borderline between hard and soft. An indication of this property is the insolubility of both $Bi(OH)_3$ and Bi_2S_3. The approximate standard potentials for bismuth in acidic aqueous solution illustrate the strongly oxidizing character of the +5 oxidation state and the mildly electropositive character of the element:

$$Bi^{3+}(aq) + 3e^- \longrightarrow Bi(s) \qquad (E^\ominus = +0.32\,V)$$
$$Bi^{5+}(aq) + 2e^- \longrightarrow Bi^{3+}(aq) \qquad (E^\ominus \approx +2\,V)$$

Bismuth(V) is prepared by heating Bi_2O_3 with sodium peroxide:

$$Bi_2O_3(s) + 2Na_2O_2(s) \longrightarrow 2NaBiO_3(s) + Na_2O(s)$$

When the sodium bismuthate product is dissolved in an aqueous solution of a non-coordinating acid, such as $HClO_4$, a poorly characterized metastable Bi(V) species is produced. This species is shown as Bi^{5+} in the half-reaction above: it may be $[Bi(OH)_6]^-$.

The coordination chemistry of Bi(III) reflects the borderline affinity for hard and soft ligands and the tendency toward apparently distorted coordination environments. In fact, the structures are often consistent with the VSEPR model, and the 'distortion' from the shape their chemical formula might suggest is actually due to the presence of a stereochemically active lone pair. The tendency for a low oxidation state p-block element to exhibit these so-called distorted structures follows the trends already discussed for Pb(II) and Sn(II). Thus:

1 Low coordination numbers favor a distorted structure. For example, in accord with VSEPR theory, BiF_3 is pyramidal in the gas phase.

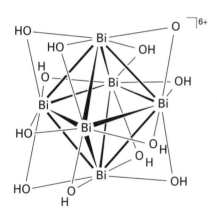

34 Bi(OR)$_3$ R = C$_2$H$_4$OCH$_3$

35 [Bi$_2$Cl$_8$]$^{2-}$

36 [Bi$_6$(OH)$_{12}$]$^{6+}$

2 Lighter *p*-block central atoms have the greater tendency toward a distorted structure. For instance, Sb(III) compounds are more often distorted from planarity than Bi(III) compounds.

3 Small ligands promote a distorted structure. Thus, fluoride or alkoxide ligands are more likely to lead to an apparently distorted structure. For example, a bismuth alkoxide with the empirical formula Bi(OC$_2$H$_4$OCH$_3$)$_3$ turns out to be an alkoxide-bridged chain in the solid state with square-bipyramidal coordination around bismuth (**34**). Similarly, the anionic complex [Bi$_2$Cl$_8$]$^{2-}$ has square-pyramidal coordination around each Bi atom (**35**).[13]

It is interesting to note that Bi(III) forms a hydroxide complex in aqueous solution, with the formula [Bi$_6$(OH)$_{12}$]$^{6+}$, which consists of an octahedral array of Bi^{3+} ions bridged with the OH$^-$ ligands bridging the edges of the octahedron (**36**).

| Three structural principles are that low coordination numbers, lighter p-block central atoms, and small ligands favor apparently distorted structures.

The *f*-block metals

Each series of fifteen elements in the *f*-block corresponds to the filling of the seven 4*f* and 5*f* orbitals, respectively (from *f*^{0} to *f*^{14}). There is a striking uniformity in the properties of the 4*f* elements, the **lanthanides**, and greater diversity in the chemistry of the 5*f* elements, the **actinides**. A general lanthanide is represented by the symbol **Ln** and an actinide by **An**.

9.19 Occurrence and recovery

Other than promethium, which has no stable isotopes, the least abundant lanthanide, thulium, is similar to iodine in crustal abundance. The principal mineral source for the early lanthanides is monazite, which contains mixtures of lanthanides and thorium (Ln, Th)PO$_4$. Another phosphate mineral, xenotime, is the principal source of yttrium and the heavier

Table 9.14 Names, symbols, and properties of the lanthanides

Z	Name	Symbol	Configuration of M^{3+}	$E^{\ominus}$/V	r(M^{3+})/Å*	O.N.†
57	Lanthanum	La	[Xe]	−2.38	1.16	**3**
58	Cerium	Ce	[Xe]4*f*^{1}	−2.34	1.14	**3**, 4
59	Praseodymium	Pr	[Xe]4*f*^{2}	−2.35	1.13	**3**, 4
60	Neodymium	Nd	[Xe]4*f*^{3}	−2.32	1.11	2(n), **3**
61	Promethium	Pm	[Xe]4*f*^{4}	−2.29	1.09	**3**
62	Samarium	Sm	[Xe]4*f*^{5}	−2.30	1.08	2(n), **3**
63	Europium	Eu	[Xe]4*f*^{6}	−1.99	1.07	2(a), **3**
64	Gadolinium	Gd	[Xe]4*f*^{7}	−2.28	1.05	**3**
65	Terbium	Tb	[Xe]4*f*^{8}	−2.31	1.04	**3**, 4
66	Dysprosium	Dy	[Xe]4*f*^{9}	−2.29	1.03	2(n), **3**
67	Holmium	Ho	[Xe]4*f*^{10}	−2.33	1.02	**3**
68	Erbium	Er	[Xe]4*f*^{11}	−2.32	1.00	**3**
69	Thulium	Tm	[Xe]4*f*^{12}	−2.32	0.99	2(n), **3**
70	Ytterbium	Yb	[Xe]4*f*^{13}	−2.22	0.99	2(a), **3**
71	Lutetium	Lu	[Xe]4*f*^{14}	−2.30	0.98	**3**

*Ionic radii for C.N. = 8 from R.D. Shannon, *Acta Crystallogr.* **A32**, 751 (1976).
†Oxidation numbers in bold type indicate the most stable states; other states that can be achieved in aqueous (a) and nonaqueous (n) solution are also included.

---

[13] Soluble and volatile alkoxides of bismuth, M.A. Matchett, M.Y. Chang, and W.E. Buhro, *Inorg. Chem.* **29**, 359 (1990).

lanthanides.[14] The common oxidation state for the lanthanides is +3 (Table 9.14). Cerium, which can be oxidized to Ce(IV), and europium, which can be reduced to Eu(II), are chemically separable from the other lanthanides. Separation of the remaining Ln^{3+} ions is accomplished on a large scale by multistep liquid–liquid extraction in which the ions distribute themselves between an aqueous phase and an organic phase containing complexing agents. Ion-exchange chromatography, which is described in greater detail later in this section, is used to separate the individual lanthanide ions when high purity is required. Pure and mixed lanthanide metals are prepared by the electrolysis of molten lanthanide halides.

A mixture of the early lanthanide metals, including cerium, is referred to in commerce as *mischmetal*. It is used in steel making to remove impurities such as oxygen, hydrogen, sulfur, and arsenic, which reduce the mechanical strength and ductility of steel.

Beyond bismuth (Z = 83) none of the elements has stable isotopes, but two of the actinides, thorium (Th, Z = 90) and uranium (U, Z = 92), have very long-lived isotopes and occur in significant quantities in Nature (Table 9.15). The primary source of the rest is synthesis by nuclear reactions, and all of them are more radioactive than thorium and uranium. One major ore for uranium, uranite (also called pitchblende), has the approximate formula UO_2. The current primary use for uranium is in nuclear reactors for electric power generation at hundreds of reactors throughout the world.

> The principal source of the lanthanides is the phosphate material monazite; the most important actinide, uranium, is recovered from uranite.

9.20 Lanthanides

The lanthanides are a family of highly electropositive metals in Period 6 that intervene between the *s*- and *d*-blocks.[15] They are sometimes referred to as the **rare earths**; however,

Table 9.15 Names, symbols, and properties of the actinides

Z	Name	Symbol	Mass number	$t_{1/2}^*$	$r(M^{3+})/\text{Å}$†	O.N.‡
89	Actinium	Ac	227	21.8 y	1.26	**3**
90	Thorium	Th	232	1.41×10^{10} y	–	**4**
91	Protactinium	Pa	231	3.28×10^4 y	1.18	4, **5**
92	Uranium	U	238	4.47×10^9 y	1.17	3, 4, 5, **6**
93	Neptunium	Np	237	2.14×10^6 y	1.15	3, 4, **5**, 6, 7
94	Plutonium	Pu	244	8.1×10^7 y	1.14	3, **4**, 5, 6
95	Americium	Am	243	7.38×10^3 y	1.12	**3**, 4, 5, 6
96	Curium	Cm	247	1.6×10^7 y	1.11	**3**, 4
97	Berkelium	Bk	247	1.38×10^3 y	1.10	**3**, 4
98	Californium	Cf	249	350 y	1.09	**3**, 4
99	Einsteinium	Es	254	277 d	(1.07)	**3**, 4
100	Fermium	Fm	257	100 d		2, **3**
101	Mendelevium	Md	258	55 d	(1.04)	2, **3**
102	Nobelium	No	259	1.0 h		**2**, 3
103	Lawrencium	Lr	260	3 min	(1.02)	**3**

*Half-life of the most long-lived isotope.
†Effective ionic radii for C.N. = 6, from R.D. Shannon, *Acta Crystallogr.* **A32**, 751 (1976). Estimates in parentheses are from W. Brüchle, M. Schädel, U.W. Sherer, J.V. Kratz, K.E. Gregorich, D. Lee, R.M. Chasteler, H.L. Hall, R.A. Henderson, and D.L. Hoffman, *Inorg. Chim. Acta* **146**, 267 (1988).
‡Oxidation states in aqueous solution; the predominant oxidation state is in bold face. From G.T. Seaborg and W.D. Loveland, *The elements beyond uranium*, p. 84. Wiley Interscience, New York (1990).

---

[14] Although yttrium is not strictly a lanthanide, its radius and chemical properties are similar to those of the heavier lanthanides.

[15] There is an ongoing controversy over whether the lanthanides should include 14 elements from La through Yb or be displaced one position to the right, Ce through Lu. We shall include 15 elements in this discussion, La to Lu. Similar controversy extends to the actinides. See W.B. Jensen, *J. Chem. Educ.* **59**, 634 (1982).

that name is inappropriate because they are not particularly rare, except for promethium, which has no stable isotopes. The lanthanides mark the first appearance of f orbitals in the ground-state configurations of the elements. In contrast to the wide variation in properties across each series of d elements, the chemical properties of the lanthanides are highly uniform.

The elements La through Yb favor the +3 oxidation state with a uniformity that is unprecedented in the periodic table. Their common adoption of oxidation state +3 can probably be traced to the high sensitivity of the $4f$ electrons to the charge outside the inner core of the atom, so any increase in charge beyond +3 results in the $4f$ electrons becoming too firmly held to be generally available in chemical reactions. It should be noted that various relevant properties of the elements vary significantly. For example, the radii of the M^{3+} ions (Table 9.14) contract greatly, from 1.16 Å for La^{3+} to 0.98 Å for Lu^{3+}, and this 18 per cent decrease in radius leads to a great increase in the hydration enthalpy across the series. A detailed analysis shows in fact that there is fortuitous cancellation of the various terms for sublimation, solvation, and ionization in the Born–Haber cycle for aqua ion formation; as a result of this coincidence, the potential for the reduction of La^{3+} to the metal, −2.38 V, is close to that for Lu^{3+}, −2.30 V, at the other end of the block.

Superimposed on this uniformity there are some atypical oxidation states that are most prevalent when the ion can attain an empty (f^0), half-filled (f^7), or filled (f^{14}) subshell (Table 9.14). Thus, Ce^{3+}, which is an f^1 ion, can be oxidized to the f^0 ion Ce^{4+}, a strong and useful oxidizing agent. The next most common of the atypical oxidation states is Eu^{2+}, which is an f^7 ion that readily reduces water.

A Ln^{3+} ion has hard acid properties, as indicated by its preference for F^- and oxygen-containing ligands and its occurrence with PO_4^{3-} in the mineral monazite. The decrease in ionic radius from La^{3+} (1.17 Å) to Lu^{3+} (1.00 Å) is attributed to in part to the increase in Z_{eff} as electrons are added to the $4f$ subshell (Section 1.8). Detailed calculations indicate that subtle relativistic effects also make a substantial contribution to the decrease in radius across the series. Most lanthanide ions are colored, and their spectra in complexes of solids generally show much narrower and more distinct absorption bands than those for d-metal complexes. These spectra are associated with weak f–f electronic transitions. Both the narrowness of the spectral features and their insensitivity to the nature of coordinated ligands indicate that the f orbitals have a smaller radial extension than the filled $5s$ and $5p$ orbitals. Similarly, the magnetic properties of the ions (which are dealt with more fully in Chapter 13) can be explained on the assumption that the f-electrons in Ln^{3+} ions are only slightly perturbed by ligands because they are buried so deeply. Ligand-field stabilization plays no part in the chemical properties of the lanthanide complexes.

Lanthanide complexes often have high coordination numbers and a wide variety of coordination environments. The variation in structure is in harmony with the view that the spatially buried f electrons have no significant stereochemical influence, and consequently that the ligands adopt positions that minimize ligand–ligand repulsion. In addition, polydentate ligands must satisfy their own stereochemical constraints, much as for the s-block and Al^{3+} complexes. For example, many lanthanide complexes have been formed with crown ether and β-diketonate ligands. The coordination numbers for $[Ln(OH_2)_n]^{3+}$ in aqueous solution are thought to be 9 for the early lanthanides and 8 for the later, smaller members of the series, but these ions are highly labile and the measurements are subject to considerable uncertainty. Similarly, a striking variation is observed for the coordination numbers and structures of lanthanide salts and complexes. For example, the small ytterbium cation, Yb^{3+}, forms the seven-coordinate complex $[Yb(acac)_3(OH_2)]$, and the larger La^{3+} is eight-coordinate in $[La(acac)_3(OH_2)_2]$. The structures of these two complexes are approximately a capped trigonal prism and a square antiprism, respectively (Fig. 9.29).

The partially fluorinated β-diketonate ligand nicknamed fod (**37**) produces complexes with Ln^{3+} that are volatile and soluble in organic solvents. Because of their volatility these

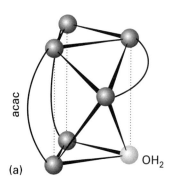

acac

(a)

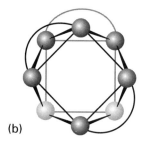

(b)

9.29 (a) Capped trigonal prism of donor atoms around ytterbium in $[Yb(acac)_3(OH_2)]$. (b) The square antiprism of donor atoms around lanthanum in $[La(acac)_3(OH_2)]$. The positions of the acac chelate rings are indicated by arcs.

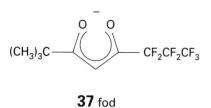

37 fod

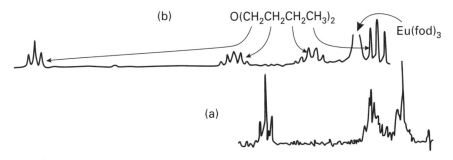

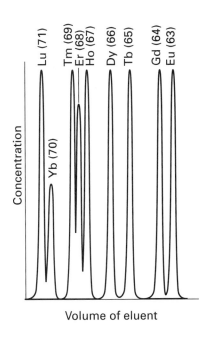

9.30 The influence of coordination to a paramagnetic Eu(III) center on the ^{1}H-NMR spectrum of an ether. (a) NMR spectrum of $O(C_4H_9)_2$. (b) The spectrum of the same ether coordinated to $[Eu(fod)_3]$. Note the greatest shift for the CH_2 groups attached to oxygen. These groups are closest to the paramagnetic Eu(III) in the complex. (From R.E. Sievers (ed.), *Nuclear magnetic resonance shift reagents*. Academic Press, New York (1973).)

9.31 Elution of the heavy lanthanide ions from a cation exchange column using ammonium 2-hydroxyisobutyrate as the eluent. Note that the higher atomic number lanthanides elute first because they have smaller radii and are more strongly complexed by the ammonium 2-hydroxyisobutyrate.

complexes are used as precursors for the synthesis of lanthanum-containing superconductors by vapor deposition (Section 18.5). Moreover, their solubility in organic solvents and presence of available coordination sites means that they are useful as NMR shift reagents, where the resonance signals for the magnetic nuclei in the attached ligand are spread out by the local magnetic field of the lanthanide ion (Fig. 9.30). This technique can be applied to a variety of molecules having a donor group that can serve as a ligand toward the lanthanide ion. The signal is most shifted for the H nuclei that are closest to the lanthanide.

Charged ligands generally have the highest affinity for the smallest Ln^{3+} ion, and the resulting increase in formation constants from large, lighter Ln^{3+} (on the left of the series) to small, heavier Ln^{3+} (on the right of the series) provides a convenient method for the chromatographic separation of these ions (Fig. 9.31 and Box 9.3). In the early days of lanthanide chemistry, before ion-exchange chromatography was developed, tedious repetitive crystallizations were used to separate the elements.

Compounds of the lanthanides find a wide range of applications, many of which are associated with their f–f electronic transitions: europium oxide or europium orthovanadate are used as red phosphors in television and computer-terminal displays, and neodymium (Nd^{3+}), samarium (Sm^{3+}), and holmium (Ho^{3+}) are employed in solid-state lasers.

Box 9.3 **Ion exchange**

To perform an ion-exchange separation, a solution of lanthanide ions is introduced at the top of a cation-exchange column in the sodium form (typically sodium polystyrene sulfonate). The Ln^{3+} ions readily undergo ion exchange displacing Na^+ ions, thus forming a band of the lanthanide ions bound to the top of the cation exchange column. To move these ions down the column and effect a separation, a solution consisting of an anionic ligand (citrate, lactate, or 2-hydroxyisobutyrate) is slowly passed through the column. These anionic chelating ligands form complexes with lanthanides. Moreover, because the complexes possess a lower positive charge than the initial Ln^{3+}, they are less tightly held by the resin than Ln^{3+}, and are displaced from the ion exchange material into the surrounding solution. The equilibria that are established between cations on the ion exchange resin (res) and neutral or anionic complexes in solution can be summarized as follows.

Initial ion exchange displacement of Na^+ by Ln^{3+} in a band at the top of column:

$$Ln^{3+}(aq) + 3\,Na^+(res) \rightleftharpoons Ln^{3+}(res) + 3\,Na^+(aq)$$

Subsequent elution with a solution of a complexing agent which leads to formation of neutral or negatively charged lanthanide complexes. To maintain electroneutrality within the ion exchange resin, sodium cations take the place of the neutral or negative lanthanide complex:

$$3\,Na^+(aq) + Ln^{3+}(res) + 3\,RCO_2^-(aq) \rightleftharpoons$$
$$3\,Na^+(res) + Ln(RCO_2)_3(aq)$$

The Ln^{3+} cations with smallest radius are most strongly bound to the anionic ligand so these ions have the greatest tendency to be eluted first (see Fig. 9.31).

Oxidation state +3 predominates for the lanthanides. Two common exceptions are Ce(IV) and Eu(II). A 15 per cent contraction in radius occurs from La^{3+} to Lu^{3+}.

9.21 Actinides

The fourteen elements from actinium (Ac, $Z = 89$) through nobelium (No, $Z = 102$) involve the filling of the 5f subshell, and in this sense are analogs of the lanthanides. However, the actinides do not exhibit the chemical uniformity of the lanthanides. A common oxidation state of the actinides (which for the present discussion will include Ac through Lr) is +3. Unlike the lanthanides, however, the early members of the series occur in a rich variety of oxidation states. The Frost diagrams (Fig. 9.32) and the data in Table 9.15 show that oxidation states higher than +3 are preferred for the early elements of the block (Th, Pa, U, and Np) and that linear or nearly linear MO^{2+} and MO_2^{2+} ions are the dominant aqua species for oxidation numbers +5 and +6.

A further general point is that the actinides have large atomic radii, and as a result often have high coordination numbers. For example, uranium in solid UCl_4 is eight-coordinate and in solid UBr_4 it is seven-coordinate in a pentagonal-bipyramidal array.

(a) Thorium and uranium

Because of their ready availability and low level of radioactivity, the chemical manipulation of thorium and uranium can be carried out with ordinary laboratory techniques. As indicated by

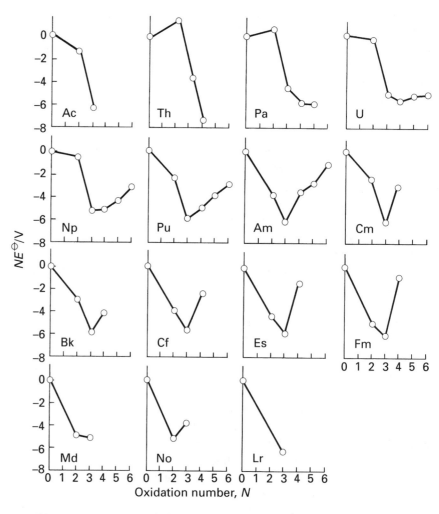

9.32 Frost diagrams for the actinides in acidic solution. (From J.J. Katz, G.T. Seaborg, and L. Morss, *Chemistry of the actinide elements*. Chapman & Hall, London (1986).)

Fig. 9.32 and Table 9.15, the only stable oxidation state of thorium in aqueous solution is $+4$. This oxidation state also dominates the solid state chemistry of the element. Eight-coordination is common in simple thorium(IV) compounds. For example, ThO_2 has the fluorite structure (in which a Th atom is surrounded by a cube of O^{2-} ions) and in $ThCl_4$ the coordination number is again 8 with dodecahedral symmetry (**38**). The coordination number of Th in $Th(NO_3)_4(OPPh_3)_2$ is 10 with the NO_3^- ions and triphenylphosphine oxide groups arranged in a capped cubic array around the metal (**39**).

The chemical properties of uranium are more varied than those of thorium because the element has access to oxidation states from $+3$ through $+6$, with $+4$ and $+6$ the most common. Uranium halides are known for the full range of oxidation states $+3$ through $+6$, with a trend toward decreasing coordination number with increase in oxidation number. The uranium atom is nine-coordinate in solid UCl_3, eight-coordinate in UCl_4, and six-coordinate for the U(V) and U(VI) chlorides U_2Cl_{10} and UCl_6, both of which are molecular compounds. The high volatility of UF_6 (it sublimes at $57\,°C$) together with the occurrence of fluorine in a single isotopic form account for the use of this compound in the separation of the uranium isotopes by gaseous diffusion or gaseous centrifugation.

Uranium metal does not form a passive oxide coating, and so it is corroded on prolonged exposure to air to give a complex mixture of oxides. The most important oxide is UO_3, which dissolves in acid to give the uranyl ion, UO_2^{2+}. In water, this bright iridescent yellow ion forms complexes with many anions, such as NO_3^- (**40**) and SO_4^{2-}. In contrast to the angular shape of the VO_2^+ ion and similar d^0 complexes, the AnO_2^{2+} moiety, with An = U, Np, Pu, and Am, maintains its linearity in all complexes. Both f-orbital bonding and relativistic effects have been invoked to explain this linearity. Unlike the lanthanides, f orbitals extend into the bonding region for the early actinides, so the complex actinide spectra are strongly affected by ligands.

The separation of uranium from most other metals is accomplished by the extraction of the neutral uranyl nitrato complex $[UO_2(NO_3)_2(OH_2)_4]$ from the aqueous phase into a polar organic phase, such as a solution of tributylphosphate dissolved in a hydrocarbon solvent. These kinds of solvent extraction processes are used to separate actinides from other fission products in spent nuclear fuel.

The fission of heavy elements, such as ^{235}U, occurs upon bombardment by neutrons. Thermal neutrons (neutrons with low velocities) bring about the fission of ^{235}U to produce two medium mass nuclides, and a large amount of energy is released because the binding energy per nucleon decreases steadily for atomic numbers beyond about 26 (see Fig. 1.2). That unsymmetrical fission of the uranium nucleus has a high probability of occurring is shown by the double-humped distribution of fission products (Fig. 9.33), which has maxima close to mass numbers 95 (Mo) and 135 (Ba). Almost all the fission products are unstable nuclides. The most troublesome are those with half-lives in the range of years to centuries: these nuclides decay fast enough to be highly radioactive but not sufficiently fast to disappear in a convenient time.

The difficult tasks of separating, immobilizing, and storing unwanted fission products have not been satisfactorily solved. One proposal is to extract uranium and plutonium and other fissionable materials, immobilize the unwanted nuclides in a glass, and deposit that glass in a geological stratum that is stable and out of communication with ground water.

> The common nuclides of thorium and uranium exhibit only low levels of radioactivity, so their chemistry has been extensively developed. The uranyl cation, with a linear OUO^{2+} array, is an important species which is associated with multiple donor atoms.

(b) The transamericium elements

For americium (Am, $Z = 95$) and beyond, the properties of the actinides begin to converge with those of the lanthanides. With increasing atomic number, M(III) becomes progressively more stable relative to higher oxidation states. Oxidation state $+3$ is dominant for curium

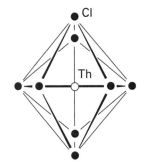

38 $ThCl_4$

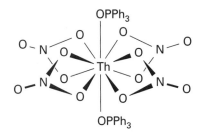

39 $[Th(NO_3)_4(OPPh_3)_2]$

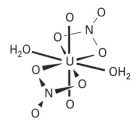

40 $[UO_2(NO_3)_2(OH_2)_2]$

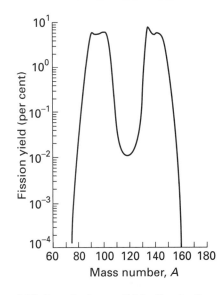

9.33 Smoothed mass distribution for the products from the thermal neutron induced fission of ^{235}U. (From G.T. Seaborg and W.D. Loveland, *The elements beyond uranium*. Wiley, New York (1990).)

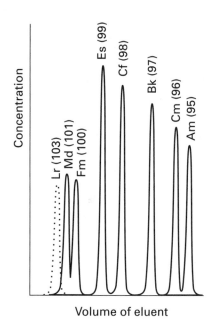

9.34 Elution of the heavy actinide ions from an ion exchange column by ammonium 2-hydroxyisobutyrate as the eluent. Note the similarity in elution sequence with Fig. 9.31. Like the Ln^{3+} ions, heavy (smaller) An^{3+} ions elute first. (From J.J. Katz, G.T. Seaborg, and L. Morss, *Chemistry of the actinide elements.* Chapman and Hall, London (1986).)

(Cm), berkelium (Bk), californium (Cf), and einsteinium (Es); these elements therefore resemble the lanthanides.

The striking difference between the chemical properties of the lanthanides and early actinides led to controversy about the most useful placement of the actinides in the periodic table. For example, before 1945, periodic tables usually showed uranium below tungsten because both elements have a maximum oxidation state of +6. The emergence of oxidation state +3 for the later actinides was a key point in the way that chemists thought about the location of the actinides in the periodic table. The similarity of the heavy actinides and the lanthanides is illustrated by their similar elution behavior in ion exchange separation (compare Fig. 9.34 and Fig. 9.31).

Because of the small quantities of material available in most cases and the intense radioactivity, most of the chemical properties of the transamericium elements have been established by experiments carried out on a microgram scale or even a few hundreds of atoms. For example, the actinide ion complexes have been absorbed on and eluted from a single bead of ion exchange material of about 0.2 mm in diameter. For the heaviest and most unstable post-actinides, such as hassium ($Z = 108$), the lifetimes are too short for chemical separation, and the identity of the element is based exclusively on the properties of the radiation it emits.

> *Oxidation state +3 becomes progressively more stable for the transamericium elements. Because of their intense radioactivity and corresponding short half-lives, their chemical properties have not been fully investigated.*

Example 9.6 Assessing the redox stability of actinide ions

Use the Frost diagram for thorium (Fig. 9.32) to describe the relative stability of Th(II) and Th(III).

Answer The initial slope in the Frost diagrams indicates that the Th^{2+} ion might be readily attained with a mild oxidant. However, Th^{2+} is above lines connecting Th(0) with the higher oxidation states, so it is susceptible to disproportionation. Thorium(III) is readily oxidized to Th(IV) and the steep negative slope indicates that it would be oxidized by water:

$$Th^{3+}(aq) + H^+(aq) \longrightarrow Th^{4+}(aq) + \tfrac{1}{2}H_2(g)$$

We can confirm from Appendix 2 that this reaction is highly favored, for $E^{\ominus} = +3.8$ V. Thus, Th(IV) will dominate in aqueous solution.

Self-test 9.6 Use the Frost diagrams and data in Appendix 2 to determine the most stable uranium ion in acid aqueous solution in the presence of air and give its formula.

Further reading

G. Wilkinson, R.D. Gillard, and J. McCleverty (ed.), *Comprehensive coordination chemistry*. Pergamon Press, Oxford (1987 *et seq.*). Volume 1 of this set provides a general introduction and succeeding volumes cover the particulars of the coordination chemistry for each metallic element.

A.F. Wells, *Structural inorganic chemistry*. Oxford University Press (1984). This book is a good single-volume reference for information on the structures of compounds of the metals.

The next two references provide general information on the metallic elements, with emphasis on mineral sources, methods of extraction, and application.

Ullmann's encyclopedia of industrial chemistry. VCH, Weinheim (1990 *et seq.*).

Kirk–Othmer encyclopedia of chemical technology. Wiley-Interscience, New York (1991 *et seq.*).

R.B. King (ed.). *Encyclopedia of inorganic chemistry*. Wiley, New York (1994). See entries for individual metals.

G.T. Seaborg and W.D. Loveland, *The elements beyond uranium*. Wiley-Interscience, New York (1990). This primer for the transuranium elements describes their discovery and provides a good overview of their separation, detection, and chemical properties.

G.L. Soloveichik, Actinide coordination chemistry. In *Encyclopedia of inorganic chemistry* (ed. R.B. King), Vol. 1, pp. 2–19. Wiley, New York (1994).

Exercises

9.1 Without using reference material, sketch the *s*-block of the periodic table, including the chemical symbols for the elements, and indicate the trends in (a) melting point, (b) radii for the common cations, (c) the tendency of peroxides to decompose thermally to simple oxides.

9.2 Which of the following pairs are most likely to form the desired compound or undergo the process mentioned? Describe the periodic trend and the physical basis for your answer in each case. (a) Cs^+ or Mg^{2+}, form an acetate complex; (b) Be or Sr, dissolve in liquid ammonia in the absence of air; (c) Li^+ or K^+, form a complex with crypt 2.2.2.

9.3 Contrast the coordination environment of a metal ion in each of the following pairs of compounds and give a plausible explanation for the difference: (a) CaF_2 versus MoS_2; (b) CdI_2 versus $MoCl_2$; (c) BeO versus CaO; (d) molybdenum(II) acetate and the beryllium acetate compound crystallized from slightly basic solution.

9.4 Without reference to a periodic table, sketch the first series of the *d*-block, including the symbols of the elements. Indicate those elements for which the group oxidation number is common by *C*, those for which the group oxidation number can be reached but the state is a powerful oxidizing agent by *O*, and those for which the group oxidation number is not achieved by *N*.

9.5 State the trend in the stability of the group oxidation number on descending a group of metallic elements in the *d*- and *p*-blocks. Illustrate the trend using standard potentials in acidic solution for Groups 5, 6, and 13/III.

9.6 For each part, give balanced chemical equations or NR (for no reaction) and rationalize your answer in terms of trends in oxidation states.

(a) $Cr^{2+}(aq) + Fe^{3+}(aq) \rightarrow$

(b) $CrO_4^{2-}(aq) + MoO_2(s) \rightarrow$

(c) $MnO_4^-(aq) + Cr^{3+}(aq) \rightarrow$

9.7 (a) Which ion, $Ni^{2+}(aq)$ or $Mn^{2+}(aq)$, is more likely to form a sulfide in the presence of H_2S? (b) Rationalize your answer with the trends in hard and soft character across Period 4. (c) Give a balanced equation for the reaction.

9.8 Preferably without reference to the text (a) write out the *d*-block of the periodic table, (b) indicate the metals that form difluorides with the rutile or fluorite structures, and (c) indicate the region of the periodic table in which metal–metal bonded halide compounds are formed, and give one example.

9.9 Write a balanced chemical equation for the reaction that occurs when *cis*-$[RuLCl(OH_2)]^+$ (see Fig. 9.12) in acidic solution at 0.2 V is made strongly basic at the same potential. Write a balanced equation for each of the successive reactions when this same complex at pH = 6 and 0.2 V is exposed to progressively more oxidizing environments up to 1.0 V. Give other examples and a reason for the redox state of the metal center affecting the extent of protonation of coordinated oxygen.

9.10 Give plausible balanced chemical reactions (or NR for no reaction) for the following combinations, and state the basis for your answer: (a) $MoO_4^{2-}(aq)$ plus $Fe^{2+}(aq)$ in acidic solution; (b) the preparation of $[Mo_6O_{19}]^{2-}(aq)$ from $K_2MoO_4(s)$; (c) $ReCl_5(s)$ plus aqueous $KMnO_4$; (d) $MoCl_2$ plus warm HBr(aq).

9.11 Speculate on the structures of the following species and present bonding models to justify your answers. (a) $[Re(O)_2(py)_4]^+$, (b) $[V(O)_2(ox)_2]^{3-}$, (c) $[Mo(O)_2(CN)_4]^{4-}$, (d) $[VOCl_4]^{2-}$.

9.12 Which of the following are likely to have structures that are typical of (a) predominantly ionic, (b) significantly covalent, (c) metal–metal bonded compounds: NiI_2, $NbCl_4$, FeF_2, PtS, and WCl_2? Rationalize the difference and speculate on the structure.

9.13 From trends in the chemical properties of the metals, write plausible balanced chemical equations (or NR for no reaction) for the following and state your reasoning: (a) TiO with aqueous HCl under an inert atmosphere; (b) $Ce^{4+}(aq)$ with $Fe^{2+}(aq)$; (c) Rb_9O_2 with water; (d) Na(am) with CH_3OH.

9.14 Indicate the probable occupancy of σ, π, and δ bonding and antibonding orbitals and the bond order for the following tetragonal prismatic complexes: (a) $[Mo_2(O_2CCH_3)_4]$, (b) $[Cr_2(O_2CC_2H_5)_4]$, (c) $[Cu_2(O_2CCH_3)_4]$.

9.15 The acid–base chemistry of liquid ammonia often parallels that of aqueous solutions. Assuming this to be the case, write balanced chemical reactions for the reaction of solid $Zn(NH_2)_2$ with (a) NH_4^+ in liquid ammonia and (b) with KNH_2 in liquid ammonia.

9.16 Give balanced chemical equations (or NR for no reaction) for the following cases and give a rationale for the reaction or lack of reaction: (a) $Hg^{2+}(aq)$ added to Cd(s); (b) $Ti^{3+}(aq)$ plus Ga(s); (c) $[AlF_6]^{3-}(aq)$ with $Tl^{3+}(aq)$.

9.17 (a) Summarize the trends in relative stabilities of the oxidation states of the elements of Groups 13/III and 14/IV, and indicate the elements that display the inert pair effect. (b) With this information in mind, write balanced chemical reactions or NR (for no reaction) for the following combinations, and explain how the answer fits the trends.

(i) $Sn^{2+}(aq) + PbO_2(s)$ (excess) $\rightarrow$ (air excluded)

(ii) $Tl^{3+}(aq) + Al(s)$ (excess) $\rightarrow$ (air excluded)

(iii) $In^+(aq) \rightarrow$ (air excluded)

(iv) $Sn^{2+} + O_2(air) \rightarrow$

(v) $Tl^+(aq) + O_2(air) \rightarrow$

9.18 Use data from Appendix 2 to determine the standard potential for each of the reactions in Exercise 9.17(b). In each case, comment on the agreement or disagreement with the qualitative assessment you gave for reactions (i) through (v).

9.19 (a) Give a balanced equation for the reaction of any of the lanthanide metals with aqueous acid. (b) Justify your answer with redox potentials and with a generalization about the most stable positive oxidation states for the lanthanides. (c) Name two lanthanides that have the greatest tendency to deviate from the usual positive oxidation state and correlate this deviation with electronic structure.

9.20 From a knowledge of their chemical properties, speculate on why cerium and europium were the easiest lanthanides to isolate before the development of ion-exchange chromatography.

9.21 Describe the general nature of the distribution of the elements formed in the thermal neutron fission of ^{235}U, and decide which of the following highly radioactive nuclides are likely to present the greatest radiation hazard in the spent fuel from nuclear power reactors: (a) ^{39}Ar, (b) ^{228}Th, (c) ^{90}Sr, (d) ^{144}Ce.

9.22 Account for the similar electronic spectra of Eu^{3+} complexes with various ligands and the variation of the electronic spectra of Am^{3+} complexes as the ligand is varied.

Problems

9.1 It is wrong to think of the periodic table as a fixed and unambiguous arrangement of the elements that was put forward by Mendeleev. In fact, there were many forerunners of Mendeleev's periodic table. One of these, put forward by William Olding in 1864, placed H, Ag, and Au in the same group. Mendeleev considered many variations on the periodic table. In one of these variations he placed Na over Cu, Ag, and Au. Discuss the arguments that might be made for and against each of these arrangements from the standpoints of: (i) chemical properties of the elements (their accessible oxidation states, and formulas and physical properties of their simple halides and oxides), and (ii) modern information on the electronic configurations of the elements and their ions.

9.2 An amateur chemist claimed the existence of a new metallic element, grubium (Gr), which has the following characteristics. Metallic Gr reacts with 1 M $H^+(aq)$ in the absence of air to produce $Gr^{3+}(aq)$ and $H_2(g)$. In the absence of air $GrCl_2(s)$ dissolves in 1 M $H^+(aq)$ and very slowly yields $H_2(g)$ plus $Gr^{3+}(aq)$. When $Gr^{3+}(aq)$ is exposed to air $GrO^{2+}(aq)$ is produced. From this information, estimate the range of potentials for (a) the reduction of $Gr^{3+}(aq)$ to Gr(s), (b) the reduction of $Gr^{3+}(aq)$ to $GrCl_2(s)$, and (c) the reduction of $GrO^{2+}(aq)$ to $Gr^{3+}(aq)$. Suggest a known element that fits the description of grubium.

9.3 The following arrangements have been used for Group 3 in the periodic table at various times:

B		
Al		
Sc	Sc	Sc
Y	Y	Lu
La	La	Lr

Discuss the relative merits of each classification assuming that the following chemical trends are the criteria for the classification: (a) aqueous oxidation state stability, (b) radii of the atoms and +3 ions, (c) electronic configuration of the atoms.

9.4 Many metal salts can be vaporized to a small extent at high temperatures and their structures studied in the vapor phase by electron diffraction. Speculate on the structures that you might expect to find for the following gas-phase species and present your reasoning: (a) TaF_5; (b) MoF_6.

9.5 The bonding in the linear uranyl ion, OUO^{2+}, is often explained in terms of significant π bonding utilizing $5f$ orbitals on the metal. Using the f orbitals illustrated in Fig. 1.16, construct a reasonable molecular orbital diagram for π-bonding with the appropriate oxygen p orbitals.

9.6 In the presence of catalytic amounts of $[Pt(P_2O_5H_2)_4]^{4-}$ and light, 2-propanol produces H_2 and acetone (E.L. Harley, E. Stiegman, A. Vlček Jr, and H.B. Gray, *J. Am. Chem. Soc.* **109**, 5233 (1987); D.C. Smith and

H.B. Gray, *Coord. Chem. Rev.* **100**, 169 (1990)). (a) Give the equation for the overall reaction. (b) Give a plausible molecular orbital scheme for the metal–metal bonding in this tetragonal-prismatic complex and indicate the nature of the excited state that is thought to be responsible for the photochemistry. (c) Indicate the metal complex intermediates and the evidence for their existence.

9.7 The initial extraction of a uranium ore with 2 M sulfuric acid contains 0.2 mol L^{-1} UO_2^{2+} and 0.2 mol L^{-1} Fe^{3+} in equilibrium with their sulfate complexes as the primary metal constituents. The equilibrium constants for coordination of UO_2^{2+} with sulfate are log $K_1 = 3.3$, log $\beta_2 = 4.3$ and for Fe^{3+} the formation constants with SO_4^{2-} as a ligand are log $K_1 = 2.2$, log $\beta_2 = 2.5$. The equilibrium mixture is adsorbed on the top of an *anion* exchange column and the column is then eluted with 2 M $HClO_4$(aq). (The perchlorate ion is an extremely weak ligand.) (a) What are the major metal-containing species in the initial solution? (b) Write chemical equations for the interaction of the initial solution with the ion exchange resin which is initially in the perchlorate form. (c) What is the effect of the 2 M perchloric acid solution on metal complexes in solution? (d) Sketch a qualitative concentration versus time profile for the two metal constituents in the eluent, and explain your reasoning.

9.8 Identify the incorrect statements in the following description and provide corrections and an explanation of the trend. (a) Sodium dissolves in ammonia and amines to produce the sodium cation and solvated electrons or the sodide ion. (b) Sodium dissolved in liquid ammonia will not react with NH_4^+ because of strong hydrogen bonding with the solvent. (c) The ion Fe^{2+}(aq) is a poorer reducing agent than V^{2+}(aq), in keeping with the trends in oxidation state stability across the 3*d* series. (d) The compound WBr_2 has a layered CdI_2 structure, in keeping with the tendency for bromides to be significantly covalent.

9.9 The existence of a maximum oxidation number of +6 for both uranium ($Z = 92$) and tungsten ($Z = 74$) prompted the placement of uranium under tungsten in early periodic tables. When the element after uranium, neptunium ($Z = 93$), was discovered in 1940 its properties did not correspond to those of rhenium ($Z = 75$), and this cast doubt on the original placement of uranium. (G.T. Seaborg and W.D. Loveland, *The elements beyond uranium*, p. 9 *et seq.* Wiley-Interscience, New York (1990).) Using standard potential data from Appendix 2, discuss the differences in oxidation state stability between Np and Re.

9.10 The extraction of uranium involves both chemical and physical separation techniques. Consult a general source (for instance, *Kirk-Othmer*) and prepare a summary of the steps involved in the separation of nuclear fuel grade uranium from its ore.

10

The boron and carbon groups

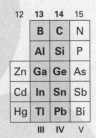

12	13	14	15
	B	C	N
	Al	Si	P
Zn	Ga	Ge	As
Cd	In	Sn	Sb
Hg	Tl	Pb	Bi
	III	IV	V

This chapter and the following two expand upon the general chemical properties of the p-block elements with emphasis on the nonmetals. The p-block is a very rich region of the periodic table, for its members show a much greater variation in properties than those of the s- and d-blocks. In contrast to the exclusively metallic character of the elements in the s- and d-blocks, the p-block elements range from metals, such as aluminum, to the highly electronegative nonmetals, such as fluorine. A single viewpoint cannot cope adequately with this great diversity, and we shall adjust our perspective as we travel across the block. Toward the right of a period, the number of oxidation states available to the elements increases, so redox properties become more important. This character is in contrast to that shown by the elements on the left of the p-block (boron, carbon, and silicon) for which redox reactions are less important. However, some of the boron and carbon group elements make up for lack of richness in their redox properties by their ability in some cases to form chains, rings, and clusters. Compounds of these elements with oxygen are important throughout the p-block, and will frequently recur in the discussion.

The elements of Groups 13/III (the boron group) and 14/IV (the carbon group) have interesting and diverse physical and chemical properties as well as being of fundamental importance in industry and nature. Carbon, of course, plays a central role in organic chemistry, but it also forms many binary compounds with metals and nonmetals and a rich range of organometallic compounds (as we see in Chapters 15 and 16). In combination with oxygen and aluminum, carbon's congener silicon is a dominant component of minerals in the Earth's crust, just as carbon in combination with hydrogen and oxygen is dominant in the biosphere. The other elements of these two groups are vital to modern high technology, particularly as semiconductors and optical waveguides.

The elements

The elements of the boron and carbon groups show a wide variation in abundance in crustal rocks, the oceans, and the atmosphere. Carbon, aluminum, and silicon are all abundant (Fig. 10.1), but the low cosmic and terrestrial abundance of boron, like that of lithium and beryllium, reflects how these light elements are sidestepped in nucleosynthesis (Section 1.1). The low abundance of heavier members of both groups is in keeping with

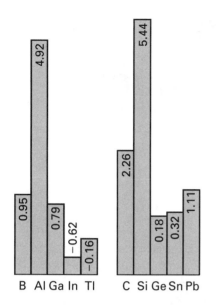

10.1 Crustal abundances of the Group 13/III and Group 14/IV elements. The numbers are logarithms of the abundance in parts per million by mass.

the progressive decrease in nuclear stability of the elements that follow iron. Except for germanium, all the carbon-group elements are more abundant than the corresponding members of the boron and nitrogen groups. This difference stems from the greater stability of nuclei that have even numbers of protons (and therefore even atomic number) compared with those having odd numbers.

There is a wide variation in chemical and physical properties on descending Groups 13 and 14. The lightest members of each group are nonmetals and the heaviest are metals. Chemical and physical similarities are particularly pronounced between boron and its diagonal neighbor silicon. In compounds, boron and silicon are chemically hard, and in their elemental form they are mechanically hard, semiconducting solids. The occurrence of two or more significantly different polymorphs is a common characteristic of *p*-block elements, and is well illustrated by elemental boron and carbon (as we describe below). The *p*-block metals, gallium, indium, thallium, and lead, were discussed in Chapter 9.

The chemical properties of boron, carbon, silicon, and germanium are distinctly those of nonmetals. Their electronegativities are similar to hydrogen's and they form many covalent hydrogen and alkyl compounds. The light elements, boron, aluminum, carbon, and silicon, are strong **oxophiles** and **fluorophiles**, in the sense that they have high affinities for oxygen and fluorine, respectively. Their oxophilic character is evident in the existence of an extensive series of oxoanions, the borates, aluminates, carbonates, and silicates. The similar oxophilicity and fluorophilicity of boron and silicon is an example of their diagonal relationship. In contrast to the behavior of the light elements, the heavy elements thallium and lead have higher affinities for soft anions, such as I^- and S^{2-} ions, than for hard anions. Thallium and lead are therefore classified as chemically soft.

Table 10.1 shows that, for most of the members of the two groups, the group oxidation number (+3 for Group 13/III and +4 for Group 14/IV) is dominant in the compounds the elements form. The major exceptions are thallium and lead, for which the most common oxidation number is 2 less than the group maximum, being +1 for thallium and +2 for lead.

Table 10.1 Properties of the boron and carbon group elements

Element	$I/(\text{kJ mol}^{-1})$	χ^*	$r_{cov}/\text{Å}$†	$r_{ion}/\text{Å}$‡	Appearance and properties	Common oxidation numbers§
Group 13/III						
B	899	2.04	0.85		Dark Semiconductor	**3**
Al	578	1.61	1.43	0.54	Metal	**3**
Ga	579	1.81	1.53	0.62	Metal m.p. 30 °C	1, **3**
In	558	1.78	1.67	0.80	Soft metal	1, **3**
Tl	589	2.04	1.71	0.89	Soft metal	**1**, 3
Group 14/IV						
C	1086	2.55	0.77		Hard insulator (diamond) Semimetal (graphite)	**4**
Si	786	1.90	1.17	0.40	Hard semiconductor	**4**
Ge	760	2.01	1.22	0.53	Metal	2, **4**
Sn	708	1.96	1.40	0.69	Metal	2, **4**
Pb	715	2.33	1.75	0.92	Soft metal	**2**, 4

* Pauling values recalculated by A.L. Allred, *J. Inorg. Nucl. Chem.* **17**, 215 (1961).
† Covalent radii from M.C. Ball and A.H. Norbury, *Physical data for inorganic chemists.* Longman, London (1974).
‡ Ionic radii from R.D. Shannon, *Acta Crystallogr.* **A32**, 751 (1975) for elements with C.N. = 6 and in their maximum group oxidation states.
§ Most common oxidation number in bold type.

This relative stability of the low oxidation state is an example of the inert pair effect (Chapter 9).

> The most abundant element in Group 13/III is aluminum and in Group 14/IV the most abundant are carbon and silicon.

The boron group (Group 13/III)

The boron group shows considerable structural diversity. For instance, boron itself exists in several hard and refractory polymorphs.[1] The three solid phases for which crystal structures are available contain the icosahedral (20-faced) B_{12} unit as a building block (Fig. 10.2). This icosahedral unit is a recurring motif in boron chemistry and we shall meet it again in the structures of metal borides and boron hydrides. All boron's congeners are metals, and their chemical properties are described in Chapter 8. Only gallium, which has one nearest neighbor in the solid (and hence resembles the structure of solid iodine), is structurally unusual.

10.1 Occurrence and recovery

The light members of Group 13/III are found in nature in combination with oxygen. The primary sources for boron are hydrated sodium borates, such as the mineral borax, $Na_2B_4O_5(OH)_4 \cdot 8H_2O$. Bauxite, the primary ore of aluminum, consists of various hydrates of aluminum oxide, such as $Al_2O_3 \cdot H_2O$. The difficulty of reducing aluminum by carbon is evident from an Ellingham diagram (Fig. 6.3), which shows that aluminum oxide has a more negative Gibbs energy of formation than its heavier congeners. Because of its abundance and wide use as a structural metal, the recovery of elemental aluminum (by the Hall–Héroult process, Section 6.1) is carried out on by far the largest scale of any element in its group. Table 10.2 summarizes the occurrence and recovery of the elements.

> In nature, boron and aluminum occur primarily as oxides and oxoanions.

Table 10.2 Mineral sources and methods of recovery of the nonmetals in Gorups 13/III and 14/IV*

Element	Natural source	Recovery
Boron	Borax	Reduction by magnesium
Carbon	Coal, hydrocarbons, graphite, charcoal	Pyrolysis
Silicon	Silica	Reduction by carbon $SiO_2 + 2C \xrightarrow{\Delta} Si + 2CO$
Germanium	Byproduct of zinc refining	Reduction of GeO_2 by hydrogen $GeO_2 + 4H_2 \xrightarrow{\Delta} Ge + 2H_2O$

*Information on the recovery of *p*-block metals is available in Table 9.13.

10.2 Compounds of boron with the electronegative elements

In the first part of this section we introduce the boron halides, which are very useful reagents and Lewis acid catalysts, and the numerous boron oxides and oxoanions. In following sections we describe the chemistry of boron with other electronegative elements.

(a) Halides

All the boron trihalides except BI_3 may be prepared by direct reaction between the elements. However, the preferred method for BF_3 is the reaction of B_2O_3 with CaF_2 in H_2SO_4. This

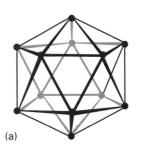

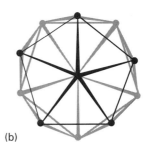

(a)

(b)

10.2 A view of the B_{12} icosahedron in α-rhombohedral boron (a) along and (b) perpendicular to the threefold axis of the crystal. The individual icosahedra are linked by 3c,2e bonds.

---

[1] However, the diversity is less than once thought, for all but three of these 'polymorphs' have been shown, in fact, to be boron-rich carbides such as $B_{50}C$ and $B_{48}Al_3C_2$.

reaction is driven in part by the reaction of the strong acid H_2SO_4 with oxides and the affinity of the hard boron atom for fluorine:

$$B_2O_3(s) + 3\,CaF_2(s) + 6\,H_2SO_4(l) \longrightarrow$$
$$2\,BF_3(g) + 3\,[H_3O][HSO_4](soln) + 3\,CaSO_4(s)$$

Boron trihalides consist of trigonal-planar BX_3 molecules. Unlike the halides of the other elements in the group, they are monomeric in the gas, liquid, and solid states. However, halogen exchange does occur, perhaps through the transient formation and dissociation of a halide-bridged dimer (**1**). Boron trifluoride and boron trichloride are gases, the tribromide is a volatile liquid, and the triiodide is a solid (Table 10.3). This trend in volatility is consistent with the increase in strength of dispersion forces with the number of electrons in the molecules.

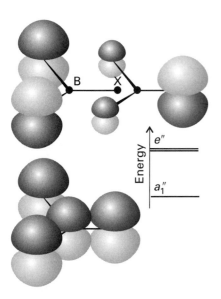

1 Br_3BBCl_3

Table 10.3 Properties representative of the boron trihalides

	Halide	m.p./°C	b.p./°C	$\Delta_fG^{\ominus}/(\text{kJ mol}^{-1})^*$
BX_3, $X =$	F	−127	−100	−1112
	Cl	−107	12	−339
	Br	−46	91	−232
	I	49	210	+21

*For the formation of the gaseous trihalide at 25 °C.

Boron trihalides are Lewis acids. We have already drawn attention to the order of their strengths in this role, which is $BF_3 < BCl_3 \leq BBr_3$ and contrary to the order of electronegativity of the attached halogens (Section 5.8). This trend stems from greater X—B π bonding for the lighter, smaller halogens giving rise to the partial occupation of the p orbital on the B atom by electrons donated by the halogen atoms (Fig. 10.3). All the boron trihalides form simple Lewis complexes with suitable bases, as in the reaction

$$BF_3(g) + :NH_3(g) \longrightarrow F_3B—NH_3(s)$$

However, boron chlorides, bromides, and iodides are susceptible to protolysis by mild proton sources such as water, alcohols, and even amines. As shown in Chart 10.1, this reaction, together with metathesis reactions, is very useful in preparative chemistry. An example is the

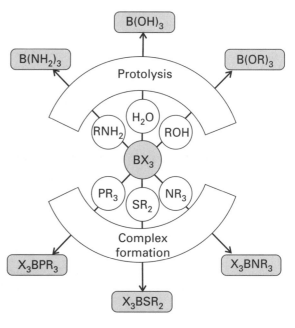

Chart 10.1 The reactions of boron–halogen compounds (X = halogen).

10.3 The bonding π orbitals of boron trihalide are largely localized on the electronegative halogen atoms, but overlap with a p orbital of boron is significant in the a_1'' orbital.

rapid hydrolysis of BCl_3:

$$BCl_3(g) + 3H_2O(l) \longrightarrow B(OH)_3(aq) + 3HCl(aq)$$

It is probable that a first step in this reaction is the formation of the complex $Cl_3B{-}OH_2$, which then eliminates HCl and reacts further with water.

Example 10.1 Predicting the products of reactions of the boron trihalides

Predict the likely products of the following reactions, and write the balanced chemical equations. (a) BF_3 and excess NaF in acidic aqueous solution; (b) BCl_3 and excess NaCl in acidic aqueous solution; (c) BBr_3 and excess $NH(CH_3)_2$ in a hydrocarbon solvent.

Answer (a) The F^- ion is a hard and fairly strong base; BF_3 is a hard and strong Lewis acid with a high affinity for the F^- ion. Hence, the reaction should result in a complex:

$$BF_3(g) + F^-(aq) \longrightarrow [BF_4]^-(aq)$$

Excess F^- and acid prevent the formation of hydrolysis products such as $[BF_3OH]^-$, which are formed at high **pH**. (b) Unlike B—F bonds, which are only mildly susceptible to hydrolysis, the other boron–halogen bonds are vigorously hydrolyzed by water. We can anticipate that BCl_3 will undergo hydrolysis rather than coordinate to aqueous Cl^-:

$$BCl_3(g) + 3H_2O(l) \longrightarrow B(OH)_3(aq) + 3HCl(aq)$$

(c) Boron tribromide will undergo protolysis with formation of a B—N bond:

$$BBr_3(g) + 6NH(CH_3)_2 \longrightarrow B(N(CH_3)_2)_3 + 3[NH_2(CH_3)_2]Br$$

In this reaction the HBr produced by the protolysis protonates excess dimethylamine.

Self-test 10.1 Write and justify balanced equations for plausible reactions between (a) BCl_3 and ethanol, (b) BCl_3 and pyridine in hydrocarbon solution, (c) BBr_3 and $F_3BN(CH_3)_3$.

The tetrafluoroborate anion, BF_4^-, which is mentioned in Example 10.1, is used in preparative chemistry when a relatively large non-coordinating anion is needed. The tetrahaloborate anions BCl_4^- and BBr_4^- can be prepared in nonaqueous solvents. However, because of the ease with which B—Cl and B—Br bonds undergo solvolysis, they are stable in neither water nor alcohols.

Boron halides are the starting point for the synthesis of many boron–carbon and boron–pseudohalogen compounds. Examples include the formation of alkylboron and arylboron compounds, such as trimethylboron by the reaction of boron trifluoride with a methyl Grignard reagent in ether solution:

$$BF_3 + 3CH_3MgI \longrightarrow B(CH_3)_3 + \text{magnesium halides}$$

When an excess of the Grignard (or organolithium) reagent is present, tetraalkyl or tetraaryl borates are formed:

$$BF_3 + Li_4(CH_3)_4 \longrightarrow Li[B(CH_3)_4] + 3LiF$$

Boron halides containing B—B bonds have been prepared. The best known of these compounds have the formula B_2X_4, with X = F, Cl, and Br, and the tetrahedral cluster compound B_4Cl_4. The B_2Cl_4 molecules are planar (**2**) in the solid state but staggered (**3**) in the gas. This difference suggests that rotation about the B—B bond is quite easy, as is expected for a single bond.

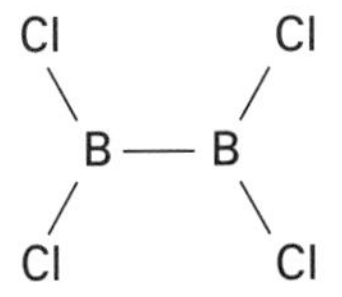

One route to B_2Cl_4 is to pass an electric discharge through BCl_3 gas in the presence of a Cl atom scavenger, such as mercury vapor. Spectroscopic data indicate that BCl is produced by electron impact on BCl_3:

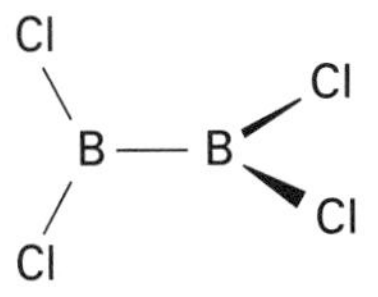

$$BCl_3(g) \xrightarrow{\text{electron impact}} BCl(g) + 2Cl(g)$$

2 D_{2h}

3 D_{2d}

The Cl atoms are scavenged by mercury vapor and removed as $Hg_2Cl_2(s)$, and the BCl fragment is thought to combine with BCl_3 to yield B_2Cl_4. Metathesis reactions can be used to make B_2X_4 derivatives from B_2Cl_4. The thermal stability of these derivatives increases with increasing tendency of the X group to form a π bond with B:

$$B_2Cl_4 < B_2F_4 < B_2(OR)_4 \ll B_2(NR_2)_4$$

It was thought for a long time that X groups with lone pairs were essential for the existence of B_2X_4 compounds, but diboron compounds with alkyl or aryl groups have been prepared. Compounds that survive at room temperature can be obtained when the groups are bulky, as in $B_2(^tBu)_4$.

Diboron tetrachloride is a highly reactive volatile molecular liquid. One of the interesting reactions of B_2Cl_4 is its addition across a C–C double bond:

$$B_2Cl_4 + C_2H_4 \xrightarrow{\text{low temperature}} Cl_2BCH_2CH_2BCl_2$$

A secondary product in the synthesis of B_2Cl_4 is B_4Cl_4, a pale yellow solid composed of molecules with the four B atoms forming a tetrahedron (4). Like B_2Cl_4, B_4Cl_4 does not have a formula analogous to those of the boranes (such as B_2H_6 discussed below). This difference may lie in the tendency of halogens to form π bonds with boron by the donation of lone electron pairs on the halide into the otherwise vacant p-orbital on boron (as in Fig 10.3).

> Boron trihalides are useful Lewis acids, with BCl_3 stronger than BF_3, and important electrophiles for the formation of boron–element bonds. Subhalides with B—B bonds, such as B_2Cl_4, are also known.

(b) Oxides and oxo compounds

Boric acid, $B(OH)_3$, is a very weak Brønsted acid in aqueous solution. However, the equilibria are more complicated than the simple Brønsted proton transfer reactions characteristic of the later p-block oxoacids. Boric acid itself is in fact primarily a weak *Lewis* acid, and the complex it forms with H_2O, $H_2OB(OH)_3$, is the actual source of protons:

$$B(OH)_3(aq) + 2\,H_2O(l) \rightleftharpoons H_3O^+(aq) + [B(OH)_4]^-(aq) \qquad pK_a = 9.2$$

As is typical for many of the lighter elements in the two groups, there is a tendency for the anion to polymerize by condensation, with the loss of H_2O. Thus, in concentrated neutral or basic solution, equilibria such as

$$3\,B(OH)_3(aq) \rightleftharpoons [B_3O_3(OH)_4]^-(aq) + H^+(aq) + 2\,H_2O(l) \qquad K = 1.4 \times 10^{-7}$$

occur to yield polynuclear anions (5).

The reaction of boric acid with an alcohol in the presence of sulfuric acid leads to the formation of simple **borate esters**, which are compounds of the form $B(OR)_3$:

$$B(OH)_3 + 3\,CH_3OH \xrightarrow{H_2SO_4} B(OCH_3)_3 + 3\,H_2O$$

Borate esters are much weaker Lewis acids than the boron trihalides, presumably because the O atom acts as an intramolecular π donor, like the F atom in BF_3 (Section 5.8), and donates electron density to the p orbital of the B atom. Hence, judging from Lewis acidity, an O atom is more effective than an F atom as a π donor toward boron. 1,2-Diols have a particularly strong tendency to form borate esters on account of the chelate effect (Section 7.7), for they produce a cyclic borate ester (6).

As with silicates and aluminates, there are many polynuclear borates, and both cyclic and chain species are known. An example is the cyclic polyborate anion $[B_3O_6]^{3-}$ (7), the conjugate base of (4). A notable feature of borate formation is the possibility of both three-coordinate B atoms, as in (6), and four-coordinate B atoms, as in $[B(OH)_4]^-$ and for one of the B atoms in (4). Polyborates form by sharing one O atom with a neighboring B atom, as in (4) and (5); structures in which two adjacent B atoms share two or three O atoms are unknown.

4 B_4Cl_4

5 $[B_3O_3(OH)_4]^-$

6

7 $[B_3O_6]^{3-}$

The rapid cooling of molten B_2O_3 or metal borates often leads to the formation of borate glasses. Although these glasses themselves have little technological significance, the fusion of sodium borate with silica leads to the formation of borosilicate glasses (such as Pyrex), which have low thermal expansivities and hence little tendency to crack when heated or cooled rapidly (Section 18.6). Borosilicate glass is used extensively for cooking- and laboratory-ware.

> Boron is oxophilic, examples of its compounds with oxygen being B_2O_3, polyborate salts, and borosilicate glasses.

(c) Compounds with nitrogen

The simplest compound of boron and nitrogen, BN, is easily synthesized by heating boron oxide with a nitrogen compound:

$$B_2O_3(l) + 2\,NH_3(g) \xrightarrow{1200\,°C} 2\,BN(s) + 3\,H_2O(g)$$

The form of boron nitride this reaction produces, and the thermodynamically stable phase under normal laboratory conditions, consists of planar sheets of atoms like those in graphite (Section 10.8). The planar sheets of alternating B and N atoms consist of edge-shared hexagons, and, as in graphite, the B–N distance within the sheet (1.45 Å) is much shorter than the distance between the sheets (3.33 Å, Fig. 10.4). The difference between the structures of graphite and boron nitride, however, lies in the register of the atoms of neighboring sheets. In BN, the hexagonal rings are stacked directly over each other, with B and N atoms alternating in successive layers; in graphite, the hexagons are staggered. Molecular orbital calculations suggest that the stacking in BN stems from a partial positive charge on B and partial negative charge on N. This charge distribution is consistent with the electronegativity difference of the two elements ($\chi_P(B) = 2.04$, $\chi_P(N) = 3.04$).

As with impure graphite, layered boron nitride is a slippery material that is used as a lubricant. Unlike graphite, however, it is a colorless electrical insulator, for there is a large energy gap between the filled and vacant π bands. The size of the band gap is consistent with its high electrical resistivity and lack of absorptions in the visible spectrum. In keeping with this large band gap, BN forms a much smaller number of intercalation compounds than graphite (Section 10.8).

Layered boron nitride changes into a denser cubic phase at high pressures and temperatures (60 kbar and 2000 °C, Fig. 10.5). This phase is a hard crystalline analog of diamond but, as it has a lower lattice enthalpy, it has a slightly lower mechanical hardness (Fig. 10.6). Cubic boron nitride is manufactured and used as an abrasive for certain high-temperature applications in which diamond cannot be used because it forms carbides with the material being ground.

There are many molecular compounds that contain BN bonds. When considering them, it is helpful to bear in mind that BN and CC are isoelectronic, which suggests that there might be analogies between these compounds and hydrocarbons. Many **amine-boranes**, the boron–nitrogen analogs of saturated hydrocarbons, can be synthesized by reaction between a nitrogen Lewis base and a boron Lewis acid:

$$\tfrac{1}{2}B_2H_6 + N(CH_3)_3 \longrightarrow H_3BN(CH_3)_3$$

However, although amine-boranes are isoelectronic with hydrocarbons, their properties are significantly different, in large part due to the difference in electronegativity of B and N. For example, whereas ammoniaborane, H_3NBH_3, is a solid at room temperature with a vapor pressure of a few torr, its analog ethane, H_3CCH_3, is a gas that condenses at $-89\,°C$. This difference can be traced to the difference in polarity of the two molecules: ethane is nonpolar, whereas ammoniaborane has a large dipole moment of 5.2 D (**8**).

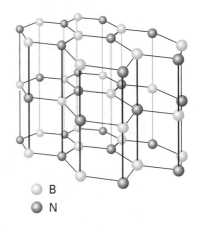

B
N

10.4 The structure of layered hexagonal boron nitride. Note that the rings are in register between layers.

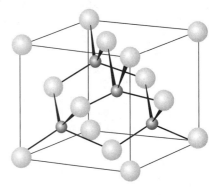

10.5 The sphalerite structure of cubic boron nitride.

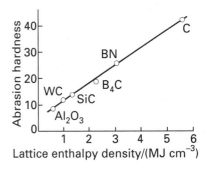

10.6 The correlation of hardness with lattice enthalpy density (the lattice enthalpy divided by the molar volume of the substance). The point for carbon represents diamond; that for boron nitride represents the diamond-like sphalerite structure.

8 NH_3BH_3

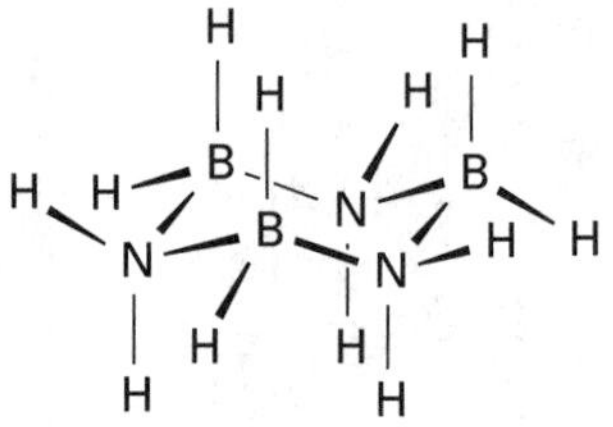

9 $N_3B_3H_{12}$

Several BN analogs of the amino acids have been prepared, including ammoniacarboxyborane, H_3NBH_2COOH, the analog of propprionic acid, CH_3CH_2COOH.[2] These compounds display significant physiological activity, including tumor inhibition and reduction of serum cholesterol.

The simplest unsaturated boron–nitrogen compound is aminoborane, H_2NBH_2, which is isoelectronic with ethene. It has only a transient existence in the gas phase because it readily forms cyclic ring compounds such as a cyclohexane analog (**9**). However, the aminoboranes do survive as monomers when the double bond is shielded from reaction by bulky alkyl groups on the N atom and by Cl atoms on the B atom (**10**). For instance, monomeric aminoboranes can be synthesized readily by the reaction of a dialkylamine and a boron halide:

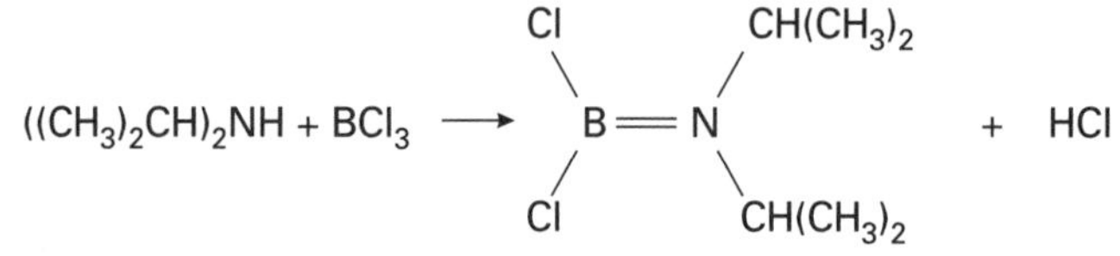

$$((CH_3)_2CH)_2NH + BCl_3 \longrightarrow$$

10 $Cl_2B=N(^iPr)_2$

The reaction also occurs with xylyl (2,4,6-trimethylphenyl) groups in place of isopropyl groups.

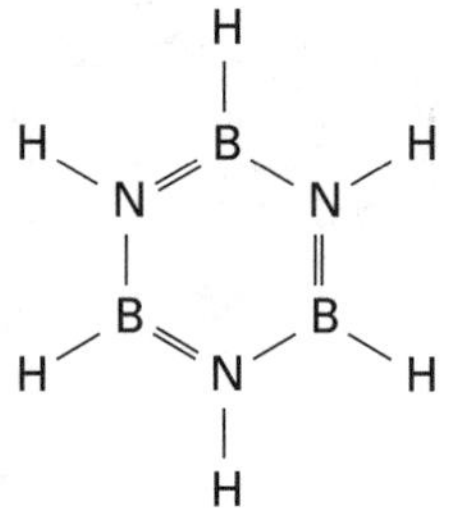

Apart from layered boron nitride, the best known unsaturated compound of boron and nitrogen is borazine, $H_3B_3N_3H_3$ (**11**), which is isoelectronic and isostructural with benzene. Borazine was first prepared in Alfred Stock's laboratory in 1926 by the reaction between diborane and ammonia. Since then many symmetrically tri-substituted derivatives have been made by procedures that depend on the protolysis of BCl bonds of BCl_3 by an ammonium salt (**12**):

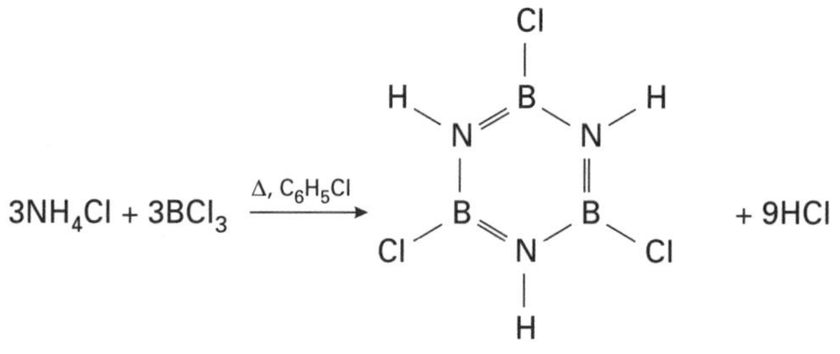

11 $H_3B_3N_3H_3$

$$3NH_4Cl + 3BCl_3 \xrightarrow{\Delta,\ C_6H_5Cl} \qquad + 9HCl$$

12 $Cl_3B_3N_3H_3$

The use of an alkylammonium chloride yields *N*-alkyl substituted *B*-trichloroborazines.

Despite their structural resemblance, there is little chemical resemblance between borazine and benzene. Once again, the difference in the electronegativities of boron and nitrogen is influential, and BCl bonds in trichloroborazine are much more labile than the CCl bonds in chlorobenzene. In the borazine compound, the π electrons are concentrated on the N atoms and there is a partial positive charge on the B atoms which leaves them open to electrophilic attack. A sign of the difference is that the reaction of a chloroborazine with a Grignard reagent or hydride source results in the substitution of Cl by alkyl, aryl, or hydride groups.

2   B.F. Spielvogel, F.U. Ahmed, and A.T. McPhail, *Inorg. Chem.* **25**, 4395 (1986).

Another example of the difference is the ready addition of HCl to borazine to produce a trichlorocyclohexane analog (**13**):

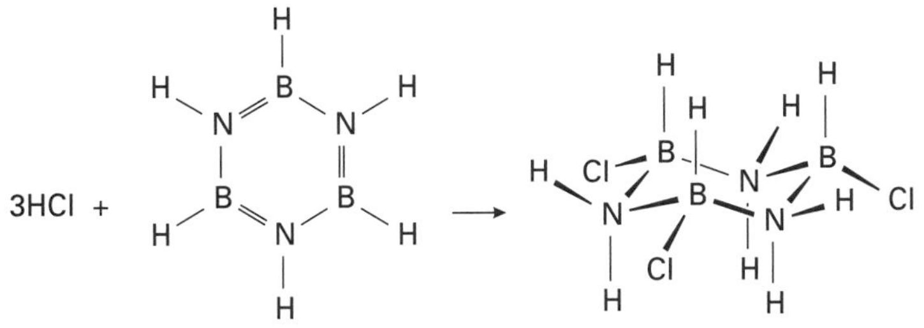

13 $Cl_3B_3N_3H_9$

The electrophile, H^+, in this reaction attaches to the partially negative N atom and the nucleophile Cl^- attaches to the partially positive B atom.

Ultraviolet spectra indicate that the HOMO–LUMO gap is greater in borazine than in benzene. This difference recalls the much greater separation of the valence and conduction bands in layered BN compared with graphite. The greater size of the gap in borazine is due to the substantial differences in the energies of the boron and nitrogen atomic orbitals, which lead to bonding orbitals that are dominated by the electronegative N atoms and excited orbitals that are predominantly those of the more electropositive boron atoms.

> Compounds containing BN, *which is isoelectronic with* CC, *include the ethane analog ammonia borane* H_3NBH_3, *the benzene analog* $H_3N_3B_3H_3$, *and BN analogs of graphite and diamond.*

Example 10.2 Preparing borazine derivatives

Give balanced chemical equations for the synthesis of borazine starting with NH_4Cl and other reagents of your choice.

Answer The reaction of NH_4Cl with BCl_3 in refluxing chlorobenzene will yield the *B*-trichloroborazine:

$$3\,NH_4Cl + 3\,BCl_3 \longrightarrow H_3N_3B_3Cl_3 + 9\,HCl$$

The Cl atoms in *B*-trichloroborazine can be displaced by hydride ions from reagents such as $LiBH_4$, to yield borazine:

$$3\,LiBH_4 + H_3N_3B_3Cl_3 \xrightarrow{\text{THF}} H_3B_3N_3H_3 + 3\,LiCl + 3\,THF{\cdot}BH_3$$

Self-test 10.2 Suggest a reaction or series of reactions for the preparation of *N,N′,N″*-trimethyl-*B,B′,B″*-trimethylborazine starting with methylamine and boron trichloride.

10.3 Boron clusters

The boron subhalide B_4Cl_4 discussed in the previous section provides a hint about boron's ability to form cluster compounds. The first recognition of these clusters was a direct outgrowth of improvements in X-ray crystallography which gave the first accurate indication of the structures of metal borides and of boranes more complex than B_2H_6. The neutral boranes and the anionic borohydrides were the first examples of molecular cluster compounds to be studied in depth.

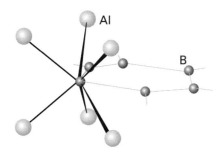

10.7 The AlB_2 structure. To give a clear picture of the hexagonal layer B atoms outside the unit cell are displayed.

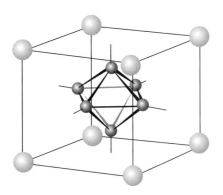

10.8 The CaB_6 structure. Note that the B_6 octahedra are connected by a bond between vertices of adjacent B_6 octahedra. The crystal is a simple cubic analog of CsCl. Thus eight Ca atoms surround the central B_6 octahedron.

14

(a) Metal borides

The direct reaction of elemental boron and a metal at high temperatures provides a useful route to many metal borides. An example is the reaction of calcium and some other highly electropositive metals with boron to produce a phase of composition MB_6:

$$Ca(l) + 6B(s) \longrightarrow CaB_6(s)$$

Metal borides are found with a wide range of compositions, for boron can occur in numerous types of structure, including isolated B atoms, chains, planar and puckered nets, and clusters. The simplest metal borides are metal-rich compounds that contain isolated B^{3-} ions. The most common examples of these compounds have the formula M_2B, where M may be one of the middle to late $3d$ metals (manganese through nickel). Another important class of metal borides contain planar or puckered hexagonal nets and have the composition MB_2 (Fig. 10.7). These compounds are formed primarily by electropositive metals, including aluminum, the early d-block metals (such as scandium through manganese), and uranium.

The boron-rich borides, typically MB_6 and MB_{12}, where M is an electropositive metal, are of even greater structural interest. In them, the B atoms link to form an intricate network of interconnecting cages. In MB_6 compounds (which are formed by metals such as sodium, potassium, calcium, barium, strontium, europium, and ytterbium) the B_6 octahedra are linked by their vertices to form a cubic framework (Fig. 10.8).[3] The linked B_6 clusters bear a charge of -1, -2, or -3 depending on the cation with which they are associated. In the MB_{12} compounds the B-atom networks are based on linked cuboctahedra (**14**) rather than the more familiar icosahedron. This type of compound is formed by some of the heavier electropositive metals, particularly those of the f-block.

> *Metal borides include boron anions as isolated B, linked closo-boron polyhedra, and hexagonal boron networks.*

(b) Bonding and structure of higher boranes and borohydrides

In Chapter 8 we introduced the simplest isolable boron hydride diborane, B_2H_6, and mentioned the existence of higher boranes. In this section we describe the structures and properties of the cage-like boranes and borohydrides, which include Alfred Stock's series B_nH_{n+4} and B_nH_{n+6} as well as the more recently discovered $[B_nH_n]^{2-}$ closed polyhedra. Boranes and borohydrides occur with a variety of shapes, some resembling nests and others (to the imaginative eye) butterflies and spiders' webs. Some of their structures are shown in Table 10.4; they are all electron-deficient in the sense that they cannot be described by simple Lewis structures.[4]

The modern interpretation of the bonding in boranes and borohydrides derives from the work of Christopher Longuet-Higgins who, as an undergraduate at Oxford, published a seminal paper in which he introduced the concept of 3c,2e bonds (Section 3.12). He later developed a fully delocalized molecular orbital treatment for boron polyhedra and predicted the stability of the icosahedral ion $[B_{12}H_{12}]^{2-}$, which was subsequently verified. William Lipscomb and his students in the USA used single-crystal X-ray diffraction to determine the structures of a large number of boranes and borohydrides, and generalized the concept of multicenter bonding to these more complex species.

---

[3] The similarity between these anionic linked octahedra and the structure of *closo*-$[B_6H_6]^{2-}$, to be described later, should be noted.

[4] An introduction to this subject is provided by C.E. Housecroft, *Boranes and metalloboranes*. Ellis Horwood, Chichester (1990), and by a series of articles in R.B. King (ed.), *Encyclopedia of inorganic chemistry*, Wiley, New York (1994): Boron hydrides by J.T. Spencer, pp. 338–57; Metallacarboranes, by C.E. Housecroft, pp. 375–89; Boron–nitrogen compounds by R.T. Paine, pp. 389–401; Orgaonoboranes, by J.A. Soderquist, pp. 401–33; and Polyhedral carboranes by R.E. Williams, pp. 433–52.

Table 10.4 Structure of some molecular and anionic boranes

closo-B$_n$H$_n^{2-}$

nido-B$_n$H$_{n+4}$

arachno-B$_n$H$_{n+6}$

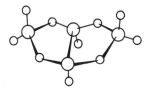

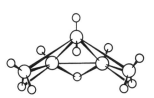

B$_4$H$_{10}$ Tetraborane(10), C_{2v}

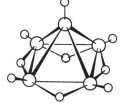

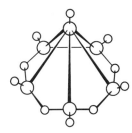

B$_5$H$_9$ Pentaborane(9), C_{4v}

B$_5$H$_{11}$ Pentaborane(11), C_s

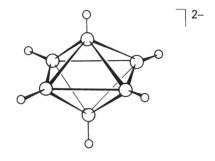

2–

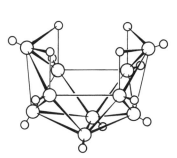

closo-[B$_6$H$_6$]$^{2-}$, O_h

B$_6$H$_{10}$ Hexaborane(10), C_s

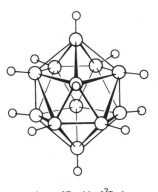

2–

closo-[B$_{12}$H$_{12}$]$^{2-}$, I_h

B$_{10}$H$_{14}$ Decaborane(14), C_{2v}

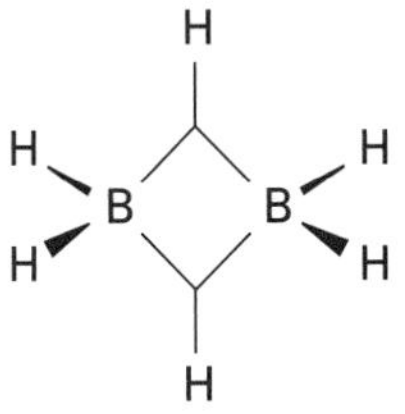

Boron cluster compounds are best considered from the standpoint of fully delocalized molecular orbitals containing electrons that contribute to the stability of the entire molecule. However, it is sometimes fruitful to identify groups of three atoms and to regard them as bonded together by versions of the 3c,2e bonds of the kind that occur in diborane itself (**15**). In the more complex boranes, the three centers of the 3c,2e bonds may be B—H—B bridge bonds, but they may also be bonds in which three B atoms lie at the corners of an equilateral

15 B$_2$H$_6$

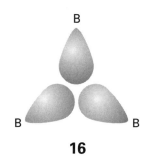

16

17

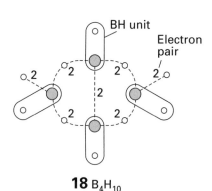

18 B_4H_{10}

triangle with their sp^3 hybrid orbitals overlapping at its center (**16**). To reduce the complexity of the structural diagrams, the illustrations that follow will not in general indicate the 3c,2e bonds in the structures.

> The bonding in boron hydrides and polyhedral boron hydride ion can be approximated by conventional 2c,2e bonds together with 3c,2e bonds.

(c) Wade's rules

A correlation between the number of electrons (counted in a specific way), the formula, and the shape of the molecule was established by the British chemist Kenneth Wade in the 1970s.[5] These so-called **Wade's rules** apply to a class of polyhedra called **deltahedra** (because they are made up of triangular faces resembling Greek deltas, Δ) and can be used in two ways. For molecular and anionic boranes, they enable us to predict the general shape of the molecule or anion from its formula. However, because the rules are also expressed in terms of the number of electrons, we can extend them to other analogous species in which there are atoms other than boron, such as carboranes and other *p*-block clusters. Here we concentrate on the boron clusters, where knowing the formula is sufficient for predicting the shape. However, so that we can cope with other clusters we shall show how to count the framework electrons too.

The building block from which the deltahedron is constructed is assumed to be one BH group (**17**). The electrons in the B—H bond are ignored in the counting procedure, but all others are included whether or not it is obvious that they help to hold the skeleton together. By the 'skeleton' is meant the framework of the cluster with each BH group counted as a unit. If a B atom happens to carry two H atoms, only one of the B—H bonds is treated as a unit.[6] For instance, in B_5H_{11}, one of the B atoms has two 'terminal' H atoms, but only one BH entity is treated as a unit, the other pair of electrons being treated as part of the skeleton and hence referred to as 'skeletal electrons'. A BH group makes two electrons available to the skeleton (the B atom provides three electrons and the H atom provides one but, of these four, two are used for the B—H bond).

Example 10.3 Counting skeletal electrons

Count the number of skeletal electrons in B_4H_{10} (Table 10.4).

Answer Four BH units contribute $4 \times 2 = 8$ electrons, and the six additional H atoms contribute a further 6 electrons, giving 14 in all. The resulting seven pairs are distributed as shown in (**18**): two are used for the additional terminal B—H bonds, four are used for the four B—H—B bridges, and one is used for the central B—B bond.

Self-test 10.3 How many skeletal electrons are present in B_5H_9?

According to Wade's rules (Table 10.5), species of formula $[B_nH_n]^{2-}$ and $n + 1$ pairs of skeletal electrons have a **closo structure**, a name derived from the Greek for 'cage', with a B atom at each corner of a closed deltahedron and no B—H—B bonds. This series of anions is known for $n = 5$ to 12, and examples include the trigonal-bipyramidal $[B_5H_5]^{2-}$ ion, the octahedral $[B_6H_6]^{2-}$ ion, and the icosahedral $[B_{12}H_{12}]^{2-}$ ion. The *closo*-borohydrides and their carborane analogs (Section 10.6) are typically thermally stable and moderately unreactive.

Boron clusters of formula B_nH_{n+4} and $n + 2$ pairs of skeletal electrons have the **nido structure**, a name derived from the Latin for 'nest'. They can be regarded as derived from a

5 More detailed accounts of Wade's rules are available in K.D. Wade, *Adv. Inorg. Chem. Radiochem.* **18**, 1 (1976), and in J.T. Spencer, In *Encyclopedia of inorganic chemistry* (ed. R.B. Knight), Vol. 1, p. 338. Wiley, New York (1994).

6 This rather odd feature is just a part of Wade's rules: it works. That is, counting electrons in this way provides a parameter that helps to correlate properties.

Table 10.5 Classification and electron count of boron hydrides

Type	Formula*	Skeletal electron pairs	Examples
Closo	$[B_nH_n]^{2-}$	$n+1$	$[B_5H_5]^{2-}$ to $[B_{12}H_{12}]^{2-}$
Nido	B_nH_{n+4}	$n+2$	B_2H_6, B_5H_9, B_6H_{10}
Arachno	B_nH_{n+6}	$n+3$	B_4H_{10}, B_5H_{11}
Hypho†	B_nH_{n+8}	$n+4$	None‡

*In some cases, protons can be removed; thus $[B_5H_8]^-$ results from the deprotonation of B_5H_9.
†The name comes from the Greek word for 'net'.
‡Some derivatives are known.

closo-borane that has lost one vertex but have B—H—B bonds as well as B—B bonds. An example is B_5H_9 (Table 10.4). In general, the thermal stability of the *nido*-boranes is intermediate between that of *closo*- and the *arachno*-boranes described below.

Clusters of formula B_nH_{n+6} and $n+3$ skeletal electron pairs have an **arachno structure**, from the Greek for 'spider' (for they resemble an untidy spider's web). They can be regarded as *closo*-borane polyhedra less two vertices (and must have B—H—B bonds). One example of an *arachno*-borane is pentaborane(11) (B_5H_{11}; Table 10.4). As with most *arachno*-boranes, pentaborane(11) is thermally unstable at room temperature and is highly reactive.[7]

> Boron hydride structures include simple polyhedral *closo* compounds and the progressively more open *nido* and *arachno* structures.

Example 10.4 Using Wade's rules

Infer the structure of $[B_6H_6]^{2-}$ from its formula and from its electron count.

Answer We note that the formula $[B_6H_6]^{2-}$ belongs to a class of borohydrides having the formula $[B_nH_n]^{2-}$, which is characteristic of a *closo* species. Alternatively, we can count the number of skeletal electron pairs and from that deduce the structural type. Assuming one B—H bond per B atom, there are six BH units to take into account and therefore twelve skeletal electrons plus two from the overall charge of -2: $6 \times 2 + 2 = 14$, or $2(n+1)$ with $n = 6$. This number is characteristic of *closo* clusters. The closed polyhedron must contain triangular faces and six vertices; therefore an octahedral structure is indicated.

Self-test 10.4 How many framework electron pairs are present in B_4H_{10} and to what structural category does it belong? Sketch its structure.

(d) The origin of Wade's rules

Wade's rules have been justified by molecular orbital calculations. We shall indicate the kind of reasoning involved by considering the first of them (the $n+1$ rule). In particular, we shall show that $[B_6H_6]^{2-}$ has a low energy if it has an octahedral *closo* structure, as predicted by the rules.

A B—H bond utilizes one electron and one orbital of the B atom, leaving three orbitals and two electrons for the skeletal bonding. One of these orbitals, which is called a **radial orbital**, can be considered to be a boron *sp* hybrid pointing toward the interior of the fragment (as in (**17**)). The remaining two boron *p* orbitals, the **tangential orbitals**, are perpendicular to the radial orbital (**19**). The shapes of the 18 symmetry-adapted linear combinations of these

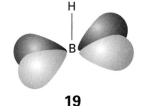

19

---

[7] A rarer set of even more open heteroborane clusters are referred to as *hypho*-boranes and a series in which polyhedra share one or more atoms are *conjuncto*-boranes.

18 orbitals in an octahedral B_6H_6 cluster can be inferred from the drawings in Appendix 4, and we show the ones with net bonding character in Fig. 10.9.

The lowest energy molecular orbital is totally symmetric (a_{1g}) and arises from in-phase contributions from all the radial orbitals. Calculations show that the next higher orbitals are the t_{1u} orbitals, each of which is a combination of four tangential and two radial orbitals. Above these three degenerate orbitals lie another three t_{2g} orbitals, which are tangential in character, giving seven bonding orbitals in all. Hence, there are seven orbitals with net bonding character delocalized over the skeleton, and they are separated by a considerable gap from the remaining eleven largely antibonding orbitals (Fig. 10.10).

There are seven electron pairs to accommodate, one pair from each of the six B atoms and one pair from the overall charge (-2). These seven pairs can all enter and fill the seven bonding skeleton orbitals, and hence give rise to a stable structure, in accord with the $n + 1$ rule. Note that the unknown neutral octahedral B_6H_6 molecule would have too few electrons to fill the t_{2g} bonding orbitals.

> The molecular orbitals in a closo-borane can be constructed from BH units each of which contributes one radial atomic orbital pointing toward the center of the cluster and two perpendicular p orbitals that are tangential to the polyhedron.

(e) Structural correlations

A very useful structural correlation between *closo*-, *nido*-, and *arachno*-species is based on the observation that clusters with the same numbers of skeletal electrons are related by removal of successive BH groups and the addition of the appropriate numbers of electrons and H atoms. This conceptual process provides a good way to think about the structures of the various boron clusters but does not represent how they are interconverted chemically.

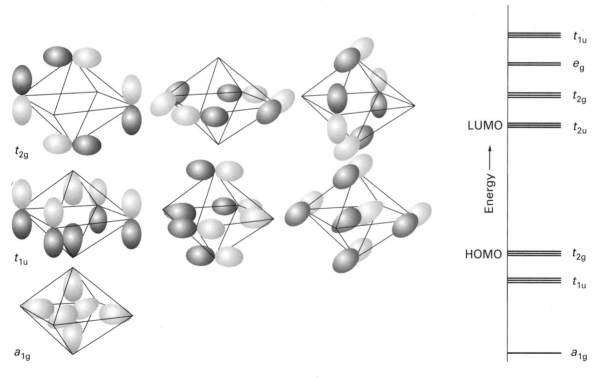

10.9 Radial and tangential bonding molecular orbitals for $[B_6H_6]^{2-}$. The relative energies are $a_{1g} < t_{1u} < t_{2g}$.

10.10 Schematic molecular orbital energy levels of the B-atom skeleton of $[B_6H_6]^{2-}$. The form of the bonding orbitals is shown in Fig. 10.9.

The idea is amplified in Fig. 10.11, where the removal of a BH unit and two electrons and the addition of four H atoms converts the octahedral *closo*-$[B_6H_6]^{2-}$ anion to the square-pyramidal *nido*-B_5H_9 borane. A similar process (removal of a BH unit and addition of two H atoms) converts *nido*-B_5H_9 into a butterfly-like *arachno*-B_4H_{10} borane. Each of these three boranes has 14 skeletal electrons but, as the number of skeletal electrons per B atom increases, the structure becomes more open. A more systematic correlation of this type is indicated for many different boranes in Fig. 10.12 (overleaf).

> Conceptually, the *closo*-, *nido*-, and *arachno*-structures are related by the successive removal of a BH fragment and addition of H or electrons.

10.4 Synthesis of higher boranes and borohydrides

As discovered by Alfred Stock and perfected by many subsequent workers, the controlled pyrolysis of B_2H_6 in the gas phase provides a route to most of the higher boranes and borohydrides, including B_4H_{10}, B_5H_9, and $B_{10}H_{14}$. A key first step in the proposed mechanism is the dissociation of B_2H_6 and the condensation of the resulting BH_3 with borane fragments. For example, the mechanism of tetraborane(10) formation by the pyrolysis of diborane appears to be

$$B_2H_6(g) \longrightarrow 2\,BH_3(g)$$
$$B_2H_6(g) + BH_3(g) \longrightarrow B_3H_7(g) + H_2(g)$$
$$BH_3(g) + B_3H_7(g) \longrightarrow B_4H_{10}(g)$$

The synthesis of tetraborane(10), B_4H_{10}, is particularly difficult because it is highly unstable, in keeping with the instability of the B_nH_{n+6} (*arachno*) series. To improve the yield, the product that emerges from the hot reactor is immediately quenched on a cold surface. Pyrolytic syntheses to form species belonging to the more stable B_nH_{n+4} (*nido*) series proceed in higher yield, without the need for a rapid quench. Thus B_5H_9 and $B_{10}H_{14}$ are readily prepared by the pyrolysis reaction. More recently, these brute force methods of pyrolysis have given way to more specific methods that are described below (see reaction 3 on p. 347).

> Pyrolysis followed by rapid quenching provides one method of converting small boranes to larger boranes.

(a) The characteristic reactions of boranes and borohydrides

The characteristic reactions of boron clusters with a Lewis base range from cleavage of BH_n from the cluster to deprotonation of the cluster, cluster enlargement, or abstraction of one or more protons.

1 Lewis base cleavage reactions have already been introduced in Section 8.9 in connection with diborane. Another example is

With the robust higher borane B_4H_{10}, cleavage may break some B—H—B bonds leading to partial fragmentation of the cluster:

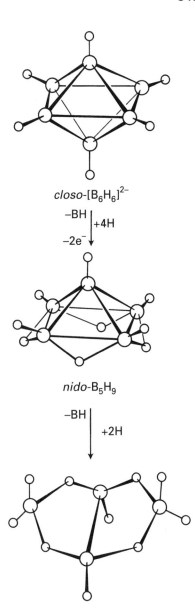

closo-$[B_6H_6]^{2-}$

$$\begin{array}{c} -BH \\ -2e^- \end{array} \Big\downarrow +4H$$

nido-B_5H_9

$$-BH \Big\downarrow +2H$$

arachno-B_4H_{10}

10.11 Structural correlations between a B_6 *closo* octahedral structure, a B_5 *nido* square pyramid, and a B_4 *arachno* butterfly.

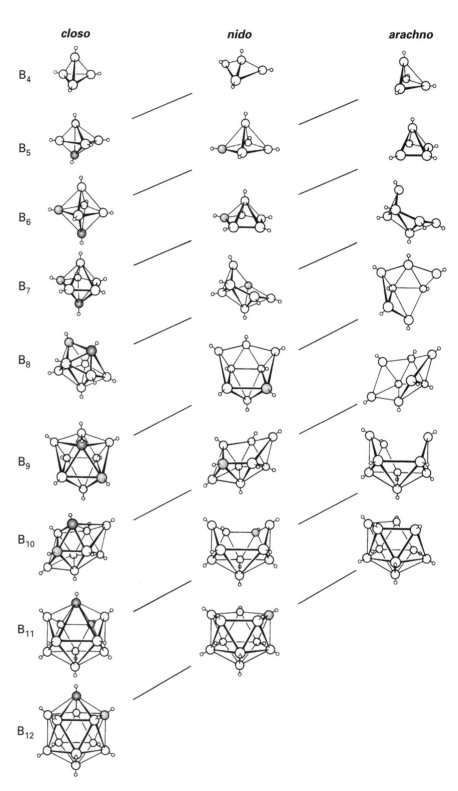

10.12 Structural relations between *closo*, *nido*, and *arachno* boranes and heteratomic boranes. Diagonal lines connect species that have the same number of skeletal electrons. Hydrogen atoms beyond those in the B—H framework and charges have been omitted. The dark tinted atom is removed first, and the pale tinted atom is removed second. (Based on R.W. Rudolph, *Acc. Chem. Res.* **9**, 446 (1976).)

2 Deprotonation, rather than cleavage, occurs readily with the large borane $B_{10}H_{14}$:

$$B_{10}H_{14} + N(CH_3)_3 \longrightarrow [HN(CH_3)_3]^+[B_{10}H_{13}]^-$$

The structure of the product anion indicates that deprotonation occurs from a 3c,2e B—H—B bridge, leaving the electron count on the boron cluster unchanged. This deprotonation of a B—H—B 3c,2e bond to yield a 2c,2e bond occurs without major disruption of the bonding:

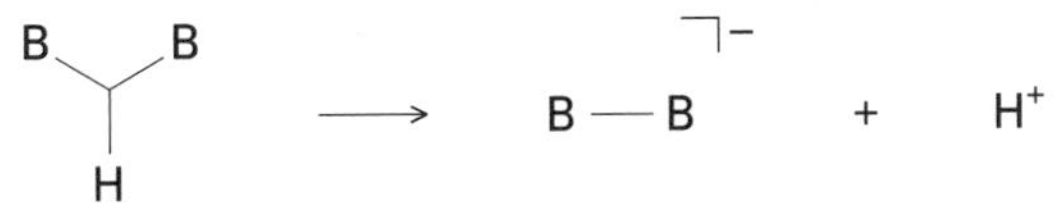

The Brønsted acidity of boron hydrides increases approximately with size:

$$B_4H_{10} < B_5H_9 < B_{10}H_{14}$$

This trend correlates with the greater delocalization of charge in the larger clusters, in much the same way that delocalization accounts for the greater acidity of phenol than methanol. The variation in acidity is illustrated by the observation that, as shown above, the weak base trimethylamine deprotonates decaborane(10), but the much stronger base methyllithium is required to deprotonate B_5H_9:

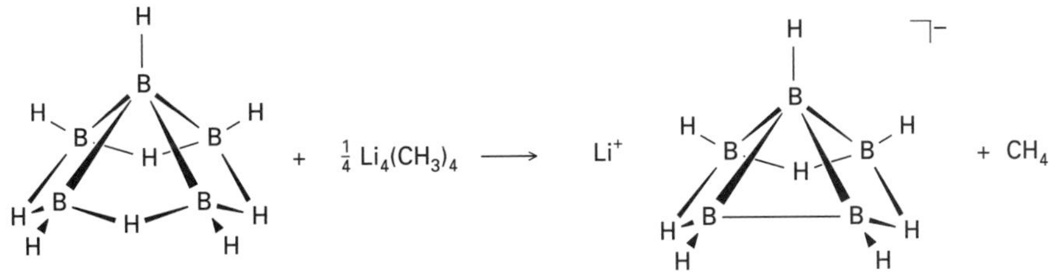

Hydridic character is most characteristic of small anionic borohydrides. As an illustration, whereas BH_4^- readily surrenders a hydride ion in the reaction

$$BH_4^- + H^+ \longrightarrow \tfrac{1}{2}B_2H_6 + H_2$$

the $[B_{10}H_{10}]^{2-}$ ion survives even in strongly acidic solution. Indeed, the hydronium salt $(H_3O)_2B_{10}H_{10}$ can even be crystallized.

3 The cluster-building reaction between a borane and a borohydride provides a convenient route to higher borohydride ions:[8]

$$5\,K[B_9H_{14}] + 2\,B_5H_9 \xrightarrow{\text{polyether, 85 °C}} 5\,K[B_{11}H_{14}] + 9\,H_2$$

Similar reactions are used to prepare other borohydrides, such as $[B_{10}H_{10}]^{2-}$. This type of reaction has been employed to synthesize a wide range of polynuclear borohydrides. Boron-11 NMR spectroscopy (Fig. 10.13) reveals that the boron skeleton in $[B_{11}H_{14}]^-$ consists of an icosahedron with a missing vertex.

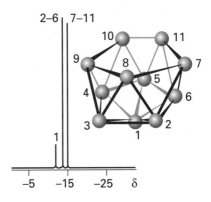

10.13 The proton-decoupled ^{11}B-NMR spectrum of $[B_{11}H_{14}]^-$. The *nido* structure (truncated icosahedron) is indicated by the 1:5:5 pattern. (Based on N.S. Hosmane, J.R. Wermer, Z. Hong, T.D. Getman, and S.G. Shore, *Inorg. Chem.* **26**, 3638 (1987).)

---

8 KB$_9$H$_{14}$ is produced from KH and B$_5$H$_9$, so by adjusting the ratio of B$_5$H$_9$ to KH a 'single pot reaction' of KH and B$_5$H$_9$ gives the desired product. The 'pot' is a reaction flask attached to a chemical high vacuum line (Box 8.1). Further details are given by N.S. Hosmane, J.R. Wermer, Z. Hong, T.D. Getman, and S.G. Shore, *Inorg. Chem.* **26**, 3638 (1987).

4 The **electrophilic displacement** of H^+ provides a route to alkylated and halogenated species. As with Friedel–Crafts reactions, the electrophilic displacement of H is catalyzed by a Lewis acid, such as aluminum chloride, and the substitution generally occurs on the closed portion of the boron clusters:

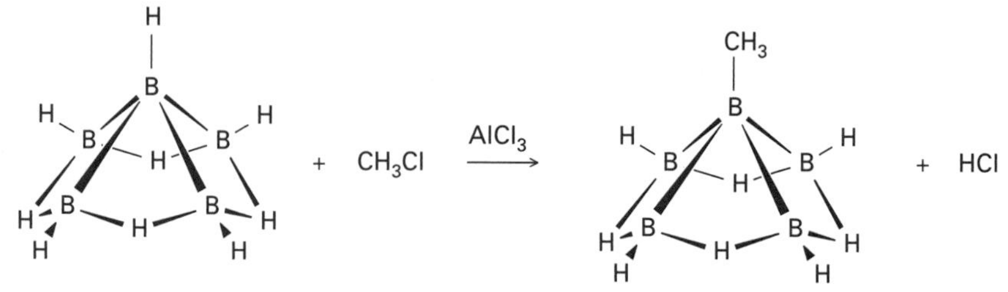

| *Characteristic reactions of boranes are cleavage of a BH_2 from diborane and tetraborane by NH_3, deprotonation of large boron hydrides by bases, reaction of a boron hydride with a borohydride ion to produce a larger borohydride anion, and Friedel–Crafts type substitution of an alkyl group for hydrogen in pentaborane and some larger boron hydrides.*

Example 10.5 Proposing a structure for a boron-cluster reaction product

Propose a structure for the product of the reaction of $B_{10}H_{14}$ with $LiBH_4$ in a refluxing polyether, $CH_3OC_2H_4OCH_3$ (which boils at $162\,°C$).

Answer The prediction of the probable outcome for the reactions of a boron cluster is difficult because several products are often plausible and the actual outcome is often sensitive to the conditions of the reaction. In the present case we note that an acidic borane, $B_{10}H_{14}$, is brought into contact with the hydridic anion BH_4^- under rather vigorous conditions. Therefore, we might expect the evolution of hydrogen:

$$B_{10}H_{14} + Li[BH_4] \xrightarrow{\text{ether, } R_2O} Li[B_{10}H_{13}] + R_2OBH_3 + H_2$$

This set of products suggests the further possibility of condensation of the neutral BH_3 complex with $B_{10}H_{13}^-$ to yield a larger borohydride. That is in fact the observed outcome under these conditions:

$$Li[B_{10}H_{13}] + R_2OBH_3 \longrightarrow Li[B_{11}H_{14}] + H_2 + R_2O$$

It turns out that, in the presence of excess $LiBH_4$, the cluster building continues to give the very stable icosahedral $B_{12}H_{12}^{2-}$ anion:

$$Li[nido\text{-}B_{11}H_{14}] + Li[BH_4] \longrightarrow Li_2[closo\text{-}B_{12}H_{12}] + 3\,H_2$$

Self-test 10.5 Propose a plausible product for the reaction between $Li[B_{10}H_{13}]$ and $Al_2(CH_3)_6$.

10.5 Metallaboranes

Many **metallaboranes**, or metal-containing boron clusters, have been characterized. In some cases the metal is appended to a borohydride ion through hydrogen bridges. A more common and generally more robust group of metallaboranes have direct metal–boron bonds. An example of a main-group metallaborane with an icosahedral framework is shown in

Fig. 10.14. It is prepared by interaction of the acidic hydrogens in $Na_2[B_{11}H_{13}]$ with trimethylaluminum:

$$2\,[B_{11}H_{13}]^{2-} + Al_2(CH_3)_6 \xrightarrow{\Delta} 2\,[B_{11}H_{11}AlCH_3]^{2-} + 4\,CH_4$$

When B_5H_9 is heated with $Fe(CO)_5$, a metallated analog of pentaborane is formed (**20**).

> Main-group and *d*-block metals may be incorporated into boron hydrides through B—H—M *bridges or more robust* B—M *bonds.*

10.6 Carboranes

Closely related to the polyhedral boranes and borohydrides are the **carboranes** (more formally, the *carbaboranes*), a large family of clusters that contain both B and C atoms. Now we begin to see the full generality of Wade's electron counting rules, for BH⁻ is isoelectronic with CH (**21**), and we can expect the polyhedral borohydrides and carboranes to be related. Thus, an analog of $[B_6H_6]^{2-}$ (**22**) is the neutral carborane $B_4C_2H_6$ (**23**).

One entry into the interesting and diverse world of carboranes is the conversion of decaborane(14) to *closo*-1,2-$B_{10}C_2H_{12}$ (**24**). The first reaction in this preparation is the displacement of an H_2 molecule from decaborane by a thioether:

$$B_{10}H_{14} + 2\,SEt_2 \longrightarrow B_{10}H_{12}(SEt_2)_2 + H_2$$

The loss of two H atoms in this reaction is compensated by the donation of electron pairs by the added thioethers, so the electron count is unchanged. The product of the reaction is then converted to the carborane by the addition of an alkyne:

$$B_{10}H_{12}(SEt_2)_2 + C_2H_2 \longrightarrow B_{10}C_2H_{12} + 2\,SEt_2 + H_2$$

The four π electrons of ethyne displace two thioether molecules (two two-electron donors) and an H_2 molecule (which leaves with two additional electrons). The net loss of two electrons correlates with the change in structure from a *nido* starting material to the *closo* product. The

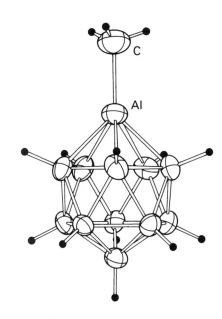

10.14 The structure of *closo*-$[B_{11}H_{11}AlCH_3]^{2-}$. (T.D. Getman and S.G. Shore, *Inorg. Chem.* **27**, 3439–40 (1988).)

20 [Fe(CO)$_3$B$_4$H$_8$]

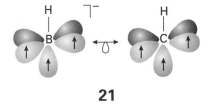

21

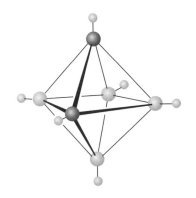

22 *closo*-[B$_6$H$_6$]$^{2-}$

23 *closo*-1, 2-B$_4$C$_2$H$_6$

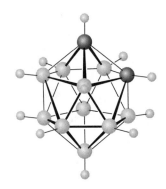

24 *closo*-1, 2-B$_{10}$C$_2$H$_{12}$

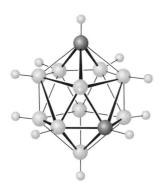

25 *closo*-1,7-B$_{10}$C$_2$H$_{12}$

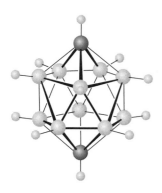

26 *closo*-1,12-B$_{10}$C$_2$H$_{12}$

C atoms are in adjacent (1,2) positions, reflecting their origin from ethyne. This *closo*-carborane survives in air and can be heated without decomposition. At 500 °C in an inert atmosphere it undergoes isomerization into 1,7-B$_{10}$C$_2$H$_{12}$ (**25**), which in turn isomerizes at 700 °C to the 1,12-isomer (**26**).

The H atoms attached to carbon in *closo*-B$_{10}$C$_2$H$_{12}$ are very mildly acidic, so it is possible to lithiate these compounds with butyllithium:

$$B_{10}C_2H_{12} + 2\,LiC_4H_9 \longrightarrow B_{10}C_2H_2Li_2 + 2\,C_4H_{10}$$

These dilithiocarboranes are good nucleophiles and undergo many of the reactions characteristic of organolithium reagents (Section 15.7). Thus, a wide range of carborane derivatives can be synthesized. For example, reaction with CO$_2$ gives a dicarboxylic acid carborane:

$$B_{10}C_2H_{10}Li_2 \xrightarrow{\text{(1) }2\,CO_2;\ (2)\ 2\,H_2O} B_{10}C_2H_{10}(COOH)_2$$

Similarly, I$_2$ leads to the diiodocarborane and NOCl yields B$_{10}$C$_2$H$_{10}$(NO)$_2$.

Although 1,2-B$_{10}$C$_2$H$_{12}$ is very stable, the cluster can be partially fragmented in strong base, and then deprotonated with NaH to yield *nido*-[B$_9$C$_2$H$_{11}$]$^{2-}$:

$$B_{10}C_2H_{12} + OEt^- + 2\,EtOH \longrightarrow [B_9C_2H_{12}]^- + B(OEt)_3 + H_2$$
$$Na[B_9C_2H_{12}] + NaH \longrightarrow Na_2[B_9C_2H_{11}] + H_2$$

The importance of these reactions is that *nido*-[B$_9$C$_2$H$_{11}$]$^{2-}$ (Fig. 10.15a) is an excellent ligand. In this role it mimics the cyclopentadienide ligand (C$_5$H$_5^-$; Fig. 10.15b) which is widely employed in organometallic chemistry:

$$2\,Na_2[B_9C_2H_{11}] + FeCl_2 \xrightarrow{\text{THF}} 2\,NaCl + Na_2[Fe(B_9C_2H_{11})_2]$$
$$2\,Na[C_5H_5] + FeCl_2 \xrightarrow{\text{THF}} 2\,NaCl + Fe(C_5H_5)_2$$

Although we shall not go into the details of their synthesis, a wide range of metal-coordinated carboranes can be synthesized. A notable feature is the ease of formation of multi-decker sandwich compounds containing carborane ligands, (**27**) and (**28**).[9] The highly negative [B$_3$C$_2$H$_5$]$^{4-}$ ligand has a much greater tendency to form stacked sandwich compounds than the less negative and therefore poorer donor C$_5$H$_5^-$.

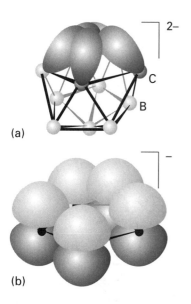

10.15 The isolobal relation between (a) [B$_9$C$_2$H$_{11}$]$^{2-}$ and (b) C$_5$H$_5^-$. The H atoms have been omitted for clarity.

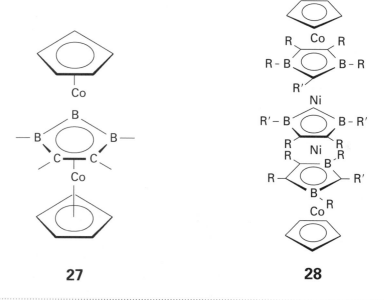

27

28

9   R.N. Grimes, Boron–carbon ring ligands in organometallic synthesis. *Chem. Rev.* **92**, 251 (1992).

When C—H is introduced in place of B—H in a polyhedral boron hydride, the charge of the resulting carboranes is one unit more positive. Carborane anions are useful precursors of boron-containing organometallic compounds.

Example 10.6 Planning the synthesis of a carborane derivative

Give balanced chemical equations for the synthesis of $1,2\text{-}B_{10}C_2H_{10}(Si(CH_3)_3)_2$ starting with decaborane(10) and other reagents of your choice.

Answer The attachment of substituents to C atoms in $1,2\text{-}closo\text{-}B_{10}C_2H_{12}$ is most readily carried out using the dilithium derivative $B_{10}C_2H_{10}Li_2$. We first prepare $1,2\text{-}B_{10}C_2H_{12}$ from decaborane:

$$B_{10}H_{14} + 2\,SR_2 \longrightarrow B_{10}H_{12}(SR_2)_2 + H_2$$
$$B_{10}H_{12}(SR_2)_2 + C_2H_2 \longrightarrow B_{10}C_2H_{12} + 2\,SR_2$$

The product is then lithiated by lithium alkyl, where the alkyl carbanion abstracts the slightly acidic hydrogen atoms from $B_{10}C_2H_{12}$, replacing them with Li^+:

$$B_{10}C_2H_{12} + 2\,LiC_4H_9 \longrightarrow B_{10}C_2H_{10}Li_2 + 2\,C_4H_{10}$$

The resulting carborane is then employed in a nucleophilic displacement on $Si(CH_3)_3Cl$ to yield the desired product:

$$B_{10}C_2H_{10}Li_2 + 2\,Si(CH_3)_3Cl \longrightarrow B_{10}C_2H_{10}(Si(CH_3)_3)_2 + 2\,LiCl$$

Self-test 10.6 Propose a synthesis for the polymer precursor $1,7\text{-}B_{10}C_2H_{10}(Si(CH_3)_2Cl)_2$ from $1,2\text{-}B_{10}C_2H_{12}$ and other reagents of your choice.

The carbon group (Group 14/IV)

We discuss carbon in many contexts throughout this text, including organometallic compounds in Chapters 15 and 16 and catalysis in Chapter 17. In this section we shall focus attention on the more classically 'inorganic' chemistry of carbon and its congeners.

All the elements of the group except lead have at least one solid phase with a diamond structure. That of tin, which is called *gray tin*, is not stable at room temperature; the more stable phase, *white tin*, has six nearest neighbors in a highly distorted octahedral array. The gap between the valence and conduction bands decreases steadily from diamond, which is commonly regarded as an insulator, to gray tin, which behaves like a metal just below its transition temperature (Table 10.6).

10.7 Occurrence and recovery

Two reasonably pure forms of carbon, diamond and graphite, are mined. There are many less pure forms, such as coke, which is made by the pyrolysis of coal, and lamp black, which is the product of incomplete combustion of hydrocarbons. As these examples imply, carbon is polymorphous, and some of its forms will be discussed in more detail in Section 10.8.

The recovery of elemental silicon from silica, SiO_2, by very high temperature reduction with carbon in an electric arc furnace (Section 6.1), is the first step in the production of pure silicon for the production of modern semiconducting devices. Germanium is low in abundance and generally not concentrated in nature. Most germanium is recovered in the processing of zinc ores. Tin is produced by the reduction of the mineral cassiterite, SnO_2, with coke in an electric furnace. Lead is obtained from its sulfide ores, which are converted to oxide and reduced by carbon in a blast furnace.

Elemental carbon is mined as graphite and diamond; elemental silicon is recovered from SiO_2 by carbon-arc reduction; the much less abundant germanium is found in zinc ores.

Table 10.6 Band gaps at 25 °C of Group 14/IV elements and compounds and some Group III/V compounds

Material	E_g/eV
C (diamond)	5.47
SiC	3.00
Si	1.12
Ge	0.66
Sn	0
BN	7.5 (approx.)
BP	2.0
GaN	3.36
GaP	2.26
GaAs	1.42
InAs	0.36

Source: S.A. Schwartz, Semiconductors, theory and applications. In *Kirk–Othmer encyclopedia of chemical technology*, Vol. 21, pp. 720–816. Wiley-Interscience, New York (1991 *et seq.*).

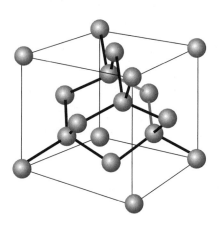

10.16 The cubic diamond structure.

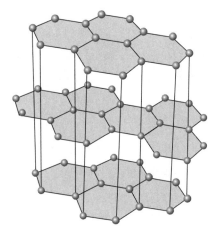

10.17 The structure of graphite. As indicated by the vertical lines, the rings are in register in alternate planes, not adjacent planes.

10.8 Diamond and graphite

Diamond and graphite, the two common crystalline forms of elemental carbon, are strikingly different. Diamond is effectively an electrical insulator; graphite is a good conductor. Diamond is the hardest known substance and hence the ultimate abrasive; impure (partially oxidized) graphite is slippery and frequently used as a lubricant. Because of its durability, clarity, and high refractive index, diamond is one of the most highly prized gemstones; graphite is soft and black with a slightly metallic luster and is neither durable nor particularly attractive. The origin of these widely different physical properties can be traced to the very different structures and bonding in the two polymorphs.

In diamond, each C atom forms single bonds of length 1.54 Å with four adjacent C atoms at the corners of a regular tetrahedron (Fig. 10.16). The result is a rigid, covalent, three-dimensional framework. On the other hand, graphite consists of stacks of planar layers within which each C atom has three nearest-neighbors at 1.42 Å (Fig. 10.17). The σ bonds between neighbors within the sheets are formed from the overlap of sp^2 hybrids, and the remaining perpendicular p orbitals overlap to form π bonds that are delocalized over the plane. The planes themselves are widely separated from each other (at 3.35 Å), which indicates that there are weaker forces between them. These forces are sometimes, but not very appropriately, called 'van der Waals forces' (because in the common impure form of graphite, graphitic oxide, they are weak, like intermolecular forces), and consequently the region between the planes is called the **van der Waals gap**. The ready cleavage of graphite parallel to the planes of atoms (which is enhanced by the presence of impurities) accounts for its slipperiness. Diamond can be cleaved, but this ancient craft requires considerable expertise since the forces in the crystal are more symmetrical.

The conversion of diamond to graphite at room temperature and pressure is spontaneous ($\Delta_{trs}G^{\ominus} = -2.90 \text{ kJ mol}^{-1}$) but does not occur at an observable rate under ordinary conditions. Diamond is the denser phase, so it is favored by high pressures, and large quantities of diamond abrasive are manufactured commercially by a d-metal catalyzed high-temperature, high-pressure process. The d metal (typically nickel) dissolves the graphite at 1800 °C and 70 kbar, and the less soluble diamond phase crystallizes from it. The synthesis of gem-quality diamonds is possible but not yet economical.

Because the high-pressure synthesis of diamond is costly and cumbersome, a low-pressure process would be highly attractive. It has in fact been known for a long time that microscopic diamond crystals mixed with graphite can be formed by depositing C atoms on a hot surface in the absence of air. The C atoms are produced by the pyrolysis of methane, and the atomic hydrogen also produced in the pyrolysis plays an important role in favoring diamond over graphite.[10] One property of the atomic hydrogen is that it reacts more rapidly with the graphite than with diamond to produce volatile hydrocarbons, so the unwanted graphite is swept away. Although the process is not fully perfected, synthetic diamond films are already finding applications ranging from the hardening of surfaces subjected to wear to the construction of electronic devices.

The electrical conductivity and many of the chemical properties of graphite are closely related to the structure of its delocalized π bonds. Its conductivity perpendicular to the planes is low (5 S cm^{-1} at 25 °C) and increases with increasing temperature, signifying that graphite is a semiconductor in that direction. The electrical conductivity is much higher parallel to the planes (3×10^4 S cm^{-1} at 25 °C) but decreases as the temperature is raised. This behavior indicates metallic conduction in that direction.[11] The anisotropy of the

[10] J.C. Anderson and C.C. Hayman, Low-pressure metastable growth of diamond and 'diamond-like' phases. *Science* **241**, 913 (1988). See also M.N. Geis and J.C. Angus, Thin diamond films. *Scientific American* **267** (4), 64 (1992).

[11] More precisely, graphite is a semimetal in that direction (Section 3.14).

conductivity is consistent with a simple band model in which the mobile electrons are in a half-full π band extending over the sheets.

A chemical consequence of the small band gap is that graphite may serve as either an electron donor or an electron acceptor toward atoms and ions that penetrate between its sheets and give rise to an **intercalation compound**. Thus, K atoms reduce graphite by donating their valence electron to the empty orbitals of the π band and the resulting K^+ ions penetrate between the layers. The electrons added to the band are mobile, and therefore alkali metal graphite intercalates have high electrical conductivity. The stoichiometry of the compound depends on the quantity of potassium and reaction conditions. The different stoichiometries are associated with an interesting series of structures, where the alkali metal ion may insert between neighboring carbon layers, every other layer, and so on (Fig. 10.18).

An example of an oxidation of graphite by removal of electrons from the π band is the formation of the substances called **graphite bisulfates** by heating graphite with a mixture of sulfuric and nitric acids. In this reaction, electrons are removed from the π band, and HSO_4^- ions penetrate between the sheets to give substances of approximate formula $(C_{24})^+SO_3(OH)^-$. In this oxidative intercalation reaction, the removal of electrons from the full π band leads to a higher conductivity than that of pure graphite. This process is analogous to the formation of p-type silicon by electron-accepting dopants (Section 3.15).

Graphite consists of stacked two-dimensional carbon sheets; oxidizing agents or reducing agents may be intercalated between these sheets with concomitant electron transfer.

(a) Carbon clusters

Metal and nonmetal cluster compounds have been known for decades, but the discovery of the soccer-ball shaped C_{60} cluster in the 1980s created great excitement in the scientific community and in the popular press. Much of this interest undoubtedly stemmed from the fact that carbon is a common element and there had seemed little likelihood that new molecular carbon structures would be found.[12]

When an electric arc is struck between carbon electrodes in an inert atmosphere, a large quantity of soot is formed together with significant quantities of C_{60} and much smaller quantities of related **fullerenes**[13] such as C_{70}, C_{76}, and C_{84}. The fullerenes can be dissolved in a hydrocarbon or halogenated hydrocarbon solvent and separated by chromatography on an alumina column. The structure of C_{60} has been determined by X-ray crystallography on the solid at low temperature and electron diffraction in the gas phase.[14] The molecule consists of five- and six-membered carbon rings, and the overall symmetry is icosahedral in the gas phase (**29**).

Fullerene reacts with alkali metals to produce solids having compositions such as K_3C_{60}. The structure of K_3C_{60} consists of a face-centered cubic array of C_{60} clusters in which K^+ ions occupy the one octahedral and two tetrahedral sites available per C_{60} (Fig. 10.19). The compound is a superconductor below 18 K.

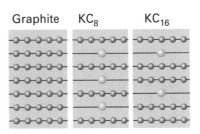

10.18 Potassium graphite compounds showing two types of alternation of intercalated atoms.

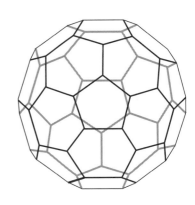

29 C_{60}

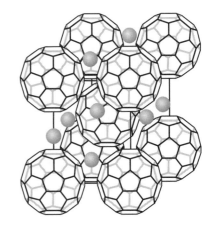

10.19 The fcc structure of K_3C_{60}. Only a fragment of the complete unit cell is shown. The full cell is face-centered cubic.

---

[12] An informative series of articles on the discovery, structure, and chemistry of C_{60} and related carbon clusters appears in *Accounts Chem. Res.* **25**, 98 *et seq.* (1992). See also J. Baggott, *Perfect symmetry*. Oxford University Press (1994) and H. Addersley-Williams, *The most beautiful molecule*. Aurum Press, London (1995).

[13] The name fullerene (or sometimes buckministerfullerene) is applied to these carbon clusters because they resemble the geodesic domes devised by the architect Buckminster Fuller.

[14] S. Lu, Y. Lu, M.M. Kappes, and J.A. Ibers, *Science* **254**, 408 (1991); K. Hedberg, L. Hedberg, D.S. Gethune, C.S. Brown, and D.R. Dorn, *Science* **254**, 410 (1991).

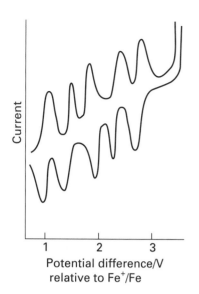

10.20 The cyclic voltammogram of C_{60} in toluene at 25 °C.

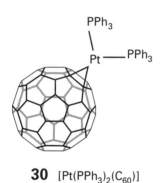

30 $[Pt(PPh_3)_2(C_{60})]$

(b) Fullerene–metal complexes

Reasonably efficient methods of synthesizing the fullerenes have been developed, and their redox and coordination chemistry is receiving serious investigation. In keeping with the formation of alkali metal fullerides described above, C_{60} undergoes five electrochemically reversible reduction and oxidation steps in nonaqueous solvents (Fig. 10.20). These observations suggest that the fullerenes might serve as either electrophiles or nucleophiles when paired with the appropriate metal.

One realization of this expectation is the attack of electron-rich platinum(0) phosphine complexes on C_{60}, yielding compounds such as (**30**), in which the Pt atom spans a pair of carbon atoms in the fullerene molecule. This reaction is analogous to the well known addition of a platinum–phosphine complex across double bonds of alkenes. An analogy with the well known η^6-benzenechromium complexes suggests that a metal atom might coordinate to a sixfold face of C_{60} but this expectation has not been realized. The lack of formation of η^6 complexes is attributed to the disposition of the $C2p\pi$ orbitals radially from each carbon vertex (**31**). This orientation of the p orbitals results in poor overlap with d orbitals of a metal atom centered above a sixfold face of the molecule.

In contrast to the poor interaction of a fullerene sixfold face ring with a single metal center, a larger metal array, the triruthenium cluster $Ru_3(CO)_{12}$, reacts with C_{60}, to form a $Ru_3(CO)_9$ cap on a sixfold face of C_{60}, and in the process three CO ligands are displaced (**32**). The relatively large triangle of three metal atoms provides a favorable geometry for overlap with radially oriented carbon orbitals.[15]

The chemistry of C_{60} is not limited to its interaction with the electron-rich metal complexes. Reaction with a strong electrophile and oxidant, OsO_4 in pyridine, yields an oxo bridge complex analogous to the adducts of OsO_4 with alkenes (**33**). These examples suggest that there is an extensive, subtle, and as yet unexplored coordination chemistry of fullerenes.

> The polyhedral fullerenes undergo reversible multielectron reduction and form complexes with d-metal organometallic compounds and with OsO_4.

(c) Carbon nanotube chemistry

Following the discovery of fullerenes, carbon nanotubes were identified. These new tubular forms of carbon are synthesized by striking an electric arc between carbon electrodes

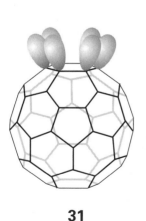

31

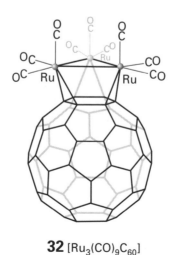

32 $[Ru_3(CO)_9C_{60}]$

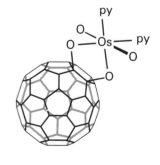

33 $[Os(O)_2(py)_2(OC_{60}O)]$

[15] H.-F. Hsu and J.R. Shapley, *J. Am. Chem. Soc.* **118**, 9192 (1993).

in an inert gas.[16] Another exciting discovery is the preparation of the isolectronic BN nanotubes.[17]

The first of the carbon samples consisted of tubes having multiple concentric walls made up of interconnected C_6 rings, and the ends of the tubes were closed (Fig. 10.21). More recently, single-walled tubes have been observed. When heated in air in the presence of either lead or bismuth the closed ends are opened and the metal is sucked into the nanotubes; however, no reaction occurs when the tubes are heated with lead or bismuth in the absence of air. Apparently, oxygen selectively attacks the tube ends and their opening is followed by the entry of the metal.

Treatment of the closed nanotubes with refluxing nitric acid followed by heating to 900 °C opens the tubes more selectively.[18] Tubes treated in this manner take up a wide variety of salts, including $AgNO_3$ and $AuCl_3$, from concentrated aqueous solutions. The metal is deposited in the interior of the tubes when these materials are heated at temperatures where the salts ordinarily decompose. A variety of organometallic compounds such as $Co(C_5H_5)_2$ have also been introduced into open tubes.[19]

High-resolution electron microscopy is an essential tool for characterizing these materials; for example, Fig. 10.22 shows an electron micrograph of Sm_2O_3 deposited in a carbon nanotube. In this example the opened nanotube was treated with a samarium(III) nitrate solution and the material was heated to 500 °C. Unlike the noble metal salts, heat treatment does not lead to reduction of samarium, so the material inside the nanotube is Sm_2O_3.

> *The closed ends of carbon nanotubes may be opened by partial oxidation. The open nanotubes imbibe solutions of metal salts or organometallic compounds.*

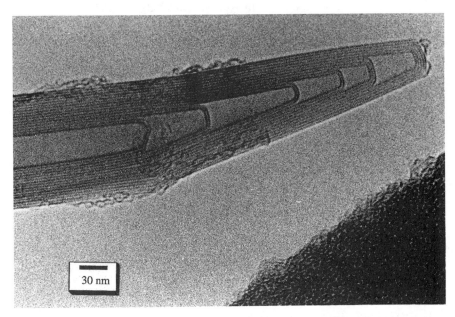

10.21 A multiwalled carbon nanotube with closed ends. (P.J.F. Harris, *Microscopy and Analysis*, Sept., 13 (1994).)

[16] S. Iijima, *Nature* **345**, 56 (1991); S. Iijima, T. Ichihashi, and Y. Ando, *Nature* **356**, 776 (1992).

[17] E.J.M. Hamilton, S.E. Dolan, C.M. Mann, H.O. Colijn, C.A. McDonald, and S.G. Shore, *Science* **260**, 659 (1993).

[18] R.M. Lago, S.C. Tsang, K.L. Lu, Y.K. Chen, and M.L.H. Green, *Chem. Commun.* 1355 (1995).

[19] J. Cook, J. Sloan, and M.L.H. Green, *Chem. and Industry* 600 (1996).

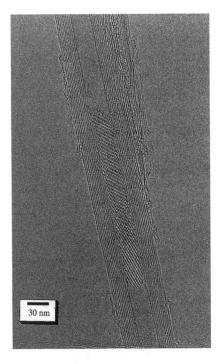

10.22 Electron micrograph of samarium(III) oxide inside a carbon nanotube. (J. Cook, J. Sloan, and M.L.H. Green, *Chem. in Industry*, 600 (1996).)

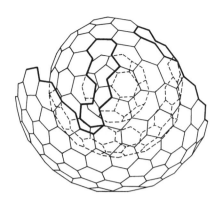

10.23 A proposed structure for soot resulting from imperfect closure of a curved C-atom network. Graphite-like structures have also been proposed.

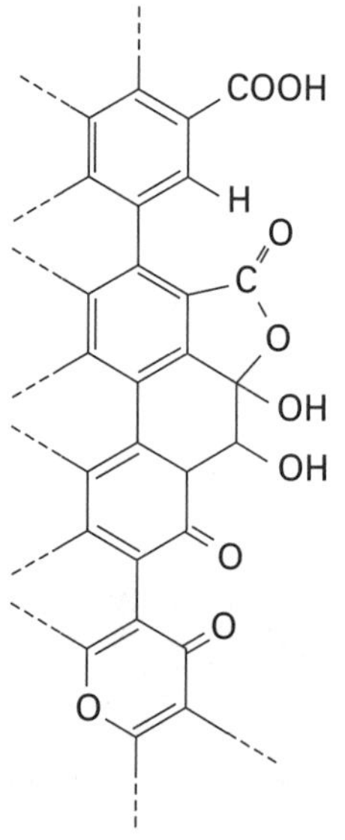

34

Polyhedral carbon clusters and carbon tubes are known and their chemistry is being developed; isoelectronic BN tubes have been prepared.

(d) Partially crystalline carbon

There are many forms of carbon that have a low degree of crystallinity. These partially crystalline materials have considerable commercial importance, for they include *carbon black, activated carbon,* and *carbon fibers.* Because single crystals suitable for complete X-ray analyses of these materials are not available, their structures are uncertain. However, what information there is suggests that their structures are similar to that of graphite, but the degree of crystallinity and shapes of the particles differ.

'Carbon black' is a very finely divided form of carbon. It is prepared (on a scale that exceeds 8×10^9 kg annually) by the combustion of hydrocarbons under oxygen-deficient conditions. Planar stacks, like those of graphite, and multilayer balls, reminiscent of the fullerenes, have both been proposed for its structure (Fig. 10.23). Carbon black is used on a huge scale as a pigment, in printer's ink (as on this page), and as a filler for rubber goods, including automobile tires, where it greatly improves the strength and wear resistance of the rubber and helps to protect it from degradation by sunlight.

'Activated carbon' is prepared from the controlled pyrolysis of organic material, including coconut shells. It has a high surface area (in some cases exceeding $1000 \text{ m}^2 \text{ g}^{-1}$), which arises from the small particle size. It is therefore a very efficient adsorbent for molecules, including organic pollutants from drinking water, noxious gases from the air, and impurities from reaction mixtures. There is evidence that the parts of the surface defined by the edges of the hexagonal sheets are covered with oxidation products, including carboxyl and hydroxyl groups (**34**). This structure may account for some of its surface activity.

Carbon fibers are made by the controlled pyrolysis of asphalt fibers or synthetic fibers and are incorporated into a variety of high-strength plastic products, such as tennis rackets and aircraft components. Their structure bears a resemblance to that of graphite, but in place of the extended sheets the layers consist of ribbons parallel to the axis of the fiber. The strong in-plane bonds (which resemble those in graphite) give the fiber its very high tensile strength.

Amorphous and partially crystalline carbon in the form of small particles are used on a large scale as adsorbents and as strengthening agents for rubber; carbon fibers impart strength to polymeric materials.

Example 10.7 Comparing bonding in diamond and boron

Each B atom in elemental boron is bonded to five other B atoms but each C atom in diamond is bonded to four nearest neighbors. Suggest an explanation of this difference.

Answer The B and C atoms both have four orbitals available for bonding (one *s* and three *p*). However, a C atom has four valence electrons, one for each orbital, and it can therefore use all its electrons and orbitals in forming 2c,2e bonds with the four neighboring C atoms. In contrast, B has one less electron; therefore, to use all its orbitals, it forms 3c,2e bonds. The formation of these three-center bonds brings another B atom into binding distance.

Self-test 10.7 Describe how the electronic structure of graphite is altered when it reacts with (a) potassium, (b) bromine.

10.9 Compounds of carbon with the electronegative elements

Carbon has a high affinity for highly electronegative elements such as O and F, with which it forms many important compounds. For example, CO_2 is essential for plant life, and carbon-fluorine compounds are widely utilized in consumer products, industry, and the laboratory.

From the viewpoint of a synthetic chemist, many of the compounds with heavier halogens and chalcogens are important starting materials for the synthesis of other compounds.

Table 10.7 Properties of tetrahalomethanes

	CF_4	CCl_4	CBr_4	CI_4
m.p./°C	−187	−23	90	171 dec
b.p./°C	−128	77	190	sub
$\Delta_f G^{\ominus}$/(kJ mol^{-1})	−879(g)	−65.2(l)	+47.7(s)	>0

dec = decomposes; sub = sublimes.

(a) Halides

The tetrahalomethanes, the simplest halocarbons, vary from the highly stable and volatile CF_4 to the thermally unstable solid CI_4 (Table 10.7). These tetrahalomethanes and analogous partially halogenated alkanes provide a route to a wide variety of derivatives, mainly by nucleophilic displacement of one or more halogen atoms. Some useful and interesting reactions from an inorganic perspective are outlined in Chart 10.2. Note in particular the metal–carbon bond-forming reactions, which take place either by complete displacement of halogen or by oxidative addition. The rates of nucleophilic displacement increases greatly from fluorine to iodine, and lie in the order $F \ll Cl < Br < I$. All tetrahalomethanes are thermodynamically unstable with respect to hydrolysis,

$$CX_4(l \text{ or } g) + 2 H_2O(l) \longrightarrow CO_2(g) + 4 HX(aq)$$

However, the rate for C–F bonds is extremely slow, and fluorocarbon polymers such as poly(tetrafluoroethylene) (e.g. Teflon) or fluorochlorocarbon polymers are highly resistant to attack by water. Halocarbons can be reduced by strong reducing agents, such as alkali metals. For example, the reaction of carbon tetrachloride with sodium is highly exoergic:

$$CCl_4(l) + 4 Na(s) \longrightarrow 4 NaCl(s) + C(s) \qquad \Delta_r G^{\ominus} = -249 \text{ kJ mol}^{-1}$$

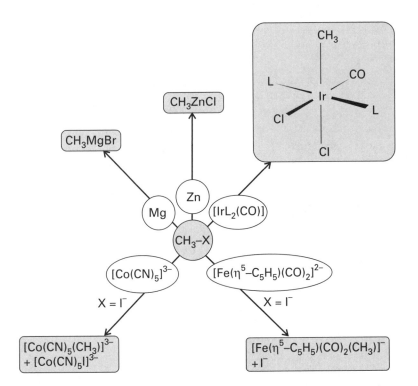

Chart 10.2 Some characteristic reactions of carbon–halogen bonds (X = halogen).

Table 10.8 Properties of carbonyl halides

	COF$_2$	COCl$_2$	COBr$_2$
m.p./°C	−114	−128	
b.p./°C	−83	8	65
$\Delta_f G^{\ominus}$/(kJ mol^{-1})	−619(g)	−205(g)	−111(g)

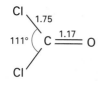

35 COCl$_2$

This reaction can occur with explosive violence with CCl$_4$ and other polyhalocarbons, so alkali metals such as sodium should never be used to dry them. Analogous reactions occur on the surface of poly(tetrafluoroethylene) when it is exposed to alkali metals or strongly reducing organometallic compounds. Fluorocarbons, together with other fluorine-containing molecules, exhibit many interesting properties, such as high volatility and strong electron-withdrawing character (Section 12.7).

The **carbonyl halides** (Table 10.8) are planar molecules and useful chemical intermediates. The simplest of these compounds (OCCl$_2$, phosgene, (**35**)) is a highly toxic gas. It is prepared on a large scale by the reaction of chlorine with carbon monoxide:

$$CO + Cl_2 \xrightarrow{200\,°C,\ charcoal} Cl_2CO$$

The utility of phosgene lies in the ease of nucleophilic displacement of chlorine to produce carbonyl compounds and isocyanates (Chart 10.3). The fact that hydrolysis leads to carbon dioxide and hydrogen chloride, rather than carbonic acid, (HO)$_2$CO, can be traced to the latter's instability.

> *Nucleophiles displace halogens in carbon–halogen bonds; organometallic nucleophiles produce new M—C bonds; mixtures of polyhalocarbons and alkali metals are explosion hazards.*

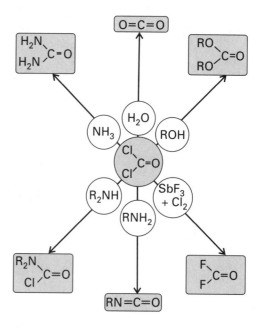

Chart 10.3 Characteristic reactions of Cl$_2$CO.

(b) Oxygen and sulfur compounds

The two familiar oxides of carbon, CO and CO$_2$, have already been mentioned in several contexts; among its less familiar oxides is carbon suboxide, O=C=C=C=O. Physical data on all three compounds are summarized in Table 10.9. It should be noted that the bond length in

Table 10.9 Properties of some oxides of carbon

Oxide	m.p./°C	b.p./°C	$\bar{\nu}(CO)/cm^{-1}$	$k(CO)/(N\ m^{-1})$	Bond length/Å	
					CC	CO
CO	−199	−191.5	2145	1860		1.13
OCO		−78*	2349, 1318	1550		1.16
OCCCO	−111	7	2200, 2290		1.28	1.16

*Sublimes; at 5 atm, CO_2 melts at −57°C.

carbon monoxide is short and the force constant high; both features are in accord with its possession of a triple bond, as in the Lewis structure :C≡O:.

The uses of CO include the reduction of metal oxides in blast furnaces (Section 6.1) and the shift reaction (Section 8.3) for the production of H_2:

$$CO(g) + H_2O(g) \longrightarrow CO_2(g) + H_2(g)$$

In Chapter 17, where we deal with catalysis, we describe the conversion of carbon monoxide to methanol, acetic acid, and aldehydes. The CO molecule has very low Brønsted basicity and negligible Lewis acidity toward neutral electron pair donors. Despite its weak Lewis acidity, however, CO is attacked by strong Lewis bases at high pressure and somewhat elevated temperatures. Thus, the reaction with hydroxide yields the formate ion, HCO_2^-:

$$CO(g) + OH^-(s) \longrightarrow HCO_2^-(s)$$

Similarly, the reaction with methoxide ions (CH_3O^-) yields the acetate ion, $CH_3CO_2^-$.

Carbon monoxide is an excellent ligand toward metal atoms in low oxidation states (Chapter 16). Its well-known toxicity is an example of this behavior, for it binds to the Fe atom in hemoglobin, so excluding the attachment of O_2, and the victim suffocates. An interesting point is that H_3BCO can be prepared from B_2H_6 and CO at high pressures in a rare example of the coordination of CO to a simple Lewis acid. A complex of similar stability is not formed by BF_3; this observation is consistent with the classification of BH_3 as a soft acid and BF_3 as a hard acid.

Carbon dioxide shows a number of subtle but significant differences from carbon monoxide. The CO bond is longer and the stretching force constants smaller in CO_2 than in CO, which is consistent with the bonds in CO_2 being double rather than triple. Carbon dioxide is only a very weak Lewis acid. For example, only a small fraction of molecules are complexed with water to form H_2CO_3 in acidic aqueous solution (Section 5.4) but, at higher pH, OH^- coordinates to the C atom, so forming the hydrogencarbonate (bicarbonate) ion, HCO_3^-.

Carbon dioxide is one of several polyatomic molecules that are implicated in the **greenhouse effect**. In this effect, a polyatomic molecule in the atmosphere permits the passage of visible light but, because of its vibrational infrared absorptions, it blocks the immediate radiation of heat from the Earth. There is strong evidence for a significant increase in atmospheric CO_2 since the industrialization of society. In the past, nature has managed to stabilize the concentration of atmospheric CO_2, in part by precipitation of calcium carbonate in the deep oceans, but it seems that the rate of diffusion of CO_2 into the deep waters is now too slow to compensate for the increased influx of CO_2 into the atmosphere.[20] Although there is convincing evidence for increasing concentrations of the greenhouse gases CO_2, CH_4, N_2O, and cholorofluorocarbons, it is not clear whether they are having an impact on global temperatures. The difficulty with detecting global warming is associated with large natural short-term and long-term variations in temperature, which may mask the effects of greenhouse gases.[21]

[20] C. Baird, *Environmental chemistry*. W.H. Freeman & Co, New York (1995).

[21] H.B. Gray, J.D. Simon, and W. Trogler, *Braving the elements*. University Science Books, Mill Valley (1995).

36 [Co(NH$_3$)$_5$(CO$_3$)]$^+$

37 [Ni(PR$_3$)$_2$(CO$_2$)]

The principal chemical properties of CO$_2$ are summarized in Chart 10.4. These properties are based on its mild Lewis acidity toward hard donors, as in the formation of CO$_3^{2-}$ ions in strongly basic solution. The ability of CO$_2$ to form limestone and thereby to moderate the concentration of atmospheric CO$_2$ has been mentioned above. Similarly, carbonato complexes of metals can be formed in which CO$_3^{2-}$ is a ligand (**36**). These complexes are often useful intermediates because CO$_3^{2-}$ can be displaced in acidic solution to yield complexes that are otherwise difficult to prepare:

$$[Co(NH_3)_5(CO_3)]^+ + 2\,HF \longrightarrow [Co(NH_3)_5F]^{2+} + CO_2 + H_2O + F^-$$

Carbonato complexes can be either monodentate, as shown above, or bidentate; in the latter case the CO$_3^{2-}$ ion has a small bite angle.

From an economic perspective, an important reaction is CO$_2$ with ammonia to yield ammonium carbonate, (NH$_4$)$_2$CO$_3$, which at elevated temperatures is directly converted to urea, CO(NH$_2$)$_2$, a useful fertilizer, feed supplement for cattle, and chemical intermediate. In organic chemistry, a common synthetic reaction is that between CO$_2$ and carbanion reagents to produce carboxylic acids.

Metal complexes of CO$_2$ are known (**37**), but they are rare and far less important than the metal carbonyls. In its interaction with a low-oxidation state, electron-rich metal center, the neutral CO$_2$ molecule acts as a Lewis acid and the bonding is dominated by electron donation from the metal atom into an antibonding π orbital of CO$_2$.

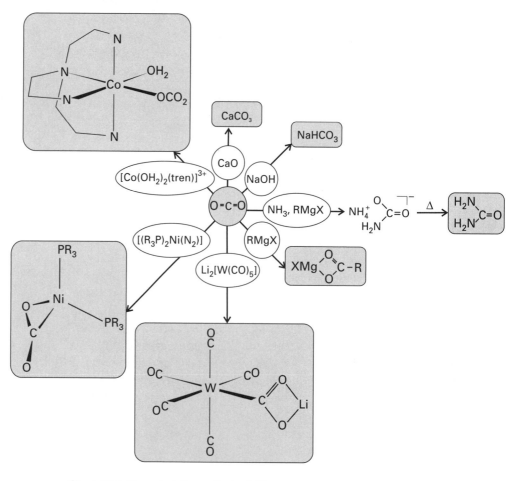

Chart 10.4 Characteristic reactions of CO$_2$.

The sulfur analogs of carbon monoxide and carbon dioxide, CS and CS_2, are known. The former is an unstable transient molecule and the latter is endoergic ($\Delta_f G^\ominus = +65$ kJ mol^{-1}). Some complexes of CS (**38**) and CS_2 (**39**) exist, and their structures are similar to those formed by CO and CO_2. In basic aqueous solution, CS_2 undergoes hydrolysis and yields a mixture of carbonate ions, CO_3^{2-}, and trithiocarbonate ions, CS_3^{2-}.

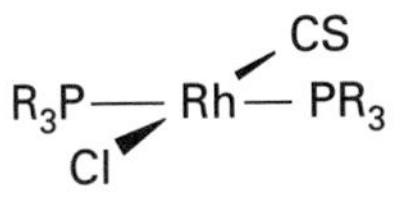

38

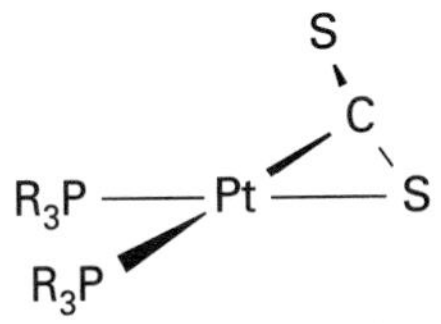

39

Example 10.8 Proposing a synthesis that uses the reactions of carbon monoxide

Propose a synthesis of $CH_3{}^{13}CO_2^-$ that uses ^{13}CO, a primary starting material for many carbon-13 labeled compounds.

Answer We first note that CO_2 is readily attacked by strong nucleophiles such as $LiCH_3$ to produce acetate ions. Therefore an appropriate procedure would be to oxidize ^{13}CO to $^{13}CO_2$ and then to react the latter with $LiCH_3$. A strong oxidizing agent such as solid MnO_2 can be used in the first step, to avoid the problem of excess O_2 in the direct oxidation.

$$^{13}CO(g) + 2MnO_2(s) \xrightarrow{\Delta} {}^{13}CO_2(g) + Mn_2O_3(s)$$
$$4\,{}^{13}CO_2(g) + Li_4(CH_3)_4(et) \longrightarrow 4Li[CH_3{}^{13}CO_2](et)$$

where et denotes solution in ether.

Self-test 10.8 Propose a synthesis of $D^{13}CO_2^-$ starting from ^{13}CO.

> CO *is a key reducing agent in the production of iron and a common ligand in d-metal chemistry. The greenhouse gas* CO_2 *is the product of combustion of fuels, and is much less important as a ligand; it is the acid anhydride of carbonic acid.*

(c) Compounds with nitrogen

Hydrogen cyanide, HCN, is produced in large amounts by the high-temperature catalytic combination of methane and ammonia, and is used as an intermediate in the synthesis of common polymers such as poly(methyl methacrylate) and polyacrylonitrile. It is highly volatile (b.p. 26 °C) and, like the CN$^-$ ion, highly poisonous. In some respects the toxicity of the CN$^-$ ion is similar to that of the isoelectronic CO molecule, because both form complexes with iron porphyrin molecules. However, whereas CO attaches to the Fe in hemoglobin and causes oxygen starvation, CN$^-$ has a higher affinity for the Fe in cytochrome-c, and blocks electron transfer.

Unlike the neutral ligand CO, the negatively charged CN$^-$ ion is a strong Brønsted base ($pK_a = 9.4$) and a much poorer Lewis acid π acceptor. Its coordination chemistry is therefore mainly associated with metal ions in positive oxidation states, as with Fe^{2+} in the hexacyanoferrate(II) complex, $[Fe(CN)_6]^{4-}$.

> CN$^-$ *forms complexes with many d-metal ions; its coordination to iron in biomolecules such as cytochrome-c accounts for its high toxicity.*

10.10 Carbides

It proves helpful to classify the numerous binary compounds of carbon with metals and metalloids, the carbides, into three main categories:

1 **Saline carbides**, which are largely ionic solids. These compounds are formed by the elements of Groups 1 and 2 and by aluminum.
2 **Metallic carbides**, which have a metallic conductivity and luster. These compounds are formed by the *d*-block elements.
3 **Metalloid carbides**, which are hard covalent solids formed by boron and silicon.

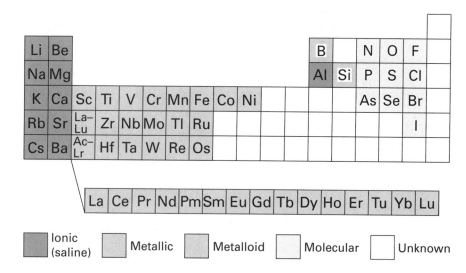

10.24 The distribution of carbides in the periodic table. Molecular compounds of carbon are included for completeness, but are not carbides.

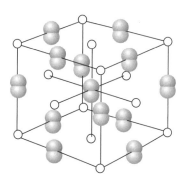

10.25 In KC_8, a graphite intercalation compound, the potassium atoms lie in a symmetrical array between the sheets. (See Fig. 10.18 for a view parallel to the sheets.)

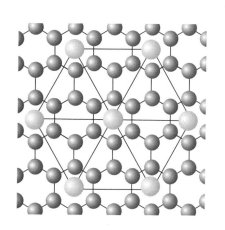

10.26 The calcium carbide structure. Note that this structure bears a similarity to the rock-salt structure. Because C_2^{2-} is not spherical, the cell is elongated along one axis. This crystal is therefore tetragonal rather than cubic.

Figure 10.24 summarizes the distribution of the different types in the periodic table; it includes binary molecular compounds of carbon with electronegative elements, which are not normally regarded as carbides. This classification is very useful for correlating chemical and physical properties, but (as so often in inorganic chemistry) the borderlines are sometimes indistinct.

(a) Saline carbides

The saline carbides of Group 1 and 2 metals may be divided into three categories: **graphite intercalation compounds**, such as KC_8, **dicarbides** (or 'acetylides'), which contain the C_2^{2-} anion, and **methides**, which contain formally the C^{4-} anion.

The graphite intercalation compounds are formed by the Group 1 metals. They are formed by a redox process, and specifically by the reaction of graphite with alkali metal vapor or with metal–ammonia solution. For example, contact between graphite and potassium vapor in a sealed tube at 300 °C leads to the formation of KC_8 in which the alkali metal ions lie in an ordered array between the graphite sheets (Fig. 10.25). A series of alkali metal–graphite intercalation compounds can be prepared with different metal:carbon ratios, including KC_8 and KC_{16}.

The dicarbides are formed by a broad range of electropositive metals, including those from Groups 1 and 2 and the lanthanides. The C_2^{2-} ion has a very short C—C distance in some dicarbides (for example, 1.19 Å in CaC_2), which is consistent with it having a triply bonded $[C{\equiv}C]^{2-}$ ion isoelectronic with $[C{\equiv}N]^-$ and $N{\equiv}N$. Some dicarbides have a structure related to rock salt, but replacement of the spherical Cl^- ion by the elongated $[C{\equiv}C]^{2-}$ ion leads to an elongation of the crystal along one axis, and a resulting tetragonal symmetry (Fig. 10.26). The C—C bond is significantly longer in the lanthanide dicarbides, which suggests that for them the simple triply bonded structure is not a good approximation.

Carbides such as Be_2C and Al_4C_3 are borderline between saline and metalloid, and the isolated C ion is only formally C^{4-}. The existence of directional bonding to the C atom in carbides (as distinct from the nondirectional character expected of purely ionic bonding) is indicated by the crystal structures of methides, which are not those expected for the simple packing of spherical ions.

The principal synthetic routes to the saline carbides and acetylides of Groups 1 and 2 are very straightforward:

1 *Direct reaction* of the elements at high temperatures:

$$Ca(l) + 2\,C(s) \xrightarrow{>2000\,°C} CaC_2(s)$$

The formation of graphite intercalation compounds is another example of a direct reaction, but is carried out at much lower temperatures. The intercalation reaction is more facile because no C—C covalent bonds are broken when an ion slips between the graphite layers.

2 *Reaction of a metal oxide and carbon* at a high temperature:

$$CaO(l) + 3\,C(s) \xrightarrow{2200\,°C} CaC_2(l) + CO(g)$$

Crude calcium carbide is prepared in electric arc furnaces by this method. The carbon serves both as a reducing agent to remove the oxygen and as a source of carbon to form the carbide.

3 *Reaction of acetylene with a metal–ammonia solution:*

$$2\,Na(am) + C_2H_2(g) \longrightarrow Na_2C_2(s) + H_2(g)$$

This reaction occurs under mild conditions and leaves the carbon–carbon bonds of the starting material intact. As the ethyne molecule is a very weak Brønsted acid, the reaction can be regarded as a redox reaction between a highly active metal and a weak acid to yield H_2 (with H^+ the oxidizing agent) and the metal dicarbide.

The saline carbides have high electron density on carbon, so they are readily oxidized and protonated. For example, calcium carbide reacts with the weak acid water to produce ethyne:

$$CaC_2(s) + 2\,H_2O(l) \longrightarrow Ca(OH)_2(s) + HC{\equiv}CH(g)$$

This reaction is readily understood as the transfer of a proton from a Brønsted acid (H_2O) to the conjugate base (C_2^{2-}) of a weaker acid (HC≡CH). Similarly, the controlled hydrolysis or oxidation of the graphite intercalation compound KC_8 restores the graphite and produces a hydroxide or oxide of the metal:

$$2\,KC_8(s) + 2\,H_2O(g) \longrightarrow 16\,C(graphite) + 2\,KOH(s) + H_2(g)$$

> Metal–carbon compounds of highly electropositive metals are saline; those of the *d*-block metals are often mechanically hard and good electrical conductors; nonmetal carbides are mechanically hard and are semiconductors.

(b) Metallic carbides

Most metallic carbides are formed by the *d* metals. They are sometimes referred to as **interstitial carbides** because their structures are often related to those of metals by the insertion of C atoms in octahedral holes. However, this nomenclature gives the erroneous impression that the metallic carbides are not legitimate compounds. In fact the hardness and other properties of metallic carbides demonstrate that strong metal–carbon bonding is present in them. Some of these carbides are economically useful materials. Tungsten carbide (WC), for example, is used for cutting tools and high-pressure apparatus such as that used to produce diamond. Cementite, Fe_3C, is a major constituent of steel and cast iron.

Metallic carbides of composition **MC** have an fcc or hcp arrangement of metal atoms with the C atoms in the octahedral holes. The fcc arrangement results in a rock-salt structure (Fig. 2.10). The C atoms in carbides of composition M_2C occupy only half the octahedral holes between the close-packed metal atoms. A C atom in an octahedral hole is formally hypercoordinate because it is surrounded by six metal atoms. However, the bonding can be expressed in terms of delocalized molecular orbitals formed from the C2*s* and C2*p* orbitals and the *d* orbitals (and perhaps other valence orbitals) of the surrounding metal atoms, and there is nothing particularly mysterious about its bonding.

It has been found empirically that the formation of simple compounds in which the C atom resides in an octahedral hole of a close-packed structure occurs when $r_C/r_M < 0.59$, where r_C

is the covalent radius of C and r_M the metallic radius of **M**. This relation, which is known as **Hägg's rule**, also applies to metal compounds containing nitrogen or oxygen.

> *d-Metal carbides are often hard materials with the carbon atom octahedrally surrounded by metal atoms.*

10.11 Silicon and germanium

The band gap and therefore conductivity of silicon are ideal for many semiconductor applications. However, germanium was the first widely used material for the construction of transistors because it was easier to purify than silicon. When methods for purifying silicon were developed in the 1960s, germanium was largely eclipsed for the construction of semiconductors.

The principal method of synthesizing semiconductor-grade silicon is to reduce high-purity silicon tetrachloride with hydrogen:

$$SiCl_4(g) + 2\,H_2(g) \longrightarrow Si(s) + 4\,HCl(g)$$

The semiconducting material used for solar cells to power calculators is called 'amorphous silicon', but it is better regarded as a solid silicon hydride SiH_x ($x \leq 0.5$; Section 8.12).

> *Elemental silicon and, to a lesser extent, germanium are produced for the construction of semiconductor devices.*

10.12 Compounds of silicon with electronegative elements

Some of the most important compounds of silicon and germanium contain the electronegative halogens, oxygen, and nitrogen. The oxides are so extensive that they deserve their own section. In this section we consider the halides and the nitrides.

(a) Compounds with halogens

The full range of tetrahalides is known for silicon and germanium; all of them are volatile molecular compounds. Lower down the group, germanium shows signs of an inert pair effect in that it also forms nonvolatile dihalides. Among the silicon tetrahalides, the most important is the tetrachloride, which is prepared by direct reaction of the elements:

$$Si(s) + 2\,Cl_2(g) \longrightarrow SiCl_4(l)$$

Silicon and germanium halides are mild Lewis acids. They display this behavior when they add one or two ligands to yield five- or six-coordinate complexes:

$$SiF_4(g) + 2\,F^-(aq) \longrightarrow [SiF_6]^{2-}(aq)$$
$$GeCl_4(l) + N\equiv CCH_3(l) \longrightarrow Cl_4GeN\equiv CCH_3(s)$$

The hydrolysis of the silicon and germanium tetrahalides is fast and it is represented schematically as follows:

$$MX_4 + 2\,H_2O \longrightarrow MX_4(OH_2)_2 \longrightarrow MO_2 + 4\,HX$$

The corresponding carbon tetrahalides are kinetically more resistant to hydrolysis on account of the lack of access to the sterically shrouded C atom to form an intermediate aqua complex.

The substitution reactions of halosilanes have been studied extensively. The reactions are more facile than for their carbon analogs because a Si atom can readily expand its coordination sphere to accommodate the incoming nucleophile. The stereochemistry of these substitution reactions[22] indicates that a five-coordinate intermediate is formed with the most electronegative substituents adopting the axial position. Moreover, substituents leave from the axial position. The H^- ion is a poor leaving group, and alkyl groups are even poorer:

----

[22] R.R. Holmes, Sterochemistry of nucleophilic substitution at tetracoordinate silicon. *Chem. Rev.* **90**, 17 (1990).

Note that in these examples, the R^4 substituent replaces H with retention of configuration.

> *Because silicon can form hypervalent transition states whereas carbon cannot, substitution reactions of silicon halides are more facile than those of carbon halides.*

(b) Compounds with nitrogen

The direct reaction of silicon and nitrogen gas at high temperatures produces Si_3N_4. This substance is very hard and inert, and may find use in high-temperature ceramic materials. Current industrial research projects focus on the use of suitable organosilicon–nitrogen compounds that might undergo pyrolysis to yield silicon nitride fibers and other shapes. When SiO_2 is heated with carbon, CO is evolved and silicon carbide, SiC, forms. This very hard material is widely used as the abrasive carborundum.

Trisilylamine, $(H_3Si)_3N$, the silicon analog of trimethylamine, has very low basicity. It has a planar structure, or is fluxional with a very low barrier to inversion.

> *Silicon nitride and silicon carbide are hard solids; silyl amines are weaker bases than their carbon analogs.*

10.13 Extended silicon–oxygen compounds

The high affinity of silicon for oxygen accounts for the existence of a vast array of silicate minerals and synthetic silicon-oxygen compounds, which are important in mineralogy, industrial processing, and the laboratory. Aside from rare high-temperature phases, the structures of silicates are confined to tetrahedral four-coordinate Si. The complicated silicate structures are often easier to comprehend if the SiO_4 unit is drawn as a tetrahedron with the Si atom at the center and O atoms at the vertices. The representation is often cut to the bone by drawing the SiO_4 unit as a simple tetrahedron with the atoms omitted. In general, these tetrahedra share vertices and (much more rarely) edges or faces. Each terminal O atom contributes -1 to the charge of the SiO_4 unit, but each shared O atom contributes zero. Thus, orthosilicate is $[SiO_4]^{4-}$ (**40**), disilicate is $[O_3SiOSiO_3]^{6-}$ (**41**), and the SiO_2 unit of silica has no net charge because all O atoms are shared.

With the above principles of charge balance in mind, it should be clear that an endless single-stranded chain or a ring of SiO_4 units, which has two shared O atoms for each Si atom, will have the formula and charge $[(SiO_3)^{2-}]_n$. An example of a compound containing such a cyclic metasilicate ion is the mineral beryl, $Be_3Al_2Si_6O_{18}$, which contains the $[Si_6O_{18}]^{12-}$ ion (**42**). Beryl is a major source of beryllium. The gemstone emerald is beryl in which Cr^{3+} ions are substituted for some Al^{3+} ions. A chain metasilicate is present in the mineral jadeite, $NaAl(SiO_3)_2$ (**43**), one of two different minerals sold as jade. In addition to other configurations for the single chain, there are double-chain silicates which include the family of minerals known commercially as asbestos.[23]

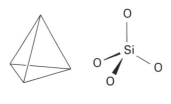

40

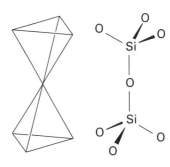

41

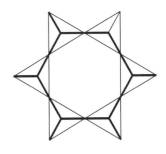

42 $[Si_6O_{18}]^{12-}$

43

---

[23] Asbestos has many commercially desirable qualities, but the fine asbestos particles are readily airborne and workers who handle the mineral are subject to a degeneration of the lung tissue which may manifest itself many years after the exposure. See L. Michaelis and S.S. Chissick (ed.), *Asbestos: properties, applications, and hazards.* Wiley, New York (1979).

Example 10.9 Determining the charge on a cyclic silicate

Draw the structure and determine the charge on the cyclic silicate anion $[Si_3O_9]^{n-}$.

Answer The ion is a six-membered ring with alternating Si and O atoms and six terminal O atoms, two on each Si atom. Because each terminal O atom contributes -1 to the charge, the overall charge is -6. From another perspective, the conventional oxidation states of silicon and oxygen, $+4$ and -2, respectively, also indicate a charge of -6 for the anion.

Self-test 10.9 Repeat the question for the cyclic anion $[Si_4O_{12}]^{n-}$.

Silica and many silicates crystallize slowly. Amorphous solids known as **glasses** can be obtained instead of crystals by cooling the melt at an appropriate rate. In some respects these glasses resemble liquids. As with liquids, their structures are ordered over distances of only a few interatomic spacings (such as within a single SiO_4 tetrahedron). Unlike liquids, their viscosities are very high, and for most practical purposes they behave like solids.

The composition of silicate glasses has a strong influence on their physical properties. For example, fused quartz (amorphous SiO_2) softens at about $1600\,°C$, borosilicate glass (which contains boron oxide, as we have already seen) softens at about $800\,°C$, and sodalime glass softens at even lower temperatures. The variation in softening point can be understood by appreciating that the Si—O—Si links in silicate glasses form the framework that imparts rigidity. When basic oxides such as Na_2O and CaO are incorporated (as in sodalime glass), they react with the SiO_2 melt and convert Si—O—Si links into terminal SiO groups and hence lower its softening temperature.

A quite different set of properties are found for the —Si—O—Si— backbone of **silicone polymers**. A typical silicone polymer has the polydimethylsiloxane repeat unit and end-capping alkyl groups (**44**). The properties of these polymers vary from mobile high-boiling liquids to flexible high polymers. The liquids are used as lubricants and the high polymers are used for a wide variety of applications, including weather- and chemical-resistant flexible sealants. The unusual property of these polymers is their very low **glass transition temperature**, the temperature at which they become rigid. The origin of the low transition temperature appears to be the highly compliant nature of the —Si—O—Si— linkage. Another desirable property is their bio-compatibility, which allows their use in devices that can be implanted in humans.

> The Si—O—Si *link is present in silica, a wide range of metal silicate minerals, and silicone polymers.*

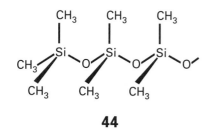

44

10.14 Aluminosilicates

Even greater structural diversity than displayed by the silicates themselves is possible when Al atoms replace some of the Si atoms. The resulting **aluminosilicates** are largely responsible for the rich variety of the mineral world. We have already seen that in γ-alumina, Al^{3+} ions are present in both octahedral or tetrahedral holes. This versatility carries over into the aluminosilicates, where Al may substitute for Si in tetrahedral sites, enter an octahedral environment external to the silicate framework, or, more rarely, occur with other coordination numbers. Because aluminum occurs as Al(III), its presence in place of Si(IV) in an aluminosilicate renders the overall charge negative by one unit. An additional cation, such as H^+, Na^+, or $\frac{1}{2}Ca^{2+}$, is therefore required for each Al atom that replaces an Si atom. As we shall see, these additional cations have a profound effect on the properties of the materials.

(a) Layered aluminosilicates

Many important minerals are varieties of layered aluminosilicates that also contain metals such as lithium, magnesium, and iron: they include clays, talc, and various micas. In one class of layered aluminosilicate, the repeating unit consists of a silicate layer with the structure shown in Fig. 10.27. An example of a simple aluminosilicate of this type (simple, that is, in the sense of there being no additional elements) is the mineral kaolinite, $Al_2(OH)_4Si_2O_5$, which is used commercially as china clay. The electrically neutral layers are held together by rather weak hydrogen bonds, so the mineral readily cleaves and incorporates water between the layers.

A larger class of aluminosilicates has Al^{3+} ions sandwiched between silicate layers (Fig. 10.28). One such mineral is pyrophyllite, $Al_2(OH)_2Si_4O_{10}$. The mineral talc, $Mg_3(OH)_2Si_4O_{10}$, is obtained when three Mg^{2+} ions replace two Al^{3+} ions in the octahedral sites. In talc (and in pyrophyllite) the repeating layers are neutral, and as a result talc readily cleaves between them. This structure accounts for its familiar slippery feel.

Muscovite mica, $KAl_2(OH)_2Si_3AlO_{10}$, has charged layers because one Al(III) atom substitutes for one Si(IV) atom in the pyrophyllite structure. The resulting negative charge is compensated by a K^+ ion that lies between the repeating layers. Because of this electrostatic cohesion, muscovite is not soft like talc, but it is readily cleaved into sheets. More highly charged layers with dipositive ions between the layers lead to greater hardness.

> The friable layered aluminosilicates are the primary constituents of clay and some common minerals.

(b) Three-dimensional aluminosilicates

There are many minerals based on a three-dimensional aluminosilicate framework. The **feldspars**, for instance, which are the most important class of rock-forming minerals (and contribute to granite), belong to this class. The aluminosilicate frameworks of feldspars are built up by sharing all vertices of SiO_4 or AlO_4 tetrahedra. The cavities in this three-dimensional network accommodate ions such as K^+ and Ba^{2+}. Two examples are the feldspars orthoclase, $KAlSi_3O_8$, and albite, $NaAlSi_3O_8$.

> The hard three-dimensional aluminosilicates are common rock-forming minerals.

(c) Molecular sieves

The **molecular sieves** are crystalline aluminosilicates having open structures with apertures of molecular dimensions. These substances represent a major triumph of solid state chemistry, for their synthesis and our understanding of their properties combine challenging determinations of structures, imaginative synthetic chemistry, and important practical applications.

The name 'molecular sieve' is prompted by the observation that these materials adsorb only molecules that are smaller than the aperture dimensions and so can be used to separate molecules of different sizes. A subclass of molecular sieves, the **zeolites**,[24] have an aluminosilicate framework with cations (typically from Groups 1 or 2) trapped inside tunnels or cages. In addition to their function as molecular sieves, zeolites can exchange their ions for those in a surrounding solution.

The cages are defined by the crystal structure, so they are highly regular and of precise size. Consequently, molecular sieves capture molecules with greater selectivity than high surface area solids such as silica gel or activated carbon, where molecules may be caught in irregular voids between the small particles. Zeolites are also used for shape-selective heterogeneous catalysis. For example, the molecular sieve ZSM-5 is used to synthesize 1,2-dimethylbenzene (o-xylene) for use as an octane booster in gasoline. The other xylenes are not produced because the catalytic process is controlled by the size and shape of the zeolite cages and

---

[24] The name zeolite is derived from the Greek words for 'boiling stone'. Geologists found that certain rocks appeared to boil when subjected to the flame of a blowpipe.

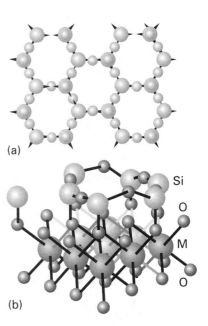

(a)

(b)

10.27 (a) A net of SiO_4 tetrahedra with one O atom on top of each Si atom projecting toward the viewer. (b) Edge view of the above net, with one O/Si incorporated into a net of MO_6 octahedra. This structure is for the mineral chrysotile, for which M is Mg. When M is Al^{3+} and each of the atoms on the bottom is replaced by an OH group, this structure is close to that of the 1:1 clay mineral kaolinite.

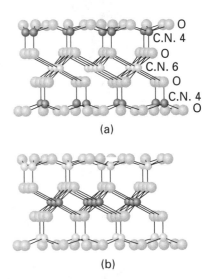

(a)

(b)

10.28 (a) The structure of 2:1 clay minerals such as muscovite mica $KAl_2(OH)_2Si_3AlO_{10}$, in which K^+ resides between the charged layers (exchangeable cation sites), Si^{4+} resides in sites of coordination number 4, and Al^{3+} in sites of coordination number 6. (b) In talc, Mg^{2+} ions occupy the octahedral sites and O atoms on the top and bottom are replaced by OH groups.

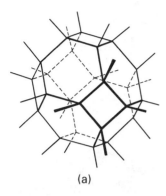

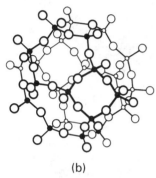

10.29 (a) Framework representation of a truncated octahedron (truncation perpendicular to the fourfold axes of the octahedron). (b) Relation of Si and O atoms to the framework. Note that an Si atom is at each vertex of the truncated octahedron and an O atom is approximately along each edge.

Table 10.10 Some uses of zeolites

Function	Application
Ion exchange	Water softeners in washing detergents
Absorption of molecules	Selective gas separation
	Industrial processes, gas chromatography, laboratory experiments
Solid acid	Cracking of high molar mass and selectivity hydrocarbons for fuel and petrochemical intermediates
	Shape selective alkylation and isomerization of aromatics for gasoline and polymer intermediates

tunnels. This and other applications are summarized in Table 10.10 and discussed in Chapter 17.

Synthetic procedures have added to the many naturally occurring zeolite varieties that have specific cage sizes and specific chemical properties within the cages. These synthetic zeolites are sometimes made at atmospheric pressure, but more often they are produced in a high-pressure autoclave. Their open structures appear to form around hydrated cations or other large cations such as NR_4^+ ions introduced into the reaction mixture. For example, a synthesis may be performed by heating colloidal silica to between $100\,°C$ and $200\,°C$ in an autoclave with an aqueous solution of tetrapropylammonium hydroxide. The microcrystalline product, which has the typical composition $\{[N(C_3H_7)_4]OH\}(SiO_2)_{48}$, is converted into the zeolite by burning away the C, H, and N of the quaternary ammonium cation at $500\,°C$ in air. Aluminosilicate zeolites are made by including high surface area alumina in the starting materials.

A wide range of zeolites has been prepared with varying cage and bottleneck sizes (Table 10.11). Their structures are based on approximately tetrahedral MO_4 units, which in the great majority of cases are SiO_4 and AlO_4. Because the structures involve many such tetrahedral units, it is common practice to abandon the polyhedral representation in favor of one that emphasizes the position of the Si and Al atoms. In this scheme, the Si or Al atom lies at the intersection of four line segments and the O atom bridge lies on the line segment (Fig. 10.29). This **framework representation** has the advantage of giving a clear impression of the shapes of the cages and channels in the zeolite.[25]

Table 10.11 Composition and properties of some molecular sieves

Molecular sieve	Composition	Bottleneck diameter/Å	Chemical properties
A	$Na_{12}[(AlO_2)_{12}(SiO_2)_{12}]\cdot xH_2O$	4	Absorbs small molecules; ion exchanger, hydrophilic
X	$Na_{86}[(AlO_2)_{86}(SiO_2)_{106}]\cdot xH_2O$	8	Absorbs medium-sized molecules; ion exchanger, hydrophilic
Chabazite	$Ca_2[(AlO_2)_4(SiO_2)_8]\cdot xH_2O$	4–5	Absorbs small molecules; ion exchanger, hydrophilic
ZSM-5	$Na_3[(AlO_2)_3(SiO_2)_{93}]\cdot xH_2O$	5.5	Moderately hydrophilic
ALPO-5	$AlPO_4\cdot xH_2O$	8	Moderately hydrophilic
Silicalite	SiO_2	6	Hydrophobic

[25] The combination of X-ray crystallography, solid state ^{29}Si-NMR and ^{27}Al-NMR with high-resolution electron microscopy has revolutionized the structural determination of zeolites and other aluminosilicates. An interesting account is given by J.M. Thomas and C.R.A. Catlow, *Prog. Inorg. Chem.* **35**, 1 (1987).

A large and important class of zeolites is based on the **sodalite cage**. This cage is a truncated octahedron (Fig. 10.30), a shape that can be formed by slicing off each vertex of an octahedron. The truncation leaves a square face in the place of each vertex and the triangular faces of the octahedron are transformed into regular hexagons. The substance known as 'zeolite type A' is based on sodalite cages that are joined by O bridges between the square faces. Eight such sodalite cages are linked in a cubic pattern with a large central cavity called an **α cage**. The α cages share octagonal faces, with an open diameter of 4.20 Å. Thus water or other small molecules can fill them and diffuse through octagonal faces. However, these faces are too small to permit the entrance of molecules with van der Waals diameters larger than 4.20 Å.

Example 10.10 Analyzing the structure of the sodalite cage

Identify the fourfold and sixfold axes in the truncated octahedral polyhedron used to describe the sodalite cage.

Answer There is one fourfold axis running through each pair of opposite square faces. This gives a total of six fourfold axes. Similarly, a set of four sixfold axes run through opposite sixfold faces.

Self-test 10.10 How many Si and Al atoms are there in one sodalite cage?

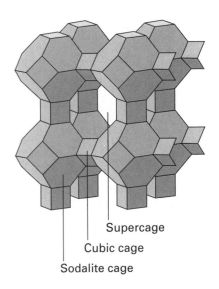

10.30 Framework representation of a type-A zeolite. Note the sodalite cages (truncated octahedra), the small cubic cages, and the central supercage.

The charge on the aluminosilicate zeolite framework is neutralized by cations lying within the cages. In the type-A zeolite, Na^+ ions are present and the formula is $Na_{12}(AlO_2)_{12}(SiO_2)_{12} \cdot xH_2O$. Numerous other ions, including d-block cations and NH_4^+, can be introduced by ion exchange with aqueous solutions. Zeolites are therefore used for water-softening and as a component of laundry detergent to remove the di- and tri-positive ions that decrease the effectiveness of the surfactant. Zeolites have largely replaced polyphosphates because the latter, which are plant nutrients, find their way into natural waters and stimulate the growth of algae.

In addition to the control of properties by selecting a zeolite with the appropriate cage and bottleneck size, the zeolite can be chosen for its affinity for polar or nonpolar molecules according to its polarity (Table 10.11). The aluminosilicate zeolites, which always contain charge-compensating ions, have high affinities for polar molecules such as H_2O and NH_3. In contrast, the nearly pure silica molecular sieves bear no net electric charge and are nonpolar to the point of being mildly hydrophobic. Another group of hydrophobic zeolites is based on the aluminum phosphate frameworks, for $AlPO_4$ is isoelectronic with Si_2O_4 and the framework is similarly uncharged.

One interesting aspect of zeolite chemistry is that large molecules can be synthesized from smaller molecules inside the zeolite cage. The result is like a ship in a bottle, because once assembled the molecule is too big to escape. For example, Na^+ ions in a Y-type zeolite may be replaced by Fe^{2+} ions (by ion exchange). The resulting Fe^{2+}-Y zeolite is heated with phthalonitrile, which diffuses into the zeolite and condenses around the Fe^{2+} to form iron phthalocyanine (**45**), which remains imprisoned in the cage.

> Zeolite aluminosilicates have large open cavities or channels giving rise to useful properties such as ion exchange and molecular absorption.

45

10.15 Silicides

Silicon, like its neighbors boron and carbon, forms a wide variety of binary compounds with metals. Some of these **silicides** contain isolated Si atoms. The structure of ferrosilicon, Fe_3Si, for instance, which plays an important role in steel manufacture, can be viewed as an fcc array of Fe atoms with some atoms replaced by Si. Compounds such as K_4Si_4 contain isolated tetrahedral cluster anions $[Si_4]^{4-}$ that are isoelectronic with P_4. Many of the f-block elements form compounds with the formula MSi_2 that have the hexagonal layers which adopt the AlB_2 structure shown in Fig. 10.7.

> Silicon–metal compounds (silicides) may contain isolated Si, tetrahedral Si_4 units or hexagonal nets of Si atoms.

Further reading

A.H.H. Stephens and M.L.H. Green, Organometallic complexes of fullerenes. *Adv. Inorg. Chem.* **44**, 1–43 (1997).

The carbon group

See especially the articles on carbon, semiconductors, and silicon in the following two references.

Ullmann's encyclopedia of industrial chemistry. VCH, Weinheim (1990 *et seq.*).

Kirk–Othmer encyclopedia of chemical technology. Wiley-Interscience, New York (1991 *et seq.*).

A series of useful reviews on fullerene chemistry is available in the following reference:

Acc. Chem. Res. **25**, 98 *et. seq.* (1992).

Aqueous chemistry, silicate minerals, and aluminosilicates

A.C.D. Newman, *Chemistry of clays and clay minerals.* Wiley, New York (1987).

W. Stumm and J.J. Morgan, *Aquatic chemistry.* Wiley, New York (1996).

C.E. Weaver and L.D. Pollard, *The chemistry of clay minerals.* Elsevier, Amsterdam (1993).

A. Dyer, *An introduction to zeolite molecular sieves.* Wiley, Chichester (1988).

Boron and carbon groups

J.T. Spencer, Boron hydrides. In *Encyclopedia of inorganic chemistry* (ed. R.B. King), Vol. 1, pp. 338–56. Wiley, New York (1994).

G.A. Olah, K. Williams, and R.E. Williams (ed.). *Electron deficient boron and carbon clusters.* Wiley-Interscience, New York (1991). This book contains a series of articles that cover major areas of borane and carborane chemistry.

C. Housecroft, *Boranes and metalloboranes: structure, bonding, and reactivity.* Ellis Horwood, Chichester; Halsted Press, New York (1990).

T.P. Fehlner, The metallic face of boron. *Adv. Inorg. Chem.* **35**, 199 (1990).

Exercises

10.1 Preferably without consulting reference material, list the elements in Groups 13/III and 14/IV and indicate (a) the metals and nonmetals, (b) those that can crystallize in the diamond structure, (c) the most oxophilic elements in these groups.

10.2 Draw the B_{12} unit that is a common motif of boron structures; take a viewpoint along a C_2 axis.

10.3 Give balanced chemical equations and conditions for the recovery of boron, silicon, and germanium from their ores. Which of these is likely to be the most energy efficient? Explain your answer.

10.4 Arrange the following in order of increasing Lewis acidity: BF_3, BCl_3, SiF_4, $AlCl_3$. (b) In the light of this order, write balanced chemical reactions (or NR) for

(i) $SiF_4N(CH_3)_3 + BF_3 \rightarrow$?

(ii) $BF_3N(CH_3)_3 + BCl_3 \rightarrow$?

(iii) $BH_3CO + BBr_3 \rightarrow$?

10.5 Using BCl_3 as a starting material and other reagents of your choice, devise a synthesis for the Lewis acid chelating agent, $F_2B—C_2H_4—BF_2$.

10.6 State the coordination numbers of boron, carbon, and silicon in their oxoanions and devise plausible explanations for the differences based on Lewis electron dot structures.

10.7 Which boron hydride would you expect to be more thermally stable, B_6H_{10} or B_6H_{12}? Give a generalization by which the thermal stability of a borane can be judged.

10.8 (a) Give a balanced chemical equation (including the state of each reactant and product) for the air oxidation of pentaborane(9). (b) Describe the probable disadvantages, other than cost, for the use of pentaborane as a fuel for an internal combustion engine.

10.9 (a) From its formula, classify $B_{10}H_{14}$ as *closo*, *nido*, or *arachno*. (b) Use Wade's rules to determine the number of framework electron pairs for decaborane(14). (c) Verify by detailed accounting of valence electrons that the number of cluster valence electrons of $B_{10}H_{14}$ is the same as that determined in (b).

10.10 Starting with $B_{10}H_{14}$ and other reagents of your choice, give the equations for the synthesis of $[Fe(nido-B_9C_2H_{11})_2]^{2-}$, and sketch the structure of this species.

10.11 (a) What are the similarities and differences in the structures of layered BN and graphite? (b) Contrast their reactivity with Na and Br_2. (c) Suggest a rationalization for the differences in structure and reactivity.

10.12 Devise a synthesis for the borazines (a) $Ph_3N_3B_3Cl_3$ and (b) $Me_3N_3B_3H_3$, starting with BCl_3 and other reagents of your choice. Draw the structures of the products.

10.13 Give the structures and names of B_4H_{10}, B_5H_9, and $1,2\text{-}B_{10}C_2H_{12}$.

10.14 Arrange the following boron hydrides in order of increasing Brønsted acidity, and draw a structure for the probable structure of the deprotonated form of one of them: B_2H_6, $B_{10}H_{14}$, B_5H_9.

10.15 (a) Describe the trend in band gap energy, E_g, for the elements carbon (diamond) through tin (gray), and for cubic BN, AlP, and GaAs. (b) Does the electrical conductivity of silicon increase or decrease when its temperature is changed from 20°C to 40°C? (c) State the functional dependence of conductivity on temperature for a semiconductor and decide whether the conductivity of AlP or GaAs will be more sensitive to temperature.

10.16 Preferably without consulting reference material, draw a periodic table and indicate the elements that form saline, metallic, and metalloid carbides.

10.17 Describe the preparation, structure, and classification of (a) KC_8, (b) CaC_2, (c) K_3C_{60}.

10.18 There are major commercial applications for semicrystalline and amorphous solids, many of which are formed by Group 14/IV elements or their compounds. List four different examples of amorphous or partially crystalline solids described in this chapter and briefly state their useful properties.

10.19 The lightest p-block elements often display different physical and chemical properties from the heavier members. Discuss the similarities and differences by comparison of:

(a) The structures and electrical properties of (i) boron and aluminum and of (ii) carbon and silicon.

(b) The physical properties and structures of the oxides of carbon and silicon.

(c) The Lewis acid–base properties of the tetrahalides of carbon and silicon.

(d) The structures of the halides of boron and aluminum.

10.20 Write balanced chemical equations for the reactions of K_2CO_3 with HCl(aq) and of Na_4SiO_4 with aqueous acid.

10.21 Describe in general terms the nature of the $[SiO_3]_n^{2n-}$ ion in jadeite and the silica-alumina framework in kaolinite.

10.22 (a) Determine the number of bridging O atoms in the framework of a single sodalite cage. (b) Describe the (supercage) polyhedron at the center of the zeolite A structure in Fig. 10.30.

10.23 Describe the physical properties of pyrophyllite and muscovite mica and explain how these properties arise from the composition and structures of these closely related aluminosilicates.

Problems

10.1 Boron-11 NMR is an excellent spectroscopic tool for inferring the structures of boron compounds. Under conditions in which the $^{11}B-^{11}B$ coupling is absent, it is possible to determine the number of attached H atoms by the multiplicity of a resonance: BH gives a doublet, BH_2 a triplet, and BH_3 a quartet. Also, B atoms on the closed side of *nido* and *arachno* clusters are generally more shielded than those on the open face. Assuming no B—B or B—H—B coupling, predict the general pattern of the ^{11}B-NMR spectra of (a) BH_3CO, (b) $[B_{12}H_{12}]^{2-}$, and (c) B_4H_{10}.

10.2 Identify the incorrect statements in the following description of Group 13/III chemistry and provide corrections along with explanations of principles or chemical generalizations that apply. (a) All the elements in Group 13/III are nonmetals. (b) The increase in chemical hardness on going down the group is illustrated by greater oxophilicity and fluorophilicity for the heavier elements. (c) The Lewis acidity increases for BX_3 from X = F to Br and this may be explained by stronger Br—B π bonding. (d) *Arachno*-boron hydrides have a $2(n+3)$ skeletal electron count and are more stable than *nido*-boron hydrides. (e) In a series of *nido*-boron hydrides acidity increases with increasing size. (f) Layered boron nitride is similar in structure to graphite and, because it has a small separation between HOMO and LUMO, is a good electrical conductor.

10.3 Correct any inaccuracies in the following descriptions of Group 14/IV chemistry. (a) None of the elements in this group is a metal. (b) At very high pressures, diamond is a thermodynamically stable phase of carbon. (c) Both CO_2 and CS_2 are weak Lewis acids and the hardness increases from CO_2 to CS_2. (d) Zeolites are layered materials exclusively composed of aluminosilicates. (e) The reaction of calcium carbide with water yields acetylene and this product reflects the presence of a highly basic C_2^{2-} ion in calcium carbide.

10.4 Acetylcholine, $[(CH_3)_3N(CH_2)_2OC(O)CH_3]^+$, is an important neurotransmitter, and the physiological properties of the neutral boron analog $(CH_3)_2(BH_3)N(CH_2)_2OC(O)CH_3$ are of interest. (B.F. Spielvogel, F.U. Ahmed, and A.T. McPhail, *Inorg. Chem.* **25**, 4395 (1986).) Devise a method of synthesis for this analog starting with $[(CH_3)_2HN(CH_2)_2OC(O)CH_3]^+$ and other reagents of your choice.

10.5 Two reactions used in commercial processes for the production of hydrogen are the reduction of $H_2O(g)$ with CO to give CO_2 and H_2, often referred to as the 'water-gas shift reaction', and the analogous reduction of water with methane. Both reactions are carried out with reactants and products in the gas phase at elevated temperatures in the presence of a catalyst. (a) Using a reliable source of thermodynamic data,[26] determine $\Delta_r G^\ominus$, and $\Delta_r S^\ominus$ per mole of H_2 for each of these reactions assuming gas-phase species at 1 bar. (b) Are these reactions

---

[26] D.D. Wagman, *et al.*, The NBS tables of thermodynamic properties. *J. Phys Chem. Ref. Data* **11**, Supplement 2 (1982).

thermodynamically favorable under the above conditions? (c) From qualitative considerations and also from the thermodynamic data determine whether each of these reactions is more, or less favored, thermodynamically, by (i) higher temperature, (ii) higher total pressure.

10.6 Use a suitable extended Hückel molecular orbital program[27] to calculate the wavefunctions and energy levels for *closo*-$[B_6H_6]^{2-}$. From that output, draw a molecular orbital energy diagram for the orbitals primarily involved in BB bonding and sketch the form of the orbitals. How do these orbitals compare qualitatively with the description for this anion in Section 10.3? Is BH bonding neatly separated from the BB bonding in the extended Hückel wavefunctions?

10.7 The layered silicate compound $CaAl_2Si_2O_8$ contains a double aluminosilicate layer, with both Si and Al in four-coordinate sites. Sketch an edge-on view of a reasonable structure for the double layer, involving only vertex sharing between the SiO_4 and AlO_4 units. Discuss the likely sites occupied by Ca^{2+} in relation to the silica–alumina double layer.

----

[27] Suitable programs include QCMP 001 from Quantum Chemistry Program Exchange, Chemistry Dept., Indiana University, Bloomington, IN; CACAO by C. Meali and D.M. Proserpio, *J. Chem. Educ.* **67**, 3399 (1990); PLOT3D, J.A. Bertrand and M.R. Johnson, School of Chemistry, Georgia Institute of Technology, Atlanta, GA.

11

The nitrogen and oxygen groups

The elements

The nitrogen group

The oxygen group

Further reading

Exercises

Problems

14	15	16	17
C	N	O	F
Si	P	S	Cl
Ge	As	Se	Br
Sn	Sb	Te	I
Pb	Bi	Po	At
IV	V	VI	VII

As in the rest of the p-block, the elements at the head of Groups 15/V and 16/VI, nitrogen and oxygen, differ significantly from their congeners. Their coordination numbers are generally lower in their compounds and they are the only members of the group to exist as diatomic molecules under normal conditions. The chemical properties of nitrogen, phosphorus, and sulfur are intricate because these elements possess a wide range of oxidation states. Moreover, their reactions are strongly influenced by kinetic factors. A good approach to the mastery of the chemical properties of the elements is to learn the structures of important species for each oxidation state and the general thermodynamic trends for their redox reactions.

The nitrogen and oxygen groups, Groups 15/V and 16/VI of the periodic table, contain some of the most important elements for geology, life, and industry. However, the heaviest elements of the groups are much less abundant than the lighter elements, for they lie at the frontier of nuclear stability. Indeed, the Group 15/V element bismuth ($Z = 83$) is the heaviest element to have stable isotopes. The most common isotope of its heavier neighbor, the Group 16/VI element polonium ($Z = 84$), ^{210}Po, is an intense α emitter with a half-life of 138 d.

The elements

All the members of the two groups other than nitrogen and oxygen are solids under normal conditions (Table 11.1) and metallic character generally increases down the groups. However, the trend is not clear cut, because the electrical conductivities of the heavier elements actually decrease from arsenic to bismuth (Fig. 11.1). The usual trend (an increase down a group) reflects the closer spacing of the atomic energy levels in heavier elements and hence a smaller separation of the valence and conduction bands. The opposite trend in conductivity in these two groups suggests that there must be a more pronounced molecular character in the solid state. Indeed, the structures of solid arsenic, antimony, and bismuth have three nearest-neighbor atoms and three more at significantly larger distances, which suggests the onset of localized covalent bonding.

In addition to their distinctive physical properties, nitrogen and oxygen are significantly different chemically from the other members of the groups. For one thing, they are among the most electronegative elements in

Table 11.1 Properties of the nitrogen and oxygen group elements

	$I/(\mathrm{kJ\ mol^{-1}})$	χ_P	Radius		Appearance and properties	Common oxidation numbers
			$r_\mathrm{cov}/\text{Å}$	$r_\mathrm{ion}/\text{Å}$		
Group 15/V						
N	1410	3.04	0.75		Gas b.p. −196°C	−**3**, 0, +1, +**3**, +**5**
P	1020	2.06	1.10		Polymorphic solid	−**3**, +3, +**5**
As	953	2.18	1.22		Dark solid	+**3**, +**5**
Sb	840	2.05	1.43		Solid metallic luster brittle	+**3**, +**5**
Bi	710	2.02	1.52		Solid metallic luster brittle	+**3**, +5
Group 16/VI						
O	1320	3.44	0.73	1.24†	Paramagnetic gas b.p. −183°C	−**2**, −1, 0
S	1005	2.44	1.02	1.70†	Yellow polymorphic solid	−**2**, 0, +4, +**6**
Se	947	2.55	1.17	1.84†	Polymorphic solid	−2, +4, +**6**
Te	875	2.10	1.35	2.07†	Solid silvery brittle	−2, +4, +**6**
Po					Solid all isotopes radioactive	−2, +2, +4, +6

†For oxidation number −2.
Sources: Ionization enthalpies from D.D. Wagman, W.H. Evans, V.B. Parker, R.H. Schumm, I. Halow, S.M. Bailey, K.L. Churney, and R.L. Nuttall, *J. Phys. Chem. Ref. Data* **11** (Suppl. 2) (1982). Electronegativities from A.L. Allred, *J. Inorg. Nucl. Chem.* **17**, 215 (1961). Covalent radii from L.C. Allen and J.E. Huheey, *J. Inorg. Nucl. Chem.* **42**, 1523 (1980). Ionic radii from R.D. Shannon, *Acta Crystallogr.* **A32**, 1751 (1976).

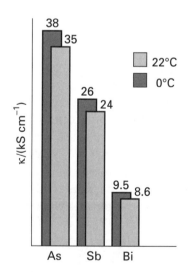

11.1 Electrical conductivities of the heavier elements of Group 15/V and their variation with temperature.

the periodic table and significantly more electronegative than their congeners. Although oxygen never achieves the group maximum oxidation state of +6, the slightly less electronegative nitrogen does achieve the maximum for its group (+5), but only under much stronger oxidizing conditions than are necessary to achieve this state for its congeners. The small radii of the N and O atoms and absence of accessible *d* orbitals also contribute to their distinctive chemical character. Thus, nitrogen and oxygen seldom have coordination numbers greater than 4 in simple molecular compounds, but their heavier congeners frequently reach coordination numbers of 5 and 6, as in PCl_5, AsF_6^-, and SeF_6. However, even this trend down the group is complicated by the inert pair effect (introduction to Section 9.15), for the heaviest member of Group 15/V, bismuth, requires a very strong oxidizing agent to achieve the group oxidation number, and in most of its compounds it has oxidation number +3.

Oxygen is the most electronegative element in Groups 15/V and 16/VI, and together with nitrogen is the only gas. The heaviest members of the groups, tellurium and bismuth, are metalloids.

The nitrogen group

The members of the nitrogen group are sometimes referred to collectively as the **pnictides** but this name is neither widely used nor officially sanctioned.

11.1 Occurrence and recovery of the elements

The substantial range in physical and chemical properties of the elements in the nitrogen and oxygen groups is reflected in the wide range of strategies for their recovery from their natural sources.

(a) Nitrogen

Nitrogen is readily available as dinitrogen, N_2, for it is the principal constituent of the atmosphere and is obtained from it on a massive scale by the distillation of liquid air. Although the bulk price of N_2 is low, the scale on which it is used is an incentive for the development of less expensive processes than liquefaction and distillation. Membrane materials that are more permeable to O_2 than to N_2 are used in laboratory-scale separations of oxygen and nitrogen at room temperature (Fig. 11.2), and an active area of research is the development of membranes that can be used for large-scale commercial separations.

The major nonchemical use of nitrogen gas is as an inert atmosphere in metal processing, petroleum refining, and food processing. Nitrogen gas is used to provide an inert atmosphere in the laboratory, and liquid nitrogen (b.p. $-196\,°C$, $77\ K$) is a convenient refrigerant in both industry and the laboratory. Nitrogen enters the chain of industrial and agricultural chemicals through its conversion into ammonia by the Haber process, which we describe later. Once 'fixed' in this way it can be converted to a wide range of compounds. The destinations of fixed nitrogen are summarized in Chart 11.1, and later in the chapter we shall see something of the chemistry involved in the reactions mentioned there.

> Nitrogen is recovered by distillation from liquid air and used as an inert gas and in the production of ammonia.

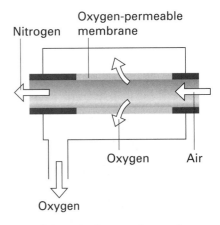

11.2 Schematic diagram of a membrane separator for nitrogen and oxygen.

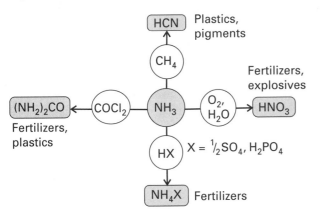

Chart 11.1 Uses of ammonia.

(b) Phosphorus

Phosphorus (along with nitrogen and potassium) is an essential plant nutrient. However, as a result of the low solubility of many metal phosphates it is often depleted in soil, and hence hydrogenphosphates are important components of balanced fertilizers.[1] Approximately 85 per cent of the phosphoric acid produced goes into fertilizer manufacture.

---

[1] The availability of soluble phosphate nutrients is slightly dependent on the pH of the soil because metal salts of PO_4^{3-} are much less soluble than those of HPO_4^{2-} and $H_2PO_4^-$.

The principal raw material for the production of elemental phosphorus and phosphoric acid is phosphate rock, the insoluble, crushed, and compacted remains of ancient organisms, which consists primarily of the minerals fluorapatite, $Ca_5(PO_4)_3F$, and hydroxyapatite, $Ca_5(PO_4)_3OH$. Phosphoric acid can be produced by an acid–base reaction between the mineral and concentrated sulfuric acid:

$$Ca_5(PO_4)_3F(s) + 5\,H_2SO_4(l) \longrightarrow 3\,H_3PO_4(l) + 5\,CaSO_4(s) + HF(aq)$$

The potential pollutant, hydrogen fluoride, from the fluoride component of the rock is scavenged by reaction with silicates to yield the less reactive $[SiF_6]^{2-}$ complex ion.

The product of the treatment of phosphate rock with acid contains d-metal contaminants that are difficult to remove completely, so its use is largely confined to fertilizers and metal treatment. However, ion-exchange processes have been developed to remove the offending ions. Most pure phosphoric acid and phosphorus compounds are still produced from the element because it can be purified by sublimation. The production of elemental phosphorus starts with crude calcium phosphate (as calcined phosphate rock) which is reduced with carbon in an electric arc furnace. (This process is another example of the high-temperature carbon reduction of a highly oxophilic element, Section 6.1.) Silica is added (as sand) to produce a slag of calcium silicate:

$$2\,Ca_3(PO_4)_2 + 6\,SiO_2 + 10\,C \xrightarrow{1500\,°C} 6\,CaSiO_3 + 10\,CO + P_4$$

The slag is molten at these high temperatures and so can be easily removed from the furnace. The phosphorus itself vaporizes and is condensed to the solid, which is stored under water to protect it from reaction with air. Most phosphorus produced in this way is burned to form P_4O_{10}, which is then hydrated to yield pure phosphoric acid.

The solid elements of Group 15/V (as well as those in Group 16/VI) exist as a number of allotropes.[2] White phosphorus, for example, is a solid consisting of tetrahedral P_4 molecules (**1**). Despite the small PPP angle, the molecules persist in the vapor up to about $800\,°C$, but above that temperature the equilibrium concentration of the allotrope P_2 becomes appreciable. As with the N_2 molecule, P_2 has a formal triple bond and a short bond length. White phosphorus is thermodynamically less stable than the other solid phases under normal conditions. However, in contrast to the usual practice of choosing the most stable (lowest Gibbs energy) phase of an element as the reference phase for thermodynamic calculations, white phosphorus is adopted because it is more accessible and better characterized than the other forms.

Red phosphorus can be obtained by heating white phosphorus at $300\,°C$ in an inert atmosphere for several days. It is normally obtained as an amorphous solid, but crystalline materials can be prepared that have very complex three-dimensional network structures. Unlike white phosphorus, red phosphorus does not ignite spontaneously in air. When phosphorus is heated under high pressure, a series of phases of *black phosphorus* are formed. One of these phases consists of puckered layers composed of pyramidal three-coordinate P atoms (Fig. 11.3).

Elemental phosphorus is recovered from the minerals fluorapatite and hydroxyapatite by carbon arc reduction. The resulting white phosphorus is a molecular solid, P_4. Treatment of apatite with sulfuric acid yields phosphoric acid, which is converted to fertilizers and other chemicals.

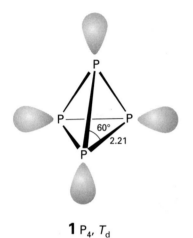

1 P_4, T_d

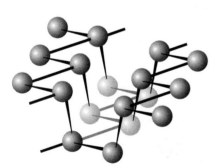

11.3 Two of the puckered layers of black phosphorus. Note the trigonal-pyramidal coordination of the atoms. Dark atoms are closest to the viewer and lighter ones are farthest away.

---

2 The term *allotrope* refers to different molecular structures of the same element; thus ozone and dioxygen are allotropes of oxygen. The word *polymorph* denotes the same substance (an element or a compound) in different crystal forms. For example, phosphorus occurs in several solid polymorphs (including white phosphorus and black phosphorus). For a discussion of the use of the terms, see B.D. Sharma, *J. Chem. Educ.* **64**, 404 (1987).

(c) Arsenic, antimony, and bismuth

The chemically softer elements arsenic, antimony, and bismuth are often found in sulfide ores. Arsenic is usually present in copper and lead sulfide ores, and most of it is obtained from the flue dust of copper and lead smelters. The solubilities of arsenates, AsO_4^{3-}, are similar to those of phosphates, so traces of these ions are present in phosphate rock. In keeping with the general trends down the *p*-block, the +3 oxidation state becomes more favorable relative to +5 on going from phosphorus to bismuth. This trend is apparent in the unfavorable standard potential for the reduction of phosphate to phosphite:

$$H_3PO_4(aq) + 2\,H^+(aq) + 2\,e^- \longrightarrow H_3PO_3(aq) + H_2O(l) \quad (E^\ominus = -0.28\ \text{V})$$

This potential is consistent with the use of phosphite as a reducing agent. Bismuth(III), on the other hand, is favored over Bi(V), so Bi(V) is a useful oxidizing agent:

$$\text{"}Bi^{5+}(aq)\text{"} + 2\,e^- \longrightarrow \text{"}Bi^{3+}(aq)\text{"} \qquad (E^\ominus = +2\ \text{V})$$

The bismuth species in this half-reaction are indicated schematically because their molecular structures are not well characterized; accordingly the impressively large potential is reported to only one significant figure.

Arsenic, antimony, and bismuth exist as several allotropes. The most stable structures at room temperature for all three elements are built from puckered hexagonal nets in which each atom has three nearest neighbors. The nets stack in a way that gives three more distant neighbors in the adjacent net (Fig. 11.4). Arsenic vapor resembles phosphorus in that it consists of tetrahedral As_4 molecules. Bismuth, in common with the α-forms of arsenic and antimony, has a metallic luster and is one of a small class of substances that, like water, expand upon solidification. The electrical conductivity of bismuth is not as high as for most metals, and its structure is not typical of the isotropic bonding normally found. The band structure of bismuth suggests a low density of conduction electrons and holes and it is best classified as a semimetal (Section 3.14) rather than as a semiconductor or a true metal.

Progressing from arsenic down to bismuth, the +5 oxidation state becomes increasingly disfavored relative to +3, so Bi(V) is a very strong oxidizing agent.

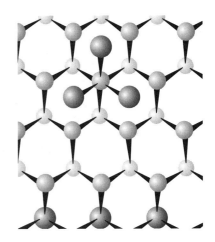

11.4 The puckered network structure of bismuth. Each Bi atom has three nearest neighbors; the dark shaded atoms indicate next-nearest neighbors from an adjacent puckered sheet. The lightest atoms are farthest from the viewer.

<hr>

Example 11.1 Examining the electronic structure and chemistry of P_4

Draw the Lewis structure of P_4, and discuss its possible role as a ligand.

Answer As shown to the right, in the Lewis structure of P_4, there is a lone pair of electrons on each P atom. This structure, together with the fact that the electronegativity of phosphorus is moderate ($\chi_P = 2.06$), suggests that P_4 might be a moderately good donor ligand. Indeed, though rare, P_4 complexes are known.

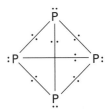

<hr>

Self-test 11.1 Consider the Lewis structure of a segment of the structure of bismuth shown in Fig. 11.4. Is this puckered structure consistent with the VSEPR model?

<hr>

11.2 Nitrogen activation

Nitrogen occurs in many compounds, but N_2 itself is strikingly unreactive. A few strong reducing agents can transfer electrons to the N_2 molecule at room temperature, leading to scission of the N—N bond, but usually the reaction needs extreme conditions. The prime example of this reaction is the slow reaction of lithium metal at room temperature, which yields Li_3N. Similarly, when magnesium (lithium's diagonal neighbor) burns in air it forms the nitride as well as the oxide.

The slowness of the reactions of N_2 appears to be the result of several factors. One is the strength of the N—N bond and hence the high activation energy for breaking it. (The strength of this bond also accounts for the lack of nitrogen allotropes.) Another factor is the size of the HOMO–LUMO gap in N_2, which makes the molecule resistant to simple electron-transfer redox processes. A third factor is the low polarizability of N_2, which does not encourage the formation of the highly polar transition states that are often involved in electrophilic and nucleophilic displacement reactions.

Cheap methods of nitrogen fixation are highly desirable because they would have a profound effect on the economy. In the Haber process for the production of ammonia (which we discuss in detail in Section 17.7), H_2 and N_2 are combined at high temperatures and pressures over an iron catalyst. Although very successful and used throughout the world, the process requires a costly high-temperature, high-pressure plant, and research continues into more economical alternatives.

Much of the recent research aimed at more economical conversion of atmospheric N_2 into useful chemicals has focused on the way in which bacteria carry out the transformation at room temperature. The dominant process for the fixation of nitrogen in soils is the catalytic conversion of nitrogen to NH_4^+ by the enzyme nitrogenase, which occurs in the root nodules of legumes. The identity of this metalloenzyme is the topic of considerable research. In this connection, dinitrogen complexes of metals were discovered in 1965, and at about the same time the partial elucidation of the structure of nitrogenase indicated that the active site contains Fe and Mo atoms (Chapter 19). These developments led to optimism that efficient homogeneous catalysts might be devised in which metal ions will coordinate to N_2 and promote its reduction. Many N_2 complexes have in fact been prepared, and in some cases the preparation is as simple as bubbling N_2 through an aqueous solution of a complex:

$$[Ru(NH_3)_5(H_2O)]^{2+}(aq) + N_2(g) \longrightarrow [Ru(NH_3)_5(N_2)]^{2+}(aq) + H_2O(l)$$

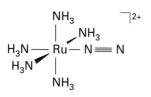

2 $[Ru(NH_3)_5N_2]^{2+}$

As with the isoelectronic CO molecule, end-on bonding is typical of N_2 when it acts as a ligand (**2**). The N—N bond in the Ru(II) complex is only slightly altered from that in the free molecule. However, when N_2 is coordinated to a more strongly reducing metal center, the nitrogen-nitrogen bond is considerably lengthened by back-donation of electron density into the π^* orbitals of N_2. Although new catalysts for N_2 reduction have not yet emerged from this research, there are high hopes of it, because it is possible to convert bound N_2 in some of these complexes into NH_4^+.[3]

> *Nitrogen is not very reactive, but bacteria manage to reduce it at room temperature. The commercial Haber process requires high temperatures and pressures to yield ammonia, which is a major ingredient in fertilizers and an important chemical intermediate. Thus, ammonia but not nitrogen is oxidized to high oxidation state oxoanions, such as nitrate and nitrite.*

11.3 Halides

Halogen compounds of phosphorus, arsenic, and antimony are numerous and important in synthetic chemistry. Trihalides are known for all the elements phosphorus through bismuth. However, whereas pentafluorides are known for phosphorus through bismuth, pentachlorides are known only for phosphorus through antimony and the pentabromide is known only for phosphorus. Nitrogen does not reach its group oxidation state in neutral binary halogen compounds, but it does achieve it in NF_4^+. Presumably, an N atom is too small for NF_5 to be sterically feasible.[4] The difficulty of oxidizing Bi(III) to Bi(V) by chlorine or bromine is an example of the inert pair effect. Bismuth pentafluoride, BiF_5 exists, but $BiCl_5$ and $BiBr_5$ do not.

> *Of the halogens, only fluorine can bring out the maximum oxidation state (+5) of bismuth (in BiF_5) and nitrogen (in NF_4^+).*

[3]   T.A. George, D.J. Rose, Y. Chang, Q. Chen, and J. Zubieta, *Inorg. Chem.* **34**, 1295 (1995).

[4]   The possible existence of NF_5 has been pursued by experiment and theory: see K.O. Christie, *J. Am. Chem. Soc.* **114**, 9934 (1992).

(a) Nitrogen halides

The most extensive series of halogen compounds of nitrogen are the fluorides, and NF_3 is in fact its only exoergic binary halogen compound. This pyramidal molecule is not very reactive. Thus, unlike NH_3, it is not a Lewis base because the strongly electronegative F atoms make the lone pair of electrons unavailable: whereas the polarity in NH_3 is $^{\delta-}NH_3^{\delta+}$, in NF_3 it is $^{\delta+}NF_3^{\delta-}$. Nitrogen trifluoride can be converted into the N(V) species NF_4^+ by the following reaction:

$$NF_3 + 2F_2 + SbF_3 \longrightarrow [NF_4^+][SbF_6^-]$$

Nitrogen trichloride, NCl_3, is a highly endoergic, explosive, and volatile liquid. It is prepared commercially by the electrolysis of a solution of ammonium chloride, and the resulting product is used directly as an oxidizing bleach for flour. The very unstable nitrogen tribromide, NBr_3, has been prepared, but NI_3 is known only in the form of the highly explosive ammoniate $NI_3 \cdot NH_3$. In complete contrast to NF_3, X-ray diffraction indicates a polymeric structure for this compound, with iodine-bridged NI_4 tetrahedra.

> *Except for NF_3, nitrogen trihalides have limited stability and nitrogen triiodide is treacherously explosive.*

(b) Other halides

The trihalides and pentahalides of nitrogen's congeners are used extensively in synthetic chemistry, and their simple empirical formulas conceal an interesting and varied structural chemistry.

The trihalides range from gases and volatile liquids, such as PF_3 (b.p. $-102\,°C$) and AsF_3 (b.p. $63\,°C$), to solids, such as BiF_3 (m.p. $649\,°C$). A common method of preparation is direct reaction of the element and halogen. For phosphorus, the trifluoride is prepared by metathesis of the trichloride and a fluoride:

$$2\,PCl_3(l) + 3\,ZnF_2(s) \longrightarrow 2\,PF_3(g) + 3\,ZnCl_2(s)$$

The trichlorides PCl_3, $AsCl_3$, and $SbCl_3$ are useful starting materials for the preparation of a variety of alkyl, aryl, alkoxy, and amino derivatives because they are susceptible to protolysis and metathesis:

$$ECl_3 + 3\,EtOH \longrightarrow E(OEt)_3 + 3\,HCl \quad (E = P, As, Sb)$$
$$ECl_3 + 6\,Me_2NH \longrightarrow E(NMe_2)_3 + 3\,[Me_2NH_2]Cl \quad (E = P, As, Sb)$$

Phosphorus trifluoride, PF_3, is an interesting ligand because in some respects it resembles CO. Like that molecule, it is a weak σ donor but a strong π acceptor, and complexes of PF_3 exist that are the analogs of carbonyls, such as $Ni(PF_3)_4$ the analog of $Ni(CO)_4$. The π-acceptor character is attributed to a P—F antibonding LUMO, which has mainly phosphorus p-orbital character. The trihalides also act as mild Lewis acids toward bases such as trialkylamines and halides. Many halide complexes have been isolated, such as the simple mononuclear species $AsCl_4^-$ (**3**) and SbF_5^{2-} (**4**). More complex dinuclear and polynuclear anions linked by halide bridges, such as the polymeric chain $([BiBr_3]^{2-})_n$, in which Bi(I) is surrounded by a distorted octahedron of Br atoms, are also known.

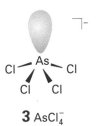

3 $AsCl_4^-$

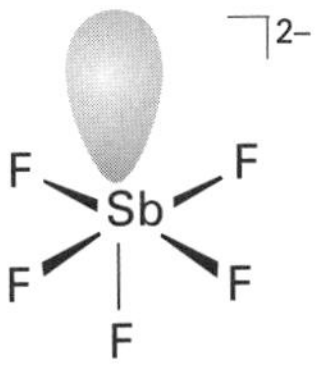

4 Sb_5^{2-}, C_{4v}

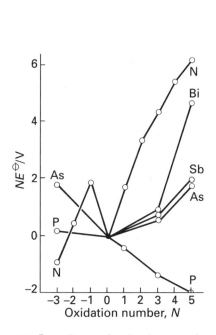

5 (SbF₅)₄

The pentahalides vary from volatile substances, such as PF₅ (b.p. −85 °C) and AsF₅ (b.p. −53 °C), to solids, such as PCl₅ (sublimes at 162 °C) and BiF₅ (m.p. 154 °C). The five-coordinate gas-phase molecules are trigonal-bipyramidal. In contrast to the volatile PF₅ and AsF₅, SbF₅ is a highly viscous liquid in which the molecules are associated through F-atom bridges. In solid SbF₅ these bridges result in a cyclic tetramer (**5**), which reflects the tendency of Sb(V) to achieve a coordination number of 6. A related phenomenon occurs with PCl₅, which in the solid state exists as $[PCl_4^+][PCl_6^-]$. In this case, the ionic contribution to the lattice enthalpy provides the driving force for the transfer of a Cl⁻ ion from one molecule to another. The pentafluorides of phosphorus, arsenic, antimony, and bismuth are strong Lewis acids. As we remarked in Section 5.9, SbF₅ is a very strong Lewis acid; it is much stronger, for example, than the aluminum halides.

The pentahalides of phosphorus and antimony are very useful in syntheses. Phosphorus pentachloride, PCl₅, is widely used in the laboratory and in industry as a starting material and some of its characteristic reactions are shown in Chart 11.2. Note, for example, that reaction of PCl₅ with Lewis acids yields PCl_4^+ salts, simple Lewis bases like F⁻ give six-coordinate complexes such as PF_6^-, compounds containing the NH₂ group lead to the formation of P—N bonds, and the interaction of PCl₅ with either H₂O or P₄O₁₀ yields O=PCl₃.

> The halides of nitrogen have limited stability, but their heavier congeners form an extensive series of compounds. Typical formulas are **EX₃** and **EX₅**. The trihalides and pentahalides are useful starting materials for the synthesis of derivatives by metathetical replacement of the halide.

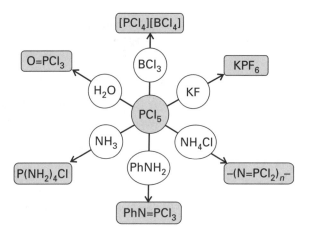

Chart 11.2 Uses of PCl₅.

11.4 Oxides and aqueous redox chemistry

We can infer the redox properties of the compounds of the elements in Group 15/V in acidic aqueous solution from the Frost diagram shown in Fig. 11.5. The steepness of the slopes of the lines on the far right of the diagram show the thermodynamic tendency for reduction of the +5 oxidation states of the elements. They show, for instance, that Bi₂O₅ is potentially a very strong oxidizing agent, which is consistent with the inert pair effect and the tendency of Bi(V) to form Bi(III). The next strongest oxidizing agent is NO₃⁻. Both As(V) and Sb(V) are milder oxidizing agents, and P(V), in the form of phosphoric acid, is a very weak oxidant.

The redox properties of nitrogen are important because of its widespread occurrence in the atmosphere, biosphere, industry, and the laboratory. Nitrogen chemistry is also quite intricate, partly on account of the large number of accessible oxidation states, but also because reactions that are thermodynamically favorable are often slow or have rates that depend critically on the identity of the reactants. In contrast to the more systematic reaction

11.5 Frost diagram for the elements of the nitrogen group in acidic solution. The species with oxidation number −3 are NH₃, PH₃, and AsH₃, and those with oxidation numbers −2 and −1 are N₂H₄ and NH₂OH respectively. The positive oxidation states refer to the most stable oxo or hydroxo species in acidic solution, and may be oxides, oxoacids, or oxoanions.

mechanisms of organic and coordination compounds, many of the reactions of N_2 and its oxides and oxoanions have turned out to be difficult to rationalize.

The main points to keep in mind include the fact that, as the N_2 molecule is kinetically inert, so redox reactions that *consume* N_2 are slow. Moreover, for mechanistic reasons that seem to be different in each case, the *formation* of N_2 is also slow and often sidestepped in aqueous solution (Chart 11.3). As with several other *p*-block elements, the barriers to reaction of high oxidation state oxoanions, such as NO_3^-, are greater than for low oxidation state oxoanions, such as NO_2^-. We should also remember that low **pH** enhances the oxidizing power of oxoanions (Section 6.3). Low **pH** also often accelerates their oxidizing reactions by protonation, for this step is thought to facilitate subsequent N—O bond breaking. Finally, it should be remembered that the reactions of the oxides and oxoanions of nitrogen commonly take place by atom or ion transfer, and outer-sphere electron transfer is rare.

Table 11.2 summarizes some of the properties of the nitrogen oxides, and Table 11.3 does the same for the nitrogen oxoanions. Both tables will help us to pick our way through the details of their properties.

> Reactions of nitrogen oxygen compounds that liberate or consume N_2 are generally very slow at normal temperatures and $pH = 7$.

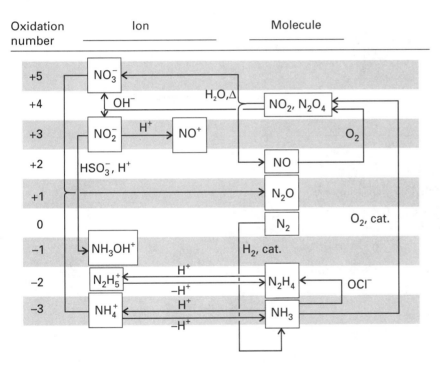

Chart 11.3 Interconversion of important nitrogen species.

(a) Nitrogen(V) oxoanions

The most common source of N(V) is nitric acid, HNO_3, which is a major industrial chemical used in the production of fertilizers, explosives, and a wide variety of nitrogen-containing chemicals. It is produced by modern versions of the *Ostwald process*, which make use of an indirect route from N_2 to the highly oxidized compound HNO_3 via the fully reduced compound NH_3. Thus, after nitrogen has been reduced to the −3 state as NH_3 by the Haber process, it is oxidized to the +4 state:

$$4\,NH_3(g) + 7\,O_2(g) \longrightarrow 6\,H_2O(g) + 4\,NO_2(g) \quad \Delta_r G^\ominus = -308.0 \text{ kJ (mol NO}_2)^{-1}$$

Table 11.2 Oxides of nitrogen

Oxidation number	Formula	Name	Structure (gas phase)	Remarks
+1	N_2O	Nitrous oxide (Dinitrogen oxide)	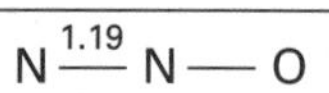 $N \overset{1.19}{\text{——}} N — O \quad C_{\infty v}$	Colorless gas, not very reactive
+2	NO	Nitric oxide (Nitrogen monoxide)	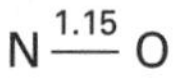 $N \overset{1.15}{\text{——}} O \quad C_{\infty v}$	Colorless, reactive paramagnetic gas
+3	N_2O_3	Dinitrogen trioxide	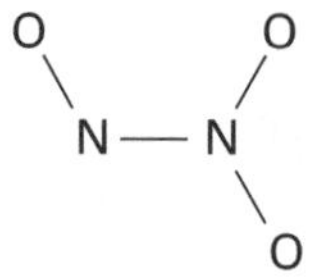 Planar, C_s	Forms blue solid (m.p. $-101\,°C$) and dissociates into NO and NO_2 in the gas phase
+4	NO_2	Nitrogen dioxide	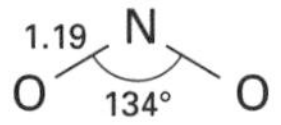 C_{2v}, angle $134°$, 1.19	Brown, reactive, paramagnetic gas
+4	N_2O_4	Dinitrogen tetroxide	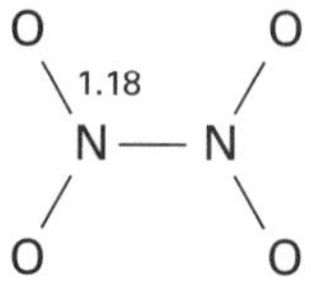 1.18, Planar, D_{2h}	Forms colorless liquid (m.p. $-11\,°C$); in equilibrium with NO_2 in the gas phase
+5	N_2O_5	Dinitrogen pentoxide (Dinitrogen oxide)	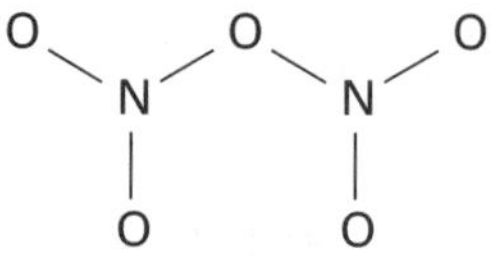 Planar, C_{2v}	Colorless ionic solid $[NO_2][NO_3]$ (m.p. $32\,°C$); unstable

Table 11.3 Nitrogen oxoions

Oxidation number	Formula	Common name	Structure	Remarks
+1	$N_2O_2^{2-}$	Hyponitrite	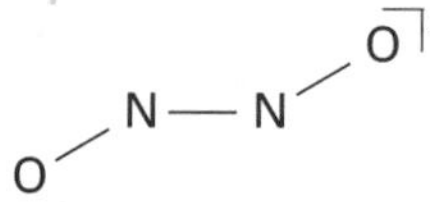 C_{2h}	Usually acts as a reducing agent
+3	NO_2^-	Nitrite	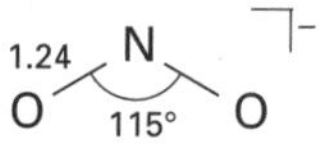 C_{2v}, 1.24, $115°$	Weak base; acts as an oxidizing and a reducing agent
+3	NO^+	Nitrosonium (nitrosyl cation)	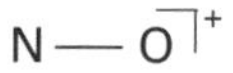 $N — O^+ \quad C_{\infty v}$	Oxidizing agent and Lewis acid; π-acceptor ligand
+5	NO_3^-	Nitrate	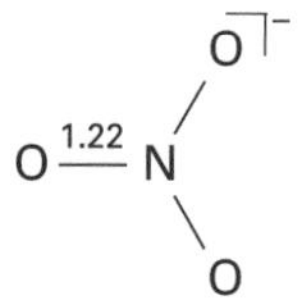 $O \overset{1.22}{\text{——}} N$ Planar, D_{3h}	Very weak base; an oxidizing agent
+5	NO_2^+	Nitronium (nitryl cation)	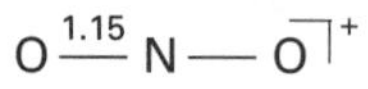 $O \overset{1.15}{\text{——}} N — O^+$ Linear, $D_{\infty h}$	Oxidizing agent, nitrating agent, and a Lewis acid

The NO_2 then undergoes disproportionation into N(II) and N(V) in water at elevated temperatures:

$$3\,NO_2(aq) + H_2O(l) \longrightarrow 2\,HNO_3^-(aq) + NO(g) \quad \Delta_r G^\ominus = -5.0\ kJ\,(mol\,HNO_3)^{-1}$$

All the steps are thermodynamically favorable. The byproduct NO is oxidized with O_2 to NO_2 and recirculated. Such an indirect route is employed because the direct oxidation of N_2 to NO_2 is thermodynamically unfavorable, with $\Delta_f G^\ominus(NO_2) = +51\ kJ\,mol^{-1}$. In part, this endoergic character is due to the great strength of the N—N bond.

Standard potential data imply that the NO_3^- ion is a fairly strong oxidizing agent; however, the kinetic aspects of its reactions are important, and they are generally slow in dilute acid solution. Because protonation of oxygen promotes N—O bond breaking, concentrated HNO_3 (in which NO_3^- is protonated) undergoes more rapid reactions than the dilute acid (in which HNO_3 is fully ionized). It is also a thermodynamically more potent oxidizing agent at low pH. A sign of this oxidizing character is the yellow color of the concentrated acid, which indicates its instability with respect to decomposition into NO_2:

$$4\,HNO_3(aq) \longrightarrow 4\,NO_2(aq) + O_2(g) + 2\,H_2O(l)$$

This decomposition is accelerated by light and heat.[5]

The reduction of NO_3^- ions rarely yields a single product, as so many lower oxidation states of nitrogen are available with substantial kinetic barriers to their interconversion. For example, a strong reducing agent such as zinc can reduce a substantial proportion of dilute HNO_3 as far as oxidation state -3:

$$HNO_3(aq) + 4\,Zn(s) + 9\,H^+(aq) \longrightarrow NH_4^+(aq) + 3\,H_2O(l) + 4\,Zn^{2+}(aq)$$

A weaker reducing agent, such as copper, proceeds only as far as oxidation state $+4$ in the concentrated acid:

$$2\,HNO_3(aq) + Cu(s) + 2\,H^+(aq) \longrightarrow 2\,NO_2(g) + Cu^{2+}(aq) + 2\,H_2O(l)$$

With the dilute acid, the $+2$ oxidation state is favored, and NO is formed:

$$2\,NO_3^-(aq) + 3\,Cu(s) + 8\,H^+(aq) \longrightarrow 2\,NO(g) + 3\,Cu^{2+}(aq) + 4\,H_2O(l)$$

The nitrate ion is a strong but sluggish oxidizing agent at room temperature; strong acid and heating accelerate the reaction.

Example 11.2 Correlating trends in the stabilities of N(V) and Bi(V)

Compounds of N(V) and Bi(V) are stronger oxidizing agents than the $+5$ oxidation states of the three intervening elements. Correlate this observation with trends in the periodic table.

Answer The light p-block elements are more electronegative than the elements immediately below them in the periodic table; accordingly, these light elements are generally less easily oxidized. Nitrogen is generally a good oxidizing agent in its positive oxidation states. Bismuth is much less electronegative, but favors the $+3$ oxidation state in preference to the $+5$ state on account of the inert pair effect.

Self-test 11.2 From trends in the periodic table, decide whether phosphorus or sulfur is likely to be the stronger oxidizing agent.

---

[5]   Useful comments on the explosion hazards of NO_3^- in combination with oxidizable cations, such as NH_4^+ in ammonium nitrate, will be found in L. Bretherick, *J. Chem. Educ.* **66**, A220 (1989).

6 N_2O_4

7 $C_2O_4^{2-}$

8 $[O_3SONO]^-$

9 ONF

(b) Nitrogen(IV) and nitrogen(III)

Nitrogen(IV) oxide, which is commonly called nitrogen dioxide, exists as an equilibrium mixture of the brown NO_2 radical and its colorless dimer N_2O_4 (dinitrogen tetroxide):

$$N_2O_4(g) \rightleftharpoons 2\,NO_2(g) \qquad K = 0.115 \text{ at } 25\,^\circ\text{C}$$

This readiness to dissociate is consistent with the N—N bond in N_2O_4 (**6**) being significantly longer and weaker than the C—C bond in the isoelectronic oxalate ion, $C_2O_4^{2-}$ (**7**), which does not equilibrate with CO_2^- radicals in the same way. The unpaired electron occupies an antibonding orbital in NO_2 and the transient CO_2^- radical, and so is less localized on the more electronegative N atom in NO_2 than it is on C in CO_2^-.[6] Hence the bond in N_2O_4 is relatively weak and N_2O_4 is more likely to dissociate than the oxalate ion.

Nitrogen(IV) oxide is a poisonous oxidizing agent that is present in low concentrations in the atmosphere, especially in photochemical smog. In basic aqueous solution it disproportionates into N(III) and N(V), forming NO_2^- and NO_3^- ions:

$$2\,NO_2(aq) + 2\,OH^-(aq) \longrightarrow NO_2^-(aq) + NO_3^-(aq) + H_2O(l)$$

In acidic solution (as in the Ostwald process) the reaction product is N(II) in place of N(III) because nitrous acid itself readily disproportionates:

$$3\,HNO_2(aq) \longrightarrow NO_3^-(aq) + 2\,NO(g) + H_3O^+(aq) \quad (E^\ominus = +0.05 \text{ V}, K = 50)$$

Nitrous acid, HNO_2, is a strong oxidizing agent:

$$HNO_2(aq) + H^+(aq) + e^- \longrightarrow NO(g) + H_2O(l) \quad (E^\ominus = +1.00 \text{ V})$$

and its reactions as an oxidizing agent are often more rapid than its disproportionation.

The rate at which nitrous acid oxidizes is increased by acid as a result of its conversion to the nitrosonium ion, NO^+:

$$HNO_2(aq) + H^+(aq) \longrightarrow NO^+(aq) + H_2O(l)$$

The nitrosonium ion is a strong Lewis acid and forms complexes rapidly with anions and other Lewis bases. The resulting species may not themselves be susceptible to oxidation (as in the case of SO_4^{2-} and F^- ions, which form $[SO_3ONO]^-$ (**8**) and ONF (**9**), respectively), but the association is often the initial step in an overall redox reaction. Thus there is good experimental evidence that the reaction of HNO_2 with I^- ions leads to the rapid formation of INO:

$$I^-(aq) + NO^+(aq) \longrightarrow INO(aq)$$

followed by the rate-determining second-order reaction between two INO molecules:

$$2\,INO(aq) \longrightarrow I_2(aq) + 2\,NO(g)$$

Nitrosonium salts containing poorly coordinating anions, such as $[NO][BF_4]$, are useful reagents in the laboratory as facile oxidizing agents and as a source of NO^+.

The intermediate oxidation states of nitrogen are often susceptible to disproportionation.

(c) Nitrogen(II) oxide

Nitrogen(II) oxide, more commonly nitric oxide, is an odd-electron molecule. However, unlike NO_2 it does not form a stable dimer in the gas phase. This difference reflects the greater delocalization of the odd electron in the π^* orbital in NO than the corresponding HOMO in NO_2. Nitric oxide reacts with O_2 to generate NO_2, but in the gas phase the rate law is second-order in NO, because a transient dimer, $(NO)_2$, is produced that subsequently collides with an

---

[6] We saw in Section 3.9 that bonding electrons are more likely to be found near the more electronegative atom in a bond, and antibonding electrons are more likely to be found near the less electronegative atom.

O_2 molecule. Because the reaction is second-order, atmospheric NO (which is produced in low concentrations by coal-fired power plants and by gasoline and diesel engines) is slow to convert to NO_2. Until the late 1980s, no beneficial biological roles were known for NO. However, since then it has been found that NO is generated *in vivo*, and that it performs functions such as the reduction of blood pressure, neurotransmission, and the destruction of microbes. Thousands of scientific papers have been published on the physiological functions of NO but our fundamental knowledge of its biochemistry is still quite meager.[7]

Because NO is endoergic, it should be possible to find a catalyst to convert the pollutant NO to the natural atmospheric gases N_2 and O_2 at its source in exhausts. It has been found that Cu^+ in a zeolite will catalyze the decomposition of NO and a reasonable understanding of the mechanism has been developed; however, this system is not yet sufficiently robust for a practical automobile exhaust catalytic converter.[8]

| *Nitric oxide is a strong π-acceptor ligand, and a troublesome pollutant in urban atmospheres; the molecule acts as a neurotransmitter.*

(d) Low oxidation states

The average oxidation number of nitrogen in dinitrogen oxide, N_2O (specifically, NNO), which is commonly called nitrous oxide, is $+1$; that of the azide ion, N_3^-, is $-\frac{1}{3}$. Both species are isoelectronic with CO_2, but the resemblance barely extends beyond their common linear shapes.

Dinitrogen oxide, a colorless unreactive gas, is produced by the comproportionation of molten ammonium nitrate. Care must be taken to avoid an explosion in this reaction, in which the cation is oxidized by the anion:

$$NH_4NO_3(l) \xrightarrow{250\,°C} N_2O(g) + 2\,H_2O(g)$$

Standard potential data suggest that N_2O should be a strong oxidizing agent in acidic and basic solutions:

$$N_2O(g) + 2\,H^+(aq) + 2\,e^- \longrightarrow N_2(g) + H_2O(l) \quad (E^{\ominus} = +1.77\ \text{V at pH} = 0)$$
$$N_2O(g) + H_2O(l) + 2\,e^- \longrightarrow N_2(g) + 2\,OH^-(aq) \quad (E = +0.94\ \text{V at pH} = 14)$$

However, kinetic considerations are paramount, and the gas is unreactive toward many reagents at room temperature. One sign of this inertness is that N_2O has been used as the propellant gas for instant whipping cream. Similarly, N_2O was used for many years as a mild anesthetic; however, this practice has been discontinued because of undesirable physiological side-effects.

Azide ions may be synthesized by the oxidation of sodium amide with either NO_3^- ions or N_2O at elevated temperatures:

$$3\,NH_2^- + NO_3^- \xrightarrow{175\,°C} N_3^- + 3\,OH^- + NH_3$$
$$2\,NH_2^- + N_2O \xrightarrow{190\,°C} N_3^- + OH^- + NH_3$$

The azide ion is a reasonably strong Brønsted base, the pK_a of its conjugate acid, hydrazoic acid, HN_3, being 4.77. It is also a good ligand toward d-block ions. However, heavy metal complexes or salts, such as $Pb(N_3)_2$ and $Hg(N_3)_2$, are shock-sensitive detonators. Ionic azides such as NaN_3 are thermodynamically unstable but kinetically inert; they can be handled at

[7] For an up-to-date assessment, see J.S. Stamler, D.J. Singel, and J. Loscalzo, Biochemistry of nitric oxide and its redox active forms. *Science* **258**, 1898 (1992); E.A. Dierks, S. Hu, K.M. Vogel, A.E. Yu, T.G. Spiro, and J.N. Burstyn, *J. Am. Chem. Soc.* **119**, 7316 (1997).

[8] M. Iwamoto, H. Furkawa, Y. Mine, F. Uemera, S. Mikuriya, and S. Kagawa, *Chem. Commun.* 1272 (1986); T. Beutel, J. Sarkany, G.-D. Lei, J.Y. Yan, and W.M.H. Sachtler, *J. Phys Chem.* **100**, 845 (1996).

10 N_2H_4

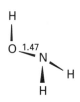

11 NH_2OH

room temperature. When alkali metal azides are heated they liberate N_2; this reaction is employed in the inflation of automobile air bags.

> The triatomic gas N_2O is unreactive, for kinetic reasons; the isoelectronic azide ion, N_3^-, forms many metal complexes.

(e) Hydrazine and hydroxylamine

Hydrazine, N_2H_4 (**10**), in which the oxidation number of nitrogen is -2, and hydroxylamine, NH_2OH (**11**), in which it is -1, are isolobal molecules that are formally related to NH_3 by the replacement of an H atom by an NH_2 group or an OH group, respectively. Both compounds are liquids at room temperature. In each case, the electronegative substituent makes the nitrogen lone pair less readily available and results in weaker Brønsted bases (and hence stronger acidity for the conjugate acids) than NH_3:

	NH_4^+	$N_2H_5^+$	NH_3OH^+
pK_a:	9.26	7.93	5.82

The Lewis base strength is reduced in the same way.

Most commercial hydrazine is prepared by the oxidation of NH_3 by ClO^- ions:

$$2\,NH_3(aq) + ClO^-(aq) \longrightarrow N_2H_4(aq) + Cl^-(aq) + H_2O(l)$$

This is formally a redox reaction. However, the mechanism is more intricate than simple electron transfer because it proceeds through the formation of the intermediate NH_2Cl. Once formed, the NH_2Cl is attacked by the Lewis base NH_3, which results in the displacement of Cl^- and the formation of an N–N bond:

$$NH_2Cl(aq) + 2\,NH_3(aq) \longrightarrow H_2NNH_2(aq) + Cl^-(aq) + NH_4^+(aq)$$

Hydrazine (m.p. 2 °C, b.p. 113 °C) is strongly associated through hydrogen bonding. It is an endoergic compound ($\Delta_f G^\ominus = +149$ kJ mol^{-1}), kinetically fairly inert, and widely used as a reducing agent. For example, it is used to reduce dissolved oxygen in boiler water to suppress corrosion. Similarly, alkylhydrazines are used as the reducing component in some rocket fuels. When hydrazine reacts with oxidizing agents, it yields a variety of nitrogen-containing products. One common product is N_2, as in the reactions

$$N_2H_4(aq) + O_2(g) \longrightarrow N_2(g) + 2\,H_2O(l)$$
$$N_2H_4(aq) + 2\,Cl_2(g) \longrightarrow N_2(g) + 4\,HCl(aq)$$

Hydrazine is a much stronger reducing agent in basic solution than in acidic solution:

$$N_2(g) + 5\,H^+(aq) + 4\,e^- \longrightarrow N_2H_5^+(aq) \quad E = -0.23\text{ V at pH} = 0$$
$$N_2(g) + 4\,H_2O(l) + 4\,e^- \longrightarrow N_2H_4(aq) + 4\,OH^-(aq)$$
$$E = -1.16\text{ V at pH} = 14$$

Hydroxylamine, NH_2OH, is an unstable endoergic compound ($\Delta_f G^\ominus = +23$ kJ mol^{-1}) which is prepared by the reduction of NO_2^- ions by HSO_3^- ions in neutral solution, followed by acidification and heating. In the first stage, SO_3^{2-} attacks NO_2^- and forms an N—S bond; in the second stage, this N—S bond is hydrolyzed:

$$NO_2^-(aq) + 2\,HSO_3^-(aq) \xrightarrow{0\,°C} N(OH)(SO_3)_2^{2-}(aq)$$
$$\xrightarrow{H^+,\,50\,°C} H_3NOH^+(aq) + 2\,SO_4^{2-}(aq)$$

Although the standard potentials indicate that hydroxylamine can serve as either an oxidizing agent or a reducing agent, the latter reactions generally occur more readily, as in

$$4\,Fe^{3+}(aq) + 2\,NH_3OH^+(aq) \longrightarrow 4\,Fe^{2+}(aq) + N_2O(g) + 6\,H^+(aq) + H_2O(l)$$

This reaction sidesteps N_2, despite the latter's great stability, and the nitrogen is carried directly from oxidation state -1 in H_2NOH to the $+1$ state in N_2O by a mechanistically complex reaction.

Example 11.3 Comparing the redox properties of nitrogen oxoanions and oxo compounds

Compare (a) NO_3^- and NO_2^- as oxidizing agents; (b) NO_2, NO, and N_2O with respect to their ease of oxidation in air; (c) N_2H_4 and H_2NOH as reducing agents.

Answer (a) NO_3^- and NO_2^- ions are both strong oxidizing agents. The reactions of the former are often sluggish but are generally faster in acidic solution. The reactions of NO_2^- ions are generally faster and become even faster in acidic solution, where the NO^+ is a common identifiable intermediate. (b) NO_2 is stable with respect to oxidation in air. N_2O and NO are thermodynamically susceptible to oxidation. However, the reaction of N_2O with oxygen is slow, and at low NO concentrations the reaction between NO and O_2 is slow because the rate law is second-order in NO. (c) Hydrazine and hydroxylamine are both good reducing agents. In basic solution hydrazine becomes a stronger reducing agent.

Self-test 11.3 Summarize the reactions that are employed for the synthesis of hydrazine and hydroxylamine. Are these reactions best described as electron-transfer processes or nucleophilic displacements?

Chart 11.3 (shown earlier) summarized the redox and acid–base reactions of some important nitrogen species. All the redox reactions shown in solution occur by atom or ion transfer rather than simple electron transfer. These reactions include the disproportionation of NO_2 in basic solution, shown at the top of the chart, and the oxidation of NH_3 to N_2H_4 and to NO_2, on the right of the chart. As already remarked, the oxidation of NH_3 by OCl^- to form N_2H_4, shown on the lower right of the chart, involves the initial formation of NH_2Cl followed by nucleophilic attack of NH_3 on NH_2Cl.

Nitrate, NO_3^-, and nitrite, NO_2^-, are the most important oxoanions of nitrogen. Their reactions can be sluggish, especially those of nitrate. The mechanisms often involve atom transfer, and rates are increased by low pH.

(f) Oxides of phosphorus, antimony, and bismuth

The complete combustion of phosphorus yields phosphorus(V) oxide, P_4O_{10}. Each P_4O_{10} molecule has a cage structure in which a tetrahedron of P atoms is held together by bridging O atoms, and each P atom has a terminal O atom (**12**). Combustion in a limited supply of oxygen results in the formation of phosphorus(III) oxide, P_4O_6; this molecule has the same O-bridged framework as P_4O_{10}, but lacks the terminal O atoms (**13**). It is also possible to isolate the intermediate compositions having one, two, or three O atoms terminally attached to the P atoms. Both principal oxides can be hydrated to yield the corresponding acids, the P(V) oxide giving phosphoric acid, H_3PO_4, and the P(III) oxide giving phosphorous acid, H_3PO_3. As remarked in Section 5.4, phosphorous acid has one H atom attached directly to the P atom; it is therefore a diprotic acid and better represented as $(HO)_2PHO$.

In contrast to the high stability of phosphorus(V) oxide, arsenic, antimony, and bismuth more readily form oxides with oxidation state $+3$, specifically As_2O_3, Sb_2O_3, and Bi_2O_3. In the gas phase, the arsenic(III) and antimony(III) oxides have the molecular formula E_4O_6, with the same tetrahedral structure as P_4O_6. Arsenic, antimony, and bismuth do form oxides with oxidation state $+5$, but Bi(V) oxide is unstable and has not been structurally characterized. Note the coordination number of 6 for antimony and 4 for the lighter elements phosphorus and arsenic, with their smaller atoms.

The oxides of phosphorus include P_4O_6 and P_4O_{10}, both of which are cage compounds with T_d symmetry. On progressing from arsenic to bismuth, the $+5$ oxidation state is more readily reduced to $+3$.

12 P_4O_{10}, T_d

13 P_4O_6, T_d

Table 11.4 Latimer diagrams for phosphorus

Acidic solution

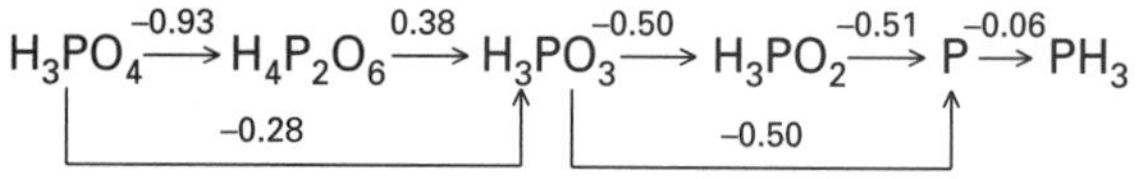

$$H_3PO_4 \xrightarrow{-0.93} H_4P_2O_6 \xrightarrow{0.38} H_3PO_3 \xrightarrow{-0.50} H_3PO_2 \xrightarrow{-0.51} P \xrightarrow{-0.06} PH_3$$

with branches labelled -0.28 (from H_3PO_4 to H_3PO_3) and -0.50 (from H_3PO_3 to P).

Basic solution

$$PO_4^{3-} \xrightarrow{-1.12} HPO_3^{2-} \xrightarrow{-1.57} H_2PO_2^{-} \xrightarrow{-2.05} P \xrightarrow{-0.89} PH_3$$

with branches labelled -1.73 (from HPO_3^{2-} to P).

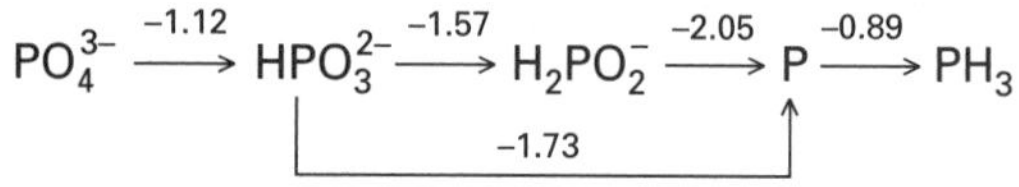

(g) Oxoanions

It can be seen from the Latimer diagram in Table 11.4 that elemental phosphorus and most of its compounds other than P(V) are strong reducing agents. White phosphorus disproportionates into phosphine, PH_3, (oxidation number -3) and hypophosphite ions (oxidation number $+1$) in basic solution:

$$P_4(s) + 3\,OH^-(aq) + 3\,H_2O(l) \longrightarrow PH_3(g) + 3\,H_2PO_2^-(aq)$$

Some common phosphorus oxoanions are listed in Table 11.5. The approximately tetrahedral environment of the P atom in their structures should be noted, as should the existence of P—H bonds in the hypophosphite and phosphite anions. The synthesis of various P(III) oxoacids and oxoanions, including HPO_3^{2-} and alkoxophosphanes, is conveniently performed by solvolysis of phosphorus(III) chloride under mild conditions, such as in cold tetrachloromethane solution:

$$PCl_3(l) + 3\,H_2O(l) \longrightarrow H_3PO_3(aq) + 3\,HCl(aq)$$
$$PCl_3(l) + 3\,ROH(sol) + 3\,N(CH_3)_3(sol) \longrightarrow P(OR)_3(sol) + 3\,[HN(CH_3)_3]Cl(sol)$$

Table 11.5 Some phosphorus oxoanions

Oxidation number	Formula	Name	Structure	Remarks
+1	$H_2PO_2^-$	Hypophosphite (dihydrodioxophosphate)	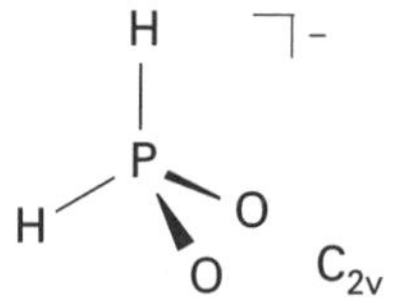 (structure, C_{2v})	Facile reducing agent
+3	HPO_3^{2-}	Phosphite (hydrotrioxophosphate)	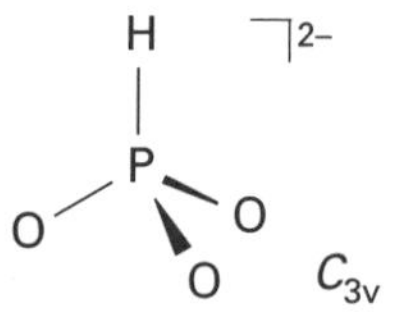 (structure, C_{3v})	Facile reducing agent
+4	$P_2O_6^{4-}$	Hypophosphate	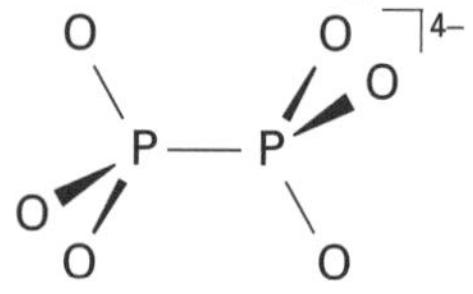 (structure)	Basic
+5	PO_4^{3-}	Phosphate	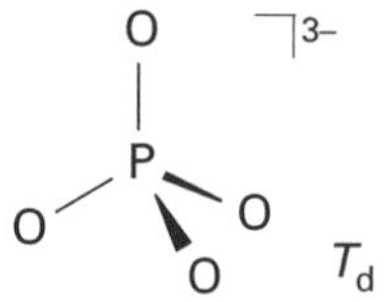 (structure, T_d)	Strongly basic
+5	$P_2O_7^{4-}$	Diphosphate	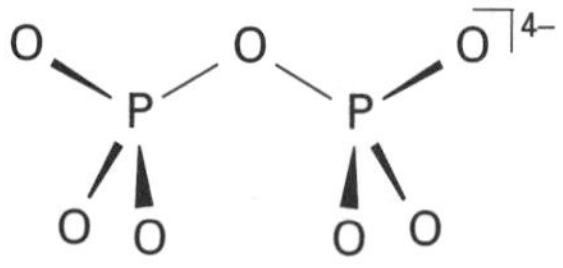 (structure)	Basic; longer-chain analogs are known

Reductions with $H_2PO_2^-$ and HPO_3^{2-} are usually fast. One of the commercial applications of this lability is the use of $H_2PO_2^-$ to reduce Ni^{2+}(aq) ions and so coat surfaces with metallic nickel in the process called 'electrodeless plating'.

The Frost diagram for the elements reveals similar trends in aqueous solution (Fig. 11.5), with oxidizing character following the order $PO_4^{3-} \approx AsO_4^{3-} < Sb(OH)_6^- \approx Bi(V)$. The thermodynamic tendency and kinetic ease of reducing AsO_4^{3-} is thought to be key to its toxicity toward animals. Thus, As(V) as AsO_4^{3-} readily mimics PO_4^{3-}, and so may be incorporated into cells. There, unlike phosphorus, it is reduced to an As(III) species, which is thought to be the real toxic agent. This toxicity may stem from the affinity of As(III) for sulfur-containing amino acids.

Important oxoanions are the P(I) species hypophosphite, $H_2PO_2^-$, the P(III) species phosphite, HPO_3^{2-}, and the P(V) species phosphate, PO_4^{3-}. The existence of P—H bonds and the highly reducing character of the two lower oxidation states is notable. Phosphorus(V) also forms an extensive series of O-bridged polyphosphates. In contrast to N(V), P(V) species are not strongly oxidizing. As(V) is more easily reduced than P(V).

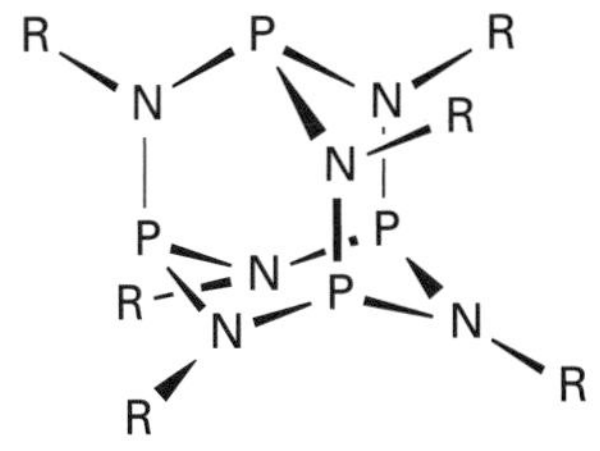

11.5 Compounds of nitrogen with phosphorus

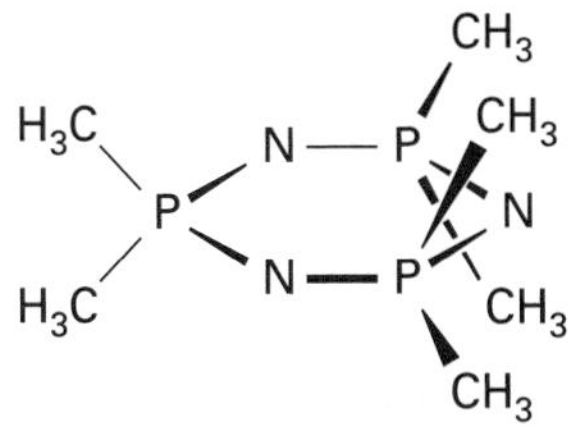

Many analogs of phosphorus–oxygen compounds exist in which the O atom is replaced by the isolobal NR or NH group, such as $P_4(NR)_6$ (**14**), the analog of P_4O_6. Other compounds exist in which OH or OR groups are replaced by the isolobal NH_2 or NR_2 groups. An example is $P(NMe_2)_3$, the analog of $P(OMe)_3$. Another indication of the scope of PN chemistry, and a useful point to remember, is that PN is isoelectronic with SiO. For example, various phosphazenes, which are chains and rings containing R_2PN units (**15**), are analogous to the siloxanes (Section 10.13) and their R_2SiO units (**16**).[9]

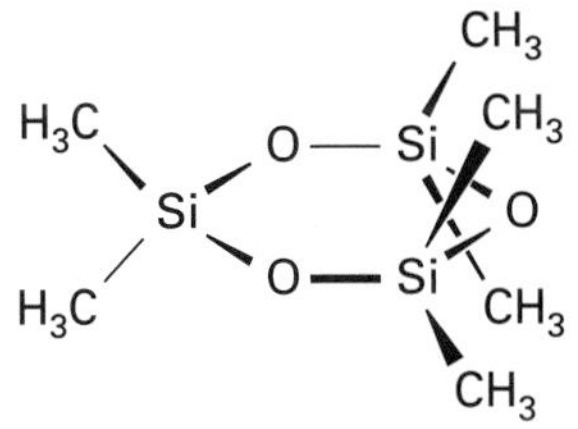

The cyclic phosphazene dichlorides are good starting materials for the preparation of the more elaborate phosphazenes. They are easily synthesized:

$$n\,PCl_5 + n\,NH_4Cl \longrightarrow (Cl_2PN)_n + 4n\,HCl \qquad n = 3 \text{ or } 4$$

A chlorocarbon solvent and temperatures near $130\,°C$ produce the cyclic trimer (**17**) and tetramer (**18**), and when the trimer is heated to about $290\,°C$ it changes to polyphosphazene. The Cl atoms in the trimer, tetramer, and polymer are readily displaced by other Lewis bases

$$(Cl_2PN)_n + 2n\,CF_3CF_2O^- \longrightarrow [(CF_3CF_2O)_2PN]_n + 2n\,Cl^-$$

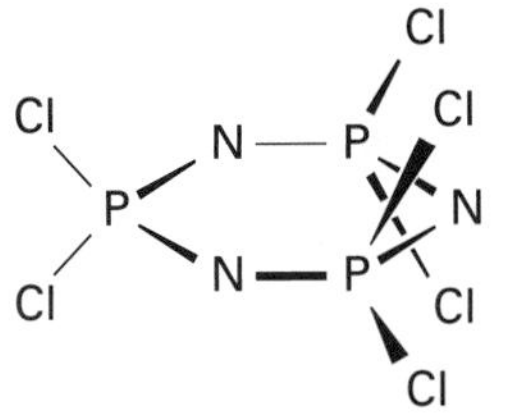

Like silicone rubber, the polyphosphazenes remain rubbery at low temperatures because, as with the isoelectronic SiOSi group, the molecules are helical and the PNP groups are highly flexible.

A phosphorus-nitrogen compound that was particularly useful in the laboratory until its carcinogenic properties were recognized was the aprotic solvent hexamethylphosphoramide, $((CH_3)_2N)_3P{=}O$, which is sometimes designated HMPA. The large bis(triphenylphosphine)imminium cation, $[Ph_3P{=}N{=}PPh_3]^+$, which is commonly abbreviated as PPN^+, is very useful. The salts of this cation are usually soluble in polar aprotic solvents such as HMPA, dimethylformamide, and even dichloromethane.

The range of PN compounds is extensive, and includes cyclic and polymeric phosphazenes, $(PX_2N)_n$. Phosphazenes form highly flexible elastomers.

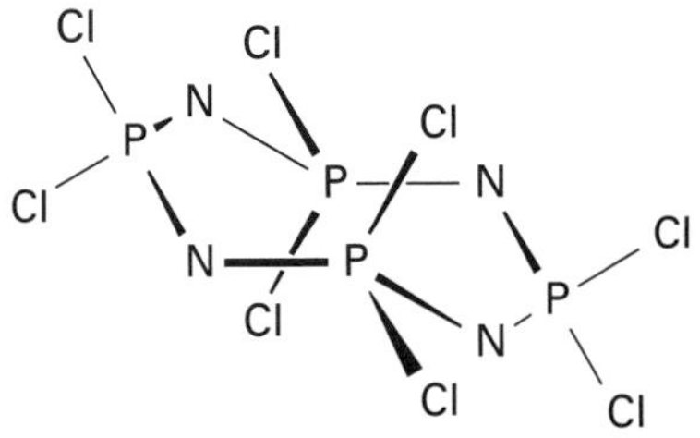

[9] H.R. Allcock, *Phosphorus–nitrogen chemistry*. Academic Press, New York (1972); R.H. Neilson, Phosphorus–nitrogen compounds. In *Encyclopedia of inorganic chemistry* (ed. R.B. King), Vol. 5, p. 3180. Wiley-Interscience, New York (1994).

14 $P_4(NR)_6$

15 $[(CH_3)_2PN]_3$

16 $[(CH_3)_2SiO]_3$

17 $[Cl_2PN]_3$, D_{3h}

18 $[Cl_2PN]_4$

Example 11.4 Devising a synthetic route to an alkoxy substituted cyclophosphazene

Suggest syntheses and give balanced equations for the preparation of $[NP(OCH_3)_2]_4$ from PCl_5, NH_4Cl, and $NaOCH_3$.

Answer The cyclic chlorophosphazene can be synthesized first:

$$4\,PCl_5 + 4\,NH_4Cl \xrightarrow{130\,°C} (Cl_2PN)_4 + 16\,HCl$$

The Cl atoms are readily replaced by strong Lewis bases, such as alkoxides, in the present case to give the desired product:

$$(Cl_2PN)_4 + 8\,NaOCH_3 \longrightarrow [(CH_3O)_2PN]_4 + 8\,NaCl$$

Self-test 11.4 Suggest a procedure and give the balanced chemical equations for the preparation of a high polymer phosphazene containing a PN backbone with two $N(CH_3)_2$ side groups attached to each P atom.

The oxygen group

The elements of the oxygen group, Group 16/VI, are often, and officially, called the **chalcogens**. The name derives from the Greek word for bronze, and refers to the association of sulfur and its congeners with copper.

11.6 Occurrence and recovery of the the elements

The principal elements of the group are oxygen and sulfur, which are both found in native form.

(a) Oxygen

Oxygen is readily available as O_2 from the atmosphere and is obtained on a massive scale by the distillation of liquid air. The main commercial motivation is to recover O_2 for steel making, in which it reacts exothermically with coke (carbon) to produce carbon monoxide and heat. The heat is necessary to achieve a fast reduction of iron oxides by CO and carbon (Sections 6.1 and 9.6). Pure oxygen, rather than air, is advantageous in this process because energy is not wasted in heating the nitrogen. About **1 tonne** of oxygen is needed to make **1 tonne** of steel.

The common allotrope of oxygen, formally dioxygen, O_2, boils at $-183\,°C$ and is very pale blue in the liquid state. This color arises from electronic transitions involving pairs of neighboring molecules. The molecular orbital description of O_2 implies the existence of a double bond; however, as we saw in Section 3.8, the outermost two electrons occupy different antibonding π orbitals with parallel spins; as a result, the molecule is paramagnetic. The molecular term symbol for the ground state is $^3\Sigma_g^-$, and henceforth the molecule will be denoted $O_2(^3\Sigma_g^-)$ when it is appropriate to specify the spin state.[10] The singlet state $^1\Delta_g$, with paired electrons in the same two π^* orbitals as in the ground state, is higher in energy by 1.61 eV, and another singlet state $^1\Delta_g$ ('singlet delta'), with electrons paired in one orbital lies between these two terms at 0.95 eV above the ground state. Of the two singlet states, the latter has much the longer excited state lifetime. This $O_2(^1\Delta_g)$ survives long enough to

---

[10] The symbols Σ, Π, and Δ are used for linear molecules such as dioxygen in place of the symbols S, P, and D used for atoms. The Greek letters represent the magnitude of the total orbital angular momentum around the internuclear axis.

participate in chemical reactions. When it is needed for reactions, $O_2(^1\Delta_g)$ can be generated in solution by energy transfer from a photoexcited molecule. Thus $[\text{Ru(bipy)}_3]^{2+}$ can be excited by absorption of blue light (452 nm) to give an electronically excited state, denoted $*[\text{Ru(bipy)}_3]^{2+}$, and this state transfers energy to $O_2(^3\Sigma_g^-)$:

$$*[\text{Ru(bipy)}_3]^{2+} + O_2(^3\Sigma_g^-) \longrightarrow [\text{Ru(bipy)}_3]^{2+} + O_2(^1\Delta_g)$$

Another efficient way to generate $O_2(^1\Delta_g)$ is through the thermal decomposition of an ozonide:

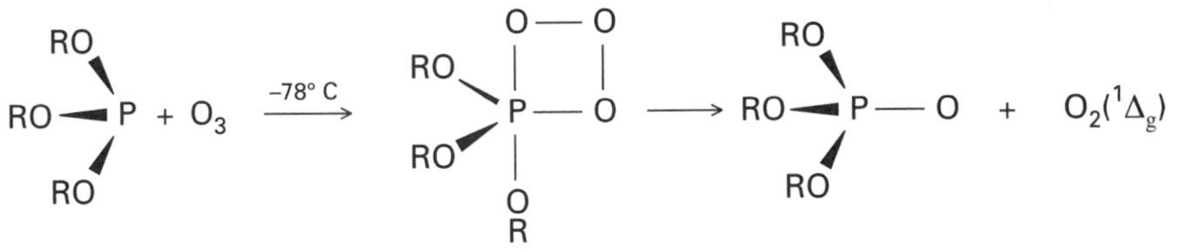

In contrast to the radical character of many $O_2(^3\Sigma_g^-)$ reactions, $O_2(^1\Delta_g)$ reacts as an electrophile. This mode of reaction is plausible because $O_2(^1\Delta_g)$ has a vacant π^* orbital. For example, $O_2(^1\Delta_g)$ adds across a diene, thus mimicking the Diels–Alder reaction of butadiene with an electrophilic alkene:

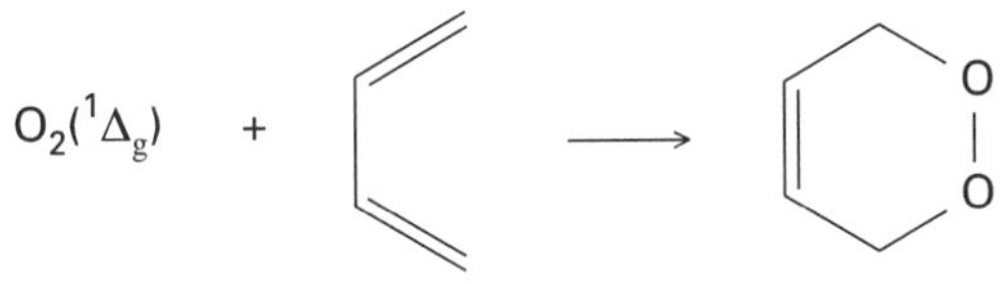

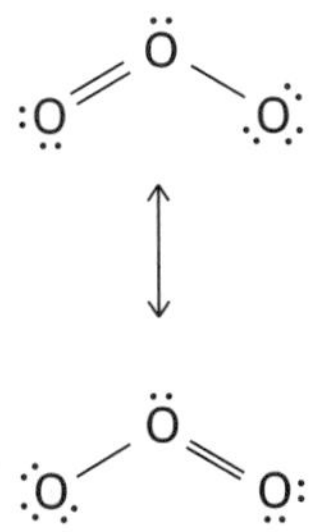

Singlet oxygen is implicated as one of the biologically hazardous products of photochemical smog.

The other allotrope of oxygen, *ozone*, O_3, boils at $-112\,°C$ and is an explosive and highly reactive endoergic blue gas $(\Delta_f G^\ominus = +163\ \text{kJ mol}^{-1})$. The O_3 molecule is angular, in accord with the VSEPR model (**19**), and has bond angle 117°; it is diamagnetic.

> Oxygen has two allotropes, dioxygen and ozone. Dioxygen has a triplet ground state that oxidizes hydrocarbons by a radical chain mechanism. Reaction with an excited state molecule can produce a fairly long-lived singlet state O_2, which is found in photochemical smog and often reacts as an electrophile. Ozone is an unstable and highly aggressive oxidizing agent.

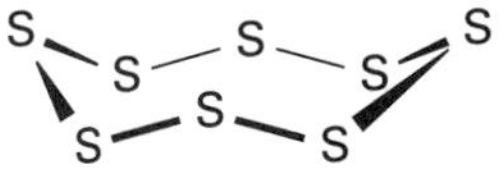

(b) Sulfur

Sulfur is obtained from deposits of the native element, metal sulfide ores, and liquid or gaseous hydrocarbons with a high sulfur content.

Unlike oxygen, sulfur (and all the heavier members of the group) tends to form single bonds with itself rather than double bonds on account of the poor π overlap of its orbitals. As a result, it aggregates into larger molecules or extended structures and hence is a solid at room temperature. Sulfur vapor, which is formed at high temperatures, consists partially of paramagnetic disulfur, S_2, molecules that resemble O_2 in having a triplet ground state and a formal double bond.

All the crystalline forms of sulfur that can be isolated at room temperature consist of S_n rings. The common orthorhombic polymorph, α-S_8, consists of crown-like eight-membered rings (**20**), but it is possible to synthesize and crystallize rings with from six to 20 S atoms. Orthorhombic sulfur melts at 113 °C; the yellow liquid darkens above 160 °C and becomes more viscous as the sulfur rings break open and polymerize. The resulting helical S_n polymers (**21**) can be drawn from the melt and quenched to form metastable rubber-like materials that slowly revert to α-S_8 at room temperature.

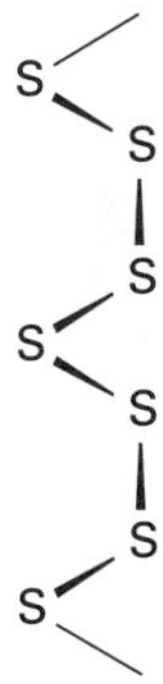

19 O_3, C_{2v}

20 S_8

21 S_n

| Sulfur is extracted as the element from underground deposits and its most stable form is the cyclic S_8 molecule. Sulfur has many allotropic forms, including a metastable polymer.

(c) Selenium, tellurium, and polonium

The chemically soft elements selenium and tellurium occur in metal sulfide ores, and their principal source is the electrolytic refining of copper.

Selenium exists as several different polymorphs. As with sulfur, a nonmetallic allotrope of selenium exists that contains Se_8 rings, but the most stable form at room temperature is *gray selenium*, a crystalline material composed of helical chains. The photoconductivity of gray selenium arises from the ability of incident light to excite electrons across its reasonably small band gap (2.6 eV in the crystalline material, 1.8 eV in the amorphous material), and is used in photocells. The common commercial form of the element is amorphous *black selenium*. Another amorphous form of selenium, obtained by deposition of the vapor, is used as the photoreceptor in the xerographic photocopying process. Selenium is an essential element for humans, but there is only a narrow range of concentration between the minimum daily requirement and toxicity.

Tellurium crystallizes in a chain structure like that of gray selenium. Polonium crystallizes in a primitive cubic structure and a closely related higher temperature form above 36 °C. We remarked in Section 2.5 that a simple cubic array represents inefficient packing of atoms, and polonium is the only element that adopts this structure under normal conditions. Tellurium and polonium are both highly toxic; the toxicity of polonium is enhanced by its intense radioactivity.

| Selenium and tellurium crystallize in helical chains; polonium crystallizes in a primitive cubic form.

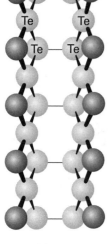

22 Te$_2$I

11.7 Halides

Oxygen forms many halogen oxides and oxoanions, and we shall discuss them in Chapter 12. The oxidation state of oxygen is −2 in all its compounds with the halogens other than fluorine. Oxygen difluoride, OF_2, is the highest fluoride of oxygen and hence contains oxygen in its highest oxidation state (+2).

Sulfur, selenium, tellurium, and polonium have a very rich halogen chemistry, and some of the most common halides are summarized in Table 11.6. The more electronegative element sulfur forms very unstable iodides, but tellurium and polonium form more robust compounds with iodine, presumably because of their bigger atomic radii. Of the halogens, the small, electronegative F atom alone brings out the maximum group oxidation state of the chalcogen elements, but it does not form stable binary compounds of selenium, tellurium, and polonium in low oxidation states (+1 and +2). A series of catenated subhalides exist for the heavy members of the group. For example, Te$_2$I and Te$_2$Br consist of ribbons of edge-shared Te hexagons with halogen bridges (**22**). The inability of the halogens other than fluorine to bring out the higher oxidation states is understandable on the basis that they are less electronegative than fluorine, and their single bond strengths to other elements are generally weaker.[11] The lack of low oxidation state fluorides may be a consequence of their instability toward disproportionation into the element and a higher oxidation state fluoride.

The structures of the sulfur halides S_2F_2, SF_4, SF_6, and S_2F_{10} (Table 11.6) are all in line with the VSEPR model. Thus, SF_4 has ten valence electrons around the S atom, two of which form a lone pair in an equatorial position of a trigonal bipyramid. We have already mentioned the theoretical evidence that the molecular orbitals bonding the F atoms to the central atom in SF_6 primarily utilize the sulfur $4s$ and $4p$ orbitals, with the $3d$ orbitals playing a relatively unimportant role (Section 3.12). The same seems to be true of SF_4 and S_2F_{10}.

[11] J. Passmore, *Acc. Chem. Res.* **22**, 234 (1989).

Table 11.6 Some halides of sulfur, selenium, and tellurium

Oxidation number	Formula	Structure	Remarks
$+\frac{1}{2}$	Te_2X (X = Br, I)	Halide bridges	Silver-gray
$+1$	S_2F_2	Two isomers:	
			Reactive
	S_2Cl_2	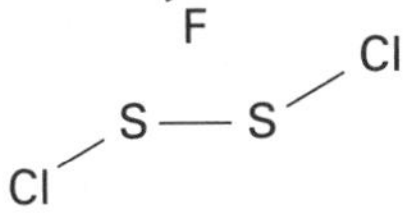	Reactive
$+2$	SCl_2	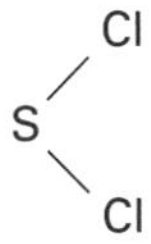	Reactive
$+4$	SF_4	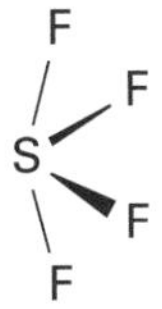	Gas
	SeX_4 (X = F, Cl, Br)		SeF_4 liquid
	TeX_4 (X = F, Cl, Br, I)		TeF_4 solid
$+5$	S_2F_{10}		Reactive
	Se_2F_{10}		
$+6$	SF_6, SeF_6		Colorless gases
	TeF_6		Liquid (b.p. 36 °C)

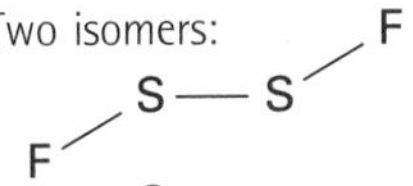

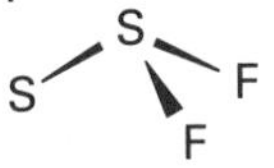

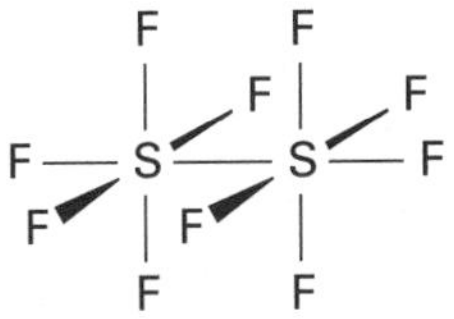

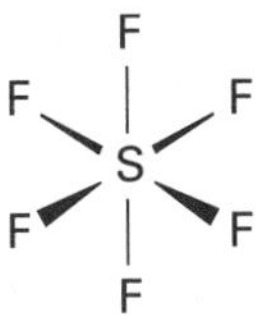

Sulfur hexafluoride is a gas at room temperature. Its inertness stems from the suppression, presumably by steric protection of the central S atom, of thermodynamically favorable reactions such as the hydrolysis

$$SF_6(g) + 4\,H_2O(l) \longrightarrow 6\,HF(aq) + H_2SO_4(aq)$$

The less sterically crowded SeF_6 molecule is easily hydrolyzed and is generally more reactive than SF_6. Similarly, the sterically less hindered molecule SF_4 is reactive and undergoes rapid partial hydrolysis:

$$SF_4 + H_2O \longrightarrow OSF_2 + 2\,HF$$

Both SF_4 and SeF_4 are very selective fluorinating agents for the conversion of –COOH into –CF_3 and C=O and P=O groups into CF_2 and PF_2 groups:

$$2\,R_2CO + SF_4 \longrightarrow 2\,R_2CF_2 + SO_2$$

Other halides of the chalcogens exist, and are summarized in Table 11.6. Sulfur chlorides are commercially important. The reaction of molten sulfur with Cl_2 yields the foul-smelling and toxic substance disulfur dichloride, S_2Cl_2, which is a yellow liquid at room temperature (b.p. 138 °C). Disulfur dichloride and its further chlorination product sulfur dichloride, SCl_2, an unstable red liquid, are produced on a large scale for use in the vulcanization of rubber. In this process, S atom bridges are introduced between polymer chains so the rubber object can retain its shape.

> *The halides of oxygen have limited stability, but its heavier congeners form an extensive series of halogen compounds. Typical formulas are EX_2, and EX_6. Binary compounds of oxygen and halogens are less extensive and less important than the halogen compounds of sulfur, selenium, and tellurium.*

11.8 Oxygen and the *p*-block oxides

Oxygen is by no means an inert molecule, and yet many of its reactions are sluggish (a point first made in connection with overpotentials in Section 6.4). For example, a solution of Fe^{2+} is only slowly oxidized by air even though the reaction is thermodynamically favored.

Several factors contribute to the appreciable activation energy of many reactions of O_2. One factor is that, with weak reducing agents, single electron transfer to O_2 is mildly unfavorable thermodynamically:

$$O_2(g) + H^+(aq) + e^- \longrightarrow HO_2(g) \quad (E^\ominus = -0.13 \text{ V at pH} = 0)$$
$$O_2(g) + e^- \longrightarrow O_2^-(aq) \quad (E = -0.33 \text{ V at pH} = 14)$$

A single-electron reducing reagent must exceed these potentials to achieve a significant reaction rate. Secondly, the ground state of O_2, with both π^* orbitals singly occupied, is neither an effective Lewis acid nor an effective Lewis base, and therefore has little tendency to undergo displacement reactions with *p*-block Lewis bases or acids. Finally, the high bond energy of O_2 (497 kJ mol^{-1}) results in a high activation energy for reactions that depend on its dissociation. Radical chain mechanisms can provide reaction paths that circumvent some of these activation barriers in combustion processes at elevated temperatures, and radical oxidations also occur in solution.

> *The reactions of dioxygen are often thermodynamically favorable but sluggish.*

(a) Hydrogen peroxide

The Frost diagram for oxygen shows that H_2O_2 is a good oxidizing agent, but it is unstable with respect to disproportionation (Fig. 11.6). In practice, however, it is not very labile and survives reasonably well at moderate temperatures unless traces of some ions are present, for these act as catalysts. Some insight into the mechanism is obtained by noting that effective catalysts have standard potentials in the range bounded by +1.76 V, the value for the reduction of H_2O_2 to H_2O, and −1.03 V, the value for the reduction of O_2 to H_2O_2. It is believed that the catalyzing ion shuttles back and forth between two oxidation states as it alternately oxidizes and reduces H_2O_2 (Fig. 11.7).

> *Hydrogen peroxide is susceptible to decomposition by disproportionation at elevated temperatures or in the presence of catalysts.*

Example 11.5 Deciding whether an ion can catalyze H_2O_2 disproportionation

Is Fe^{3+} thermodynamically capable of catalyzing the decomposition of H_2O_2?

Answer The standard potential for the reduction of Fe^{3+} to Fe^{2+} is +0.77 V. This value falls between the potentials for the reduction of H_2O_2 to H_2O (+1.76 V) and for the reduction of O_2 to H_2O_2 (−1.03 V), so catalytic decomposition is expected. We can verify that the potentials are favorable. Subtraction of the equations and potentials for reduction of O_2 to H_2O_2 from that for reduction of Fe^{3+} gives:

$$2\,Fe^{3+}(aq) + H_2O_2(aq) \longrightarrow 2\,Fe^{2+}(aq) + O_2(g) + 2\,H^+(aq) \quad (E^\ominus = +1.80 \text{ V})$$

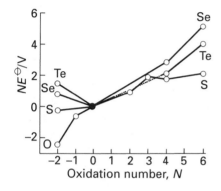

11.6 Frost diagram for the elements of the oxygen group in acidic solution. The species with oxidation number −2 are H_2E and for oxidation number −1 the compound is H_2O_2. The positive oxidation states refer to oxides or oxoacids. The species for S(II) is thiosulfate, $S_2O_3^{2-}$, and that for S(III) is hydrogendithionite, $HO_2SSO_2^-$.

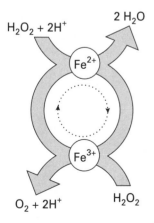

11.7 The catalytic cycle for the reduction of oxygen to hydrogen peroxide in the presence of iron(III) ions.

Because $E^\ominus > 0$, the reaction is thermodynamically favorable. Next we subtract the equation and potential for the reduction of Fe^{3+} from those for the reduction of H_2O_2 to H_2O and obtain

$$2\,Fe^{2+}(aq) + H_2O_2(aq) + 2\,H^+(aq) \longrightarrow 2\,Fe^{3+}(aq) + 2\,H_2O(l)$$
$$(E^\ominus = +0.99\ \text{V})$$

This reaction is also spontaneous, so catalytic decomposition is thermodynamically favored. In fact, the rates also are high, so Fe^{3+} is a highly effective catalyst for the decomposition of H_2O_2, and in its manufacture great pains are taken to minimize contamination by iron.

....

Self-test 11.5 Determine whether either Br^- or Cl^- is a candidate for catalysis of the decomposition of H_2O_2.

....

(b) Sulfur oxides and oxohalides

The molecules of the two common oxides of sulfur, SO_2 (b.p. $-10\,°C$, (**23**)) and SO_3 (b.p. $44.8\,°C$, (**24**)), are angular- and trigonal-planar in the gas phase, respectively. They are both Lewis acids, with the S atom the acceptor site, but SO_3 is the much the stronger and harder acid. The high Lewis acidity of SO_3 accounts for its occurrence as a polymeric O-bridged solid at room temperature and pressure (**25**).

Sulfur dioxide forms weak complexes with simple *p*-block Lewis bases. For example, although it does not form a stable complex with H_2O, it does form stable complexes with stronger Lewis bases, such as trimethylamine and F^- ions. Sulfur dioxide is a useful solvent for acidic substances (see Box 11.1).

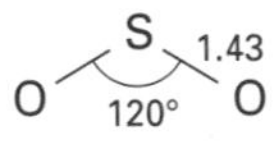

....

Example 11.6 Deducing the structures and chemistry of SO_2 complexes

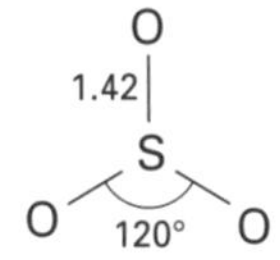

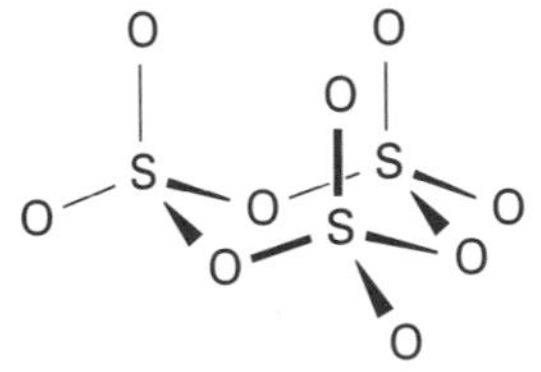

Suggest the probable structures of SO_2F^- and $(CH_3)_3NSO_2$, and predict their reactions with OH^-.

Answer Although the Lewis structure of SO_2 (**26**) has an electron octet around the S atom, that atom can act as a Lewis acid (Section 5.7). Both resulting complexes have a lone pair on S, and the four electron pairs form a trigonal pyramid around S in both complexes, (**27**) and (**28**). The OH^- ion is a stronger Lewis base than either F^- or $N(CH_3)_3$, so exposure of either complex to OH^- will yield the hydrogensulfite ion, HSO_3^-, which has been found to exist in two isomers, (**29**) and (**30**).

....

Self-test 11.6 Give the Lewis structures and point groups of (a) $SO_3(g)$ and (b) SO_3F^-.

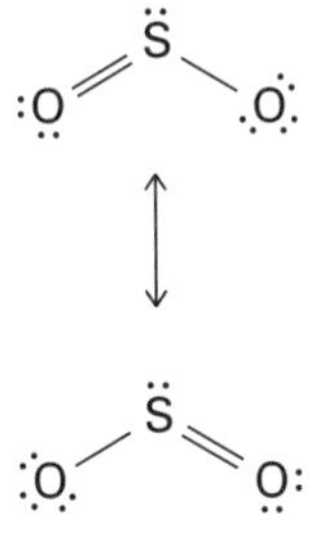

....

23 SO_2, C_{2v}

24 SO_3, D_{3h}

25 $(SO_3)_3$, C_{3v}

26 SO_2

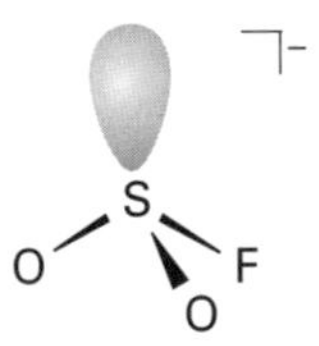

27 $[SO_2F]^-$

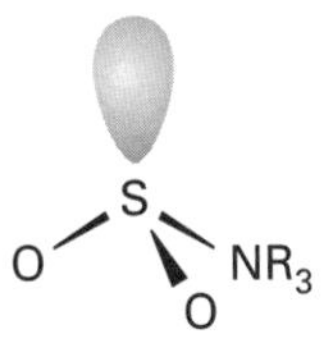

28 NR_3SO_2

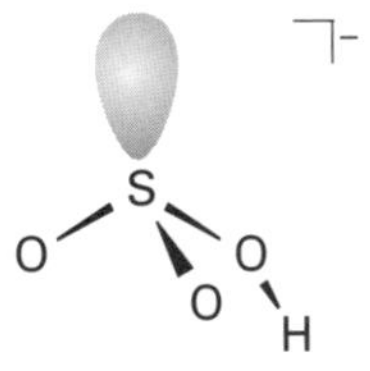

29 HSO_3^-

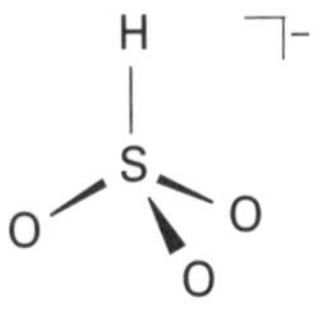

30 HSO_3^-, C_{3v}

Box 11.1 Synthesis in nonaqueous solvents

The nonaqueous solvents liquid ammonia, liquid sulfur dioxide, and sulfuric acid show an interesting set of contrasts as they range from a good Lewis base with a resistance to reduction (ammonia) to a strong Brønsted acid with resistance to oxidation (sulfuric acid).

Liquid ammonia (b.p. $-33\,°C$) and the solutions it forms may be handled in an open Dewar flask in an efficient hood (Fig. B11.1), in a vacuum line when the liquid is kept below its boiling point, or—with caution—in sealed heavy-walled glass tubes at room temperature (the vapor pressure of ammonia is about $10\,atm$ at room temperature). Liquid ammonia solutions of the s-block metals are excellent reducing agents. One example of their application is in the preparation of a complex of nickel in its unusual $+1$ oxidation state:

$$2\,K(am) + 2\,[Ni(CN)_4]^{2-}(am) \longrightarrow$$
$$[Ni_2(CN)_6]^{4-}(am) + 2\,KCN(am)$$

Liquid sulfur dioxide (b.p. $-10\,°C$) can be handled like liquid ammonia. An example of its use is the reaction

$$NH_2OH + SO_2(l) \longrightarrow H_2NSO_2OH$$

This reaction is performed in a sealed thick-walled borosilicate glass tube, which is sometimes called a 'Carius tube' (Fig. B11.2). The hydroxylamine is introduced into the tube and the sulfur dioxide is condensed into the tube after it has been cooled to about $-45\,°C$. After the tube has been sealed, it is allowed to warm to room temperature behind a blast shield in a hood. The reaction is complete after several days at room temperature; the tube is then cooled (to reduce the pressure) and broken open.

Sulfuric acid and solutions of SO_3 in sulfuric acid (fuming sulfuric acid, $H_2S_2O_7$) are used as oxidizing acidic media for the preparation of polychalcogen cations:

$$8\,Se + 5\,H_2SO_4 \longrightarrow Se_8^{2+} + 2\,H_3O^+ + 4\,HSO_4^- + SO_2$$

Further reading

W.M. Burgess and J.W. Estes, *Inorg. Synth.* **5**, 197 (1957).

G. Pass and H. Sutcliffe, *Practical inorganic chemistry*, p. 150. Chapman & Hall, London (1974).

R.J. Gillespie and J. Passmore, *Acc. Chem. Res.* **4**, 413 (1971).

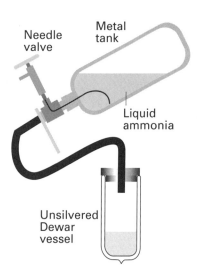

B11.1 The transfer of liquid ammonia from a pressurized tank. (Based on W.L. Jolly, *The synthesis and characterization of inorganic compounds.* Waveland Press, Prospect Heights (1991).)

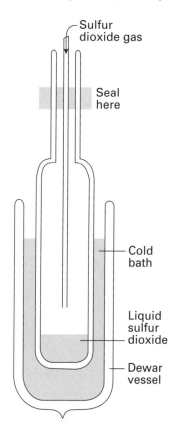

B11.2 The condensation of sulfur dioxide into a Carius tube.

Many chalcogen oxohalides are known. The most important are the thionyl dihalides, OSX_2, and the sulfuryl dihalides, O_2SX_2. One laboratory application of thionyl dichloride is the dehydration of metal chlorides:

$$MgCl_2 \cdot 6H_2O + 6\,OSCl_2 \longrightarrow MgCl_2 + 6\,SO_2 + 12\,HCl$$

The compound $F_5TeOTeF_5$ and its selenium analog are known, and the $OTeF_5^-$ ion, which is known informally as 'teflate', is one of the largest anions that combines very low basicity and high resistance to oxidation. Accordingly, it has been employed to stabilize large unstable cations. For example a complex between Ag^+ and 1,2-dichloroethene can be isolated as a teflate salt, $[Ag(CH_2Cl_2)][OTeF_5]$.[12]

> *Sulfur dioxide is a mild Lewis acid toward p-block bases, and $OSCl_2$ is a useful drying agent. The anion $OTeF_5^-$ coordinates only very weakly and has little tendency to undergo redox reactions.*

(c) Redox properties of sulfur oxoanions

A wide range of sulfur oxoanions exist (Table 11.7), and many of them are important in the

Table 11.7 Some sulfur oxoanions

Oxidation number	Formula	Name	Structure	Remarks
One S atom				
+4	SO_3^{2-}	Sulfite	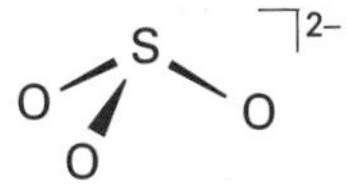	Basic, reducing agent
+6	SO_4^{2-}	Sulfate	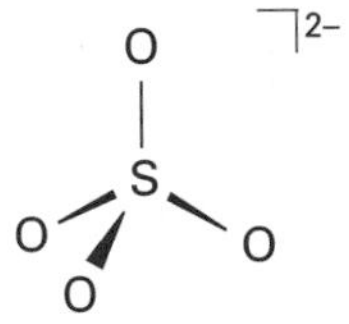	Weakly basic
Two S atoms				
+2	$S_2O_3^-$	Thiosulfate	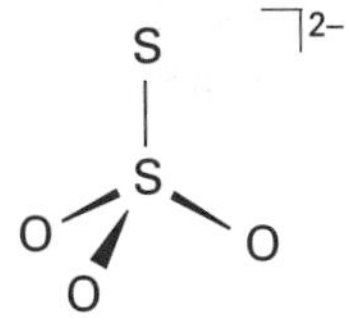	Mild reducing agent
+3	$S_2O_4^{2-}$	Dithionite	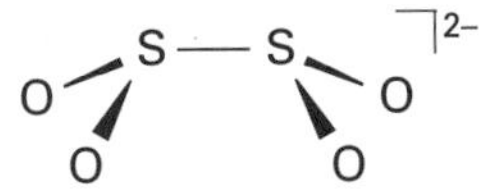	Strong and facile reducing agent
+5	$S_2O_6^{2-}$	Dithionate	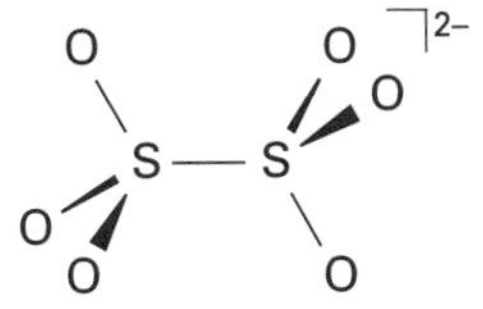	Resists oxidation and reduction
Polysulfur oxanions				
Variable	$S_nO_6^{2-}$ $3 \leq n \leq 20$	$n = 3$, Trithionate	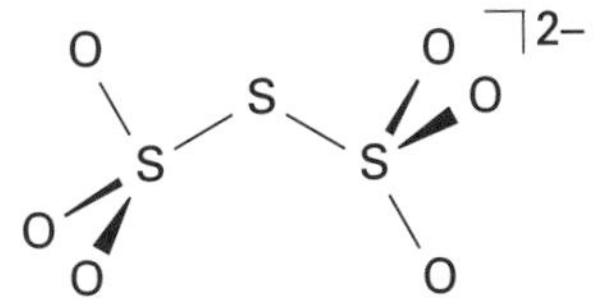	

[12] M.R. Colsman, T.D. Newbound, L.J. Marshall, M.D. Foirot, M.M. Miller, G.P. Wulfsburg, J.S. Frey, O.P. Anderson, and S.H. Strauss, *J. Am. Chem. Soc.* **112**, 2349 (1990).

laboratory and in industry. The peroxodisulfate anion, $O_3SOOSO_3^{2-}$, for example, is a powerful and useful oxidizing agent:

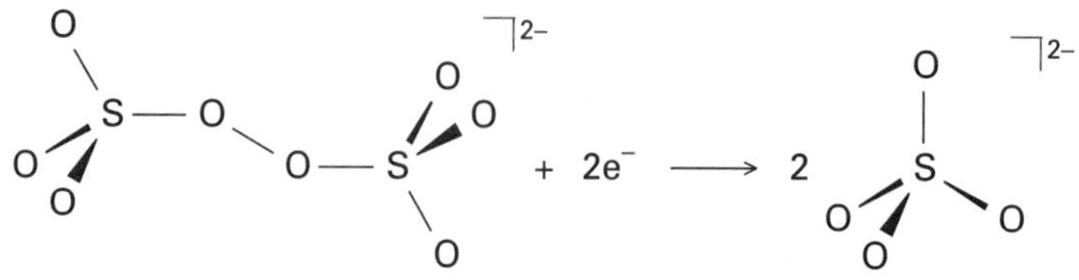

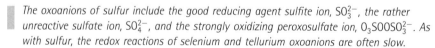

Sulfur's common oxidation numbers are -2, 0, $+2$, $+4$, and $+6$, but there are also many S—S bonded species that are assigned odd and fractional average oxidation numbers. A simple example is the thiosulfate ion, $S_2O_3^{2-}$, in which the average oxidation number of S is $+2$, but in which the environments of the two S atoms are quite different. The thermodynamic relations between the oxidation states are summarized by the Frost diagram (Fig. 11.6). As with many other p-block oxoanions, many of the thermodynamically favorable reactions are slow when the element is in its maximum oxidation state ($+6$), as in SO_4^{2-}. Another kinetic factor is suggested by the fact that oxidation numbers of compounds containing a single S atom generally change in steps of 2, which alerts us to look for an O atom transfer path for the mechanism. In some cases a radical mechanism operates, as in the oxidation of thiols and alcohols by peroxodisulfate, in which O—O bond cleavage produces the transient radical anion SO_4^-.

We saw in Section 6.3 that the **pH** of a solution has a marked effect on the redox properties of oxoanions. This strong dependence is true for SO_2 and SO_3^{2-}, because the former is easily reduced in acidic solution and is therefore a good oxidizing agent, whereas the latter in basic solution is primarily a reducing agent:

$$SO_2(aq) + 4H^+(aq) + 4e^- \longrightarrow S(s) + 2H_2O(l) \quad (E^{\ominus} = +0.50\ V)$$
$$SO_4^{2-}(aq) + H_2O(l) + 4e^- \longrightarrow SO_3^{2-}(aq) + 2OH^-(aq) \quad (E^{\ominus} = -0.94\ V)$$

The principal species present in acidic solution is SO_2, not H_2SO_3 (Section 5.9), but in more basic solution HSO_3^- exists in equilibrium with H—SO_3^- and H—OSO_2^-. The oxidizing character of SO_2 accounts for its use as a mild disinfectant and bleach for foodstuffs, such as dried fruit and wine.

The oxoanions of selenium and tellurium are a much less diverse and extensive group. Selenic acid is thermodynamically a strong oxidizing acid:

$$SeO_4^{2-}(aq) + 4H^+(aq) + 2e^- \longrightarrow H_2SeO_3(aq) + H_2O(l) \quad (E^{\ominus} = +1.1\ V)$$

However, like SO_4^{2-} and in common with the behavior of oxoanions of other elements in high oxidation states, the reduction of SeO_4^{2-} is generally slow. Telluric acid exists as $Te(OH)_6$ and also as $(HO)_2TeO_2$ in solution. Again, its reduction is thermodynamically favorable but kinetically sluggish.

> The oxoanions of sulfur include the good reducing agent sulfite ion, SO_3^{2-}, the rather unreactive sulfate ion, SO_4^{2-}, and the strongly oxidizing peroxosulfate ion, $O_3SOOSO_3^{2-}$. As with sulfur, the redox reactions of selenium and tellurium oxoanions are often slow.

11.9 Metal oxides

The O_2 molecule readily takes on electrons from metals to form a variety of metal oxides containing the anions O^{2-} (oxide), O_2^- (superoxide), or O_2^{2-} (peroxide). Even though these solids can be expressed in these terms, and the existence of O^{2-} can be rationalized in terms of a closed-shell noble gas electron configuration, the formation of $O^{2-}(g)$ from $O(g)$ is highly endothermic, and the ion is stabilized in the solid state by favorable lattice energies (Section 2.12).

Alkali metals and alkaline earth metals often form peroxides or superoxides (see Section 9.3), but peroxides and superoxides of other metals are rare. Among the metals, only some of the noble metals do not form thermodynamically stable oxides. However, even where no bulk oxide phase is formed, an atomically clean metal surface (which can be prepared only in an ultra-high vacuum) is quickly covered with a surface layer of oxide when it is exposed to traces of oxygen.

An important trend in the chemical properties of metal oxides is the high Brønsted basicity of oxides with metal ions of low charge and large radius (low electrostatic parameter, ξ) and the progression through amphoteric to acidic oxides as the charge-to-radius ratio increases (high ξ). These trends were examined in Section 5.5.

Structural trends in the metal oxides are not so readily summarized but, for oxides in which the metal has oxidation state +1, +2, or +3, the oxide ion is generally in a site of high coordination number. Thus, oxides of M^{2+} ions usually have the rock-salt structure (6:6 coordination, Section 2.9), oxides of M^{3+} ions (of formula M_2O_3) often have 6:4 coordination, and oxides with the formula MO_2 often occur in the rutile or fluorite structures (6:3 and 8:4 coordination, respectively). At the other extreme, MO_4 compounds are molecular: the tetrahedral compound osmium tetraoxide, OsO_4, is an example. The structures of the oxides with metals in high oxidation states and the oxides of nonmetallic elements often have multiple bond character. Deviations from these simple structures are common with p-block metals, where the less symmetric packing of oxide around the metal can often be rationalized in terms of the existence of a stereochemically active lone pair, as in PbO (Section 9.17). Another common structural motif for nonmetals and some metals in high oxidation states is a bridging oxygen atom, as in E—O—E, in angular and linear structures.

The oxides formed by metals include the basic oxides with high oxygen coordination number that are formed with most M^+ and M^{2+} ions. Oxides of metals in intermediate oxidation states often have more complex structures and are amphoteric. Metal peroxides and superoxides are formed between O_2 and alkali metals and alkaline earth metals. Terminal E=O linkages and E—O—E bridges are common with nonmetals and with metals in high oxidation states.

11.10 Metal sulfides, selenides, and tellurides

Sulfide, selenide, and telluride ions are soft ligands and often occur together in nature in copper and zinc deposits. As with oxygen, Se and Te atoms can serve as bridging ligands between metal centers.[13] For example, S, Se, and Te atoms are found to bridge both two or three metal atoms (**31**). The tendency to form multiple bonds with a metal atom in a low oxidation state is greater for sulfur, selenium, and tellurium than for oxygen, for which, as remarked above, M=O bond formation is generally confined to metals in high oxidation states.

We saw in Section 9.11 that sulfur has a rich coordination chemistry as the formal S^{2-} anion and higher polysulfide anions.[14] The extensive series of polysulfide ions, S_n^{2-}, with $n = 2$ to 22, is not matched by either selenium or tellurium, for which only the isolated ions Se_3^{2-} and Te_3^{2-} and d-metal complexes of larger polyselenides and polytellurides are known, such as $[TiCp_2Se_5]$ (**32**).

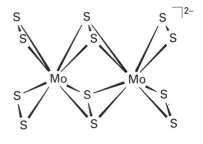

31 $[Mo_2(S_2)_6]^{2-}$

32 $[TiCp_2Se_5]$

[13] For a review see W.A. Herrmann, *Angew. Chem., Int. Ed. Engl.* **25**, 56 (1986).

[14] Additional examples are given by A. Muller, *Coord. Chem. Rev.* **46**, 245 (1982).

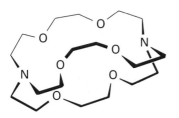

33 [WS(Se$_4$)$_2$]

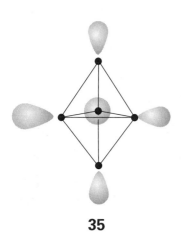

34 2,2,2–crypt

35

It appears that in polysulfides, polyselenides, and polytellurides electron density is concentrated at the ends of an E$_n^{2-}$ chain, which accounts for coordination through the terminal atoms, as shown in (**32**) and (**33**).

> Monatomic and polyatomic sulfide, selenide, and telluride ions are known as discrete anions and as ligands.

11.11 *p*-Block ring and cluster compounds

Cluster compounds are known for elements from across the periodic table.[15] We have already discussed boron clusters (the boranes and carboranes, Chapter 10) and clusters formed by *d*-block metals (Chapter 7); the latter will be met again in Chapter 16. In this section we concentrate on the clusters formed by the heavier *p*-block elements. These clusters are frequently called 'naked clusters' because they generally have no attached groups, and hence are of the form E$_n$, as in P$_4$. Many of the *p*-block clusters are ions.

(a) Anionic clusters

Some *p*-block metals and metalloids react with alkali metals to form compounds such as K$_2$Pb$_5$. These compounds were investigated by Eduard Zintl in Germany in the 1930s, who found that some could be dissolved in liquid ammonia, a hospitable solvent to strong reducing agents. These polyatomic clusters are often referred to as **Zintl ions**. It has been found that by complexing the alkali metal cation with 2,2,2-crypt (**34**), it is possible to prepare species in ethylenediamine solution that can be crystallized and investigated by X-ray diffraction. The crypt ligand encapsulates the alkali metal ion and thus creates a large cation complex:

$$K_2Pb_5 + 2 \text{ crypt} \xrightarrow{\text{en, 50°C}} [K(\text{crypt})]_2[Pb_5]$$

The stability of the product stems from the stabilizing effect that large cations have on large anions. Similar techniques have led to a variety of reduced species that have been characterized in the solid state by X-ray diffraction and in some cases in solution by NMR.

The electron-counting correlations (Wade's rules) introduced for boranes and carboranes in Section 10.3 are successful with many *p*-block naked cluster anions, and we can use Table 10.5 to predict their shapes. As an example, the number of valence electrons available in the Pb$_5^{2-}$ ion is $5 \times 4 = 20$ from the Pb atoms and two more from the charge, giving 22 electrons, or 11 pairs. Of these 11 pairs, one pair on each Pb atom, and so five pairs in all, are unavailable for bonding since, like the electron pair of a B—H unit in boranes, they are assumed to be directed away from the skeleton (**35**). Thus the total number of skeletal electron pairs is $11 - 5 = 6$. This number agrees with the count expected for a five-atom ($n = 5$) *closo* cluster, for $5 + 1 = 6$.

> Cyclic and cluster compounds are known for many of the heavier *p*-block elements. The valence electron count and structures of Group 14/IV and 15/V anionic clusters can often be correlated by Wade's rules.

....

Example 11.7 Correlating the electron count and structure of Zintl ions

Determine the electron count and the structural classification for (a) the Sn$_9^{4-}$ ion and (b) the diamond-shaped Bi$_4^{2-}$ ion.

....

[15] J.D. Corbett, *Chem. Rev.* **85**, 383 (1985); H.G. von Schnering, *Angew. Chem. Intl. Ed. Engl.* **20**, 33 (1981); A.H. Cowley (ed.), *Rings, polymers, and clusters of main group elements*, ACS Symposium Series, no. 232. American Chemical Society, Washington DC (1983); S.M. Kauzlarich (ed.), *Chemistry, structure, and bonding of Zintl phases and ions*. VCH, Weinheim (1996).

36 Sn_9^{4-}

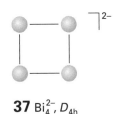

37 Bi_4^{2-}, D_{4h}

38 Ge_9^{2-}

Answer (a) The Sn_9^{4-} ion contains $4 \times 9 + 4 = 40$ valence electrons, and hence 20 pairs. Subtracting one nonbonding pair on each atom (a total of 9) leaves 11 pairs, which is consistent with the number expected for a *nido* cluster. In agreement with this conclusion, the structure is a truncated M_{10} deltahedron characteristic of a *nido* cluster (**36**). (b) The Bi_4^{2-} anion contains $(4 \times 5) + 2 - (4 \times 2) = 14$ skeletal electrons, or 7 pairs. The count suggests an *arachno* cluster, but it is square-planar rather than the expected butterfly (**37**); see the isoelectronic Se_4^{2+} below.

....

Self-test 11.7 Determine the classification from electron count for Ge_9^{2-} and determine whether your conclusion is consistent with the structure (**38**).

....

(b) Polycations

Many cationic chain, ring, and cluster compounds of the *p*-block elements have been prepared. The majority of them contain sulfur, selenium, or tellurium, but they may also contain elements from mercury in Group 12 to the halogens in Group 17/VII. Because these cations are oxidizing agents and Lewis acids, the preparative conditions are quite different from those used to synthesize the highly reducing polyanions. For example, S_8 is oxidized by AsF_5 in liquid sulfur dioxide to yield the S_8^{2+} ion:

$$S_8 + 2\,AsF_5 \xrightarrow{SO_2} [S_8][AsF_6]_2 + AsF_3$$

A strong acid solvent, such as fluorosulfuric acid, is used and the strongly oxidizing peroxide compound FO_2SOOSO_2F (that is, $S_2O_6F_2$) oxidizes Se to Se_4^{2+}:

$$4\,Se + S_2O_6F_2 \xrightarrow{HSO_3F} [Se_4][SO_3F]_2$$

The Se_4^{2+} cation has a square-planar structure (**39**). In the molecular orbital model of the bonding, this square cation has a closed-shell configuration in which the delocalized π-bonding a_{1g} and nonbonding e_g orbitals are filled, and a higher energy nonbonding orbital is vacant (Fig. 11.8). In contrast, most of the larger ring systems can be understood in terms of localized 2c,2e bonds. For these larger rings, the removal of two electrons brings about the formation of an additional 2c,2e bond, thereby preserving the local electron count on each element. This change is readily seen for the oxidation of S_8 to S_8^{2+} (**40**). An X-ray single-crystal structure determination shows that the transannular bonds in S_8^{2+} are long compared with the other bonds. Long transannular bonds are common in these types of compounds.

> Polyatomic cations of S and Se can be produced by the action of mild oxidizing agents on the elements in strong acid media.

(c) Neutral heteroatomic ring and clusters

The P_4 molecule is a good example of a cluster that can be described by local 2c,2e bonds. It turns out that several chalcogen derivatives are structurally related to P_4 by the insertion of

39 Se_4^{2+}

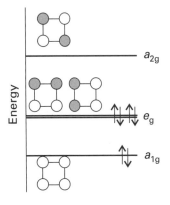

40 S_8^{2+}

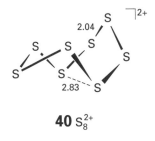

11.8 A schematic molecular orbital energy level diagram for the Se_4^{2+} cation.

41 P_4S_3

42 S_4N_4

43 $S_4N_4SO_3$

atoms into P–P bonds to give analogs of the P_4O_n compounds we have already described. Some sulfur compounds in this series are P_4S_3 (**41**), which apparently does not have an oxygen analog, and P_4S_{10}, the analog of P_4O_{10} (**12**).

Sulfur–nitrogen compounds have structures that can be related to the polycations discussed above. The oldest known, and easiest to prepare, is the pale yellow–orange tetrasulfurtetranitride, S_4N_4 (**42**), which is made by passing ammonia through a solution of S_2Cl_2:

$$6\,S_2Cl_2 + 16\,NH_3 \longrightarrow S_4N_4 + S_8 + 12\,NH_4Cl$$

Tetrasulfurtetranitride is endoergic ($\Delta_f G^{\ominus} = +536\ \text{kJ mol}^{-1}$) and may decompose explosively. The molecule is an eight-membered ring with the four N atoms in a plane and bridged by S atoms that project above and below the plane. The short S—S distance (2.58 Å) suggests that there is a weak interaction between pairs of S atoms. Lewis acids such as BF_3, SbF_5, and SO_3 form 1:1 complexes with one of the nitrogen atoms, and in the process the S_4N_4 ring rearranges (**43**).

Disulfurdinitride, S_2N_2 (**44**), is formed (together with Ag_2S and N_2) when S_4N_4 vapor is passed over hot silver wool. It is even more touchy than its precursor, for it explodes above room temperature. When allowed to stand at $0\,°C$ for several days, disulfurdinitride transforms into a bronze-colored polymer of composition $(SN)_n$ (**45**). The chains have a zig-zag shape and the compound exhibits metallic conductivity along the chain axis. The polymer is superconducting below $0.3\ \text{K}$. The discovery of this superconductivity was important because it was the first example of a superconductor that had no metal constituents.

> Neutral heteroatomic ring and cluster compounds of the p-block elements include P_4S_{10} and cyclic S_4N_4. Disulfurdinitride transforms into a polymer, $(SN)_n$, which is superconducting at very low temperatures.

44 S_2N_2 1.66   89.6°   90.4°

45 $(SN)_n$

Further reading

J. Emsley and D. Hall, *The chemistry of phosphorus.* Harper & Row, New York (1976).

D.E.C. Corbridge, *Phosphorus.* Elsevier, New York (1980). Vol. 1, Phosphorus compounds; Vol. 2, Organophosphorus compounds.

H.G. Heal, *The inorganic heterocyclic chemistry of sulfur, nitrogen, and phosphorus.* Academic Press, New York (1980).

A.J. Bard, R. Parsons, and J. Jordan, *Standard potentials in aqueous solution.* Dekker, New York (1985). See Chapters 4, 6, and 9 particularly. In addition to a discussion of thermodynamic data, these chapters give qualitative information on rates.

A.F. Wells, *Structural inorganic chemistry.* Oxford University Press (1984). See especially Chapters 11, 12, and 16–20.

A.E. Martell and D.T. Sawyer (ed.), *Oxygen complexes and oxygen activation by transition metals.* Plenum, New York (1988). This volume presents a range of informative review papers on O_2 complexes and O_2 activation by complexes.

D.T. Sawyer, *Oxygen chemistry.* Oxford University Press, New York (1991).

Kirk–Othmer encyclopedia of chemical technology. Wiley-Interscience, New York (1991 *et seq.*). Because of the great industrial importance of N, O, P, and S compounds, they receive extensive coverage in these volumes. Vol. 15: nitric acid, nitrides, nitrogen, and nitrogen fixation; Vol. 16: oxygen and ozone; Vol. 17: phosphoric acid, phosphates, phosphorus, phosphides, phosphorus compounds, and peroxides; Vol. 22; sulfur, sulfur recovery, sulfuric acid, sulfur trioxide, and sulfur compounds.

Exercises

11.1 List the elements in Groups 15/V and 16/VI and indicate the ones that are (a) diatomic gases, (b) nonmetals, (c) metalloids, (d) true metals. Also indicate those elements that do not achieve the maximum group oxidation number and indicate the elements that display the inert pair effect.

11.2 (a) Give complete and balanced chemical equations for each step in the synthesis of H_3PO_4 from hydroxyapatite to yield (a) high purity phosphoric acid and (b) fertilizer grade phosphoric acid. (c) Account for the large difference in costs between these two methods.

11.3 Ammonia can be prepared by (a) the hydrolysis of Li_3N or (b) the high-temperature high-pressure reduction of N_2 by H_2. Give balanced chemical equations for each method starting with N_2, Li, and H_2, as appropriate. (c) Account for the lower cost of the second method.

11.4 Compare and contrast the formulas and stabilities of the oxidation states of the common nitrogen chlorides with the phosphorus chlorides.

11.5 Use Lewis structures and the VSEPR model to predict the probable structure of (a) PCl_4^+, (b) PCl_4^-, (c) $AsCl_5$.

11.6 Give balanced chemical equations for each of the following reactions: (a) oxidation of P_4 with excess oxygen; (b) reaction of the product from part (a) with excess water; (c) reaction of the product from part (b) with a solution of $CaCl_2$ and name the product.

11.7 Select isoelectronic species from O_2^+, O_2, O_2^-, O_2^{2-}, N_2, NO, NO^+, CN^-, and $N_2H_5^+$. Within these isoelectronic groups, describe the probable relative oxidizing strengths and Lewis basicities.

11.8 Give the formula and name of a carbon-containing molecule or molecular ion that is isoelectronic and isostructural with (a) NO_3^-, (b) NO_2^-, (c) N_2O_4, (d) N_2, (e) NH_3.

11.9 Starting with $NH_3(g)$ and other reagents of your choice, give the chemical equations and conditions for the synthesis of (a) HNO_3, (b) NO_2^-, (c) NH_2OH, (d) N_3^-.

11.10 Write the balanced chemical equation corresponding to the standard enthalpy of formation of $P_4O_{10}(s)$. Specify the structure, physical state (s, l, or g), and allotrope of the reactants. Do either of these reactants differ from the usual practice of taking as reference state the most stable form of an element?

11.11 Without reference to the text, sketch the general form of the Frost diagrams for phosphorus (oxidation states 0 to +5) and bismuth (0 to +5) in acidic solution and discuss the relative stabilities of the +3 and +5 oxidation states of both elements.

11.12 Are reactions of NO_2^- as an oxidizing agent generally faster or slower when pH is lowered? Give a mechanistic explanation for the pH dependence of NO_2^- oxidations.

11.13 When equal volumes of nitric oxide (NO) and air are mixed at atmospheric pressure a rapid reaction occurs, to form NO_2 and N_2O_4. However, nitric oxide from an automobile exhaust, which is present in the parts per million concentration range, reacts slowly with air. Give an explanation for this observation in terms of the rate law and the probable mechanism.

11.14 Give balanced chemical equations for the reactions of the following reagents with PCl_5 and indicate the structures of the products: (a) water (1:1), (b) water in excess, (c) $AlCl_3$, (d) NH_4Cl.

11.15 Use standard potentials (Appendix 2) to calculate the standard potential of the reaction of H_3PO_2 with Cu^{2+}. Are HPO_3^{2-} and $H_2PO_2^{2-}$ useful as oxidizing or reducing agents?

11.16 (a) Use standard potentials (Appendix 2) to calculate the standard potential of the disproportionation of H_2O_2 and HO_2 in acid solution. (b) Is Cr^{2+} a likely catalyst for the disproportionation of H_2O_2? (c) Given the Latimer diagram

$$O_2 \xrightarrow{-0.13} HO_2 \xrightarrow{1.51} H_2O_2$$

in acidic solution, calculate $\Delta_r G^{\ominus}$ for the disproportionation of hydrogen superoxide (HO_2^-) into O_2 and H_2O_2, and compare the result with its value for the disproportionation of H_2O_2.

11.17 Which of the solvents ethylenediamine (which is basic and reducing) or SO_2 (which is acidic and oxidizing) might not react with (a) Na_2S_4, (b) K_2Te_3, (c) $Cd_2(Al_2Cl_7)_2$. Explain your reasons.

11.18 Rank the following species from the strongest reducing agent to the strongest oxidizing agent and give a plausible balanced reaction illustrating the strongest reducing agent and strongest oxidizing agent: SO_4^{2-}, SO_3^{2-}, $O_3SO_2SO_3^{2-}$.

11.19 (a) Give the formula for Te(VI) in acidic aqueous solution and contrast it with the formula for S(VI). (b) Offer a plausible explanation for this difference. (c) If there is precedent for this trend in Group 15/V, identify and describe it.

11.20 Use isoelectronic analogies to infer the probable structures of (a) Sb_4^{2-} and (b) P_7^{3-}. Describe the bonding in these ions.

Problems

11.1 The tetrahedral P_4 molecule may be described in terms of localized 2c,2e bonds. Determine the number of skeletal valence electrons and from this decide whether P_4 is *closo*, *nido*, or *arachno*. If it is not *closo*, determine the parent *closo* polyhedron from which the structure of P_4 could be formally derived by the removal of one or more vertices.

11.2 On account of their slowness at electrodes, the potentials of most redox reactions of nitrogen compounds cannot be measured in an electrochemical cell. Instead, the values must be determined from other theromodynamic data. Illustrate such a calculation using $\Delta_f G^{\ominus}(NH_3, aq) = -26.5$ kJ mol^{-1} to calculate the standard potential of the N_2/NH_3 couple in basic aqueous solution.

11.3 Correct any inaccuracies in the following statements and, after correction, provide examples to illustrate each statement. (a) Elements in the middle of Groups 15/V and 16/VI are easier to oxidize to the group oxidation state than are the lightest and heaviest members. (b) In its ground state, O_2 is a triplet and it undergoes Diels–Alder electrophilic attack on dienes. (c) The diffusion of ozone from the stratosphere into the troposphere poses a major environmental problem. (d) The starting material for the production of sodium nitrite is ammonia.

11.4 A mechanistic study of reaction between chloramine and sulfite has been reported (B.S. Yiin, D.M. Walker, and D.W. Margerum, *Inorg. Chem.* **26**, 3435 (1987)). Summarize the observed rate law and the proposed mechanism. Accepting the proposed mechanism, why should $SO_2(OH)^-$ and HSO_3^- display different rates of reaction? Explain why it was not possible to distinguish the reactivity of $SO_2(OH)^-$ from that of HSO_3^-.

11.5 A compound containing pentacoordinate nitrogen has been characterized (A. Frohmann, J. Riede, and H. Schmidbaur, *Nature* **345**, 140 (1990)). Describe (a) the synthesis and (b) the structure of the compound and (c) give a description of the bonding.

11.6 Tetramethyltellurium, $Te(CH_3)_4$ was prepared in 1989 (R.W. Gedrige, D.C. Harris, K.R. Higa, and R.A. Nissan, *Organometallics* **8**, 2817 (1989)), and its synthesis was soon followed by the preparation of the hexamethyl compound (L. Ahmed and J.A. Morrison, *J. Am. Chem. Soc.* **112**, 7411 (1990)). Explain why these compounds are so unusual, give equations for their syntheses, and speculate on why these synthetic procedures are successful. In the latter connection, speculate on why reaction of TeF_4 with a methyllithium does not yield the tetramethyltellurium.

11.7 The bonding in the square-planar Se_4^{2+} ion is briefly described in Section 11.12. Explore this description in more detail by carrying out an extended Hückel calculation on Se_4^{2+} with S—S bond distances of 2.0 Å (sulfur is recommended because its extended Hückel parameters are more reliable than those of Se).[16] From the output (a) draw the molecular orbital energy level diagram, (b) assign the symmetry of each level, and (c) sketch the highest energy molecular orbital. Is a closed-shell molecule predicted?

---

[16] Y. Jean, F. Volatron, and J. Burdett, *An introduction to molecular orbitals*, p. 60. Oxford University Press (1993).

12

The halogens and the noble gases

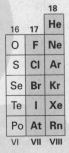

We now turn to the last two groups in the p-block, and see that many of the systematic themes that were helpful for discussing the two preceding groups will again prove useful. For instance, we see that the VSEPR model can be used to predict the shapes of the wide range of molecules that the halogens form among themselves, with oxygen, and with the noble gases. As with the elements in Groups 15/V and 16/VII, we shall see that the oxoanions of the halogens and of xenon are oxidizing agents that often react by atom transfer. Another similarity between the groups is the useful correlation between the oxidation number of the central atom and the rates of redox reactions. A wide range of oxidation states is observed for most of the halogens; moreover, as a result of this diversity and the relative weakness of halogen–halogen bonds in the elemental dihalogen molecules, the reactions of the dihalogens are often fast. The range of compounds of the noble gases is much less extensive than that of the halogens, but there are many similarities between the two groups in the structures of their compounds.

The **halogens**, the members of Group 17/VII, are among the most reactive nonmetallic elements; the **noble gases**, their neighbors in Group 18/VIII, are the least. Despite this contrast, there are resemblances between the two groups, particularly in the structures of their compounds. The two groups are also related in the sense that the first compounds of xenon to be prepared were fluorides, and these fluorides are the most common starting materials for other compounds of the noble gases.

The chemical properties of the halogens are extensive and have been mentioned many times already. Therefore, in this chapter we highlight their systematic features and pay particular attention to trends within the group.

The elements

The atomic properties of the halogens and the noble gases are listed in Table 12.1. The features to note include their high ionization energies and (for the halogens) their high electronegativities and electron affinities. The halogens have high electron affinities because the incoming electron can occupy an orbital of an incomplete valence shell and experience a strong nuclear attraction: recall that Z_{eff} increases progressively across the period (Section 1.6). The noble gases have negative electron affinities because their valence shells are full and an incoming electron occupies an orbital of a new shell.

Table 12.1 Atomic properties of the halogens and noble gases

Element	Ionization enthalpy/ (kJ mol^{-1})	Electron affinity/ (kJ mol^{-1})	χ_P	Ionic radius/ Å	Common oxidation numbers
Halogens					
F	1687	334	3.98	1.17	−1
Cl	1257	355	3.16	1.67	−1, 1, 3, 5, 7
Br	1146	325	2.96	1.82	−1, 1, 3, 5
I	1015	295	2.66	2.06	−1, 1, 3, 5, 7
At		270			
Noble gases					
He	2378	−48			0
Ne	2087	−120			0
Ar	1527	−96			0
Kr	1357	−96			0, 2
Xe	1177	−77	2.6		0, 2, 4, 6, 8
Rn	1043				

Sources: As for Table 11.1. Electronegativity of Xe from L.C. Allen and J.E. Huheey, *J. Inorg. Nucl. Chem.* **42**, 1523 (1980).

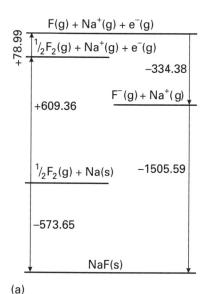

(a)

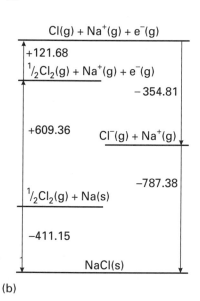

(b)

12.1 Thermochemical cycles for (a) sodium fluoride and (b) sodium chloride (values are in kilojoules per mole).

We have seen when discussing the earlier groups in the *p*-block that the element at the head of each group has structures and properties that are quite different from those of its heavier congeners. The anomalies are much less striking for the halogens, and the most notable difference is that fluorine has a lower electron affinity than chlorine. Intuitively, this feature seems to be at odds with the high electronegativity of fluorine, but it stems from the larger electron–electron repulsion in the compact F atom as compared with the larger Cl atom. Despite this difference in electron affinity, however, the enthalpies of formation of metal fluorides are generally much greater than those of metal chlorides. The explanation of this observation is that the low electron affinity of fluorine is more than offset by the high lattice enthalpies of ionic compounds containing the small F$^-$ ion (Fig. 12.1) and the strengths of bonds in covalent species (for example, the fluorides of metals in high oxidation states).

Because fluorine is the most electronegative of all elements, it is never found in a positive oxidation state (except in the transient gas-phase species F$_2^+$). With the possible exception of astatine, the other halogens occur with oxidation numbers ranging from −1 to +7. The dearth of chemical information on astatine stems from its lack of any stable isotopes and the relatively short half-life (8.3 h) of its most long-lived isotope. Because of this short half-life, astatine solutions are intensely radioactive and may be studied only in high dilution. Astatine appears to exist as the anion At$^-$ and as At(I) and At(III) oxoanions; no evidence for At(VII) has been obtained.

The noble gas with the most extensive chemical properties is xenon. The most important nonzero oxidation numbers of xenon are +2, +4, and +6 and compounds with Xe—F, Xe—O, Xe—N, and Xe—C bonds are known. The chemical properties of xenon's lighter neighbor, krypton, are much more limited. The study of radon chemistry, like that of astatine, is inhibited by the high radioactivity of the element.

Except for fluorine and the highly radioactive astatine, the halogens exist in oxidation states ranging from −1 to +7. The small and highly electronegative fluorine atom is effective in oxidizing many elements to high oxidation states. Of the noble gases, xenon forms a range of compounds with fluorine and oxygen.

The halogens (Group 17/VII)

As a consequence of the high electronegativities and abundances of the lighter halogens, their compounds are important in practically every area of chemistry and are encountered throughout the text. Our attention here will be on the halogens themselves, halogen-rich compounds (the interhalogens), and oxides of the halogens.

12.1 Occurrence and recovery

The halogens are so reactive that they are found naturally only as compounds. Their abundances in the Earth's crust decrease steadily with atomic number from fluorine to iodine. All the dihalogens (except the radioactive At_2) are produced commercially on a large scale, with chlorine production by far the greatest, followed by fluorine. Chlorine is widely used in industry to make chlorinated hydrocarbons and in applications in which a strong and effective oxidizing agent is needed (including bleaching and water purification). These applications are under close scrutiny, however, because some organic chlorine compounds are carcinogenic and chlorocarbons are implicated in the destruction of ozone in the stratosphere.

The elements are found mainly as halides in nature, but the most easily oxidized element, iodine, is also found as sodium or potassium iodate, KIO_3, in alkali metal nitrate deposits. Because many chlorides, bromides, and iodides are soluble, these anions occur in the oceans and in brines. The primary source of fluorine is calcium fluoride, which is highly insoluble and often found in sedimentary deposits (as *fluorite*, CaF_2).

The principal method of production of the elements is by oxidation of the halides (Section 6.2). The strongly positive standard potentials $E^\ominus(F_2, F^-) = +2.87$ V and $E^\ominus(Cl_2, Cl^-) = +1.36$ V indicate that the oxidation of F^- and Cl^- ions requires a strong oxidizing agent. Only electrolytic oxidation is commercially feasible. An aqueous electrolyte cannot be used for fluorine production because water is oxidized at a much lower potential (+1.23 V) and any fluorine produced would react rapidly with water. The isolation of elemental fluorine eluded chemists for most of the nineteenth century until in 1886 the French chemist Henri Moisson prepared it by the electrolysis of a solution of potassium fluoride in liquid hydrogen fluoride using a cell very much like the one still used today (Fig. 12.2).

Most commercial chlorine is produced by the electrolysis of aqueous sodium chloride solution in a *chloralkali cell* (Fig. 12.3). The half-reactions are

Anode half-reaction: $2\,Cl^-(aq) \longrightarrow Cl_2(g) + 2\,e^-$

Cathode half-reaction: $2\,H_2O(l) + 2\,e^- \longrightarrow 2\,OH^-(aq) + H_2(g)$

The oxidation of water at the anode is suppressed by employing an electrode material that has a higher overpotential for O_2 evolution than for Cl_2 evolution. The best anode material appears to be RuO_2 (see Chapter 17).

In modern chloralkali cells, the anode and cathode compartments are separated by a polymer ion-exchange membrane. The membrane can exchange cations, and hence permits Na^+ ions to migrate from the anode to the cathode compartment. The flow of cations maintains electroneutrality in the two compartments because, during electrolysis, negative charge is removed at the anode (by conversion of $2Cl^-$ to Cl_2) and supplied at the cathode (by formation of OH^-). The migration of OH^- in the opposite direction would also maintain electroneutrality, but OH^- would react with Cl_2 and spoil the process. Hydroxide ion

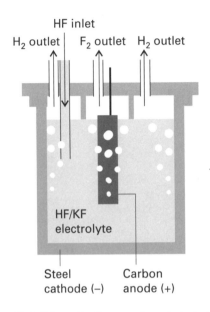

12.2 Schematic diagram of an electrolysis cell for the production of fluorine from potassium fluoride dissolved in liquid hydrogen fluoride.

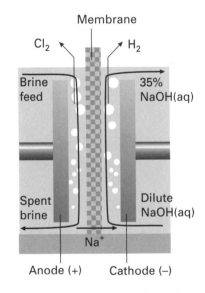

12.3 Schematic diagram of a modern chloralkali cell using a cation-exchange membrane, which has a high permeability to Na^+ ions and low permeabiliity to OH^- and Cl^- ions.

migration is suppressed because the membrane does not exchange anions.[1] A final detail of the membrane material is that the C atoms in the polymer backbone are fluorinated: fluorination protects the membrane because the C—F bonds are resistant both to the strong oxidizing agent Cl_2 and to the strong nucleophile OH^-.

Bromine is obtained by the chemical oxidation of Br^- ions in sea water. A similar process is used to recover iodine from certain natural brines that are rich in I^-. The more strongly oxidizing halogen, chlorine, is used as the oxidizing agent in both processes, and the resulting Br_2 and I_2 are driven from the solution in a stream of air:

$$Cl_2(g) + 2\,X^-(aq) \xrightarrow{\ \text{air}\ } 2\,Cl^-(aq) + X_2(g) \qquad (X = \text{Br or I})$$

Fluorine, chlorine, and bromine are prepared by electrochemical oxidation of halide salts; chlorine is used to oxidize Br^- and I^- to the corresponding dihalogen.

Example 12.1 Recovery of Br_2 from brine

Give the chemical equation and potential for the commercial conversion of Br^- in brine to Br_2. Show that from a thermodynamic standpoint bromide could be oxidized to Br_2 by O_2, and suggest a reason why O_2 is not used for this purpose.

Answer Chlorine is used to oxidize Br^-:

$$Cl_2(g) + 2\,Br^-(aq) \longrightarrow 2\,Cl^-(aq) + Br_2(g) \qquad (E^\ominus = +0.26\ \text{V})$$

and the resulting volatile Br_2 is removed in a steam–air mixture. Oxygen would be thermodynamically capable of carrying out this reaction in acidic solution:

$$O_2(g) + 4\,Br^-(aq) + 4\,H^+(aq) \longrightarrow 2\,H_2O(l) + 2\,Br_2(l) \qquad (E^\ominus = +0.16\ \text{V})$$

but the reaction is not favorable at $pH = 7$, when $E^\ominus = -0.15$ V. Even though the reaction is thermodynamically favorable in acidic solution, it is doubtful that the rate would be adequate, because an overpotential of about 0.6 V is associated with the reactions of O_2 (Section 6.4). Even if the oxidation by O_2 in acidic solution were kinetically favorable, the process would be unattractive because of the cost of acidifying large quantities of brine and then neutralizing the effluent.

Self-test 12.1 One source of iodine is sodium iodate, $NaIO_3$. Which of the reducing agents $SO_2(aq)$ or $Sn^{2+}(aq)$ would seem practical from the standpoints of thermodynamic feasibility and plausible judgements about cost? Standard potentials are given in Appendix 2.

12.2 Trends in properties

Unlike the structures of the elements of the preceding groups in the p-block, the molecular structures of the halogens are strikingly similar. They are all diatomic, and many properties change smoothly down the group. For instance, there is a trend toward metallic character on descending the group.

---

[1] From the standpoint of electrostatic interactions it might seem unlikely that the OH^- ion would migrate away from the positive electrode and toward the negative electrode. It must be remembered however that a solute will diffuse from regions of higher concentration to regions of lower concentration; see P.W. Atkins, *Physical chemistry.* Oxford University Press and W.H. Freeman (1998). Even the cation exchange nature of the membrane material does not totally suppress the diffusion of Cl^- into the negative electrode compartment.

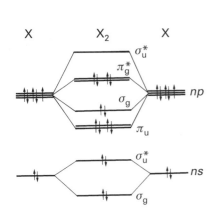

12.4 Schematic molecular orbital energy diagram for Cl_2, Br_2, and I_2. For F_2, the order of the π_u and the upper σ_g orbitals is reversed.

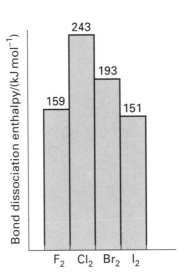

12.5 Bond dissociation enthalpies of the halogens (in kilojoules per mole).

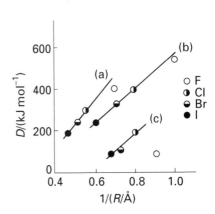

12.6 Dissociation enthalpies of (a) carbon–halogen, (b) hydrogen–halogen, and (c) halogen–halogen bonds plotted against the reciprocal of the bond length. Note the weakness of the X—F bond. (From P. Politzer, *J. Am. Chem. Soc.* **91**, 6235 (1969).)

(a) Molecular structure and properties

Among the most striking physical properties of the halogens are their colors. In the vapor they range from the almost colorless F_2, through yellowish-green Cl_2 and red-brown Br_2, to purple I_2. The progression of the maximum absorption to longer wavelengths reflects the decrease in the HOMO–LUMO gap on descending the group. In each case, the optical absorption spectrum arises primarily from transitions in which an electron is promoted from the highest filled σ and π^* orbitals into the vacant antibonding σ^* orbital (Fig. 12.4).

Except for F_2, the analysis of the UV absorption spectra gives precise values for the dihalogen bond dissociation energies (Fig. 12.5). It is found that bond strengths decrease down the group from Cl_2. The UV spectrum of F_2, however, is a broad continuum that lacks structure because absorption is accompanied by dissociation of the F_2 molecule. The lack of discrete absorption bands makes it difficult to estimate the dissociation energy spectroscopically, and thermochemical methods are complicated by the highly corrosive nature of this reactive halogen. When these corrosion problems were solved, the F—F bond enthalpy was found to be less than that of Br_2 and thus out of line with the trend in the group. However, fluorine's low bond enthalpy is consistent with the low single-bond enthalpies of N—N, O—O, and various combinations of N, F, and O (Fig. 12.6). The simplest explanation (like the explanation of the low electron affinity of fluorine) is that the bond is weakened by the strong repulsions between nonbonding electrons in the small F_2 molecule. In molecular orbital terms, the molecule has numerous electrons in strongly antibonding orbitals.

Chlorine, bromine, and iodine all crystallize in lattices of the same symmetry (Fig. 12.7), so it is possible to make a detailed comparison of distances between bonded and nonbonded neighboring atoms (Table 12.2). The important conclusion is that nonbonded distances do not

Table 12.2 Bonding and shortest nonbonding distances for solid dihalogens

Element	Temperature/ °C	Bond length/ Å	Nonbonding distance/ Å	Ratio
Cl_2	−160	1.98	3.32	1.68
Br_2	−106	2.27	3.32	1.46
I_2	−163	2.72	3.50	1.29

Data from J. Donohue, *The structure of the elements.* Wiley, New York (1974).

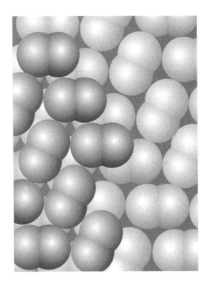

12.7 Solid chlorine, bromine, and iodine have similar structures; however, the closest nonbonded interactions are relatively less compressed in Cl_2 and Br_2 than in I_2. (Based on J. Donohue, *The structures of the elements.* Wiley, New York (1974).)

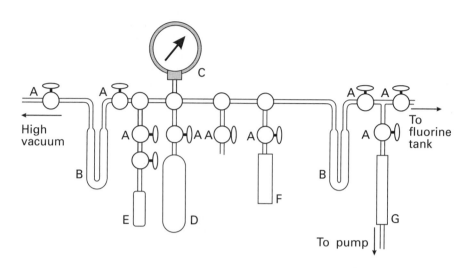

12.8 A typical metal vacuum system for handling fluorine and reactive fluorides. Nickel tubing is used throughout. (A) Monel valves, (B) nickel U-traps for the condensation of gases, (C) monel pressure gauge, (D) nickel container for gas storage, (E) polytetrafluoroethylene reaction tube, (F) nickel reaction vessel, (G) nickel canister filled with soda lime (a mixture of sodium and calcium hydroxides) to neutralize HF and react with F_2 and oxidizing fluorine compounds.

increase as rapidly as the bond lengths. This observation suggests the presence of weak intermolecular bonding interactions that strengthen on going from Cl_2 to I_2. The interaction also leads to a weakening of the I—I bond within the I_2 molecules, as is demonstrated by the lower I—I stretching frequency and greater I—I bond length in the solid as compared with in the gas phase. Moreover, solid iodine is a semiconductor, and under high pressure exhibits metallic conductivity.

> The F—F bond is weak relative to the Cl—Cl bond; in the solid state I_2 molecules cohere to their nearest neighbors more than the other dihalogens do.

(b) Reactivity trends

Fluorine, F_2, is the most reactive nonmetal and is the strongest oxidizing agent among the halogens. The rapidity of many reactions of fluorine with other elements may in part be due to a low kinetic barrier associated with the weak F—F bond. Despite the thermodynamic stability of most metal fluorides, fluorine can be handled in metal containers because a substantial number of common metals form a passive metal fluoride surface film upon contact with fluorine gas. Nickel and a nickel–molybdenum alloy (monel) are particularly resistant to fluorine and fluorides once the passivating film is formed. Fluorocarbon polymers, such as polytetrafluoroethylene, are also useful materials for the construction of apparatus to contain fluorine and oxidizing fluorine compounds (Fig. 12.8). Relatively few laboratories have the equipment and expertise for research on fluorine chemistry performed with elemental F_2.

The standard potentials for the halides indicate that F_2 $(E^\ominus = +2.87\ \mathrm{V})$ is a much stronger oxidizing agent than Cl_2 $(E^\ominus = +1.36\ \mathrm{V})$. The decrease in oxidizing strength continues in more modest steps from Cl_2 through Br_2 $(+1.07\ \mathrm{V})$ to I_2 $(+0.54\ \mathrm{V})$. Although the half-reaction

$$X_2(g) + 2\,e^- \longrightarrow 2\,X^-(aq)$$

is favored by a high electron affinity (which suggests that fluorine should have a lower reduction potential than chlorine), the process is favored by the low bond enthalpy of F_2 and by the highly exothermic hydration of the small F^- ion (Fig. 12.9). The net outcome of these

Figure (a)

F(g) + Na⁺(g) + e⁻(g)

+78.99

½F₂(g) + Na⁺(g) + e⁻(g)

−334.38

F⁻(g) + Na⁺(g)

+609.36

½F₂(g) + Na(s)

−926.72

−572.75

Na⁺(aq) + F⁻(aq)

(a)

Figure (b)

Cl(g) + Na⁺(g) + e⁻(g)

+121.68

½Cl₂(g) + Na⁺(g) + e⁻(g)

−354.81

+609.36   Cl⁻(g) + Na⁺(g)

½Cl₂(g) + Na(s)

−783.50

−407.27

Na⁺(aq) + Cl⁻(aq)

(b)

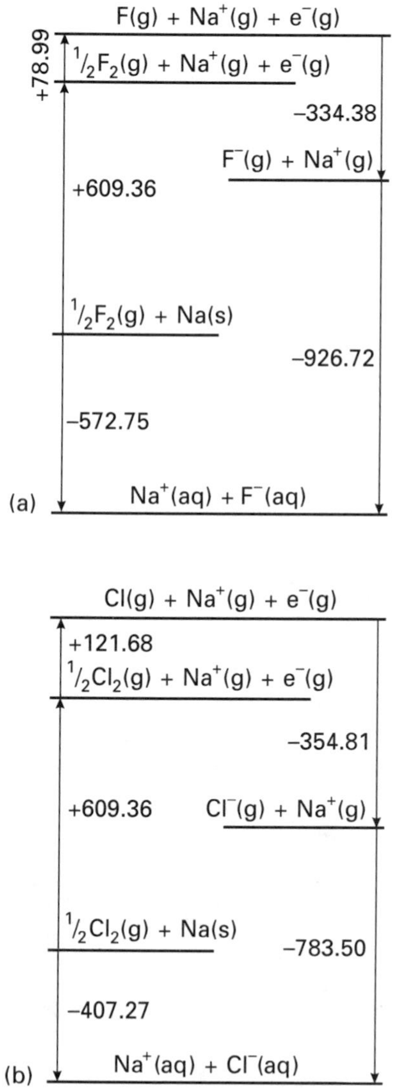

12.9 Thermochemical cycles for the enthalpy of formation of (a) aqueous sodium fluoride and (b) aqueous sodium chloride. Note that the hydration is much more exothermic for F^- than for Cl^-.

Table 12.3 Normal boiling points (in °C) of compounds of fluorine and their analogs*

F_2	−188.2	H_2	−252.8	Cl_2	−34.0
CF_4	−127.9	CH_4	−161.5	CCl_4	76.7
PF_3	−101.5	PH_3	−87.7	PCl_3	75.5

*Hydrogen-bonded molecules such as HF and NH_3 and their analogs have been omitted.

three competing effects of size is that fluorine is the most strongly oxidizing element of the group.

> *Fluorine is the most oxidizing halogen; the oxidizing power of the halogens decreases down the group.*

(c) Special properties of fluorine compounds

The boiling points in Table 12.3 demonstrate that molecular fluorine compounds tend to be highly volatile, in some cases even more volatile than the corresponding hydrogen compounds (compare, for example, PF_3, b.p. −101.5 °C, and PH_3, b.p. −87.7 °C) and in all cases much more volatile than the chlorine analogs. The volatilities of the compounds are a result of variations in the strength of the dispersion interaction (the interaction between instantaneous transient electric dipole moments), which is strongest for highly polarizable molecules. The electrons in the small F atoms are held tightly by the nuclei, and consequently fluorine compounds have low polarizabilities and hence weak dispersion interactions. In contrast to the volatility trend just discussed, the volatility of HF is low (b.p. 19.5 °C) for such a light molecule. As with water, this low volatility stems from extensive hydrogen bonding in liquid HF.

Another characteristic of fluorine is the ability of an F atom in a compound to withdraw electrons from the atoms to which it is attached. This withdrawal leads to enhanced Brønsted acidity. An example of this effect is the increase by three orders of magnitude in the acidity of trifluoromethanesulfonic acid, HSO_3CF_3 ($pK_a = 3.0$ in nitromethane) over that of methanesulfonic acid, HSO_3CH_3 ($pK_a = 6.0$ in nitromethane). The presence of F atoms in a molecule also results—for the same reason—in high Lewis acidity. For example, we saw in Section 5.9 that SbF_5 is one of the strongest Lewis acids of its type (it is much stronger than $SbCl_5$).

The ability of fluorine to stabilize high oxidation states is second only to that of oxygen. Some examples of high oxidation state fluorine compounds are IF_7, PtF_6, BiF_5, and $KAgF_4$. All these compounds are examples of the highest oxidation state attainable for these elements, the rare oxidation state Ag(III) being perhaps the most notable. Another example is the existence of the stable Pb(IV) fluoride, PbF_4, and the instability of all other Pb(IV) halides. A related phenomenon is the tendency of fluorine to disfavor low oxidation states. Thus solid copper(I) fluoride, CuF, is unknown but CuCl, CuBr, and CuI are well known. These trends were discussed in Section 2.12 in terms of a simple ionic model in which the small size of the F^- ion in combination with a small, highly charged cation results in a high lattice enthalpy. As a result, there is a thermodynamic tendency for CuF to disproportionate and form copper metal and CuF_2 (because Cu^{2+} is doubly charged and its ionic radius is smaller than that of Cu^+).

> *Fluorine substituents promote volatility, increase the strengths of Lewis and Brønsted acids, and stabilize high oxidation states.*

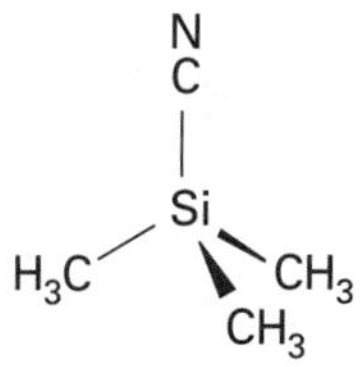

N
C
|
Si
H₃C   CH₃
      CH₃

1 (CH₃)₃SiCN

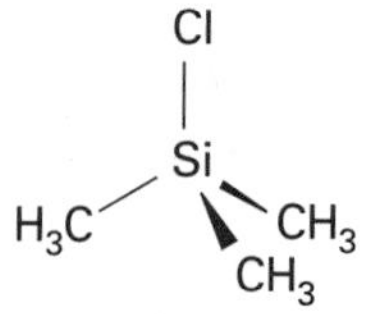

Cl
|
Si
H₃C   CH₃
      CH₃

2 (CH₃)₃SiCl

12.3 Pseudohalogens

A number of compounds have properties so similar to those of the halogens that they are called **pseudohalogens** (Table 12.4). For example, like the dihalogens, cyanogen, $(CN)_2$, undergoes thermal and photochemical dissociation in the gas phase; the resulting CN radicals are isolobal with halogen atoms and undergo similar reactions, such as a chain reaction with hydrogen:

$$NC—CN \xrightarrow{\text{heat or light}} 2\,CN\cdot$$
$$H_2 + CN\cdot \longrightarrow HCN + H\cdot$$
$$H\cdot + NC—CN \longrightarrow HCN + CN\cdot$$

Overall: $H_2 + C_2N_2 \longrightarrow 2\,HCN$

Table 12.4 Pseudohalides, pseudohalogens, and corresponding acids

Pseudohalide	Pseudohalogen	$E^{\ominus}/V$	Acid	pK_a^*
CN⁻ cyanide	NCCN cyanogen	+0.27	HCN hydrogen cyanide	9.2
SCN⁻ thiocyanate	NCSSCN dithiocyanogen	+0.77	HNCS hydrogen isothiocyanate	−1.9
OCN⁻ cyanate			HNCO isocyanic acid	3.5
CNO⁻ fulminate			HCNO fulminic acid	
NNN⁻ azide			HNNN hydrazoic acid	4.92

*A. Albert and E.R. Serjeant, *The determination of ionization constants.* Chapman & Hall, London (1984).

Another similarity is the reduction of a pseudohalogen:

$$(CN)_2(aq) + 2e^- \longrightarrow 2\,CN^-(aq)$$

The anion formally derived from a pseudohalogen is called a **pseudohalide ion**. An example is the cyanide anion, CN⁻. We say 'formally derived' because pseudohalide ions are known in many cases even though the neutral dimer (the pseudohalogen) is not. Covalent pseudohalides similar to the covalent halides of the *p*-block elements are also common. They are often structurally similar to the corresponding covalent halides (compare (**1**) and (**2**)), and undergo similar metathesis reactions.

As with all analogies, the concepts of pseudohalogen and pseudohalide have many limitations. For example, pseudohalogen ions are not spherical, so structures of ionic compounds often differ (for example, NaCl is fcc but NaSCN is not). The pseudohalides are generally less electronegative than the lighter halides and some pseudohalides have more versatile donor properties. The thiocyanate ion, SCN⁻, for instance, acts as an ambidentate ligand with a soft base site, S, and a hard base site, N (see Section 7.2).

> *Pseudohalogens and pseudohalides mimic halogens and halides, respectively. The pseudohalogens form dimers and form molecular compounds with nonmetals and ionic compounds with alkali metals.*

12.4 Interhalogens

The halogens form compounds among themselves. These binary **interhalogens** are molecular compounds with formulas XY, XY_3, XY_5, and XY_7, where the heavier, less electronegative halogen **X** is the central atom. They are of special importance as highly reactive intermediates and for providing useful insights into bonding.

(a) Physical properties and structure

The diatomic interhalogens, **XY**, have been made for all combinations of the elements, but many of them do not survive for long. The least labile is ClF, but ICl and IBr can also be obtained in pure crystalline form. Their physical properties are intermediate between those of their component molecules. For example, the deep red ICl (m.p. 27 °C, b.p. 97 °C) is intermediate between yellowish-green Cl_2 (m.p. −101 °C, b.p. −35 °C) and dark purple I_2 (m.p. 114 °C, b.p. 184 °C). Photoelectron spectra indicate that the molecular orbital energy levels in the mixed dihalogen molecules lie in the order $3\sigma^2 < 1\pi^4 < 2\pi^4$, which is the same as in the homonuclear dihalogen molecules (Fig. 12.10). An interesting historical note is that ICl was discovered before Br_2 in the early nineteenth century, and, when later the first samples of the dark red–brown Br_2 (m.p. −7 °C, b.p. 59 °C) were prepared, they were mistaken for ICl.

Most of the higher interhalogens are fluorides (Table 12.5). The only neutral interhalogen with the central atom in a +7 oxidation state is IF_7, but the cation ClF_6^+, a compound of Cl(VII), is known. The absence of a neutral ClF_7 reflects the destabilizing effect of nonbonding repulsions between fluorine atoms (indeed, coordination numbers greater than six are not observed for other p-block central atoms in Period 3). The lack of BrF_7 might be rationalized in a similar way, but in addition we shall see later that bromine is reluctant to achieve its maximum oxidation state. In this respect, it resembles some other Period 4 p-block elements, notably arsenic and selenium.

The shapes of interhalogen molecules (**3**), (**4**), and (**5**) are largely in accord with the VSEPR model. For example, the XY_3 compounds (such as ClF_3) have five valence electron pairs around the X atom, and they adopt a trigonal-bipyramidal arrangement. The Y atoms attach to the two axial pairs and one of the three equatorial pairs, and then the two axial bonding pairs move away from the two equatorial lone pairs. As a result, XY_3 molecules have a C_{2v} drooping-T shape. There are some discrepancies: for example, ICl_3 is a Cl-bridged dimer.

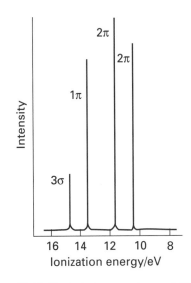

12.10 Photoelectron spectrum of ICl. The 2π levels give rise to two peaks because of spin–orbital interaction in the positive ion. (Based on S. Evans and A.F. Orchard, *Inorg. Chem. Acta* **5**, 81 (1971).)

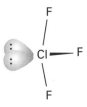

3 ClF_3, C_{2v}

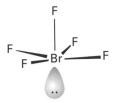

4 BrF_5, C_{4v}

Table 12.5 Representative interhalogens

XY	XY₃	XY₅	XY₇
ClF colorless b.p. −100 °C	ClF₃ colorless b.p. 12 °C	ClF₅ colorless b.p. −13 °C	
BrF* light brown b.p. c. −20 °C	BrF₃ yellow b.p. 126 °C	BrF₅ colorless b.p. 41 °C	
IF*	(IF₃)ₙ yellow dec. −28 °C	IF₅ colorless b.p. 105 °C	IF₇ colorless subl. 5 °C
BrCl*† red brown b.p. c. 5 °C			
ICl red solid	I₂Cl₆ bright yellow m.p. 101 °C(16 atm)		
IBr black solid			

dec., decomposes; subl., sublimes.
*Very unstable.
†The pure solid is known at low temperatures.

5 IF_7, $\approx D_{5h}$

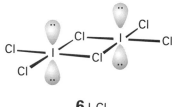

6 I_2Cl_6

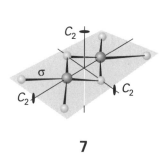

7

Example 12.2 Predicting the shape of an interhalogen molecule

Predict the shape of the chlorine-bridged I_2Cl_6 molecule by using the VSEPR model, and assign the point group.

Answer Each ICl_3 moiety has five electron pairs, and the bridging Cl atom increases this number to six (**6**). According to the VSEPR model, these pairs adopt an octahedral configuration, with the lone pairs in nonadjacent *trans* positions. Thus the bonds to surrounding Cl atoms take on a square-planar configuration and the molecule should be planar. This shape is verified by X-ray diffraction on single crystals. To assign the point group, we note that there are three perpendicular twofold axes (**7**), so the molecule has D_2 symmetry. The presence of a mirror plane identifies the group as D_{2h}.

Self-test 12.2 Predict the structure and identify the point group of ClO_2F.

The Lewis structure of XF_5 puts five bonding pairs and one lone pair on the central halogen atom and, as expected from the VSEPR model, the XF_5 molecules are square-pyramidal. As already mentioned, the only known XY_7 compound is IF_7, which is predicted to be pentagonal-bipyramidal. The experimental evidence for its actual structure is inconclusive, for it has been both claimed and denied that electron diffraction data support the prediction. As with other hypervalent molecules, the bonding in IF_7 can be explained without invoking *d*-orbital participation by adopting a molecular orbital model in which bonding and nonbonding orbitals are occupied but antibonding orbitals are not.

| *Molecular interhalogen compounds include diatomic species and polyatomic species with structures in accord with the VSEPR model; fluorine brings out high oxidation states of other halogens.*

(b) Chemical properties

All the interhalogens are oxidizing agents. As with all the known interhalogen fluorides, ClF_3 is an exoergic compound, so thermodynamically it is a weaker fluorinating agent than F_2. However, the rate at which it fluorinates substances generally exceeds that of fluorine, so it is in fact an aggressive fluorinating agent toward many elements and compounds. In general, the rates of oxidation of interhalogens do not have a simple relation to their thermodynamic stabilities. Thus, ClF_3 and BrF_3 are much more aggressive fluorinating agents than BrF_5, IF_5, and IF_7; iodine pentafluoride, for instance, is a convenient mild fluorinating agent that can be handled in glass apparatus. One use of ClF_3 as a fluorinating agent is in the formation of a passivating metal fluoride film on the inside of nickel apparatus used in fluorine chemistry.

Both ClF_3 and BrF_3 react vigorously (often explosively) with organic matter, burn asbestos, and expel oxygen from many metal oxides:

$$2\,Co_3O_4(s) + 6\,ClF_3(g) \longrightarrow 6\,CoF_3(s) + 3\,Cl_2(g) + 4\,O_2(g)$$

They are produced on a significant scale for the production of UF_6 for use in the enrichment of uranium:

$$UF_4(s) + ClF_3(g) \longrightarrow UF_6(s) + ClF(g)$$

Bromine trifluoride autoionizes in the liquid state:

$$2\,BrF_3(l) \rightleftharpoons BrF_2^+ + BrF_4^-$$

This Lewis acid–base behavior is shown by its ability to dissolve a number of halide salts:

$$CsF(s) + BrF_3(l) \longrightarrow CsBrF_4(soln)$$

Table 12.6 Representative interhalogen cations

$[XF_2]^+$	$[XF_4]^+$	$[XF_6]^+$
$[ClF_2]^+$	$[ClF_4]^+$	$[ClF_6]^+$
$[BrF_2]^+$	$[BrF_4]^+$	$[BrF_6]^+$
	$[IF_4]^+$	$[IF_6]^+$

Bromine trifluoride is a useful solvent for ionic reactions that must be carried out under highly oxidizing conditions. The Lewis acid character of BrF_3 is shared by other interhalogens, which react with alkali metal fluorides to produce anionic fluoride complexes.

Fluorine-containing interhalogens are typically Lewis acids and strong oxidizing agents.

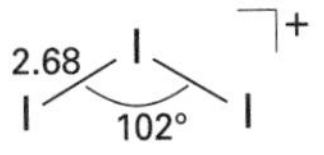

(c) Cationic polyhalogens

Under special strongly oxidizing conditions, such as in fuming sulfuric acid, I_2 is oxidized to the blue paramagnetic diiodinium cation, I_2^+. The dibrominium cation, Br_2^+, is also known. The bonds of these cations are shorter than those of the corresponding neutral dihalogens, which is the expected result for loss of an electron from a π^* orbital and the accompanying increase in bond order from 1 to 1.5 (see Fig. 12.4). Three higher polyhalogen cations, Br_5^+, I_3^+, and I_5^+, are known, and X-ray diffraction studies of the iodine species have established the structures shown in (**8**) and (**9**). The angular shape of I_3^+ is in line with the VSEPR model because the central I atom has two lone pairs of electrons.

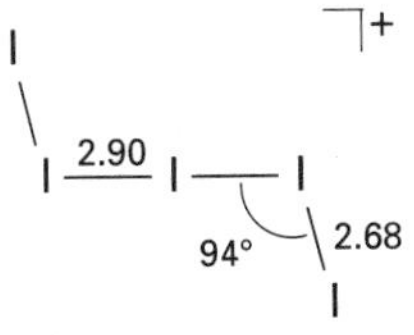

Another class of polyhalogen cations of formula XF_n^+ is obtained when a strong Lewis acid, such as SbF_5, abstracts F^- from interhalogen fluorides:

$$CIF_3 + SbF_5 \longrightarrow [CIF_2^+][SbF_6^-]$$

Table 12.6 lists a variety of interhalogen cations that are prepared in the same manner. X-ray diffraction of solid compounds that contain these cations indicates that the F^- abstraction from the cations is incomplete and that the anions remain weakly associated with the cations by fluorine bridges (**10**).

Cationic interhalogen compounds have structures in accord with the VSEPR model; with fluorine substituents they are Lewis acids.

12.5 Halogen complexes and polyhalides

The Lewis acidity of diiodine and the other heavy dihalogens toward electron pair donor molecules was mentioned in Section 5.10. This Lewis acidity is also evident in the interaction of halogen molecules with ion donors to give the range of ions known as **polyhalides**.

(a) Polyiodides

A deep brown color develops when I_2 is added to a solution of I^- ions. This color is characteristic of the homoatomic polyiodides, which include triiodide ions, I_3^-, and pentaiodide ions, I_5^-. These polyiodides are Lewis acid-base complexes in which I^- and I_3^- act as the bases and I_2 acts as the acid (Fig. 12.11). The Lewis structure of I_3^- has three equatorial lone pairs on the central I atom and two axial bonding pairs in a trigonal-bipyramidal arrangement. This hypervalent Lewis structure is consistent with the observed linear structure of I_3^-, which is described in more detail below.

An I_3^- ion can interact with other I_2 molecules to yield larger mononegative polyiodides of composition, $[(I_2)_n(I^-)]$. The I_3^- ion is the most stable member of this series. In combination with a large cation, such as $[N(CH_3)_4]^+$, it is symmetrical and linear with a longer I—I bond than in I_2. However, the structure of the triiodide ion, like that of the polyiodides in general, is highly sensitive to the identity of the counterion. For example Cs^+, which is smaller than the tetramethylammonium ion, distorts the I_3^- ion and produces one long and one short I—I bond (**11**). The ease with which the ion responds to its environment is a reflection of the weakness of bonds that just manage to hold the atoms together. A more extreme example of sensitivity to the cation is provided by NaI_3, which can be formed in aqueous solution but decomposes when the water is evaporated:

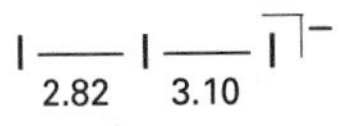

$$Na^+(aq) + I_3^-(aq) \xrightarrow{\text{remove water}} NaI(s) + I_2(s)$$

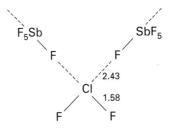

8 I_3^+

9 I_5^+

10 $(CIF_2)(SbF_6)$

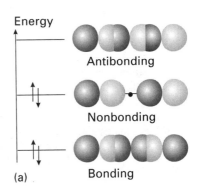

12.11 Some representations of the polyiodide ion. (a) The σ interaction in an I_3^- ion, (b) Lewis and VSEPR rationalization of the linear structure, where the five electron pairs around the central atom are arranged in a trigonal-bipyramidal array with the two bonding pairs along the threefold axis and the three nonbonding pairs in the equatorial plane.

11 I_3^- in CsI_3

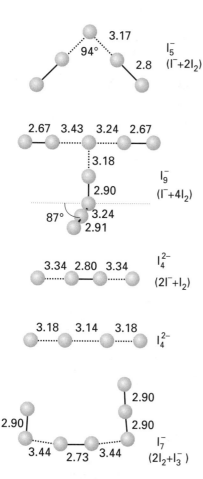

12.12 Some representative polyiodide structures and their approximate description in terms of I^-, I_3^-, and I_2 building blocks. The bond lengths and angles vary with the identity of the cation.

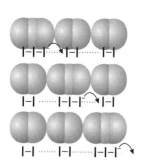

12.13 One possible mode of charge transport along a polyiodide chain is the shift of long and short bonds resulting in the effective migration of an I^- ion along the chain. Three successive stages of the migration are shown. As with the transport of the proton in ice or water, the iodide ion from the I_3^- on the left is not the same as the one emerging on the right.

This behavior is another example of the instability of large anions in combination with small cations which, as we saw in Section 2.12, can be rationalized by the ionic model.

The existence and structures of the higher polyiodides are sensitive to the counterion for similar reasons, and large cations are necessary to stabilize them in the solid state. In fact, entirely different shapes are observed for polyiodide ions in combination with various large cations, for the structure of the anion is determined in large measure by the manner in which the ions pack together in the crystal. The bond lengths in a polyiodide ion often suggest that it can be regarded as a chain of associated I^-, I_2, I_3^-, and sometimes I_4^{2-} units (Fig. 12.12). Solids containing polyiodides exhibit electrical conductivity, which may arise from the hopping of electrons (or holes) along the polyiodide chain or by an ion relay along the polyiodide chain (Fig. 12.13).

Some dinegative polyiodides are known. These contain an even number of I atoms and their general formula is $[I^-(I_2)_n I^-]$. They have the same sensitivity to the cation as their mononegative counterparts.

> *Polyiodides, such as I_3^- are formed by adding I_2 to I^-; they are stabilized by large cations.*

(b) Other polyhalides

Although polyhalide formation is most pronounced for iodine, other polyhalides are also known. They include Cl_3^-, Br_3^-, and BrI_2^-, which are known in solution and (in partnership with large cations) as solids too. Even F_3^- has been detected spectroscopically at low temperatures in an inert matrix. This technique, which is known as *matrix isolation*, employs the codeposition of the reactants with a large excess of noble gas at very low temperatures (in the region of 4 K to 14 K). The solid noble gas forms an inert matrix within which the F_3^- ion can sit in chemical isolation.

In addition to complex formation between dihalogens and halide ions, some interhalogens can act as Lewis acids toward halide ions. The reaction results in the formation of polyhalides which, in contrast to the chain-like polyiodides, are assembled around a central halogen acceptor atom in a high oxidation state. As mentioned earlier, for instance, BrF_3 reacts with CsF to form $CsBrF_4$, which contains the square-planar BrF_4^- anion (**12**). Many of these interhalogen anions have been synthesized (Table 12.7). Their shapes generally agree with the

Table 12.7 Representative interhalogen anions

Compound	Shape
IF_8^-	Square antiprism[a]
ClF_6^-	Octahedral[b]
BrF_6^-	Octahedral[c]
IF_6^-	Trigonally distorted octahedron[d]

[a]*Chem. Commun.*, 837 (1991): *Angew. Chem., Int. Ed. Engl.* **30**, 876 (1991).
[b]*Inorg. Chem.* **29**, 3506 (1990).
[c]*Angew. Chem., Int. Ed. Engl.* **28**, 1526 (1989).
[d]*Angew. Chem., Int. Ed. Engl.* **30**, 323 (1991).

VSEPR model, but there are some interesting exceptions. Two such exceptions are ClF_6^- and BrF_6^-, in which the central halogen has a lone pair of electrons, but the apparent structure is octahedral. The ions IF_6^- participates in an extended array through I—F⋯I interactions.

> *Some of the most stable polyhalogen anions contain fluorine as the substituent. Their structures usually conform to the VSEPR model.*

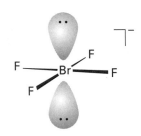

12 [BrF$_4$]$^-$

Example 12.3 A bonding model for I$^+$ complexes

In some cases the interaction of I_2 with strong donor ligands leads to the formation of cationic complexes such as bis(pyridine)iodine(+1), [py—I—py]$^+$. Propose a bonding model for this linear complex from (a) the standpoint of the VSEPR model and (b) simple molecular orbital considerations.

Answer (a) The Lewis electron structure places 10 electrons around the central I$^+$ in [py—I—py]$^+$, six from the iodine cation and four from the lone pairs on the two pyridine ligands. According to the VSEPR model, these pairs should form a trigonal bipyramid. The lone pairs will occupy the equatorial positions and consequently the complex should be linear. (b) From a molecular orbital perspective, the orbitals of the N—I—N array can be pictured as formed from an iodine $5p$ orbital and an orbital of σ symmetry from each of the two ligand atoms. Three orbitals can be constructed: 1σ (bonding), 2σ (nearly nonbonding), and 3σ (antibonding). There are four electrons to accommodate (two from each ligand atom; the iodine $5p$ orbital is empty). The resulting configuration is $1\sigma^2 2\sigma^2$, which is net bonding.

Self-test 12.3 From the perspective of structure and bonding, indicate several polyhalides that are analogous to [py—I—py]$^+$, and describe their bonding.

12.6 Compounds of the halogens with oxygen

In contrast to the rather simple formulas and structures of most halogen fluorides, the molecular compounds of the halogens with oxygen are a diverse group. The aqueous chemistry of the halogen oxoanions is the most uniform and important, so we shall concentrate on them after a brief discussion of the neutral oxides.

(a) Halogen oxides

Many binary compounds of the halogens and oxygen are known, but most are unstable and not commonly encountered in the laboratory. We shall mention only a few of the most important.

Oxygen difluoride (FOF; m.p. $-224\,°C$, b.p. $-145\,°C$), the most stable binary compound of oxygen and fluorine, is prepared by passing fluorine through dilute aqueous hydroxide solution:

$$2\,F_2(g) + 2\,OH^-(aq) \longrightarrow OF_2(g) + 2\,F^-(aq) + H_2O(l)$$

The pure difluoride survives in the gas phase above room temperature and does not react with glass. It is a strong fluorinating agent, but less so than fluorine itself. As suggested by the VSEPR model (or by a Walsh diagram, by analogy with H_2O), the OF_2 molecule is angular.

Dioxygen difluoride (FOOF; m.p. $-154\,°C$, b.p. $-57\,°C$) can be synthesized by photolysis of a liquid mixture of the two elements. It is unstable in the liquid state and decomposes rapidly above $-100\,°C$, but may be transferred (with some decomposition) as a low-pressure gas in a metal vacuum line. Dioxygen difluoride is an even more aggressive fluorinating agent than ClF_3. For example, it oxidizes plutonium metal and its compounds to PuF_6 in a reaction that ClF_3 cannot accomplish:

$$Pu(s) + 3\,O_2F_2(g) \longrightarrow PuF_6(g) + 3\,O_2(g)$$

The interest in this reaction is in removing plutonium as its volatile hexafluoride from spent nuclear fuel.

Chlorine occurs with many different oxidation numbers in its oxides, for they include:

Oxidation number	+1	+4	+6	+7
Formula	Cl_2O	ClO_2	Cl_2O_6	Cl_2O_7
Color	brown–yellow	yellow	dark red	colorless
State	gas	gas	liquid	liquid

Some of these oxides are odd-electron species, including ClO_2, in which chlorine has the unusual oxidation number +4, and Cl_2O_6, which exists as a mixed oxidation state ionic solid, $[ClO_2][ClO_4]$.

Chlorine dioxide is the only halogen oxide produced on a large scale. The reaction used is the reduction of ClO_3^- with HCl or SO_2 in strongly acidic solution:

$$2\,ClO_3^-(aq) + SO_2(g) \xrightarrow{\text{acid}} 2\,ClO_2(g) + SO_4^{2-}(aq)$$

It is a strongly endoergic compound ($\Delta_f G^{\ominus} = +121\ kJ\,mol^{-1}$) and must be kept dilute to avoid explosive decomposition; it is therefore consumed at the site of production. Its major uses are to bleach paper pulp and to disinfect sewage and drinking water. Some controversy surrounds these applications because the action of chlorine (or its product of hydrolysis, HClO) and chlorine dioxide on organic matter produces low concentrations of chlorocarbon compounds, some of which are potential carcinogens. On the other hand, the disinfection of water undoubtedly saves many more lives than the carcinogenic byproducts may take.

Chlorine oxides are known for Cl oxidation numbers from +1, +4, +6, and +7. The strong and facile oxidizing agent ClO_2 is the most commonly used halogen oxide.

(b) Oxoacids and oxoanions

The wide range of oxoanions and oxoacids of the halogens presents a challenge to those who devise systems of nomenclature. We shall use the most commonly employed names, such as chlorate for ClO_3^-, rather than the systematic names, such as trioxochlorate(V). Table 12.8 is a brief dictionary for converting between the two systems of nomenclature.

Table 12.8 Halogen oxoanions

Oxidation number	Formula	Name*	Point group	Shape	Remarks
+1	ClO^-	Hypochlorite [monoxochlorate(I)]	$C_{\infty v}$	$[Cl-O]^-$	Good oxidizing agent
+3	ClO_2^-†	Chlorite [dioxochlorate(III)]	C_{2v}		Strong oxidizing agent, disproportionates
+5	ClO_3^-	Chlorate [trioxochlorate(V)]	C_{3v}		Oxidizing agent
+7	ClO_4^-‡	Perchlorate [tetraoxochlorate(VII)]	T_d		Oxidizing agent and very weak ligand

*IUPAC names in square brackets.
†Iodite has not been isolated.
‡For iodine, IO_4^- and $H_3IO_6^{2-}$ are present in basic solution, and H_5IO_6 is present in acidic solution.

The strengths of the oxoacids vary systematically with the number of O atoms on the central atom (Table 12.9; see Pauling's rules in Section 5.4, specifically $pK_a = 8 - 5p$ for the acid $O_pE(OH)_q$). At first sight, periodic acid—the I(VII) analog of perchloric acid—appears to be out of line, for it is weak ($pK_{a1} = 3.29$). However, as soon as we note that its formula is $(HO)_5IO$ with $p = 1$, we see it is its structure, not its strength, that is anomalous. The O atoms in the conjugate base $H_4IO_6^-$ are very labile on account of the rapid equilibration

$$H_4IO_6^- (aq) \rightleftharpoons IO_4^- (aq) + 2 H_2O(l) \qquad K = 40$$

and in basic solution IO_4^- is the dominant ion. The tendency to have an expanded coordination shell is shared by the oxoacids of the neighboring Group 16/VI element tellurium, which in its maximum oxidation state forms the weak acid $Te(OH)_6$.

The halogen oxoanions, like many oxoanions, form metal complexes, including the metal perchlorates and periodates to be discussed here. In this connection we note that, because $HClO_4$ is a very strong acid and H_5IO_6 is a weak acid, it follows that ClO_4^- is a very weak base and $H_4IO_6^-$ is a relatively stronger base.

In view of the low Brønsted basicity and single negative charge of the perchlorate ion, ClO_4^-, it is not surprising that it is a weak Lewis base with little tendency to form complexes with cations in aqueous solution. Therefore, metal perchlorates are often used to study the properties of hexaaqua ions in solution. The ClO_4^- ion is used as a weakly coordinating ion that can readily be displaced from a complex by other ligands, or as a medium-sized anion that might stabilize solid salts containing large cationic complexes with easily displaced ligands.

However, the ClO_4^- ion is a treacherous ally. Because it is a powerful oxidizing agent, solid compounds of perchlorate should be avoided whenever there are oxidizable ligands or ions present (which is commonly the case). In some cases the danger lies in wait, for the reactions of ClO_4^- are generally slow and it is possible to prepare many metastable perchlorate complexes or salts that may be handled with deceptive ease. However, once reaction has been initiated by mechanical action, heat, or static electricity, these compounds can detonate with disastrous consequences.[2] Such explosions have injured chemists who may have handled a compound many times before it unexpectedly exploded. Some readily available and more docile weakly basic anions may be used in place of ClO_4^-; they include trifluoromethane-sulfonate $[SO_3CF_3]^-$, tetrafluoroborate $[BF_4]^-$, and hexafluorophosphate $[PF_6]^-$.

In contrast to perchlorate, periodate is a rapid oxidizing agent and a stronger Lewis base than perchlorate. These properties lead to the use of periodate as an oxidizing agent and stabilizing ligand for metal ions in high oxidation states. Some of the high oxidation states it can be used to form are very unusual: they include Cu(III) in a salt containing the $[Cu(HIO_6)_2]^{5-}$ complex and Ni(IV) in an extended complex containing the $[Ni(IO_6)]^-$ unit. The periodate ligand is bidentate in these complexes, and in the last example it forms a bridge between Ni(IV) ions.

The halogen oxoanions are thermodynamically strong oxidizing agents, but the high activation energy for oxidation by perchlorate provides a false sense of security and perchlorates of oxidizable cations should be avoided.

(c) Thermodynamic aspects of redox reactions

The thermodynamic tendencies of the halogen oxoanions and oxoacids to participate in redox reactions are well understood. As we shall see, we can summarize their behavior with a Frost diagram that is quite easy to rationalize. It is a very different story with the rates of the reactions, for these vary widely. Their mechanisms are only partly understood despite many years of investigation. Recent progress in the understanding of some of these mechanisms

Table 12.9 Acidities of chlorine oxoacids

Acid	q/p	pK_a
HOCl	0	7.53 (weak)
HOClO	1	2.00
HOClO$_2$	2	−1.2
HOClO$_3$	3	−10 (strong)

---

[2] Two examples are the explosion in a rocket-fuel plant of ammonium perchlorate with the intensity of an earthquake, *Chem. and Eng. News*, p. 5, May 16 (1988) and the injury of a chemist by the explosion of the perchlorate salt of a lanthanide complex, *Chem. and Eng. News*, p. 4, December 5 (1983).

stems from advances in techniques for fast reactions and interest in oscillating reactions (Box 12.1).

We saw in Section 6.9 that, if a species lies above the line joining its two neighbors of higher and lower oxidation state in a Frost diagram, then it is unstable with respect to disproportionation into them. From the Frost diagram for the halogen oxoanions and oxoacids in Fig. 12.14 we can see that many of oxoanions in intermediate oxidation states are susceptible to disproportionation. Chlorous acid, $HClO_2$, for instance, lies above the line joining its two neighbors, and is liable to disproportionation:

$$2\,HClO_2(aq) \longrightarrow ClO_3^-(aq) + HClO(aq) + H^+(aq) \qquad (E^{\ominus} = +0.52\ V)$$

Although BrO_2^- is well characterized, the corresponding I(III) species is so unstable that it does not exist in solution, except perhaps as a transient intermediate.

We also saw in Section 6.9 that the more positive the slope for the line from a lower to higher oxidation state species in a Frost diagram, the stronger the oxidizing power of the couple. A glance at Fig. 12.14 shows that all three diagrams have steep positively sloping lines, which immediately shows that all the oxidation states except the lowest (Cl^-, Br^-, and I^-) are strongly oxidizing.

Finally, basic conditions decrease reduction potentials for oxoanions as compared with their conjugate acids (Section 6.10). This decrease is evident in the less steep slopes of the lines in the Frost diagrams for the oxoanions in basic solution. The numerical comparison for ClO_4^- ions in 1 M acid compared with 1 M base makes this clear:

$$At\ pH = 0:\ ClO_4^-(aq) + 2\,H^+(aq) + 2\,e^- \longrightarrow ClO_3^-(aq) + H_2O(l)$$
$$(E^{\ominus} = +1.20\ V)$$

$$At\ pH = 14:\ ClO_4^-(aq) + H_2O(l) + 2\,e^- \longrightarrow ClO_3^-(aq) + 2\,OH^-(aq)$$
$$(E_B^{\ominus} = +0.37\ V)$$

Box 12.1 Oscillating reactions

Clock reactions and oscillating reactions are an active topic of research and provide fascinating lecture demonstrations. Most examples of oscillating reactions are based on reactions of halogen oxoanions, apparently because of the variety of oxidation states and their sensitivity to changes in pH. In 1895, H. Landot discovered that a mixture of sulfite, iodate, and starch in acidic solution remains nearly colorless for an initial period and then suddenly switches to the dark purple of the I_2-starch complex. When the concentrations are properly adjusted the reaction oscillates between nearly colorless and opaque blue. The reactions leading to this oscillation are the reduction of periodate to iodide by sulfite, where all reactants are colorless:

$$IO_4^-(aq) + 4\,SO_3^{2-}(aq) \longrightarrow I^-(aq) + 4\,SO_4^{2-}(aq)$$

A comproportionation reaction between I^- and IO_3^- then produces I_2, which forms an intensely colored complex with starch:

$$IO_3^-(aq) + 6\,H^-(aq) + 5\,I^-(aq) \longrightarrow$$
$$3\,H_2O(l) + 3\,I_2(starch)$$

Under some conditions this is the final state, but adjustment of concentrations may lead to bleaching of the I_2-starch complex by sulfite reduction of iodine to the colorless $I^-(aq)$ ion:

$$3\,I_2(starch) + 3\,SO_3^{2-}(aq) + 3\,H_2O(l) \longrightarrow$$
$$6\,I^-(aq) + 6\,H^+(aq) + 3\,SO_3^{2-}(aq)$$

The reaction may then oscillate between colorless and blue as the I_2/I^- ratio changes.

The detailed analysis of the kinetic conditions for oscillating reactions is pursued by chemists and chemical engineers. In the former case the challenge is to employ kinetic data determined separately for the individual steps to model the observed oscillations with a view to testing the validity of the overall scheme. Since oscillating reactions have been observed in commercial catalytic processes, the concern of the chemical engineer is to avoid large fluctuations or even chaotic reactions that might degrade the process.

Further reading

R.J. Field and F.W. Schneider, *J. Chem. Educ.* **66**, 195 (1989), and other papers in that issue.

P. Gray and S.K. Scott, *Chemical oscillations and instabilities.* Clarendon Press, Oxford (1990).

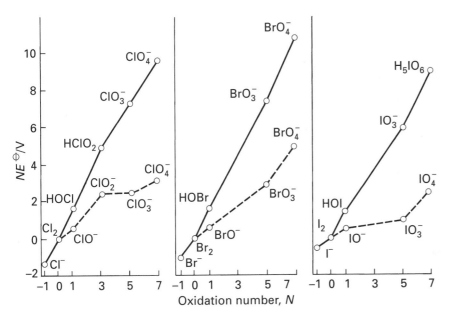

12.14 Frost diagrams for chlorine, bromine, and iodine in acidic solution (solid line) and basic solution (dashed line).

The reduction potentials show that perchlorate is thermodynamically a weaker oxidizing agent in basic solution than in acidic solution.

> *The oxoanions of halogens are strong oxidizing agents, especially in acidic solution.*

(d) Trends in rates of redox reactions

Mechanistic studies show that the redox reactions of halogen oxoanions are intricate. Nevertheless, despite this complexity there are a few discernible patterns that help to correlate the trends in rates of reaction. These correlations have practical value and give some clues about the mechanisms that may be involved.

The oxidation of many molecules and ions by halogen oxoanions becomes progressively faster as the oxidation number of the halogen decreases. Thus the rates observed are often in the order

$$ClO_4^- < ClO_3^- < ClO_2^- \approx ClO^- \approx Cl_2$$
$$BrO_4^- < BrO_3^- \approx BrO^- \approx Br_2$$
$$IO_4^- < IO_3^- < I_2$$

For example, aqueous solutions containing Fe^{2+} and ClO_4^- are stable for many months in the absence of dissolved oxygen, but an equilibrium mixture of aqueous $HClO$ and Cl_2 rapidly oxidizes Fe^{2+}.

Oxoanions of the heavier halogens tend to react most rapidly, particularly for the elements in their highest oxidation states:

$$ClO_4^- < BrO_4^- < IO_4^-$$

As we have remarked, perchlorates in dilute aqueous solution are usually unreactive, but periodate oxidations are fast enough to be employed for titrations. The mechanistic details are often complex, but the fast reaction of periodate appears to be due to the ease with which the reductant can reach the large I atom. The existence of both 4-coordinate and 6-coordinate periodate ions shows that the I atom in periodate is accessible to nucleophiles. Thus H_2O may add to the I atom in IO_4^- to yield $IO_2(OH)_4^-$ and protons. This reaction is facile, as can be seen from the rapid exchange of labeled water ($H_2^{18}O$) with periodate:

$$IO_2(OH)_4^- + H_2^{18}O \rightleftharpoons IO_2(OH)_3(^{18}OH) + H_2O$$

We have already seen that the thermodynamic tendency of oxoanions to act as oxidizing agents increases as the **pH** is lowered. It is also found that their rates are increased too. Thus, kinetics and equilibria unite to bring about otherwise difficult oxidations. The oxidation of halides by BrO_3^- ions, for instance, is second-order in H^+:

$$Rate = k[BrO_3^-][X^-][H^+]^2$$

and so the rate increases as the **pH** is decreased. The acid is thought to act by protonating the oxo group in the oxoanion, so aiding oxygen–halogen bond scission. Another role of protonation is to increase the electrophilicity of the halogen. An example is HClO where, as described below, the Cl atom may be viewed as an electrophile toward an incoming reducing agent (**13**). An illustration of the effect of acidity on rate is the use of a mixture of H_2SO_4 and $HClO_4$ in the final stages of the oxidation of organic matter in certain analytical procedures.

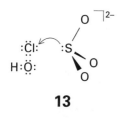

13

> Oxidation by halogen oxoanions is faster for the lower oxidation states. Rates and thermodynamics of oxidation are both enhanced by an acidic medium.

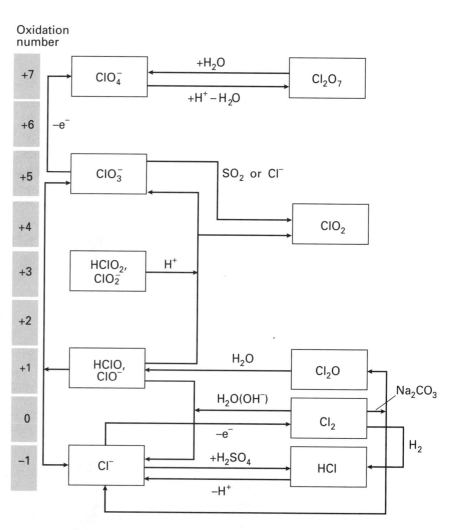

Chart 12.1 The interconversion of oxidation states of some important chlorine species.

(e) Reactions of individual oxidation states

With the general redox properties of the halogens now established, we can consider the characteristic properties and reactions of specific oxidation states, with emphasis on the mechanism of reaction. Although we are dealing here with halogen oxides, it is convenient to mention, for the sake of completeness, the redox properties of halogen(−1) and halogen(0) species. Chart 12.1 summarizes some of the reactions that interconvert the oxoanions and oxoacids of chlorine in its various oxidation states. One point to note is the major role of disproportionation and electrochemical reactions in the scheme. For example, the chart includes the production of Cl_2 by the electrochemical oxidation of Cl^-, which was discussed in Section 12.1. Once Cl_2 is available, the chart indicates that two successive disproportionations lead to the important Cl(V) species, ClO_3^-. The highest oxidation state Cl(VII) is achieved by the electrochemical oxidation of ClO_3^- to ClO_4^-.

Halogen(−1)

The halides are familiar ions found in natural waters. They also occur as crystalline minerals, such as rock salt (NaCl), sylvite (KCl), and fluorite (CaF_2). As starting materials they are often needed in their anhydrous form. The removal of residual water is not straightforward for metal ions in the +3 or higher oxidation state, because simple heating leads to hydrolysis:

$$FeCl_3 \cdot H_2O(s) \xrightarrow{\Delta} FeCl_2(OH)(s) + HCl(g) \xrightarrow{\Delta} FeOCl(s) + 2\,HCl(g)$$

Several methods are available to dehydrate metal salts without hydrolysis (see Box 12.2). The halides become thermodynamically easier to oxidize to their elemental form on descending the group from fluorine to iodine.

Box 12.2 **Preparation of anhydrous metal halides**

Because anhydrous halides are very common starting materials for inorganic syntheses, their preparation and the removal of water from impure commercial halides is of considerable practical importance. The principal synthetic methods are the direct reaction of a metal and a halogen and the reaction of halogen compounds with metals or metal oxides.

An example of direct reaction is shown in Fig. B12.1, in which the halogen in the gas phase reacts with the metal contained in a heated tube. When the halide is volatile at the temperature of the reaction, it sublimes to the exit end of the tube.

It is sometimes advantageous to use a halogen compound for the synthesis of a metal halide. For example, the reaction of ZrO_2 with CCl_4 vapor in a heated tube is a good method for the preparation of $ZrCl_4$. A contribution to the Gibbs energy of this reaction comes from the production of the C=O bond in phosgene, $COCl_2$:

$$ZrO_2(s) + 2\,CCl_4(g) \longrightarrow ZrCl_4(s) + 2\,COCl_2(g)$$

Anhydrous halides can often be prepared by the removal of water from hydrated metal chlorides.* Simply heating the sample in a dry gas stream is usually not satisfactory for metals with high charge-to-radius ratio because significant hydrolysis may lead to the oxide or oxohalide:

$$2\,CrCl_3 \cdot 6H_2O(s) \longrightarrow Cr_2O_3(s) + 6\,HCl(g) + 9\,H_2O(g)$$

This hydrolysis can be suppressed by carrying out the dehydration in a heated tube in a stream of hydrogen halide, or by dehydration with thionyl chloride (b.p. 79 °C), which produces the desired anhydrous chloride and volatile HCl and SO_2 byproducts:

$$FeCl_3 \cdot 6H_2O(s) + 6\,SOCl_2(l) \longrightarrow$$
$$FeCl_3(s) + 6\,SO_2(g) + 12\,HCl(g)$$

....

* The procedures for dehydration of metal chloride hydrates by thionyl chloride and references to other methods are given by A.R. Pray, *Inorg. Synth.* **28**, 321 (1990).

B12.1 A tube furnace used for the synthesis of anhydrous gallium bromide.

> Metal halides are often soluble and therefore they are common in natural waters. In the
> laboratory, simple heating to drive off water may lead to OH^- contamination of a metal
> halide.

Halogen(0)

One of the most characteristic redox reactions of the halogens is disproportionation, which is thermodynamically favorable for basic solutions of Cl_2, Br_2, and I_2. The equilibria in basic aqueous solution are:

$$X_2(aq) + 2\,OH^-(aq) \rightleftharpoons XO^-(aq) + X^-(aq) + H_2O(l) \qquad K = \frac{[XO^-][X^-]}{[X_2][OH^-]^2}$$

with $K = 7.5 \times 10^{15}$ for X = Cl, 2×10^8 for Br, and 30 for I.[3]

Disproportionation is much less favorable in acidic solution, as would be expected from the fact that H^+ is a product of the reaction:

$$Cl_2(aq) + H_2O(l) \longrightarrow HClO(aq) + H^+(aq) + Cl^-(aq) \qquad K = 3.9 \times 10^{-4}$$

The mechanism of this reaction appears to involve attack of H_2O on the Cl_2 molecule:

$$H_2O + Cl_2 \longrightarrow [H_2O{-}Cl{-}Cl] \longrightarrow H^+ + [HO{-}Cl{-}Cl]^-$$
$$\longrightarrow H^+ + HOCl + Cl^-$$

It is not known if the Lewis complex $H_2O{-}Cl_2$ is actually formed. According to one interpretation, proton transfer is so rapid that the reaction yields $HOCl_2^-$ directly. The $[HO{-}Cl{-}Cl]^-$ complex is at most transitory, for Cl^- is rapidly displaced by the strong nucleophile OH^-. The important insight gained from this mechanism is that this reaction, which is formally a redox reaction (because the oxidation number of chlorine changes), may be mechanistically a nucleophilic displacement.

Because the redox reactions of HClO and Cl_2 are often fast, Cl_2 in water is widely used as an inexpensive and powerful oxidizing agent. A part of the reason for its usefulness is that the rapid equilibration between ClO^- and Cl_2 gives rise to a variety of possible reaction pathways. For example, the nonlabile complex $[Fe(phen)_3]^{2+}$ reacts more rapidly with Cl_2 than with HClO because the former can more readily undergo an outer-sphere electron transfer, whereas ClO^- or HClO appear to require dissociation of phen prior to an inner-sphere redox process. In other environments HClO may be a more facile oxidizing agent than Cl_2. In general, labile aqua complexes, such as $[Cr(OH_2)_6]^{2+}$, react faster with ClO^- than with Cl_2.

The equilibrium constants for the hydrolysis of Br_2 and I_2 in acid solution are smaller than for Cl_2, and both elements are unchanged when dissolved in slightly acidified water.[4] Because F_2 is a much stronger oxidizing agent than the other halogens, it produces mainly O_2 and H_2O_2 when in contact with water. As a result, hypofluorous acid, HFO, was discovered only long after the other hypohalous acids. The first indication of its existence came from IR spectroscopic studies of H_2O and F_2 trapped in a matrix of frozen N_2 at about 20 K. With this encouragement, HFO was finally isolated and chemically characterized in the early 1970s by the controlled reaction of F_2 with ice at $-40\,°C$, which leads to the highly unstable HFO in approximately 50 per cent yield:

$$F_2(g) + H_2O(s) \longrightarrow HFO(g) + HF(g)$$

The product decomposes to O_2 and HF with a half-life of 30 min at $20\,°C$.[5]

---

[3] When writing the associated equilibrium expressions in aqueous solution, we conform to the usual convention that the activity of water is 1, so $[H_2O]$ does not appear in the equilibrium constant.

[4] The value for $Br_2 + H_2O \rightleftharpoons HOBr + HBr$ is 3.5×10^{-9} at $25\,°C$. See R.C. Beckwith, T.X. Wang, and D.W. Margerum, *Inorg. Chem.* **90**, 995 (1996).

[5] E.H. Appleman, *Acc. Chem. Res.* **6**, 113 (1973). This interesting paper describes the discovery of two simple compounds that had long eluded chemists.

> Disproportionation of dihalogen molecules in aqueous solution occurs with the formation of the +1 (OX^-) and −1 (X^-) species.

Halogen(I)

The aqueous Cl(I) species hypochlorous acid, HClO (the angular molecular species H—O—Cl)[6] and hypochlorite ions, ClO^-, are facile oxidizing agents that are used as household bleach and disinfectant and as useful laboratory oxidizing agents. The mechanism invoked for oxidation reactions of HClO was thought for many years to involve O atom transfer, but mechanistic studies strongly implicate Cl^+ transfer in the reactions of ClO^- and HClO.[7] For example, kinetic data indicate that the first step in the mechanism of reaction of HOCl with SO_3^{2-} is Cl^+ transfer to produce the chlorosulfate anion:

$$HClO(aq) + SO_3^{2-}(aq) \longrightarrow OH^-(aq) + ClSO_3^-(aq)$$

The activated complex is pictured as a Cl-bridged species $[HO—Cl—SO_3]^{2-}$ (as in **12**). To account for the outcome of this reaction, the Cl atom is thought to carry only six electrons with it as it transfers to SO_3^{2-}; so in effect a Cl^+ ion is transferred even though an isolated Cl^+ cation is not involved. This step is followed by the hydrolysis of chlorosulfate, to yield the sulfate and chloride ions:

$$ClSO_3^-(aq) + H_2O(l) \longrightarrow SO_4^{2-}(aq) + Cl^-(aq) + 2H^+(aq)$$

A similar Cl^+ ion transfer is indicated for the reactions of HClO with other nucleophilic anions and the rate of reaction, which lies in the order

$$Cl^- < Br^- < I^- < SO_3^{2-} < CN^-$$

correlates with the nucleophilicity of the anion. The ready access to the unobstructed, electrophilic Cl atom in HClO appears to be one feature that leads to the very fast redox reactions of this compound. These rates contrast with the much slower redox reactions of perchlorate ions, in which access of nucleophiles to the Cl atom is blocked by the surrounding O atoms.

Hypohalite ions undergo disproportionation. For instance, ClO^- disproportionates into Cl^- and ClO_3^-:

$$3\,ClO^-(aq) \longrightarrow 2\,Cl^-(aq) + ClO_3^-(aq) \qquad K = 1.5 \times 10^{27}$$

This reaction (which is used for the commercial production of chlorates) is slow at or below room temperature for ClO^- but is much faster for BrO^-. It is so fast for IO^- that this ion has been detected only as a reaction intermediate.

> Hypochlorite is a facile oxidizing agent; hypohalite ions undergo disproportionation.

Halogen(III)

Chlorite ions, ClO_2^-, and bromite ions, BrO_2^-, are both susceptible to disproportionation. However, the rate is strongly dependent on **pH**, and ClO_2^- (and to a lesser extent BrO_2^-) can be handled in basic solution with only slow decomposition. In contrast, chlorous acid, $HClO_2$, and bromous acid, $HBrO_2$, both disproportionate rapidly. As the following potentials indicate, disproportionation is thermodynamically favorable in acidic solution.

$$2\,HClO_2(aq) \longrightarrow H^+(aq) + HClO(aq) + ClO_3^-(aq) \qquad (E^{\ominus} = +0.51\ \text{V})$$

$$2\,HBrO_2(aq) \longrightarrow H^+(aq) + HBrO(aq) + BrO_3^-(aq) \qquad (E^{\ominus} = +0.50\ \text{V})$$

---

[6] Hypohalous acids are widely denoted HOX to emphasize their structure. We have adopted the formula HXO to emphasize their relationship to the other oxoacids, HXO_n.

[7] C. Gerritsen and D.W. Margerum, *Inorg. Chem.* **29**, 2757 (1990) and references therein.

Iodine(III) is even more elusive, and HIO_2 was only recently identified as a transient species in aqueous solution.

These primary reactions are usually accompanied by further reaction. For example, the disproportionation of $HClO_2$ under a broad range of conditions is quite well described by the equation

$$4\,HClO_2(aq) \longrightarrow 2\,H^+(aq) + 2\,ClO_2(aq) + ClO_3^-(aq) + Cl^-(aq) + H_2O(l)$$

This complex disproportionation is shown in the center of Chart 12.1. Because two oxidized products (ClO_2 and ClO_3^-) are produced, the equation can be balanced differently by changing their ratio. The ratio actually observed reflects the relative rates of the alternative reactions.

| Halite ions typically undergo disproportionation.

Halogen(V)

The Frost diagram for chlorine shown in Fig. 12.14 indicates that chlorate ions, ClO_3^-, are unstable with respect to disproportionation in both acidic and basic solution:

$$4\,ClO_3^-(aq) \longrightarrow 3\,ClO_4^-(aq) + Cl^-(aq) \qquad (E^\ominus = +0.25 \text{ V})$$

This value of $E^\ominus$ corresponds to $K = 1.4 \times 10^{25}$; however, as $HClO_3$ is a strong acid, and this reaction is slow at both low and high pH, ClO_3^- ions can be handled readily in aqueous solution. Bromates and iodates are thermodynamically stable with respect to disproportionation.

| Chlorate ions undergo disproportionation in solution but bromates and iodates do not.

Example 12.4 The decomposition of HFO

It was remarked that HFO is unstable in water. Write plausible equations for the decomposition of HFO(aq) and discuss why it is different from that for HClO.

Answer Oxidation states 0, +1, and +3 are accessible for all the halogens except fluorine. In part, this distinction is simply an artifact of the rule for assigning the most negative oxidation number to the most electronegative element, but the higher electronegativity of fluorine is also reflected in the chemical properties of HFO. We might for example expect this thermodynamically unstable compound to yield more stable compounds, such as HF, O_2, H_2O_2, and possibly OF_2. In fact, some of these products are observed:

$$2\,HFO(aq) \longrightarrow 2\,HF(aq) + O_2(g)$$
$$HFO(aq) + H_2O(l) \longrightarrow H_2O_2(aq) + HF(aq)$$

In the presence of excess fluorine a further reaction occurs:

$$F_2(aq) + HFO(aq) \longrightarrow OF_2(aq) + HF(aq)$$

Self-test 12.4 Use standard potentials to decide which dihalogens are thermodynamically capable of oxidizing H_2O to O_2. Which of these reactions are facile?

Halogen(VII)

The perchlorate ion, ClO_4^-, and periodate ion, IO_4^-, were known in the nineteenth century, but the perbromate ion BrO_4^- was not discovered until the late 1960s (see footnote 5). This difference is another example of the instability of the maximum oxidation state of the Period 4 elements arsenic, selenium, and bromine. Although perbromate is not readily available commercially, it may be synthesized by the action of fluorine on bromate ions in basic aqueous solution:

$$BrO_3^-(aq) + F_2(g) + 2\,OH^-(aq) \longrightarrow BrO_4^-(aq) + 2\,F^-(aq) + H_2O(l)$$

Of the three XO_4^- ions, BrO_4^- is the strongest oxidizing agent. That perbromate is out of line with its adjacent halogen congeners fits a general pattern for anomalies in the chemistry of p-block elements of Period 4.

The reduction of BrO_4^- is faster than that of ClO_4^- because perbromate ions undergo electron transfer more rapidly (by an outer-sphere mechanism). However, reduction of periodate ions in dilute acid solution is by far the fastest. Periodates are therefore used in analytical chemistry as oxidizing titrants and also in syntheses, such as the oxidative cleavage of diols. As we remarked earlier, the speed of periodate reactions appear to stem from the ease with which the reducing agent can coordinate to the big, electrophilic I(VII) atom, and thereby undergo inner-sphere reaction, as in the diol oxidation:

$$HIO_4(aq) + HOC(CH_3)_2C(CH_3)_2OH \longrightarrow$$

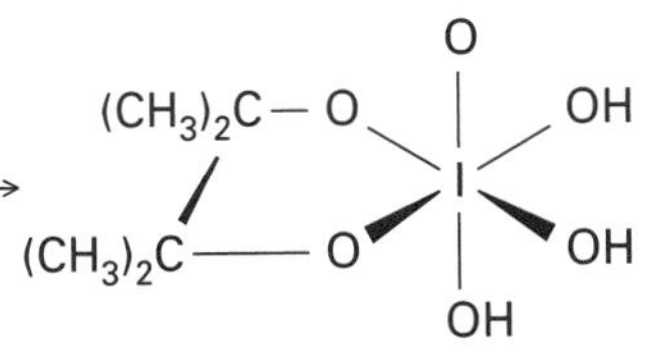

$$\longrightarrow 2(CH_3)_2CO + HIO_3 + H_2O$$

We have already commented on the low Lewis basicity of ClO_4^-, which leads to very weak coordination with metal aqua-ions, and the fact that ClO_4^- in combination with many reducing cations forms metastable compounds that do not react under mild conditions but explode when provoked. With certain reducing agents, ClO_4^-, however, reacts smoothly in aqueous solution. The order of reactivity observed for a series of metal ion reducing agents is

$$Ru^{2+} > Ti^{3+} > Mo^{3+} > V^{2+} \approx V^{3+} > Cr^{2+} \gg Fe^{2+}$$

For the most part, this series correlates with the ligand-field splitting parameter Δ_0 but not with the values of $E^\ominus$ for the individual ions. The correlation of rate and Δ_0 indicates that stronger bonding between the metal ion and a perchlorate O atom may facilitate O atom transfer from Cl(VII) to the metal ion, thereby reducing the chlorine to Cl(V) (**14**).

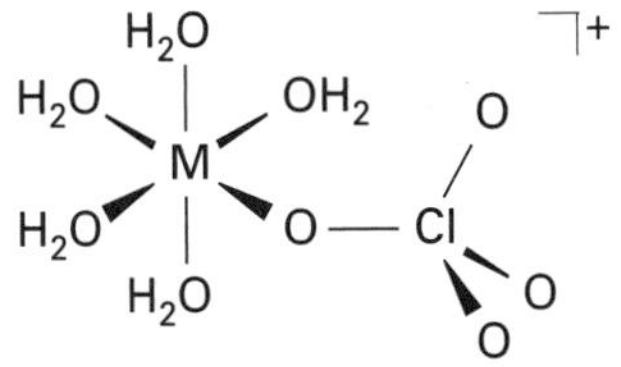

Anhydrous perchloric acid is a particularly aggressive oxidizing medium, which must not be allowed to come into accidental contact with organic matter or similar reducing agents. The presence of Cl_2O_7 as a result of the equilibrium

$$3\, HClO_4 \rightleftharpoons Cl_2O_7 + H_3O^+ + ClO_4^-$$

and at elevated temperatures of the radical species $\cdot OH$, $\cdot ClO_3$, and $\cdot ClO_4$, appears to be responsible for its rapid oxidation reactions.

> The rates of reactions in which ClO_4^- acts as an oxidizing agent are often slow in basic solution but rapid in acidic solution.

14 $[(H_2O)_5MOClO_3]^+$

12.7 Fluorocarbons

The synthesis of fluorocarbon compounds is of great technological importance because they are useful in applications ranging from coatings for non-sticking cooking ware and halogen-resistant laboratory vessels to the volatile chlorofluorocarbons used as refrigerants in air conditioners and refrigerators. Fluorocarbon derivatives also have been the topic of considerable exploratory synthetic research because their derivatives often have unusual properties.

The direct reaction of an aliphatic hydrocarbon with an oxidizing metal fluoride leads to the formation of C—F bonds and produces HF as a byproduct:

$$RH + 2\,CoF_3 \longrightarrow RF + 2\,CoF_2 + HF \qquad (R = alkyl\ or\ aryl)$$

When R is an aryl, CoF_3 yields the cyclic saturated fluoride:

$$C_6H_6 + 18\,CoF_3 \longrightarrow C_6F_{12} + 18\,CoF_2 + 6\,HF$$

The strongly oxidizing fluorinating agent used in these reactions, CoF_3, is regenerated by the reaction of CoF_2 with fluorine:

$$2\,CoF_2 + F_2 \longrightarrow 2\,CoF_3$$

Another important method of C—F bond formation is halogen exchange by the reaction of a nonoxidizing fluoride, such as HF, with a chlorocarbon in the presence of a catalyst, such as SbF_3:

$$CHCl_3 + 2\,HF \longrightarrow CHClF_2 + 2\,HCl$$

This process is performed on a large scale to produce the chlorofluorocarbons (CFCs) that are used as refrigerant fluids, the propellant in spray cans, and in the blowing agent in plastic foam products. Unfortunately, the CFCs find their way up into the stratosphere, where they catalyze the destruction of the ozone layer and open the way for dangerous ultraviolet radiation to reach the ground. They may also contribute to global warming.[8]

When heated, dichlorofluoromethane is converted to the useful monomer, C_2F_4:

$$2\,CHClF_2 \xrightarrow{600\text{-}800\,°C} C_2F_4 + 2\,HCl$$

The polymerization of tetrafluoroethylene is carried out with a radical initiator:

$$n\,C_2F_4 \xrightarrow{ROO·} (—CF_2—CF_2—)_n$$

Polytetrafluoroethylene is sold under many trade names, one of which is Teflon (DuPont). It can be depolymerized at high temperatures and this is the most convenient method of preparing tetrafluoroethylene in the laboratory:

$$(—CF_2—CF_2—)_n \xrightarrow{600\,°C} n\,C_2F_4$$

Although tetrafluoroethylene is not highly toxic, a byproduct, perfluoroisobutylene, is toxic and its presence dictates care in handling the crude tetrafluoroethylene.

Fluorocarbon molecules and polymers are resistant to oxidation but are attacked by alkali metals.

The noble gases (Group 18/VIII)

The elements in Group 18/VIII of the periodic table have been given and then lost various collective names over the years as different aspects of their properties have been identified and then disproved. Thus they have been called the *rare gases* and the *inert gases*, and are currently called the **noble gases**. The first name is inappropriate because argon is by no means rare (it is more abundant than CO_2 in the atmosphere). The second has been inappropriate since the discovery of compounds of xenon. The appellation 'noble gases' is now accepted because it gives the sense of low but significant reactivity.

[8] C. Baird, *Environmental chemistry.* W.H. Freeman & Co, New York (1995); H.B. Gray, J.D. Simon, and W.C. Trogler, *Braving the elements.* University Science Books, Mill Valley (1995).

12.8 Occurrence and recovery

Because they are so unreactive and are rarely concentrated in nature, the noble gases eluded recognition until the end of the nineteenth century. Indeed, Mendeleev did not make a place for them in his periodic table because the chemical regularities of the other elements, upon which his table was based, did not suggest their existence. However, in 1868 a new spectral line observed in the spectrum of the Sun did not correspond to a known element. This line was eventually attributed to a new element, helium, and in due course the element itself and its congeners were found on Earth.

All the noble gases occur in the atmosphere. The crustal and atmospheric abundances of argon (0.94 per cent by volume) and neon (1.5×10^{-3} per cent) make these two elements more plentiful than many familiar elements, such as arsenic and bismuth, in the Earth's crust (Fig. 12.15). Xenon and radon are the rarest elements of the group. Radon is a product of radioactive decay and, since it lies beyond bismuth, is itself unstable. The elements neon through xenon are extracted from liquid air by low-temperature distillation. Helium atoms are too light, and hence move too fast on average, to be retained by the Earth's gravitational field, so most of the helium on Earth is the product of α-emission in the decay of radioactive elements and has been trapped by rock formations. High helium concentrations are found in certain natural gas deposits (mainly in the USA and eastern Europe) from which it can be recovered by low-temperature distillation. Some helium arrives from the Sun as a solar wind of α particles.

The low boiling points of the lighter noble gases (He, 4.2 K; Ne, 27.1 K; Ar, 87.3 K) follow from the weakness of the dispersion forces between the atoms and the absence of other forces. Helium has the lowest boiling point of any substance and is widely used in cryogenics as a very low temperature refrigerant; it is the coolant for superconducting magnets used for NMR spectroscopy and magnetic resonance imaging. When ^{4}He is cooled below 2.178 K it undergoes a transformation into a second liquid phase known as **helium-II**. This phase is classed as a superfluid, for it flows without viscosity. Solid helium is formed only under pressure.

On account of its low density and nonflammability, helium is used in balloons and lighter-than-air craft. The greatest use of argon is to provide an inert atmosphere for the production of air-sensitive compounds and as an inert gas blanket to suppress the oxidation of metals when they are welded. The noble gases are also extensively used in various light sources, including conventional sources (neon signs, fluorescent lamps, xenon flash lamps) and lasers (helium–neon, argon-ion, and krypton-ion lasers). In each case, an electric discharge through the gas ionizes some of the atoms and promotes both ions and neutral atoms into excited states which then emit electromagnetic radiation upon return to a lower state.

Radon, which is a product of nuclear power plants and of the radioactive decay of naturally occurring uranium and thorium, is a health hazard because of the ionizing nuclear radiation it produces. It is usually a minor contributor to the background radiation arising from cosmic rays and terrestrial sources. However, in regions where the soil, underlying rocks, or building materials contain significant concentrations of uranium, excessive amounts of the gas have been found in buildings.

The noble gases are monatomic. Helium is used as an inert gas and as light sources in lasers and electric discharge lamps. Liquid helium is a very low temperature refrigerant. Radon is radioactive and highly mobile because it is a gas.

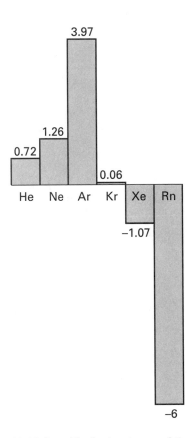

12.15 Logarithmic abundances of the noble gases in the Earth's crust (atmospheric ppm by volume). (Source: J. Emsley, *The elements*. Oxford University Press (1998).)

12.9 Compounds

The reactivity of the noble gases has been investigated sporadically ever since their discovery, but all early attempts to coerce them into compound formation were unsuccessful. However, in March 1962, Neil Bartlett, then at the University of British Columbia, observed the reaction of a noble gas.[9] Bartlett's report, and another from Rudolf Hoppe's group in the University of Munster a few weeks later, set off a flurry of activity throughout the world. Within a year, a series of xenon fluorides and oxo compounds had been synthesized and characterized.[10] However, the field does appear to be somewhat limited, and little significant progress has been made in recent years.

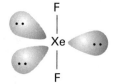

15 XeF_2

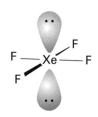

16 XeF_4

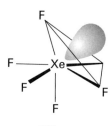

17 XeF_6

(a) Synthesis and structure of xenon fluorides

Bartlett's motivation for studying xenon was based on the observations that PtF_6 can oxidize O_2 to form a solid, and that the ionization energy of xenon is similar to that of molecular oxygen. Indeed, reaction of xenon with PtF_6 did give a solid, but the reaction is complex and the complete formulation of the product (or products) is not clear. The direct reaction of xenon and fluorine leads to a series of compounds with oxidation numbers +2 (XeF_2), +4 (XeF_4), and +6 (XeF_6).

The molecules XeF_2, XeF_4, and XeF_6 are linear (**15**), square-planar (**16**), and distorted octahedral (**17**), respectively. Solid XeF_6 is more complex, and contains fluoride ion-bridged XeF_5^+ cations. In solution, it forms Xe_4F_{24} tetramers. The first two structures are consistent with the VSEPR model for a Xe atom with five and six electron pairs, respectively, and they bear a molecular and electronic structural resemblance to the isoelectronic polyhalide anions I_3^- and ClF_4^- (Section 12.5). The structures of XeF_2 and XeF_4 are well established over a wide timescale by diffraction and spectroscopic methods. Similar measurements on XeF_6 in the gas phase, however, lead to the conclusion that this molecule is fluxional. Infrared spectra and electron diffraction on XeF_6 show that a distortion occurs about a threefold axis, suggesting that a triangular face of F atoms opens up to accommodate a lone pair of electrons (as in **14**). One interpretation is that the fluxional process arises from the migration of the lone pair from one triangular face to another.

The xenon fluorides are synthesized by direct reaction of the elements, usually in a nickel reaction vessel that has been passivated by exposure to F_2 to form a thin protective NiF_2 coating. This treatment also removes surface oxide, which would react with the xenon fluorides. The synthetic conditions indicated in the following equations show that formation of the higher halides is favored by a higher proportion of fluorine and higher total pressure:

$$Xe(g) + F_2(g) \xrightarrow{400\,°C,\ 1\ atm} XeF_2(g) \qquad \text{(Xe in excess)}$$

$$Xe(g) + 2\,F_2(g) \xrightarrow{600\,°C,\ 6\ atm} XeF_4(g) \qquad (Xe{:}F_2 = 1{:}5)$$

$$Xe(g) + 3\,F_2(g) \xrightarrow{300\,°C,\ 60\ atm} XeF_6(g) \qquad (Xe{:}F_2 = 1{:}20)$$

A simple 'windowsill' synthesis also is possible. Xenon and fluorine are sealed in a glass bulb (previously rigorously dried to prevent the formation of HF and the attendant etching of the glass) and the bulb is exposed to sunlight, whereupon beautiful crystals of XeF_2 slowly form in the bulb. It will be recalled that fluorine undergoes photodissociation (Section 12.1), and in this synthesis the photochemically generated F atoms react with Xe atoms.

Xenon reacts with fluorine to form a series of fluorides: XeF_2, XeF_4, and XeF_6.

[9]   For an enjoyable account of the early failure and final success in the quest for compounds of the noble gases, see P. Lazlo and G.J. Schrobilgen, *Angew. Chem., Int. Ed. Engl.* **27**, 479 (1988).

[10]   John Holloway provides a nice summary of the development of this field in 'Twenty-five years of noble gas chemistry', in *Chem. in Brit.*, 658 (1987). For detailed coverage, see B. Zemva, In *Encyclopedia of inorganic chemistry* (ed. R.B. King), Vol. 5, p. 2660. Wiley-Interscience, New York (1994).

(b) Reactions of xenon fluorides

The reactions of the xenon fluorides are similar to those of the high oxidation number interhalogens, and redox and metathesis reactions dominate. One important reaction of XeF_6 is double-displacement with oxides:

$$XeF_6(s) + 3 H_2O(l) \longrightarrow XeO_3(aq) + 6 HF(g)$$
$$2 XeF_6(s) + 3 SiO_2(s) \longrightarrow 2 XeO_3(s) + 3 SiF_4(g)$$

Xenon trioxide presents a serious hazard since this endoergic compound is highly explosive. In basic aqueous solution the Xe(VI) oxoanion $HXeO_4^-$ slowly decomposes in a coupled disproportionation and water oxidation to yield a Xe(VIII) perxenate ion, XeO_6^{4-}, and xenon. Another important chemical property of the xenon halides is their strong oxidizing power, as illustrated by the following examples:

$$2 XeF_2(s) + 2 H_2O(l) \longrightarrow 2 Xe(g) + 4 HF(g) + O_2(g)$$
$$XeF_4(s) + Pt(s) \longrightarrow Xe(g) + PtF_4(s)$$

As with the interhalogens, the xenon fluorides react with strong Lewis acids to form xenon fluoride cations:

$$XeF_2(s) + SbF_5(l) \longrightarrow [XeF]^+[SbF_6]^-(s)$$

These cations are associated with the counterion by F^- bridges. Compounds with Xe—N (**18**) and Xe—C bonds, as in $[Xe(C_6F_5)][C_6F_5BF_3]$, have also been synthesized.[11] Evidence also exists for the formation of the paramagnetic species Xe_2^+.

Another similarity with the interhalogens is the reaction of XeF_4 with the Lewis base F^- in cyanomethane solution to produce the XeF_5^- ion:

$$XeF_4 + [N(CH_3)_4]F \longrightarrow [N(CH_3)_4]XeF_5$$

The XeF_5^- ion has a planar-pentagonal shape (**19**), and in the VSEPR model the two electron pairs on Xe occupy axial positions.[12] Similarly, it has been known for many years that reaction of XeF_6 with an F^- source produces the XeF_7^- or XeF_8^{2-} ions depending on the proportion of fluoride. Only the shape of XeF_8^{2-} is known; it is a square antiprism (**20**), which is difficult to reconcile with the simple VSEPR model because this shape does not provide a site for the lone pair on Xe.

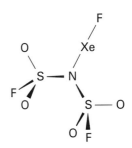

18 $FXeN(SO_2F)_2$

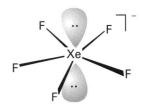

19 XeF_5^-

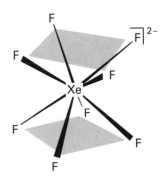

20 XeF_8^{2-}

Example 12.5 Synthesis and a structure of a noble gas compound

(a) Describe a procedure for the synthesis of potassium perxenate, starting with xenon and other reagents of your choice. (b) Using VSEPR theory, determine the probable structure of the perxenate ion.

Answer (a) Xe—O compounds are endoergic, so they cannot be prepared by direct reaction of xenon and oxygen. The hydrolysis of XeF_6 yields XeO_3 and, as described in the text, the latter in basic solution undergoes disproportionation yielding perxenate, XeO_6^{4-}. Thus the reaction of xenon and excess F_2 would be carried out at 300 °C and 60 atm in a stout nickel container. The resulting XeF_6 might then be converted to the perxenate in one step by exposing it to aqueous KOH solution. The resulting potassium perxenate (which turns out to be a hydrate) could then be crystallized. (b) A Lewis structure of the perxenate ion is shown in (**21**). With six electron pairs around the Xe atom, the VSEPR model predicts an octahedral arrangement of bonding electron pairs and an octahedral overall structure.

21 XeO_6^{4-}

[11] H.J. Frohn and S. Jakobs, *J. Chem. Soc. Chem. Commun.*, 625 (1989). In each of these cases, strongly electron withdrawing groups are attached to the C or N, but this is not the case for $HCNXeF^+$; A.A.A. Emara and G.J. Schrobilgen, *J. Chem. Soc. Chem. Commun.*, 1644 (1987).

[12] A.-R. Mahjoub and K. Seppelt, The structure of $[IF_8]^-$. *Angew. Chem.* **30**, 876 (1991).

Self-test 12.5 Write a balanced equation for the decomposition of xenate ions in basic solution for the production of perxenate ions, xenon, and oxygen.

The xenon fluorides are the gateway to the preparation of compounds of the noble gases with elements other than fluorine and oxygen. The reaction of nucleophiles with a xenon fluoride is one useful strategy for the synthesis of such bonds. For instance, the reaction

$$XeF_2 + HN(SO_2F) \longrightarrow FXeN(SO_2F) + HF$$

is driven forward by the stability of the product HF and the energy of formation of the Xe—N bond. A strong Lewis acid such as AsF_5 can extract F^- from the product of this reaction to yield the cation $[XeN(SO_2F)_2]^+$. Another route to Xe—N bonds is the reaction of one of the fluorides with a strong Lewis acid:

$$XeF_2 + AsF_5 \longrightarrow [XeF][AsF_6]$$

followed by the introduction of a Lewis base, such as CH_3CN, to yield $[CH_3CNXeF][AsF_6]$.

> Xenon fluorides are strong oxidizing agents and they form complexes with F^-, such as XeF_5^-, XeF_7^-, and XeF_8^{2-}. They are used in the preparation of compounds containing Xe—O and Xe—N bonds.

(c) Compounds of other noble gases

Radon has a lower ionization energy than xenon, so it can be expected it to form compounds even more readily. Evidence exists for the formation of RnF_2, and cationic compounds, such as $[RnF^+][SbF_6^-]$, but detailed characterization is frustrated by their radioactivity.[13] Krypton has a much higher ionization energy than xenon (Table 12.1) and its ability to form compounds is much more limited. Krypton difluoride, KrF_2, is prepared by passing an electric discharge or ionizing radiation through a fluorine-krypton mixture at low temperatures ($-196\,°C$). As with XeF_2, the krypton compound is a colorless volatile solid and the molecule is linear. It is an endoergic and highly reactive compound that must be stored at low temperatures.

> Krypton and radon fluorides are known but their known chemistry is much less extensive than that of xenon.

---

[13] L. Stein, *Inorg. Chem.* **23**, 3670 (1984).

Further reading

A.J. Bard, R. Parsons, and J. Jordan, *Standard potentials in aqueous solution*. Dekker, New York (1985). See the chapters on the halogens and the 'inert' gases.

J.H. Canterford and R. Colton, *Halides of the transition elements: halides of the second and third row transition metals*. Wiley, New York (1968).

D. Brown, *Halides of the transition elements: halides of the lanthanides*. Wiley, New York (1968).

H. Selig and J.H. Holloway, Cationic and anionic complexes of the noble gases. *Topics in Current Chemistry* **124**, 33 (1984).

R.C. Thompson, Reaction mechanisms of the halogens and oxohalogen species in acidic, aqueous solution. In *Advances in*

inorganic and bioinorganic mechanisms (ed. A.G. Sykes), Vol. 4. Academic Press, New York (1986).

K. Seppelt and D. Lentz, Novel developments in noble gas chemistry. *Phys. Inorg Chem.* **29**, 167 (1982).

Kirk–Othmer encyclopedia of chemical technology. Wiley-Interscience, New York (1991 *et seq.*). The series of volumes contains many reviews of the individual halogens and their compounds with emphasis on practical application. Useful information on the noble gases is also provided.

G.J. Miller, Halides: solid state chemistry. In *Encyclopedia of inorganic chemistry* (ed. R.B. King), Vol. 3, pp. 1343–57. Wiley, New York (1994).

Exercises

12.1 Preferably without consulting reference material, write out the halogens and noble gases as they appear in the periodic table, and indicate the trends in (a) physical state (s, l, or g) at room temperature and pressure, (b) electronegativity, (c) hardness of the halide ion, and (d) color.

12.2 Describe how the halogens are recovered from their naturally occurring halides and rationalize the approach in terms of standard potentials. Give balanced chemical equations and conditions where appropriate.

12.3 Sketch the chloroalkali cell. Show the half-cell reactions and indicate the direction of diffusion of the ions. Give the chemical equation for the unwanted reaction that would occur if OH^- migrated through the membrane and into the anode compartment.

12.4 Sketch the form of the vacant σ^* orbital of a dihalogen molecule and describe its role in the Lewis acidity of the dihalogens.

12.5 Nitrogen trifluoride, NF_3, boils at $-129\,°C$ and is devoid of Lewis basicity. By contrast the lower molar mass compound NH_3 boils at $-33\,°C$ and is well known as a Lewis base. (a) Describe the origins of this very large difference in volatility. (b) Describe the probable origins of the difference in basicity.

12.6 Based on the analogy between halogens and pseudohalogens: (a) Write the balanced equation for the probable reaction of cyanogen $(CN)_2$ with aqueous sodium hydroxide. (b) Write the equation for the probable reaction of excess thiocyanate with the oxidizing agent $MnO_2(s)$ in acidic aqueous solution. (c) Write a plausible structure for trimethylsilyl cyanide.

12.7 (a) Use the VSEPR model to predict the probable shapes of IF_6^+ and IF_7. (b) Give a plausible chemical equation for the preparation of $[IF_6][SbF_6]$.

12.8 Predict whether each of the following solutes is likely to make liquid BrF_3 a stronger Lewis acid or a stronger Lewis base: (a) SbF_5, (b) SF_6, (c) CsF.

12.9 Predict whether each of the following compounds is likely to be dangerously explosive in contact with BrF_3, and explain your answer: (a) SbF_5, (b) CH_3OH, (c) F_2, (d) S_2Cl_2.

12.10 The formation of Br_3^- from a tetraalkylammonium bromide and Br_2 is only slightly exoergic. Write an equation (or NR for no reaction) for the interaction of $[NR_4][Br_3]$ with excess I_2 in CH_2Cl_2 solution, and give your reasoning.

12.11 Explain why $CsI_3(s)$ is stable with respect to the elements but $NaI_3(s)$ is not.

12.12 Write plausible Lewis structures for (a) ClO_2 and (b) I_2O_5 and predict their shapes and point groups.

12.13 (a) Give the formulas and the probable relative acidities of perbromic acid and periodic acid. (b) Which is the more stable?

12.14 (a) Describe the expected trend in the standard potential of an oxoanion in a solution with decreasing pH. (b) Demonstrate this phenomenon by calculating the reduction potential of ClO_4^- at $pH = 7$ and comparing it with the tabulated value at $pH = 0$.

12.15 With regard to the general influence of pH on the standard potentials of oxoanions, explain why the disproportionation of an oxoanion is often promoted by low pH.

12.16 Which oxidizing agent reacts more readily in dilute aqueous solution, perchloric acid or periodic acid? Give a mechanistic explanation for the difference.

12.17 (a) For which of the following anions is disproportionation thermodynamically favorable in acidic solution: ClO^-, ClO_2^-, ClO_3^-, and ClO_4^-? (If you do not know the properties of these ions, determine them from a table of standard potentials.) (b) For which of the favorable cases is the reaction very low at room temperature?

12.18 Which of the following compounds present an explosion hazard? (a) NH_4ClO_4, (b) $Mg(ClO_4)_2$, (c) $NaClO_4$, (d) $[Fe(OH_2)_6][ClO_4]_2$. Explain your reasoning.

12.19 Explain why helium is present in low concentration in the atmosphere even though it is the second most abundant element in the universe.

12.20 Which of the noble gases would you choose as (a) the lowest temperature liquid refrigerant, (b) an electric discharge light source requiring a safe gas with the lowest ionization energy, (c) the least expensive inert atmosphere?

12.21 By means of balanced chemical equations and a statement of conditions, describe a suitable synthesis of (a) xenon difluoride, (b) xenon hexafluoride, (c) xenon trioxide.

12.22 Give the formula and describe the structure of a noble gas species that is isostructural with (a) ICl_4^-, (b) IBr_2^-, (c) BrO_3^-, (d) ClF.

12.23 (a) Give Lewis formulas and formal charges for ClO^-, Br_2, and XeF^+. (b) Are these species isolobal? (c) Describe the chemical similarities as judged by their reactions with nucleophiles. (d) Rationalize the trends in electrophilicity and basicity.

12.24 (a) Give a Lewis structure for XeF_7^-. (b) Speculate on its possible structures by using the VSEPR model and analogy with other xenon fluoride anions.

Problems

12.1 Many of the acids and salts corresponding to the positive oxidation numbers of the halogens are not listed in the catalog of a major international chemical supplier: (a) $KClO_4$ and KIO_4 are available but $KBrO_4$ is not; (b) $KClO_3$, $KBrO_3$, and KIO_3 are all available; (c) $NaClO_2$ and $NaBrO_2 \cdot 3H_2O$ are available but salts of IO_2^- are not; (d) ClO^- salts are available but the bromine and iodine analogs are not. Describe the probable reasons for the missing salts of oxoanions.

12.2 Determine the incorrect statements among the following descriptions and provide correct statements. (a) Oxidation of the halides is the only commercial method of preparing the halogens F_2 through I_2. (b) ClF_4^- and I_5^- are isolobal and isostructural. (c) Atom-transfer processes are common in the mechanisms of oxidations by the halogen oxoanions, and an example is the O atom transfer in the oxidation of SO_3^{2-} by ClO^-. (d) Periodate appears to be a more facile oxidizing agent than perchlorate because the former can coordinate to the reducing agent at the I(VII) center, whereas the Cl(VII) center in perchlorate is inaccessible to reducing agents.

12.3 The reaction of I^- ions is often used to titrate ClO^-, giving deeply colored I_3^- ions, along with Cl^- and H_2O. Although never proved, it was thought that the initial reaction proceeded by O atom transfer from Cl to I. However, it is now believed that the reaction proceeds by Cl atom transfer to give ICl as the intermediate (K. Kumar, R.A. Day, and D.W. Margerum, *Inorg. Chem.* **25**, 4344 (1986)). Summarize the evidence for Cl atom transfer.

12.4 Given the bond lengths and angles in I_5^+ (**9**), describe the bonding in terms of two-center and three-center σ bonds and account for the structure in terms of the VSEPR model.

12.5 Until the work of K.O. Christe (*Inorg. Chem.* **25**, 3721 (1986)), F_2 could be prepared only electrochemically. Give chemical equations for Christe's preparation and summarize the reasoning behind it.

12.6 The first compound containing an Xe–N bond was reported by R.D. LeBlond and D.D. DesMarteau (*J. Chem. Soc., Chem. Commun,* 554 (1974)). Summarize the method of synthesis and characterization. (The proposed structure was later confirmed in an X-ray crystal structure determination.)

Part 3

Advanced topics

Our final set of topics deals with more advanced aspects of inorganic chemistry. The first two chapters (on spectroscopy and reaction mechanisms) develop topics that have been introduced earlier, and introduce some more advanced material to show how a detailed understanding of structure and mechanism can be obtained. The remaining four chapters (on main-group and d-block organometallic compounds, catalysis, solid state chemistry, and bioinorganic chemistry) span areas that are currently being pursued vigorously in laboratories throughout the world. Research in these fields often combines the talents of inorganic chemists with those of researchers in other disciplines, such as condensed matter physics, biology, and engineering. Chapter 17 (catalysis) follows naturally from the two preceding chapters (organometallic compounds) because it shows how certain classes of organometallic compound can be used as catalysts. Chapter 18 extends the discussion of the solid state that we began in Chapter 2 to include new structural types, nonstoichiometry, and the role of defects. Chapter 19, the final chapter of the book, brings inorganic chemistry face-to-face with biology, and we see how metal ions and their complexes are involved in some vital biological processes, such as oxygen transport and storage and electron transfer.

The electronic spectra of complexes

The aim of this chapter is to demonstrate how to interpret the origins of the electronic spectra of coordination compounds and to correlate these spectra with bonding. Because atoms and complexes are so compact, the mutual repulsions of their electrons have a strong influence on the energies of electronic transitions. The role of electron–electron repulsion was originally determined by the analysis of atoms and ions in the gas phase, and much of that information can be utilized in the analysis of the spectra of metal complexes. The main differences between gas-phase metal atoms and ions and metal atoms in complexes are changes in symmetry and the resultant changes in the degeneracy of energy levels. This chapter also extends the material in Chapter 7 by describing how to apply molecular orbital techniques to systems of lower symmetry than treated there. In particular, we see how to build up a description of the molecular orbitals of complexes from simpler fragments.

Figure 13.1 sets the stage for the topics covered in this chapter by showing the electronic absorption spectrum of the d^3 complex $[Cr(NH_3)_6]^{3+}$ in aqueous solution. The band at lowest energy (longest wavelength) is very weak, and later we shall see that it is an example of a 'spin-forbidden' transition. Next are two bands with intermediate intensities, which will turn out to be 'spin-allowed' transitions between the t_{2g} and e_g orbitals of the complex $(t_{2g}^2 e_g^1 \leftarrow t_{2g}^3)$,[1] which are mainly derived from the metal d orbitals. The third feature in the spectrum is the intense band at short wavelength labeled CT, of which only the low-energy tail is evident in the illustration.

One problem that immediately confronts us is why two absorptions can be ascribed to the apparently single transition $t_{2g}^2 e_g^1 \leftarrow t_{2g}^3$. This splitting of a single transition into two bands is in fact an outcome of the electron–electron repulsions that will be one focus of this chapter. The intense band labeled CT has an entirely different origin from the other three. Later in the chapter we shall see that it is an example of a transition in which an electron is transferred from the ligands to the central metal atom.

---

[1] Throughout this discussion we employ the convention of indicating spectroscopic transitions with the lower energy state on the right and upper energy state on the left of the arrow.

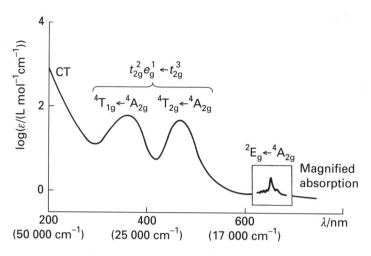

13.1 The spectrum of the d^3 complex $[Cr(NH_3)_6]^{3+}$, which illustrates the features studied in this chapter, and the assignments of the transitions as explained in the text.

The electronic spectra of atoms

In Part 1 we described atoms and molecules by giving their electronic configurations, the designation of the number of electrons in each orbital (as in $1s^2 2s^1$ for Li). However, a configuration is an incomplete description of the arrangement of electrons in atoms. In the configuration $2p^2$, for instance, the two electrons might occupy orbitals with different orientations of their orbital angular momenta (that is, with different values of m_l from among the possibilities $+1$, 0, and -1 that are available when $l = 1$). Similarly, the designation $2p^2$ tells us nothing about the spin orientations of the two electrons ($m_s = +\frac{1}{2}$ or $-\frac{1}{2}$). The atom may in fact have several different states of total orbital and spin angular momenta, each one corresponding to the occupation of orbitals with different values of m_l by electrons with different values of m_s. The different ways in which the electrons can occupy the orbitals specified in the configuration are called the **microstates** of the configuration. For example, one microstate of a $2p^2$ configuration is $(1^+, 1^-)$; this notation signifies that both electrons occupy an orbital with $m_l = +1$ but do so with opposite spins, the superscript $+$ indicating $m_s = +\frac{1}{2}$ and $-$ indicating $m_s = -\frac{1}{2}$). Another microstate of the same configuration is $(-1^+, 0^+)$. In this microstate, both electrons have $m_s = +\frac{1}{2}$ but one occupies the $2p$ orbital with $m_l = -1$ and the other occupies the orbital with $m_l = 0$.

13.1 Spectroscopic terms

The microstates of a given configuration have the same energy only if interelectronic repulsions are negligible. However, because atoms and most molecules are compact, interelectronic repulsions are strong and cannot always be ignored. As a result, microstates that correspond to different relative spatial distributions of electrons have different energies. If we group together the microstates that have the same energy when electron repulsions are taken into account, we obtain the spectroscopically distinguishable energy levels called **terms**.

For light atoms and the $3d$ and $4f$ (lanthanide) series, it turns out that the most important property of a microstate for helping us to decide its energy is the relative orientation of the spins of the electrons.[2] Next in importance is the relative orientation of the orbital angular momenta of the electrons. It follows that we can identify the terms of light atoms and put

---

[2] This point is taken up again in Section 13.7.

them in order of energy by sorting the microstates according to their total spin quantum number S (which is determined by the relative orientation of the individual spins) and then according to their total orbital angular momentum quantum number L (which is determined by the relative orientation of the individual orbital angular momenta of the electrons). For heavy atoms, such as those of the $4d$, $5d$, and $5f$ series, the relative orientations of orbital momenta or of spin momenta are less important. In these atoms the spin and orbital angular momenta of individual electrons are strongly coupled together into a resultant by the magnetic interaction known as **spin–orbit coupling**, so the relative orientation of the spin and orbital angular momenta of each electron is the most important feature for determining the energy. The terms of heavy atoms are therefore sorted on the basis of the values of the total angular momentum quantum number j for an electron in each microstate.

The process of combining electron angular momenta by summing first the spins, then the orbital momenta, and finally the two resultants is called **Russell–Saunders coupling**. This coupling scheme is used to identify the terms of light atoms. The coupling scheme most appropriate to heavy atoms is called **jj-coupling**.

To develop the Russell–Saunders scheme, we need to know what values of L and S can arise in an atom. Suppose we have two electrons with quantum numbers l_1, s_1 and l_2, s_2. Then, according to the **Clebsch–Gordan series**, the possible values of L and S are

$$L = l_1 + l_2, l_1 + l_2 - 1, \cdots, |l_1 - l_2| \qquad S = s_1 + s_2, s_1 + s_2 - 1, \cdots, |s_1 - s_2| \tag{1}$$

(The modulus signs appear because neither L nor S can be negative.) For example, an atom with configuration d^2 can have the following values of L:

$$L = 2 + 2, 2 + 2 - 1, \cdots, |2 - 2| = 4, 3, 2, 1, 0$$

The total spin can be

$$S = \tfrac{1}{2} + \tfrac{1}{2}, \tfrac{1}{2} + \tfrac{1}{2} - 1, \cdots, |\tfrac{1}{2} - \tfrac{1}{2}| = 1, 0$$

To find the values of L and S for atoms with three electrons, we continue the process by combining l_3 with the value of L just obtained, and likewise for s_3.

Once L and S have been found, we can write down the allowed values of the quantum numbers M_L and M_S,

$$M_L = L, L - 1, \cdots, -L \qquad M_S = S, S - 1, \cdots, -S$$

These quantum numbers give the orientation of the angular momentum relative to an arbitrary axis: there are $2L + 1$ values of M_L for a given value of L and $2S + 1$ values of M_S for a given value of S. The actual values of M_L and M_S for a given microstate can be found very easily by adding together the values of m_l or m_s for the individual electrons. Therefore, if one electron has the quantum number m_{l1} and the other has m_{l2}, then

$$M_L = m_{l1} + m_{l2}$$

The same applies to the total spin:

$$M_S = m_{s1} + m_{s2}$$

Thus $(0^+, -1^-)$ is a microstate with $M_L = -1$ and $M_S = 0$ and may contribute to any term for which these two quantum numbers apply.

By analogy with the notation $s, p, d, \cdots$ for orbitals with $l = 0, 1, 2, \cdots$, the total orbital angular momentum of an atomic term is denoted by the equivalent upper case letter:

$L =$	0	1	2	3	4	$\cdots$
	S	P	D	F	G	then alphabetical (omitting J)

The total spin is normally reported as the value of $2S + 1$, which is called the **multiplicity** of the term:

$$
\begin{array}{cccccc}
S = & 0 & \tfrac{1}{2} & 1 & \tfrac{3}{2} & 2 & \cdots \\
2S + 1 = & 1 & 2 & 3 & 4 & 5 & \cdots
\end{array}
$$

The multiplicity is written as a left superscript on the letter representing the value of L, and the entire label of a term is called a **term symbol**. Thus, the term symbol ^{3}P denotes a term (a collection of degenerate states) with $L = 1$ and $S = 1$, and is called a 'triplet' term.

> The terms of atoms are specified by symbols in which the value of L is indicated by one of the letters $\mathrm{S, P, D}, \cdots$, and the value of $2S + 1$ is given as a left superscript.

Example 13.1 Deriving term symbols

Give the term symbols for an atom with the configurations (a) s^1, (b) p^1, and (c) $s^1 p^1$.

Answer (a) The single s electron has $l = 0$ and $s = \tfrac{1}{2}$. Because there is only one electron, $L = 0$ (an S term), $S = \tfrac{1}{2}$, and $2S + 1 = 2$ (a doublet term). The term symbol is therefore ^{2}S. (b) For a single p electron, $l = 1$ so $L = 1$ and the term is ^{2}P. (These terms arise in the spectrum of an alkali metal atom, such as Na.) With one s and one p electron, $L = 0 + 1 = 1$, a P term. The electrons may be paired ($S = 0$) or parallel ($S = 1$). Hence both ^{1}P and ^{3}P terms are possible.

Self–test 13.1 What terms arise from a $p^1 d^1$ configuration?

13.2 Terms of a d^2 configuration

The Pauli principle restricts the microstates that can occur in a configuration, and consequently it affects the terms that can occur. For example, two electrons cannot both have the same spin and be in a d orbital with $m_l = +2$. Therefore, the microstate $(2^+, 2^+)$ is forbidden and so are the values of L and S to which such a microstate might contribute. We shall illustrate how to determine what terms are allowed by considering a d^2 configuration, for the outcome will be useful in the discussion of the complexes encountered later in the chapter. An example of a species with a d^2 configuration is a Ti^{2+} ion.

(a) The classification of microstates

We start the analysis by setting up a table of microstates of the d^2 configuration (Table 13.1); only the microstates allowed by the Pauli principle have been included. Next, we note the largest value of M_L, which for a d^2 configuration is $+4$. This state must belong to a term with $L = 4$ (a G term). Table 13.1 shows that the only value of M_S that occurs for this term is $M_S = 0$, so the G term must be a singlet. Moreover, since there are nine values of M_L when $L = 4$, one of the microstates in each of the boxes in the column below $(2^+, 2^-)$ must belong to this term.[3] We can therefore strike out one microstate from each row in the central column of Table 13.1, which leaves 36 microstates to classify.

The next largest value is $M_L = +3$, which must stem from $L = 3$ and hence belong to an F term. That row contains one microstate in each column (that is, each box contains one unassigned combination for $M_S = +1, 0$, and -1), which signifies a triplet term. Hence the microstates belong to a ^{3}F term. The same is true for one microstate in each of the rows down to $M_L = -3$, which accounts for a further $3 \times 7 = 21$ microstates. If we strike out one state in each of the 21 boxes, we are left with 15 to be assigned.

There is one unassigned microstate in the row with $M_L = +2$ ($L = 2$) and the column under $M_S = 0$ ($S = 0$), which must therefore belong to a ^{1}D term. This term has five values of M_L, which removes one microstate from each row (in the column headed $M_S = 0$) down to

---

[3]   In fact, it is unlikely that one of the microstates itself will correspond to one of these states: in general, a state is a linear combination of microstates. However, as N linear combinations can be formed from N microstates, each time we cross off one microstate, we are taking one linear combination into account, so the bookkeeping is correct even though the detail may be wrong.

Table 13.1 Microstates of the d^2 configuration

M_L	-1	M_S 0	$+1$
$+4$		$(2^+, 2^-)$	
$+3$	$(2^-, 1^-)$	$(2^+, 1^-)(2^-, 1^+)$	$(2^+, 1^+)$
$+2$	$(2^-, 0^-)$	$(2^+, 0^-)(2^-, 0^+)$	$(2^+, 0^+)$
		$(1^+, 1^-)$	
$+1$	$(2^-, -1^-)(1^-, 0^-)$	$(2^+, -1^-)(2^-, -1^+)$	$(2^+, -1^+)(1^+, 0^+)$
		$(1^+, 0^-)(1^-, 0^+)$	
0	$(1^-, -1^-)(2^-, -2^-)$	$(1^+, -1^-)(1^-, -1^+)$	$(1^+, -1^+)(2^+, -2^+)$
		$(2^+, -2^-)(2^-, -2^+)$	
		$(0^+, 0^-)$	
-1 to -4*			

*The lower half of the diagram is a reflection of the upper half.

$M_L = -2$, leaving 10 microstates unassigned. Because these unassigned microstates include one with $M_L = +1$ and $M_S = +1$, nine of these microstates must belong to a ^3P term. There now remains only one microstate in the central box of the table, with $M_L = 0$ and $M_S = 0$. This microstate (or, in fact, a linear combination of all the microstates in that box; see footnote 3) must be the one and only state of a ^1S term (which has $L = 0$ and $S = 0$).

At this point we can conclude that the terms of a $3d^2$ configuration are ^1G, ^3F, ^1D, ^3P, and ^1S. These terms account for all 45 permitted states (see table in the margin).

Term	Number of states
^1G	$9 \times 1 = 9$
^3F	$7 \times 3 = 21$
^1D	$5 \times 1 = 5$
^3P	$3 \times 3 = 9$
^1S	$1 \times 1 = 1$
	Total: 45

> The allowed terms of a configuration can be found by identifying the values of L and S to which the microstates of an atom can contribute. For a d^2 configuration the 45 microstates are distributed over five terms, two being triplets and three singlets.

(b) The energies of the terms

Once the values of L and S that can arise from a given configuration are known, it is possible to identify the term of lowest energy by using **Hund's rules**. The first of these empirical rules was introduced in Section 1.7. There it was expressed as 'the lowest energy configuration is achieved if the electrons are parallel'. Because a high value of S stems from parallel electron spins, an alternative statement is

1 For a given configuration, the term with the greatest multiplicity lies lowest in energy.

The rule implies that a triplet term of a configuration (if one is permitted) has a lower energy than a singlet term of the same configuration. For the d^2 configuration, this rule predicts that the ground state will be either ^3F or ^3P.

By inspecting spectroscopic data, Hund also identified a second rule for the relative energies of the terms of a given multiplicity:

2 For a term of given multiplicity, the greater the value of L, the lower the energy.

The physical justification for this rule is that (in a classical picture of the atom), when L is high, the electrons are orbiting in the same direction like cars going round a traffic circle. As a result, the electrons can stay clear of one another and hence experience a lower repulsion. If L is low, the electrons are orbiting in different directions, meet more often, and hence repel one another more strongly. The second rule implies that, of the two triplet terms of a d^2 configuration, the ^3F term is lower in energy than the ^3P term. It follows that the ground term of a d^2 species such as Ti^{2+} is expected to be ^3F.

The spin multiplicity rule is fairly reliable for predicting the ordering of terms, but the 'greatest L' rule is reliable only for predicting the ground term, the term of lowest energy; there is generally little correlation of L with the order of the higher terms. Thus, for d^2 the rules predict the order

$$^3F < {}^3P < {}^1G < {}^1D < {}^1S$$

but the order observed for Ti^{2+} from spectroscopy is

$$^3F < {}^1D < {}^3P < {}^1G < {}^1S$$

All the previous work can be shortened considerably if, as is normally the case, all we want to know is the identity of the ground term of an atom or ion. The procedure may then be summarized as follows:

1 Identify the microstate that has the highest value of M_S.

This step tells us the highest multiplicity of the configuration.

2 Identify the highest permitted value of M_L for that multiplicity.

This step tells us the highest value of L consistent with the highest multiplicity.

> *Hund's rules indicate the likely ground term of a gas-phase atom or ion.*

Example 13.2 Identifying the ground term of a configuration

What is the ground term for the configurations (a) $3d^5$ of Mn^{2+} and (b) $3d^3$ of Cr^{3+}?

Answer (a) The d^5 configuration permits occupation of each d orbital singly, so the maximum value of S is $\frac{5}{2}$ and the maximum multiplicity is $2 \times \frac{5}{2} + 1 = 6$, a sextet term. If each of the electrons is to have the same spin quantum number, all must occupy different orbitals and hence have different m_l values. Thus, the m_l values of the occupied orbitals will be $+2, +1, 0, -1,$ and -2. The sum of these values is 0, so $L = 0$ and the term is ^6S. (b) For the configuration d^3, the maximum multiplicity corresponds to all three electrons having the same spin quantum number, so $S = \frac{3}{2}$. The multiplicity is therefore $2 \times \frac{3}{2} + 1 = 4$, a quartet. The three m_l values must be different if all spins are the same, which allows a maximum value of $M_L = 2 + 1 + 0 = +3$, indicating $L = 3$, an F term. Hence, the ground term of d^3 is ^4F.

Self-test 13.2 Identify the ground terms of (a) $2p^2$ and (b) $3d^9$. (*Hint:* Because d^9 is one electron short of a closed shell with $L = 0$ and $S = 0$, treat it on the same footing as a d^1 configuration.)

(c) Racah parameters

Different terms of a configuration have different energies on account of the repulsion between electrons. If we wanted to calculate the energies of the terms we would have to evaluate these electron–electron repulsion energies as complicated integrals over the orbitals occupied by the electrons. Mercifully, though, all the integrals for a given configuration can be collected together in three specific combinations, and the repulsion energy of any term of a configuration can be expressed as a sum of these three quantities. The three combinations of integrals are called the **Racah parameters** and denoted A, B, and C. We do not even need to know the theoretical values of the parameters or the theoretical expressions for them, because it is more reliable to use A, B, and C as empirical quantities obtained from gas-phase atomic spectroscopy.

Each term stemming from a given configuration has an energy that may be expressed as a linear combination of the three Racah parameters. For a d^2 configuration a detailed analysis shows that

$$E(^1S) = A + 14B + 7C$$
$$E(^1G) = A + 4B + 2C$$
$$E(^1D) = A - 3B + 2C \qquad\qquad (2)$$
$$E(^3P) = A + 7B$$
$$E(^3F) = A - 8B$$

The values of A, B, and C can be determined by fitting these expressions to the observed energies of the terms. Note that A is common to all the terms; therefore, if we are interested only in their relative energies, we do not need to know the value of A. Likewise, if we are interested only in the relative energies of the two triplet terms, we also do not need to know the value of C.

All three Racah parameters are positive (they represent electron–electron repulsions). Therefore, provided $C > 5B$, the energies of the terms of the d^2 configuration lie in the order

$$^3F < {}^3P < {}^1D < {}^1G < {}^1S$$

This order is nearly the same as obtained by using Hund's rules. However, if $C < 5B$, the advantage of having an occupation of orbitals that corresponds to a high orbital angular momentum is greater than the advantage of having a high multiplicity, and the ^3P term lies above ^1D (as is in fact the case for Ti^{2+}). Some experimental values of B and C are given in Table 13.2. The values in parenthesis indicate that $C \approx 4B$, so the ions listed there are in the region where Hund's rules are not reliable for predicting anything more than the ground term of a configuration.

> The Racah parameters summarize the effects of electron–electron repulsion on the energies of the terms that arise from a single configuration. The parameters are the quantitative expression of the ideas underlying Hund's rules, and account for deviations from them.

Table 13.2 Racah parameters for some d-block ions*

	1+	2+	3+	4+
Ti		720(3.7)		
V		765(3.9)	860(4.8)	
Cr		830(4.1)	1030(3.7)	1040(4.1)
Mn		960(3.5)	1130(3.2)	
Fe		1060(4.1)		
Co		1120(3.9)		
Ni		1080(4.5)		
Cu	1220(4.0)	1240(3.8)		

*The table gives the B parameter with the value of C/B in parentheses; B is in cm^{-1}.

The electronic spectra of complexes

We now return to the spectrum of $[Cr(NH_3)_6]^{3+}$ in Fig. 13.1 and put the preceding discussion to work. The two central bands with intermediate intensities are HOMO–LUMO transitions with energies that differ on account of the electron–electron repulsions (as we shortly explain). Because both the HOMO and LUMO of an octahedral complex are predominantly metal d orbital in character, with a separation characterized by the strength of the ligand-field splitting parameter Δ_O, these two transitions are called **d–d transitions** or **ligand-field transitions**.

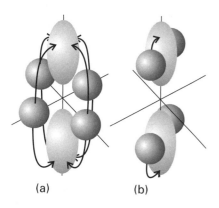

(a)   (b)

13.2 The shifts in electron density that accompany the two transitions discussed in the text. There is a considerable relocation of electron density toward the ligands on the z-axis in (a), but a much less substantial relocation in (b).

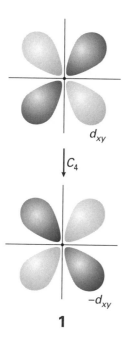

d_{xy}

$\downarrow C_4$

$-d_{xy}$

1

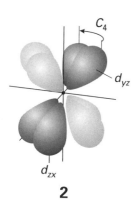

C_4

d_{yz}

d_{zx}

2

13.3 Ligand-field transitions

According to the discussion in Chapter 7, we would expect the octahedral d^3 complex $[\text{Cr}(\text{NH}_3)_6]^{3+}$ to have the ground configuration t_{2g}^3. The absorption in the region near $25\,000$ cm^{-1} can be identified as arising from the excitation $t_{2g}^2 e_g^1 \leftarrow t_{2g}^3$ because that wavenumber is typical of ligand-field splittings in complexes.

Before we embark on a Racah-like analysis of the transition, it may be helpful to see qualitatively from the viewpoint of molecular orbital theory why the transition gives rise to two bands. First, note that a $d_{z^2} \leftarrow d_{xy}$ transition, which is one way of achieving $e_g \leftarrow t_{2g}$, promotes an electron from the xy-plane into the already electron-rich z direction: that axis is electron-rich because both d_{yz} and d_{zx} are occupied (Fig. 13.2). However, a $d_{z^2} \leftarrow d_{zx}$ transition, which is another way of achieving $e_g \leftarrow t_{2g}$, merely relocates an electron that is already largely concentrated along the z-axis. In the former case, but not in the latter, there is a distinct increase in electron repulsion and, as a result, the two $e_g \leftarrow t_{2g}$ transitions lie at different energies. All six $e_g \leftarrow t_{2g}$ transitions resemble one or other of these two cases: three of them fall into one group and the other three fall into the second group.

> The presence of electron–electron repulsion splits ligand-field transitions into components with different energies.

(a) The spectroscopic terms

The two bands we are discussing in Fig. 13.1 are labeled $^4T_{2g} \leftarrow {}^4A_{2g}$ (at $21\,550$ cm^{-1}) and $^4T_{1g} \leftarrow {}^4A_{2g}$ (at $28\,500$ cm^{-1}). The labels are **molecular term symbols** and serve a purpose similar to that of the atomic term symbols we have already encountered. The left superscript denotes the multiplicity, so the superscript 4 denotes a quartet state with $S = \frac{3}{2}$, as expected when there are three unpaired electrons. However, L is not a good quantum number in an octahedral environment because orbital angular momentum is not well defined unless a system has spherical symmetry. Therefore, we need a different notation.

The letters in the term symbols are the symmetry species of the overall electronic orbital state of the complex. For example, the nearly totally symmetric ground state of a d^3 complex (with an electron in each of the three t_{2g} orbitals) is denoted A_{2g}. We say *nearly* totally symmetric, because close inspection of the behavior of the three occupied t_{2g} orbitals shows that the C_3 rotation of the O_h point group transforms the product $t_{2g} \times t_{2g} \times t_{2g}$ into itself, which identifies the complex as an A symmetry species (see the character table in Appendix 3). Moreover, because each orbital has even parity (g), the overall parity is also g. However, each C_4 rotation transforms one t_{2g} orbital into the negative of itself (**1**) and the other two t_{2g} orbitals into each other (**2**), so overall there is a change of sign under this operation and its character is -1. Thus the term is A_{2g} rather than the totally symmetrical A_{1g} of a closed shell.

It is more difficult to establish that the term symbols that can arise from the quartet $t_{2g}^2 e_g^1$ excited configuration are $^4T_{2g}$ and $^4T_{1g}$ and we shall not consider this aspect here. The superscript 4 implies that the upper configuration continues to have the same number of unpaired spins as in the ground state, and the subscript g stems from the even parity of all the contributing orbitals.

> The terms of an octahedral metal complex are labeled by the symmetry species of the overall orbital state; a superscript prefix shows the multiplicity of the term.

(b) The energies of the terms

In a free atom, where all the d orbitals are degenerate, we needed to consider only the electron–electron repulsions in order to arrive at the relative ordering of the terms of a given d^n configuration. In an octahedral complex, where the d orbitals are not all degenerate, it is necessary to take into account the difference in energy between the t_{2g} and e_g orbitals as well as the electron–electron repulsions.

We can illustrate the considerations involved by treating first the simplest case of an atom or ion with a single valence electron. Because a totally symmetrical orbital in one environment becomes a totally symmetrical orbital in another environment, an s orbital in a free atom becomes an a_{1g} orbital in an octahedral field. We express the change by saying that the s orbital of the atom 'correlates' with the a_{1g} orbital of the complex. Similarly, the five d orbitals of a free atom correlate with the triply degenerate t_{2g} and doubly degenerate e_g sets in an octahedral complex.

Now consider a many-electron atom. In exactly the same way as for a single electron, the totally symmetrical overall S term of a many-electron atom correlates with the totally symmetrical A_{1g} term of an octahedral complex. Likewise, an atomic D term splits into a T_{2g} term and an E_g term in O_h symmetry. The same kind of analysis can be applied to other states, and Table 13.3 summarizes the correlations between free atom terms and terms in an octahedral complex.

> The free atom terms split in the ligand field of an octahedral complex and are then labeled by their symmetry species as enumerated in Table 13.3.

Table 13.3 The correlation of spectroscopic terms for d electrons in O_h complexes

Atomic term	Number of states	Terms in O_h symmetry
S	1	A_{1g}
P	3	T_{1g}
D	5	$T_{2g} + E_g$
F	7	$T_{1g} + T_{2g} + A_{2g}$
G	9	$A_{1g} + E_g + T_{1g} + T_{2g}$

Example 13.3 Identifying correlations between terms

What terms in a complex with O_h symmetry correlate with the ^3P term of a free ion with a d^2 configuration?

Answer The three p orbitals of a free atom become the triply degenerate t_{1u} orbitals of an octahedral complex. Therefore, if we disregard parity for the moment, a P term of a many-electron atom becomes a T_1 term in the point group O_h. Because d orbitals have even parity, the term overall must be g. The multiplicity is unchanged in the correlation, so the ^3P term becomes the $^3T_{1g}$ term.

Self-test 13.3 What terms in a d^2 complex of O_h symmetry correlate with the ^3F and ^1D terms of a free ion?

(c) Weak and strong field limits

Electron–electron repulsions are difficult to take into account, but the discussion is simplified by considering two extreme cases. In the **weak field limit** the ligand field, as measured by Δ_O, is so weak that only electron–electron repulsions are important and the relative energies of the terms can be expressed as combinations of Racah parameters B and C. In the **strong field limit** the ligand field is so strong that electron–electron repulsions can be ignored and the energies of the terms can be expressed solely in terms of Δ_O. Then, with the two extremes established, we can consider intermediate cases by drawing a correlation diagram between the two. We shall illustrate what is involved by considering two simple cases, namely, d^1 and d^2. Then we show how the same ideas are used to treat more complicated cases.

The only term arising from the d^1 configuration of a free atom is ^2D. In an octahedral complex the configuration is either t^1_{2g}, which gives rise to a $^2T_{2g}$ term, or e^1_g, which gives rise to a ^2E term. Because there are no electron–electron repulsions to worry about, the separation of the $^2T_{2g}$ and ^2E_g terms is the same as the separation of the t_{2g} and e_g orbitals, or Δ_O. The correlation diagram for the d^1 configuration will therefore resemble that shown in Fig. 13.3.

For a d^2 configuration, the triplet terms are ^3F and ^3P in the free atom, as we have seen, and their energies relative to the lower term (^3F) are $E(^3F) = 0$ and $E(^3P) = 15B$. These two energies are marked on the left of Fig. 13.4. Now consider the very strong field limit. A d^2 ion has the configurations:

$$t^2_{2g} < t^1_{2g}e^1_g < e^2_g$$

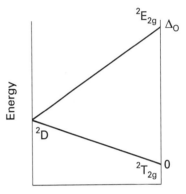

13.3 Correlation diagram for a free ion (left) and the strong-field terms (right) of a d^1 configuration.

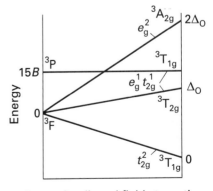

13.4 Correlation diagram for a free ion (left) and the strong-field terms (right) of a d^2 configuration.

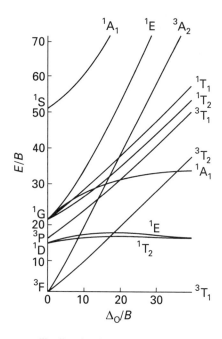

13.5 The Tanabe–Sugano diagram for the d^2 configuration. A complete collection of diagrams for d^n configurations is given in Appendix 5 at the back of the book. The parity subscript g has been omitted from the term symbols for clarity.

From the information in Fig. 7.9, we can write their energies as

$$E(t_{2g}^2, T_{1g}) = 2(-\tfrac{2}{5}\Delta_O) = -0.8\Delta_O$$

$$E(t_{2g}^1 e_g^1, T_{2g}) = (-\tfrac{2}{5} + \tfrac{3}{5})\Delta_O = +0.2\Delta_O$$

$$E(e_g^2, A_{2g}) = 2(\tfrac{3}{5})\Delta_O = +1.2\Delta_O$$

Therefore, relative to the energy of the lowest term, their energies are

$$E(t_{2g}^2, T_{1g}) = 0 \qquad E(t_{2g}^1 e_g^1, T_{2g}) = +\Delta_O \qquad E(e_g^2, A_{2g}) = +2\Delta_O$$

These energies are marked on the right in Fig. 13.4.

Our problem now is to account for the energies of the more realistic cases, where neither the ligand-field nor the electron repulsion terms are dominant. We do this by correlating the terms between the two extreme cases by identifying the term that each free atom term becomes in the very high field limit. Thus, the triplet t_{2g}^2 configuration gives rise to a $^3T_{1g}$ term, and so it correlates with the ^3F term of the free ion. The remaining correlations can be established similarly, and are shown in Fig. 13.4, which is a simplified version of an **Orgel diagram**.

> For a given metal ion, the energies of the individual terms respond differently to ligands of increasing field strength and the correlation between free atom terms and terms of a complex can be displayed on an Orgel diagram.

(d) Tanabe–Sugano diagrams

Diagrams showing the correlation of terms can be constructed for any symmetry and strength of ligand field. The most widely used versions are called **Tanabe–Sugano diagrams**, after the Japanese scientists who devised them, and an example for d^2 is given in Fig. 13.5. In these diagrams the term energies, E, are expressed as E/B and plotted against Δ_O/B, where B is a Racah parameter. Because $C \approx 4B$, terms with energies that depend on both B and C can be plotted on the same diagrams. Some lines in Tanabe–Sugano diagrams are curved on account of the mixing of terms of the same symmetry type. Terms of the same symmetry obey the **noncrossing rule**, which states that, if the increasing ligand field causes two weak field terms of the same symmetry to approach, they do not cross but bend apart from each other (Fig. 13.6). The effect of the noncrossing rule can be seen for the two ^1E terms, the two $^1T_2$ terms, and the two $^1A_1$ terms in Fig. 13.5.

Tanabe–Sugano diagrams for O_h complexes with configurations d^2 through d^8 are given in Appendix 5. The zero of energy in a Tanabe–Sugano diagram is always taken as that of the lowest term. Hence the lines in the diagrams have abrupt changes of slope (see that for d^4, for instance) when there is a change in the identity of the ground term.

> Tanabe–Sugano diagrams are correlation diagrams that depict the energies of electronic states of complexes as a function of the strength of the ligand field.

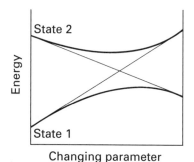

13.6 The noncrossing rule states that, if two states of the same symmetry are likely to cross as a parameter is changed (as shown by the thin lines), they will in fact mix together and avoid the crossing (as shown by the heavy lines).

Example 13.4 Calculating Δ_O and B using a Tanabe–Sugano diagram

Deduce the values of Δ_O and B for $[Cr(NH_3)_6]^{3+}$ from the spectrum in Fig. 13.1 and a Tanabe–Sugano diagram.

Answer The relevant diagram (for d^3) is shown in Fig. 13.7. We have seen that Fig. 13.1 contains two low-energy ligand-field transitions at 21 550 cm^{-1} ($^4T_{2g} \leftarrow {}^4A_{2g}$) and 28 500 cm^{-1} ($^4T_{1g} \leftarrow {}^4A_{2g}$), a ratio of 1.32. The only point in Fig. 13.7 where this energy ratio is satisfied is on the far right. Hence, we can read off the value of $\Delta_O/B = 32.8$ from the location of this point. The tip of the arrow representing the lower energy transition lies at 32.8B vertically, so equating 32.8B and 21 550 cm^{-1} gives $B = 657$ cm^{-1} and $\Delta_O = 21\,550$ cm^{-1}.

Self-test 13.4 Use the same Tanabe–Sugano diagram to predict the wavenumbers of the first two spin-allowed quartet bands in the spectrum of $[Cr(OH_2)_6]^{3+}$ for which $\Delta_O = 17\,600\text{ cm}^{-1}$ and $B = 700\text{ cm}^{-1}$.

(e) The nephelauxetic series

In Example 13.4 we found that $B = 657\text{ cm}^{-1}$ for $[Cr(NH_3)_6]^{3+}$. This value is only 64 per cent of the value for a free Cr^{3+} ion in the gas phase, which indicates that electron repulsions are weaker in the complex than in the free ion. This weakening occurs because the occupied molecular orbitals are delocalized over the ligands and away from the metal. The delocalization increases the average separation of the electrons and hence reduces their mutual repulsion.

The reduction of B from its free ion value is normally reported in terms of the **nephelauxetic parameter**,[4] β:

$$\beta = \frac{B(\text{complex})}{B(\text{free ion})} \tag{3}$$

The values of β depend on the identity of the metal ion and the ligand. They vary along the **nephelauxetic series**:

$$Br^- < Cl^-, CN^- < NH_3 < H_2O < F^-$$

A small value of β indicates a large measure of d-electron delocalization on to the ligands and hence a significant covalent character in the complex. Thus the series shows that a Br^- ligand results in a greater reduction in electron repulsions in the ion than an F^- ion, which is consistent with a greater covalent character in bromo complexes than in analogous fluoro complexes. Another way of expressing the trend represented by the series is that, the softer the ligand, the smaller the nephelauxetic parameter.

The nephelauxetic character of a ligand is different for electrons in t_{2g} and e_g orbitals. Because the σ overlap of e_g is usually larger than the π overlap of t_{2g}, the cloud expansion is larger in the former case. The measured nephelauxetic parameter of an $e \leftarrow t$ transition is an average of the effects on both types of orbital.

> The nephelauxetic parameter is a measure of the extent of d-electron delocalization on to the ligands of a complex; the softer the ligand, the smaller the nephelauxetic parameter.

13.4 Charge-transfer bands

Another feature in the spectrum of $[Cr(NH_3)_6]^{3+}$ that remains to be explained is the very intense shoulder of an absorption that appears to have a maximum well above $50\,000\text{ cm}^{-1}$. The apparently high intensity suggests that this is not a simple ligand-field d–d transition, but is consistent with a charge-transfer transition. In a **charge-transfer transition** (a CT transition) an electron is moved between orbitals that are predominantly ligand in character and orbitals that are predominantly metal in character. The transition is classified as a **ligand-to-metal charge-transfer transition** (LMCT transition) if the migration of the electron is from the ligand to the metal. In some complexes the charge migration occurs in the opposite direction, in which case it is classified as a **metal-to-ligand charge-transfer transition** (MLCT transition). An example of an MLCT transition is the one responsible for the red color of tris(bipyridyl)iron(II), the complex used for the colorimetric analysis of Fe(II). In this case, an electron makes a transition from a d orbital of the central metal into a π^* orbital of the ligand. Figure 13.8 summarizes the transitions we classify as charge transfer.

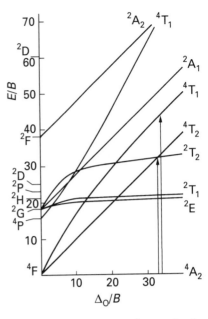

13.7 The Tanabe–Sugano diagram for the d^3 configuration. The parity subscript g has been omitted from the term symbols for clarity.

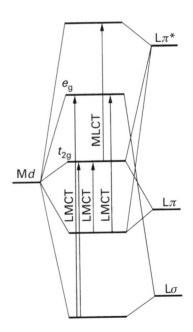

13.8 A summary of the charge-transfer transitions in an octahedral complex.

---

[4] The name is from the Greek words for 'cloud expanding'.

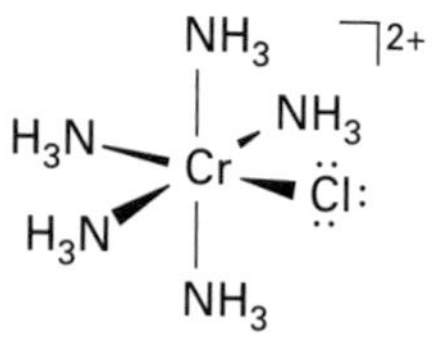

3 $[CrCl(NH_3)_5]^{2+}$

Several lines of evidence are used to identify a band as due to a CT transition. The high intensity of the band, which is evident in Fig. 13.9, is one strong indication of a CT transition, particularly if such a band occurs when a new complex is formed with a particular ligand. The CT character is most often identified (and distinguished from $\pi^* \leftarrow \pi$ transitions on ligands) by demonstrating **solvatochromism**, the variation of the transition frequency with changes in solvent permittivity. Solvatochromism indicates that there is a large shift in electron density as a result of the transition.[5]

Figure 13.9 shows the visible and UV spectrum of the less symmetric complex $[CrCl(NH_3)_5]^{2+}$ (**3**). This spectrum resembles that in Fig. 13.1, and we can recognize the two ligand-field, d–d, bands in the visible region. The replacement of one NH_3 ligand by a weaker field Cl^- ligand moves the lowest energy d–d bands to lower energy than those of $[Cr(NH_3)_6]^{3+}$ in Fig. 13.1. Also, a shoulder appears on the high-energy side of one of the d–d bands as a result of the reduction in symmetry from O_h to C_{4v}. Another new feature in the spectrum is the strong absorption maximum in the ultraviolet, near $42\,000$ cm^{-1}. This band is at lower energy than the corresponding band in the spectrum of $[Cr(NH_3)_6]^{3+}$ and has been assigned as an LMCT transition from the Cl^- ligand to the metal. The LMCT character of similar bands in $[CoX(NH_3)_6]^{2+}$ is confirmed by the decrease in wavenumber in steps of about 8000 cm^{-1} as X is varied from Cl to Br to I. In this LMCT transition a lone-pair electron of the halide ligand is promoted into a predominantly metal orbital.

> *Charge-transfer transitions are identified by their high intensity and the sensitivity of their energies to solvent polarity.*

(a) LMCT transitions

As in Fig. 13.9, CT transitions are generally intense compared with ligand-field transitions. Charge-transfer bands in the visible region of the spectrum (and hence contributing to the intense colors of complexes) may occur if the ligands have lone pairs of relatively high energy (as in sulfur and selenium) or if the metal has low-lying empty orbitals. The color of the artists' pigment 'cadmium yellow', CdS, for instance, is due to the transition $Cd^{2+}(5s) \leftarrow S^{2-}(\pi)$. Similarly, HgS is red as a result of the transition $Hg^{2+}(6s) \leftarrow S^{2-}(\pi)$ and ochers are iron oxides that are colored red and yellow by the transition $Fe(3d) \leftarrow O^{2-}(\pi)$.

The tetraoxoanions of metals with high oxidation numbers (such as MnO_4^-) provide what are probably the most familiar examples of LMCT bands. In them, an O lone-pair electron is

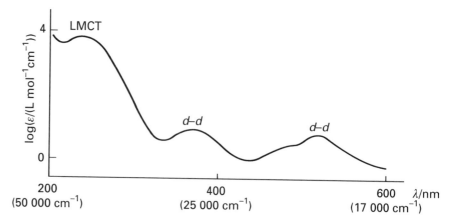

13.9 The absorption spectrum of $[CrCl(NH_3)_5]^{2+}$ in water in the visible and ultraviolet regions. The peak corresponding to the transition $^2E \leftarrow {}^4A$ is not visible at this magnification.

5   A useful discussion of solvatochromism is provided in A.B.P. Lever, *Inorganic electronic spectroscopy*. Elsevier, Amsterdam (1984).

promoted into a low-lying empty e metal orbital. High metal oxidation states correspond to a low d-orbital population (many are d^0), so the acceptor level is available and low in energy. The trend in LMCT energies is:

Oxidation number	
+7	$MnO_4^- < TcO_4^- < ReO_4^-$
+6	$CrO_4^{2-} < MoO_4^{2-} < WO_4^{2-}$
+5	$VO_4^{3-} < NbO_4^{3-} < TaO_4^{3-}$

The energies of the transitions correlate with the order of the electrochemical series, with the lowest energy transitions taking place to the most easily reduced metal ions. This correlation is consistent with the transition being the transfer of an electron from the ligands to the metal, corresponding, in effect, to the reduction of the metal ion by the ligands.

Polymeric and monomeric oxoanions follow the same trends, with the oxidation number of the metal the determining factor.[6] The similarity suggests that these LMCT transitions are localized processes that take place on discrete molecular fragments.

The variation in the position of LMCT bands can be expressed in terms of the **optical electronegativities** of the metal, χ_{metal}, and the ligands, χ_{ligand}. The wavenumber of the transition is then written as the difference between the two electronegativities:

$$\tilde{\nu} = |\chi_{ligand} - \chi_{metal}|\tilde{\nu}_0 \tag{4}$$

Optical electronegativities have values comparable to Pauling electronegativities if we take the constant $\tilde{\nu}_0$ to be $3.0 \times 10^4 \text{ cm}^{-1}$ (Table 13.4). If the LMCT transition terminates in an e_g orbital, Δ_O must be added to the energy predicted by this equation. Electron pairing energies must also be taken into account if the transition results in the population of an orbital that already contains an electron. The values for metals are different in complexes of different symmetry, and the ligand values are different if the transition originates from a π orbital rather than a σ orbital.

> Ligand-to-metal transitions are observed when the metal is in a high oxidation state and ligands contain lone-pair electrons; the variation in the position of LMCT bands can be parametrized in terms of optical electronegativities.

Table 13.4 Optical electronegativities

Metal	O_h	T_d	Ligand	π	σ
Cr(III)	1.8–1.9		F^-	3.9	4.4
Co(III)*	2.3		Cl^-	3.0	3.4
Ni(II)		2.0–2.1	Br^-	2.8	3.3
Co(II)		1.8–1.9	I^-	2.5	3.0
Rh(III)*	2.3		H_2O	3.5	
Mo(V)	2.1		NH_3	3.3	

*Low-spin complexes.

(b) MLCT transitions

Charge-transfer transitions from metal to ligand are most commonly observed in complexes with ligands that have low-lying π^* orbitals, especially aromatic ligands. The transition will occur at low energy if the metal ion has a low oxidation number, for its d orbitals will then be relatively high in energy.

The family of ligands most commonly involved in MLCT transitions are the diimines, which have two N donor atoms: two important examples are 2,2'-bipyridine (**4**) and 1,10-phenanthroline (**5**). Complexes of diimines with strong MLCT bands include tris(diimine)

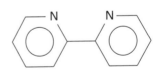

4 2,2'–Bipyridine (bipy)

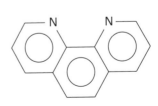

5 1,10–Phenanthroline (phen)

[6] E.S. Dodsworth and A.B.P. Lever, Solvatochromism of dinuclear complexes. *Inorg. Chem.* **29**, 499 (1990).

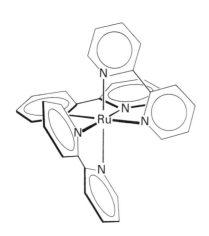

6 [Ru(bipy)₃]²⁺

7 Dithiolene

species such as tris(2,2′-bipyridyl)ruthenium(II) (**6**), which is orange on account of its MLCT band. A diimine ligand may also be easily substituted into a complex with other ligands that favor a low oxidation state. Two examples are $[W(CO)_4(phen)]$ and $[Fe(CO)_3(bipy)]$. However, the occurrence of MLCT transitions is by no means limited to diimine ligands. Another important ligand type that shows typical MLCT transitions is dithiolene, $S_2C_2R_2$ (**7**). Resonance Raman spectroscopy (Box 13.1) is a powerful technique for the study of MLCT transitions, particularly of diimine complexes.

The MLCT excitation of tris(2,2′-bipyridyl)ruthenium(II) has been the subject of intense research efforts because the excited state that results from the charge transfer has a lifetime of microseconds, and the complex is a versatile photochemical redox reagent (Section 14.15). The photochemical behavior of a number of related complexes has also been studied on account of their relatively long excited-state lifetimes.

> *Charge-transfer transitions from metal to ligand are most commonly observed in complexes with ligands that have low-lying π^* orbitals, especially aromatic ligands.*

13.5 Selection rules and intensities

The contrast in intensity between typical charge-transfer bands and typical ligand-field bands raises the question of the factors that control the intensities of absorption bands. In an octahedral, nearly octahedral, or square-planar complex, the maximum molar absorption coefficient ε_{max} (which measures the strength of the absorption)[7] is typically less than or close to $100 \, L \, mol^{-1} \, cm^{-1}$ for ligand-field transitions. In tetrahedral complexes, which have no center of symmetry, ε_{max} might exceed $250 \, L \, mol^{-1} \, cm^{-1}$. In many cases, charge-transfer bands have an ε_{max} of between 1000 and $50\,000 \, L \, mol^{-1} \, cm^{-1}$.

To understand the intensities of transitions in complexes we have to explore the strength with which the complex couples with the electromagnetic field. Intense transitions indicate strong coupling; weak transitions indicate feeble coupling. The strength of coupling when an electron makes a transition from a state with wavefunction ψ_i to one with wavefunction ψ_f is measured by the **transition dipole moment**, which is defined as the integral

$$\boldsymbol{\mu}_{fi} = \int \psi_f^* \boldsymbol{\mu} \psi_i \, d\tau \tag{5}$$

where $\boldsymbol{\mu}$ is the electric dipole moment operator $-e\boldsymbol{r}$. The transition dipole moment can be regarded as a measure of the impulse that a transition imparts to the electromagnetic field: a large impulse corresponds to an intense transition; zero impulse corresponds to a forbidden transition. The intensity of the transition is proportional to the square of the transition dipole moment.

A spectroscopic **selection rule** is a statement about which transitions are allowed and which are forbidden. An **allowed transition** is a transition with a nonzero transition dipole moment, and hence nonzero intensity. A **forbidden transition** is a transition for which the transition dipole moment is calculated as zero. Forbidden transitions may occur in a spectrum if the assumptions on which the transition dipole moment were calculated are invalid, such as the complex having a lower symmetry than assumed.

(a) Spin selection rules

The electromagnetic field of the incident radiation cannot change the relative orientations of the spins of the electrons in a complex. For example, an initially antiparallel pair of electrons cannot be converted to a parallel pair, so a singlet ($S = 0$) cannot undergo a transition to a

---

[7] The molar absorption coefficient is the constant in the Beer–Lambert law for the transmittance $\tau = I_f/I_i$ when light passes through a length l of solution of molar concentration $[X]$ and is attenuated from an intensity I_i to an intensity I_f: $\log \tau = -\varepsilon l[X]$. Its earlier name is the 'extinction coefficient'.

Box 13.1 Resonance Raman spectroscopy

Resonance Raman spectroscopy is a powerful technique for the assignment and characterization of charge-transfer bands. The conventional Raman technique described in Box 4.1 involves the inelastic scattering of radiation by a molecule. The frequency of the scattered radiation differs from that of the incident radiation by an amount corresponding to a vibrational excitation energy of the molecule, so the observation of the frequency shift can be interpreted as a vibrational spectrum. Conventional Raman spectroscopy depends upon the variation of the polarizability of the molecule as it vibrates, and the incident radiation is typically not at an absorption frequency of the molecule. Resonance Raman spectroscopy uses incident radiation that has a frequency within an electronic absorption band of the molecule. Under such circumstances, there is an enhancement of the Raman effect for the vibration modes that are affected by the electronic excitation.

Resonance Raman bands are observed when the electronic transition results in a change in the electron distribution that causes a shift in nuclear positions (bond lengths, bond angles). Thus, the vibrations that give intense resonance Raman bands at a particular incident frequency are vibrations that occur in the part of the molecule that is strongly affected by the electronic transition. As such, resonance Raman enhancements aid in the identification of the electronic transition.

Figure B13.1 shows the resonance Raman spectrum of $[W(CO)_4(phen)]$.[†] The incident 488.0 nm laser radiation is close to the maximum of the absorption band of the complex that has been assigned (on the basis of solvatochromism) as the $\pi^*(phen) \leftarrow d(W)$ MLCT transition. Notice that three stretching vibrations of the phen ligand give strong Raman lines. This observation provides clear confirmation of the assignment of the electronic transition.

A surprise in this spectrum is the strong resonance enhancement of the CO stretch (marked eq). The substantial intensity of these signals suggests that the π^* levels of both phen and the coplanar CO(eq) participate in the electronic transition.

The resonance enhancement of the phen ring vibrations decreases as the metal d- and phen ligand π^*-orbitals approach each other in energy along a series of analogous complexes. This decrease indicates a reduction in the extent of charge transfer in the transition as both the initial and final orbital become more mixed in character. There is a very good correlation between the extent of resonance Raman enhancement and solvatochromism, the simplest indicator of charge transfer.

....

[†] R.W. Balk, Th.L. Snoeck, D.J. Stufkens, and A. Oskam, *Inorg. Chem.* **19**, 3015 (1980).

B13.1 The resonance Raman spectrum of $[W(CO)_4(phen)]$ using 488.0 nm (blue) laser light. Only the symmetric stretch of the equatorial CO ligands (near 2000 cm^{-1}) and the phen modes (between 1400 cm^{-1} and 1600 cm^{-1}) are enhanced. That observation indicates that the electronic transition is confined to the equatorial plane.

triplet $(S = 1)$. This restriction is summarized by the rule $\Delta S = 0$ for **spin-allowed transitions**.

The coupling of spin and orbital angular momenta can relax the spin selection rule, but such **spin-forbidden**, $\Delta S \neq 0$, transitions are generally much weaker than spin-allowed transitions. The intensity of spin-forbidden bands increases as the atomic number increases because the strength of the spin–orbit coupling is greater for heavy atoms than for light atoms. The breakdown of the spin selection rule by spin–orbit coupling is often called the **heavy-atom effect**. In the first d series, in which spin–orbit coupling is weak, spin-forbidden bands have ε_{max} less than about $1\ \text{L mol}^{-1}\ \text{cm}^{-1}$; however, spin-forbidden bands are a significant feature in the spectra of heavy d-metal complexes.

> *Intensities of spin-forbidden transitions are greater for 4d or 5d metal complexes than for comparable 3d complexes.*

(b) The Laporte selection rule

The **Laporte selection rule** is a statement about the change in parity that accompanies a transition: *in a centrosymmetric molecule or ion, the only allowed transitions are those accompanied by a change in parity*. That is, transitions between g and u terms are permitted, but a g term cannot combine with another g term and a u term cannot combine with another u term:

$$\text{g} \longleftrightarrow \text{u} \qquad \text{g} \longleftrightarrow\!\!\!| \ \text{g} \qquad \text{u} \ |\!\!\!\longleftrightarrow \text{u}$$

The Laporte selection rule is based on the properties of the transition dipole moment. Because r changes sign under inversion (and is therefore u), the entire integral in eqn 5 will also change sign under inversion if ψ_i and ψ_f have the same parity, because

$$\text{g} \times \text{u} \times \text{g} = \text{u} \qquad \text{and} \qquad \text{u} \times \text{u} \times \text{u} = \text{u} \tag{6}$$

Therefore, because the value of an integral cannot depend on the choice of coordinates,[8] it vanishes if ψ_i and ψ_f have the same parity. However, if they have opposite parity, the integral does not change sign under inversion of the coordinates because

$$\text{g} \times \text{u} \times \text{u} = \text{g}$$

and therefore need not vanish.

In a centrosymmetric complex, ligand-field d–d transitions are g–g and are therefore forbidden. Their forbidden character accounts for the relative weakness of these transitions in octahedral complexes compared with those in tetrahedral complexes, on which the Laporte rule is silent (as they have no center of symmetry).

The question remains why d–d transitions occur at all, albeit weakly. The Laporte selection rule may be relaxed in two ways. First, a complex may depart slightly from perfect centrosymmetry, perhaps on account of the intrinsic asymmetry in the structure of polyatomic ligands or a distortion imposed by the environment of a complex packed into a crystal. Alternatively, the complex might undergo an asymmetrical vibration, which also destroys its center of inversion. In either case, a Laporte-forbidden d–d band tends to be much more intense than a spin-forbidden transition.

Table 13.5 summarizes typical intensities of electronic transitions of complexes of the first series of d elements. The width of spectroscopic absorption bands is due principally to the simultaneous excitation of vibration when the electron is promoted from one distribution to another. According to the **Franck–Condon principle**, the electronic transition takes place within a stationary nuclear framework. As a result, after the transition has occurred, the nuclei experience a new force field and the molecule bursts into vibration.

---

[8] An integral is an area, and areas are independent of the coordinates used for their evaluation.

Table 13.5 Intensities of spectroscopic bands in $3d$ complexes

Band type	$\varepsilon_{max}/(\text{L mol}^{-1}\,\text{cm}^{-1})$
Spin-forbidden	< 1
Laporte-forbidden d–d	20–100
Laporte-allowed d–d	c. 500
Symmetry-allowed (e.g. CT)	1000–50 000

> Electronic transitions must involve an electric dipole moment change to absorb light so transitions between d orbitals are forbidden; however, asymmetric vibrations relax this restriction.

Example 13.5 Assigning a spectrum using selection rules

Assign the bands in the spectrum in Fig. 13.9 by considering their intensities.

Answer Like the hexaammine spectrum in Fig. 13.1, which has d–d bands at similar wavenumbers, the complex is moderately strong field ($\Delta_O/B > 20$). If we assume that the complex is approximately octahedral, examination of the Tanabe–Sugano diagram again reveals that the ground term is $^4A_{2g}$. The next higher term is ^2E_g. The next two are $^2T_{1g}$ and $^2T_{2g}$. Transitions to these three will not have ε_{max} greater than 1 $\text{L mol}^{-1}\,\text{cm}^{-1}$ because they are spin-forbidden. Very weak bands at low energy are predicted. They may be too high in energy to be resolved from the more intense bands. The next two higher terms are $^4T_{2g}$ and $^4T_{1g}$. These are reached by spin-allowed but Laporte-forbidden d–d transitions, and have ε_{max} near 100 $\text{L mol}^{-1}\,\text{cm}^{-1}$. In the near UV, the band with ε_{max} near 10 000 $\text{L mol}^{-1}\,\text{cm}^{-1}$ corresponds to the LMCT transitions in which an electron from a chlorine π lone pair is promoted into a molecular orbital that is principally metal d orbital.

Self-test 13.5 The spectrum of $[\text{Cr(NCS)}_6]^{3-}$ has a very weak band near 16 000 cm^{-1}, a band at 17 700 cm^{-1} with $\varepsilon_{max} = 160$ $\text{L mol}^{-1}\,\text{cm}^{-1}$, a band at 23 800 cm^{-1} with $\varepsilon_{max} = 130$ $\text{L mol}^{-1}\,\text{cm}^{-1}$, and a very strong band at 32 400 cm^{-1}. Assign these transitions using the d^3 Tanabe–Sugano diagram and selection rule considerations. (*Hint:* NCS$^-$ has low-lying π^* orbitals.)

13.6 Luminescence

A complex is **luminescent** if it emits radiation after it has been electronically excited by the absorption of radiation. Luminescence competes with nonradiative decay by thermal degradation of energy through heat loss to the surroundings. Relatively fast radiative decay is not especially common at room temperature for d-metal complexes, so strongly luminescent systems are comparatively rare. Nevertheless, they do occur, and we can distinguish two types of process. Traditionally, rapidly decaying luminescence was called 'fluorescence' and luminescence that persists after the exciting illumination is extinguished was called 'phosphorescence'. However, because the lifetime criterion is not reliable, the modern definitions of the two kinds of luminescence are based on the distinctive mechanisms of the processes. **Fluorescence** is radiative decay from an excited state of the same multiplicity as the ground state. The transition is spin-allowed and is fast; fluorescence half-lives are a matter of nanoseconds. **Phosphorescence** is radiative decay from a state of different multiplicity from the ground state. It is a spin-forbidden process, and hence is often slow.

(a) Phosphorescent complexes

The initial excitation of a complex usually populates a state by a spin-allowed transition, so the mechanism of phosphorescence involves **intersystem crossing**, the nonradiative

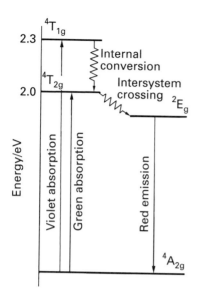

13.10 The transitions responsible for the absorption and luminescence of Cr^{3+} ions in ruby.

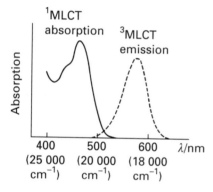

13.11 The absorption and phosphorescence spectra of $[Ru(bipy)_3]^{2+}$.

conversion of the initial excited state into another excited state of different multiplicity. This second state acts as an energy reservoir because radiative decay to the ground state is spin-forbidden. However, just as spin–orbit coupling allows the intersystem crossing to occur, it also breaks down the spin selection rule, so the radiative decay can occur. Radiative decay back to the ground state is slow, so a phosphorescent state of a d-metal complex may survive for microseconds or even longer.

An important example of phosphorescence is provided by ruby, which consists of a low concentration of Cr^{3+} ions in place of Al^{3+} in alumina. Each Cr^{3+} ion is surrounded octahedrally by six O^{2-} ions and the initial excitations are the spin-allowed processes

$$t_{2g}^2 e_g^1 \leftarrow t_{2g}^3 : \qquad {}^4T_{2g} \leftarrow {}^4A_{2g} \text{ and } {}^4T_{1g} \leftarrow {}^4A_{2g}$$

These absorptions occur in the green and violet regions of the spectrum and are responsible for the red color of the gem (Fig. 13.10). Intersystem crossing to a ${}^2E$ term of the t_{2g}^3 configuration occurs in a few picoseconds or less, and red 627 nm phosphorescence occurs as this doublet decays back into the quartet ground state. This red emission adds to the red perceived by the subtraction of green and violet light from white light, and adds luster to the gem's appearance. When the optical arrangement is such that the emission is stimulated by 627 nm photons reflecting back and forth between two mirrors, it grows strongly in intensity. This stimulated emission in a resonant cavity is the process used by Theodore Maiman in the first successful laser (in 1960) based on ruby.

A similar ${}^2E \rightarrow {}^4A$ phosphorescence can be observed from a number of Cr(III) complexes in solution. The emission is always in the red and close to the wavelength of ruby emission. The ${}^2E$ term belongs to the t_{2g}^3 configuration, which is the same as the ground state, and the strength of the ligand field is not important. If the ligands are rigid, as in $[Cr(bipy)_3]^{3+}$, the ${}^2E$ term may live for several microseconds in solution.

Another interesting example of a phosphorescent state is found in $[Ru(bipy)_3]^{2+}$. The excited singlet term produced by a spin-allowed MLCT transition of this d^6 complex undergoes intersystem crossing to the lower energy triplet term of the same configuration, $t_{2g}^5 \pi^{*1}$. Bright orange emission then occurs with a lifetime of about 1 μs (Fig. 13.11). The effects of other molecules (quenchers) on the lifetime of the emission may be used to monitor the rate of electron transfer from the excited state.

> Phosphorescence occurs when an excited state undergoes intersystem crossing to a state of different multiplicity, and then undergoes radiative decay.

13.7 Spectra of f-block complexes

The partially filled f shell can also participate in transitions in the visible region of the spectrum. In one respect, their description is more complex than for the analogous transitions in the d-block because there are seven f orbitals. However, it is simplified by the fact that f orbitals are relatively deep inside the atom and overlap only weakly with ligand orbitals. Hence, as a first approximation, their spectra can be discussed in the free-ion limit.

One systematizing feature is that the colors of the aqua ions from La^{3+} (f^0) to Gd^{3+} (f^7) tend to repeat themselves in the reverse sequence Lu^{3+} (f^{14}) to Gd^{3+} (f^7):

f electrons:	0	1	2	3	4	5	6	7
Color:	colorless	colorless	green	red	pink	yellow	pink	colorless
f electrons:	14	13	12	11	10	9	8	

This sequence suggests that the absorption maxima are related in a simple manner to the number of unpaired f electrons and largely independent of the specific stereochemistry of the complex. Unfortunately, though, the apparent simplicity of this correlation is deceptive and is not fully borne out by detailed analysis.

Figure 13.12 shows the spectrum of f^2 Pr^{3+}(aq) from the infrared to the ultraviolet regions. The four bands are labeled with the free-ion term symbols, which is appropriate when the interaction with ligands is very weak. The Russell–Saunders coupling scheme remains a good approximation despite the elements having high atomic numbers, because the f electrons penetrate only slightly through the inner shells. As a result, they do not sample the high electric field at the nucleus and hence their spin–orbit coupling is weak.

The other noteworthy feature of the spectrum is the narrowness of the bands, which indicates that the electronic transition does not excite much molecular vibration when it occurs. The narrowness implies that there is little difference in the molecular potential energy surface when an electron is excited, which is consistent with an f electron interacting only weakly with the ligands. Because the excited electron interacts only weakly with its environment, the nonradiative lifetime of the excited state is quite long and luminescence can be expected. These complexes do tend to be strongly luminescent, and for that reason are used as phosphors on TV screens.

> The lanthanides typically display weak but sharp absorption spectra because the f orbitals are shielded from the ligands.

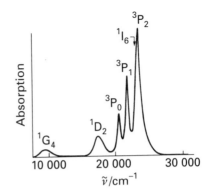

13.12 The spectrum of the $f^2(^3H)$ Pr^{3+}(aq) ion in the visible region.

13.8 Circular dichroism

The photon is a particle with spin 1 (that is, it carries one unit of spin angular momentum), and in a state of circular polarization it has a definite **helicity**, or component of angular momentum along its direction of propagation. Left-circularly polarized light consists of photons of one helicity and right-circularly polarized light of photons of the opposite helicity (Fig. 13.13).

A chiral molecule (one lacking an improper rotation axis, Sections 4.1 and 4.4) displays **circular dichroism**. That is, the molecule has different absorption coefficients for right- and left-circularly polarized radiation at any given wavelength. A **circular dichroism spectrum** (a CD spectrum) is a plot of the difference of the molar absorption coefficients for right- and left-circularly polarized light against wavelength. As we see in Fig. 13.14, enantiomers have CD spectra that are mirror-images of each other.

The usefulness of CD spectra can be appreciated by comparing the CD spectra of two enantiomers of $[Co(en)_3]^{3+}$ with their conventional absorption spectra. The latter show two ligand-field bands, just as though the complex were an octahedral complex, and give no sign

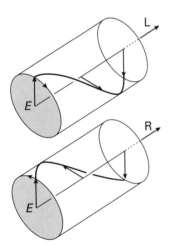

13.13 Left (upper) and right (lower) circularly polarized component of electromagnetic radiation. Only the electric field is shown.

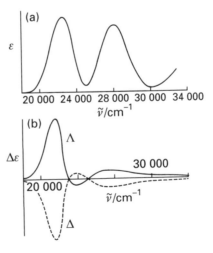

13.14 (a) The absorption spectrum of $[Co(en)_3]^{3+}$ and (b) the CD spectra of the two optical isomers.

that we are dealing with a complex of lower symmetry (D_3). That difference, however, shows up in the CD spectra, where an additional band is seen near $24\,000$ cm^{-1}.

The absolute configuration of chiral complexes may be identified in favorable cases, for the same absolute configuration of two complexes with similar electronic configurations gives CD spectra of the same sign. Hence CD spectra can be used to relate large families of chiral complexes to the small number of primary cases, including [Co(en)$_3$]$^{3+}$, for which absolute configurations have been established by X-ray diffraction.

Circular dichroism spectra can be used to study geometrical as well as optical isomerism. For example, the amino acid alanine CH$_3$CH(NH$_2$)COOH forms *fac* and *mer* tris complexes with Co(III), and each form can have Λ or Δ optical isomers. The problem of identifying the arrangement of ligands is first to decide between the *fac* and *mer* geometrical isomers, and then between their Λ and Δ optical isomers. The two geometrical isomers have different absorption spectra because the immediate symmetry of ligands about the Co(III) center differs (Fig. 13.15). In the *mer* isomer, there is no axis of rotational symmetry; the *fac* isomer has a C_3 axis passing through the two faces surrounded by N and the O atoms. The lower symmetry of the *mer* isomer is likely to lead to a somewhat broader absorption band. The corresponding CD spectra are shown in Fig. 13.15(b). The two pairs of curves with peaks of opposite sign clearly indicate which isomers are enantiomers of the same geometrical isomer. Furthermore, we can assign the Λ and Δ absolute configuration labels by comparing these spectra with the CD spectrum of [Co(en)$_3$]$^{3+}$, for its absolute configuration is known from X-ray diffraction.

> Circular dichroism is observed for chiral complexes; it can be used to infer the absolute configuration of enantiomers.

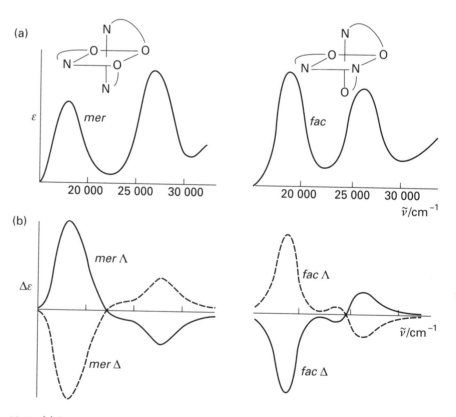

13.15 (a) The absorption spectra of the *fac* and *mer* isomers of [Co(ala)$_3$], where ala is alanine (specifically, its conjugate base), and (b) the CD spectra of the isomers and the assignment of their absolute configuration by comparison with the CD spectrum of [Co(en)$_3$]$^{3+}$.

Example 13.6 Using CD spectra to assign a configuration

The enantiomer commonly labeled $(-)546\,[\mathrm{Co(edta)}]^-$, because it rotates the plane of linearly polarized 546 nm light counterclockwise (to an observer looking toward the oncoming beam), has the CD spectrum shown in Fig. 13.16. Assign its absolute configuration by comparison with $[\mathrm{Co(en)_3}]^{3+}$.

Answer The complex is similar to the alanine complex with one more O atom bound to Co. Therefore it seems reasonable to continue to correlate the spectra with the two d–d transitions of $[\mathrm{Co(en)_3}]^{3+}$. The lower energy band has a positive sign followed by a weak negative peak at higher energy. The higher energy absorption band corresponds to weak CD absorption with a larger positive than negative area. These features match the Λ absolute configuration of $[\mathrm{Co(en)_3}]^{3+}$.

Self–test 13.6 What evidence do you find in Fig. 13.16 that $[\mathrm{Co(edta)}]^-$ has lower symmetry at Co than $[\mathrm{Co(en)_3}]^{3+}$?

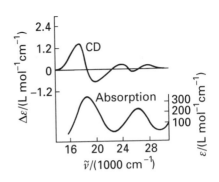

13.16 The absorption and CD spectra of $(-)546[\mathrm{Co(edta)}]^-$ referred to in Example 13.6.

13.9 Electron paramagnetic resonance

The electronic analog of nuclear magnetic resonance is called **electron paramagnetic resonance** (EPR; the technique is also called *electron spin resonance*, ESR). It is a powerful technique for studying complexes with unpaired electrons (those in other than singlet states). EPR can be used to map the distribution of an unpaired electron in a molecule and help to decide the extent to which electrons are delocalized over the ligands. The technique can also provide some information about the energy levels of complexes.

(a) The *g* value

Electron paramagnetic resonance relies on the fact that an electron spin can adopt two orientations along the direction defined by an applied magnetic field $\mathscr{B}$. The energy difference between the states $m_s = +\frac{1}{2}$ and $-\frac{1}{2}$ is

$$\Delta E = g\mu_B \mathscr{B} \tag{7}$$

where μ_B is the Bohr magneton and g is a constant characteristic of the complex. For a free electron, g has the 'free-spin' value $g_e = 2.0023$. If the sample is exposed to electromagnetic radiation of frequency ν, a strong absorption—a resonance—occurs when the magnetic field satisfies the condition

$$h\nu = g\mu_B \mathscr{B} \tag{8}$$

In a typical spectrometer operating near 0.3 T, resonance occurs in the 3 cm band of the microwave region of the electromagnetic spectrum (at about 9 GHz).

The g-factor in a complex differs from g_e by an amount that depends on the ability of the applied field to induce local magnetic fields in the complex. If $g > g_e$, the local field is larger than that applied; if $g < g_e$, the local field is smaller. The sign and magnitude of the induced local fields depend on the separation of the energy levels of the complex. The closer they are together, the easier it is for the applied field to induce orbital circulation of the electrons and hence produce a local magnetic field.

The value of g can be determined by noting the applied field needed to achieve resonant absorption for a given microwave frequency. If the complex is part of a single crystal, g may be measured along different directions and hence used to deduce the symmetry of the complex. For example, the tetragonally distorted D_{4h} complex $[\mathrm{Cu(OH_2)_6}]^{2+}$ has $g = 2.08$ in the direction parallel to the C_4 axis and $g = 2.40$ perpendicular to it. The fact that two different g values are measured confirms the existence of the distortion, because an octahedral complex has an isotropic g value (one that is the same in all directions). The two numerical values of g give information about the energy levels of the complex.

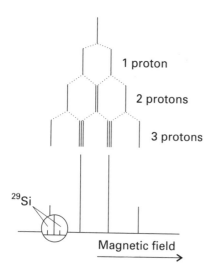

13.17 The analysis of the hyperfine structure of the EPR spectrum of the silyl radical, SiH_3. The origin of the lines is described in the text.

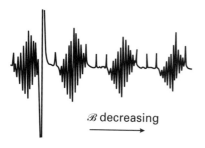

8 Bis(salicylaldehyde–imine) copper(II)

13.18 The EPR spectrum of copper(II) bis(salicylaldehyde-imine) diluted 1:200 with the isomorphous nickel(II) chelate. The copper is isotopically pure copper-63. The strong absorption at applied field (on the left) is a radical marker (DPPH) used to calibrate the spectrum. As is common in EPR, the spectrum is displayed as the first derivative of the absorption with respect to applied field, so each line has a shape like that of the strong DPPH absorption.

> The g value for a paramagnetic complex indicates the symmetry of the ligand environment and gives an indication of the availability of excited states.

(b) Hyperfine structure

Magnetic nuclei—those with nonzero spin—give rise to an additional magnetic field in a complex, and any unpaired electrons respond to this field. Because a magnetic nucleus with spin I can adopt $2I + 1$ different orientations, it can give rise to $2I + 1$ different contributions to the local field. As a result, the resonance condition will be fulfilled at $2I + 1$ different values of the applied field and the spectrum is split into multiplets called its **hyperfine structure**. In the presence of a single nucleus of spin I, the $2I + 1$ resonance conditions can be expressed as

$$h\nu = g\mu_B(\mathscr{B} + Am_I) \tag{9}$$

where A is the **hyperfine splitting constant**.

When there are several magnetic nuclei present, the hyperfine structure is the outcome of all possible combinations of the nuclear spin orientations. As an illustration, the EPR spectrum of the SiH_3 radical consists of four lines with an intensity pattern 1:3:3:1. This pattern is due to the coupling of the electron spin to the three protons ($I = \frac{1}{2}$), as shown in Fig. 13.17. A further splitting occurs in the 4.7 per cent of radicals that have a ^{29}Si nucleus ($I = \frac{1}{2}$): the four lines are split into doublets by this interaction.

The size of the hyperfine splitting constant depends on the probability that the unpaired electron will be found close to the magnetic nucleus responsible for the splitting. Therefore, the measurement of A provides a means of mapping the wavefunction of the molecular orbital occupied by the unpaired electron. In particular, the observation of hyperfine structure due to nuclei in the ligands shows that the unpaired electron is delocalized from the metal on to them. The value of A for the interaction with either the ligand nuclei or with the metal nucleus is a very good indication of the extent to which electrons have migrated from the central metal atom to the surrounding ligands. For instance, for a Cu(II) complex, the hyperfine splitting of the Cu nucleus varies as follows:

Ligand	F^-	H_2O	O^{2-}	S^{2-}
A/mT	9.8	9.8	8.5	6.9

These values show that, as the energy of the ligand orbitals approaches that of the metal, the orbital occupied by the unpaired electron becomes more ligand-like and interacts more weakly with the metal nucleus.

A classic study of the delocalization of electrons on to ligands is the work on bis(salicylaldehyde-imine)copper(II) with ^{63}Cu (**8**).[9] There are four main regions of absorption in the EPR spectrum of the complex, and each one is split into eleven lines (Fig. 13.18). Substitution of H atoms of the NH group by deuterium (for which $I = 1$) does not alter the spectrum. However, replacement of the H of the neighboring CH group by CH_3 groups (the spin of ^{12}C is $I = 0$) results in the collapse of each of the four groups into lines with only five components. The original $4 \times 11 = 44$ hyperfine lines correspond to coupling to ^{63}Cu ($I = \frac{3}{2}$, to give four lines of equal intensity) and then to two equivalent ^{14}N nuclei ($I = 1$, jointly giving five lines) and two equivalent protons (jointly giving three lines). The fact that only eleven lines are seen whereas this analysis requires $5 \times 3 = 15$ is a consequence of partial overlap of the lines. When the proton coupling is removed by substitution with CH_3 groups, only the five lines of the ^{14}N splitting survive. The results indicate that electron delocalization is important to the H of C—H but not as far as the H of N—H.

9   A.H. Maki and B.R. McGarvey, Electron spin resonance in transition metal chelates. II. Copper(II) bis-salicaldehyde-imine. *J. Chem. Phys.* **29**, 35 (1958).

The hyperfine structure in the EPR spectrum of a complex can indicate the extent of delocalization of unpaired electrons on to ligands with magnetic nuclei.

Bonding and spectra of M—M bonded compounds

In this section we discuss the molecular orbitals of some complexes that we meet again in later chapters, particularly when we discuss the d-block organometallic compounds (Chapter 16) and catalysis (Chapter 17). In particular, we deal here with complexes of lower symmetry than those considered in Chapter 7. We shall show that one very useful technique for constructing molecular orbitals of complicated molecules is to build them up from the valence orbitals of subunits. The technique is especially useful for compounds containing metal–metal bonds and which can be regarded as composed of two or more ML_n subunits.

13.10 The ML₅ fragment

Our aim here is to infer the molecular orbitals for a square-pyramidal ML_5 structure from the MOs of the octahedral molecular orbital diagram. Once we have established the orbitals of the ML_5 fragment with C_{4v} symmetry, we can interpet some aspects of the spectra of C_{4v} and D_{4h} complexes. We shall also be able to use the fragment orbitals to construct an orbital diagram of an M_2L_{10} binuclear metal–metal bonded complex.

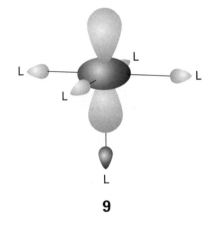

9

(a) ML₅ molecular orbitals

The HOMO and LUMO of an ML_6 octahedral complex are shown on the left of Fig. 13.19. Suppose we remove one ligand from the z-axis to give ML_5. The principal change occurs to the d_{z^2} orbital, which becomes a_1 in C_{4v} and falls sharply in energy as it loses its antibonding character toward the departing ligand (**9**). Moreover, as the symmetry of d_{z^2} changes from e_g in O_h and becomes a_1 in C_{4v}, it can mix with s and p_z, which are also a_1 in C_{4v}. We can therefore expect the resulting hybrid orbital to have a large amplitude in the vacant coordination position, which makes it an ideal orbital for acting as a σ acceptor to a new donor ligand.

In C_{4v} the three t_{2g} orbitals are split into a doubly degenerate e pair (d_{zx}, d_{yz}) and one b_2 orbital (d_{xy}). The splitting between e and b_2 is smaller than that between d_{z^2} and $d_{x^2-y^2}$ because the three orbitals are less directly affected by the ligand that is being removed. However, if the d_{zx} and d_{yz} orbitals are involved in π bonding with the departing ligand there may be an additional effect. Thus, if the ligand is a π donor (like Cl⁻), then the antibonding effect will be reduced and the new e pair will lie below the b_2 orbital (as we have shown in

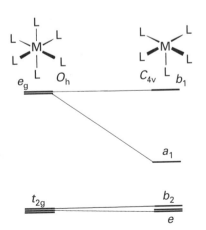

13.19 An orbital correlation diagram for the conversion of an ML_6 octahedral complex into an ML_5 fragment. Only the frontier orbitals are shown.

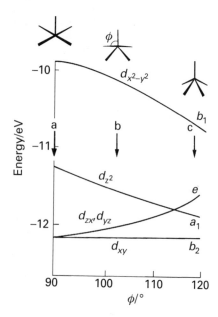

13.20 The correlation diagram for the metal d orbitals in an ML_5 fragment as the basal ligands bend away from the metal and the angle ϕ increases from $90°$ to $120°$. The arrows show the location of (a) low-spin d^6 oxyhemoglobin, (b) low-spin d^8 $[Ni(CN)_5]^{3-}$, and (c) a d^{10} tetradentate macrocyclic complex of Cu(I) with an axial CO ligand.

10 $[Re_2Cl_8]^{2-}$

11 $[Re_2(CO)_{10}]$

Fig. 13.19). On the other hand, if the departing ligand is a π acceptor (like CO), then its removal will lead to an increase in the energy of the e orbitals above the b_2 orbital.

> The d_{z^2} orbital in an ML_6 complex falls in energy as one of the ligands is removed and hybridizes with the metal s and p orbitals.

(b) Square-pyramidal complexes

A square-pyramidal five-coordinate complex can have its central **M** atom above the plane defined by the four basal ligands.

Figure 13.20 shows an example of the energy variations of the HOMO and LUMO as the metal atom is moved above the basal plane and the angle denoted ϕ increases from $90°$ to $120°$. This increase in angle reduces the overlap of the ligands with the d_{z^2} and $d_{x^2-y^2}$ orbitals and the net antibonding character is reduced. At the same time, σ overlap with the d_{zx} and d_{yz} orbitals increases and they become more antibonding.

Figure 13.20 can be used to correlate the structures of square-pyramidal complexes with their electron counts. A low-spin d^6 complex should be expected to have ϕ close to $90°$ because d_{zx} and d_{yz}, which are filled, rise in energy as ϕ increases from $90°$. In contrast, a low-spin d^8 pyramid should have ϕ greater than $90°$, for then d_{z^2} (which is now occupied) is stabilized. In support of this conclusion, the d^6 oxyhemoglobin system has $\phi \approx 90°$ whereas the d^8 complex $[Ni(CN)_5]^{3-}$ has $\phi = 101°$.

> The four basal atoms of a five-coordinate complex bend away from the axial ligand as the d-electron count on the central metal changes from six to eight.

13.11 Binuclear complexes

In Section 9.9 we saw that it is possible to synthesize metal–metal bonded compounds in which the metal–metal bond order ranges from 1 to 4. In many cases, these binuclear complexes can be regarded as formed from the union of two ML_n units. Not too far removed conceptually from these complexes are species in which two metal atoms are in different oxidation states, in some cases insulated from each other by a bridging ligand.

(a) Metal–metal multiple bonding

X-ray diffraction shows that $[Re_2Cl_8]^{2-}$ has the structure shown in (**10**). The striking feature is the eclipsed arrangement of the two sets of equatorial Cl ligands rather than the staggered conformation that would minimize van der Waals repulsions between the two sets of ligands. The eclipsed configuration is attributed to the confacial interaction of d_{xy} orbitals on each of the two Re atoms. The short Re—Re bond distance of only $2.24 Å$ can also be traced to the presence of δ bonds between the two atoms (see Fig. 9.20).

Another line of evidence for the presence of a δ bond is provided by the electronic transition at $15\,000\,\text{cm}^{-1}$ in the near IR, which shows an extensive vibrational structure with an average spacing of $351\,\text{cm}^{-1}$ (Fig. 13.21). The structure has been assigned to a $\delta^* \leftarrow \delta$ transition.[10] The vibrational splitting is almost identical to the Re—Re stretching wavenumber determined by Raman spectroscopy, which confirms that the absorption band is due to the metal–metal region of the complex. The stretching wavenumber is in fact much higher than that observed for Re_2CO_{10} ($125\,\text{cm}^{-1}$), which has only a single Re—Re bond (**11**).

> Spectroscopic and structural studies indicate the presence of a quadruple rhenium-rhenium bond in $[Re_2Cl_8]^{2-}$.

[10] C.D. Cowman and H.B. Gray, *J. Am. Chem. Soc.* **95**, 8177 (1973). This is probably the only spectroscopic transition to have been featured in a detective novel, Joseph Wambaugh's *The delta star*, Bantam Books, New York (1974).

(b) Mixed-valence complexes

Another type of binuclear complex with interesting spectroscopic properties is a **mixed-valence complex** with identical metal atoms in different oxidation states. One of the earlier and more important examples is the 'Creutz–Taube ion', $[(NH_3)_5Ru\text{-}pyz\text{-}Ru(NH_3)_5]^{5+}$ **(12)**, where pyz is a bridging pyrazine ligand. The question this formula poses is whether the two Ru atoms are a localized Ru(II) and a localized Ru(III) or whether the electrons are sufficiently delocalized to give two equivalent Ru atoms in a +2.5 oxidation state. The problem can be expressed (but not solved) by writing a wavefunction ψ' for Ru(II)—Ru(III) and another wavefunction ψ'' for Ru(III)—Ru(II) and expressing the overall ground state by the wavefunction $\psi = c'\psi' + c''\psi''$. The values $c' = 0$ or $c'' = 0$ correspond to complete localization and $|c'|^2 = |c''|^2$ corresponds to complete delocalization.

A classification of mixed-valence compounds that reflects the differences in the strength of the interaction has been proposed by M.B. Robin and Peter Day:

Class I: Fully localized electrons
Class II: Intermediate
Class III: Fully delocalized electrons

Class III is subdivided into IIIA for pairs or clusters of metals and IIIB for indefinitely delocalized solids. The classification is summarized in Table 13.6.

Class I includes a large number of metal oxide and sulfide solids that seem curious from the standpoint of oxidation numbers. For example, the apparent oxidation state of lead in Pb_3O_4 may be regarded as a mixture of the two common oxidation states Pb(IV) and Pb(II). As outlined in Table 13.6, compounds in this class contain metal ions in ligand fields of different symmetry or strength, and there is definite structural evidence for localized oxidation states

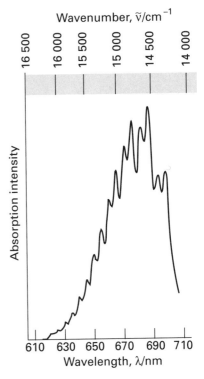

13.21 The polarized visible absorption spectrum of a single crystal of $[(C_4H_9)_4N]_2[Re_2Cl_8]$. (C.D. Cowman and H.B. Gray, *J. Am. Chem. Soc.* **95**, 8177 (1973).)

12 $[Ru_2(NH_3)_{10}(pyz)]^{5+}$

Table 13.6 The Robin and Day classification of intervalence transition (IT) bands

Class I	Class II	Class III
M and **M′** very different, different ligand fields, etc.	**M** and **M′** in similar environments, but not equivalent	**M** and **M′** not distinguishable
Orbital 'trapped'	Orbital distinguishable but not fully localized	Orbital delocalized
IT bands at high energy	IT in visible or near IR	Bands in visible or near IR
Electronic spectra of constituent ions seen	Electronic spectra of constituent ions seen but modified	Electronic spectra of constituent ions not distinguishable
Insulator	Semiconductor	IIIA: Usually insulator IIIB: Usually metallic conductor
Magnetic properties of complex	Magnetic properties of isolated complex except at low temperature	IIIA: Magnetic properties of isolated complex IIIB: Ferromagnetic or paramagnetic

(corresponding to $|c'|^2 \gg |c''|^2$, or vice versa). Spectroscopic and magnetic properties near room temperature are characteristic of the coordination environment of the individual ions.

Class III includes cases in which formation of metal–metal bonds leads to the delocalization of charge ($|c'|^2 \approx |c''|^2$). The bronze metallic conductor Ag_2F, for example, contains sheets of metal–metal bonded Ag atoms bearing a positive charge, and is a member of Class IIIB. In contrast, Ta_6Cl_{15} contains cluster cations of the formula $[Ta_6Cl_{12}]^{3+}$ containing six equivalent Ta atoms. Discrete clusters of this kind belong to Class IIIA. The solid is an electrical insulator and does not have a metallic appearance.

Class II includes the compounds for which **intervalence transitions** are observed in which an electron is excited from the valence shell of one atom into the valence shell of the other atom. Thus, they correspond to metal-to-metal charge-transfer transitions. As is the case in the Creutz–Taube ion, intervalence transitions are strong if there is an appropriate bridging ligand between the two centers. If the metal ions do not share a bridging ligand, the transition may be too weak to be observed. The classic example of this type of compound is the pigment Prussian blue. Its structure is a cubic lattice with Fe^{2+} octahedrally coordinated by the C atom of the CN^- ligands and Fe^{3+} octahedrally coordinated by the N atom of the ligand. The bridging cyanide provides the conduit for the strong intervalence transition and the resulting blue color. The magnetic properties of Class II compounds at room temperature give little indication of differences from localized complexes but magnetic interactions between sites become evident at low temperatures.

Mixed oxidation state compounds are classified as Class I, with distinct spectra for the different metal ions, Class II, which have metal ions of similar energies and are highly colored, and Class III, which have fully delocalized metal–metal bonding.

Further reading

A.B.P. Lever, *Inorganic electronic spectroscopy*. Elsevier, Amsterdam (1984). A comprehensive, reasonably up-to-date treatment with a good collection of data.

D.W. Smith, Ligand field theory and spectra. In *Encyclopedia of inorganic chemistry* (ed. R.B. King), Vol. 4, pp. 1965–82. Wiley, New York (1994).

T.A. Albright, J.K. Burdett, and M.H. Whangbo, *Orbital interactions in chemistry*. Wiley, New York (1985). A good source for the treatment of construction of molecular orbitals by combination of fragments.

I. Bersuker, *Electronic structure and properties of transition metal complexes*. Wiley Interscience, New York (1996).

B.N. Figgis, In *Comprehensive coordination chemistry* (ed. G. Wilkinson, R.D. Gillard, and J.A. McCleverty), Vol. 1, p. 213. Pergamon, Oxford (1987).

T.J. Meyer, Excited state electron transfer. *Prog. Inorg. Chem.* **30**, 389 (1983).

S.F.A. Kettle, *Physical inorganic chemistry: a coordination approach*. Oxford University Press (1996).

Exercises

13.1 Write the Russell–Saunders term symbols for states with the angular momentum quantum numbers (L, S): (a) $(0, \frac{5}{2})$, (b) $(3, \frac{3}{2})$, (c) $(2, \frac{1}{2})$, (d) $(1, 1)$.

13.2 Identify the ground term from each set of terms: (a) ^1G, ^3F, ^3P, ^1P; (b) ^3H, ^3P, ^5D, ^1I, ^1G; (c) ^6S, ^4G, ^4P, ^2I.

13.3 Give the Russell–Saunders terms of the configurations: (a) $4s^1$, (b) $3p^2$. Identify the ground term.

13.4 Identify an atom or an ion corresponding to the configuration $3p^5$ and $4d^9$. What are the ground terms for each?

13.5 Identify the ground terms of the gas-phase species B^+, Na, Ti^{2+}, and Ag^+.

13.6 The free gas-phase ion V^{3+} has a ^3F ground term. The ^1D and ^3P terms lie respectively $10\,642\ cm^{-1}$ and $12\,920\ cm^{-1}$ above it. The energies of the terms are given in terms of Racah parameters as $E(^3F) = A - 8B$, $E(^3P) = A + 7B$, $E(^1D) = A - 3B + 2C$. Calculate the values of B and C for V^{3+}.

13.7 Write the d orbital configurations and use the Tanabe–Sugano diagrams (Appendix 5) to identify the ground term of (a) low-spin $[Rh(NH_3)_6]^{3+}$, (b) $[Ti(OH_2)_6]^{3+}$, (c) high-spin $[Fe(OH_2)_6]^{3+}$.

13.8 Using the Tanabe–Sugano diagrams in Appendix 5, estimate Δ_O and B for (a) $[Ni(OH_2)_6]^{2+}$ (absorptions at 8500, 15\,400, and 26\,000 cm^{-1}) and (b) $[Ni(NH_3)_6]^{2+}$ (absorptions at 10\,750, 17\,500, and 28\,200 cm^{-1}).

13.9 If an octahedral Fe(II) complex has a large paramagnetic susceptibility, what is the ground-state label according to the Tanabe–Sugano diagram? What is the description of the states involved in the spin-allowed electronic transition?

13.10 The spectrum of $[Co(NH_3)_6]^{3+}$ has a very weak band in the red and two moderate intensity bands in the visible to near-UV. How should these transitions be assigned?

13.11 Explain why $[FeF_6]^{3-}$ is colorless whereas $[CoF_6]^{3-}$ is colored but exhibits only a single band in the visible.

13.12 The Racah parameter B is $460\ cm^{-1}$ in $[Co(CN)_6]^{3-}$ and $615\ cm^{-1}$ in $[Co(NH_3)_6]^{3+}$. Consider the nature of bonding with the two ligands and explain the difference in nephelauxetic effect.

13.13 An approximately 'octahedral' complex of Co(III) with ammine and chloro ligands gives two bands with ε_{max} between 60 and 80 L $mol^{-1}cm^{-1}$, one weak peak with $\varepsilon_{max} = 2$ L $mol^{-1}\ cm^{-1}$ and a strong band at higher energy with $\varepsilon_{max} = 2 \times 10^4$ L $mol^{-1}\ cm^{-1}$. What do you suggest for the origins of these transitions?

13.14 Ordinary bottle glass appears nearly colorless when viewed through the wall of the bottle but green when viewed from the end so that the light has a long path through the glass. The color is associated with the presence of Fe^{3+} in the silicate matrix. Describe the transitions.

13.15 $[Cr(OH_2)_6]^{3+}$ ions are pale blue-green but the chromate ion CrO_4^{2-} is an intense yellow. Characterize the origins of the transitions and explain the relative intensities.

13.16 Classify the symmetry type of the d_{z^2} orbital in a tetragonal C_{4v} symmetry complex, such as $[CoCl(NH_3)_5]^+$ where the Cl is on the z-axis. (a) Which orbitals will be displaced from their position in the octahedral molecular orbital diagram by π interactions with the lone pairs of the Cl^- ligand? (b) Which orbital will move because the Cl^- ligand is not as strong a σ-donor as NH_3? (c) Sketch the qualitative molecular orbital diagram for the C_{4v} complex.

13.17 Consider the molecular orbital diagram for a tetrahedral complex (based on Fig. 7.13) and the relevant d orbital configuration. Show that the purple color of MnO_4^- ions cannot arise from a ligand-field transition.

13.18 Given that the wavenumbers of the two transitions in MnO_4^- are $18\,500\ cm^{-1}$ and $32\,200\ cm^{-1}$, explain how to estimate Δ_T from an assignment of the two charge-transfer transitions, even though Δ_T cannot be observed directly.

13.19 Prussian blue was one of the first synthetic pigments. It has the overall composition $K[Fe_2(CN)_6]$ in which each Fe(II) is surrounded by six carbon coordinated cyanide ligands which bridge to Fe(III) bound to the nitrogen. Assign the intense absorption bands at 15\,000 and 25\,000 cm^{-1}.

13.20 The M(II) oxidation state is not common among the lanthanides but a 'normal' M(II) chemistry does exist for Sm^{2+}, Eu^{2+}, and Yb^{2+}. Write the f-electron configurations for these species. By comparison with Table 13.1, identify the ground terms.

Problems

13.1 Consider the trigonal-prismatic six-coordinate ML_6 complex with D_{3h} symmetry. Use the D_{3h} character table to divide the d orbitals of the metal atom into sets of defined symmetry type. Assume that the ligands are at the same angle relative to the xy-plane as in a tetrahedral complex.

13.2 The absorption spectra of the two isomers of $[CoCl_2(en)_2]^+$ are shown in Fig. 13.22. Considering the high symmetry of the *trans* isomer, suggest which is *cis* and which is *trans*. Supposing that this assignment is correct, predict the CD spectrum of the Λ isomer of *cis*-$[CoCl_2(en)_2]^+$ by comparison with $[Co(en)_3]^{3+}$ shown in Fig. 13.22.

(a)

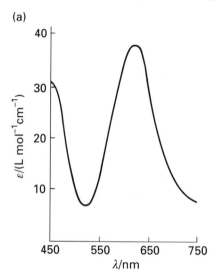

(b)

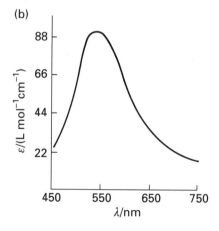

Fig. 13.22 The absorption spectra of two isomers of $[CoCl_2(en)_2]^+$.

13.3 Vanadium(IV) species that have the V=O group have quite distinct spectra. What is the d-electron configuration of V(IV)? The most symmetrical of such complexes are VOL_5 with C_{4v} symmetry with the O atom on the z-axis. What are the symmetry species of the five d orbitals in VOL_5 complexes? How many d–d bands are expected in the spectra of these complexes? A band near $24\,000\ cm^{-1}$ in these complexes shows vibrational progressions of the V=O vibration, implicating an orbital involving V=O bonding. Which d–d transition is a candidate? (See C.J. Ballhausen and H.B. Gray, *Inorg. Chem.* **1**, 111 (1962).)

13.4 Figure 13.23 shows the EPR spectra of the anion produced by reduction of the square-planar iron tetraphenylporphyrin $[Fe(TPP)]^-$, of labeled $[^{57}Fe(TPP)]^-$, and of $[Fe(TPP)]^-$ in the presence of excess pyridine. Account for the form of the observed spectra. (G.S. Srivatsa, D.T. Sawyer, N.J. Boldt, and D.F. Bocian, *Inorg. Chem.* **24**, 2123 (1985).)

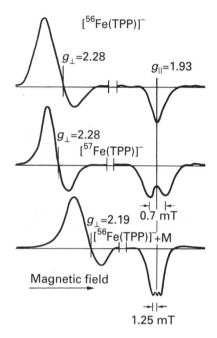

Fig. 13.23 The EPR spectra of the anion produced by reduction of the square-planar iron tetraphenylporphyrin $[Fe(TPP)]^-$, of labeled $[^{57}Fe(TPP)]^-$, and of $[Fe(TPP)]^-$ in the presence of excess pyridine.

13.5 The compound Ph_3Sn-$Re(CO)_3(t\text{-butDAB})$ (t-butDAB = R—N=CH—CH=N—R where R is a t-butyl group) has a metal-to-diimine ligand MLCT band as the lowest energy transition in its spectrum. Irradiation of this band gives an ·$Re(CO)_3(t\text{-butDAB})$ radical as a photochemical product. The EPR spectrum shows extensive hyperfine splitting by ^{14}N and ^1H. The radical is assigned as a complex of the DAB radical anion with Re(I). Explain the argument (see D.J. Stufkens, *Coord. Chem. Rev.* **104**, 39 (1990)).

13.6 It was remarked that MLCT bands can be recognized by the fact that the energy is a sensitive function of the polarity of the solvent (because the excited state is more polar than the ground state). Two simplified molecular orbital diagrams are shown in Fig. 13.24. In (a) is a case with a ligand π level higher than the metal d orbital. In (b) is a case in which the metal d orbital and the ligand level are at the same energy. Which of the two MLCT bands should be more solvent-sensitive? These two cases are realized by $[W(CO)_4(phen)]$ and $[W(CO)_4(iPr\text{-DAB})]$, where DAB = 1,4-diaza-1,3-butadiene, respectively. (See P.C. Servas, H.K. van Dijk, T.L. Snoeck, D.J. Stufkens, and A. Oskam, *Inorg. Chem.* **24**, 4494 (1985).) Comment on the CT character of the transition as a function of the extent of back-donation by the metal.

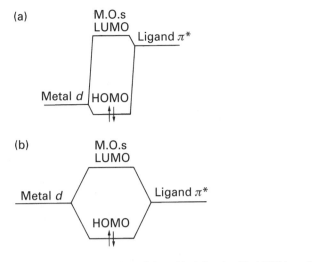

(a)

M.O.s
LUMO
Ligand π^*

Metal d | HOMO

(b)

M.O.s
LUMO

Metal d

Ligand π^*

HOMO

Fig. 13.24 Representation of the orbitals involved in MLCT transitions for cases in which the energy of the ligand π^* orbital varies with respect to the energy of the metal d orbital. See Problem 13.7.

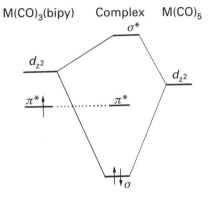

M(CO)$_3$(bipy)   Complex   M(CO)$_5$

σ^*

d_{z^2}

d_{z^2}

π^* .... π^*

σ

Fig. 13.25 Simplified MO diagram for a complex with an M—M bond formed from the combination of a fragment M(CO)$_5$ with a fragment with a π acceptor ligand, M'(CO)$_3$(bipy). See Problem 13.7.

13.7 Figure 13.25 shows a simplified scheme for the interaction of a fragment **M(CO)$_5$** with **M'(CO)$_3$**(bipy). The **M—M** bond is simplified to the interaction of the d_{z^2} orbital of **M** and **M'**. The spectrum of [(CO)$_5$ReRe(CO)$_3$(phen)] has a strong band in the visible (at 18 940 cm^{-1}) and a second of similar intensity in the near ultraviolet (at 28 570 cm^{-1}). Propose assignments. (See D.J. Stufkens, *Coord. Chem. Rev.* **104**, 39 (1990).)

13.8 Predict the number of lines in the EPR spectrum of bis(benzene)-vanadium(0) given that the abundance of the isotope ^{51}V ($I = \frac{7}{2}$) is 99.8 per cent. Comment on the interpretation of the spectrum given in S.M. Mattar and G.A. Ozin, *J. Chem. Phys.* **90**, 1037 (1986).

14

Reaction mechanisms of d-metal complexes

In Chapter 6 we encountered some of the factors that influence the rates of electron transfer, and in Chapter 7 we met the distinction between labile and inert complexes. Since then, many aspects of reactivity have been interpreted in terms of reaction mechanisms. Now we look in more detail at the evidence and experiments that are used in the analysis of reaction pathways, and so develop a deeper understanding of the mechanisms of the reactions of d-metal complexes. Because the mechanism is rarely known finally and completely, the nature of the evidence for a mechanism should always be kept in mind in order to recognize what other possibilities might also be consistent with it. In the first part of this chapter we describe how reaction mechanisms are classified, and distinguish between the steps by which the reaction takes place and the details of the formation of the activated complex. Then these concepts are used to describe the currently accepted mechanisms for the substitution and redox reactions of complexes.

We have seen the central importance of the interplay of equilibrium and kinetics for determining the outcome of inorganic reactions. It is helpful to understand the mechanisms of the reactions in cases where kinetic considerations are dominant, and in this chapter we see how to achieve at least partial understanding of the reactions of d-metal complexes. Ligand substitution and redox reactions were the first two classes of reactions to be closely analyzed, and we concentrate mainly on them. As we shall see, there are many ambiguities in the detailed understanding of inorganic reaction mechanisms. However, kinetic data, together with stereochemical and other experimental evidence, provide strong indications of the mechanisms of many reactions.

Ligand substitution reactions

A **ligand substitution reaction** is a reaction in which one Lewis base displaces another from a Lewis acid:

$$Y + M\text{—}X \longrightarrow M\text{—}Y + X$$

This class of reactions includes complex formation reactions, in which the **leaving group**, the displaced base X, is a solvent molecule and the **entering group**, the displacing base Y, is some other ligand. An example is the replacement of a water ligand by Cl^-:

$$[Co(OH_2)_6]^{2+} + Cl^- \longrightarrow [CoCl(OH_2)_5]^+ + H_2O$$

The majority of the reactions we consider in this chapter take place in water, and we shall not write the phase explicitly.

14.1 Patterns of reactivity

We saw in Section 7.7 that the equilibrium constants of displacement reactions can be used to rank ligands in order of their strength as Lewis bases. However, a different order may be found if bases are ranked according to the *rates* at which they displace a ligand from the central metal ion in a complex. Therefore, for kinetic considerations we replace the equilibrium concept of basicity by the kinetic concept of **nucleophilicity** (from the Greek for 'nucleus loving'). By nucleophilicity we mean the rate of attack on a complex by a given Lewis base relative to the rate of attack by a reference base. The shift from equilibrium to kinetic considerations is emphasized by referring to ligand displacement as **nucleophilic substitution**.

We have already seen that the rates of substitution reactions span a very wide range, and that to some extent there is a pattern in their behavior (Section 7.8). For instance, aqua complexes of Group 1, 2, and 12 metal ions, lanthanide ions, and some $3d$-metal ions, particularly those in low oxidation states, have half-lives as short as nanoseconds. In contrast, half-lives are years long for complexes of heavier d metals in high oxidation states, such as those of Ir(III) or Pt(IV). Intermediate half-lives are also observed. Two examples are $[Ni(OH_2)_6]^{2+}$, which has a half-life of the order of milliseconds, and $[Co(NH_3)_5(OH_2)]^{3+}$, in which H_2O survives for several minutes as a ligand before it is replaced by a stronger base. Table 14.1 puts the timescales of reactions into perspective.

Ligands other than the entering and leaving groups may play a significant role in controlling the rate of reaction. A **spectator ligand** is a ligand that is present in the complex but not lost in the reaction. It is observed for square-planar complexes that most good nucleophiles are good reaction accelerators if they occupy a *trans* position (T) as a spectator ligand. Thus X, Y, and T play related roles, and knowing how a ligand affects the rate when it is Y usually allows us to predict how it affects the rate when it is X or T.

> The nucleophilicity of an entering group is a measure of its relative rate of attack on a complex in a nucleophilic substitution reaction.

Table 14.1 Representative timescales of chemical and physical processes

Timescale*	Process	Example
10^2 s	Ligand exchange (inert complex)	$[Cr(OH_2)_6]^{3+} - H_2O$ (ca. 10^8 s)
60 s	Ligand exchange (inert complex)	$[V(OH_2)_6]^{3+} - H_2O$ (50 s)
1 ms	Ligand exchange (labile complex)	$[Pt(OH_2)_4]^{2+} - H_2O$ (0.4 ms)
1 μs	Intervalence charge transfer	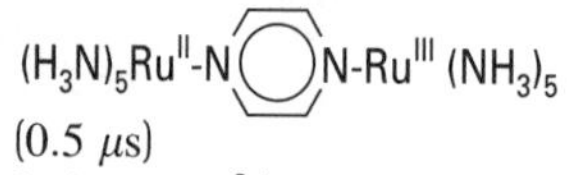 $(H_3N)_5Ru^{II}\text{-}N\bigcirc N\text{-}Ru^{III}(NH_3)_5$ (0.5 μs)
1 ns	Ligand exchange (labile complex)	$[Ni(OH_2)_5py]^{2+}$ (1 ns)
10 ps	Ligand association	$Cr(CO)_5 + THF$
1 ps	Rotation time in liquid	CH_3CN (10 ps)
1 fs	Molecular vibration	$Sn-Cl$ stretch (300 fs)

*Approximate time at room temperature.

14.2 The classification of mechanisms

The **mechanism** of a reaction is the sequence of elementary steps by which the reaction takes place. Once the mechanism has been unraveled, attention turns to the details of the activation process of the rate-determining step.[1] In some cases the overall mechanism is not fully resolved, and the only information available is the rate-determining step.

(a) Association, dissociation, and interchange

The first stage in the kinetic analysis of a reaction is to study how its rate changes as the concentrations of reactants are varied. This type of investigation leads to the identification of **rate laws**, the differential equations governing the rate of change of the concentrations of reactants and products. For example, the observation that the rate of formation of $[Ni(OH_2)_5(NH_3)]^{2+}$ from $[Ni(OH_2)_6]^{2+}$ is proportional to the concentration of both NH_3 and $[Ni(OH_2)_6]^{2+}$ implies that the reaction is first-order in each of these two reactants, and that the overall rate law is[2]

$$\text{rate} = k[Ni(OH_2)_6^{2+}][NH_3] \qquad (1)$$

In simple reaction schemes, the slowest elementary step of the reaction dominates the overall reaction rate and the overall rate law, and is called the **rate-determining step**. However, in general, all the steps in the reaction contribute to the rate law and affect its rate. Therefore, in conjunction with stereochemical and isotope tracer studies, the determination of the rate law is the route to the elucidation of the mechanism of the reaction.

Three main classes of reaction mechanism have been identified. A **dissociative mechanism**, denoted D, is a reaction sequence in which an intermediate of reduced coordination number is formed by the departure of the leaving group:

$$ML_nX \longrightarrow ML_n + X \quad \text{followed by} \quad ML_n + Y \longrightarrow ML_nY$$

Here ML_n (the metal atom and any spectator ligands) is a true intermediate that can, in principle, be isolated. For example, it has been proposed that the substitution of hexacarbonyltungsten (a d^6, 18-electron complex) by phosphine takes place by dissociation of CO from the complex

$$W(CO)_6 \longrightarrow W(CO)_5 + CO$$

followed by coordination of phosphine:

$$W(CO)_5 + PPh_3 \longrightarrow W(CO)_5PPh_3$$

Under the conditions in which this reaction is usually performed in the laboratory, the intermediate $W(CO)_5$ is rapidly captured by an ether solvent, such as tetrahydrofuran, to form $[W(CO)_5(thf)]$. That complex in turn is converted to the phosphine product, presumably by a second dissociative process. The generalized reaction profile is shown in Fig. 14.1a.

An **associative mechanism**, denoted A, involves a step in which an intermediate is formed with a higher coordination number than the original complex (Fig. 14.1c):

$$ML_nX + Y \longrightarrow ML_nXY \quad \text{followed by} \quad ML_nXY \longrightarrow ML_nY + X$$

Once again, the intermediate ML_nXY can, in principle at least, be isolated. This mechanism plays a role in many reactions of square-planar Pt(II), Pd(II), and Ir(I) d^8 complexes. A specific example is the exchange of $^{14}CN^-$ with the ligands in the square-planar complex $[Ni(CN)_4]^{2-}$. The first step is the coordination of a ligand to the complex

$$[Ni(CN)_4]^{2-} + {}^{14}CN^- \longrightarrow [Ni(CN)_4({}^{14}CN)]^{3-}$$

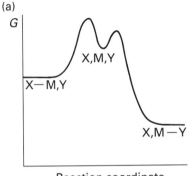

(a)

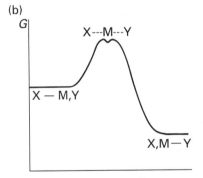

(b)

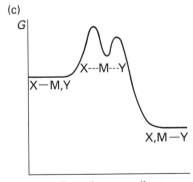

(c)

14.1 Reaction profiles for (a) dissociative D, (b) interchange I, and (c) associative A mechanisms. A true intermediate exists in (a) and (c) but not in (b).

---

[1] The latter is sometimes called the 'intimate mechanism' of the reaction.

[2] In rate equations, we omit the brackets that are part of the chemical formula of the complex; the surviving brackets denote molar concentration.

Subsequently, a ligand is discarded:

$$[Ni(CN)_4(^{14}CN)]^{3-} \longrightarrow [Ni(CN)_3(^{14}CN)]^{2-} + CN^-$$

The radioactivity of carbon-14 provides a means of monitoring this reaction. The intermediate $[Ni(CN)_5]^{3-}$ has been detected and isolated.

The rates of the associative substitution reactions of square-planar Pt(II), Pd(II), and Ir(I) complexes lie in the order

$$\mathbf{Y}: \qquad H_2O < Cl^- < I^- < H^- < PR_3 < CO, CN^-$$

Good entering groups (good nucleophiles) are usually poor (nonlabile) leaving groups. Therefore, to summarize the dependence of the rate on the leaving group all we need do is to reverse the order:

$$\mathbf{X}: \qquad CN^-, CO < PR_3 < H^- < I^- < Cl^- < H_2O$$

An **interchange mechanism**, denoted I (Fig. 14.1b), takes place in one step:

$$ML_nX + Y \longrightarrow X \cdots ML_n \cdots Y \longrightarrow ML_nY + X$$

The leaving and entering groups exchange in a single step by forming an activated complex but not a true intermediate. The interchange mechanism is common for many reactions of six-coordinate complexes.

The distinction between the A and I mechanisms hinges on whether or not the intermediate persists long enough to be detectable. One type of evidence is the isolation of an intermediate in another related reaction or under different conditions. If an argument by extrapolation to the actual reaction conditions suggests that a moderately long-lived intermediate might exist during the reaction in question, then the A path is indicated. For example, the synthesis of the first trigonal-bipyramidal Pt(II) complex, $[Pt(SnCl_3)_5]^{3-}$, indicates that a five-coordinate platinum complex may be plausible in substitution reactions of square-planar Pt(II) ammine complexes. Similarly, the fact that $[Ni(CN)_5]^{3-}$ is observed spectroscopically in solution, and that it has been isolated in the crystalline state, provides support for the view that it is involved when CN^- exchanges with the square-planar tetracyanonickelate(II) ion.

A second indication of the persistence of an intermediate is the observation of a stereochemical change, which implies that the intermediate has lived long enough to undergo rearrangement. *Cis* to *trans* isomerization is observed in the substitution reactions of certain square-planar phosphine Pt(II) complexes, which is in contrast to the retention of configuration usually observed. This difference implies that the trigonal bipyramid lives long enough for an exchange between the axial and equatorial ligand positions to occur.

Direct spectroscopic detection of the intermediate, and hence an indication of A rather than I, may be possible if a sufficient amount accumulates. Such direct evidence, however, requires an unusually stable intermediate with favorable spectroscopic characteristics and has not yet been obtained for Pt(II) complexes.

> *The mechanism of a nucleophilic substitution reaction is the sequence of elementary steps by which the reaction takes place. It is classified as associative, dissociative, or interchange. An associative mechanism is distinguished from an interchange mechanism by demonstrating that the intermediate has a relatively long life. Good nucleophiles are usually nonlabile leaving groups.*

(b) The rate-determining step

Now we consider the rate-determining step of a reaction and the details of its formation. The step is called **associative** and denoted a if its rate depends strongly on the identity of the incoming group. Examples are found among reactions of the d^8 square-planar complexes of Pt(II), Pd(II), and Au(III), including[3]

$$[PtCl(dien)]^+ + I^- \longrightarrow [PtI(dien)]^+ + Cl^-$$

It is found, for instance, that the replacement of I^- by Br^- decreases the rate constant by an order of magnitude. Experimental observations on the substitution reactions of square-planar complexes support the view that the rate-determining step is associative.

The strong dependence on **Y** of an associative rate-determining step indicates that the activated complex must involve significant bonding to **Y**. A reaction with an associative mechanism (A) will be associatively activated (a) if the attachment of **Y** to the initial reactant ML_nX is the rate-determining step. A reaction with a dissociative mechanism (D) is associatively activated (a) if the attachment of **Y** to the intermediate ML_n is the rate-determining step. An interchange mechanism (I) is associatively activated if the rate of formation of the activated complex depends on the rate at which the new $Y \cdots M$ bond forms.

The rate-determining step is called **dissociative** and denoted d if its rate is largely independent of the identity of **Y**. This category includes some of the classic examples of ligand substitution in octahedral d-metal complexes, including

$$[Ni(OH_2)_6]^{2+} + NH_3 \longrightarrow [Ni(OH_2)_5(NH_3)]^{2+} + H_2O$$

It is found that replacement of NH_3 by pyridine in this reaction changes the rate by at most a few per cent.

The weak dependence on **Y** of a dissociatively activated process indicates that the rate of formation of the activated complex is determined largely by the rate at which the bond to the leaving group **X** can break. A reaction with an associative mechanism (A) will be dissociatively activated (d) provided the loss of **X** from the intermediate YML_nX is the rate-determining step. A reaction with a dissociative mechanism (D) is dissociatively activated (d) if the initial loss of **X** from the reactant ML_nX is the rate-determining step. An interchange mechanism (I) is dissociatively activated if the rate of formation of the activated complex depends on the rate at which the $M \cdots X$ bond breaks.

> The rate-determining step is classified as associative or dissociative according to the dependence of its rate on the identity of the entering group.

The substitution of square-planar complexes

The elucidation of the mechanism of the substitution of square-planar complexes is often complicated by the occurrence of alternative pathways. For instance, if a reaction of the form

$$[PtCl(dien)]^+ + I^- \longrightarrow [PtI(dien)]^+ + Cl^-$$

is first-order in the complex and independent of the concentration of I^-, then the rate of reaction will be equal to $k_1[PtCl(dien)^+]$. However, if there is a pathway in which the rate law is overall second-order, first-order in the complex, and first-order in the incoming group, then the rate would be given by $k_2[PtCl(dien)^+][I^-]$. If both reaction pathways occur at comparable rates, the rate law would have the form

$$\text{rate} = (k_1 + k_2[I^-])[PtCl(dien)^+] \tag{2}$$

Experimental evidence for this type of rate law comes from a study of substitution in the presence of excess I^- ions, when the rate law becomes pseudofirst-order:

$$\text{rate} = k_{obs}[PtCl(dien)^+] \qquad k_{obs} = k_1 + k_2[I^-] \tag{3}$$

---

[3]   Dien is diethylenetriamine, $NH_2CH_2CH_2NHCH_2CH_2NH_2$.

A plot of the observed pseudofirst-order rate constant against $[I^-]$ gives k_2 as the slope and k_1 as the intercept.

14.3 The nucleophilicity of the entering group

We deal first with the path that leads to second-order kinetics and consider the variation of the rate of the reaction as the entering group Y is varied. The reactivity of Y (for instance, I^- in the reaction above) can be expressed in terms of a **nucleophilicity parameter**, n_{Pt}:

$$n_{Pt} = \log \frac{k_2(Y)}{k_2^{\circ}} \qquad (4)$$

where $k_2(Y)$ is the second-order rate constant for the reaction

$$\textit{trans-}[PtCl_2(py)_2] + Y \longrightarrow \textit{trans-}[PtClY(py)_2]^+ + Cl^-$$

and k_2° is the rate constant for the same reaction with a reference nucleophile, methanol. The entering group is highly nucleophilic, or has a high nucleophilicity, if n_{Pt} is large.

Table 14.2 gives some values of n_{Pt}. One striking feature of the data is that, although the entering groups in the table are all quite simple, the rate constants span nearly nine orders of magnitude. Another feature is that the nucleophilicity of the entering group toward Pt appears to correlate with soft Lewis basicity, with $Cl^- < I^-$, $O < S$, and $NH_3 < PR_3$.

The nucleophilicity parameter is defined in terms of the reaction rates of a specific platinum complex. When the complex itself is varied we find that the reaction rates show a range of different sensitivities toward changes in the entering group. To express this range of sensitivities we rearrange eqn 4 into

$$\log k_2(Y) = n_{Pt}(Y) + C \qquad (5)$$

where $C = \log k_2^{\circ}$. Now consider the analogous substitution reactions for the general complex $[PtL_3X]$:

$$[PtL_3X] + Y \longrightarrow [PtL_3Y] + X$$

The relative rates of these reactions can be expressed in terms of the same nucleophilicity parameter n_{Pt} provided we replace eqn 5 by

$$\log k_2(Y) = S n_{Pt}(Y) + C \qquad (6)$$

The parameter S, which characterizes the sensitivity of the rate constant to the nucleophilicity parameter, is called the **nucleophilic discrimination factor**. We see from Fig. 14.2 that the straight line obtained by plotting $\log k_2(Y)$ against n_{Pt} for reactions of Y with *trans*-$[PtCl_2(PEt_3)_2]$ is steeper than that for reactions with *cis*-$[PtCl_2(en)]$. Hence, S is larger for the former reaction, which indicates that rate of the reaction is more sensitive to changes in the nucleophilicity of the entering group.

Some values of S are given in Table 14.3. Notice that S is close to 1 in all cases, so all the complexes are quite sensitive to n_{Pt}. This sensitivity is what we expect for associative reactions. Another feature to note is that larger values of S are found for complexes of platinum with softer base ligands.

> The nucleophilicity of an entering group is expressed in terms of the nucleophilicity parameter defined in terms of the substitution reactions of a specific square-planar platinum complex. The sensitivity of other platinum complexes to changes in the entering group is expressed in terms of the nucleophilic discrimination factor.

Table 14.2 A selection of n_{Pt} values for a range of nucleophiles

Nucleophile	Donor atom	n_{Pt}
Cl^-	Cl	3.04
C_6H_5SH	S	4.15
CN^-	C	7.14
$(C_6H_5)_3P$	P	8.93
CH_3OH	O	0
I^-	I	5.46
NH_3	N	3.07

Table 14.3 Nucleophilic discrimination factors

	S
trans-$[PtCl_2(PEt_3)_2]$	1.43
trans-$[PtCl_2(py)_2]$	1.00
$[PtCl_2(en)]$	0.64
$[PtCl(dien)]^+$	0.65

··

Example 14.1 Using the nucleophilicity parameter

The second-order rate constant for the reaction of I^- with *trans*-$[PtCl(CH_3)(PEt_3)_2]$ in methanol at $30\,^{\circ}C$ is $40\ L\,mol^{-1}\,s^{-1}$. The corresponding reaction with N_3^- has

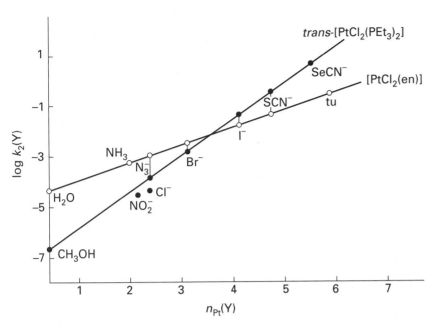

14.2 The slope of the straight line obtained by plotting $\log k_2(Y)$ against the nucleophilicity parameter $n_{Pt}(Y)$ for a series of ligands is a measure of the responsiveness of the complex to the nucleophilicity of the entering group. (The abbreviation tu denotes thiourea.)

$k_2 = 7.0 \text{ L mol}^{-1}\text{s}^{-1}$. Estimate S and C for the reaction given the n_{Pt} values of 5.42 and 3.58, respectively, for the two nucleophiles.

Answer Substituting the two values of n_{Pt} into eqn 5 gives

$$1.60 = 5.42S + C \qquad \text{(for I}^-)$$
$$0.85 = 3.58S + C \qquad \text{(for N}_3^-)$$

Solving these two simultaneous equations gives $S = 0.41$ and $C = -0.61$. The value of S is rather small, showing that the discrimination of this complex among different nucleophiles is not great. This lack of sensitivity is related to the rather large value of C, which corresponds to the rate constant being large and hence to the complex being reactive. It is commonly found that high reactivity correlates with low selectivity.

Self-test 14.1 Calculate the second-order rate constant for the reaction of the same complex with NO_2^-, for which $n_{Pt} = 3.22$.

14.4 The shape of the activated complex

Careful studies of the variation of the reaction rates of square-planar complexes with changes in the composition of the reactant complex and the conditions of the reaction shed light on the general shape of the activated complex. They also confirm that substitution almost invariably has an associative rate-determining stage.

(a) The *trans* effect

The spectator ligands T that are *trans* to the leaving group in square-planar complexes influence the rate of substitution. This phenomenon is called the **trans effect**. If T is a strong σ-donor ligand or π-acceptor ligand, then it greatly accelerates substitution of a ligand that lies *trans* to itself (Table 14.4):

For T a σ-donor: $OH^- < NH_3 < Cl^- < Br^- < CN^-, CO, CH_3^- < I^- < SCN^- < PR_3 < H^-$

For T a π-acceptor: $Br^- < I^- < NCS^- < NO_2^- < CN^- < CO, C_2H_4$

Table 14.4 The effect of the *trans* ligand in reactions of *trans*-[PtCl(PEt$_3$)$_2$L]

L	k_1/s^{-1}	$k_2/(\text{L mol}^{-1}\text{ s}^{-1})$
CH_3^-	1.7×10^{-4}	6.7×10^{-2}
$C_6H_5^-$	3.3×10^{-5}	1.6×10^{-2}
Cl^-	1.0×10^{-6}	4.0×10^{-4}
H^-	1.8×10^{-2}	4.2
PEt_3	1.7×10^{-2}	3.8

These orders are approximately the order of increasing overlap of the ligand orbitals with either a σ or a π Pt$5d$ orbital, and, the greater the overlap, the stronger the *trans* effect. It should be noticed that the same factors contribute to a large ligand-field splitting.

Example 14.2 Using the *trans* effect synthetically

Use the *trans* effect series to suggest synthetic routes to *cis*- and *trans*-[PtCl$_2$(NH$_3$)$_2$] from [Pt(NH$_3$)$_4$]$^{2+}$ and [PtCl$_4$]$^{2-}$.

Answer Reaction of [Pt(NH$_3$)$_4$]$^{2+}$ with HCl leads to [PtCl(NH$_3$)$_3$]$^+$. Because the *trans* effect of Cl$^-$ is greater than that of NH$_3$, substitution reactions will occur preferentially *trans* to Cl$^-$, and further action of HCl gives *trans*-[PtCl$_2$(NH$_3$)$_2$]. When the starting complex is [PtCl$_4$]$^{2-}$, reaction with NH$_3$ leads first to [PtCl$_3$(NH$_3$)]$^-$. A second step should substitute one of the two mutually *trans* Cl$^-$ ligands with NH$_3$ to give *cis*-[PtCl$_2$(NH$_3$)$_2$].

Self–test 14.2 Given the reactants PPh$_3$, NH$_3$, and [PtCl$_4$]$^-$, propose efficient routes to *cis*- and *trans*-[PtCl$_2$(NH$_3$)(PPh$_3$)].

The significant degree of correlation of the strength of the *trans* effect with the nucleophilicity of the entering group and the lability of the leaving group provides a strong clue to the structure of the activated complex. First we note that the two *cis* spectator ligands have little influence on the rate of substitution. Then we note that, if the *trans*, entering, and leaving ligands have complementary influences on the reaction rate, we can suspect that the activated complex has a trigonal-bipyramidal structure. In this model there are two locations of one kind (the axial locations, occupied by the two *cis* ligands) and three of another kind (the equatorial positions, occupied by **X**, **Y**, and **T**).

The process that might occur when **Y** attacks from above the plane of the initial complex and **X** is destined to leave from below it, is shown in (**1**).[4] The π-acceptor ligands may facilitate nucleophilic attack on a *d*-metal atom by removing some *d*-electron density from it (**2**).

> A strong σ-donor ligand or π-acceptor ligand greatly accelerates substitution of a ligand that lies in the trans position.

(b) Steric effects

Steric crowding at the reaction center usually inhibits associative reactions and facilitates dissociative reactions. Bulky groups can block the approach of attacking nucleophiles, and the reduction of coordination number in a dissociative reaction can relieve overcrowding.

The rate constants for the hydrolysis of *cis*-[PtClL(PEt$_3$)$_2$] complexes (the replacement of Cl$^-$ by H$_2$O) at 25 °C illustrate the point:

L =	pyridine	2-methylpyridine	2,6-dimethylpyridine
k/s^{-1}	8×10^{-2}	2.0×10^{-4}	1.0×10^{-6}

The methyl groups adjacent to the N donor atom greatly decrease the rate. In the 2-methylpyridine complex they block positions either above or below the plane. In the 2,6-dimethylpyridine complex they block positions both above and below the plane (**3**). Thus, along the series the methyl groups increasingly hinder attack by H$_2$O. The effect is smaller if L is *trans* to Cl$^-$. This difference is explained by the methyl groups then being further from the entering and leaving groups in the trigonal-bipyramidal activated complex if the pyridine ligand is in the trigonal plane (**4**).

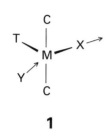

1

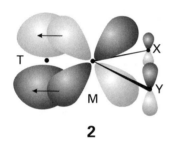

2

3

4

---

4   Note that the stereochemistry is quite different from that of *p*-block central atoms, such as Si(IV) and P(V), where the leaving group departs from an axial position.

Steric crowding at the reaction center usually inhibits associative reactions and facilitates dissociative reactions.

(c) Stereochemistry

Further insight into the nature of the activated complex is obtained from the observation that substitution of a square-planar complex preserves the original geometry. That is, a *cis* complex gives a *cis* product and a *trans* complex gives a *trans* product. This behavior is explained by the formation of an approximately trigonal-bipyramidal activated complex with the entering, leaving, and *trans* groups in the trigonal plane. The steric course of the reaction is shown in Fig. 14.3. We can expect a *cis* ligand to exchange places with the T ligand in the trigonal plane only if the intermediate lives long enough to be stereomobile. That is, it must be a long-lived associative (A) intermediate.

Substitution of a square-planar complex preserves the original geometry, which supports a trigonal-pyramidal activated complex.

(d) Temperature and pressure dependence

Another clue to the nature of the activated complex comes from determining the entropies and volumes of activation for reactions of Pt(II) and Au(III) complexes (Table 14.5). The entropy of activation is obtained from the temperature dependence of the rate constant, and effectively indicates the change in disorder (of reactants and solvent) when the activated complex forms. Likewise, the volume of activation, which is obtained (with considerable difficulty) from the pressure dependence of the rate constant, is the change in volume that occurs on formation of the activated complex. The two striking aspects of the data in the table are the consistently strongly negative values of both quantities. The simplest explanation of

Table 14.5 Activation parameters for substitution in square-planar complexes (in methanol as solvent)

Reaction	k_1			k_2		
	$\Delta^\ddagger H$	$\Delta^\ddagger S$	$\Delta^\ddagger V$	$\Delta^\ddagger H$	$\Delta^\ddagger S$	$\Delta^\ddagger V$
trans-[PtCl(NO$_2$)(py)$_2$] + py				50	−100	−38
trans-[PtBrP$_2$(mes)]* + SC(NH$_2$)$_2$	71	−84	−46	46	−138	−54
cis-[PtBrP$_2$(mes)]* + I$^-$	84	−59	−67	63	−121	−63
cis-[PtBrP$_2$(mes)]* + SC(NH$_2$)$_2$	79	−71	−71	59	−121	−54
[AuCl(dien)]$^{2+}$ + Br$^-$				54	−17	

*[PtBrP$_2$(mes)] is [PtBr(PEt$_3$)$_2$ (2,4,6-(CH$_3$)$_3$C$_6$H$_2$)].
Enthalpy in kJ mol^{-1}, entropy in J K^{-1} mol^{-1}, and volume in cm^3 mol^{-1}.

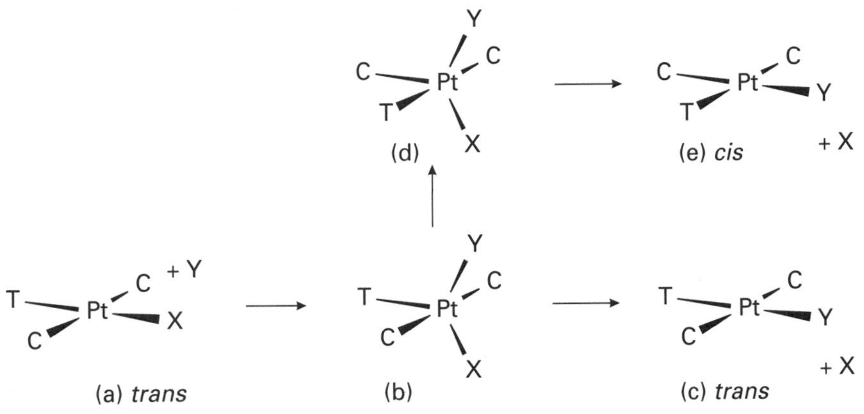

14.3 The stereochemistry of substitution in a square-planar complex. The normal path (resulting in retention) is from (a) to (c). However, if the intermediate (b) is sufficiently long-lived, it can undergo pseudorotation to (d), which lends to the isomer (e).

the reduction in disorder and the reduction in volume is that the entering ligand is being incorporated into the activated complex without release of the leaving group. That is, we can conclude that the rate-determining step is associative.

> Negative volumes and entropies of activation support the view that the rate-determining step of square-planar Pt(II) complexes is associative.

14.5 The k_1 pathway

Now we consider the first-order pathway for the substitution of square-planar complexes. The first issue we must address is the first-order pathway in the rate equation and decide whether k_1 in the rate law in eqn 2 and its generalization

$$\text{rate} = (k_1 + k_2[\text{Y}])[\text{PtL}_4] \tag{7}$$

does indeed represent the operation of an entirely different reaction mechanism. It turns out that it does not, and k_1 represents an associative reaction involving the solvent. Thus, the substitution of Cl^- by pyridine in methanol as solvent proceeds in two steps, with the first rate-determining:

$$[\text{PtCl(dien)}]^+ + \text{CH}_3\text{OH} \longrightarrow [\text{Pt(dien)(CH}_3\text{OH)}]^{2+} + \text{Cl}^- \quad \text{(slow)}$$
$$[\text{Pt(dien)(CH}_3\text{OH)}]^{2+} + \text{py} \longrightarrow [\text{Pt(dien)(py)}]^{2+} + \text{CH}_3\text{OH} \quad \text{(fast)}$$

The evidence for this two-step mechanism comes from a correlation of the rates of these reactions with the nucleophilicity parameters of the solvent molecules and the observation that reactions of entering groups with solvent complexes are rapid compared to the step in which the solvent displaces a ligand.

> The first-order contribution to the rate law in eqn 7 is in fact a pseudofirst-order process in which the solvent participates.

Substitution in octahedral complexes

The interchange mechanism plays a central role in the discussion of octahedral substitution. But is the rate-determining step associative or dissociative? The analysis of rate laws for reactions that take place by such a mechanism helps to formulate the precise conditions for distinguishing these two possibilities and identifying the substitution as I_a (interchange with an associative rate-determining stage) or I_d (interchange with a dissociative rate-determining stage). The difference between the two classes of reaction hinges on whether the rate-determining step is the formation of the new $\text{Y} \cdots \text{M}$ bond or the breaking of the old $\text{M} \cdots \text{X}$ bond.

14.6 Rate laws and their interpretation

The rate law of an octahedral substitution by an interchange mechanism is usually consistent with an **Eigen–Wilkins mechanism**. In this mechanism, an encounter complex is formed in a pre-equilibrium step. This complex then rearranges to form the products.

(a) The Eigen–Wilkins mechanism

The first step in the Eigen–Wilkins mechanism is an encounter in which the complex ML_6 and the entering group Y diffuse together and come into contact. They may also separate at diffusion-limited rates (that is, at a rate governed by their ability to migrate by diffusion through the solvent), and the equilibrium

$$\text{ML}_6 + \text{Y} \rightleftharpoons \text{ML}_6\text{Y} \qquad K_E = \frac{[\{\text{ML}_6, \text{Y}\}]}{[\text{ML}_6][\text{Y}]} \tag{8}$$

is established, where $\{ML_6, Y\}$ denotes the encounter complex. Because in ordinary solvents the lifetime of a diffusional encounter is approximately 1 ns, the formation of an encounter complex can be treated as a pre-equilibrium in all reactions that take longer than a few nanoseconds. A special case arises if Y is the solvent. Then the encounter equilibrium is 'saturated' in the sense that, because the complex is always surrounded by solvent, a solvent molecule is always available to take the place of one that leaves the complex.

The second step is the rate-determining reaction of the encounter complex to give products:

$$\{ML_6, Y\} \longrightarrow \text{Product} \qquad \text{rate} = k[\{ML_6, Y\}] \tag{9}$$

When the equilibrium equation is solved for the concentration of ML_6Y and the result substituted into the rate law for the rate-determining step, the overall rate law is found to be

$$\text{rate} = \frac{kK_E[C]_{\text{tot}}[Y]}{1 + K_E[Y]} \tag{10}$$

where $[C]_{\text{tot}}$ represents the sum of the molar concentrations of all complex species. An example is the reaction

$$[Ni(OH_2)_6]^{2+} + NH_3 \rightleftharpoons \{[Ni(OH_2)_6]^{2+}, NH_3\}$$

with equilibrium constant K_E and

$$\{[Ni(OH_2)_6]^{2+}, NH_3\} \longrightarrow [Ni(OH_2)_5NH_3]^{2+} + H_2O$$

with rate constant k; then

$$\text{rate} = \frac{kK_E[Ni^{2+}]_{\text{tot}}[NH_3]}{1 + K_E[NH_3]} \tag{11}$$

It is rarely possible to conduct experiments over a range of concentrations wide enough to test the rate law in eqn 11 exhaustively (see Problem 14.4). However, at such low concentrations of the entering group that $K_E[Y] \ll 1$, the rate law reduces to the second-order form

$$\text{rate} = k_{\text{obs}}[C]_{\text{tot}}[Y] \qquad k_{\text{obs}} = kK_E \tag{12}$$

Because k_{obs} can be measured and K_E can be either measured or estimated as we describe below, the rate constant k can be found from k_{obs}/K_E. The results for reactions of Ni(II) hexaaqua complexes with various nucleophiles are shown in Table 14.6. The very small variation in k indicates a model I_d reaction with very slight response to the nucleophilicity of the entering group.

In the special case that Y is the solvent and the encounter equilibration is saturated, $K_E[Y] \gg 1$ and $k_{\text{obs}} = k$. Thus, reactions with the solvent can be directly compared to reactions with other entering ligands without needing to estimate the value of K_E.

Table 14.6 Complex formation by the $[Ni(OH_2)_6]^{2+}$ ion

Ligand	$k_{\text{obs}}/(\text{L mol}^{-1}\,\text{s}^{-1})$	$K_E/(\text{L mol}^{-1})$	$(k_{\text{obs}}/K_E)/\text{s}^{-1}$
$CH_3CO_2^-$	1×10^5	3	3×10^4
F^-	8×10^3	1	8×10^3
HF	3×10^3	0.15	2×10^4
H_2O^*			3×10^3
NH_3	5×10^3	0.15	3×10^4
$[NH_2(CH_2)_2NH_3]^+$	4×10^2	0.02	2×10^3
SCN^-	6×10^3	1	6×10^3

*The solvent is always in encounter with the ion so that K_E is undefined and all rates are inherently first-order.

> In the Eigen–Wilkins mechanism for the nucleophilic substitution of an octahedral complex, an encounter complex is formed in a pre-equilibrium step, and the encounter complex forms products in a subsequent rate-determining step.

(b) The Fuoss–Eigen equation

The equilibrium constant K_E for the encounter complex can be estimated theoretically by using a simple equation proposed independently by R.M. Fuoss and M. Eigen. Both sought to take the particle size and charge into account, expecting that larger, oppositely charged particles would meet more frequently than small ions of the same charge. Fuoss used an approach based on statistical thermodynamics and Eigen used one based on kinetics. Their result, which is called the **Fuoss–Eigen equation**, is

$$K_E = \tfrac{4}{3}\pi a^3 N_A e^{-V/RT} \tag{13}$$

In this expression a is the distance of closest approach, V the Coulombic potential energy $(z_1 z_2 e^2 / 4\pi\varepsilon a)$ of the ions at that distance, and N_A is the Avogadro constant. Although the value predicted by this equation depends strongly on the details of the charges and radii of the ions, typically it favors the encounter if the reactants are large (so a is large) and oppositely charged (V negative).

> The Fuoss–Eigen equation, eqn 13, provides an estimate of the pre-equilibrium constant in the Eigen–Wilkins mechanism.

14.7 The activation of octahedral complexes

Many studies of substitution in octahedral complexes support the view that the rate-determining step is dissociative, and we summarize these studies first. However, the reactions of octahedral substitutions can acquire a distinct associative character in the case of large central ions (as in the second and third series of the *d*-block) or where the *d*-electron population at the metal is low (the early members of the *d*-block). More room for attack or lower π^* electron density appears to facilitate nucleophilic attack and hence permit association.

(a) Leaving-group effects

We can expect the identity of the leaving group X to have a large effect in dissociatively activated reactions because their rate depends on the scission of the $M \cdots X$ bond. When X is the only variable, as in the reaction

$$[CoX(NH_3)_5]^{2+} + H_2O \longrightarrow [Co(NH_3)_5(OH_2)]^{3+} + X^-$$

it is found that there is a linear relation between the logarithms of the rate constants and equilibrium constants of the reaction (Fig. 14.4), and that specifically

$$\ln k = \ln K + c \tag{14}$$

Because both logarithms are proportional to Gibbs energies ($\ln k$ is approximately proportional to the activation Gibbs energy, $\Delta^{\ddagger}G$, and $\ln K$ is proportional to the standard reaction Gibbs energy, $\Delta_r G^{\ominus}$), we can write

$$\Delta^{\ddagger}G = p\Delta_r G^{\ominus} + b \tag{15}$$

with p and b constants (and $p \approx 1$). The existence of this **linear free energy relation** (LFER) of unit slope shows that changing X has the same effect on $\Delta^{\ddagger}G$ for the conversion of Co—X to the activated complex as it has on $\Delta_r G^{\ominus}$ for the complete elimination of X^- (Fig. 14.5). This observation in turn suggests that in an I_d reaction the leaving group (an anionic ligand) has become a solvated ion in the activated complex.

A similar LFER (with a slope smaller than 1.0, indicating some associative character) is observed for the corresponding complexes of Rh(III). However, as we should expect for the

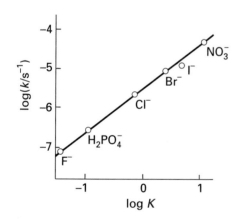

14.4 The straight line obtained when the logarithm of a rate constant is plotted against the logarithm of an equilibrium constant shows the existence of a linear free energy relation. This graph is for the reaction $[CoX(NH_3)_5]^{2+} + H_2O \rightarrow [Co(NH_3)_5(OH_2)]^{3+} + X^-$ with different leaving groups X.

softer Rh(III) center, the order of leaving groups is reversed, and the reaction rates are in the order $I^- < Br^- < Cl^-$ because the softer acid, Rh(III), forms stronger bonds to I^-.

> A large effect of the leaving group X is expected in I_d reactions. A linear relation is found between the logarithms of the rate constants and equilibrium constants.

(b) The effects of spectator ligands

In Co(III), Cr(III), and related octahedral complexes, both *cis* and *trans* ligands affect rates of substitution in proportion to the strength of the bonds they form with the metal. There is no important *trans* effect. For instance, hydrolysis reactions such as

$$[NiXL_5]^+ + H_2O \longrightarrow [NiL_5(OH_2)] + X^-$$

are much faster when L is NH_3 than when it is H_2O. This difference can be explained on the grounds that, as NH_3 is a stronger σ-donor than H_2O, it increases the electron density at the metal and hence facilitates the scission of the $M—X$ bond and the formation of X^-. In the activated complex, the good donor stabilizes the reduced coordination number.

> There is no important *trans* effect in octahedral complexes. Both *cis* and *trans* ligands affect rates of substitution in proportion to the strength of the bonds they form with the metal.

(c) Steric effects

Steric effects on reactions with dissociative rate-determining steps can be illustrated by considering the rate of hydrolysis of the first Cl^- ligand in two complexes of the type $[CoCl_2(bn)_2]^+$:

$$[CoCl_2(bn)_2]^+ + H_2O \longrightarrow [CoCl(bn)_2(OH_2)]^{2+} + Cl^-$$

The ligand bn is 2,3-butanediamine, and may be either chiral (**5**) or achiral (**6**). The important observation is that the complex formed with the chiral form of the ligand hydrolyzes 30 times more slowly than the complex of the achiral form. The two ligands have very similar electronic effects, but the CH_3 groups are on the opposite sides of the chelate ring in (**5**) and adjacent and crowded in (**6**). The latter arrangement is more reactive because the strain is relieved in the dissociative activated complex with its reduced coordination number. In general, steric crowding favors an I_d process because the five-coordinate activated complex can relieve strain.

Quantitative treatments of the steric effects of ligands have been developed using molecular modeling computer software that takes into account van der Waals interactions.[5] However, a more pictorial semiquantitative approach was introduced by C.A. Tolman. It assesses the extent to which various ligands (especially phosphines) crowd each other by approximating the volume occupied by a ligand by a cone with an angle determined from a space-filling model and, for phosphine ligands, an $M—P$ bond length of 2.28 Å (Fig. 14.6 and

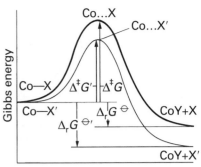

14.5 The existence of a linear free energy relation with unit slope shows that changing X has the same effect on $\Delta^{\ddagger}G$ for the conversion of $M—X$ to the activated complex as it has on $\Delta_rG^{\ominus}$ for the complete elimination of X^-. This reaction profile shows the effect of changing the leaving group from X to X'.

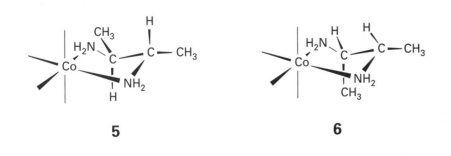

5   **6**

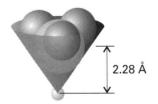

14.6 The determination of ligand cone angles from space-filling molecular models of the ligand and an assumed $M—P$ bond length of 2.28 Å.

5   J.F. Endicott, K. Kumar, C.L. Schwarz, W. Perkovic, and W. Lin, *J. Am. Chem. Soc.* **111**, 7411 (1989).

Table 14.7 Tolman cone angle for various ligands

Ligand	$\theta/^\circ$	Ligand	$\theta/^\circ$
CH_3	90	$P(OCH_3)_3$	128
CO	95	$P(OC_2H_5)_3$	134
Cl, Et	102	$\eta^5\text{-}C_5H_5$ (Cp)	136
PF_3	104	PEt_3	137
Br, Ph	105	PPh_3	145
I, $P(OCH_3)_3$	107	$C_5(CH_3)_5$ (Cp*)	165
$P(CH_3)_3$	118	$2,4\text{-}(CH_3)_2C_5H_3$	180
t-butyl	126	$P(t\text{-}Bu)_3$	182

Source: C.A. Tolman, *Chem. Rev.* **77**, 313 (1975); L. Stahl and R.D. Ernst, *J. Am. Chem. Soc.* **109**, 5673 (1987).

Table 14.7). The ligand CO is small in the sense of having a small cone angle; $P(^tBu)_3$ is regarded as bulky because it has a large cone angle. Bulky ligands have considerable steric repulsion with each other when packed around a metal center. They favor dissociative activation and inhibit associative activation.

The rate of the reaction of $[Ru(SiCl_3)_2(CO)_3(PR_3)]$ with **Y** to give $[Ru(SiCl_3)_2(CO)_2Y(PR_3)]$ is independent of the identity of **Y**, which suggests that the rate-determining step is dissociative. Furthermore, it has been found that there is only a small variation in rate for substituents with similar cone angles but significantly different values of pK_a. This observation supports the assignment of the rate changes to steric effects because changes in pK_a should correlate with changes in electron distributions in the ligands.

> *Steric crowding favors dissociative activation because the formation of the activated complex can relieve strain.*

(d) Associative activation

As we have seen, activation volumes reflect the changes in compactness (including that of the surrounding solvent) when the activated complex forms from the reactants. The last column in Table 14.8 gives $\Delta^\ddagger V$ for some H_2O ligand exchange reactions. We see that $\Delta^\ddagger V$ increases from -4.1 $cm^3\,mol^{-1}$ for V^{2+} to $+7.2$ $cm^3\,mol^{-1}$ for Ni^{2+}. Because the negative volume of activation can be interpreted as the result of shrinkage when an H_2O molecule enters the activated complex, we can infer that the activation has significant associative character. The increase in $\Delta^\ddagger V$ follows the increase in the number of nonbonding d electrons from d^3 to d^8 across the first row of the d-block. In the earlier part of the d-block, associative reaction appears to be favored by a low population of d electrons.

Table 14.8 Activation parameters for the H_2O exchange reactions $[M(OH_2)_6]^{2+} + H_2^{17}O \rightarrow [M(OH_2)_5(^{17}OH_2)]^{2+} + H_2O$

	$\Delta^\ddagger H/$ (kJ mol^{-1})	LFSE*/ Δ_O	LFSE†/ Δ_O	LFAE/ Δ_O	$\Delta^\ddagger V/$ (cm^3 mol^{-1})
$Ti^{2+}(d^2)$		8	9.1	-1.1	
$V^{2+}(d^3)$	68.6	12	10	2	-4.1
$Cr^{2+}(d^4$, hs)		6	9.1	-3.1	
$Mn^{2+}(d^5$, hs)	33.9	0	0	0	-5.4
$Fe^{2+}(d^6$, hs)	31.2	4	4.6	-0.6	$+3.8$
$Co^{2+}(d^7$, hs)	43.5	8	9.1	-1.1	$+6.1$
$Ni^{2+}(d^8)$	58.1	12	10	2	$+7.2$

*Octahedral.
†Square-pyramidal.
hs: High spin.

Table 14.9 Kinetic parameters for anion attack on Cr(III)*

| X | $L = H_2O$ | | | $L = NH_3$ |
	$k/(10^{-8} \text{ mol}^{-1} \text{ s}^{-1})$	$\Delta^{\ddagger}H$	$\Delta^{\ddagger}S$	$k/(10^{-4} \text{ L mol}^{-1} \text{ s}^{-1})$
Br^-	0.46	122	8	3.7
Cl^-	1.15	126	38	0.7
NCS^-	48.7	105	4	4.2

*Enthalpy in kJ mol^{-1}, entropy in $\text{J K}^{-1} \text{ mol}^{-1}$.

Negative volumes of activation are also observed in the second and third rows of the block, such as for Rh(III), and also indicate association of the entering group into the activated complex in the reactions of these complexes. Associative activation begins to dominate when the metal center is more accessible to nucleophilic attack, either because it is large or has a low (nonbonding or π^*) d-electron population, and the mechanism shifts from I_d toward I_a. Table 14.9 shows some data for the formation of Br^-, Cl^-, and NCS^- complexes from $[Cr(NH_3)_5(OH_2)]^{3+}$ and $[Cr(OH_2)_6]^{3+}$. The pentaammine complex shows only a weak nucleophile dependence, in contrast to the strong dependence of the hexaaqua complex. The two complexes probably mark the transition from I_d to I_a.

The rate constants for the replacement of H_2O in $[Cr(OH_2)_6]^{3+}$ by Cl^-, Br^-, or NCS^- are smaller than those for the analogous reactions of $[Cr(NH_3)_5(OH_2)]^{3+}$ by a factor of about 10^4 (see Table 14.9). This difference suggests that the NH_3 ligands, which are stronger σ-donors than H_2O, promote dissociation of the sixth ligand more effectively. As we saw above, this is to be expected in dissociatively activated reactions.

> Negative volumes of activation indicate association of the entering group into the activated complex.

Example 14.3 Interpreting kinetic data in terms of a mechanism

The second-order rate constants for formation of $[VX(OH_2)_5]^+$ from $[V(OH_2)_6]^{2+}$ and X^- for $X^- = Cl^-$, NCS^-, and N_3^- are in the ratio 1:2:10. What do the data suggest about the rate-determining step for the substitution reaction?

Answer Because all three ligands are singly charged anions of similar size, the encounter equilibrium constants are similar, and second-order rate constants (which are equal to $K_E k_2$) are proportional to first-order rate constants for substitution in the encounter complex, k_2. The greater rate constants for NCS^- than for Cl^- and especially the fivefold difference of NCS^- from its close structural analog N_3^-, suggest some contribution from nucleophilic attack and an associative reaction. There is no such systematic pattern for the same anions reacting with Ni(II), for which the reaction is believed to be dissociative.

Self-test 14.3 Using the data in Table 14.8, estimate an appropriate value for K_E and calculate k_2 for the reactions of V(II) with Cl^- if the observed second-order rate constant is $1.2 \times 10^2 \text{ L mol}^{-1} \text{ s}^{-1}$.

14.8 Stereochemistry

Classic examples of octahedral substitution stereochemistry are provided by Co(III) complexes, and in Table 14.10 we give some data for the hydrolysis of *cis*- and *trans*-$[CoAX(en)_2]^{2+}$ where X is the leaving group (either Cl^- or Br^-) and A is OH^-, NCS^-, or Cl^-. The stereochemical consequences of substitution of octahedral complexes are more involved than

Table 14.10 Stereochemical course of hydrolysis reactions of $[CoAX(en)_2]^+$

	A	X	Percentage *cis* in product
cis	OH^-	Cl^-	100
	Cl^-	Cl^-	100
	NCS^-	Cl^-	100
	Cl^-	Br^-	100
trans	NO_2^-	Cl^-	0
	NCS^-	Cl^-	50–70
	Cl^-	Cl^-	35
	OH^-	Cl^-	75

X is the leaving group.

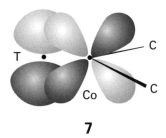

7

those of square-planar complexes. The *cis* complexes do not undergo isomerization when substitution occurs. The *trans* complexes show a tendency to isomerize to *cis* that increases in the order

$$A = NO_2^- < Cl^- < NCS^- < OH^-$$

The data can be understood in terms of an I_d mechanism by recognizing that the five-coordinate metal center in the activated complex may resemble either of the two stable geometries for five-coordination, namely square-pyramidal or trigonal-bipyramidal. As we can see from Fig. 14.7, reaction through the square-pyramidal complex results in retention of the original geometry, but reaction through the trigonal-bipyramidal complex can lead to isomerization. The results suggest that a good π-donor ligand *trans* to the leaving group Cl^- favors isomerization, because it can stabilize a trigonal bipyramid by participating in π bonding (**7**).

> Reaction through the square-pyramidal complex results in retention of the original geometry, but reaction through the trigonal-bipyramidal complex can lead to isomerization.

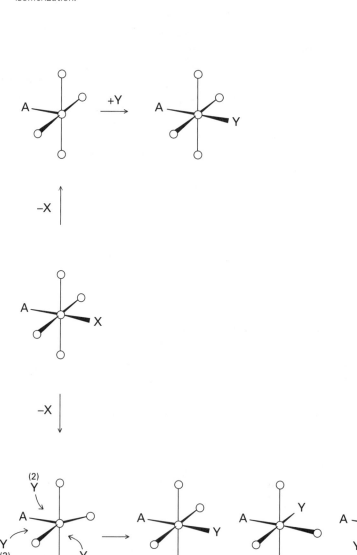

14.7 Reaction through a square-pyramidal complex (top path) results in retention of the original geometry but reaction through a trigonal-bipyramidal complex (bottom path) can lead to isomerization.

14.9 Base hydrolysis

Octahedral substitution is commonly greatly accelerated by OH^- ions when ligands with acidic protons are present. A typical rate law for such reactions is

$$\text{rate} = k[\text{CoCl(NH}_3)_5^{2+}][\text{OH}^-] \tag{16}$$

However, an extended series of studies has shown that the mechanism is not a simple bimolecular attack by OH^- on the complex. The stereochemistry of the reaction indicates that a strong π-donor ligand is present and that the steric effects observed are characteristic of dissociative activation.

There is a considerable body of indirect evidence relating to the problem, but one elegant experiment makes the essential point. This conclusive evidence comes from a study of ^{18}O isotope distribution in the product $[\text{Co(OH)(NH}_3)_5]^{2+}$, where it is found that the $^{18}O/^{16}O$ ratio differs between water and hydroxide ion at equilibrium. The isotope ratio in the product matches that for water, not that for the OH^- ions. Therefore, an H_2O molecule and not an OH^- ion appears to be the entering group.

The mechanism that takes these observations into account supposes that the role of OH^- is to act as a Brønsted base, not an entering group:

$$[\text{CoCl(NH}_3)_5]^{2+} + \text{OH}^- \rightleftharpoons [\text{CoCl(NH}_2)(\text{NH}_3)_4]^+ + \text{H}_2\text{O}$$

$$[\text{CoCl(NH}_2)(\text{NH}_3)_4]^+ \longrightarrow [\text{Co(NH}_2)(\text{NH}_3)_4]^{2+} + \text{Cl}^- \quad \text{(slow)}$$

$$[\text{Co(NH}_2)(\text{NH}_3)_4]^{2+} + \text{H}_2\text{O} \longrightarrow [\text{Co(OH)(NH}_3)_5]^{2+} \quad \text{(fast)}$$

In the first step, an NH_3 ligand acts as a Brønsted acid, resulting in the formation of its conjugate base, the NH_2^- ion, as a ligand. Because NH_2^- is a strong π-donor, it greatly accelerates Cl^- ion loss (see Example 14.4).

> Octahedral substitution is commonly greatly accelerated by OH^- ions when ligands with acidic protons are present.

14.10 Isomerization reactions

Isomerization reactions are closely related to substitution reactions; indeed, a major pathway for isomerization is often via substitution. The square-planar Pt(II) and octahedral Co(III) complexes we have discussed can both form five-coordinate trigonal-bipyramidal activated complexes. The interchange of the axial and equatorial ligands in a trigonal-bipyramidal complex can be pictured as occurring by a pseudorotation through a square pyramid (Fig. 14.8). As we saw above, when a trigonal bipyramid adds a ligand to produce a six-coordinate complex, a new direction of attack of the entering group can lead to an isomerization.

If a chelate ligand is present, isomerization can occur as a consequence of metal–ligand bond breaking, and substitution need not occur. An example is the exchange of the

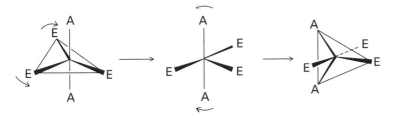

14.8 The exchange of axial and equatorial ligands by a twist through a square-pyramidal conformation of the complex.

'outer' CD_3 group with the 'inner' CH_3 group during the isomerization of tris(acetylacetylacetonato)cobalt(III), (**8**)→ (**9**).

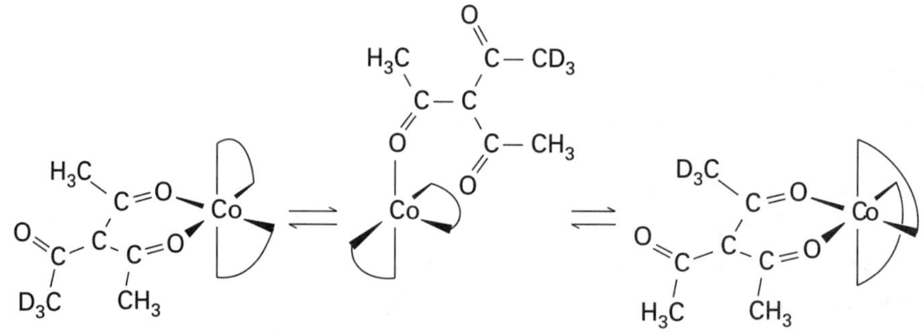

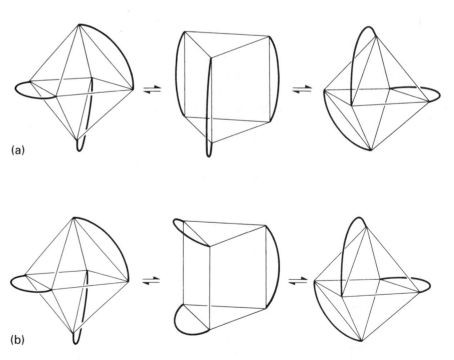

An octahedral complex can also undergo isomerization by an intramolecular twist without loss of a ligand or breaking of a bond. There is evidence, for example, that racemization of $[Ni(en)_3]^{2+}$ occurs by such an internal twist. Two possible paths are the **Bailar twist** (Fig. 14.9a) and the **Ray–Dutt twist** (Fig. 14.9b).

> *Isomerization of a chelate ligand can take place by mechanisms that involve substitution, bond-cleavage and reformation, and a variety of twists.*

Example 14.4 Interpreting the stereochemical course of a reaction

Substitution of Co(III) complexes of the type $[CoAX(en)_2]^+$ results in *trans* to *cis* isomerization, but only when the reaction is catalyzed by a base. Show that this observation is consistent with the model in Fig. 14.9.

(a)

(b)

14.9 (a) The Bailar twist and (b) the Ray–Dutt twist by which an octahedral complex can undergo isomerization without losing a ligand or breaking a bond.

Answer For base hydrolysis, the conjugate base is the strong π-donor ligand $:\ddot{N}HR^-$. A trigonal bipyramid of the type shown in Fig. 14.9b is possible, and it may be attacked in the way shown there.

Self-test 14.4 If the directions of attack on the above trigonal bipyramid are random, what cis-trans isomer distribution does the analogy with Fig. 14.9 suggest for products of the type $[CoA(NH_3)(OH_2)]^{2+}$?

14.11 Involvement of the ligand: alkyl migration and CO insertion

Substitution reaction pathways are more complex when an atom belonging to a ligand other than an entering or leaving group becomes directly involved. For instance, in substitution reactions of alkyl carbonyl complexes such as

$$[CH_3Mn(CO)_5] + PPh_3 \longrightarrow [CH_3Mn(CO)_4PPh_3] + CO$$

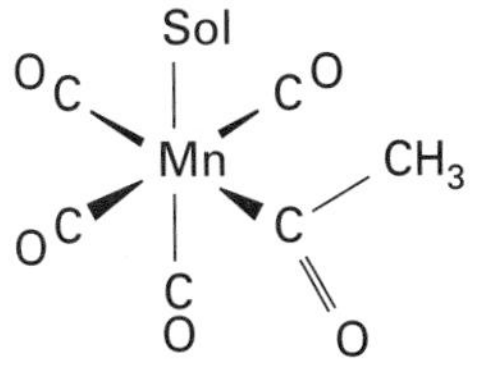

there is evidence that an acyl intermediate (**10**) is involved, where **Sol** is a solvent molecule. This intermediate arises from the cleavage of the **M—CH₃** bond and its replacement by an $H_3CC(O)—$ bond. The substitution reaction is commonly completed by replacement of **Sol** by a more strongly binding ligand and restoration of the CH₃—M bond with loss of CO. When the entering ligand is CO itself, the net outcome of the reaction is **CO insertion** of the form

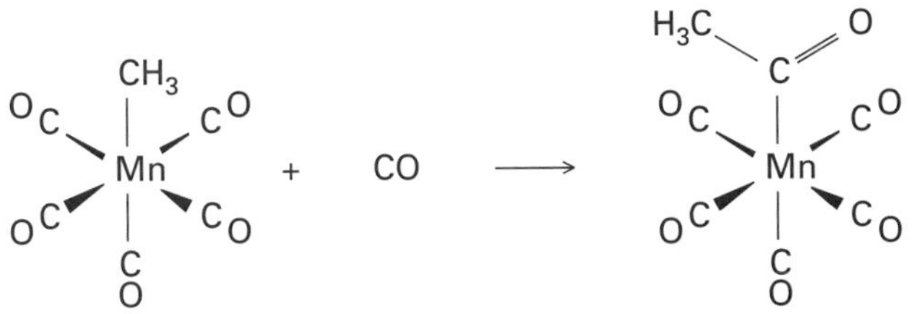

10

11

The mechanism of the reaction can be similar to that of the PPh₃ substitution above, up to the formation of the acyl intermediate. Labeling studies using IR and ^{13}C-NMR show for this complex that the stereochemistry at the metal atom requires a methyl group to move. It is proposed that the formation of the acyl intermediate is by **methyl migration**, in which the CH₃ group engages in nucleophilic attack on the C atom of a neighboring CO ligand (**11**). This interpretation of the reaction is consistent with the activation parameters for the reaction, especially $\Delta^{\ddagger}S = -88.2 \text{ J K}^{-1} \text{ mol}^{-1}$, the negative value indicating incorporation of an additional ligand into the activated complex. Moreover, electron-withdrawing groups substituted into CH₃— dramatically slow the reaction. This effect would be expected if CH₃ functioned as a nucleophile. For example, $[Mn(CH_2NO_2)(CO)_5]$ is much less reactive than $[MnCH_3(CO)_5]$.

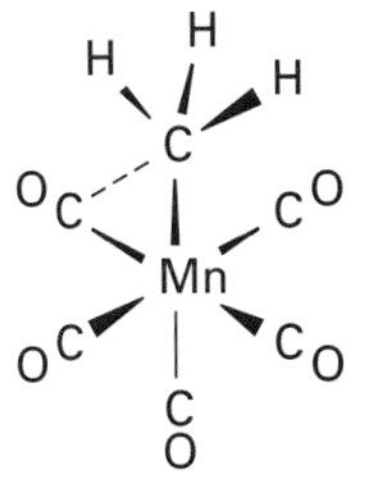

Substitution reactions of alkyl carbonyl complexes can involve an acyl intermediate formed by alkyl migration.

Redox reactions

As remarked in Chapter 6, redox reactions can occur by the direct transfer of electrons (as in some electrochemical cells). Alternatively, they may occur by the transfer of atoms and ions, as in the transfer of O atoms in reactions of oxoanions. Because redox reactions involve both an oxidizing and a reducing agent, they are usually bimolecular in character. The rare exceptions are reactions in which one molecule has both oxidizing and reducing centers.

14.12 The classification of redox reactions

In the 1950s, Henry Taube[6] identified two mechanisms of redox reactions. One is the **inner-sphere mechanism**, which includes atom transfer processes. In an inner-sphere mechanism, the coordination spheres of the reactants share a ligand transitorily and form a bridged intermediate activated complex. The other is an **outer-sphere mechanism**, which includes many simple electron transfers. In an outer-sphere mechanism, the complexes come into contact without sharing a bridging ligand.

(a) Inner-sphere mechanisms

The inner-sphere mechanism was first confirmed for the reduction of the nonlabile complex $[CoCl(NH_3)_5]^{2+}$ by $Cr^{2+}(aq)$, because the products of the reaction included both Co^{2+} and $[CrCl(OH_2)_5]^{2+}$. Moreover, addition of $^{36}Cl^-$ to the solution did not lead to the incorporation of any of the isotope in the Cr(III) product. Furthermore, the reaction is faster than reactions that remove Cl^- from Co(III) or introduce Cl^- into the nonlabile $[Cr(OH_2)_6]^{3+}$ complex. These observations must mean that Cl^- has moved directly from the coordination sphere of one complex to that of the other during the reaction.

The Cl^- attached to Co(III) can easily enter into the labile coordination sphere of $[Cr(OH_2)_6]^{2+}$ to produce a bridged intermediate (**12**). Good bridging ligands are those with more than one pair of electrons available to donate to the two metal centers; they include:

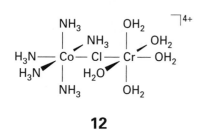

12

$$:\!\overset{\cdot\cdot}{\underset{\cdot\cdot}{Cl}}\!:^- \qquad :S\!-\!C\!\equiv\!N\!:^- \qquad :N\!\equiv\!N\!: \qquad :\overset{\cdot\cdot}{N}\!=\!N\!=\!\overset{\cdot\cdot}{N}\!:^- \qquad :C\!\equiv\!N\!:^-$$

> The original evidence for an inner-sphere mechanism of a redox reaction as set out above was the observation of accompanying ligand transfer.

(b) Outer-sphere mechanism

An example of an outer-sphere mechanism is the reduction of $[Fe(phen)_3]^{3+}$ by $[Fe(CN)_6]^{4-}$. Both reactants have nonlabile coordination spheres, and no ligand substitution can occur on the very short timescale of the redox reaction. The only possible mechanism is an electron transfer between the two complex ions in outer-sphere contact.

An understanding of ligand substitution is a prerequisite to the full analysis of a redox reaction mechanism. If a redox reaction is faster than ligand substitution, then the reaction has an outer-sphere mechanism. Similarly, there is no difficulty in assigning an inner-sphere mechanism when the reaction involves ligand transfer from an initially nonlabile reactant to a nonlabile product. Unfortunately, it is difficult to make unambiguous assignments when the complexes are labile. Much of the study of well defined examples is directed toward the identification of the parameters that differentiate the two paths in the hope of being able to making correct assignments in more difficult cases.

> If a redox reaction is faster than ligand substitution, then the reaction has an outer-sphere mechanism.

14.13 The theory of redox reactions

The theory of the mechanisms of redox reactions is well established. This is especially so for outer-sphere reactions where in favorable cases it is even possible to calculate rate constants.

---

6 Taube has given an excellent retrospective account in his 1983 Nobel Prize lecture reprinted in *Science* **226**, 1028 (1984). Informative chapters on the mechanisms of the redox reactions of coordination compounds are also available in *Prog. Inorg. Chem.* **30** (1983).

(a) The concepts needed to discuss outer-sphere processes

The analysis of the outer-sphere mechanism depends on two concepts. One is the **Born–Oppenheimer approximation**, which implies that electron distributions can be calculated by assuming that the nuclei are stationary at a given location. This separation of the motion of electrons and nuclei is based on the great difference of mass of electrons and nuclei, and hence the former's ability to respond almost instantaneously to rearrangements of nuclei and, conversely, for the nuclei to respond only sluggishly to movement of the electrons. If we assume that the nuclei are fixed in the intermediate (transition-state) configuration, we can picture the wavefunction of the migrating electron as distributed over both centres. It is also energetically more economical for ion–ligand bond lengths to adjust to intermediate values, and then for electron transfer to take place, than for electron transfer to occur at reactant bond lengths (as would be the case in photo-effected electron transfer, which is subject to the Franck–Condon principle, Section 13.5). The second concept we need is that electron transfer is most facile when the nuclei in the two complexes have positions that ensure that the electron has the same energy on each site. It follows from these two points that the rate of electron transfer, and the activation energy for the process, is governed by the ability of nuclei to adopt arrangements that achieve this matching of energies.

To understand how an outer-sphere reaction occurs, we consider the exchange reaction

$$[Fe(OH_2)_6]^{2+} + [Fe^*(OH_2)_6]^{3+} \longrightarrow [Fe(OH_2)_6]^{3+} + [Fe^*(OH_2)_6]^{2+}$$

where the * indicates a radioactive isotope of iron that acts as a tracer. The second-order rate constant is $3.0\ L\,mol^{-1}\,s^{-1}$ at $25\,°C$ and the activation energy is $32\ kJ\,mol^{-1}$. With the two theoretical points in mind we must consider the following three factors. First, bond lengths in Fe(II) are longer than those in Fe(III). Hence a part of the activation energy will arise from their adjustment to a common value in both complexes. This adjustment requires a change in Gibbs energy that is called the **inner-sphere rearrangement energy**, $\Delta^{\ddagger}G_{IS}$. Secondly, the solvent immediately outside the coordination sphere must be reorganized, which results in a change in Gibbs energy called the **outer-sphere reorganization energy**, $\Delta^{\ddagger}G_{OS}$. Thirdly, there is the electrostatic interaction energy between the two reactants, $\Delta^{\ddagger}G_{ES}$. The total activation Gibbs energy is therefore

$$\Delta^{\ddagger}G = \Delta^{\ddagger}G_{IS} + \Delta^{\ddagger}G_{OS} + \Delta^{\ddagger}G_{ES} \tag{17}$$

> The theoretical model of an outer-sphere reaction is based on the Born–Oppenheimer approximation and the greatest likelihood of electron transfer when the initial and final states of the electron have the same energy.

(b) The potential energy curves for reaction

The reactants initially have their normal bond lengths for Fe(II) and Fe(III), respectively, and the reaction corresponds to motion in which the Fe(II) bonds shorten and the Fe(III) bonds simultaneously lengthen (Fig. 14.10). The potential energy curve for the products of this symmetrical reaction is the same as for the reactants, the only difference being the interchange of the roles of the two Fe atoms. We have assumed that the metal–ligand stretching motions resemble a harmonic vibration and so have drawn them as parabolas.

The activated complex is located at the intersection of the two curves. However, the noncrossing rule (Section 13.3) states that molecular potential energy curves of states of the same symmetry do not cross but instead split into an upper and a lower curve (as seen in Fig. 14.10). The noncrossing rule implies that, if the reactants in their ground states slowly distort, then they follow the path of minimum energy and transform into products in their ground states.

More general redox reactions correspond to a nonzero reaction Gibbs energy, so the parabolas representing the reactants and the products lie at different heights. If the product

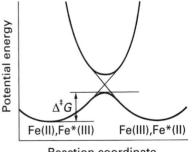

14.10 A simplified reaction profile for electron exchange in a symmetrical reaction. On the left of the graph, the nuclear coordinates correspond to Fe(II) and Fe*(III); on the right, the ligands and solvent molecules have adjusted locations and the nuclear coordinates correspond to Fe(III) and Fe*(II), where * denotes the isotope label.

surface is higher (Fig. 14.11a), then the crossing point moves up and the reaction has a higher activation energy. Conversely, moving the product curve down (Fig. 14.11c) leads to lower crossing points and lower activation energy, at least until the crossing begins to occur at the left. At the extreme of exergonic reaction (Fig. 14.11d), the crossing point rises and rates may become slower again.[7]

The noncrossing rule implies that, if the reactants in their ground states slowly distort, then they follow the path of minimum potential energy and transform into products in their ground states.

(c) The Marcus equation

The diagrams in Fig. 14.11 suggest that there are two factors that determine the rate of electron transfer. The first is the shape of the potential curves. If the parabolas rise steeply, indicating a rapid increase in energy with increasing bond extension, then their crossing points will be high and activation energies will be high too. Shallow potential curves, in contrast, imply low activation energies. Similarly, large changes in the equilibrium internuclear distance mean that the equilibrium points are far apart, and the crossing point will not be reached without large distortions. The second factor is the standard reaction Gibbs energy: the more negative it is, the lower the activation energy of the reaction.

These considerations were expressed quantitatively by R.A. Marcus. He derived an equation for predicting the rate constant for an outer-sphere reaction from the exchange rate constants for each of the redox couples involved and the equilibrium constant for the overall reaction. The Marcus equation for the rate constant k is

$$k^2 = fk_1k_2K \tag{18}$$

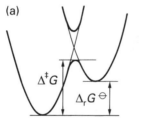
(a)

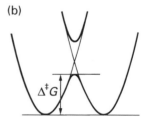
(b)

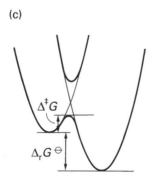
(c)

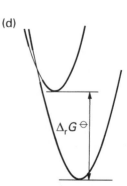
(d)

14.11 The effect on the activation energy of a change of reaction Gibbs energy for electron transfer when the shape of the potential surfaces remains constant (corresponding to equal self-exchange rates).

[7] We shall not consider this last point further here. The slowing of the rate at high exothermicity is referred to as the 'Marcus inverted region'. In an experimental *tour de force*, appropriate highly energetic oxidizing agents have been prepared photochemically and direct evidence for this inverted region has been obtained. See, for example, L.S. Fox, M. Kozik, J.R. Winkler, and H.B. Gray, *Science* **297**, 1069 (1990).

where k_1 and k_2 are the rate constants for the two exchange reactions and K is the equilibrium constant for the overall reaction. The factor f is a parameter composed of the rate constants and the encounter rate; it may be taken as near unity for approximate calculations. The idea of a weighted average of the two rates of self-exchange is emphasized by the name 'Marcus cross relation', which is sometimes given to eqn 18.

Example 14.5 Using the Marcus equation

Calculate the rate constant at $0\,°C$ for the outer-sphere reduction of $[Co(bipy)_3]^{3+}$ by $[Co(terpy)_2]^{2+}$.

Answer The required exchange reactions are

$$[Co(bipy)_3]^{2+} + [Co^*(bipy)_3]^{3+} \xrightarrow{k_1} [Co(bipy)_3]^{3+} + [Co^*(bipy)_3]^{2+}$$

$$[Co(terpy)_2]^{2+} + [Co^*(terpy)_2]^{3+} \xrightarrow{k_2} [Co(terpy)_2]^{3+} + [Co^*(terpy)_2]^{2+}$$

where $k_1 = 9.0\,\mathrm{mol^{-1}\,s^{-1}}$, $k_2 = 48\,\mathrm{L\,mol^{-1}\,s^{-1}}$, and $K = 3.57$, all at $0\,°C$. Setting $f = 1$ gives $k = (9.0 \times 48 \times 3.57)^{1/2}\,\mathrm{L\,mol^{-1}\,s^{-1}}$, or $39\,\mathrm{L\,mol^{-1}\,s^{-1}}$. This result compares reasonably well with the experimental value, which is $64\,\mathrm{L\,mol^{-1}\,s^{-1}}$.

Self-test 14.5 For reactions with k_1 and k_2 as given in the example, what would be the rate of electron transfer if the overall reaction had $E^{\ominus} = +1.00\,\mathrm{V}$?

The Marcus equation can be expressed as a linear free energy relation because the logarithms of the rate constants are proportional to the Gibbs energies of activation. Thus,

$$2\ln k = \ln k_1 + \ln k_2 + \ln K \tag{19}$$

implies that

$$2\Delta^{\ddagger}G = \Delta^{\ddagger}G_1 + \Delta^{\ddagger}G_2 + \Delta_r G^{\ominus} \tag{20}$$

Figure 14.12 shows the correlation between rates and overall reaction Gibbs energy for the oxidation of a series of substituted complexes of Fe(II) by Ce(IV). The fit to the straight line is a sign of the success of the approximate Marcus equation.[8]

Departures from the Marcus equation are taken as a sign of special features in outer-sphere reactions. These features may include a barrier created by change from high to low spin during electron transfer, or a change in symmetry, such as from octahedral to distorted tetragonal. An important example is that of cobalt ammine complexes, in which electron exchange between Co(II) and Co(III) is quite slow. The spin factor may play a role because Co(II) complexes are high-spin d^7 and Co(III) are low-spin d^6. However, transfer of an electron into the σ antibonding LUMO of a low-spin d^6 complex also requires a substantial tetragonal distortion. Alternatively, departures from the Marcus region could indicate simply that the reaction is inner-sphere rather than outer-sphere.

> The Marcus equation, eqn 18, is an approximate equation for the rate constant of outer-sphere electron transfer. Departures from the Marcus equation are taken as a sign of special features in outer-sphere reactions, such as spin and symmetry constraints.

(d) Inner-sphere reactions

Inner-sphere reactions are much more difficult to analyze because, whereas an outer-sphere reaction must be an electron transfer, an inner-sphere reaction may be either electron or group transfer. In many cases, it is very difficult to distinguish the two.

8   Tests of the Marcus equation are summarized in D.A. Pennington, In *Coordination chemistry* (ed. A.E. Martell), ACS Monograph 174, Vol. 2. American Chemical Society, Washington, DC (1978).

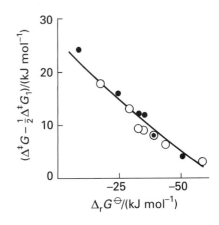

14.12 The correlation between rates (as expressed by the activation Gibbs energy) and reaction Gibbs energy for the oxidation of a series of phenanthroline complexes of Fe(II) by Ce(IV) in sulfuric acid (open circles) and $Fe^{2+}(aq)$ in perchloric acid (closed circles).

An inner-sphere reaction can be regarded as the outcome of three steps:

1 Formation of the bridged (μ) complex:

$$M^{II}L_6 + XM'^{III}L'_5 \longrightarrow L_5M^{II}-X-M'^{III}L'_5 + L$$

2 Electron transfer:

$$L_5M^{II}-X-M'^{III}L'_5 \longrightarrow L_5M^{III}-X-M'^{II}L'_5$$

3 Decomposition of the successor complex into final products:

$$L_5M^{III}-X-M'^{II}L'_5 \longrightarrow \text{products}$$

The rate-determining step of the overall reaction may be any one of these processes, but the most common one is the electron-transfer step. Reactions in which break-up of the successor complex is rate-determining are common only if the configuration of both metal ions after electron transfer results in substitutional inertness—but then an inert binuclear complex is a characteristic product. A good example is the reduction of $[RuCl(NH_3)_5]^{2+}$ by $[Cr(OH_2)_6]^{2+}$ in which the rate-determining step is the dissociation of $[Ru^{II}(NH_3)_5(\mu\text{-Cl})Cr^{III}(OH_2)_5]^{4+}$. Reactions in which the formation of the bridged complex is rate-determining tend to be quite similar for a series of partners of a given species. For example, the oxidation of $V^{2+}(aq)$ proceeds at more or less the same rate for a long series of Co(III) oxidants with different bridging ligands. The explanation is that the rate-determining step is the substitution of an H_2O molecule from the coordination sphere of V(II).

The numerous reactions in which electron transfer is rate-determining do not display such simple regularities. Rates vary over a wide range as metal ions and bridging ligands are varied. Because the change of oxidation states requires reorganization of ligands and solvent just as is the case for outer-sphere reactions, the trends indicated by the Marcus relation will also find some application to inner-sphere reactions. A very interesting way to study the role of the barriers to reorganization is the synthesis of bridged complexes of the type shown in (**13**) as models for inner-sphere precursor complexes. The data in Table 14.11 show some typical variations as bridging ligand, oxidizing metal, and reducing metal are changed.

$$[(bipy)_2ClRu^{II}-N\bigcirc N-Ru^{III}(bipy)_2Cl]^{3+}$$

13

All the reactions in Table 14.11 result in the change of oxidation number by ± 1. Such reactions are still often called **one-equivalent processes**, the name reflecting the largely outmoded term 'chemical equivalent'. Their mechanisms require transfer of electrons or radicals. Similarly, reactions that result in the change of oxidation number by ± 2 are often called **two-equivalent processes** and may resemble nucleophilic substitutions. This resemblance can be seen by considering the reaction

$$[Pt^{II}Cl_4]^{2-} + [^*Pt^{IV}Cl_6]^{2-} \longrightarrow [Pt^{IV}Cl_6]^{2-} + [^*Pt^{II}Cl_4]^{2-}$$

which occurs through a Cl^- bridge (**14**). The reaction depends on the transfer of a Cl^- ion in the break-up of the successor complex.

> The rate-determining step of the overall reaction may be any one of the three component processes enumerated above, but a common one is the electron-transfer step.

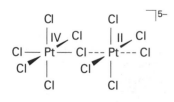

14

Table 14.11 Second-order rate constants for selected inner-sphere reactions with variable bridging ligands

Oxidant	Reductant	Bridging ligand	$k/(\text{L mol}^{-1}\text{ s}^{-1})$
$[\text{Co(NH}_3)_6]^{3+}$	$[\text{Cr(OH}_2)_6]^{2+}$		8×10^{-5}
$[\text{CoF(NH}_3)_5]^{2+}$	$[\text{Cr(OH}_2)_6]^{2+}$	F^-	2.5×10^5
$[\text{CoCl(NH}_3)_5]^{2+}$	$[\text{Cr(OH}_2)_6]^{2+}$	Cl^-	6.0×10^5
$[\text{CoI(NH}_3)_5]^{2+}$	$[\text{Cr(OH}_2)_6]^{2+}$	I^-	3.0×10^6
$[\text{Co(NCS)(NH}_3)_5]^{2+}$	$[\text{Cr(OH}_2)_6]^{2+}$	NCS^-	1.9×10
$[\text{Co(SCN)(NH}_3)_5]^{2+}$	$[\text{Cr(OH}_2)_6]^{2+}$	SCN^-	1.9×10^5
$[\text{Co(NH}_3)_5(\text{OH}_2)]^{3+}$	$[\text{Cr(OH}_2)_6]^{2+}$	H_2O	1×10^{-1}
$[\text{CrF(OH}_2)_5]^{2+}$	$[\text{Cr(OH}_2)_6]^{2+}$	F^-	7.4×10^{-3}

$[\text{Co(NH}_3)_5\text{O}_2\text{C}-\bigcirc-\text{N}-\text{Ru(NH}_3)_4(\text{OH}_2)]$		$^-\text{O}_2\text{C}-\bigcirc-\text{N}$	1×10^2
$[\text{Co(NH}_3)_5\text{O}_2\text{C}-\bigcirc-\text{N}-\text{Ru(NH}_3)_4(\text{OH}_2)]$		$^-\text{O}_2\text{C}-\bigcirc-\text{N}$	1.6×10^{-3}

14.14 Oxidative addition

A reaction of the type

$$[\text{IrCl(CO)L}_2] + \text{RI} \longrightarrow [\text{Ir(R)(I)(Cl)(CO)L}_2]$$

where $\text{L} = \text{PPh}_3$ and R is an alkyl group, is called an **oxidative addition**.[9] In the reaction, the d^8 square-planar complex uses the nonbonding axial electron pair to bond to the incoming Lewis acid (R^+ or H^+ in this case) and the lone pair of the base X or H coordinates to the metal acting as an acid. The net effect is a metal oxidized to the extent of two electrons, from d^8 to d^6. Because the alkyl group and the anion are both more electronegative than the metal, they formally acquire all four electrons involved in $\text{M}-\text{R}$ and $\text{M}-\text{I}$ bonding, so the oxidation number of the metal changes by $+2$ (corresponding to oxidation). Similar reactions are exhibited by the d^{10} species $[\text{Pd(PPh}_3)_4]$ and $[\text{Ni(P(OEt)}_3)_4]$.

There are three important mechanisms of oxidative addition. The first is addition of H_2 followed by scission of the $\text{H}-\text{H}$ bond:

Oxidative addition of H_2 in this way always produces a *cis* dihydride.[10] The activation enthalpy for the reaction, which typically lies in the range 20 to 40 kJ mol^{-1}, is much smaller than the bond enthalpy of H_2 (436 kJ mol^{-1}), which suggests that little bond breaking has occurred in the formation of the activated complex. This inference is confirmed by replacing H by D. The fact that dihydrogen (as distinct from dihydrido) complexes have been prepared in which H_2 is bonded side-on to a metal atom suggests that the activated complex for oxidative addition is also a side-on $\text{M}-\text{H}_2$ arrangement that subsequently collapses to the *cis*-dihydrido product.

An alternative mechanism involves nucleophilic attack on a C atom by the lone-pair electrons of the metal atom:

[9] H_2 or HX, where X is a halogen, may replace the iodoalkane.

[10] A.J. Kunin, C.E. Johnson, J.A. Maguire, W.D. Jones, and R. Eisenberg, *J. Am. Chem. Soc.* **109**, 2963 (1987).

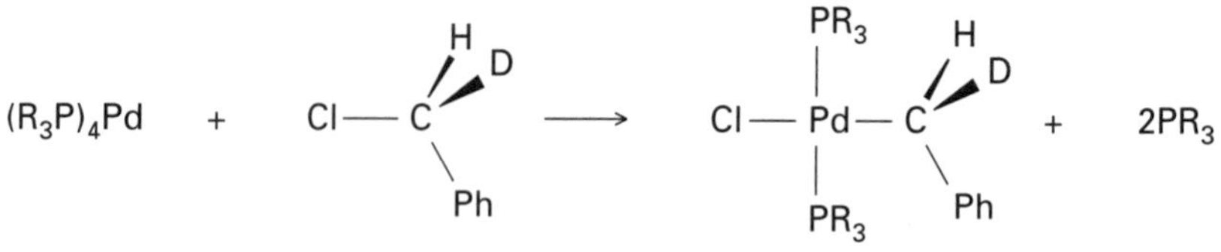

The oxidative addition of haloalkanes often proceeds in this way, by nucleophilic attack of the metal atom on the halogen-bearing C atom. In this example, attack by Pd displaces Cl^- in the rate-determining step; the subsequent addition of Cl^- is then rapid.[11] The fact that these reactions are not very stereoselective at the metal atom is consistent with this mechanism. In contrast, inversion does occur at the C atom, which demonstrates that nucleophilic displacement is indeed occurring. The reactivity order for haloalkanes indicates that steric hindrance does inhibit attack, for reaction rates lie in the order

$$CH_3X > CH_3CH_2X > CHR_2X > \text{cyclohexyl-}X \qquad X = \text{halogen}$$

The third mechanism involves radicals; it is accelerated by O_2 and light and occurs one electron at a time:

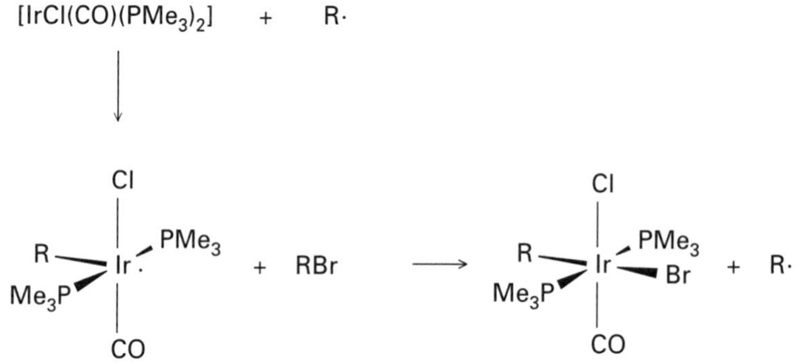

Loss of stereochemistry, inhibition by radical scavengers, and sensitivity to O_2 are commonly indicators of radical pathways. Haloalkenes, haloarenes, and α-haloesters show evidence of this radical pathway in reactions with $[IrCl(CO)(PMe_3)_2]$.[12]

The reverse of oxidative addition is **reductive elimination**. As we shall see in Chapter 17, the combination of oxidative addition and reductive elimination can be used to modify organic compounds in ways that are useful in the construction of catalytic cycles. A simple example is the oxidative addition

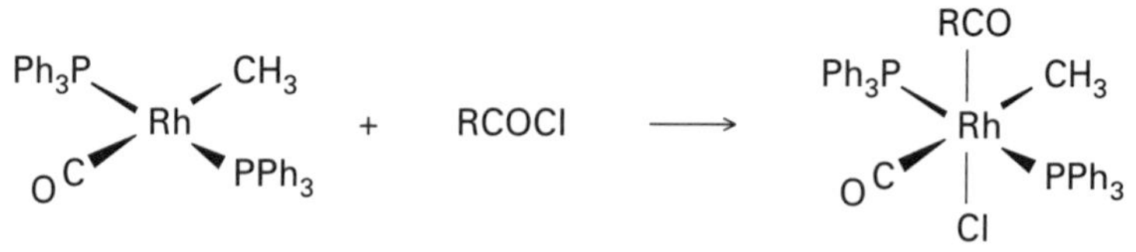

followed by the reductive elimination

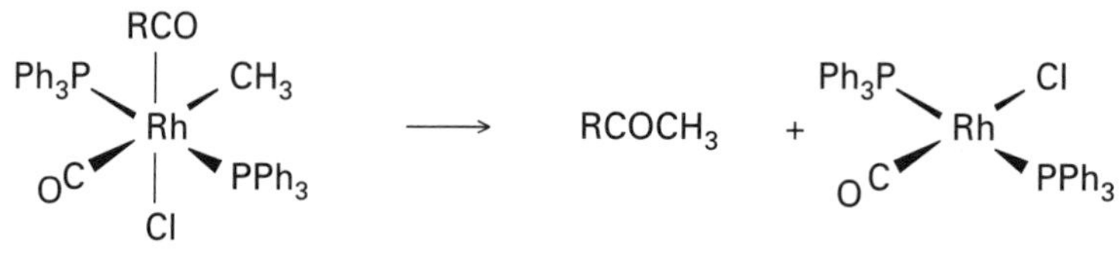

[11] J. Halpern, *Acc. Chem. Res.* **3**, 386 (1970).

[12] J.A. Labinger, J.A. Osborne, and N.J. Coville, *Inorg. Chem.* **19**, 3236 (1980).

The overall reaction is the conversion of an acyl chloride to a ketone.

> The three important mechanisms of oxidative addition are (1) nonpolar addition, (2) initial nucleophilic attack on a carbon atom by the lone pairs on a metal atom, (3) stepwise addition by radicals.

Photochemical reactions

The absorption of a photon increases the energy of a complex, typically by between 170 and 600 kJ mol^{-1}. Because these energies are larger than typical activation energies it should not be surprising that new reaction channels are opened. However, when the high energy of a photon is used to provide the energy of the primary forward reaction, the back reaction is almost always very favorable, and much of the design of efficient photochemical systems lies in trying to avoid the back reaction.

14.15 Prompt and delayed reactions

In some cases, the excited states dissociate almost immediately they are formed. Examples include formation of the pentacarbonyl intermediates that initiate ligand substitution in carbonyls

$$Cr(CO)_6 \xrightarrow{h\nu} Cr(CO)_5 + CO$$

and the scission of Co—Cl bonds:

$$[Co^{III}Cl(NH_3)_5]^{2+} \xrightarrow{h\nu \, (\lambda < 350 \text{ nm})} [Co^{II}(NH_3)_5]^{2+} + Cl\cdot$$

Both these processes occur in less than 10 ps and hence are called **prompt reactions**.

In the second reaction, the **quantum yield**, the amount of reaction per mole of photons absorbed, increases as the wavelength of the radiation is decreased (and the photon energy correspondingly increased). The energy in excess of the bond energy is available to the newly formed fragments and increases the probability that they will escape from each other through the solution before they have an opportunity to recombine.

Some excited states have long lifetimes. They may be regarded as energetic isomers of the ground state that can participate in **delayed reactions**. The excited state of $[Ru^{II}(bipy)_3]^{2+}$ created by photon absorption in the metal-to-ligand charge-transfer band (Section 13.4) may be regarded as a Ru(III) complexed to a radical anion of the ligand. Its redox reactions can be explained by adding the excitation energy (expressed as a potential by using $-FE = \Delta G$ and equating ΔG to the molar excitation energy) to the ground-state reduction potential:[13]

$$[Ru(bipy)_3]^{3+} + e^- \longrightarrow [Ru(bipy)_3]^{2+}$$
$$E^{\ominus} = +1.26 \text{ V} \; (-122 \text{ kJ mol}^{-1})$$

$$[Ru(bipy)_3]^{2+} + h\nu(590 \text{ nm}) \longrightarrow [^{\ddagger}Ru(bipy)_3]^{2+}$$
$$\Delta G = +202 \text{ kJ mol}^{-1}$$

$$[Ru(bipy)_3]^{3+} + e^- \longrightarrow [^{\ddagger}Ru(bipy)_3]^{2+}$$
$$E^{\ominus\ddagger} = -0.84 \text{ V} \; (+80 \text{ kJ mol}^{-1})$$

> Electronically excited species may dissociate promptly or after a delay.

[13] For a discussion, see T.J. Meyer, Excited state electron transfer. *Prog. Inorg. Chem.* **30**, 389 (1983).

14.16 *d–d* and charge-transfer reactions

We have seen (Sections 13.3 and 13.4) that there are two main types of spectroscopically observable electron promotion in metal complexes, namely *d–d* transitions and charge-transfer transitions. A *d–d* transition corresponds to the essentially angular redistribution of electrons within a *d* shell. In octahedral complexes, this redistribution often corresponds to the occupation of M—L antibonding e_g orbitals. An example is the ${}^4T_{1g} \leftarrow {}^4A_{2g}$ ($t_{2g}^2 e_g^1 \leftarrow t_{2g}^3$) transition in $[Cr(NH_3)_6]^{3+}$. The occupation of the antibonding e_g orbital results in a quantum yield close to 1 (specifically 0.6) for the photosubstitution

$$[Cr(NH_3)_6]^{3+} + H_2O \xrightarrow{h\nu} [Cr(NH_3)_5(OH_2)]^{3+} + NH_3$$

This is a prompt reaction, for it occurs in less than 5 ps.

Charge-transfer transitions correspond to the *radial* redistribution of electron density. They correspond to the promotion of electrons into more predominantly ligand orbitals if the transition is metal-to-ligand or into orbitals of greater metal character if the transition is ligand-to-metal. The former process corresponds to metal oxidation and the latter to metal reduction. These excitations commonly initiate photoredox reactions of the kind already mentioned in connection with Co(III) and Ru(II).

Although it is a very useful first approximation to associate photosubstitution and photoisomerization with *d–d* transitions and photoredox with charge-transfer transitions, the rule is not absolute. For example, it is not uncommon for a charge-transfer band to result in photosubstitution by an indirect path:

$$[Co^{III}Cl(NH_3)_5]^{2+} + H_2O \xrightarrow{h\nu} [Co^{II}(NH_3)_5(OH_2)]^{2+} + Cl\cdot$$
$$[Co^{III}(NH_3)_5(OH_2)]^{2+} + Cl\cdot \longrightarrow [Co^{III}(NH_3)_5(OH_2)]^{3+} + Cl^-$$

In this case, the aqua complex formed after the homolytic fission of the Co—Cl bond is reoxidized by the Cl atom. The net result leaves the Co oxidized and substituted. Conversely, the long-lived ${}^2E$ state of $[Cr(bipy)_3]^{3+}$ is a pure *d–d* state which lacks substitutional reactivity in acidic solution. Its lifetime of several microseconds allows the excess energy to enhance its redox reactions. The reduction potential (+1.3 V), calculated by adding the excitation energy to the ground-state value, accounts for its function as a good oxidizing agent, in which it undergoes reduction to $[Cr(bipy)_3]^{2+}$.

> A useful first approximation is to associate photosubstitution and photoisomerization with *d–d* transitions and photoredox with charge-transfer transitions, but the rule is not absolute.

14.17 Transitions in metal–metal bonded systems

We encountered the δ^*–δ transition of multiple metal-metal bonds in Chapter 13, and it should not be surprising that population of an antibonding orbital of the metal–metal system can sometimes initiate photodissociation. It is more interesting that such excited states have been shown to initiate multielectron redox photochemistry.

One of the best characterized systems is the dinuclear platinum complex $[Pt_2(\mu\text{-}P_2O_5H_2)_4]^{4-}$ called informally 'PtPOP' (**15**).[14] There is no metal-metal bonding in the ground state of this Pt(II)–Pt(II) d^8–d^8 species. The HOMO–LUMO pattern indicates that excitation populates a bonding orbital between the metals (Fig. 14.13). The lowest-lying excited state has a lifetime of 9 μs. It is a powerful reducing agent, and reacts by both electron and halogen atom transfer. The most interesting oxidation products are Pt(III)—Pt(III) which contain X^- ligands (where X is halogen or pseudohalogen) at both ends and a metal-

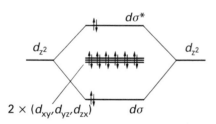

15

Pt	Pt–Pt	Pt
	$p\sigma^*$	
p_z	$2 \times d_{x^2-y^2}$	p_z

	$d\sigma^*$	
d_{z^2}		d_{z^2}
$2 \times (d_{xy}, d_{yz}, d_{zx})$	$d\sigma$	

14.13 The dinuclear complex $[Pt_2(\mu\text{-}P_2O_5H_2)_4]^{4-}$. The structure consists of two face-to-face square-planar complexes held together by a bridging pyrophosphito ligand. The metal p_z–p_z and d_{z^2}–d_{z^2} orbitals interact along the Pt-Pt axis. The other *p* and *d* orbitals are considered to be nonbonding.

---

[14] The chemistry of this interesting complex is reviewed in D.M. Roundhill, H.B. Gray, and C.-M. Chi, *Acc. Chem. Res.* **22**, 55 (1989).

metal single bond. Irradiation in the presence of $(Bu)_3SnH$ gives a dihydrido product that can eliminate H_2.

Irradiation of the quadruply bonded dinuclear cluster $Mo_2(O_2P(OC_6H_5)_2)_4$ (16) at 500 nm in the presence of $ClCH_2CH_2Cl$ results in production of ethene and the addition of two Cl atoms to the two Mo atoms, with a two-electron oxidation.[15] The reaction proceeds in steps of one electron, and requires a complex with the metals blocked by sterically crowding ligands. If smaller ligands are present, the reaction that occurs instead is a photochemical oxidative addition of the organic molecule.

> *Population of a metal–metal antibonding orbital can sometimes initiate photodissociation; such excited states have been shown to initiate multielectron redox photochemistry.*

16

Further reading

R.B. Jordan, *Reaction mechanisms of inorganic and organometallic systems.* Oxford University Press, New York (1998).

R.G. Wilkins, *Kinetics and mechanism of reactions of transition metal complexes.* VCH, Weinheim (1991).

J.D. Atwood, *Inorganic and organometallic reaction mechanisms.* VCH, Weinheim (1997).

J.D. Atwood, Mechanisms of reactions of organometallic complexes. In *Encyclopedia of inorganic chemistry* (ed. R.B. King), Vol. 4, pp. 2119–35. Wiley, New York (1994).

Important older books include:

F. Basolo and R.G. Pearson, *Mechanisms of inorganic reactions.* Wiley, New York (1967). This book is the classic in the field.

M.L. Tobe, *Inorganic reaction mechanisms.* Nelson, London (1972). This is a very readable treatment.

C.H. Langford and H.B. Gray, *Ligand substitution processes.* W.A. Benjamin, New York (1966). This book introduced the associative/dissociative notation.

A series of reviews summarizing current work is:

M.V. Twigg (ed.), *Mechanisms of inorganic and organometallic reactions,* Vols 1–8. Plenum Press, New York (1983–94).

W. Preetz, G. Peters, and D. Bublitz, Preparation and investigation of mixed octahedral complexes. *Chem. Rev.,* **96**, 977 (1996).

Exercises

14.1 Classify (a) NH_3, (b) Cl^-, (c) Ag^+, (d) S^{2-}, (e) Al^{3+} as nucleophiles or electrophiles.

14.2 The reactions of $Ni(CO)_4$ in which phosphines or phosphites replace CO to give the family $Ni(CO)_3L$ occur at the same rate for different phosphines or phosphites. Is the reaction d or a?

14.3 In experiment (a), $^{36}Cl^-$ is added to the solution during the study of the *cis*-to-*trans* isomerism of $[CoCl_2(en)_2]^+$ to determine whether isomerization is accompanied by substitution. In experiment (b), D replaces H in the ammine ligands of $[Cr(NCS)_4(NH_3)_2]^-$ and reduces the rate of replacement of NCS^- by water. Which experiment studies the stoichiometric mechanism and which studies the intimate mechanism?

14.4 Write the rate law for formation of $[MnX(OH_2)_5]^+$ from the aqua ion and X^-. How would you undertake to determine whether the reaction is d or a?

14.5 Octahedral complexes of metal centers with high oxidation numbers or of d metals of the second and third series are less labile than those of the low oxidation number and d metals of the first series of the block. Account for this observation on the basis of a dissociative rate-determining step.

14.6 A Pt(II) complex of tetraethyldiethylenetriamine is attacked by Cl^- 10^5 times less rapidly than the diethylenetriamine analog. Explain this observation in terms of an associative rate-determining step.

14.7 The rate of loss of chlorobenzene, PhCl, from $[W(CO)_4L(PhCl)]$ increases with increase in the cone angle of L. What does this observation suggest about the mechanism? When PhCl is replaced by a phosphine ligand, the rate of reaction depends upon the concentration of the phosphine in experiments conducted at low phosphine concentration. Is this observation in conflict with the mechanistic conclusion drawn from cone angle effects?

---

[15] C.M. Partigianoni, I.J. Chang, and D. Nocera, *Coord. Chem. Rev.* **105**, 97 (1990).

14.8 Does the fact that $[Ni(CN)_5]^{3-}$ can be isolated help to explain why substitution reactions of $[Ni(CN)_4]^{2-}$ are very rapid?

14.9 Design two-step syntheses of *cis*- and *trans*-$[PtCl_2(NO_2)(NH_3)]^-$ starting from $[PtCl_4]^{2-}$.

14.10 How does each of the following modifications affect the rate of a square-planar complex substitution reaction? (a) Changing a *trans* ligand from H to Cl. (b) Changing the leaving group from Cl to I. (c) Adding a bulky substituent to a *cis* ligand. (d) Increasing the positive charge on the complex.

14.11 The rate of attack on Co(III) by an entering group **Y** is nearly independent of **Y** with a spectacular exception of the rapid reaction with OH^-. Explain the anomaly. What is the implication of your explanation for the behavior of a complex lacking Brønsted acidity on the ligands?

14.12 Predict the products of the following reactions:

(a) $[Pt(PR_3)_4]^{2+} + 2Cl^-$

(b) $[PtCl_4]^{2-} + 2PR_3$

(c) *cis*-$[Pt(NH_3)_2(py)_2]^{2+} + 2Cl^-$

14.13 Put in order of increasing rate of substitution by H_2O the complexes (a) $[Co(NH_3)_6]^{3+}$, (b) $[Rh(NH_3)_6]^{3+}$, (c) $[Ir(NH_3)_6]^{3+}$, (d) $[Mn(OH_2)_6]^{2+}$, (e) $[Ni(OH_2)_6]^{2+}$.

14.14 State the effect on the rate of dissociatively activated reactions of Rh(III) complexes of (a) an increase in the overall charge on the complex; (b) changing the leaving group from NO_3^- to Cl^-; (c) changing the entering group from Cl^- to I^-; (d) changing the *cis* ligands from NH_3 to H_2O; (e) changing an ethylenediamine ligand to propylenediamine when the leaving ligand is Cl^-.

14.15 The mechanism of CO insertion is thought to be

$RMn(CO)_5 \rightleftharpoons (RCO)Mn(CO)_4$

$(RCO)Mn(CO)_4 + L \rightleftharpoons (RCO)MnL(CO)_4$

In the first step, the acyl intermediate is formed, leaving a vacant coordination position. In the second step, a ligand enters. At high ligand concentrations, the rate is independent of [L]. In this limit, what rate constant can be extracted from rate data?

14.16 The pressure dependence of the replacement of chlorobenzene (PhCl) by piperidine (pip) in the complex $[W(CO)_4(PPh_3)(PhCl)]$ has been studied. The volume of activation is found to be $+11.3$ $cm^3 mol^{-1}$. What does this value suggest about the mechanism?

14.17 Write out the inner-and outer-sphere pathways for reduction of azidopentaamminecobalt(III) ion with $V^{2+}(aq)$. What experimental data might be used to distinguish between the two pathways?

14.18 The intermediate $[Fe(SCN)(OH_2)_5]^{2+}$ can be detected in the reaction of $[Co(NCS)(NH_3)_5]^{2+}$ with $Fe^{2+}(aq)$ to give $Fe^{3+}(aq)$ and $Co^{2+}(aq)$. What does this observation suggest about the mechanism?

14.19 The photochemical substitution of $[W(CO)_5(py)]$ (py $=$ pyridine) with triphenylphosphine gives $W(CO)_5(P(C_6H_5)_3)$. In the presence of excess phosphine, the quantum yield is approximately 0.4. A flash photolysis study reveals a spectrum that can be assigned to the intermediate $W(CO)_5$. What product and quantum yield do you predict for substitution of $[W(CO)_5(py)]$ in the presence of excess triethylamine? Is this reaction expected to be initiated from the ligand field or MLCT excited state of the complex?

14.20 From the spectrum of $[CrCl(NH_3)_5]^{2+}$ shown in Fig. 13.9, propose a wavelength for transient photoreduction of Cr(III) to Cr(II) accompanied by oxidation of a ligand.

14.21 Reactions of $[Pt(Ph)_2(SMe_2)_2]$ with the bidentate ligand, 1,10-phenanthroline (phen) give $[Pt(Ph)_2phen]$. There is a kinetic pathway with activation parameters $\Delta^{\ddagger}H = +101$ kJ mol^{-1} and $\Delta^{\ddagger}S = +42$ J K^{-1} mol^{-1}. Propose a mechanism.

Problems

14.1 Solutions of $[PtH_2(P(CH_3)_3)_2]$ exist as a mixture of *cis* and *trans* isomers. Addition of excess $P(CH_3)_3$ led to formation of $[PtH_2(P(CH_3)_3)_3]$ at a concentration that could be detected using NMR. This complex exchanged phosphine ligands rapidly with the *trans* isomer but not the *cis*. Propose a pathway. What are the implications for the *trans* effect of H versus $P(CH_3)_3$? (See D.L. Packett and W.G. Trogler, *Inorg. Chem.* **27**, 1768 (1988).)

14.2 Given the following mechanism for the formation of a chelate complex

$[Ni(OH_2)_6]^{2+} + L\text{—}L \rightleftharpoons [Ni(OH_2)_6]^{2+}, L\text{—}L$ K_E, rapid

$[Ni(OH_2)_6]^{2+}, L\text{—}L \rightleftharpoons [Ni(OH_2)_5L\text{—}L]^{2+} + H_2O$ k_a, k_a'

$[Ni(OH_2)_5L\text{—}L]^{2+} \rightleftharpoons [Ni(OH_2)_4LL]^{2+} + H_2O$ k_b, k_b'

Derive the rate law for the formation of the chelate. Discuss the step that is different from that for two monodentate ligands. It is found that the formation of chelates with strongly bound ligands occurs at the rate of formation of the analogous monodentate complex, but that the formation of chelates of weakly bound ligands is often significantly slower. Assuming an I_d mechanism, explain this observation. (See R.G. Wilkins, *Acc. Chem. Res.* **3**, 408 (1970).)

14.3 The complex $[PtH(PEt_3)_3]^+$ was studied in deuterated acetone in the presence of excess PEt_3, where Et is CH_3CH_2. In the absence of excess ligand the ^{1}H-NMR spectrum in the hydride region exhibits two triplets. As excess PEt_3 ligand is added the triplets begin to collapse, the line shape depending on the ligand concentration. Give a mechanism to account for the effects of excess PEt_3.

14.4 The substitution reactions of the bridged dinuclear Rh(II) complex $[Rh_2(\mu\text{-}O_2CCH_3)_4XY]$ (**17**) have been studied by M.A.S. Aquino and D.H. Macartney (*Inorg. Chem.* **26**, 2696 (1987)). Reaction rates show little dependence on the choice of the entering group. The table below shows the dependence on the leaving group, **X**, and the ligand on the opposite Rh, (*trans*) **Y** at 298K. What conclusions can you draw concerning the mechanism?

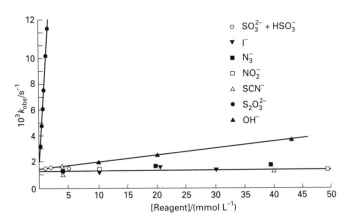

17

X	Y	$k/(\text{L mol}^{-1}\,\text{s}^{-1})$
H_2O	H_2O	10^5–10^7
CH_3OH	CH_3OH	2×10^6
CH_3CN	CH_3CN	1.1×10^5
PPh_3	PPh_3	1.5×10^5
CH_3CN	PR_3	10^7–10^9
PR_3	CH_3CN	10^{-1}–10^2
N-donor	H_2O	10^2–10^3

Note that the complex has a d^7 configuration at each Rh and a single Rh—Rh bond.

14.5 Figure 14.14 (which is from J.B. Goddard and F. Basolo, *Inorg. Chem.* **7**, 936 (1968)) shows the observed first-order rate constants for the reaction of $[PdBrL]^+$ with various Y^- to give $[PdYL]^+$ where L is $Et_2NCH_2CH_2NHCH_2CH_2NH_2$. Notice the steep slope for $S_2O_3^{2-}$ and zero slopes for $X = N_3^-$, I^-, NO_2^-, and SCN^-. Propose a mechanism.

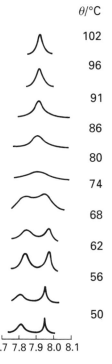

14.6 The activation enthalpy for the reduction of *cis*-$[CoCl_2(en)_2]^+$ by Cr^{2+}(aq) is -24 kJ mol^{-1}. Explain the negative value. (See R.C. Patel, R.E. Ball, J.F. Endicott, and R.G. Hughes, *Inorg. Chem.* **9**, 23 (1970).)

14.7 The rate of reduction of $[Co(NH_3)_5(OH_2)]^{3+}$ by Cr(II) is seven orders of magnitude slower than reduction of its conjugate base, $[Co(NH_3)_5(OH)]^{2+}$ by Cr(II). For the corresponding reductions with $[Ru(NH_3)_6]^{2+}$ the two differ by less than a factor of 10. What do these observations suggest about mechanisms? Comment on H_2O and OH^- as bridging ligands.

14.8 Using the NMR spectra in Fig. 14.15, estimate the rate of exchange between free and coordinated ammonia for the alkyl Co(III) complex represented. (See R.J. Guschl and T.L. Brown, *Inorg. Chem.* **13**, 959 (1974).)

$\theta/°C$

102

96

91

86

80

74

68

62

56

50

7.7 7.8 7.9 8.0 8.1

Fig. 14.15 The 100 MHz ^{1}H-NMR spectra required for Problem 14.8.

14.9 Calculate the rate constants for outer-sphere reactions from the following data. Compare your results to the measured values in the last column.

Reaction	$k_1/(\text{L mol}^{-1}\,\text{s}^{-1})$	$k_2/(\text{L mol}^{-1}\,\text{s}^{-1})$	$E^\ominus/\text{V}$	$k_{obs}/(\text{L mol}^{-1}\,\text{s}^{-1})$
$Cr^{2+} + Fe^{2+}$	2×10^{-5}	4.0	$+1.18$	2.3×10^3
$[W(CN)_8]^{4-}$ $+ Ce(IV)$	$> 4 \times 10^4$	4.4	$+0.90$	$> 10^8$
$[Fe(CN)_6]^{4-}$ $+ MnO_4^-$	7.4×10^2	3×10^3	$+0.20$	1.7×10^5
$[Fe(phen)_3]^{2+}$ $+ Ce(IV)$	$> 3 \times 10^7$	4.4	$+0.36$	1.4×10^5

Fig. 14.14 The data required for Problem 14.5.

14.10 Oxidative addition of CH$_3$I to the Rh(I) complex (**18**) has the rate law:

$$k_{obs} = k_s + k_2[CH_3I]$$

The term k_s is sensitive to solvent nucleophilicity. It is suggested (S.J. Basson, J.G. Leipoldt, A. Roodt, and J.A. Venter, *Inorg. Chim. Acta* **128**, 31 (1987)) that the rate law implies competing pathways for the initial step: one involving electrophilic attack by the CH$_3$ group to give a five-coordinate species and a second in which the solvent forms the five-coordinate species prior to CH$_3$I attack. Write pathways for the oxidative addition and identify the electrophilic and nucleophilic steps.

18

14.11 Figure 14.16 (which is from W.R. Muir and C.H. Langford, *Inorg. Chem.* **7**, 1032 (1968)) shows plots of observed first-order rate constants against [X$^-$] for the reactions

$$[Co(NO_2)(dmso)(en)_2]^{2+} + X^- \rightarrow [Co(NO_2)X(en)_2]^{2+} + dmso$$

Assume that all three reactions have the same mechanism. (a) If the mechanisms were D, to what would the limiting rate constants correspond? (b) If the mechanisms were I_d, to what would the limiting rate correspond? (c) What is the significance of the fact that the dmso exchange rate constant is larger than the limits for anion attack? (d) The limiting rate constants are 5×10^{-4} s^{-1}, 1.2×10^{-4} s^{-1}, and about 1×10^{-4} s^{-1} for Cl$^-$, NO$_2^-$, and NCS$^-$, respectively. Does the evidence favor a D or an I_d mechanism?

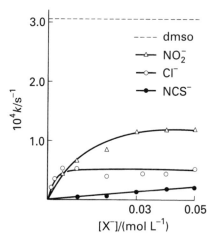

Fig. 14.16 The data required for Problem 14.11. Rate of reaction of *cis*-[Co(en)$_2$(NO$_2$)(dmso)]$^{2+}$ ion as a function of the concentration of the entering anion X$^-$ for X$^-$ = NO$_2^-$, Cl$^-$, and NCS$^-$. The broken line shows the rate of the dmso exchange reaction.

15

Main-group organometallic compounds

There are many similarities between the chemical properties of hydrogen compounds and the alkyl derivatives of the main-group elements. This is due in part to the similar electronegativities of hydrogen and carbon and the resulting similar strengths and polarities of element–carbon and element–hydrogen bonds. Here we consider the organometallic compounds of the main-group elements and include the zinc group, which is very similar. The organometallic compounds of the d- and f-blocks are treated in Chapter 16 and the role of organometallic compounds in catalysis is treated in Chapter 17. We shall see that most organometallic compounds can be synthesized by using one of four M—C bond-forming reactions. Moreover, many of their properties can be rationalized in terms of a small number of classes of reaction. One feature that we shall constantly emphasize is the important role of steric congestion around the central atom. This congestion is believed to be responsible for the ability of some organometallic compounds to withstand hydrolysis, and has led to the synthesis of compounds containing multiple bonds between heavy elements.

One of the pioneers of main-group organometallic chemistry was the English chemist E.C. Frankland, who learned about organoarsenic compounds during his doctoral studies in Germany. In 1848 he was the first person to synthesize dimethylzinc, and over the next 14 years he discovered $Zn(C_2H_5)_2$, $Hg(CH_3)_2$, $Sn(C_2H_5)_4$, and $B(CH_3)_3$. He also developed the use of organozinc reagents in organic synthesis and introduced the term 'organometallic' into the vocabulary of chemistry.

The organic derivatives of lithium, magnesium, boron, aluminum, and silicon received considerable attention in the early part of the twentieth century. These and other developments led to several major industrial applications for organometallic compounds, such as catalysts for alkene polymerization and the production of silicone polymers. Since the 1960s, exploratory synthetic research in organometallic compounds has been dominated by studies of d-block compounds, but a revival of interest in main-group organometallic compounds has produced new classes of compounds and has broadened our knowledge of the range of types of bonding and reactions of the s- and p-block elements.

Classification, nomenclature, and structure

A compound is regarded as 'organometallic' if it contains at least one metal–carbon (M—C) bond. In this context the suffix 'metallic' is interpreted broadly to include metalloids such as boron, silicon, and arsenic as well as true metals. To emphasize the distinctions between the members of this broad class of compounds, we write E—C for bonds to carbon where E is usually a p-block metalloid, and M—C when the element M is specifically an electropositive metal.

15.1 Nomenclature

The s-block organometallic compounds are commonly named according to the substituent names used in organic chemistry (Table 15.1), as in methyllithium for $Li_4(CH_3)_4$. Compounds that have considerable ionic character are named as salts, as for sodium naphthalide, $Na[C_{10}H_8]$. The p-block organometallic compounds are also often named as simple organic species, such as either boron trimethyl or trimethylboron, $B(CH_3)_3$. They may also be named as derivatives of their hydrogen counterparts, as in trimethylborane for $B(CH_3)_3$, tetramethyl-silane for $Si(CH_3)_4$, and trimethylarsane for $As(CH_3)_3$.

The nomenclature we shall use for main-group organometallic compounds does not include an explicit statement of oxidation number, but it is sometimes useful to take note of the oxidation number of the metallic element when considering the properties of a compound. As elsewhere in inorganic chemistry, the oxidation number of an element is useful for keeping track of electrons and inferring the type of reaction it is likely to undergo, but it must not be interpreted as the actual charge on the atom. The convention employed in assigning an oxidation number to a metal atom in an organometallic compound is that the organic moiety is considered to be present as an anion. For example, in dimethylmercury, $Hg(CH_3)_2$, each CH_3 group is assigned a charge of -1, which (because the sum of the oxidation numbers of all the atoms in a neutral compound is zero) implies that the oxidation number of Hg is $+2$. In bis(cyclopentadienyl)magnesium, $Mg(\eta^5\text{-}C_5H_5)_2$,[1] the C_5H_5 group is assigned a charge of -1, so the oxidation number of Mg is $+2$.

> The oxidation numbers of organic groups attached to metal atoms are assigned as though the groups were present as anions.

Table 15.1 Names of common organic substituents in organometallic chemistry*

Formula	Systematic name	Alternative name (and abbreviation)
CH_3—	Methyl	(Me)
CH_3CH_2—	Ethyl	(Et)
$(CH_3)_2CH$—	1-Methylethyl	Isopropyl (^{i}Pr)
CH_2=$CHCH_2$—	2-Propenyl	Allyl
$(CH_3)_2CHCH_2$—	2-Methylpropyl	Isobutyl (^{i}Bu)
$(CH_3)_3C$—	1,1-Dimethylethyl	$tert$-Butyl (^{t}Bu)
C_5H_5	Cyclopentadienyl	(Cp)
C_6H_5—	Phenyl	(Ph)
H_2C=	Methylene	
HC≡	Methylidyne	

*For a longer list, see *Further information 1*.

----

[1] The notation η^5 indicates that the ligand has five points of attachment to the metal atom; this notation is explained in more detail in Section 15.8.

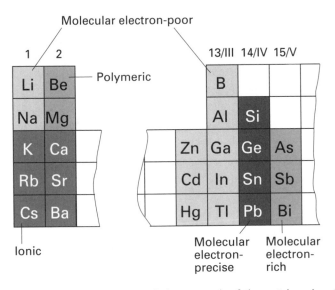

Molecular electron-poor

Polymeric

Ionic

13/III 14/IV 15/V

Molecular electron-precise

Molecular electron-rich

15.1 The classification of methyl compounds of the metals and metalloids.

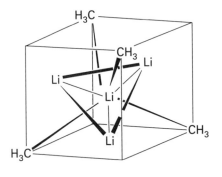

1

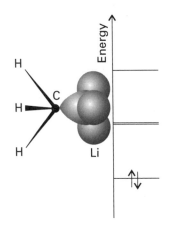

2 $Li_4(CH_3)_4$

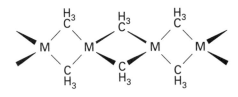

3 M = Be or Mg

15.2 Comparison with hydrogen compounds

The polarities and strengths of many E—C bonds are similar to those of E—H bonds. As a result, there are structural and chemical similarities between alkyl compounds and their simple hydrogen analogs, as may be seen by comparing Fig. 15.1 and Fig. 8.2. As expected from electronegativity considerations, the alkyl compounds of the s-block elements have highly polar $M^{\delta+}$—$C^{\delta-}$ bonds. The electron-precise compounds of Group 14/IV and the electron-rich organometallic compounds of Groups 15/V and 16/VI have M—C bonds of relatively low polarity.

> The polarities of element–carbon bonds are in accord with electronegativity considerations.

15.3 Structure and bonding

There are some subtle differences between organometallic compounds and binary hydrogen compounds, which stem in part from the tendency of alkyl groups to avoid ionic bonding. For example, the molecular structures of trimethylaluminum (**1**) and methyllithium (**2**) are in sharp contrast to the saline character of their solid hydrides, AlH_3 and LiH. Even the more ionic methylpotassium crystallizes in the nickel-arsenide structure rather than the rock-salt structure adopted by potassium chloride. It should be recalled that the nickel-arsenide structure is typical of soft-cation, soft-anion combinations (Section 2.9). There is a series of electron-deficient compounds, such as trimethylaluminum, that may contain 3c,2e bonds analogous to the B—H—B bridges in diborane (Section 8.9).

(a) s-Block organometallic compounds

Methyllithium in nonpolar solvents consists of a tetrahedron of Li atoms with each face bridged by a methyl group (**2**). As with $Al_2(CH_3)_6$, it is convenient to describe the bonding in terms of a set of localized molecular orbitals. Thus, a totally symmetric combination of three Li2s orbitals on each face of the Li_4 tetrahedron and one sp^3 hybrid orbital from CH_3 gives an orbital that can accommodate one electron pair (Fig. 15.2), leading to a **4-center, 2-electron bond** (a 4c,2e bond). The lower energy of the C orbital compared with the Li orbitals indicates that the bonding pair of electrons will be associated primarily with the CH_3 group, in agreement with the observed carbanionic character of the molecule. Some analyses of the bonding in these compounds indicate about 90 per cent ionic character for the Li—CH_3 interaction. Dimethylberyllium and dimethylmagnesium exist in a polymeric structure with two 3c,2e-bonding CH_3 bridges between each metal atom (**3**).

15.2 The interaction between an sp^3 orbital from a face-bridging methyl group and the 2s orbitals of the Li atoms in a triangular face of $Li_4(CH_3)_4$ to form a totally symmetric 4c,2e bonding orbital. The next higher orbital is doubly degenerate and nonbonding. The uppermost orbital is antibonding.

Compounds with greater ionic character are known in which the alkali metal is combined with cyclic or polycyclic arenes (which can accept an electron into their π^* orbitals). One such compound, sodium naphthalide, has already been mentioned: it contains the radical anion $[\cdot C_{10}H_8]^-$ in which a single unpaired electron occupies an antibonding π orbital of naphthalene. Because of the delocalization of the charge, the effective radius of the anion is large and the Coulombic interaction between cation and anion is weak. As a result, salts that contain such aromatic radical anions are slightly dissociated in polar aprotic solvents, such as tetrahydrofuran or tris(dimethylamine)phosphine oxide, $((CH_3)_2N)_3PO$.

> The methyl compounds of lithium, sodium, beryllium, magnesium, and aluminum are associated through alkyl bridges, and multi-center two-electron bonding may be invoked.

(b) The zinc group

A general feature to keep in mind is that the metals of Group 12 form linear molecular compounds, such as $Zn(CH_3)_2$, $Cd(CH_3)_2$, and $Hg(CH_3)_2$, that are not associated in the solid, liquid, or gaseous states or in hydrocarbon solution.

The elements in Group 12 have two valence electrons, and their linear monomeric structures indicate that they employ these electrons to form molecules with localized 2c,2e bonds (**4**). Unlike their beryllium and magnesium analogs, the elements do not complete their valence shells by association through alkyl bridges. To put these monomeric linear compounds into perspective we should note the general tendency toward linear coordination of metals with configuration d^{10}, such as Cu(I), Ag(I), and Au(I) in Group 11 (for example, $[N\equiv C-M-C\equiv N]^-$, M = Ag, Au). This tendency is sometimes rationalized by invoking pd hybridization in the M^+ ion, which leads to orbitals that favor linear attachment of ligands (much like the analogous spd hybridization illustrated in Fig. 9.23).

(c) The boron group

Unlike BH_3, which exists as the dimer, trimethylboron is a monomeric molecular compound, probably on account of the steric repulsion that would result between methyl groups in the bridged structure. In support of this explanation, the bigger Al atoms in alkylaluminum compounds do allow dimerization: they form 3c,2e bonds composed of orbitals supplied by the two Al atoms and bridging CH_3 groups (as in **1**). However, $Ga(CH_3)_3$ is monomeric even though Ga is larger than Al, so size is not the only factor.

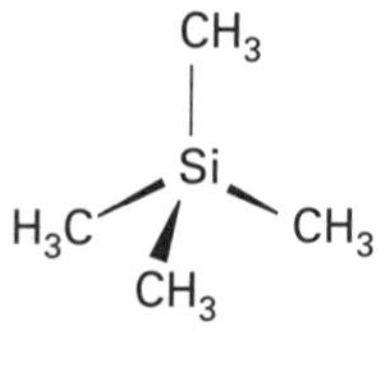

> Size is an important factor, but not the only one, in deciding whether alkyl compounds of Group 13/III form dimers.

(d) The carbon and nitrogen groups

The electron-precise compounds of the carbon group, such as tetramethylsilane (**5**), and the electron-rich compounds of the nitrogen group, such as trimethylarsane (**6**), are adequately described by conventional, localized 2c,2e bonds, and their shapes can be rationalized by the VSEPR model. However, that the simplest form of the VSEPR model cannot account for everything is shown by the fact that C—E—C bond angles for trimethyl compounds of the heavy elements (arsenic and antimony) are close to 90°. This angle is reminiscent of the bond angles of the corresponding hydrogen compounds, such as AsH_3 and SbH_3 (Table 8.3).

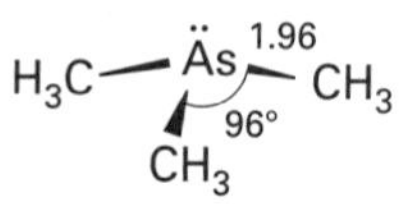

> The electron-precise compounds of Group 14/IV and the electron-rich compounds of Group 15/V, are described by localized 2c,2e bonds, and in most cases their shapes are in accord with the VSEPR model.

15.4 Stability

As reliable Gibbs energies of formation of organometallic compounds are not widely available, we have to rely on the enthalpies of formation in Table 15.2 for an approximate indication of

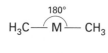

$$H_3C \overset{180°}{\underset{}{\frown}} M \text{—} CH_3$$

4 M = Zn,Cd,Hg

5

$$H_3C \overset{\overset{\overset{\displaystyle CH_3}{|}}{Si}}{\underset{CH_3}{\diagup}} CH_3$$

6

$$H_3C \overset{\overset{\displaystyle \ddot{A}s}{}}{\underset{CH_3}{\diagup}} {}^{1.96}_{96°} CH_3$$

Table 15.2 Standard molar enthalpies of formation of gaseous methyl derivatives of the *p*-block elements ($\Delta_f H^{\ominus}/\text{kJ mol}^{-1}$)

Period	12	13/III	14/IV	15/V
2		$B(CH_3)_3$	$C(CH_3)_4$	$N(CH_3)_3$
		−124	−167	−24
3		$Al(CH_3)_3$	$Si(CH_3)_4$	$P(CH_3)_3$
		−74	−239	−101
4	$Zn(CH_3)_2$	$Ga(CH_3)_3$	$Ge(CH_3)_4$	$As(CH_3)_3$
	+53	−45	−71	+13
5	$Cd(CH_3)_2$	$In(CH_3)_3$	$Sn(CH_3)_4$	$Sb(CH_3)_3$
	+101	> 0	+21	+32
6	$Hg(CH_3)_2$	$Tl(CH_3)_3$	$Pb(CH_3)_4$	$Bi(CH_3)_3$
	+94	> 0	+136	+194

Sources: D.D. Wagman, W.H. Evans, V.B. Parker, R.H. Shumm, I. Halow, S.M. Baiky, K.L. Churney, and R.L. Nuttall, *J. Phys. Chem. Ref. Data*, **11**, Supplement 2 (1982); M.E. O'Neill and K. Wade in *Comprehensive organometallic chemistry* (ed. G. Wilkinson, F.G.A. Stone, and E.W. Abel). Pergamon, Oxford (1982).

the stability of the compounds with respect to their elements. The trends are similar to those encountered for hydrogen compounds of the *p*-block: the methyl compounds of the light elements of Groups 13/III to 15/V are exothermic with respect to the elements, but the exothermic character decreases down the groups, and the compounds of the heaviest elements are endothermic.

The relative instability of the heavy members of the group is a consequence in part of the decrease in M—C bond strength (Fig. 15.3). The trend toward decreasing E—C bond strengths down a main group is also encountered for other types of bonds, such as E—H bonds (recall Fig. 8.11). As with the analogous hydrogen compounds, the endothermic heavy metal organometallic compounds are susceptible to homolytic cleavage of the M—C bond. For example, when $Pb(CH_3)_4$ is heated in the gas phase, it undergoes homolytic Pb—C bond cleavage to produce CH_3 radicals. Similarly, dimethylcadmium can decompose explosively.

> The methyl compounds of the light elements of Groups 13/III to 15/V are exothermic, but the exothermic character decreases down the groups, and the compounds of the heaviest elements are endothermic.

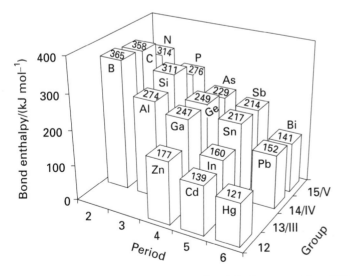

15.3 Average M—CH_3 bond enthalpies (in kilojoules per mole) at 298 K. From the data in M.E. O'Neill and K. Wade, in *Comprehensive organometallic chemistry* (ed. G. Wilkinson, F.G.A. Stone, and E.W. Abel), Vol. 1, p. 5. Pergamon Press, Oxford (1982).

15.5 Synthesis

Most organometallic compounds can be synthesized by using one of four M—C bond-forming reactions: reaction of a metal with an organic halide, metal displacement, metathesis, and hydrometallation.

(a) Reaction with metal and transmetallation

The net reaction of an electropositive metal M and a halogen-substituted hydrocarbon is

$$2\,M + RX \longrightarrow MR + MX$$

where RX is an alkyl or aryl halide. For example, methyllithium is produced commercially by the reaction

$$8\,Li + 4\,CH_3Cl \longrightarrow Li_4(CH_3)_4 + 4\,LiCl$$

With other active metals, such as magnesium, aluminum, and zinc, the reaction generally yields the organometal halide. A familiar example is the synthesis of a Grignard reagent, an alkylmagnesium halide:

$$Mg + CH_3Br \xrightarrow{\text{diethyl ether}} CH_3MgBr$$

In a **transmetallation reaction**, one metal atom takes the place of another, as in

$$M + M'R \longrightarrow M' + MR$$

This reaction is favorable when the metal M is higher in the electrochemical series than M' (which is typically Hg), so the reaction is analogous to a simple inorganic reaction in which one metal is displaced from a compound by a metal with a more negative standard potential. The probable outcome of a transmetallation reaction correlates with the electrochemical series even though the conditions for the reaction are very far from those used to determine standard potentials. With somewhat lower reliability, the outcome of the reaction can also be predicted from relative electronegativities: a metal with low electronegativity tends to displace a metal of higher electronegativity.

Because all the metals of Groups 1, 2, and 13/III have more negative standard potentials than mercury, transmetallation can be carried out between them and dimethylmercury, as in

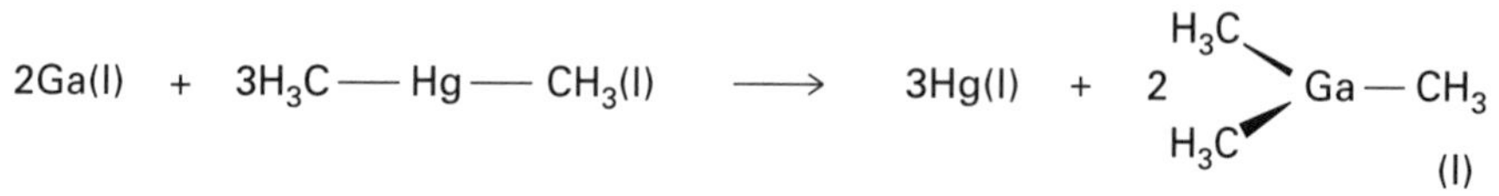

$$2Ga(l) \; + \; 3H_3C-Hg-CH_3(l) \; \longrightarrow \; 3Hg(l) \; + \; 2 \; \begin{matrix} H_3C \\ H_3C \end{matrix}\!\!\!>\!Ga-CH_3$$

(1)

In this reaction, gallium $(E^{\ominus}(Ga^{3+}/Ga) = -0.53 \text{ V})$ drives mercury $(E^{\ominus}(Hg^{2+}/Hg) = +0.85 \text{ V})$ out of dimethylmercury.

Transmetallation is favorable when the displacing metal is higher in the electrochemical series than the displaced metal.

(b) Synthesis by metathesis

The metathesis of an organometallic compound MR and a binary halide EX provides a widely used synthetic procedure in organometallic chemistry:

$$MR + EX \longrightarrow MX + ER$$

An approximate but instructive view of the reaction is that it takes place by the exchange of a formal carbanion (R^-) and a halide (X^-). Metathesis is an effective way of preparing a large number of organoelement compounds: the most common reagents are alkyllithium, alkylmagnesium, and alkylaluminum compounds and the halides of the Group 13/III to 15/V elements:

$$Li_4(CH_3)_4 + SiCl_4 \longrightarrow 4\,LiCl + Si(CH_3)_4$$
$$Al_2(CH_3)_6 + 2\,BF_3 \longrightarrow 2\,AlF_3 + 2B(CH_3)_3$$

The tendency for reaction can frequently be predicted from electronegativity or hard and soft acid–base considerations. From the standpoint of electronegativity, hydrocarbon groups form stronger covalent than ionic bonds, so the hydrocarbon group tends to bond to the more electronegative element; the halogen favors the formation of ionic compounds with the more electropositive metal. In brief, the alkyl or aryl group tends to migrate from the less to the more electronegative element:

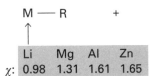

$$M - R \quad + \quad E - X \longrightarrow MX + R - E$$

	Li	Mg	Al	Zn		Si	B	As	P
χ:	0.98	1.31	1.61	1.65		1.90	2.04	2.18	2.19

When the electronegativities are similar, the correct outcome can usually be predicted as the outcome of the combination of the softer element with the organic group and harder element with fluoride or chloride. However, predictions from electronegativities and hard–soft characters must be used with some care. For example, an insoluble product or reactant may change the outcome, as in the reaction

$$SnPh_4(thf) + HgBr_2(thf) \longrightarrow HgPhBr(s) + Ph_3SnBr(thf)$$

where HgPhBr turns out to be insoluble in tetrahydrofuran.

Metathesis reactions involving the same central element are often referred to as **redistribution reactions**. For example, silicon tetrachloride and tetramethylsilicon (tetramethylsilane) undergo redistribution when heated and produce a variety of chloromethylsilanes:

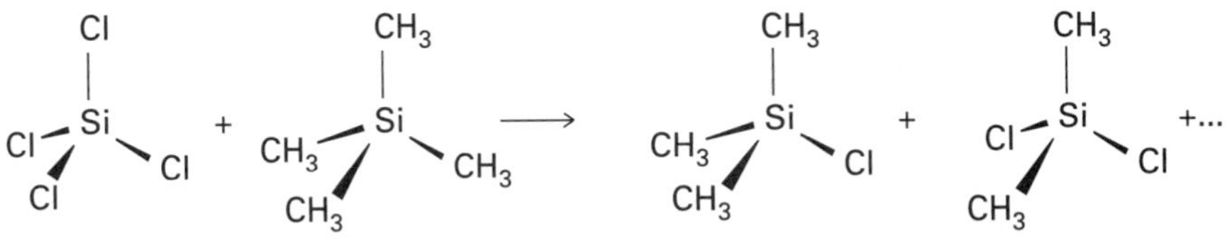

The formation of compounds with mixed substituents is favored by entropy considerations, for a random distribution of ligands is a more disorderly arrangement than the ligand-specific starting compounds.

> Metathesis can be regarded as taking place by the exchange of a formal carbanion (R^-) and a halide (X^-). The alkyl or aryl group tends to migrate from the less to the more electronegative element.

Example 15.1 Classifying and predicting a potential M—C bond-forming reaction

Decide on the type of reaction that might occur between $Al_2(CH_3)_6$ and $GeCl_4$.

Answer We are presented with an organometallic compound and a metal halide, so we should explore the possibility that these compounds can undergo a metathesis reaction. The organometallic compound has a metal (aluminum) that is more electropositive than the central atom in the halide (germanium); therefore the reaction

$$3\,GeCl_4 + 2\,Al_2(CH_3)_6 \longrightarrow 3\,Ge(CH_3)_4 + 4\,AlCl_3$$

should be thermodynamically favorable. This expectation is borne out in practice and it provides a convenient synthetic procedure.

Self-test 15.1 What is the likely outcome of the reaction between magnesium and dimethylmercury?

(c) Synthesis by hydrometallation

The net outcome of the addition of a metal hydride to an alkene is an alkylmetal compound:

$$EH + H_2C{=}CH_2 \longrightarrow ECH_2{-}CH_3$$

This reaction is driven mainly by the high strength of the C—H bond relative to that of most E—H bonds, and occurs with a wide variety of compounds that contain E—H bonds. We have already encountered several examples in Chapter 8, the most important being

1 *Hydroboration* (Section 8.9):

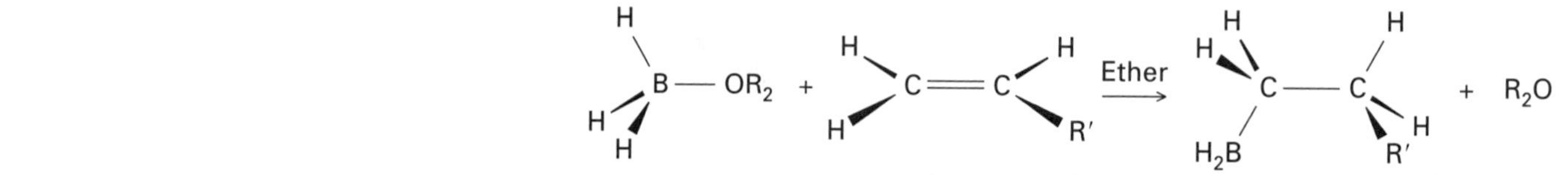

2 *Hydrosilylation* (Section 8.12):

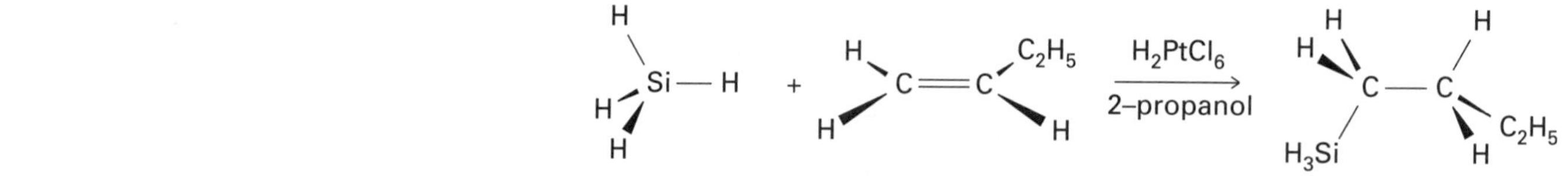

In the hydroboration and hydrosilylation of unsymmetrical alkenes, the bulkier B or Si group adds to the sterically less hindered C atom, and the smaller H atom adds to the more hindered C atom, as illustrated above.

> *Hydroboration and hydrosilylation of alkenes are two routes to the formation of boron–carbon and carbon–silicon bonds.*

15.6 Reaction patterns

The reactions of organometallic compounds of electropositive elements are dominated by factors such as the carbanion character of the organic moiety and the availability of a coordination site on the central metal atom.

(a) Oxidation

All organometallic compounds are potentially reducing agents; those of the electropositive elements are in fact very strong reducing agents. It is important to be mindful of the potential fire hazard this character implies, for many organometallic compounds are pyrophoric (ignite spontaneously on contact with air). The strong reducing character also presents a potential explosion hazard if the compounds are mixed with large amounts of oxidizing agents.

All organometallic compounds of the electropositive metals that have unfilled valence orbitals, or that readily dissociate into fragments with unfilled orbitals, are pyrophoric. These compounds include $Li_4(CH_3)_4$, $Zn(CH_3)_2$, $B(CH_3)_3$, and $Al_2(CH_3)_6$. Volatile pyrophoric compounds, such as $B(CH_3)_3$, may be handled in a vacuum line (Box 8.1) and inert atmosphere techniques are used for less volatile but air-sensitive organometallic compounds (Box 15.1). Compounds such as $Si(CH_3)_4$ and $Sn(CH_3)_4$, which do not have low-lying empty orbitals, require elevated temperatures to initiate combustion, and can be handled in air. The combustion of many organometallic compounds takes place by a radical chain mechanism.

> *All organometallic compounds are potentially reducing agents; those of the electropositive elements are in fact very strong reducing agents. All organometallic compounds of the electropositive metals that have unfilled valence orbitals, or that readily dissociate to fragments with unfilled orbitals, are pyrophoric.*

(b) Nucleophilic (carbanion) character

The partial negative charge of an organic group attached to an electropositive metal results in the group being a strong nucleophile and Lewis base. This is frequently referred to as its **carbanion character** even though the compound itself is not ionic.

Alkyllithium and alkylaluminum compounds and Grignard reagents are the most common carbanion reagents in laboratory-scale synthetic chemistry; carbanion character is greatly diminished for the less metallic elements boron and silicon. We described this nucleophilic character of organometallic compounds in Section 15.5 in connection with the formation of M—C bonds. As may be inferred from the summary in Chart 15.1, carbanion character has many other synthetic applications. For instance, an $R^{\delta-}$ group from the organometallic reagent attacks the carbonyl C atom of a ketone and upon hydrolysis a tertiary alcohol results. Similarly, aldehydes can be converted to secondary alcohols by reaction with an

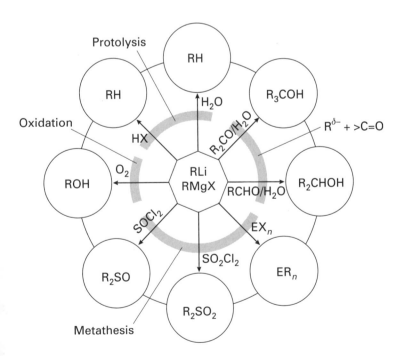

Chart 15.1 Typical protolysis, attack of carbonyl, metathesis, and oxidation reactions of alkyllithium compounds and Grignard reagents; X = halide; E = B, Si, Ge, Sn, Pb, As, and Sb.

Box 15.1 Inert atmosphere techniques

Because they react so readily with oxygen, moisture, and carbon dioxide, many organometallic compounds are handled in an inert atmosphere. The simplest of these techniques (conceptually, at least) is the **inert-atmosphere glove box** (Fig. B15.1). This apparatus allows manipulations to be carried out in a large metal enclosure filled with nitrogen, argon, or helium. The enclosure has a window, and a tightly attached pair of rubber gloves are used to manipulate chemicals inside the box. Two other important components of the box are a source of pure inert gas and an antechamber—an airlock—which permits items to be brought into or out of the box while avoiding the influx of air. Because air slowly diffuses into the box through the rubber gloves, it is common to maintain the purity of atmosphere inside the chamber by a recirculating gas purifier or a constant flush of inert gas.

Another common method of handling air-sensitive compounds uses standard glassware constructed so that an inert atmosphere can be maintained inside it. Typically this apparatus is equipped with sidearms for pumping out the air and introducing inert gas (Fig. B15.2). When apparatus containing air-sensitive compounds must be reconfigured, it is opened under a flush of inert gas. This type of apparatus can be employed for all the standard synthetic operations such as reactions in solution, filtration, and crystallization. The apparatus is often referred to as **Schlenk ware** in recognition of the German chemist Wilhelm Schlenk, who introduced this general design during his pioneering research in organometallic chemistry during the first quarter of the twentieth century.

A variation on the above method for transferring air-sensitive solutions utilizes syringes or metal cannulae (flexible, small-diameter metal tubing). For these operations, a flask is fitted with a rubber serum bottle cap (a septum) which can be pierced with a syringe needle or sharpened cannula. Figure B15.3 illustrates the transfer of a solution from one flask to the other by a cannula and a pressure differential between the two flasks.

Further reading

J.J. Eisch (ed.), *Organometallic synthesis*, Vol. 2. Academic Press, New York (1981).

A.L. Wayda and M.Y. Darensbourg (ed.), *Experimental organometallic chemistry: a practicum in synthesis and characterization*, ACS Symposium Series 357. American Chemical Society, Washington, DC (1988).

D.F. Shriver and M.A. Drezdzon, *The manipulation of air-sensitive compounds*. Wiley, New York (1986).

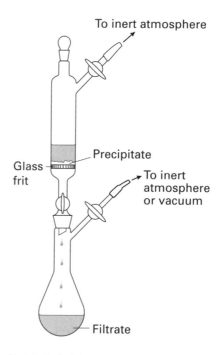

B15.2 Typical inert-atmosphere Schlenk ware. In this example, a reaction mixture has been filtered in the upper part of the apparatus and the filtrate collected in the Schlenk tube shown in the lower part of the illustration.

B15.1 An inert-atmosphere glove box. The door to the airlock is on the right. A vacuum pump is used to evacuate the transfer lock, which is refilled with inert gas before items are brought into the main chamber. The system also contains a circulation pump and columns, which remove water and oxygen from the glove box atmosphere. (Reproduced by permission of LabConco, Kansas City, MO.)

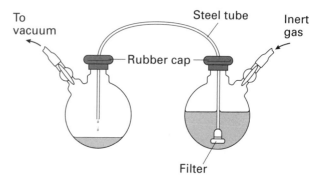

B15.3 Filtration and liquid transfer by means of a pressure differential and stainless steel tube (a cannula).

organometallic compound followed by hydrolysis. There are many examples of the use of metathesis reactions for the preparation of p-block organometallic compounds and sulfones and sulfoxides by treatment of SO_2Cl_2 or $SOCl_2$ with an alkyllithium or Grignard reagent. We amplify on many of these reactions later in the chapter.

One noteworthy consequence of high carbanion character is the protolysis reaction that takes place with very weak Brønsted acids, including water and alcohols. The criteria for judging when this reaction can be rapid are similar to those for judging whether reaction with O_2 (combustion) is likely.[2] Thus, organometallic compounds with low-lying vacant orbitals have low kinetic barriers for the formation of an initial complex, which can then undergo proton transfer. According to this mechanism, the lone pairs on the O atom of an alcohol coordinate to the Ga atom in triethylgallium. This complex formation is followed by transfer of the proton to the ethyl group and hence leads to the evolution of ethane:

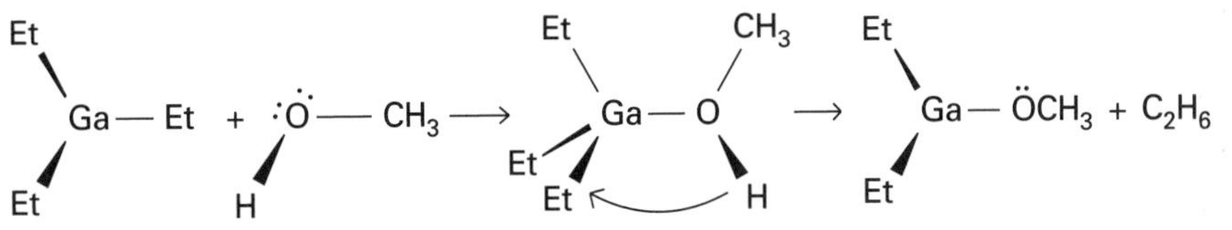

Similarly, alkylaluminum compounds react vigorously with excess alcohol to produce alkoxyaluminum compounds:

$$Al_2(CH_3)_6 + 6\,C_2H_5OH \longrightarrow 2\,Al(OC_2H_5)_3 + 6\,CH_4$$

Again, it appears that protolysis occurs by prior coordination:

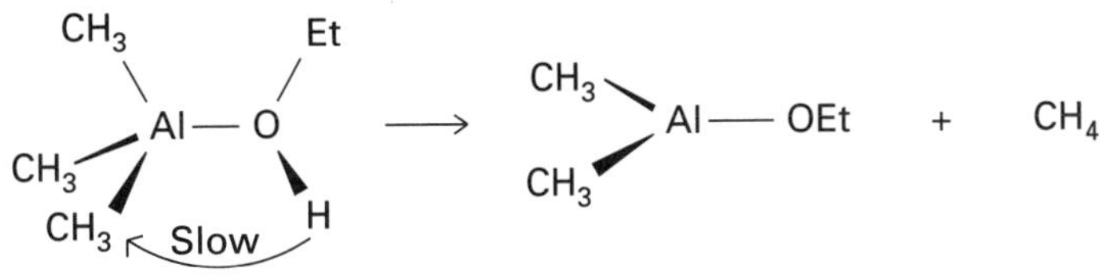

In keeping with the lengthening of the O—H bond implied by this mechanism, the reaction shows a significant kinetic isotope effect on deuteration, and the reaction is much slower when H is replaced by an alkyl group. For example, $k(CH_3O-H)/k(CH_3O-R) = 2.9$ when R is an octyl group.

Because the protolysis of organometallic compounds liberates hydrocarbons, water should not be used to extinguish a fire involving trialkylaluminum compounds or any other organometallic compounds of electropositive metals. Protolysis of electropositive organo-metallic compounds with sterically hindered alcohols, such as 2-propanol or *tert*-butanol (2-methyl-2-propanol) proceeds at a moderate rate and therefore provides a convenient way of destroying reactive organometallic wastes.

Organometallic compounds with low-lying vacant orbitals have low kinetic barriers for the formation of an initial complex, which can then undergo proton transfer.

(c) Lewis acidity

As a result of the presence of unoccupied orbitals on the metal atom, electron-deficient organometallic compounds are observed to be Lewis acids. An illustration of this Lewis acidity is the synthesis of organometallic anions, such as the tetraphenylborate ion:

[2] As with most chemical generalizations, this statement has limitations. For example, $B(CH_3)_3$ is oxidized rapidly by air but is resistant to hydrolysis.

$$B(C_6H_5)_3 + Li(C_6H_5) \longrightarrow Li[B(C_6H_5)_4]$$

This reaction may be viewed as the transfer of the strong base $C_6H_5^-$ from the weak Lewis acid Li^+ to the stronger acid B(III).

Organometallic species that are bridged by organic groups can also serve as Lewis acids and, in the process, undergo bridge cleavage. For example, $Al_2(CH_3)_6$ is cleaved by tertiary amines to form a simple Lewis acid–base complex:

$$Al_2(CH_3)_6 + 2\,N(C_2H_5)_3 \longrightarrow 2\,(CH_3)_3AlN(C_2H_5)_3$$

This reaction illustrates once again the weakness of the 3c,2e Al—CH_3—Al bond. Solvents such as THF coordinate to the Li atoms in $Li_4(CH_3)_4$, but such a mild base does not disrupt the metal cluster.

▌ *Electron-deficient organometallic compounds are Lewis acids.*

(d) β-hydrogen elimination

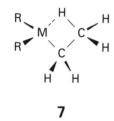

7

In the process known as **β-hydrogen elimination**, a metal atom extracts an H atom from the next-nearest neighbor (β) carbon atom. An example is the reaction

$$M—CH_2—CH_3 \longrightarrow M—H + H_2C{=}CH_2$$

The reaction is the reverse of the addition of an **M**—H bond to an alkene (hydrometallation) and under some conditions significant equilibrium concentrations of both reactants and products are observed. The reaction mechanism is believed to involve the formation of β-hydrogen bridges to an open coordination site on the metal (**7**). As might be expected from the proposed mechanism, compounds in which the central atom has a low coordination number tend to undergo this reaction. For example, it occurs with trialkylaluminum compounds but not with tetraalkylgermanium compounds.

▌ *Compounds in which the central atom has a low coordination number tend to undergo β-elimination.*

Example 15.2 Predicting the products of thermal decomposition

Summarize the stabilities of (a) $Bi(CH_3)_3$ and (b) $Al_2({}^iBu)_6$, with respect to their thermal decomposition and give chemical equations for their decomposition.[3]

Answer (a) As with the other heavy *p*-block elements, bismuth–carbon bonds are weak and readily undergo homolytic cleavage. The resulting methyl radicals will react with other radicals or form ethane:

$$2\,Bi(CH_3)_3 \xrightarrow{\Delta} 2\,Bi(s) + 3\,H_3C—CH_3$$

(b) The $Al_2({}^iBu)_6$ dimer readily dissociates. At elevated temperatures dissociation is followed by β-hydrogen elimination. This type of elimination is common for organometallic compounds that have alkyl groups with β-hydrogens, can form stable **M**—H bonds, and can provide a coordination site on the central metal. The decomposition reaction is

$$Al_2({}^iBu)_6 \xrightarrow{\Delta} 2\,Al({}^iBu)_3 \xrightarrow{\Delta\Delta} Al({}^iBu)_2H + (CH_3)_2C{=}CH_2$$

In this equation, Δ implies moderate heating and $\Delta\Delta$ implies stronger heating.

Self-test 15.2 Describe the probable mode of thermal decomposition of $Pb(CH_3)_4$.

---

[3] ${}^iBu$ denotes the $(CH_3)_2CHCH_2—$ group.

Ionic and electron-deficient compounds of Groups 1, 2, and 12

In this and the following sections we discuss the chemical properties of certain organometallic compounds in more detail. We concentrate on the compounds formed by lithium, magnesium, zinc, and mercury because of their interesting properties and utility in syntheses.

15.7 Alkali metals

Organometallic derivatives have been made for all the Group 1 metals. Of the simple alkyl compounds the alkyllithium compounds are by far the most thoroughly studied and useful synthetic reagents.

(a) Organolithium compounds

Organolithium compounds are available commercially as solutions. Methyllithium is generally handled in ether solution, but alkyllithium compounds with longer chains are soluble in hydrocarbons. Because the commercial preparation of alkyllithium compounds is by the reaction of the metal with the organic halide, they are often contaminated by halide. This contamination can be avoided by a preparation of the form

$$HgR_2 + 2\,Li \longrightarrow 2\,LiR + Hg$$

Although methyllithium exists as a tetrahedral cluster in the solid state and in solution, many of its higher homologs exist in solution as hexamers or equilibrium mixtures of aggregates ranging up to hexamers. These larger aggregates can be broken down by strong Lewis bases, such as chelating amines. For example, the interaction of TMEDA[4] with phenyllithium produces a complex containing two Li atoms bridged by phenyl groups with each Li atom coordinated by the chelating diamine (**8**).

In addition to the common organolithium compounds with one Li atom per organic group, a variety of **polylithiated** organic molecules, or organometallic compounds containing several Li atoms per molecule, are known.[5] The simplest example is dilithiomethane, Li_2CH_2, which can be prepared by the pyrolysis of methyllithium. This compound crystallizes in a distorted antifluorite structure (see Fig. 2.13), but the finer details of the orientation of the CH_2 groups are as yet unknown.

> Organolithium compounds with one lithium atom per organic group and a variety of polylithiated organic molecules are known.

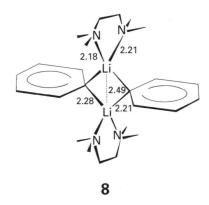

8

(b) Radical anion salts

Sodium naphthalide is an example of an organometallic salt that contains a delocalized radical anion, $C_{10}H_8^-$. Such compounds are readily prepared by the direct reaction of the aromatic compound with an alkali metal in a polar aprotic solvent. Thus, naphthalene dissolved in THF reacts with sodium metal to produce a dark green solution of sodium naphthalide:

$$Na(s) + C_{10}H_8(thf) \longrightarrow Na[C_{10}H_8](thf)$$

EPR spectra show that the odd electron is delocalized in an antibonding orbital of $C_{10}H_8$.

---

[4] TMEDA is *N,N,N′,N′*-tetramethylethylenediamine. It is a favorite chelating ligand in *p*-block organometallic chemistry because it lacks the mildly acidic N—H bonds of ethylenediamine, and therefore does not undergo protolysis with carbanionic organometallic compounds such as the alkyllithium compounds.

[5] A description of X-ray structures of lithium compounds is given by W.N. Setzer and P. von R. Schleyer, *Adv. Organomet. Chem.* **24**, 1385 (1985).

The formation of the radical anion is more favorable when the π LUMO of the arene is low in energy. Simple molecular orbital theory predicts that the LUMO occurs at progressively lower energies on going from benzene to more extensively conjugated hydrocarbons. This is analogous to the lowering of energy levels of an electron in a box as the length of the box is increased. The prediction is borne out by the standard potentials of aromatic hydrocarbons (Table 15.3). For this reason, the radical anion of benzene cannot be formed in most solvents, but naphthalene and more extensively conjugated arenes readily yield alkali metal salts.

Sodium naphthalide and similar compounds are highly reactive reducing agents. They are often preferred to sodium because—unlike sodium itself—they are readily soluble in ethers. The resulting homogeneous reaction is generally faster and easier to control than a heterogeneous reaction between one component in solution and pieces of sodium metal, which are often coated with unreactive oxide or with insoluble reaction products. As indicated in Table 15.3, an additional advantage of the radical anion reagents is that by proper choice of the aromatic group the reduction potential of the reagent can be chosen to match the requirements of a particular synthetic task.

Another route to delocalized anions is the reductive cleavage of acidic C—H bonds by an alkali metal or alkylmetallic compound. Thus, in the presence of a good coordinating ligand such as TMEDA, butyllithium reduces dihydronaphthalene to produce a diamagnetic dinegative anion:

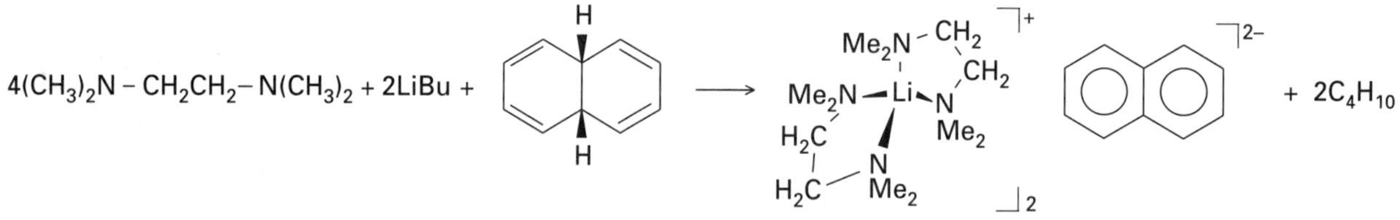

Table 15.3 Standard potentials of some conjugated hydrocarbons*

Compound		$E^{\ominus}/\text{V}$
Biphenyl	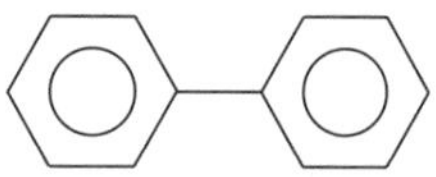	+0.00
Naphthalene	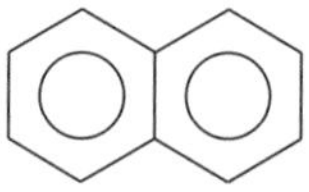	+0.09
Phenanthrene	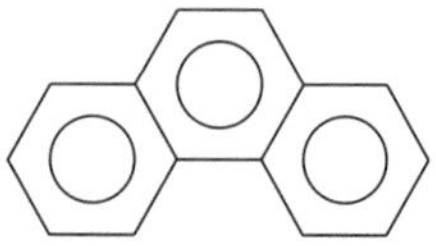	+0.17
Anthracene	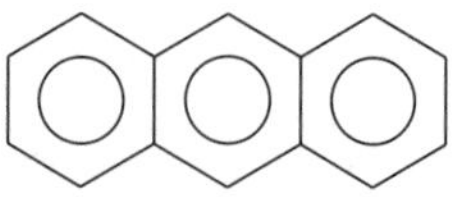	+0.78

*Relative to the value for biphenyl in 1,2-dimethoxyethane. *Source:* E. de Boer, *Adv. Organometal. Chem.* **2**, 115 (1964).

TMEDA contributes to the favorable reaction Gibbs energy through its affinity for the Li^+ produced in the reaction.

The planar aromatic cyclopentadienide ion $C_5H_5^-$ can be prepared by the reductive cleavage of a C—H bond in cyclopentadiene, by using either sodium metal or NaH:

$$+ \quad Na \quad \xrightarrow{THF} \quad Na^+ \quad \left[\begin{array}{c} - \end{array}\right] \quad + \quad {}^1/_2\,H_2(g)$$

The product, sodium cyclopentadienide, is an important reagent in organometallic synthesis. It readily participates in double replacement with a variety of halides of p-block elements to produce either σ- or π-bonded cyclopentadienyl compounds (which we discuss later in this chapter). It is also used to synthesize a wide variety of d-block organometallic compounds (Chapter 16).

> The formation of the radical anion is more favorable when the π LUMO of the arene is low in energy, which it is in more extensively conjugated systems.

15.8 Alkaline earth metals

Organoberyllium and organomagnesium compounds have significant covalent character, whereas the analogous compounds of the heavier congeners are more ionic. The latter have not yet been thoroughly investigated.

A characteristic of organoberyllium and organomagnesium compounds is the tendency of the metal atoms to adopt four-coordination; three-coordination is also observed for beryllium. The dimethyl compounds, for instance, appear to be bridged species (as in **3**). The fact that more bulky groups lead to decreased association is shown by the dimeric structure of diethylberyllium in benzene solution compared with the monomeric character of $Be(^tBu)_2$.

(a) Synthesis and structure

Because beryllium is more electronegative than magnesium, ether-solvated dialkylberyllium compounds can be prepared by transmetallation with a Grignard reagent:

$$BeCl_2 + 2\,RMgX + (C_2H_5)_2O \xrightarrow{\text{diethyl ether}} R_2Be{-}O(C_2H_5)_2 + 2\,MgXCl$$

When an ether-free product is needed, transmetallation with dialkylmercury can be used because beryllium is higher than mercury in the electrochemical series:

$$Be + Hg(CH_3)_2 \longrightarrow Be(CH_3)_2 + Hg$$

(Remember that $Be(CH_3)_2$ has the polymeric structure shown in (**3**).) Bis(cyclopentadienyl)beryllium is readily prepared by a metathesis reaction between beryllium halide and $Na[C_5H_5]$. In agreement with the usual trend, the $C_5H_5^-$ ion transfers to the more electronegative Be(II) atom:

$$BeCl_2 + 2\,NaC_5H_5 \longrightarrow Be(C_5H_5)_2 + 2\,NaCl$$

Data from X-ray crystal structure determinations demonstrate that in the solid state the compound has a mixed η^5-C_5H_5, η^1-C_5H_5 coordination (**9**). The symbol η^1 indicates the attachment of a C_5H_5 group through only one C atom and denotes **monohapto bonding**. The notation η^5 for the other C_5H_5 ring indicates that all five C atoms bond to beryllium; this arrangement is called **pentahapto bonding**.[6] The structure in the gas phase is less clear. It

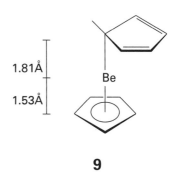

9

....

[6] The term 'hapto', is derived from the Greek word 'to fasten'. More examples of this type are given in Chapter 16.

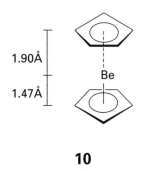

1.90Å

1.47Å

Be

10

may be the same unsymmetrical sandwich (**9**) with the molecule undergoing rearrangements on a timescale too fast to resolve by electron diffraction, or it may be an antisymmetric sandwich (**10**), where both C_5H_5 groups are pentahapto, but one ring is further than the other from the Be atom. This ambiguity illustrates the difficulty of obtaining precise structural data on large, possibly fluxional molecules in the gas phase.

> Organometallic compounds of the Group 2 elements may be formed by transmetallation and metathesis.

(b) Reactions

Simple organoberyllium compounds react with oxygen, water, and other weak Brønsted acids:

$$Be(CH_3)_2 + CH_3OH \longrightarrow (CH_3)Be(OCH_3) + CH_4$$

Dimethylberyllium also readily forms complexes with Lewis bases:

$$Be(CH_3)_2 + N(CH_3)_3 \longrightarrow (CH_3)_2BeN(CH_3)_3$$

$$Be(CH_3)_2 \quad + \quad (CH_3)_2NCH_2CH_2N(CH_3)_2 \quad \longrightarrow$$

As beryllium compounds are highly toxic, and there appear to be no compelling reasons to use them as reagents, organoberyllium compounds do not have significant commercial applications. Their use as synthetic intermediates in the laboratory is confined to the synthesis of other beryllium compounds.

> Simple organoberyllium compounds react with oxygen, water, and other weak Brønsted acids, and form complexes with Lewis bases.

(c) Grignard reagents

Organomagnesium compounds are familiar to any student of organic chemistry as useful carbanion reagents. The most common of these compounds are the alkylmagnesium halides, or **Grignard reagents**, prepared by the reaction of a haloalkane with magnesium metal.[7] This reaction is carried out in ether and, because a coating of oxide on the magnesium acts as a kinetic barrier, a trace of iodine is often added to initiate the reaction.

From the early work carried out by Wilhelm Schlenk, it is known that redistribution equilibria occur in ether solution. The simplest of these, often called the **Schlenk equilibrium**, is

$$2\,RMgX \rightleftharpoons MgR_2 + MgX_2$$

The addition of dioxane to the equilibrium mixture in diethyl ether leads to the precipitation of a dioxane complex of the magnesium halide, $MgX_2 \cdot (C_4H_8O_2)$, and the filtrate can be evaporated to yield the dialkylmagnesium. More recent spectroscopic studies indicate a rather complex set of equilibria between alkylmagnesium halides in ether solution. In line with these observations, the species crystallized from ether solution include a wide variety of structures, such as monomeric four-coordinate magnesium complexes (**11**) and larger clusters (**12**). An important feature of the latter structure is that halide bridges, where conventional 2c,2e bonds can be formed by donation of halide lone-electron pairs, are preferred over alkyl

Et

Br — Mg — OEt₂

OEt₂

11

12

---

[7] A translation of Grignard's first full-length paper on this subject is available in *J. Chem. Educ.* **47**, 290 (1970).

bridges, which would require two-electron multicenter bonds. We shall encounter this preference for 2c,2e bridge bonds in aluminum chemistry.

Because Grignard reagents are produced in ether solution, their use is limited to reactions in which that mildly Lewis basic solvent is not objectionable. The ether may form an unwanted complex with Lewis acid products, such as $(CH_3)_3BO(C_2H_5)_2$ in the reaction of BF_3 with CH_3MgBr in diethyl ether. In these cases, alkylaluminum or alkyllithium compounds are preferred because they can be used in hydrocarbon solution.

> Grignard reagents are prepared by the reaction of a haloalkane with magnesium metal; the species crystallized from ether solution include a wide variety of structures, such as monomeric magnesium complexes and larger clusters.

15.9 The zinc group

As mentioned earlier in the chapter, the dialkyl compounds of zinc, cadmium, and mercury are remarkable for their lack of association through alkyl bridges. Another feature to bear in mind is that dialkylzinc compounds are only weak Lewis acids, organocadmium compounds are even weaker, and organomercury compounds do not act as Lewis acids except under special circumstances.

(a) Organozinc and organocadmium compounds

A convenient synthesis of organometallic compounds of zinc is metathesis with alkylaluminum or alkyllithium compounds. With alkyllithium compounds as reactants this reaction conforms to the correlation of electronegativity that is often characteristic of metathesis reactions. However, the correlation is not decisive for alkylaluminum reactions because the electronegativities of aluminum and zinc are so similar (1.61 and 1.65, respectively). In this case, hardness considerations correctly predict the formation of the softer $ZnCH_3$ and harder $AlCl$ pairs:[8]

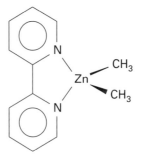

13

14

$$ZnCl_2(s) + \text{[aluminum alkyl bridge]} \longrightarrow H_3C-Zn-CH_3 + \text{[aluminum chloride bridge]}$$

Alkylzinc compounds are pyrophoric and readily hydrolyzed, whereas alkylcadmium compounds react more slowly with air. Because of their mild Lewis acidity, dialkylzinc and dialkylcadmium compounds form stable complexes with amines, especially chelating amines (**13**). The C—Zn bond has greater carbanionic character than the C—Cd bond. One example is the addition of alkylzinc compounds across the carbonyl group of a ketone:

$$Zn(CH_3)_2 + (CH_3)_2C{=}O \longrightarrow (CH_3)_3C{-}O{-}ZnCH_3$$

This reaction does not take place when the less polar alkylcadmium and alkylmercury compounds are used instead. The reaction also occurs for organolithium, organomagnesium, and organoaluminum compounds, all of which contain metals of even lower electronegativity than zinc.

Yet again, the cyclopentadienyl compounds are structurally unusual. (Cyclopentadienyl)methylzinc is monomeric in the gas phase with a pentahapto C_5H_5 group (**14**). In the solid it is associated in a zig-zag chain (**15**), each C_5H_5 group being pentahapto with respect to two Zn atoms.

15

8   For examples where hard and soft acid–base principles are more reliable for the prediction of the outcome of a reaction than are electronegativities, see R.G. Pearson, *J. Chem. Educ.* **45**, 643 (1968). The great convenience of using electronegativity as a first approximation to reactivity is the availability of numerical values that follow fairly simple trends in the periodic table.

> *Alkylzinc compounds are pyrophoric and readily hydrolyzed, whereas alkylcadmium compounds react more slowly with air. Because of their mild Lewis acidity, dialkylzinc and dialkylcadmium compounds form stable complexes with amines, especially chelating amines.*

(b) Organomercury compounds

Organomercury compounds are readily prepared by metathesis reactions between mercury(II) halides and strong carbanion reagents, such as Grignard reagents or trialkylaluminum compounds:

$$2\,RMgX + HgX_2 \longrightarrow HgR_2 + MgX_2$$

This reaction is in line with both electronegativity and hardness considerations. As we have already remarked, dialkylmercury compounds are versatile starting materials for the synthesis of many organometallic compounds of more electropositive metals by transmetallation. However, owing to the high toxicity of alkylmercury compounds (Box 15.2), other syntheses are often preferred. In striking contrast to the high sensitivity of dimethylzinc to oxygen, dimethylmercury survives exposure to air.

> *Dialkylmercury compounds can be used as starting materials for the synthesis of organometallic compounds of more electropositive metals by transmetallation.*

Electron-deficient compounds of the boron group

We illustrate two general points in this section. One point is that electron-deficient organometallic compounds of Group 13/III are molecular. The second point is that, for compounds in which the metallic element has oxidation number +3, only the organoaluminum compounds are significantly associated through 3c,2e bonded organic bridges.

Box 15.2 The toxicity of organomercury compounds

The toxicity of mercury arises from the very high affinity of the soft Hg atom for sulfhydryl (—SH) groups in enzymes. Simple mercury–sulfur compounds have been studied as potential analogs of natural systems. The Hg atoms are most commonly four-coordinated, as in $[Hg_2(SMe_2)_6]^{2-}$ (**B1**).

Mercury poisoning was a serious problem for early scientists, including Isaac Newton in the eighteenth century and Alfred Stock in the early twentieth century, both of whom worked with mercury in poorly ventilated laboratories. More recently, it became a major public concern following the incidence of brain damage and death it caused among the inhabitants in Minamata, Japan. This incident arose because mercury from a plastics factory was allowed to escape into a bay where it found its way into fish that were later eaten. Research since that time has shown that bacteria found in sediments are capable of methylating mercury, and that species such as $Hg(CH_3)_2$ and $[HgCH_3]^+$ enter the food chain because they readily penetrate cell walls. The bacteria appear to produce $Hg(CH_3)_2$ as a means of eliminating toxic mercury ions through their cell walls and into the environment.

Further reading

P.J. Craig (ed.), *Organometallic compounds in the environment.* Wiley, New York (1986).

J.G. Wright, M.J. Natan, F.M. MacDonnell, D. Ralston, and T.V. O'Halloran, Mercury(II) thiolate chemistry and the mechanism of the heavy metal biosensor. *Prog. Inorg. Chem.* **38**, 323 (1990).

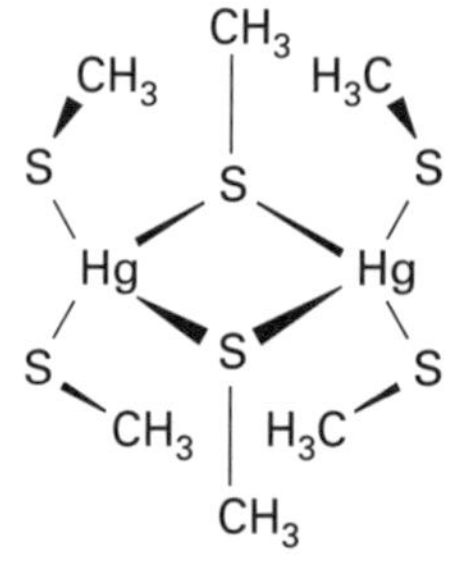

B1

15.10 Organoboron compounds

Trimethylboron is colorless, gaseous (b.p. $-22\,°C$), and monomeric. It is pyrophoric but is not rapidly hydrolyzed by water. The alkylboranes can be synthesized by metathesis between boron halides and organometallic compounds of metals with low electronegativity, such as Grignard reagents or organoaluminum compounds:

$$BF_3 + 3\,CH_3MgBr \xrightarrow{\text{dibutyl ether}} B(CH_3)_3 + 3\,MgBrF$$

Dibutyl ether is used here rather than diethyl ether because it has a much lower vapor pressure than trimethylboron, which facilitates separation by trap-to-trap distillation on a vacuum line. Another key to the success of this separation is the very weak association of the dibutyl ether–trimethylboron complex.

Although trialkyl- and triarylboron compounds are mild Lewis acids, strong carbanion reagents lead to anions of the type $[BR_4]^-$. The best known of these anions is the tetraphenylborate ion, in which $R = C_6H_5$. This bulky anion hydrolyzes very slowly in neutral or basic water and is useful for the precipitation of large monopositive cations. For example, the addition of an aqueous solution of $Na[BPh_4]$ to a solution containing K^+ ions leads to the formation of insoluble $K[BPh_4]$. This precipitation reaction, which can be used for the gravimetric determination of potassium, is an example of the low solubility of large-cation large-anion salts in water (Section 2.12).

The incorporation of boron into heterocycles is common, as in the protolysis of a triarylboron with 1,2-dihydroxybenzene at elevated temperatures:

Organohaloboron compounds are more reactive than simple trialkylboron compounds. One preparative route is by the reaction of boron trichloride with a stoichiometric amount of an alkylaluminum in a hydrocarbon solvent:

$$3\,BCl_3 + 6\,AlR_3 \longrightarrow 3\,R_2BCl + 6\,AlR_2Cl$$

Another procedure is the redistribution of boron trihalide and trialkylborane in the presence of diborane as catalyst:

$$2\,BCl_3 + BMe_3 \xrightarrow{\text{diborane}} 3\,BMeCl_2$$

The products may be subjected to the full range of protolysis reactions with ROH, R_2NH, and other reagents,

$$B(CH_3)_2Cl + 2\,HNR_2 \longrightarrow B(CH_3)_2NR_2 + [NR_2H_2]Cl$$

and to displacement of the halide by a carbanion reagent:

$$B(CH_3)_2Cl + Li(C_4H_9) \longrightarrow B(CH_3)_2(C_4H_9) + LiCl$$

Among the many other organoboranes, an especially interesting series contains B—N linkages. A BN fragment is isoelectronic with a CC fragment and B—N and C—C analogs generally have the same structures. However, despite this formal similarity, their chemical and physical properties are often quite different (see Chapters 10 and 11).

The alkylboranes can be synthesized by metathesis between boron halides and organometallic compounds of metals with low electronegativity, such as Grignard reagents or organoaluminum compounds. Organohaloboron compounds are more reactive than simple trialkylboron compounds.

16

17 Al(mes)$_3$

18

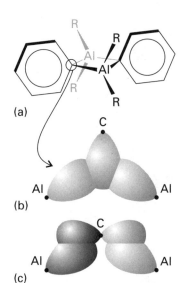

(a)

(b)

(c)

15.4 Structure and bonding of the phenyl bridge in Ph$_6$Al$_2$. (a) Structure illustrating the perpendicular orientation of the bridging phenyl relative to the AlCAlC plane. (b) The 3c,2e bond formed by a symmetric combination of C and Al orbitals. (c) An additional interaction between the $p\pi$ orbital on C and an antisymmetric combination of Al orbitals.

15.11 Organoaluminum compounds

One of the distinguishing features of the methyl bridge bond in alkylaluminum compounds (as in **1**) is the small Al—C—Al angle, which is approximately 75° (**16**). The weakness of these 3c,2e bridge bonds is indicated by the dissociation of trialkylaluminum compounds in the pure liquid to an extent that increases with the bulkiness of the alkyl group:[9]

$$Al_2(CH_3)_6 \rightleftharpoons 2\,Al(CH_3)_3 \qquad K = 1.52 \times 10^{-8}$$
$$Al_2(C_4H_9)_6 \rightleftharpoons 2\,Al(C_4H_9)_3 \qquad K = 2.3 \times 10^{-4}$$

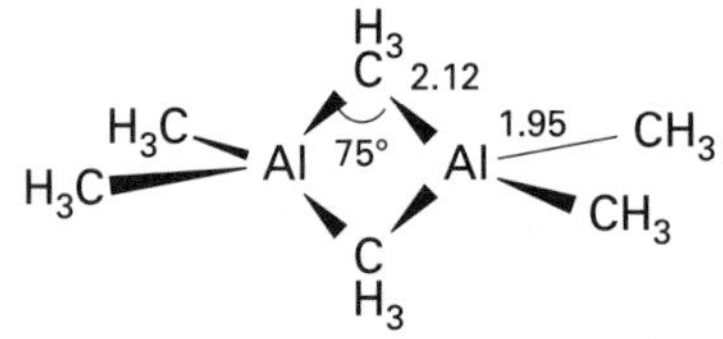

With very bulky groups, dissociation is virtually complete: trimesitylaluminum (**17**), for instance, is a monomer.[10] These examples provide clear evidence for the powerful role of steric effects on the structures of the alkylaluminum compounds.

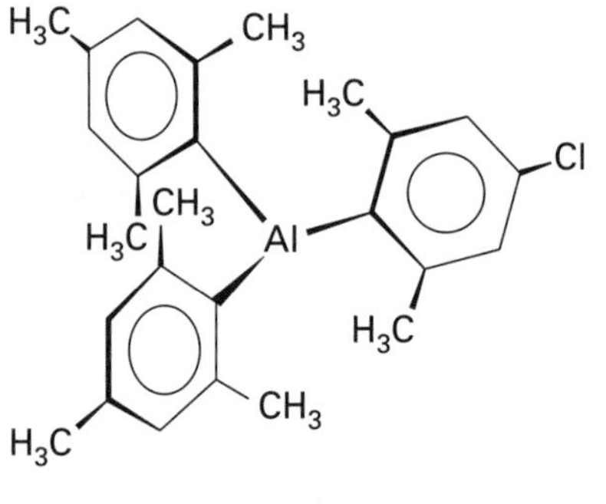

With bridging halides, alkoxides, and amides the Al—X—Al angle is close to 90° (**18**). In contrast to a bridging alkyl group, a halogen atom has more orbitals and electrons available and can form a bridge utilizing two 2c,2e bonds. Similarly, triphenylaluminum exists as a dimer with bridging η^1-phenyl groups lying in a plane perpendicular to the line joining the two Al atoms (Fig. 15.4). This structure is favored partly on steric grounds and partly by supplementation of the Al—C—Al bond by electron donation from the phenyl π orbitals to the Al atoms. These bonding arguments are consistent with the preference for the bridging site, which falls in the order (OR, halide) > phenyl > alkyl.

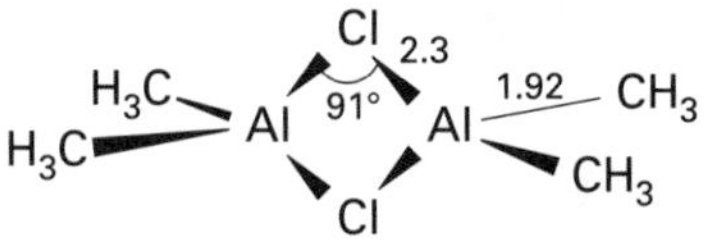

(a) Synthesis

Alkylaluminum compounds have been extensively investigated on account of their use as alkene polymerization catalysts and chemical intermediates. They are relatively inexpensive carbanion reagents for the replacement of halogens with organic groups by metathesis. A general laboratory-scale preparation of trimethylaluminum is transmetallation with dimethylmercury:

$$2\,Al + 3\,Hg(CH_3)_2 \longrightarrow Al_2(CH_3)_6 + 3\,Hg$$

The commercial synthesis uses the reaction of aluminum metal with chloromethane to produce Al$_2$Cl$_2$(CH$_3$)$_4$, which is then reduced with sodium:

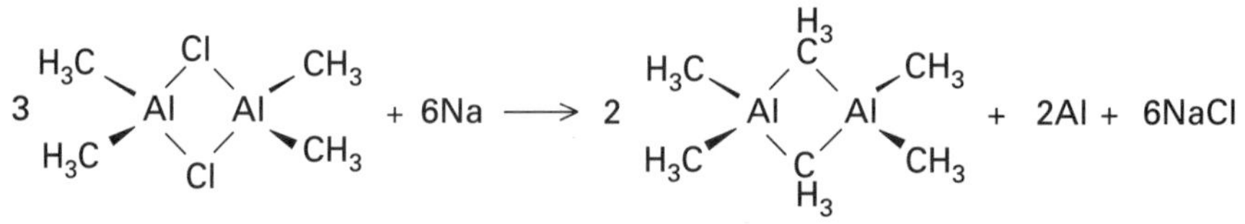

The commercial synthesis of triethylaluminum and higher homologs is by the addition of H$_2$ and the appropriate alkene to aluminum metal at elevated pressures and temperatures:

$$2\,Al + 3\,H_2 + 6\,RHC{=}CH_2 \xrightarrow{60-110\,°C,\ 100-200\ atm} Al_2(CH_2CH_2R)_6$$

It is probable that this reaction proceeds by the formation of a surface Al—H species that adds across the double bond of the alkene in a hydrometallation reaction. The commercial use of alkylaluminum compounds would be very limited without this relatively economical synthesis.

Organoaluminum compounds are carbanion reagents for the replacement of halogens with organic groups by metathesis. They are prepared by transmetallation or, commercially, by reactions with aluminum metal.

---

9. The equilibrium constants refer to 25 °C and are expressed in mole fractions.

10. The mesityl (2,4,6-trimethylphenyl) group is commonly used on account of its bulk. It plays a prominent role in the discussion of organosilicon chemistry in Section 15.13.

(b) Reactions

As might be expected from the more electropositive nature of aluminum, alkylaluminum compounds have much greater carbanionic character than alkylboron compounds. As a result, the former are very sensitive to water and oxygen and most are pyrophoric; both the pure liquid and solutions must be handled using inert atmosphere or vacuum line techniques. However, this reactivity can be turned to advantage, for the susceptibility of the Al—R bond to protolysis provides a simple method for the preparation of aluminum alkoxides and amides:

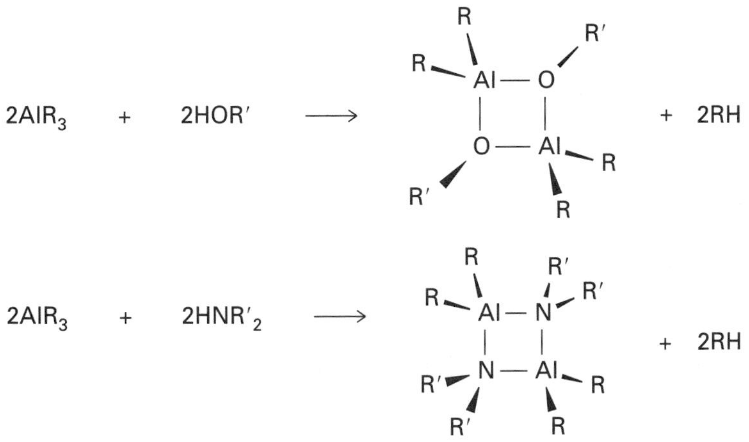

$$2AlR_3 \quad + \quad 2HOR' \quad \longrightarrow \quad \qquad + \quad 2RH$$

$$2AlR_3 \quad + \quad 2HNR'_2 \quad \longrightarrow \quad \qquad + \quad 2RH$$

Alkylaluminum compounds are mild Lewis acids and form complexes with ethers, amines, and anions.

β-Hydrogen elimination (Section 15.6) yields dialkylaluminum hydride when triethylaluminum and higher alkylaluminum compounds are heated. Tri(isobutyl)aluminum has a strong tendency to undergo this reaction:

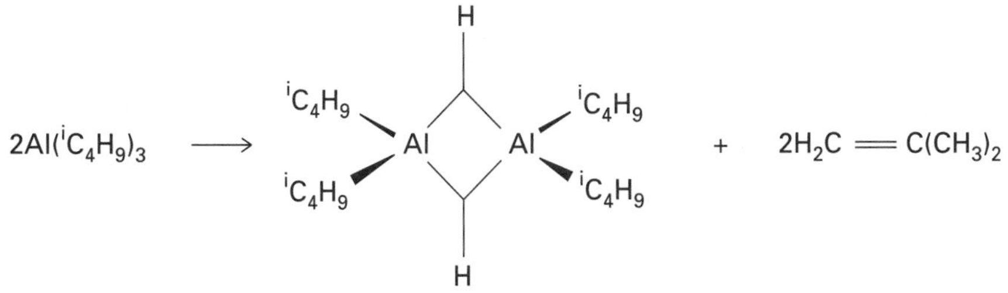

$$2Al(^iC_4H_9)_3 \quad \longrightarrow \quad \qquad + \quad 2H_2C = C(CH_3)_2$$

The general preference of hydride for the bridging position indicates that the 3c,2e bond is stronger for H than an alkyl group, probably because the small H atom can lie more readily between the atoms, and its orbital, being spherical, has less stringent directional requirements.

> Organoaluminum compounds have much greater carbanionic character than their boron analogs; as a result, they are very sensitive to water and oxygen and most are pyrophoric.

Example 15.3 Proposing structures for some boron and aluminum organometallic compounds

Based on your knowledge of the bonding in organoboron and organoaluminum compounds, propose structures for compounds having the empirical formulas (a) B(^{i}Pr)$_3$, (b) Al(C$_2$H$_5$)$_2$Ph, (c) Al(C$_2$H$_5$)$_2$P(CH$_3$)$_2$.

Answer (a) Triisopropylboron, as with all the simple alkylboranes, will be monomeric and the B atom and the three C atoms bonded to it should lie in a plane. (b) If the attached groups have moderate bulk, the alkyl and arylaluminum compounds are associated into dimers

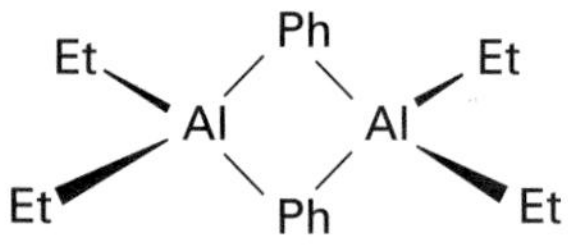

19

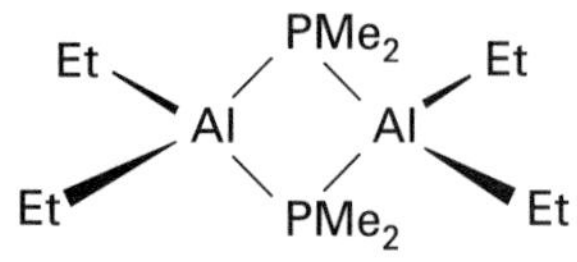

20

through multicenter bonds. Recalling that the tendency toward bridge structures is $PR_2^- > X^- > H^- > Ph^- > R^-$ (where R is alkyl), the structures will be as in (19) and (20).

Self-test 15.3 Propose a structure for $Al_2(^tBu)_4Cl_2$.

15.12 Organometallic compounds of gallium, indium, and thallium

There is a striking alternation of structures for the alkyl compounds of the first three elements in Group 13/III. As we have seen, the first member of the series, trimethylboron, is a monomer with three-coordinate B; the next, trimethylaluminum, is a dimer with four-coordinate Al; all the succeeding members (trimethylgallium, trimethylindium, and trimethylthallium) are monomeric in solution and in the gas phase. Carbanion character, as indicated by the tendency to hydrolyze and undergo double replacement reactions, follows the order indicated by aqueous potentials, namely Al > Ga > In > Tl. Alkylaluminum compounds, for instance, are rapidly and completely hydrolyzed by water:

$$Al_2(CH_3)_6 + 6 H_2O \longrightarrow 2 Al(OH)_3 + 6 CH_4$$

Hydrolysis of the analogous gallium, indium, and thallium compounds under mild conditions yields $M(CH_3)_2^+$ which itself is susceptible to hydrolysis in acidic solution in the order Ga > In > Tl. As we have seen (Section 15.6), trialkylboranes require much more forcing conditions to undergo hydrolysis.

> Carbanion character is greatest for aluminum and decreases in the order $Al_2R_6 > GaR_3 > InR_3 > TlR_3 > BR_3$.

(a) Gallium

Trialkylgallium compounds can be synthesized by the reaction of alkyllithium compounds with gallium chloride in a hydrocarbon solvent:

$$3 Li_4(C_2H_5)_4 + 4 GaCl_3 \longrightarrow 12 LiCl + 4 Ga(C_2H_5)_3$$

Trialkylgallium compounds are mild Lewis acids, and so the corresponding metathesis in ether produces the complex $(C_2H_5)_2OGa(C_2H_5)_3$. Similarly, the use of excess methyllithium leads to the uptake of a fourth alkyl group by the Ga atom to form a salt:

$$Li_4(C_2H_5)_4 + GaCl_3 \longrightarrow 3 LiCl + Li[Ga(C_2H_5)_4]$$

Trialkylgallium compounds are pyrophoric and react with weak Brønsted acids such as water, alcohols, and thiols. In keeping with the decrease in carbanion character with the more electronegative metal atom, their reactivity toward Brønsted acids is somewhat less than that of alkylaluminum compounds. The general (unbalanced) form of the reactions they undergo is

$$Ga(CH_3)_3 + H_2O \longrightarrow [Ga(CH_3)_2OH]_n + CH_4 \qquad n = 2, 3, \text{ or } 4$$

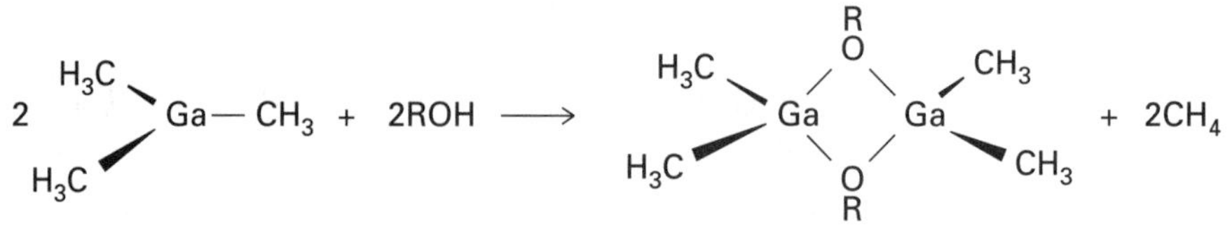

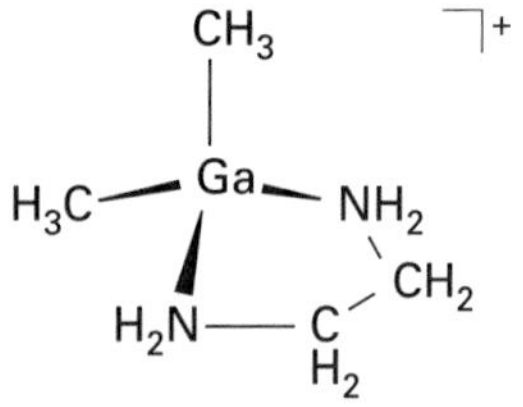

21

The GaR_2 group is somewhat resistant to protolysis, and complexes such as $[Ga(CH_3)_2(en)]^+$ (21) and even $[Ga(CH_3)_2(OH_2)_2]^+$ can be handled in aqueous solution.

> Trialkylgallium compounds are pyrophoric and react with weak Brønsted acids, but less vigorously than alkylaluminum compounds.

(b) Indium and thallium and group trends

Alkylindium and alkylthallium compounds may be prepared by reactions analogous to those used to make the alkylgallium compounds. Trimethylindium is monomeric in the gas phase and in the solid the bond lengths indicate that association is weak (if present at all). Partial hydrolysis of $Tl(CH_3)_3$ yields the linear $[CH_3TlCH_3]^+$ ion, which is isoelectronic and isostructural with CH_3HgCH_3.

A new aspect of organometallic chemistry becomes evident this deep in the boron group, for now the inert-pair effect can play a role and give rise to stable indium(I) and thallium(I) compounds. Two examples of these compounds are $(\eta^5\text{-}C_5H_5)In$ and $(\eta^5\text{-}C_5H_5)Tl$, which exist as monomers in the gas phase but are associated as solids. Cyclopentadienylthallium is useful as a synthetic reagent in organometallic chemistry because it is not as highly reducing as $Na[C_5H_5]$ and the insolubility of TlCl provides an added driving force for metathesis reactions. A drawback is that thallium is even more poisonous than mercury, so the disposal of the reaction byproducts must be done with care.

> *Trimethylindium is monomeric in the gas phase. The inert-pair effect is displayed for indium and thallium and gives rise to stable indium(I) and thallium(I) compounds.*

Electron-precise compounds of the carbon group

In Group 14/IV we meet organometallic compounds that are formed by carbon with its own congeners. Because the electronegativity of carbon is similar to that of its congeners, the bonds these elements form to one another are not very polar. This low polarity in conjunction with the greater steric protection of the four-coordinate central atom and the lack of a low-energy LUMO appear to be responsible for their greater resistance to hydrolysis compared with the organometallic compounds of the boron group.

15.13 Organosilicon compounds

The chemistry of organosilicon compounds is very extensive. This is partly because the compounds have been studied for a long time but also because they have wide commercial use as water repellents, lubricants, and sealants.

(a) Structures and properties

Many oxo-bridged organosilicon compounds can be synthesized; one example is hexamethyldisiloxane $(CH_3)_3Si$—O—$Si(CH_3)_3$. As with all simple organosilicon compounds, this compound is resistant to moisture and air. Two properties of these materials, the very weak Lewis basicity of the O atom and ready deformation of the Si—O—Si bond angle, have been rationalized by a model in which lone pairs on O are partially delocalized into vacant σ^* or d orbitals of Si (**22**). The delocalization reduces the directionality of the Si—O single bond and hence makes the structure more flexible. This flexibility permits silicone elastomers to remain rubber-like down to very low temperatures. Delocalization also accounts for the low basicity of an O atom attached to silicon, as in $(CH_3)_3Si$—O—$Si(CH_3)_3$, because the electrons that are needed for the O atom to act as a base are partially removed. The planarity of trisilylamine, $(SiH_3)_3N$, is also explained by the delocalization of the lone pair on N; moreover, this compound is only very weakly basic.[11]

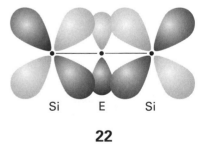

Si   E   Si

22

....

[11] The interesting shapes of silyl molecules containing Si—N and Si—O bonds have been reviewed by E.A.V. Ebsworth, *Acc. Chem. Res.* **20**, 295 (1987).

Another closely related observation is the relative ease of deprotonation of tetra-methylsilane by very strong bases, such as strong carbanion reagents:

$$LiBu + Si(CH_3)_4 \longrightarrow Li[CH_2Si(CH_3)_3] + BuH$$

Once again, the CH_3— group is thought to exhibit this mild Brønsted acidity because the resulting conjugate base, the —CH_2— group, can delocalize electron density on to the neighboring Si atom.

The general feature that will be illustrated in this section is that in contrast to the **M—C** *bonds of more electropositive elements, such as aluminum,* Si—C *bonds are resistant both to hydrolysis and to air oxidation.*

Example 15.4 Predicting the properties of silyl ethers and amines

Predict the extent of hydrogen bonding of ethanol with (a) $(H_3Si)_2O$ compared with $(H_3CCH_2)_2O$ and (b) $(H_3Si)_3N$ compared with $(H_3C)_3N$.

Answer Strong Lewis bases form the stronger hydrogen bonds with a given reference hydrogen donor, such as ethanol. We have seen that O or N attached to silicon has reduced Lewis basicity as compared with the carbon analogs. So we expect that for (a) diethyl ether and for (b) trimethyl amine will form the stronger hydrogen bonds.

Self-test 15.4 For the compounds in (a), which do you expect to have the lower force constant, Si—O—Si bending or C—O—C bending?

(b) The formation of Si—C bonds

A convenient way of linking alkyl groups to Group 14/IV elements is by metathesis reactions between E—Cl and Grignard or alkyllithium reagents:

$$Li_4(CH_3)_4 + SiCl_4 \longrightarrow 4\,LiCl + Si(CH_3)_4$$

Large quantities of chloromethylsilanes are needed in industry to synthesize silicone rubber and oils, but Grignard reagents and alkyllithium reagents are too expensive to be used on such a scale. It was the desire to satisfy this need more cheaply that led Eugene Rochow (in the early 1940s at General Electric in the USA) to develop a process for their direct synthesis from elemental silicon and an alkyl or aryl halide in the presence of copper as a catalyst. The reaction has the (unbalanced) form

$$Si + RX \xrightarrow{\text{250-550 °C, Cu}} R_nSiX_{4-n}$$

The conditions are usually adjusted to favor the formation of dimethyldichlorosilane, but other useful halosilanes are also produced. This relatively inexpensive direct process transformed silicone polymers from expensive laboratory curiosities into widely used materials.[12]

Metathesis reactions between **E—Cl** *and Grignard or alkyllithium reagents are used to link alkyl groups to Group 14/IV elements.*

(c) Redistribution and substitution reactions

Redistribution reactions (Section 15.5) are useful for the synthesis of a variety of Group 14/IV compounds. The reaction is readily performed in the laboratory and is employed commercially to prepare halosilanes:

$$2\,SiCl_4 + Si(CH_3)_4 \longrightarrow SiCl(CH_3)_3, \ SiCl_2(CH_3)_2, \ \text{and} \ SiCl_3(CH_3)$$

---

[12] An interesting personal account of his discovery of the direct process is given by E.G. Rochow in his book *Silicon and silicones.* Springer-Verlag, Berlin (1987).

Traces of Lewis acids such as $AlCl_3$ are effective catalysts for this reaction. It might be expected that, because Si—Cl and Si—CH_3 bonds are being broken and then reformed, the product distribution from an initial 1:1 mixture of $SiCl_4$ and $Si(CH_3)_4$ should yield a statistical mixture of products: $SiCl_4$, $SiCl_3(CH_3)$, $SiCl_2(CH_3)_2$, $SiCl(CH_3)_3$, and $Si(CH_3)_4$ in the ratio 1:4:6:4:1. However, subtle aspects of the bonding and steric interactions often lead to nonstatistical mixtures.

Metathesis reactions that utilize haloorganosilane, germane, and stannane compounds are very useful for laboratory-scale syntheses, especially for preparing compounds with mixed organic substituents:

$$4\,SiCl_3(C_6H_5) + 3\,Li_4(CH_3)_4 \longrightarrow 4\,Si(CH_3)_3(C_6H_5) + 12\,LiCl$$

Studies of the kinetics of substitution reactions on silicon suggest that the rate-determining step of these reactions is associative (Section 14.2). Thus, the reaction is generally second-order overall

$$\text{Rate} = k[SiXL_3][Y]$$

and k depends strongly on the identity of the entering group Y. That redistribution reactions on silicon and heavier members of the group are more facile than those on carbon also suggests that an Si atom forms a five-coordinate activated complex much more readily than a C atom does. This difference is in line with the existence of many more five-coordinate silicon inorganic compounds than five-coordinate carbon compounds.[13]

Stereochemical investigations reveal that substitution reactions may proceed with either inversion or retention of configuration. A mechanism that accounts for inversion is very similar to that proposed for associative substitutions on C (that is, a mechanism resembling an S_N2 substitution in organic chemistry), with the entering and leaving groups occupying the axial positions in a trigonal-bipyramidal activated complex:

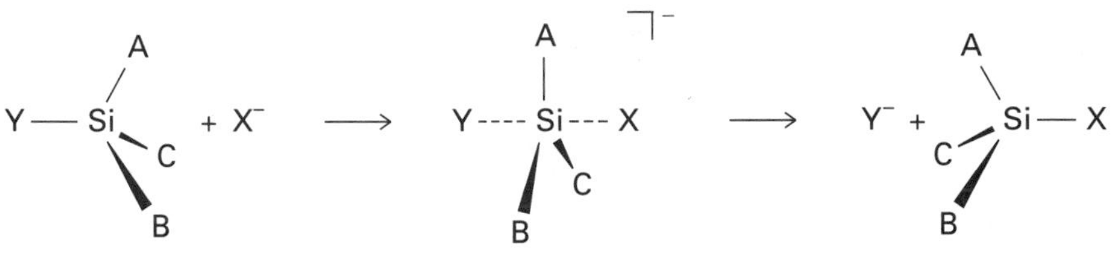

Retention is common when Y is a poor leaving group, such as H^- or OR^-. This kind of reaction is believed to involve a pseudorotation of a five-coordinate intermediate that places Y on the threefold axis and *trans* to one of the original substituents:

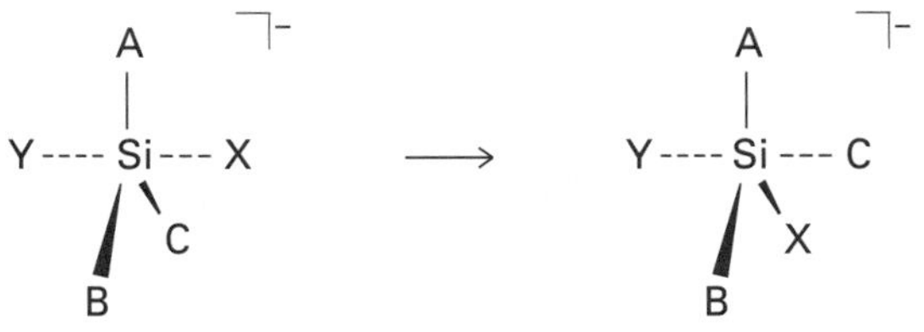

The pseudorotation is then followed by departure of the leaving group:

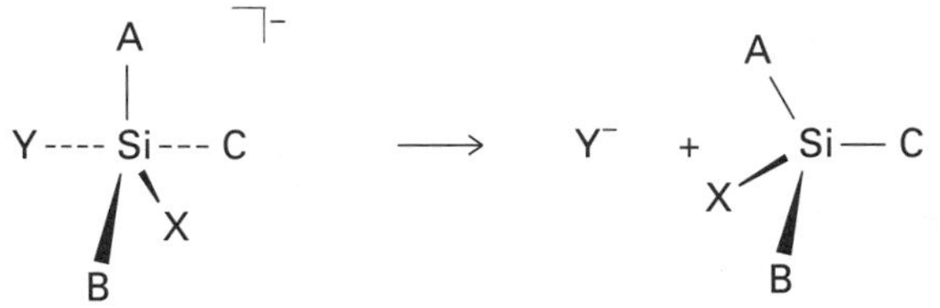

....
[13] The latter are largely confined to carboranes, metal cluster compounds, and solid metal carbides.

The ability of the Si atom to form a five-coordinate intermediate with sufficiently long lifetime to rearrange results in retention of configuration in some nucleophilic displacement reactions even though they proceed by an associative mechanism. On the other hand, the smaller C atom, which has a much less stable five-coordinate activated complex, immediately undergoes inversion.

> Redistribution reactions on silicon and heavier members of the group are more facile than those on carbon; substitution reactions may proceed with either inversion or retention of configuration.

(d) Protolysis of halosilanes

The protolysis of Si—X bonds is a convenient route to a myriad of compounds, particularly **siloxanes**, which are compounds with Si—O bonds. Thus, polar Si—Cl bonds are susceptible to protolysis by species with H—O, H—N, and H—S bonds, but the much less polar Si—C bonds are more resistant. This difference accounts, for instance, for the formation of hexamethyldisiloxane in the reaction of water with chlorotrimethylsilane. The initial reaction is the hydrolysis of the Si—Cl bond:

$$2\,(CH_3)_3SiCl + 2\,H_2O \longrightarrow 2\,(CH_3)_3SiOH + 2\,HCl$$

This step is followed by a slower reaction, the elimination of water to form the Si—O—Si link:

$$2(CH_3)_3SiOH \longrightarrow (CH_3)_3Si-O-Si(CH_3)_3 + H_2O$$

The condensation of the SiOH compound is analogous to the transformation of metal hydroxide complexes to polycations (Section 5.6) and to the polymerization of $Si(OH)_4$ in aqueous solution to produce silica gel. These reactions demonstrate the general tendency of Si—OH groups to eliminate water.

When dimethyldichlorosilane is hydrolyzed, both cyclic (**23**) and long-chain compounds are produced:

$$(CH_3)_2SiCl_2 + H_2O \longrightarrow HO[Si(CH_3)_2O]_nH + [(CH_3)_2SiO]_4 + etc.$$

When $RSiCl_3$ is used as a starting material and contains bulky organic groups, more elaborate structures are possible, including cage compounds (**24**). Similar reactions occur with ammonia and with primary and secondary amines to produce a variety of **silazanes**, compounds with Si—N bonds. For example, excess $(CH_3)_3SiCl$ reacts with NH_3 to produce $((CH_3)_3Si)_2NH$. In this case, the protolysis is incomplete on account of the protection afforded by the bulky groups around the Si atom; the less hindered $HSi(CH_3)_2Cl$ yields $((CH_3)_2HSi)_3N$.

Siloxanes undergo redistribution reactions to form polymers, the **silicones**, in the presence of sulfuric acid as a catalyst:

$$n[(CH_3)_2SiO]_4 + (CH_3)_3SiOSi(CH_3)_3 \xrightarrow{H_2SO_4} (CH_3)_3SiO[Si(CH_3)_2O]_{4n}Si(CH_3)_3$$

In this reaction, the hexamethyldisiloxane provides the $Si(CH_3)_3$ end groups; so, the higher the proportion of it present, the lower the molar mass of the resulting polymer. Silicones, including poly(dimethylsiloxane), are fluids, waxes, and, when cross-linked, elastomers. They are produced on a large scale. In general, the polysiloxanes are highly pliable materials that retain their flexibility at low temperatures on account of the low bending force constants of Si—O—Si bonds. Their useful properties include their flexibility at low temperatures, their ability to repel water, and their resistance to oxidation by air. Additionally, their low toxicity leads to their use in medical and cosmetic implants.

> Polar Si—Cl bonds are susceptible to protolysis by species with H—O, H—N, and H—S bonds, but the much less polar Si—C bonds are more resistant.

23

24

Example 15.5 Contrasting the ease of hydrolytic cleavage of Al—CH_3 and Si—CH_3 bonds

Give chemical equations for the possible reactions of trimethylaluminum and tetramethylsilane with water, and identify the difference in bonding in the two compounds that accounts for the difference in behavior.

Answer Organoaluminum compounds have considerable carbanion character, which correlates with the highly electropositive character of aluminum. This bonding description is consistent with the experimental observation that trimethylaluminum hydrolyzes rapidly on contact with water:

$$Al_2(CH_3)_6 + 6 H_2O \longrightarrow 2 Al(OH)_3 + 6 CH_4$$

As mentioned in Section 15.6, this reaction probably occurs by initial coordination of the O atom in H_2O to the relatively low-energy acceptor orbitals of the Al atom of the alkylaluminum compound. In contrast, silicon and carbon have similar electronegativity and they form covalent bonds of low polarity. Their low polarities, relative steric congestion, and weak Lewis acidities prevent coordination prior to hydrolysis. This resistance to hydrolysis is illustrated by the use of silicone polymers as waterproofing agents on cloth or leather and the use of silicone cements to isolate electrical and electronic devices from water.

Self-test 15.5 Illustrate how the difference in reactivity between Al—C and Si—C bonds with O—H groups leads to the choice of different strategies for the synthesis of aluminum and silicon alkoxides.

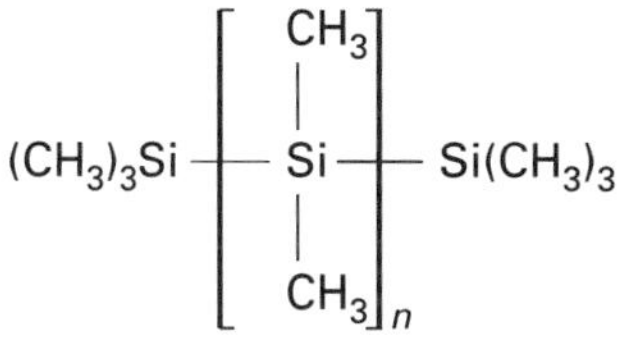

(e) Single bonds between Si atoms

Although not as extensive as the simple carbon–carbon bonded organic compounds, there are many catenated silicon organometallic compounds.[14] Considerable progress has been made in the synthesis of compounds containing the Si—Si bond, which is weaker than the C—C bond but not dramatically so. Acyclic, cyclic, bicyclic, and cage alkylsilicon compounds are known (25), (26). These compounds are prepared by reductive halide elimination reaction:[15]

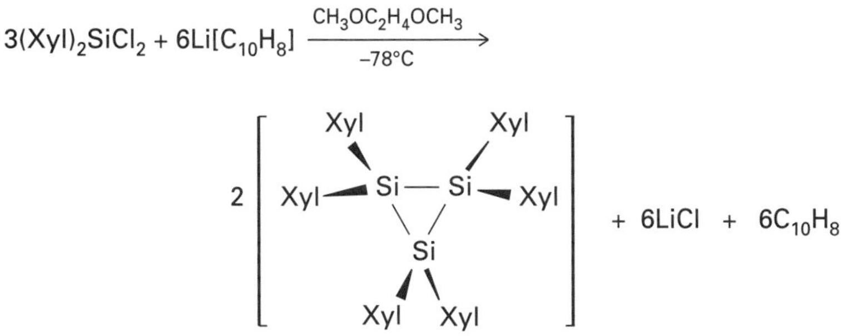

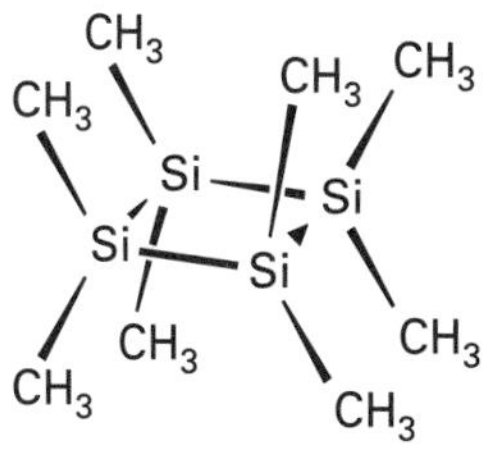

25

$$3(Xyl)_2SiCl_2 + 6Li[C_{10}H_8] \xrightarrow[-78°C]{CH_3OC_2H_4OCH_3}$$

$+ 6LiCl + 6C_{10}H_8$

26

The strong reducing agent lithium naphthalide reduces an Si—X halogen bond with the expulsion of the X^- halide ion and the formation of Si—Si bonds. The product of this reaction will be important in our discussion of Si=Si bonds in the following section.

Spectroscopic and chemical evidence suggests that the catenated compounds of silicon and germanium have fairly low-lying vacant orbitals that are delocalized over the chain or ring. Thus, they have near-UV absorption bands that decrease in energy with increasing chain length. This absorption is attributed to the promotion of an electron from a σ orbital of the

---

[14] Catenation means the formation of chains, such as $R_3Si—SiR_3$ and $R_3Si—SiR_2—SiR_3$.

[15] The abbreviation Xyl represents the bulky xylyl group (2,6-dimethylphenyl).

Si—Si or Ge—Ge backbone into a vacant σ^* orbital that is delocalized along the Si—Si or Ge—Ge chain. In keeping with this interpretation, silane chains undergo photolysis when exposed to UV radiation. For example, polydimethylsilane, $H_3C(—Si(CH_3)_2—)_nCH_3$, is cleaved when exposed to UV radiation.[16]

Another aspect of the chemical properties of silicon that emphasizes the delocalized nature of the σ orbitals is the formation, by one-electron reduction of a saturated silicon compound, of a radical anion in which the odd electron occupies a delocalized Si—Si σ^* orbital (represented by the ring, as in benzene, but here signifying σ delocalization, not π delocalization):

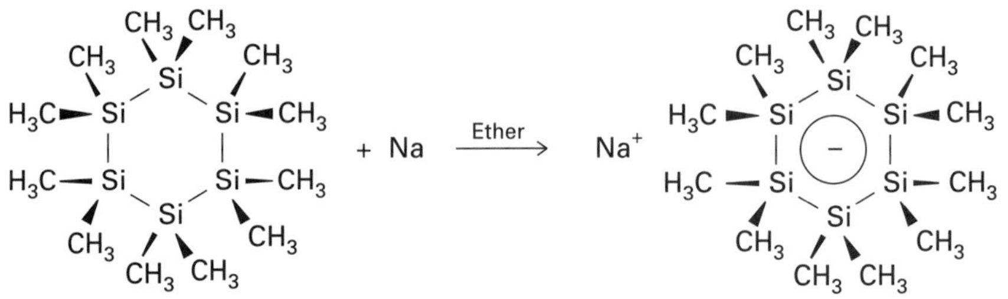

The EPR spectrum of the anion shows that the unpaired electron occurs with equal probability on all six Si atoms, proving that the electron is fully delocalized (at least on the timescale of this experiment, which is about 1 μs).

Acyclic, cyclic, bicyclic, and cage alkylsilicon compounds are prepared by reductive halide elimination. Low-energy σ^* orbitals are present in polyalkyl silanes.

(f) Multiple bonds between Si atoms

As is the case for carbon, two single Si—Si bonds are stronger than one Si=Si double bond, so there is an energy advantage for Si=C or Si=Si groups to couple together. Carbon C=C bonds are protected from coupling by the high activation energy of the reaction, but there is much less kinetic protection for multiple bonds involving silicon or germanium.

A clue that multiply bonded silicon compounds might exist came from the detection of transient compounds such as $(CH_3)_2Si=CH_2$ in the gas phase and in low-temperature inert-gas matrices. However, few of the chemical properties of the molecule were established because it rapidly forms dimers and polymers:

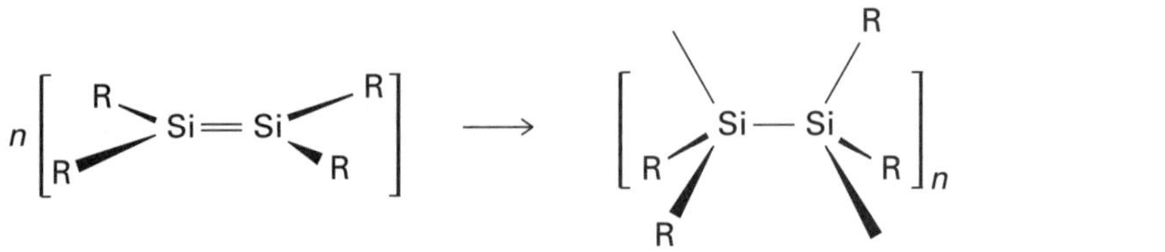

The key to success in the formation of Si=Si and double-bonded compounds of germanium, tin, phosphorus, and arsenic was the discovery that bulky substituents block dimerization and polymerization.

---

[16] This photolysis forms the basis for the potential use of polydimethylsilane in the photolithographic production of integrated circuits. In this application the polydimethylsilane appears to be superior to the photosensitive polymers that are currently used. See: J. Michl, J.W. Downing, T. Karatsu, A.J. McKinley, G. Poggi, G.W. Wallraff, R. Sooriyakumaran, and R.D. Miller, *Pure Appl. Chem.* **60**, 959 (1988).

Compounds containing Si═C bonds, the **silaethenes**, and Si═Si bonds, the **disilaethenes**, are now known. These new silaethenes and disilaethenes not only extend our understanding of chemical bonding but also exhibit interesting chemical properties.[17] Photochemical cleavage of Si—Si bonds, following excitation into an antibonding Si—Si σ^* orbital, provides a convenient route to disilaethenes. An example is the photolysis of a cyclic trisilane with bulky substituents:

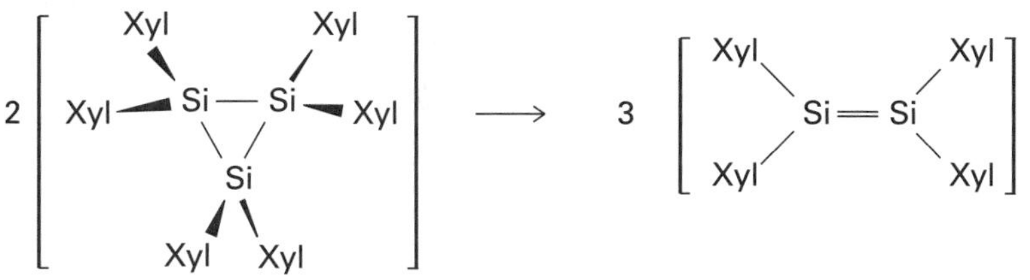

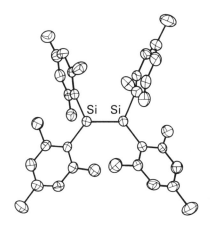

15.5 An ORTEP diagram of the structure of $(Mes)_2Si=Si(Mes)_2$ as determined by X-ray single-crystal diffraction. (From M.J. Fink, M.J. Michalczyk, K.J. Haller, R. West, and J. Michl, *Organometallics* **3**, 793 (1983).)

The bulky groups on the Si atoms prevent cyclization and polymerization; they also obstruct reaction of the disilaethene product.

X-ray structure determinations show the expected bond length decrease from Si—Si to Si═Si. For example, tetramesityldisilaethene has a silicon–silicon bond length of $2.16 \, \text{Å}$ (Fig. 15.5), which is about $0.20 \, \text{Å}$ shorter than a typical Si—Si single bond. A significant difference between the disilaethenes and alkenes, however, is the greater ease of distorting the $R_2Si═SiR_2$ group from planarity. We shall see that this tendency is even more pronounced in heavier Group 14/IV analogs, and an explanation in terms of bonding will be given there.

The electronic absorption spectra of the disilaethenes contain a band in the visible or near-UV region, and the compounds are often brightly colored. This coloration indicates that the π and π^* orbitals are closer in energy than in the alkenes, in which the corresponding absorption is well into the UV (Fig. 15.6).

The energies of the bonding and antibonding π orbitals can be inferred from reduction potentials in solution. It is found that the oxidation of disilaethenes occurs at less positive potentials and the reduction at less negative potentials than for corresponding alkenes. As depicted in Fig. 15.6, this observation indicates that the energies of π and π^* orbitals of the disilaethenes are between those of the corresponding alkenes. As a result, disilaethenes are better π donors (using electron donation from the filled π orbital) and better π acceptors (accepting electrons into the π^* orbital).

Molecules that are not sterically encumbered can penetrate a molecule's stereochemical defences and have access to the Si═Si bond in disilaethenes. Some of the resulting reactions are very similar to those found for alkenes. For example, hydrogen halides and halogens add across the Si═Si double bond:

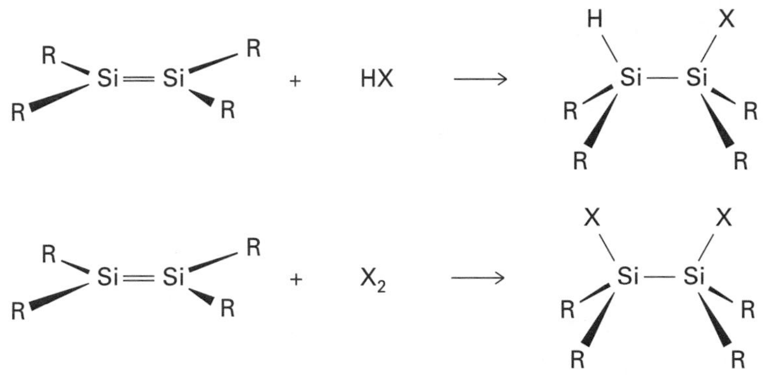

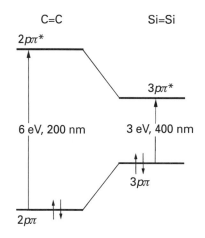

15.6 Approximate energy levels for alkenes and disilaethenes. The energy separations are obtained from the analysis of UV absorption spectra. (From R. West, *Angew. Chem. Int. Ed. Engl.* **26**, 1201 (1987).)

---

[17] Silaethene chemistry is reviewed by A.G. Brook and K.M. Baines, *Adv. Organometal. Chem.* **25**, 1 (1987). A helpful survey of disilaethenes is given by R. West in *Angew. Chem. Int. Ed. Engl.* **26**, 1201 (1987).

where $X = Cl$ or Br. More interesting, perhaps, are the contrasts with the properties of alkenes. Unlike alkenes, disilaethenes undergo addition of ROH across the $Si{=}Si$ bond:

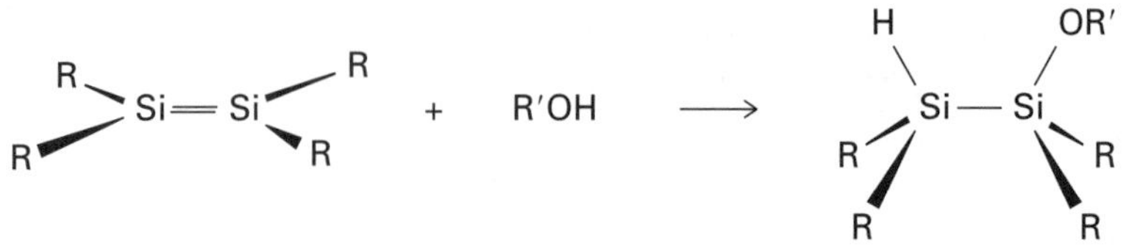

They also undergo $2 + 2$ additions with some alkynes:

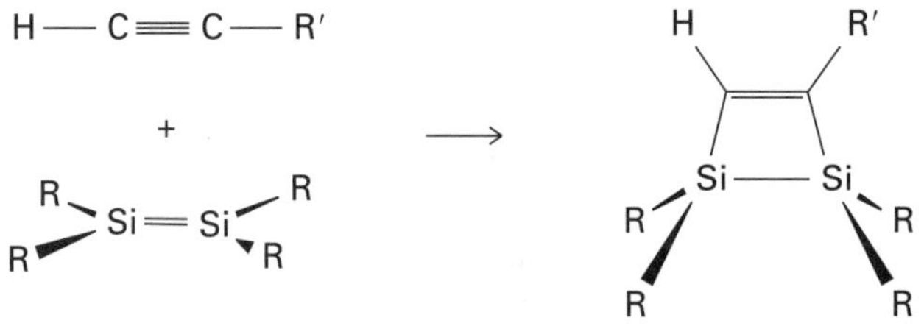

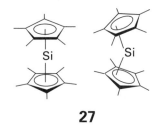

27

It is possible to attach a pentamethylcyclopentadienyl ligand to a Si atom. This possibility has led to the synthesis of the first formal Si(II) organometallic compound, bis(pentamethylcyclopentadienyl)silicon.[18] An X-ray crystal structure determination of the compound reveals the presence of two different configurations (**27**). Apparently the energy difference between the two configurations is small enough for packing forces in the solid to influence the structure.

> *Bulky substituents block dimerization and polymerization and silaethenes and disilaethenes with bulky substituents are now known.*

15.14 Organometallic compounds of germanium, tin, and lead

Many of the reactions of organotin compounds (organostannanes) and organolead compounds (organoplumbanes) are similar to those already mentioned for silicon and germanium. One major difference from organosilicon compounds, however, is the existence (because of the inert-pair effect) of some Ge(II), Sn(II), and Pb(II) compounds. Another contrast with organosilicon compounds results from the rapid decrease in E—C bond strength down the group (see Fig 15.3). Because of the latter, organolead compounds generally decompose when heated above $100\,^{\circ}C$.

The gas-phase decomposition of alkyllead compounds occurs with the formation of alkyl radicals:

$$Pb(CH_3)_4(l \text{ or } g) \xrightarrow{\Delta} Pb(s) + 4 \cdot CH_3(g)$$

This decomposition was the reason why, for many years, tetramethyllead and tetraethyllead were added to gasoline to improve its octane rating (a measure of its smoothness of combustion). The alkyl radicals generated by the decomposition of the tetraalkyllead in the hot combustion chamber of the engine are radical chain terminators that reduce the likelihood of explosive combustion. The toxicity of lead, however, and its deactivation of catalytic converters for pollution control, has prompted the elimination of organolead additives from gasoline in many parts of the world.

....

[18] P. Jutzi, Main-group metallocenes: recent developments. *Pure Appl. Chem.* **61**, 1731 (1989).

Organotin compounds find many different applications, ranging from stabilizers for poly(vinyl chloride) plastics to fungicides and antifouling paints for the hulls of boats. Recently, though, some of these applications have come under close scrutiny because they may harm benign and desirable organisms: organotin compounds kill not only barnacles but oysters too.

Germaethenes ($R_2Ge=CR_2$), stannaethenes ($R_2Sn=CR_2$), digermaethenes ($R_2Ge=GeR_2$), and distannaethenes ($R_2Sn=SnR_2$) have all been prepared. As with the corresponding silicon compounds, it is necessary for the R groups to be bulky to prevent association. The digermaethenes and distannaethenes are distinctly nonplanar and several bonding explanations have been proposed. In one of them, the nonplanarity is attributed to an unconventional pattern of multiple bonding, which is between the $sp^2\sigma$ orbital on one atom with a $p\pi$ orbital on the other (**28**).

Germanium–germanium and tin–tin multiple bonds are long, and dissociation has been observed in solution to form the germylene and stannylene divalent compounds, GeR_2 and SnR_2. As expected from the increased stability of the divalent state going down the group, the SnR_2 species are more stable than their GeR_2 counterparts. The divalent state is also seen in $Sn(\eta^5\text{-}C_5H_5)_2$ and $Pb(\eta^5\text{-}C_5H_5)_2$ which have angular structures in the gas phase, indicating that a stereochemically active lone pair may be present on the metal atom (**29**). The very bulky pentaphenylcyclopentadienyl ligands in $Sn(C_5Ph_5)_2$ suppress the stereochemical activity of the electron pair on the Sn atom, and X-ray structure determination reveals that the two C_5 rings are coplanar.

In addition to the simple monomeric organogermanium and organotin compounds, catenated compounds (**30**) and cyclic compounds (**31**) are known. The synthetic routes to these compounds resemble those used for their silicon analogs, but the reactions are often more facile with organogermanium and organotin compounds. One possible reason for this greater reactivity is that these larger atoms may be more accessible to reactants. We have seen already that the heavier members of this group more readily participate in radical reactions.

Polyhedral compounds of carbon, such as cubane and its derivatives, R_4C_4, are thermodynamically less stable than arenes, but the same is not true for analogous compounds of carbon's congeners, and these heavier members of Group 14/IV have never been found to form benzene-like structures. Instead, a series of closed polyhedral compounds has been prepared recently, including trigonal-prismatic Ge_6R_6 (**32**), square-prismatic Si_8R_8 and Sn_8R_8, (**33**), and pentagonal-prismatic $Sn_{10}R_{10}$ (**34**).[19] It appears that the bulkiness of the R group plays a role in determining the particular structure, but the general systematics of its role have not been identified. The preparation of a germanium prismane compound is representative of the simple reactions that lead to these polyhedral compounds:

$$6\,RGeCl_3 + 18\,Li \longrightarrow Ge_6R_6 + 18\,LiCl \qquad R = -CH_2Si(CH_3)_3$$

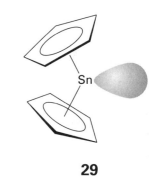

28

29

$$(CH_3)_3Ge \left[\begin{array}{c} CH_3 \\ | \\ Ge \\ | \\ CH_3 \end{array} \right]_n Ge(CH_3)_3$$

30

31 R = CH_3, CH_2CH_3

32 R = $CH(Si(CH_3)_3)_2$

33

34 R = $(C_2H_5)_2C_6H_3$

[19] A. Sekiguchi, C. Kabuto, and H. Sakuri, *Angew. Chem. Int. Ed. Engl.* **28**, 55 (1989) and L.R. Sita and I. Kinoshita, *J. Am. Chem. Soc.* **113**, 1856 (1991).

We can view this reaction as the reduction of the Ge(IV) compound $RGeCl_3$ to a formal Ge(I) cluster compound.

> *Many of the reactions of organostannanes and organoplumbanes are similar to those of their silicon and germanium analogs, but the inert-pair effect leads to the existence of some Ge(II), Sn(II), and Pb(II) compounds.*

Electron-rich compounds of the nitrogen group

Some new features of organometallic compounds are encountered in Group 15/V, including the Lewis basicity arising from the lone pair on the central atom (as in $:AsR_3$). Moreover, the central atom may exist in oxidation states +3 or +5, as in AsR_3 and $AsPh_4^+$, respectively.

15.15 Organometallic compounds of arsenic, antimony, and bismuth

Because the first two members of the nitrogen group (N and P) are unambiguously nonmetals, we do not consider them here (see Chapter 11). Work on the organic derivatives of the heavier elements in the group has resulted in interesting new compounds with unusual bonding patterns.

Oxidation states +3 and +5 are encountered in many of the organometallic compounds of arsenic, antimony, and bismuth. An example of a compound with an element in the +3 oxidation state is $As(CH_3)_3$ (**6**) and an example of the +5 state is $As(C_6H_5)_5$ (**35**). Compounds in which the elements have oxidation number +3 contain lone pairs and, therefore, can be considered electron-rich; those with elements in the +5 oxidation state are electron-precise.

Organoarsenic compounds were once widely used to treat bacterial infections and as herbicides and fungicides. However, because of their high toxicity arsenic, antimony, and bismuth organometallic compounds no longer have major commercial applications. Nevertheless, they can be handled in the laboratory with proper techniques, and many interesting compounds have been prepared.

(a) Oxidation state +3

Here we consider a broad class of compounds of these elements with the general formula ER_3, and related organohalides and cyclopentadiene compounds.

The metathesis reaction of a Grignard or organolithium reagent with EX_3 (E = As, Sb, Bi; X = Cl, Br) provides a convenient route to the trialkyl and triaryl compounds:

$$ECl_3 \quad + \quad 3MgCH_3Br \quad \xrightarrow{THF} \quad \underset{H_3C}{\overset{CH_3}{E:}}{\overset{|}{\underset{CH_3}{\diagdown}}} \quad + \quad 3MgClBr$$

The $E(CH_3)_3$ molecules and their analogs are trigonal-pyramidal as expected in view of the lone pair on the central atom. Alkylarsanes, such as trimethylarsane, are volatile, malodorous, and toxic. The arylarsanes are more air-stable and less volatile.

The aryl- and alkylarsanes, such as trimethylarsane (**6**), are encountered as ligands in d-metal complexes. The order of affinity of these soft Lewis bases for a d-metal ion generally follows the order $PR_3 > AsR_3 > SbR_3 \gg BiR_3$. Many complexes of alkyl- and arylarsanes have been prepared but fewer stibane complexes are known. A useful ligand, for example, is the bidentate compound known as diars (**36**). Because of their soft-donor character, many aryl-

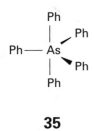

35

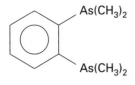

36 $C_6H_4(As(CH_3)_2)_2$, diars

and alkylarsane complexes of the soft species Rh(I), Ir(I), Pd(II), and Pt(II) have been prepared. However, hardness criteria are only approximate, so we should not be surprised to see phosphine and arsane complexes of some metals in higher oxidation states. For example, the unusual +4 oxidation state of palladium is stabilized by the diars ligand (**37**).

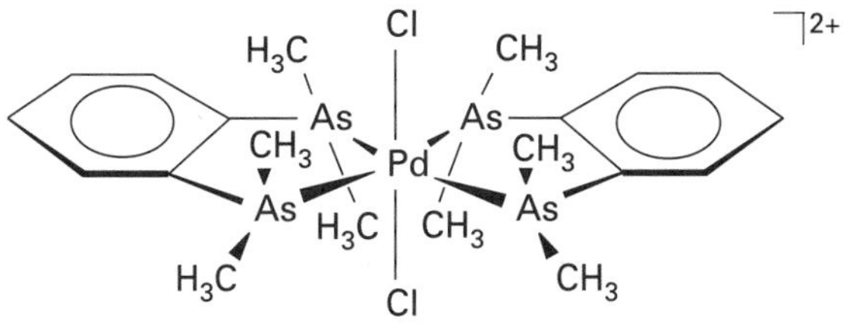

37

The synthesis of diars provides a good illustration of some common reactions in the synthesis of organoarsenic compounds. The starting material is $(CH_3)_2AsI$. This compound is not conveniently prepared by metathesis reaction between AsI_3 and a Grignard or similar carbanion reagent because that reaction is not selective to partial substitution on the As atom when the organic group is compact. Instead, the compound can be prepared by the direct action of a haloalkane, CH_3I, on the metallic allotrope of arsenic:

$$4\,As + 6\,CH_3I \longrightarrow 3\,(CH_3)_2AsI + AsI_3$$

In the next step, the action of sodium on $(CH_3)_2AsI$ is used to produce $[(CH_3)_2As]^-$:

$$(CH_3)_2AsI + 2\,Na \longrightarrow Na[(CH_3)_2As] + NaI$$

The resulting powerful nucleophile $[(CH_3)_2As]^-$ is then employed to displace chlorine from 1,2-dichlorobenzene:

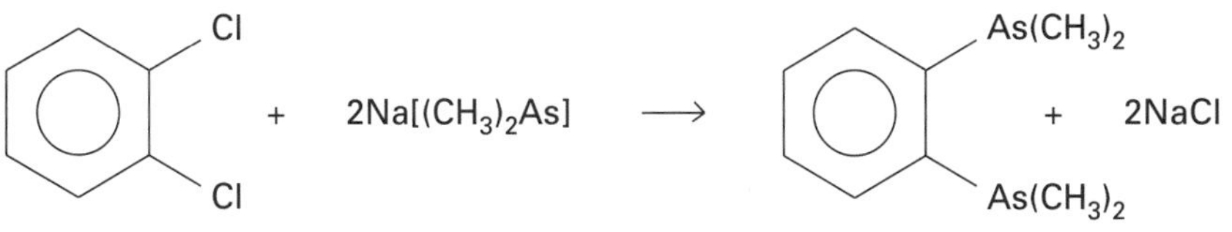

Another interesting compound of As(III), arsabenzene (**38**), is an analog of pyridine. It lacks the σ-donor ability of pyridine but forms π-complexes with d-block metals (**39**) analogous to those formed by benzene.[20]

> The metathesis reaction of a Grignard or organolithium reagent with EX_3 is used to prepare trialkyl and triaryl compounds of arsenic, antimony, and bismuth. The order of affinity of these soft Lewis bases for a d-metal ion is as specified above.

38 Arsabenzene

(b) Oxidation state +5

The trialkylarsanes act as nucleophiles toward haloalkanes to produce tetraalkylarsonium salts, which contain As(V):

$$As(CH_3)_3 + CH_3Br \longrightarrow [As(CH_3)_4]Br$$

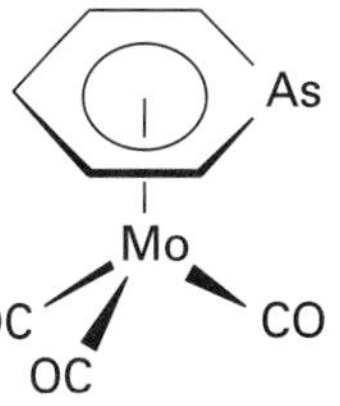

[20] A.J. Ashe, The Group 5 heterobenzenes. *Acc. Chem. Res.* **11**, 153 (1978). Volume 39 of *Advances in Organometallic Chemistry* (1996) is devoted to multiple bonds between main-group elements.

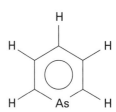

39

This type of reaction cannot be used for the preparation of the tetraphenylarsonium ion, $[AsPh_4]^+$, because triphenylarsane is a much weaker nucleophile than trimethylarsane. Instead, a suitable synthetic reaction is:

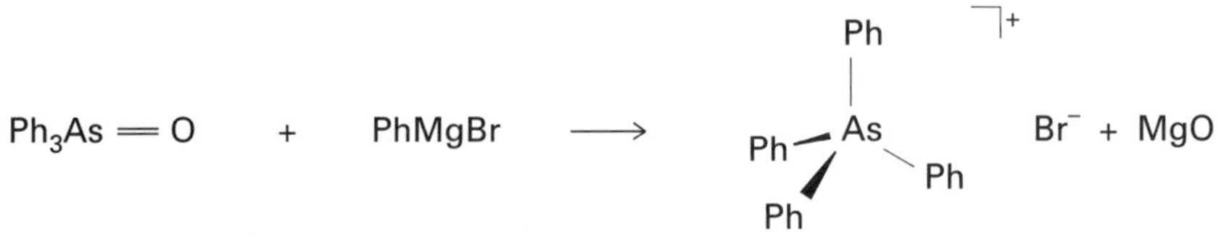

$$Ph_3As = O \quad + \quad PhMgBr \quad \longrightarrow \quad \left[\begin{array}{c} Ph \\ | \\ Ph - As \\ | \\ Ph \end{array} \right]^+ Ph \quad Br^- + MgO$$

This reaction may look unfamiliar, but it is simply a metathesis in which the Ph^- anion replaces the formal O^{2-} ion attached to the As atom, resulting in a compound in which the arsenic retains its +5 oxidation state. The formation of the highly exoergic compound MgO also contributes to the Gibbs energy of this reaction, and its formation drives the reaction forward.

The tetraphenylarsonium, tetraalkylammonium, and tetraphenylphosphonium cations are used in synthetic inorganic chemistry as bulky cations to stabilize bulky anions. The tetraphenylarsonium ion is also a starting material for the preparation of other As(V) organometallic compounds. For instance, the action of phenyllithium on a tetraphenylarsonium salt produces pentaphenylarsenic (**35**), a compound of As(V):

$$[AsPh_4]Br + LiPh \longrightarrow AsPh_5 + LiBr$$

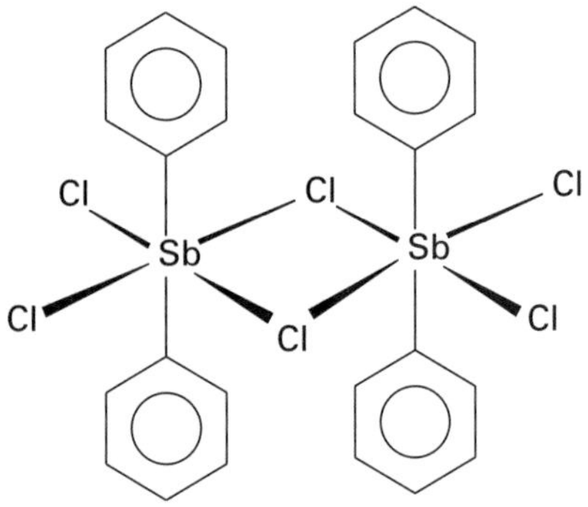

Pentaphenylarsenic is trigonal-bipyramidal, as expected from VSEPR considerations. We have seen (Section 3.3) that a square-pyramidal structure is often close in energy to the trigonal-bipyramidal structure, and the antimony analog, $SbPh_5$, is in fact square-pyramidal. A similar reaction under carefully controlled conditions yields the unstable compound $As(CH_3)_5$.

Mixed-substituent compounds are common for oxidation state +5, and we have already seen one example in $OAsR_3$; similarly, halides such as $Sb_2Ph_4Cl_6$ are also known. Molecules of the latter compound have an edge-shared structure with octahedral coordination around each Sb atom (**40**).

40

> Tetraalkylammonium, tetraphenylphosphonium, and tetraphenylarsonium cations are used in synthetic inorganic chemistry as bulky cations to stabilize bulky anions. The tetraphenylarsonium ion is also a starting material for the preparation of other As(V) organometallic compounds.

Example 15.6 Correlating oxidation numbers and stabilities

Describe the stability of the alkyl compounds of the adjacent elements germanium and arsenic with their group oxidation number and the group oxidation number −2.

Answer The most common oxidation number for germanium in organogermanium compounds is +4, the group oxidation number. Only a few organogermanium(II) compounds are known. In contrast, As(V) compounds such as $As(CH_3)_5$ are unstable with respect to organoarsenic(III) compounds, such as $As(CH_3)_3$. One way of explaining this trend is that, on going to the right along a period, the elements become more electronegative and it becomes progressively more difficult to bring out their high oxidation states. This difficulty is particularly strong with a substituent, such as CH_3, with a low electronegativity. These trends in the stability of oxidation states follow the trends for simple inorganic compounds of germanium and arsenic.

Self-test 15.6 Compare formulas of the most stable hydrogen compounds of germanium and arsenic (Chapter 8) with those of their methyl compounds. Can the differences be explained in terms of the relative electronegativities of C and H?

15.16 Catenated and multiply bonded compounds

Tetramethyldiarsane, $(CH_3)_2AsAs(CH_3)_2$, a catenated compound, was one of the first organometallic compounds to be made. One convenient synthesis is the reaction of $As(CH_3)_2Br$ with zinc:

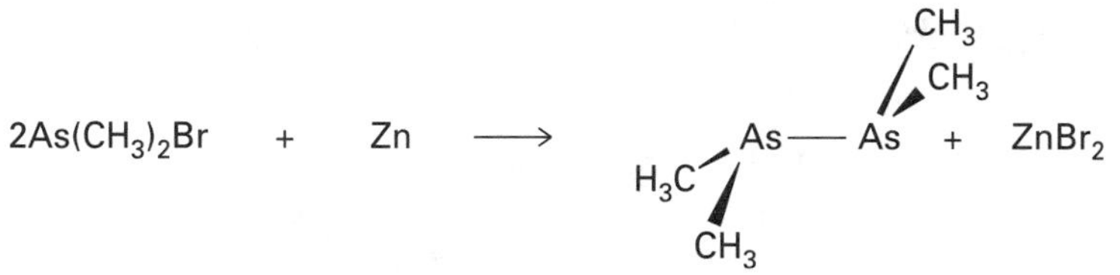

$$2As(CH_3)_2Br \quad + \quad Zn \quad \longrightarrow \quad \text{[structure]} \quad + \quad ZnBr_2$$

Cyclic arsanes are also known, including $(PhAs)_6$. The synthesis of $(PhAs)_6$ is accomplished by abstracting iodine from phenylarsenic diiodide:

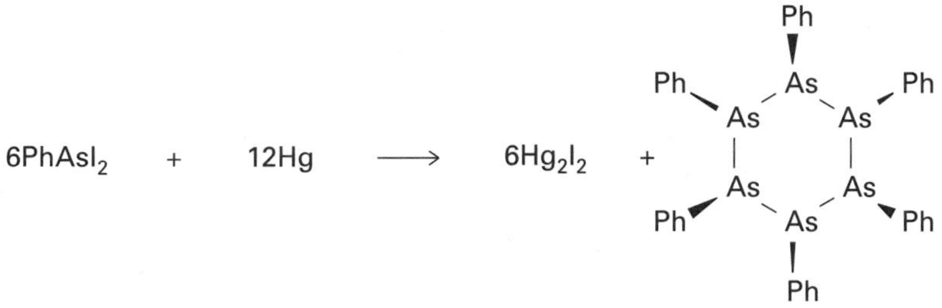

$$6PhAsI_2 \quad + \quad 12Hg \quad \longrightarrow \quad 6Hg_2I_2 \quad + \quad \text{[structure]}$$

There are also many other arsenic–arsenic bonded compounds. Among them is one containing a ladder structure (**41**), which is obtained by reducing CH_3AsI_2 with $Sb(C_4H_9)_3$:

$$n\,CH_3AsI_2 + Sb(C_4H_9)_3 \longrightarrow [AsCH_3]_n + n\,Sb(C_4H_9)_3I_2$$

As with the disilaethenes, it has been found that bulky substituents make it possible to prevent the polymerization and cyclization of E—R groups. This strategy leads to formal double-bonded systems of formula RE=ER, where E may be P, As, or Sb, including the series of compounds that contain P=As and P=Sb bonds (**42**).

Catenated and cyclic arsanes with arsenic–arsenic bonds are known. There are also many other arsenic–arsenic bonded compounds and some compounds with double bonds to arsenic and antimony.

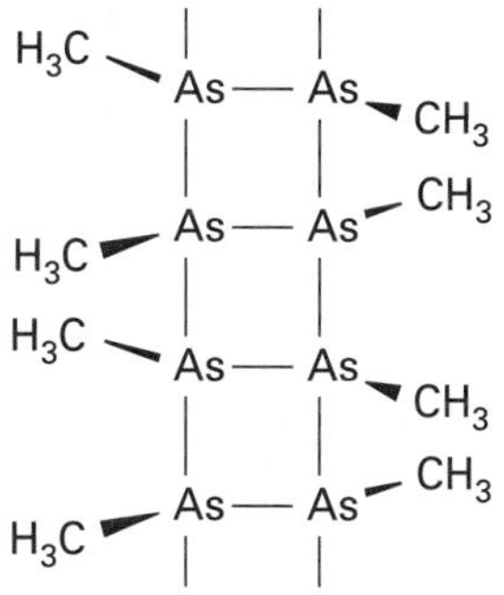

41

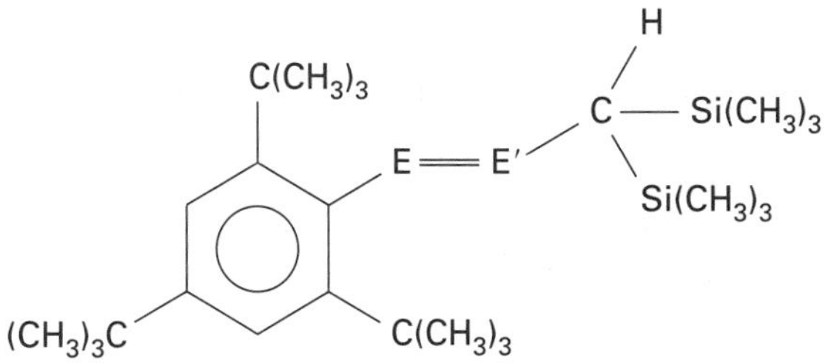

42 (E,E′) = (P,As),(P,Sb),(As,P)

Further reading

Ch. Elschenbroich and A. Salzer, *Organometallics*. VCH, Weinheim and New York (1992). This concise and well illustrated book provides an excellent introduction to the chemistry of both main-group and *d*-metal organometallic compounds.

G. Wilkinson, F.G.A. Stone, and E.W. Abel (ed.), *Comprehensive organometallic chemistry*. Pergamon, Oxford (1982). Volumes 1 and 2 are devoted to the *s*- and *p*-blocks. Chapter 1, Main group structure and bonding relationships (Vol. 1, p. 1) by M.E. O'Neill and K. Wade, provides a good introduction, and later chapters provide thorough coverage. Volumes 1 and 2 of the second edition: *Comprehensive organometallic chemistry*. E.W. Abel,

F.G.A. Stone, G. Wilkinson and C.E. Housecroft (ed.), Vols. 1 and 2, (1995).

B.J. Aylett, *Organometallic compounds*. Chapman & Hall, London (1979). Group 14/IV and 15/V organometallic compounds.

J.J. Eisch (ed.), *Organometallic syntheses*, Vols. 1–4. Academic Press, New York (1965–88). This series provides a thorough description of methods and detailed procedures for the synthesis of important *p*-block organoelement compounds.

J.L. Atwood, J.E. Davies, D.D. MacNicol, and F. Vögtle (ed.), *Comprehensive supramolecular chemistry*, Vols 1–11. Pergamon Press, Oxford (1996).

Exercises

15.1 Explain why certain of the following compounds qualify as organometallic compounds whereas others do not: (a) $B(CH_3)_3$, (b) $B(OCH_3)_3$, (c) $Na_4(CH_3)_4$, (d) $SiCl_3(CH_3)$, (e) $N(CH_3)_3$, (f) sodium acetate, (g) $Na[B(C_6H_5)_4]$.

15.2 Preferably without consulting reference material, construct the periodic table for the *s*- and *p*-block elements and the Group 12 elements. Head each group with the formula of a representative methyl compound and indicate (a) the positions of the salt-like methyl compounds, (b) the positions of the electron-deficient, electron-precise, and electron-rich methyl compounds, (c) trends in $\Delta_f H^{\ominus}$ in the *p*-block.

15.3 Write formulas for each of the following compounds. For nonionic compounds give alternative names as derivatives of their hydrogen compounds if they are named as organometallic compounds of the element, and vice versa: (a) trimethylbismuth, (b) tetraphenylsilane, (c) tetraphenylarsonium bromide, (d) potassium tetraphenylborate.

15.4 Name each of the following compounds and classify them: (a) $SiH(C_2H_5)_3$, (b) $BCl(C_6H_5)_2$, (c) $Al_2Cl_2(C_6H_5)_4$, (d) $Li_4(C_2H_5)_4$, (e) $RbCH_3$.

15.5 Sketch the structures of: (a) methyllithium, (b) trimethylboron, (c) hexamethyldialuminum, (d) tetramethylsilane, (e) trimethylarsane, and (f) tetraphenylarsonium.

15.6 Describe the tendency toward association through methyl bridges for the trimethyl compounds of boron, aluminum, gallium, and indium. Give a plausible explanation for the difference between boron and aluminum.

15.7 For each of the following compounds, indicate those that may serve as (1) a good carbanion nucleophile reagent, (2) a mild Lewis acid, (3) a mild Lewis base at the central atom, (4) a strong reducing agent. (A compound may have more than one of these properties.) (a) $Li_4(CH_3)_4$, (b) $Zn(CH_3)_2$, (c) $(CH_3)MgBr$, (d) $B(CH_3)_3$, (e) $Al_2(CH_3)_6$, (f) $Si(CH_3)_4$, (g) $As(CH_3)_3$.

15.8 For an appropriate compound from Exercise 15.7, give balanced chemical equations for: (a) the reactions of one of the carbanion

reagents with $AsCl_3$ and $SiPh_2Cl_2$, (b) the reaction of a Lewis acid with NH_3, (c) the reaction of a Lewis base with the Lewis acid $[HgCH_3][BF_4]$.

15.9 Give examples of the synthesis of organometallic compounds by each of the following reaction types and describe the factors that favor reaction in each case: (a) reaction of a metal with an organic halide, (b) transmetallation, (c) double replacement.

15.10 Determine which compound in each pair is likely to be the stronger reducing agent and explain the physical basis for your answer: (a) $Na[C_{10}H_8]$ and $Na[C_{14}H_{10}]$, (b) $Na[C_{10}H_8]$ and $Na_2[C_{10}H_8]$ (where $C_{10}H_8$ is naphthalene and $C_{14}H_{10}$ is anthracene).

15.11 Determine the likely reaction type, write a reasonable balanced chemical equation (or NR for no reaction), and explain the systematics of the reaction type that led you to decide whether or not a reaction is likely.

   (a) Calcium with dimethylmercury.
   (b) Mercury with diethylzinc.
   (c) Methyllithium with triphenylchlorosilane in ether.
   (d) Tetramethylsilane with zinc chloride in ether.
   (e) Trimethylsilane with ethene in a solution of chloroplatinic acid in 2-propanol.

15.12 Give balanced chemical equations that illustrate: (a) direct synthesis of $SiCl_2(CH_3)_2$; (b) redistribution of $SiCl_2(CH_3)_2$.

15.13 Using silicon and a chloromethane as the primary starting materials, give equations and conditions for the synthesis of a poly(dimethylsiloxane).

15.14 Disregarding compounds with metal–metal bonds, describe the trends in the oxidation states for the organometallic compounds of Groups 13/III through 15/V.

15.15 For the simple organometallic compounds, briefly describe the periodic trends in (a) metal–carbon bond enthalpies, (b) Lewis acidity, (c) Lewis basicity for Groups 1, 2, 12, 13/III, and 14/IV.

15.16 Summarize the trend in each process listed below, giving your reasoning:

(a) The relative ease of pyrolysis of $Si(CH_3)_4$ and $Pb(CH_3)_4$ at $300°C$.

(b) The relative Lewis acidity of $Li_4(CH_3)_4$, $B(CH_3)_3$, $Si(CH_3)_4$, and $Si(CH_3)Cl_3$.

(c) The relative Lewis basicity of $Si(CH_3)_4$ and $As(CH_3)_3$.

(d) The tendency of $Li_4(CH_3)_4$ and $Hg(CH_3)_2$ to displace halide from $GeCl_4$.

15.17 Taking into account sensitivity to oxygen and moisture and the volatility of the compound, indicate the general techniques (vacuum line, inert atmosphere, Schlenk ware, or open flasks) that are most appropriate for handling the following organometallic compounds: (a) $Li_4(CH_3)_4$ in ether, (b) trimethylboron, (c) triisobutylaluminum, (d) $AsPh_3$ (solid), (e) $(CH_3)_3SiOSi(CH_3)_3$ (liquid).

15.18 The 'saturated' cyclic compound $Si_6(CH_3)_{12}$, which has a near-UV absorption band, can be reduced by sodium to produce a cyclic radical anion $[Si_6(CH_3)_{12}]^-$. What types of molecular orbitals are thought to be involved in these processes?

15.19 Give equations and conditions for the synthesis of $R_2Si{=}SiR_2$ from R_2SiCl_2 and indicate the type of R group that is necessary to yield a stable product.

Problems

15.1 The bulk price per mole of trimethylaluminum is an order of magnitude greater than that of triethylaluminum. Describe the methods of synthesis available for these two compounds and rationalize the price difference.

15.2 Correct any errors and explain the nature of the errors in the following statements. (a) The methyl compounds of lithium, zinc, and germanium are associated through multicenter $\mathbf{M}-\mathbf{C}-\mathbf{M}$ bonds because they each contain fewer methyl groups than valence orbitals on the metal. (b) Substitution reactions of organometallic compounds of silicon such as chloroethylmethylphenylsilane generally occur by an associative mechanism with complete loss of stereochemistry. (c) The tendency toward radical reaction by $\mathbf{ER}_4$ compounds increases on going down the group from $\mathbf{E}=Si$ to Pb, which correlates with decreasing $\mathbf{E}-\mathbf{C}$ bond enthalpies.

15.3 The compound $(C_6H_5)_3Pb-Pb(C_6H_5)_3$ survives up to about $300°C$. Write plausible reactions for this compound with (a) sodium in liquid ammonia, (b) HBr, and (c) $[Pt(PPh_3)_4]$ (this compound readily loses up to two triphenylphosphine ligands). Give your reasoning in each case.

15.4 Based on general trends in atomic energy levels, explain the probable order of the π-π^* separation in $R_2Si{=}SiR_2$ and $R_2Ge{=}GeR_2$.

Give specific examples of how this energy order might be reflected in the reactions of digermene in comparison with disilaethenes.

15.5 Although tetraalkyltin compounds are not Lewis acids, the haloalkylstannanes are Lewis acids. C. Yoder and coworkers have determined thermodynamic parameters of complex formation of triphenylphosphine oxide with a series of halotrialkystannanes (*Organometallics* **5**, 118 (1986)). Summarize the trends in enthalpies of complex formation and discuss reasonable explanations of these trends.

15.6 The synthesis of the first stable germaethene was reported by C. Couret, J. Escudie, J. Satge, and M. Lazraq, *J. Am. Chem. Soc.* **109**, 4411 (1987). Summarize the reactions involved in this synthesis and speculate on why each reaction is favorable. Describe the interaction of Lewis bases with this germaethene and propose reasons for the interaction of the germaethene with a Lewis base and lack of similar interaction for a simple alkene.

15.7 Given $Ge(CH_3)_3Cl$, $H_2C{=}CH(C_6H_5)$, and other reagents of your choice, and drawing on analogies with the chemical properties of silicon, give balanced chemical equations and conditions for the synthesis of $(CH_3)_3GeCH_2CH_2(C_6H_5)$.

16

d- and f-block organometallic compounds

Bonding

d-Block carbonyls

Other organometallic compounds

Metal–metal bonding and metal clusters

Further reading

Exercises

Problems

Much of the basic organometallic chemistry of the p-block metals was understood by the early part of the twentieth century, but that of the d- and f-blocks has been developed much more recently. Since the mid-1950s the latter field has grown into a thriving area that spans interesting new types of reactions, unusual structures, and practical applications in organic synthesis and industrial catalysis. We begin by considering the structures, bonding, and reactions of metal carbonyls, which form the foundation of much of d-block organometallic chemistry. Then we consider hydrocarbon ligands, and their variety of structures, bonding modes, and reactions. Finally, we deal with the structures and reactions of metal cluster compounds, which has been one of the most recent areas of the subject to develop. Chapter 17 will take the story further by describing how d-block organometallic compounds are used in catalysis.

A few d-block organometallic compounds were synthesized and partially characterized in the nineteenth century. The first of them (**1**), an ethene complex of platinum(II), was prepared by W.C. Zeise in 1827. The first metal carbonyl, $[Pt(Cl)_2(CO)]_2$, was made by P. Schützenberger in 1868. The next major discovery was the metal carbonyl, tetracarbonylnickel (**2**), which was synthesized by Ludwig Mond, Carl Langer, and Friedrich Quinke in 1890. Beginning in the 1930s, Walter Hieber in Munich synthesized a wide variety of metal carbonyl cluster compounds, many of which are anionic, including $[Fe_4(CO)_{13}]^{2-}$ (**3**). It was clear from this work that metal carbonyl chemistry was potentially a very rich field. However, as the structures of these and other d- and f-block organometallic compounds are difficult or impossible to deduce by chemical means alone, fundamental advances had to await

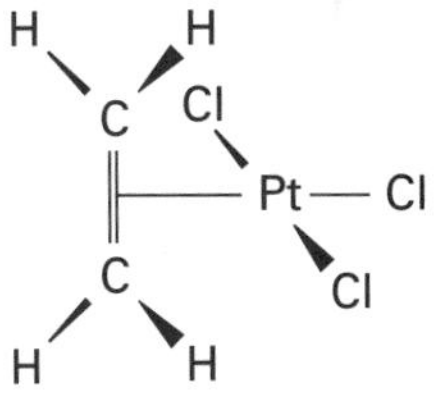

1 $[PtCl_3(C_2H_4)]^-$

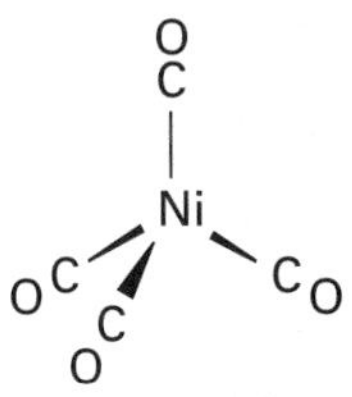

2 $Ni(CO)_4$

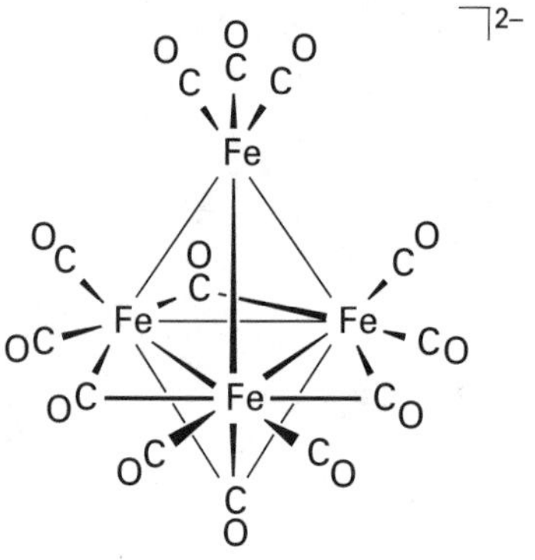

3 $[Fe_4(CO)_{13}]^{2-}$

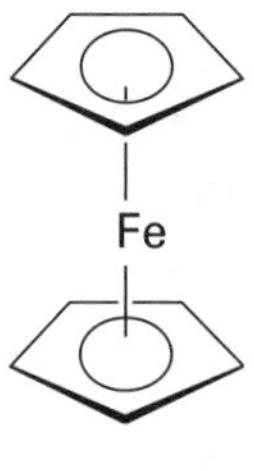

4 $Fe(C_5H_5)_2$

the development of X-ray diffraction for precise structural data on solid samples and of IR and NMR spectroscopy for structural information in solution. The discovery of the remarkably stable organometallic compound ferrocene, $Fe(C_5H_5)_2$, occurred at a time (in 1951) when these techniques were becoming widely available. The 'sandwich' structure of ferrocene (**4**) was soon correctly inferred from its IR spectrum and then determined in detail by X-ray crystallography.

The stability, structure, and bonding of ferrocene defied the classical Lewis description and therefore captured the imagination of chemists. This puzzle in turn set off a train of synthesizing, characterizing, and theorizing that led to a rapid development of *d*-block organometallic chemistry. Two highly productive research workers in the formative stage of the subject, Ernst-Otto Fischer in Munich and Geoffrey Wilkinson in London, were awarded the Nobel Prize in 1973 for their contributions.[1] Similarly, *f*-block organometallic chemistry blossomed soon after the discovery in the late 1970s that the pentamethylcyclopentadienyl ligand, C_5Me_5, forms stable *f*-block compounds (**5**).

As in Chapter 15, we adhere to the convention that an organometallic compound contains at least one metal–carbon (**M**—**C**) bond. Thus, compounds (**1**) to (**5**) clearly qualify as organometallic, whereas a complex such as $[Co(en)_3]^{3+}$, which contains carbon but has no **M**—**C** bonds, does not. Cyano complexes, such as hexacyanoferrate(II) ions, do have **M**—**C** bonds, but as their properties are more akin to those of conventional Werner complexes they are generally not considered as organometallic. In contrast, complexes of the isoelectronic ligand CO are considered to be organometallic. The justification for this somewhat arbitrary distinction is that many metal carbonyls are significantly different from Werner complexes both chemically and physically.

Following the recommended convention, we use the same nomenclature as for Werner complexes (Section 7.2). Ligands are listed in alphabetical order followed by the name of the metal, all of which is written as one word. The name of the metal may be followed by its oxidation number in parentheses. The nomenclature employed in research journals, however, does not always obey these rules strictly, and it is common to find the name of the metal buried in the middle of the name of the compound and the oxidation number omitted. For example, (**6**) is sometimes referred to as benzenemolybdenumtricarbonyl, rather than the preferred name benzene(tricarbonyl)molybdenum(0). All ligands that are radicals when neutral have the suffix -yl, as in methyl, cyclopentadienyl, and allyl.

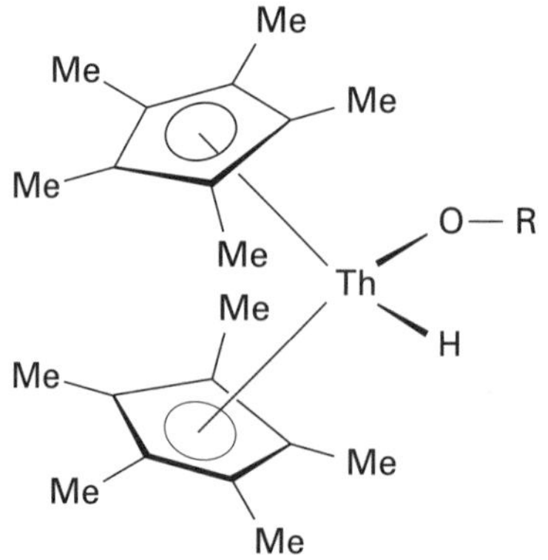

5

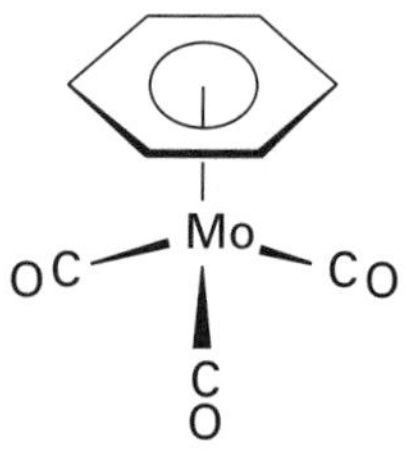

6 $Mo(\eta^6-C_6H_6)(CO)_3$

....

[1] Wilkinson's personal account captures the excitement of these developments: *J. Organometal. Chem.*
     100, 273 (1975).

The IUPAC recommendation for formulas is to write them in the same form as for Werner complexes: metal first followed by formally anionic ligands, listed in alphabetical order. The neutral ligands are then listed in alphabetical order based on their chemical symbol. We shall follow these conventions unless a different order of ligands helps to clarify a particular point.

A further notational point is that organic ligands possess a special kind of versatility: a single ligand may attach to a central metal atom using several of its atoms simultaneously (ferrocene is an example). As mentioned in Section 15.1, the hapticity, η, of a ligand is the number of its atoms that are within bonding distance of the metal atom. The definition is independent of a model of the bonding itself. For example, a CH_3 group attached by a single M—C bond is monohapto, η^1. The ethene ligand is dihapto, η^2, if its two C atoms are both within bonding distance of the metal (**1**). The bonding of a C_5H_5 to the Fe atom in ferrocene (**4**) is pentahapto, η^5. The C_5H_5 ligand, which is so common that it is represented by the symbol Cp, can also be monohapto (**7**) and trihapto (**8**).

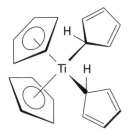

7 $Ti(\eta^1-C_5H_5)_2(\eta^5-C_5H_5)_2$

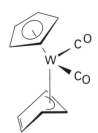

8 $W(\eta^3-C_5H_5)(\eta^5-C_5H_5)(CO)_2$

Bonding

In organometallic chemistry, as in most areas of chemistry, the concepts of closed electronic shell and (to a lesser extent) oxidation state help to rationalize the structures and reactions of the compounds.

16.1 Valence electron count

In the 1920s, the British chemist N.V. Sidgwick recognized that the metal atom in a simple metal carbonyl, such as $Ni(CO)_4$, has the same valence electron count (18) as the noble gas that terminates the long period to which the metal belongs. Sidgwick coined the term 'inert gas rule' for this indication of stability, but it is now usually referred to as the **18-electron rule**. It should be noted, however, that the 18-electron rule is not as uniformly obeyed for *d*-block organometallic compounds as the octet rule is obeyed for compounds of Period 2 elements. We shall see, however, that the exceptions to the rule can usually be rationalized.

When electrons are counted, each metal atom and ligand is treated as neutral. If the complex is charged, we simply add or subtract the appropriate number of electrons to the total. We must include in the count all valence electrons of the metal atom and all the electrons donated by the ligands. For example, $Fe(CO)_5$ acquires 18 electrons from the eight valence electrons on the Fe atom and the 10 electrons donated by the five CO ligands. Table 16.1 lists the maximum number of electrons available for donation to a metal for several common ligands.

The 18-electron rule recognizes the special stability of electron configurations corresponding to the noble gas atom that terminates the long period to which a metal belongs.

Table 16.1 Some organic ligands

Available electrons	Hapticity	Ligand	Metal–ligand structure
1	η^1	Methyl, alkyl ·CH$_3$, ·CH$_2$R	M — CH$_3$
2	η^1	Alkylidene (carbene)	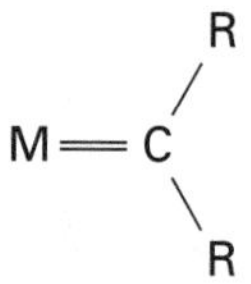
2	η^2	Alkene H$_2$C=CH$_2$	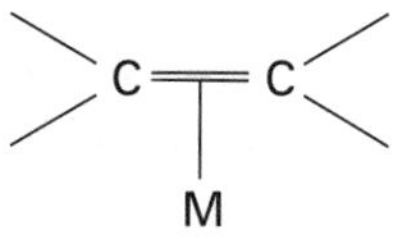
3	η^3	π-Allyl C$_3$H$_5$	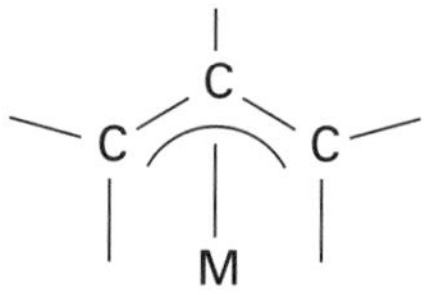
3	η^1	Alkylidyne (carbyne) C—R	M≡C—R
4	η^4	1,3-Butadiene C$_4$H$_6$	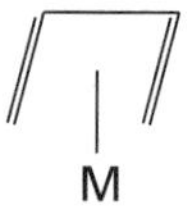
4	η^4	Cyclobutadiene C$_4$H$_4$	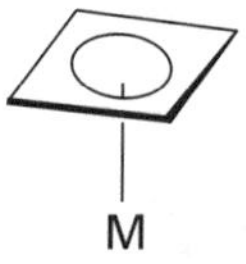
5 (3) (1)	η^5 η^3 η^1	Cyclopentadienyl C$_5$H$_5$ (Cp)	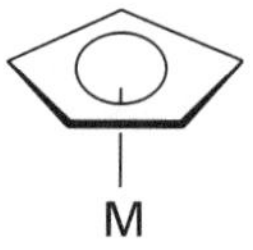
6	η^6	Benzene C$_6$H$_6$	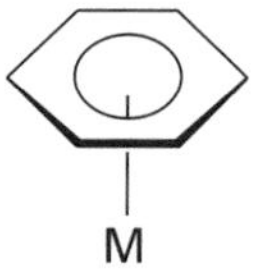
6	η^7	Cycloheptatrienyl (tropylium) C$_7$H$_7$	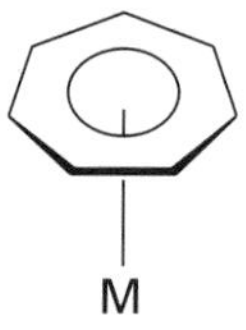
6	η^6	Cycloheptatriene C$_7$H$_8$	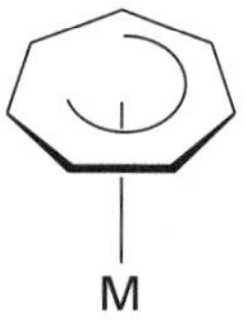
8 (6) (4)	η^8 η^6 η^4	Cyclooctatetraenyl C$_8$H$_8$ (cot)	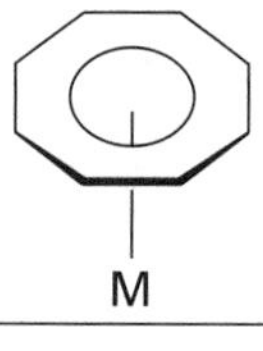

(a) The 18-electron rule and formulas of metal carbonyls

The 18-electron rule helps to systematize the formulas of metal carbonyls. As shown in Table 16.2, the carbonyls of the Period 4 elements of Groups 6 to 10 have alternately one and two metal atoms and a decreasing number of CO ligands. The binuclear carbonyls are formed by elements of the odd-numbered groups, which have an odd number of valence electrons and therefore dimerize by forming metal–metal (M—M) bonds. The decrease in the number of CO ligands from left to right across a period matches the need for fewer CO ligands to achieve 18 valence electrons.

The same principles apply when other soft ligands occupy the metal coordination shell. Two examples are benzene(tricarbonyl)molybdenum(0) (**6**) and bis(cyclopentadienyl)(tetracarbonyl)diiron(0) (**9**), both of which obey the 18-electron rule. The 18-electron rule can be extended to simple polynuclear carbonyls by adding one electron to the valence electron count of a specific metal atom in the cluster for each M—M bond to that metal atom. However, we shall see later that the 18-electron rule does not apply to M_6 and larger clusters; nor does it apply to some clusters smaller than M_6 ([$Fe_4(CO)_{13}$]$^{2-}$, for instance).

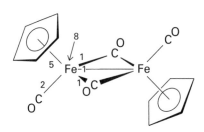

9 $Fe_2(CO)_4(C_5H_5)_2$

> The carbonyls of the Period 4 elements of Groups 6 to 10 have alternately one and two metal atoms and a decreasing number of CO ligands.

Table 16.2 Formulas and electron count for some Period 4 carbonyls

Group	Formula	Valence electrons		Structure
6	$Cr(CO)_6$	Cr	6	
		6(CO)	$\frac{12}{18}$	
7	$Mn_2(CO)_{10}$	Mn	7	
		5(CO)	10	
		M—M	$\frac{1}{18}$	
8	$Fe(CO)_5$	Fe	8	
		5(CO)	$\frac{10}{18}$	
9	$Co_2(CO)_8$	Co	9	
		4(CO)	8	
		M—M	$\frac{1}{18}$	
10	$Ni(CO)_4$	Ni	10	
		4(CO)	$\frac{8}{18}$	

Table 16.3 Scope of the 16/18-electron rule for *d*-block organometallic compounds

Usually less than 18			Usually 18 electrons			16 or 18 electrons	
Sc	Ti	V	Cr	Mn	Fe	Co	Ni
Y	Zr	Nb	Mo	Tc	Ru	Rh	Pd
La	Hf	Ta	W	Re	Os	Ir	Pt

(b) Sixteen-electron complexes

Organometallic compounds with 16 valence electrons are common on the right of the *d*-block, particularly in Groups 9 and 10 (Table 16.3). Examples of such complexes (which are generally square-planar) include $[IrCl(CO)(PPh_3)_2]$ (**10**) and the anion of Zeise's salt, $[PtCl_3(C_2H_4)]^-$ (**1**). Square-planar 16-electron complexes are particularly common for the d^8 metals of the heavier elements in Groups 9 and 10, especially for Rh(I), Ir(I), Pd(II), and Pt(II). It should be recalled from Section 7.5 that the ligand-field stabilization energy of d^8 complexes favors a low-spin square-planar configuration when Δ is large, as is typical of Period 5 and 6 *d*-metal atoms and ions. The $d_{x^2-y^2}$ orbital is then empty and d_{z^2}, which is doubly occupied, is stabilized.

> Organometallic compounds with 16 valence electrons are common on the right of the *d*-block, particularly in Groups 9 and 10.

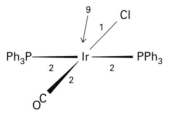

10 *trans*–$[IrCl(CO)(PPh_3)_2]$

Example 16.1 Counting the valence electrons on a metal atom

Do (a) $[IrBr_2(CH_3)(CO)(PPh_3)]$ and (b) $[Cr(\eta^5\text{-}C_5H_5)(\eta^6\text{-}C_6H_6)]$ obey the 18-electron rule?

Answer (a) An Ir atom (Group 9) has nine valence electrons; the two Br atoms and the CH_3 radical are one-electron donors; CO and PPh_3 are two-electron donors. Thus, the number of valence electrons on the metal atom is $9 + 3 \times 1 + 2 \times 2 = 16$, in accord with the common occurrence of either 16- or 18-electron complexes of Group 9 metals. (b) A Cr atom (Group 6) has six valence electrons, the $\eta^5\text{-}C_5H_5$ ligand donates five electrons, and the $\eta^6\text{-}C_6H_6$ ligand donates six; so the number of metal valence electrons is $6 + 5 + 6 = 17$. This complex does not obey the 18-electron rule and is not stable. A related but stable 18-electron compound is $[Cr(\eta^6\text{-}C_6H_6)_2]$.

Self-test 16.1 Is $Mo(CO)_7$ likely to exist?

(c) Exceptions to the 16/18-electron rule

Exceptions to the 16/18-electron rule are common on the left of the *d*-block.[2] Here steric and electronic factors are in competition, and it is not possible to crowd enough ligands around the metal either to satisfy the rule or to permit dimerization. For example, the simplest carbonyl in Group 5, which is $[V(CO)_6]$, is a 17-electron complex. Other examples include $[W(CH_3)_6]$, which has 12 electrons, and $[Cr(\eta^5\text{-}Cp)(CO)_2(PPh_3)]$ with 17 electrons. The latter compound provides a good example of the role of steric crowding. When the compact CO ligand is present in place of bulky triphenylphosphine, a dimeric compound with a long but definite Cr—Cr bond is observed in the solid state and in solution. The formation of the Cr—Cr bond in $[Cr(\eta^5\text{-}Cp)(CO)_3]_2$ raises the electron count on each metal to 18.

Deviations from the 16/18-electron rule are common in neutral bis(cyclopentadienyl) complexes, which are known for most of the *d*-block metals. Since two η^5-cyclopentadienyl ligands jointly contribute 10 valence electrons, the 18-electron rule can be satisfied only for

---

[2] For a series of articles on 17- and 19-electron organometallic compounds see: W.C. Trogler (ed.), *Organometallic radical processes.* Elsevier, Amsterdam (1990); and D.R. Tyler, 19-Electron organometallic adducts. *Acc. Chem. Res.* **24**, 325 (1991).

neutral compounds of the Group 8 metals. Complexes of this kind that do obey the 18-electron rule (such as ferrocene itself) are the most stable. Their stability is suggested by their bond lengths and their redox reactions. For example, the 19-electron complex $[Co(\eta^5\text{-}Cp)_2]$ is readily oxidized to the 18-electron cation $[Co(\eta^5\text{-}Cp)_2]^+$.

Exceptions to the 16/18-electron rule are common on the left of the d-block.

16.2 Oxidation numbers and formal ligand charges

Oxidation numbers are of less importance in the organometallic chemistry of the d-block metals than in many other areas of inorganic chemistry. However, oxidation numbers do help to systematize reactions such as oxidative addition (Section 8.3). They also bring out analogies between the chemical properties of organometallic complexes and Werner complexes. For instance, U(IV), Fe(II), and Fe(III) are known for both types of complex. The principal difference between organometallic complexes and Werner complexes with hard ligands is that the soft organic ligands in the former help to stabilize elements in low oxidation states. In organometallic complexes, as in halides and oxides, there is a decreasing ability for the metal to achieve high oxidation states on moving from Group 6 to the right of the d-block.

The rules for calculating the oxidation number of an element in an organometallic complex are the same as for conventional compounds. Ligands such as $\cdot H$, $\cdot CH_3$, and $\cdot C_5H_5$ are formally considered to take an electron from the metal atom, and hence are assigned oxidation number -1. The ligand CO is formally assigned an oxidation number of 0, as are substituted phosphine ligands, such as PMe_3. Thus, ferrocene can be considered to be a compound of Fe(II) and the ferrocenium ion, $[Fe(\eta^5\text{-}Cp)_2]^+$, a compound of Fe(III). It follows that the formal name of ferrocene is bis(η^5-cyclopentadienyl)iron(II). The oxidation number of the metal atom is often omitted when no ambiguity results. When planar, the η^8-cyclooctatetraene ligand is considered to extract two electrons and to become (formally) the aromatic dinegative ligand $[C_8H_8]^{2-}$. With two such dinegative ligands in uranocene $[U(\eta^8\text{-}C_8H_8)_2]$, the U atom must be assigned an oxidation number of $+4$, and the formal name of uranocene is bis(η^8-cyclooctatetraenyl)uranium(IV).

In addition to the high oxidation state of uranium in uranocene, some other examples of compounds with metals in high oxidation state are the unstable complex $[W(CH_3)_6]$, which contains formal W(VI). Another example is Re(VII) in $[Re(O)_3(\eta^5\text{-}Cp)]$ (11).

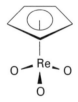

11 $Re(\eta^5\text{-}C_5H_5)O_3$

Example 16.2 Assigning oxidation states

Determine the oxidation states of iridium and chlorine in $[IrCl(CO)(PPh_3)_2]$.

Answer The ligands CO and PPh_3 are neutral two-electron donors and thus do not influence the oxidation state of iridium. A Cl atom is a one-electron ligand, and in combination with metals is assigned oxidation number -1. Because the complex as a whole is electrically neutral (implying that the oxidation numbers sum to zero), the oxidation state of the Ir atom is $+1$.

Self-test 16.2 Assign the oxidation state of cobalt in $[Co(\eta^5\text{-}C_5H_5)(CO)_2]$.

As remarked in Section 16.1, the valence electron count on a metal in an organometallic compound can be determined without assigning oxidation numbers. Therefore, it is sensible to ignore oxidation number when counting the number of valence-shell electrons on a metal atom in an organometallic compound. To establish the electron count, it is necessary to consider only the number of electrons on the neutral metal atom, the electrons donated by neutral ligands, and the overall charge of the compound.

In organometallic complexes there is a decreasing ability for the metal to achieve high oxidation numbers on moving from Group 6 to the right of the d-block.

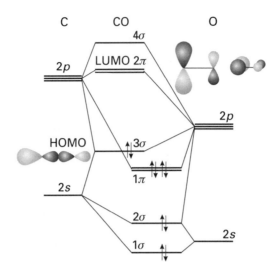

16.1 The molecular orbital energy level diagram for CO. The filled 3σ and vacant 2π orbitals are important in metal complex formation.

d-Block carbonyls

A **homoleptic carbonyl** is a carbonyl complex with only one kind of ligand. Simple homoleptic metal carbonyls can be prepared for most of the *d* metals, but those of palladium and platinum are so unstable that they exist only at low temperatures. No simple neutral metal carbonyls are known for copper, silver, and gold or for the members of Group 3.[3] The metal carbonyls are useful synthetic precursors for other organometallic compounds and are used in organic syntheses and as industrial catalysts.

16.3 Carbon monoxide as a ligand

Carbon monoxide is the most common π-acceptor ligand in organometallic chemistry. Its primary mode of attachment to metal atoms is through the C atom.

(a) The molecular orbitals of CO

The molecular orbital scheme for CO (Fig. 16.1) shows that the HOMO has σ symmetry and is essentially a lobe that projects away from the C atom. The orbitals themselves were depicted in Fig. 3.23. When CO acts as a ligand, this σ orbital serves as a very weak donor to a metal atom, and forms a σ bond with the central metal atom (**12**). The LUMOs of CO are the π^* orbitals. These two orbitals play a crucial role because they can overlap with metal *d* orbitals that have local π symmetry (such as the t_{2g} orbitals in an O_h complex (**13**)). The π interaction leads to the delocalization of electrons from filled *d* orbitals on the metal atom into the empty π^* orbitals on the CO ligands, so the ligand also acts as a π acceptor.

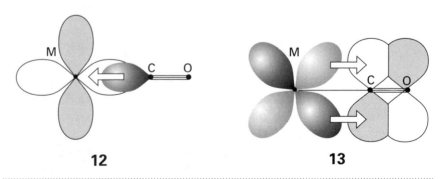

12	**13**

---

[3]   The silver carbonyl cation $[OCAgCO]^+$ is described by P.K. Hurlburt, J.J. Rack, S.F. Dec, O.P. Anderson, and S.H. Strauss, *Inorg. Chem.* **32**, 373 (1993).

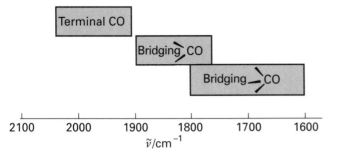

16.2 Approximate ranges for the CO stretching wavenumber in neutral metal carbonyls. Note that high wavenumbers (and hence high frequencies) are on the left, in keeping with the way in which infrared spectrometers generally plot spectra.

Trends in CO bond lengths and stretching frequencies obtained from IR spectra are in general agreement with the bonding model just described. Thus, strong σ-donor ligands attached to a metal carbonyl and a formal negative charge on a metal carbonyl anion both result in slightly greater CO bond lengths and significantly lower CO stretching frequencies (Table 16.4). This behavior is consistent with the bonding model because an increase in the electron density on the metal atom is delocalized over the CO ligands by populating the carbonyl π^* orbital (**13**), so weakening the CO bond.

Carbon monoxide is versatile as a ligand because it can bridge two (**14**) or three (**15**) metal atoms. Although the description of the bonding is now more complicated, the concepts of σ-donor and π-acceptor ligands remain useful. The CO stretching frequencies generally follow the order $MCO > M_2CO > M_3CO$ (Fig. 16.2), which suggests an increasing occupation of the π^* orbital as the CO molecule bonds to more metal atoms.

> The σ orbital of CO serves as a very weak donor while the π^* orbitals act as acceptors.

Table 16.4 The influence of coordination and charge on CO stretching wavenumbers

Compound	$\tilde{\nu}/\mathrm{cm}^{-1}$
$CO(g)$	2143
$[Mn(CO)_6]^+$	2090
$Cr(CO)_6$	2000
$[V(CO)_6]^-$	1860
$[Ti(CO)_6]^{2-}$	1750

Source: K. Nakamoto, *Infrared and Raman spectra of inorganic and coordination compounds.* Wiley, New York (1997); data for $[Ti(CO)_6]^{2-}$ are from S.R. Frerichs, B.K. Stein, and J.E. Ellis, *J. Am. Chem. Soc.* **109**, 5558 (1987).

14

15

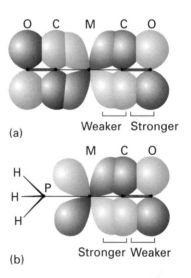

16.3 The influence of metal–ligand π bonding on CO bonds in metal carbonyls. (a) An electron-poor metal center that has d orbitals lowered in energy because of the positive charge on the metal atom or competing π bonding ligands. In this case, the CO bond length is reduced and the CO stretching frequency is raised. (b) An electron-rich center, which occurs with strong σ-donor ligands or negative charge. Now the CO π antibonding orbital is more highly populated, so the CO bond is weakened. As a result, the CO bond lengthens slightly and the CO stretching frequency decreases.

(b) Related π-acceptor ligands

Many other π-acceptor ligands are found in organometallic complexes, but not all of them contain carbon. Some of them (most importantly, those isoelectronic with CO) can be arranged according to increasing π-acceptor strength:

$$C{\equiv}N^- < N{\equiv}N < C{\equiv}NR < C{\equiv}O < C{\equiv}S < N{\equiv}O^+$$

Carbonyl stretching frequencies are often used to determine the order of π-acceptor strengths for the other ligands present in a complex. The basis of the approach is that the CO stretching frequency is decreased when it serves as a π acceptor. However, as other π acceptors in the same complex compete for the d electrons of the metal, they cause the CO frequency to increase (Fig. 16.3). This behavior is opposite to that observed with donor ligands, which cause the CO stretching frequency to decrease as they supply electrons to the metal and hence, indirectly, to the CO π^* orbitals.

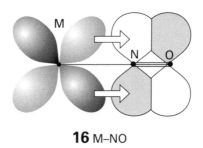

16 M–NO

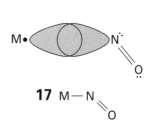

17 M—N

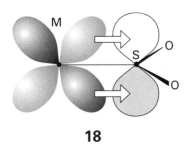

18

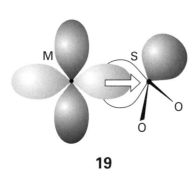

19

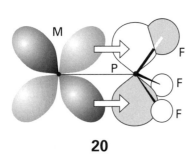

20

The ligands at the ends of the π-acceptor series, CN^- and NO^+, differ significantly from CO and are not always good analogs. For example, CN^-, which is electron-rich, is a good σ donor and a weak π acceptor and therefore forms many classical Werner complexes with metal atoms in high oxidation states. At the other extreme, NO^+ is very strongly electron withdrawing. It forms a linearly linked M—NO ligand in complexes in which it is considered to be NO^+ (**16**) but, when NO takes on an additional electron to form NO^-, the M—NO link is bent (**17**). For example, in $[IrCl(CO)(NO)(PPh_3)_2]^+$ the IrNO angle is 124°. The other ligands in the series (N_2 through CS) are generally found with metals having low oxidation numbers, and their compounds resemble each other quite closely.

Another useful perspective is obtained by regarding linearly coordinated NO as a neutral three-electron ligand. Thus, two NO ligands can replace three CO ligands and preserve the same 18-electron count on the metal. The tetrahedral nitrosyl compound $[Cr(NO)_4]$, for instance, has the same electron count as the octahedral carbonyl compound $[Cr(CO)_6]$.

Two other important π-acceptor ligands are SO_2 and PF_3. The former can serve as a fairly strong π-acceptor ligand. Like NO, it binds in various ways depending on the number of electrons available. One common case is the coplanar M—SO_2 fragment, in which the bonding is σ donation of an electron pair from the S atom to the metal atom in conjunction with π back-bonding from the metal to the SO_2 (**18**). In another extreme, the plane of the SO_2 is tilted with respect to the M—S link (**19**). This type of bonding is generally explained in terms of σ donation to the SO_2 ligand of a lone pair from an electron-rich metal center. In the latter case, SO_2 acts as a simple Lewis acid in its interaction with the metal.[4]

As judged from spectroscopic data, the π-acceptor character of PF_3 is comparable to that of CO. This similarity was originally explained in terms of the vacant P3d orbitals acting as electron acceptors. Molecular orbital calculations on PF_3, however, indicate that the acceptor orbital is primarily the P—F antibonding orbitals formed from P3p orbitals (**20**), and that the P3d orbitals play only a minor role. Experimental data on complexes and calculations both indicate that $P(OR)_3$ ligands are somewhat weaker π acceptors than PF_3. When the attached groups are less electronegative, as in PH_3 and alkylphosphines, the P atom becomes a weaker π acceptor but a stronger σ donor.[5]

> *The* CO *stretching frequency is decreased when it serves as a π acceptor, but stronger π acceptors in the same complex cause the* CO *frequency to increase. Donor ligands cause the* CO *stretching frequency to decrease as they supply electrons to the metal.*

[4]   For a review of the reactions of SO_2 with metal coordination compounds, including organometallic compounds see G.J. Kubas and R.R. Ryan, *Polyhedron* **1/2**, 473 (1986).

[5]   A theoretical interpretation of phosphine–metal bonding is given by G. Pacchioni and P.S. Bagus, *Inorg. Chem.* **31**, 4391 (1992).

16.4 Synthesis of carbonyls

The two principal methods for the synthesis of monometallic metal carbonyls are direct combination of carbon monoxide with a finely divided metal and the reduction of a metal salt in the presence of carbon monoxide under pressure. Many polymetallic carbonyls are synthesized from monometallic carbonyls.

(a) Direct combination

Mond, Langer, and Quinke discovered that the direct combination of nickel and carbon monoxide produced nickel carbonyl, $Ni(CO)_4$:[6]

$$Ni(s) + 4 CO(g) \xrightarrow{30\,°C,\ 1\ atm\ CO} Ni(CO)_4(l)$$

Tetracarbonylnickel(0) is in fact the metal carbonyl that is most readily synthesized in this way.

Other metal carbonyls, such as $Fe(CO)_5$, are formed more slowly. They are therefore synthesized at high pressures and temperatures (Fig. 16.4):

$$Fe(s) + 5 CO(g) \xrightarrow{200\,°C,\ 200\ atm} Fe(CO)_5(l)$$

$$2 Co(s) + 8 CO(g) \xrightarrow{150\,°C,\ 35\ atm} Co_2(CO)_8(s)$$

▌ *Some metal carbonyls are formed by direct reaction, but of those that can be formed in this way most require high pressures and temperatures.*

(b) Reductive carbonylation

Direct reaction is impractical for most of the remaining d metals and **reductive carbonylation**, the reduction of a salt or metal complex in the presence of CO, is normally employed instead. Reducing agents vary from active metals such as aluminum and sodium, to alkylaluminum compounds, H_2, and CO itself:

$$CrCl_3(s) + Al(s) + 6 CO(g) \xrightarrow{AlCl_3,\ benzene} AlCl_3(soln) + Cr(CO)_6(soln)$$

$$3 Ru(acac)_3(soln) + H_2(g) + 12 CO(g) \xrightarrow{150\,°C,\ 200\ atm,\ CH_3OH} Ru_3(CO)_{12}(soln) + \cdots$$

$$Re_2O_7(s) + 17 CO(g) \xrightarrow{250\,°C,\ 350\ atm} Re_2(CO)_{10}(s) + 7 CO_2(g)$$

▌ *Metal carbonyls are commonly formed by reductive carbonylation.*

16.5 Structure

Simple metal carbonyl molecules often have well defined, simple, symmetrical shapes that correspond to the CO ligands taking up the most-distant locations, like electron pairs in the VSEPR model. Thus, the Group 6 hexacarbonyls are octahedral (**21**), pentacarbonyliron(0) is trigonal-bipyramidal (**22**), and tetracarbonylnickel(0) is tetrahedral (as shown in **2**). Decacarbonyldimanganese(0) consists of two square-pyramidal $Mn(CO)_5$ groups joined by a metal–metal bond. In one isomer of octacarbonyldicobalt(0) the metal–metal bond is bridged by CO.

....

[6] Mond, Langer, and Quinke discovered $Ni(CO)_4$ in the course of studying the corrosion of nickel valves in process gas containing CO. They quickly applied their discovery to develop a new industrial process for the separation of nickel from cobalt. The centennial of this important discovery is celebrated in a special volume of *J. Organometal. Chem.* **383** (1990), which is devoted to metal carbonyl chemistry.

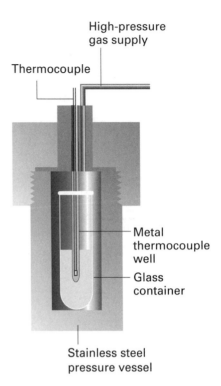

High-pressure gas supply

Thermocouple

Metal thermocouple well

Glass container

Stainless steel pressure vessel

16.4 A high-pressure reaction vessel. The reaction mixture is in the glass container.

21 $Cr(CO)_6, O_h$

22 $Fe(CO)_5, D_{3h}$

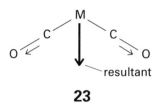

23

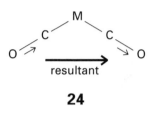

24

Infrared and ^{13}C-NMR spectroscopy are widely used to determine the arrangement of atoms in metal carbonyl compounds, as separate signals are observed for inequivalent CO ligands. The terminal ligand ^{13}C nucleus is the more shielded. NMR spectra generally contain more detailed structural information than IR spectra provided the molecule is not fluxional. However, IR spectra are often simpler to obtain, and are particularly useful for following reactions. Most CO stretching bands occur in the range 2100 to 1700 cm^{-1}, a region that is generally free of bands due to organic groups. The range of CO stretching frequencies (see Fig. 16.2) and the number of CO bands (Table 16.5) are both important for making structural inferences.

If the CO ligands are not related by a center of inversion or a threefold or higher axis of symmetry, a molecule with N CO ligands will have N CO stretching absorption bands. Thus a bent OC—M—CO group (with only a twofold symmetry axis) will have two infrared absorptions because both the symmetric (**23**) and antisymmetric (**24**) stretches cause the electric dipole moment to change and are IR active. Highly symmetric molecules have fewer bands than CO ligands. Thus, in a linear OC—M—CO group, only one IR band is observed in the CO stretching region because the symmetric stretch leaves the overall electric dipole moment unchanged. As shown in Fig. 16.5, the positions of CO ligands in a metal carbonyl compound may be more symmetrical than the point group of the whole compound suggests, and then fewer bands will be observed than are predicted on the basis of the overall point group.

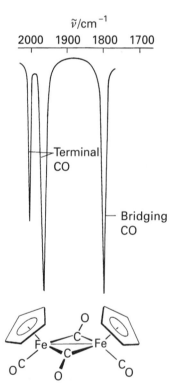

16.5 The infrared spectrum of $[Fe_2(\eta^5\text{-}Cp)_2(CO)_4]$. Note the two high-frequency terminal CO stretches and the lower frequency absorption of the bridging CO ligands. Although two bridging CO bands would be expected on account of the low symmetry of the complex, a single band is observed because the two bridging CO groups are nearly collinear.

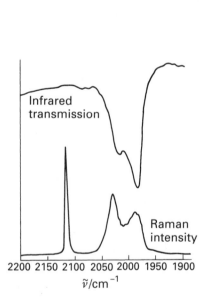

16.6 Infrared and Raman vibrational spectra of liquid $[Fe(CO)_5]$ in the CO stretching region. Note that, in accord with Table 16.5, only two CO stretching bands are observed for this D_{3h} complex. The totally symmetric band at 2115 cm^{-1} is absent in the infrared spectrum but strong in the Raman spectrum because the selection rules are different in the two cases.

Infrared spectroscopy is also useful for distinguishing terminal CO (MCO) from two-metal bridging CO (μ_2-CO) and three-metal face-bridging CO (μ_3-CO). As explained earlier, the CO stretches occur at lower frequency in the more highly bridging structures (as in Figs 16.2 and 16.5).

Example 16.3 Determining the structure of a carbonyl from IR data

The complex $[Cr(CO)_4(PPh_3)_2]$ has one very strong IR absorption band at 1889 cm^{-1} and two other very weak bands in the CO stretching region. What is the probable structure of this compound? (The CO stretching wavenumbers are lower than in the corresponding hexacarbonyl because the phosphine ligands are better σ-donors and poorer π-acceptors than CO.)

Answer A disubstituted hexacarbonyl may exist with either *cis* or *trans* configurations. In the *cis* isomer the four CO ligands are in a low-symmetry (C_{2v}) environment and therefore four IR bands should be observed, as indicated in Table 16.5. The *trans* isomer has a square-planar array of four CO ligands (D_{4h}), for which only one band in the CO stretching region is expected (Table 16.5). The *trans* CO arrangement is indicated by the data, since it is reasonable to assume that the weak bands reflect a small departure from D_{4h} symmetry imposed by the PPh$_3$ ligands.

Self-test 16.3
The IR spectrum of $[Ni_2(\eta^5\text{-Cp})_2(CO)_2]$ has a pair of CO stretching bands at 1857 cm^{-1} (strong) and 1897 cm^{-1} (weak). Does this complex contain bridging or terminal CO ligands, or both? (Substitution of η^5-C$_5$H$_5$ ligands for CO ligands leads to small shifts in the CO stretching frequency for a terminal CO.)

A motionally averaged NMR signal is observed when a molecule undergoes changes in structure more rapidly than the technique can resolve. Although this phenomenon is quite common in the NMR spectra of organometallic compounds, it is not observed in their IR or Raman spectra. An example of this difference is $[Fe(CO)_5]$, for which the NMR signal shows a single line at $\delta = 210$, whereas IR and Raman spectra are consistent with a trigonal-bipyramidal structure (Fig. 16.6). We shall meet more NMR evidence for fluxional behavior when we discuss cyclic polyene ligands and metal carbonyl clusters.

Simple metal carbonyl molecules commonly have well defined, symmetrical shapes which correspond to the CO ligands taking up the most-distant locations, like electron pairs in the VSEPR model.

16.6 Properties and reactions

Iron and nickel carbonyls are liquids at room temperature and pressure, but all other common carbonyls are solids. All the mononuclear carbonyls are volatile; their vapor pressures at room temperature range from approximately 350 **Torr** for tetracarbonylnickel(0) to approximately 0.1 **Torr** for hexacarbonyltungsten(0). The high volatility of Ni(CO)$_4$ coupled with its extremely high toxicity require unusual care in handling it. Although the other carbonyls appear to be less toxic, they too must not be inhaled or allowed to touch the skin.

Since they are nonpolar, all the mononuclear and many of the polynuclear carbonyls are soluble in hydrocarbon solvents. The most striking exception among the common carbonyls is nonacarbonyldiiron(0), $[Fe_2(CO)_9]$, which has a very low vapor pressure and is insoluble in solvents with which it does not react. In contrast, $[Mn_2(CO)_{10}]$ and $[Co_2(CO)_8]$ are soluble in hydrocarbon solvents and sublime readily.

Most of the mononuclear carbonyls are colorless or lightly colored. Polynuclear carbonyls are colored, the intensity of the color increasing with the number of metal atoms. For example, pentacarbonyliron(0) is a light straw-colored liquid, nonacarbonyldiiron(0) forms golden-yellow flakes, and dodecacarbonyltriiron(0) is a deep green compound which looks

Table 16.5 Relation of the number of CO stretching bands in the IR spectrum to structure

Complex	Isomer	Structure*	Point group	Number of bands†
$M(CO)_6$		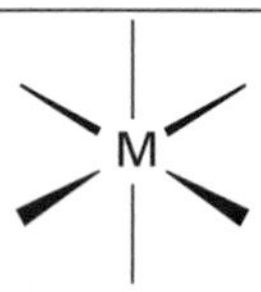	O_h	1
$M(CO)_5L$		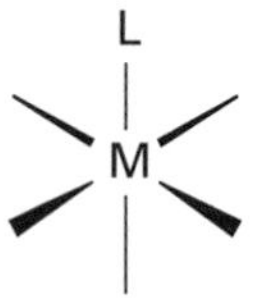	C_{4v}	3‡
$M(CO)_4L_2$	*trans*	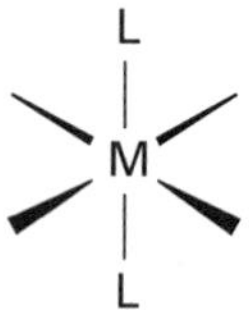	D_{4h}	1
	cis	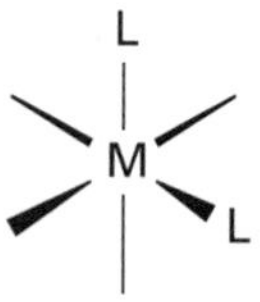	C_{2v}	4‡‡
$M(CO)_3L_3$	*mer*	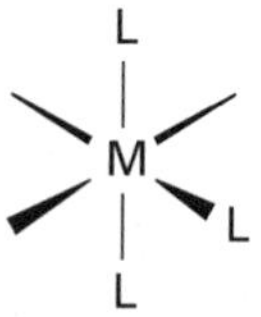	C_{2v}	3‡‡
	fac	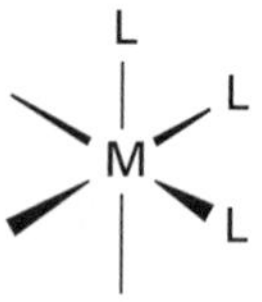	C_{3v}	2
$M(CO)_5$		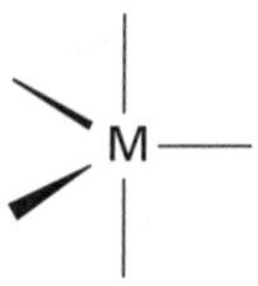	D_{3h}	2
$M(CO)_4L$	*ax*	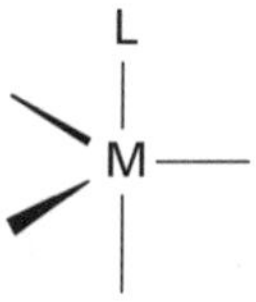	C_{3v}	3**
	eq	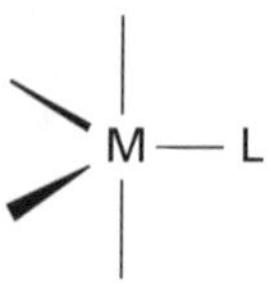	C_{2v}	4

Table 16.5 (continued)

Complex	Isomer	Structure*	Point group	Number of bands†
$M(CO)_3L_2$		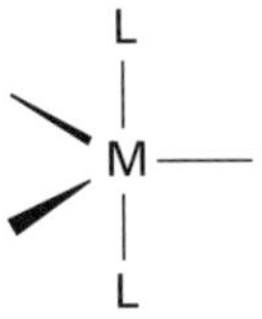	D_{3h}	1
		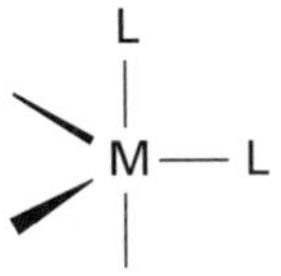	C_s	3
$M(CO)_4$			T_d	1

*Each bonding line that is not terminated by an L is a CO.
†The number of IR bands expected in the CO stretching region is based on formal selection rules. In some cases, fewer bands are observed, as explained in the footnotes.
‡If the fourfold array of CO ligands lies in the same plane as the metal atom, two bands will be observed.
‡‡If the trans CO ligands are nearly collinear, one fewer band will be observed.
**If the threefold array of CO ligands is nearly planar, only two bands will be observed.

black in the solid state. The colors of polynuclear carbonyls arise from electronic transitions between orbitals that are largely localized on the metal framework.

The principal reactions of the metal center of simple metal carbonyls are substitution, condensation into clusters, reduction, and oxidation. In certain cases, the CO ligand itself is also subject to attack by nucleophiles or electrophiles.

All the mononuclear carbonyls are volatile; all the mononuclear and many of the polynuclear carbonyls are soluble in hydrocarbon solvents; polynuclear carbonyls are colored. The principal reactions are as set out above.

(a) Substitution

Substitution reactions provide a route to organometallic compounds containing a variety of other ligands. The simplest examples involve the replacement of CO by another electron-pair donor, such as a phosphine. Studies of the rates at which alkylphosphines and other ligands replace CO in $Ni(CO)_4$, $Fe(CO)_5$, and the hexacarbonyls of the chromium group indicate that a dissociative mechanism (D or I_d, Section 14.2) is in operation. In the D mechanism, an intermediate of reduced coordination number, or more probably a solvated intermediate such as $[Cr(CO)_5(THF)]$, is produced. This intermediate then combines with the entering group in a bimolecular process:[7]

$$Cr(CO)_6 + Sol \rightleftharpoons Cr(CO)_5(Sol) + CO$$
$$Cr(CO)_5(Sol) + PR_3 \longrightarrow Cr(CO)_5PR_3 + Sol$$

In keeping with this interpretation, solvated metal carbonyls have been observed by IR spectroscopy as the photolysis products of metal carbonyls in polar solvents such as THF. The evidence for an I_d mechanism is that the rate laws for reactions of strong nucleophiles at high concentrations with Group 6 carbonyls have a second-order term. Thus, the proposed reaction pathways are

[7] Sol denotes a solvent molecule.

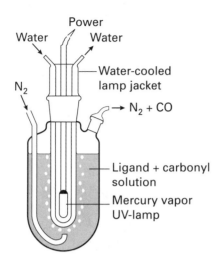

16.7 Apparatus for photochemical ligand substitution of metal carbonyls.

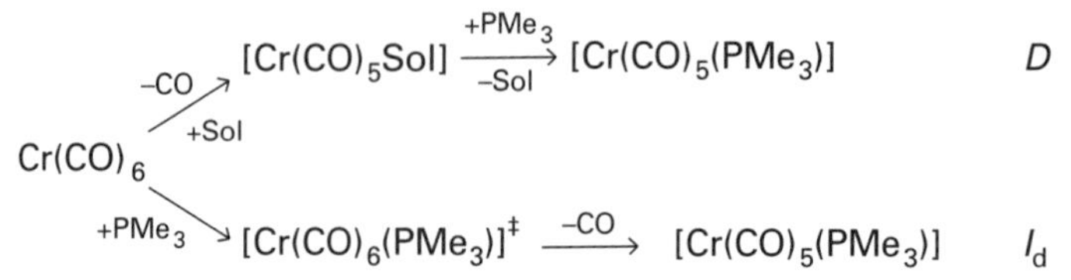

$$Cr(CO)_6 \begin{cases} \xrightarrow[+Sol]{-CO} [Cr(CO)_5Sol] \xrightarrow[-Sol]{+PMe_3} [Cr(CO)_5(PMe_3)] \qquad D \\ \xrightarrow{+PMe_3} [Cr(CO)_6(PMe_3)]^{\ddagger} \xrightarrow{-CO} [Cr(CO)_5(PMe_3)] \qquad I_d \end{cases}$$

Loss of the first CO group from $Ni(CO)_4$ occurs easily, and substitution is fast at room temperature. The CO ligands are much more tightly bound in the Group 6 carbonyls, and loss of CO often needs to be promoted thermally or photochemically. For example, the substitution of CH_3CN for CO is carried out in refluxing acetonitrile, using a stream of nitrogen to sweep away the carbon monoxide and hence drive the reaction to completion. To achieve photolysis, mononuclear carbonyls (which do not absorb strongly in the visible region) are exposed to near-UV radiation in an apparatus like that shown in Fig. 16.7. As with the thermal process, there is strong evidence that the photo-assisted substitution reaction leads to the formation of a labile intermediate complex with the solvent (Sol), which is then displaced by the entering group:

$$Cr(CO)_6 \xrightarrow{h\nu,\ -CO,\ +Sol} [Cr(CO)_5(Sol)] \xrightarrow{+L,\ -Sol} Cr(CO)_5L$$

> Substitution reactions provide a route to organometallic compounds containing a variety of other ligands.

(b) Influence of steric repulsion

As in the reactions of conventional complexes, we can expect steric crowding between ligands to accelerate dissociative processes and to decrease the rates of associative processes (Section 14.2). The extent to which various ligands crowd each other is approximated by a cone with an angle determined from a space-filling model (Fig 14.6 and Table 14.7). Ligands with large cone angles exert considerable steric repulsion on one another when packed around a metal center.

We can see how the cone angle of a ligand influences the equilibrium constant for ligand binding by examining the dissociation constant of $[Ni(PR_3)_4]$ complexes (Table 16.6). These complexes are slightly dissociated in solution if the phosphine ligands are compact, such as PMe_3, with a cone angle of $118°$. However, a complex such as $[Ni\{P(^tBu)_3\}_4]$, where the cone angle is huge ($182°$), is highly dissociated. The ligand $P(^tBu)_3$ is so bulky that the 14-electron complex $[Pt(P(^tBu)_3)_2]$ can be isolated.

> Steric crowding between ligands increases the rate of dissociative processes and decreases the rates of associative processes.

Table 16.6 Cone angle θ and dissociation constant K_d for some Ni complexes*

L	$\theta/°$	K_d
$P(CH_3)_3$	118	$< 10^{-9}$
$P(C_2H_5)_3$	137	1.2×10^{-5}
$P(CH_3)Ph_2$	136	5.0×10^{-2}
PPh_3	145	Large
$P(t\text{-}Bu)_3$	182	Large

*Data are for $NiL_4 \rightleftharpoons NiL_3 + L$ (in benzene, at $25°C$); from C.A. Tolman, *Chem. Rev.* **77**, 313 (1977).

....

Example 16.4 Preparing substituted metal carbonyls

Starting with MoO_3 as a source of Mo, and CO and PPh_3 as the ligand sources, plus other reagents of your choice, give equations and conditions for the synthesis of $Mo(CO)_5PPh_3$.

Answer A typical procedure is to synthesize $Mo(CO)_6$ first and then carry out a ligand substitution. Reductive carbonylation of MoO_3 might be performed using $Al(CH_2CH_3)_3$ as a reducing agent in the presence of carbon monoxide under pressure. The temperature and pressure required for this reaction are less than those for the direct combination of molybdenum and carbon monoxide.

$$MoO_3 + Al(CH_2CH_3)_3 + 6\,CO \xrightarrow{50\ atm,\ 150°C,\ heptane} Mo(CO)_6$$
$$+ \text{oxidation products of } Al(CH_2CH_3)_3$$

The subsequent substitution could be carried out photochemically using the apparatus illustrated in Fig. 16.7.

$$Mo(CO)_6 + PPh_3 \xrightarrow{THF, h\nu} Mo(CO)_5PPh_3 + CO$$

The progress of the reaction can be followed by IR spectroscopy in the CO stretching region using small samples that are periodically removed from the reaction vessel.

Self–test 16.4 If the highly substituted complex $[Mo(CO)_3L_3]$ is desired, which of the ligands $P(CH_3)_3$ or $P({}^tBu)_3$ would be preferred? Give reasons for your choice.

(c) Influence of electronic structure on CO substitution

Extensive studies of CO substitution reactions of simple carbonyl complexes have revealed systematic trends in mechanisms and rates.[8]

The 18-electron metal carbonyls generally undergo dissociatively activated substitution reactions. As we saw in Section 14.2, this class of reaction is characterized by insensitivity of its rate to the identity of the entering group. Dissociative activation of an 18-electron complex is usually favored because the alternative, associative activation, would require the formation of an energetically unfavorable 20-electron activated complex.

The rates of substitution of 16-electron complexes are sensitive to the identity and concentration of the entering group, which indicates associative activation. For example, the reactions of $[IrCl(CO)(PPh_3)_2]$ with triethylphosphine are associatively activated. As with Werner coordination compounds, 16-electron organometallic compounds appear to undergo associatively activated substitution reactions because the 18-electron activated complex is energetically more favorable than the 14-electron activated complex that would occur in dissociative activation.

Although these generalizations apply to a wide range of reactions, some exceptions are observed, especially if cyclopentadienyl or nitrosyl ligands are present. In these cases it is common to find evidence of associatively activated substitution even for 18-electron complexes. The common explanation is that NO may switch from being a three-electron donor when M—NO is linear (as in **16**), to being a one-electron donor when it is angular, since two of the electrons can then be regarded as localized on NO (as in **17**). Similarly, the η^5-Cp five-electron donor can slip relative to the metal and become an η^3-Cp three-electron donor. In this case, the C_5H_5 ligand is regarded as having a three-carbon interaction with the metal while the remaining two electrons form a simple C=C bond that is not engaged with the metal (**25**), and the relatively electron-depleted central metal atom becomes susceptible to substitution.

25

One final general observation is that the rate of CO substitution in six-coordinate metal carbonyls often decreases as more strongly basic ligands replace CO, and two or three alkylphosphine ligands often represent the limit of substitution. With bulky phosphine ligands, further substitution may be thermodynamically unfavorable on account of ligand crowding. Increased electron density on the metal center, which arises when a π-acceptor ligand is replaced by a net donor ligand, appears to bind the remaining CO ligands more tightly and therefore reduce the rate of CO dissociative substitution. The donor character of P(III) ligands as judged by ease of protonation and other data lie in the order[9]

$$PF_3 \ll P(OAr)_3 < P(OR)_3 < PAr_3 < PR_3$$

where Ar is aryl and R is alkyl.

[8] For an informative article on the development of the major principles see F. Basolo, *Inorg. Chim. Acta* **50**, 65 (1981).

[9] For a discussion of various measures of phosphine basicity, see R.C. Bush and R.J. Angelici, *Inorg. Chem.* **27**, 681 (1988).

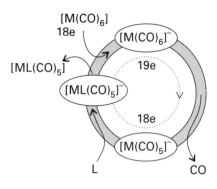

16.8 Schematic diagram of an electron transfer catalyzed CO substitution. After addition of a small amount of a reducing initiator, the cycle continues until the limiting reagent $[M(CO)_6]$ or L has been consumed.

The explanation of the influence of σ-donor ligands on CO bonding is that the increased electron density contributed by the phosphine leads to stronger π back-bonding to the remaining CO ligands and therefore strengthens the M—CO bond. This stronger M—C bond decreases the tendency of CO to leave the metal and therefore decreases the rate of dissociative substitution. In some instances these bonding effects may lead to greater thermodynamic stability for the metal carbonyl than its substitution product.

It has been found that, for some metal carbonyls, the displacement of CO can be catalyzed by electron transfer processes that create anion or cation radicals and modify the valence electron count adversely. A typical process of this type is illustrated in Fig. 16.8. As can be seen, the key features are the lability of CO in the 19-electron anion radical and the smaller reduction potential of the phosphine-substituted product than for the metal carbonyl starting material. Similarly, the less common 19- and 17-electron metal compounds are labile with respect to substitution.

> The 18-electron metal carbonyls generally undergo dissociatively activated substitution reactions, whereas 16-electron compounds are commonly associatively activated.

(d) Reduction to form metal carbonylates

Most metal carbonyls can be reduced to metal carbonylate anions. In monometallic carbonyls, two-electron reduction is generally accompanied by loss of the two-electron donor CO ligand:

$$Fe(CO)_5 \xrightarrow{\text{Na, THF}} [Fe(CO)_4]^{2-} + CO$$

The metal carbonylate contains Fe with oxidation number -2, and it is rapidly oxidized by air. That much of the negative charge is delocalized over the CO ligands is confirmed by the observation of a low CO stretching frequency in the IR spectrum, corresponding to about 1730 cm^{-1}. Polynuclear carbonyls containing M—M bonds are generally cleaved by strong reducing agents. Once again the 18-electron rule is obeyed and a mononegative mononuclear carbonyl results:

$$2\,Na + (OC)_5Mn{-}Mn(CO)_5 \xrightarrow{\text{THF}} 2\,Na^+[Mn(CO)_5]^-$$

Some metal carbonyls disproportionate in the presence of a strongly basic ligand, producing the ligated cation and a carbonylate. Much of the driving force for this reaction is the stability of the metal cation when it is surrounded by strongly basic ligands. Octacarbonyldicobalt(0) is highly susceptible to this type of reaction when exposed to a good Lewis base such as pyridine (py):

$$3\,Co_2^{(0)}(CO)_8 + 12\,py \longrightarrow 2\,[Co^{(+2)}(py)_6][Co^{(-1)}(CO)_4]_2 + 8\,CO$$

It is also possible for the CO ligand to be oxidized in the presence of the strongly basic ligand OH^-, the net outcome being the reduction of a metal center:

$$3\,Fe^{(0)}(CO)_5 + 4\,OH^- \longrightarrow [Fe_3^{(-2/3)}(CO)_{11}]^{2-} + CO_3^{2-} + 2\,H_2O + 3\,CO$$

> Most metal carbonyls can be reduced to metal carbonylates. Some metal carbonyls disproportionate in the presence of a strongly basic ligand, producing the ligated cation and a carbonylate anion.

(e) Metal basicity

Many organometallic compounds can be protonated at the metal center. Metal carbonylates provide many examples of this basicity:

$$[Mn(CO)_5]^-(aq) + H^+(aq) \longrightarrow HMn(CO)_5(s)$$

The affinity of metal carbonylates for the proton varies widely (Table 16.7). It is observed that, the more negative the anion, the higher its Brønsted basicity and hence the lower the acidity of its conjugate acid (the metal carbonyl hydride).

The d-block M—H complexes are commonly referred to as hydrides. This name reflects the assignment of oxidation number -1 to an H atom attached to a metal atom. Nevertheless, most of the carbonyl hydrides of metals to the right of the d-block are Brønsted acids. The Brønsted acidity of a metal CO hydride is a reflection of the π-acceptor strength of CO, which stabilizes the conjugate base. Thus, $[CoH(CO)_4]$ is acidic whereas $[CoH(PMe_3)_4]$ is strongly hydridic. This behavior underlines yet again the formal nature of oxidation numbers. In striking contrast to p-block hydrogen compounds, the Brønsted acidity of d-block M—H compounds decreases on descending a group.

Neutral metal carbonyls (such as pentacarbonyliron) can be protonated in air-free concentrated acid. Compounds having metal–metal bonds are even more easily protonated. The Brønsted basicity of a metal atom with oxidation number 0 is associated with the presence of nonbonding d electrons. Similarly, the tendency to protonate metal–metal bonds can be explained by the transformation of an M—M bond into M—H—M (**26**). A characteristic feature of the Groups 6 to 10 organometallic hydrides is the appearance of a highly shielded ^{1}H-NMR signal ($\delta = -8$ to -60); the hydrogen signal is significantly more shielded than one attached to most elements (see *Further information 2*).

Metal hydrides play an important role in inorganic chemistry and not all of them are generated by protonation. For example, in Section 8.3 we encountered complexes in which an η^2-H$_2$ is coordinated to a metal center. Even when H$_2$ adds to a metal center giving conventional H—M—H complexes, it is likely that a transitory η^2-H$_2$ complex is formed. We discuss the addition of H$_2$ to metal complexes in Section 16.7 and the role of hydrogen complexes in homogeneous catalysis in Section 17.4.

Metal basicity is turned to good use in the synthesis of a wide variety of organometallic compounds. For example, alkyl and acyl groups can be attached to metal atoms by the reaction of an alkyl or acyl halide with an anionic metal carbonyl:

$$[Mn(CO)_5]^- + CH_3I \longrightarrow [Mn(CH_3)(CO)_5] + I^-$$
$$[Co(CO)_4]^- + CH_3COI \longrightarrow [Co(COCH_3)(CO)_4] + I^-$$

A similar reaction with organometallic halides may be used to form M—M bonds:

$$[Mn(CO)_5]^- + ReBr(CO)_5 \longrightarrow (OC)_5Mn—Re(CO)_5 + Br^-$$

| Most organometallic metal compounds can be protonated at the metal center; the more
negative the anion, the higher its Brønsted basicity and hence the lower the acidity of its
conjugate acid.

Table 16.7 Acidity constant of d-metal hydrides in acetonitrile at 25 °C

Hydride	pK_a(MH)
HCo(CO)$_4$	8.3
HCo(CO)$_3$P(OPh)$_3$	11.3
H$_2$Fe(CO)$_4$	11.4
HCpCr(CO)$_3$	13.3
HCpMo(CO)$_3$	13.9
HMo(CO)$_5$	15.1
HCo(CO)$_3$PPh$_3$	15.4
HCpW(CO)$_3$	16.1
HCp*Mo(CO)$_3$	17.1
H$_2$Ru(CO)$_4$	18.7
HCpFe(CO)$_2$	19.4
HCpRu(CO)$_2$	20.2
H$_2$Os(CO)$_4$	20.8
HRe(CO)$_5$	21.1
HCp*Fe(CO)$_2$	26.3
HCpW(CO)$_2$(PMe$_3$)	26.6

Source: E.J. Moore, J.M. Sullivan, and J.R. Norton, *J. Am. Chem. Soc.* **108**, 2257 (1986). Cp* is the pentamethylcyclopentadienyl ligand, C$_5$Me$_5$.

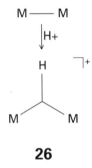

26

(f) Oxidation to form metal carbonyl halides

Metal carbonyls are susceptible to oxidation by air. Although uncontrolled oxidation produces the metal oxide and CO or CO$_2$, of more interest in organometallic chemistry are the controlled reactions that give rise to organometallic halides. One of the simplest of these is the oxidative cleavage of an M—M bond:

$$(OC)_5Mn^{(0)}—Mn^{(0)}(CO)_5 + Br_2 \longrightarrow 2\,Mn^{(+1)}Br(CO)_5$$

In keeping with the loss of electron density from the metal when a halogen atom is attached, the CO stretching frequencies of the product are significantly higher than those of $[Mn_2(CO)_{10}]$.

| Metal carbonyls are susceptible to oxidation by air; the metal–metal bond undergoes
oxidative cleavage.

(g) Reactions of the CO ligand

The C atom of CO is susceptible to attack by nucleophiles if it is attached to a metal atom that is not electron-rich. Thus, terminal carbonyls with high CO stretching frequencies are liable to attack by nucleophiles. The *d* electrons in these neutral or cationic metal carbonyls are not extensively delocalized on to the carbonyl C atom, so that atom can be attacked by electron-rich reagents. For example, strong nucleophiles (such as methyllithium, Section 15.3) attack the CO in many neutral metal carbonyl compounds:

$$\tfrac{1}{4}Li_4(CH_3)_4 + Mo(CO)_6 \longrightarrow Li[Mo(COCH_3)(CO)_5]$$

The resulting anionic carbene reacts with carbocation reagents to produce a stable and easily handled neutral product:

$$Li[Mo(\overset{\overset{\textstyle O}{\textstyle \|}}{C}Me)(CO)_5] + [Et_3O][BF_4] \longrightarrow Mo(=\overset{\overset{\textstyle OEt}{\textstyle |}}{C}Me)(CO)_5 + LiBF_4 + Et_2O$$

Compounds of this type, with a direct **M=C** bond, are often called **Fischer carbenes** because they were discovered in E.O. Fischer's laboratory. The attack of a nucleophile on the C atom is also important for the mechanism of the hydroxide-induced dissociation of metal carbonyls:

$$[(OC)_nM(CO)] + OH^- \longrightarrow [(OC)_nM(COOH)]^-$$

$$\xrightarrow{3\,OH^-} [M(CO)_n]^{2-} + CO_3^{2-} + 2\,H_2O$$

In electron-rich metal carbonyls, considerable electron density is delocalized on to the CO ligand. As a result, in some cases the O atom of a CO ligand is susceptible to attack by electrophiles. Once again, IR data give an indication of when this type of reaction should be expected, for a low CO stretching frequency indicates significant back-donation to the CO ligand, and hence appreciable electron density on the O atom. Thus a bridging carbonyl is particularly susceptible to attack at the O atom:

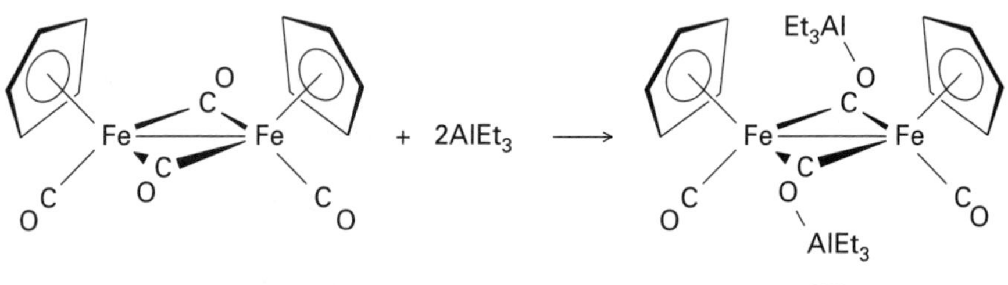

The attachment of an electrophile to the oxygen of a CO ligand, as in (**27**), promotes migratory insertion reactions and C—O cleavage reactions.[10]

Some alkyl-substituted metal carbonyls undergo a **migratory insertion** reaction in which an alkyl ligand and a CO ligand are converted into an acyl ligand, —(CO)**R**. As discussed in Section 14.11, there is good evidence that the reaction of [Mn(CH₃)(CO)₅] occurs by initial migration of the CH₃ group to CO, producing a low concentration of the solvent-coordinated

[10] These and other reactions of C- and O-bonded metal carbonyls are described by C.P. Horwitz and D.F. Shriver, *Adv. Organometal. Chem.* **23**, 219 (1984).

acyl compound. That intermediate then takes on a ligand to produce the nonlabile acyl product:

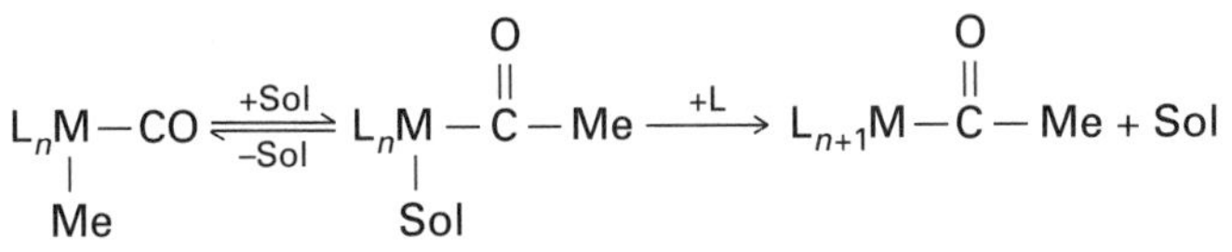

$$L_nM-CO \underset{-Sol}{\overset{+Sol}{\rightleftharpoons}} \underset{\substack{|\\Sol}}{L_nM} \overset{\substack{O\\||}}{-C}-Me \overset{+L}{\longrightarrow} L_{n+1}M \overset{\substack{O\\||}}{-C}-Me + Sol$$

(with Me below the first L_nM)

We shall see the importance of this reaction in catalytic reactions in Sections 17.3 and 17.4.

The C atom of CO is susceptible to attack by nucleophiles if it is attached to a metal atom that is not electron-rich; the O atom of CO is susceptible to attack by electrophiles in electron-rich carbonyls.

Example 16.5 Converting CO to carbene and acyl ligands

Propose a set of reactions for the formation of $W(CO)_5(C(OCH_3)Ph)$, starting with hexacarbonyltungsten(0) and other reagents of your choice.

Answer As with its neighbor in Group 6, the CO ligands in hexacarbonyltungsten(0) are susceptible to attack by nucleophiles, and the reaction with phenyllithium should give a C-phenyl intermediate:

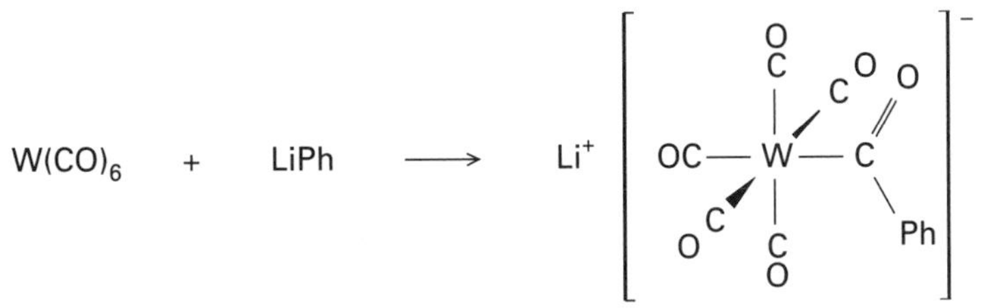

The anion can then be alkylated on the O atom of the CO ligand by an electrophile:

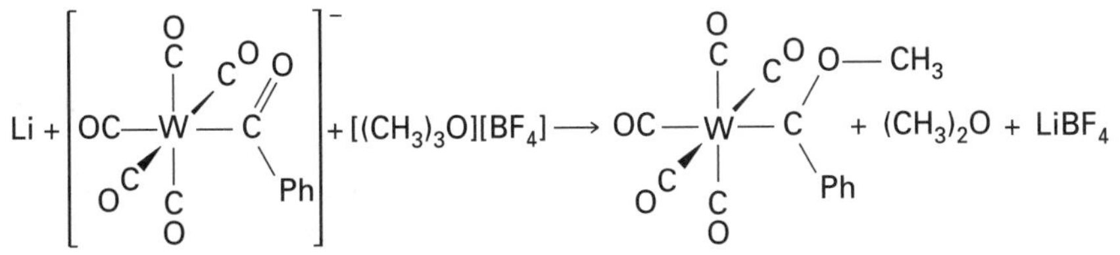

Self-test 16.5 Propose a synthesis for $[Mn(CO)_4(PPh_3)(COCH_3)]$ starting with $[Mn_2(CO)_{10}]$, PPh_3, Na, and CH_3I.

Other organometallic compounds   •

A large number of *d*-block complexes containing hydrocarbon ligands have been discovered and investigated since the mid-1950s. We shall discuss the basic structures, bonding, and reactions here and continue the discussion of reactions in Chapter 17 in the context of catalysis.

16.7 Hydrogen and open-chain hydrocarbon ligands

It will be useful to keep an eye on Table 16.1, which summarized the number of electrons that an organic ligand may donate to a metal, as we work through the following examples.

(a) Hydrogen

The one-electron ligand H is intimately involved in organometallic chemistry. We have already seen how an M—H bond can be produced by protonation of neutral and anionic metal carbonyls (Section 16.6); in some cases a similar reaction gives a stable product with other organometallic compounds. For example, ferrocene can be protonated in strong acid to produce an Fe—H bond. As a consequence of the way in which negative oxidation states are assigned to nonmetal ligands, such as H or halides, this protonation reaction leads to a change in oxidation state of the metal:

$$Cp_2Fe^{(+2)} + HBF_4 \longrightarrow [Cp_2Fe^{(+4)}-H]^+[BF_4]^-$$

One of the most interesting and important methods for introducing hydrogen into a metal complex is by oxidative addition (Section 8.3). In this reaction an XY molecule adds to a 16-electron complex $[ML_4]$ with cleavage of the $X-Y$ bond to give $[M(X)(Y)L_4]$. Thus, the addition of either H_2 or HX to a 16-electron complex may produce an 18-electron hydrido complex:

$$IrCl(CO)(PPh_3)_2 + H_2 \rightarrow IrCl(H)_2(CO)(PPh_3)_2$$
$$\uparrow \qquad\qquad\qquad\qquad \uparrow$$
16e Ir(I)   18e Ir(III)
$$\downarrow \qquad\qquad\qquad\qquad \downarrow$$
$$IrCl(CO)(PPh_3)_2 + HCl \rightarrow IrCl_2(H)(CO)(PPh_3)_2$$

The upper reaction may proceed through a dihydrogen complex in which η^2-H_2 forms transiently.

> An **M**—H *bond can be produced by protonation of neutral and anionic metal carbonyls and by oxidative addition.*

(b) Alkyl ligands

Table 16.8 Bond dissociation enthalpies

	$B/(\text{kJ mol}^{-1})$
$(OC)_5Mn-CH_2Ph$	87
$(OC)_5Mn-Mn(CO)_5$	94
$(OC)_5Mn-CH_3$	153
$(OC)_5Mn-H$	213

An alkyl ligand forms an M—C single bond, and in doing so the alkyl group acts as a one-electron monohapto ligand. Many such compounds are known, but they are less common in the *d*-block than in the *s*- and *p*-blocks. This difference may in part be a result of the modest M—C bond strengths (Table 16.8). Of greater importance are the kinetically facile reactions, such as β-hydrogen elimination, CO insertion, and reductive elimination, which lead to the transformation of the alkyl ligand into other groups.

As we saw in Section 15.6, in a β-hydrogen elimination reaction, an H atom on the β-C atom of an alkyl group is transferred to the metal atom and an alkene is eliminated

$$[L_nM(CH_2CH_3)] \longrightarrow [L_nMH] + H_2C{=}CH_2$$

We discuss the reverse of this reaction, alkene insertion into the M—H bond, in Chapter 17. Both reactions are thought to proceed through a cyclic intermediate involving a 3c,2e M—H—C bond, which is called an **agostic interaction**:[11]

---

[11] These agostic interactions have been identified spectroscopically and by diffraction, and they are often invoked in organometallic mechanisms, see: M. Brookhart and M.L.H. Green, *J. Organometal. Chem.* **205**, 395 (1983) and A.J. Schulz, *et al., Science* **220**, 197 (1983).

$$M—CH_2CH_3 \longrightarrow M \overset{H}{\underset{CH_2}{\diamondsuit}} CH_2 \longrightarrow M—H + H_2C=CH_2$$

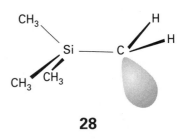

28

Because β-hydrogen elimination is blocked if there are no H atoms on the β-C atom (or if there is no β-C atom), benzyl ($CH_2(C_6H_5)$) and trimethylsilylmethyl ($CH_2Si(CH_3)_3$, (**28**)) ligands are much more robust than ethyl ligands. Similarly, the lack of β-H atoms on the methyl group accounts for the greater stability of complexes containing methyl ligands rather than ethyl ligands. The β-hydrogen elimination reaction is also seen with some p-block organometallic compounds, most notably alkylaluminum compounds (Section 15.11).

> Metal–alkyl compounds undergo β-hydrogen elimination, CO insertion, and reductive elimination, which lead to the transformation of the alkyl ligand into other groups.

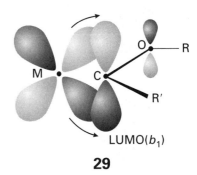

29

(c) Alkylidene ligands

The methylidene groups, CH_2, CHR, or CR_2, are monohapto, two-electron ligands that form M=C double bonds, such as the Fischer carbenes (**29**), (**30**). Fischer carbenes are attacked at the C atom by nucleophiles. For example, the attack of an amine on the electrophilic C atom of a Fischer carbene results in the displacement of the —OR group to yield a new carbene ligand:[12]

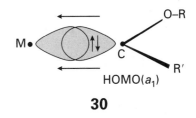

30

$$(OC)_5Cr=C\overset{OMe}{\underset{Ph}{<}} + :NHR_2 \longrightarrow (OC)_5Cr—\overset{OMe}{\underset{Ph}{\overset{|}{C}}}—NHR_2 \longrightarrow (OC)_5Cr=C\overset{NR_2}{\underset{Ph}{<}} + MeOH$$

This type of reaction is useful synthetically for the preparation of a wide range of Fischer carbenes, which are formed by middle to late d-block metals. The electrophilicity of the C atom bound to the metal atom in these carbenes is attributed to the significant electronegativity of the middle to late d-block metals. The corresponding molecular orbital explanation, which is illustrated in Fig. 16.9, is that in the late d-block the metal $d\pi$ orbitals are at lower energy than the carbon p orbitals. As a result, the electron density of the lower energy π orbital is largely on the metal atom, where it is subject to attack by electrophiles. Conversely, the vacant π orbital is mainly located on C, which is therefore susceptible to attack by nucleophiles.

The reactions of Fischer carbene complexes with alkynes have considerable utility in organic synthesis.[13] For example, naphthyl compounds can be synthesized by the reaction of methoxy phenyl Fischer carbenes with an alkyne:

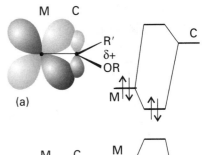

(a)

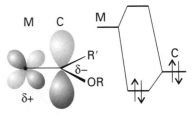

(b)

16.9 In the Fischer carbene (a), the vacant p_z orbital of CR'(OR) is higher in energy than the metal $d\pi$ orbitals, so the π-electron density is largely concentrated on the metal atom and the carbene C is electrophilic. For the Schrock carbene (b), the p_z orbital of CR_2 is lower in energy than the $d\pi$ orbitals, so π-electron density is concentrated in the carbon and the carbene C is nucleophilic.

$$(OC)_5Cr=C\overset{OMe}{\underset{\bigcirc}{<}} + RC\equiv CR' \longrightarrow \text{[naphthyl product with OH, R, R', OMe, }(OC)_3Cr\text{]} + CO$$

[12] E.O. Fischer, *Adv. Organometal. Chem.* **14**, 1 (1976).

[13] K.H. Dotz, *Angew. Chem. Int. Ed. Engl.* **23**, 587 (1984).

The ten C atoms in the naphthalene rings are contributed by a CO ligand (one atom), the carbene ligand (seven atoms), and the ethyne reagent (two atoms). The chromium can be removed from the organic product by mild oxidation.

The metals on the left of the *d*-block form another series of methylidene compounds that complement the Fischer carbenes formed largely by the metals toward the right of the block. These compounds are called **Schrock carbenes** after their discoverer.[14] The chemistry and spectroscopy of the Schrock carbenes indicate that the $M{=}C$ bond is polarized so as to put negative charge on the C atom bound to the metal. This polarization is consistent with the low electronegativity of the metal atoms on the left of the *d*-block. The chemical evidence for the higher electron density on that atom is the reaction of the C atom bound to the metal atom with electrophiles such as Me_3SiBr:

$$Cp_2Ti(CH_2)Me + Me_3SiBr \longrightarrow [Cp_2Ti(CH_2SiMe_3)Me]^+ + Br^-$$

An interesting reaction of the Schrock carbenes is the **alkene metathesis reaction**:

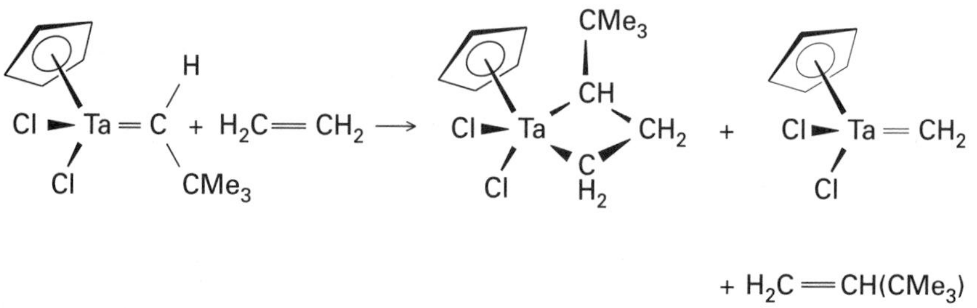

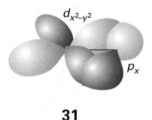

31

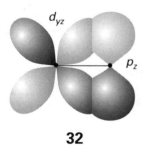

32

$$+ \; H_2C = CH(CMe_3)$$

This reaction proceeds through a four-membered ring that includes the metal atom.

> The Fischer carbenes are formed by middle to late *d*-block metals and Schrock carbenes are formed by metals on the left of the *d*-block.

(d) Alkylidyne ligands

The alkylidyne ligands have the general formula CH or CR.[15] They are monohapto three-electron ligands and are bound to the metal by an $M{\equiv}C$ triple bond involving one σ bond and two (d,p)-overlap π bonds (**31**), (**32**). The simplest member of this series is methylidyne, CH; the next simplest is ethylidyne, CCH_3. There are several routes to these compounds, and we have already described their preparation by the cleavage of $M{\equiv}M$ bonds (Section 9.9). The first synthesis of a metal alkylidyne involved the abstraction of an alkoxide group from a Fischer carbene by BBr_3:

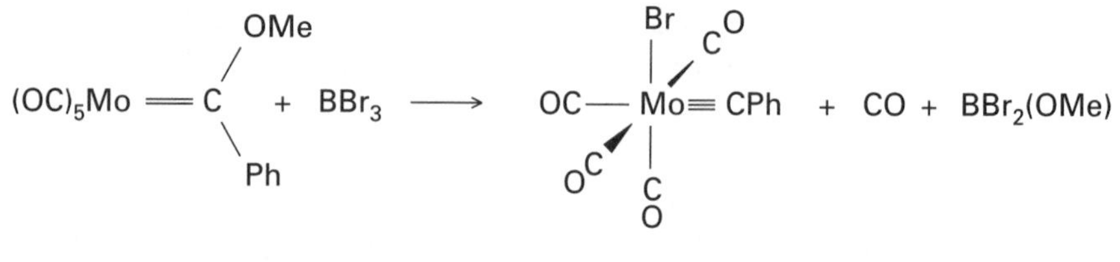

$$(OC)_5Mo = C \;\; \binom{OMe}{Ph} + BBr_3 \longrightarrow OC{-}Mo{\equiv}CPh + CO + BBr_2(OMe)$$

This reaction was originally carried out in an attempt to convert the carbene ligand to its bromo analog, $M{=}CBrPh$. As with many exploratory synthetic projects, the outcome was different from that intended, and in this case much more interesting.

> The alkylidyne ligands CH or CR are monohapto three-electron ligands bound to the metal by an $M{\equiv}C$ triple bond.

[14] Alkylidene complexes of niobium and tantalum; R.R. Schrock, *Acc. Chem. Res.* **12**, 98 (1984).

[15] H. Fischer, P. Hoffmann, F.R. Kreissl, R.R. Schrock, U. Schubert, and K. Weiss, *Carbyne complexes.* VCH, Weinheim (1988).

(e) Alkene ligands

A C=C π bond can act as a source of electron pairs in complex formation, as in Zeise's salt (**1**). This salt is prepared by bubbling ethene through an aqueous solution of tetrachloroplatinate(II) ions in the presence of Sn(II), which aids the removal of Cl$^-$ ions from the Pt(II) coordination spheres:

$$K_2[PtCl_4] + H_2C{=}CH_2 \xrightarrow{\text{SnCl}_2} K[PtCl_3(\eta^2\text{-}C_2H_4)] + KCl$$

Another route to alkene complexes is the removal of H$^-$ from a coordinated alkyl ligand:

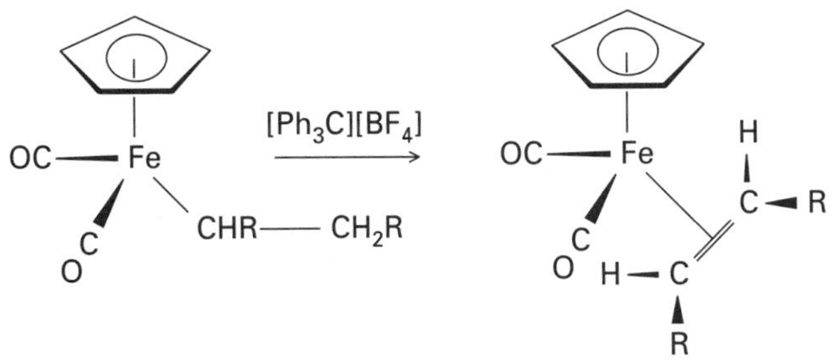

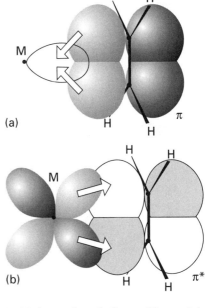

(a)

(b)

16.10 Interaction of ethene with a metal atom. (a) Donation of electron density from the filled π molecular orbital of ethene to a vacant metal σ orbital. (b) Acceptance of electron density from a filled $d\pi$ orbital into the vacant π^* orbital of ethene.

Simple alkene ligands are dihapto two-electron donors because the filled π orbital projects toward the metal and donates electrons to suitably oriented metal orbitals (Fig. 16.10a). In addition, the π^* orbital of the ligand can accept electron density from filled metal orbitals of the appropriate symmetry (Fig. 16.10b). This description of metal–alkene bonding in terms of electron donation from the filled bonding π orbital of the ligand and simultaneous electron acceptance into the empty antibonding π^* orbital is called the **Dewar–Chatt–Duncanson model**.

Electron donor and acceptor character appear to be fairly evenly balanced in most ethene complexes of the d metals, but the degree of donation and back-donation can be altered by substituents. An extreme example is tetracyanoethene (**33**), which is an abnormally strong electron acceptor ligand on account of its electron-withdrawing CN groups. Tetracyanoethene qualifies as a π-acceptor ligand because its role as an acceptor dominates its role as a donor. When the degree of donation of electron density from the metal atom to the alkene ligand is small, substituents on the ligand are bent only slightly away from the metal,:and the C—C bond length is only slightly greater than in the free alkene. With electron-rich metals or electron-withdrawing substituents on the alkene, the back-donation is greater. Substituents on the alkene are then bent away from the metal, and the C—C bond length approaches that of a single C—C bond. These differences lead some chemists to depict the first group as simple π complexes (**34**) and the second as metallocycles (**35**) with M—C single bonds.

Electron donor and acceptor character are evenly balanced in most ethene complexes of the d metals, but the degree of donation and back-donation can be altered by substituents on the alkene and the oxidation state of the metal.

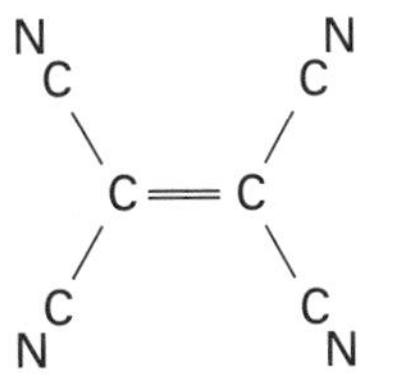

33 TCNE

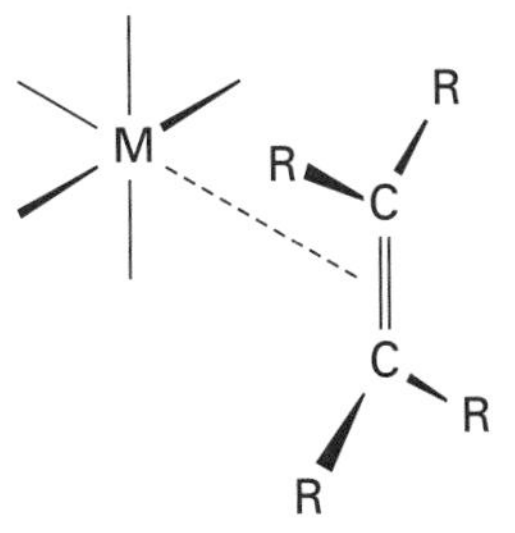

34

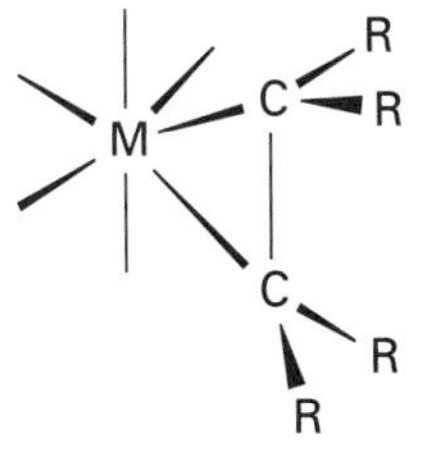

35

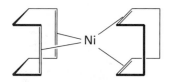

36 Ni(cod)$_2$

37 Fe(η^1–C$_3$H$_5$)(CO)$_2$(η^5–Cp)

38 Co(η^3–C$_3$H$_5$)(CO)$_3$

39 anti

40 syn

(f) Diene and polyene ligands

Diene (—C=C—C=C—) and polyene ligands present the possibility of polyhapto bonding. As with the chelate effect in Werner complexes (Section 7.7), the resulting polyene complexes are usually more stable than the equivalent complex with individual ligands because the entropy of dissociation of the complex is much smaller than when the liberated ligands can move independently. For example, bis(η^4-cycloocta-1,5-diene)nickel(0) (**36**) is more stable than the corresponding complex containing four ethene ligands. Cycloocta-1,5-diene is a fairly common ligand in organometallic chemistry, where it is referred to engagingly as 'cod', and is introduced into the metal coordination sphere by simple ligand displacement reactions. An example is:

Metal complexes of cod are often used as starting materials because they often have intermediate stability. Many of them are sufficiently stable to be isolated and handled, but cod can be displaced by many stronger ligands. For example, if the highly toxic Ni(CO)$_4$ molecule is needed in a reaction, then it may be generated from Ni(cod)$_2$ directly in the reaction flask:

$$[\text{Ni(cod)}_2](\text{soln}) + 4\,\text{CO(g)} \longrightarrow \text{Ni(CO)}_4(\text{soln}) + 2\,\text{cod(soln)}$$

> *Metal complexes of cycloocta-1,5-diene are often used as starting materials in syntheses; many of them can be isolated, but the ligand can be displaced by stronger ligands.*

(g) The π-allyl ligand

The allyl ligand can bind to a metal atom in either of two configurations. As an η^1-ligand (**37**) it is a one-electron donor; as an η^3-ligand (**38**) it is a three-electron donor. In the latter case the terminal substituents are bent slightly out of the plane of the three-carbon backbone and are either anti (**39**) or syn (**40**) relative to the metal. It is common to observe anti and syn group exchange, which in some cases is fast on an NMR timescale. A mechanism that involves the transformation $\eta^3 \rightarrow \eta^1 \rightarrow \eta^3$ is often invoked to explain this exchange. Because of this flexibility in type of bonding, η^3-allyl complexes are often highly reactive.

There are many routes to allyl complexes. One is the nucleophilic attack of an allyl Grignard reagent on a metal halide:

$$2\,\text{C}_3\text{H}_5\text{MgBr} + \text{NiCl}_2 \xrightarrow{\text{Ether}} [\text{Ni}(\eta^3\text{-C}_3\text{H}_3)_2] + 2\,\text{MgBrCl}$$

Conversely, electrophilic attack of a haloalkane on a metal atom in a low oxidation state yields allyl complexes:

If the metal center is not basic, the protonation of a butadiene complex leads to a π-allyl complex:

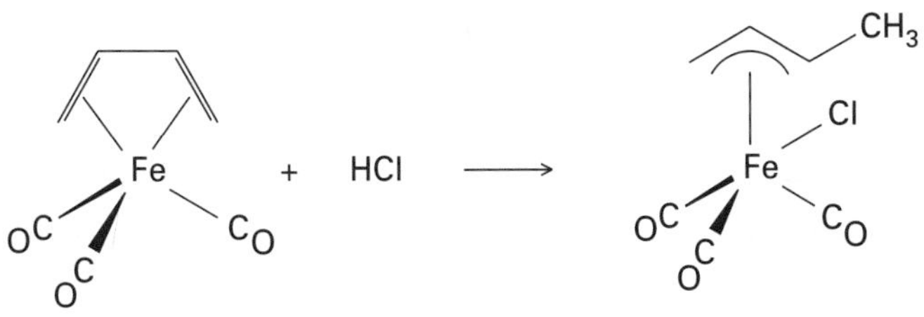

+ HCl ⟶

Because of the variability of the allyl ligand in its type of bonding, η^3-allyl complexes are often highly reactive.

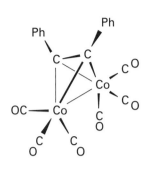

41 $Co_2(PhC_2Ph)(CO)_6$

(h) Alkyne ligands

Ethyne (acetylene) has two π bonds and hence is a potential four-electron donor. However, it is not always clear from experimental information that it acts in this way. Substituted ethynes form very stable polymetallic complexes in which the ethyne can be regarded as a four-electron donor. An example is η^2-diphenylethyne(hexacarbonyl)dicobalt(0) (**41**), in which we can view one π bond as donating to one of the Co atoms and the second π bond as overlapping with the other Co atom (**42**), (**43**). As in this example, the alkyl or aryl groups present on the ethyne impart stability. They do so by reducing the tendency toward secondary reactions of the coordinated ethyne, such as loss of the slightly acidic acetylenic H atom to the metal.

Substituted ethynes form very stable polymetallic complexes in which the ethyne can be regarded as a four-electron donor.

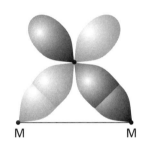

42 View along CC

16.8 Cyclic polyene complexes

Cyclic polyene ligands are among the most important in organometallic chemistry. In this section we shall consider cyclic polyene ligands ranging from cyclobutadiene to cyclooctatetraene.[16] These ligands are capable of forming the metallocenes ferrocene (**4**), bis(benzene)chromium(0) (**44**), and uranocene (**45**). A **metallocene** consists of a metal atom between two planar polyhapto rings (as in ferrocene) and they are informally called 'sandwich compounds'. Some cyclic polyenes are known to form complexes in which they are bound to a metal atom through some but not all of their C atoms. In these cases, the ring is nonplanar. Besides these homoleptic cyclopolyene complexes there are many other organometallic compounds in which other ligands share the coordination sphere of a metal with a cyclic polyene.

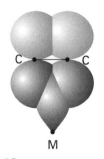

43 View along MM

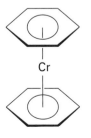

44 $Cr(\eta^6-C_6H_6)_2$

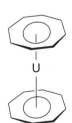

45 $U(\eta^8-C_8H_8)_2$

[16] The cyclopropenyl cation is the simplest member in the series; some η^3-cyclopropenyl complexes are known, but are not common.

46 Ru(C$_4$H$_4$)(CO)$_3$

(a) Metallo cyclobutadiene compounds

The simplest common ligand of the group, cyclobutadiene, is a four-electron donor. It is unstable as the free molecule, but stable complexes are known, including [Ru(CO)$_3$(η^4-C$_4$H$_4$)] (**46**). This species is one of many in which coordination to a metal atom stabilizes an otherwise unstable molecule. Because of the instability of cyclobutadiene, the ligand must be generated in the presence of the metal to which it is coordinated. This synthesis can be accomplished in a variety of ways. One method is the dehalogenation of a halogenated cyclobutene:

Another synthesis is the dimerization of a substituted ethyne:

Cyclobutadiene is a four-electron donor that forms some stable complexes.

(b) Metallo cyclopentadienyl compounds

The cyclopentadienyl ligand C$_5$H$_5$ has played a major role in the development of organometallic chemistry and continues to be the archetype of cyclic polyene ligands. A huge number of metal cyclopentadienyl compounds are known. Some have C$_5$H$_5$ as a monohapto ligand, in which case it contributes one electron to a σ bond to the metal (**7**); others contain C$_5$H$_5$ as a trihapto ligand, when it donates three electrons. Usually, though, C$_5$H$_5$ is present as a pentahapto ligand contributing five electrons. Structure (**8**) shows a compound that contains both η^3 and η^5 ligands.

Sodium cyclopentadienide is a common starting material for the preparation of cyclopentadienyl compounds. It can be prepared by the action of Na on cyclopentadiene in tetrahydrofuran solution:

$$2\,\text{Na} + 2\,\text{C}_5\text{H}_6 \xrightarrow{\text{THF}} 2\,\text{Na}[\text{C}_5\text{H}_5] + \text{H}_2$$

Sodium cyclopentadienide reacts with *d*-metal halides to produce metallocenes:

$$2\text{Na}[\text{C}_5\text{H}_5] + \text{MnCl}_2 \xrightarrow{\text{THF}} \quad\text{Mn} + 2\text{NaCl}$$

The overall result is simple ligand displacement, but the mechanism may be more complicated than this suggests. The bis(cyclopentadienyl) complexes of iron, cobalt, and nickel are also readily prepared by this method.

> Many metal cyclopentadienyl compounds are known. Some have C_5H_5 as a monohapto ligand, others contain C_5H_5 as a trihapto ligand, but usually C_5H_5 is present as a pentahapto ligand contributing five electrons.

(c) Reactions

Because of their great stability, the 18-electron Group 8 compounds ferrocene, ruthenocene, and osmocene maintain their ligand–metal bonds under rather harsh conditions, and it is possible to carry out a variety of transformations on the cyclopentadienyl ligands. For example, they undergo reactions similar to those of simple aromatic hydrocarbons, such as Friedel–Crafts acylation:

$$CH_3COCl \; + \; Fe(\eta^5\text{–}C_5H_5)_2 \; \xrightarrow{\;AlCl_3\;}$$

It also is possible to replace H on a C_5H_5 ring by Li:

$$LiBu \; + \; Fe(\eta^5\text{–}C_5H_5)_2 \; \longrightarrow \qquad + \; C_4H_{10}$$

As might be imagined, the lithiated product is an excellent starting material for the synthesis of a wide variety of ring-substituted products and in this respect resembles simple organolithium compounds (Section 15.7).

> Because of their great stability, the 18-electron Group 8 metallocenes maintain their ligand–metal bonds under rather harsh conditions, and it is possible to carry out a variety of transformations on the cyclopentadienyl ligands.

(d) Related metallocenes

There are many interesting related structures in addition to the simple bis(cyclopentadienyl) and bis(arene) complexes. In the jargon of this area, these species are referred to as 'bent sandwich compounds' (Table 16.9), 'half-sandwich' or 'piano stool' compounds (**6**), and, inevitably, 'triple deckers' (**47**). We have already met triple deckers and multidecker sandwich compounds that incorporate $[B_3C_2H_5]^{4-}$, which is isoelectronic with $[C_5H_5]^-$ but has a greater tendency to form stacked sandwich compounds than the cyclopentadienyl ligand.

Bent sandwich compounds play a major role in the organometallic chemistry of the early and middle d-block elements, and we have already encountered one example, the Schrock carbene $[Ta(\eta^5\text{-}Cp)_2(CH_3)(CH_2)]$ (**48**). Other examples include $[TiCl_2(\eta^5\text{-}Cp)_2]$, $[Re(\eta^5\text{-}Cp)_2(Cl)]$, $[W(\eta^5\text{-}Cp)_2(H)_2]$, and $[Nb(\eta^5\text{-}Cp)_2(Cl)_3]$.

As shown in Table 16.9, bent sandwich compounds occur with a variety of electron counts and stereochemistries. Their structures can be systematized in terms of a model in which three metal atom orbitals project toward the open face of the bent Cp_2M moiety. According to this

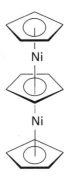

47 $[Ni_2(C_5H_5)_3]^+$

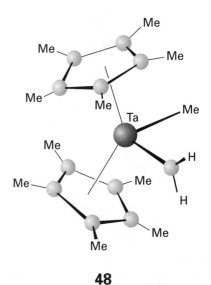

48

Table 16.9 Bent-sandwich compounds

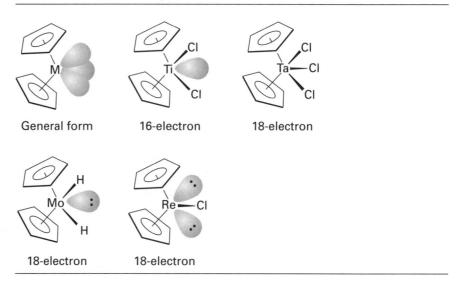

General form   16-electron   18-electron

18-electron   18-electron

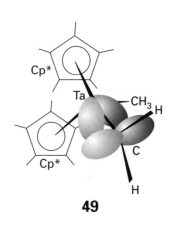

49

model, the metal atom often satisfies its electron deficiency when the electron count is less than 18 by interaction with lone pairs or C—H groups on the ligands. For example, this model explains why the sterically disfavored orientation of CH_2 exists in the Schrock carbene, because by adopting this conformation the filled $p\pi$ orbital on CH_2 donates electron density to an orbital on the Ta atom (**49**).

> The structures of bent sandwich compounds can be systematized in terms of a model in which three metal atom orbitals project toward the open face of the bent Cp_2M moiety.

(e) Bonding in ferrocene

Although details of the bonding in ferrocene are not settled, the molecular orbital energy level diagram shown in Fig. 16.11 accounts for a number of experimental observations. This diagram refers to the eclipsed (D_{5h}) form of the complex, which in the gas phase is about 4 kJ mol^{-1} lower in energy than the staggered conformation; almost all metallocenes have low staggered–eclipsed conversion barriers. We shall focus our attention on the frontier orbitals. As shown in Fig. 16.12, the e_1'' symmetry-adapted linear combinations of ligand orbitals have the same symmetry as the d_{zx} and d_{yz} orbitals of the metal atom. The lower energy frontier orbital (a_1') is composed of d_{z^2} and the corresponding SALC of ligand orbitals. However, there is little interaction between the ligands and the metal orbitals because the ligand p orbitals happen to lie—by accident—in the conical nodal surface of the metal's d_{z^2} orbital. In ferrocene and the other 18-electron bis(cyclopentadienyl) complexes, the a_1' frontier orbital and all lower orbitals are full but the e_1'' frontier orbital and all higher orbitals are empty.

The frontier orbitals are neither strongly bonding nor strongly antibonding. This characteristic permits the possibility of the existence of metallocenes that diverge from the 18-electron rule, such as the 17-electron complex [Fe(η^5-Cp)$_2$]$^+$ and the 20-electron complex [Ni(η^5-Cp)$_2$]. Deviations from the 18-electron rule, however, do lead to significant changes in M—C bond lengths that correlate quite well with the molecular orbital scheme (Table 16.10). Similarly, the redox properties of the complexes can be understood in terms of the electronic structure. Thus, we have already mentioned that ferrocene is fairly readily oxidized to the ferrocinium ion, [Fe(η^5-Cp)$_2$]$^+$. From an orbital viewpoint, this oxidation corresponds to the removal of an electron from the nonbonding a_1' orbital. The 19-electron complex [Co(η^5-Cp)$_2$] is much more readily oxidized than ferrocene because the electron is lost from the antibonding e_1'' orbital to give the 18-electron [Co(η^5-Cp)$_2$]$^+$ ion.

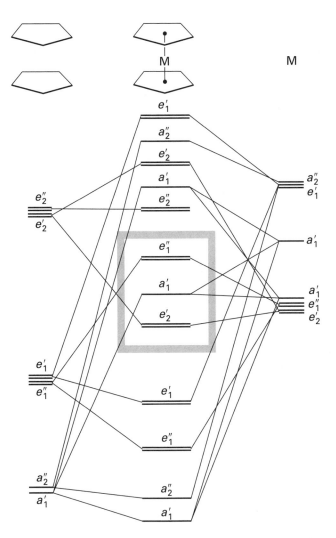

16.11 Molecular orbital energy diagram of a metallocene on the assumption of D_{5h} symmetry. The energies of the symmetry-adapted π orbitals of the C_5H_5 ligands are shown on the left, relevant d orbitals of the metal are on the right, and the resulting molecular orbital energies are in the center. Eighteen electrons can be accommodated by filling the molecular orbitals up to and including the a'_1 orbital in the box. The box denotes the orbitals typically regarded as frontier orbitals in these molecules.

Another useful comparison can be made with octahedral complexes. The e''_1 frontier orbital of a metallocene is the analog of the e_g orbital in an octahedral complex, and the a' orbital plus the e'_2 pair of orbitals are analogous to the t_{2g} orbitals of an octahedral complex. This formal similarity extends to the existence of high- and low-spin bis(cyclopentadienyl)

Table 16.10 Electronic configuration and M—C bond length in $M(\eta^5\text{-}C_5H_5)_2$ complexes

Complex	Valence electrons	Electron configuration	$R(\text{M—C})/\text{Å}$
$V(C_5H_5)_2$	15	$e'^2_2 a'^1_1$	2.28
$Cr(C_5H_5)_2$	16	$e'^3_2 a'^1_1$	2.17
$Mn(C_5H_4CH_3)^*_2$	17	$e'^3_2 a'^2_1$	2.11
$Fe(C_5H_5)_2$	18	$e'^4_2 a'^2_1$	2.06
$Co(C_5H_5)_2$	19	$e'^4_2 e''^1_1 a'^2_1$	2.12
$Ni(C_5H_5)_2$	20	$e'^4_2 e''^2_1 a'^2_1$	2.20

*Data are quoted for this complex because $Mn(C_5H_5)_2$ has a high-spin configurations a'^2_1, e'^2_1, e''^2_1 and hence an anomalously long M—C bond length (2.38 Å).

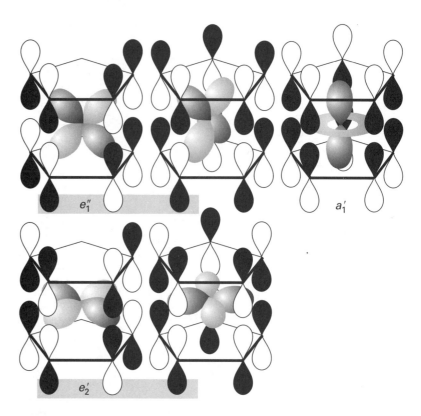

16.12 Symmetry-adapted orbital combinations giving rise to the three types of metallocene frontier orbitals. The accidental coincidence of the nodal surface of d_{z^2} and the C_5H_5 π orbitals results in negligible interaction despite its being symmetry-allowed.

complexes (Table 16.10). The redox chemistry of metallocenes is permeated by similarities to that of simple octahedral complexes.

The bonding scheme in ferrocene is shown in Figs 16.11 and 16.12.

Example 16.6 Identifying metallocene electronic structure and stability

Refer to Fig. 16.11. Discuss the occupancy and nature of the HOMO in $[Co(\eta^5\text{-Cp})_2]^+$ and the change in metal–ligand bonding relative to neutral cobaltocene.

Answer The $[Co(\eta^5\text{-Cp})_2]^+$ ion contains 18 valence electrons (nine from Co, ten from the two Cp ligands, less one for the single positive charge). Assuming that the molecular orbital energy level diagram for ferrocene applies, the 18-electron count leads to double occupancy of the orbitals up through a'_{1g}. The 19-electron cobaltocene molecule has an additional electron in the e''_g orbital, which is antibonding with respect to the metal and ligand. Therefore the metal–ligand bond should be stronger and shorter in $[Co(\eta^5\text{-Cp})_2]^+$ than in $[Co(\eta^5\text{-Cp})_2]$. This is borne out by structural data.

Self-test 16.6 Using the same molecular orbital diagram, comment on whether the removal of an electron from $[Fe(\eta^5\text{-Cp})_2]$ to produce $[Fe(\eta^5\text{-Cp})_2]^+$ should produce a substantial change in M—C bond length relative to neutral ferrocene.

(f) Metallo–arene compounds

Benzene and its derivatives are most often hexahapto six-electron donors (as in (**44**)) in which all six π electrons are shared with the metal. Photochemical or thermal activation of metal

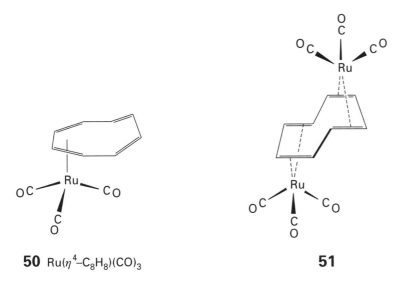

50 $Ru(\eta^4-C_8H_8)(CO)_3$

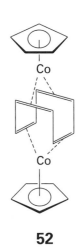

51

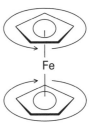

52

carbonyls is often an effective route to introduce neutral cyclic polyene ligands into the metal coordination sphere. For example, hexacarbonylchromium(0) can be refluxed with an arene to produce the arene tricarbonyl:

$$Cr(CO)_6 + C_6H_6 \xrightarrow{\Delta} [Cr(\eta^6-C_6H_6)(CO)_3] + 3\,CO$$

Benzene and its derivatives are most often hexahapto.

(g) Metallo–cyclooctatetraene compounds

Cyclooctatetraene is a large ligand that is found in a wide variety of bonding arrangements. It uses some of its π electrons for bonding to metals when it is an η^2, an η^4 (**50**), or an η^6 ligand, and in these cases the ring is puckered. Once again, photochemistry provides a good route to cyclooctatetraene carbonyl complexes:

$$Fe(CO)_5 + C_8H_8 \xrightarrow{h\nu} [Fe(\eta^4-C_8H_8)(CO)_3] + 2\,CO$$

The bonding versatility of cyclooctratetraene is also shown by its ability to act as a bridge between two metal atoms, (**51**) and (**52**). As already mentioned, when C_8H_8 is a planar η^8 ligand (**45**), it is best regarded as the planar aromatic $C_8H_8^{2-}$ group.

Cyclooctatetraene may be dihapto, tetrahapto, hexahapto, or octahapto as a ligand; in the last case, the ring is planar.

(h) Fluxional cyclic polyene complexes

One of the most remarkable aspects of many cyclic polyene complexes is their stereochemical nonrigidity. For example, at room temperature the two rings in ferrocene rotate rapidly relative to each other (**53**). This type of fluxional process is called **internal rotation**, and is similar to the process in ethane.

Of greater interest is the stereochemical nonrigidity that is often seen when a conjugated cyclic polyene is attached to a metal atom through some but not all of its C atoms. In such complexes the metal–ligand bonding may hop around the ring; in the informal jargon of organometallic chemists this internal rotation is called 'ring whizzing'. A simple example is found in $[Ge(\eta^1-Cp)(CH_3)_3]$, in which the single site of attachment of the Ge atom to the cyclopentadiene ring hops around the ring in a series of **1,2-shifts**, or motion in which a

53

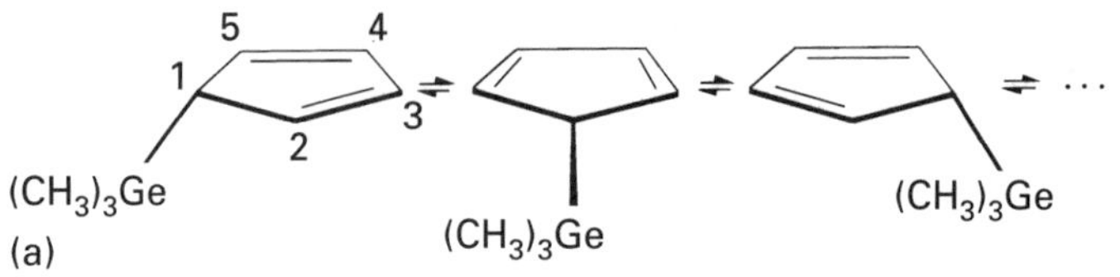

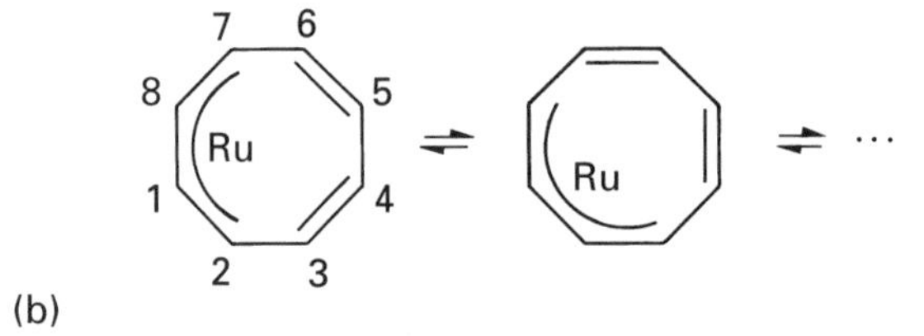

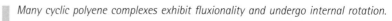

16.13 (a) The fluxional process in $[Ge(\eta^1\text{-}Cp)(CH_3)_3]$ occurs by a series of 1,2-shifts. (b) The fluxionality of $[Ru(\eta^4\text{-}C_8H_8)(CO)_3]$ can be described similarly. We need to imagine that the Ru atom is out of the plane of the page; the CO ligands have been omitted.

C—M bond is replaced by a C—M bond to the next C atom around the ring (Fig. 16.13a). The great majority of fluxional conjugated polyene complexes that have been investigated migrate by 1,2-shifts, but it is not known whether these shifts are controlled by a principle of least motion or by some aspect of orbital symmetry.

NMR provides the primary evidence for the existence and mechanism of these fluxional processes, for they occur on a timescale of 10^{-2} to 10^{-4} s, and can be studied by ^{1}H- and ^{13}C-NMR. The compound $[Ru(\eta^4\text{-}C_8H_8)(CO)_3]$ (**50**) is an illustration of the approach. At room temperature, its ^{1}H-NMR spectrum consists of a single, sharp line that could be interpreted as arising from a symmetrical $\eta^8\text{-}C_8H_8$ ligand (Fig. 16.14). However, X-ray diffraction studies of single crystals show unambiguously that the ligand is tetrahapto. This conflict is resolved by ^{1}H-NMR spectra at lower temperatures, for as the sample is cooled the signal broadens and then separates into four peaks. These peaks are expected for the four pairs of protons of a $\eta^4\text{-}C_8H_8$ ligand (**54**). The interpretation is that at room temperature the ring is whizzing around the metal atom rapidly compared with the timescale of the NMR experiment, so an averaged signal is observed. At lower temperatures the motion of the ring is slower, and the distinct conformations exist long enough to be resolved. A detailed analysis of the line shape of the NMR spectra can be used to measure the activation energy of the migration.[17]

Many cyclic polyene complexes exhibit fluxionality and undergo internal rotation.

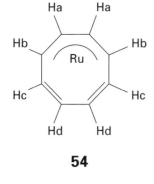

54

16.9 Reactivity of early *d-*block and *f-*block organometallic compounds

There are strong similarities between the chemical properties of the early *d-*block organometallic compounds (those of Groups 3 to 5) and those of the *f-*block.[18] Specifically, early *d-*block organometallic compounds and those of the *f-*block are oxophilic and halophilic, highly reactive toward C—H bond cleavage, and have little tendency to form M—M bonds without bridging ligands. Under normal laboratory conditions, neutral metal carbonyl compounds are unknown for the *f-*block elements and are very rare for Groups 3

[17] L.M. Jackman and F.A. Cotton (ed.), *Dynamic nuclear magnetic resonance spectroscopy.* Academic Press, New York (1975).

[18] For a review of organolanthanide chemistry see W.J. Evans, *Adv. Organometal. Chem.* **24**, 131 (1985). For organoactinides see T.J. Marks and A. Streitwieser, p. 1547, and T.J. Marks, p. 1588, in G. Seaborg, J. Katz, and L.R. Morss, *Chemistry of the actinide elements,* Vol. 2. Chapman & Hall, London (1986).

and 4 in the early *d*-block; one example is $[Ti(CO)_5(dmpe)]$. We have already encountered some of these factors in the discussion of Schrock carbenes. Many of these properties can be rationalized in terms of the hard, electropositive character, low electron count, and limited number of accessible oxidation states that characterize these *d*- and *f*-block elements.

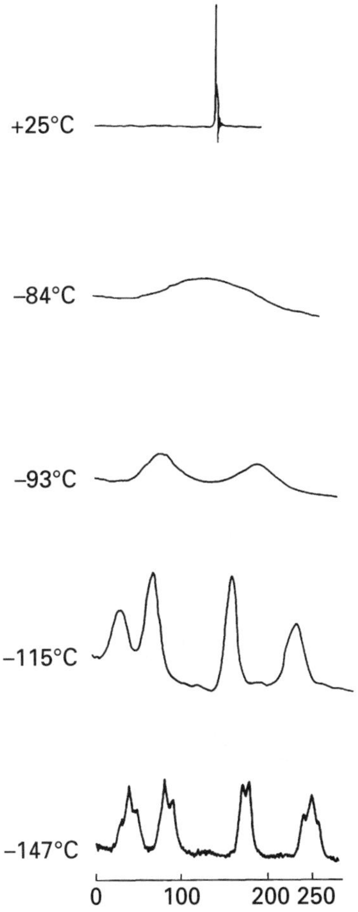

(a) Oxophilicity

That the early *d*- and *f*-block elements are hard is shown by their affinity for hard ligands such as O and Cl. The affinity for O is evident in the properties of the carbonyls of these elements. For example, the reaction of carbon monoxide with $[Zr(\eta^5\text{-Cp})_2(CH_3)_2]$ gives a C- and O-bonded η^2-acetyl (instead of a simple C-bonded η^1-acetyl typical of late *d*-metals):

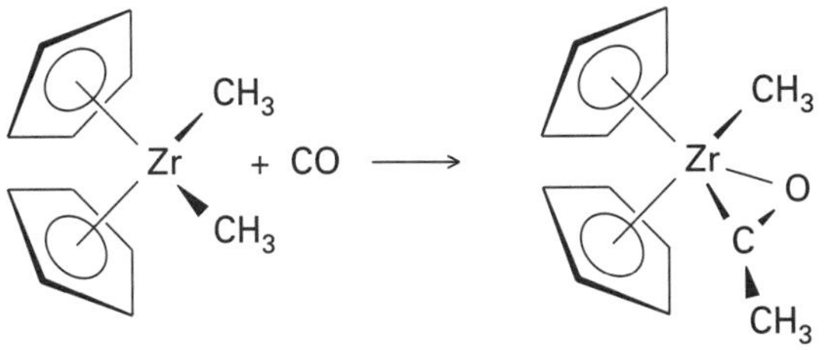

In this reaction, the vacant central orbital on the starting material $[Zr(\eta^5\text{-Cp})_2(CH_3)_2]$ is utilized in the product by the η^2-acetyl ligand. Similarly, the reaction between a zirconium dihydride complex and CO in the presence of H_2 results in reduction and migration of the oxygen into the coordination sphere of the Zr atom:[19]

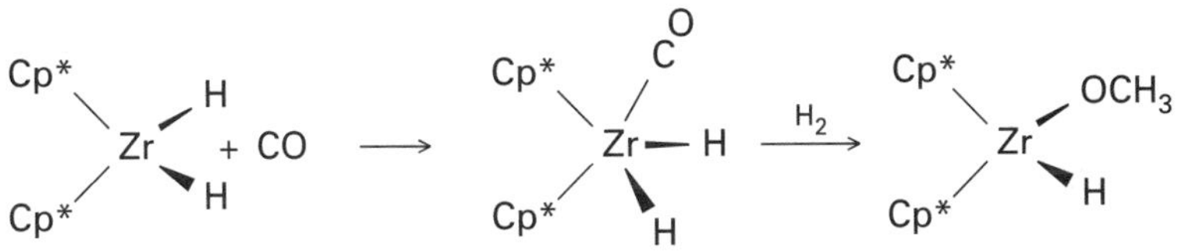

As we explain below, pentamethylcyclopentadienyl confers greater stability than the unsubstituted cyclopentadienyl ligand on early *d*-block and *f*-block organometallic compounds.

> Early *d*-block organometallic compounds and those of the *f*-block are oxophilic.

(b) C—H cleavage

The early *d* metals have a marked tendency to activate C—H bonds. Intermolecular C—H cleavage is widespread for Ru, Rh, and Ir. This tendency is so pronounced, for instance, that attempts to prepare $[Ti(\eta^5\text{-Cp})_2]$ in fact produce the dimer (**55**) in which C—H bonds have oxidatively added to the Ti atom. To suppress this and other unwanted reactions, the pentamethylcyclopentadienyl ligand is often used. This ligand has no H atoms on the ring carbons, so C—H cleavage is avoided. It is also more electron-rich than Cp and therefore forms somewhat stronger metal–ligand bonds. Moreover, the bulk of the Cp* ligand blocks access to the metal and thus hinders associative reactions.

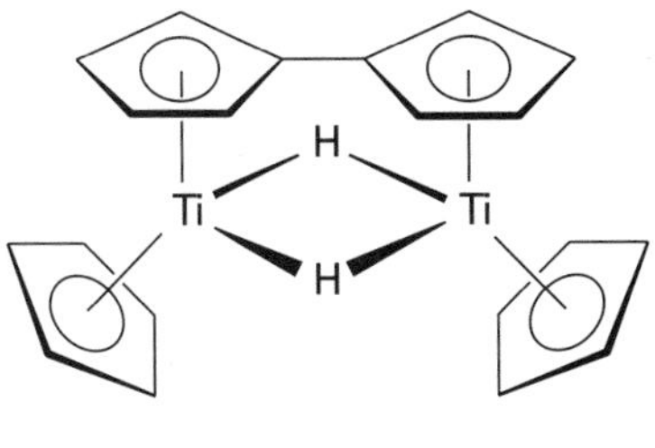

+25°C

−84°C

−93°C

−115°C

−147°C

0   100   200 250

Spectrometer frequency, ν/MHz

16.14 Observed ^{1}H-NMR spectra for $[Ru(\eta^4\text{-C}_8\text{H}_8)(CO)_3]$ at various temperatures. (From F.A. Cotton, In *Dynamic nuclear magnetic resonance spectroscopy* (ed. L.M. Jackman and F.A. Cotton), Chapter 10. Academic Press, New York (1975).)

55

----

[19] Cp* is pentamethylcyclopentadienyl.

A striking example of C—H activation, the activation of C—H bonds in methane by a lanthanide organometallic compound, was discovered by Patricia Watson. The discovery was based on the observation that $^{13}CH_4$ exchanges ^{13}C with the CH_3 group attached to Lu:

$$[LuCp^*_2(CH_3)] + {}^{13}CH_4 \longrightarrow [LuCp^*_2({}^{13}CH_3)] + CH_4$$

This reaction can be carried out in deuterated cyclohexane with no evidence for activation of the cyclohexane C—D bond, presumably because cyclohexane is too bulky to gain access to the metal center. Methane activation has also been observed with organoactinides and late *d*-block complexes. Electrophilic metal centers promote this reaction, and a four-center intermediate has been proposed:

$$M{-}R + R'{-}H \longrightarrow \begin{array}{c} R'{-}{-}{-}H \\ | \quad\quad | \\ M{-}{-}{-}R \end{array}^{\ddagger} \longrightarrow \begin{array}{c} R' \quad H \\ | \quad | \\ M \quad R \end{array}$$

> The early *d*-block metals have a marked tendency to activate C—H bonds; the late *d*-block metals share this ability in the right environment.

Metal–metal bonding and metal clusters

Organic chemists must go to heroic efforts to synthesize their cluster molecules, such as cubane (**56**). In contrast, one of the distinctive characteristics of inorganic chemistry is the large number of closed polyhedral molecules, such as the tetrahedral P_4 molecule (Section 11.1), the octahedral halide-bridged early *d*-block clusters (Section 9.9), the polyhedral carboranes (Section 10.6), and the organometallic cluster compounds that we discuss here. We shall include some non-organometallic compounds in this section so as to permit a fuller discussion of metal–metal multiple bonds.

16.10 Structure

We define **metal clusters** to be molecular complexes with metal–metal bonds that form triangular or larger closed structures. This definition excludes linear M—M—M compounds. It also excludes cage compounds, in which several metal atoms are held together exclusively by ligand bridges.[20] The presence of bridging ligands in a cluster (such as (**57**)), raises the possibility that the atoms are held together by M—L—M interactions rather than M—M bonds. Bond lengths are of some help in resolving this issue. If the M—M distance is much greater than twice the metallic radius, it is reasonable to conclude that the M—M bond is either very weak or absent. However, if the metal atoms are within a reasonable bonding distance, the proportion of the bonding that is attributable to direct M—M interaction is ambiguous. For example, there has been debate in the literature over the extent of Fe—Fe bonding in $[Fe_2(CO)_9]$ (**58**).

Metal–metal bond strengths in metal complexes cannot be determined with great precision, but a variety of evidence–such as the stability of compounds and M—M force constants–indicates that there is an increase in M—M bond strengths down a group in the *d*-block. This trend contrasts with that in the *p*-block where element–element bonds are usually weaker for the heavier members of a group. As a consequence of this trend, metal–metal bonded systems are most numerous for the Period 4 and 5 *d* metals.

56 C_8H_8

57 $Co_4(CO)_{12}$

58 $Fe_2(CO)_9$

---

[20] This definition is sometimes relaxed. Occasionally any M—M bonded system is referred to as a cluster, and quite often chemists do not make the distinction between cage and cluster compounds.

(a) Electron count and structure of clusters

Organometallic cluster compounds are rare for the early d metals and unknown for the f metals, but a large number of metal carbonyl clusters exist for the elements of Groups 6 to 10. The bonding in the smaller clusters can be readily explained in terms of a local **M—M** and **M—L** electron pair bonding and the 18-electron rule, but octahedral M_6 and larger clusters do not conform to this pattern.[21]

A semiempirical correlation of electron count and structure of the larger organometallic clusters was introduced by K. Wade and refined by D.M.P. Mingos and J. Lauher. These **Wade–Mingos–Lauher rules** are summarized in Table 16.11; they apply most reliably to metal clusters in Groups 6 to 9. In general (and as in the boron hydrides where similar considerations apply), more open structures (which have fewer metal–metal bonds) occur when there is a higher cluster valence electron (CVE) count.

Table 16.11 Correlation of cluster valence electron count and structure*

Number of metal atoms	Structure of metal framework	Cluster valence electron count	Example
1	Single metal	18	$Ni(CO)_4$ (**2**)
2	Linear	34	$Mn_2(CO)_{10}$†
3	Closed triangle	48	$Co_3(CO)_9CH$ (**59**)
4	Tetrahedron	60	$Co_4(CO)_{12}$ (**57**)
	Butterfly	62	$[Fe_4(CO)_{12}C]^{2-}$ (**60**)
	Square	64	$Os_4(CO)_{16}$
5	Trigonal bipyramid	72	$Os_5(CO)_{16}$
	Square pyramid	74	$Fe_5C(CO)_{15}$
6	Octahedron	86	$Ru_6C(CO)_{17}$
	Trigonal prism	90	$[Rh_6C(CO)_{15}]^{2-}$

*For a more extensive table, see J.W. Lauher, *J. Am. Chem. Soc.* **100**, 5305 (1978).
†Table 16.2.

[21] D.M.P. Mingos and D.J. Wales, *Introduction to cluster chemistry.* Prentice Hall, Englewood Cliffs (1990); J.W. Lauher, *J. Am. Chem. Soc.* **100**, 5305 (1978).

> The Wade–Mingos–Lauher rules identify a correlation between the valence electron count and the structures of large organometallic complexes.

Example 16.7 Correlating spectroscopic data, cluster valence electron count, and structure

The reaction of chloroform with $[Co_2(CO)_8]$ yields a compound of formula $[Co_3(CH)(CO)_9]$. NMR and IR data indicate the presence of only terminal CO ligands and the presence of a CH group. Propose a structure consistent with the spectra and the correlation of CVE with structure.

Answer Electrons available for the cluster are: 27 for three Co atoms, 18 for nine CO ligands, and 3 for CH (assuming the latter is simply C-bonded: one C electron is used for the C—H bond, so three are available for bonding in the cluster). The resulting total CVE of 48 indicates a triangular cluster (see Table 16.11). A structure consistent with this conclusion and the presence of only terminal CO ligands and a capping CH ligand is illustrated in (**59**).

59 $Co_3(CH)(CO)_9$

Self-test 16.7 The compound $[Fe_4Cp_4(CO)_4]$ is a dark-green solid. Its IR spectrum shows a single CO stretch at 1640 cm^{-1}. The ^1H-NMR is a single line even at low temperatures. From this spectroscopic information and the CVE propose a structure for $[Fe_4Cp_4(CO)_4]$.

(b) Structure and isolobal analogies

A good way to picture the incorporation of hetero atoms into a metal cluster is to use isolobal analogies (Section 3.6).[22] These analogies let us draw a parallel between $[Co_3(CO)_9(CH)]$ (**59**) and $[Co_4(CO)_{12}]$ (**57**), both of which can be regarded as triangular $Co_3(CO)_9$ fragments capped on one side either by $Co(CO)_3$ or by CH. The CH and $Co(CO)_3$ groups are isolobal because each one has three orbitals and three electrons available for participation in framework bonding. A minor complication in this comparison is the occurrence of $Co(CO)_2$ groups together with bridging CO ligands in $[Co_4(CO)_{12}]$ because bridging and terminal ligands often have similar energies.

Isolobal analogies allow us to draw many parallels between metal–metal bonded systems containing only *d* metals and mixed systems containing *d*-block and *p*-block metals. Additional examples are given in Table 16.12, where it will be noted that a P atom is isolobal with CH; accordingly, a cluster similar to (**59**) is known, but with a capping P atom. Similarly, the ligands CR_2 and $Fe(CO)_4$ are both capable of bonding to two metal atoms in a cluster; CH_3 and $Mn(CO)_5$ can bond to one metal atom.

> Isolobal analogies allow us to draw many parallels between metal–metal bonded systems containing only *d* metals and mixed systems containing *d*-block and *p*-block metals.

16.11 Syntheses

One of the oldest methods for the synthesis of metal clusters is the thermal expulsion of CO from a metal carbonyl.[23] The pyrolytic formation of metal cluster compounds can be viewed from the standpoint of electron count: a decrease in valence electrons around the metal

22 The application of isolobal analogies to metal cluster compounds is described in the Nobel Prize lecture by R. Hoffmann, *Angew. Chem. Int. Ed. Eng.* **21**, 711 (1982). Structures of organometallics and clusters also are nicely systematized by isoelectronic relations: J. Ellis, *J. Chem. Educ.* **53**, 2 (1976).

23 R.D. Adams, In *The chemistry of metal cluster complexes* (ed. D.F. Shriver, H.D. Kaesz, and R.D. Adams), Chapter 3. VCH, Weinheim (1990).

Table 16.12 Some isolobal *p*-block and *d*-block fragments*

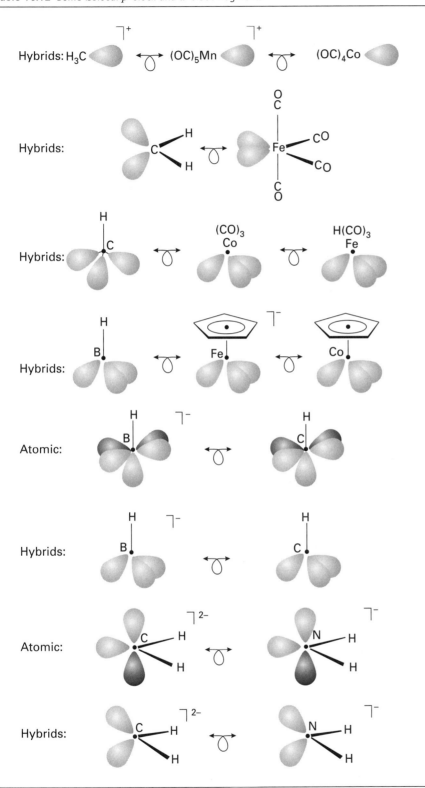

*Showing both atomic orbital and hybrid orbital representations.

resulting from loss of CO is compensated by the formation of **M—M** bonds. One example is the synthesis of $[Co_4(CO)_{12}]$ by heating $[Co_2(CO)_8]$:

$$2\,Co_2(CO)_8 \xrightarrow{\Delta} Co_4(CO)_{12} + 4\,CO$$

This reaction proceeds slowly at room temperature, so samples of octacarbonyldicobalt are usually contaminated with dodecacarbonyltetracobalt.[24]

A widely employed and more controllable reaction is based on the condensation of a carbonyl anion and a neutral organometallic complex:

$$[Ni_5(CO)_{12}]^{2-} + Ni(CO)_4 \longrightarrow [Ni_6(CO)_{12}]^{2-} + 4\,CO$$

The Ni_5 complex has a CVE of 76 whereas the N_6 complex has a count of 86. The descriptive name **redox condensation** is often given to reactions of this type, which are very useful for the preparation of anionic metal carbonyl clusters. In this example, a trigonal-bipyramidal cluster containing Ni with oxidation number $-\frac{2}{5}$ and $Ni(CO)_4$ containing Ni(0) is converted into an octahedral cluster having Ni with oxidation number $-\frac{1}{3}$. The $[Ni_5(CO)_{12}]^{2-}$ cluster, which has four electrons in excess of the 72 expected for a trigonal bipyramid, illustrates a fairly common tendency for the Group 10 metal clusters to have an electron count in excess of that expected from the Wade–Mingos–Lauher rules.

A third method, pioneered by F.G.A. Stone,[25] is based on the condensation of an organometallic complex containing displaceable ligands with an unsaturated organometallic compound. The unsaturated complex may be a metal alkylidene, $L_nM{=}CR_2$, a metal alkylidyne, $L_nM{\equiv}CR$, or a compound with multiple metal–metal bonds:

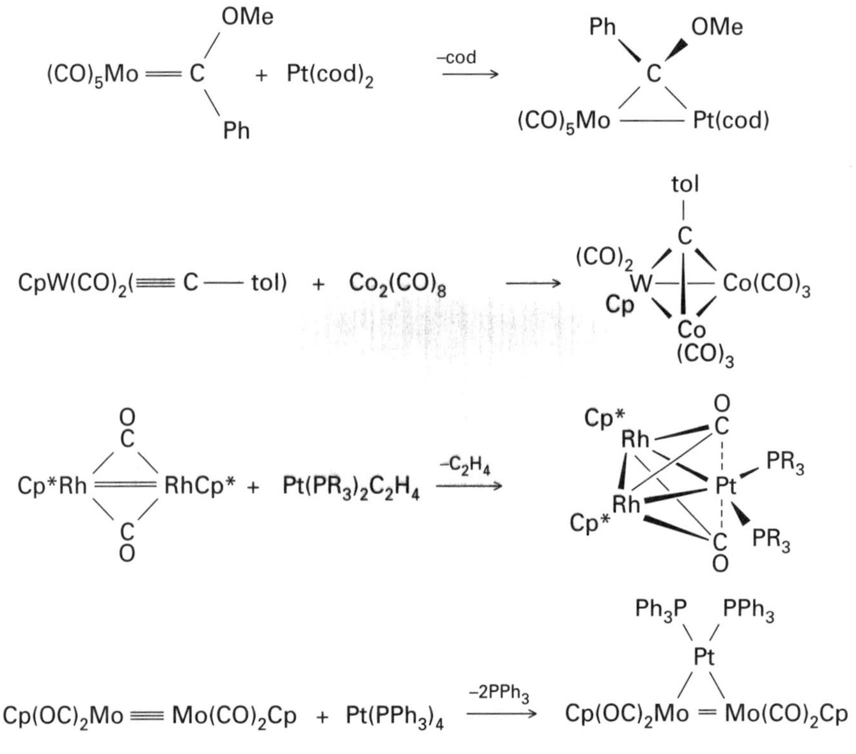

[24] A wide range of osmium and ruthenium cluster compounds with interesting structures have been prepared by the pyrolysis method: J. Lewis and B.F.G. Johnson, *Pure Appl. Chem.* **54**, 97 (1982).

[25] F.G.A. Stone, *Pure Appl. Chem.* **23**, 89 (1984).

> *Three methods are commonly used to prepare metal clusters: thermal expulsion of CO*
> *from a metal carbonyl, the condensation of a carbonyl anion and a neutral organometallic*
> *complex, and the condensation of an organometallic complex with an unsaturated*
> *organometallic compound.*

16.12 Reactions

We have already seen some examples of the reactivity of metal clusters in the above discussion of metal cluster synthesis. Some other common reactions of clusters are ligand substitution, fragmentation, and protonation.

(a) Substitution versus fragmentation

Because $M—M$ bonds are generally comparable in strength to $M—L$ bonds, there is often a delicate balance between ligand substitution and cluster fragmentation. For example, dodecacarbonyltriiron(0), $[Fe_3(CO)_{12}]$, reacts with triphenylphosphine under mild conditions to yield simple mono- and disubstituted products as well as some cluster fragmentation products

$$Fe_3(CO)_{12} + PPh_3 \longrightarrow Fe_3(CO)_{11}(PPh_3)$$
$$+ Fe_3(CO)_{10}(PPh_3)_2 + Fe(CO)_5$$
$$+ Fe(CO)_4(PPh_3) + Fe(CO)_3(PPh_3)_2 + CO$$

However, for somewhat longer reaction times or elevated temperatures, only monoiron cleavage products are obtained. Since the strength of $M—M$ bonds increases down a group, substitution products of the heavier clusters such as $[Ru_3(CO)_{10}(PPh_3)_2]$ or $[Os_3(CO)_{10}(PPh_3)_2]$ can be prepared without significant fragmentation into mononuclear complexes.

> *In organometallic clusters there is often a delicate balance between ligand substitution*
> *and cluster fragmentation.*

(b) Protonation

The tendency of protons to attach to metal atoms in clusters is even more pronounced than that of mononuclear carbonyls. This Brønsted basicity of clusters is associated with the ready protonation of $M—M$ bonds to produce a formal 3c,2e bond like that in diborane:

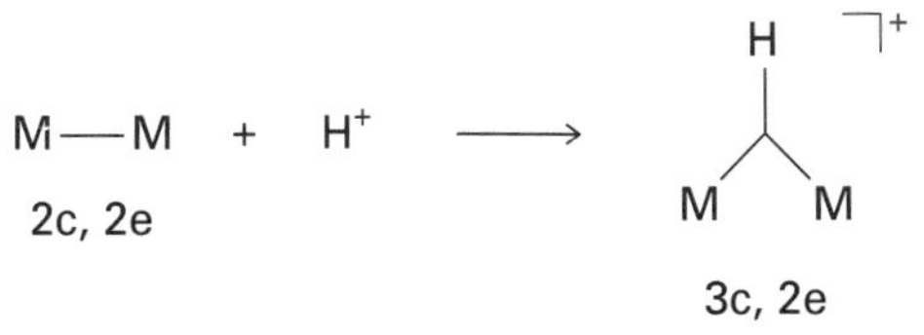

$$
M—M \quad + \quad H^+ \quad \longrightarrow \quad
\begin{bmatrix}
& H & \\
& | & \\
M & & M
\end{bmatrix}^+
$$

2c, 2e   3c, 2e

For example:

$$[Fe_3(CO)_{11}]^{2-} + H^+ \longrightarrow [Fe_3H(CO)_{11}]^-$$

As expected, metal cluster anions are much stronger bases than their neutral analogs. The $M—H—M$ bridge is by far the most common bonding mode of hydrogen in clusters, but hydrogen also is known to bridge a triangular three-atom face and to reside inside a metal polyhedron.

> *The tendency of protons to attach to metal atoms in clusters is more pronounced than in*
> *mononuclear carbonyls.*

(c) Cluster-assisted ligand transformations

Transformations of ligands on metal clusters sometimes mimic those on single metal centers. However, a cluster of metal atoms often facilitates ligand reactions because the presence of several metal atoms near a ligand provides additional opportunities for interactions. Two examples will be given, one involving the CH_3 ligand and the other CO.[26]

The osmium cluster $[Os_3(H_2)(CH_3)(CO)_{10}]$ undergoes a transfer of H atoms to the metal-atom framework:

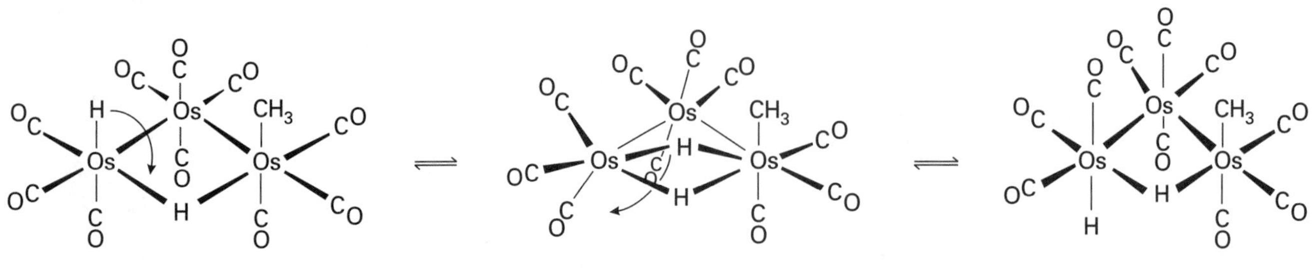

The resulting equilibrium mixture of products can lose a CO ligand. A similar set of transformations may occur on catalytic metal surfaces. However, the chemistry of metal surfaces cannot be studied in the detail that was possible in this study of a metal cluster compound. In solution, NMR proved to be very useful and, in the solid state, X-ray and neutron diffraction measurements provided precise structural information.

Both CO migratory insertion and the attack on CO by nucleophiles have also been observed in clusters. Of greater interest, however, are reactions in which the metal atoms work in concert to bring about ligand transformations. One example is the greater ease of cleaving the CO ligand in a cluster:

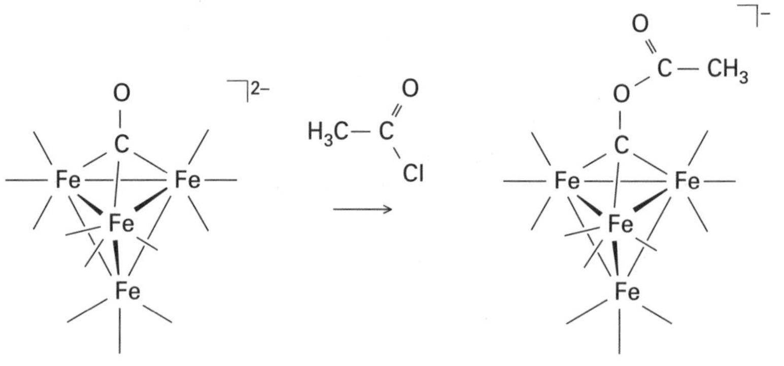

$[Fe_4(CO)_{12}(\mu_3\text{-CO})]^{2-}$ $[Fe_4(CO)_{12}(\mu_3\text{-COAc})]^-$

where Ac denotes CH_3CO. This acylation of the O atom of CO has been observed only in clusters. The acetate group so produced is a good leaving group and it can be removed by electron transfer to the cluster, which maintains the electron count on the cluster when the acetate has departed. This process yields a metal carbide cluster with the carbon in a four-coordinate environment but far from tetrahedral:

[26] G. Lavigne, Cluster-assisted ligand transformations, in *The chemistry of metal cluster complexes* (ed. D.F. Shriver, H.D. Kaesz, and R.D. Adams), p. 201. VCH, Weinheim (1990).

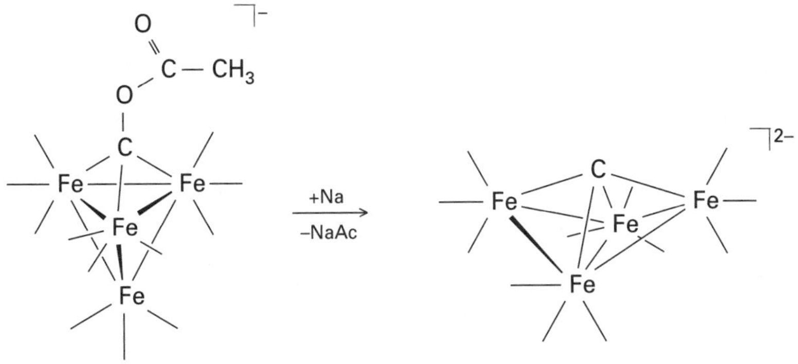

$[Fe_4(CO)_{12}(\mu_3-COAc)]^-$ $[Fe_4(CO)_{12}(\mu_4-C)]^{2-}$

60

In penta- and hexairon clusters, the C atom is found in square-pyramidal and octahedral coordination environments. From these examples and other areas of inorganic chemistry—such as carboranes (Section 10.6) and solid metal carbides (Section 10.10)—we see that carbon participates in a wide range of bonding arrangements that are unprecedented in organic chemistry.

Further reading

Some excellent introductory texts are available:

Ch. Elschenbroich and A. Salzer, *Organometallics*. VCH, Weinheim (1992).

P. Powell, *Principles of organometallic chemistry*. Chapman & Hall, London (1988).

R.H. Crabtree, *The organometallic chemistry of the transition metals*. Wiley-Interscience, New York (1994).

A. Yamamoto, *Organotransition metal chemistry*. Wiley-Interscience, New York (1986).

C.M. Lukehart, *Fundamental transition metal organometallic chemistry*. Brooks Cole, Belmont (1985).

More detailed coverage is available in:

J.P. Collman, L.S. Hegedus, J.R. Norton, and R.G. Finke, *Principles and applications of organotransition metal chemistry*. University Science Books, Mill Valley (1987).

G. Wilkinson, F.G.A. Stone, and E.W. Abel (ed.), *Comprehensive organometallic chemistry*. Pergamon Press, Oxford (1982 *et seq.*). This multivolume set provides excellent introductions and details for all areas of organometallic chemistry. More recent developments appear in the second edition (1995).

D.F. Shriver, H.D. Kaesz, and R.D. Adams (ed.), *The chemistry of metal cluster complexes*. VCH, New York (1990).

M. Bochman, *Organometallics*. Oxford University Press (1994).

Exercises

16.1 Name the species and draw the structures of (a) $Fe(CO)_5$, (b) $Ni(CO)_4$, (c) $Mo(CO)_6$, (d) $Mn_2(CO)_{10}$, (e) $V(CO)_6$, (f) $[PtCl_3(C_2H_4)]^-$.

16.2 (a) Sketch an η^2 interaction of 1,3-butadiene with a metal atom, and (b) do the same for an η^4 interaction.

16.3 Assign oxidation numbers to the metal atom in (a) $[Fe(\eta^5-C_5H_5)_2][BF_4]$, (b) $Fe(CO)_5$, (c) $[Fe(CO)_4]^{2-}$, and (d) $Co_2(CO)_8$.

16.4 Count the number of valence electrons for each metal atom in the complexes listed in Exercises 16.1 and 16.3. Do any of them deviate from the 18-electron rule? If so, is this reflected in their structure or chemical properties?

16.5 State the two common methods for the preparation of simple metal carbonyls and illustrate your answer with chemical equations. Is the selection of method based on thermodynamic or kinetic considerations?

16.6 Suggest a sequence of reactions for the preparation of $Fe(diphos)(CO)_3$, given iron metal, CO, diphos ($Ph_2PCH_2CH_2PPh_2$), and other reagents of your choice.

16.7 Suppose that you are given a series of metal tricarbonyl compounds having the respective symmetries C_{2v}, D_{3h}, and C_s. Without consulting reference material, which of these should display the greatest number of CO stretching bands in the IR spectrum? Check your answer and give the number of expected bands for each by consulting Table 16.5.

16.8 The compound $Ni_3(C_5H_5)_3(CO)_2$ has a single CO stretching absorption at 1761 cm^{-1}. The IR data indicate that all C_5H_5 ligands are pentahapto and probably in identical environments. (a) On the basis of these data, propose a structure. (b) Does the electron count for each metal in your structure agree with the 18-electron rule? If not, is nickel in a region of the periodic table where deviations from the 18-electron rule are common?

16.9 Decide which of the two complexes (a) $W(CO)_6$ or (b) $IrCl(PPh_2)_2(CO)$ should undergo the fastest exchange with ^{13}CO. Justify your answer.

16.10 Which metal carbonyl in each of (a) $[Fe(CO)_4]^{2-}$ or $[Co(CO)_4]^-$, (b) $[Mn(CO)_5]^-$ or $[Re(CO)_5]^-$ should be the most basic toward a proton? What are the trends upon which your answer is based?

16.11 For each of the following mixtures of reactants, give (i) a plausible chemical equation and (ii) structure for the organometallic product, and (iii) general reason for the course of the reaction: (a) methyllithium and $W(CO)_6$, (b) $Co_2(CO)_8$ and $AlBr_3$.

16.12 What hapticities are possible for the interaction of each of the following ligands with a single *d-*block metal atom such as cobalt? (a) C_2H_4, (b) cyclopentadienyl, (c) C_6H_6, (d) butadiene, (e) cyclooctatetraene.

16.13 Draw plausible structures and give the electron count of (a) $Ni(\eta^3\text{-}C_3H_5)_2$, (b) η^4-cyclobutadiene-η^5-cyclopentadienylcobalt, (c) $(\eta^3\text{-}C_3H_5)Co(CO)_2$. If the electron count deviates from 18, is the deviation explicable in terms of periodic trends?

16.14 Write out the *d-*block of the periodic table. (By this stage, you should not have to consult a periodic table.) Indicate on this table (a) the elements that form 18-electron neutral Cp_2M compounds, (b) the Period 4 elements for which the simplest carbonyls are dimeric, (c) the Period 4 elements that form neutral carbonyls with six, five, and four carbonyl ligands, and (d) the elements that most commonly obey the 18-electron rule.

16.15 Using the 18-electron rule as a guide, indicate the probable number of carbonyl ligands in (a) $W(\eta^6\text{-}C_6H_6)(CO)_n$, (b) $Rh(\eta^5\text{-}C_5H_5)(CO)_n$, and (c) $Ru_3(CO)_n$.

16.16 Give plausible equations, different from those in the text, to demonstrate the utility of metal carbonyl anions in the synthesis of M—C, M—H, and M—M′ bonds.

16.17 Propose a synthesis for $MnH(CO)_5$, starting with $Mn_2(CO)_{10}$ as the source of Mn and other reagents of your choice.

16.18 Give the probable structure of the product obtained when $Mo(CO)_6$ is allowed to react first with LiPh and then with a strong carbocation reagent, $CH_3OSO_2CF_3$.

16.19 Provide plausible reasons for the differences in IR wavenumbers between each of the following pairs: (a) $Mo(PF_3)_3(CO)_3$ 2040, 1991 cm^{-1}, versus $Mo(PMe_3)_3(CO)_3$ 1945, 1851 cm^{-1}, (b) $MnCp(CO)_3$ 2023, 1939 cm^{-1}, versus $MnCp^*(CO)_3$ 2017, 1928 cm^{-1}.

16.20 Which compound would you expect to be more stable, $Rh(\eta^5\text{-}C_5H_5)_2$ or $Ru(\eta^5\text{-}C_5H_5)_2$? Give a plausible explanation for the difference in terms of simple bonding concepts.

16.21 Give the equation for a workable reaction that will convert $Fe(\eta^5\text{-}C_5H_5)_2$ into $Fe(\eta^5\text{-}C_5H_5)(\eta^5\text{-}C_5H_4COCH_3)$.

16.22 Sketch the a_1' symmetry-adapted orbitals for the two eclipsed C_5H_5 ligands stacked together with D_{5h} symmetry. Identify the *s*, *p*, and *d* orbitals of a metal atom lying between the rings that may have nonzero overlap, and state how many a_1' molecular orbitals may be formed.

16.23 The compound $Ni(\eta^5\text{-}C_5H_5)_2$ readily reacts with HF to yield $[Ni(\eta^5\text{-}C_5H_5)(\eta^4\text{-}C_5H_6)]^+$ whereas $Fe(\eta^5\text{-}C_5H_5)_2$ reacts with strong acid to yield $[Fe(\eta^5\text{-}C_5H_5)_2H]^+$. In the latter compound the H atom is attached to the Fe atom. Provide a reasonable explanation for this difference.

16.24 Write a plausible mechanism, giving your reasoning, for the reactions (a) $[Mn(CO)_5(CF_2)]^+ + H_2O \rightarrow [Mn(CO)_6]^+ + 2HF$ and (b) $Rh(CO)(C_2H_5)(PR_3)_2 \rightarrow RhH(CO)(PR_3)_2 + C_2H_4$.

16.25 Contrast the general chemical characteristics of the organometallic complexes of *d-*block elements in Groups 6 to 8 with those in Groups 3 and 4 with respect to (a) stability of the $\eta^5\text{-}C_5H_5$ ligand, (b) hydridic or protonic character of M—H bonds, and (c) adherence to the 18-electron rule.

16.26 (a) What cluster valence electron (CVE) count is characteristic of octahedral and trigonal prismatic complexes? (b) Can these CVE values be derived from the 18-electron rule? (c) Determine the probable geometry (octahedral or trigonal-prismatic) of $[Fe_6(C)(CO)_{16}]^{2-}$ and $[Co_6(C)(CO)_{15}]^{2-}$. (The C atom in both cases resides in the center of the cluster and can be considered to be a four-electron donor.)

16.27 Based on isolobal analogies, choose the groups that might replace the group in boldface in

(a) $Co_2(CO)_9\textbf{CH}$ OCH_3, $N(CH_2)_2$, or $SiCH_3$

(b) $(OC)_5Mn\textbf{Mn(CO)}_5$ I, CH_2, or CCH_3

16.28 Ligand substitution reactions on metal clusters are often found to occur by associative mechanisms, and it is postulated that these occur by initial breaking of a M—M bond, thereby providing an open coordination site for the incoming ligand. If the proposed mechanism is applicable, which would you expect to undergo the fastest exchange with added ^{13}CO: $Co_4(CO)_{12}$ or $Ir_4(CO)_{12}$? Suggest an explanation.

Problems

16.1 Propose the structure of the product obtained by the reaction of $[Re(CO)(\eta^5\text{-}C_5H_5)(PPh_3)(NO)]^+$ with $Li[HBEt_3]$. The latter contains a strongly nucleophilic hydride. (For full details see: W. Tam, G.Y. Lin, W.K. Wong, W.A. Kiel, V. Wong, and J.A. Gladysz, *J. Am. Chem. Soc.* **104**, 141 (1982).

16.2 Develop a qualitative molecular orbital diagram for $Cr(CO)_6$ based on the symmetry-adapted linear combinations in Appendix 4. First consider the influence of σ bonding and then the effects of π bonding. (For help see T.A. Albright, J.K. Burdett, and M.H. Whangbo, *Orbital interactions in chemistry.* Wiley-Interscience, New York (1985).)

16.3 When several CO ligands are present in a metal carbonyl, an indication of the individual bond strengths can be determined by means of force constants derived from the experimental IR frequencies. In $Cr(CO)_5(PPh_3)$ the *cis*-CO ligands have the higher force constants, whereas in $Ph_3SnCo(CO)_4$ the force constants are higher for the *trans*-CO. Explain which carbonyl C atoms should be susceptible to nucleophiles to attack in these two cases. (For details see D.J. Darensbourg and M.Y. Darensbourg, *Inorg. Chem.* **9**, 1691 (1970).)

16.4 The complex $Ru_2Cp_2(CO)_4$ is known to exist as three isomers in hydrocarbon solution, one of which contains no CO bridges and presumably has nearly free rotation about the Ru—Ru bond. The two other isomers both have two CO bridges; one has a *cis* disposition of the Cp and terminal CO ligands and the other has a *trans* disposition of these ligands. The IR spectrum of this equilibrium mixture of isomers in solution contains CO stretching bands at 2030, 2024, 2021*, 1993, 1988, 1967*, 1945*, 1834 and 1795 cm^{-1}. The bands marked with * have been assigned to the nonbridged isomer. The variation of the IR spectra with $AlEt_3$ concentration indicates that, at an intermediate concentration, a species is produced with the simpler spectrum 2030, 2024, 1993, 1988, 1834, and 1679 cm^{-1}. When the concentration of $AlEt_3$ is increased further the IR spectrum is even simpler: 2046, 2011, 2006, and 1679 cm^{-1}. Propose a chemical basis for the simplification of the IR spectrum at high $AlEt_3$ concentrations. Accompany this answer with equations for the probable reactions and an interpretation of the IR data for the two reaction products. (For further details see A.A. Alich, N.J. Nelson, D. Strope, and D.F. Shriver, *Inorg. Chem.* **11**, 2976 (1972).)

16.5 The dinitrogen complex $Zr_2(\eta^5\text{-}Cp^*)_4(N_2)_3$ has been isolated and its structure determined by single-crystal X-ray diffraction. Each Zr atom is bonded to two Cp* and one terminal N_2. The third N_2 bridges between the Zr atoms in a nearly linear ZrN≡NZr array. Before consulting the reference, write a plausible structure for this compound that accounts for the ^{1}H-NMR spectrum obtained on a sample held at $-7\,^\circ C$. This spectrum is a doublet, indicating that the Cp* rings are in two different environments. At somewhat above room temperature these rings become equivalent on the NMR timescale and ^{15}N-NMR indicates that N_2 exchange between the terminal ligands and dissolved N_2 is correlated with the process that interconverts the Cp* ligand sites. Propose a way in which this equilibration could interconvert the sites of the Cp* ligands. (For further details see J.M. Manriquez, D.R. McAlister, E. Rosenberg, H.M. Shiller, K.L. Williamson, S.I. Chan, and J.E. Bercaw, *J. Am. Chem. Soc.* **108**, 3078 (1978).)

16.6 Devise a qualitative molecular orbital scheme for $Ni(\eta^5\text{-}C_5H_5)_2$ assuming D_{5h} symmetry and making use of the symmetry-adapted orbitals in Appendix 4.

16.7 If any of the following statements are incorrect, provide a corrected counterpart and an example that bears out your correction. (a) A high-field NMR signal is characteristic of the hydride ligand in Groups 8 and 9 organometallic complexes. (b) Fischer carbenes are synthesized from carbonyl compounds of the early *d*-block metals. (c) The appearance of a single feature in the ^{1}H-NMR spectrum of a cyclooctatetraene metal complex is proof that the ligand is octahapto. (d) Square-planar 16-electron compounds generally undergo ligand substitution by a dissociative mechanism. (e) Metal–metal bonds in cluster compounds generally become weaker going down a group in the *d*-block.

17

Catalysis

In this chapter we apply the concepts of organometallic chemistry and coordination chemistry to catalysis. We emphasize general principles, such as the nature of catalytic cycles, in which a catalytic species is regenerated in a reaction, and the delicate balance of reactions required for a successful cycle. We shall see that there are numerous requirements for a successful catalytic process: the reaction being catalyzed must be thermodynamically favorable and fast enough when catalyzed; the catalyst must have an appropriate selectivity toward the desired product and a lifetime long enough to be economical. We then survey the reaction types commonly encountered in the homogeneous catalysis of hydrocarbon interconversion and show how these reactions are invoked in proposals about mechanisms. The final part of the chapter develops a similar theme in heterogeneous catalysis, and we shall see that many parallels between homogeneous and heterogeneous catalysis lurk beneath differences in terminology. In neither type of catalysis are mechanisms finally settled, and there is still considerable scope for making new discoveries.

A **catalyst** is a substance that increases the rate of a reaction but is not itself consumed. Catalysts are widely used in nature, in industry, and in the laboratory, and it is estimated that they contribute to one-sixth of the value of all manufactured goods in industrialized countries. As shown in Table 17.1, 13 of the top 20 synthetic chemicals are produced directly or indirectly by catalysis in the USA. For example, a key step in the production of a dominant industrial chemical, sulfuric acid, is the catalytic oxidation of SO_2 to SO_3. Ammonia, another chemical essential for industry and agriculture, is produced by the catalytic reduction of N_2 by H_2. Inorganic catalysts are also used for the production of the major organic chemicals and petroleum products, such as fuels, petrochemicals, and polyalkene plastics. Catalysts play a steadily increasing role in achieving a cleaner environment, both through the destruction of pollutants (as with the automotive catalytic exhaust converters) and through the development of cleaner industrial processes with less abundant byproducts. Enzymes, a class of elaborate biochemical catalysts, are discussed in Chapter 19.

In addition to their economic importance and contribution to the quality of life, catalysts are interesting for the subtlety with which they go about their business. The understanding of the mechanisms of catalytic reactions has improved greatly in recent years because of the availability of isotopically labeled molecules, improved methods for determining reaction

Table 17.1 The top 20 synthetic chemicals in the USA

Synthetic chemical	Rank†	Catalytic process
Ethene	1	Hydrocarbon cracking, heterogeneous
Sulfuric acid	2	SO_2 oxidation, heterogeneous
Propene	3	Hydrocarbon cracking, heterogeneous
1,2-Dichloroethane	4	$C_2H_4 + Cl_2$; heterogeneous
Calcium hydroxide	5	Not catalytic
Ammonia	6	$N_2 + H_2$; heterogeneous
Urea	7	NH_3 precursor catalytic
Phosphoric acid	8	Not catalytic
Chlorine	9	Electrolysis
Ethylbenzene	10	Alkylation of benzene; homogeneous
Sodium carbonate	11	Not catalytic
Sodium hydroxide	12	Electrolysis
Styrene	13	Dehydrogenation of ethylbenzene; heterogeneous
Nitric acid	14	$NH_3 + O_2$; heterogeneous
Ammonium nitrate	15	Precursors catalytic
Hydrogen chloride	16	Precursors catalytic
Acrylonitrile	17	$HCN + C_2H_2$, homogeneous
Ammonium sulfate	18	Precursors catalytic
Potassium oxide	19	Not catalytic
Titanium dioxide	20	Not catalytic

† Based on mass, from *Chemical and Engineering News* survey of US industrial chemicals, June 29 (1998) for production in 1997.

rates, improved spectroscopic and diffraction techniques, and much more reliable molecular orbital calculations.

General principles

Catalysts are classified as 'homogeneous' if they are present in the same phase as the reagents; this normally means that they are present as solutes in a liquid reaction mixture. Catalysts are 'heterogeneous' if they are present in a different phase from that of the reactants. Both types of catalysis are discussed in this chapter, and it will be seen that they are fundamentally similar. Of the two, heterogeneous catalysis has a much greater economic impact.

The term 'negative catalyst' is sometimes applied to substances that retard reactions. We shall not use the term because these substances are best considered to be **catalyst poisons** that block one or more elementary steps in a catalytic reaction.

17.1 Description of catalysts

A catalyzed reaction is faster (or, in some cases, more specific) than an uncatalyzed version of the same reaction because the catalyst provides a different reaction pathway with a lower activation energy. To describe these characteristics we develop in this section some of the terminology used to express the speed of a catalytic reaction and its mechanism.

(a) Catalytic efficiency

The **turnover frequency**, N (formerly 'turnover number'), is often used to express the efficiency of a catalyst. For the conversion of A to B catalyzed by Q and with a rate v,

$$A \xrightarrow{\text{Q}} B \qquad v = \frac{d[B]}{dt} \tag{1}$$

the turnover frequency is given by

$$N = \frac{v}{[Q]} \tag{2}$$

if the rate of the uncatalyzed reaction is negligible. A highly active catalyst, one that results in a fast reaction even in low concentrations, has a large turnover frequency.

In heterogeneous catalysis, the reaction rate is expressed in terms of the rate of change in the amount of product (in place of concentration) and the concentration of catalyst is replaced by the amount present. The determination of the number of active sites in a heterogeneous catalyst is particularly challenging, and often the denominator $[Q]$ in eqn 2 is replaced by the surface area of the catalyst.

A highly active catalyst, one that results in a fast reaction even in low concentrations, has a large turnover frequency.

(b) Catalytic cycles

The essence of catalysis is a cycle of reactions that consumes the reactants, forms products, and regenerates the catalytic species. A simple example of a catalytic cycle involving a homogeneous catalyst is the isomerization of allyl alcohol (3-hydroxypropene, $CH_2=CHCH_2OH$)) to propanal (CH_3CH_2CHO) with the catalyst $[CoH(CO)_3]$. The first step is the coordination of the reactant to the catalyst. That complex isomerizes in the coordination sphere of the catalyst and goes on to release the product and reform the catalyst.

As with all mechanisms, the cycle has been proposed on the basis of a range of information like that summarized in Chart 17.1. Many of the components shown in the chart were encountered in Chapter 14 in connection with the determination of mechanisms of substitution reactions. However, the elucidation of catalytic mechanisms is complicated by

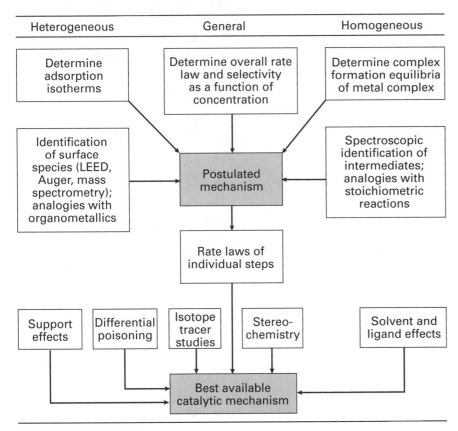

Chart 17.1 Determination of catalytic mechanisms.

the occurrence of several delicately balanced reactions, which often cannot be studied in isolation.

Two stringent tests of any proposed mechanism are the determination of rate laws and the elucidation of stereochemistry. If intermediates are postulated, their detection by magnetic resonance and IR spectroscopy also provides support. If specific atom transfer steps are proposed, isotope tracer studies may serve as a test. The influences of different ligands and different substrates are also sometimes informative. Although rate data and the corresponding laws have been determined for many *overall* catalytic cycles, it is also necessary to determine rate laws for the individual steps in order to have reasonable confidence in the mechanism. However, because of experimental complications, it is rare that catalytic cycles are studied in this detail.

> The essence of catalysis is a cycle of reactions that consumes the reactants, forms products, and regenerates the catalytic species.

(c) Energetics

A catalyst increases the rates of processes by introducing new pathways with lower Gibbs energies of activation, $\Delta^{\ddagger}G$. It is important to focus on the Gibbs energy profile of a catalytic reaction, not just the enthalpy or energy profile, because the new elementary steps that occur in the catalyzed process are likely to have quite different entropies of activation.[1] A catalyst does not affect the Gibbs energy of the overall reaction, $\Delta_r G^{\ominus}$, because G is a state function.[2] The difference is illustrated in Fig. 17.1, where the overall reaction Gibbs energy is the same in both energy profiles. Reactions that are thermodynamically unfavorable cannot be made favorable by a catalyst.

Figure 17.1 also shows that the Gibbs energy profile of a catalyzed reaction contains no high peaks and no deep troughs. The new pathway introduced by the catalyst changes the mechanism of the reaction to one with a very different shape and with lower maxima. However, an equally important point is that stable or nonlabile catalytic intermediates do not occur in the cycle. Similarly, the product must be released in a thermodynamically favorable step. If, as shown by the gray line in Fig. 17.1, a stable complex is formed with the catalyst, it would turn out to be the product of the reaction and the cycle would terminate. Similarly, impurities may suppress catalysis by coordinating strongly to catalytically active sites and act as catalyst poisons.

> A catalyst increases the rates of processes by introducing new pathways with lower Gibbs energies of activation; the reaction profile contains no high peaks and no deep troughs.

17.2 Properties of catalysts

Although the increased rate of a catalytic reaction is important, it is not the only criterion. Of similar importance are a minimum of side products and a long catalyst lifetime.

(a) Selectivity

A **selective catalyst** yields a high proportion of the desired product with minimum amounts of side products. In industry, there is considerable economic incentive to develop selective catalysts. For example, when metallic silver is used to catalyze the oxidation of ethene with oxygen to produce ethylene oxide (**1**), the reaction is accompanied by the more thermodynamically favored but undesirable formation of CO_2 and H_2O. This lack of selectivity increases the consumption of ethene, so chemists are constantly trying to devise a more selective catalyst for ethylene oxide synthesis. Selectivity can be ignored in only a very few

17.1 Schematic representation of the energetics in a catalytic cycle. The uncatalyzed reaction (a) has a higher Gibbs energy of activation $\Delta^{\ddagger}G$ than any step in the catalyzed reaction (b). The Gibbs energy of reaction, $\Delta_r G^{\ominus}$, for the overall reaction is unchanged from (a) to (b).

1 Ethylene oxide

---

[1]   For further discussion of this point see A. Haim, Catalysis: new reaction pathways, not just lowering of the activation energy. *J. Chem. Educ.* **66**, 731 (1989).

[2]   That is, G depends only on the current state of the system and not on the path that led to the state.

simple inorganic reactions, where there is essentially only one thermodynamically favorable product, as in the formation of NH_3 from H_2 and N_2.

A selective catalyst yields a high proportion of the desired product with minimum amounts of side products.

(b) Lifetime

A small amount of catalyst must survive through a large number of cycles if it is to be economically viable. However, a catalyst may be destroyed by side reactions to the main catalytic cycle or by the presence of small amounts of impurities in the starting materials (the feedstock). For example, many alkene polymerization catalysts are destroyed by O_2, so in the synthesis of polyethylene and polypropylene the concentration of O_2 in the ethene or propene feedstock should be no more than a few parts per billion.

Some catalysts can be regenerated quite readily. For example, the supported metal catalysts used in the reforming reactions that convert hydrocarbons to high octane gasoline become covered with carbon because the catalytic reaction is accompanied by a small amount of dehydrogenation. These supported metal particles can be cleaned by periodically interrupting the catalytic process and burning off the accumulated carbon.

Table 17.2 Some homogeneous catalytic processes

Hydroformylation of alkenes (Oxo process)

$$RCH{=}CH_2 + CO + H_2 \xrightarrow[\text{or Pt(II)}]{\text{Co(I), Rh(I),}} RCH_2CH_2C({=}O)H + RCH(Me){-}C({=}O)H$$

Oxidation of alkenes (Wacker process)

$$H_2C{=}CH_2 + O_2 \xrightarrow{\text{Pd(II) or Cu(II)}} H_3CC({=}O)H$$

Carbonylation of methanol to acetic acid (Monsanto process)

$$CH_3OH + CO \xrightarrow{[RhI_2(CO)_2]^-} H_3CC({=}O)OH$$

Hydrocyanation of butadiene to adiponitrile

$$H_2C{=}CHCH{=}CH_2 + 2HCN \xrightarrow{Ni(P(OR)_3)_4} NCCH_2CH_2CH_2CH_2CN$$

Oligomerization of ethene

$$nH_2C{=}CH_2 \xrightarrow{Ni} H_2C{=}CH(CH_2CH_2)_{n-2}CH_2CH_3$$

Alkene dismutation (olefin metathesis)

$$2H_2C{=}CHCH_3 \xrightarrow{WOCl_4/AlCl_2Et} H_2C{=}CH_2 + H_3CCH{=}CHCH_3$$

Asymmetric hydrogenation of prochiral alkenes

$$(H)(R)C{=}C(COOR)(NHCOR) + H_2 \xrightarrow{[Rh(DiPAMP)_2]^+} RCH_2C^*(COOR)(NHCOR){-}H$$

90 per cent L

Cyclotrimerization of acetylene

$$3CH{\equiv}CH \xrightarrow{Ni(acac)_2}$$

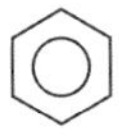

Adapted from J. Halpern, *Inorg. Chim. Acta* **50**, 11 (1981).

Homogeneous catalysis

In this section we describe some important homogeneous catalytic reactions and describe their currently favored mechanisms (Table 17.2). As with nearly all mechanistic chemistry, catalytic mechanisms are subject to refinement or change as more detailed experimental information becomes available. It should also be borne in mind that the mechanism of homogeneous catalysis is more accessible to detailed investigation than that of heterogeneous catalysis because the interpretation of rate data is frequently easier. Moreover, species in solution are often easier to characterize than those on a surface.

From a practical standpoint, homogeneous catalysis is attractive because it is often highly selective toward the formation of a desired product. In large-scale industrial processes homogeneous catalysts are preferred for exothermic reactions because it is easier to dissipate heat from a solution than from the solid bed of a heterogeneous catalyst.

17.3 Catalytic steps

In this section we review five types of reaction (and in some cases their reverse) which, in combination, account for most of the homogeneous catalytic cycles that have been proposed for hydrocarbon transformations. This list will undoubtedly alter and grow as our knowledge of catalysis deepens through further research.

(a) Ligand coordination and dissociation

The catalysis of molecular transformations generally requires facile coordination of reactants to metal ions and equally facile loss of products from the coordination sphere. Both processes must occur with low activation Gibbs energy, so highly labile metal complexes are required. These labile complexes are **coordinatively unsaturated** in the sense that they contain an open coordination site or, at most, a site that is only weakly coordinated.

Square-planar 16-electron complexes are coordinatively unsaturated and are often employed to catalyze the reactions of organic molecules. In Sections 7.8 and 14.2 we met a number of examples of associative reactions between square-planar complexes and entering groups. These facile reactions are quite common for catalytic systems involving ML_4 complexes of Pd(II), Pt(II), and Rh(I), such as the hydrogenation catalyst $[RhCl(PPh_3)_3]$.

> *The catalysis of molecular transformations generally requires facile coordination of reactants to metal ions and equally facile loss of products from the coordination sphere, so coordinatively unsaturated complexes are commonly used.*

(b) Insertion and elimination

The migration of alkyl and hydride ligands to unsaturated ligands, as in the reaction

$$L + \underset{}{\overset{R}{\underset{|}{M}}}-CO \longrightarrow \underset{}{\overset{L}{\underset{|}{M}}}-\underset{\overset{\|}{O}}{C}\overset{R}{}$$

is an example of a migratory insertion reaction (or simply 'insertion reaction') discussed in Section 16.6. Another example of the same kind is the migration of an H ligand to a coordinated alkene to produce a coordinated alkyl ligand:

$$\underset{}{\overset{H}{\underset{|}{M}}}\underset{\overset{\|}{CH_2}}{\overset{CH_2}{}} \longrightarrow \underset{}{M}-CH_2CH_3$$

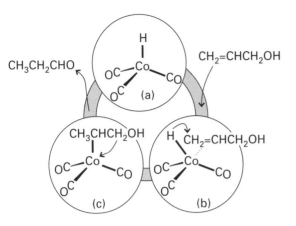

Cycle 17.1

The conversion of (b) to (c) in Cycle 17.1 is a hydrogen migration of this type.

The reverse of insertion is elimination. Elimination reactions include the important β-hydrogen elimination (Section 15.6):

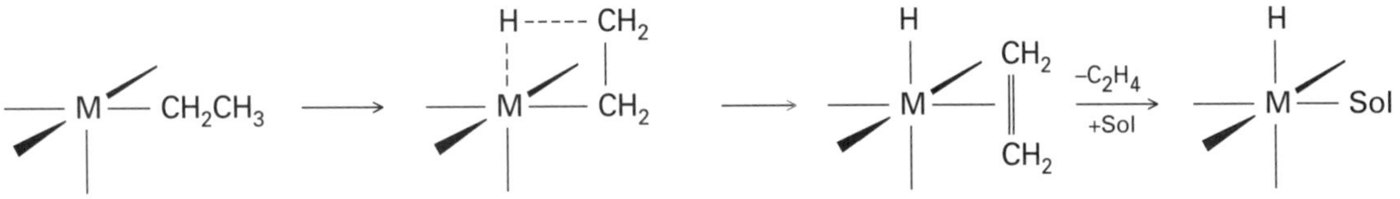

(c) Nucleophilic attack on coordinated ligands

The coordination of ligands such as CO and alkenes to metal ions in positive oxidation states results in the activation of the coordinated C atoms toward attack by nucleophiles. Such reactions are useful in catalysis as well as in organometallic chemistry (Chapter 16).

The hydration of ethene bound to Pd(II) is a good example of catalysis by nucleophilic activation. The stereochemical evidence indicates that the reaction occurs by direct attack on the most highly substituted C atom of the coordinated alkene:

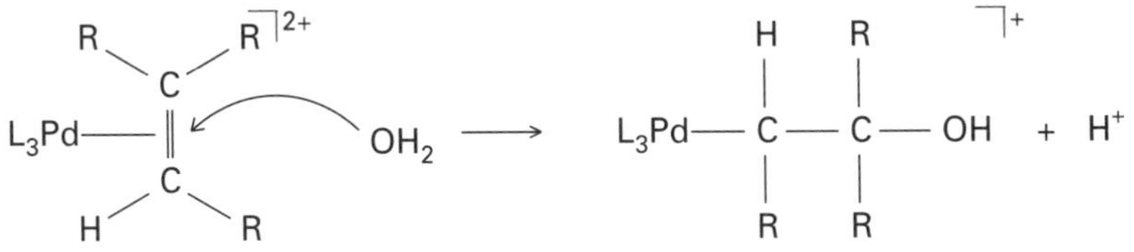

It also is possible for the hydroxylation of a coordinated alkene to take place by the prior coordination of H_2O to a metal complex, followed by an insertion reaction:

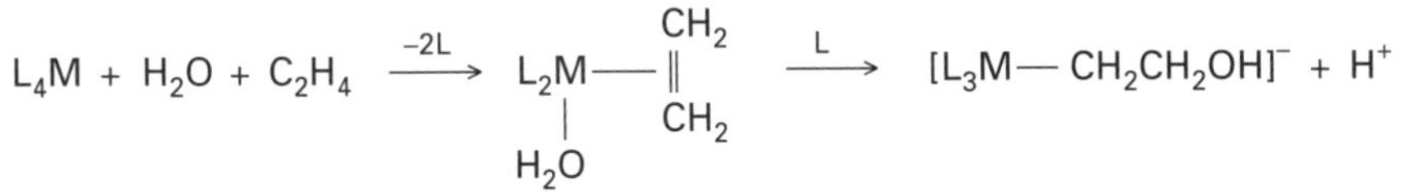

Similarly, a coordinated CO ligand is attacked by an OH^- ion at the C atom, forming a —CO(OH) ligand, which subsequently loses CO_2:

$$L_5M-CO \ + \ OH^- \ \longrightarrow \ L_5M-\overset{\displaystyle O}{\overset{\|}{C}}-OH \ \longrightarrow \ [L_5M-H]^- \ + \ CO_2$$

We discussed this reaction in Section 16.6 in connection with the formation of metal carbonylates. It is thought to be a critical step in the shift reaction catalyzed by metal carbonyl complexes or metal ions on solid surfaces:

$$CO + H_2O \ \xrightarrow{\text{catalyst}} \ CO_2 + H_2$$

> *The coordination of ligands such as CO and alkenes to metal ions in positive oxidation states results in the activation of the coordinated C atoms toward attack by nucleophiles.*

(d) Oxidation and reduction

Metal complexes are often used for the catalytic oxidation of organic substrates. In the catalytic cycle the metal atom alternates between two oxidation states.[3] Some common catalytic one-electron couples are Cu^{2+}/Cu^+, Co^{3+}/Co^{2+}, and Mn^{3+}/Mn^{2+}; a few examples of catalytic two-electron processes are also known, such as with the couple Pd^{2+}/Pd.

Catalysts containing metal ions are used in large-scale processes for the oxidation of hydrocarbons, as in the oxidation of p-xylene (1,4-dimethylbenzene) to terephthalic acid (1,4-benzenedicarboxylic acid). The metal ion can play various roles in these radical oxidations, as we can see by considering a mechanism of the following general form:

Initiation (In· = radical initiator):

$$In \cdot + RH \longrightarrow InH + R\cdot$$

Propagation:

$$R \cdot + O_2 \longrightarrow R-O-O\cdot$$
$$R-O-O \cdot + R-H \longrightarrow R-O-O-H + R\cdot$$

Termination:

$$R \cdot + \cdot R \longrightarrow R_2$$
$$R \cdot + R-O-O\cdot \longrightarrow R-O-O-R$$
$$R-O-O \cdot + \cdot O-O-R \longrightarrow R-O-O-O-O-R \longrightarrow O_2 + \text{nonradicals}$$

The metal ions control the reaction by contributing to the formation of the $R-O-O\cdot$ radicals:

$$Co(II) + R-O-O-H \longrightarrow Co(ROOH) \longrightarrow Co(III)OH + R-O\cdot$$
$$Co(III) + R-O-O-H \longrightarrow Co(II) + R-O-O \cdot + H^+$$

The metal atom shuttles back and forth between oxidation states in this pair of reactions. A metal ion can also act as an initiator. For example, when arenes are involved it is believed that initiation occurs by a simple redox process:

$$ArCH_3 + Co(III) \longrightarrow \cdot ArCH_3^+ + Co(II)$$

> *Metal complexes are often used for the catalytic oxidation of organic substrates, and in the catalytic cycle the metal atom alternates between two oxidation states.*

....

[3] Good general references are: R.S. Sheldon and J.K. Kochi, *Metal-catalyzed oxidations of organic compounds.* Academic Press, Boca Raton (1981); P.M. Henry, *Palladium-catalyzed oxidation of hydrocarbons.* Reidel, Dordrecht (1980).

(e) Oxidative addition and reductive elimination

We saw in Sections 15.6 and 16.6 that the oxidative addition of a molecule AX to a complex brings about dissociation of the A—X bond and coordination of the two fragments:

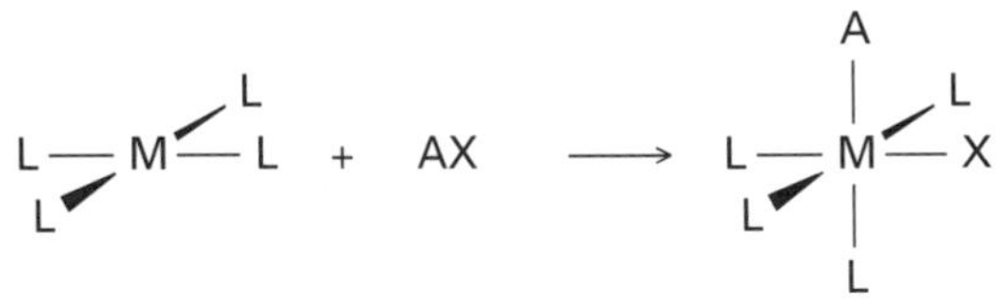

Reductive elimination is the reverse of oxidative addition, and often follows it in a catalytic cycle.

The mechanisms of oxidative addition reactions vary. Depending upon reaction conditions and the nature of the reactants, there is evidence for oxidative addition by simple concerted reaction, heterolytic (ionic) addition of A^+ and X^-, or radical addition of A· and X·. Despite this diversity of mechanism it is found that the rates of oxidative addition of alkyl halides generally follow the orders

primary alkyl < secondary alkyl < tertiary alkyl

$F \ll Cl < Br < I$

17.4 Examples

We shall now see how several individual elementary reactions combine forces and contribute to catalytic cycles. The following examples illustrate our current understanding of the mechanisms of some important types of catalytic reactions and give some insight into the manner in which catalytic activity and selectivity might be altered. As we have already remarked, there is usually even more uncertainty associated with catalytic mechanisms than with mechanisms of simpler kinds of reactions. Unlike simple reactions, a catalytic process frequently contains many steps over which the experimentalist has little control. Moreover, highly reactive intermediates are often present in concentrations too low to be detected spectroscopically. The best attitude to adopt toward these catalytic mechanisms is to learn the pattern of transformations and appreciate their implications, but to be prepared to accept new mechanisms that might be indicated by future work.

The scope of homogeneous catalysis can be appreciated from the examples given in Tables 17.1 and 17.2. The reactions cited there include hydrogenation, oxidation, and a host of other processes. Often the complexes of all metal atoms in a group will exhibit catalytic activity in a particular reaction, but the $4d$-metal complexes are often superior as catalysts to their lighter and heavier congeners. In some cases the difference may be associated with the greater substitutional lability of $4d$ organometallic compounds in comparison with their $3d$ and $5d$ analogs. It is often the case that the complexes of costly metals must be used on account of their superior performance compared with the complexes of cheaper metals.

(a) Hydrogenation of alkenes

The addition of hydrogen to an alkene to form an alkane is favored thermodynamically ($\Delta_r G^\ominus = -101$ kJ mol^{-1} for the conversion of ethene to ethane). However, the reaction rate is negligible at ordinary conditions in the absence of a catalyst. Efficient homogeneous and heterogeneous catalysts are known for the hydrogenation of alkenes and are used in such diverse areas as the manufacture of margarine, pharmaceuticals, and petrochemicals.

One of the most studied catalytic systems is the Rh(I) complex $[RhCl(PPh_3)_3]$, which is often referred to as 'Wilkinson's catalyst'. This useful catalyst hydrogenates a wide variety of alkenes at pressures of hydrogen close to 1 atm or less. The dominant cycle (Cycle 17.2) for the hydrogenation of cyclohexene by Wilkinson's catalyst involves the oxidative addition of H_2 to the 16-electron complex $[RhCl(PPh_3)_3]$ (a), to form the 18-electron dihydrido complex (b). The dissociation of the phosphine ligands from (b) results in the formation of the coordinatively unsaturated complex (c), which then forms the alkene complex (d). Hydrogen transfer from the rhodium atom in (d) to the coordinated alkene yields a transient 16-electron alkyl complex (e). This complex takes on a phosphine ligand to produce (f), and hydrogen migration to carbon results in the reductive elimination of the alkene and the formation of (a), which is set to repeat the cycle. A parallel but slower cycle (which is not shown) reverses the order of H_2 and alkene addition.

Wilkinson's catalyst is highly sensitive to the nature of the phosphine ligand and the alkene substrate. Analogous complexes with alkylphosphine ligands are inactive, presumably because they are more strongly bound to the metal and do not readily dissociate. Similarly, the alkene must be just the right size: highly hindered alkenes or the sterically unencumbered ethene are not hydrogenated by the catalyst. It is presumed that the sterically crowded alkenes do not coordinate and that ethene forms a strong complex, which does not react further. These observations emphasize the point made earlier that a catalytic cycle is usually a delicately poised sequence of reactions, and anything that upsets their flow may block catalysis or alter the mechanism.

Wilkinson's catalyst is used in laboratory-scale organic synthesis and in the production of fine chemicals. Related Rh(I) phosphine catalysts that contain a chiral phosphine ligand have been developed to synthesize optically active products in **enantioselective** reactions. The alkene to be hydrogenated must be **prochiral**, which means that it must have a structure that

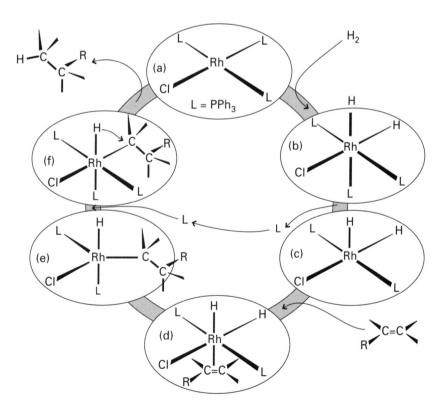

Cycle 17.2

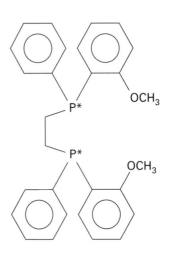

2 DiPAMP

17.2 The diasteromeric complexes that may form from a complex with chiral phosphine ligands (*) and a prochiral alkene.

leads to R or S chirality when complexed to the metal (Fig. 17.2).[4] The resulting complex will thus have two diastereomeric forms depending on the mode of coordination of the alkene. In general, diastereomers have different stabilities and labilities, and in favorable cases one or the other of these effects leads to product enantioselectivity.

An enantioselective hydrogenation catalyst containing a chiral phosphine ligand referred to as DiPAMP (**2**) is used by Monsanto to synthesize L-dopa (**3**).[5] An interesting detail of the process is that the minor diastereomer in solution leads to the major product. The explanation of the greater turnover frequency of the minor isomer lies in the difference in activation Gibbs energies (Fig. 17.3). Spurred by clever ligand design, this field is growing rapidly and providing clinically useful compounds.[6]

> *Wilkinson's catalyst, [RhCl(PPh₃)₃], and related Rh(I) complexes are used for the hydrogenation of a wide variety of alkenes at pressures of hydrogen close to 1 atm or less.*

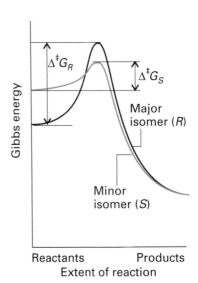

3 L-dopa

(b) Hydroformylation

In a **hydroformylation reaction**, an alkene, CO, and H_2 react to form an aldehyde containing one more C atom than in the original alkene:

$$RCH=CH_2 + CO + H_2 \longrightarrow RCH_2CH_2CHO$$

Both cobalt and rhodium complexes are employed as catalysts. Aldehydes produced by hydroformylation are normally reduced to alcohols that are used as solvents, plasticizers, and in the synthesis of detergents. The scale of production is enormous, amounting to millions of tonnes per year.

The term 'hydroformylation' derived from the idea that the product resulted from the addition of formaldehyde to the alkene, and the name has stuck even though experimental data indicate a different mechanism. A less common but more appropriate name is **hydrocarbonylation**. The general mechanism of cobalt-carbonyl-catalyzed hydroformylation

17.3 Kinetically controlled stereoselectivity. Note that $\Delta^{\ddagger}G_S < \Delta^{\ddagger}G_R$, so the minor isomer reacts faster than the major isomer.

---

[4] The designations R and S for a chiral center are determined as follows. With the element with the lowest atomic number (Z) away from the viewer, the center is R if the sequence of Z from highest to lowest for the remaining three atoms decreases in a clockwise manner. The center is S if the decrease in Z occurs in a counterclockwise manner. There are additional rules in the event of atoms having identical atomic numbers.

[5] L-dopa is a chiral amino acid used to treat Parkinson's disease. The development of this catalyst and the origin of stereoselectivity in the catalytic synthesis are described by W.S. Knowles, *Acc. Chem. Res.* **16**, 106 (1983), and J. Halpern, *Pure Appl. Chem.* **55**, 99 (1983), respectively.

[6] W.A. Nugent, T.V. RajanBabu, and M.J. Burk, Beyond nature's chiral pool: enantioselective catalysis in industry. *Science* **259**, 479 (1993).

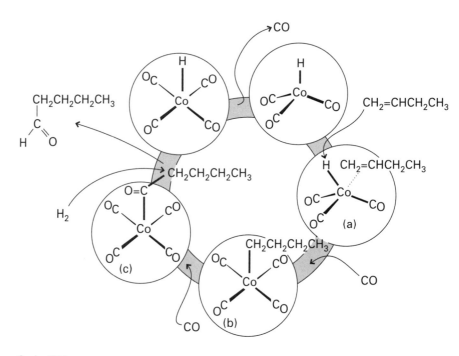

Cycle 17.3

was proposed in 1961 by Heck and Breslow by analogy with reactions familiar from organometallic chemistry (Cycle 17.3). Their general mechanism is still invoked, but has proved difficult to verify in detail.

In the proposed mechanism, a pre-equilibrium is established in which octacarbonyldicobalt combines with hydrogen at high pressure to yield the known tetracarbonylhydridocobalt complex:

$$[Co_2(CO)_8] + H_2 \rightleftharpoons 2\,[CoH(CO)_4]$$

This complex, it is proposed, loses CO to produce the coordinatively unsaturated complex $[CoH(CO)_3]$:

$$[CoH(CO)_4] \rightleftharpoons [CoH(CO)_3] + CO$$

It is then thought that $[CoH(CO)_3]$ coordinates to an alkene, producing (a) in Cycle 17.3, which undergoes an insertion reaction with the coordinated hydrido ligand. The product at this stage is a normal alkane complex (b). In the presence of CO at high pressure, (b) undergoes migratory insertion yielding the acyl complex (c) which has been observed by IR spectroscopy under catalytic reaction conditions. The formation of the aldehyde product is thought to occur by attack of either H_2 (as depicted in Cycle 17.3) or by the strongly acidic complex $[CoH(CO)_4]$ to yield an aldehyde and regenerate the coordinatively unsaturated $[CoH(CO)_3]$.

A significant portion of branched aldehyde is also formed in the cobalt-catalyzed hydroformylation. This product may result from a 2-alkylcobalt intermediate formed when isomerization of (a) is followed by insertion of CO:

Hydrogenation then yields a branched aldehyde:

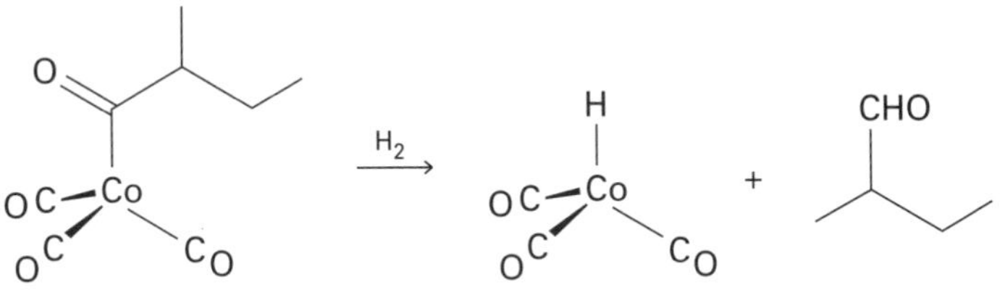

When the linear aldehyde is preferred, such as for the synthesis of biodegradable detergents, it is desirable to suppress the isomerization. It is found that addition of an alkylphosphine to the reaction mixture gives much higher selectivity for the linear product. One plausible explanation is that the replacement of CO by a bulky ligand disfavors the formation of complexes of sterically crowded 2-alkenes:

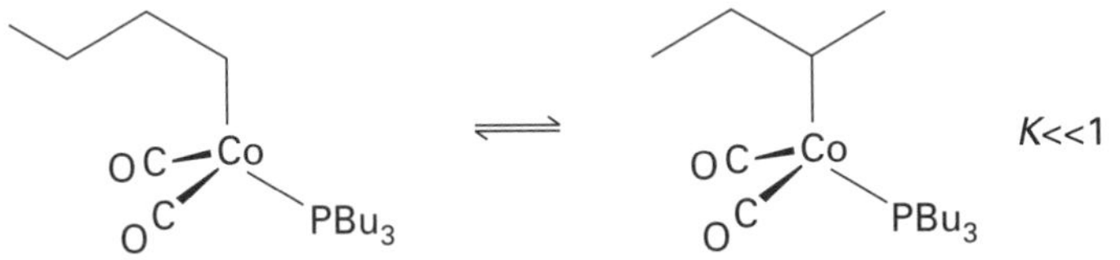

Here again we see an example of the powerful influence of ancillary ligands on catalysis.

In keeping with our comments about the high catalytic activity of 4d-metal complexes, rhodium–phosphine complexes are even more active hydroformylation catalysts than cobalt complexes. One effective catalyst precursor, $[RhH(CO)(PPh_3)_3]$ (4), loses a phosphine ligand to form the coordinatively unsaturated 16-electron complex $[RhH(CO)(PPh_3)_2]$, which promotes hydroformylation at moderate temperatures and 1 atm. This behavior contrasts with the cobalt carbonyl catalyst, which typically requires 150 °C and 250 atm. The rhodium catalyst is useful in the laboratory as it is effective under convenient conditions. Because it favors linear aldehyde products, it competes with the phosphine-modified cobalt catalyst in industry.

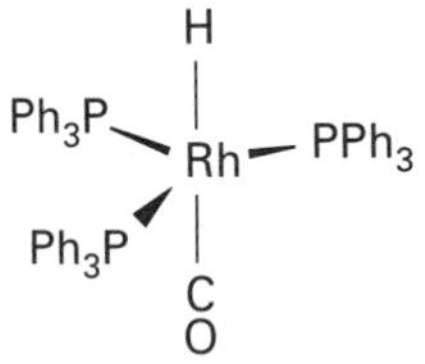

4 $[RhH(CO)(PPh_3)_3]$

> The mechanism of hydrocarbonylation is thought to involve a pre-equilibrium in which octacarbonyldicobalt combines with hydrogen at high pressure.

Example 17.1 Interpreting the influence of chemical variables on a catalytic cycle

An increase in CO partial pressure above a certain threshold decreases the rate of the cobalt-catalyzed hydroformylation of 1-pentene. Suggest an interpretation of this observation.

Answer The decrease in rate with increasing partial pressure suggests that CO suppresses the concentration of one of the catalytic species. An increase in CO pressure will lower the concentration of $CoH(CO)_3$ in the equilibrium

$$[CoH(CO)_4] \rightleftharpoons [CoH(CO)_3] + CO$$

This type of evidence was used as the basis for postulating the existence of $[CoH(CO)_3]$, which is not detected spectroscopically in the reaction mixture.

Self-test 17.1 Predict the influence of added triphenylphosphine on the rate of hydroformylation catalyzed by $[RhH(CO)(PPh_3)_3]$.

(c) Monsanto acetic acid synthesis

The time-honored method for synthesizing acetic acid is by aerobic bacterial action on dilute aqueous ethanol, which produces vinegar. However, this process is uneconomical as a source of concentrated acetic acid for industry. A highly successful commercial process is based on the rhodium-catalyzed carbonylation of methanol:

$$CH_3OH + CO \xrightarrow{[RhI_2(CO)_2]^-} CH_3COOH$$

The reaction is catalyzed by all three members of Group 9 (cobalt, rhodium, and iridium), but complexes of the 4d metal rhodium are the most active. Originally a cobalt complex was employed, but the rhodium catalyst developed at Monsanto greatly reduced the cost of the process by allowing lower pressures to be used. As a result, the **Monsanto process** is used throughout the world.

The principal catalytic cycle in the Monsanto process is illustrated in Cycle 17.4. Under normal operating conditions, the rate-determining step is oxidative addition of iodomethane to the four-coordinate, 16-electron complex $[RhI_2(CO)_2]^-$ (a), producing the six-coordinate 18-electron complex $[(H_3C)RhI_3(CO)_2]^-$ (b). This step is followed by CO migratory insertion, yielding a 16-electron acyl complex (c). Coordination of CO restores an 18-electron complex (d), which is then set to undergo reductive elimination of acetyl iodide with the regeneration of $[RhI_2(CO)_2]^-$. Water then hydrolyzes the acetyl iodide to acetic acid and regenerates HI:

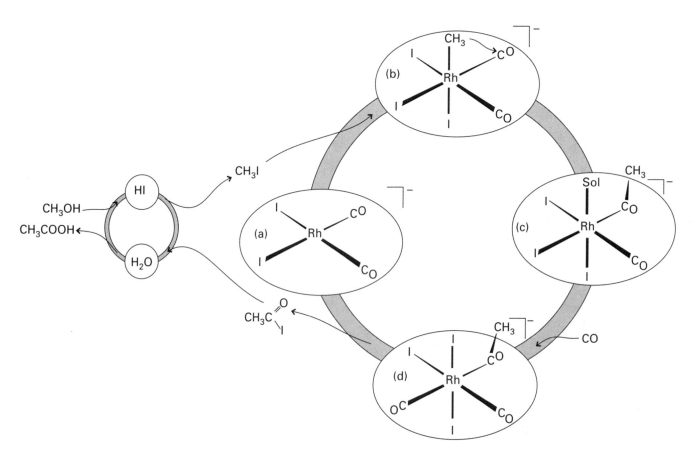

Cycle 17.4

No other anion works as well as iodide in this catalytic system. Its special ability arises from several factors. Among them is the greater rate of oxidative addition of iodomethane relative to the other haloalkanes in the rate-determining step. In addition, the soft I^- ion, which is a good ligand for the soft Rh(I), appears to form a five-coordinate complex, $[RhI_3(CO)_2]^{2-}$, which undergoes oxidative addition with iodomethane more rapidly than $[RhI_2(CO)_2]^-$.

> The Monsanto process is the rhodium-catalyzed carbonylation of methanol to form acetic acid.

(d) Wacker oxidation of alkenes
The **Wacker process** is primarily used to produce acetaldehyde from ethene and oxygen:

$$C_2H_4 \ + \ \tfrac{1}{2}O_2 \ \longrightarrow \ CH_3C\!\!\overset{\displaystyle O}{\underset{\displaystyle H}{\big\langle}} \qquad \Delta_r G^{\ominus} = -197 \text{ kJ mol}^{-1}$$

Its invention at the Wacker Consortium für Elektrochemische Industrie in the late 1950s marked the beginning of an era of production of chemicals from petroleum feed-stock.[7] Although this is no longer a major industrial process, it has some interesting new mechanistic features that are worth noting. An initial step is thought to be

$$C_2H_4 \ + \ PdCl_2 \ + \ H_2O \ \longrightarrow \ CH_3C\!\!\overset{\displaystyle O}{\underset{\displaystyle H}{\big\langle}} \ + \ Pd(0) \ + \ 2HCl$$

The exact nature of the Pd(0) is unknown, but it probably is present as a complex. The slow oxidation of Pd(0) back to Pd(II) by oxygen is catalyzed by the addition of Cu(II), which shuttles back and forth to Cu(I):

$$Pd(0) + 2\,[CuCl_4]^{2-} \ \longrightarrow \ Pd^{2+} + 2\,[CuCl_2]^- + 4\,Cl^-$$
$$2\,[CuCl_2]^- + \tfrac{1}{2}O_2 + 2\,H^+ + 4\,Cl^- \ \longrightarrow \ 2\,[CuCl_4]^{2-} + H_2O$$

The overall catalytic cycle is shown in Cycle 17.5. Detailed stereochemical studies on related systems indicate that the hydration of the alkene–Pd(II) complex (a) occurs by the attack of H_2O from the solution on the coordinated ethene rather than the insertion of coordinated OH.[8] Hydration, to form (b), is followed by two steps that isomerize the coordinated alcohol. First, β-hydrogen elimination occurs with the formation of (c), and then migratory insertion results in the formation of (d). Elimination of the acetaldehyde and a hydrogen ion then leaves Pd(0). The Pd(0) is converted back to Pd(II) by the auxiliary copper(II)-catalyzed air oxidation cycle mentioned above.

Alkene ligands coordinated to Pt(II) are also susceptible to nucleophilic attack, but only palladium leads to a successful catalytic system. The principal reason for palladium's unique behavior appears to be the greater lability of the $4d$ Pd(II) complexes in comparison with their $5d$ Pt(II) counterparts. Furthermore, the potential for the oxidation of Pd(0) to Pd(II) is more favorable than for the corresponding platinum couple.

---

[7] Ethene and other alkenes, the starting materials in the Wacker process and many other petrochemical syntheses, are produced by the thermal cracking of saturated hydrocarbons.

[8] This and related details of the mechanism are still being debated, see G.W. Parshall and S.D. Ittle, *Homogeneous catalysis*. Wiley, New York (1992), and references therein.

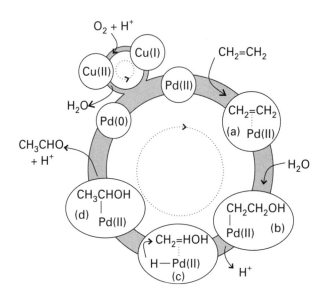

Cycle 17.5

The Wacker process is used to produce acetaldehyde from ethene and oxygen; alkene ligands coordinated to Pt(II) are also susceptible to nucleophilic attack, but only palladium leads to a successful catalytic system.

(e) Alkene polymerization

Polyalkenes, which are among the most common and useful class of synthetic polymers, are most often prepared by use of organometallic catalysts, either in solution or supported on a solid surface. The latter are heterogeneous catalysts but we mention them here both for convenience and because they have important homogeneous analogs.

In the 1950s Karl Ziegler, working in Germany, developed a catalyst for ethene polymerization based on a catalyst formed from $TiCl_4$ and $Al(C_2H_5)_3$, and soon thereafter G. Natta in Italy utilized this type of catalyst for the stereospecific polymerization of propene. When the catalyst is prepared, the $TiCl_4$ and the $Al(C_2H_5)_3$ react to give polymeric $TiCl_3$, which is used in the form of a fine powder. These developments ushered in a revolution in packaging materials, fabrics, and constructional materials. A **Ziegler–Natta catalyst** and another widely used chromium polymerization catalyst are solid particles on which the polymerization takes place at crystal defects where the titanium is coordinatively unsaturated.

The full details of the mechanism of Ziegler–Natta catalysts are still uncertain, but the **Cossee–Arlman mechanism** is regarded as highly plausible (Cycle 17.6). The alkylaluminum alkylates a titanium atom on the surface of the solid, and an ethene molecule coordinates to the neighboring vacant site (represented by the open circle). In the propagation steps for the polymerization the coordinated alkene undergoes a migratory insertion reaction. This migration opens up another neighboring vacancy, and so the reaction can continue and the polymer chain can grow. The release of the polymer from the metal atom occurs by β-hydrogen elimination, and the chain is terminated. The stereospecific character of the polymerization reflects the constraints on the orientation in which the monomer can attach to the metal atom next to the growing chain. Some catalyst remains in the polymer, but the process is so efficient that the amount is negligible.

Homogeneous catalysts related to Ziegler–Natta catalysts provide additional hints on the course of reaction and are of considerable industrial significance in their own right. One example is the tipped ring complex $[(\eta^5\text{-}Cp)_2Zr(CH_3)L]^+$ (**5**), which catalyzes alkene polymerization by successive insertion steps that are thought to involve prior coordination of

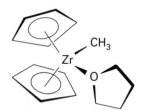

5 $[Zr(Cp)_2(CH_3)(thf)]^+$

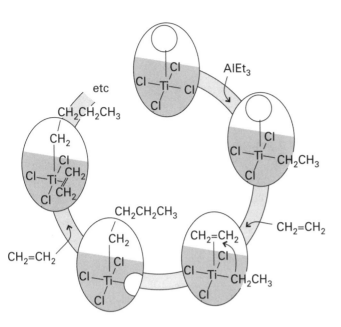

Cycle 17.6

the alkene to the electrophilic Zr center.[9] This type of homogeneous catalyst is now employed commercially for the synthesis of specialized polymers. With a chiral zirconium catalyst (**6**), it is possible to produce optically active polypropylene.

Cycloalkenes other than cyclohexene are susceptible to **ring-opening polymerization** (ROMP), which is catalyzed by certain organometallic complexes:[10]

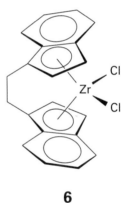

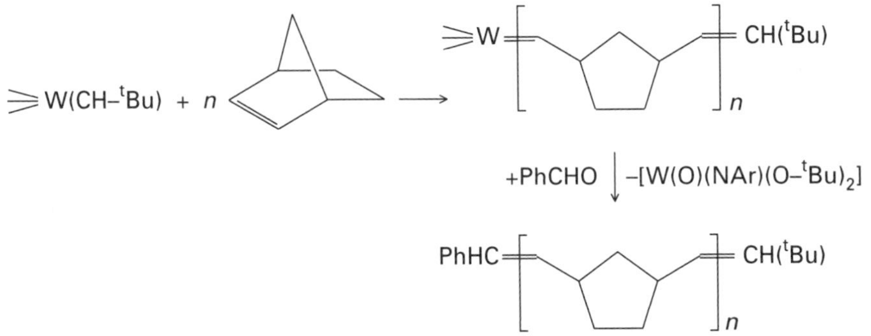

6

In this polymerization reaction an alkene inserts into the metal carbene, giving rise to a new carbene complex with a longer side chain on the carbene. Successive insertions lead to longer chains. The polymerization reaction is thermodynamically favored by the relief of ring strain when the cyclic compound breaks open. A wide variety of metal carbenes are known to catalyze this reaction and many of these result in a **living polymerization**. One characteristic of a living polymerization is that the polymerization continues until all of the monomer is consumed and then can be restarted by simply adding more monomer. The polymer isolated from a living polymerization is often highly monodisperse (that is, has a very narrow range of molar masses). These characteristics result from the absence of chain-transfer reactions, so each catalytic metal center has a chain of similar length attached to it during the polymerization process. The polymerization is terminated by hydrolysis.

[9] R.F. Jordan, *J. Chem. Educ.* **65**, 285 (1988).

[10] R.H. Grubbs and W. Tumas, *Science* **243**, 907 (1989); R.H. Schrock, *Acc. Chem. Res.* **23**, 158 (1990).

> *Ziegler–Natta catalysts are employed in alkene polymerization; the Cossee–Arlman mechanism describes their function. Cycloalkenes other than cyclohexene are susceptible to ring-opening polymerization which is catalyzed by certain organometallic complexes.*

Heterogeneous catalysis

Heterogeneous catalysts are used very extensively in industry. One attractive feature is that many of these solid catalysts are robust at high temperatures and therefore make available a wide range of operating conditions. Another reason for their widespread use is that extra steps are not needed to separate the product from the catalyst. Typically, gaseous or liquid reactants enter a tubular reactor at one end, pass over a bed of the catalyst, and products are collected at the other end. This same simplicity of design applies to the catalytic converter employed to oxidize CO and hydrocarbons and reduce nitrogen oxides in automobile exhausts (Fig. 17.4).

17.5 The nature of heterogeneous catalysts

Information on the structure of a heterogeneous catalyst can sometimes be determined by electron diffraction and related studies of reactive molecules adsorbed on single crystal surfaces. However these studies are generally carried out in ultra-high vacuum conditions, which are very different from those used in practice. Practical heterogeneous catalysts are high-surface-area materials that may contain several different phases and which operate at 1 atm and higher pressures. In some cases the bulk of a high-surface-area material serves as the catalyst, and such a material is called a **uniform catalyst**. One example is the catalytic zeolite ZSM-5, which contains channels through which reacting molecules diffuse. More often **multiphasic catalysts** are used. These consist of a high-surface-area material that serves as a support on to which an active catalyst is deposited (Fig. 17.5).[11]

(a) Surface area and porosity

An ordinary dense solid is unsuitable as a catalyst because its surface area is quite low. Thus α-alumina, which is a dense material with a low specific surface area, is used much less as a catalyst support than the microcrystalline solid γ-alumina, which can be prepared with small particle size, and therefore a high specific surface area. The high surface area results from the many small but connected particles like those shown in Fig. 17.5, and a gram or so of a typical catalyst support has an internal surface area equal to that of a tennis court. Similarly, quartz is not used as a catalyst support but the high specific surface area versions of SiO_2, the silica gels, are widely employed.

Both γ-alumina and silica gel are metastable materials, but under ordinary conditions they do not convert to their more stable phases (α-alumina and quartz, respectively). The preparation of γ-alumina involves the dehydration of an aluminum oxide hydroxide:

$$2\,AlO(OH) \xrightarrow{\ \Delta\ } \gamma\text{-}Al_2O_3 + H_2O$$

Similarly, silica gel is prepared from the acidification of silicates to produce $Si(OH)_4$, which rapidly forms a hydrated silica gel from which much of the adsorbed water can be removed by gentle heating. When viewed with an electron microscope, the texture of silica gel or alumina appears to be that of a rough gravel bed with irregularly shaped voids between the interconnecting particles (as in Fig. 17.5).

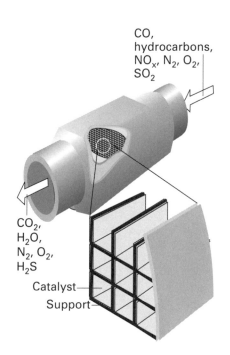

CO,
hydrocarbons,
NO_x, N_2, O_2,
SO_2

CO_2,
H_2O,
N_2, O_2,
H_2S

Catalyst
Support

17.4 A reactor for heterogeneous catalysis. This automobile catalytic converter oxidizes CO and hydrocarbons and reduces nitrogen and sulfur oxides. The metal catalyst is supported on a ceramic honeycomb, which is more robust in this application than a bed of loose particles.

Pt particle   Silica gel

|← 600 Å →|

17.5 Schematic diagram of metal particles supported on silica gel.

---

[11] The distinctions between uniform and multiphase catalysts are developed by J.M. Thomas, *Angew. Chem. Int. Ed. Eng.* **12**, 1673 (1988).

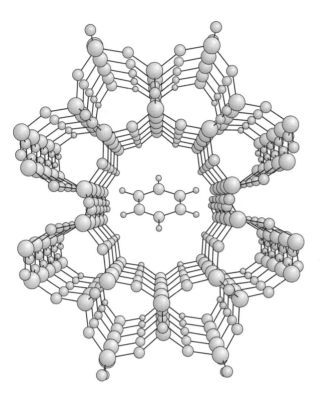

17.6 A view through the channels in Theta-1 zeolite with an adsorbed benzene molecule in one of the channels. (Reproduced by permission from A. Dyer, *An introduction to molecular sieves*. Wiley, Chichester (1988).)

Zeolites (Section 10.14) are examples of uniform catalysts.[12] They are prepared as very fine crystals that contain large regular channels and cages defined by the crystal structure (Fig. 17.6). The openings in these channels vary from one crystalline form of the zeolite to the next, but are typically 3 to 10 Å in their smallest diameter. The zeolite absorbs molecules small enough to enter the channels and excludes larger molecules. This selectivity, in combination with catalytic sites inside the cages, provides a degree of control over catalytic reactions that is unattainable with silica gel or γ-alumina. The synthesis of new zeolites and similar shape-selective solids and the introduction of catalytic sites into them is a vigorous area of research (see Section 10.14).

> The shape selectivity of zeolites in combination with catalytic sites inside the cages, provides a degree of control over catalytic reactions that is unattainable with silica gel or γ-alumina.

(b) Surface acidic and basic sites

When exposed to atmospheric moisture, the surface of γ-alumina is covered with adsorbed water molecules. Dehydration at 100 to 150 °C leads to the desorption of water, but surface OH groups remain and act as weak Brønsted acids:

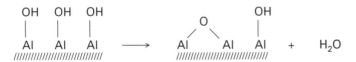

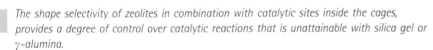

---

[12] An introductory discussion of zeolite structure and catalysis is given by A. Dyer, *An introduction to zeolite molecular sieves*. Wiley, Chichester (1988).

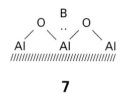

7

At even higher temperatures adjacent OH groups condense to liberate more H_2O and generate exposed Al^{3+} Lewis acid sites as well as O^{2-} Lewis base sites (**7**). The rigidity of the surface permits the coexistence of these strong Lewis acid and base sites, which would otherwise immediately combine to form Lewis acid–base complexes. Surface acids and bases are highly active for catalytic reactions such as the dehydration of alcohols and isomerization of alkenes. Similar Brønsted and Lewis acid sites exist on the interior of certain zeolites.

> *Surface acids and bases are highly active for catalytic reactions such as the dehydration of alcohols and isomerization of alkenes.*

Example 17.2 Using IR spectra to probe molecular interaction with surfaces

Infrared spectra of hydrogen-bonded and Lewis acid complexes of pyridine (py), such as $Cl_3Al(py)$, show that bands near 1540 cm^{-1} can be ascribed to hydrogen-bonded pyridine and that bands near 1465 cm^{-1} are due to Al-py Lewis acid–base interactions. A sample of γ-alumina that had been pre-treated by heating to 200 °C and then cooled and exposed to pyridine vapor had absorption bands near 1540 cm^{-1} and none near 1465 cm^{-1}. Another sample that was heated to 500 °C, cooled, and then exposed to pyridine had bands near 1540 cm^{-1} and 1465 cm^{-1}. Correlate these results with the statements made in the text concerning the effect of heating γ-alumina. (Much of the evidence for the chemical nature of the γ-alumina surface comes from experiments like these.)

Answer At about 200 °C, surface H_2O is lost but OH^- bound to Al^{3+} remains. These groups appear to be mildly acidic as judged by color indicators and bands around 1540 cm^{-1} that indicate the presence of hydrogen-bonded pyridine generated in the reaction

$$
\begin{array}{ccc}
\text{OH} & & \text{O} \text{---} \text{H} \text{----} \text{py} \\
| & & | \\
\text{Al} \quad + \quad \text{py} \longrightarrow & & \text{Al} \\
/\!/\!/\!/\!/ & & /\!/\!/\!/\!/
\end{array}
$$

When heated to 500 °C, much but not all of the OH is lost as H_2O, leaving behind O^{2-} and exposed Al^{3+}. The evidence is the appearance of the 1465 cm^{-1} bands, which are indicative of Al^{3+}—NC_5H_5, as well as the 1540 cm^{-1} band.

Self-test 17.2 If the statements in this section about the effect of heating γ-alumina are correct, what would be the intensities of the diagnostic IR bands if the γ-alumina sample were pre-treated at 900 °C, cooled in the absence of water, exposed to pyridine vapor, and a spectrum determined?

(c) Surface metal sites

Metal particles are often deposited on supports to provide a catalyst. For example, finely divided platinum–rhenium alloys distributed on the surface of γ-alumina particles are used to interconvert hydrocarbons, and finely divided platinum–rhodium alloy particles supported on γ-alumina are employed in automobile catalytic converters to promote the combination of O_2 with CO and hydrocarbons to CO_2 and reduction of nitrogen oxides to nitrogen. A supported metal particle about 25 Å in diameter has about 40 per cent of its atoms on the surface, and the particles are protected from fusing together into bulk metal by their separation. The high proportion of exposed atoms is a great advantage for these small supported particles, particularly for metals such as platinum and the even more expensive rhodium.

The metal atoms on the surface of metal clusters are capable of forming bonds such as M—CO, M—CH_2R, M—H, and M—O (Table 17.3). Often the nature of surface ligands is inferred by comparison of IR spectra with those of organometallic or inorganic complexes. Thus, both terminal and bridging CO can be identified on surfaces by IR spectroscopy, and the IR spectra of many hydrocarbon ligands on surfaces are similar to those of discrete organometallic complexes. The case of the N_2 ligand is an interesting contrast, because coordinated N_2 was identified by IR spectroscopy on metal surfaces before dinitrogen complexes were prepared and characterized by inorganic chemists.

The development of new techniques for studying single crystal surfaces has greatly expanded our knowledge of the surface species that may be present in catalysis. For example, the desorption of molecules from surfaces (thermally, or by ion or atom impact) combined with mass spectrometric analysis of the desorbed substance, provides insight into the chemical identity of surface species. Similarly, Auger and X-ray photoelectron spectroscopy (XPS) provide information on the elemental composition of surfaces.

Low-energy electron diffraction (LEED) provides information about the structure of single crystal surfaces and, when adsorbate molecules are present, their arrangement on the surface. One important finding from LEED is that the adsorption of small molecules on a surface may bring about a structural modification of the surface. This surface reconstruction is often observed to reverse when desorption occurs. Scanning tunneling microscopy (STM) provides an unrivaled method for locating adsorbates on surfaces. This striking technique provides a contour map of single crystal surfaces at or close to atomic resolution.[13]

Although most of these modern surface techniques cannot be applied to the study of supported multiphasic catalysts, they are very helpful for revealing the range of probable

Table 17.3 Chemisorbed ligands on surfaces

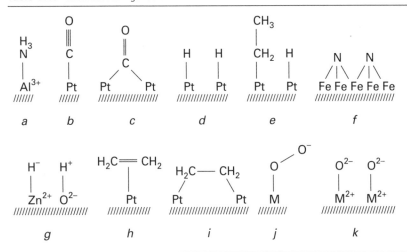

a Ammonia adsorbed on the Lewis and Al^{3+} sites of γ-alumina.
b,c CO coordinated to platinum metal.
d Hydrogen dissociatively chemisorbed on platinum metal.
e Ethane dissociatively chemisorbed on platinum.
f Nitrogen dissociatively chemisorbed on iron.
g H_2 dissociatively chemisorbed on ZnO.
h Ethene η^2 coordinated to a Pt atom.
i Ethene bonded to two Pt atoms.
j O_2 bound as a superoxide to a metal surface.
k O_2 dissociatively chemisorbed on a metal surface.
Adapted from: R.L. Burwell, Jr., Heterogeneous catalysis. *Survey of Progress in Chemistry* **8**, 2 (1977).

[13] These and other modern surface techniques and their help in the elucidation of catalysis are summarized by G.A. Samorjai in *Surface chemistry and catalysis.* Wiley, New York (1994).

surface species and circumscribing the structures that may plausibly be invoked in a mechanism of heterogeneous catalysis. The application of these techniques to heterogeneous catalysis is similar to the use of X-ray diffraction and spectroscopy for the characterization of organometallic homogeneous catalyst precursors and model compounds.

> Terminal and bridging CO can be identified on surfaces by IR spectroscopy; the IR spectra of many hydrocarbon ligands on surfaces are similar to those of discrete organometallic complexes.

17.6 Catalytic steps

There are many parallels between the individual reaction steps encountered in heterogeneous and homogeneous catalysis.[14]

(a) Chemisorption and desorption

The adsorption of molecules on surfaces often activates molecules just as coordination activates molecules in complexes. The desorption of product molecules that is necessary to refresh the active sites in heterogeneous catalysis is analogous to the dissociation of a complex in homogeneous catalysis.

Before a heterogeneous catalyst is used it is usually 'activated'. Activation is a catch-all term; in some instances it refers to the desorption of adsorbed molecules such as water from the surface, as in the dehydration of γ-alumina. In other cases it refers to the preparation of the active site by a chemical reaction, such as by reduction of metal oxide particles to produce active metal particles.

An activated surface can be characterized by the adsorption of various inert and reactive gases. The adsorption may be either **physisorption**, when no new chemical bond is formed, or **chemisorption**, when surface–adsorbate bonds are formed (Fig. 17.7). Low-temperature physisorption of a gas like nitrogen is useful for the determination of the total surface area of a solid, whereas chemisorption is used to determine the number of exposed reactive sites. For example, the dissociative chemisorption of H_2 on supported platinum particles reveals the number of exposed surface Pt atoms.

The interaction of small molecules with metal surfaces is similar to their interaction with low-oxidation-state metal complexes. Table 17.4 shows that a wide range of metals chemisorb CO, and that many fewer are capable of chemisorbing N_2, just as there is a much

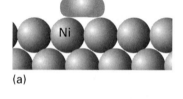

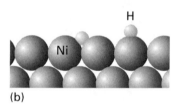

17.7 Schematic representation of (a) physisorption and (b) chemisorption of hydrogen on a nickel metal surface.

Table 17.4 The ability of metals to chemisorb simple molecules

	Gases						
	O_2	C_2H_2	C_2H_4	CO	H_2	CO_2	N_2
Ti, Zr, Hf, V, Nb, Ta, Cr, Mo, W, Fe, Ru, Os	+	+	+	+	+	+	+
Ni, Co	+	+	+	+	+	+	+
Rh, Pd, Pt, Ir	+	+	+	+	+	+	+
Mn, Cu	+	+	+	+	±	+	+
Al, Au	+	+	+	+	−	−	−
Na, K	+	+	−	−	−	−	−
Ag, Zn, Cd, In, Si, Ge, Sn, Pb, As, Sb, Bi	+	−	−	−	−	−	−

+ Strong chemisorption, ± weak, − unobservable.
Adapted from G.C. Bond, *Heterogeneous catalysis*, p. 29. Oxford University Press (1987).

----

[14] For an entertaining and informative account of the development of mechanistic concepts in heterogeneous catalysis see R.L. Burwell, Jr, *Chemtech* **17**, 586 (1987).

wider variety of metals that form carbonyls than form dinitrogen complexes. Furthermore, just as with metal carbonyl complexes, both bridging and terminal CO surface species have been identified by IR spectroscopy. The dissociative chemisorption of H_2 is analogous to the oxidative addition of H_2 to metal complexes.

Even though adsorption is essential for heterogeneous catalysis to occur, it must not be so strong as to block the catalytic sites and prevent further reaction. This factor is in part responsible for the limited number of metals that are effective catalysts. The catalytic decomposition of formic acid on metal surfaces

$$HCOOH \xrightarrow{M} CO + H_2O$$

provides a good example of this balance between adsorption and catalytic activity. It is observed that the catalysis is fastest on metals for which the metal formate is of intermediate stability (Fig. 17.8). The plot in Fig. 17.8 is an example of a 'volcano diagram', and is typical of many catalytic reactions. The implication is that the earlier d-block metals form very stable surface compounds whereas the later noble metals such as silver and gold form very weak surface compounds, both of which are detrimental to a catalytic process. Between these extremes the metals in Groups 8 to 10 have high catalytic activity, and this is especially true of the platinum metals (Group 10). In Section 17.4 we saw a similar high activity of these metal complexes in the homogeneous catalysis of hydrocarbon transformations.

The active sites of heterogeneous catalysts are not uniform and many diverse sites are exposed on the surface of a poorly crystalline solid such as γ-alumina or a noncrystalline solid such as silica gel. However, even highly crystalline metal particles are not uniform. A crystalline solid has typically more than one type of exposed plane, each with its characteristic pattern of surface atoms (Fig. 17.9). In addition, single crystal metal surfaces have irregularities such as steps that expose metal atoms with low coordination numbers (Fig. 17.10). These highly exposed, coordinatively unsaturated sites appear to be particularly reactive. As a result, the different sites on the surface may serve different functions in catalytic reactions. The variety of sites also accounts for the lower selectivity of many heterogeneous catalysts in comparison with their homogeneous analogs.

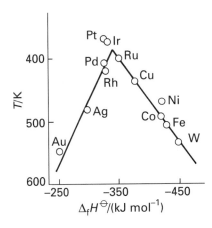

17.8 A volcano diagram. In this case the reaction temperature for a set rate of formic acid decomposition is plotted against the stability of the corresponding metal formate as judged by enthalpy of formation. (W.J.M. Rootsaert and W.M.H. Sachtler, *Z. Physik. Chem.* **26**, 16 (1960).)

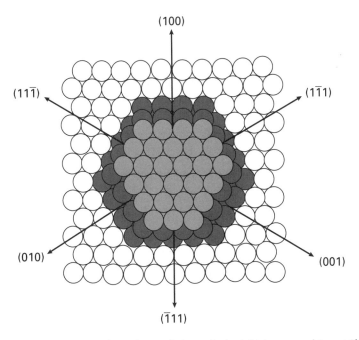

17.9 A collection of metal crystal planes that might be exposed to reactive gases. ($\bar{1}$11), ($1\bar{1}$1), etc. are close-packed hexagonal planes. The planes represented by (100), (010), etc. have square arrays of atoms.

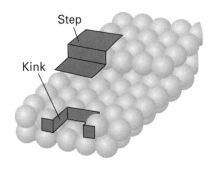

17.10 Schematic representation of surface irregularities, steps and kinks.

| Even though adsorption is essential for heterogeneous catalysis to occur, it must not be so strong as to block the catalytic sites and prevent further reaction.

(b) Surface migration

We saw in Chapter 16 that ligands in organometallic clusters are often fluxional. The surface analog of stereochemical mobility in clusters is diffusion, and there is abundant evidence for the diffusion of chemisorbed molecules or atoms on metal surfaces. For example, adsorbed H atoms and adsorbed CO molecules are known to move over the surface of a metal particle. This mobility is important in catalytic reactions for it allows atoms or molecules to approach one another. Surface diffusion rates are difficult to measure and reliable data have been obtained only recently.

| Adsorbed atoms and molecules migrate over metal surfaces.

17.7 Examples

A huge range of industrial processes are facilitated by heterogeneous catalysis. We give a few examples here that illustrate some of this range. Despite some superficial similarities the first two examples, hydrogenation of alkenes and ammonia synthesis (the hydrogenation of nitrogen), employ quite different catalysts and conditions. The example of the isomerization of aromatics on zeolites illustrates acid catalysis. Finally, we illustrate catalysis at the electrode of an electrochemical cell.

(a) Hydrogenation of alkenes

A milestone in heterogeneous catalysis was Paul Sabatier's observation that nickel catalyzes the hydrogenation of alkenes (1900). He was actually attempting to synthesize $Ni(C_2H_4)_4$ in response to Mond, Langer, and Quinke's synthesis of $Ni(CO)_4$ (Section 16.4). However, when he passed ethene over heated nickel he detected ethane. His curiosity was sparked, so he included hydrogen with the ethene, whereupon he observed a good yield of ethane. Major industrial applications soon followed.

The hydrogenation of alkenes on supported metal particles is thought to proceed in a manner very similar to that in metal complexes. As pictured in Fig. 17.11, H_2, which is dissociatively chemisorbed on the surface, is thought to migrate to an adsorbed ethene molecule, giving first a surface alkyl and then the saturated hydrocarbon. When ethene is hydrogenated with D_2 over platinum, the simple mechanism depicted in Fig. 17.11 indicates that CH_2DCH_2D should be the product. In fact, a complete range of $C_2H_nD_{6-n}$ ethane molecules is observed. It is for this reason that the central step is written as reversible; the rate of the reverse reaction must be greater than the rate at which the ethane molecule is formed and desorbed in the final step.

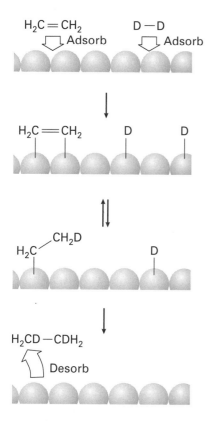

17.11 Schematic diagram of the hydrogenation of ethene by deuterium on a metal surface.

| In the hydrogenation of alkenes on supported metal particles H_2 is dissociatively chemisorbed on the surface, migrates to an adsorbed ethene molecule, and gives first a surface alkyl and then the saturated hydrocarbon.

(b) Ammonia synthesis

The synthesis of ammonia has already been discussed from several different viewpoints (Sections 8.14 and 11.2). Here we concentrate on details of the catalytic steps. The formation of ammonia is exoergic and exothermic at 25 °C, the relevant thermodynamic data being $\Delta_f G^\ominus = -16.5 \text{ kJ mol}^{-1}$, $\Delta_f H^\ominus = -46.1 \text{ kJ mol}^{-1}$, and $\Delta_f S^\ominus = -99.4 \text{ J K}^{-1} \text{ mol}^{-1}$. The negative entropy of formation reflects the fact that two NH_3 molecules form in place of four reactant molecules.

The great inertness of N_2 (and to a lesser extent H_2) requires that a catalyst be used to effect the reaction. Iron metal together with small quantities of alumina and potassium salts and other promoters is used as the catalyst. Extensive studies on the mechanism of ammonia synthesis indicate that the rate-determining step under normal operating conditions is the dissociation of N_2 coordinated to the catalyst surface. The other reactant, H_2, undergoes much more facile dissociation on the metal surface and a series of insertion reactions between adsorbed species leads to the production of NH_3:

$N_2(g) \longrightarrow$ $N_2 \atop /////$ $\longrightarrow$ $2N \atop //////$

$H_2(g)$ $\longrightarrow$ $2H \atop //////$

$N \atop ////$ $+$ $H \atop ////$ $\longrightarrow$ $NH \atop //////$

$NH \atop //////$ $+$ $H \atop ////$ $\longrightarrow$ $NH_2 \atop ///////$

$NH_2 \atop ///////$ $+$ $H \atop ////$ $\longrightarrow$ $NH_3 \atop ////////$

$NH_3 \atop ////////$ $\longrightarrow$ $NH_3(g)$

Because of the slowness of the N_2 dissociation, it is necessary to run the ammonia synthesis at high temperatures, typically $400\,°C$. However, since the reaction is exothermic, high temperature greatly reduces the equilibrium constant of the reaction: from the van't Hoff equation

$$\frac{\mathrm{d}\ln K}{\mathrm{d}T} = \frac{\Delta_f H^{\ominus}}{RT^2} < 0 \tag{3}$$

we see that K decreases as T is increased. To recover some of this reduced yield, pressures on the order of $100\,\mathrm{bar}$ are employed. A catalyst that would operate at room temperature would give good equilibrium yields of NH_3, but such catalysts have not yet been discovered (Section 11.2).

In the course of developing the original ammonia synthesis process, Haber and his coworkers investigated the catalytic activity of most of the metals in the periodic table, and found that the best are iron, ruthenium, and uranium promoted (rendered more active) by small amounts of alumina and potassium salts. Cost and toxicity considerations led to the choice of iron as the basis of the commercial catalyst. Other metals, such as lithium, are more active for the critical N≡N cleavage step, but in these cases the succeeding steps are not favorable on account of the great stability of the metal nitride. This observation once again emphasizes the detrimental nature of a highly stable intermediate in a catalytic cycle.

Promoted iron metal is used as the catalyst for the synthesis of ammonia; the rate-determining step is the dissociation of N_2 coordinated to the catalyst surface.

(c) SO_2 oxidation

The oxidation of SO_2 to SO_3 is a key step in the production of sulfuric acid (Section 11.8). The reaction of sulfur with oxygen to produce SO_3 gas is exoergic ($\Delta_r G^{\ominus} = -371\ \mathrm{kJ\,mol^{-1}}$) but very slow, and the principal product of combustion is SO_2:

$$S(s) + O_2(g) \longrightarrow SO_2(g)$$

The combustion is followed by the catalytic oxidation of SO_2:

$$SO_2(g) + \tfrac{1}{2}O_2(g) \longrightarrow SO_3(g)$$

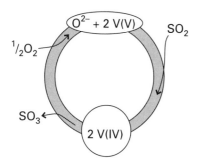

Cycle 17.7

This step is also exothermic and, as with ammonia synthesis, has a less favorable equilibrium constant at elevated temperatures. The process is therefore generally run in stages. In the first stage, the combustion of sulfur raises the temperature to about $600\,^\circ$C, but by cooling before the catalytic stage the equilibrium is driven to the right and high conversion of SO_2 to SO_3 is achieved.

Several quite different catalytic systems have been used to catalyze the combination of SO_2 with O_2. The most widely used catalyst at present is potassium vanadate supported on a high-surface-area silica (diatomaceous earth). One interesting aspect of this catalyst is that the vanadate is a molten salt under operating conditions. The current view of the mechanism of the reaction is that the rate-determining step is the oxidation of V(IV) to V(V) by O_2 (Cycle 17.7). In the melt the vanadium and oxide ions are part of a polyvanadate complex (Section 5.6), but little is known about the evolution of the oxo species.

> The most widely used catalyst for the oxidation of sulfur dioxide is potassium vanadate supported on a high-surface-area silica.

(d) Interconversion of aromatics by zeolites

The zeolite-based heterogeneous catalysts play an important role in the interconversion of hydrocarbons and the alkylation of aromatics as well as in oxidation and reduction.

The synthetic zeolite ZSM-5 is widely used by the petroleum industry for a variety of hydrocarbon interconversions.[15] ZSM-5 is an aluminosilicate zeolite with a higher silica content and a much lower aluminum content than the zeolites discussed in Section 10.14. Its channels consist of a three-dimensional maze of intersecting tunnels (Fig. 17.12). As with other aluminosilicate catalysts, the aluminum sites are strongly acidic. The charge imbalance of Al^{3+} in place of tetrahedrally coordinated Si(IV) requires the presence of an added positive ion. When this ion is H^+ (Fig. 17.13) the Brønsted acidity of the aluminosilicate can be as great as that of concentrated H_2SO_4 and the turnover frequency for hydrocarbon reactions at these sites can be very high.[16]

Reactions such as xylene isomerization and toluene disproportionation are illustrative of the selectivity that can be achieved with acidic zeolite catalysis. The shape selectivity of these catalysts has been attributed to faster diffusion of product molecules that have dimensions

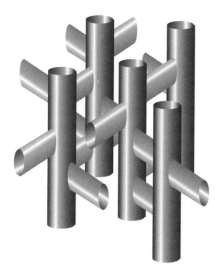

17.12 The structure of ZSM-5 has intersecting channels defined by the crystal structure. The tubes in this diagram represent the channels with the three-dimensional structure.

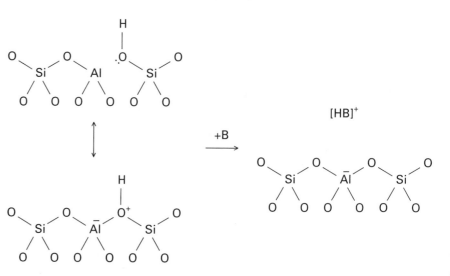

17.13 The Brønsted acid site in HZSM-5 and its interaction with a base. (From W.O. Haag, R.M. Lago, and P.B. Weisz, *Nature* **309**, 589 (1984).)

[15] The catalyst was developed in the research laboratories of Mobil Oil; the initials stand for Zeolite Socony–Mobil.

[16] W.O. Haag, R.M. Lago, and P.B. Weisz, *Nature* **309**, 589 (1984).

compatible with the channels. According to this idea, molecules that do not fit the channels diffuse slowly and, because of their long residence in the zeolite, have ample opportunity to be converted to the more mobile isomers that can escape rapidly. A currently more favored view, however, is that the orientation of reactive intermediates within the zeolite channels favors specific products, such as 1,4-dialkylbenzenes.

Aside from their important shape selectivity, acidic zeolite catalysts appear to promote reactions by standard carbonium ion mechanisms. For example, the isomerization of m-xylene to p-xylene may occur by the following steps:

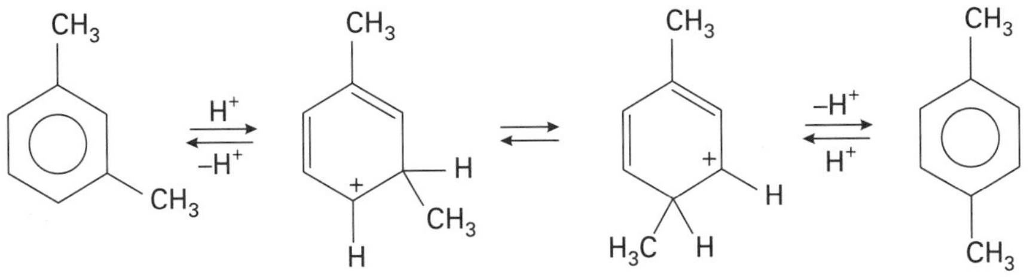

Another common reaction in zeolites is the alkylation of aromatics with alkenes.

The shape selectivity of zeolite catalysts has been attributed to faster diffusion of product molecules that have dimensions compatible with the channels.

Example 17.3 Proposing a mechanism for the alkylation of benzene

In its protonated form, ZSM-5 catalyzes the reaction of ethene with benzene to produce ethylbenzene. Write a plausible mechanism for the reaction.

Answer The acidic form of ZSM-5 is strong enough to generate carbocations:

$$CH_2{=}CH_2 + H^+ \longrightarrow [CH_2CH_3]^+$$

As we saw in Section 5.8, the carbocation can attack benzene. Subsequent deprotonation of the product yields ethylbenzene:

$$[CH_2CH_3]^+ + C_6H_6 \longrightarrow C_6H_5CH_2CH_3 + H^+$$

Self-test 17.3 A pure silica analog of ZSM-5 can be prepared. Would you expect this compound to be an active catalyst for benzene alkylation? Explain your reasoning.

(e) Electrocatalysis

The kinetic barrier to electrochemical reactions involving H_2 and O_2 was discussed in Chapter 6. Kinetic barriers are quite common for electrochemical reactions at the interface between a solution and an electrode and, as we saw in Section 6.4, it is common to express these barriers as overpotentials.

The overpotential, η, of an electrolytic cell is the potential in addition to the zero-current cell potential that must be applied in order to bring about an otherwise slow reaction within the cell. The overpotential is empirically related to the current density, j (the current divided by the area of the electrode) that passes through the cell by[17]

$$j = j_0 e^{a\eta} \tag{4}$$

[17] The exponential relation between the current and the overpotential is accounted for by the Butler–Volmer equation, which is derived by applying activated complex theory to dynamical processes at electrodes. See Chapter 29 of P.W. Atkins, *Physical chemistry*. Oxford University Press and W.H. Freeman & Co (1998).

where j_0 and a are empirical constants. The constant j_0, the **exchange current density**, is a measure of the rates of the forward and reverse electrode reactions at dynamic equilibrium in the absence of an overpotential. For systems obeying these relations, the reaction rate (as measured by the current density) increases rapidly with increasing applied potential when $a\eta > 1$. If the exchange current density is high, an appreciable reaction rate may be achieved for only a small overpotential. If the exchange current density is low, a high overpotential is necessary. There is therefore considerable interest in controlling the exchange current density. In an industrial process an overpotential in a synthetic step is very costly because it represents wasted energy. However, a low exchange current density may block undesirable pathways.

A catalytic electrode surface can increase the exchange current density and hence dramatically decrease the overpotential required for sluggish electrochemical reactions, such as H_2, O_2, or Cl_2 evolution and consumption. For example, 'platinum black', a finely divided form of platinum, is very effective for increasing the exchange current density and hence decreasing the overpotential of reactions involving the consumption or evolution of H_2. The role of platinum is to dissociate the strong H—H bond and thereby to reduce the large barrier that this strength imposes on reactions involving H_2. Palladium metal also has a high exchange current density (and hence requires only a low overpotential) for H_2 evolution or consumption.

The effectiveness of metals can be judged from Fig. 17.14, which also gives insight into the process. The volcano-like plot of exchange current density against **M—H** bond enthalpy suggests that **M—H** bond formation and cleavage are both important in the catalytic process. It appears that an intermediate **M—H** bond energy leads to the proper balance for the existence of a catalytic cycle and the most effective metals for electrocatalysis are clustered around Group 10.

Ruthenium dioxide is an effective catalyst for both O_2 and Cl_2 evolution and it also is a good electrical conductor. It turns out that at high current densities RuO_2 is more effective for the catalysis of Cl_2 evolution than for O_2 evolution. Therefore RuO_2 is extensively used as an electrode material in commercial chorine production. The electrode processes that contribute to this subtle catalytic effect do not appear to be well understood.

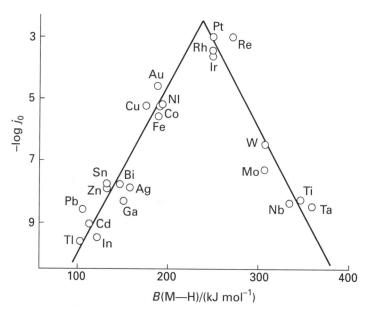

17.14 The rate of H_2 evolution expressed as the logarithm of the exchange current density versus M—H bond energies.

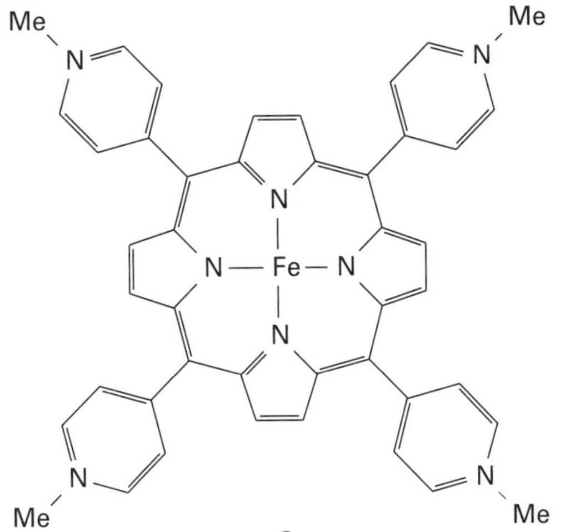

8

There is great interest in devising new catalytic electrodes, particularly ones that decrease the O_2 overpotential on surfaces such as graphite. Thus, tetrakis(4-N-methylpyridyl) tetraphenylporphyriniron(II), [Fe(TMPyP)] (**8**), has been deposited on the exposed edges of graphite electrodes (on which O_2 reduction requires a high overpotential) and the resulting electrode surface was found to catalyze the electrochemical reduction of O_2. A plausible explanation for this catalysis is that [Fe^{III}(TMPyP)] attached to the electrode (indicated below by *) is first reduced electrochemically:

$$[Fe^{III}(TMPyP)]^* + e^- \longrightarrow [Fe^{II}(TMPyP)]^*$$

The resulting [Fe^{II}(TMPyP)] forms an O_2 complex:

$$[Fe^{II}(TMPyP)]^* + O_2 \longrightarrow [Fe^{II}(O_2)(TMPyP)]^*$$

This iron(II) porphyrin oxygen complex is then susceptible to reduction to both water and hydrogen peroxide:

$$[Fe^{II}(O_2)(TMPyP)]^* + ne^- + mH^+ \longrightarrow [Fe^{II}(TMPyP)]^* + (H_2O_2, H_2O)$$

Although details of the mechanism are still elusive, the general set of reactions given above is in harmony with electrochemical measurements and the known properties of iron porphyrins. The investigation of porphyriniron(III) complex as a catalyst was undoubtedly motivated by Nature's use of metalloporphyrins for oxygen activation, a subject we explore in Chapter 19.

Further reading

Homogeneous catalysis

G.W. Parshall and S.D. Ittel, *Homogeneous catalysis*. Wiley, New York (1992). This extensive revision of the highly successful first edition, documents the chemistry of homogeneous catalysts that are of importance in industrial and laboratory-scale organic synthesis.

R.S. Dixon, *Homogeneous catalysis with compounds of rhodium and iridium*. D. Reidel, Dordrecht (1985).

L.H. Pignolet, *Homogeneous catalysis with metal phosphine complexes*. Plenum, New York (1983).

C. Masters, *Homogeneous transition-metal catalysis: a gentle art*. Chapman & Hall, London (1981).

B.R. James, *Homogeneous hydrogenation*. Wiley, New York (1973).

R.A. Shelton and J.K. Kochi, *Metal catalyzed oxidations of organic compounds*. Academic Press, New York (1981).

J.P. Collman, L.S. Hegedus, J.R. Norton, and R.G. Finke, *Principles and applications of organotransition metal chemistry*. University Science Books, Mill Valley (1987). This important book covers both stoichiometric and catalytic organometallic chemistry.

Heterogeneous catalysis

F.H. Ribeiro and G.A. Somorjai, Heterogeneous catalysis by metals. In *Encyclopedia of inorganic chemistry* (ed. R.B. King), Vol. 3, pp. 1359–71. Wiley, New York (1994).

V. Ponec and G.C. Bond, *Catalysis by metals and alloys*. Elsevier, Amsterdam (1995).

R.D. Srivatava, *Heterogeneous catalytic science*. CRC Press, Boca Raton (1988). A survey of experimental methods and several major catalytic processes, including the oxidation of hydrocarbons, the hydrogenation of CO, and hydrocarbon reforming.

J.M. Thomas and W.J. Thomas, *Principles and practice of heterogeneous catalysis*. VCH, Weinheim (1997).

B.C. Gates, *Catalytic chemistry*. Wiley-Interscience, New York (1991). This text discusses both homogeneous and heterogeneous catalysis.

Exercises

17.1 Which of the following constitute genuine examples of catalysis and which do not? Present your reasoning. (a) The addition of H_2 to C_2H_4 when the mixture is brought into contact with finely divided platinum. (b) The reaction of an H_2/O_2 gas mixture when an electric arc is struck. (c) The combination of N_2 gas with lithium metal to produce Li_3N, which then reacts with H_2O to produce NH_3 and LiOH.

17.2 Define the terms (a) turnover frequency, (b) selectivity, (c) catalyst, (d) catalytic cycle, and (e) catalyst support.

17.3 Classify the following as homogeneous or heterogeneous catalysis and present your reasoning. (a) The increased rate in the presence of NO(g) of SO_2(g) oxidation by O_2(g) to SO_3(g). (b) The hydrogenation of liquid vegetable oil using a finely divided nickel catalyst. (c) The conversion of an aqueous solution of D-glucose to a D,L mixture catalyzed by HCl(aq).

17.4 You are approached by an industrialist with the proposition that you develop catalysts for the following processes at $80\,°C$ with no input of electrical energy or electromagnetic radiation.

    (a) The splitting of water into H_2 and O_2.
    (b) The decomposition of CO_2 into C and O_2.
    (c) The combination of N_2 with H_2 to produce NH_3.
    (d) The hydrogenation of the double bonds in vegetable oil.

The industrialist's company will build the plant to carry out the process and the two of you will share equally in the profits. Which of these would be easy to do, which are plausible candidates for investigation, and which are unreasonable? Describe the chemical basis for the decision in each case.

17.5 Addition of PPh_3 to a solution of Wilkinson's catalyst, $[RhCl(PPh_3)_3]$, reduces the turnover frequency for the hydrogenation of propylene. Give a plausible mechanistic explanation of this observation.

17.6 The rates of H_2 gas absorption (in $L\ mol^{-1}\ s^{-1}$) by alkenes catalyzed by $[RhCl(PPh_3)_3]$ in benzene at $25°C$ are: hexene, 2910; *cis*-4-methyl-2-pentene, 990; cyclohexene, 3160; 1-methylcyclohexene, 60. Suggest the origin of the trends and identify the affected reaction step in the proposed mechanism (Cycle 17.2).

17.7 Infrared spectroscopic investigation of a mixture of CO, H_2, and 1-butene under conditions that bring about hydroformylation indicate the presence of compound (c) in Cycle 17.3 in the reaction mixture. The same reacting mixture in the presence of added tributylphosphine was studied by infrared spectroscopy and neither (c) nor an analogous phosphine-substituted complex was observed. What does the first observation suggest as the rate-limiting reaction in the absence of phosphine? Assuming the sequence of reactions remains unchanged, what are the possible rate-limiting reactions in the presence of tributylphosphine?

17.8 (a) Starting with the alkene complex shown in Cycle 17.5(a) with *trans*-DHC=CHD in place of C_2H_4, assume dissolved OH^- attacks from the side opposite the metal. Give a stereochemical drawing of the resulting compound. (b) Assume attack on the coordinated *trans*-DHC=CHD by an OH^- ligand coordinated to Pd, and draw the stereochemistry of the resulting compound. (c) Does the stereochemistry differentiate these proposed steps in the Wacker process?

17.9 Formic acid is thermodynamically unstable with respect to CO_2 and H_2. Give a plausible mechanism for the catalysis of formic acid decomposition by $[IrCl(H)_2(PPh_3)_3]$.

17.10 At elevated temperatures, finely divided titanium readily reacts with N_2(g) to form a stable nitride. The rate-determining step in ammonia synthesis over iron is $N≡N$ bond breaking. Why then is titanium ineffective whereas iron is effective as a catalyst for ammonia synthesis?

17.11 Aluminosilicate surfaces were described in the text as strong Brønsted acids, whereas silica gel is a very weak acid. (a) Give an explanation for the enhancement of acidity by the presence of Al^{3+} in a silica lattice. (b) Name three other ions that might enhance the acidity of silica gel.

17.12 Indicate the difference between each of the following: (a) The Lewis acidity of γ-alumina that has been heated to $900\,°C$ versus γ-alumina that has been heated to $100\,°C$. (b) The Brønsted acidity of silica gel versus γ-alumina. (c) The structural regularity of the channels and voids in silica gel versus that for ZSM-5.

17.13 Why is the platinum-rhodium in automobile catalytic converters dispersed on the surface of a ceramic rather than used in the form of thin foil?

17.14 Describe the concept of shape-selective catalysts and how this shape selectivity is attained.

17.15 Alkanes are observed to exchange hydrogen atoms with deuterium gas over some platinum metal catalysts. When 3,3-dimethylpentane in the presence of D_2 is exposed to a platinum catalyst and the gases are observed before the reaction has proceeded very far the main product is $CH_3CH_2C(CH_3)_2CD_2CD_3$ plus unreacted 3,3-dimethylpentane. Devise a plausible mechanism to explain this observation.

Problems

17.1 Carbon monoxide is known to undergo dissociative chemisorption on nickel at elevated temperatures, which results in surface carbide and oxide. With this as an initial step, propose a series of plausible reactions for the nickel-catalyzed conversion of CO and H_2 to CH_4 and H_2O.

17.2 Consider the truth of each of the following statements and provide corrections where required. (a) A catalyst introduces a new reaction pathway with lower enthalpy of activation. (b) Since the Gibbs energy is more favorable for a catalytic reaction, yields of the product are increased by catalysis. (c) An example of a homogeneous catalyst is the Ziegler–Natta catalyst made from $TiCl_4(l)$ and $Al(C_2H_5)_3(l)$. (d) Highly favorable Gibbs energies for the attachment of reactants and products to a homogeneous or heterogeneous catalyst are the key to high catalytic activity.

17.3 When direct evidence for a mechanism is not available, chemists frequently invoke analogies with similar systems. Describe how J.E. Bäckvall, B. Åkermark, and S.O. Ljunggren (*J. Am. Chem. Soc.* **101**, 2411 (1979)) inferred the attack of uncoordinated water on η^2-C_2H_4 in the Wacker process.

17.4 The removal of the sulfur atom (as H_2S) from organosulfur compounds in crude oil (hydrodesulfurization) is an important process in petroleum refining. Metal sulfides are used as catalysts in hydrodesulfurization and both the **4d** and **5d** metal sulfides display catalytic activity that is greatest in Group 8 (Ru and Os). The **3d** metal sulfides have relatively low catalytic activity and do not display maximum activity for the Group 8 element, iron. After reviewing the paper by S. Harris and R.R. Chianelli (*J. Catal.* **86**, 400 (1984)) describe the feature of the electronic structures of these sulfides that correlates with the catalytic activity and how that parameter might be relevant to the catalysis.

17.16 The effectiveness of platinum in catalyzing the reaction $2H^+(aq) + 2e^- \rightarrow H_2(g)$ is greatly decreased in the presence of CO. Suggest an explanation.

18

Structures and properties of solids

The physical and chemical properties of solids are a pervasive aspect of inorganic chemistry, and have been discussed throughout the text. As part of the current enthusiasm for 'materials chemistry' there has been a great increase in the synthesis and study of new inorganic solids. In this chapter we touch on some of the areas of current investigation. Some basic concepts, including the structures of some simple protypical solids, lattice enthalpies, and both ionic and covalent bonding models, have been discussed in Chapter 2. In intervening chapters we have only touched on defects, and here this important feature is developed more thoroughly. We see how interruptions in the uniformity of a crystal structure influence properties such as the migration of ions. In addition, we extend the discussion of intercalation compounds, and see how their formation, structures, and properties correlate with the crystal structure and the electronic band structure of the parent compound.

Solid-state inorganic chemistry is concerned with the synthesis, structures, and properties of solids. Many of these solids exhibit new phenomena or possess desirable properties, such as high-temperature superconductivity or ferromagnetism. It is a vigorous and exciting area of inorganic research, partly on account of the technological applications of these properties but also because they are challenging to understand. We shall draw on some of the concepts developed in earlier chapters for discussing the solid state, such as lattice enthalpies and band structure, and introduce some additional concepts that are needed to discuss the dynamics of events that occur in the interior of solids. We also need to go beyond the view that solids have a well defined stoichiometry, for interesting phenomena arise from nonstoichiometry and atom mobility.

Some general principles

Chapter 2 provided a summary of prototypical structures adopted by many metals and ionic solids; it also explored some general rules for the stability of ionic solids. Most of that discussion centered on the structures of simple compounds with well defined composition. In the present chapter, we describe a wider range of solids, including those where atomic-scale defects and deviations from simple whole-number ratios of the constituent elements influence their physical and chemical properties.

Much of the current research in solid-state chemistry is motivated by the search for commercially useful materials, such as components of batteries and fuel cells, catalysts for hydrocarbon interconversion, and improved electronic and photonic devices for information processing and storage. The scope for the synthesis of new inorganic solids is enormous; for example, although it is known that 100 structural types account for 95 per cent of the known binary or ternary intermetallic compounds,[1] there are plenty of opportunities for extending these studies to the synthesis and characterization of four-, five-, and six-component systems. Among the many other areas that are ripe for exploration are new mesoporous solids for use in molecular separations and heterogeneous catalysis; these materials are discussed in Chapters 10 and 17.

18.1 Defects

All solids contain **defects**, or imperfections of structure or composition. Defects are important because they influence properties such as mechanical strength, electrical conductivity, and chemical reactivity. We need to consider both **intrinsic defects**, which are defects that occur in the pure substance, and **extrinsic defects**, which stem from the presence of impurities. It is also common to distinguish **point defects**, which occur at single sites, from **extended defects**, which are ordered in one, two, and three dimensions. Point defects are random errors in a periodic lattice, such as the absence of an atom at its usual site or the presence of an atom at a site that is not ordinarily occupied. Extended defects involve various irregularities in the stacking of planes.

(a) Why crystals have defects

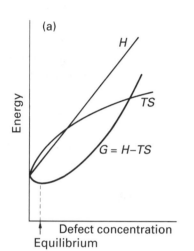

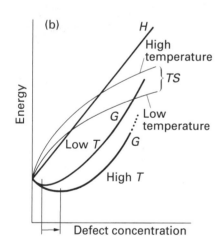

All solids have a thermodynamic tendency to acquire defects, because defects introduce disorder into an otherwise perfect structure and hence increase its entropy. The Gibbs energy, $G = H - TS$, of a solid with defects has contributions from the enthalpy and the entropy of the sample. It follows that, through their effect on the entropy, defects contribute a negative term to the Gibbs energy of the solid. The formation of defects may be endothermic (so H is higher in the presence of defects). Provided $T > 0$, however, the Gibbs energy will have a minimum at a nonzero concentration of defects and their formation will be spontaneous (Fig. 18.1a). Moreover, as the temperature is raised, the minimum in G shifts to higher defect concentrations (Fig. 18.1b).

> All solids have a thermodynamic tendency to acquire defects, because defects introduce disorder into an otherwise perfect structure and hence increase its entropy.

(b) Intrinsic point defects

Point defects are difficult to detect directly. X-ray diffraction, for example, samples the periodic structure of a solid over thousands of ångströms, so small random deviations from periodicity often go unnoticed. Occasionally, spectroscopic measurements indicate the presence of an ion in an unusual position or an electron trapped at a particular site in the lattice. The presence of an appreciable number of vacancies or excess atoms may sometimes be inferred from the difference between the measured and calculated densities of a sample (as we illustrate in Example 18.1 below). The electrical conductivity of a sample has also been used to infer the presence of defects. Modern electron microscopy has improved our ability to detect defects because its resolution is so high that defects can be observed virtually directly.

18.1 (a) The variation of the enthalpy and entropy of a crystal as the number of defects increases. The resulting Gibbs energy $G = H - TS$ has a minimum at a nonzero concentration, and hence defect formation is spontaneous. (b) As the temperature is increased, the minimum in the Gibbs energy moves to higher defect concentrations, so more defects are present at equilibrium at higher temperatures than at low.

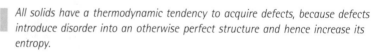

[1] J.R. Rodgers and P. Villars, *MRS Bulletin*, February, 27 (1993).

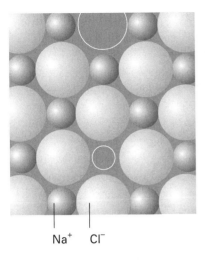

18.2 A Shottky defect is the absence of an ion from a site. For charge neutrality overall, there must be equal numbers of cation and anion absences in a 1:1 compound.

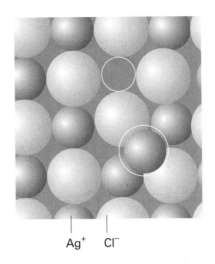

18.3 A Frenkel defect forms when an ion moves to an interstitial site.

In the 1930s, two solid state physicists—W. Schottky in Germany and J. Frenkel in Russia—used conductivity and density data to identify specific types of point defect. A **Schottky defect** (Fig. 18.2) is a vacancy in an otherwise perfect lattice. That is, it is a point defect in which an atom or ion is missing from its normal site in the lattice. The overall stoichiometry of a solid is not usually affected by the presence of Schottky defects because there are equal numbers of vacancies at cation and anion sites. A **Frenkel defect** (Fig. 18.3) is a point defect in which an atom or ion has been displaced into an interstitial site. For example, in silver chloride, which has the rock-salt structure, a small number of Ag^+ ions reside in tetrahedral sites (**1**). The stoichiometry of the compound is unchanged when a Frenkel defect forms. A useful generalization is that Frenkel defects are most often encountered in the more open structures such as wurtzite and sphalerite, where the coordination numbers are low and the open structure provides sites that can accommodate interstitial atoms.

The concentration of Schottky defects varies considerably from one type of compound to the next. The concentration of vacancies is very low in the alkali metal halides, being of the order of 10^6 cm^{-3} at 130 °C. That concentration corresponds to about one defect per 10^{14} formula units. On the other hand, some d-metal oxides, sulfides, and hydrides have very high concentrations of vacancies. An extreme example is the high-temperature form of TiO, which has vacancies on both the cation and anion sites at a concentration corresponding to about one defect per 10 formula units.

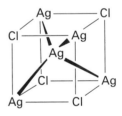

1 Interstitial Ag^+

Example 18.1 Inferring the type of defect from density measurements

Titanium monoxide has a rock-salt structure. X-ray diffraction data show that the length of one edge of the cubic unit cell for TiO with a 1:1 ratio of Ti to O is 4.18 Å, and the density determined by volume and mass measurements is 4.92 g cm^{-3}. Do the data indicate that defects are present? If so, are they vacancy or interstitial defects?

Answer The presence of vacancies (Schottky defects) at the Ti and O sites should be reflected in a lower measured density than that calculated from the size of the unit cell and the assumption that every Ti and O site is occupied. Interstitial (Frenkel) defects would give little if any difference between the measured and theoretical densities. There are four TiO formula units per unit cell (Fig. 2.10). The mass of one formula unit is 63.88 u, so the corresponding theoretical mass in one unit cell is $4 \times 63.88\ \text{u} = 255.52\ \text{u}$, which corresponds to 4.24×10^{-22} g. The corresponding theoretical density is

$$\rho = \frac{4.24 \times 10^{-22}\ \text{g}}{(4.18 \times 10^{-8}\ \text{cm})^3} = 5.81\ \text{g cm}^{-3}$$

which is significantly greater than the measured density. Therefore, the crystal must contain numerous vacancies. Because the overall composition of the solid is TiO, there must be equal numbers of vacancies on the cation and anion sites.

Self-test 18.1 The measured density of VO (1:1 stoichiometry) is $5.92\ \text{g cm}^{-3}$ and the theoretical density is $6.49\ \text{g cm}^{-3}$. Do the data indicate the presence of vacancies or interstitials?

Schottky and Frenkel defects are only two of the many possible types of defect. Another example is an **atom interchange defect**, which consists of an interchanged pair of M and X atoms or ions. This type of defect is common in metal alloys. For example, a copper–gold alloy of overall composition CuAu has extensive disorder at high temperatures and a significant fraction of Cu and Au atoms are interchanged (Fig. 18.4).

> *Schottky defects are lattice site vacancies. Frenkel defects are most often encountered in the more open structures where the coordination numbers are low and the open structure provides sites that can accommodate interstitial atoms.*

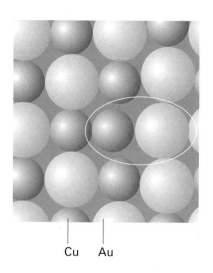

Cu   Au

18.4 Atom interchange can also give rise to a point defect.

(c) Extrinsic point defects

Extrinsic defects are inevitable because perfect purity is unattainable in crystals of any significant size. In Section 3.15 we discussed impurities that had been introduced intentionally, where a pure substance was deliberately doped with another, such as when As atoms are substituted for Si atoms in a single crystal of silicon in order to modify its semiconducting properties. We can begin to see how the electronic structure of the solid influences the type of defect it is likely to form. Thus, when As replaces Si the added electron from each As atom finds its way into the conduction band. In the more ionic substance ZrO_2, the introduction of Ca^{2+} impurities in place of Zr^{4+} ions is accompanied by the formation of an O^{2-} ion vacancy to maintain charge neutrality (Fig. 18.5). In some cases a change in oxidation state (if one is possible) may be induced by an impurity. For example the introduction of Li_2O into NiO places Li^+ in Ni^{2+} sites, and the system balances charge by oxidation of one Ni^{2+} ion to Ni^{3+} for each Li^+ ion present. As with the doping of silicon by boron, this process introduces holes in the valence band of NiO and the conductivity increases greatly.

Another example of an extrinsic point defect is a **color center**, a generic term for defects responsible for modifications to the IR, visible, and UV absorption characteristics of solids that have been irradiated or exposed to chemical treatment. One type of color center is produced by heating an alkali halide crystal in the vapor of the alkali metal. The process results in an alkali metal cation at a normal cation site, but the electron the atom brings into the crystal occupies a halide ion vacancy. A color center consisting of an electron in a halide ion vacancy (Fig. 18.6) is called an **F-center**.[2] The color results from the excitation of the electron in the

---

[2] The name comes from the German word for color center, *Farbenzenter.*

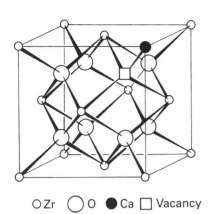

18.5 Introduction of a calcium ion into the ZrO_2 lattice produces a vacancy on the oxide sublattice and helps to stabilize the fluorite structure.

○ Zr ◯ O ● Ca ☐ Vacancy

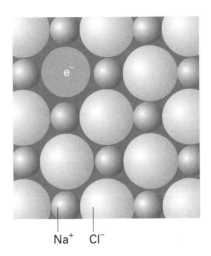

Na⁺ Cl⁻

18.6 An F-center is an electron that occupies a halide ion vacancy. The energy levels of the electron resemble those of a particle in a three-dimensional square well.

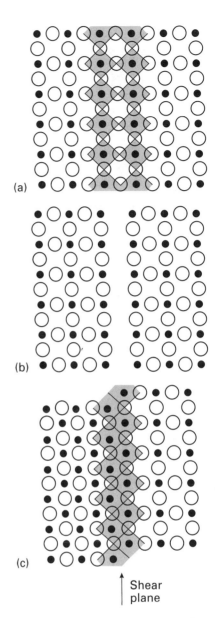

(a)

(b)

(c)

↑ Shear plane

18.7 The concept of a crystallographic shear plane illustrated by the (100) plane of the ReO_3 structure. (a) A plane of metal, ●, and oxygen, ◯, atoms. The octahedron around each metal atom is completed by a plane of O atoms above and below the plane illustrated here. Some of the octahedra are shaded to clarify the processes that follow. (b) Oxygen atoms in the plane perpendicular to the page are removed, leaving two planes of metal atoms that lack their sixth oxygen ligand. (c) The octahedral coordination of the two planes of metal atoms is restored by translating the right slab as shown. This creates a plane (labeled the shear plane) vertical to the paper in which the MO_6 octahedra share edges.

localized environment of its surrounding ions, and its quantized energy levels resemble those of an electron in a box.

The structure of a solid influences the type of defect it is likely to accommodate.

(d) Extended defects

All the point defects discussed so far entail a significant local distortion of the lattice and in some instances localized charge imbalances too. Therefore it should not be surprising that defects may cluster together and sometimes form lines and planes.

Tungsten oxides illustrate the formation of planes of defects. As illustrated in Fig. 18.7a, the idealized structure of WO_3 (which is usually referred to as the 'ReO_3 structure'; see below) consists of WO_6 octahedra sharing all vertices. To picture the formation of the defect plane we should imagine the removal of shared O atoms along a diagonal (Fig. 18.7b). Then adjacent slabs slip past each other in a motion that results in the completion of the vacant coordination sites around each W atom (Fig. 18.7c). This shearing motion creates edge-shared octahedra along a diagonal direction. The resulting structure was named a **crystallographic shear plane** by the Australian crystallographer A.D. Wadsley, who first devised this way of describing extended planar defects.

Crystallographic shear planes randomly distributed in the solid are called **Wadsley defects**. Such defects lead to a continuous range of compositions, as in tungsten oxide, which ranges from WO_3 to $WO_{2.93}$. If, however, the crystallographic shear planes are distributed in a non-random, periodic manner, we should regard the material as a new stoichiometric phase. Thus, when even more O^{2-} ions are removed from tungsten oxide, a series of discrete phases having ordered crystallographic shear planes and compositions W_nO_{3n-2} ($n = 20, 24, 25,$ and 40) are observed. Closely spaced shear plane phases are known for oxides of tungsten, molybdenum, titanium, and vanadium. They can be identified by X-ray diffraction, but electron microscopy

(a)

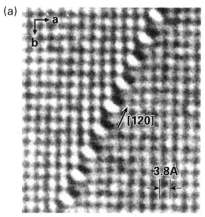

(b)

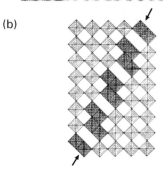

18.8 (a) High-resolution electron micrograph lattice image of a crystallographic shear plane ($1\bar{2}0$) in WO_{3-x}. (b) Drawing of the oxygen octahedral polyhedra that surround the tungsten atoms imaged in the electron micrograph. Note the edge-shared octahedra along the crystallographic shear plane. (Reproduced by permission from S. Iijima, *J. Solid State Chem.* **14**, 52 (1975).)

provides a more general and direct method because it reveals both ordered and random arrays of shear planes (Fig. 18.8).

| *Wadsley defects are crystallographic shear planes randomly distributed in the solid; such defects lead to a continuous range of compositions.*

18.2 Nonstoichiometric compounds

A **nonstoichiometric compound** is a substance with variable composition but which retains essentially the same basic structure. For example, at 1000 °C the composition of wüstite (which is nominally FeO) can vary from $Fe_{0.89}O$ to $Fe_{0.96}O$. Apart from gradual changes in the size of the unit cell as the composition is varied, the main features of the rock-salt structure are retained throughout this composition range. Some representative nonstoichiometric hydrides, oxides, and sulfides are listed in Table 18.1. Deviations from stoichiometry are common with *d*-, *f*-, and some *p*-block metals in combination with soft anions, such as S^{2-} and H^- ions, as well as with the harder O^{2-} anion. In contrast, hard anions such as fluorides, chlorides, sulfates, and nitrates form fewer nonstoichiometric compounds.

Nonstoichiometric compounds can be regarded as solid solutions in the sense that, as with liquid solutions, the chemical potentials of the components vary continuously as the composition changes. For example, as the partial pressure of oxygen is varied while equilibrium is maintained in the presence of a metal oxide, the composition of the oxide changes continuously (Fig. 18.9).

It may be very difficult to distinguish a nonstoichiometric compound from a series of discrete phases if the Gibbs energies of the individual phases are similar. For example, the partial pressure of oxygen over a series of closely spaced phases may appear to vary continuously rather than in steps (Fig. 18.10), and the presence of separate phases would not be apparent. Discrete stoichiometric phases can also be detected by X-ray diffraction, and often by optical or electron microscopy.

| *Deviations from stoichiometry are common with d-, f-, and some p-block metals in combination with soft anions as well as with the harder O^{2-} anion. In contrast, hard anions form fewer nonstoichiometric compounds.*

Table 18.1 Representative composition ranges† of nonstoichiometric binary hydrides, oxides, and sulfides

d-block			*f*-block		
Hydrides					
TiH_x	1–2			Fluorite type	Hexagonal
ZrH_x	1.5–1.6		GdH_x	1.8–2.3	2.85–3.0
HfH_x	1.7–1.8		ErH_x	1.95–2.31	2.82–3.0
NbH_x	0.64–1.0		LuH_x	1.85–2.23	1.74–3.0
Oxides					
	Rock-salt type	Rutile type			
TiO_x	0.7–1.25	1.9–2.0			
VO_x	0.9–1.20	1.8–2.0			
NbO_x	0.9–1.04				
Sulfides					
ZrS_x	0.9–1.0				
YS_x	0.9–1.0				

†Expressed as the range of values that x may take.

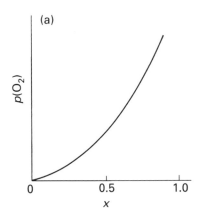

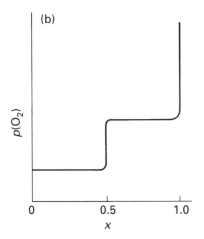

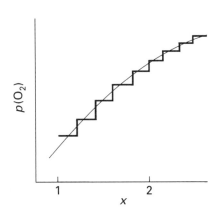

18.9 Schematic representation of the variation of the partial pressure of oxygen with composition at constant pressure for (a) a nonstoichiometric oxide; (b) a stoichiometric pair of metal oxides MO and MO_2. The axis x is the atom ratio in MO_x.

18.10 The stepped line indicates the variation in pressure of O_2 in equilibrium with a series of closely spaced discrete phases at constant temperature. Two solid phases are present in the horizontal regions. The smooth curve indicates the observed pressure when the reaction is so slow that equilibrium is not established. The axis x is the atom ratio in MO_x.

18.3 Atom and ion diffusion

One reason why diffusion in solids is much less familiar than diffusion in gases and liquids is that at room temperature it is generally very much slower. However, there are some striking exceptions to this generalization. Diffusion in solids is in fact very important in many areas of solid state technology, such as semiconductor manufacture, the synthesis of new solids, metallurgy, and heterogeneous catalysis.

(a) General principles of diffusion

The diffusion of particles through any medium is governed by the **diffusion equation**:

$$\frac{\partial c}{\partial t} = D \frac{\partial^2 c}{\partial x^2} \tag{1}$$

where c is the concentration of the diffusing species and D is the **diffusion coefficient**. The diffusion equation summarizes the fact that the rate of gain or loss of particles in a region is proportional to the spatial inhomogeneity of the concentration there (Fig. 18.11). Where

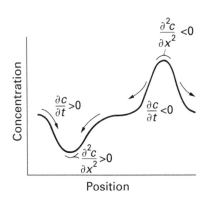

18.11 A schematic representation of the content of the diffusion equation. Where there is a trough in the concentration, particles tend to accumulate to remove it; where there is a peak, particles tend to spread away.

there are more particles than average (in the sense that $\partial^2 c/\partial x^2 < 0$, corresponding to a bump in the distribution), the particles will tend to disperse and lower the concentration locally ($\partial c/\partial t < 0$). If there is a dip in the concentration locally (so $\partial^2 c/\partial x^2 > 0$), particles will tend to diffuse into the region ($\partial c/\partial t > 0$).[3]

The rate of approach to a uniform distribution of particles is high when D is large. In a solid, D may depend more strongly than in a liquid on the concentration of particles, for the presence of foreign particles affects the local structure of the solid and has a marked effect on the ability of atoms to migrate. The diffusion coefficient also increases with temperature, because particle migration is an activated process, and we can write the Arrhenius-like expression

$$D = D_0 e^{-E_a/RT} \tag{2}$$

where E_a is the activation energy for diffusion. Figure 18.12 shows the temperature dependence of the diffusion coefficients for some representative solids at high temperature. According to eqn 2, the slopes of the lines are proportional to the activation energy for atom or ion transport. Thus Na^+ is highly mobile and has a low activation energy for motion through β-alumina, whereas Ca^{2+} in CaO is much less mobile and has a much higher activation energy for diffusion.

> The diffusion equation summarizes the fact that the rate of gain or loss of particles in a region is proportional to the spatial inhomogeneity of the concentration there.

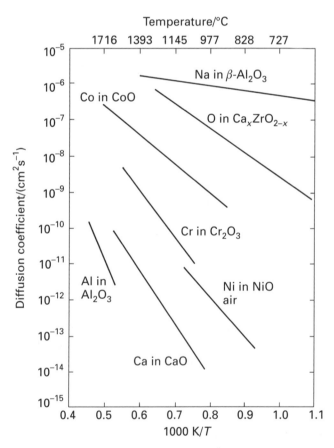

18.12 Arrhenius plot of the diffusion coefficient (on a logarithmic scale) of the mobile ion in a series of solids versus $1/T$.

....

[3]   The diffusion equation can be regarded as a mathematical form of the remark that 'Nature abhors a wrinkle' (see P.W. Atkins, *Physical chemistry.* Oxford University Press and W.H. Freeman & Co (1998)). Peaks in distributions tend to disperse, and troughs tend to fill.

(b) Mechanisms of diffusion

The diffusion coefficients of ions can often be understood in terms of the mechanism for their migration and the activation barriers the ions encounter as they move. In particular, diffusion coefficients in solids are markedly dependent on the presence of defects. The role defects play is summarized in Fig. 18.13, which shows some commonly postulated mechanisms for atom or ion motion in solids.

As with the mechanisms of chemical reactions, the evidence for the operation of a particular diffusion process is circumstantial. The individual events are never directly observed but are inferred from the influence of experimental conditions on atom or ion diffusion rates. In addition, a detailed analysis of the thermal motion of ions in crystals based on X-ray and neutron diffraction provides strong hints about the ability of ions to move through the crystal, including their most likely paths. Theoretical models, which are often elaborations of the ionic model (Chapter 2), give very useful guidance to the feasibility of these migration mechanisms.

Diffusion is an activated process; diffusion coefficients in solids are markedly dependent on the presence of defects.

(c) Solid electrolytes

Any electrochemical cell such as a battery, fuel cell, electrochromic display, or electrochemical sensor, requires an electrolyte. There is currently considerable interest in the use of solid electrolytes in these applications. Silver tetraiodomercurate(II), Ag_2HgI_4, and sodium β-alumina, a sodium-doped version of alumina of composition $Na_{1+x}Al_{11}O_{17+x/2}$, provide examples of two quite different types of solids in which cations are highly mobile. The former is a soft material with a lattice of low rigidity and the latter is a hard refractory solid.

Below 50 °C, Ag_2HgI_4 has an ordered crystal structure in which Ag^+ and Hg^{2+} ions are tetrahedrally coordinated by I^- ions and there are unoccupied tetrahedral holes (Fig. 18.14a). At this temperature its electrical conductivity is low. Above 50 °C, the Ag^+

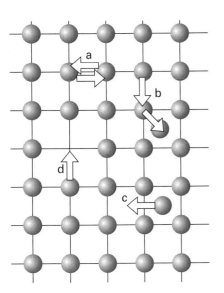

18.13 Some diffusion mechanisms for ions or atoms in a solid: (a) exchange, (b) interstitialcy, (c) interstitial, and (d) vacancy.

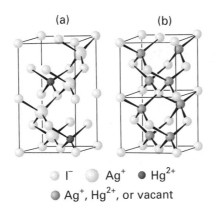

I^-　Ag^+　Hg^{2+}
Ag^+, Hg^{2+}, or vacant

18.14 (a) Low-temperature ordered structure of Ag_2HgI_4. (b) High-temperature disordered structure. Ag_2HgI_4 is an Ag^+ ion conductor in the high-temperature form.

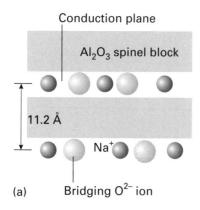

Conduction plane

Al$_2$O$_3$ spinel block

11.2 Å

Na$^+$

(a)   Bridging O^{2-} ion

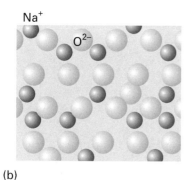

Na$^+$

O^{2-}

(b)

18.15 (a) Schematic side view of β-alumina showing the Na$_2$O conduction planes between Al$_2$O$_3$ slabs. The O atoms in these planes bridge the two slabs. (b) A view of the conduction plane. Note the abundance of mobile ions and vacancies in which they can move.

and Hg^{2+} cations of Ag$_2$HgI$_4$ are randomly distributed over the tetrahedral sites (Fig. 18.14b). At this temperature the material is a good electrical conductor largely on account of the mobility of the Ag$^+$ ions. The weak and easily deformed close-packed array of I$^-$ ions is responsible for low activation energy for ion migration from one lattice site to the next. This example illustrates the general observation that solid inorganic electrolytes have a low-temperature phase in which the ions are ordered on a subset of lattice sites and that at higher temperatures the ions become disordered over the sites and the ionic conductivity increases. There are many similar solid electrolytes having soft lattices, such as AgI and RbAg$_4$I$_5$, both of which are Ag$^+$ conductors. The mobility of Ag$^+$ ions is high, for the conductivity of RbAg$_4$I$_5$ at room temperature is greater than that of an aqueous sodium chloride solution.

Sodium β-alumina is an example of a mechanically hard material that is a good ionic conductor. In this case the rigid and dense Al$_2$O$_3$ slabs are bridged by a sparse array of O^{2-} ions (Fig. 18.15). The plane containing these bridging ions also contains Na$^+$ ions which can move from site to site because there are no major bottlenecks to hinder their motion. Many similar rigid materials having planes or channels through which ions can move are known and are called **framework electrolytes**. Another closely related material, sodium β″-alumina, has even less restricted motion of ions than β-alumina, and it has been found possible to substitute dipositive cations such as Mg^{2+} or Ni^{2+} for Na$^+$. Even the large lanthanide cation Eu^{2+} can be introduced into β″-alumina, although the diffusion of such ions is slower than that of their smaller counterparts.

These examples illustrate a fairly common strategy for the formation of new solids by diffusion processes much lower than the melting point of the host material. Such **low-temperature syntheses** provide routes to a wide variety of solids that are not thermodynamically stable and therefore cannot be prepared from the melt. The lanthanide β″-aluminas prepared in this manner are of interest for possible applications in lasers.

Example 18.2 Correlating conductivity and ion size in a framework electrolyte

Conductivity data on β-alumina containing monopositive ions of various radii show that Ag$^+$ and Na$^+$ ions, both of which have radii close to 1.0 Å, have activation energies for conductivity close to 17 kJ mol^{-1} whereas that for Tl$^+$ (radius 1.4 Å) is about 35 kJ mol^{-1}. Suggest an explanation of the difference.

Answer In sodium β-alumina and related β-aluminas a fairly rigid framework provides a two-dimensional network of passages that permit ion migration. Judging from the experimental results, the bottlenecks for ion motion appear to be large enough to allow Na$^+$ or Ag$^+$ to pass quite readily (with a low activation energy) but too small to let the larger Tl$^+$ pass through as readily.

Self-test 18.2 Why does increased pressure reduce the conductivity of K$^+$ in β-alumina more than that of Na$^+$ in β-alumina?

The common characteristics of solid electrolytes are high concentrations of mobile ions and vacancies and the existence of pathways between the vacancies that provide routes for ion motion with low activation energy. In both Ag$_2$HgI$_4$ and sodium β-alumina there are high concentrations of carrier ions (Ag$^+$ and Na$^+$, respectively) as well as of vacant sites into which these carriers may diffuse.

> *Solid inorganic electrolytes often have a low-temperature phase in which the ions are ordered on a subset of lattice sites; at higher temperatures the ions become disordered over the sites and the ionic conductivity increases.*

Prototypical oxides and fluorides

In this section we explore compounds of the hard oxide and fluoride anions with metals. These compounds provide chemical insight into defects, nonstoichiometry, and ion diffusion, and into the influence of these characteristics on physical properties. This section also describes the structures of solids that are of interest in the synthesis of magnetic and superconducting materials.

18.4 Monoxides of the 3d metals

The monoxides of most of the 3d metals have the rock-salt structure, so it might at first appear that there would be little more to say about them and that their properties should be simple. In fact, the structures and properties of these oxides are so interesting that they have been repeatedly investigated (Table 18.2). In particular, the compounds provide examples of how mixed oxidation states and defects lead to nonstoichiometry, for TiO, VO, FeO, CoO, and NiO can all be prepared with significant deviations from the stoichiometry MO. The range of character of the d-metal monoxides, from mild to aggressive reducing agents, is reflected in the various ways in which they are prepared.

(a) Defects and nonstoichiometry

The origin of the nonstoichiometry in FeO has been studied in more detail than that in most other compounds. The consensus is that it arises from the creation of vacancies on the Fe^{2+} octahedral sites, and that each vacancy is charge-compensated by the conversion of two neighboring Fe^{2+} ions to two Fe^{3+} ions. The relative ease of oxidizing Fe(II) to Fe(III) accounts for the rather broad range of compositions of Fe. At high temperatures the interstitial Fe^{3+} ions associate with the Fe^{2+} vacancies to form clusters distributed throughout the structure (Fig. 18.16). A similar clustering of defects appears to occur with all other 3d monoxides, with the possible exception of NiO. It is more difficult to study NiO because its range of nonstoichiometry is extremely narrow, but conductivity and the rate of atom diffusion vary with oxygen partial pressure in a manner that suggests the presence of isolated point defects.

Both CoO and NiO occur in a metal-deficient state, although their range of stoichiometry is not as broad as that of FeO. As indicated by standard potentials in aqueous solution (Section 6.3), Fe(II) is more easily oxidized than is either Co(II) or Ni(II), and this solution redox chemistry correlates well with the much smaller range of oxygen deficiency in NiO and CoO. Chromium(II) oxide does not exist because it spontaneously disproportionates:

$$3\,CrO(s) \longrightarrow Cr_2O_3(s) + Cr(s)$$

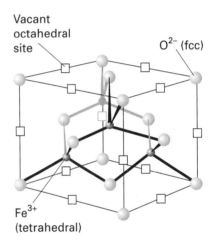

18.16 Defect sites proposed for FeO. Note that the tetrahedral Fe^{3+} interstitials and octahedral Fe^{2+} vacancies have clustered together.

Table 18.2 Monoxides of the Period 4 metals

Compound	Structure	Stoichiometry MO_x	Electrical properties
CaO_x	Rock salt	1	Insulator
TiO_x	Rock salt	0.65–1.25	Metallic
VO_x [a]	Rock salt	0.79–1.29	Metallic
MnO_x	Rock salt	1–1.15	Semiconductor
FeO_x	Rock salt	1.04–1.17	Semiconductor
CoO_x	Rock salt	1–1.01	Semiconductor
NiO_x	Rock salt	1–1.001	Semiconductor
CuO_x	PtS	1	Semiconductor
ZnO_x	Wurtzite	Slight Zn excess	Wide gap n-type semiconductor

[a] At low temperatures, VO adopts a less symmetrical structure.

It is possible, however, to stabilize CrO as a solid solution in CuO. Significant deviation from the stoichiometry **MO** is out of the question for Group 2 metal oxides such as CaO because for these metals M^{3+} ions are chemically inaccessible.

> The nonstoichiometry of FeO arises from the creation of vacancies on the Fe^{2+} octahedral sites, with each vacancy charge-compensated by the conversion of two Fe^{2+} ions to two Fe^{3+} ions.

(b) Electrical conductivity

The $3d$ monoxides MnO, FeO, CoO, and NiO have low electrical conductivities that increase with temperature; hence they are semiconductors. The electron or hole migration in these oxide semiconductors is attributed to a hopping mechanism. In this model the electron or hole hops from one localized metal site to the next. When it lands on a new site it causes the surrounding ions to adjust their locations and the electron or hole is trapped temporarily in the potential well produced by this atomic polarization. The electron resides at its new site until it is thermally activated to migrate into another nearby site. Another aspect of this charge hopping mechanism is that the electron or hole tends to associate with local defects, so the activation energy for charge transport may also include the energy of freeing the hole from its position next to a defect. The conduction process in MnO, FeO, CoO, and probably NiO occurs by such a hopping mechanism, but it is generally not clear whether site-to-site hopping or dissociation from a defect is rate-limiting.

Hopping contrasts with the band model for semiconductivity discussed in Section 3.15, where the conduction and valence electrons occupy orbitals that spread through the whole crystal. The difference stems from the less diffuse d orbitals in the monoxides of the mid to late $3d$ metals, which are too compact to form the broad bands necessary for metallic conduction. We have already mentioned that when NiO is doped with Li_2O in an O_2 atmosphere a solid solution $Li_xNi^{(II)}_{1-x}Ni^{(III)}_xO$ is obtained, which has greatly increased conductivity for reasons similar to the increase in conductivity of silicon when doped with boron. The characteristic pronounced increase in electrical conductivity with increasing temperature of metal oxide semiconductors is used in 'thermistors' to measure temperature.

In contrast to the semiconductivity of the monoxides in the center and right of the $3d$ series, TiO and VO have high electric conductivities that decrease with increasing temperature. This metallic conductivity persists over a broad composition range from highly oxygen-rich to metal-rich. In these compounds a conduction band is formed by the overlap of t_{2g} orbitals of metal ions in neighboring octahedral sites that are oriented toward each other (Fig. 18.17). The radial extension of the d orbitals of these early d-block elements is greater than for elements later in the period, and a band results from their overlap (Fig. 18.18); this band is in general only partly filled. The widely varying compositions of these monoxides also appears to be associated with the electronic delocalization: the conduction band serves as a rapidly accessible source and sink of electrons that can readily compensate for the formation of vacancies.

> The $3d$ monoxides MnO, FeO, CoO, and NiO are semiconductors; TiO and VO are metallic conductors.

18.5 Higher oxides

Now we turn to the higher oxides and see something of the elaborate structures and interesting properties of these compounds.

(a) M_2O_3 corundum structure

Although α-alumina (the mineral corundum) is best regarded as a highly polar covalent compound, it adopts a structure that can be modeled as a hexagonal close-packed array of

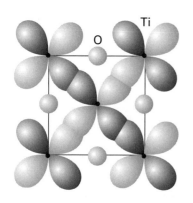

18.17 Overlap of d_{zx} orbitals in TiO to give a t_{2g} band.

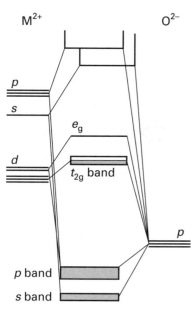

18.18 Molecular orbital energy level diagram for early d-block metal monoxides. The t_{2g} band is only partly filled and metallic conductivity results.

O^{2-} ions with the cations in two-thirds of the octahedral holes. The details of the structure are complex and we need not go into them. The corundum structure is adopted by the oxides of titanium, vanadium, chromium, iron, and gallium in their +3 oxidation states. Two of these oxides, Ti_2O_3 and V_2O_3, exhibit metallic-to-semiconducting transitions below 410 and 150 K, respectively (Fig. 18.19). In V_2O_3, the transition is accompanied by antiferromagnetic ordering of the spins. The two insulators Cr_2O_3 and Fe_2O_3 also display antiferromagnetic ordering.

Another interesting aspect of the M_2O_3 compounds is the formation of solid solutions of dark-green Cr_2O_3 and colorless Al_2O_3 to form brilliant red ruby. As we remarked in Section 13.6, this shift in the ligand-field transitions of Cr^{3+} stems from the compression of the O^{2-} ions around Cr^{3+} in the Al_2O_3 host lattice. (In Al_2O_3, $a = 4.75$ Å and $c = 13.00$ Å; in Cr_2O_3 the lattice constants are 4.93 Å and 13.56 Å, respectively.) The compression shifts the absorption toward the blue and the solid appears red in white light. The responsiveness of the absorption (and fluorescence) spectrum of Cr^{3+} ions to compression is sometimes used to measure pressure in high-pressure spectroscopic experiments. In this application, a tiny crystal of ruby in one segment of the sample can be interrogated by visible light and the shift in its fluorescence spectrum provides an indication of the pressure inside the cell.

> The corundum structure is adopted by the oxides of titanium, vanadium, chromium, iron, and gallium in their +3 oxidation states, as well as by alumina itself.

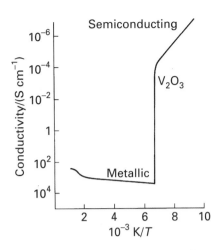

18.19 The temperature dependence of the electrical conductivity of V_2O_3 showing the metal-to-semiconductor transition.

(b) MO_2 and fluorite

The fluorite structure contains cations that are surrounded by eight anions (Fig. 2.13). As with the cesium-chloride structure, this high coordination number is favored by relatively large cations. The structure is found for fluorides such as PbF_2 and oxides such as ZrO_2.

Michael Faraday reported in 1834 that red-hot solid PbF_2 is a good conductor of electricity. Much later it was recognized that the conductivity arises from the mobility of F^- ions through the solid. The property of anion conductivity is shared by other crystals having the fluorite structure. Ion transport in these solids is thought to be by an interstitial mechanism in which an F^- ion first migrates from its normal position into an interstitial site and then moves to a vacant F^- site.

Certain oxides that have the fluorite structure conduct by O^{2-} ion migration. The best known of these compounds is ZrO_2 with added CaO to stabilize the fluorite phase (which in its absence is stable only at high temperature). For each Ca^{2+} substituted for Zr^{4+}, a vacancy on an O^{2-} site is created to maintain electroneutrality, and these vacancies improve the mobility of the O^{2-} ions.

A CaO-doped ZrO_2 electrolyte is used in a solid-state electrochemical sensor for the oxygen partial pressure in automobile exhaust systems (Fig. 18.20).[4] The platinum electrodes in this cell adsorb O atoms and, if the partial pressure of oxygen is different between the sample and reference side, there is a thermodynamic tendency for oxygen to migrate through the electrolyte as the O^{2-} ion. The thermodynamically favored processes are:

High p_{O_2} side:

$$O_2(g) + Pt(s) \longrightarrow 2\,O(Pt,\ surface)$$
$$O(Pt,\ surface) + 2\,e^- \longrightarrow O^{2-}(ZrO_2)$$

Low p_{O_2} side:

$$O^{2-}(ZrO_2) \longrightarrow O(Pt,\ surface) + 2\,e^-$$
$$O(Pt,\ surface) \longrightarrow O_2(g) + Pt(s)$$

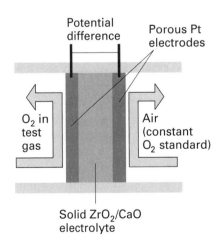

18.20 An oxygen sensor based on the solid electrolyte ZrO_2/CaO.

---

4   The signal from this sensor is used to adjust the air/fuel ratio and thereby the composition of the exhaust gas being fed to the catalytic converter.

The cell potential is related to the oxygen partial pressures through the Nernst equation (Section 6.3), so a simple measurement of the potential provides a measure of the oxygen partial pressure in the exhaust gases.

(c) Spinels

The simple d-block oxides Fe_3O_4, Co_3O_4, and Mn_3O_4 and many related mixed-metal compounds have interesting properties. They have structures related to spinel, $MgAl_2O_4$, and may be given the general formula AB_2O_4. The spinel structure was described briefly in Section 2.9(i), where we saw that it consists of an fcc array of O^{2-} ions in which the A ions reside in one-eighth of the tetrahedral holes and the B ions inhabit half the octahedral holes (Fig. 18.21); this structure is commonly denoted $A[B_2]O_4$. In the inverse spinel structure the cation distribution is $B[AB]O_4$. Lattice enthalpy calculations based on a simple ionic model indicate that for A^{2+} and B^{3+} the normal spinel structure, $A[B_2]O_4$, should be the more stable. The observation that many d-metal spinels do not conform to this expectation has been traced to the effect of ligand-field stabilization energies on the site preferences of the ions.

The **occupation factor**, λ, of a spinel is the fraction of B atoms in the tetrahedral sites. Thus $\lambda = 0$ for a normal spinel and 0.5 for an inverse spinel, $B[AB]O_4$. The distribution of cations in (A^{2+}, B^{3+}) spinels (Table 18.3) illustrates that for d^0 A and B ions the normal structure is preferred ($\lambda = 0$) as predicted by electrostatic considerations. Table 18.3 shows that, when A^{2+} is a d^6, d^7, d^8, or d^9 ion and B^{3+} is Fe^{3+}, the inverse structure is generally favored. This preference can be traced to the lack of ligand-field stabilization of the high-spin d^5 Fe^{3+} ion in either the octahedral or the tetrahedral site and the ligand-field stabilization of the other d^n ions in the octahedral site. It is important to note that simple ligand-field stabilization appears to work over this limited range of cations, but more detailed analysis is necessary when cations of different radius are introduced. Moreover, λ is often found to depend on the temperature.

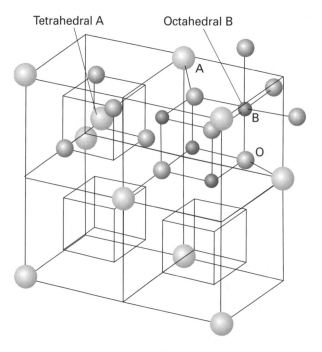

18.21 A segment of the spinel (AB_2O_4) unit cell showing the tetrahedral environment of A ions and the octahedral environment of B ions. (Compare with Fig. 2.19.)

Table 18.3 Occupation factor, λ, in some spinels[a]

B	A	Mg^{2+} d^0	Mn^{2+} d^5	Fe^{2+} d^6	Co^{2+} d^7	Ni^{2+} d^8	Cu^{2+} d^9	Zn^{2+} d^{10}
Al^{3+}	d^0	0	0	0	0	0.38	0	
Cr^{3+}	d^3	0	0	0	0	0	0	0
Fe^{3+}	d^5	0.45	0.1	0.5	0.5	0.5	0.5	0
Mn^{3+}	d^4	0						0
Co^{3+}	d^6					0		0

[a] $\lambda = 0$ corresponds to a normal spinel; $\lambda = 0.5$ corresponds to an inverse spinel.

The inverse spinels of formula AFe_2O_4 are sometimes called **ferrites**. When $RT > J$, where J is the energy of interaction of the spins on different ions, ferrites are paramagnetic. However, when $RT < J$, a ferrite may be either ferromagnetic or antiferromagnetic (see Box 18.1). The antiparallel alignment of spins characteristic of antiferromagnetism is illustrated by $ZnFe_2O_4$, which has the cation distribution $Fe[ZnFe]O_4$. In this compound the Fe^{3+} ions (with $S = \frac{5}{2}$) in the tetrahedral and octahedral sites are antiferromagnetically coupled below 9.5 K to give nearly zero net magnetic moment to the solid as a whole.

> The observation that many d-metal spinels do not have the normal spinel structure is related to the effect of ligand-field stabilization energies on the site preferences of the ions.

Example 18.3 Predicting the structures of spinel compounds

Is $MnCr_2O_4$ likely to have a normal or inverse spinel structure?

Answer A normal spinel structure is expected because Cr^{3+} will have a ligand-field stabilization energy in the octahedral site whereas the d^5 Mn^{2+} ion will not. Table 18.3 shows that this prediction is verified experimentally.

Self-test 18.3 Table 18.3 indicates that $FeCr_2O_4$ is a normal spinel. Rationalize this observation.

(d) Perovskites and related phases

To introduce the structure of perovskites we first examine the related but simpler ReO_3 structure, which consists of vertex-shared octahedra (Fig. 18.22a). The illustration shows that the unit cell of the ReO_3 structure contains M atoms at the corners of a cube and O atoms on the edges. The presence of the vertex-shared octahedra is clarified by extending the structure to more than one unit cell (Fig. 18.22b). Note in particular that the ReO_3 structure is very open and that at the center there is a large hole of coordination number 12.

The perovskites have the general formula ABX_3, in which the 12-coordinate hole of BX_3 is occupied by a large A ion (Fig. 18.23; a different view of this structure was given in Fig. 2.17). The X ion is generally O^{2-} or F^-; perovskite itself is $CaTiO_3$ and an example of a fluoride perovskite is $NaFeF_3$. The structure is often observed to be distorted in such a manner that the unit cell is no longer centrosymmetric and the crystal has an overall permanent electric polarization. Some polar crystals are **ferroelectric** in the sense that they resemble ferromagnets, but instead of a large number of spins being aligned over a region the electric polarizations of many unit cells are aligned. As a result the relative permittivities for a ferroelectric material often exceed 1×10^3 and can be as high as 15×10^3; for comparison the relative permitivity of liquid water is about 80 at room temperature.

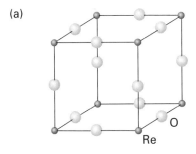

(a)

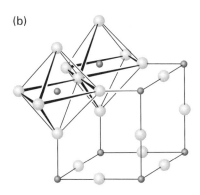

(b)

18.22 (a) The ReO_3 unit cell. (b) The ReO_3 unit cell together with some oxygen atoms from adjacent unit cells to show the corner-shared octahedra.

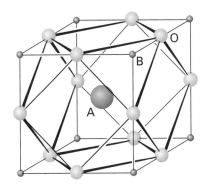

18.23 A view of the perovskite (ABO_3) structure emphasizing the dodecahedral coordination of the large A ion. The octahedral environment of the B ion is shown in Fig. 18.22.

Box 18.1 Cooperative magnetism

The diamagnetic and paramagnetic properties of compounds were covered in Section 7.4; they are characteristic of individual atoms or complexes. In contrast, properties such as ferromagnetism and antiferromagnetism depend on interactions between electron spins on many atoms and arise from the **cooperative** behavior of many unit cells in a crystal.

In a **ferromagnetic** substance the spins on different metal centers are coupled into a parallel alignment (Fig. B18.1a) that is

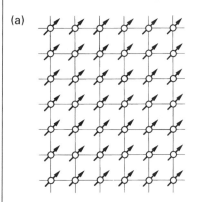

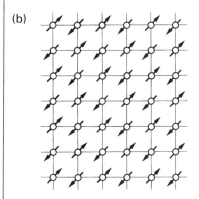

B18.1 (a) The parallel alignment of individual magnetic moments in a ferromagnetic material and (b) the antiparallel arrangement of spins in an antiferromagnetic material.

sustained over thousands of atoms in a magnetic **domain**. The net magnetic moment may be very large because the magnetic moments of individual spins augment each other. Moreover, once established and with the temperature maintained below the **Curie temperature** (T_C), the magnetization persists because the spins are locked together. Ferromagnetism is exhibited by materials containing unpaired electrons in d or f orbitals that couple with unpaired electrons in similar orbitals on surrounding atoms. The key feature is that this interaction is strong enough to align spins but not so strong as to form covalent bonds, in which the electrons would be paired.

The magnetization, M, of a ferromagnet is not linearly proportional to the applied field strength H. Instead, a hysteresis loop (Fig. B18.2) is observed. For **hard ferromagnets** the loop is broad (Fig. B18.2a) and M is large when the applied field has been reduced to zero. Hard ferromagnets are used for permanent magnets. A **soft ferromagnet** has a narrower hysteresis loop and is therefore much more responsive to the applied field (Fig. B18.2b). Soft ferromagnets are used in transformers, where they must respond to a rapidly oscillating field.

In an **antiferromagnetic** substance, neighboring spins are locked into an antiparallel alignment (Fig. B18.1b) and the sample has a low magnetic moment. Antiferromagnetism is often observed when a paramagnetic material is cooled to a low temperature and is signaled by a decrease in magnetic susceptibility with decreasing temperature (Fig. B18.3). The critical temperature for the onset of antiferromagnetism is called the **Néel temperature**, T_N. The spin coupling responsible for antiferromagnetism generally occurs through intervening ligands by a mechanism called **superexchange**. As indicated in Fig. B18.4, the spin on one metal atom induces a small spin polarization in an occupied orbital of a ligand. This polarization results in an antiparallel alignment of the spin on the adjacent metal atom. Many compounds exhibit antiferro-

(continued)

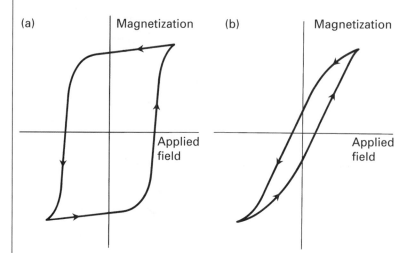

B18.2 Magnetization curves for ferromagnetic materials. A hysteresis loop results because the magnetization of the sample with increasing field (→) is not retraced as the field is decreased (←). (a) A hard ferromagnet, (b) a soft ferromagnet.

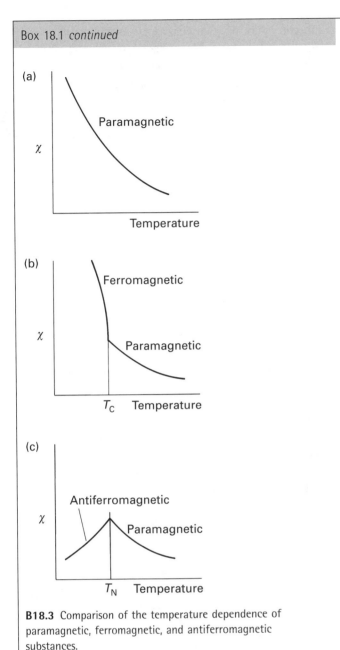

B18.3 Comparison of the temperature dependence of paramagnetic, ferromagnetic, and antiferromagnetic substances.

magnetic behavior in the appropriate temperature range; for example MnO is antiferromagnetic below −151 °C, and Cr_2O_3 is antiferromagnetic below 37 °C. Coupling of spins through intervening ligands is frequently observed in molecular complexes containing two ligand-bridged metal ions.

In a third type of collective magnetic interaction, **ferrimagnetism**, net magnetic ordering is observed below the Curie temperature. However, ferrimagnets differ from ferromagnets because ions with different local moments are present. These ions order with opposed spins, as in antiferromagnetism, but because of the different magnitudes of the individual spin moments there is incomplete cancellation and the sample has a net overall moment. As with antiferromagnetism, these interactions are generally transmitted by the ligand.

Further reading

R.L. Carlin, *Magnetochemistry*. Springer-Verlag, Berlin (1986).

C. Kittel, *Introduction to solid state physics*. Wiley, New York (1996).

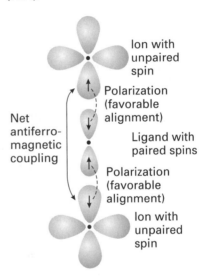

B18.4 Antiferromagnetic coupling between two metal centers created by spin polarization of a bridging ligand.

Another characteristic of many crystals that lack a center of symmetry is **piezoelectricity**, the ability to generate an electrical field when the crystal is under stress or to change dimensions when an electrical field is applied to the crystal. Piezoelectric materials are used for a variety of applications such as pressure transducers, ultramicromanipulators, as sound detectors, and as the probe support in scanning tunneling microscopy. Some important examples are $BaTiO_3$, $NaNbO_3$, $NaTaO_3$, and $KTaO_3$. Although a non-centrosymmetric structure is required for both ferroelectric and piezoelectric behavior, the two phenomena do not necessarily occur for the same crystal. For example, quartz is piezoelectric but not ferroelectric. Quartz is widely used to set the clock rate of microprocessors and watches because a thin sliver oscillates at a specific frequency to produce a small oscillating electrical field.

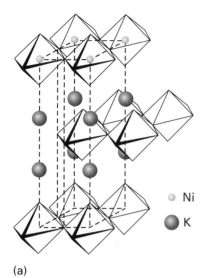

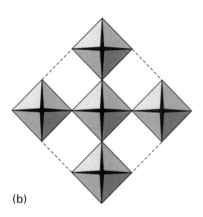

○ Ni

● K

(a)

(b)

18.24 The K_2NiF_4 structure. (a) The displaced layers of NiF_6 octahedra and (b) a view of the NiF_6 octahedra showing the corner-sharing of F.

Another prototypical structure, that of potassium tetrafluoronickelate(II), K_2NiF_4 (Fig. 18.24), is related to perovskite and has a bearing on the mechanism of high-temperature superconduction. The compound can be thought of as containing layers from the perovskite structure that share four F atom vertices within the layer and have terminal vertices above and below the layer. These layers are displaced relative to each other, so the terminal vertex in one layer is capped by a K^+ ion in the next layer.

Compounds with the K_2NiF_4 structure have come under renewed investigation because some high-temperature superconductors, such as $La_{1.85}Sr_{0.15}CuO_4$, crystallize with this structure. Apart from their importance in superconductivity, compounds with the K_2NiF_4 structure also provide an opportunity to investigate two-dimensional magnetic domains.

> The perovskites have the general formula ABX_3, in which the 12-coordinate hole of a ReO_3-type BX_3 structure is occupied by a large A ion.

(e) Superconductors

The versatility of the perovskites extends to superconductivity, because some of the high-temperature superconductors first reported in 1986 can be viewed as variants of the perovskite structure. Superconductors have two striking characteristics. Below a critical temperature, T_c, they enter the superconducting state and have zero electrical resistance. They also exhibit the **Meissner effect**, the exclusion of a magnetic field.

Following the discovery in 1911 that mercury is a superconductor below 4.2 K, physicists and chemists made slow but steady progress in the discovery of superconductors with higher values of T_c, at a rate of about 3 K per decade. After 75 years, T_c had been edged up to 23 K. Then, in 1986, **high-temperature superconductors** (HTSC) were discovered.[5] Several materials are now known with T_c well above 77 K, the boiling point of the inexpensive refrigerant liquid nitrogen, and in a few years T_c has increased by more than a factor of five.

Two types of superconductors are known. Those classed as **Type I** show abrupt loss of superconductivity when an applied magnetic field exceeds a value characteristic of the material. **Type II** superconductors, which include high-temperature materials, show a gradual loss of superconductivity above a critical field, H_c.[6] Figure 18.25 shows that there is a degree of periodicity in the elements that exhibit superconductivity. Note in particular that the ferromagnetic metals iron, cobalt, and nickel do not display superconductivity; nor do the alkali metals, nor the coinage metals copper, silver, and gold. For simple metals, ferromagnetism and superconductivity never coexist, but in some of the oxide superconductors ferromagnetism and superconductivity appear to coexist on different sublattices of the same solid.

One of the most widely studied oxide superconductor materials $YBa_2Cu_3O_7$ (informally called '123', from the proportions of metal atoms in the compound) has a structure similar to perovskite but with missing O atoms. In terms of the structure shown in Fig. 18.26, the A atoms of the original perovskite are yttrium and barium in $YBa_2Cu_3O_7$ and the B atoms are copper. The square-planar CuO_4 units are arranged in chains. Sheets and chains of CuO_4 or CuO_5 units can also be detected in other oxocuprate high-temperature superconductors, and it is thought that they are an important component of the mechanism of superconduction.

There is as yet no settled explanation of high-temperature superconductivity. It is believed that the Cooper pairs responsible for conventional superconductivity (Section 3.16) are important in the high-temperature materials, but the mechanism for pairing is hotly debated.[7]

[5]   For a good survey of the range of superconducting materials see J. Etourneau, in *Solid state chemistry, compounds* (ed. A.K. Cheetham and P. Day), pp. 60–111. Oxford University Press (1992).

[6]   The Chevrel phases discussed in Section 18.8 have the highest observed values of H_c.

[7]   For an entertaining account of the people investigating superconductivity see R.M. Hazen, *The breakthrough. The race for the superconductor.* Summit Books, New York (1988).

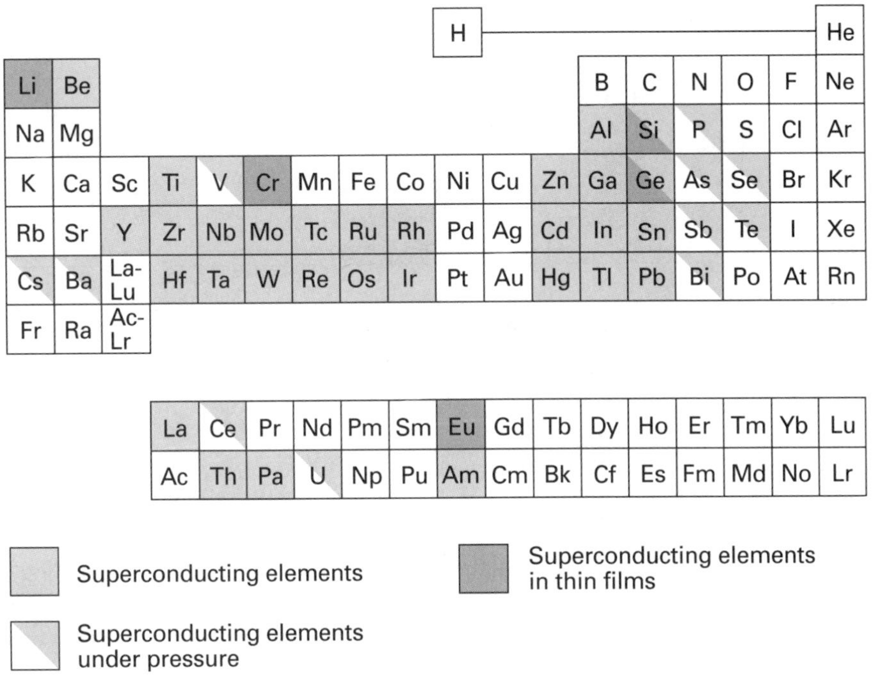

H																	He
Li	Be											B	C	N	O	F	Ne
Na	Mg											Al	Si	P	S	Cl	Ar
K	Ca	Sc	Ti	V	Cr	Mn	Fe	Co	Ni	Cu	Zn	Ga	Ge	As	Se	Br	Kr
Rb	Sr	Y	Zr	Nb	Mo	Tc	Ru	Rh	Pd	Ag	Cd	In	Sn	Sb	Te	I	Xe
Cs	Ba	La-Lu	Hf	Ta	W	Re	Os	Ir	Pt	Au	Hg	Tl	Pb	Bi	Po	At	Rn
Fr	Ra	Ac-Lr															

La	Ce	Pr	Nd	Pm	Sm	Eu	Gd	Tb	Dy	Ho	Er	Tm	Yb	Lu
Ac	Th	Pa	U	Np	Pu	Am	Cm	Bk	Cf	Es	Fm	Md	No	Lr

☐ Superconducting elements

■ Superconducting elements in thin films

◪ Superconducting elements under pressure

18.25 Superconducting elements. (From J. Etourneau, Superconducting materials, in *Solid state chemistry: compounds* (ed. A.K. Cheetham and P. Day), Chapter 3. Oxford University Press (1992).)

The perovskite-like structure of $YBa_2Cu_3O_7$ is shown in Fig. 18.26. Copper ions occupy the **B** sites and Y or Ba ions occupy the **A** sites of the layered lattice. However, unlike in a true perovskite structure, the **B** sites are not surrounded by an octahedron of O atoms. Some have five O atom neighbors and others have only four. Similarly, Ba^{2+} and Y^{3+} in the **A** site have less than 12-coordination. Another point to note is that, if we assign the usual oxidation

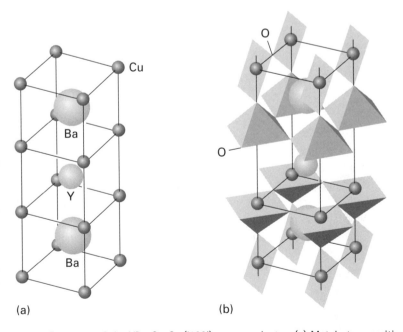

(a)   (b)

18.26 Structure of the $YBa_2Cu_3O_7$ ('123') superconductor. (a) Metal atom positions. (b) Oxygen polyhedra around the metal atoms. Unlike the octahedral environment in perovskite, metal ions in 123 are in square-planar and square-pyramidal coordination environments.

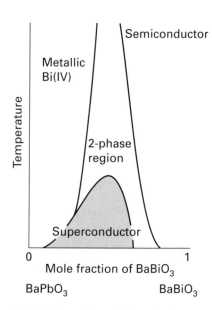

18.27 Schematic correlation of phase instability and superconductivity.

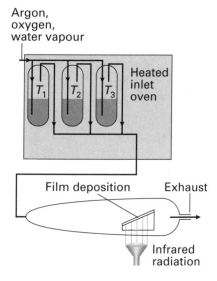

18.28 Schematic diagram of the chemical vapor deposition of superconductor thin films. A reactive carrier gas mixture (Ar, O_2, and H_2O) is passed through traps of the volatile metal precursors held at temperatures T_1, T_2, and T_3 which provide the desired vapor pressure of each reactant. Deposition occurs on the wedge-shaped block, which is heated to a high temperature by an IR lamp. After the deposition, the resulting film is annealed at high temperature to improve the crystallinity of the film.

Table 18.4 Some superconductors

Elements	T_c/K	Compounds	T_c/K
Zn	0.88	Nb_3Ge	23.2
Cd	0.56	Nb_3Sn	18.0
Hg	4.15	$LiTi_2O_4$	13
Pb	7.19	$K_{0.4}Ba_{0.6}BiO_3$	29.8
Nb	9.50	$YBa_2Cu_3O_7$	95
		$Tl_2Ba_2Ca_2Cu_3O_{10}$	122

numbers (Y $+3$, Ba $+2$, and O -2), then the average oxidation number of copper turns out to be $+2.33$, so it is inferred that $YBa_2Cu_3O_7$ is a mixed oxidation state material that contains Cu^{2+} and Cu^{3+}.

A few of the new superconductors are listed in Table 18.4 along with other superconducting materials. As illustrated in Fig. 18.27 for $BaPb_{1-x}Bi_xO_3$, superconductivity is usually observed only for materials that are not highly conducting above the critical temperature.

The synthesis of high-temperature superconductors has been guided by a variety of qualitative considerations, such as the demonstrated success of the layered structures and of mixed oxidation state copper in combination with heavy *p*-block elements. Additional considerations are the radii of ions and their preference for certain coordination environments. Some of these materials are prepared simply by heating an intimate mixture of the metal oxides to 800–900 °C in an open alumina crucible or a sealed gold tube.

Thin films are needed for the use of superconductors in electronic devices. Their preparation is an active area of research and a promising strategy is **chemical vapor deposition**.[8] The general strategy is to form the thin film by decomposing a thermally unstable compound on a hot solid substrate material (Fig. 18.28). Fluorinated acetylacetonato complexes of the metals (using ligands such as fod, (**2**)) are sometimes used because they are more volatile than simple acetylacetonato complexes. These complexes, such as $Cu(acac)_2$, $Y(dmpm)_3$, and $Ba(fod)_2$, are swept into the reaction chamber by slightly moist oxygen gas. When conditions are properly controlled this gaseous mixture reacts on the hot substrate to produce the desired $YBa_2Cu_3O_{7-x}$ film. The film may be amorphous and require subsequent heating to form a crystalline product.

High-temperature superconductors have structures related to perovskite.

18.6 Glasses

The term **ceramic** is often applied to all inorganic nonmetallic, non-molecular materials, including both amorphous and crystalline materials, but the term is commonly reserved for compounds or mixtures that have undergone heat treatment. The term **glass** is used in a variety of contexts, but for present purposes it implies an amorphous ceramic with a viscosity so high that it can be considered rigid. A substance in its glassy form is said to be in its **vitreous state**. Although ceramics and glasses have been utilized since antiquity, their development is currently an area of rapid scientific and technological progress. This enthusiasm stems from interest in the scientific basis of their properties and the development of novel synthetic routes to new high-performance materials. We confine our attention here to glasses. The most familiar glasses are alkali metal or alkaline earth metal silicates or borosilicates.

8   L.M. Tonge, D.S. Richeson, T.J. Marks, J. Zhao, J. Zhang, B.W. Wessels, H.O. Marcy, and C.R. Kannewurf, Organometallic chemical vapor deposition. In *Electron transfer in biology and solid state* (ed. M.K. Johnson, R.B. King, D.M. Kurtz Jr, C. Kutal, M.L. Norton, and R.A. Scott), Advances in Chemistry Series no. 226, pp. 351–68. American Chemical Society, Washington, DC (1990).

(a) Glass formation

A glass is prepared by cooling a melt more quickly than its rate of crystallization. Thus, cooling molten silica gives vitreous quartz. Under these conditions the solid mass has no long-range periodicity as judged by the lack of X-ray diffraction peaks, but spectroscopic and other data indicate that each Si atom is surrounded by a tetrahedral array of O atoms. The lack of long-range order results from variations of the Si—O—Si angles. Figure 18.29 illustrates in two dimensions how a local coordination environment can be preserved but long-range order lost by variation of the bond angles around oxygen. Silicon dioxide readily forms a glass because the three-dimensional network of strong covalent Si—O bonds in the melt does not readily break and reform upon cooling. The lack of strong directional bonds in metals and simple ionic substances makes it much more difficult to form glasses from these materials. Recently, however, techniques have been developed for ultrafast cooling and, as a result, a wide variety of metals and simple inorganic materials can now be frozen into a vitreous state.

The concept that the local coordination sphere of the glass-forming element is preserved but that bond angles around O are variable was originally proposed by the American crystallographer W.H. Zachariasen in 1932. He reasoned that these conditions would lead to similar Gibbs energies and molar volumes for the glass and its crystalline counterpart. Zachariasen also proposed that the vitreous state is favored by corner-shared O atoms, rather than edge- or face-shared, which would enforce greater order. These and other **Zachariasen rules** hold for common glass-forming oxides, but exceptions are known.

An instructive comparison between vitreous and crystalline materials is seen in their change in volume with temperature (Fig. 18.30). When a molten material crystallizes an abrupt change in volume (usually a decrease) occurs. In contrast, a glass-forming material that is cooled sufficiently rapidly persists in the liquid state to form a metastable supercooled liquid. When cooled below the **glass transition temperature**, T_g, the supercooled liquid becomes rigid, and this change is accompanied by only an inflection in the cooling curve.

> Silicon dioxide readily forms a glass because the three-dimensional network of strong covalent Si—O bonds in the melt does not readily break and reform upon cooling. The Zachariasen rules summarize the properties likely to lead to glass formation.

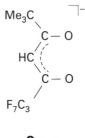

2 fod

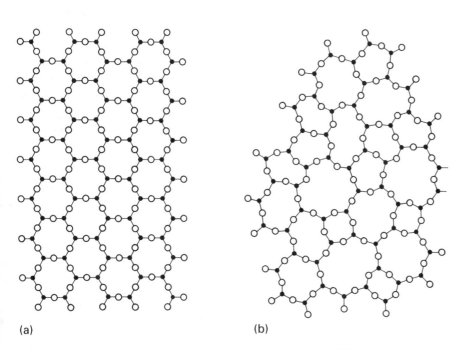

(a)   (b)

18.29 Schematic representation of (a) two-dimensional crystal and (b) two-dimensional glass. (Reproduced with permission from W.H. Zachariasen, *J. Am. Chem. Soc.* **54**, 3841 (1932).)

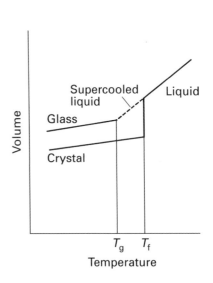

18.30 Comparison of the volume change for supercooled liquids and glasses with that for a crystalline material.

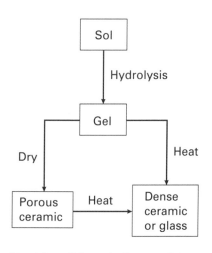

18.31 The role of a modifier is to introduce oxide ions and cations that can pin the negative charges together.

Chart 18.1 Schematic diagram of the sol–gel process. When the gel is dried at high temperatures dense ceramics or glasses are formed. Drying at low temperatures above the critical pressure of water produces microporous solids known as xerogels or aerogels.

(b) Glass production

Although vitreous silica is a strong glass that can withstand rapid cooling or heating without cracking, it has a high glass transition temperature and therefore must be worked at inconveniently high temperatures. Therefore a **modifier**, such as Na_2O or CaO, is commonly added to SiO_2. A modifier disrupts some of the Si—O—Si linkages and replaces them with terminal S—O^- links that associate with the cation (Fig. 18.31). The consequent partial disruption of the Si—O network leads to glasses that have lower softening points. The common glass used in bottles and windows is called 'sodalime glass' and contains Na_2O and CaO as modifiers. When B_2O_3 is used as a modifier, the resulting 'borosilicate glasses' have lower thermal expansion coefficients than sodalime glass and are less likely to crack when heated. Borosilicate glass is therefore widely used for ovenware and laboratory glassware.

Glass formation is a property of many oxides, and practical glasses have been made from sulfides, fluorides, and other anionic constituents. Some of the best glass formers are the oxides of elements near silicon in the periodic table (B_2O_3, GeO_2, and P_2O_5), but the solubility in water of most borate and phosphate glasses and the high cost of germanium limit their usefulness.

A current technological revolution is the replacement of electrical signals for voice and data transmission by optical signals. Similarly, optical circuit elements are being developed that may eventually replace electrical integrated circuits. These developments are stimulating considerable research in transparent crystalline and vitreous materials for light transmission and processing. For example, optical fibers for light transmission are currently being produced with a composition gradient from the interior to the surface. This composition gradient modifies the index of refraction and thereby decreases light loss. Fluoride glasses are also being investigated as possible substitutes for oxide glasses, because oxide glasses contain small numbers of OH groups, which absorb near-infrared radiation and attenuate the signal.

Most glasses are made simply by melting the component oxides and cooling faster than the rate of crystallization, which is very slow for many complex metal silicates, phosphates, and borates. Another route to glasses is the **sol–gel process**, which is described schematically in Chart 18.1.[9] As shown there, the sol-gel process is also used to produce crystalline ceramic materials and high-surface-area compounds such as silica gel. A typical process involves the addition of a metal alkoxide precursor to an alcohol, followed by the addition of water to hydrolyze the reactants. This hydrolysis leads to a thick gel that can be dehydrated and **sintered** (heated below its melting point to produce a compact solid). For example, ceramics containing TiO_2 and Al_2O_3 can be prepared in this way at much lower temperatures than required to produce the ceramic from the simple oxides. Often a special shape can be fashioned at the gel stage. Thus, the gel may be shaped into a fiber and then heated to expel water and produce a glass or ceramic fiber at much lower temperatures than would be necessary if the fiber were made from the melt of the components.

> *Most glasses are made by melting the component oxides and cooling faster than the rate of crystallization; another route to glasses and ceramic composites is the sol–gel process.*

Prototypical sulfides and related compounds

The soft chalcogens sulfur, selenium, and tellurium form binary compounds with metals that often have quite different structures from the corresponding oxides. As we saw in Section 2.10, this difference is consistent with the greater covalency of the compounds of sulfur and its heavier congeners. For example, we noted there that MO compounds generally adopt the rock-salt structure whereas ZnS and CdS adopt the sphalerite or the wurtzite

9　R.C. Mehrotra, Present status and future potential of the sol-gel process. *Structure and bonding* **77**, 1 (1992); R. Roy, *Science* **238**, 1665 (1987).

structures. Similarly, the d-block monosulfides generally adopt the more characteristically covalent nickel-arsenide structure rather than the rock-salt structure of alkaline earth oxides such as MgO. Even more striking are the layered MS$_2$ compounds of many d-block elements in contrast to the fluorite or rutile structures of many d-block dioxides.

Before we discuss these compounds, we should note that there are many metal-rich compounds that disobey simple valence rules and do not conform to an ionic model. Some examples are Ti$_2$S, Pd$_4$S, V$_2$O, and Fe$_3$N, and even the alkali metal suboxides, such as Cs$_3$O (Section 9.5). The occurrence of metal-rich phases is generally associated with the formation of M—M bonds. Many other intermetallic compounds display stoichiometries that cannot be understood in terms of conventional valence rules. Included among them are the important permanent magnet materials Nd$_2$Fe$_{17}$B and SmCo$_5$.

18.7 Layered MS$_2$ compounds and intercalation

We introduced the layered sulfides and their intercalation compounds in Section 11.10. Here we develop a broader picture of their structures and properties.

(a) Synthesis and crystal growth

The chalcogens react with most metals at or above room temperature, and a wide range of compounds are known. The preparation of crystalline disulfides suitable for chemical and structural studies is often performed by **chemical vapor transport**, which is a generally useful technique in solid-state chemistry. It is possible sometimes simply to sublime a compound, but the vapor transport technique can also be applied to a wide variety of nonvolatile compounds.

Typically, the crude material is loaded into one end of a borosilicate or fused quartz tube. After evacuation, a small amount of a chemical transport agent is introduced, the tube is sealed, and placed in a furnace with a temperature gradient. The polycrystalline and possibly impure metal chalcogenide is vaporized at one end and redeposited as pure crystals at the other (Fig. 18.32). The technique is called chemical vapor transport rather than sublimation because the chemical transport agent, which may be a halogen, produces an intermediate volatile species such as a metal halide. Generally, only a small amount of transport agent is needed because upon crystal formation it is released and diffuses back to pick up more reactant. For example, TaS$_2$ can be transported with I$_2$ in a thermal gradient. At 850 °C the reaction with I$_2$ to produce gaseous products is endothermic:

$$TaS_2(s) + 2\,I_2(g) \longrightarrow TaI_4(g) + S_2(g)$$

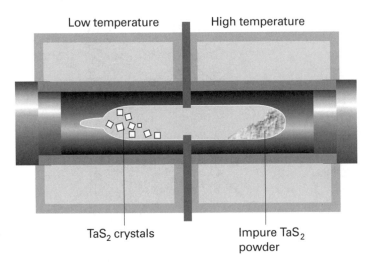

Low temperature   High temperature

TaS$_2$ crystals   Impure TaS$_2$ powder

18.32 Vapor transport crystal growth and purification of TaS$_2$. A small quantity of I$_2$ is present to serve as a transport agent.

The reaction is favored at this temperature. Therefore, the partial pressure of volatile products is greater at 850 °C than at 750 °C, so TaS_2 is deposited at the lower temperature. If, as occasionally is the case, the transport reaction is exothermic, the solid is carried from the cooler to the hotter end of the tube.

> *In the presence of iodine, polycrstalline metallic chalcogens form compounds that are volatile above room temperature, and single crystals of the chalcogenide deposit at lower temperatures.*

(b) Structure

As we saw in Section 11.10, the *d*-block disulfides fall into two classes. There are the layered materials formed by metals on the left of the *d*-block, and in the middle and toward the right of the block there are compounds containing formal S_2^{2-} ions, such as pyrite, FeS_2. We shall concentrate on the layered materials.

In TaS_2 and many other layered disulfides, the *d*-metal ion is located in octahedral holes between close-packed AB layers (Fig. 18.33a). The Ta ions form a close-packed layer denoted c, so the metal and adjoining sulfide layers can be portrayed as an AcB sandwich. These sandwich-like slabs form a three-dimensional crystal by stacking in sequences such as AcB AcB AcB . . ., where the strongly bound AcB slabs are held to their neighbors by weak dispersion forces. The Mo atoms in MoS_2 reside in the trigonal-prismatic holes between sulfide layers that are in register with one another (AA, Fig. 18.30(b)). The Mo atoms, which are strongly bonded to the adjacent sulfide layers, form a close-packed array, and so we can represent each slab as AbA or CbC. These slabs form a three-dimensional crystal by stacking in a pattern AbA CbC AbA CbC . . . . Weak dispersion forces also contribute to holding these AbA and CbC slabs together.

Polytypes can occur, which differ only in the stacking arrangement along a direction perpendicular to the plane of the slabs. Thus, MoS_2 forms several polytypes including one with the sequence CbC AbA and another with the sequence CaC AbA BcB.

In addition to variations in stacking sequences, the structures of many *d*-block dichalcogenides show subtle periodic distortions within the individual slabs. These distortions are called **charge density waves** and they will be described after we have established some background information on the band model of metal dichalcogenides.

In the discussion above it was convenient to describe—somewhat simplistically—the layered structures in terms of cations and anions. Now we shall use a band model, which is necessary if we are to account for their significant covalence, electrical conductivities, and chemical properties. Some approximate band structures derived primarily from molecular orbital

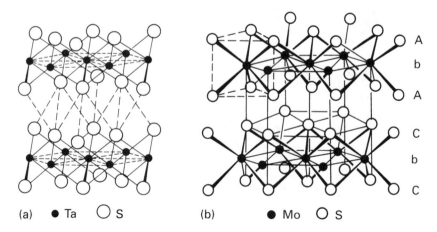

(a) ● Ta ○ S   (b) ● Mo ○ S

18.33 (a) The structure of TaS_2 (CdI_2-type). The Ta atoms reside in octahedral, c, sites between the AB layers of sulfur. (b) The MoS_2 structure; the Mo atoms reside in trigonal-prismatic sites, b, between AA or CC sulfide layers.

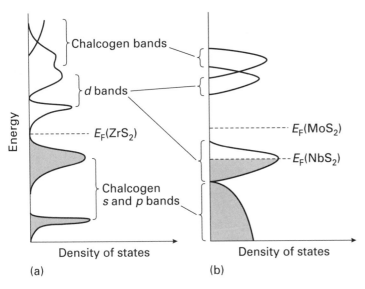

18.34 Approximate band structures of dichalcogenides. (a) Octahedral MS$_2$ compounds; for ZrS$_2$ (which is d^0), only the sulfur s and p bands are full and the compound is a semiconductor. (b) Trigonal-prismatic MS$_2$ compounds. NbS$_2$ (d^1, as illustrated) is metallic whereas MoS$_2$ (d^2) is a semiconductor.

calculations and photoelectron spectra are shown in Fig. 18.34. They show that the dichalcogenides with octahedral and trigonal-prismatic metal sites have low-lying bands primarily composed of chalcogen s and p orbitals, higher energy bands derived primarily from metal d orbitals, and above these metal and chalcogen bands.

When the metal atom resides in an octahedral site, its d orbitals are split into a lower t_{2g} set and a higher e_g set, just as in localized complexes. These atomic orbitals combine to give a t_{2g} band and an e_g band. The broad t_{2g} band may accommodate up to six electrons per metal atom. Thus TaS$_2$, with one d electron per metal, has an only partly filled t_{2g} band, and is a metallic conductor.

A trigonal-prismatic ligand field leads to a low-energy a_1' level and two doubly degenerate higher energy orbitals designated e' and e'' respectively. Now the lowest band needs only two electrons per atom to fill it. Thus MoS$_2$, with trigonal-prismatic Mo sites and two d electrons, has a filled a_1' band and is an insulator (more precisely, a large band-gap semiconductor).

The formation of charge density waves in some metal dichalcogenides is similar to the onset of the Peierls distortion of one-dimensional structures (Section 3.14). For a partly filled conduction band in one- and two-dimensional systems, a distortion of the lattice will split the band so that the resulting lower component may be filled with electrons and the upper component emptied. Since the energy advantage of this distortion is not large, a static modulation of the structure is usually only evident at low temperatures. Charge density waves are observed at low temperatures in the Group 5 dichalcogenides VS$_2$, NbS$_2$, and TaS$_2$. In these compounds four of the five electrons from each metal are required to fill the s and p band formed from the sulfur orbitals, and the remaining electron occupies but does not fill the d band. At sufficiently low temperatures the lattice distorts and the d band splits.

> Metals on the left of the d-block form layered sulfides (S^{2-}), whereas those in the middle and toward the right of the block contain formal S$_2^{2-}$ ions.

(c) Intercalation and insertion

We have already introduced the idea that alkali metal ions may insert between graphite sheets (Section 10.8) and metal disulfide slabs (Section 9.11) to form intercalation compounds. For a reaction to qualify as an intercalation, or as an **insertion reaction**, the basic structure of the

Table 18.5 Some alkali metal intercalation compounds of chalcogenides

Compound	$\Delta d/\text{Å}$†
$KZrS_2$	1.60
$NaTiS_2$	1.17
$KTiS_2$	1.92
$RbSnS_2$	2.24
$KSnS_2$	2.67
$Na_{0.6}MoS_2$ (2H)	1.35
$K_{0.4}MoS_2$ (2H)	2.14
$Rb_{0.3}MoS_2$ (2H)	2.45
$Cs_{0.3}MoS_2$ (2H)	3.66

†Change in interlayer spacing in comparison with the parent MS_2. (2H signifies a particular layer sequence.)

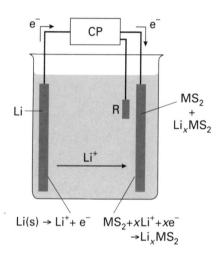

18.35 Schematic experimental arrangement for electrointercalation. A polar organic solvent (e.g. propylene carbonate) containing a lithium salt is used as the electrolyte. R is a reference electrode and CP is a coulometer (to measure the charge passed) and a potential controller.

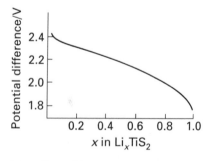

18.36 Potential versus composition diagram for the electrointercalation of lithium into titanium disulfide. The composition, x, in Li_xTiS_2 is calculated from the charge passed in the course of electrointercalation.

host should not be altered when it occurs.[10] Insertion compound formation is the basis of 'high-energy cells', which consist, for instance, of a lithium anode, a chalcogenide cathode, and an electrolyte dissolved in an inert polar organic solvent. Such a cell can produce well over 2 V and have over twice the energy density of a Ni/Cd cell. The recharging of such cells is fast because no solid phase transition is involved.

The π conduction and valence bands of graphite are contiguous in energy (we have seen in fact that graphite is formally a semimetal, Section 3.14) and the favorable Gibbs energy for intercalation arises from the transfer of an electron from the alkali metal atom to the graphite conduction band. The insertion of an alkali metal atom into a dichalcogenide involves a similar process: the electron is accepted into the d band and the charge-compensating alkali metal ion diffuses to positions between the slabs. Some representative alkali metal insertion compounds are listed in Table 18.5.

The insertion of alkali metal ions into host lattices can be achieved by direct combination of the alkali metal and the disulfide:

$$TaS_2 + xNa \xrightarrow{800\,°C} Na_xTaS_2 \qquad x = 0.4 \text{ to } 0.7$$

or by utilizing a highly reducing alkali metal compound or the electrochemical technique of **electrointercalation** (Fig. 18.35). One advantage of electrointercalation is that it is possible to measure the amount of alkali metal incorporated by monitoring the amount of electrons (as calculated from It/F) passed during the synthesis. It also is possible to distinguish solid solution formation from discrete phase formation. As illustrated in Fig. 18.36, the formation of a solid solution is characterized by a gradual change in potential as intercalation proceeds, whereas the formation of a new discrete phase yields a steady potential over the range in which one solid phase is being converted into the other, followed by an abrupt change in potential when that reaction is complete.

Insertion compounds are examples of mixed ionic and electronic conductors. In general the insertion process can be reversed either chemically or electrochemically. This reversibility makes it possible to recharge a lithium cell by removal of lithium from the compound. In a clever synthetic application of these concepts, the previously unknown layered disulfide VS_2 was prepared by first making the known layered compound $LiVS_2$ in a high-temperature process. The lithium was then removed by reaction with I_2 to produce the metastable layered VS_2, which was found to have the TiS_2 structure:

$$2\,LiVS_2 + I_2 \longrightarrow 2\,LiI + 2\,VS_2$$

Insertion compounds also can be formed with molecular guests. Perhaps the most interesting are the metallocenes $Co(\eta^5\text{-Cp})_2$ and $Cr(\eta^5\text{-Cp})_2$, where $Cp = C_5H_5$, which can be incorporated into a variety of hosts, such as TiS_2, $TiSe_2$, and TaS_2, to the extent of about $0.25\ M(\eta^5\text{-Cp})_2$ per MS_2 or MSe_2. This limit appears to correspond to the space available for forming a complete layer of $M(\eta^5\text{-Cp})_2^+$. The organometallic compound appears to undergo oxidation upon intercalation, so the favorable Gibbs energy in these reactions arises in the same way as in alkali metal intercalation. In agreement with this interpretation, $Fe(\eta^5\text{-Cp})_2$, which is harder to oxidize than its chromium or cobalt analogs, does not intercalate.

We can imagine the insertion of ions into one-dimensional channels, two-dimensional planes of the type we have been discussing, or channels that intersect to form three-dimensional networks (Fig. 18.37). Aside from the availability of a site for a guest to enter, the host must provide a conduction band of suitable energy to take up electrons reversibly (or, in some cases, be able to donate electrons to the host). Table 18.6 illustrates that a wide variety

---

[10] Reactions in which the structure of one of the solid starting materials is not radically altered are called 'topotactic' reactions. They are not limited to the type of insertion chemistry we are discussing here. For example, hydration, dehydration, and ion exchange reactions may be topotactic.

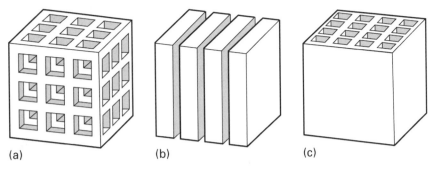

18.37 Schematic representation of host materials for intercalation reactions. (a) A three-dimensional host with intersecting channels, (b) a two-dimensional layered compound, and (c) one-dimensional channels.

of hosts are possible, including metal oxides and various ternary and quaternary compounds. We see that intercalation chemistry is by no means limited to graphite and layered disulfides.

> *Insertion compounds can be formed by direct reaction or electrochemically; insertion compounds can also be formed with molecular guests.*

18.8 Chevrel phases

We close this chapter with a brief discussion of an interesting class of ternary compounds first reported by the French solid-state inorganic chemist R. Chevrel in 1971.[11] These compounds, which illustrate three-dimensional intercalation, have formulas such as Mo_6X_8 and $M_xMo_6S_8$; Se or Te may take the place of S and the intercalated M atom may be a variety of metals such as Li, Mn, Fe, Cd, or Pb. The parent Mo_6Se_8 and Mo_6Te_8 are prepared by heating the elements at about $1000\,°C$. A structural unit common to this series is M_6S_8, which may be viewed as an octahedron of M atoms face-bridged by S atoms, or alternatively as an octahedron of M atoms in a cube of S atoms (Fig. 18.38). This type of cluster is also observed for some halides of the Period 4 and 5 early d-block elements, such as the $[M_6X_8]^{4-}$ cluster found in Mo and W dichlorides, bromides, and iodides.

Figure 18.39 shows that in the three-dimensional solid the Mo_6S_8 clusters are canted relative to each other and relative to the sites occupied by intercalated ions. The Mo_6S_8 are canted to allow secondary donor–acceptor interaction between vacant Mo d_{z^2} orbitals (which project outward from the faces of the Mo_6S_8 cube) and a filled donor orbital on S atoms of adjacent clusters.

One of the physical properties that has drawn attention to the Chevrel phases is their superconductivity. Superconductivity persists up to $14\,K$ in $PbMo_6S_8$, and it also persists to very high magnetic fields, which is of considerable practical interest because many applications involve high fields. In this respect the Chevrel phases appear to be significantly superior to the newer oxocuprate high-temperature superconductors.

> *A Chevrel phase has a formula such as Mo_6X_8 or $M_xMo_6S_8$ where Se or Te may take the place of S and the intercalated M atom may be a variety of metals such as Li, Mn, Fe, Cd, and Pb.*

Table 18.6 Some three-dimensional intercalation compounds

Phase	Composition (x)
$Li_x[Mo_6S_8]$	0.65–2.4
$Na_x[Mo_6S_8]$	3.6
$Ni_x[Mo_6Se_8]$	1.8
H_xWO_3	0–0.6
Li_xWO_3	0–1.0
Li_xNiPS_3	0–1.5

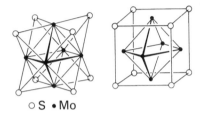

○ S • Mo

18.38 Two views of the Mo_6S_8 unit in a Chevrel phase.

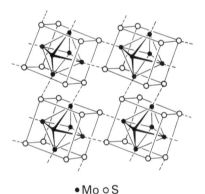

• Mo ○ S

18.39 Structure of a Chevrel phase showing the canted Mo_6S_8 unit in the slightly distorted cube of Pb atoms. An Mo atom in one cube can act as the acceptor for an electron pair donated by an S atom in a neighboring cube.

[11] T. Hughbanks and R. Hoffman, *J. Am. Chem. Soc.* **105**, 1150 (1983); R. Schöllhorn, *Angew. Chem. Int. Ed. Eng.* **19**, 983 (1980).

Further reading

J.K. Burdett, *Chemical bonding in solids*. Oxford University Press, New York (1995). A concise discussion of bonding in solids.

C.N.R. Rao and J. Gopalakrishnan, *New directions in solid state chemistry*. Cambridge University Press (1997). This book provides an up-to-date account of methods for the preparation and characterization of solids, preparative strategies, properties of solids, and their reactivity.

A.R. West, *Solid state chemistry and its applications*. Wiley, New York (1984). This book provides a very good introduction to structures and bonding in solids.

J.R. Rodgers and P. Villars, *Trends in advanced materials data: regularities and predictions*, MRS Bulletin, Feb. 1983. A global view of structure types for intermetallic compounds.

A.K. Cheetham and P. Day (ed.), *Solid state chemistry: compounds*. Oxford University Press (1992). A good discussion of inorganic solids.

A. Müller, *Inorganic structural chemistry*. Wiley, New York (1993). A well illustrated text with emphasis on solids.

M. O'Keefe and A. Navrotsky (ed.), *Structure and bonding in crystals*. Academic Press, Boca Raton (1981).

A. Wold and K. Dwight, *Solid state chemistry*. Chapman & Hall, London (1993). An introduction to the preparation and structures of metal oxides and sulfides.

A.F. Wells, *Structural inorganic chemistry*. Oxford University Press (1984). This handy book provides a wealth of information on structures and structural correlations.

L.V. Interrante, L.A. Casper, and A.B. Ellis (ed.), *Materials chemistry: an emerging discipline*, Advances in Chemistry Series no. 245. American Chemical Society, Washington, DC (1995). A series of chapters on a broad range of inorganic and organic solids.

A. Feltz, *Amorphous inorganic materials and glasses*. VCH, Weinheim (1993). This volume describes the properties of a wide range of glasses.

R.B. King (ed.), *The encyclopedia of inorganic chemistry*. Wiley, New York (1994). The following articles on solids are of interest: J. Livage, Solids: sol–gel synthesis, Vol. 7, pp. 3836–51; A.R. West, Solids: characterization by powder diffraction, Vol. 7, pp. 3851–68; C.R.A. Catlow, Solids: computer modeling, Vol. 7, pp. 3868–99. Also of interest are the following articles: J.E. Greedan, Magnetic oxides, Vol. 4, pp. 2017–48; W.E. Hatfield, Magnetism of transition metal ions, Vol. 4, pp. 2066–75; W.E. Hatfield, Magnetism of extended arrays in inorganic solids, Vol. 4, pp. 2049–65.

Exercises

18.1 Describe the nature of Frenkel and Schottky defects, and some experimental tests for their occurrence. Do either of these defects by themselves give rise to nonstoichiometry?

18.2 From each of the pairs of compounds (a) NaCl and NiO, (b) CaF_2 and PbF_2, and (c) Al_2O_3 and Fe_2O_3, pick the one most likely to have a high concentration of defects. Describe the possible defects and give your reasoning.

18.3 Distinguish intrinsic from extrinsic defects and give an example of each in actual compounds.

18.4 Draw one unit cell in the ReO_3 structure showing the **M** and O atoms. Does this structure appear to be sufficiently open to undergo Na^+ ion intercalation? If so, where might the Na^+ ions reside?

18.5 Is a crystallographic shear plane a defect, a way of describing a new structure, or both? Explain your answer.

18.6 Sketch the rock-salt crystal structure and locate in the sketch the interstitial sites to which a cation might migrate. What is the nature of the bottleneck between the normal and interstitial sites?

18.7 Contrast the nature of the defects in TiO with those in FeO.

18.8 How might you distinguish experimentally the existence of a solid solution from a series of crystallographic shear plane structures for a material that appears to have variable composition?

18.9 Bearing in mind that the redox chemistry of metal oxides often can be correlated with the redox behavior of the ions in solution, explain which of TiO, MnO, and NiO is (a) the most difficult to oxidize with air, (b) the easiest to reduce.

18.10 Describe the electrical conductivity of TiO and NiO and describe the interpretation in terms of the trends in electronic structure across the $3d$ series.

18.11 Above 146 °C, AgI is an Ag^+ ion conductor. Suppose a pellet of this compound is sandwiched between two silver electrodes at 165 °C and a potential difference of 0.1 V is imposed between the electrodes. Sketch the apparatus and indicate the changes that occur with time at the two electrodes.

18.12 What will be the quantitative changes in the mass of the cathode, the AgI, and the anode after 1.0×10^{-3} mol of electrons have been passed through the cell described in Exercise 18.11.

18.13 Contrast the mechanism of interaction between unpaired spins in a ferromagnetic material with that for an antiferromagnet.

18.14 Magnetic measurements on the ferrite $CoFe_2O_4$ indicate 3.4 spins per formula unit. Suggest a distribution of cations between octahedral and tetrahedral sites that would satisfy this observation.

18.15 The superconducting compound $YBa_2Cu_3O_7$ is described as having a perovskite-like structure. Ignoring problems of charge on the ions, describe the difference between this structure and that of perovskite.

Problems

18.1 Given the ligand-field Δ_O and Δ_T values that follow, determine the site preference for $A = Ni^{2+}$ and $B = Fe^{3+}$ in normal compared with inverse spinel, assuming that ligand-field stabilization is dominant. Δ_O: Fe^{3+} (1400 cm^{-1}), Ni^{2+} (860 cm^{-1}); Δ_T: Fe^{3+} (620 cm^{-1}), Ni^{2+} (380 cm^{-1}).

18.2 Magnetite, Fe_3O_4, has the spinel structure and is thus a member of the ferrite group of compounds. In a normal spinel unit cell eight tetrahedral holes would be filled with Fe^{2+} and 16 octahedral holes would contain Fe^{3+}. At room temperature magnetite has high conductivity (about 2×10^2 S cm^{-1}). Are the conductivity data consistent with the normal structure? If not, propose one possible variant of the spinel structure that would lead to high conductivity.

18.3 TiO adopts the NaCl structure, but it contains a high concentration of Schottky defects at both the cation and anion sites. LiCl also has the NaCl structure, but the concentration of defects is very low. Describe the probable influence of redox behavior of the cations on this difference in defect density.

18.4 Fe_xO generally has $x < 1$; describe the probable metal ion defect that leads to x being less than 1.

18.5 Ag_2HgI_4 is a good electrical conductor above $50\,°C$. Speculate on why Ag^+ rather than Hg^{2+} is the mobile species.

18.6 When NiO is doped with small quantities of Li_2O the electronic conductivity of the solid increases. Provide a plausible chemical explanation for this observation. (*Hint.* Li^+ occurs on Ni^{2+} sites.)

18.7 Cr_2O_3 is dark-green but, when a small quantity of Cr_2O_3 is doped into the isostructural Al_2O_3, the resulting solid solution is bright red. What is the origin of this color change?

18.8 Solid PbF_2 and ZrO_2 at elevated temperatures are good anion conductors. Describe their crystal structure and why it may be conducive to ion transport.

18.9 Describe the issue of occupation factors in Fe_3O_4, Cr_3O_4, and Mn_3O_4 and how the occupation factor is related to ligand-field stabilization factors.

18.10 The perovskites, such as ABO_3, often have the dipositive ion in the A site and an ion of higher oxidation state in the B site. What are the factors that lead to this site preference?

18.11 Superconductors are often classified as type I or II. Describe the physical characteristic that determines the classification of a superconductor into one or the other of these two classes.

18.12 State Zachariasen's two generalizations that favor glass formation and apply these to the observation that cooling molten CaF_2 leads to a crystalline solid whereas cooling molten SiO_2 at a similar rate produces a glass.

18.13 Starting with $TiCl_4$ and $AlCl_3$, give a plausible procedure for the formation of a ceramic containing polycrystalline Al_2O_3 and TiO_2.

18.14 Describe the interactions between Mo_6S_8 units in a Chevrel phase.

19

Bioinorganic chemistry

The elements of living systems

Enzymes exploiting acid catalysis

Redox catalysis

Metals in medicine

Further reading

Exercises

Problems

This chapter surveys the rapidly growing area of bioinorganic chemistry, which largely focuses on the roles of metal ions in biology. We begin with a general survey of the utilization of metal ions in biology and their general functions and then turn to the ligands that Nature employs, most of which are far more elaborate than the ligands considered in earlier chapters. The subtle interplay between macromolecular ligands and metal ions is then illustrated by considering biological functions such as oxygen storage and release. This discussion takes us into the realm of enzymes, where metal ions are often employed at the active site to effect important processes such as redox reactions, including nitrogen fixation. We close with a summary of our current knowledge of photosynthesis, the process that provides the primary energy source for living organisms, and a brief look at pharmaceutical applications of metal ions.

In contrast to the simpler and more static mineral world, living matter is based on much more intricate structures, such as cells, that persist in states far from equilibrium by maintaining a steady flow of nutrients and energy. Organisms use many of the elements to maintain this steady state, so the principles of inorganic chemistry are relevant to the understanding of biological systems. As a result, there is a lively exchange of ideas between inorganic chemists and biochemists, which has led to the growth and recognition of the interdisciplinary area of 'bioinorganic chemistry'. Just as a knowledge of inorganic reactions can be brought to bear on biochemical processes, the structures that have emerged in Nature have stimulated the synthesis and characterization of new inorganic compounds designed to test, mimic, and perhaps improve our understanding of the role of metals and ligands in living systems.

Although main-group elements play a major role in biochemistry, current research in bioinorganic chemistry stresses the role of d-metal ions, and this outlook will dominate the present chapter. We shall encounter interesting coordination environments for metal ions as well as the reaction types introduced earlier in the text, including Brønsted and Lewis acid–base reactions, redox reactions that depend on electron and group transfer, and photochemically induced electron transfer.

The elements of living systems

Table 19.1 Descending mass abundance of the elements

Earth		Humans
Crust	Oceans	
O	O	H
Si	H	O
Al	Cl	C
Fe	Na	N
Ca	Mg	Na
Mg	S	K
Na	Ca	Ca
K	K	Mg
Ti	C	P
H	Br	S
P	B	Cl

Living organisms do not employ the elements in the same order as that of their abundance in the Earth's crust. For example, Table 19.1 shows a high crustal abundance for aluminum and silicon, but these elements are not abundant in mammals. Some elements that are present only in trace amounts in the crust play a central role in an organism's struggle to avoid equilibrium: copper and selenium are two examples. There is a striking example of one element, cobalt, having a single known occurrence in biochemistry, in coenzyme B_{12}. Many elements of low terrestrial abundance, such as beryllium, thallium, and uranium, are not known to be used in any biochemical processes.

19.1 The biological roles of metal ions

Table 19.2 is a summary of biomolecules that utilize metal ions; many of these molecules are proteins. About 30 per cent of enzymes have a metal atom at the active site. These **metalloenzymes** facilitate a variety of reactions, which include acid-catalyzed hydrolysis (carried out by the hydrolases), redox reactions (oxidases and oxygenases), and the rearrangement of carbon–carbon bonds (synthases and isomerases). Metal ions also play

Table 19.2 The classification of some biomolecules containing metal ions

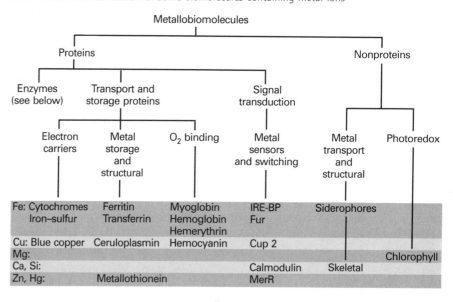

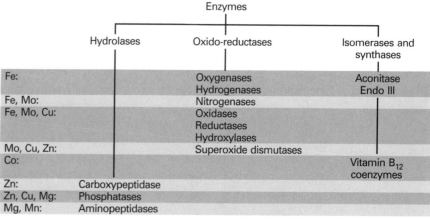

Augmented version of J.A. Ibers and R.H. Holm, *Science* **209**, 223 (1980)

structural roles. For example Ca^{2+} ions are involved in protein folding. The table shows that metal-ion containing molecules in general (not just enzymes) serve important functions as electron carriers, metal storage sites, O_2 binding and storage sites, and in signal transduction. Metal ions are also involved in sites of photoredox reactions, as in the energy-harvesting molecule chlorophyll.

Although some of the functions we have mentioned are shared by nonmetal species, it is clear that metal complexes play a major biological role. The manner in which they act is also highly varied. For instance, metal ions exert an inductive effect by coordination to the site of reaction, and serve as redox sites that function by either electron or atom transfer. Selectivity is achieved by the deployment of ions having the appropriate size, stereochemical preference, hard–soft character, or reduction potential to perform a given task.

Much of this chapter will deal with polypeptides. These macromolecules are formed from α-amino acids linked by peptide bonds (**1**). We shall sometimes use the term 'peptide residue', which denotes the component of an amino acid that remains in the chain once the peptide link has formed by elimination of H_2O. The 20 different α-amino acids that occur naturally in proteins are listed with their abbreviations in Table 19.3. Note that the different side chains in

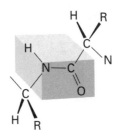

1 Peptide bond

Table 19.3 The classification of amino acids $NH_2CHRCOOH$

Type	Name	Abbreviation		R—
Hydrophobic R	Glycine	gly	G	H—
	Alanine	ala	A	CH_3—
	Valine	val	V	$(CH_3)_2CH$—
	Leucine	leu	L	$(CH_3)_2CHCH_2$—
	Isoleucine	ile	I	$CH_3CH_2CH(CH_3)$—
	Phenylalanine	phe	F	⬡—CH_2—
Inert heteroatom R	Tryptophan	trp	W	(indole)—CH_2—
Hydroxylic R	Serine	ser	S	$HOCH_2$—
	Threonine	thr	T	$HOCH(CH_3)$—
	Tyrosine	tyr	Y	HO—⬡—CH_2—
Carboxylic R	Aspartic acid	asp	D	$HOOCCH_2$—
	Glutamic acid	glu	E	$HOOCCH_2CH_2$—
Amine R	Lysine	lys	K	$H_2NCH_2CH_2CH_2CH_2$—
	Arginine	arg	R	$H_2N-\underset{\underset{NH}{\parallel}}{C}-NHCH_2CH_2CH_2$—
Amide R	Aspargine	asn	W	$H_2N\underset{\underset{O}{\parallel}}{C}CH_2$—
	Glutamine	gln	Q	$H_2N\underset{\underset{O}{\parallel}}{C}CH_2CH_2$—
Imidazole R	Histidine	his	H	(imidazole)
Sulfur-containing R	Cysteine	cys	C	$HSCH_2$—
	Methionine	met	M	$CH_3SCH_2CH_2$—
Others	Proline	pro	P	(proline structure)

Table 19.4 Minerals in biological structural materials

Mineral	Formula	Organism	Location
Calcite	$CaCO_3$	Birds	Eggshells
Aragonite	$CaCO_3$	Mollusks	Shell
Hydroxyapatite	$Ca_5(PO_4)_3(OH)$	Vertebrates	Bone
		Mammals	
Silicon dioxide	$SiO_2 \cdot nH_2O$	Diatoms	Cell wall
		Limpets	Teeth
		Plants	Leaves

these amino acids possess a range of functional groups, such as alkyl, carboxyl (—COOH), amino (—NH_2), hydroxyl (—OH), and thiol (—SH). The alkyl groups confer hydrophobic character and the more polar groups confer hydrophilic character to varying degrees. Most of these polar groups serve as Brønsted acids or bases, or Lewis bases in their complexation with metal ions. One very important role for the functional groups of the peptide residues is to modify the immediate environment of metal ions in metalloenzymes, and we shall concentrate on this subtle aspect of their behavior.

In addition to their role in the dynamics of biological processes, metal ions in the form of crystalline minerals or amorphous solids are important as structural materials in many organisms (Table 19.4).[1] In Section 11.1 we saw that the principal commercial source of phosphates and phosphorus is the calcium-containing mineral hydroxyapatite. Hydroxyapatite is widespread in the animal kingdom as a principal constituent of bones, teeth, and shells. The minerals usually exist in conjunction with other substances; for example, in bones and teeth hydroxyapatite coexists with the protein collagen, which controls the morphology of hydroxyapatite growth and influences its mechanical properties. Silicon is less widely used in Nature's structural materials, but one of the major exceptions to this generalization is the use of hydrated silicon dioxide in tiny diatoms that live in the sea. Deposits of their dead cell walls form 'diatomaceous earth', which is a useful filtering agent in the laboratory and in industry.

> *Metal ions are utilized for catalysis, signaling, energy storage, and the maintenance of the hard and soft structures of biological materials.*

19.2 Calcium biochemistry

As well as occurring in hard structural biomaterials, Ca^{2+} serves many biochemical roles, such as a messenger for hormonal action, a trigger for muscle contraction, in the initiation of blood clotting, and in the stabilization of protein structures. X-ray diffraction and NMR results illustrate how these functions are controlled by conformational changes induced by Ca^{2+} when it binds to calmodulin, troponin C, and related proteins: these proteins are involved in the activation of membrane channels and receptors on cell surfaces.[2]

The many roles of the Ca^{2+} ion appear to arise from its affinity for the hard ligand oxygen in conjunction with the intermediate lability of its complexes, which fall between the alkali metal ions and the *d*-metal ions, and the less labile ions of its lighter congeners in Group 2 (Be^{2+} and Mg^{2+}). The differences between Ca^{2+} and Mg^{2+} are illuminating and can be traced to three factors. First, as a consequence of its low selectivity, Ca^{2+} can bind neutral oxygen donor ligands (carbonyls and alcohols) in competition with water. Second, Ca^{2+} resembles Na^+ and K^+ in favoring high coordination numbers and an irregular coordination geometry. With its charge number of +2, Ca^{2+} can bind anions that alkali metal ions do not. Third, the rates of binding to and dissociation from Ca^{2+} are high: the rate of attachment is diffusion-

[1] For a useful and brief review see S. Mann, *J. Chem. Soc. Dalton*, 3953 (1997).

[2] W.J. Chazin, *Nature Structural Biology* **2**, 707 (1995).

limited and the rate of dissociation correlates with stability. The water exchange rate at Mg^{2+} is 10^5 s^{-1}, nearly three orders of magnitude slower than that of Ca^{2+}. This rapid exchange rate, corresponding to a highly responsive system, accounts for the use of Ca^{2+} when rapid responses are needed, such as the opening or closing of membrane channels, the regulation of ion transport, and muscle contraction.

Calcium binding proteins are typically rich in aspartate and glutamate, both of which have carboxylate groups as side chains and hence can act as hard anionic ligands. As an illustration of the kind of structure that occurs, one Ca^{2+} ion binding site that has been characterized by X-ray diffraction consists of four carboxylate O atoms in a distorted tetrahedral array (2). The Ca^{2+} ion commonly functions as a bridge between different protein segments, binding to anionic side groups of different amino acids or even carbonyl groups (Fig. 19.1). The utilization of this property for the control of chain folding is illustrated schematically in (2), which shows how the peptide chain folds to allow four CO_2^- groups to coordinate to Ca^{2+}. The outcome of this movement is a shift in the locations of the side groups on one helical protein region relative to the side groups on the second helical region. The changes in protein folding control cell structure and function, such as the rate of cell growth and energy metabolism.

> The dipositive ions Ca^{2+} and Mg^{2+} influence protein folding and thereby alter cell structure and function; the lability of Ca^{2+} permits rapid structural change.

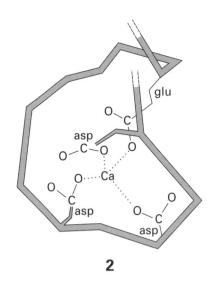

2

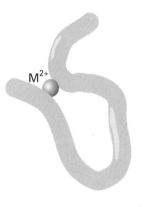

19.1 A metal ion, such as Ca^{2+}, may act as an important influence on the conformation adopted by a polypeptide chain.

19.3 Oxygen transport and storage

Oxygen is a recurring theme in this chapter because its role in respiration and its production in photosynthesis provide examples of a range of biologically important redox reactions, electron transfers, atom transfers, and photochemical processes. To ensure a supply of O_2, three types of proteins have evolved that bind and transport oxygen. They are commonly referred to as 'oxygen carriers'. Each of these proteins makes use of a metal atom or pair of metal atoms in proteins that have evolved to serve a specific type of organism.

The most familiar and most widely distributed oxygen carrier, hemoglobin (Fig. 19.2), is present in red blood cells and is utilized by vertebrates to carry oxygen from lungs or gills to the tissues where the O_2 is to be used and in the process reduced to CO_2. Vertebrates also employ the oxygen-storage protein myoglobin. This molecule has an O_2 binding site that resembles those in hemoglobin. Myoglobin provides oxygen to muscle tissue to sustain a burst of activity.

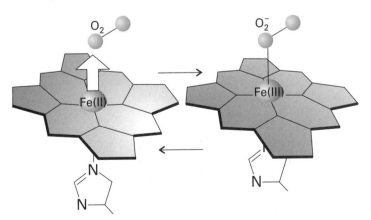

19.2 When an O_2 molecule arrives at the heme center in hemoglobin, the iron(II) atom moves up toward it as an electron is transferred; the iron is oxidized to iron(III) and the dioxygen molecule becomes an O_2^- ion. Note that the porphyrin ring bulges down toward the histidine (left), but becomes flatter when oxygen attaches (right).

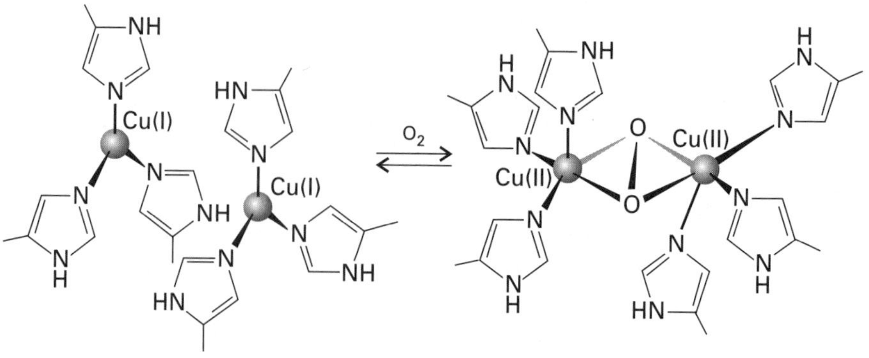

19.3 A fragment of the oxygen carrier hemocyanin found in mollusks and arthropods. When an O_2 molecule attaches, two copper(I) atoms are oxidized to copper(II) and the oxygen is reduced to O_2^{2-}.

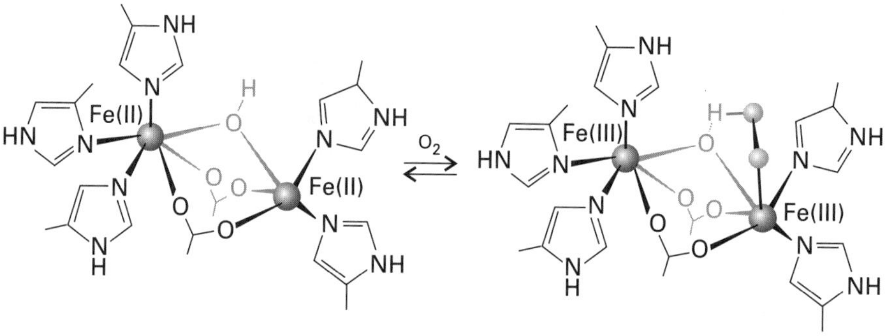

19.4 A fragment of the oxygen carrier hemerythrin found in some sea worms. When an O_2 molecule attaches, the iron(II) atoms are oxidized to iron(III) and the oxygen molecule is bound end-on as an HO_2^- ion.

A second type of oxygen carrier, hemocyanin (Fig. 19.3), is found in snails (mollusks) and crabs (arthropods). The O_2 binding site contains two Cu^+ ions attached directly to a protein. Interaction with O_2 causes the metal ions to become Cu^{2+} and the oxygen is reduced to O_2^{2-}.

The third oxygen carrier, hemerythrin (Fig. 19.4), is found in a limited range of sea worms. The oxygen binding site in hemerythrin consists of two Fe^{2+} ions in close proximity: they are bound directly to donor atoms on the protein rather than to a heme macrocycle. The O_2 is bound in an end-on fashion as a hydroperoxide ion (HO_2^-) to one of the two iron atoms. Oddly, some of the organisms that use hemerythrin as an O_2 carrier employ the entirely different protein myoglobin to store O_2.

> O_2 binds to an iron-porphyrin in hemoglobin and myoglobin, to a pair of protein-bound copper ions in hemocyanin, and to one of the two protein-bound iron atoms in hemerythrin.

(a) Hemoglobin and myoglobin

The helical protein in hemoglobin (Hb) plays an interesting regulatory role in the function of this oxygen carrier. The coils in this protein act like springs that can respond to the strain generated when oxygen binds at one iron–porphyrin site and the protein conveys that strain to other sites.[3] The net effect of this interaction is to increase the O_2 affinity of the second site. This phenomenon is called **cooperativity**, and enables hemoglobin to bind and release O_2 more effectively. We see that O_2 attachment can be a combination of complexation at the metal site and of alteration of the protein environment.

---

[3] J.A. Cowan, *Inorganic biochemistry*. Wiley–VCH, New York (1997).

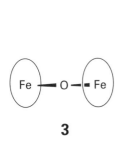

3

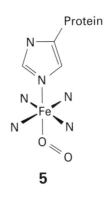

4

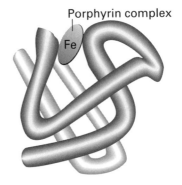

5

The O_2 molecule serves as a powerful π-acceptor in its interaction with a Fe^{2+} center, so it should not come as a surprise that other π-acceptor ligands are capable of bonding to the iron in hemoglobin and myoglobin. Complexes with NO, CO, CN^-, RNC, N_3^-, and SCN^- have been studied. These strongly bonded ligands can shut down the function of hemoglobin completely, with fatal results.

The polypeptide in hemoglobin plays an important role. Naked 'heme', the Fe–porphyrin complex without the accompanying polypeptide, is oxidized irreversibly to Fe(III) by molecular O_2, ultimately giving a stable μ-O^{2-} product (**3**) that cannot function as an O_2 carrier. This unproductive oxidative dimerization of iron porphyrins is potentially a fatal flaw for their biological function. However, it is prevented in hemoglobin or myoglobin by the polypeptide chains that surround and protect the heme site.

The structure of deoxy-myoglobin, the oxygen-free form of myoglobin (Mb), is shown in outline in Fig. 19.5. Note that the porphyrin active site is entwined with the polypeptide chain. The protein pocket in which the heme is carried is formed from amino acid residues with nonpolar side groups, and is hydrophobic. These same groups block access of larger molecules to the Fe atom neighborhood and so prevent the formation of an Fe—O_2—Fe bridged species. The hydrophobic groups also deter the solvation of ions produced in the oxidation of the heme complex. The result is that the Fe(II) complex can survive long enough to bind and release O_2. This is a typical example of the delicate control of reaction environments that proteins can exercise.

Deoxy-Mb is a five-coordinate high-spin Fe(II) complex with four of the coordination positions occupied by the porphyrin ring N atoms. The fifth position is occupied by an N atom of an imidazole ligand of a histidine residue which couples the heme to the protein (Fig. 19.6). Such five-coordinate heme complexes of Fe(II) are always high-spin. The high-spin $t_{2g}^4 e_g^2$ Fe(II) ion is larger than the central hole in the porphyrin ring, and so it lies about $0.40\,\text{Å}$ above the plane of the ring (**4**). High-spin Fe(II) porphyrin complexes, including the active site in hemoglobin and myoglobin, involve puckering and twisting of the porphyrin macrocycle, whereas a low-spin Fe(II) ion is smaller and can fit comfortably into the ring. When O_2 completes the six-coordination environment, the complex converts to low-spin (corresponding to the t_{2g}^6 configuration of Fe(II)) and the Fe(II) ion shrinks a little and moves into the plane (**5**). This structural change is clearly evident in X-ray single crystal structure determinations.

> The helical proteins associated with hemoglobin and myoglobin promote cooperative O_2 binding and thwart the formation of nonfunctional oxo dimers.

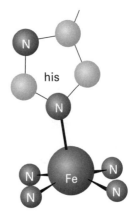

19.5 An outline of the structure of myoglobin. The tubular structure represents the polypeptide chain, and straight sections indicate helical regions.

19.6 The coordination environment of iron in myoglobin.

(b) Cobalt-containing models of O_2 binding

Many investigations of the interaction of O_2 with metal complexes have aimed to understand the function of oxygen carriers. The results of the work have contributed not only to

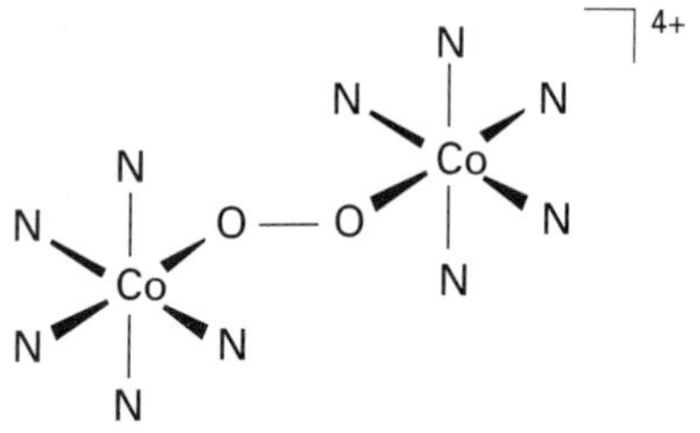

6

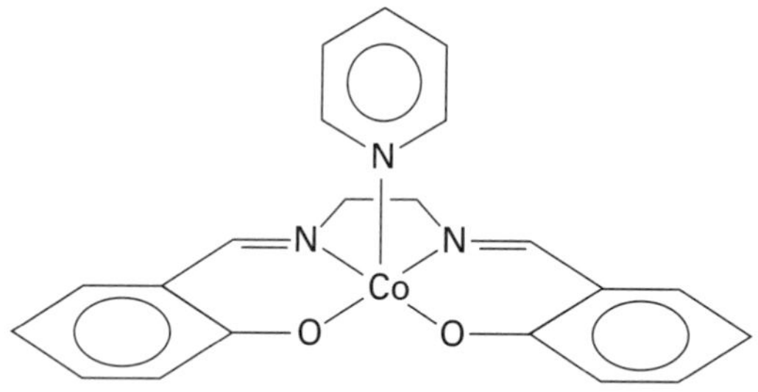

7 [Co(salen)(py)]

unraveling the biological problem but also to an understanding of the pathways by which O_2 acts as an oxidizing agent. Among the d-block metals that bind O_2 and act as helpful model systems, it is probably cobalt that has provided the best general picture.

Like simple Fe(II) complexes, Co(II) complexes react with O_2 by electron transfer:

$$[LCo]^{2+} + O_2 \longrightarrow [LCo^{3+}O_2^-]$$

The product is formally a Co(III) complex of the superoxide ion, O_2^-. It reacts very readily with a second Co(II) complex to give a bridged complex of the peroxide ion, O_2^{2-}:

$$[LCo^{3+}O_2^-] + [LCo]^{2+} \longrightarrow [LCo^{3+}-(O_2^{2-})-Co^{3+}L]$$

The structure of $[(NH_3)_5Co-O_2-Co(NH_3)_5]$ has been determined (**6**), and the O—O bond distance of 1.47 Å is satisfactorily close to the typical peroxide bond length of 1.49 Å.

Cobalt(II) complexes of ligands such as salen, the Schiff base chelating ligand in (**7**), and a base such as pyridine react with O_2 in a rapid, reversible reaction which is suggestive of the reactions of Mb or Hb:

$$[Co(salen)(py)] + O_2 \rightleftharpoons [Co(O_2)(py)(salen)]$$

An X-ray study of the complex reveals an O—O length of 1.26 Å, which is between the values for O_2 (1.21 Å) and O_2^- (1.34 Å).

EPR spectra are consistent with a single unpaired electron interacting with the ^{59}Co nucleus. That electron cannot be on a Co(II) atom because a high-spin d^7 complex has three unpaired electrons, and studies of complexes with comparable ligands show that the ligand field is not strong enough to force the complex to have the low-spin configuration (which would have one unpaired electron). A low-spin d^7 configuration is also not supported by an interpretation of the EPR hyperfine coupling constants. These observations suggest that an O_2^- ligand is coordinated to a low-spin d^6 Co(III) center.

An alternative proposal for O_2 bonding to Co accounts for the observed long O—O bond length in terms of potential σ-donor and a π-acceptor character of O_2. The suggestion is that O_2 donates a lone pair of electrons to form a σ bond and that the π^* orbitals of O_2 accept electrons from the d_{zx} or d_{yz} orbitals of Co. Such back-donation weakens and lengthens the O—O bond. The π-acceptor character of the ligand *trans* to O_2 should exert a substantial influence by competing for or supplying electrons to the π system.

> Cobalt is not found in natural O_2 carriers, but synthetic cobalt complexes have been useful models for natural oxygen carriers.

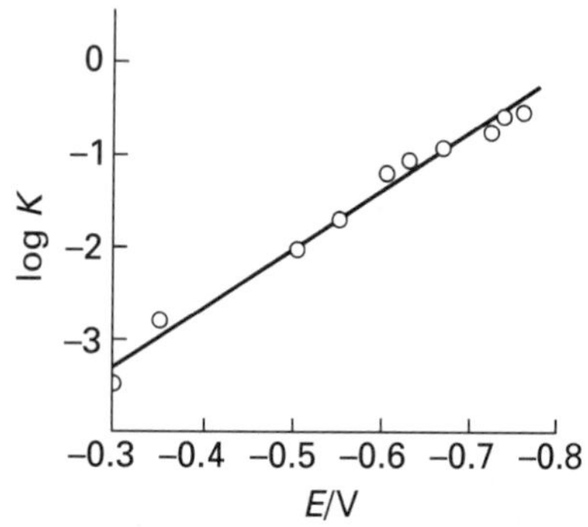

19.7 The linear correlation of $\log K$ and E for a series of Schiff-base complexes.

Example 19.1 Diagnosing the extent of electron transfer to O_2

Figure 19.7 shows that the logarithms of the equilibrium constants for O_2 binding by cobalt Schiff base complexes, [Co(Schiff base)(B)], with various axial ligands, B, correlate linearly with standard potentials for the Co(III)/Co(II) complex couples. What can you conclude from this relation?

Answer Because $\log K$ is proportional to the standard reaction Gibbs energy, and the standard potential is also proportional to this $\Delta_r G^\ominus$, the correlation is a linear free energy relation (Section 14.7). Because the Gibbs energies of reduction and O_2 binding increase together, it is reasonable to suggest that electron transfer from Co (and reduction of O_2) is involved in O_2 binding. However, an increase of 0.4 V (corresponding to 40 kJ mol^{-1}) in $E^\ominus$ leads to an increase in $\log K$ of only about 2.1 (corresponding to 12 kJ mol^{-1}). The difference suggests that electron transfer to O_2 is incomplete and that Co(II)—O_2 and Co(III)—O_2^- are idealized extremes.

Self-test 19.1 The O—O bond lengths in O_2, KO_2, and BaO_2 are 1.21, 1.34, and 1.49 Å, respectively. These values provide reference data on the relation between bond length and oxidation state. For the complexes $[Co(CN)_5(O_2)]^{3-}$, $[Co(bzacen)(py)(O_2)]$ (where bzacen is an acyclic tetradentate ligand similar to salen with two N and two O donor atoms), $[(NH_3)_5Co(O_2)Co(NH_3)_5]^{4+}$, and $[(NH_3)_5Co(O_2)Co(NH_3)_5]^{5+}$, the O—O bond lengths are 1.24, 1.26, 1.47, and 1.30 Å, respectively. Comment on the extent of Co to O_2 electron transfer in each complex.

(c) Iron-containing model O_2 carriers

Early attempts to synthesize O_2-carrying iron-porphyrin models were thwarted by the formation of oxidized porphyrin dimers having a μ-O bridge between the iron atoms. This difficulty led to the synthesis of substituted porphyrins that would be resistant to the μ-O dimer formation and thereby exhibit reversible O_2 coordination. Three approaches have been successful. One is to introduce bulky groups on the porphyrin ring, which prevent the close approach needed for μ-O dimer formation (as in the biological system itself).[4] A second is the use of low temperatures to slow the dimerization reaction, and the third is to anchor the Fe–porphyrin complex to a surface (for example, silica gel) so that dimerization is prevented. The first is the approach most closely related to the behavior of Mb and Hb.

Steric hindrance of dimer formation is achieved with so-called **picket-fence porphyrins**, in which the 'pickets' are a group of blocking substituents projecting from one side of the planar ring. Coordination of a bulky ligand, such as an N-alkylimidazole, can occur only on the unhindered side. An imidazole (Im) is an effective σ-donor that favors coordination of a π-acceptor lying *trans* to itself. It gives the complex an affinity for O_2 similar to that of Mb, and the blocking substituents create a pocket for O_2 and prevent formation of $[Fe(porph)(Im)_2]$. The pickets also prevent reaction with a second Fe center which would give the inactive μ-O_2 species.

One such picket-fence complex is shown in (**8**). It binds O_2 to give a structure very similar to (**5**); the Fe—O—O angle is $136°$ and the O—O bond length is 1.25 Å. This model system provides guidance on the structure of Mb and Hb oxygen complexes. That the complex is diamagnetic (low-spin) might be regarded as evidence of a low-spin d^6 Fe(II) complex of singlet O_2, which would mean that Fe oxidation and O_2 reduction is less important than in the Co complexes. However, we must be cautious with this interpretation because other evidence points in a different direction.

The O—O stretch lies at 1107 cm^{-1}, which is closer to the O_2^- value of 1145 cm^{-1} than the O_2 value of 1550 cm^{-1}. This difference suggests the formulation O_2^-, which is a spin-$\frac{1}{2}$ ion, in combination with low-spin Fe^{3+}, which is also a spin-$\frac{1}{2}$ ion. The low-spin character observed for the complex could arise from spin pairing of Fe(III) and O_2^-, in which case the complex would be similar to the cobalt model compounds. That such a possibility is very real is underlined by the observation that a d^3 Cr(III)–porphyrin complex of oxygen has been synthesized that has only

8

---

[4] Other strategies for blocking one face of the porphyrin ring are discussed by J.P. Collman, X. Zhang, K. Wong, and J. Brauman, *J. Am. Chem. Soc.* **116**, 6245 (1994).

two unpaired electrons. Because d^3 Cr(III) has three unpaired electrons, the only explanation for the spin observed is spin pairing with an electron from O_2^-. These conflicting indications demonstrate once again that assigning precise charges to the central metal ion and to the ligands is an oversimplification. It is employed only because it often helps in the interpretation of reactions and structures, but must never be taken too literally.

> Iron-containing models for natural O_2 carriers include the picket fence porphyrins, in which steric hindrance suppresses oxo-dimer formation.

(d) Cooperativity

The function of hemoglobin is to bind O_2 at the high oxygen partial pressures found in lung tissue, to carry it without loss through the blood, and then to release it to myoglobin in cellular tissues. The sequence requires Mb to have a greater oxygen affinity than Hb at low partial pressures, as is observed (Fig. 19.8).

The shape of the Mb curve in the illustration is reproduced by a simple equilibrium:

$$Mb + O_2 \rightleftharpoons MbO_2 \qquad K = \frac{[MbO_2]}{[Mb]p} \tag{1}$$

where p is the partial pressure of oxygen. The **fractional oxygen saturation**, α, is the ratio of the concentration of the Mb present as MbO_2 to the total concentration of Mb:

$$\alpha = \frac{[MbO_2]}{[Mb] + [MbO_2]} \tag{2}$$

It follows from the equilibrium expression that

$$\alpha = \frac{Kp}{1 + Kp} \tag{3}$$

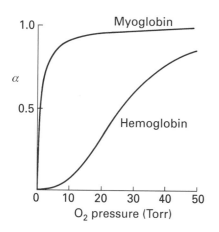

19.8 The oxygen saturation curves for Mb and Hb showing the fractional oxygen saturation α as a function of the oxygen partial pressure at pH $= 7.2$.

The Mb curve in the illustration conforms to this equation but the Hb curve does not.

The Hb curve is described by a more complex function where the dependence of α on the partial pressure of oxygen is changed to p^n, with n between 2 and 3. Another feature of the binding site is that O_2 binding by Hb depends on the **pH** despite the absence of an acid group at the heme site, and it is observed that O_2 is released more readily at lower **pH**. Thus, O_2 is released more readily in cells where metabolism is active, resulting in a high concentration of CO_2, which lowers the **pH**. A final detail to accommodate is that the binding of organic phosphates to the protein of Hb far from the heme site also affects O_2 binding.

The structural difference between Mb and Hb is that, whereas the former has a single heme group, the latter is essentially a tetramer of Mb, with four heme groups (Fig. 19.9). The difference is crucial, because it allows the four heme units of Hb to bind O_2 cooperatively: once one O_2 is bound to Hb, the affinity for subsequent O_2 molecules is greater. The origin of this cooperativity was suggested by M.F. Perutz, who determined the crystal structure of Hb, and argued that oxygenation of one Fe atom in Hb leads to structural changes in its partners.[5] When the high-spin Fe(II) atom that lies 0.4 Å above the porphyrin plane coordinates O_2, it changes to low-spin, moves into the plane, and pulls the histidine residue of the protein along with it. As a result, the shape of the protein is adjusted and the binding characteristics of the other sites are modified. The **pH** and phosphate dependence are ascribed to conformational influences stemming from points moderately distant from the metal atom. The latter change underlines yet again the sensitivity of structures to influences at distant sites.

> The phenomenon of cooperativity promotes O_2 transfer from the oxygen-uptake protein hemoglobin to the oxygen-storage protein myoglobin.

19.9 A schematic structure of Hb showing the relationship among the four subunits that constitute the quaternary structure of the protein.

5   M.F. Perutz, G. Fermi, B. Luisi, B. Shaanan, and R.C. Liddington summarize the link between stereochemistry and mechanism in *Acc. Chem. Res.* **20**, 309 (1987); see also M.F. Perutz, *Mechanism of cooperativity and allosteric regulation in proteins.* Cambridge University Press (1990) and J.A. Cowan, *Inorganic biochemistry.* Wiley–VCH, New York (1997).

Enzymes exploiting acid catalysis

We saw in Section 9.7 that the Brønsted and Lewis acidity of aquated metal ions increase with increasing oxidation number of the central metal. This fundamental property is exploited in metalloenzymes, in which the metalloenzyme can act as an acid catalyst through Lewis acidity of the metal ion itself, or Brønsted acidity of a ligand exhibiting enhanced acidity as a result of coordination to the metal ion. The metal ion in a metalloenzyme may also be able to adopt roles denied to it as a part of a small complex in solution. For instance, the active site of the enzyme may be protected from the bulk aqueous environment in such a way that hydration is disrupted or the solution permittivity is modified. Some of these features have already been seen in the discussion of oxygen transport by Hb.

The Zn^{2+} ion is a very common Lewis acid in biochemical systems, and it is of some interest to explore why zinc is employed rather than Mg^{2+} and Cu^{2+}, which are also small dipositive ions. One difference between these ions is that Zn^{2+} is a softer acceptor than Mg^{2+} so Zn^{2+} is a stronger Lewis acid toward many biomolecules. The Cu^{2+} ion is also soft; however, unlike zinc, copper has an accessible +1 oxidation state, which would allow redox reactions that complicate its behavior.

A further point is that catalysis in general requires reorganization of atoms. It follows that fast reactions will be more characteristic of a metal ion that, while affecting ligands structurally, can take up and release ligands rapidly. In fact, Zn^{2+} complexes are much more labile than the corresponding complexes of Mg^{2+} and Ni^{2+}. Moreover, geometrical rearrangement (an aspect of fluxional behavior) is usually facile for zinc complexes but not for the corresponding complexes of Mg^{2+} and Cu^{2+}. The only ion similar enough to Zn^{2+} to be a competitor for use by biochemical systems is Cd^{2+}, but cadmium is a much rarer element than zinc, and so is less available for incorporation. The unique features of zinc are summarized in Table 19.5.

Table 19.5 Special characteristics of zinc

1. More available than Ni, Cd, Fe, Cu
2. More strongly complexed than Mn(II), Fe(II)
3. Faster ligand exchange than Ni(II), Mg(II)
4. Not redox active compared with Cu(II), Fe(II), Mn(II)
5. More flexible coordination geometry than Ni(II), Mg(II)
6. Good Lewis acid; among M^{2+} ions, only Cu(II) is stronger

Adapted from J.J.R. Fraústo da Silva and R.J.P. Williams, *The biological chemistry of the elements*, p. 300. Oxford University Press (1991).

19.4 Carbonic anhydrase

We turn now to the Zn^{2+}-containing enzyme carbonic anhydrase, because it is well characterized and representative of a family of important hydrolases (see Table 19.2).[6] Carbonic anhydrase serves the simple and important function of catalyzing the hydration-dehydration equilibrium of CO_2:

$$CO_2(aq) + H_2O(l) \rightleftharpoons H_2CO_3(aq)$$

[6] D.W. Christianson and C.A. Fierke, Carbonic anhydrase: evolution of the zinc binding site by nature and design. *Acc. Chem. Res.* **29**, 331 (1996).

As with most enzyme mechanisms, the identification of the correct path is tricky and depends on clever inferences; to be safe, more than one possibility must be evaluated. This zinc(II)-containing enzyme exists as **isozymes**, or genetically distinct forms, designated by Roman numerals I–VII. Carbonic anhydrase is one of a set of hydrolytic enzymes; two others are carboxy peptidase, which promotes the removal of amino acids from proteins, and alkaline phosphatase, a catalyst of phosphate ester hydrolysis.

A variety of tools have been brought to bear on the characterization of carbonic anhydrase. The crystal structures of several different isozymes of carbonic anhydrase have been determined in the presence of a kinetic inhibitor. They display the common four-coordinate environment of Zn^{2+}, in which three of the ligands are the imidazole nitrogens of three histidines and the fourth is a water molecule or hydroxide ion.

The rates of the forward and reverse reactions in the CO_2 hydration equilibrium increase as the **pH** is raised. This observation led to the proposal that the H_2O ligand is rapidly interconverting with a zinc-coordinated OH^- ligand, which serves as the nucleophile toward CO_2. This mechanistic cycle is shown as Cycle 19.1: it invokes attack on the carbon atom of CO_2 by the rapidly equilibrating OH^- ligand on Zn^{2+} followed by formation of a transient five-coordinate Zn^{2+} ion in which a carbonato oxygen from HCO_3^- coordinates to the Zn^{2+} ion. After rearrangement, the HCO_3^- ligand is displaced by H_2O. Deprotonation of the Zn^{2+}-coordinated H_2O then regenerates the OH^- ligand, which can attack another CO_2 molecule, with the repetition of the catalytic cycle. This proposed mechanism relies in part on the X-ray crystal structure of carbonic anhydrase, which indicates the presence of a hydrophobic pocket (which should be a hospitable site for the nonpolar CO_2 molecule) adjacent to the proposed site of CO_2 association. Furthermore, IR spectra, taken upon initial association of CO_2 with an inhibited form of the enzyme, indicate that the CO_2 resides in a hydrophobic region.

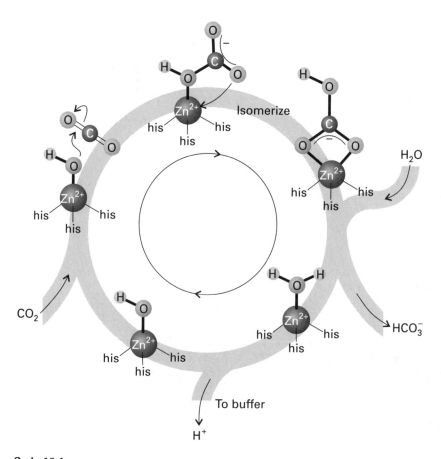

Cycle 19.1

The rapid CO_2 hydration/dehydration by carbonic anhydrase appears to occur at a hydrophobic site near Zn^{2+}.

19.5 Carboxypeptidases

The carboxypeptidases are another zinc-containing enzyme system. Their function is the catalytic hydrolysis of peptide bonds in proteins:[7]

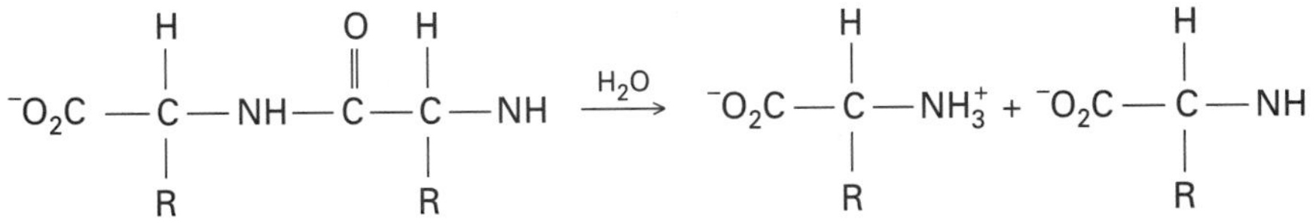

$$^-O_2C - \underset{R}{\overset{H}{C}} - NH - \underset{}{\overset{O}{C}} - \underset{R}{\overset{H}{C}} - NH \xrightarrow{H_2O} {}^-O_2C - \underset{R}{\overset{H}{C}} - NH_3^+ + {}^-O_2C - \underset{R}{\overset{H}{C}} - NH$$

A well studied example is bovine carboxypeptidase A, which is composed of a single polypeptide chain bound to one Zn^{2+}. The X-ray crystal structure indicates the coordination environment of the Zn^{2+} ion (Fig. 19.10). As with the more thoroughly characterized carbonic anhydrase, coordination of water to Zn^{2+} may enhance the rate of equilibration between H_2O and the OH^- nucleophile. Alternatively, Zn^{2+} may serve as a Lewis acid catalyst by coordination to the peptide carbonyl group and thereby reduce the electron density at its C atom (**9**) and promote hydrolysis.

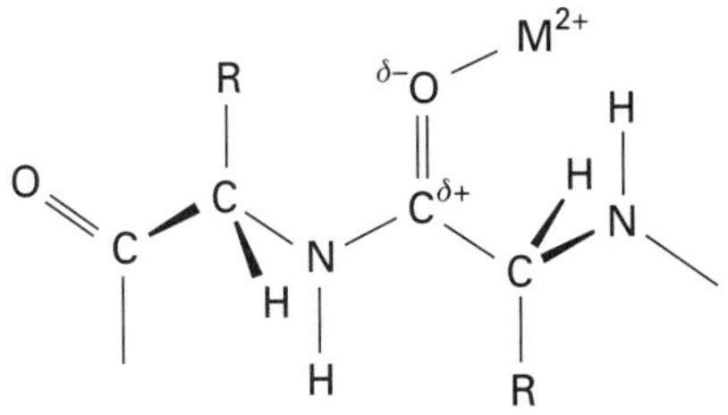

Structural data indicate a considerable movement of the tyrosine residue 248 toward the active site during substrate binding, suggesting that this tyrosine is likely to be involved in the mechanism. X-ray studies of a model substrate led to the development of an alternative hydroxo mechanism in which the carbonyl group to be attacked is hydrogen bonded to arginine residue 127; H_2O remains in the coordination sphere of the Zn^{2+} ion and attacks the substrate by the hydroxo mechanism.[8] The primary intermediate in the reaction is a diol

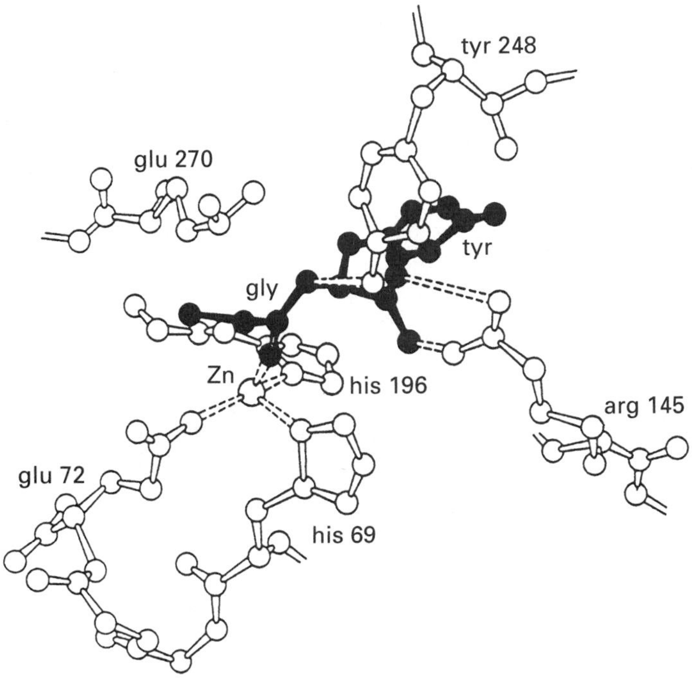

9

19.10 X-ray structure of glycyltyrosine bound to active site in carboxypeptidase A. (From D.M. Blow and T.A. Steitz, *Ann. Rev. Biochem.* **39**, 79 (1970).)

---

[7] S.J. Lippard and J.M. Berg, *Principles of bioinorganic chemistry.* University Science Books, Mill Valley (1994).

[8] W.N. Lipscomb and N. Sträter, *Chem. Rev.* **96**, 2375 (1996).

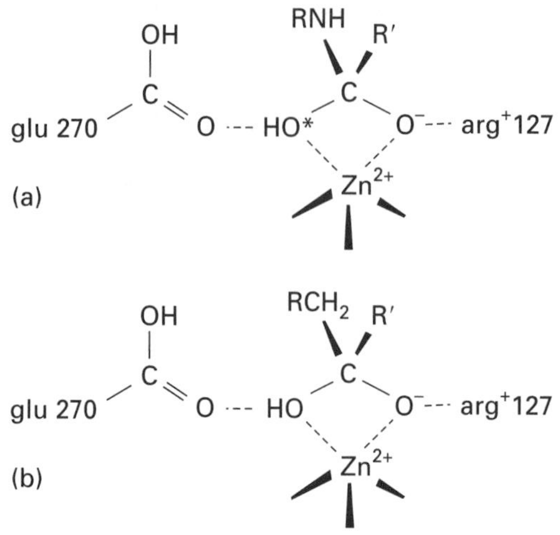

19.11 (a) The diol primary product of the reaction of a substrate catalyzed by carboxypeptidase according to the hydroxo mechanism. The starred O atom marks the hydroxo group from the attacking Zn^{2+} ion. (b) An analog of the reaction that has been studied by X-ray diffraction.

structure shown in Fig. 19.11. This proposal is consistent with the kinetics of the reaction. The mechanism implies the existence of transient five-coordination at the Zn^{2+} ion.

> *Lewis acid and hydroxo mechanisms are both plausible for the hydrolytic breakdown of proteins catalyzed by carboxypeptidase.*

Example 19.2 Proposing a role for Ca^{2+}

We have stressed the role of Zn^{2+} at the active site of enzymes. Another dipositive ion utilized by enzymes is Ca^{2+}. Taking into account factors such as lability and Lewis acidity, propose the potential advantages such as kinetics or ligand selectivity that Ca^{2+} might present at the active site of some enzymes.

Answer A Ca^{2+} ion is the harder ion and therefore it is more selective toward binding to hard donors such as oxygen rather than softer nitrogen or sulfur. Because of its higher charge to radius ratio, Ca^{2+} has a higher affinity for most hard ligands than alkali metal ions, which also are hard acceptors, so Ca^{2+} should be advantageous when especially high reaction rates are needed along with relatively high affinity for oxygen ligands. These properties may account for Nature's frequent use of Ca^{2+} to trigger nerve action, often in concert with alkali metal ions.

Self-test 19.2 Propose conditions under which Nature might select Cu^{2+} for the active site in an enzyme.

Redox catalysis

Because photosynthesis and respiration, the two primary energy conversion processes, involve redox reactions, it is clear that a major metabolic role must be played by the enzymes that catalyze oxidation and reduction. As we saw in Chapters 6 and 14, such reactions may occur by electron transfer, atom transfer, and group transfer. Metalloenzymes can function in all three of these ways. Here we shall concentrate on iron, the one metallic element that is common to all forms of life.

Reduction and oxidation in a living cell do not occur in a single step but usually involve a series of steps that act like a series of locks on a canal, allowing oxidation to occur in stages.

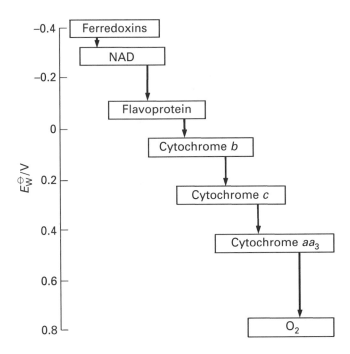

19.12 Reduction potentials of some important electron transfer mediators in biological cells at pH $= 7$.

Once again, control is of paramount importance, for uncontrolled oxidation by oxygen is combustion.

Figure 19.12 illustrates the range of standard potentials of some of the compounds that are involved in oxidation in mitochondria (the organelles in which oxygen utilization is accomplished). They include the heme-containing cytochromes and the flavoproteins, which also participate in the reactions that insert oxygen into organic molecules. Nicotine adenine dinucleotide, NAD (**10**), is a widely distributed substance that participates in reductions by achieving the equivalent of H^- ion transfer. The most strongly reducing members of the chain in the illustration are the ferredoxins, a very interesting group of iron–sulfur proteins.

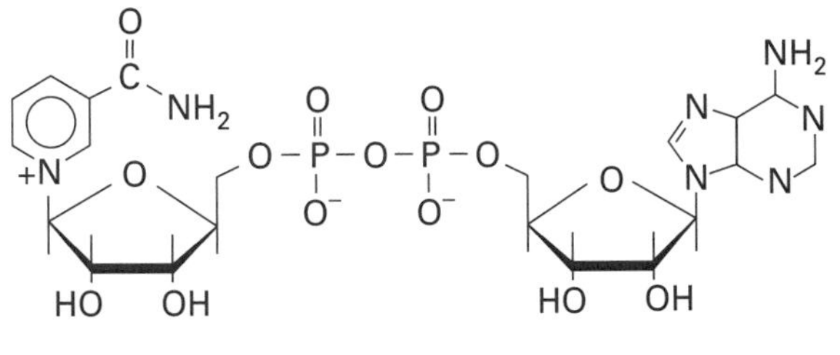

10 NAD

19.6 Iron–sulfur proteins and non-heme iron

The porphyrin ligand environment of iron that occurs in Hb and Mb is also important in redox enzymes. Thus the large class of biochemically important heme proteins has Fe coordinated to a porphyrin ligand. All other iron proteins are defined as non-heme; those containing iron in a tetrahedral environment of four S atoms are particularly important. Iron–sulfur clusters were not familiar to inorganic chemists prior to the recognition of their biochemical importance. They display reduction potentials from 0.0 to -0.5 V and are involved in reducing NAD to NADH and N_2 to NH_3; they can generate H_2 from acidic solutions.

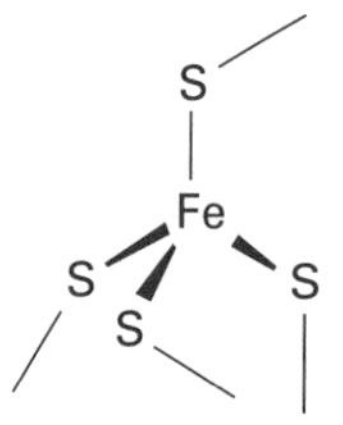

11

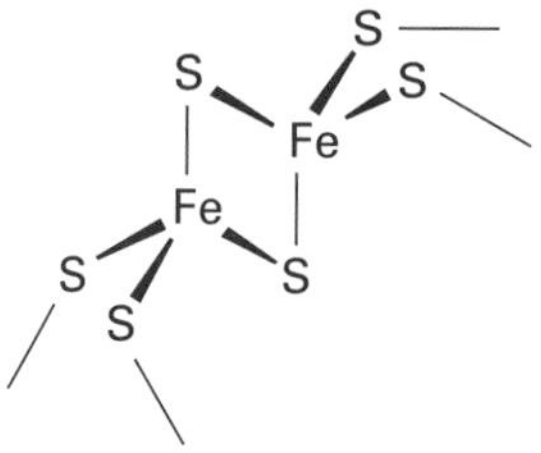

12

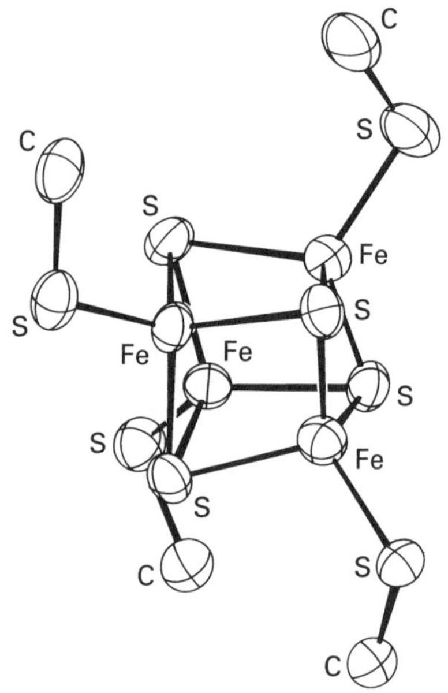

13 $[Fe_4S_4(SCH_2Ph)_4]^{2-}$

One or more Fe atoms may participate at the active site in iron–sulfur proteins. Examples are the isolated Fe atom in a tetrahedral environment of RS^- from cysteine (**11**), two Fe atom clusters formed with S^{2-} bridging atoms (**12**), which are called 'two-iron ferredoxins', and the 4Fe,4S ferredoxin shown in the model structure (**13**).

The fact that the Fe—S cluster is always held in a fold of the protein enables control to be exercised over access by substrates. The Fe—S clusters show little change in bond length on gain of an electron despite the conversion of high-spin Fe(III) to high-spin Fe(II). As a result, there is little barrier to electron transfer arising from reorganization, and the common function of these proteins is one-electron transfer. Various side groups can render the whole protein water-soluble or lipid-soluble to adapt to the cytoplasm (the cell interior) or membrane, respectively.

The synthesis of models for these naturally occurring Fe—S clusters and their structural characterization is an active area of inorganic chemistry.[9] For instance, it has been found that the reaction of $FeCl_3$, $NaOCH_3$, NaHS, and benzylthiol in methanol gives $[Fe_4S_4(SCH_2Ph)_4]^{2-}$ (**13**). The magnetic susceptibility, electronic spectrum, redox properties, and ^{57}Fe Mössbauer spectrum are analogous to those of ferredoxins. Oxidation number rules suggest two Fe(II) atoms and two Fe(III) atoms, but all spectroscopic methods, including X-ray photoelectron spectroscopy, suggest that all four Fe atoms are equivalent. Consequently, a delocalized electron description is most appropriate and the electron transfer reactions involve orbitals that are delocalized over the cluster with the S and Fe atoms both sharing in the change of oxidation state.

 Iron–sulfur clusters such as Fe_4S_4 provide catalytic redox sites in enzymes of plants and animals; synthetic models have been prepared, and their structures suggest charge delocalization.

9   L. Cai and R.H. Holm, *J. Am. Chem. Soc.* **118**, 7177 (1994).

Example 19.3 Writing redox reactions that involve thiolate ligands used in cluster synthesis

One of the problems that has plagued synthetic chemists in their attempts to prepare model compounds for cysteine-complexed metal ions in metalloproteins is the easy oxidation of thiolate anions (RS^-) to RS—SR. Simple complexes with Cu^{2+}—SR and Fe^{3+}—SR bonds that might serve as models for cytochrome P-450 and the ferredoxins (and, in addition, blue copper proteins) are labile because of this reaction. Write balanced equations for the decomposition of $[Cu^{(II)}L_n(SR)]$ and $[Fe^{(III)}L_n(SR)]$.

Answer If thiolate ligands are oxidized to disulfides, the Cu(II) or Fe(III) must be reduced to Cu(I) and Fe(II), respectively. The balanced equations, ignoring the possible charges on the complexes, are

$$2\,[Cu^{(II)}L_n(SR)] \longrightarrow 2\,[Cu^{(I)}L_n] + RSSR$$

$$2\,[Fe^{(III)}L_n(SR)] \longrightarrow 2\,[Fe^{(II)}L_n] + RSSR$$

Self-test 19.3 Early attempts to prepare models of the 2Fe,2S ferredoxins with simple thiolate ligands did not lead to $[Fe_2S_2(RS)_4]^{2-}$ clusters but instead to $[Fe_4S_4(SR)_4]^{2-}$. By considering the average oxidation state of Fe in these clusters, suggest a reason for the difficulty in their preparation.

19.7 Cytochromes of the electron transport chain

Oxygen is a powerful and potentially dangerous oxidizing agent. If the cell is to use it in metabolism, a safe distance must be kept between the sites of oxidation by O_2 and the variety of metabolic reactions. This isolation is accomplished by the series of proteins of decreasing standard potential, which pass electrons toward oxygen in the mitochondria of the cell. A schematic version of the mitochondrial electron transport chain is shown in Fig. 19.13.[10] The most prominent participants are the cytochromes, a group of heme proteins with Fe in a porphyrin-type ligand environment.

There are large reaction Gibbs energies at two stages in Fig. 19.13, one at the reaction between the pair of cytochromes denoted cyt b and cyt c_1 and the second where cytochrome a reacts with oxygen (at cyt aa_3 and O_2). The energy released in these stages can be stored for use elsewhere in the cell by coupling the reaction to the formation of ATP from ADP and

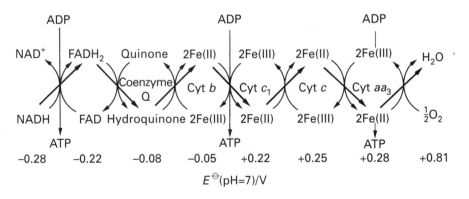

19.13 The sequence of reactions in the electron transport chain in the mitochondria of cells where oxygen is utilized.

[10] G.R. Moore and G.W. Pettigrew, *Cytochrome c: structural and physicochemical aspects.* Springer, Berlin (1990).

HPO_4^{2-}. The chain might seem unnecessarily complex, but it breaks the reaction of O_2 with NADH into cautious, discrete steps. This sequence allows coupling to the synthesis of several ATP molecules as well as organizing the process spatially from one side of the membrane (where the cytochromes are located) to the other.

| Cytochromes of the electron transport chain are heme proteins that reduce O_2 stepwise.

(a) Outer-sphere character

Cytochromes function by shuttling iron between Fe(II) and Fe(III) at the active site, and hence they are one-electron transfer agents. The Fe atoms are in porphyrin ring coordination environments buried in the middle of the protein. By using computer modeling, it is possible to generate plausible models of the complex that is formed when two of these proteins meet to exchange an electron. Because the Fe atoms of the two encountering complexes always remain far apart, the reaction cannot be a case of inner-sphere electron transfer. The reaction must take place by long-distance outer-sphere electron transfer.[11]

Table 19.6 shows some rate constants, standard potentials, and distances of closest approach of the edge of the porphyrin ring containing the Fe atom and its electron transfer partner. It is instructive to compare these protein–protein parameters with similar parameters for reactions of a protein with simple metal complexes, and some of the latter are included in the table too. Of particular interest is the variation of the reaction rate with distance and energy.

| The oxidation and reduction of iron in cytochromes appear to be controlled by outer-sphere electron transfer.

(b) Distance dependence and tunneling

The distance dependence of the rate constant is usually interpreted in terms of quantum mechanical tunneling, in which a particle may escape through a barrier despite having insufficient energy to surmount it. In simple models of tunneling, the probability of finding the particle outside the barrier declines exponentially with the width of the barrier. The data in Table 19.6 confirm that the distance between the metal centers has the expected effect if comparisons are limited to similar structures.

Table 19.6 Some rates of electron transfer reactions among redox proteins†

Couple	$E^{\ominus}/\text{V}$	$R/\text{Å}$‡	k/s^{-1}
$Fe^{II}cytb_5/Fe^{III}cytc$	0.2	8	1.5×10^3
$Znapocytc^*/Fe^{II}b_5$	0.8	8	3×10^5
$H_2porfc^*/Fe^{III}b_5$	0.4	8	1×10^4
$Zncytb_5Fe^{III}cytc$	1.1	8	1×10^3
$RuHis33/cytc$	0.15	11	40
$RuHis/azurin$	0.2	10	2.5
$RuHis/Mb$	0.05	13	0.02
$Fe^{II}ccp/Fe^{III}cytc$	0.4	16	0.025
$Fe^{III}ccp/porfcytc$	1.0	16	180

†Most of the arcane abbreviations are designations of proteins. They are elaborated in biochemical texts. For our purposes we do not need to know the details. All that the different abbreviations mean here is that different members of the family of electron transfer proteins are associated with the different values of $E^{\ominus}$ and RuHis refers to the group $(NH_3)_5Ru$ attached to a histidine (number indicated if necessary) on the enzyme. Distances can be varied by changing the histidine unit to which the Ru is bound. The notation * indicates a reaction using a photoexcited state to make the reaction more exoergic.

‡Distance between electron donor and acceptor metal centers.

[11] An interesting question is whether the pathway is along specific directions of the peptide chain or through space. T.P. Karpishin, M.W. Grinstaff, S. Jomar-Panicucci, G. McLendon, and H.B. Gray, *Structure* **2**, 415 (1994).

> Electron transfer rates between redox sites decrease approximately exponentially with distance.

(c) The energy dependence and Marcus theory

The variation of reaction rate with energy can be considered in the light of Marcus theory (Section 14.13). It can also be used to explore whether the theories developed to explain outer-sphere electron transfer between small molecules also work for large molecules that make contact only at large distances.

We saw in Section 14.13 that the Marcus theory allows us to express the rate constant k_r of any outer-sphere electron transfer as $k_r^2 = k_1 k_2 K f$, with $f \approx 1$. This expression may be written

$$\ln k_r^2 = \ln(k_1 k_2) - \frac{\Delta_r G^{\ominus}}{2RT} \qquad (4)$$

The product $k_1 k_2$ of self-exchange rate constants reflects the intrinsic barrier to electron transfer of the two complexes. The equilibrium constant K is a measure of the overall reaction Gibbs energy. Unfortunately, the intrinsic barrier is not easily estimated for many proteins because it is not feasible to study self-exchange between two oxidation states of a protein.

The best approach seems to be to consider the outer-sphere reaction of a protein and a small molecule. In such cases the reaction Gibbs energy and the self-exchange rate of the small molecule couple can be known. The difficulty lies only in finding the self-exchange rate constant for the protein couple. If we suppose that the Marcus theory does indeed apply to the reaction, then we can calculate a self-exchange rate constant for the protein couple using the known values of k_r and k_1. Then we can use the value of k_2 so obtained to discuss other reactions of the protein. In general, this approach is quite successful and supports the view that the Marcus relation is valid.

The variation of k_r with the overall reaction Gibbs energy (and, equivalently, with the standard potentials of the proteins through $\Delta_r G^{\ominus} = -\nu F E^{\ominus}$ with $\nu = 1$) is also as predicted by Marcus theory for proteins with similar structures. However, the structure also plays a role in determining k_r, and different types of protein structures have rate constants that may differ by orders of magnitude. For instance, the first three entries in Table 19.6 form a series with the dependence on $E^{\ominus}$ predicted by the Marcus equation. Similarly, the last two entries may also represent a structurally similar pair. The role of the enzyme structure in determining the intrinsic barrier is clearly indicated by comparing the fourth entry to the first three. The potential is more favorable than any of the first three yet the rate constant is smaller.[12] The general conclusion appears to be that the cytochromes and several other redox proteins are simple outer-sphere one-electron transfer reagents.

> For redox sites with similar structure, redox rates decrease with decreasing standard reaction Gibbs energy.

....

Example 19.4 Testing the Marcus theory

Test the assertion that the first three points in Table 19.6 show the dependence on Gibbs energy expected from Marcus theory.

Answer To test the theory, we plot the logarithm of the rate constants against the overall reaction Gibbs energy for the electron transfer. Because $E^{\ominus}$ is proportional to the standard reaction Gibbs energy, we may use the standard potentials. The logarithms of the first three rate constants are 7.31, 12.61, and 9.21. The corresponding potentials are 0.2, 0.8, and

....

[12] Factors controlling long-range electron transfer are analyzed by J.M. Nocek, J.S. Zhou, S. de Forest, S. Priyadarshy, D.N. Beratan, J.N. Onuchic, and B.M. Hoffman, *Chem. Rev.* **96**, 2549 (1996). Two excellent surveys of long-range electron transfer in enzymes are given by G. McLendon, *Acc. Chem. Res.* **21**, 160 (1988) and B.E. Bowler, A.L. Raphael, and H.B. Gray, *Prog. Inorg. Chem.* **38**, 259 (1990).

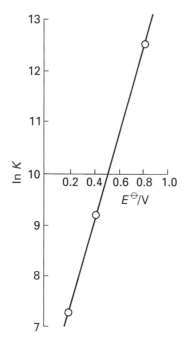

19.14 The plot of the data in Example 19.4.

0.4 V, respectively. The straight line in Fig. 19.14 confirms the approximately linear relationship predicted by the theory.

···

Self-test 19.4 Compare the first and the sixth, the third and eighth, and the fourth and the ninth entries in Table 19.6. The $E^{\ominus}$ values are nearly the same but rates and distances differ. Compare the change in distance to change in $\ln k_r$. Do these results confirm an exponential distance dependence of the intrinsic barriers to electron transfer or must another factor be invoked?

···

19.8 Cytochrome P-450 enzymes

'Cytochrome P-450' is the designation of a family of enzymes with iron porphyrin active sites that catalyze the addition of oxygen to a hydrocarbon substrate.[13] The designation 'P-450' is taken from the position of the characteristic blue to near-ultraviolet absorption band of the porphyrin; the band is called the 'Soret band'; it is red-shifted to 450 nm in carbonyl complexes of these molecules. The most important representative of this class of reactions is the insertion reaction:

$$R{-}H + \tfrac{1}{2}O_2 \longrightarrow R{-}O{-}H$$

The insertion of O into an R—H bond—which is just the sort of redox reaction likely to occur by an atom transfer mechanism—is a part of the body's defense against hydrophobic compounds such as drugs, steroid precursors, and pesticides. The hydroxylation of RH to ROH renders the target compounds more soluble in water and thereby aids their elimination.

The proposed catalytic cycle for P-450 is shown as Cycle 19.2. The sequence begins at (a) with the enzyme in a resting state with iron present as Fe(III). The hydrocarbon substrate then

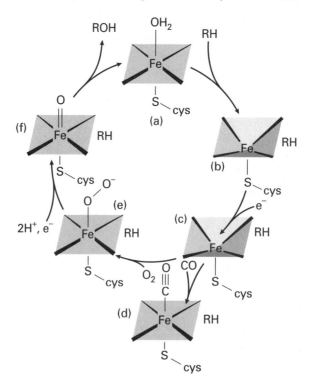

Cycle 19.2

···

[13] G.R. Moore and G.W. Pettigrew, *Cytochrome c: structural and physicochemical aspects.* Springer, Berlin (1990); D. Ostovic and T.C. Bruce, Mechanism of alkene epoxidation by iron, chromium, and manganese higher valent oxo-metalloporphyrins. *Acc. Chem. Res.* **25**, 314 (1992).

binds (b), followed by one-electron transfer to the iron porphyrin yielding (c). This Fe(II) complex with bound substrate proceeds to bind O_2 with the production of (e). (At this point in the cycle, a competing reaction with CO to give (d) leads to a species which is easily identified and is responsible for the absorption at **450 nm** that gives the enzyme its name.) A key reaction is the reduction of the porphyrin ring of the oxygen complex (e) by a second electron, which produces the ring radical anion. Uptake of two H^+ ions then leads to the formation of the Fe(IV) oxo-complex (f) which attacks the hydrocarbon substrate to insert oxygen. Loss of ROH and uptake of an H_2O molecule at the vacated coordination position bring the cycle back to the resting state. The key to the cycle is the formation of the Fe(IV) oxo complex (f).

(a) Structure of the active site

The amino acid sequence of the protein creates a folded structure around the heme site and its Fe atom, so forming an environment shielded from the solution and with a low permittivity. The hydrocarbon substrate is bound nearby, and the C—H bond under attack may be as little as 5 Å from the Fe site. The ligand on the Fe atom lying *trans* to the site of oxygen binding is a thiolate side chain of a cysteine residue (**14**). When O_2 coordinates, the outer end of the O_2 molecule projects into the solution, but the substrate is bound to the enzyme within the hydrophobic pocket.

> *The iron atom at the active site of P-450 is shielded by hydrocarbon substituents, but O_2 projects into the aqueous phase.*

(b) Oxygenation mechanism

The precise mechanism of oxygenation by P-450 remains an intense subject of research. There are thought to be two possibilities, one involving the generation of an oxygen radical species that can attack the C—H bond, and the other the transfer of an O atom to the C—H bond. There are no good inorganic precedents for oxidation of an organic compound by $Fe^{(IV)}O$ and the current debate about the mechanism of the P-450 reaction will almost certainly enlarge the range of possible oxidation mechanisms known to inorganic chemists.[14] The proposal of a radical mechanism entails some unconventional features. Radical oxidations are usually relatively unselective and not stereospecific; P-450 oxidations are quite selective and preserve the optical activity of chiral substrates.

The proposal that oxygenation is accomplished by attack of a positive oxygen center on the C—H bond is just one example of a novel mechanism proposed to explain an enzyme reaction. It is supported by evidence of retention of stereochemical configuration and the reduction of reactivity for cases in which the substrate has substituents that block the approach of the O atom to the C—H bond. The transfer of what is effectively a neutral O unit leaves the iron porphyrin reduced from Fe(IV) to Fe(III) while the organic substrate is oxidized.

The role of the S atom in this pathway can be expressed in terms of molecular orbitals by focussing attention on the atomic orbitals of the S—Fe—O fragment. If we limit our attention to the sulfur $3p_z$, the iron $3d_{z^2}$, $3d_{zx}$, $3d_{yz}$, and oxygen $2p$ orbitals, then we obtain the orbital scheme shown in Fig. 19.15. The 13 electrons available in the fragment occupy the bonding levels, the weakly antibonding levels, and the last electron occupies the strongly antibonding 3σ orbital. The high formal oxidation number of the iron in the complex can therefore be rationalized in terms of its stabilization by considerable charge donation from the readily polarized S atom into the 2σ and 2π orbitals of the fragment. The special role of the *trans* S atom is to control the charge on the Fe atom so that an O atom can be transferred.

> *The coordinated sulfur atom in P-450 appears to control the charge on iron, thereby promoting O-atom transfer.*

....
[14] A.E. Martell and D.T. Sawyer (ed.), *Oxygen complexes and oxygen activation by transition metals.* Plenum, New York (1988).

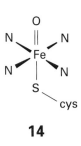

14

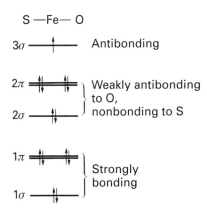

19.15 The molecular orbital scheme of the linear S—Fe—O fragment in cytochrome P-450.

15

Example 19.5 Judging the electron configuration of a metalloenzyme

The Fe=O oxidizing center in P-450 is characterized as an Fe(IV) complex with the porphyrin also oxidized by one electron. Such a picture receives strong support from the ^{57}Fe Mössbauer spectrum of the species.[15] The complex has a magnetic susceptibility that indicates the presence of three unpaired electrons. To which orbitals are they assigned?

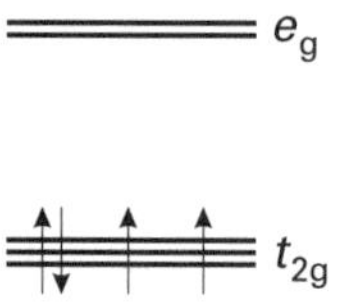

Answer The Fe(IV) oxidation state corresponds to a d^4 configuration. Assuming it to be approximately octahedral in a strong field environment it should be t_{2g}^4 with two unpaired electrons (**15**). The metal ion is in a triplet state ($S = 1$). The oxidation of the porphyrin ring removes one electron from the π HOMO leaving the third unpaired electron there. This configuration could be considered to be a doublet state of the oxidized free ligand.

Self-test 19.5 In the resting state of P-450, the complex is treated as five-coordinate Fe(III) complexed to four porphyrin N atoms and one cysteine S atom, with overall C_{4v} symmetry. Assign electrons to the iron d orbitals.

19.9 Coenzyme B$_{12}$

Cobalt is the 30th most abundant element in the Earth's crust, ahead of lithium, bromine and iodine, all of which participate in a variety of biochemical environments; however, the only known cobalt-containing biological system is coenzyme B$_{12}$, which contains a cobalt bonded to five nitrogen atoms and a carbon atom of an adenosine ligand (**16**). The carbon–cobalt bond in this molecule qualifies coenzyme B$_{12}$ as the first example of a biological organometallic compound. The closely related vitamin B$_{12}$ contains a cyano ligand in place of the adenosine ligand and other cobalt-bound organic substituents are encountered. The name **cobalamin** is used for this class of B$_{12}$ derivatives. The vitamin was first detected in

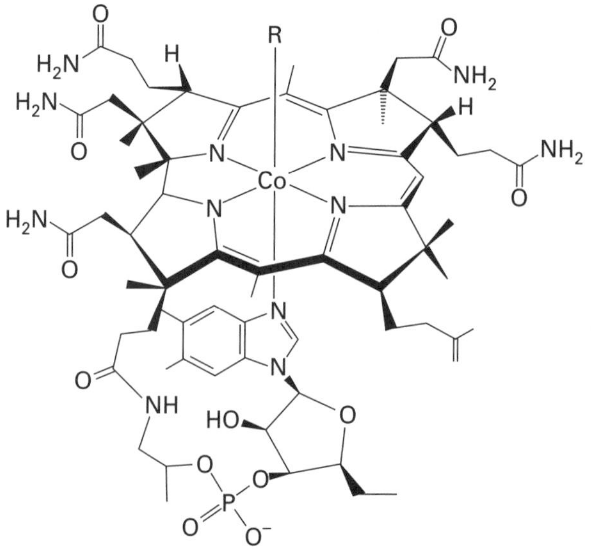

16 Coenzyme B$_{12}$ (R = adenosyl)

---

[15] Mössbauer spectroscopy measures the resonant absorption of a γ-ray by a nucleus held rigidly in a solid framework so that the emission and absorption of γ-ray photons are recoil-free. The chemical environment of the nucleus affects the absorption frequency and so the spectrum is a probe of the electron distribution. The elements most commonly studied are Fe, Sn, and I. The technique may be used to identify the oxidation state of iron.

1929 from liver extracts, and subsequently it was observed that the lack of vitamin B_{12} or coenzyme B_{12} leads to pernicious anemia in humans. The Co—CN bond was observed in an X-ray single crystal determination of the vitamin. This determination represented an extraordinary achievement at a time when automatic X-ray diffractometers and high-speed computers were not available and earned Dorothy Hodgkin the 1964 Nobel Prize.[16] In addition to its interesting structure, coenzyme B_{12} displays an unusually versatile chemistry, some of which is described below.

The reactions in Table 19.7 illustrate the chemical versatility of coenzyme B_{12}, for the special ligand environment supports cobalt in three different oxidation states despite the

Table 19.7 Reactions catalyzed by enzymes related to coenzyme B_{12}

Substrate	Enzyme	Product
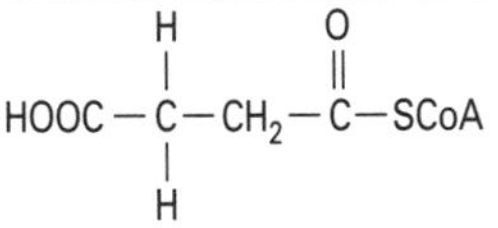 $HOOC-CH(H)-CH_2-C(=O)-SCoA$	Methylmalonyl–CoA mutase	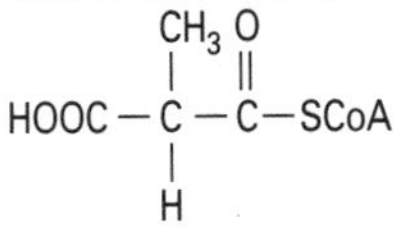 $HOOC-CH(CH_3)-C(=O)-SCoA$
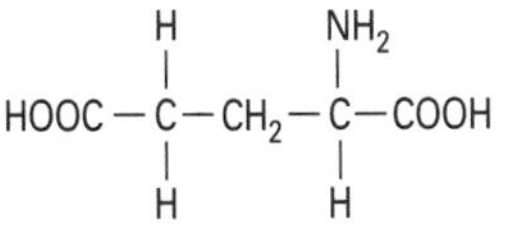 $HOOC-CH(H)-CH_2-CH(NH_2)-COOH$	Glutamate mutase	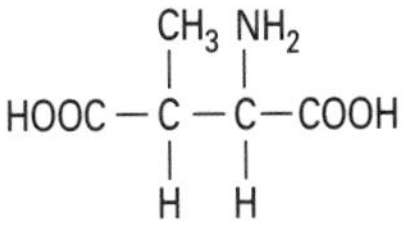 $HOOC-CH(CH_3)-CH(NH_2)-COOH$
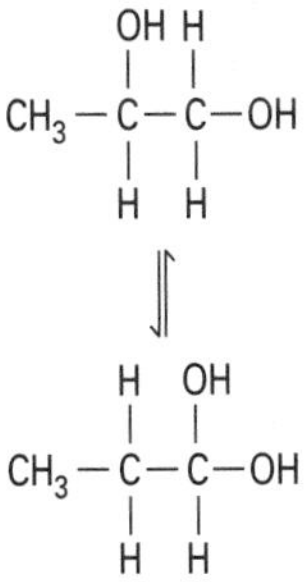 $CH_3-C(OH)(H)-CH(H)-OH$ ⇕ $CH_3-CH(H)-CH(OH)-OH$	Diol dehydrase ($-H_2O$)	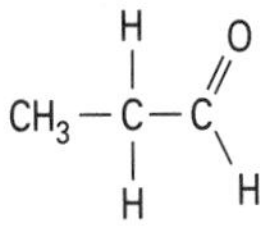 CH_3-CH_2-CHO
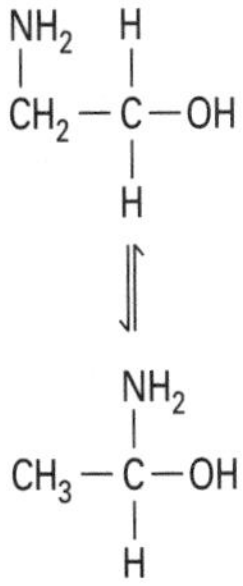 $CH_2(NH_2)-CH(H)-OH$ ⇕ $CH_3-C(NH_2)(H)-OH$	Ethanolamine ammonia lyase ($-NH_3$)	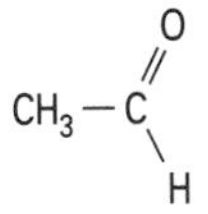 CH_3-CHO
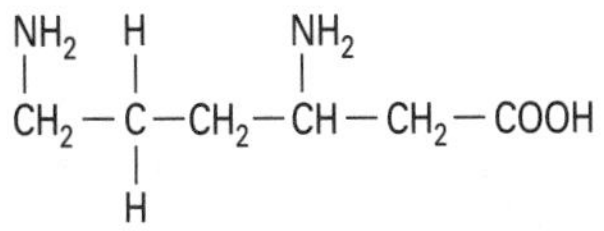 $CH_2(NH_2)-CH(H)-CH_2-CH(NH_2)-CH_2-COOH$	L-β-Lysine mutase	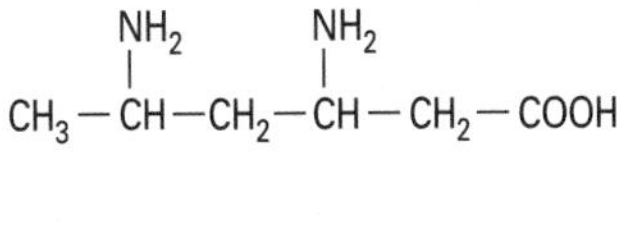 $CH_3-CH(NH_2)-CH_2-CH(NH_2)-CH_2-COOH$
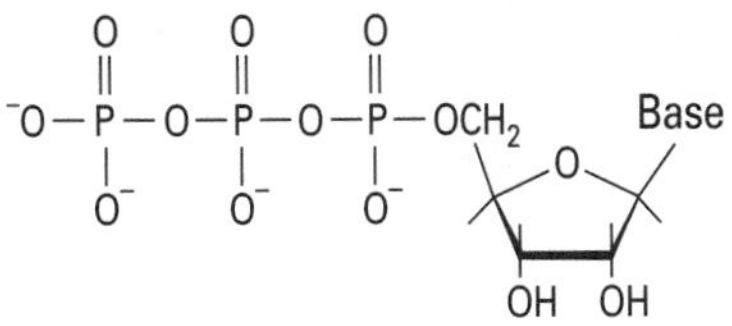 $^-O-P(=O)(O^-)-O-P(=O)(O^-)-O-P(=O)(O^-)-OCH_2$ — Base (ribose, OH OH)	Ribonucleotide reductase	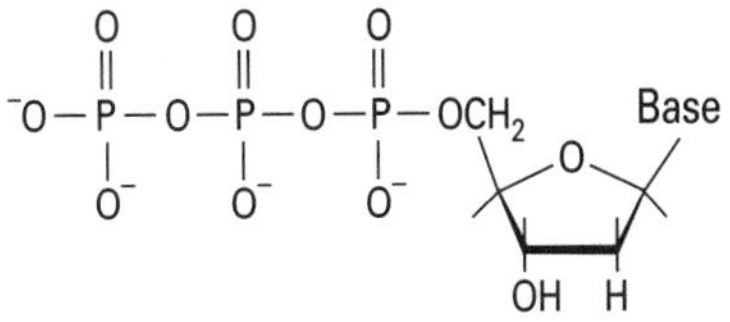 $^-O-P(=O)(O^-)-O-P(=O)(O^-)-O-P(=O)(O^-)-OCH_2$ — Base (ribose, OH H)

[16] D.C. Hodgkin, J. Kamper, M. Mackay, J. Pickworth, K.N. Trueblood, and J.G. White, *Nature* **178**, 64 (1956).

presence of an aqueous medium. The Co(III) form of the enzyme is oxidizing and electrophilic. Thus Co(III) readily undergoes reduction to Co(II). Further reduction leads to Co(I), a rare oxidation state for cobalt in aqueous media. Cobalt(I) is highly nucleophilic, so it readily undergoes methylation. As well as its beneficial functions in methylation reactions, such as methionine synthetase, 'methylcobalamin' is implicated in the methylation of heavy metal ions in aqueous solution to yield such highly toxic species as $Hg(CH_3)_2$ and $Pb(CH_3)_4$. The ligand environment around cobalt that supports this unusual chemistry includes a porphyrin-like corrin ring, which provides four nitrogen donor atoms around the equator of the cobalt. The fifth site also consists of a conjugated nitrogen donor, and the sixth is occupied by a carbon ligand (see **16**). This Co—C bond makes coenzyme B_{12} a rare example of an organometallic molecule in a biology. The Co—C bond is weak and readily undergoes homolytic cleavage to produce a carbon-based radical that can abstract hydrogen from C—H bonds. The radical-based hydrogen abstraction by coenzyme B_{12} is employed in the interconversion of C—H and C—C bonds in substrate molecules.

> Coenzyme B_{12} is a versatile enzyme; its cobalt atom can be reduced to Co(I) and when methylated it serves as a methylating agent.

19.10 Nitrogen fixation

In Section 11.2 we referred to the high-pressure and high-temperature catalysis of the reaction of H_2 with N_2 to produce NH_3. Without the benefit of high temperatures and pressures in the biosphere, Nature has evolved an entirely different and much more elaborate route to NH_3. The reducing agent used by Nature is ATP and the reduction half-reaction can be represented as

$$N_2 + 16\,MgATP + 8\,e^- + 8\,H^+ \longrightarrow 2\,NH_3 + 16\,MgADP + 16\,P_i + H_2$$

where P_i is inorganic phosphate. This process is probably much less efficient than the high-pressure, high-temperature industrial Haber process because the biological system puts considerable energy into the production of hydrogen and the protection of the highly reducing biological system from atmospheric oxygen. The appealing feature, however, is that the process occurs at room temperature and pressure in the *Rhizobium* organisms, which live in the root nodules of various legumes (for example, clover, alfalfa, beans, and peas) as well as several bacteria and blue-green algae. The enzyme nitrogenase carries out the reaction under essentially anaerobic conditions. The problem the enzyme has solved is how to overcome the great inertness of the N≡N molecule.

The mechanistic details of the fixation of N_2 are unclear, but it is known that nitrogenase employs an iron–sulfur protein and a molybdenum–iron–sulfur protein as indicated in Fig. 19.16.[17]

The metal-containing cofactors involved in this catalysis have been isolated and bioinorganic chemists have prepared many potential structural models of the active site. A major breakthrough in this area was the successful crystallization and X-ray structure determination of the $MoFe_7S_8$ cofactor and an associated 'P' cluster which contains two 4Fe,4S clusters joined by sulfur bridges (**17**).[18] As discussed in Section 19.6, the latter type of cluster is common in electron transfer systems and remains intact during sequential electron transfer reactions.

19.16 The coupling between the dephosphorylation of ATP to ADP and the reduction of dinitrogen to NH_4^+.

17

[17] W.H. Orme-Johnson, in *Molybdenum enzymes, cofactors, and model systems* (ed. E.I. Stiefel, D. Coucouvanis, and W.E. Newton), Advances in Chemistry Symposium Series no. 535, p. 257. American Chemical Society, Washington, DC (1993).

[18] M.K. Chan, J. Kim, and D.C. Rees, *Science* **260**, 792 (1993).

The X-ray structure of the Mo—Fe—S cluster at which N_2 reduction appears to take place has an open site. It is speculated that in a more reduced form of this cluster the N_2 might associate with the central cavity and then undergo reduction and protonation. We are far from knowing the details of this process or indeed if the N_2 is bound in this unprecedented manner, but the new structural data provide a more definite model for the active site than had existed previously.

Dinitrogen complexes are well known in inorganic chemistry (Section 11.2), but they are quite different from the proposed biological structure. In the presence of strong acid, the N_2 in some of the inorganic complexes undergoes a proton-induced reduction to ammonium ions:[19]

$$[Mo(PR_3)(N_2)_2] + 8H^+ \longrightarrow 2NH_4^+ + N_2 + Mo(VI) + \cdots$$

The idea that multiple metal binding sites might facilitate the reduction of N_2 is echoed in organometallic chemistry, where it has been demonstrated that metal clusters facilitate the proton-induced reduction of the isoelectronic molecule CO.[20]

A molybdenum–iron–sulfur cluster in the nitrogenase enzyme reduces N_2 to NH_4^+.

19.11 Photosynthesis

One of the most remarkable redox reactions is the thermodynamically 'uphill' conversion of water and carbon dioxide into carbohydrates and oxygen by using solar energy. This photosynthetic process is the underpinning for life, as it is the primary source of the food that we eat and the oxygen that we breathe.[21] Formally, the production of carbohydrates involves the reduction of CO_2 and the oxidation of H_2O to O_2. There are two photochemical reaction centers, 'photosystem I' and 'photosystem II' (PSI and PSII). The former is utilized by anaerobic bacteria. We will focus on the more important PSII, which is employed by all green plants and algae. In plants the photosynthetic system is organized in the green leaf organelles called chloroplasts.[22]

(a) Functions of photosystem II

The energy for the photosynthetic process is absorbed from sunlight by the magnesium-macrocycle chlorophyll a_1, a magnesium dihydroporphyrin complex (**18**). After photoexcitation PSII can act as a reducing agent for an iron–sulfur complex, and the electrons it transfers to the complex are ultimately transferred to CO_2. The oxidized form of PSII that remains after the electron-transfer step is not strong enough to oxidize water to O_2. The return of PSII to the reduced form is used to drive the conversion of two ADP molecules to two ATP molecules through a sequence of mediating species that include several iron-based redox couples and a quinone called plastoquinone.

Chlorophyll a_1 harvests light energy which PSII utilizes for the production of O_2.

(b) Proposed structures and mechanism of PSII

The light harvested by chlorophyll provides energy to PSII, which utilizes oxo-bridged clusters containing four manganese ions. These clusters are designated as S_0 through S_4 in

18 Chlorophyll a_1

[19] D.A. Hall and G.J. Leigh, *J. Chem. Soc. Dalton* 3539 (1996).

[20] M.A. Drezdzon, K.H. Whitmire, A.A. Battacharryya, W.L. Hsu, C.C. Nagel, S.G. Shore, and D.F. Shriver, *J. Am. Chem. Soc.* **104**, 6247 (1982).

[21] We do not discuss the anaerobic photosynthetic bacteria, which produce H_2S rather than oxygen.

[22] PSII has been reviewed by G.T. Babcock, *Acc. Chem. Res.* **31**, 18 (1998) and G.W. Brudvig, H.H. Thorp, and R.W. Crabtree, *Acc. Chem. Res.* **24**, 311 (1991).

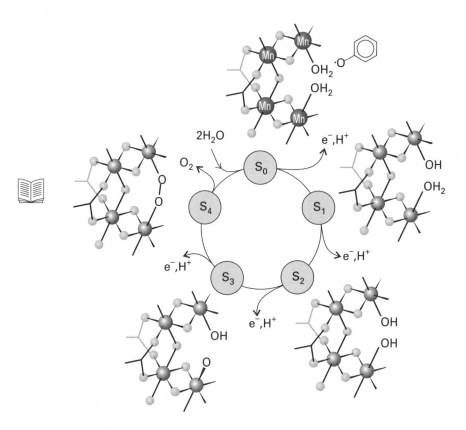

Cycle 19.3

Cycle 19.3.[23] The subscripts indicate the degree of photo-oxidation of each cluster. Owing to the lack of single crystals the detailed structures of the clusters are unclear, but a C-shaped oxo-bridged Mn_4 arrangement is indicated by X-ray spectroscopy (specifically EXAFS). The evidence for the sequential formation of four clusters rests on the observation that pulsed photoexcitation leads to the evolution of O_2 only after the fourth pulse of light. As shown in the illustration, it is postulated that the non-bridging oxo ligands on the open side of the cluster are the sites for the formation of the new O—O bond and the subsequent evolution of the newly formed O_2. This evolution results in a reduced cluster, which undergoes hydration and is then set to repeat the cycle.

> PSII employs energy from photoexcited chlorophyll for the stepwise oxidation of an oxo-bridged manganese cluster, which releases O_2.

(c) Chlorophyll

The chromophore in the photosystem, chlorophyll a_1, has the absorption spectrum shown in Fig. 19.17. The strong absorption band in the red (called the 'Q band') and the band in the blue to near ultraviolet (the Soret band) are characteristic of porphyrins. Both arise from promotion of electrons from the porphyrin π HOMO to the π^* LUMO. The non-absorbing region between these two bands accounts for the characteristic green color of vegetation. Other pigments—called appropriately 'antenna' pigments—are also present in leaves to harvest some light in the absorption gap and transfer its energy to the reaction center.

The initial absorption is from the singlet ground state to a singlet excited state of the chlorophyll Q band (^1Q). Isolated chlorophyll molecules in solution undergo rapid intersystem crossing into the lower lying spin triplet (^3Q), and this state is responsible for the photochemical electron transfer reactions of the molecule in solution. In contrast, the PS reaction center shortcircuits the intersystem crossing by immediately transferring an electron

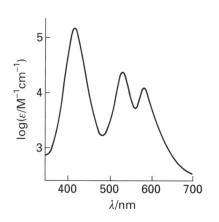

19.17 The absorption spectrum of chlorophyll a_1 in the visible region.

23 C.W. Hoganson and T.T. Babcock, *Science* **227**, 1953 (1997).

from ^1Q to electron acceptors that are adjacent to the chlorophyll in the chloroplast. We see again that spatial organization is a key aspect of biochemistry. A simplified model of such a reaction is obtained by using the porphyrin–amide–quinone molecule (**19**), which exhibits fluorescence properties, suggesting charge transfer from the singlet state of the porphyrin to the quinone that is 10^4 times faster than relaxation to the triplet.

> Chlorophyll absorbs light, and the absorbed energy is transferred to a reaction center.

19

(d) Reaction center organization

The spatial organization in the molecule allows the rapid transfer (within 10 ps) of the energy of the singlet excited state. The singlet state is at higher energy than the triplet and its use results in increased energy efficiency: intersystem crossing into a triplet squanders energy as heat. At this stage, clever spatial organization plays another essential role. The products of reactions that store energy could revert to reactants, so wasting energy; however, the system is so organized that electrons are transferred along the reaction sequence faster than the reverse reactions can take place.

A reaction center of a bacteriochlorophyll has been crystallized and the structure determined by X-ray diffraction.[24] A 'special pair' of chlorophyll molecules, which has long been recognized to be essential to the rapid initial photoelectron transfer, lies at the site of light absorption. Around this pair is organized a double chain of the acceptor molecules. One puzzle is that the series of acceptors seems to be duplicated, but the chemical evidence suggests that one component of the structure is dormant and that only one of the pair is active in electron transport.

> In the reaction center a special pair of chlorophyll molecules rapidly transfer energy to an acceptor site.

Metals in medicine

In view of the extensive variety of metal ions in biochemical systems, it is logical that metal ions are often administered as mineral supplements to maintain the needed daily intake for certain elements, such as calcium, iron, zinc, and copper. The use of inorganic compounds for therapy also has a long-standing tradition.[25]

The development of a new set of antitumor therapeutic agents followed from the observation that the simple square-planar Pt(II) complex *cis*-[PtCl$_2$(NH$_3$)$_2$] inhibits bacterial division. This observation led to the inference that the platinum complex might suppress the otherwise rapid division of tumor cells.[26] The platinum complex, called cisplatin in medical circles, attacks DNA of a fast-growing tumor. It is most effective against testicular cancer. Before the use of cisplatin a 5 per cent survival rate for this cancer was the norm but a survival rate over 80 per cent is now achieved with cisplatin treatment. This success led to the testing of a large number of platinum(II) and platinum(IV) complexes with the objective of increasing and broadening the effectiveness of platinum chemotherapy.

Other inorganic therapeutic agents include bismuth compounds for peptic ulcers and gastritis, and gold compounds for arthritis. Gallium nitrate, Ga(NO$_3$)$_3$, has been found to be effective against hypercalcemia, the rapid loss of calcium from the bones of cancer patients.

> Medicinal metal systems include the antitumor square-planar *cis*-Pt(II) complexes, gallium nitrate to reduce calcium loss, and gold compounds for arthritis.

[24] J. Deisenhofer and J.R. Norris, *The photosynthetic reaction center*, Vols 1 and 2. Academic Press, San Diego (1993).

[25] M.J. Abrams and B.A. Murrer, Metal compounds in therapy and diagnosis. *Science* **261**, 725 (1993).

[26] R. Gust, R. Krauser, B. Schmid, and H. Schonenberger, *Inorg. Chim. Acta* **250**, 203 (1996).

Further reading

D.E. Fenton, *Biocoordination chemistry*. Oxford University Press (1995). A concise introduction to the subject.

J.A. Cowan, *Inorganic biochemistry*. Wiley–VCH, New York (1997). A clear discussion of major topics in bioinorganic chemistry.

S.J. Lippard and J.M. Berg, *Principles of bioinorganic chemistry*. University Science Books, Mill Valley (1994). An unusually well illustrated and authoritative text.

I. Bertini, H.B. Gray, S.J. Lippard, and J.S. Valentine (ed.), *Bioinorganic chemistry*. University Science Books, Mill Valley (1994). A clear presentation of bioinorganic topics.

J.J.R. Fraústo da Silva and R.J.P. Williams, *The biological chemistry of the elements: the inorganic chemistry of life*. Oxford University Press (1991). Emphasis is placed on the way the chemical properties of the elements determine their role in biology.

R.J.P. Williams and J.J.R Fraústo da Silva, *The natural selection of the chemical elements*. Oxford University Press (1996). This text opens with an introduction to principles of inorganic chemistry and physical chemistry. It progresses to the selection of the elements in abiotic and biotic systems. The result is an appealing integration of topics that are generally treated in isolation.

Bioinorganic enzymology, *Chem. Rev.* **7** (1996). This thick volume provides detailed reviews of research topics at the forefront of bioinorganic chemistry.

M. Nicolini and L. Sindellarei (ed.), *Lectures in bioinorganic chemistry*. Cortina Press, Verona and Raven Press, New York (1991). Inorganic pharmaceuticals, the functions of some metal-containing enzymes, and the inorganic species used in NMR imaging of diseased tissue are the topics of this concise book.

P.J. Sadler, Inorganic chemistry and drug design. *Adv. Inorg. Chem.* **36**, 1 (1991).

Science **261**, August 1993. Provides a series of four articles that review metalloenzymes, the role of metal ions in ribozymes, metalloregulatory proteins, and metals in diagnostic reagents.

Exercises

19.1 Consider the elements O, N, K, and Ca. Identify where they are concentrated in animals and describe a major function for each one.

19.2 Describe the characteristics of the ligands that are adopted for binding Ca^{2+} to proteins and those used to bind Fe^{2+} in the oxygen-carrying protein hemoglobin. Suggest some reasons for the differences.

19.3 In the discussion of O_2 coordination, O_2, O_2^-, and O_2^{2-} were considered as limiting forms. Consider the molecular orbital energy level diagram for O_2 from Chapter 3. What is the implication for bond length and net spin of the choice of each of these O_2 species as a model of the ligand?

19.4 (a) Does the equilibrium position of the following reaction lie to the left or the right?

$$Hb + Hb(O_2)_4 \rightleftharpoons 2Hb(O_2)_2$$

Use Structure 19.8 to explain your reasoning. (b) Does the equilibrium position of the reaction

$$Hb(O_2)_4 + 4Mb \rightleftharpoons Hb + 4Mb(O_2)$$

lie to the left or the right? Does the equilibrium depend on the partial pressure of O_2?

19.5 Oxygen is a σ-donor and a π-acceptor. Carbon monoxide is also an excellent example of this type of ligand. Can you use these facts to propose a mechanism for CO poisoning?

19.6 The use of Co(III) centers for synthesis of peptides depends on the reaction

In this scheme, **A** stands for a peptide of arbitrary length. What is the nucleophile and what center is subject to catalyzed nucleophilic attack?

19.7 The diameter of a high-spin Fe(II) ion is larger than that of the 'hole' at the center of the porphyrin ring, whereas a low-spin Fe(II) ion is smaller than the hole. (a) Give the electron configurations for the two spin states in an octahedral environment. Why is the high-spin ion larger? (b) Give examples of ligands that might lead to six-coordinate high- and low-spin $[Fe(porph)L_2]$ complexes.

19.8 Why are d metals such as manganese, iron, cobalt, and copper used in redox enzymes in preference to zinc, gallium, and calcium?

19.9 By considering the specific character of enzyme–substrate binding, can you suggest why it is difficult to measure the self-exchange rate constants of redox enzymes?

19.10 Suggest two ways in which the protein parts of the enzymes contribute to making an outer-sphere electron transfer highly specific to one oxidant with one reductant partner.

19.11 Identify one significant role in biological processes for the elements iron, manganese, molybdenum, copper, and zinc.

19.12 What prevents simple iron porphyrins from functioning as O_2 carriers?

19.13 Sketch the steps illustrating a metal complex functioning as (a) a Brønsted acid and (b) a Lewis acid in an enzyme-catalyzed reaction.

19.14 What is the average change in oxidation number of each of the four Mn atoms in the proposed reaction producing O_2 in the S_4 to S_0 step in the PSII cycle? See Cycle 19.3.

19.15 Describe the differences between the source of an electron in the reaction of the excited state of chlorophyll and its source in the Fe(II) state of a cytochrome. Could magnesium be used as the metal ion in a cytochrome?

19.16 Consider the energy available from a photon at a wavelength of 700 nm. If the photochemical reaction at PSI generates a potential difference of 1 V, what is the efficiency of harvest of energy?

19.17 Metal ions in animals are often coordinated by nitrogen donor atoms. Give some examples of Nature's nitrogen ligands and name the ions with which they are often associated.

19.18 Provide several plausible reasons for the emergence of Ca^{2+}, Zn^{2+}, and Cu^{2+} in the control of protein folding rather than Be^{2+}, Al^{3+} and Cr^{3+}.

19.19 Why do sterically encumbered ligands, such as hemoglobin, occur naturally in O_2 transport systems?

19.20 Oxygen coordinates to both hemoglobin and myoglobin. What is the advantage of employing these different dioxygen complexes?

19.21 Describe the origin of CO toxicity in mammals, including a consideration of the nature of metal–CO bonding.

19.22 Describe the characteristics of Zn^{2+} that make it suitable as the metal ion in many hydrolytic enzymes.

19.23 Why are iron–sulfur proteins employed in redox catalysis?

19.24 What is the shape and makeup of the manganese complexes utilized in PSII?

19.25 Which features of manganese suit it to function as a redox center in PSII, as opposed to metals such as copper or nickel?

Problems

19.1 Correct any of the following statements that are erroneous. Calcium carbonate is the main mineral constituent of bones. The attachment of one O_2 to hemoglobin increases the affinity for the next O_2 because of simple entropy considerations. Carboxypeptidase A utilizes a Zn^{2+} site to bring about the hydrolysis of peptides and two different mechanisms may apply.

19.2 A topic of current interest, the simulation of selective ion transport across membranes, is summarized by H. Tsukube in *J. Coord. Chem. B* **16**, 101 (1987). (a) Describe the experimental arrangement for these measurements and contrast it with the nature of a cell immersed in extracellular fluids. (b) The 'armed' crown ethers described in the review display interesting selectivity. Describe the probable origin of that selectivity.

19.3 The attachment of Fe(TPP), where TPP is tetraphenylporphyrin, to a rigid silica gel support containing a 3-imidazoylpropyl group as a nitrogen donor ligand produces a good reversible oxygen carrier. How does this system prevent irreversible oxidation without a 'picket fence'? (See *Acc. Chem. Res.* **8**, 384 (1975).)

19.4 The structures of Hb may be classified as 'relaxed' (R) or 'tense' (T) as alternative terms to oxygenated and deoxygenated. The R and T structures differ in both the relation among the four subunits (the quaternary structure) and the conformation within a subunit (the tertiary structure). Explain how these structural differences relate to the difference in the oxygen binding curve of Hb as compared to Mb.

19.5 Discuss the probable difference in the pockets present in carboxypeptidase and carbonic anhydrase.

19.6 The substitution of Co(II) for Zn(II) gives a 'spectral probe'. What Co(II) spectral features are exploited? Why does Zn(II) lack them? If Co(II) is to serve as a useful probe for the zinc site, what assumption must be satisfied?

19.7 Chelating ligands that are specific for a particular metal ion are useful for the removal of toxic concentrations of metal ions from the body. See K.N. Raymond and W.L. Smith, *Structure and Bonding* **43**, 159 (1981) or F.L. Weith and K.N. Raymond, *J. Am. Chem. Soc.* **102**, 2289 (1980). After reviewing either paper, summarize the way in which coordination number, charge, and hard–soft character can be utilized to remove the highly toxic Pu(IV) from mammals.

19.8 The substituted carboxypeptidase enzymes show a reactivity order Mn < Zn < Ni < Co. To what extent does this order follow the order of LFSE for tetrahedral complexes and thus support the notion that the determining factor is the bond of the metal to the carbonyl oxygen? What order would you predict if the coordination geometry were octahedral?

19.9 Consider the Frost diagram for oxygen (Fig. 6.10). Using a table of standard potentials, add a point for superoxide. Discuss the biological significance of these potentials. Keep in mind that a biological system is a reduced system that is threatened by oxidizing agents.

19.10 When photosynthesis uses the combination of PSI and PSII to accomplish the difficult process of water oxidation, is the arrangement the analog of batteries connected in parallel or in series? When the system is organized by stacking chlorophyll in grana, which are in turn stacked in the chloroplasts, is this increasing efficiency by the analog of parallel or series connection?

Further information

Further information 1

Nomenclature

We present in this section a concise set of rules and examples for naming inorganic compounds according to IUPAC conventions. With minor exceptions these rules are followed in this text. For information on detailed structural descriptors and on alternative schemes, consult the IUPAC official nomenclature book, G.J. Leigh (ed.), *Nomenclature of inorganic chemistry*, Blackwell Scientific Publications, Oxford (1990); CRC Press, Boca Raton (1990). This publication is widely referred to as the *Red book*. Another useful presentation is given by, B.P. Block, W.H. Powell, and W.C. Fernelius, *Inorganic chemical nomenclature: principles and practice*, American Chemical Society, Washington, DC (1990). In the following sections, examples are given in parentheses after a rule is introduced.

Chemical formulas

1.1 Simple ionic compounds

For ionic compounds, the cation (more electropositive element) should always be first (KCl, Na_2S). If several cations are present, then they are listed in alphabetical order, followed by the anions in alphabetical order ($KMgClF_2$). An exception is the proton, which is listed last in the sequence of cations ($RbHF_2$).

1.2 Sequence of atoms in polyatomic ions and molecules

There is significant latitude in writing the formulas of many compounds. Examples of acceptable formulas for acids, simple salts, and coordination compounds are given in Tables F1.1 through F1.4, respectively. For neutral molecules or polyatomic ions with a central atom, the central atom is generally listed first followed by the attached atoms or attached polyatomic groups in alphabetical order (SO_4^{2-}, $NClH_2$, PCl_3O, SO_3, CF_3). In the last example, CF_3 is a polyatomic attached group; its location in a formula is determined by the letter C. This scheme is similar to that used for coordination compounds. The chemical symbols are sometimes listed in an order that represents the structure, particularly for linear species (SCN^-, OCN^-, and CNO^-).

1.3 Coordination compounds

The ligand–metal array is written with the central metal atom given first, followed by anionic ligands in alphabetical order, and then neutral ligands in alphabetical order of the first symbol of their formulas. Ligand abbreviations may be used in place of complete formulas (en, for $H_2NC_2H_4NH_2$). A list of ligand abbreviations is given inside the back cover. The formula for a metal–ligand entity, which in this text we call a complex (a usage frowned on by IUPAC), is enclosed in brackets whether it is charged or uncharged ($[Co(Cl)_3(NH_3)_3]$). When a compound consists of charged complexes, the cation is listed first ($K_2[Ni(CN)_4]$, $[CoCl_2(NH_3)_4]Cl$).

Descriptors that indicate the spatial arrangement of ligands (cis-, trans-, mer-, fac-) may be included as prefixes (cis-$[Co(Cl)_2(NH_3)_4]^+$). Further information on stereochemical descriptors and examples is given in Section 7.2 and more detailed descriptors are given in the Red book.

Chemical names

1.4 Homoatomic species

The prefix catena is used for chains, and cyclo for rings (Table F1.1).

Table F1.1 Systematic and traditional names for some chain and ring compounds

Formula	Systematic name	Traditional name
O_2	dioxygen	oxygen
O_3	catena-trioxygen	ozone
S_8	cyclo-octasulfur	sulfur
P_4	tetraphosphorus	white phosphorus
Hg_2^{2+}	dimercury(2+)	mercurous ion
O_2^{2-}	dioxide(2−)	peroxide ion
O_2^-	dioxide(1−)	superoxide ion
C_2^{2-}	dicarbide(2−)	acetylide ion*
N_3^-	trinitride(1−)	azide ion
I_3^-	triiodide(1−)	triiodide ion

*Confusingly, this traditional name also applies to HC_2^-.

1.5 Heteroatomic species

(a) Acids

The IUPAC recommendation is to use trivial names for only the most common acids and systematic names for the rest (Table F1.2). We indicate in the table the acids that can be considered common. A point worth noting is the generally accepted distinction between hydrogen halides and hydrohalic acids. For example, molecular HCl, as in the gas phase or hydrocarbon solvent, should be named hydrogen chloride and not hydrochloric acid. The latter name is reserved for the aqueous solution.

The traditional names of oxoacids and their systematic IUPAC names based on the scheme or the acid nomenclature scheme are displayed in Table F1.2. Note that, for neutral molecular acids, hydrogen is left as a separate word (dihydrogen trioxocarbonate); in anions derived from an oxoacid, the attached hydrogen is named as one word (HCO_3^-, hydrogencarbonate ion).

(b) Salts

The principles of nomenclature for salts should be clear from the representative names for simple salts given in Table F1.3. Note that cations are named first. When multiple cations are

Table F1.2 Traditional and systematic names for acids

Formula	Traditional name	Hydrogen nomenclature	Acid nomenclature
H_3BO_3	boric acid	trihydrogen trioxoborate	trioxoboric acid
H_4SiO_4	orthosilicic acid	tetrahydrogen tetraoxosilicate	tetraoxosilicic acid
H_2CO_3	carbonic acid	dihydrogen trioxocarbonate	trioxocarbonic acid
HNO_3	nitric acid*	hydrogen trioxonitrate(1−)	trioxonitric acid
HNO_2	nitrous acid*	hydrogen dioxonitrate(1−)	dioxonitric acid
HPH_2O_2	phosphinic acid	hydrogen dihydridodioxophosphate(1−)	dihydridodioxophosphoric acid
H_3PO_3	phosphorous acid*	trihydrogen trioxophosphate(3−)	trioxophosphoric(3−) acid
H_2PHO_3	phosphonic acid	dihydrogen hydridotrioxophosphate(2−)	hydridotrioxophosphoric (2−) acid
H_3PO_4	phosphoric acid* orthophosphoric acid	trihydrogen tetraoxophosphate(3−)	tetraoxophosphoric acid
$H_4P_2O_7$	diphosphoric acid	tetrahydrogen μ-oxo-hexaoxodiphosphate	μ-oxo-hexaoxo-phosphoric acid
$(HPO_3)_n$	metaphosphoric acid	poly[hydrogen trioxophosphate(1−)]	polytrioxophosphoric acid
H_3AsO_4	arsenic acid*	trihydrogen tetraoxoarsenate	tetraoxoarsenic acid
H_3AsO_3	arsenous acid*	trihydrogen trioxoarsenate(3−)	trioxoarsenic acid
H_2SO_4	sulfuric acid*	dihydrogen tetraoxosulfate	tetraoxosulfuric acid
$H_2S_2O_3$	thiosulfuric acid	dihydrogen trioxothiosulfate (S—S)	trioxothiosulfuric acid
$H_2S_2O_6$	dithionic acid	dihydrogen hexaoxodisulfate (S—S)	hexaoxodisulfuric acid
$H_2S_2O_4$	dithionous acid	dihydrogen tetraoxodisulfate	tetraoxodisulfuric acid
H_2SO_3	sulfurous acid*	dihydrogen trioxosulfate	trioxosulfuric acid
H_2CrO_4	chromic acid*		tetraoxochromic acid
$H_2Cr_2O_7$	dichromic acid*		μ-oxohexadichromic acid
$HClO_4$	perchloric acid*	hydrogen tetraoxochlorate	tetraoxochloric acid
$HClO_3$	chloric acid*	hydrogen trioxochlorate	trioxochloric acid
$HClO_2$	chlorous acid*	hydrogen dioxochlorate	dioxochloric acid
$HClO$	hypochlorous acid*	hydrogen monoxochlorate	monooxochloric acid
HIO_3	iodic acid*	hydrogen trioxoiodate	trioxoiodic acid
HIO_4	periodic acid*	hydrogen tetraoxoiodate	tetraoxoiodic acid
H_5IO_6	orthoperiodic acid	pentahydrogen hexaoxoiodate(5−)	hexaoxoiodic(5−) acid
$HMnO_4$	permanganic acid	hydrogen tetraoxomanganate(1−)	tetraoxomanganic(1−) acid
H_2MnO_4	manganic acid	dihydrogen tetraoxomanganate(2−)	tetraoxomanganic(2−) acid

*Common oxoacids for which traditional names are widely understood.

Table F1.3 Illustrative names for some salts

Formula	Name
$KMgF_3$	magnesium potassium fluoride
$NaTl(NO_3)_2$	sodium thallium(I) nitrate, or sodium thallium dinitrate
$MgNH_4PO_4 \cdot 6H_2O$	ammonium magnesium phosphate hexahydrate
$NaHCO_3$	sodium hydrogencarbonate
LiH_2PO_4	lithium dihydrogenphosphate
$CsHSO_4$	cesium hydrogensulfate, or cesium hydrogentetraoxosulfate(VI), or cesium hydrogentetraoxosulfate(1−)
$NaCl \cdot NaF \cdot 2Na_2SO_4$, $Na_6ClF(SO_4)_2$	hexasodium chloride fluoride bis(sulfate)
$Ca_5F(PO_4)_3$	pentacalcium fluoride tris(phosphate)

present they are listed in alphabetical order, disregarding prefixes such as di-, tri-, etc. The same rule holds for anions.

(c) Mononuclear coordination compounds

Ligands are named in alphabetical order followed by the name of the central metal atom. Note that, unlike in formulas, the names of charged and neutral ligands are intermingled as dictated by alphabetization. When two or more simple ligands are present, their number is indicated by a prefix (di-, tri-, tetra-, penta-, hexa-, etc.). However, when the ligand name itself contains one of these prefixes (as in ethylenediamine), then the multiplicataive prefix is bis-, tris-, tetrakis-, pentakis-, hexakis-, etc. ($[Co(NH_3)_6]^{3+}$, hexaamminecobalt(III) or hexa(ammine)cobalt(III); $[Co(en)_3]^{3+}$, tris(ethylenediamine)cobalt(III)). Note that when a multiplicative prefix is present, the ligand is enclosed in parentheses to improve the ease of reading.

The oxidation state of the metal atom is indicated by Roman numerals in parentheses following the name of the metal. It is also permissible to indicate the oxidation state by giving the overall charge on the complex ion by Arabic numerals and sign, enclosed in parentheses following the name of the complex. As a third alternative the number of counterions present may given (see Table F1.4). The suffix -ate is appended to the metal name in all anionic complexes. Note that the Latin name of the metal is sometimes used in conjunction with the -ate suffix (ferrate, argentate, aurate). The Latin names are all suggested by the symbols for the element (Fe, Ag, Au). Mercury is an exception, presumably because hydragyrate is a tongue twister.

Table F1.4 Illustrative systematic names of coordination compounds

Formula	Name
$K_4[Fe(CN)_6]$	potassium hexacyanoferrate(II)*
	potassium hexacyanoferrate(4−)
	tetrapotassium hexacyanoferrate
$[Pt(Cl)_2(C_5H_5N)(NH_3)]$	amminedi(chloro)pyridineplatinum(II)

*Preferred in this text.

As in some of the above examples, anionic ligands end in -o rather than -e of the free anion, and NH_3 as a ligand is designated ammine, with a double m. A few common ligand names are illustrated in Table F1.5.

Stereochemical descriptors, such as *cis* or *trans*, are given as prefixes in italics (when writing, underlines are used to indicate italics) and connected to the name of the complex by a hyphen (*cis*-diamminedichloroplatinum(II)). The Δ and Λ descriptors for chirality (Section 7.3) may be similarly appended (Δ-tris(ethylenediamine)cobalt(III)).

When there is ambiguity, the site of attachment of a ligand may be indicated by the element symbol surrounded by hyphens and placed after the ligand name ($[RhNO_2(NH_3)_5]^{2+}$

Table F1.5 Illustrative ligand names

Formula	Name in a complex
CN^-	cyano
H^-	hydrido
$CH_3CO_2^-$	acetato
$(CH_3)_2N^-$	dimethylamido
O^{2-}	oxo
$(O_2)^{2-}$	peroxo
NH_3	ammine
$NH_2C_2H_4NH_2$	ethylenediamine

may exist in either of two isomers: penta(ammine)nitro-*O*-rhodium(III) cation, containing a Rh—O—N—O link, or penta(ammine)nitro-*N*-rhodium(III) cation, containing Rh—NO$_2$).

(d) Polynuclear coordination compounds

Bridging ligands are indicated by means of the prefix μ ([{Cr(NH$_3$)$_5$}$_2$(μ-OH)]Cl$_5$, μ-hydroxo-bis(penta(ammine)chromium)(III) pentachloride).

(e) Organometallic compounds

The nomenclature rules follow those given for metal complexes. The hapticity, the number of sites, n, of attachment, of a ligand is commonly specified by η^n (as in bis(η^6-cyclopentadienyl)iron). Examples of η^n organic ligands are given in Table 16.1 and many examples may be seen throughout that chapter.

(f) Hydrogen compounds and their derivatives

The common names for *p*-block hydrogen compounds (which IUPAC continues to refer to as *hydrides* despite the inappropriateness of the name in many cases) were given in Table 8.2. The systematic names given in Table F1.6 form the basis for a *substitutive system* of nomenclature in which the groups replacing hydrogen are indicated. This substitutive nomenclature is generally confined to boron, aluminum, and the elements from Groups 14/IV through 16/VI.

Table F1.6 Illustrative substitutive names

Formula	Substitutive		Other
	Systematic	Traditional	
PH$_2$CH$_3$	methylphosphane	methylphosphine	
B(C$_2$H$_5$)$_3$	triethylborane		borontriethyl
S(C$_6$H$_5$)$_2$	diphenylsulfane		diphenyl sulfide
Sb(C$_2$H$_3$)$_3$	trivinylstibane	trivinylstibine	

Further information 2

Nuclear magnetic resonance

Nuclear magnetic resonance (NMR) is the most powerful and widely used spectroscopic method for the determination of molecular structures in solution and pure liquids. In many cases, it provides information about shape and symmetry with greater certainty than is possible with other spectroscopic techniques, such as infrared and Raman spectroscopy. However, unlike X-ray diffraction (Box 2.1), NMR studies of molecules in solution generally do not provide detailed bond distance and angle information. NMR also provides information about the rate and nature of the interchange of ligands in fluxional molecules.

NMR can be observed only for compounds containing elements with magnetic nuclei (those with nonzero nuclear spin). The sensitivity is dependent on several parameters, including the abundance of the isotope and the size of its nuclear magnetic moment. For example, ^1H, with 99.98 per cent natural abundance and a large magnetic moment, is easier to observe than ^{13}C, which has a smaller magnetic moment and only 1.1 per cent natural abundance. With modern multinuclear NMR techniques it is easy to observe spectra for approximately 20 different nuclei, including many elements that are important in inorganic chemistry, such as ^1H, ^7Li, ^{11}B, ^{13}C, ^{15}N, ^{19}F, ^{23}Na, ^{27}Al, ^{29}Si, ^{31}P, ^{195}Pt, and ^{199}Hg. With more effort, useful spectra can also be obtained using many other nuclei.

Determination of the spectrum

A nucleus with spin I can take up $2I + 1$ distinct orientations relative to the direction of an applied magnetic field. Each orientation has a different energy, with the lowest level (marginally) the most highly populated. The energy of the transition between these nuclear spin states is measured by exciting nuclei in the sample with a radiofrequency pulse or pulse sequence and then observing the return of the nuclear magnetization back to equilibrium. After data processing (Fourier transformation), the data are displayed as an absorption spectrum (Fig. F2.1), with peaks at frequencies corresponding to transitions between the different nuclear energy levels.

Chemical shifts

The frequency of an NMR transition depends on the *local* magnetic field the nucleus experiences. The position of an NMR signal is expressed in terms of the **chemical shift**, δ, which is defined in terms of the difference between the resonance frequency of nuclei in the sample and that of a reference compound:

$$\delta = \frac{\nu - \nu_{\text{ref}}}{\nu_{\text{ref}}} \times 10^6$$

A common standard for ^{1}H, ^{13}C, or ^{29}Si NMR spectra is tetramethylsilane, $Si(CH_3)_4$ (TMS). When δ for a signal is negative, the nucleus involved is said to be **shielded** relative to the standard. Conversely, a positive δ corresponds to nucleus that is **deshielded** with respect to the reference. An H atom bound to a closed-shell, low-oxidation-state, d-block element from Groups 6 through 10 (such as $[HCo(CO)_4]$) is generally found to be highly shielded whereas in an oxoacid, such as H_2SO_4, it is deshielded (in each case, relative to TMS). From these examples it might be supposed that the higher the electron density around a nucleus the greater its shielding. However, as several factors contribute to the shielding, a simple physical interpretation of chemical shifts in terms of electron density is generally not possible.

The chemical shifts of ^{1}H and other nuclei in various chemical environments are tabulated, so empirical correlations can often be used to identifying compounds or the element to which the resonant nucleus is bound. For example, the ^{1}H chemical shift in CH_4 is only 0.1 because the H nuclei are in an environment similar to that in tetramethylsilane, but the ^{1}H chemical shift is $\delta = 3.1$ for H bonded to Ge in GeH_4. Chemical shifts are different for the same element in inequivalent positions within a molecule. Thus in ClF_3 the chemical shift of the equatorial ^{19}F nucleus is separated by $\delta = 120$ from that of the axial F nuclei.

Spin–spin coupling

Structural assignment is often helped by the observation of the **spin–spin coupling** of nuclei, which gives rise to a multiplet of lines in the spectrum. The strength of spin–spin coupling, which is reported as the **spin–spin coupling constant**, J (in hertz, Hz), decreases rapidly with distance through chemical bonds, and in many cases is greatest when the two atoms are directly bonded to each other. For simple, so-called **first-order spectra**, which are being considered here, the coupling constant is equal to the separation of adjacent lines in a multiplet. As can be seen in Fig. F2.1, $J(^1H-^{73}Ge) \approx 100$ Hz. The chemical shift is measured at the center of the multiplet.

The allowed transitions contributing to a multiplet all occur at the same frequency when the nuclei are related by symmetry. Thus, a single ^{1}H signal is observed for the H_3Cl molecule because the three H nuclei are related to each other by a threefold axis. Similarly, in the

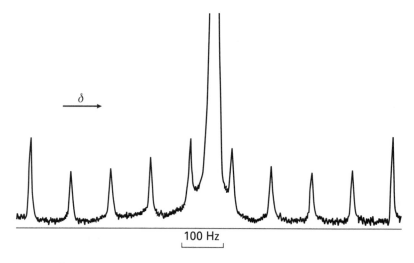

F2.1 The ^{1}H-NMR spectrum of GeH_4. (Source: E.A.V. Ebsworth, D.W.H. Rankin, and S. Cradock, *Structural methods in inorganic chemistry.* Blackwell, Oxford (1991).)

spectrum of GeH_4 the single central line arises from the four equivalent H atoms in GeH_4 molecules that contain germanium isotopes of zero nuclear spin. This strong central line is flanked by 10 evenly spaced but less intense lines that arise from a small fraction of GeH_4 that contains ^{73}Ge, for which $I = \frac{9}{2}$. The properties of spin–spin coupling are such that a multiplet of $2I + 1$ lines results in the spectrum of a spin-$\frac{1}{2}$ nucleus when that nucleus (or a set of symmetry-related spin-$\frac{1}{2}$ nuclei) is coupled to a nucleus of spin I. In the present case, the ^1H nuclei are coupled to the ^{73}Ge nucleus to yield a $2 \times \frac{9}{2} + 1 = 10$ line multiplet.

The coupling of the nuclear spins of different elements is called **heteronuclear coupling**: the Ge—H coupling discussed above is an example. **Homonuclear coupling** between nuclei of the same element is detectable when the nuclei are unrelated by the symmetry operations of the molecule, as in the ^{19}F-NMR spectrum of ClF_3 (Fig. F2.2). The signal ascribed to the two axial F nuclei is split into a doublet by the single equatorial F nucleus, and the latter is split into a triplet by the two axial F nuclei. Thus, the pattern of ^{19}F resonances readily distinguishes this unsymmetrical structure from two more symmetric possibilities, trigonal-planar and trigonal-pyramidal, both of which would have equivalent F nuclei and hence a single ^{19}F resonance.

The sizes of coupling constants are often related to the geometry of a molecule by noting empirical trends. In square-planar Pt(II) complexes, $J(Pt—H)$ is sensitive to the group trans to a phosphine ligand and the value of $J(Pt—P)$ increases in the following order of trans ligands:

$$PR_3 < H^- < R^- < NH_3 < Br^- \leq Cl^-$$

For example, cis-$[PtCl_2(PEt_3)_2]$, where Cl^- is trans to P, has $J(Pt—H) = 3.5$ kHz, whereas trans-$[PtCl_2(PEt_3)_2]$, with P trans to P, has $J(Pt—P) = 2.4$ kHz. These systematics permit us to differentiate cis and trans isomers quite readily.

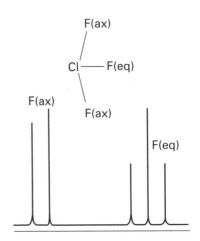

F2.2 The ^{19}F-NMR spectrum of ClF_3. (Source: R.S. Drago, *Physical methods in chemistry*. Saunders, Philadelphia (1992).)

Intensities

Two aspects of line intensities are useful in the analysis of simple NMR spectra. The first is the intensity pattern within a multiplet. The second is the relative integrated intensity for signals of magnetically inequivalent nuclei.

The general rule for a multiplet that arises from coupling to spin-$\frac{1}{2}$ nuclei is that the ratios of the line intensities are given by Pascal's triangle (**1**). This pattern is different from those generated by nuclei with higher spin moments. For example, the ^1H-NMR spectrum of HD is a triplet as a result of coupling with the ^2H nucleus ($I = 1$, so $2I + 1 = 3$), but the intensities of the components in this triplet are equal (Fig. F2.3)

The integrated intensity under a signal arising from a set of symmetry-related nuclei is proportional to the number of nuclei in the set. As an example we return to ClF_3 (Fig. F2.2) where the relative integrated intensities are 2 (doublet) to 1 (triplet). This pattern is consistent with the structure and the splitting pattern, for it indicates the presence of two symmetry-related F nuclei and one magnetically inequivalent F nucleus.

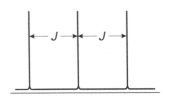

F2.3 Schematic ^1H-NMR spectrum of HD.

Intensity ratio	Coupling to
1	0 other nuclei
1 1	1 spin-$\frac{1}{2}$ nucleus
1 2 1	2 spin-$\frac{1}{2}$ nuclei
1 3 3 1	3 spin-$\frac{1}{2}$ nuclei
1 4 6 4 1	4 spin-$\frac{1}{2}$ nuclei

1

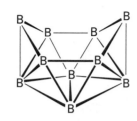

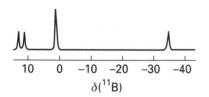

F2.4 The ^{11}B-NMR spectrum of $B_{10}H_{14}$. The ^{11}B—^{1}H coupling has been suppressed and no B—B coupling is observed. The most intense peak belongs to the four equivalent B nuclei; the rest correspond to signals for each of the three equivalent pairs of B nuclei.

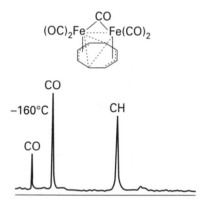

F2.5 The CPMAS spectrum of solid $[Fe_2(C_8H_8)(CO)_5]$. (From C.A. Fyfe, *Solid state NMR for chemists*. CFC Press, Guelph, Ontario (1983).)

Sometimes it is advantageous to eliminate spin–spin coupling by special electronic techniques. Figure F2.4 shows the ^{11}B-NMR spectrum of $B_{10}H_{14}$ in solution, collected under conditions in which proton spin coupling has been eliminated. Each set of symmetry-related B nuclei gives rise to a signal that has an intensity approximately proportional to the number of B atoms it contains.

Solid-state NMR

Developments of far-reaching importance in inorganic chemistry have made possible the observation of high-resolution NMR spectra on solids. One of these high-resolution techniques, referred to as **CPMAS–NMR**, involves high-speed sample spinning at a 'magic angle' with respect to the field axis; the technique is a combination of magic angle spinning (MAS) with cross-polarization (CP). Compounds containing ^{13}C, ^{31}P, and ^{29}Si and many other nuclei have been studied in the solid state using these techniques. An example is the use of ^{29}Si-MAS-NMR spectra to determine the positions of Si atoms in natural and synthetic aluminosilicates. It also is used to study molecular compounds in the solid state. For example the ^{13}C-CPMAS spectrum of $[Fe_2(C_8H_8)(CO)_5]$, at $-160\,°C$ (Fig. F2.5) indicates that all C atoms in the C_8 ring are equivalent on the timescale of the experiment. The interpretation of this observation is that the molecule is fluxional in the solid state.

Group theory

The origin of character tables is the representation of the effects of symmetry operations by matrices. As an illustration, consider the C_{2v} molecule SO_2 and the valence p_x orbitals on each atom (**1**), which we shall denote p_S, p_A, and p_B.

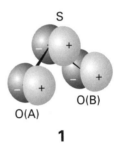

S

O(B)

O(A)

1

Under σ_v, the change

$$(p_S, p_A, p_B) \rightarrow (p_S, p_B, p_A)$$

takes place. We can express this transformation under a reflection by using matrix multiplication:

$$(p_S, p_A, p_B) \begin{pmatrix} 1 & 0 & 0 \\ 0 & 0 & 1 \\ 0 & 1 & 0 \end{pmatrix} = (p_S, p_B, p_A)$$

This relation can be expressed more succinctly as

$$(p_S, p_A, p_B) \mathbf{D}(\sigma_v) = (p_S, p_B, p_A) \quad \text{where } \mathbf{D}(\sigma_v) = \begin{pmatrix} 1 & 0 & 0 \\ 0 & 0 & 1 \\ 0 & 1 & 0 \end{pmatrix}$$

The matrix $\mathbf{D}(\sigma_v)$ is called a **representative** of the operation σ_v. Representatives take different forms according to the basis (the set of orbitals) that has been adopted.

We can use the same technique to find matrices that reproduce the other symmetry operations. For instance, C_2 has the effect

$$(p_S, p_A, p_B) \rightarrow (-p_S, -p_B, -p_A)$$

and its representative is

$$\mathbf{D}(C_2) = \begin{pmatrix} -1 & 0 & 0 \\ 0 & 0 & -1 \\ 0 & -1 & 0 \end{pmatrix}$$

The effect of σ_v' is

$$(p_S, p_A, p_B) \rightarrow (-p_S, -p_A, -p_B)$$

and its representative is

$$\mathbf{D}(\sigma_v') = \begin{pmatrix} -1 & 0 & 0 \\ 0 & -1 & 0 \\ 0 & 0 & -1 \end{pmatrix}$$

The identity operation has no effect on the basis and so its representative is the unit matrix:

$$\mathbf{D}(E) = \begin{pmatrix} 1 & 0 & 0 \\ 0 & 1 & 0 \\ 0 & 0 & 1 \end{pmatrix}$$

The set of matrices that represents all the operations of the group is called a **matrix representation** of the group for the particular basis we have chosen. We denote this three-dimensional representation by the symbol $\Gamma^{(3)}$. The discovery of a matrix representation of the group means that we have found a link between the symbolic manipulations of the operations and algebraic manipulations involving numbers. It may readily be verified that the matrices, when multiplied together, reproduce the group multiplication table.

The **character**, χ, of an operation in a particular matrix representation is the sum of the diagonal elements of the representative of the operation. Thus, in the basis we are illustrating, the characters of the representatives are

$\mathbf{D}(E)$	$\mathbf{D}(C_2)$	$\mathbf{D}(\sigma_v)$	$\mathbf{D}(\sigma_v')$
3	−1	1	−3

The **character** of an operation depends on the basis.

The representatives in the basis we have chosen are three-dimensional (i.e., they are 3×3 matrices), but inspection shows that they are all of the form

$$\begin{pmatrix} \blacksquare & 0 & 0 \\ 0 & & \\ 0 & & \end{pmatrix}$$

and that the symmetry operations never mix p_S with the other two functions. This suggests that the basis can be cut into two parts, one consisting of p_S alone and the other of (p_A, p_B). It is readily verified that the p_S orbital itself is a basis for the one-dimensional representation

$$\mathbf{D}(E) = 1 \qquad \mathbf{D}(C_2) = -1 \qquad \mathbf{D}(\sigma_v) = 1 \qquad \mathbf{D}(\sigma_v') = -1$$

which we shall call $\Gamma^{(1)}$. The remaining two basis functions are a basis for the two-dimensional representation $\Gamma^{(2)}$:

$$\mathbf{D}(E) = \begin{pmatrix} 1 & 0 \\ 0 & 1 \end{pmatrix} \qquad \mathbf{D}(C_2) = \begin{pmatrix} 0 & -1 \\ -1 & 0 \end{pmatrix}$$

$$\mathbf{D}(\sigma_v) = \begin{pmatrix} 0 & 1 \\ 1 & 0 \end{pmatrix} \qquad \mathbf{D}(\sigma_v') = \begin{pmatrix} -1 & 0 \\ 0 & -1 \end{pmatrix}$$

These matrices are the same as those of the original three-dimensional representation, except for the loss of the first row and column. We say that the original three-dimensional representation has been **reduced** to the **direct sum** of a one-dimensional representation **spanned** by p_S and a two-dimensional representation spanned by (p_A, p_B). This reduction is consistent with the common sense view that the central orbital plays a role different from the other two. The reduction is denoted symbolically by

$$\Gamma^{(3)} = \Gamma^{(1)} + \Gamma^{(2)}$$

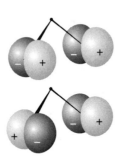

The one-dimensional representation cannot be reduced any further, and is called an **irreducible representation** of the group. We can demonstrate that the two-dimensional representation is reducible (for this basis) by switching attention to the linear combinations $p_1 = p_A + p_B$ and $p_2 = p_A - p_B$. These combinations are sketched in (2).

The representatives in the new basis can be constructed from the old. For example, because under σ_v

$$(p_A, p_B) \rightarrow (p_B, p_A)$$

it follows (by applying these transformations to the linear combinations) that

$$(p_1, p_2) \rightarrow (p_1, -p_2)$$

The transformation is achieved by writing

$$(p_1, p_2)\mathbf{D}(\sigma_v) = (p_1, -p_2) \qquad \text{with} \qquad \mathbf{D}(\sigma_v) = \begin{pmatrix} 1 & 0 \\ 0 & -1 \end{pmatrix}$$

which gives us the representative $\mathbf{D}(\sigma_v)$ in the new basis. The remaining three representatives may be found similarly, and the complete representation is

$$\mathbf{D}(E) = \begin{pmatrix} 1 & 0 \\ 0 & 1 \end{pmatrix} \qquad \mathbf{D}(C_2) = \begin{pmatrix} -1 & 0 \\ 0 & -1 \end{pmatrix}$$

$$\mathbf{D}(\sigma_v) = \begin{pmatrix} 1 & 0 \\ 0 & -1 \end{pmatrix} \qquad \mathbf{D}(\sigma_v') = \begin{pmatrix} -1 & 0 \\ 0 & -1 \end{pmatrix}$$

The new representatives are all in block diagonal form

$$\mathbf{D} = \begin{pmatrix} \blacksquare & 0 \\ 0 & \blacksquare \end{pmatrix}$$

and the two combinations are not mixed with one another by any operation of the group. We have therefore achieved the reduction of $\Gamma^{(2)}$ to the direct sum of two one-dimensional representations. Thus, p_1 spans

$$\mathbf{D}(E) = 1 \qquad \mathbf{D}(C_2) = -1 \qquad \mathbf{D}(\sigma_v) = 1 \qquad \mathbf{D}(\sigma_v') = -1$$

which is the same one-dimensional representation as that spanned by p_S, and p_2 spans

$$\mathbf{D}(E) = 1 \qquad \mathbf{D}(C_2) = 1 \qquad \mathbf{D}(\sigma_v) = -1 \qquad \mathbf{D}(\sigma_v') = -1$$

which is a different one-dimensional representation; we shall denote it $\Gamma^{(1)}$. It is easy to check that either set of 1×1 matrices is a representation by multiplying pairs together and seeing that they reproduce the original group multiplication table.

Now we can make the final link to the material in Chapter 4. The **character table** of a group is the list of the characters of all its irreducible representations. At this point we have found two irreducible representations of the group C_{2v}. Their characters are

	E	C_2	σ_v	σ_v'
$\Gamma^{(1)}$	1	-1	1	-1
$\Gamma^{(1)'}$	1	1	-1	-1

The two irreducible representations are normally labelled B_1 and A_2, respectively. An A or a B is used to denote a one-dimensional representation; A is used if the character under the principal rotation is $+1$ and B is used if the character is -1. (The letter E denotes a two-dimensional irreducible representation and a T a three-dimensional irreducible representation: all the irreducible representations of C_{2v} are one-dimensional.) There are in fact only two more species of irreducible representations of this group, for it is a surprising theorem of group theory that

$$\text{Number of symmetry species} = \text{number of classes}$$

In C_{2v} there are four classes (four columns in the character table), and so there are only four species of irreducible representation.

The most important applications in chemistry of group theory, the construction of symmetry-adapted orbitals and the analysis of selection rules, are based on the **little orthogonality theorem**:

$$\sum_C g(C)\chi^{(\Gamma)}(C)\chi^{(\Gamma')}(C) = 0$$

The sum is over the classes, C, of the operations of the group (the columns in the character table), g is the number of operations in each class (such as the 2 in $2C_3$), and Γ and Γ' are two *different* irreducible representations. If Γ and Γ' are the same irreducible representation, then

$$\sum_C g(C)\chi^{(\Gamma)}(C)\chi^{(\Gamma)}(C) = h$$

where h is the order of the group (the number of elements).

To find out whether a reducible representation contains a given irreducible representation, we use an expression derived from the little orthogonality theorem. Thus, for a given operation, the character of a reducible representation is a linear combination of the characters of the irreducible representations of the group:

$$\chi(C) = \sum_\Gamma c_\Gamma \chi^{(\Gamma)}(C)$$

To find the coefficients for a given irreducible representation Γ', we multiply both sides by $g(C)\chi^{(\Gamma')}(C)$ and sum over all the classes of operation:

$$\sum_C g(C)\chi^{(\Gamma')}(C)\chi(C) = \sum_C \sum_\Gamma c_\Gamma g(C)\chi^{(\Gamma')}(C)\chi^{(\Gamma)}(C)$$

When the right-hand side is summed over C, the little orthogonality theorem gives 0 for all terms for which Γ is not equal to Γ'. However, because we are summing over all Γ, a term with $\Gamma = \Gamma'$ is guaranteed to be present. Only that term contributes, and the right-hand side is then equal to $hc_{\Gamma'}$. It then follows that

$$c_{\Gamma'} = \frac{1}{h}\sum_C g(C)\chi^{(\Gamma')}(C)\chi(C)$$

This formula is exceptionally important for finding the decomposition of a reducible representation, because Γ' can be set equal to each irreducible representation in turn, and the coefficients C determined. It takes an even easier form if we only want to know if the totally symmetric irreducible representation (which for convenience we denote A_1) is present, because all the characters of that representation are 1, so

$$c_{A_1} = \frac{1}{h}\sum_C g(C)\chi(C)$$

To form an orbital of symmetry species Γ we form $P\psi$, where

$$P\psi = \sum_R \chi^{(\Gamma)}(R)R\psi$$

and R is an operation of the group. Note that the actual operations occur in the formula, not the classes as in the earlier expressions. The quantity P is called a **projection operator**. As an example of its form, to project out a B_1 symmetry-adapted linear combination in the group C_{2v} we would use

$$P = \chi^{(B_1)}(E)E + \chi^{(B_1)}(C_2)C_2 + \chi^{(B_1)}(\sigma_v)\sigma_v + \chi^{(B_1)}(\sigma_v')\sigma_v'$$
$$= E - C_2 + \sigma_v - \sigma_v'$$

Appendix 1

Electronic properties of the elements

Ground-state electron configurations of atoms are determined experimentally from spectroscopic and magnetic measurements. The results of these determinations are listed below. They can be rationalized in terms of the building-up principle, in which electrons are added to the available orbitals in a specific order in accord with the Pauli exclusion principle. Some variation in order is encountered in the d- and f-block elements to accommodate the effects of electron–electron interaction more faithfully. The closed shell configuration $1s^2$ characteristic of helium is denoted [He] and likewise for the other noble-gas element configurations. The ground-state electron configurations and term symbols listed below have been taken from S. Fraga, J. Karwowski, and K.M.S. Saxena, *Handbook of atomic data*. Elsevier, Amsterdam (1976).

The first three ionization energies of an element E are the energies required for the following processes:

$I_1:$ $E(g) \longrightarrow E^+(g) + e^-(g)$

$I_2:$ $E^+(g) \longrightarrow E^{2+}(g) + e^-(g)$

$I_3:$ $E^{2+}(g) \longrightarrow E^{3+}(g) + e^-(g)$

The electron affinity E_{ea} is the energy *released* when an electron attaches to a gas-phase atom:

$E_{ea}:$ $E(g) + e^-(g) \longrightarrow E^-(g)$

The values given here are taken from various sources, particularly C.E. Moore, *Atomic energy levels*, NBS Circular 467, Washington (1970) and W.C. Martin, L. Hagan, J. Reader, and J. Sugar, *J. Phys. Chem. Ref. Data* **3**, 771 (1974). Values for the actinides are taken from J.J. Katz, G.T. Seaborg, and L.R. Morss (ed.), *The chemistry of the actinide elements*. Chapman & Hall, London (1986). Electron affinities are from H. Hotop and W.C. Lineberger, *J. Phys. Chem. Ref. Data* **14**, 731 (1985).

For conversions to kilojoules per mole and reciprocal centimeters, use:

$1\ eV = 96.485\ kJ\ mol^{-1}$ $1\ eV = 8065.5\ cm^{-1}$

Atom	Ionization energy/eV			Electron affinity, E_a/eV
	I_1	I_2	I_3	
1 H $1s^1$	13.60			+0.754
2 He $1s^2$	24.59	54.51		−0.5
3 Li [He]$2s^1$	5.320	75.63	122.4	+0.618
4 Be [He]$2s^2$	9.321	18.21	153.85	≤ 0
5 B [He]$2s^2 2p^1$	8.297	25.15	37.93	+0.277
6 C [He]$2s^2 2p^2$	11.257	24.38	47.88	+1.263
7 N [He]$2s^2 2p^3$	14.53	29.60	47.44	−0.07
8 O [He]$2s^2 2p^4$	13.62	35.11	54.93	+1.461
9 F [He]$2s^2 2p^5$	17.42	34.97	62.70	+3.399
10 Ne [He]$2s^2 2p^6$	21.56	40.96	63.45	−1.2
11 Na [Ne]$3s^1$	5.138	47.28	71.63	+0.548
12 Mg [Ne]$3s^2$	7.642	15.03	80.14	≤ 0
13 Al [Ne]$3s^2 3p^1$	5.984	18.83	28.44	+0.441
14 Si [Ne]$3s^2 3p^2$	8.151	16.34	33.49	+1.385
15 P [Ne]$3s^2 3p^3$	10.485	19.72	30.18	+0.747
16 S [Ne]$3s^2 3p^4$	10.360	23.33	34.83	+2.077
17 Cl [Ne]$3s^2 3p^5$	12.966	23.80	39.65	+3.617
18 Ar [Ne]$3s^2 3p^6$	15.76	27.62	40.71	−1.0
19 K [Ar]$4s^1$	4.340	31.62	45.71	+0.502
20 Ca [Ar]$4s^2$	6.111	11.87	50.89	+0.02
21 Sc [Ar]$3d^1 4s^2$	6.54	12.80	24.76	
22 Ti [Ar]$3d^2 4s^2$	6.82	13.58	27.48	
23 V [Ar]$3d^3 4s^2$	6.74	14.65	29.31	
24 Cr [Ar]$3d^5 4s^1$	6.764	16.50	30.96	
25 Mn [Ar]$3d^5 4s^2$	7.435	15.64	33.67	
26 Fe [Ar]$3d^6 4s^2$	7.869	16.18	30.65	
27 Co [Ar]$3d^7 4s^2$	7.876	17.06	33.50	
28 Ni [Ar]$3d^8 4s^2$	7.635	18.17	35.16	
29 Cu [Ar]$3d^{10} 4s^1$	7.725	20.29	36.84	
30 Zn [Ar]$3d^{10} 4s^2$	9.393	17.96	39.72	
31 Ga [Ar]$3d^{10} 4s^2 4p^1$	5.998	20.51	30.71	+0.30
32 Ge [Ar]$3d^{10} 4s^2 4p^2$	7.898	15.93	34.22	+1.2
33 As [Ar]$3d^{10} 4s^2 4p^3$	9.814	18.63	28.34	+0.81
34 Se [Ar]$3d^{10} 4s^2 4p^4$	9.751	21.18	30.82	+2.021
35 Br [Ar]$3d^{10} 4s^2 4p^5$	11.814	21.80	36.27	+3.365
36 Kr [Ar]$3d^{10} 4s^2 4p^6$	13.998	24.35	36.95	−1.0
37 Rb [Kr]$5s^1$	4.177	27.28	40.42	+0.486
38 Sr [Kr]$5s^2$	5.695	11.03	43.63	+0.05
39 Y [Kr]$4d^1 5s^2$	6.38	12.24	20.52	
40 Zr [Kr]$4d^1 5s^2$	6.84	13.13	22.99	
41 Nb [Kr]$4d^4 5s^1$	6.88	14.32	25.04	
42 Mo [Kr]$4d^5 5s^1$	7.099	16.15	27.16	
43 Tc [Kr]$4d^5 5s^2$	7.28	15.25	29.54	
44 Ru [Kr]$4d^7 5s^1$	7.37	16.76	28.47	
45 Rh [Kr]$4d^8 5s^1$	7.46	18.07	31.06	
46 Pd [Kr]$4d^{10}$	8.34	19.43	32.92	
47 Ag [Kr]$4d^{10} 5s^1$	7.576	21.48	34.83	
48 Cd [Kr]$4d^{10} 5s^2$	8.992	16.90	37.47	
49 In [Kr]$4d^{10} 5s^2 5p^1$	5.786	18.87	28.02	+0.3
50 Sn [Kr]$4d^{10} 5s^2 5p^2$	7.344	14.63	30.50	+1.2
51 Sb [Kr]$4d^{10} 5s^2 5p^3$	8.640	18.59	25.32	+1.07
52 Te [Kr]$4d^{10} 5s^2 5p^4$	9.008	18.60	27.96	+1.971
53 I [Kr]$4d^{10} 5s^2 5p^5$	10.45	19.13	33.16	+3.059
54 Xe [Kr]$4d^{10} 5s^2 5p^6$	12.130	21.20	32.10	−0.8
55 Cs [Xe]$6s^1$	3.894	25.08	35.24	
56 Ba [Xe]$6s^2$	5.211	10.00	37.51	
57 La [Xe]$5d^1 6s^2$	5.577	11.06	19.17	
58 Ce [Xe]$4f^1 5d^1 6s^2$	5.466	10.85	20.20	

Atom	Ionization energy/eV			Electron affinity, E_a/eV
	I_1	I_2	I_3	
59 Pr [Xe]$4f^3 6s^2$	5.421	10.55	21.62	
60 Nd [Xe]$4f^4 6s^2$	5.489	10.73	20.07	
61 Pm [Xe]$4f^5 6s^2$	5.554	10.90	22.28	
62 Sm [Xe]$4f^6 6s^2$	5.631	11.07	23.42	
63 Eu [Xe]$4f^7 6s^2$	5.666	11.24	24.91	
64 Gd [Xe]$4f^7 5d^1 6s^2$	6.140	12.09	20.62	
65 Tb [Xe]$4f^9 6s^2$	5.851	11.52	21.91	
66 Dy [Xe]$4f^{10} 6s^2$	5.927	11.67	22.80	
67 Ho [Xe]$4f^{11} 6s^2$	6.018	11.80	22.84	
68 Er [Xe]$4f^{12} 6s^2$	6.101	11.93	22.74	
69 Tm [Xe]$4f^{13} 6s^2$	6.184	12.05	23.68	
70 Yb [Xe]$4f^{14} 6s^2$	6.254	12.19	25.03	
71 Lu [Xe]$4f^{14} 5d^1 6s^2$	5.425	13.89	20.96	
72 Hf [Xe]$4f^{14} 5d^2 6s^2$	6.65	14.92	23.32	
73 Ta [Xe]$4f^{14} 5d^3 6s^2$	7.89	15.55	21.76	
74 W [Xe]$4f^{14} 5d^4 6s^2$	7.89	17.62	23.84	
75 Re [Xe]$4f^{14} 5d^5 6s^2$	7.88	13.06	26.01	
76 Os [Xe]$4f^{14} 5d^6 6s^2$	8.71	16.58	24.87	
77 Ir [Xe]$4f^{14} 5d^7 6s^2$	9.12	17.41	26.95	
78 Pt [Xe]$4f^{14} 5d^9 6s^1$	9.02	18.56	29.02	
79 Au [Xe]$4f^{14} 5d^{10} 6s^1$	9.22	20.52	30.05	
80 Hg [Xe]$4f^{14} 5d^{10} 6s^2$	10.44	18.76	34.20	
81 Tl [Xe]$4f^{14} 5d^{10} 6s^2 6p^1$	6.107	20.43	29.83	
82 Pb [Xe]$4f^{14} 5d^{10} 6s^2 6p^2$	7.415	15.03	31.94	
83 Bi [Xe]$4f^{14} 5d^{10} 6s^2 6p^3$	7.289	16.69	25.56	
84 Po [Xe]$4f^{14} 5d^{10} 6s^2 6p^4$	8.42	18.66	27.98	
85 At [Xe]$4f^{14} 5d^{10} 6s^2 6p^5$	9.64	16.58	30.06	
86 Rn [Xe]$4f^{14} 5d^{10} 6s^2 6p^6$	10.75			
87 Fr [Rn]$7s^1$	4.15	21.76	32.13	
88 Ra [Rn]$7s^2$	5.278	10.15	34.20	
89 Ac [Rn]$6d^1 7s^2$	5.17	11.87	19.69	
90 Th [Rn]$6d^2 7s^2$	6.08	11.89	20.50	
91 Pa [Rn]$5f^2 6d^1 7s^2$	5.89	11.7	18.8	
92 U [Rn]$5f^3 6d^1 7s^2$	6.19	11.9	19.1	
93 Np [Rn]$5f^4 6d^1 7s^2$	6.27	11.7	19.4	
94 Pu [Rn]$5f^6 7s^2$	6.06	11.7	21.8	
95 Am [Rn]$5f^7 7s^2$	5.99	12.0	22.4	
96 Cm [Rn]$5f^7 6d^1 7s^2$	6.02	12.4	21.2	
97 Bk [Rn]$5f^9 7s^2$	6.23	12.3	22.3	
98 Cf [Rn]$5f^{10} 7s^2$	6.30	12.5	23.6	
99 Es [Rn]$5f^{11} 7s^2$	6.42	12.6	24.1	
100 Fm [Rn]$5f^{12} 7s^2$	6.50	12.7	24.4	
101 Md [Rn]$5f^{13} 7s^2$	6.58	12.8	25.4	
102 No [Rn]$5f^{14} 7s^2$	6.65	13.0	27.0	
103 Lr [Rn]$5f^{14} 6d^1 7s^2$	4.6	14.8	23.0	

Appendix 2

Standard potentials

The standard potentials quoted here are presented in the form of Latimer diagrams (Section 6.8) and are arranged according to the blocks of the periodic table in the order s, p, d, f. Data and species in parentheses are uncertain. Most of the data, together with occasional corrections, come from A.J. Bard, R. Parsons, and J. Jordan (ed.), *Standard potentials in aqueous solution*. Marcel Dekker, New York (1985). Data for the actinides are from L.R. Morss, *The chemistry of the actinide elements*, Vol. 2 (ed. J.J. Katz, G.T. Seaborg, and L.R. Morss). Chapman & Hall, London (1986). The value for $[Ru(bipy)_3]^{3+/2+}$ is from B. Durham, J.L. Walsh, C.L. Carter, and T.J. Meyer, *Inorg. Chem.* **19**, 860 (1980). Potentials for carbon species and some d-block elements are taken from S.G. Bratsch, *J. Phys. Chem. Ref. Data* **18**, 1 (1989). For further information on standard potentials of unstable radical species see D.M. Stanbury, *Adv. Inorg. Chem.* **33**, 69 (1989). Potentials in the literature are occasionally reported relative to the standard calomel electrode (SCE) and may be converted to the H^+/H_2 scale by adding 0.2412 V. For a detailed discussion of other reference electrodes, see D.J.G. Ives and G.J. Janz, *Reference electrodes*. Academic Press, New York (1961).

s Block · Group 1/I

Acidic solution   **Basic solution**

+1   0   +1   0

H^+ ——————— 0 ————————→ H_2 H_2O ——————— −0.828 ————————→ H_2

Li^+ ——————— −3.040 ————————→ Li

Na^+ ——————— −2.714 ————————→ Na

K^+ ——————— −2.936 ————————→ K

Rb^+ ——————— −2.923 ————————→ Rb

Cs^+ ——————— −3.026 ————————→ Cs

s Block · Group 2/II

Acidic solution   **Basic solution**

+2   0   +2   0

Be^{2+} ——————— −1.97 ————————→ Be

Mg^{2+} ——————— −2.356 ————————→ Mg $Mg(OH)_2$ ——————— −2.687 ————————→ Mg

Ca^{2+} ——————— −2.87 ————————→ Ca

Sr^{2+} ——————— −2.90 ————————→ Sr

Ba^{2+} ——————— −2.91 ————————→ Ba

Ra^{2+} ——————— −2.916 ————————→ Ra

p Block · Group 13/III

Acidic solution

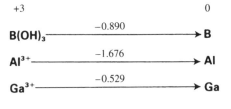

+3   0

B(OH)₃ ——————−0.890——————→ B

Al³⁺ ——————−1.676——————→ Al

Ga³⁺ ——————−0.529——————→ Ga

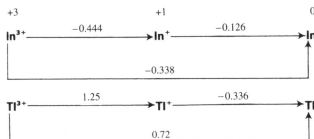

+3   +1   0

In³⁺ ——−0.444——→ In⁺ ——−0.126——→ In

          ——————−0.338——————

Tl³⁺ ——1.25——→ Tl⁺ ——−0.336——→ Tl

          ——————0.72——————

Basic solution

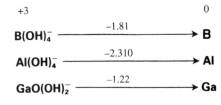

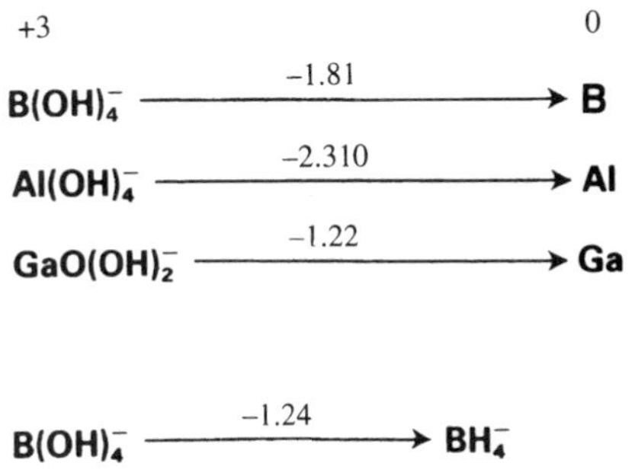

+3   0

B(OH)₄⁻ ——————−1.81——————→ B

Al(OH)₄⁻ ——————−2.310——————→ Al

GaO(OH)₂⁻ ——————−1.22——————→ Ga

B(OH)₄⁻ ——————−1.24——————→ BH₄⁻

p Block · Group 14/IV

Acidic solution

+4   +2   0   −2   −4

CO₂ ——−0.114——→ HCOOH ——−0.029——→ HCHO ——0.237——→ CH₃OH ——0.583——→ CH₄

        ——−0.104——→ CO ——0.517——→ C ——————0.132——————

Basic solution

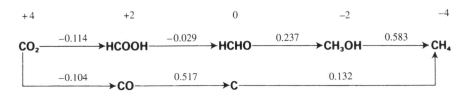

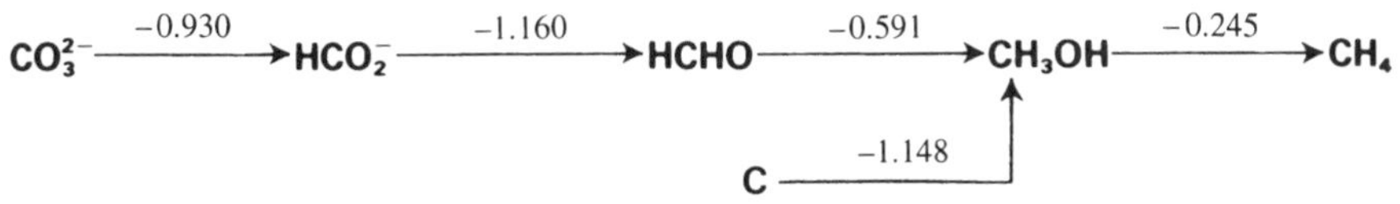

CO₃²⁻ ——−0.930——→ HCO₂⁻ ——−1.160——→ HCHO ——−0.591——→ CH₃OH ——−0.245——→ CH₄

                                    C ——−1.148——→

p Block · Group 14/IV (continued)

Acidic solution   **Basic solution**

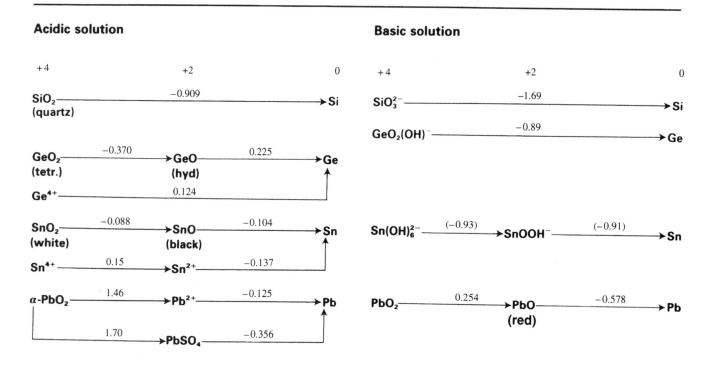

p Block · Group 15/V

Acidic solution

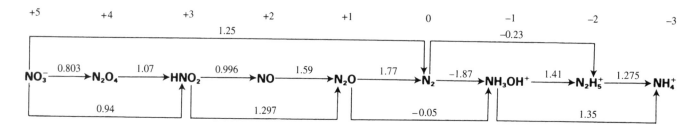

Basic solution

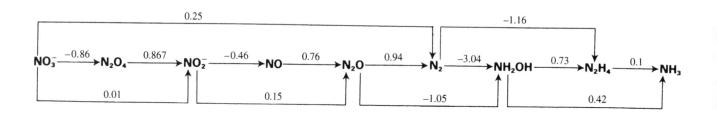

p Block · Group 15/V (continued)

Acidic solution

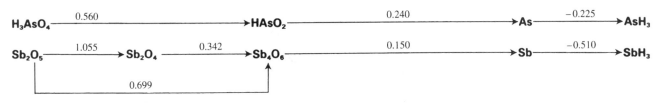

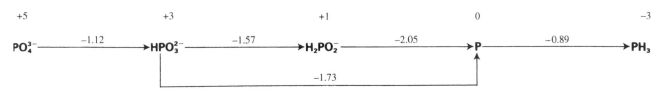

Basic solution

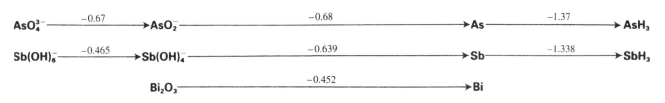

p Block · Group 16/VI

Acidic solution

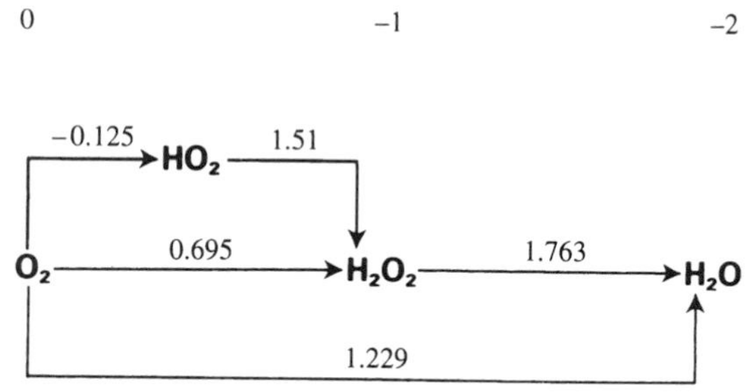

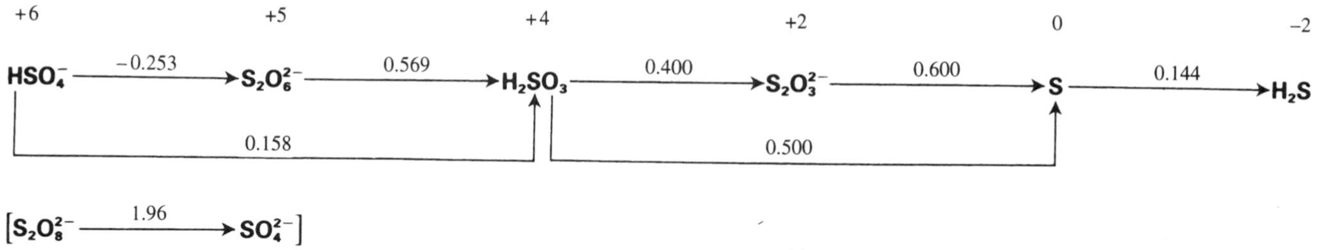

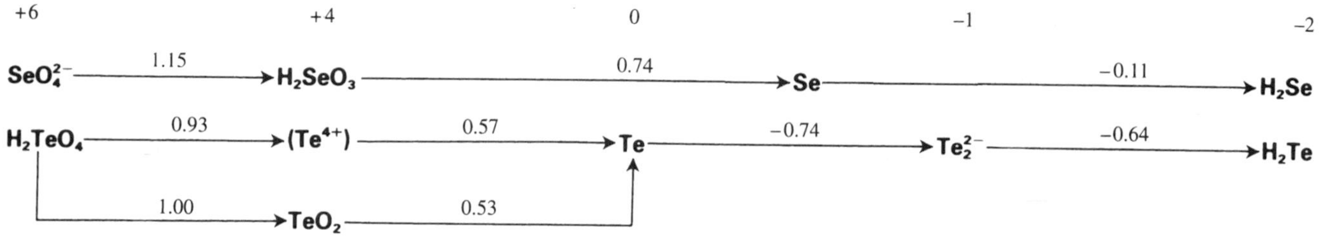

0   −1   −2

$$O_2 \xrightarrow{-0.125} HO_2 \xrightarrow{1.51} H_2O_2 \xrightarrow{1.763} H_2O$$

$$O_2 \xrightarrow{0.695} H_2O_2$$

$$O_2 \xrightarrow{1.229} H_2O$$

+6   +5   +4   +2   0   −2

$$HSO_4^- \xrightarrow{-0.253} S_2O_6^{2-} \xrightarrow{0.569} H_2SO_3 \xrightarrow{0.400} S_2O_3^{2-} \xrightarrow{0.600} S \xrightarrow{0.144} H_2S$$

$$HSO_4^- \xrightarrow{0.158} H_2SO_3$$

$$H_2SO_3 \xrightarrow{0.500} S$$

$$\left[S_2O_8^{2-} \xrightarrow{1.96} SO_4^{2-} \right]$$

+6   +4   0   −1   −2

$$SeO_4^{2-} \xrightarrow{1.15} H_2SeO_3 \xrightarrow{0.74} Se \xrightarrow{-0.11} H_2Se$$

$$H_2TeO_4 \xrightarrow{0.93} (Te^{4+}) \xrightarrow{0.57} Te \xrightarrow{-0.74} Te_2^{2-} \xrightarrow{-0.64} H_2Te$$

$$H_2TeO_4 \xrightarrow{1.00} TeO_2 \xrightarrow{0.53} Te$$

Basic solution

0   −1   −2

$$O_2 \xrightarrow{-0.33} O_2^- \xrightarrow{0.20} HO_2^- \xrightarrow{0.867} OH^-$$

$$O_2 \xrightarrow{-0.0649} HO_2^-$$

$$O_2 \xrightarrow{0.401} OH^-$$

+6   +4   +2   0   −1   −2

$$SO_4^{2-} \xrightarrow{-0.936} SO_3^{2-} \xrightarrow{-0.576} S_2O_3^{2-} \xrightarrow{-0.742} S \xrightarrow{-0.476} HS^-$$

$$SO_3^{2-} \xrightarrow{-0.659} S$$

$$SeO_4^{2-} \xrightarrow{0.03} SeO_3^{2-} \xrightarrow{-0.36} Se \xrightarrow{-0.67} Se^{2-}$$

$$TeO_4^{2-} \xrightarrow{0.07} TeO_3^{2-} \xrightarrow{-0.42} Te \xrightarrow{-0.84} Te_2^{2-} \xrightarrow{-1.445} Te^{2-}$$

$$Te \xrightarrow{-1.143} Te^{2-}$$

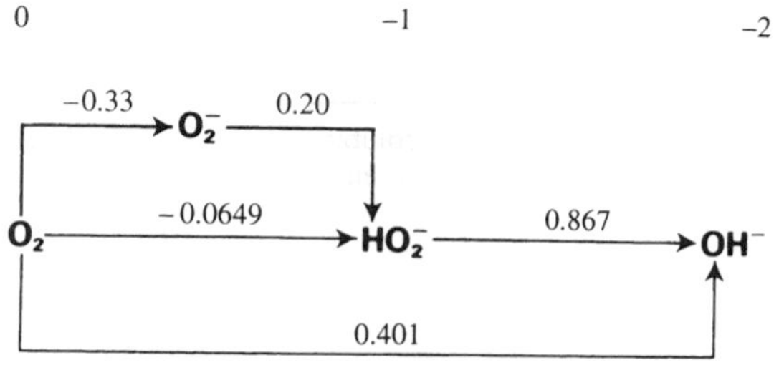

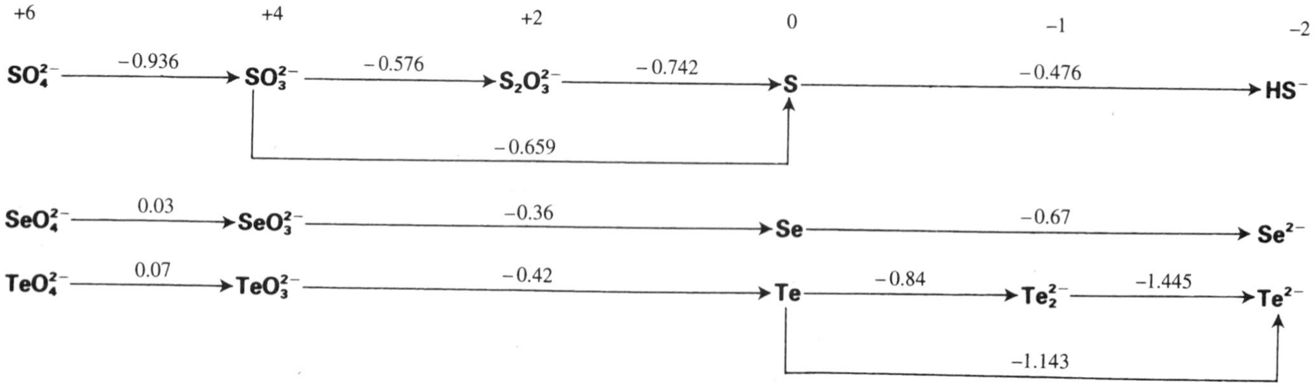

p Block · Group 17/VII

Acidic solution

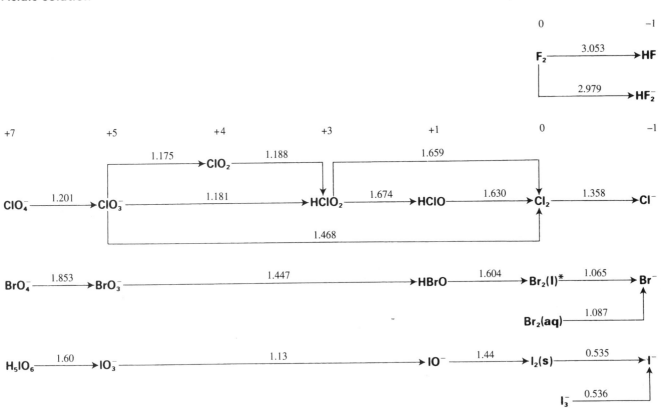

*Bromine is not sufficiently soluble in water at room temperature to achieve unit activity. Therefore the value for a saturated solution in contact with $Br_2(l)$ should be used in all practical calculations.

p **Block** · **Group 17/VII** (continued)

Basic solution

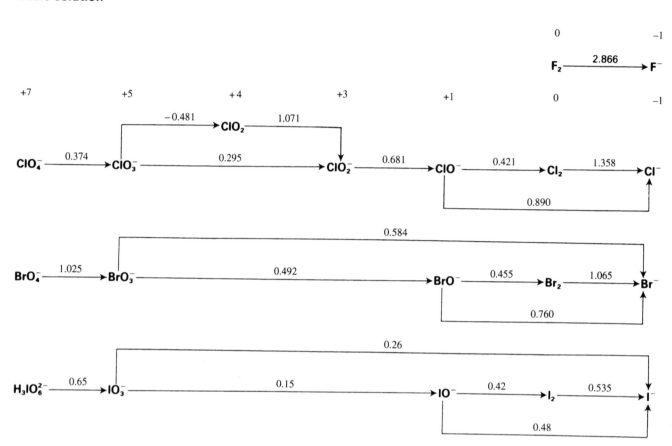

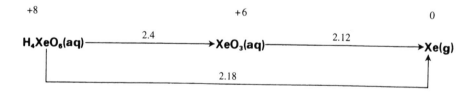

p **Block** · **Group 18/VIII**

Acidic solution

Basic solution

d Block · Group 3

Acidic solution

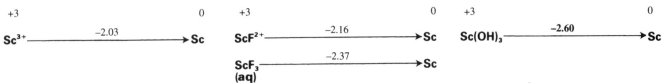

+3   0   +3   0

Basic solution

+3   0

Sc^{3+} —— -2.03 —→ Sc ScF^{2+} —— -2.16 —→ Sc $Sc(OH)_3$ —— **-2.60** —→ Sc

ScF_3 (aq) —— -2.37 —→ Sc

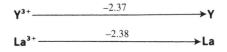

Y^{3+} —— -2.37 —→ Y

La^{3+} —— -2.38 —→ La

d Block · Group 4

Acidic solution

+4   +3   +2   0

Basic solution

+4   +3   +2   0

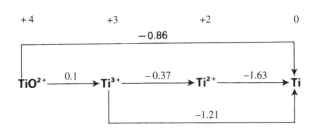

-0.86

TiO^{2+} —— 0.1 —→ Ti^{3+} —— -0.37 —→ Ti^{2+} —— -1.63 —→ Ti

-1.21

TiO_2 —— -1.38 —→ Ti_2O_3 —— -1.95 —→ TiO —— -2.13 —→ Ti

TiO_2 —— -0.56 —→ Ti_2O_3 —— -1.23 —→ TiO —— -1.31 —→ Ti

Zr^{4+} —— -1.55 —→ Zr

Hf^{4+} —— -1.70 —→ Hf

d Block · Group 5

Acidic solution

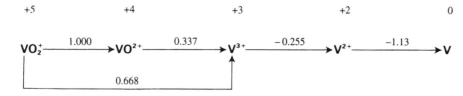

Weakly acidic solution, pH about 3.0–3.5

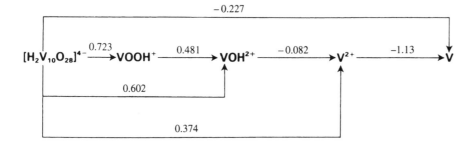

Basic solution

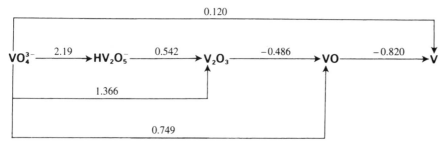

Acidic solution

+5	+3	0

Nb$_2$O$_5$ $\xrightarrow{-0.1}$ Nb^{3+} $\xrightarrow{-1.1}$ Nb

$\xrightarrow{-0.65}$

+5	0

Ta$_2$O$_5$ $\xrightarrow{-0.81}$ Ta

TaF$_7^{2-}$ $\xrightarrow{-0.45}$ Ta

d Block · Group 6

Acidic solution

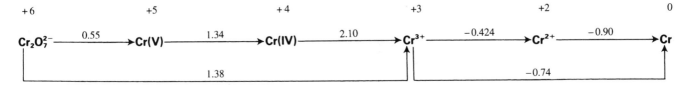

+6   +5   +4   +3   +2   0

$Cr_2O_7^{2-}$ —0.55→ $Cr(V)$ —1.34→ $Cr(IV)$ —2.10— Cr^{3+} —−0.424→ Cr^{2+} —−0.90→ Cr

—————1.38—————

—————−0.74—————

Neutral solution

+3   +2

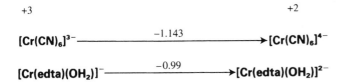

$[Cr(CN)_6]^{3-}$ ————−1.143————→ $[Cr(CN)_6]^{4-}$

$[Cr(edta)(OH_2)]^-$ ————−0.99————→ $[Cr(edta)(OH_2)]^{2-}$

Basic solution

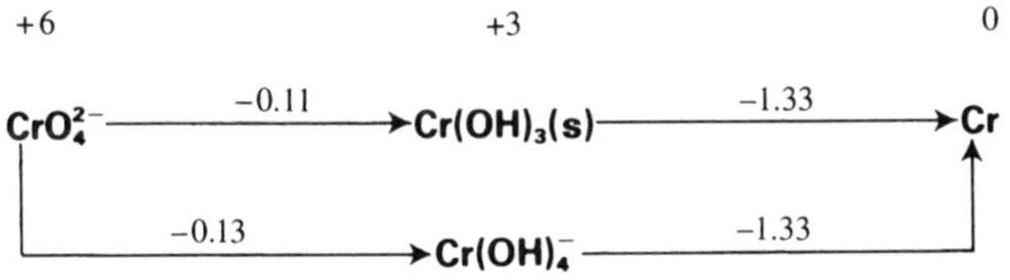

+6   +3   0

CrO_4^{2-} —−0.11→ $Cr(OH)_3(s)$ —−1.33→ Cr

—−0.13→ $Cr(OH)_4^-$ —−1.33—

Acidic solution

+6   +5   +4   +3   0

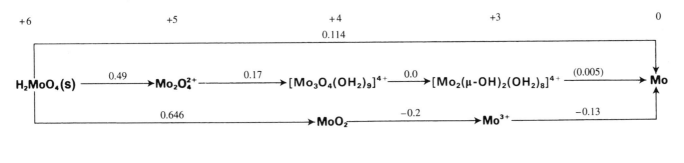

0.114

$H_2MoO_4(s)$ —0.49→ $Mo_2O_4^{2+}$ —0.17→ $[Mo_3O_4(OH_2)_9]^{4+}$ —0.0→ $[Mo_2(\mu\text{-}OH)_2(OH_2)_8]^{4+}$ —(0.005)→ Mo

———0.646———→ MoO_2 —−0.2→ Mo^{3+} —−0.13—

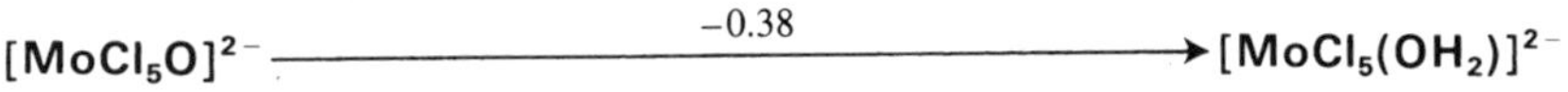

$[MoCl_5O]^{2-}$ ————−0.38————→ $[MoCl_5(OH_2)]^{2-}$

Neutral solution

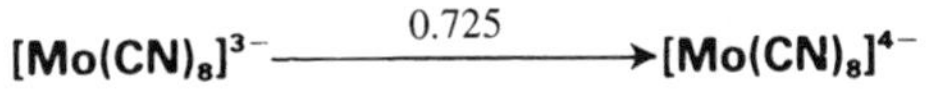

$[Mo(CN)_8]^{3-}$ —0.725→ $[Mo(CN)_8]^{4-}$

d **Block** · **Group 6** (continued)

Basic solution

+6   +4   0

MoO_4^{2-} $\xrightarrow{\;-0.780\;}$ MoO_2 $\xrightarrow{\;-0.980\;}$ Mo

$\xrightarrow{\;\;\;-0.913\;\;\;}$

Acidic solution

+6   +5   +4   0

WO_3 $\xrightarrow{\;-0.029\;}$ W_2O_5 $\xrightarrow{\;-0.031\;}$ $WO_2^{\,*}$ $\xrightarrow{\;-0.119\;}$ W

$\xrightarrow{\;\;\;-0.090\;\;\;}$

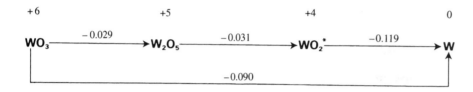

$[CoW_{12}O_{40}]^{6-}$ $\xrightarrow{\;-0.046\;}$ $[H_2CoW_{12}O_{40}]^{6-}$

$[PW_{12}O_{40}]^{3-}$ $\xrightarrow{\;0.22\;}$ $[PW_{12}O_{40}]^{4-}$

Neutral solution

+5   +4

$W(CN)_8^{3-}$ $\xrightarrow{\;0.457\;}$ $W(CN)_8^{4-}$

Basic solution

+6   +4   0

WO_4^{2-} $\xrightarrow{\;-1.259\;}$ WO_2 $\xrightarrow{\;-0.982\;}$ W

$\xrightarrow{\;\;\;-1.074\;\;\;}$

$[W(CN)_4(OH)_4]^{2-}$ $\xrightarrow{\;-0.702\;}$ $[W(CN)_4(OH)_4]^{4-}$

*Probably $[W_3(\mu_3\text{-}O)(\mu\text{-}O)_3(OH_2)_9]^{4+}$. See S.P. Gosh and E.S. Gould, *Inorg. Chem.* **30**, 3662 (1991).

d **Block · Group 7**

Acidic solution

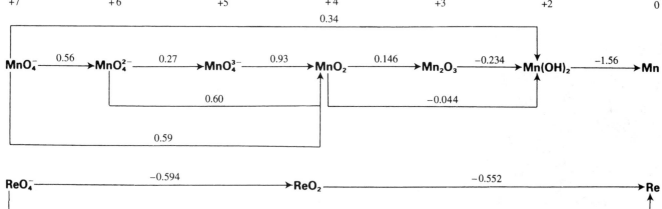

Basic solution

d Block · Group 8

Acidic solution

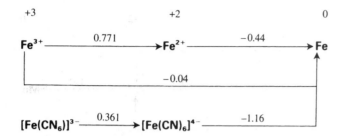

+3   +2   0

Fe^{3+} —0.771→ Fe^{2+} —−0.44→ Fe

Fe^{3+} —−0.04→ Fe

$[Fe(CN)_6]^{3-}$ —0.361→ $[Fe(CN)_6]^{4-}$ —−1.16→

Basic solution

+6   +3   +2   0

FeO_4^{2-} —0.81→ Fe_2O_3 —−0.86→ $Fe(OH)_2$ —−0.89→ Fe

$[Fe(ox)_3]^{3-}$ —0.005→ $[Fe(ox)_3]^{4-}$ (excess ox⁻)

Acidic solution

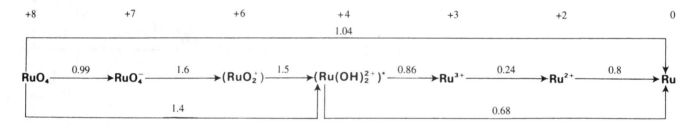

+8   +7   +6   +4   +3   +2   0

                                        1.04

RuO_4 —0.99→ RuO_4^- —1.6→ (RuO_2^+) —1.5→ $(Ru(OH)_2^{2+})^*$ —0.86→ Ru^{3+} —0.24→ Ru^{2+} —0.8→ Ru

              1.4   0.68

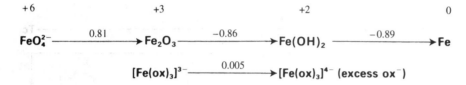

+3   +2

$[Ru(NH_3)_6]^{3+}$ —0.10→ $[Ru(NH_3)_6]^{2+}$

$[Ru(CN)_6]^{3-}$ —0.86→ $[Ru(CN)_6]^{4-}$

$[Ru(bipy)_3]^{3+}$ —1.53→ $[Ru(bipy)_3]^{2+}$

*Likely to be $H_n[Ru_4O_6(OH_2)_{12}]^{(4+n)+}$. See A. Patel and D.T. Richen, *Inorg. Chem.* **30**, 3792 (1991).

d Block · **Group 8** (continued)

Acidic solution

+8   +4   0

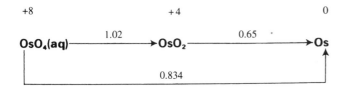

$OsO_4(aq) \xrightarrow{1.02} OsO_2 \xrightarrow{0.65} Os$

$\xrightarrow{0.834}$

+4   +3   +3   +2

$[OsCl_6]^{2-} \xrightarrow{0.45} [OsCl_6]^{3-}$ $[Os(CN)_6]^{3-} \xrightarrow{0.634} [Os(CN)_6]^{4-}$

$[OsBr_6]^{2-} \xrightarrow{0.45} [OsBr_6]^{3-}$ $[Os(bipy)_3]^{3+} \xrightarrow{0.885} [Os(bipy)_3]^{2+}$

d Block · **Group 9**

Acidic solution

+4   +3   +2   0

$CoO_2 \xrightarrow{1.4} Co^{3+} \xrightarrow{1.92} Co^{2+} \xrightarrow{-0.282} Co$

Basic solution

+4   +3   +2   0

$CoO_2 \xrightarrow{0.7} Co(OH)_3 \xrightarrow{0.42} Co(OH)_2 \xrightarrow{-0.733} Co$

Neutral solution

+3   +2

$[Co(NH_3)_6]^{3+} \xrightarrow{0.058} [Co(NH_3)_6]^{2+}$

$[Co(phen)_3]^{3+} \xrightarrow{0.33} [Co(phen)_3]^{2+}$

$[Co(ox)_3]^{3-} \xrightarrow{0.57} [Co(ox)_3]^{4-}$

Acidic solution

+3   0

$Rh^{3+} \xrightarrow{0.76} Rh$

Neutral solution

+3   +2

$[Rh(CN)_6]^{3-} \xrightarrow{0.9} [Rh(CN)_6]^{4-}$

d Block · **Group 9** (continued)

Acidic solution

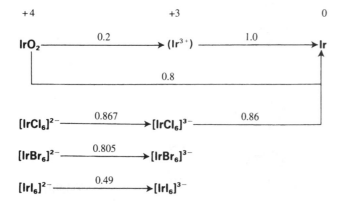

+4   +3   0

IrO$_2$ —— 0.2 —→ (Ir^{3+}) —— 1.0 —→ Ir

0.8

[IrCl$_6$]$^{2-}$ —— 0.867 —→ [IrCl$_6$]$^{3-}$ —— 0.86

[IrBr$_6$]$^{2-}$ —— 0.805 —→ [IrBr$_6$]$^{3-}$

[IrI$_6$]$^{2-}$ —— 0.49 —→ [IrI$_6$]$^{3-}$

d Block · **Group 10**

Acidic solution

+4   +3   +2   0

NiO$_2$ —————— 1.5 —————→ Ni^{2+} —— -0.257 —→ Ni

Basic solution

NiO$_2^*$ — 0.7 → NiOOH — 0.52 → Ni(OH)$_2$ — -0.72 → Ni

Neutral solution

[Ni(NH$_3$)$_6$]$^{2+}$ — -0.49 → Ni

Acidic solution

+4   +2   0

PdO$_2$ —— 1.194 —→ Pd^{2+} —— 0.915 —→ Pd

[PdCl$_6$]$^{2-}$ —— 1.47 —→ [PdCl$_4$]$^{2-}$ —— 0.60 —→ Pd

[PdBr$_4$]$^{2-}$ —— 0.49 —→ Pd

* Formulation uncertain.

---

d Block · **Group 10** (continued)

Basic solution

+4   +2   0

$PdO_2 \xrightarrow{\;\;1.47\;\;} PdO \xrightarrow{\;\;0.897\;\;} Pd$

Acidic solution

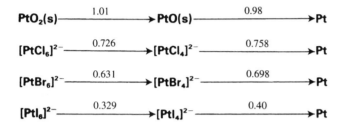

$PtO_2(s) \xrightarrow{\;\;1.01\;\;} PtO(s) \xrightarrow{\;\;0.98\;\;} Pt$

$[PtCl_6]^{2-} \xrightarrow{\;\;0.726\;\;} [PtCl_4]^{2-} \xrightarrow{\;\;0.758\;\;} Pt$

$[PtBr_6]^{2-} \xrightarrow{\;\;0.631\;\;} [PtBr_4]^{2-} \xrightarrow{\;\;0.698\;\;} Pt$

$[PtI_6]^{2-} \xrightarrow{\;\;0.329\;\;} [PtI_4]^{2-} \xrightarrow{\;\;0.40\;\;} Pt$

---

d Block · **Group 11**

Acidic solution

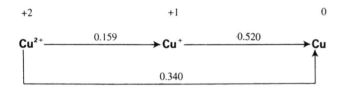

+2   +1   0

$Cu^{2+} \xrightarrow{\;\;0.159\;\;} Cu^+ \xrightarrow{\;\;0.520\;\;} Cu$

$\xrightarrow{\;\;0.340\;\;}$

$[Cu(NH_3)_4]^{2+} \xrightarrow{\;\;0.10\;\;} [Cu(NH_3)_2]^+ \xrightarrow{\;\;-0.10\;\;} Cu$

$Cu^{2+} \xrightarrow{\;\;1.12\;\;} [Cu(CN)_2]^- \xrightarrow{\;\;-0.44\;\;} Cu$

Basic solution

$Cu(OH)_2 \xrightarrow{\;\;-1.11\;\;} Cu_2O \xrightarrow{\;\;-1.36\;\;} Cu$

d **Block** · **Group 11** (continued)

Acidic solution

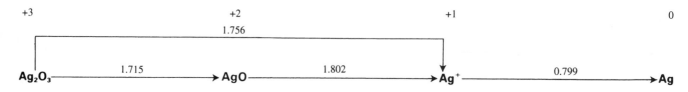

Basic solution

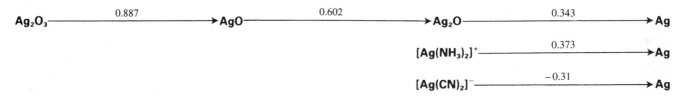

Acidic solution

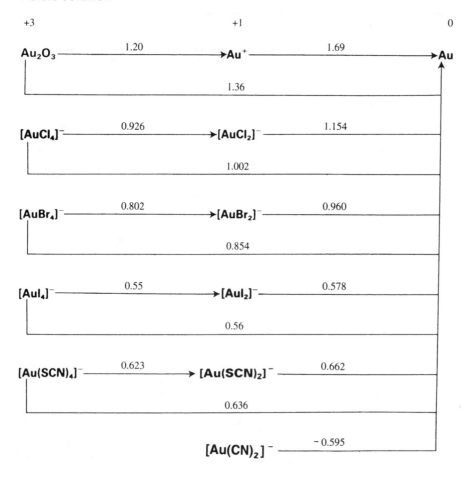

d Block · Group 12

Acidic solution

+2   0

Zn^{2+} ———— -0.762 ————→ Zn

Basic solution

$[Zn(OH)_4]^{2-}$ ———— -1.199 ————→ Zn

$Zn(OH)_2$ ———— -1.246 ————→ Zn

Acidic solution

Cd^{2+} ———— -0.402 ————→ Cd

Basic solution

$Cd(OH)_2(s)$ ———— -0.824 ————→ Cd

Acidic solution

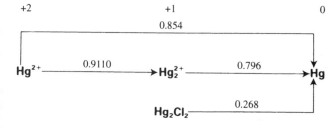

+2   +1   0

0.854

Hg^{2+} ——— 0.9110 ——→ Hg_2^{2+} ——— 0.796 ——→ Hg

Hg_2Cl_2 ——— 0.268 ———

Basic solution

HgO ———— 0.0977 ————→ Hg

f Block · Lanthanides

Acidic solution

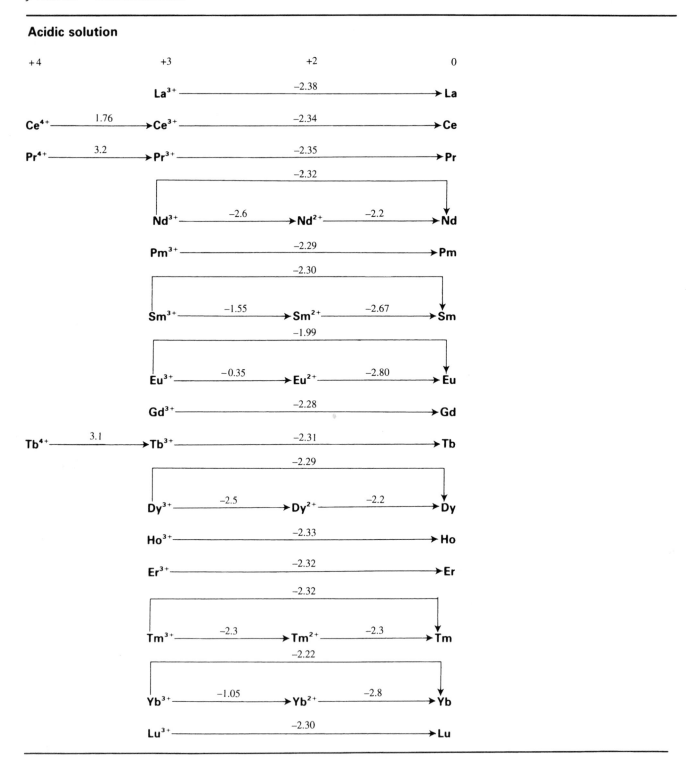

+4   +3   +2   0

La^{3+} ————————— -2.38 ————————→ La

Ce^{4+} —— 1.76 ——→ Ce^{3+} ——— -2.34 ———→ Ce

Pr^{4+} —— 3.2 ——→ Pr^{3+} ——— -2.35 ———→ Pr

Nd^{3+} —— -2.6 ——→ Nd^{2+} —— -2.2 ——→ Nd
(—— -2.32 ——)

Pm^{3+} ————— -2.29 —————→ Pm

Sm^{3+} —— -1.55 ——→ Sm^{2+} —— -2.67 ——→ Sm
(—— -2.30 ——)

Eu^{3+} —— -0.35 ——→ Eu^{2+} —— -2.80 ——→ Eu
(—— -1.99 ——)

Gd^{3+} ————— -2.28 —————→ Gd

Tb^{4+} —— 3.1 ——→ Tb^{3+} ——— -2.31 ———→ Tb

Dy^{3+} —— -2.5 ——→ Dy^{2+} —— -2.2 ——→ Dy
(—— -2.29 ——)

Ho^{3+} ————— -2.33 —————→ Ho

Er^{3+} ————— -2.32 —————→ Er

Tm^{3+} —— -2.3 ——→ Tm^{2+} —— -2.3 ——→ Tm
(—— -2.32 ——)

Yb^{3+} —— -1.05 ——→ Yb^{2+} —— -2.8 ——→ Yb
(—— -2.22 ——)

Lu^{3+} ————— -2.30 —————→ Lu

f Block · Actinides

Acidic solution

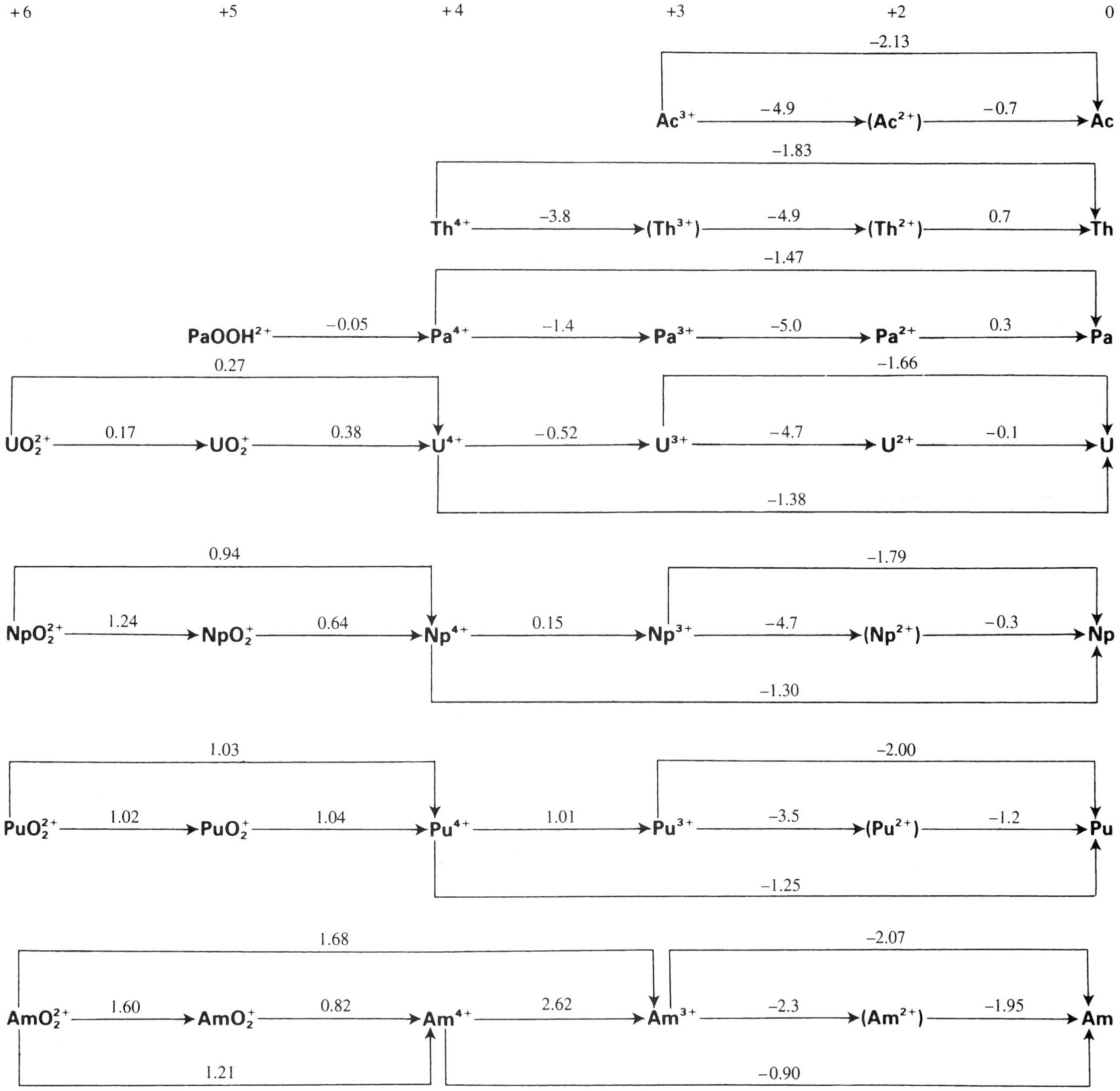

+6	+5	+4	+3	+2	0

$$
Ac^{3+} \xrightarrow{-4.9} (Ac^{2+}) \xrightarrow{-0.7} Ac
$$
$$
Ac^{3+} \xrightarrow{-2.13} Ac
$$

$$
Th^{4+} \xrightarrow{-3.8} (Th^{3+}) \xrightarrow{-4.9} (Th^{2+}) \xrightarrow{0.7} Th
$$
$$
Th^{4+} \xrightarrow{-1.83} Th
$$

$$
PaOOH^{2+} \xrightarrow{-0.05} Pa^{4+} \xrightarrow{-1.4} Pa^{3+} \xrightarrow{-5.0} Pa^{2+} \xrightarrow{0.3} Pa
$$
$$
Pa^{4+} \xrightarrow{-1.47} Pa
$$

$$
UO_2^{2+} \xrightarrow{0.17} UO_2^{+} \xrightarrow{0.38} U^{4+} \xrightarrow{-0.52} U^{3+} \xrightarrow{-4.7} U^{2+} \xrightarrow{-0.1} U
$$
$$
UO_2^{2+} \xrightarrow{0.27} U^{4+}
$$
$$
U^{3+} \xrightarrow{-1.66} U
$$
$$
U^{4+} \xrightarrow{-1.38} U
$$

$$
NpO_2^{2+} \xrightarrow{1.24} NpO_2^{+} \xrightarrow{0.64} Np^{4+} \xrightarrow{0.15} Np^{3+} \xrightarrow{-4.7} (Np^{2+}) \xrightarrow{-0.3} Np
$$
$$
NpO_2^{2+} \xrightarrow{0.94} Np^{4+}
$$
$$
Np^{3+} \xrightarrow{-1.79} Np
$$
$$
Np^{4+} \xrightarrow{-1.30} Np
$$

$$
PuO_2^{2+} \xrightarrow{1.02} PuO_2^{+} \xrightarrow{1.04} Pu^{4+} \xrightarrow{1.01} Pu^{3+} \xrightarrow{-3.5} (Pu^{2+}) \xrightarrow{-1.2} Pu
$$
$$
PuO_2^{2+} \xrightarrow{1.03} Pu^{4+}
$$
$$
Pu^{3+} \xrightarrow{-2.00} Pu
$$
$$
Pu^{4+} \xrightarrow{-1.25} Pu
$$

$$
AmO_2^{2+} \xrightarrow{1.60} AmO_2^{+} \xrightarrow{0.82} Am^{4+} \xrightarrow{2.62} Am^{3+} \xrightarrow{-2.3} (Am^{2+}) \xrightarrow{-1.95} Am
$$
$$
AmO_2^{2+} \xrightarrow{1.68} Am^{3+}
$$
$$
AmO_2^{+} \xrightarrow{1.21} Am^{4+}
$$
$$
Am^{4+} \xrightarrow{-0.90} Am
$$
$$
Am^{3+} \xrightarrow{-2.07} Am
$$

f Block · **Actinides** (continued)

Acidic solution

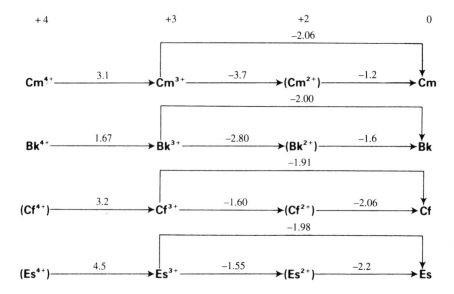

Appendix 3

Character tables

The character tables that follow are for the most common point groups encountered in inorganic chemistry. Each one is labeled with the symbol adopted in the *Schoenflies system* of nomenclature (such as C_{3v}). Point groups that qualify as crystallographic point groups (because they are also applicable to unit cells) are also labeled with the symbol adopted in the *International system* (or the Hermann–Mauguin system, such as $2/m$). In the latter system, a number n represents an n-fold axis and a letter m represents a mirror plane. A diagonal line indicates that a mirror plane lies perpendicular to the symmetry axis and a bar over the number indicates that the rotation is combined with an inversion.

The symmetry species of the p and d orbitals are shown on the right of the tables. Thus, in C_{2v}, a p_x orbital (which is proportional to x) has $\mathbf{B_1}$ symmetry. The functions $x, y,$ and z also show the transformation properties of translations and of the electric dipole moment. The set of functions that span a degenerate representation (such as x and y, which jointly span $\mathbf{E}$ in C_{3v}) are enclosed in parentheses. The transformation properties of rotation are shown by the letters R on the right of the tables.

The groups C_1, C_s, C_i

C_1 (**1**)	E	$h = 1$
A	1	

$C_s = C_h$ (*m*)	E	σ_h	$h = 2$	
A'	1	1	x, y, R_z	x^2, y^2, z^2, xy
A''	1	−1	z, R_x, R_y	yz, zx

$C_i = S_2$ (**$\bar{1}$**)	E	i	$h = 2$	
A_g	1	1	R_x, R_y, R_z	$x^2, y^2, z^2, xy, zx, yz$
A_u	1	−1	x, y, z	

The groups C_n

C_2 (**2**)	E	C_2	$h = 2$	
A	1	1	z, R_z	x^2, y^2, z^2, xy
B	1	−1	x, y, R_x, R_y	yz, zx

C_3 (**3**)	E	C_3	C_3^2	$\varepsilon = \exp(2\pi i/3)$	$h = 3$
A	1	1	1	z, R_z	$x^2 + y^2, z^2$
E	$\begin{cases} 1 \\ 1 \end{cases}$ $\begin{matrix} \varepsilon \\ \varepsilon^* \end{matrix}$ $\begin{matrix} \varepsilon^* \\ \varepsilon \end{matrix}$			$(x, y)\,(R_x, R_y)$	$(x^2 - y^2, xy)\,(yz, zx)$

C_4 (**4**)	E	C_4	C_2	C_4^3	$h = 4$	
A	1	1	1	1	z, R_z	$x^2 + y^2, z^2$
B	1	−1	1	−1		$x^2 - y^2, xy$
E	$\begin{cases} 1 \\ 1 \end{cases}$ $\begin{matrix} i \\ -i \end{matrix}$ $\begin{matrix} -1 \\ 1 \end{matrix}$ $\begin{matrix} i \\ i \end{matrix}$				$(x, y)\,(R_x, R_y)$	(yz, zx)

The groups C_{nv}

C_{2v} (**2mm**)	E	C_2	$\sigma_v(xz)$	$\sigma_v'(yz)$	$h = 4$	
A_1	1	1	1	1	z	x^2, y^2, z^2
A_2	1	1	−1	−1	R_z	xy
B_1	1	−1	1	−1	x, R_y	zx
B_2	1	−1	−1	1	y, R_x	yz

C_{3v} (**3m**)	E	$2C_3$	$3\sigma_v$	$h = 6$	
A_1	1	1	1	z	$x^2 + y^2, z^2$
A_2	1	1	−1	R_z	
E	2	−1	0	$(x, y)\,(R_x, R_y)$	$(x^2 - y^2, xy)\,(zx, yz)$

C_{4v} (4mm)	E	$2C_4$	C_2	$2\sigma_v$	$2\sigma_d$	$h = 8$	
A_1	1	1	1	1	1	z	$x^2 + y^2, z^2$
A_2	1	1	1	-1	-1	R_z	
B_1	1	-1	1	1	-1		$x^2 - y^2$
B_2	1	-1	1	-1	1		xy
E	2	0	-2	0	0	$(x, y)\,(R_x, R_y)$	(zx, yz)

C_{5v}	E	$2C_5$	$2C_5^2$	$5\sigma_v$	$h = 10,\ \alpha = 72°$	
A_1	1	1	1	1	z	$x^2 + y^2, z^2$
A_2	1	1	1	-1	R_z	
E_1	2	$2\cos\alpha$	$2\cos 2\alpha$	0	$(x, y)\,(R_x, R_y)$	(zx, yz)
E_2	2	$2\cos 2\alpha$	$2\cos\alpha$	0		$(x^2 - y^2, xy)$

C_{6v} (6mm)	E	$2C_6$	$2C_3$	C_2	$3\sigma_v$	$3\sigma_d$	$h = 12$	
A_1	1	1	1	1	1	1	z	$x^2 + y^2, z^2$
A_2	1	1	1	1	-1	-1	R_z	
B_1	1	-1	1	-1	1	-1		
B_2	1	-1	1	-1	-1	1		
E_1	2	1	-1	-2	0	0	$(x, y)\,(R_x, R_y)$	(zx, yz)
E_2	2	-1	-1	2	0	0		$(x^2 - y^2, xy)$

$C_{\infty v}$	E	C_2	$2C_\phi$	$\infty\sigma_v$	$h = \infty$	
$A_1\ (\Sigma^+)$	1	1	1	1	z	$x^2 + y^2, z^2$
$A_2\ (\Sigma^-)$	1	1	1	-1	R_z	
$E_1\ (\Pi)$	2	-2	$2\cos\phi$	0	$(x, y)\,(R_x, R_y)$	(zx, yz)
$E_2\ (\Delta)$	2	2	$2\cos 2\phi$	0		$(xy, x^2 - y^2)$
$\vdots$	$\vdots$	$\vdots$	$\vdots$	$\vdots$		

The groups D_n

D_2 (222)	E	$C_2(z)$	$C_2(y)$	$C_2(x)$	$h = 4$	
A	1	1	1	1		x^2, y^2, z^2
B_1	1	1	-1	-1	z, R_z	xy
B_2	1	-1	1	-1	y, R_y	zx
B_3	1	-1	-1	1	x, R_x	yz

D_3 (32)	E	$2C_3$	$3C_2$	$h = 6$	
A_1	1	1	1		$x^2 + y^2, z^2$
A_2	1	1	-1	z, R_z	
E	2	-1	0	$(x, y)\,(R_x, R_y)$	$(x^2 - y^2, xy)(zx, yz)$

The groups $D_{n\text{h}}$

$D_{2\text{h}}$ (mmm)	E	$C_2(z)$	$C_2(y)$	$C_2(x)$	i	$\sigma(xy)$	$\sigma(xz)$	$\sigma(yz)$	$h = 8$	
A_g	1	1	1	1	1	1	1	1		x^2, y^2, z^2
B_{1g}	1	1	-1	-1	1	1	-1	-1	R_z	xy
B_{2g}	1	-1	1	-1	1	-1	1	-1	R_y	zx
B_{3g}	1	-1	-1	1	1	-1	-1	1	R_x	yz
A_u	1	1	1	1	-1	-1	-1	-1		
B_{1u}	1	1	-1	-1	-1	-1	1	1	z	
B_{2u}	1	-1	1	-1	-1	1	-1	1	y	
B_{3u}	1	-1	-1	1	-1	1	1	-1	x	

$D_{3\text{h}}$ ($\overline{6}m2$)	E	$2C_3$	$3C_2$	σ_h	$2S_3$	$3\sigma_\text{v}$	$h = 12$	
A_1'	1	1	1	1	1	1		$x^2 + y^2, z^2$
A_2'	1	1	-1	1	1	-1	R_z	
E'	2	-1	0	2	-1	0	(x, y)	$(x^2 - y^2, xy)$
A_1''	1	1	1	-1	-1	-1		
A_2''	1	1	-1	-1	-1	1	z	
E''	2	-1	0	-2	1	0	(R_x, R_y)	(zx, yz)

$D_{4\text{h}}$ ($4/mmm$)	E	$2C_4$	C_2	$2C_2'$	$2C_2''$	i	$2S_4$	σ_h	$2\sigma_\text{v}$	$2\sigma_\text{d}$	$h = 16$	
A_{1g}	1	1	1	1	1	1	1	1	1	1		$x^2 + y^2, z^2$
A_{2g}	1	1	1	-1	-1	1	1	1	-1	-1	R_z	
B_{1g}	1	-1	1	1	-1	1	-1	1	1	-1		$x^2 - y^2$
B_{2g}	1	-1	1	-1	1	1	-1	1	-1	1		xy
E_g	2	0	-2	0	0	2	0	-2	0	0	(R_x, R_y)	(zx, yz)
A_{1u}	1	1	1	1	1	-1	-1	-1	-1	-1		
A_{2u}	1	1	1	-1	-1	-1	-1	-1	1	1	z	
B_{1u}	1	-1	1	1	-1	-1	1	-1	-1	1		
B_{2u}	1	-1	1	-1	1	-1	1	-1	1	-1		
E_u	2	0	-2	0	0	-2	0	2	0	0	(x, y)	

The groups $D_{n\text{h}}$ (continued)

$D_{5\text{h}}$	E	$2C_5$	$2C_5^2$	$5C_2$	σ_h	$2S_5$	$2S_5^3$	$5\sigma_\text{v}$	$h = 20$, $\alpha = 72°$	
A_1'	1	1	1	1	1	1	1	1		$x^2 + y^2$, z^2
A_2'	1	1	1	-1	1	1	1	-1	R_z	
E_1'	2	$2\cos\alpha$	$2\cos 2\alpha$	0	2	$2\cos\alpha$	$2\cos 2\alpha$	0	(x, y)	
E_2'	2	$2\cos 2\alpha$	$2\cos\alpha$	0	2	$2\cos 2\alpha$	$2\cos\alpha$	0		$(x^2 - y^2, xy)$
A_1''	1	1	1	1	-1	-1	-1	-1		
A_2''	1	1	1	-1	-1	-1	-1	1	z	
E_1''	2	$2\cos\alpha$	$2\cos 2\alpha$	0	-2	$-2\cos\alpha$	$-2\cos 2\alpha$	0	(R_x, R_y)	(zx, yz)
E_2''	2	$2\cos 2\alpha$	$2\cos\alpha$	0	-2	$-2\cos 2\alpha$	$-2\cos\alpha$	0		

$D_{6\text{h}}$ $(6/mmm)$	E	$2C_6$	$2C_3$	C_2	$3C_2'$	$3C_2''$	i	$2S_3$	$2S_6$	σ_h	$3\sigma_\text{d}$	$3\sigma_\text{v}$	$h = 24$	
A_{1g}	1	1	1	1	1	1	1	1	1	1	1	1		$x^2 + y^2$, z^2
A_{2g}	1	1	1	1	-1	-1	1	1	1	1	-1	-1	R_z	
B_{1g}	1	-1	1	-1	1	-1	1	-1	1	-1	1	-1		
B_{2g}	1	-1	1	-1	-1	1	1	-1	1	-1	-1	1		
E_{1g}	2	1	-1	-2	0	0	2	1	-1	-2	0	0	(R_x, R_y)	(zx, yz)
E_{2g}	2	-1	-1	2	0	0	2	-1	-1	2	0	0		$(x^2 - y^2, xy)$
A_{1u}	1	1	1	1	1	1	-1	-1	-1	-1	-1	-1		
A_{2u}	1	1	1	1	-1	-1	-1	-1	-1	-1	1	1	z	
B_{1u}	1	-1	1	-1	1	-1	-1	1	-1	1	-1	1		
B_{2u}	1	-1	1	-1	-1	1	-1	1	-1	1	1	-1		
E_{1u}	2	1	-1	-2	0	0	-2	-1	1	2	0	0	(x, y)	
E_{2u}	2	-1	-1	2	0	0	-2	1	1	-2	0	0		

$D_{\infty\text{h}}$	E	$\infty C_2'$	$2C_\phi$	i	$\infty\sigma_\text{v}$	$2S_\phi$	$h = \infty$	
A_{1g} (Σ_g^+)	1	1	1	1	1	1		z^2, $x^2 + y^2$
A_{1u} (Σ_u^+)	1	-1	1	-1	1	-1	z	
A_{2g} (Σ_g^-)	1	-1	1	1	-1	1	R_z	
A_{2u} (Σ_u^-)	1	1	1	-1	-1	-1		
E_{1g} (Π_g)	2	0	$2\cos\phi$	2	0	$-2\cos\phi$	(R_x, R_y)	(zx, yz)
E_{1u} (Π_u)	2	0	$2\cos\phi$	-2	0	$2\cos\phi$	(x, y)	
E_{2g} (Δ_g)	2	0	$2\cos 2\phi$	2	0	$2\cos 2\phi$		$(xy, x^2 - y^2)$
E_{2u} (Δ_u)	2	0	$2\cos 2\phi$	-2	0	$-2\cos 2\phi$		
$\vdots$	$\vdots$	$\vdots$	$\vdots$	$\vdots$	$\vdots$	$\vdots$		

The groups $D_{n\mathrm{d}}$

$D_{2\mathrm{d}} = V_\mathrm{d}$ ($\bar{4}2m$)	E	$2S_4$	C_2	$2C_2'$	$2\sigma_\mathrm{d}$		$h = 8$	
A_1	1	1	1	1	1		$x^2 + y^2,\ z^2$	
A_2	1	1	1	−1	−1	R_z		
B_1	1	−1	1	1	−1		$x^2 - y^2$	
B_2	1	−1	1	−1	1	z	xy	
E	2	0	−2	0	0	(x, y)	(zx, yz)	
						(R_x, R_y)		

$D_{3\mathrm{d}}$ ($\bar{3}m$)	E	$2C_3$	$3C_2$	i	$2S_6$	$3\sigma_\mathrm{d}$		$h = 12$	
A_{1g}	1	1	1	1	1	1		$x^2 + y^2,\ z^2$	
A_{2g}	1	1	−1	1	1	−1	R_z		
E_g	2	−1	0	2	−1	0	(R_x, R_y)	$(x^2 - y^2,\ xy)$	
								(zx, yz)	
A_{1u}	1	1	1	−1	−1	−1			
A_{2u}	1	1	−1	−1	−1	1	z		
E_u	2	−1	0	−2	1	0	(x, y)		

$D_{4\mathrm{d}}$	E	$2S_8$	$2C_4$	$2S_8^3$	C_2	$4C_2'$	$4\sigma_\mathrm{d}$		$h = 16$	
A_1	1	1	1	1	1	1	1		$x^2 + y^2,\ z^2$	
A_2	1	1	1	1	1	−1	−1	R_z		
B_1	1	−1	1	−1	1	1	−1			
B_2	1	−1	1	−1	1	−1	1	z		
E_1	2	$\sqrt{2}$	0	$-\sqrt{2}$	−2	0	0	(x, y)		
E_2	2	0	−2	0	2	0	0		$(x^2 - y^2,\ xy)$	
E_3	2	$-\sqrt{2}$	0	$\sqrt{2}$	−2	0	0	(R_x, R_y)	(zx, yz)	

The cubic groups

T_d ($\bar{4}3m$)	E	$8C_3$	$3C_2$	$6S_4$	$6\sigma_\mathrm{d}$		$h = 24$	
A_1	1	1	1	1	1		$x^2 + y^2 + z^2$	
A_2	1	1	1	−1	−1			
E	2	−1	2	0	0		$(2z^2 - x^2 - y^2,$ $x^2 - y^2)$	
T_1	3	0	−1	1	−1	(R_x, R_y, R_z)		
T_2	3	0	−1	−1	1	(x, y, z)	(xy, yz, zx)	

The cubic groups (continued)

O_h ($m3m$)	E	$8C_3$	$6C_2$	$6C_4$	$3C_2$ ($= C_4^2$)	i	$6S_4$	$8S_6$	$3\sigma_h$	$6\sigma_d$	$h = 48$
A_{1g}	1	1	1	1	1	1	1	1	1	1	$x^2 + y^2 + z^2$
A_{2g}	1	1	−1	−1	1	1	−1	1	1	−1	
E_g	2	−1	0	0	2	2	0	−1	2	0	$(2z^2 - x^2 - y^2,$ $x^2 - y^2)$
T_{1g}	3	0	−1	1	−1	3	1	0	−1	−1	(R_x, R_y, R_z)
T_{2g}	3	0	1	−1	−1	3	−1	0	−1	1	(xy, yz, zx)
A_{1u}	1	1	1	1	1	−1	−1	−1	−1	−1	
A_{2u}	1	1	−1	−1	1	−1	1	−1	−1	1	
E_u	2	−1	0	0	2	−2	0	1	−2	0	
T_{1u}	3	0	−1	1	−1	−3	−1	0	1	1	(x, y, z)
T_{2u}	3	0	1	−1	−1	−3	1	0	1	−1	

The icosahedral group

I	E	$12C_5$	$12C_5^2$	$20C_3$	$15C_2$	$h = 60$
A_1	1	1	1	1	1	$x^2 + y^2 + z^2$
T_1	3	$\frac{1}{2}(1 + \sqrt{5})$	$\frac{1}{2}(1 - \sqrt{5})$	0	−1	(x, y, z) (R_x, R_y, R_z)
T_2	3	$\frac{1}{2}(1 - \sqrt{5})$	$\frac{1}{2}(1 + \sqrt{5})$	0	−1	
G	4	−1	−1	1	0	
H	5	0	0	−1	1	$(2z^2 - x^2 - y^2,$ $x^2 - y^2, xy, yz, zx)$

Further information: P.W. Atkins, M.S. Child, and C.S.G. Phillips, *Tables for group theory*. Oxford University Press (1970).

Appendix 4

Symmetry-adapted orbitals

Table A4.1 gives the symmetry classes of the s, p, and d orbitals of the central atom of an AB_n molecule of the specified point group. In most cases, the z-axis is the principal axis of the molecule; in C_{2v} the x-axis lies perpendicular to the molecular plane.

The orbital diagrams that follow show the linear combinations of atomic orbitals on the peripheral atoms of AB_n molecules of the specified point groups. Where a view from above is shown, the dot representing the central atom is either in the plane of the paper (for the D groups) or above the plane (for the corresponding C groups). Different phases of the atomic orbitals ($+$ or $-$ amplitudes) are shown by tinting. Where there is a large difference in the magnitudes of the orbital coefficients in a particular combination, the atomic orbitals have been drawn large or small to represent their relative contributions to the linear combination. In the case of degenerate linear combinations (those labeled E or T), any linearly independent combination of the degenerate pair is also of suitable symmetry. In practice, these different linear combinations look like the ones shown here, but their nodes are rotated by an arbitrary axis around the z-axis.

Molecular orbitals are formed by combining an orbital of the central atom (as in Table A4.1) with a linear combination of the same symmetry.

Table A4.1

	$D_{\infty h}$	C_{2v}	D_{3h}	C_{3v}	D_{4h}	C_{4v}	D_{5h}	C_{5v}	D_{6h}	C_{6v}	T_d	O_h
s	Σ	A_1	A_1'	A_1	A_{1g}	A_1	A_1'	A_1	A_{1g}	A_1	A_1	A_{1g}
p_x	Π	B_1	E'	E	E_u	E	E_1'	E_1	E_{1u}	E_1	T_2	T_{1u}
p_y	Π	B_2	E'	E	E_u	E	E_1'	E_1	E_{1u}	E_1	T_2	T_{1u}
p_z	Σ	A_1	A_2''	A_1	A_{2u}	A_1	A_2''	A_1	A_{2u}	A_1	T_2	T_{1u}
d_{z^2}	Σ	A_1	A_1'	A_1	A_{1g}	A_1	A_1'	A_1	A_{1g}	A_1	E	E_g
$d_{x^2-y^2}$	Δ	A_1	E'	E	B_{1g}	B_1	E_2'	E_2	E_{2g}	E_2	E	E_g
d_{xy}	Δ	A_2	E'	E	B_{2g}	B_2	E_2'	E_2	E_{2g}	E_2	T_2	T_{2g}
d_{yz}	Π	B_2	E''	E	E_g	E	E_1''	E_1	E_{1g}	E_1	T_2	T_{2g}
d_{zx}	Π	B_1	E''	E	E_g	E	E_1''	E_1	E_{1g}	E_1	T_2	T_{2g}

$D_{\infty h}$	C_{2v}	
σ_g	A_1	
π_g	A_2	
π_u	B_1	
σ_u	B_2	
	A_1	
	B_2	

D_{3h}	C_{3v}	
A_1'	A_1	
A_2'	A_2	
E'	E	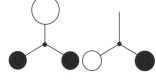
E'	E	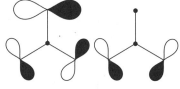
A_2''	A_1	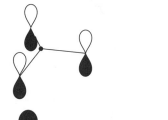
E''	E	

D_{4h} C_{4v}

A_{1g} A_1

A_{2g} A_2

B_{1g} B_1

B_{2g} B_2

E_u E

A_{2u} A_1

E_g E

B_{2u} B_2

D_{5h} C_{5v}

A_1' A_1

A_2' A_2

E_1' E_1

E_2' E_2

A_2'' A_1

E_1'' E_1

E_2'' E_2

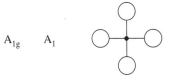
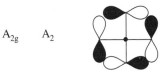
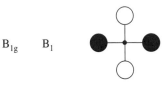

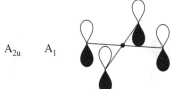

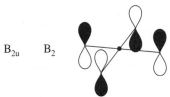
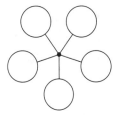

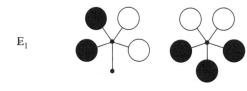

D_{6h} **C_{6v}**

A_{1g} A_1

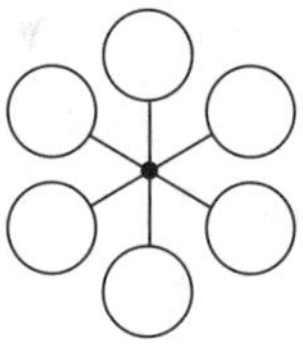

A_{2g} A_2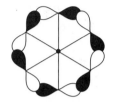

B_{1u} B_1

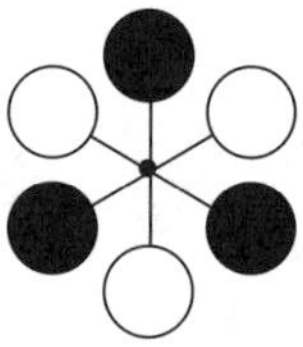

B_{2u} B_2

E_{1u} E_1

E_{2g} E_2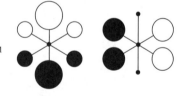

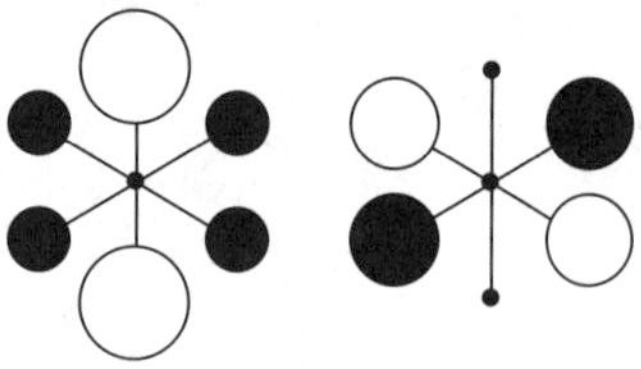

A_{2u} A_1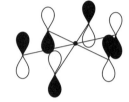

B_{2g} B_1

E_{1g} E_1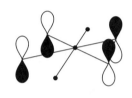

E_{2u} E_2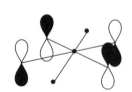

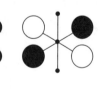

$T_\mathbf{d}$

A$_1$

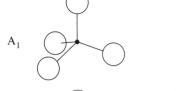

T$_2$

O_h

A$_{1g}$

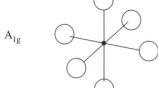

E$_g$

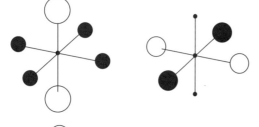

T$_{1u}$

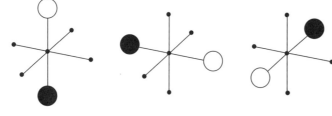

T$_{1u}$

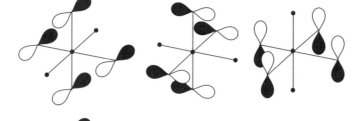

T$_{2g}$

T$_{1g}$

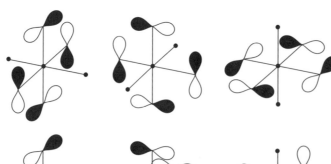

T$_{2u}$

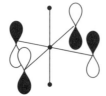

Appendix 5 Tanabe–Sugano diagrams

This appendix collects together the Tanabe–Sugano diagrams for octahedral complexes with electron configurations d^2 to d^8. The diagrams, which were introduced in Section 13.3, show the dependence of the term energies on ligand-field strength. The term energies E are expressed as the ratio E/B, where B is a Racah parameter, and the ligand-field splitting Δ_O is expressed as Δ_O/B. Terms of different multiplicity are included in the same diagram by making specific, plausible choices about the value of the Racah parameter C, and these choices are given for each diagram. The term energy is always measured from the lowest energy term, and so there are discontinuities of slope where a low-spin term displaces a high-spin term at sufficiently high ligand-field strengths for d^4 to d^8 configurations. Moreover, the noncrossing rule requires terms of the same symmetry to mix rather than to cross, and this mixing accounts for the curved rather than the straight lines in a number of cases. The term labels are those of the point group O_h.

The diagrams were first introduced by Y. Tanabe and S. Sugano, *J. Phys. Soc. Japan* **9**, 753 (1954). They may be used to find the parameters Δ_O and B by fitting the ratios of the energies of observed transitions to the lines. Alternatively, if the ligand-field parameters are known, then the ligand-field spectra may be predicted.

1. d^1 with $C = 4.42B$

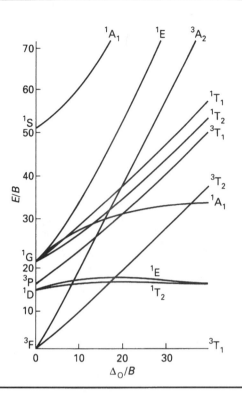

3. d^4 with $C = 4.61B$

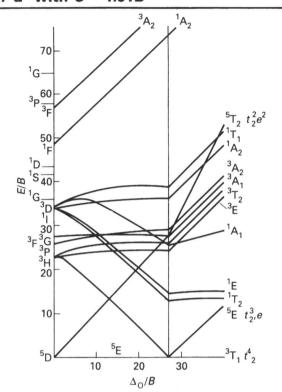

2. d^3 with $C = 4.5B$

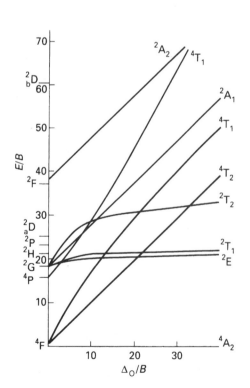

4. d^5 with $C = 4.477B$

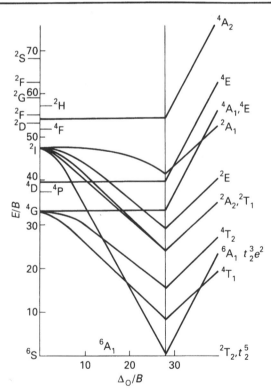

5. d^6 with $C = 4.8B$

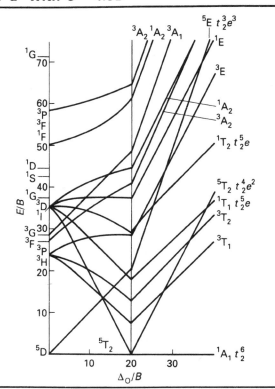

7. d^8 with $C = 4.709B$

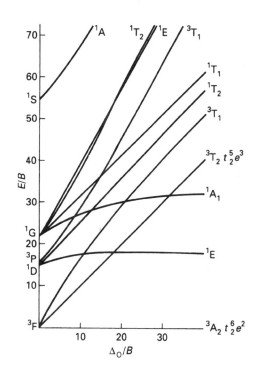

6. d^7 with $C = 4.633B$

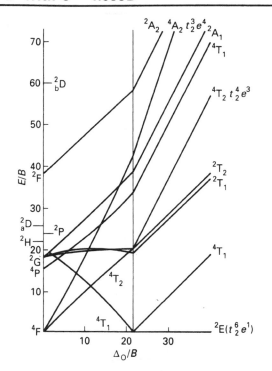

Answers to self-tests and exercises

Chapter 1

Self-tests

1.1 $^{80}_{35}Br + n \rightarrow ^{81}_{35}Br + \gamma$

1.2 $3p$.

1.3 Added p electron is in a different (p) orbital so it is less shielded.

1.4 Ni : $[Ar]3d^8 4s^2$, Ni^{2+}: $[Ar]3d^8$

1.5 Going down a group the atomic radius increases and the first ionization energy generally decreases.

1.6 For N the p orbitals are all singly occupied. On going to C the added electron must go into a half-filled orbital with added electron–electron repulsion.

Exercises

1.1 (a) $^{14}_{7}N + ^{4}_{2}He \rightarrow ^{17}_{8}O + ^{1}_{1}p + \gamma$
(b) $^{12}_{6}C + ^{1}_{1}p \rightarrow ^{13}_{7}N + \gamma$
(c) $^{14}_{7}N + ^{1}_{0}n \rightarrow ^{12}_{6}C + ^{3}_{1}H$

1.2 $^{22}_{10}Ne + ^{4}_{2}\alpha \rightarrow ^{25}_{12}Mg + n$

1.3 Check inside front cover

1.4 0.25.

1.5 $- 13.2$ eV.

1.6 0 up to $n - 1$.

1.7 n^2.

1.8 See Figs 1.9 through 1.14.

1.9 See Table 1.6 and discussion on p. 28.

1.10 Table 1.6 shows Sr > Ba < Ra. Ra is anomalous because of higher Z_{eff} due to lanthanide contraction.

1.11 Anomalously high value for Cr is associated with the stability of 1/2-filled d shell.

1.12 (a) $[He]2s^2 2p^2$; (b) $[He]2s^2 2p^5$;
(c) $[Ar]4s^2$; (d) $[Ar]3d^{10}$;
(e) $[Xe]4f^{14}5d^{10}6s^2 3p^3$;
(f) $[Xe]4f^{14}5d^{10}6s^2$.

1.13 (a) $[Ar]3d^{10}s^2$; (b) $[Ar]3d^2$;
(c) $[Ar]3d^5$; (d) $[Ar]3d^4$;
(e) $[Ar]3d^6$; (f) $[Ar]$;
(g) $[Ar]3d^{10}4s^1$;
(h) $[Xe]4f^7$.

1.14 (a) $[Xe]4f^{14}5d^4 6s^2$; (b) $[Kr]4d^6$;
(c) $[Xe]4f^6$; (d) $[Xe]4f^7$; (e) $[Ar]$;
(f) $[Kr]4d^2$.

1.15 (a) I_1 increases across group except for a dip at S; (b) A_e tends to increase except for Mg (filled subshell), P (half-filled subshell), and Ar (filled shell).

1.16 Radii of period 4 and 5 d-block metals are similar because of lanthanide contraction.

1.17 χ increases steadily in most scales. For χ_M anomalies in I are generally offset by those in A.

1.18 $2s^2$ and $2p^0$.

1.19 With some exceptions generally associated with half-full or full subshells I and χ increase and r decreases across a period.

Chapter 2

Self-tests

2.1 $r_h = ((3/2)^{1/2} - 1)r = 0.225r$.

2.2 2.

2.3 6.

2.4 6.

2.5 2421 kJ mol^{-1}.

2.6 2.63 MJ mol^{-1}.

2.7 T (decomp):
$MgSO_4 < CaSO_4 < SrSO_4 < BaSO_4$.

2.8 $NaClO_4$.

Exercises

2.1 Close-packed: (a), (b), (d), (e).

2.2 Use Fig. 2.2 as a guide.

2.3 (a) see pp. 37 and 41; (b) polymorphs: diamond and graphite; polytypes often occur in layered compounds, such as TaS_2.

2.4 Close-packed W with C in octahedral holes.

2.5 (a) Rock salt 6:6, cesium chloride 8:8; (b) cesium chloride.

2.6 (a) Re: C.N = 6, O: C.N. = 2; (b) perovskite.

2.7 Cations: Groups 1 and 2 (except Be); anions: Group 7 and oxygen.

2.8 (a) 6:6; (b) 12; (c) use Figs 2.3–2.5 as a guide.

2.9 (a) 8:8; (b) 6.

2.10 (a) One of each; (b) four of each.

2.11 Use Fig. 2.16 taking into account the sharing of peripheral atoms with the surrounding unit cells.

2.12 Use Fig. 2.17 and the suggestion in the above answer.

2.13 Fluorite (CaF_2) structure.

2.14 $r(Ca^{2+}) = 1.08$ Å, $r(Sr^{2+}) = 1.19$ Å, $r(Ba^{2+}) = 1.38$ Å, $r(Mg^{2+}) \leqslant 0.80$ Å.

2.15 (a) 6:6; (b) 6:6; (c) 6:6; (d) 6:6 or 4:4. The small Be may lead to covalency which favors 4:4 coordination.

2.16 (a) 0.65 MJ mol^{-1}; (b) large second ionization energy for K.

2.17 (b) < (c) < (a), based on size and charge.

2.18 (a) $MgCO_3$; (b) CsI_3.

2.19 (a) $MgSO_4$; (b) $NaBF_4$.

Chapter 3

Self-tests

3.1

$$: \overset{..}{Cl} : \overset{..}{P} : \overset{..}{Cl} :$$
$$: \overset{..}{Cl} :$$

3.2

$$
\left[\overset{..}{N} \overset{}{\underset{\underset{..}{O.}}{\overset{..}{\underset{.}{O}}}} \right]^{-} \longleftrightarrow \left[\overset{..}{N} \right]^{-}
$$

3.3 $+ 1/2, +5$.

3.4 24 kJ mol^{-1}.

3.5 Linear.

3.6 S_2^{2-}: $1\sigma_g^2 1\sigma_u^2 2\sigma_g^2 1\pi_u^4 2\pi_g^4$; Cl_2^-: $[S_2^{2-}] 2\sigma_u^1$.

3.7 Similar to Fig. 3.24 with Cl $3s$ and $3p$ on left and O $2s$ and $2p$ on right.

3.8 If it contains 4 or fewer electrons.

3.9 $0.724S$.

Exercises

3.1 (a)

$$: \overset{..}{Cl} : \qquad]^{-}$$
$$: \overset{..}{Ge} : \overset{..}{Cl} :$$
$$: \overset{..}{Cl} :$$

(b)

$$: \overset{..}{F} : \overset{..}{C} \qquad \longleftrightarrow \qquad : \overset{..}{F} : \overset{..}{C}$$

(c)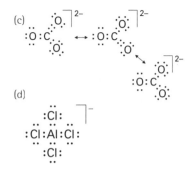

(d)

(e)

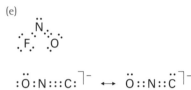

3.2 :Ö:N:::C:⁻ ↔ Ö::N::C:⁻

3.3 (a) and (b):

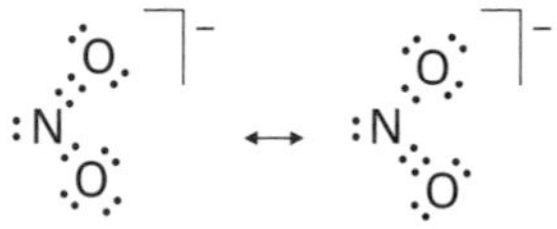

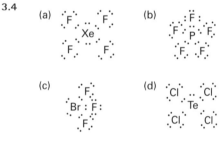

(c) N(+3), O(−2);
(i) formal charge, (ii) oxidation number, (iii) neither.

3.4 (a) ... Xe ... (b) ... P ...

(c) ... Br:F: ... (d) ... Te ...

(e)  :Cl:I:Cl:⁻

3.5 (a) Trigonal-planar; (b) trigonal-pyramidal; (c) square-pyramidal.

3.6 PCl_4^+ is tetrahedral; PCl_6^- is octahedral.

3.7 (a) 1.76 Å, (b) 2.17 Å, (c) 2.21 Å.

3.8 2(Si—O) = 932 kJ > Si=O = 640 kJ; therefore two Si—O are preferred and SiO_2 should (and does) have four single Si—O bonds.

3.9 Bond enthalpy data (Table 3.5) indicate $2N\equiv N \rightarrow N_4 \; \Delta H^{\ominus} \approx$ +912 kJ mol⁻¹, $2P_2 \rightarrow P_4 \Delta H^{\ominus} \approx$ −246 kJ mol⁻¹. Multiple bonds are much stronger for 2nd period elements than heavier elements.

3.10 − 483 kJ difference is smaller than expected because bond energies are not accurate.

3.11 (a) 0; (b) 205 kJ mol⁻¹.

3.12 (a) One; (b) one; (c) none; (d) two.

3.13 (a) $1\sigma_g^2 1\sigma_u^2$; (b) $1\sigma_g^2 2\sigma_u^2 1\pi_u^2$;
(c) $1\sigma_g^2 1\sigma_u^2 1\pi_u^4 2\sigma_g^1$;
(d) $1\sigma_g^2 1\sigma_u^2 2\sigma_g^2 1\pi_u^4 1\pi_g^3$.

3.14 (a) 2; (b) 1; (c) 2.

3.15 (a) + 0.5; (b) − 0.5; (c) + 0.5.

3.16 (a) 4; (b, c)

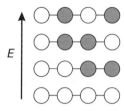

3.17 (a–c)

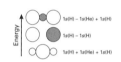

(d) possibly stable in isolation (only bonding and nonbonding orbitals are filled); not stable in solution because solvents would have higher proton affinity than He.

3.18 1.

3.19 HOMO exclusively F; LUMO mainly S.

3.20 (a) Electron-deficient; (b) electron-precise.

3.21 (a) NH_3; (b) BH_3; (c) NH_3.

3.22 (a) Compare Fig. 3.51 for a metal with Fig. 3.57 for a semiconductor; (b) for metal σ (conductivity) decreases moderately with increasing T; for semiconductor σ increases strongly with increasing T; (c) similar qualitatively but σ for insulator is lower.

3.23 (a) n; (b) p; (c) neither.

3.24 3.55 eV.

3.25 n.

3.26 n.

3.27 518 nm.

3.28 590.

Chapter 4

Self-tests

4.1 Four such axes, one along each of the NH bonds.

4.2 (a) D_{3h}; (b) T_d.

4.3 No.

4.4 Yes (but the two chiral forms rapidly interconvert).

4.5 Triple.

4.6 B_{1g}.

4.7 B_{1g}.

4.8 Yes.

4.9 See the first row in the D_{2h} character table.

4.10 T_{1u}.

Exercises

4.1 (a) (N), C_3 through N and the center of the H_3 triangle, three σ_v through N and one of each of the H atoms; (b) C_4

through Pt and perpendicular to plane of the complex; σ_h is in the plane of the complex.

4.2 (1) CO_2, C_2H_2; (2) SO_4^{2-}.

4.3 (a) C_s; (b) D_{3h}; (c) T_d; (d) $C_{\infty v}$; (e) C_1; (f) T_d.

4.4 (a) ∞C_n axes; $\infty\sigma$; i, (b) C_∞, $\infty\sigma$; (c) 3σ, $3C_2$; (d) σ; C_∞, ∞C_2, S_∞.

4.5 (a) $\sigma \perp$ to C_n axis, i, multiple non-collinear C_n axes; (b) a, d, and e may be polar.

4.6 (a) Lack of S_n symmetry element (s); (b) only SiFClBrI.

4.7 (a) C_{3v}; (b) double degeneracy; (c) p_x and p_y (where C_3 is along z-axis of SO_3^{2-}).

4.8 (a) D_{3h}; (b) doubly degenerate; (c) p_x and p_y (where C_3 is along z-axis of PF_5).

4.9 (a)  (A₁g); + (Eᵤ);

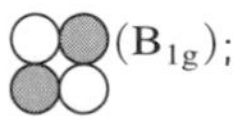

(B₁g);

(b) D_{4h} symmetry labels given in part a. (c) $d_z^2(A_{1g})$, and $d_{x^2-y^2}(B_{1g})$.

4.10 (a) 5; (b) 1.

4.11 (a) none; (b) E'.

4.12 A_2, B_1, B_2.

Chapter 5

Self-tests

5.1 Acid/conjugatebase :
(a) HNO_3/NO_3^-; (b) H_2O/OH^-;
(c) H_2S/HS^-.

5.2 $[Na(OH_2)_6]^+$ $< [Mn(OH_2)_6]^{2+} < [Ni(OH_2)_6]^{2+}$ $< [Sc(OH_2)_6]^{3+}$.

5.3 pK_a estimated by Pauling's rules: (a) 3; (b) 8; (c) 13.

5.4 Aq. NH_3 precipitates TiO_2, which dissolves in NaOH.

5.5 Acid/base: (a) $FeCl_3/Cl^-$; (b) I_2/I^-; (c) $Mn(CO)_5/SnCl_3^-$.

5.6 $(H_3Si)_3N$, trigonal-planar; $(H_3C)_3N$, pyramidal.

5.7 Aluminosilicates: Rb, Cr, Sr; sulfides: Cd, Pb, Pd.

5.8 B coordinated to O; pyramidal C_2OB array.

Exercises

5.1 Check element symbols inside cover. Check acidity and basicity in Fig. 5.4.

5.2 $[Co(NH_3)_5(OH)]^{2+}$; SO_4^{2-}; CH_3O^-; HPO_4^{2-}; $[Si(OH)_3O]^-$; S^{2-}.

5.3 $C_5H_5NH^+$; $H_2PO_4^-$; OH^-; $CH_3CO_2H_2^+$; $HCo(CO)_4$; HCN.

5.4 $I^- < F^- < HS^- < NH_2^-$.

5.5 (a) In water too strong: O^{2-}; too weak: ClO_4^-, NO_3^-; measurable: CO_3^{2-};

(b) (in H_2SO_4) too weak: ClO_4^-; measurable: NO_3^-, HSO_4^-.

5.6 CN is electron withdrawing.

5.7 From Pauling's rules, $pK_a = 13$.

5.8 $HSiO_4^{3-} < HPO_4^{2-} < HSO_4^- < HClO_4$.

5.9 (a) $[Fe(OH_2)_6]^{3+}$, higher ξ;
(b) $[Al(OH_2)_6]^{3+}$, higher ξ;
(c) $Si(OH)_4$, higher ξ; (d) $HClO_4$,
(higher ox. no. or Pauling's rules);
(e) $HMnO_4$ (higher ox. no.); (f) H_2SO_4
(higher ox. no. or Pauling's rules).

5.10 $Cl_2O_7 < SO_3 < CO_2 < B_2O_3 < Al_2O_3 < BaO$.

5.11 $NH_3 < CH_3GeH_3 < H_4SiO_4 < HSO_4^- < H_3O^+ < HSO_3F$.

5.12 Ag^+.

5.13 Polyanions: As, B, Si, Mo;
polyoxocations: Al, Ti, Cu.

5.14 Reduction of one.

5.15 $4PO_4^{3-} + 8H^+ \rightarrow P_4O_{12}^{4-} + 4H_2O$,
$2[Fe(OH_2)_6]^{3+} \rightarrow$
$[(H_2O)_4Fe(\mu\text{-}OH)_2Fe(OH_2)_4]^{4+}$
$+2H_3O^+$.

5.16 (a) $H_3PO_4 + HPO_4^{2-} \rightarrow 2H_2PO_4^-$;
(b) $CO_2 + CaCO_3 + H_2O \rightarrow$
$Ca^{2+} + 2HCO_3^-$.

5.17 Check symbols inside cover; Lewis
acids: BF_3, $AlCl_3$, M(I) or M(II) halides
of Ga, In, Td; Cl_2; M(IV) halides of Si,
Ge, Sn; $PbCl_2$; M(V) halides of P, Sb,
As; $BiCl_3$; dioxides of S, Se, Te;
Br_2, I_2, IF_7.

5.18 Acids: SO_3; Hg^{2+}; $SnCl_2$; SbF_5;
hydrogen bond.

5.19 (a) BBr_3, BCl_3, $B(n\text{-}Bu)_3$;
(b) Me_3N, $4\text{-}CH_3C_5H_4N$.

5.20 $K > 1$: (b), (d).

5.21 BH_3 binds to P, BF_3 binds to N.

5.22 Steric repulsion.

5.23 (a) DMSO stronger base to hard
and soft acid. (b) SMe_2 stronger
base toward soft acids than
DMSO.

5.24 $SiO_2 + 4HF \rightarrow 2H_2O + SiF_4$.

5.25 $Al_2S_3 + 3H_2O \rightarrow Al_2O_3 + 3H_2S$.

5.26 (a) and (b) Hard hydrogen bonding
solvent; (c) hard donor e.g. H_2O;
(d) softer base than Cl^-, possibly
PR_3.

5.27 I_2^+ and Se_8^+ disproportionate in or
coordinate to bases; S_4^{2-} and Pb_9^{4-}
disproportionate in or coordinate to
acids.

5.28 Al_2O_3 acid sites coordinate Cl^-.

5.29 HgS soft-soft, ZnO hard-hard.

5.30 (a) $EtOH + HF \rightarrow EtOH_2^+ + F^-$;
(b) $NH_3 + HF \rightarrow NH_4^+ + F^-$;
(c) $PhCOOH + HF \rightleftharpoons$
$PhCOO^- + H_2F^+$.

5.31 Both.

5.32 Hard.

5.33 SiO_2 stronger.

5.34 Base O^{2-}, acid $S_2O_7^{2-}$.

5.35 AsF_5 trigonal-bipyramidal (D_{3h});
AsF_6^{2-} octahedral (O_h); C_{3v} and
D_{2h}.

Chapter 6

Self-tests

6.1 Above c. 1750°C.

6.2 Yes; $E^\ominus$ for reaction with Cl^- only
slightly favorable (reaction is very
slow).

6.3 0.85 V.

6.4 No.

6.5 No.

6.6 Oxidation by O_2 plus H_2O to
H_2SO_4.

6.7 1.43 V.

6.8 Tl^+ at a minimum ($NE^\ominus = -0.34$ V), Tl^{3+} the highest point
($NE^\ominus = +2.19$ V).

6.9 Mn^{2+}.

6.10 Much stronger in acid solution.

6.11 Fe(II) favored.

6.12 $Ni(en)_3^{2+}/Ni$ potential more negative
(less favored).

Exercises

6.1 Above 1400 °C.

6.2 Higher overpotential for O_2
formation because of a more
demanding mechanism.

6.3 Thermodynamically suitable in acid
solution: (a) HClO, $\alpha\text{-}PbO_2$ etc.;
(b) Fe, Zn, etc.;
(c) Al, Fe, Zn etc.;
(d) same as c.

6.4 (a) $4Cr^{2+} + O_2 + 4H^+ \rightarrow 4Cr^{3+} + 2H_2O$;
(b) $4Fe^{2+} + O_2 + 4H^+ \rightarrow 4Fe^{3+} + 2H_2O$; (c) NR; (d) NR;
(e) $2Zn + O_2 + 4H^+ \rightarrow 2Zn^{2+} + 2H_2O$.

6.5 (a) $4Fe^{2+} + O_2 + 4H^+ \rightarrow 4Fe^{3+} + 2H_2O$;
(b) $4Ru^{2+} + O_2 + 4H^+ \rightarrow 4Ru^{3+} + 2H_2O$ and/or
$3Ru^{2+} \rightarrow Ru(s) + 2Ru^{3+}$;
(c) $3HClO_2 + H_2O \rightarrow ClO_3^- + 2HClO + H^+$
(followed by ClO_3^-
disproportionation)
$HClO_2 + O_2 \rightarrow ClO_3^-$ (slow due to
O_2 overpotential).
(d) NR

6.6 (a) $E = E^\ominus - (0.059\,\text{V}/4)[\log p_{O_2}^{-1} + 4\,\text{pH}]$; (b) $E = E^\ominus - (0.059\,\text{V}/6)(6\,\text{pH})$.

6.7 For $Cr_2O_4^{2-}$: $\Delta G^\ominus = +31.8\,\text{kJ mol}^{-1}$, $K = 2.7 \times 10^{-6}$
For $[Cu(NH_3)_2]^+$: $\Delta G^\ominus = +9.65\,\text{kJ mol}^{-1}$, $K = 2.0 \times 10^{-2}$.
Despite similar $E^\ominus$, $\Delta G^\ominus$ and K
differ because of differing n.

6.8 (a) Disproportionation:
$Cl_2 + 2OH^- \rightarrow Cl^- + ClO^- + H_2O$;
(b) very little disproportionation;
(c) in acid kinetic; in base
thermodynamic.

6.9 (a) $5N_2O + 2OH^-$
$\rightarrow 2NO_3^- + 4N_2 + H_2O$;
(b) $Zn + I_3^- \rightarrow Zn^{2+} + 3I^-$;
(c) $3I_2 + 5ClO_3^- + 3H_2O \rightarrow 6IO_3^- + 5Cl^- + 6H^+$.

6.10 Added acid (lower pH) (a) disfavored;
(b) favored; (c) disfavored; (d) no
effect.

6.11 (a) Atom transfer; (b) outer-
sphere electron transfer;
(c) multistep.

6.12 $+1.39$ V; $2ClO_4^- + 16H^+ + 14e^- \rightarrow Cl_2 + 8H_2O$.

6.13 3.7×10^{38}.

6.14 (a) Fe_2O_3; (b) Mn_2O_3; (c) HSO_4^-.

6.15 c. -0.1 V.

6.16 Potential less positive.

6.17 $Fe^{2+}/Fe(OH)_2$; $Fe^{3+}/Fe(OH)_3$.

6.18 If potential for reduction shifts upon
addition of ligand, complex formation
is indicated.

6.19 CO_2 lowers pH and favors
oxidation.

Chapter 7

Self-tests

7.1 A is *trans*, B is *cis*.

7.2 (a) *cis*-$[PtCl_2(H_2O)_2]$;
(b) $[Cr(NCS)_4(NH_3)_2]^-$;
(c) $[Rh(en)_3]^{3+}$.

7.3 Replace Co for Cr in **39a** and **39b**.

7.4 Only (a).

7.5 $t_{2g}^3 e_g^2$.

7.6 See discussion of Fig. 7.12.

7.7 The differences in the 6–8 eV region
can be attributed to the lack of d
electrons for Mg (II).

7.8 Increased LFSE for high to low spin
state.

Exercises

7.1 Check element positions inside front
cover, tetrahedral halide complexes:
Fe^{2+}, Co^{2+}, Ni^{2+}, Cu^{2+} and Zn^{2+}.

7.2 (a) Rh^+, Ir^+, Ni^{2+}, Pd^{2+}, Pt^{2+}, Au^{3+};
(b) $PtCl_4^{2-}$, $Pt(NH_3)_4^{2+}$,
$Ni(CN)_4^{2-}$.

7.3 (a) See **10** and **11**; (b) the trigonal
prism; (c) $[Cr(OH_2)_6]^{3+}$,
$[Co(NH_3)_6]^{3+}$, $[Fe(CN)_6]^{4-}$.

7.4 (a) Tetracarbonylnickel(0), tetrahedral;
(b) tetracyanonickelate(2−), square-
planar; (c) tetrachlorocobaltate(2−);
(d) hexaamminenickel(2+).

7.5 (a) See **34**; (b) see ox^{2-} formula and
description in Table 7.1 and Fig. 7.7;
(c), (d), and (e) see ligands in Table 7.1.

7.6 (a) Fig. 7.1;
(b) **12**;
(c) $[NCAgCN]^-$.

7.7 (a) $[CoCl(NH_3)_5]Cl_2$;

(b) $[Fe(OH_2)_6]NO_3$;
(c) $[cis\text{-}RuCl_2(en)_2]$;
(d) $[Cr_2(NH_3)_5(\mu\text{-}OH)]Cl_5$.

7.8 (a) cis-tetraamminedichlorochromium($+1$), $trans$-tetraisocyanatodiamminechromium(-1); (b) $trans$-diamminetetraisothiocyanatochromium(III); (c) bis(ethylenediamine)oxalatocobalt(III).

7.9 (a) cis and $trans$; (b) square-planar with $trans$ phosphanes; (c) one isomer; (d) two isomers; (e) three isomers.

7.10 $trans$-carbonylchlorobis(triphenylphosphane)iridium(I); dicarbonylchlorobis(triphenylphosphane)iridium(I); dicarbonylhydridobis(triphenylphosphane)iridium(I).

7.11 Chiral: (a) (d) (e).

7.12 Pink $[Co(NH_3)_5(OH_2)]Cl_3$; pentaammineaquacobalt(III) chloride; purple $[CoCl(NH_3)_5]Cl_2$, pentaammine chlorocobalt(III) chloride.

7.13 Violet: $[Cr(OH_2)_6]Cl_3$; green: $[Cr(OH_2)_5Cl]Cl_2$; both complexes are octahedral.

7.14 $trans$-diamminediaquaplatinum(II), square-planar.

7.15 $[Pt(NH_3)_4][PtCl_4]$ tetraammineplatinum(II) tetrachloroplatinate(II); $[Pt(NH_3)_4][NO_3]_2$, tetraammineplatinum(II)nitrate; $Ag_2[PtCl_4]$, silver(I) tetrachloroplatinate(II).

7.16 Both are square-planar; the β isomer is $trans$ and the other is cis.

7.17 The PX_3 ligands are equivalent within each isomer, but the chemical shifts will differ between the isomers. (^{31}P coupling will be quite different between the isomers).

7.18 (a) For ^{31}P NMR see 7.17; the ^{13}C NMR will show equivalent CO ligands for the cis isomer and two inequivalent sets for the $trans$ isomer.
(b) Only one P chem. shift and one ^{13}C shift for the axial phosphane isomer; a single chemical shift for ^{31}P in the phosphane isomer but two separate ^{13}C shifts in a 1:2 intensity ratio.

7.19 The bonding orbital consists of the two Pσ orbitals in phase with Ptd_{z^2}, and in the antibonding orbital the Pσ orbitals are out of phase with d_{z^2}.

7.20 (a) and (e) t_{2g}^6, $2.4\Delta_o$;
(b) $t_{2g}^4 e_g^2$, $0.4\Delta_o$; (c) t_{2g}^5, $2\Delta_o$;
(d) t_{2g}^3, $1.2\Delta_o$; (f) $e^3 t^2$,
$0.6\Delta_T$; (g) $e^4 t^6$, 0.

7.21 No; H^-, strong σ donor no π interaction; PPh_3 σ donor $+\pi$ acceptor.

7.22 (a) 0; (b) 4.9; (c) 1.7; (d) 3.9; (e) 0; (f) 4.9; (g) 0.

7.23 Yellow, pink, and blue, respectively.

7.24 See Fig. 7.12 and associated discussion.

7.25 Perchlorate: square-planar $(d_{xz}, d_{yz})^4 (d_{xy})^2 (d_{z^2})^2$ diamagnetic; thiocyanate approximately octahedral $t_{2g}^6 e_g^2$ with unpaired electron in e_g.

7.26 Distorted octahedral.

7.27 Jahn–Teller distortion in excited state.

7.28 Dissociative mechanism.

7.29 Rate of associative process depends on identity of entering ligand and, therefore, it isn't an inherent property of $[M(OH_2)_6]^{n+}$.

Chapter 8

Self-tests

8.1 (a) $Ca(s) + H_2(g) \rightarrow CaH_2(s)$;
(b) $NH_3(g) + BF_3(g) \rightarrow$ $H_3N\text{—}BF_3(s)$; (c) $LiOH(s) +$ $H_2(g) \rightarrow NR$.

8.2 $Et_3SnH + CH_3Br \rightarrow$ $Et_3SnCH_3 + HBr$.

8.3 $H_3B\text{—}OR_2 + H_2C\text{=}CH_2 \rightarrow$ $CH_3CH_2BH_2 + OR_2$,
$BH_4^- + CH_3CH\text{=}CH_2 \rightarrow$ $CH_3CH_2CH_2BH_3^-$,
$BH_4^- + NH_4^+ \rightarrow H_2 + H_3B\text{—}NH_3$.

Exercises

8.1 (a) $H(+1), S(-2)$; (b) $K(+1)$, $H(-1)$; (c) $Re(+7), H(-1)$;
(d) $H(+1), S(+6), O(-2)$;
(e) $H(+1), P(+1), O(-2)$.

8.2 Industry at high temperature:
$CH_4 + H_2O \rightarrow CO + 3H_2$
$CO + H_2O \rightarrow CO_2 + H_2$
$C + 2H_2O \rightarrow CO_2 + 4H_2$
Laboratory at ambient temperature:
$2HCl + Zn \rightarrow ZnCl_2 + H_2$
$NaH + CH_3OH \rightarrow NaOCH_3 + H_2$.

8.3 (a) Check with Fig. 8.2;
(b) check with Table 8.6;
(c) Group 13, Group 14, Groups 15 through 17.

8.4 (a) Barium hydride, saline; (b) silane, electron-precise, molecular;
(c) ammonia electron-rich, molecular;
(d) arsane, same as NH_3;
(e) palladium hydride, metallic;
(f) hydrogen iodide, electron-rich.

8.5 (a) $BaH_2 + 2H_2O \rightarrow 2H_2 + Ba(OH)_2$;
(b) $HI + NH_3 \rightarrow NH_4I$;
(c) $PdH_x \xrightarrow{\Delta} PdH_{x-\delta} + (\delta/2)H_2$;
(d) $NH_3 + BMe_3 \rightarrow H_3NBMe_3$.

8.6 Solids: BaH_2, $PdH_{0.9}$; liquids, none; gases: SiH_4, NH_3, AsH_3, HI; PdH_x electrical conductor.

8.7 H_2Se, bent, C_{2v}; P_2H_4, pyramidal around each P, C_2; H_3O^+, pyramidal, C_{3v}.

8.8 (b)

8.9 Me_3SnH, weak Sn—H bond.

8.10 (a) $H_2O < H_2S < H_2Se$;
(b) $H_2Se < H_2S < H_2O$.

8.11 (i) Direct combination: $Pd + x/2$ $H_2 \rightarrow PdH_x$;
(ii) protonation of a salt: $NaCl + H_2SO_4 \rightarrow HCl + NaHSO_4$;
(iii) metathesis : $4BCl_3 + 3LiAlH_4 \rightarrow$ $2B_2H_6 + LiAlCl_4$.

8.12 (a) $CaSe + HCl(aq) \rightarrow CaCl_2$ $+H_2Se$; (b) $SiCl_4 + LiAlD_4 \rightarrow$ $SiD_4 + LiAlCl_4$;
(c) $2GeMe_2Cl_2 + LiAlH_4 \rightarrow$ $2GeMe_2H_2 + LiAlCl_4$;
(d) $Si + 2HCl \rightarrow SiH_2Cl_2$ (+ other chlorosilanes),
$2SiH_2Cl_2 \rightleftharpoons SiH_4 + SiCl_4$.

8.13 (i) No, $B_2H_6 + 3O_2 \rightarrow B_2O_3 + 3H_2O_2$,
(ii) See Box 8.1.

8.14 $BH_4^- \ll AlH_4^- > GaH_4^-$; AlH_4^- strongest reducing agent;
$GaH_4^- + 4HCl \rightarrow GaCl_4^- + 4H_2$.

8.15 (a) $2NaBH_4 + 2H_3PO_4 \rightarrow$ $2NaHPO_4 + B_2H_6$,
$B_2H_6 + THF \rightarrow 2H_3BTHF$,
$H_3BTHF + 3C_2H_4 \rightarrow 3BEt_3 + THF$;
(b) BH_3THF (from above) $+Et_3N \rightarrow Et_3NBH_3 +$ THF or $NaBH_4 + NH_4Br$ $\xrightarrow{THF} H_3 + H_2 + NaBr$.

8.16 $Si + 2Cl_2 \rightarrow SiCl_4$ (distill); $SiCl_4 + 2H_2 \rightarrow Si + 4HCl$.

8.17 Period 2 hydrides are volatile except for BeH_2; Period 3 CaH_2 and AlH_3 are not volatile, the rest are volatile. Period 2 hydrides generally more thermodynamically stable (B_2H_6 vs AlH_3 is an exception).

8.18 Clathrate hydrates in which the krypton is contained in a hydrogen-bonded $(H_2O)_n$ cage.

8.19 See Fig. 8.6.

Chapter 9

Self-tests

9.1 (a) 2.2.2 crypt, $Rb^+ > Na^+$ $\approx Cs^+ > Li$; (b) EDTA, $Fe^{3+} > Cu^{2+}$.

9.2 V(V), VO_2^+.

9.3 $Re_3Cl_9(PPh_3)_3$; Re_3 triangle with bridging and terminal Cl and one PPh_3 on each Re.

9.4 Layered structure with weak van der Waals interaction between layers.

9.5 (a) $(Me)_2SAlCl_3 + GaBr_3 \rightarrow$ $Me_2SGaBr_3 + AlCl_3$

(Ga − S soft/soft), (b) $2TlCl_3$
$+CH_2O + H_2O \rightarrow 2TlCl$
$+CO_2 + 4H^+ + 4Cl^-$.

9.6 UO_2^{2+} which contains U(VI).

Exercises

9.1 Check position inside cover;
(a) decrease down a group;
(b) increase down a group;
(c) increase up a group.

9.2 (a) Mg^{2+} (larger q/r); (b) Sr; (c) K^+.

9.3 (a) CaF_2 8 coord. (cubic), MoS_2
6 coord. (trigonal prism); (b) CdI_2
layered, $MoCl_2$ contains Mo_6 clusters;
(c) BeO 4:4 coord., CaO 6:6 coord.;
(d) acetate-bridged Mo_2 dimer **(27)**;
central O surrounded by 4 Be^{2+} **(5)**.

9.4 See inside cover, **C**: Sr^{3+}, Ti^{4+}, V^{5+};
O: Cr, Mn; **N**: Fe, Co, Ni, Cu, Zn.

9.5 d block: higher ox. states more stable,
p block higher ox. states less stable;
see Appendix 2.

9.6 (a) $Cr^{2+} + Fe^{3+} \rightarrow Cr^{3+} + Fe^{2+}$ (+2
ox. state more stable to the right of a
d-block series; (b) $2CrO_4^{2-} + 2MoO_2$
$+2OH^- \rightarrow Cr_2O_3 + 2MoO_4^{2-} + H_2O$
(stability of highest oxidation state
increases down a group); (c) $6MnO_4^- +$
$10Cr^{3+} +11H_2O \rightarrow 6Mn^{2+} +$
$5Cr_2O_7^{2-} + 22H^+$ (maximum ox. state
less stable from right to left in $3d$
series).

9.7 (a) Ni^{2+}; (b) for the same ox. no. metal
ions become softer going toward the
right of a d series; (c) $Ni^{2+} + H_2S \rightarrow$
$NiS + 2H^+$.

9.8 (a) Check inside front cover; (b) TiF_2
through NiF_2; (c) Sc, Y and lanthanides,
Zr and Hf, Nb and Ta, Mo and W, Tc
and Re; one example is Mo_6Cl_{12}.

9.9 cis-$[Ru^{II}LCl(OH_2)]^+ + OH^- \rightarrow$
cis-$[Ru^{III}LCl(OH)]^+ + H_2O + e^-$ (e^-
might be transferred to an electrode
which maintains the potential on the
system). Higher ox. state promotes
acidity thus $[Fe(OH_2)_6]^{3+}$ is more
acidic than $[Fe(OH_2)_6]^{2+}$.

9.10 (a) NR; (b) $6MoO_4^{2-} + 10H^+ \rightarrow$
$[Mo_6O_{19}]^{2-} + 5H_2O$;
(c) $3ReCl_5 + 2MnO_4^- + 8H_2O \rightarrow$
$3ReO_4^- + 2MnO_2 + 16H^+$;
(d) $6MoCl_2 + 2HCl + 2H_2O \rightarrow$
$[H_3O]_2[Mo_6Cl_{14}]$.

9.11 $trans$-octahedral; cis-octahedral (d^0),
$trans$-octahedral (d^2), sq. pyramid.

9.12 Ionic or borderline: NiI_2, ionic; $NbCl_4$,
FeF_2; more covalent PtS; M—M
bonded: WCl_2.

9.13 (a) $2TiO + 6HCl \rightarrow$
$2Ti^{3+} + H_2 + 2H_2O + 6Cl^-$;
(b) $Ce^{4+} + Fe^{2+} \rightarrow Fe^{3+} + Ce^{3+}$;
(c) $2Rb_9O_2 + 14H_2O \rightarrow 18Rb^+$
$+18OH^- + 5H_2$; (d) $2Na+$
$2CH_3OH \rightarrow 2Na(OCH_3) + H_2$.

9.14 (a) $\sigma^2\pi^4\delta^2$, B.O. = 4;
(b) $\sigma^2\pi^4\delta^2$, B.O. = 4;
(c) $\sigma^2\pi^4\delta^2\delta^{*2}\pi^{*2}\sigma^{*2}$, B.O. = 0.

9.15 (a) $Zn(NH_2)_2 + 2NH_4^+ \rightarrow$
$Zn(NH_3)_6^{2+}$;
(b) $Zn(NH_2)_2 + 2KNH_2 \rightarrow$
$Zn(NH_2)_4 + 2K^+$.

9.16 (a) $Hg^{2+} + Cd \rightarrow Hg + Cd^{2+}$;
(b) $Tl^{3+} + Ga \rightarrow Tl + Ga^{3+}$;
(c) NR.

9.17 (a) Group 13 max. ox. state 3+;
ignoring element–element bonded
systems such as $B_{10}H_{10}^{2-}$ +3 most
stable for light elements B; Al. The +1
state is progressively more stable
from Ga to Tl. Similarly in Group 14
ignoring element–element bonded
system +4 is most stable for lighter
elements and +2 is most stable for
Pb. (b) (i) $Sn^{2+} + PbO_2 + 4H^+$
$\rightarrow Sn^{4+} + Pb^{2+} + 2H_2O$ (fits);
(ii) $Tl^{3+} + Al \rightarrow Al^{3+} + Tl$;
(iii) $3In^+ \rightarrow In^{3+} + 3In$;
(iv) $2Sn^{2+} + O_2 + 4H^+ \rightarrow 2Sn^{4+}$
$+2H_2O$; (v) NR.

9.18 (i) 1.32 V; (ii) 2.40 V; (iii) 0.32 V;
(iv) 1.08 V; (v) −0.02 V.

9.19 (a) $2Ln(s) + 6H^+ \rightarrow 2Ln^{3+} + 3H_2$.
(b) The + 3 state is by far the
most stable. (c) Cerium (+4) and
europium (+2).

9.20 $Ce(IO_3)_4$ and $EuSO_4$ have low
solubility compared with the Ln^{3+}
salts of these anions.

9.21 See Fig. 9.33; ^{90}Sr and ^{144}Ce have high
yields and present serious hazards.

9.22 The $4f$ orbitals in lanthanides do not
have significant probability density at
the ionic radii. The $5f$ orbitals are
more exposed to ligands.

Chapter 10

Self-tests

10.1 (a) $BCl_3 + 3EtOH \rightarrow B(OEt)_3$
$+3HCl$, protolysis;
(b) $BCl_3 + NC_5H_5 \rightarrow Cl_3BNC_5H_5$,
Lewis acid–base;
(c) $BBr_3 + F_3BNMe_3 \rightarrow$
$Br_3BNMe_3 + BF_3$, Lewis acidity
$BF_3 < BBr_3$.

10.2 $3NMeH_2 + 3BCl_3 \rightarrow$
$Me_3N_3B_3Cl_3 + 6HCl$.
$Me_3N_3B_3Cl_3 + 3MgMeBr \rightarrow$
$Me_3N_3B_3Me_3 + 3MgBrCl$.

10.3 7 pairs of skeletal electrons.

10.4 Framework electron pairs; $nido$; see
Table 10.4.

10.5 $2[B_{10}H_{13}]^- + Al_2Me_6 \rightarrow$
$2[B_{10}H_{11}AlMe]^- + 4CH_4$.

10.6 $1,2$-$B_{10}C_2H_2 + 2LiBu \rightarrow$
$1,2$-$B_{10}C_2H_{10}Li_2 + 2C_4H_{10}$.
$1,2$-$B_{10}C_2H_{10}Li_2 + 2SiMe_2Cl_2 \rightarrow$
$1,2$-$B_{10}C_2H_{10}(SiMe_2Cl)_2 + 2LiCl$.

10.7 (a) Electrons are added to the π bond;
(b) electrons are withdrawn from the
π bond.

10.8 $^{13}CO + 2MnO_2 \rightarrow {}^{13}CO_2 + Mn_2O_3$.
$^{13}CO_2 + LiBMe_3H \rightarrow Li[^{13}CHO_2]$

$+BMe_3$.

10.9 Cyclic $Si_4O_{12}^{n-}$ (with one bridging O
between each Si and two terminal O
on each Si)
$4Si^{4+} \rightarrow 16+, 12O^{2-} \rightarrow 24-$;
therefore $n = 8$.

10.10 Fig. 10.29 (a) has 24 vertices;
therefore the total number of Si and
Al is 24.

Exercises

10.1 Check inside front cover; (a) Group 13
metals Al through Tl, Group 14 metals
Sn and Pb; (b) C through Sn in Group
14; (c) highly oxophilic: B, Al, C, Si.

10.2 Work from Fig. 10.2; two B atoms are
closest to viewer in this orientation.

10.3 $B_2O_3 + 3Mg \rightarrow 2B + 3MgO$;
$SiO_2 + 2C \rightarrow Si + 2CO$;
$GeO_2 + H_2 \rightarrow Ge + 2H_2O$;
Even though C is the cheapest
reducing agent a high temperature
arc furnace is used to make Si, and the
Ge recovery is probably the most
efficient.

10.4 $SiF_4 < BF_3 < BCl_3 < AlCl_3$;
(i) $F_4SiNMe_3 + BF_3 \rightarrow$
$F_3BNMe_3 + SiF_4$;
(ii) $F_3BNMe_3 + BCl_3 \rightarrow$
$Cl_3BNMe_3 + BF_3$;
(iii) NR (reaction is controlled by
hard–soft).

10.5 $BCl_3 + Hg \rightarrow B_2Cl_4 + HgCl_2$
(electric discharge);
$B_2Cl_4 + C_2H_4 \rightarrow$
$Cl_2BCH_2CH_2BCl_2$;
$3Cl_2BCH_2CH_2BCl_2 + 4AsF_3 \rightarrow$
$3F_2BCH_2CH_2BF_2 + 4AsCl_3$.

10.6 B: C.N. = 3 or 4, C: C.N. = 3; Si:
C.N. = 4; partly determined by size of
the central atom: C < B << Si, also
multiple bonding in CO and BO bonds.

10.7 B_6H_{10} $nido$ boranes more stable than
$arachno$ (B_6H_{12}).

10.8 (a) $2B_5Hg(l) + 12O_2(g) \rightarrow$
$5B_2O_3(s) + 9H_2O(g)$;
(b) the solid B_2O_3 would deposit
inside the engine.

10.9 (a) $B_{10}H_{14}$ is $nido$ (b) 12 skeletal
(framework) electron pairs.

10.10 Check against $closo$-$B_{10}C_2H_{12}$
synthesis, p. 350, follow by base
cleavage and reaction with $FeCl_2$,
p. 350.

10.11 (a) Hexagonal array of atom in
stacked sheets; different stacking B
over N in BN, C offset from C in
graphite; (b) graphite reacts with Na
and with Br_2, BN does not; (c) large
HOMO–LUMO gap in BN does not
provide source and sink for e^-.

10.12 (a) $3PhNH_2 + 3BCl_3 \rightarrow$
$Ph_3N_3B_3Cl_3 + 6HCl$;
(b) $Ph_3N_3B_3Cl_3 + 3NaBH_4 \rightarrow$
$Ph_3N_3B_3H_3 + 3NaBH_3Cl$.

10.13 Check against Table 10.4, and
structure **24**.

10.14 $B_2H_6 < B_5H_9 < B_{10}H_{14}$.

10.15 (a) E_g: C > Si > Sn ≈ 0; BN > AlP > GaAs; (b) increase; (c) $\sigma = \sigma^0 \exp[E_g/2kT]$, so σ for AlP with a larger E_g is more temperature sensitive.

10.16 Check against Fig. 10.24.

10.17 (a) Graphite + K(am) → KC_8; (b) C heated with Cu → CuC_2; (c) K(vapor) + C_{60} → K_8C_{60}.

10.18 Carbon black (paint pigment); activated carbon (adsorbent), silica gel (adsorbent), amorphous silicon (photovoltaic cells).

10.19 (a) (i) B hard semiconductor, Al malleable metal; (ii) diamond harder and larger E_g than Si; (b) CO_2 linear molecule, gas at room T and p; SiO_2 hard 3-dimensional solid; (c) CX_4 are not simple Lewis acids; $SiCl_4$ mild Lewis acid; (d) BX_3 molecular for X = F, Cl, Br, I; AlX_3 various extended structures in solid, Al_2Cl_6 and Al_2Br_6 in non-coordinating solvents.

10.20 $K_2CO_3 + 2HCl \rightarrow$ $2KCl + H_2O + CO_2$. $Na_2SiO_4 + 2H^+ \rightarrow$ $2Na^+ + SiO_2 \cdot H_2O$ (silica gel)

10.21 Jadeite linear chains of $(SiO_3^{2-})_n$; kaolinite layered see Fig. 10.27. Si is 4-coordinate in both materials.

10.22 (a) 48 bridging O; (b) see Fig. 10.29.

10.23 See Fig. 10.28, 2:1 layers consisting of SiO tetrahedra, AlO_6 octahedra, SiO tetrahedra; pyrophyllite has neutral layers; muscovite has charged layers held together by K^+ ions and is more brittle.

Chapter 11

Self-tests

11.1 According to the VSEPR interpretation the 3-coordinate Bi may have a lone pair on Bi.

11.2 Sulfur (Group 16)/VI should be a stronger oxidant than P (Group 15) since electronegativity increases with group number.

11.3 The reaction employed to synthesize hydroxylamine, which produces the intermediate $N(OH)(SO_3)_2$, probably involves the attack of HSO_3^- (acting as a Lewis base) on NO_2^- (acting as a Lewis acid) although, since the formal oxidation state of the N atom changes from +3 in NO_2^- to +1 in $N(OH)(SO_3)_2$, this can also be seen as a redox reaction.

11.4 Polydichlorophoshazene can be converted to other polyphashazenes by nucleophilic substitution. Equation on p. 389, substituting $[Na^+][N(CH_3)_2^-]$ for $C_2F_5O^-$.

11.5 Both Br^- and Cl^- are potential catalysts.

11.6 (a) SO_3: see resonance forms for isoelectronic NO_3^- (**6** on p. 68), D_{3h}; (b) SO_3F^-: see isoelectronic XO_4^{n-} Table 3.1, p. 66; C_{3v}.

11.7 $38e^-$, (19 pairs), excluding 9 radial lone pairs gives 10 skeletal e^- pairs, *closo*, consistent with **38**.

Exercises

11.1 Check inside front cover; (a), (b), (c), (d): Bi is a metalloid the rest are nonmetals; all but oxygen can achieve max. group ox. no.; Bi strong inert pair effect.

11.2 (a) Second equation, p. 376, followed by $P_4 + 5O_2 \rightarrow P_4O_{10}$, then $P_4O_{10} + 6H_2O \rightarrow 4H_3PO_4$; (b) fertilizer grade, first equation p. 376; (c) high energy cost for P_4 preparation.

11.3 (a) $6Li + N_2 \rightarrow 2Li_3N$, $Li_3N + 3H_2O \rightarrow 3LiOH + NH_3$; (b) $3H_2 + N_2 \rightarrow NH_3$; (c) H_2 cheaper than Li to produce.

11.4 NCl_3 highly unstable, no NCl_5; PCl_3 and PCl_5 stable.

11.5 (a) PCl_4^+ tetrahedral; (b) PCl_4^- see-saw; (c) $AsCl_5$ trigonal bipyramid.

11.6 (a), (b) see 11.2(a); (c) $2H_3PO_4 + 3CaCl_2 \rightarrow$ $Ca_3(PO_4)_2 + 6HCl$.

11.7 $10e^-$: $N_2, NO^+, (CN^-)$; $11e^-$: $O_2^+, (NO)$; $11e^-$: $O_2^+, (NO)$; $14e^-$: (O_2^{2-}); $N_2H_5^+$. Strongest oxidant is italic; strongest base in parentheses.

11.8 (a) CO_3^{2-}; (b) CO_3^{2-}; (c) $C_2O_4^{2-}$; (d) C_2^{2-}; (e) CH_3^-.

11.9 (a) $4NH_3 + 7O_2 \rightarrow 6H_2O + 2NO_2$; $NO_2 + H_2O \rightarrow 2HNO_3 + NO$; (b) $2NO_2 + 2OH^- \rightarrow NO_2^-$ $+NO_3^- + H_2O$; (c) $NO_3^- + 2HSO_3^- + H_2O \rightarrow$ $NH_3OH^+ + SO_4^{2-}$, $NH_3OH^+ + OH^- \rightarrow NH_2OH + H_2O$; (d) $NaNH_2 + N_2O \rightarrow$ $NaN_3 + NaOH + NH_3$.

11.10 $P_4(s) + 5O_2(g) \rightarrow P_4O_{10}(s)$; P_4 is not the most stable allotrope.

11.11 Check against Fig. 11.5; Bi(V) is readily reduced; the Bi(III), P(III), and P(V) oxo species are much less prone to redox.

11.12 Faster: the good electrophile NO^+ is formed and the mechanisms generally proceed through attack on NO^+ by nucleophiles.

11.13 The rate law contains a $p(NO)^3$ term. Thus at low partial pressure NO is not rapidly oxidized.

11.14 (a) $PCl_5 + H_2O \rightarrow POCl_3 + 2HCl$; (b) $PCl_5 + 8H_2O \rightarrow 2H_3PO_4 + 10HCl$; (c) $PCl_5 + AlCl_3 \rightarrow [PCl_4][AlCl_4]$; (d) $3PCl_5 + 3NH_4Cl \rightarrow$ $N_3P_3Cl_6 + 12HCl$.

11.15 $H_3PO_2 + Cu^{2+} + H_2O \rightarrow$ $H_3PO_4 + Cu + 3H^+, E^\ominus =$

+ 0.839 V, H_3PO_4 is a useful reducing agent.

11.16 (a) + 1.068 V ; (b) Cr^{2+} is not a good catalyst for H_2O_2 disproportionation.

11.17 (a) en; (b) en; (c) possibly SO_2; basic solvents are compatible with highly reducing anions. Cd_2^{2+} would disproportionate with en to form a Ca^{2+} complex with en and Cd.

11.18 Most reducing SO_3^{2-}, SO_4^{2-}, $S_2O_8^{2-}$ most oxidizing.

11.19 (a) $Te(OH)_6$; SO_4^{2-}, (b) The large Te(VII) accommodates more ligands and is less acidic; (c) $PO(OH)_3$ and $Sb(OH)_5$ show the same trend in Group 15.

11.20 (a) Square; (b) possibly P_4S_3 structure, **41**.

Chapter 12

Self-tests

12.1 Both SO_2 and Sn^{2+} are thermodynamically suitable, but the oxidation products of the former are easier to separate from the product and S is much cheaper than Sn.

12.2 Pyramidal, C_s.

12.3 I_3^-, IBr_2^-, Br_3^-, central atom hypervalent.

12.4 In acid solution: F_2 and Cl_2. F_2 reaction is fast.

12.5 $2XeO_4^{2-} \rightarrow XeO_6^{4-} + Xe + O_2$, $2XeO_4^{2-} + 2H_2O \rightarrow$ $2Xe + 4OH^- + 3O_2$, two independent reactions (disproportionation of xenate and oxidation of water); their ratio will depend on conditions.

Exercises

12.1 Check with Table 12.1, F_2 light green-yellow, Cl_2 green-yellow, Br_2 brown, I_2 purple in gas or dilute hydrocarbon solution; Noble gases are colorless.

12.2 Br_2 or I_2 by oxidation of halide solution with Cl_2; Cl_2 aqueous electrolysis, F_2 nonaqueous electrolysis. See pp. 407–8.

12.3 Check with Fig. 12.3; $Cl_2 + OH^- \rightarrow HClO + Cl^-$.

12.4 ⬭×⬭ ⬭×⬭ , vacant lobes on each end accept electrons from donor molecules.

12.5 (a) NH_3 is hydrogen bonded; (b) high electronegativity of F decreases basicity of N.

12.6 (a) $NCCN + 2OH^- \rightarrow$ $NCO^- + CN^- + H_2O$; (b) $2SCN^- + MnO_2 + H^+ \rightarrow$ $Mn^{2+} + (SCN)_2 + 2H_2O$; (c) $+ C_{3v}$.

12.7 (a) Octahedral, pentagonal bipyramid; (b) $IF_7 + SbF_5 \rightarrow [IF_5][SbF_6]$.

12.8 (a) Stronger; (b) no change; (c) weaker.

12.9 (a) No, Sb in maximum oxidation state; (b) explosive CH_3OH is a reducing agent, BrF_5 a strong oxidizing agent; (c) BrF_3 can be oxidized, but probably not explosively; (d) explosive, S_2Cl_2 easily oxidized.

12.10 $Br_3^- + I_2 \rightarrow 2IBr + Br^-$.

12.11 Large cations stabilize large unstable anions.

12.12 ClO_2 angular radical; I_2O_5 has oxo bridge between two IO_2.

12.13 $HBrO_4$ is more acidic but less stable than H_5IO_6.

12.14 (a) Reduction potential increases with decreasing pH; (b) at $E^{\ominus}(\text{pH} = 7) = 0.78$ V vs. $E^{\ominus}(\text{pH} = 0) = +1.20$ V.

12.15 Reduction of the oxoanion is more favorable because protons are consumed in that half reaction.

12.16 Periodic I—O bonds are more labile.

12.17 (a) ClO_2^-; (b) ClO^-.

12.18 (a) and (d) because they have reducible cations associated with a strongly oxidizing anion, ClO_4^-.

12.19 He has sufficient speed to escape from the atmosphere.

12.20 (a) He; (b) Xe; (c) Ar.

12.21 (a) $Xe + F_2 \rightarrow XeF_2$ sunlight; (b) $Xe + 3F_2 \rightarrow XeF_6$ high pressure of F_2, $300\,°C$; (c) $XeF_6 + 3H_2O \rightarrow XeO_3 + 6HF$.

12.22 (a) XeF_4 (square-planar); (b) XeF_2 linear; (c) XeO_3 pyramidal; (d) IF^+.

12.23 (a) All three share a single electron pair and have an octet around each atom. Formal charges: Cl(0) O(−1); Br(0); Xe(+1), F(0); (b) They are isolobal; (c) All three display some electrophilicity in the order $ClO^- < Br_2 < XeF^+$. Positive formal charge on XeF^+ correlates with its Lewis acidity and electrophilicity.

12.24 (a) Eight electron pairs around Xe; one is a lone pair. (b) The configuration of these pairs may lead to a square antiprism (**17**, p. 218) or dodecahedron (**18**, p. 218) with one unoccupied vertex. If the lone pair is not stereochemically active typical 7-coordinate structures may result: a pentagonal bipyramid (**13**, p. 218) or capped octahedron (**14**, p. 218). The latter two structures seem most probable.

Chapter 13

Self-tests

13.1 $^1F, ^3F$.

13.2 (a) ^3P; (b) ^2D.

13.3 $^3T_{1g}, ^3T_{2g}, ^3A_{2g}$ and $^1T_{2g}, ^1E_g$.

13.4 $17\,500\,\text{cm}^{-1}$ and $22\,400\,\text{cm}^{-1}$.

13.5 $^4A_{2g}$ to $^2E_g, ^4T_{1g}, ^4T_{2g}$, ligand.

13.6 An extra band.

Exercises

13.1 (a) ^6S; (b) ^3F; (c) ^2D; (d) ^3P.

13.2 (a) ^3F; (b) ^5D; (c) ^6S.

13.3 (a) ^2S; (b) ^3P (ground), $^1D, ^1S$.

13.4 $3p^5$: F; $4d^{10}$; Cd^{2+}.

13.5 B^+: ^1S; Na: ^2S; Ti^{2+}: ^3F; Ag^+: ^1S.

13.6 $B = 861\,\text{cm}^{-1}$, $C = 3168\,\text{cm}^{-1}$.

13.7 $d^6, ^1A_{1g}; d^1, ^2T_{2g}; d^5, ^6A_{1g}$.

13.8 From T-S diagram: (a) $B \approx 770\,\text{cm}^{-1}$, $\Delta_O \approx 8500\,\text{cm}^{-1}$; (b) $B \approx 720\,\text{cm}^{-1}$, $\Delta_O \approx 10\,750\,\text{cm}^{-1}$.

13.9 $^5T_{2g}$, charge transfer, LMCT.

13.10 Weak: spin forbidden LF; medium: spin allowed LF transitions; strong: CT transition.

13.11 $[FeF_3]^{3-}$ has only one spin-forbidden LF band; $[CoF_6]^{3-}$ has $^5E_g \leftarrow {}^5T_{2g}$ which is spin-allowed.

13.12 CN^- is more covalent.

13.13 A spin-forbidden LF band, two spin-allowed LF bands, and a LMCT band respectively.

13.14 High spin d^5 has only spin-forbidden LF bands.

13.15 $[Cr(OH_2)_6]^{3+}$: LF bands which are weak; CrO_4^{2-}: allowed LMCT bands.

13.16 a_{1g}; (a) d_{xz}, d_{yz}; (b) d_{z^2}; (c) lowest xy; $(xz, yz), z^2, x^2 - y^2$ highest energy.

13.17 Mn(VII) is d^0 and has no LF bands.

13.18 The two LMCT bands are to t_{2g} and e_g. The difference is Δ_T.

13.19 Metal–metal CT.

13.20 $Sm^{2+}, ^7F$; $Eu^{2+}, ^8S$; $Yb^{2+}, ^1S$.

Chapter 14

Self-tests

14.1 $5.1\,\text{L mol}^{-1}\,\text{s}^{-1}$.

14.2 $[PtCl_4]^{2-} \overset{PH_3}{\rightarrow} [PtCl_3PPH_3]^- \overset{NH_3}{\rightarrow} trans\text{-}[PtCl_2NH_3PPH_3]$; $[PtCl_4]^{2-} \overset{NH_3}{\rightarrow} [PtCl_3NH_3]^- \overset{PPH_3}{\rightarrow} cis\text{-}[PtCl_2NH_3PPh_3]$.

14.3 $K_E \approx 1, k = 1.2 \times 10^2\,\text{s}^{-1}$.

14.4 2 *cis* to 1 *trans*.

14.5 $K = 3 \times 10^{18}$, $k = 3.6 \times 10^{10}\,\text{L mol}^{-1}\,\text{s}^{-1}$.

Exercises

14.1 Nucleophiles; NH_3, Cl^-, S^{2-}; electrophiles; Ag^+, Al^{3+}.

14.2 d.

14.3 (a) Stoichiometric; (b) intimate.

14.4 Rate $= k[Mn(OH_2)_6^{2+}]$ $[X^-]/(1 + K_a[X^-])$, vary X^-.

14.5 Stronger covalent bonds to be broken for higher ox. states and heavier metals.

14.6 Steric hindrance to nucleophilic attack.

14.7 The rate-controlling step may be dissociative. The influence of phosphane is consistent with this conclusion.

14.8 Yes, a favorable A intermediate.

14.9 $[PtCl_4]^{2-} \overset{NH_3}{\rightarrow} [PtCl_3NH_3]^-$ $\overset{NO_2^-}{\rightarrow} cis\text{-}[PtCl_2(NO_2)(NH_3)]$, $[PtCl_4]^{2-} \overset{NO_2^-}{\rightarrow} [PtCl_3NO_2]^- \overset{NH_3}{\rightarrow} trans\text{-}[PtCl_2(NO_2)(NH_3)]$.

14.10 (a) Decrease rate; (b) decrease rate; (c) decrease rate; (d) increase rate.

14.11 Conjugate base path; Brønsted acidity is required.

14.12 (a) $cis\text{-}[PtCl_2(PR_3)_2]$; (b) $trans\text{-}[PtCl_2(PR_3)_2]$; (c) $trans\text{-}[PtCl_2(NH_3)(py)]$.

14.13 $[Ir(NH_3)_6]^{3+} < [Rh(NH_3)_6]^{3+}$ $< [Co(NH_3)_6]^{3+} < [Ni(OH_2)_6]^{2+}$ $< [MN(OH_2)_6]^{2+}$.

14.14 (a) Decreases; (b) decreases; (c) little effect; (d) decreases; (e) increases.

14.15 The rate constant for the initial CO insertion step.

14.16 Dissociative.

14.17 Inner sphere goes through N_3^- bridged complex. Outer sphere has no ligand bridge.

14.18 Inner-sphere mechanism.

14.19 $[W(CO)_5(NEt_3)]0.4$; ligand field

14.20 250 nm.

14.21 Rate-determining step is SMe_2 dissociation.

Chapter 15

Self-tests

15.1 $Mg + HgMe_2 \rightarrow MgMe_2 + Hg$.

15.2 Homolytic Hg—CH_3 bond cleavage.

15.3 Bridging chlorine atoms and terminal Bu groups.

15.4 Lower force constant for Si-O-Si bend.

15.5 $Al_2Me_6 + 6MeOH \rightarrow$ $2Al(OMe)_3 + 6CH_4$; no reaction of $SiMe_4 + MeOH$; $SiCl_4 +$ $4MeOH \rightarrow Si(OMe)_4 + 4HCl$.

15.6 $GeH_4, GeR_4; AsH_3, AsR_3$. Hydrogen and alkyl compounds are similar for each element. This may reflect similar H and C electronegativity.

Exercises

15.1 Organometallic: (a), (c), (d), and (g) because each contains a metal (or metalloid)–carbon bond.

15.2 Check against Fig. 15.1 and Table 15.2.

15.3 (a) $BiMe_3$, trimethylbismuthane; (b) $SiPh_4$, tetraphenylsilicon (IV); (c) $[AsPh_4][Br]$, tetraphenylarsenic(1+) bromide; (d) $K[BPh_4]$, potassium tetraphenylboron(1−).

15.4 (a) Triethylsilane (tetrahedral monomer, electron-precise); (b) diphenylchloroborane (trigonal monomer, electron-deficient); (c) tetraphenyldichloroaluminum (two Al—Cl—Al bridges, in this structural form it is electron-precise); (d) ethyllithium or, more precisely, tetraethyltetralithium (tetrahedral Li$_4$ array with a phenyl carbon bridging each face, electron-deficient); methylrubidium, salt-like.

15.5 (a) Li tetrahedron with each face capped by CH$_3$ (structure **2**);
(b) planar triangular array of B and C;
(c) four terminal CH$_3$ and two CH$_3$ bridges in a diborane-like structure;
(d) tetrahedral;
(e) pyramidal;
(f) pseudotetrahedral.

15.6 The small B atom bonds to three CH$_3$ groups. The larger Al is a dimer with each Al bonded to four carbons. Although similar in size to Al, GaMe$_3$ does not associate into dimers. GeMe$_4$ is a simple tetrahedral molecule, InMe$_3$ is a monomer in solution.

15.7 (1): (a), (b), (c), (e);
(2): (a), (c), (d), (e); (3): AsMe$_3$;
(4): (a), (b), (c), (e).

15.8 (a) 2AsCl$_3$ + 3ZnMe$_2$ →
2AsMe$_3$ + 3ZnCl$_2$;
SiPh$_2$Cl$_2$ + ZnMe$_2$→ZnCl$_2$;
(b) NH$_3$ + BMe$_3$→H$_3$NBMe$_3$;
(c) AsMe$_3$ + [HgMe][BF$_4$] →
[Me$_3$AsHgMe][BF$_4$].

15.9 (a) Mg + MeBr→MeMgBr, active metal (Me);
(b) 3HgMe$_2$ + 2Al→3Hg + Al$_2$Me$_6$, organometallic of less active metal (Me) plus a more active metal; (c) 3Li$_4$Me$_4$ + 4BCl$_3$→4BMe + 12LiCl, organometallic of more active metal with a metal halide.

15.10 (a) Na[C$_{10}$H$_8$], less delocalized (stabilized) anion; (b) Na$_2$[C$_{10}$H$_8$], the anion is more highly reduced.

15.11 (a) Ca + HgMe$_2$→CaMe$_2$ + Hg, Ca more electropositive than Hg;
(b) NR, converse of (a);
(c) Li$_4$Me$_4$ + 4SiPh$_3$Cl→
4LiCl + 4SiPh$_3$Me, the Me$^-$ prefers the more electronegative element;
(d) NR, converse of (c);
(e) SiMe$_3$H + C$_2$H$_4$→SiMeEt; hydrosilylation.

15.12 (a) Si + 2MeCl→SiMe$_2$Cl$_2$ (elevated temp and Cu catalyst);
(b) 2SiCl$_2$Me$_2$ ⇌ SiClMe$_3$ + SiCl$_3$Me (heat).

15.13 Reaction (a) above followed by hydrolysis: 4SiMe$_2$Cl + 4H$_2$O→
(SiMe$_2$O)$_4$ + 8HCl,
n(SiMe$_2$O)$_4$→4(−Si−O−)$_n$,
H$_2$SO$_4$ catalyzed.

15.14 Group 13: + 3 predominates, some Tl(I) organometallics; Group 14: + 4 predominates, some Sn(II) and Pb(II); Group 15: + 3 predominates some As(II) and Sb(V).

15.15 (a) **M**—C enthalpy decreases down a group; (b) Lewis acidity decreases from AlR$_3$ to TlR$_3$; (c) Basicity of central metal only in Group 15, :ER$_3$, compounds, order of basicity:
As > Sb > Bi.

15.16 (a) PbMe$_4$ > SiMe$_4$;
(b) BMe$_3$ > Li$_4$Me$_4$ > SiMeCl$_3$ >>
SiMe$_4$; (c)AsMe$_3$ >>> SiMe$_4$;
Li$_4$Me$_4$ >> HgMe$_2$.

15.17 (a) Schlenk ware; (b) BMe$_3$ is highly volatile, vacuum line or solutions in Schlenk ware; (c) Schlenk ware;
(d) open vessels; (e) open vessels.

15.18 Delocalized σ antibonding orbitals.

15.19 3SiR$_2$Cl$_2$ + 6Na→Si$_3$R$_6$ + 6NaCl,
Si$_3$R$_6$ $\xrightarrow{h\nu}$ R$_2$Si═SiR$_2$, R must be bulky.

Chapter 16

Self tests

16.1 No (20 e$^-$).

16.2 +1.

16.3 Bridging only.

16.4 P(CH$_3$)$_3$, less steric repulsion.

16.5 Prepare: [Mn(CO)$_5$]$^-$ with Na (p. 554), then Mn(CO)$_5$CH$_3$ with CH$_3$I (p. 555), migratory insertion with L = PPh$_3$ (p. 554).

16.6 Small change.

16.7 CVE = 60; tetrahedral Fe$_4$ with Cp on each vertex, CO on each face.

Exercises

16.1 (a) Pentacarbonyliron(0);
(b) tetracarbonylnickel(0);
(c) hexacarbonylmolybdenum(0);
(d) decacarbonyldimanganese(0);
(e) hexcarbonylvanadium(0);
(f) trichloroethyleneplatinate(II).

16.2 (a)

M

(b)

M

16.3 (a) +3; (b) 0; (c) −2; (d) 0.

16.4 (a) to (d) 18e$^-$ in both 16.1 and 16.3;
V(CO)$_6$17e$^-$, no influence on structure, easily reduced;
[PtCl$_3$(C$_2$H$_4$)]$^-$16e$^-$, square-planar;
[(η^5-C$_5$H$_5$)$_2$Fe][BF$_4$], 17e$^-$ slightly longer Fe—C distances; easily reduced.

16.5 Direct combination, reductive carbonylation.

16.6 Fe + 5CO $\xrightarrow{\text{high pressure}}$ Fe(CO)$_5$,
diphos + Fe(CO)$_5$→Fe(CO)$_{3-}$
(diphos) + 2CO.

16.7 C_s(3); C_{3v}(2); D_{3h}(1).

16.8 Ni$_3$ triangle each Ni capped by Cp, CO above and below nickel triangle and collinear; 18$\frac{1}{3}$e$^-$ per Ni atom, deviations from 18-e$^-$ rule are common in Group 10.

16.9 (b) Reacts by an associative process.

16.10 (a) [Fe(CO)$_4$]$^{2-}$ more negative, complex is more basic; (b) [Re(CO)$_5$]$^-$ heavier metal, more basic.

16.11

Products: (a) Li[W(CO)$_5$—C$\overset{\displaystyle O}{\underset{\displaystyle CH_3}{\|}}$],

(b) AlBr$_3$ coordinated to bridging CO ligands.

16.12 (a) η^2; (b) η^1, η^3, η^5; (c) η^2, η^4, η^6; in practice the last is observed;
(d) η^2, η^4; (e) η^2, η^4, η^6, η^8.

16.13 (a) 16e sandwich; (b) 18e; (c) 18e.

16.14 (a) Fe, Ru, Os; (b) Mn, Co; (c) Cr, Fe, Ni; (d) Table 16.3.

16.15 (a) 3; (b) 2; (c) 12.

16.16 See pp. 557, 558, and 562–4.

16.17 Prepare Mn(CO)$_5^-$ (p. 554), protonate with HCl.

16.18

Mo(CO)$_5$C$\overset{\displaystyle OCH_3}{\underset{\displaystyle Ph}{\diagup}}$

16.19 (a) PF$_3$ π-acceptor ligand (increases ν(CO)), P(CH$_3$)$_3$ σ donor ligand (decreases ν(CO)); (b) Cp* stronger donor, and therefore lower ν(CO).

16.20 RuCp$_2$ (18e) more stable than RhCp$_2$ (19e); (η^5-C$_5$H$_5$)$_2$Fe +
CH$_3$C(O)Cl $\xrightarrow{\text{AlCl}_3}$ (η^5-C$_5$H$_5$)
(η^5-C$_5$H$_4$COCH$_3$)Fe + HCl.

16.21 (η^5-C$_5$H$_5$)$_2$Fe + LiBu→
(η^5-C$_5$H$_5$)(η^5 − C$_5$H$_4$Li)Fe
+BuH, (η^5-C$_5$H$_5$)−
(η^5-C$_5$H$_4$Li)Fe + ClC(O)CH$_3$ →
(η^5-C$_5$H$_5$)(η^5-C$_5$H$_4$COOH)Fe
+LiCl.

16.22 See Appendix 4.

16.23 C$_5$H$_6$ is a tetrahapto four-electron ligand. Its formation reduces the metal electron count to 18 on Ni; protonation of Fe in FeCp$_2$ leaves the 18e count unchanged.

16.24 (a) Attack of H$_2$O on the CF$_2$ carbon followed by elimination of HF; (b) β-H elimination.

16.25 (a) Groups 6–9; stable Cp compounds, Groups 3 and 4; C—H insertion; (b) Groups 6 to 8; **M**H more acidic than in Groups 3 and 4. (c) 18e rule violation for many Group 3 and 4 and some Group 9 metals.

16.26 (a) Octahedral (86), trigonal prismatic, CVE = 90; (b) no;
(c) [Fe$_6$C(CO)$_{16}$]$^{2-}$, CVE = 86, octahedral;
[Co$_6$C(CO)$_{16}$]$^{2-}$, CVE = 92, trigonal prismatic.

16.27 (a) Si(CH$_3$); (b) I.

16.28 Co$_4$(CO)$_{12}$ has weaker **M**M bonds.

Chapter 17

Self-tests

17.1 Decrease rate.

17.2 Single 1465 cm^{-1} band expected.

17.3 No, it lacks acidic sites.

Exercises

17.1 (a) Catalytic; (b) noncatalytic; (c) noncatalytic.

17.2 See (a) p. 584; (b) p. 586; (c) p. 583; (d) p. 585; (e) p. 600.

17.3 (a) Homogeneous; (b) heterogeneous; (c) homogeneous.

17.4 Accomplished with existing technology: (d); thermodynamically prohibited: (a) and (b); need research: (c).

17.5 Added PPh_3 suppresses equilibrium concentration of $RhCiL_2(Sol)$.

17.6 Steric effects reduce equilibrium formation of (c) in Cycle 17.2.

17.7 Formation of (a) or (c) may be rate-limiting with added phosphane.

17.8 Attack of coordinated OH^- gives one enantiomer, attack of solution OH^- gives the other.

17.9 Coordination of formic acid with loss of L, reductive elimination of H_2, ox. addn. HCO_2H, and CO_2 elim.

17.10 Titanium nitride is too stable.

17.11 (a) An extra proton is required to charge-compensate Al^{3+} substituted for Si^{4+}. (b) Ga(III), Sc(III), and Fe(III).

17.12 (a) 900 °C treatment Lewis acid; (b) silica gel more Brønsted acidic; (c) ZSM-5 is highly regular, silica gel is not.

17.13 To achieve high surface area.

17.14 See p. 608.

17.15 Initial chemisorption as $C^*H_2C^*HC(CH_3)_2CH_2CH_3$, exchange of H for D at surface attached carbon atoms (*), dissociation (via reductive elimination).

17.16 CO strongly chemisorbs and blocks sites for H_2.

Chapter 18

Self-tests

18.1 Vacancies on both anion and cation sites.

18.2 Reduction in thickness of conduction plane impedes the larger cation more.

18.3 $Cr^{[III]}$ in octahedral holes provides maximum LFSE.

End-of-chapter exercises

18.1 Frenkel, interstitial, Schottky, holes; no.

18.2 (a) NiO; (b) PbF_2; (c) Fe_2O_3.

18.3 Intrinsic in pure substance, extrinsic (due to impurities).

18.4 Fig. 18.22; Na^+ in center of cell.

18.5 Both. It is a defect when disordered, and describes a new phase when ordered.

18.6 See Fig. 18.16 for interstitial sites; triangular oxygen arrays.

18.7 TiO vacancies on both Ti and O sites; FeO; Fe^{3+} interstitials with vacancies on Fe^{2+} sites.

18.8 Electron microscopy, electron diffraction.

18.9 (a) NiO (MnO is much more easily oxidized than the aqueous potentials suggest); (b) NiO.

18.10 TiO metallic, NiO semiconductor.

18.11 Positive Ag electrode: $Ag(s) \rightarrow Ag^+(in\ AgI) + e^-$ (external circuit); AgI electrolyte: Ag^+ to − electrode; negative Ag electrode: $Ag^+(in\ AgI) + e^- \rightarrow Ag(s)$.

18.12 Positive electrode mass loss 108 mg, AgI no change, negative electrode mass gain 108 mg.

18.13 Ferromagnet: domains of parallel spins on atoms; antiferromagnet: domains of antiparallel spins on atoms.

18.14 In the inverse spinel structure, Fe[CoFe] with Co^{2+} and Fe^{3+}; three unpaired electrons are expected from the d^7 octahedral Co^{2+}. The structure appears to be close to the inverse spinel formulation.

18.15 See p. 633.

Chapter 19

Self-tests

19.1 1.24 Å is just slightly above O_2, suggesting little electron transfer; 1.26 Å is just less than O_2^- suggesting nearly one electron transferred. 1.47 Å is near O_2^- suggesting transfer of mainly one electron from each Co, and 1.30 Å is just over O_2^- suggesting transfer of just over one electron between the two Co centers.

19.2 Cu^{2+} is redox-active, binds ligands strongly, undergoes rapid ligand exchange, and has a flexible coordination geometry.

19.3 The average oxidation number of Fe in the latter is 2.5. The former is 3. Reduction and disulfide formation can convert the Fe_2 clusters to Fe_4 clusters.

19.4 First and sixth entries have same $E^{\ominus}$ but change of R by 2 Å correlates with greater than 10^2 rate change. The comparison of third and eighth entries is similar. Fourth and ninth show doubling R increases rate by a factor of 10. All imply qualitative agreement with Marcus theory.

19.5 C_{4v} has the order of d orbitals $xz, yz < xy < z^2 < x^2 - y^2$. The configuration is $(xz)^1(yz)^1(xy)^1(z^2)^1(x^2-y^2)^1$.

Exercises

19.1 O: water, amino acids, phosphates in bone. N: proteins and nucleic acids. K: inside cells, promotes the hydrolysis of phosphorylated biomolecules. Ca: bones, shells, and other structural materials.

19.2 Carboxylate side chains of polypeptides for calcium; imidazole side chains of histidine amino acids of polypeptides and porphyrin macrocycles for iron

19.3 Neutral O_2 with net double bond, triplet; O_2^- longer; 1 1/2 order bond, doublet; O_2^{2-} longer single bond, singlet.

19.4 Left for HbO_2; right for Mb at low $p(O_2)$, left at high $p(O_2)$.

19.5 CO can bind to, and block, the oxygen transport sites of hemoglobin.

19.6 Nucleophile is N end of peptide; site of attack is coordinated carbonyl group.

19.7 (a) High spin $t_{2g}^4 e_g^2$, low spin t_{2g}^6; (b) high H_2O, low $(C_6H_5)_3P$:.

19.8 Variable oxidation number.

19.9 The two oxidation states are not configured to conform to each other. They bind substrates.

19.10 Specificity in binding to allow close approach, control of inner-sphere reorganization energies.

19.11 Consult text.

19.12 Fe—O—O—Fe bridge formation.

19.13 (a) $ZnOH_2$ transfers a proton to promote–OH attack; (b) Zn—O=C promotes nucleophilic attack on C.

19.14 O_2 evolution requires $4e^-$, over four Mn atoms, Mn charge changes by −1.

19.15 Chlorophyll–porphyrin π^*, cytochrome Fe(II). Mg could not be used; Mg(III) inaccessible.

19.16 56.5%.

19.17 Imidazole rings, porphyrins, dihydroporphyrins, and corrins.

19.18 Ca^{2+}, Zn^{2+}, and Cu^{2+} form more labile complexes than Be^{2+}, Al^{3+}, and Cr^{3+}.

19.19 To prevent the formation of a μ-O^{2-} diporphyrin complex.

19.20 Hemoglobin contains four subunits to myoglobin's one. Hemoglobin exhibits cooperative and pH-dependent O_2 binding.

19.21 CO binds to the iron atom and blocks coordination, and therefore transport, of O_2.

19.22 Borderline Lewis acidity, strong ligand binding, rapid ligand substitution reactions.

19.23 Small structural reorganization leads to rapid outer-sphere electron transfer.

19.24 Octahedral complexes with a predominance of oxygen-containing ligands.

19.25 Manganese ions are redox-active and are harder Lewis acids than nickel or copper ions.

Copyright acknowledgements

Formula Index

E: Electronic structure; L: Lewis structure; P: preparation; S: structure.

Ag

Ag^{2+} 194
$[Ag(CH_2Cl_2)][OTeF_5]$ 397
Ag_2HgI_4 623–4
AgI 47S
$[Ag(NH_3)_2]^+$ 308S
$Ag(NO_3)$ 355
Ag-containing formulas
  $[C_6H_6Ag]^+$ 159S
  $KAgF_4$ 411
  $RbAg_4I_5$ 624

Al

AlB_2 340S, 370
$[Al(BH_4)_3]$ 274S
Al_4C_3 362
$Al_2(CH_3)_6$ 509, 518P
$Al(C_2H_5)_3$ 598
$Al(C_4H_9)_3$ 519
$Al_2(C_2H_5)_4H_2$ 275S
Al_2Cl_6 161S
$AlCl_3$ 316P
AlF_3 316
Al_2O_3 154, 181, 183, 316, 600, 624, 627
$[AlO_4]^{5-}$ 156S
$[AlO_4(Al(OH)_2)_{12}]^{7+}$ 156
$Al_2(OH)_2Si_4O_{10}$ 367
$AlPO_4$ 369
$Al(mes)_3$ 518S
Al-containing formulas
  $(C_2H_5)NAlCl_3$ 318P
  $[(CH_3)_3N]_2AlH_3$ 275S
  $Cl_3Al(N(CH_3)_2)_2$ 316
  $KAl_2(OH)_2Si_3AlO_{10}$ 367S
  $KAlSi_3O_8$ 367
  $LiAlCl_4$ 267P, 276P
  $LiAlF_4$ 270P
  $LiAlH_4$ 267, 269–70, 274–6
  $Li[Al(OEt)_4](thf)$ 269P
  $MgAl_2O_4$ 49S, 628
  Na_3AlF_6 183, 316
  $Na_{12}(AlO_2)_{12}(SiO_2)_{12}$ 369
  $NaAlSi_3O_8$ 367

As

$As(CH_3)_3$ 502S
$As(C_6H_5)_3$ 278
$As_2(CH_3)_4Br$ 531P
$AsCl_4^-$ 379S
As_2O_3 387
As-containing formulas
  $GaAs$ 278

Au

$[Au(CN)_2]^-$ 184
$AuCl_3$ 355
$[AuI_2]^-$ 308S
Au-containing formulas
  Cu_3Au 44

B

B_{12} 333S
$B_2(Bu)_4$ 336
$B_{10}C_2H_{12}$ 349S
$[B_3C_2H_5]^{4-}$ 350S, 565
$B(CH_3)_3$ 159–60S, 335P, 500, 517P
$B_4C_2H_6$ 349S
$B(C_6H_5)_3$ 510
$B(CH_3)_2(C_4H_9)$ 517P
$B_{10}C_2H_{10}(COOH)_2$ 350P
$B(CH_3)_3$ 517P
$B_{10}C_2H_{10}(NO)_2$ 350P
$B_{10}C_2H_{10}(Si(CH_3)_3)_2$ 351P
$B(C_2H_5)_3(et)$ 264
B_2Cl_4 335S
B_4Cl_4 335–6S
BF_3 68L, 75S, 121S, 123S, 128S, 161, 165L, 269, 333–4P
$[BF_4]^-$ 67S, 335S, 419
$B_{10}H_{14}$ 341S, 347
$[B_{12}H_{12}]^{2-}$ 341–2S
B_2H_6 100S, 121S, 260S, 265, 270P, 272–3, 341S
BH_3 272–3, 502
BH_4^- 273
B_4H_{10} 341–2S, 345–7S
B_5H_9 341S, 343, 345S, 347
$[B_6H_6]^{2-}$ 341S, 343, 344E, 349S
B_5H_{11} 341–2S
B_6H_{10} 341S
$[B_{11}H_{11}AlCH_3]^{2-}$ 349S
BN 337P
BO 73
B_2O_3 333–4
$[B_3O_6]^{3-}$ 336S
$B(OCH_3)_3$ 336P
$B(OH)_3$ 271P
$[B_3O_3(OH)_4]^-$ 336S
B-containing formulas
  CaB_6 340
  $CH_3CH_2BH_2$ 273
  Cl_4B_4 339
  $Cl_2BCH_2CH_2BCl_2$ 336
  $Cl_3B_3N_3H_3$ 338–9P
  $Cl_3B_3N_3H_9$ 339
  $F_3BS(CH_3)_2$ 272
  $H_3BN(CH_3)_3$ 272, 337
  $H_3B_3N_3H_3$ 338S
  H_2NBH_2 338
  H_3NBH_3 337
  H_3NBH_2COOH 338
  $H_2OB(OH)_3$ 336
  KBH_4 274
  $K[B_{11}H_{14}]$ 347P
  $Li[B(CH_3)_4]$ 335
  $Li[B(C_6H_5)_4]$ 510P
  $LiBH_4$ 270, 273
  $Li[BH_4](et)$ 269
  $Li[B_{11}H_{14}]$ 348
  $Li_2[B_{12}H_{12}]$ 348P
  $Li[BH(CH_3)_3](et)$ 267P
  $NaBH_4$ 274
  $Na_2B_4O_5(OH)_4$ 323
  $Na[HB(C_2H_5)_3](et)$ 264P
  $N_3B_3H_{12}$ 338
  $[NO][BF_4]$ 384S
  $[Zr(BH_4)_4]$ 274S

Ba

$BaBiO_3$ 634
$BaPbO_3$ 634
$BaSO_4$ 288
Ba-containing formulas
  $YBa_2Cu_3O_7$ 632–3

Be

Be 287
$Be_3Al_2Si_6O_{18}$ 365
Be_2C 70, 362
$Be(CH_3)_2$ 513–14
$Be(C_5H_5)_2$ 513–14S
$BeCO_3$ 289
BeH_2 103E
BeO 288
$Be_4O(O_2CCH_3)_6$ 289S

Bi

Bi 377S
Bi_4^{2-} 400–1S
$([BiBr_3]^{2-})_n$ 379
$[Bi_2Cl_8]^{2-}$ 320S
BiF_3 319S
BiF_5 378, 411
Bi_2O_3 387
$Bi(OC_2H_4OCH_3)_3$ 320S
$[Bi_6OH_{12}]^{6+}$ 320S
Bi-containing formulas
  $NaBiO_3$ 319P

Br

Br_2 259, 409E
BrF_3 414
$[BrF_4]^-$ 417S
BrF_5 413S
BrO_3^- 425
BrO_4^- 426–7
Br-containing formulas
  $(CH_3)_2COBr_2$ 164S
  $CsBrF_4$ 414P
  $SiBrClFI$ 121S

C

C_{60} 123, 353
CBr_4 357
CCl_4 357, 423
CF_4 357
C_2F_4 428P
$[(CF_3CF_2O)_2PN]_n$ 389P
CF_3SO_3H 152S
CH_4 75S, 120S, 121S, 125S, 261S, 273
C_2H_2 121S
C_2H_6 120S
C_3H_8 275
C_8H_8 572S
$[C_6H_6Ag]^+$ 159S
$(CH_3)_3AlN(C_2H_5)_2$ 510P
$C_6H_4(As(CH_3)_2)_2$ 530S
$(CH_3)_2AsI$ 531P
CH_2BrCl 121–2
$CHBrClF$ 120–1S
$CH_3CH_2BH_2$ 273
$CH_3CH_2SiH_3$ 277P
$[CH_3(CN)]$ 170L, 171
$(CH_3)_2CO$ 427P

Subject Index

(T) signifies tabular matter. Bold type signifies key entries.

Useful relations

At 298.15 K, $RT = 2.4790$ kJ mol^{-1} and $RT/F = 25.693$ mV

1 atm $= 101.325$ kPa $= 760$ Torr (exactly)

1 bar $= 10^5$ Pa

1 eV $= 1.602\ 18 \times 10^{-19}$ J $= 96.485$ kJ mol^{-1} $= 8065.5$ cm^{-1}

1 cm^{-1} $= 1.986 \times 10^{-23}$ J $= 11.96$ J mol^{-1} $= 0.1240$ meV

1 cal $= 4.184$ J (exactly)

1 D (debye) $= 3.335\ 64 \times 10^{-30}$ C m

1 T $= 10^4$ G

1 Å (ångström) $= 100$ pm

1 M $= 1$ mol dm^{-3}

General data and fundamental constants

Quantity	Symbol	Value
Speed of light	c	$2.997\ 925 \times 10^8$ m s^{-1}
Elementary charge	e	$1.602\ 177 \times 10^{-19}$ C
Faraday constant	$F = eN_A$	9.6485×10^4 C mol^{-1}
Boltzmann constant	k	$1.380\ 66 \times 10^{-23}$ J K^{-1}
		8.6174×10^{-5} eV K^{-1}
Gas constant	$R = kN_A$	$8.314\ 51$ J K^{-1} mol^{-1}
		$8.205\ 78 \times 10^{-2}$ dm^3 atm K^{-1} mol^{-1}
Planck constant	h	$6.626\ 08 \times 10^{-34}$ J s
	$\hbar = h/2\pi$	$1.054\ 57 \times 10^{-34}$ J s
Avogadro constant	N_A	$6.022\ 14 \times 10^{23}$ mol^{-1}
Atomic mass unit	u	$1.660\ 54 \times 10^{-27}$ kg
Mass of electron	m_e	$9.109\ 39 \times 10^{-31}$ kg
Vacuum permittivity	ε_0	$8.854\ 19 \times 10^{-12}$ J^{-1} C^2 m^{-1}
	$4\pi\varepsilon_0$	$1.112\ 65 \times 10^{-10}$ J^{-1} C^2 m^{-1}
Bohr magneton	$\mu_B = e\hbar/2m_e$	$9.274\ 02 \times 10^{-24}$ J T^{-1}
Bohr radius	$a_0 = 4\pi\varepsilon_0\hbar^2/m_e e^2$	$5.291\ 77 \times 10^{-11}$ m
Rydberg constant	$R_\infty = m_e e^4/8h^3 c\varepsilon_0^2$	$1.097\ 37 \times 10^5$ cm^{-1}

Prefixes

f	p	n	μ	m	c	d	k	M	G
femto	pico	nano	micro	milli	centi	deci	kilo	mega	giga
10^{-15}	10^{-12}	10^{-9}	10^{-6}	10^{-3}	10^{-2}	10^{-1}	10^3	10^6	10^9

Glossary of chemical abbreviations

Ac	acetyl, CH_3CO
acac	acetylacetonato (Table 7.1)
ADP	adenosine diphosphate (158)
amino acids	see Table 19.3
aq	aqueous solution species
ATP	adenosine triphosphate (158)
bpy	2,3-bipyridine (Table 7.1)
bn	2,3-butanediamine (477)
cod	1,5-cyclooctadiene (560)
cot	cyclooctatetraene (Table 16.1)
Cp	cyclopentadienyl (Table 16.1)
Cp*	pentamethylcyclopentadienyl (569)
cyclam	tetraazacyclotetradecane (Table 7.1)
2.2.2-crypt	a bicyclic octadentate ligand (288)
diars	$1,2-C_6H_4(AsMe_2)_2$ (528)
dien	diethylenetriamine (Table 7.1)
DMSO	dimethylsulfoxide (149)
E	generally, a p-block element
η	hapticity (511)
edta	ethylenediaminetetraacetato (Table 7.1)
en	ethylenediamine (Table 7.1)
Et	ethyl
fod	partially fluorinated β-diketone (322)
gly	glycinato (Table 7.1)
Hal	halide
Hb	hemoglobin (649)
i-Pr or ^{i}Pr	isopropyl
KCP	$K_2Pt(CN)_4Br_{0.3} \cdot 3H_2O$ (109)
L	a ligand
μ	signifies a bridging ligand
Mb	myoglobin (651)
Me	methyl
mes	mesityl, 2,4,6-trimethylphenyl (516)
mnt	maleonitrilethiolato (Table 7.1)
NAD	nicotinamide adenine dinucleotide (659)
nta	nitritotriacetato (Table 7.1)
Ox	an oxidized species
ox	oxalato (Table 7.1)
Ph	phenyl
phen	phenanthroline (447)
py	pyridine
pyz	pyrazine (459)
salen	(652)
Sol	solvent, or a solvent molecule
soln	nonaqueous solution species
TCNE	tetracyanoethene (160)
THF	tetrahydrofuran (171)
TMEDA	N,N,N',N'-tetramethylethylenediamine (509)
TMPyP	tetrakis(4-N-methylpyridyl)teraphenylporphyrin (611)
trien	$2,2',2''$-triaminotriethylene (Table 7.1)
X	generally halogen, also a leaving group
Xyl	xylyl, 2,6-dimethylphenyl (523)
Y	an entering group
Zeolites:	A, ALPO-5, X, ZSM-5 (Table 10.11)